HETEROGENEOUS CATALYSIS

HETEROGENEOUS CATALYSIS

Experimental and Theoretical Studies

ELI RUCKENSTEIN

HAIYOU WANG

WILEY

Published by John Wiley & Sons, Inc., Hoboken, New Jersey
Published simultaneously in Canada

Library of Congress Cataloging-in-Publication Data:

Ruckenstein, Eli, 1925–
Heterogeneous catalysis : experimental and theoretical studies / Eli Ruckenstein, Department of Chemical and Biological Engineering, State University of New York at Buffalo, Haiyou Wang, Buffalo, NY.
pages cm
Includes index.
ISBN 978-1-118-54690-1 (hardback)
1. Heterogeneous catalysis. I. Wang, Haiyou, 1964– II. Title.
QD505.R83 2014
541′.395–dc23

2014009463

Printed in the United States of America

10 9 8 7 6 5 4 3 2 1

CONTENTS

PREFACE

This book contains a collection of papers that Professor Eli Ruckenstein and his coworkers have published in the area of heterogeneous catalysis during the past four decades. It consists of the following eight chapters:

1 Catalytic conversion of methane to synthesis gas by CO_2 reforming
2 Catalytic conversion of methane to synthesis gas by partial oxidation
3 Catalytic combustion of clean as well as nitrogen-bound fuels over transitional metal oxides
4 Zeolites and their applications as catalysts and/or catalyst supports
5 Synthesis of mesoporous V–Mg–O oxides and their applications as catalysts
6 Synthesis of polymer-supported catalysts and polymer-coated silica supports and their applications in catalysis
7 Metal sintering during heating in various atmospheres
8 Heterogeneous catalysis: a theoretical approach.

Chapters 1–3 deal with a number of catalysts involved in various catalytic reactions, including, methane CO_2 reforming, methane partial oxidation, and catalytic combustion; Chapters 4–6 are concerned with applications of various materials, such as zeolites, mesoporous V–Mg–O oxides, polymers, and polymer-coated SiO_2 as catalysts or catalyst supports; Chapter 7 is devoted to theoretical and experimental investigations of metal sintering in model metal catalysts during heating in various atmospheres; and Chapter 8 deals with some theoretical approaches relevant to heterogeneous catalysis. Each chapter starts with an introduction that summarizes the papers included.

This book will be of interest to catalytic scientists and to material scientists working in universities and industry, and to graduate students.

Eli Ruckenstein
Haiyou Wang

ABOUT THE AUTHORS

Professor Eli Ruckenstein is SUNY Distinguished Professor at State University of New York at Buffalo and Member of the National Academy of Engineering and American Academy of Arts and Sciences. His pioneering contributions to various areas of physical chemistry and applied sciences have brought him numerous distinctions, among which are the National Medal of Science and the Founders Award of the National Academy of Engineering. In addition, from the American Institute of Chemical Engineers, he won the Alpha Chi Sigma Award for his work in transport phenomena, the Walker Award for his work in the area of catalysis, and the Founders Award for his overall contribution to Science. From the American Chemical Society, he received the Kendall Award for his contributions to colloids and interfaces, the Langmuir Lecture Award for his achievement in the area of polymers, the Schoellkoph Medal for his work in catalysis, and the Murphree Award for his work in heterogeneous catalysis. He was given the Humboldt Award by Germany for his work in surfactants and the Creativity Award by the National Science Foundation for his work on biomolecules. He is a fellow of the American Institute of Chemical Engineers, which with the occasion of its 100th anniversary designated him as one of 50 Eminent Chemical Engineers of the Foundation Age. He is also a fellow of the American Nano Society.

Dr. Haiyou Wang was born in 1964 in Lanxi, Zhejiang Province, P.R. China. He is currently a principal engineer of Performance Materials and Technologies, Honeywell International. He was a senior scientist of Catalytic Materials LLC from 2001 to 2005, a visiting research scholar of Hong Kong Baptist University from 1994 to 1996, and a faculty member of Xiamen University from 1991 to 1997. He received a BS in 1985, a MS in 1988, and a DSc. in 1991, all in physical chemistry from Xiamen University, as well as a PhD in chemical engineering in 2001 from the State University of New York at Buffalo. He has authored or coauthored 50+ papers and 50+ patents/patent applications. He has been active within heterogeneous catalysis for over two decades.

1

CATALYTIC CONVERSION OF METHANE TO SYNTHESIS GAS BY CO_2 REFORMING

INTRODUCTION

CO_2 reforming of methane,

$$CH_4 + CO_2 \rightarrow 2CO + 2H_2, \qquad \Delta H^0_{298} = 247 \text{ kJ/mol.}$$

is an endothermic reaction; it produces synthesis gas with a H_2/CO ratio around 1, which is suitable for the Fischer–Tropsch synthesis of long-chain hydrocarbons. In addition, this process may have important environmental implications, since it consumes CO_2 and CH_4, two greenhouse gases. Unfortunately, no industrial technology for CO_2 reforming of methane has yet been developed, because no effective and commercially viable catalysts have been discovered [1]. When the conventional Ni-based catalyst for steam reforming is used for CO_2 reforming, it deactivates rapidly mainly because of carbon deposition on the active sites. A high molar ratio of CO_2 to CH_4 ($\geq$3) can be used to reduce the carbon deposition, but this decreases the selectivity to synthesis gas. Therefore, the inhibition of carbon deposition without extra cost and loss of catalyst performance posses a major challenge for CO_2 reforming of methane.

In CO_2 reforming of methane, carbon formation can take place through two possible routes: CH_4 decomposition and CO disproportionation (Boudouard reaction). While the former is an endothermic reaction, the latter is an exothermic one. Carbon formation by CH_4 decomposition is a structure-sensitive reaction [2, 3]. Specifically, the Ni(100) and Ni(110) surfaces are more active in CH_4 decomposition to carbon than the Ni(111) surface [2]. The CO disproportionation is favored at temperatures below 973 K due to its exothermic nature. Carbon deposition with measurable rates occurs over Co, Fe, and Ni catalysts at temperatures higher than 623 K [3]. Given that CO_2 reforming of methane often takes place at a temperature $\geq$973 K, CH_4 decomposition to carbon constitutes the main concern. The diffusion and segregation of carbon are dependent on the structure of metal surface as well. For instance, the carbon on Ni(110) diffuses more quickly into the bulk of the metal than carbon on Ni(100) [3]. In addition, it is more difficult for the carbon adsorbed on a smaller metal particle to diffuse than on a larger metal particle [4]. The structure-sensitivity nature of carbon formation and its diffusion provides an opportunity to avoid, or at least to reduce carbon deposition through the

Heterogeneous Catalysis: Experimental and Theoretical Studies, First Edition. Edited by Eli Ruckenstein and Haiyou Wang.

modification of the metal surface structure. One method is to control the size of the ensembles of metal atoms on surface since the ensembles required for carbon formation are larger than those needed for CH_4 reforming [5]. Accordingly, by controlling the metal particle size, one can control carbon formation. In this regard, strong adsorption of sulfur (sulfur passivation) was used to control the size of ensembles, and the degree of carbon deposition on Ni catalysts, as in the successfully commercialized SPARG process [6, 7]. Sulfur passivation is presumably responsible for the control of the size of the active metal ensembles since sulfur can preferentially eliminate the larger ensembles.

Besides the surface structure, the surface acidity of a catalyst is another major property that affects carbon deposition. It has been noted that carbon deposition can be attenuated or even completely suppressed when the metal is supported on a metal oxide with a strong Lewis basicity [8–11]. This suppression occurs because the high Lewis basicity of the support increases the ability of the catalyst to chemisorb CO_2 in the CO_2 reforming of methane and water in the steam reforming of methane, and these species react with carbon to form CO, resulting in decreased net carbon formation.

Almost all the catalysts used for CO_2 reforming of methane are supported metal catalysts. Based on the catalytic metal component, catalysts can be divided into two groups, that is, supported noble metal (such as Pd, Rh, Ir, Pt, Ru, etc.) catalysts and supported nonnoble metal (such as Fe, Co, and Ni) catalysts. Inui [12] and Rostrup-Nielsen [7] reported that the amount of carbon deposited on metal catalysts decreases in the order Ni $\gg$ Rh $>$ Ir $=$ Ru $>$ Pt $\approx$ Pd at 773 K and Ni $>$ Pd $=$ Rh $>$ Ir $>$ Pt $\gg$ Ru at 923 K. Thus, noble metal catalysts, in general, provide greater likelihood of carbon-free operation than Ni and other non-noble metal catalysts. For the same metal component, the rate of carbon deposition is also dependent on the nature of the support [7, 13–16].

Among the noble metal catalysts, Pt has been the most widely studied one, probably because of its lower rate of carbon formation and its greater availability than other noble metals. Regarding the support for Pt, ZrO_2 has been the most widely used because of its lower rate of carbon formation compared with other supports [17–22]. Bitter et al. [20] found that the rate of carbon formation decreases in the order $Pt/Al_2O_3 > Pt/TiO_2 > Pt/ZrO_2$ and that carbon deposition from methane decomposition instead of metal sintering is mainly responsible for the deactivation of supported Pt catalysts. The higher stability of Pt/ZrO_2 is probably due to the strong $Pt\text{-}Zr^{n+}$ interactions, which reduce carbon formation during reaction by promoting CO_2 dissociation [21]. The promoter, which can be a metal component or a metal oxide, has a significant effect on carbon deposition as well. It was reported that the bimetallic $Pt\text{-}Au/SiO_2$, $Pt\text{-}Sn/SiO_2$, and $Pt\text{-}Sn/ZrO_2$ catalysts generate less deposited carbon during CO_2 reforming of CH_4 than the corresponding monometallic Pt catalysts [23], probably because of the formation of metal alloys. It was also reported that the addition of cerium or lanthanum greatly improves the stability of a Pt/ZrO_2 catalyst without decreasing the CH_4 or CO_2 conversions [24]. Although the total amount of carbon deposition remained comparable for a Ce-promoted Pt/ZrO_2 catalyst and an unpromoted Pt/ZrO_2 catalyst, TPO (temperature programmed oxidation) studies suggested that the carbon species on Ce-promoted Pt/ZrO_2 catalyst could be removed at much lower temperatures. On the other hand, the La-promoted catalyst experienced a much lower carbon deposition than the unpromoted one.

More research work has been devoted to non-noble metal catalysts, particularly Ni-based catalysts due to their low cost with comparable activity and selectivity as the noble metal catalysts. Nevertheless, thermodynamic studies indicate that Ni catalysts are prone to carbon deposition in CO_2 reforming of CH_4, causing unstable activity [25]. Therefore, it poses an even bigger challenge to develop a stable Ni-based catalyst. Alumina is one of the most commonly used supports for Ni catalysts [10, 26–36]. The rate of carbon deposition over Ni/Al_2O_3 catalysts is dependent on the catalyst structure, composition, and preparation conditions [12]. Chen and Ren [37] reported that carbon deposition was significantly suppressed when $NiAl_2O_4$ was formed during pretreatment. The Ni–O bond in $NiAl_2O_4$ is stronger than that in NiO [38]. Consequently, it is more difficult to reduce Ni^{2+} to Ni^0, resulting in smaller nickel crystallites on the surface of catalyst. Such nickel particles, which are smaller than the size needed for carbon deposition, decrease the carbon deposition [28]. The catalyst preparation method exerts an influence on the dispersion of metal particles and ultimately on the rate of carbon deposition. Kim et al. [27, 39] and Osaki et al. [40] reported that Ni/Al_2O_3 catalysts prepared from aerogel alumina exhibited remarkably lower coking rates compared with alumina-supported Ni catalysts prepared by the conventional impregnation method. Osaki et al. [40] suggested that the Ni–O–Al bonds formed in aerogels are responsible for the formation of fine nickel particle after hydrogen reduction and ultimately for the high activity and low carbon deposition. A water-in-oil (w/o) micromulsion method was also found to be effective in the preparation of a Ni/Al_2O_3 catalyst possessing low carbon deposition and good stability [41].

Various promoters have been used to improve the performance of Ni/Al_2O_3 catalysts. The addition of basic oxides of Na, K, Mg, and Ca to Ni/Al_2O_3 catalysts was found to greatly reduce carbon deposition [10, 35, 42–44]. Kinetic results indicated that these added metal oxides changed the reaction order in CH_4 from negative to positive and the order in CO_2 from positive to negative. This observation implies that the surface of a basic metal oxide doped Ni catalyst is rich in adsorbed CO_2, whereas the surface of a Ni catalyst free of such a basic metal oxide is rich in adsorbed CH_4 [10]. The coverage of nickel with CO_2 is most likely unfavorable to CH_4 decomposition and, consequently, the carbon deposition is decreased. Choi et al. [45] examined the effect of Co, Cu, Zr, and Mn as promoters. They found that, compared with unpromoted Ni/Al_2O_3 catalyst, those promoted with Co, Cu, and Zr provided slightly improved activities, while the Mn-promoted catalyst exhibited a significant reduction in coke formation with only a small decrease in catalytic activity. Moreover, Seok et al. [46] found that the addition of Mn to Ni/Al_2O_3 caused a partial coverage of the surface of nickel particles with patches of MnO_x, which promoted the adsorption of CO_2. Both the partial coverage of the nickel surface with MnO_x, and the promoted CO_2 adsorption appear to contribute to decreased carbon deposition on Mn-promoted Ni/Al_2O_3 catalyst. Mo was also found to improve the stability of Ni/Al_2O_3 catalyst by reducing carbon deposition [47]. Rare earth metals or metal oxides were also investigated as promoters [48–50]. Nd [48] and CeO_2 [49, 50] were found to promote the Ni/Al_2O_3 catalyst, by reducing carbon deposition.

Our work was mainly focused on the effect of various supports on the reactivity of supported metal catalysts and the structure–property relationships. Efforts were also made to identify the mechanism of CO_2 reforming of CH_4 to synthesis gas. In brief, Chapter 1 consists of four parts: the first is concerned with Ni-based catalysts (**1.1**, **1.2**, **1.3**, and **1.4**); the second with Co-based catalysts (**1.5**, **1.6**, and**1.7**); the third with Rh-based catalysts (**1.8**); and the fourth with mechanistic studies of CO_2 reforming of CH_4 (**1.9**).

The nature of the support was found to play a very important role in determining the performance (activity and stability) of supported metal catalysts for CO_2 reforming of methane (**1.1**, **1.2**, **1.5**, and **1.8**). Among all supports investigated, MgO was the best support for both Ni and Co catalysts (**1.1** and **1.5**). The reduced NiO/MgO catalyst with a NiO content in the range of 9.2–28.6 wt% prepared by impregnation provided both high CO yield (94%) and excellent stability, because it generated NiO-MgO solid solution catalysts (**1.2** and **1.3**). The strong interactions between Ni and Mg induced by the formation of a NiO–MgO solid solution help to inhibit not only the sintering of NiO, but also the formation of large clusters of Ni, which generate coke (**1.4**). Similarly, the reduced 12 wt% CoO/MgO solid solution catalyst provided a CO yield of 93% and excellent stability (**1.5**). Since Co has two oxidation states, Co^{2+} and Co^{3+}, depending upon the calcination temperature and Co loading, one, two, or three Co-containing species, that is, Co_3O_4, $MgCo_2O_4$, and CoO–MgO solid solution, can be generated. This affects the reduction behavior and ultimately the catalytic performance of MgO-supported Co catalysts. Indeed, the Co/MgO solid solution catalysts with a loading between 8 and 36 wt%, precalcined at 500 and 800°C, respectively, provided high and stable activities; a too-high calcination temperature (e.g., 900°C) resulted in low activity, whereas too-high Co loading (e.g., 48 wt%) combined with a low calcination temperature $\leq$800°C, resulted in an unstable activity (**1.6**). For supported Rh catalysts, most of the reducible oxides, such as CeO_2, Nb_2O_5, Ta_2O_5, TiO_2, and ZrO_2, provided much lower yields of CO and H_2 than the irreducible oxides, such as γ-Al_2O_3, La_2O_3, MgO, SiO_2, and Y_2O_3; among the irreducible oxides, MgO and γ-Al_2O_3 provided high and stable activities (**1.8**).

Numerous mechanistic studies have suggested that during CO_2 reforming of methane, CH_4 is decomposed on metallic sites to reactive carbon species (CH_x, with $x = 0$–3), which are oxidized to CO by the oxygen-containing species that originate from CO_2. In this regard, a transient response analysis via a broadened pulse combined with a step change or an isotopic pulse was developed and applied to CO_2 reforming of methane over Ni/SiO_2 (**1.9**). It was found that Ni^0 is the active site for methane decomposition and that the reaction between C and O species on Ni^0 is the rate-determining step. Consequently, the rate of carbon accumulation on the catalyst surface is determined by the relative rates of formation of carbon species and their oxidative removal. When the former dominates, an excess of carbon is deposited, whereas when the latter dominates, the number of metallic sites is decreased because of their oxidation and subsequent restructuring of the catalyst. Both result in unstable activity. By optimizing the metal loading and the calcination temperature, T_c, for example, 6% wt% for $T_c = 500$°C and 9 wt% for $T_c = 1000$°C, highly effective and stable Co/γ-Al_2O_3 catalysts were obtained (**1.7**).

REFERENCES

[1] Puskas, I, *CHEMTECH*, **24**, 43 (1995).

[2] Schouten, S.C., Gijzeman, O.L.J., and Bootsma, G.A., *Bull. Soc. Chim. Belg.*, **88**, 418 (1996).

[3] Bartholomew, C.H., *Catal. Rev. Sci. Eng.*, **24**, 67 1982).

[4] Eizenberg, M., and Blakely, J.M., *Surf. Sci.*, **82**, 228 (1979).

[5] Rostrup-Nielsen, J.R., *Stud. Surf. Sci. Catal.*, **68**, 85 (1991).

[6] (a) Udengaar, N.R., Hansen, J.H.B., Hanson, D.C., and Stal, J.A., *Oil Gas J.*, **90**, 62 (1992); (b) Dibbern, H.C., Olesen, P., Rostrup-Nielsen, J.R., Totrrup, P.B., and Udengaar, N.R., *Hydrocarbon Process*, **65**, 71 (1986).

[7] Rostrup-Nielsen, J.R., and Hansen, J.H.B., *J. Catal.*, **144**, 38 (1993).

[8] Yamazaki, O., Nozaki, T., Omata, K., and Fujimoto, K., *Chem. Lett.*, 1953 (1992).

[9] Zhang, Z.L., and Verykios, X.E., *Catal. Today*, **21**, 589 (1994).

[10] Horiuchi, T., Sakuma, K., Fukui, T., Kubo, Y., Osaki, T., and Mori, T., *Appl. Catal. A: Gen.*, **144**, 111 (1996).

[11] Kim, G.J., Cho, D.S., Kim, K.H., and Kim, J.H., *Catal. Lett.*, **28**, 41 (1994).

[12] Inui, T., *Catalysis*, Vol. **16**, p. 133, The Royal Society of Chemistry, Cambridge, 2002.

[13] Perera, J.S.H., Couves, J.W., Sankar, G., and Thomas, J.M., *Catal. Lett.*, **11**, 219 (1991).

[14] Qin, D., and Lapszewicz, J., *Catal. Today*, **21**, 551 (1994).

[15] Solymosi, F., Kustan, Gy., and Erohelyi, A., *Catal. Lett.*, **11**, 149 (1991).

[16] Erohelyi, A., Fodor, K., and Solymosi, F., *Stud. Surf. Sci. Catal.*, **107**, 525 (1997).

[17] Bitter, J.H., Hally, W., Seshan, K., van Ommen, J.G., and Lercher, J.A., *Catal. Today*, **29**, 349 (1996).

[18] Bitter, J.H., Seshan, K., and Lercher, J.A., *J. Catal.*, **176**, 93 (1998).

[19] O'Connor, A.M., and Ross, J.R.H., *Catal. Today*, **46**, 203 (1998).

[20] Bitter, J.H., Seshan, K., and Lercher, J.A., *J. Catal.*, **183**, 336 (1999).

[21] Souza, M.M.V.M., Aranda, D.A.G., and Schmal, M., *Ind. Engn. Chem. Res.*, **41**, 4681(2002).

[22] Chen, Y.Z., Liaw, B.J., and Lai, W.H., *Appl. Catal. A: Gen.*, **230**, 73 (2002).

[23] Stagg, S.M., and Resaco, D.E., *Stud. Surf. Sci. Catal.*, **111**, 543 (1997).

[24] Stagg-Williams, S.M., Noronha, F.B., Fendley, G., and Resasco, D.E., *J. Catal.*, **194**, 240 (2000).

[25] Gadalla, A.M., and Bower, B., *Chem. Engng Sci.*, **43**, 3049 (1988).

[26] Tsipouriari, V.A., Zhang, Z., and Verykios, X.E., *J. Catal.*, 179, 283 (1998).

[27] Kim, J.H., Suh, D.J., Park, T.J., and Kim, K.L., *Appl. Catal. A: Gen.*, 197, 191 (2000).

[28] Bhattacharyya, A., and Chang, V.W., *Stud. Surf. Sci. Catal.*, **88**, 207 (1994).

[29] Halliche, D., Bouarab, R., Cherifi, O., and Bettahar, M.M., *Catal. Today*, **29**, 373 (1996).

[30] Wang, S., and Lu, G.O.M., *Appl. Catal. A: Environ.*, **16**, 269 (1998).

[31] Yan, Z.F., Ding, R.G., Song, L.H., and Qian, L., *Energy Fuels*, **12**, 1114 (1998).

[32] Wang, S., and Lu, G.O.M., *Appl. Catal. A: Gen.*, **169**, 271 (1998).

[33] Wang, S., and Lu, G.O.M., *Ind. Engng Chem. Res.*, **38**, 2615 (1999).

[34] Ito, M., Tagawa, T., and Goto, S., *Appl. Catal. A: Gen.*, **177**, 15 (1999).

[35] Xu, Z., Li, Y.M., Zhang, J.Y., Chang, L., Zhou, R.Q., and Duan, Z.T., *Appl. Catal. A: Gen.*, **210**, 45 (2001).

[36] Wang, S., and Lu, G.O.M., *Energy Fuels*, **12**, 1235 (1998).

[37] Chen, Y.G., and Ren, J., *Catal. Lett.*, **29**, 39 (1994).

[38] Sridhar, S., Sichen, D., and Seetharaman, S.Z., *Metallkd.*, **85**, 9 (1994).

[39] Kim, J.H., Suh, D.J., Park, T.J., and Kim, K.L., *Natural Conversion V. Stud. Surf. Sci. Catal.*, **119**, 771 (1998).

[40] Osaki, T., Horiuchi, T., Sugiyama, T., Suzuki, K., and Mori, T., *Catal. Lett.*, **52**, 171 (1998).

[41] Hayashi, H., Murata, S., Tago, T., Kishida, M., and Wakabayashi, K., *Sekiyu Gakkaishi – J. Jpn Petrol. Inst.*, **44**, 334 (2001).

[42] Moris, T., Osaki, T., Horiuchi, T., Sugiyama, T., and Suzuki, K., Catalyst Deactivation 1999, *Stud. Surf. Sci. Catal.*, 365 (1999).

[43] Lemonidou, A.A., Goula, M.A., and Vasalos, I.A., *Catal. Today*, **46**, 175 (1998).

[44] Osaki, T., and Mori, T., *J. Catal.*, **204**, 89 (2001).

[45] Choi, J.S., Moon, K.I., Kim, Y.G., Lee, J.S., Kim, C.H., and Trimm, D.L., *Catal. Lett.*, **52**, 43 (1998).

[46] Seok, S.H., Han, S.H., and Lee, J.S., *Appl. Catal. A: Gen.*, **215**, 31 (2001).

[47] Quincoces, C.E., de Vargas, S.P., Grange, P., and Gonzalez, M.G., *Mater. Lett.*, **56**, 698 (2002).

[48] Li, W.Y., Feng, J., and Xie, K.C., *React. Kinet. Catal. Lett.*, **64**, 381 (1998).

[49] Wang, S.B., and Lu, G.Q., *Appl. Catal. B: Environ.*, **19**, 267 (1998)

[50] Xu, G.L., Shi, K.Y., Gao, Y., Xu, H.Y., and Wei, Y.D., *J. Mol. Catal. A: Chem.*, **147**, 47 (1999).

Role of Support in CO_2 Reforming of CH_4 to Syngas over Ni Catalysts

Eli Ruckenstein[1] and Yun Hang Hu[2]

Department of Chemical Engineering, State University of New York at Buffalo, Amherst, New York 14260

Received November 29, 1995; revised April 5, 1996

The dynamic changes in activity and selectivity for the CO_2 reforming of CH_4 over unsupported and supported Ni catalysts were investigated. The reduced NiO has low initial conversions of CH_4 and CO_2, which decrease rapidly with time. The mixture of powders of NiO and Al_2O_3 has low initial conversions of CH_4 and CO_2 and a low yield to CO, which, however, increase with reaction time, due most likely to Ni migration to Al_2O_3. In contrast, the reduced 1 wt% Ni/Al_2O_3 catalyst (reduced from NiO/Al_2O_3) has high initial conversions and CO yield, which slowly decrease with time due to carbon deposition. The reduced 13.6 wt% Ni/Al_2O_3 catalyst has high initial conversions and CO yield, but, because of carbon deposition, the reactor was completely plugged in 6 h. The reduced 1 wt% Ni/SiO_2 catalyst has moderate initial activity and CO yield, which decay fast with reaction time. The reduced 13.6 wt% Ni/SiO_2 has a high initial activity and CO yield, which decay slowly with reaction time. The reduced 13.6 wt% Ni/TiO_2 has a moderate initial activity and CO yield, which decay with reaction time. Carbon deposition was found to occur over all these catalysts, and the sequence of carbon deposition was: 13.6 wt% Ni/Al_2O_3 > 13.6 wt% Ni/SiO_2 > the reduced mixture of NiO and Al_2O_3(13.6 wt% Ni) > 13.6 wt% Ni/TiO_2 > 1 wt% Ni/Al_2O_3 > 1 wt% Ni/SiO_2.

1. INTRODUCTION

Syngas constitutes a very important feed in petrochemistry. While the dominant commercial method employed to produce synthesis gas is the steam reforming of methane,

$$CH_4 + H_2O \rightarrow CO + 3H_2, \quad \Delta H = 225.4 \text{ kJ/mol}, \quad [1]$$

this process has poor selectivity for CO and a too high H_2/CO product ratio, unsuitable for the methanol and the Fischer–Tropsch syntheses. Two other processes, namely, the catalytic partial oxidation of methane (1–7) and CO_2 reforming (8–14) can be employed to produce syngas from methane:

$$CH_4 + 1/2O_2 \rightarrow CO + 2H_2, \quad \Delta H = -38 \text{ kJ/mol}. \quad [2]$$

$$CH_4 + CO_2 \rightarrow 2CO + 2H_2, \quad \Delta H = 247 \text{ kJ/mol}. \quad [3]$$

Reaction [3] provides a high CO selectivity and a more suitable H_2/CO ratio. In the CO_2 reforming of CH_4, the carbon deposition sequence for various metals, reported by Rostrup–Nielsen (15), is: Ni ≫ Pt > Ru. Ashcroft *et al.* (8) and Rostrup–Nielsen (16) demonstrated that, over nickel-supported catalysts and for a CO_2/CH_4 molar ratio of 1 : 1, carbon deposition cannot be avoided. The effect of the support on carbon deposition was also investigated by Tang *et al.* (17). Ashcroft *et al.* (8) suppressed the carbon deposition in CO_2 reforming by replacing Ni with platinum-group catalysts; this provided a CO yield between 69% for Ru and 89% for Ir from a stoichiometric feed of CO_2 and CH_4. It is, however, worthwhile to develop improved stable and effective nickel-based catalysts because of the high cost of the noble metals. Rostrup–Nielsen (18) noted that carbon-free steam reforming of methane can be achieved using a partly sulfur-passivated nickel catalyst. However, although sulfur inhibits carbon formation, it also reduces the activity of the catalyst. Recently, we found that the reduced NiO/MgO catalysts have high activities and selectivities, as well as excellent stability (19, 20). In those cases the support (MgO), with which NiO forms a solid solution, inhibits carbon deposition.

In this article, we present a number of observations regarding the dynamic changes in activity and CO yield, during the CO_2 reforming of methane over Ni unsupported and supported catalysts with the emphasis on the effect of the nature of the support.

2. EXPERIMENT

2.1. Catalyst Preparation

NiO catalyst preparation. NiO was prepared by the decomposition of Ni nitrate in air at 800°C for 1.5 h.

NiO-supported catalyst preparation. NiO/Al_2O_3 and NiO/SiO_2 (containing after reduction 1 wt% Ni) and NiO/Al_2O_3, NiO/SiO_2, and NiO/TiO_2 (containing after reduction 13.6 wt% Ni) catalysts were prepared by impregnating the support with an aqueous solution of nickel nitrate. The obtained paste was dried at room temperature in air, then decomposed and calcined at 800°C in air for 1.5 h.

[1] To whom correspondence should be addressed.

[2] Permanent address: Department of Chemistry, Xiamen University, Xiamen 361005, People's Republic of China.

This paper, which was originally published in **Journal of Catalysis**: 162, 230–238 (1996), is republished with permission.

2.2. *Catalytic Reaction*

The catalytic reaction was carried out under atmospheric pressure, at 790°C, in a flow system, using a vertical quartz tube (2 mm inside diameter) as reactor. The catalyst powder (weight: 0.02 g) was held on quartz wool. The molar ratio of the reactant gases, with a GHSV = 60000 cm^3/g/h, was $CO_2/CH_4 = 1:1$. The catalyst was reduced in H_2 (20 ml/min) at 500°C for 14 h, followed by an increase in temperature to 790°C at a rate of 20°C/min, and also in H_2 (20 ml/min). The analysis of the reactants/products mixtures was performed, with an *in situ* gas chromatograph equipped with a Porapak Q column. The yield to CO was calculated from the expression: $[CO/(CO + CO_2 + CH_4)] \times 100\%$.

2.3. *CO Chemisorption and Temperature-Programmed Desorption (TPD)*

The exposed Ni metal surface area and hence the dispersion of the catalyst were measured by CO pulse adsorption at room temperature, assuming a stoichiometry of 1/1. The catalyst (0.25 g), held on quartz wool in a vertical quartz tube (6 mm inside diameter), was reduced in H_2 (40 ml/min) at 500°C for 3 h, followed by an increase in temperature to 790°C at a rate of 20°C/min, also in H_2 (40 ml/min). At 790°C, the reduced catalyst was purged with He (40 ml/min) for 20 min. Then, the catalyst was allowed to cool at room temperature. The analysis of gases during the adsorption of CO at room temperature was carried out with an on-line mass spectrometer (HP Quadrupole, 5971 series mass selective detector) equipped with a fast-response inlet capillary system. For the mechanical mixture of NiO and Al_2O_3, the dispersion was measured both before and after the CO_2 reforming reaction of CH_4.

After the CO adsorption, a TPD experiment was carried out at a rate of 10°C/min. The analysis of gases during the desorption of CO was performed continuously with the on-line mass spectrometer.

2.4. *BET Surface Area and Pore Size Distribution Measurements*

The surface area and the pore-size distribution were determined by nitrogen adsorption, using a Micromeritics ASAP 2000 instrument. The surface area was calculated using the BET method, while the pore-size distribution curve was obtained from the adsorption branch of the N_2 isotherm by the BJH method (21). The sample was degassed at 200°C in high vacuum before measurements.

2.5. *The Determination of Carbon Deposition*

The deposited carbon was determined by O_2 pulse–MS. The catalyst (0.02 g) was reduced in H_2 and the reaction carried out for 6 h, both as described in Section 2.2. Then, the catalyst was purged with He at 790°C for 1 h. The analysis of gases during the O_2 pulses at 790°C was carried out with an on-line mass spectrometer. The amount of carbon deposited was calculated from the amount of CO and CO_2 produced.

3. RESULTS

3.1. *CO_2 Reforming Reaction of CH_4 over Unsupported NiO Catalysts*

Figure 1 shows that the reduced NiO provides very low initial conversions and CO yield, which decrease rapidly with time. The unreduced NiO has high initial conversion and CO yield, which decrease catastrophically with time.

The reduced mixture of powders of NiO and γ-alumina (0.2 wt ratio NiO/Al_2O_3) has very low initial conversions of CH_4 (15%) and CO_2 (19%) and a low yield to CO (17%) (Fig. 2). As the reaction time increases, the conversions and yield increase. After 10 h, they reach maximum values (about 45, 42, and 40%), which do not change at longer times (Fig. 2).

In contrast, the unreduced NiO catalyst has higher initial conversions of CH_4 and CO_2 yield to CO, which first rapidly decrease and then increase somewhat with time (Fig. 2).

3.2. *CO_2 Reforming of CH_4 over 1 wt% Ni/Al_2O_3 and 1 wt% Ni/SiO_2 Catalysts*

Figure 3 shows that the reduced 1 wt% Ni/Al_2O_3 catalyst has high initial conversions of CH_4 and CO_2 (78 and 90%) and yield to CO (84%), which decrease slowly as the reaction time increases. After about 7 h, the conversions of CH_4 and CO_2 and the yield of CO become 73, 90, and 81%, respectively; after about 30 h, they decrease to 60, 75, and 70%, respectively. Compared with the 1 wt% Ni/Al_2O_3, the reduced 1 wt% Ni/SiO_2 has lower initial conversions of CH_4 and CO_2 (52 and 72%) and yield to CO (62%). The conversions and the yield to CO decrease with time. After 7 h, they become 35, 40, 37%, respectively. As shown in Fig. 3, the deactivation of the 1 wt% Ni/SiO_2 is faster than that of the 1 wt% Ni/Al_2O_3.

The unreduced 1 wt% Ni/Al_2O_3 and 1 wt% Ni/SiO_2 catalysts provide initial conversions of CH_4 and CO_2 and yield to CO of about 10%, which increase with time, the increase being faster for the 1 wt% Ni/Al_2O_3 catalyst than for the 1 wt% Ni/SiO_2 catalyst (Fig. 4).

3.3. *CO_2 Reforming of CH_4 over Reduced 13.6 wt% Ni Supported Catalysts*

The 13.6% Ni/Al_2O_3 reduced catalyst has very high initial conversions of CH_4 and CO_2 (91 and 98%) and yield to CO (95%) (Table 1). It was, however, impossible to obtain the relationship between the CO yield and reaction time due to the difficulty of maintaining constant feed flow. This happened because the pressure drop in reactor increased, due

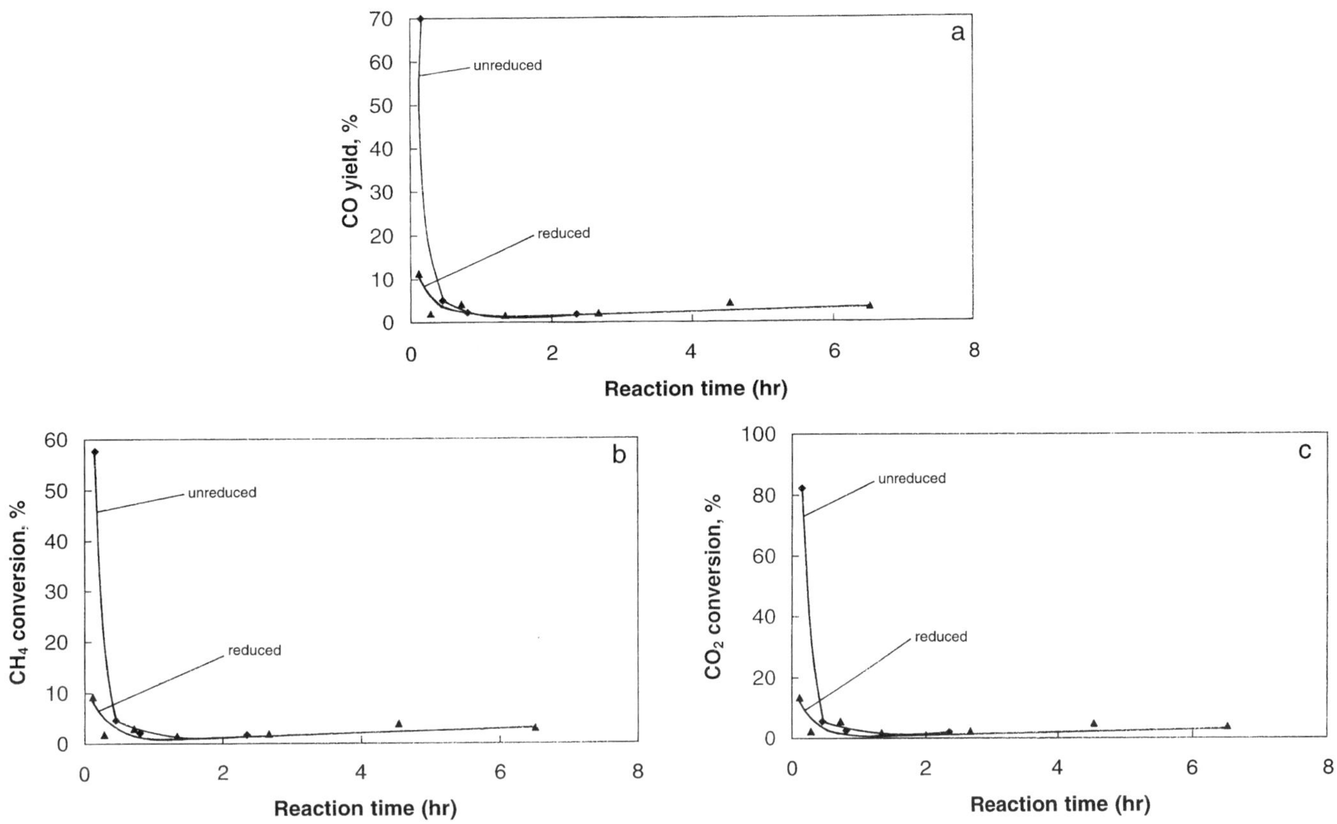

FIG. 1. The relationship between reaction time and (a) CO yield, (b) CH_4 conversion, and (c) CO_2 conversion over NiO. Reaction conditions: total pressure, 1 atm; $T = 790°C$; $CH_4/CO_2 = 1:1$; GHSV = 60000 $cm^3/g/h$.

to the fast carbon deposition, with reaction time. After 6 h, the reactor was completely plugged. In contrast, this did not occur over the reduced 13.6 wt% Ni/SiO_2 or Ni/TiO_2. The Ni/SiO_2 provides very high initial conversions of CH_4 and CO_2 (89 and 97%) and yield to CO (93%) (Table 1), which decrease slowly with reaction time (Fig. 5). The Ni/TiO_2 catalyst has relatively low initial conversions of CH_4 and CO_2 (41 and 67%) and yield to CO (53%) (Table 1), which decrease with reaction time faster than for the Ni/SiO_2 catalyst (Fig. 5).

3.4. BET Surface Area and Pore Size Distribution

The surface areas of the unreduced catalysts are listed in Table 1. The 1 wt% and 13.6 wt% Ni/SiO_2 have the high surface areas of 447 and 362 m^2/g, respectively, the 1 and 13.6 wt% Ni/Al_2O_3 catalysts have the intermediate surface areas of 75 and 63 m^2/g, respectively, and the 13.6 wt% NiO/TiO_2 and pure NiO catalysts have the very low surface areas of 6 and 3 m^2/g, respectively. The pore size varied for 13.6 wt% Ni/SiO_2 between 20 and 150 Å with an average of 52 Å and for 13.6 wt% Ni/Al_2O_3 between 20 and 1000 Å with an average of 130 Å. NiO and 13.6 wt% Ni/TiO_2 have very few pores (Fig. 6).

3.5. CO Turnover Number and Amount of Carbon Deposited

The initial CO turnover numbers (TON) are listed in Table 1. The 1 wt% Ni/SiO_2 has the very high initial turnover number of 1721 s^{-1}. (Because the amount of CO chemisorped is very small, there is, however, a large error in the determination of the number of active sites.) The sequence of TON is: 1 wt% $Ni/SiO_2 >$ 1 wt% $Ni/Al_2O_3 >$ 13.6 wt% $Ni/TiO_2 >$ 13.6 wt% $Ni/SiO_2 >$ 13.6 wt% Ni/Al_2O_3.

Both the 1 wt% Ni/Al_2O_3 and Ni/SiO_2 catalysts exhibit low carbon depositions (Table 1). For 13.6 wt% Ni-based catalysts, the sequence of carbon deposition is: $Ni/Al_2O_3 > Ni/SiO_2 > Ni/TiO_2$ (Table 1).

3.6. CO TPD on Reduced Ni Unsupported and Supported Catalysts

Figure 7 shows that the reduced mixture of NiO and Al_2O_3 (0.2 wt ratio NiO/Al_2O_3) has a sharp CO peak at 80°C and two wide CO overlapped peaks around 300°C. At about 420°C, a CO_2 peak appears. The reduced 1 wt% Ni/Al_2O_3 catalyst has a similar TPD curve: one sharp CO peak at 80°C and two wide CO overlapped peaks at a temperature over 150°C; no CO_2 peak was, however, detected

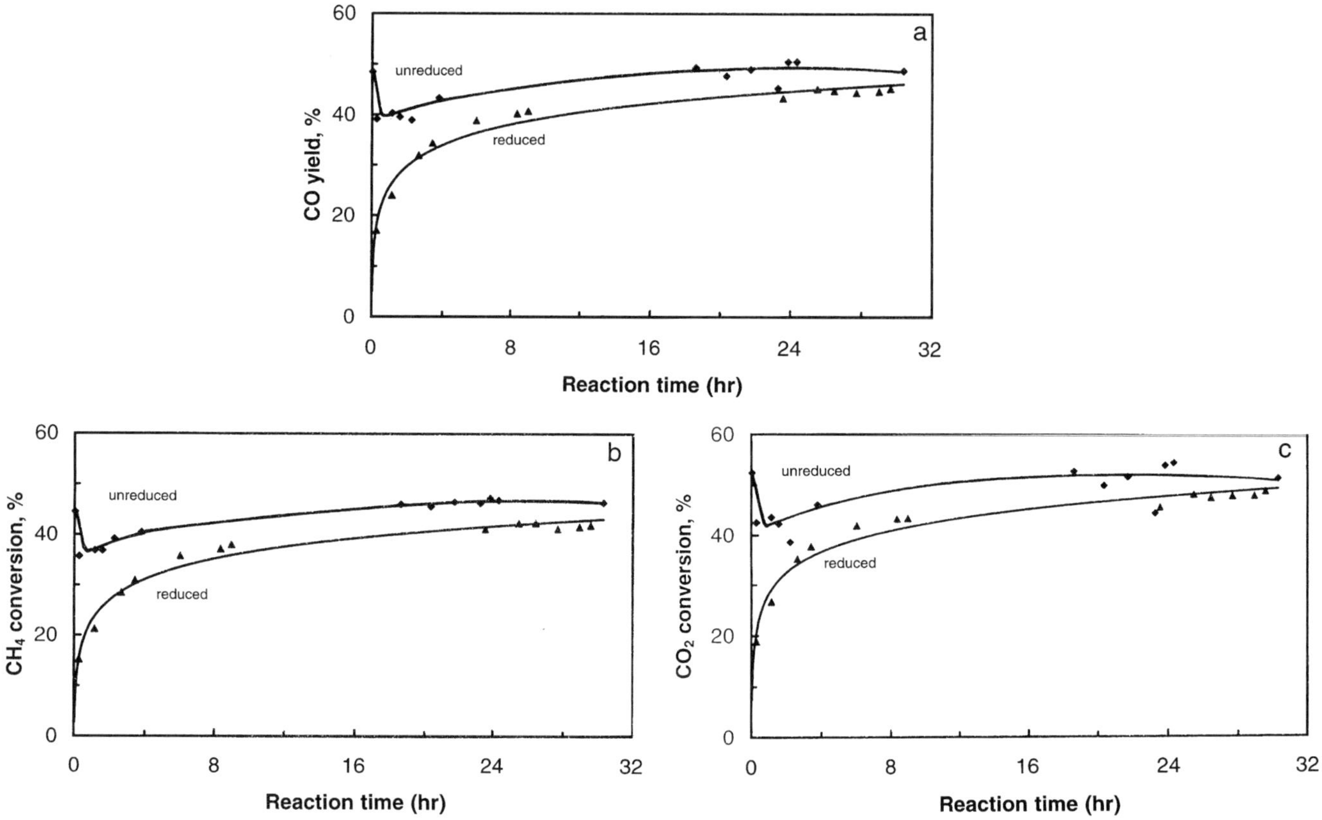

FIG. 2. The relationship between reaction time and (a) CO yield, (b) CH_4 conversion, and (c) CO_2 conversion over NiO catalyst diluted with Al_2O_3 via mechanical mixing (0.2 wt ratio NiO/Al_2O_3). Reaction conditions: total pressure, 1 atm; $T = 790°C$; $CH_4/CO_2 = 1:1$; GHSV = 60000 $cm^3/g/h$.

(Fig. 7). In contrast, very small amounts of CO and CO_2 were detected for the 1 wt% Ni/SiO_2 catalyst (Fig. 7). The TPD results also show that the curves for CO desorption from 13.6 wt% Ni/Al_2O_3 and 13.6 wt% Ni/SiO_2 (Figs. 8 and 9) are very different from those for 1 wt% Ni/Al_2O_3 and 1 wt% Ni/SiO_2. There are five overlapping CO peaks and four overlapping CO_2 peaks in the TPD curves for the 13.6 wt% Ni/Al_2O_3. This indicates that the increase in Ni content results in the increase of the CO disproportionation. The 13.6 wt% Ni/SiO_2 catalyst has a TPD curve similar to that of Ni/Al_2O_3. The TPD curve of the 13.6 wt% Ni/TiO_2 catalyst has two CO peaks at 98 and 200°C, respectively, and one CO_2 peak at 240°C (Fig. 10).

4. DISCUSSION

4.1. Effect of Al_2O_3 in Mechanical Mixture with NiO on CO_2 Reforming of CH_4

For the reduced mixture of NiO and Al_2O_3, the initial conversions of CH_4 and CO_2 and yield to CO are very low; however, they increase fast with reaction time, attain maximum values, and then remain constant (Fig. 1). The number of Ni sites increased after the reforming reaction. Before reaction, but after reduction, the dispersion was 0.065%, while after reaction it was 0.18%. The increase in activity with reaction time is obviously due to the increase in the number of active sites. Two explanations are possible: either new sites, which consist of Ni and C, are generated during reaction (22), or Ni atoms are migrating to the surface of Al_2O_3 particles. Because the experiments carried out with reduced NiO particles alone (Fig. 1) show that the activity decreases with time, the second explanation is the more likely one. It is possible that the carbon deposited on Ni during reaction stimulates the migration of Ni atoms to the Al_2O_3 surface. Indeed, the deposited carbon has the tendency to diffuse into the bulk of NiO (23), thus weakening the bonds among the Ni atoms and making the latter more mobile.

For both the unreduced pure NiO and the unreduced mixture of NiO and Al_2O_3, the initial conversions of CH_4 and CO_2 and yield to CO are high; they first decrease fast and then slowly with time (Figs. 1 and 2). This can be explained as follows: The reduction of NiO by CH_4 provides Ni^0 active sites for the reforming process. For this reason, the initial activity is high. The rapid sintering decreases, however, the number of active sites, and hence the activity decreases with time. A more rapid sintering occurs at

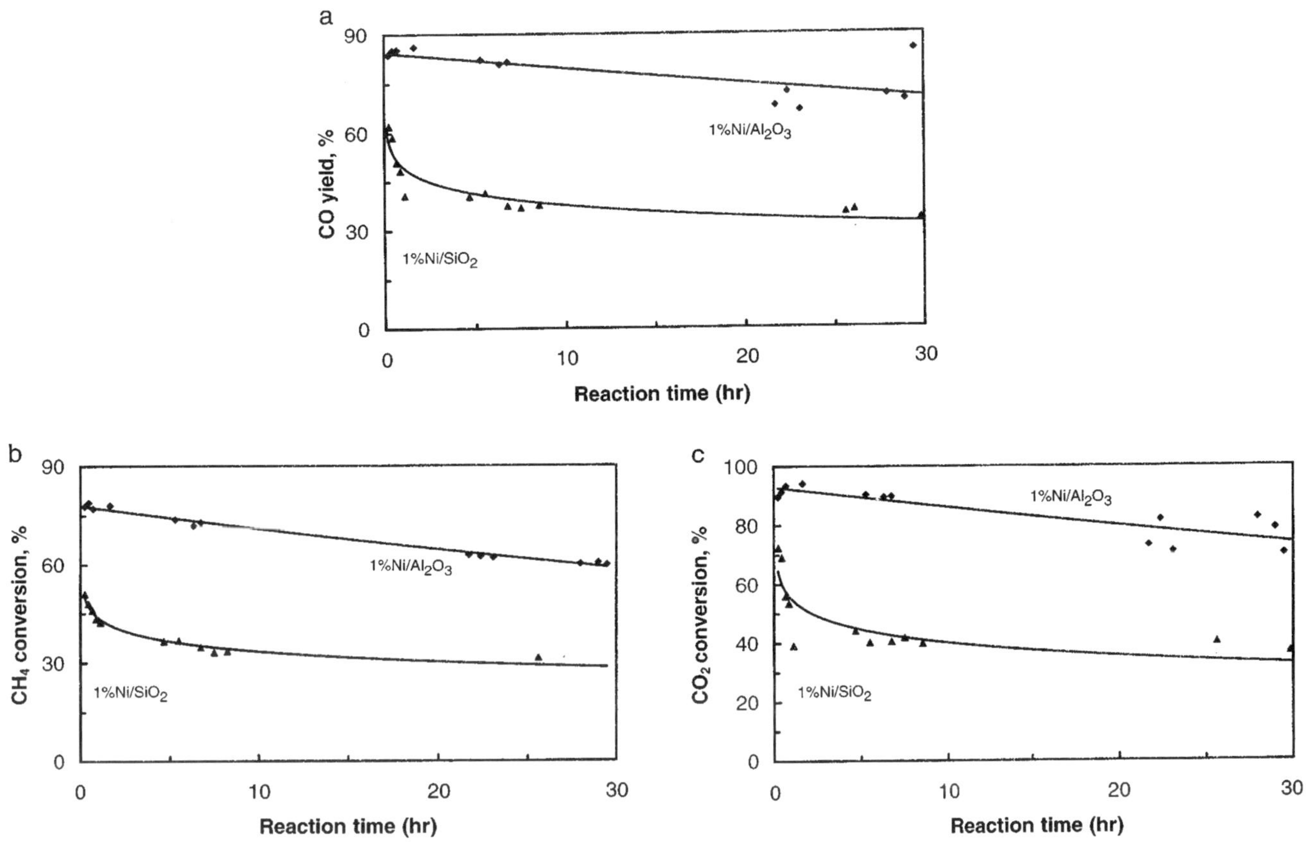

FIG. 3. The relationship between reaction time and (a) CO yields, (b) CH_4 conversion, and (c) CO_2 conversion over reduced 1% Ni/Al_2O_3 and 1% Ni/SiO_2 catalysts. Reaction conditions: total pressure, 1 atm; $T = 790°C$; $CH_4/CO_2 = 1:1$; GHSV = 60000 $cm^3/g/h$.

shorter times because then the Ni particles are smaller; this explains the rapid decay of activity at shorter times.

4.2. *The Interactions between NiO and Al_2O_3 or SiO_2 Support*

In contrast to the mechanical mixture of NiO and Al_2O_3, the unreduced 1 wt% Ni/Al_2O_3 catalyst has low initial activity, which increases slowly with reaction time (Fig. 4). This occurs because, during reaction, the unreduced 1 wt% Ni/Al_2O_3 is reduced by CH_4, and the number of Ni^0 active sites increases. Comparing Figs. 1, 2, and 4, one can conclude that the alumina-supported NiO catalyst is reduced during reaction with greater difficulty than the unsupported one. This can be attributed to the formation of a nickel aluminate (24).

While the initial activities are comparable, the increase of activity is much slower over the unreduced 1 wt% Ni/SiO_2 catalyst than over the unreduced 1 wt% Ni/Al_2O_3 (Fig. 4). This indicates that the reduction of NiO is more difficult over 1 wt% Ni/SiO_2 than over 1 wt% Ni/Al_2O_3. Although the BET measurements show that the unreduced 1 wt% Ni/SiO_2 has a surface area (447 m^2/g) much larger than the unreduced 1 wt% Ni/Al_2O_3 (71 m^2/g), the CO chemisorption indicates that the reduced 1 wt% Ni/SiO_2 has a much smaller Ni metal surface area (0.01 m^2/g) than the reduced 1 wt% Ni/Al_2O_3 (0.11 m^2/g) (Table 1). This suggests that chemical interactions occur between NiO and SiO_2 in the NiO/SiO_2 (containing 1 wt% Ni) calcined at high temperatures. de Lange and Vissar (25) suggested as early as 1946 that chemical interactions between silica support and NiO, with the formation of silicate, is the reason for the difficult reducibility of silica-supported NiO catalysts. This was later supported by several investigations (26–28). The small number of active sites in 1 wt% Ni/SiO_2 catalyst can be explained as follows: SiO_2 has a very high surface area over which a small amount of NiO is dispersed. As a result, many of the NiO molecules are in direct contact with SiO_2. The high temperature (800°C) employed during the catalyst preparation leads to the reaction between the two, with the generation of a nickel silicate, which is difficult to reduce. This hinders the reduction of Ni^{2+} to Ni^0. When the NiO content increases, large NiO particles are formed, and the fraction of NiO that does not contact SiO_2 can be reduced easily to Ni. Indeed, the Ni metal surface area of the 13.6 wt% Ni/SiO_2 is much greater than that of 1 wt% Ni/SiO_2 (Table 1).

Table 1 shows that the reduced 1 wt% Ni/SiO_2 has a very high turnover number (1721 s^{-1}), much larger than that

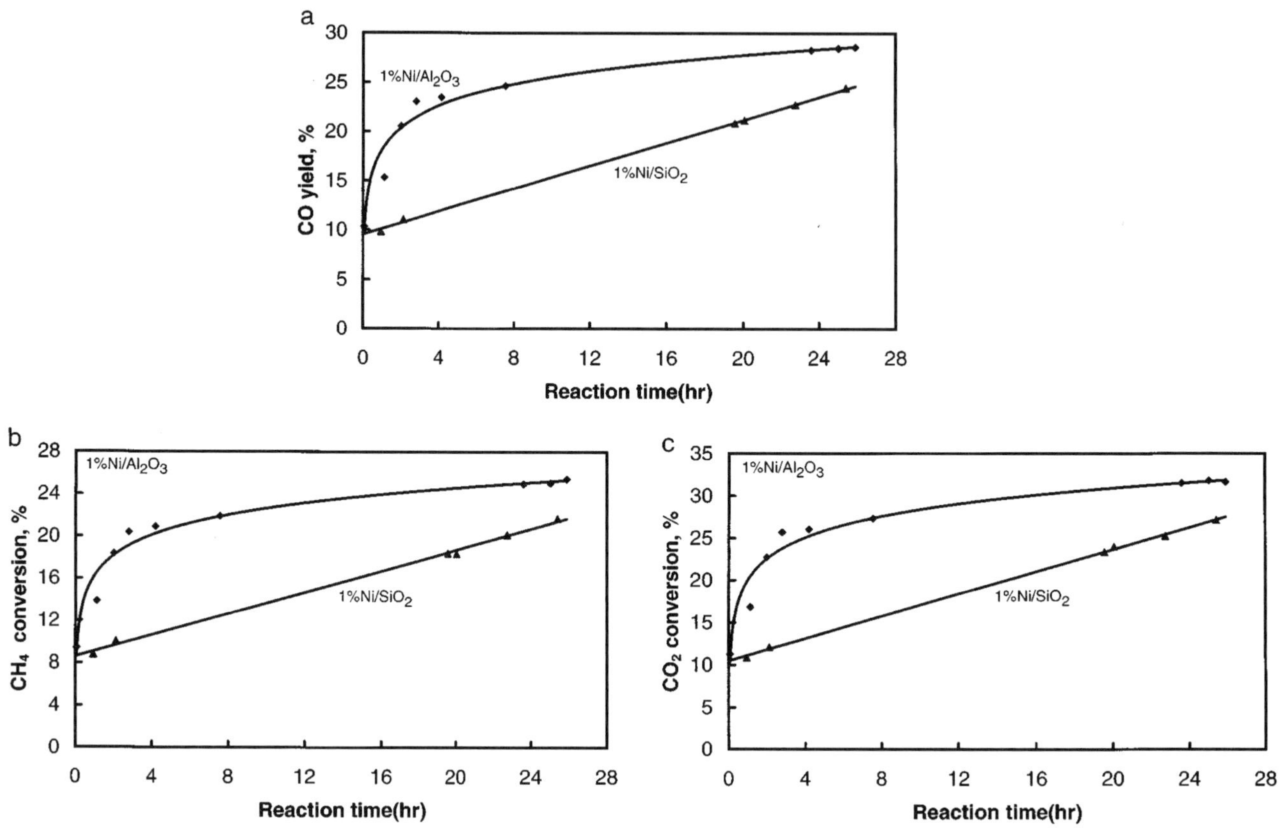

FIG. 4. The relationship between reaction time and (a) CO yield, (b) CH_4 conversion, and (c) CO_2 conversion over unreduced 1% Ni/Al_2O_3 and 1% Ni/SiO_2 catalysts. Reaction conditions: total pressure, 1 atm; $T = 790°C$; $CH_4/CO_2 = 1:1$; GHSV = 60000 $cm^3/g/h$.

of 1 wt% Ni/Al_2O_3 (230 s^{-1}). Because the reduced 1 wt% Ni/SiO_2 has a very small number of active sites with very high turnover number, a small decrease in their number during reaction, due to carbon deposition or sintering, results in large decrease in activity. This explains why the deactivation is faster over the reduced 1 wt% Ni/SiO_2 than over the reduced 1 wt% Ni/Al_2O_3 catalyst, although the carbon deposition is greater over 1 wt% Ni/Al_2O_3 than over 1 wt% Ni/SiO_2.

4.3. *Effect of Ni Content on the Activity of Ni-Supported Catalysts for CO_2 Reforming of CH_4*

The reduced 1 wt% Ni/Al_2O_3 catalyst has high initial conversions of CH_4 and CO_2 and yield to CO, which decrease

TABLE 1

The Surface Area, Carbon Deposition, CO Yield, and CO TON of Various Catalysts

Catalyst	BET surface area (m^2/g)	Average pore diameter (Å)	Ni surface area (m^2/g)	CH_4 initial conver. (%)	CO_2 initial conver. (%)	CO initial yield[a] (%)	CO TON[b] (s^{-1})	C[c] (g/gcat)
$NiO + Al_2O_3$[d]	—	—	0.06	15	18	17	83	3.3×10^{-3}
1% Ni/SiO_2	447	53	0.01[e]	52	72	62	1721[e]	2.3×10^{-4}
1% Ni/Al_2O_3	75	130	0.11	78	90	84	230	3.9×10^{-4}
13.6% Ni/Al_2O_3	63	129	1.76	91	98	95	15	[f]
13.6% Ni/SiO_2	362	52	1.57	89	97	93	18	1.9×10^{-1}
13.6% Ni/TiO_2	8	—	0.19	41	67	53	80	8.0×10^{-4}

[a] The initial yield for GHSV = 60,000 $cm^3/g/h$, $T = 790°C$, $CH_4/CO_2 = 1:1$.
[b] CO turnover number.
[c] Amount of carbon deposited over the catalyst during CO_2 reforming of methane at 790°C for 6 h.
[d] The mixture of NiO and Al_2O_3 (containing 13.6 wt% Ni).
[e] The error is very large because the amount of CO chemisorbed is extremely small.
[f] The carbon deposition was so fast that the reactor was completely plugged in 6 h.

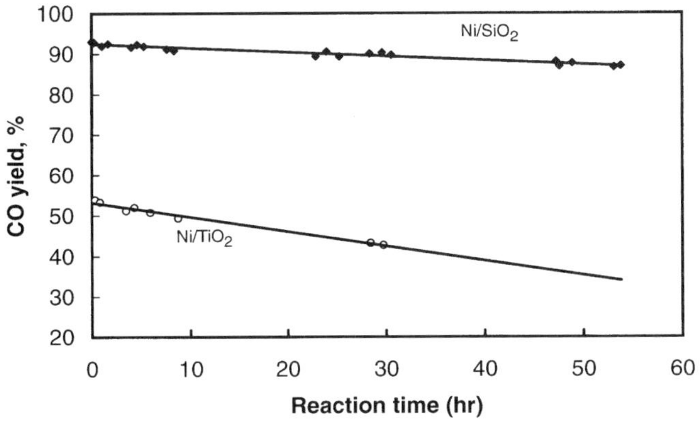

FIG. 5. The relationship between reaction time and CO yield over reduced 13.6% Ni/SiO_2 and Ni/TiO_2 catalysts. Reaction conditions: total pressure, 1 atm; $T = 790°C$; $CH_4/CO_2 = 1:1$; GHSV = 60000 $cm^3/g/h$.

slowly with reaction time (Fig. 3). The relatively small amount of carbon deposited during reaction (Table 1) may explain the slow decrease in activity with time. Instead, over the reduced 13.6 wt% Ni/Al_2O_3 catalyst, the carbon deposition was so fast that the reactor was plugged in 6 h. As shown by the TPD curves, the decomposition of CO to CO_2 is much higher over the reduced 13.6 wt% Ni/Al_2O_3 than over the reduced 1 wt% Ni/Al_2O_3 (Figs. 7 and 8). Therefore, the increase of CO disproportionation is partly responsible for the increase in carbon deposition with the increase of Ni content in Ni/Al_2O_3 catalysts. In addition, the increase of Ni content also increases the decomposition of CH_4, resulting in an increase in carbon deposition. As shown later, the latter mechanism is the main one at 790°C.

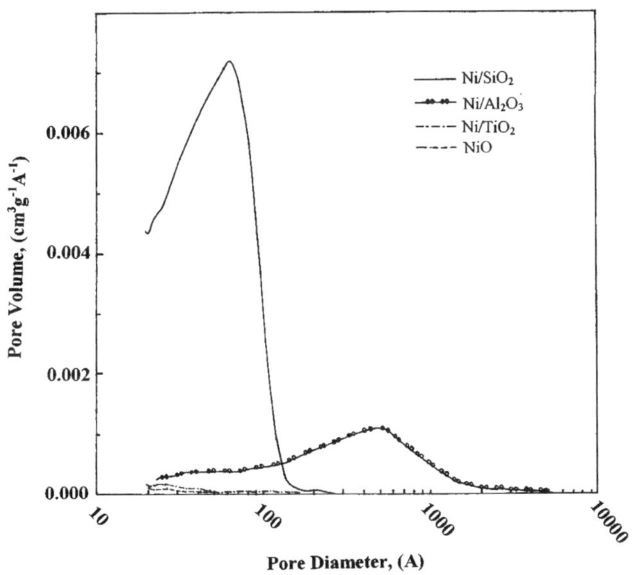

FIG. 6. The pore size distribution of 13.6 wt% Ni-based catalysts Ni/SiO_2, Ni/Al_2O_3, and Ni/TiO_2.

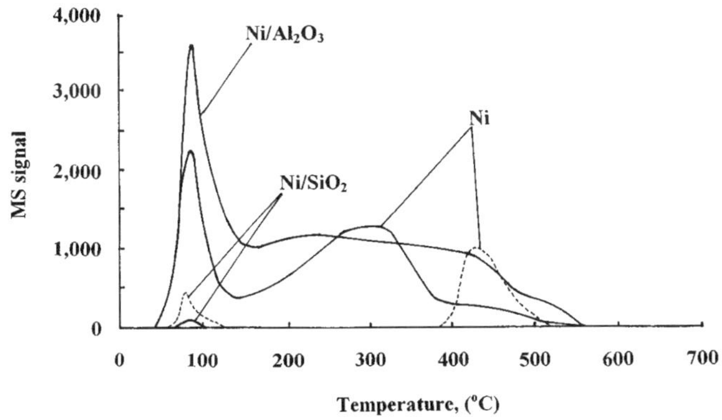

FIG. 7. Temperature-programmed desorption spectrum of CO over Ni from the reduced mixture of NiO and Al_2O_3, reduced 1 wt% Ni/Al_2O_3, and Ni/SiO_2 catalysts (—) CO, (- - -) CO_2.

Even though sintering occurs with some ease, the large amount of Ni present in the reduced 13.6 wt% Ni/SiO_2 provides a larger number of Ni sites than the 1 wt% Ni/SiO_2 catalyst. Indeed, both the CO chemisorption and TPD curve show that the 13.6 wt% Ni/SiO_2 adsorbs a much larger amount of CO than the 1 wt% Ni/SiO_2 (Figs. 7 and 9). Similar to the 13.6 wt% Ni/Al_2O_3 catalyst, the 13.6 wt% Ni/SiO_2 catalyst exhibits a much higher carbon deposition than the 1 wt% Ni/SiO_2 (Table 1). The activity of the 13.6 wt% Ni/SiO_2 catalyst does not decay, however, as fast as that of the 1 wt% Ni/SiO_2 catalyst. This happens probably because the initial number of sites is much larger for the 13.6 wt% catalyst, and the large amount of carbon deposited partially inhibits sintering.

Although the TPD CO curves show that the reduced 13.6 wt% Ni/Al_2O_3 and 13.6 wt% Ni/SiO_2 have comparable CO disproportionation abilties, the former catalyst exhibits at 790°C a much higher carbon deposition than the latter. This indicates that the CO dispropotionation does not constitute the main mechanism for carbon deposition on Ni-based catalysts at 790°C. This observation

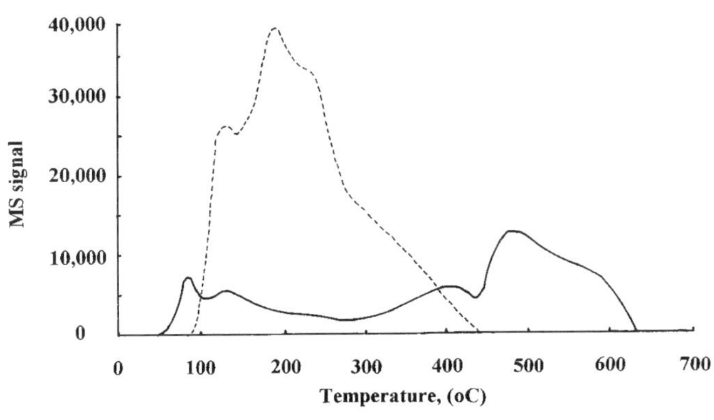

FIG. 8. Temperature-programmed desorption spectrum of CO over the reduced 13.6 wt% Ni/Al_2O_3 catalyst (—) CO, (- - -) CO_2.

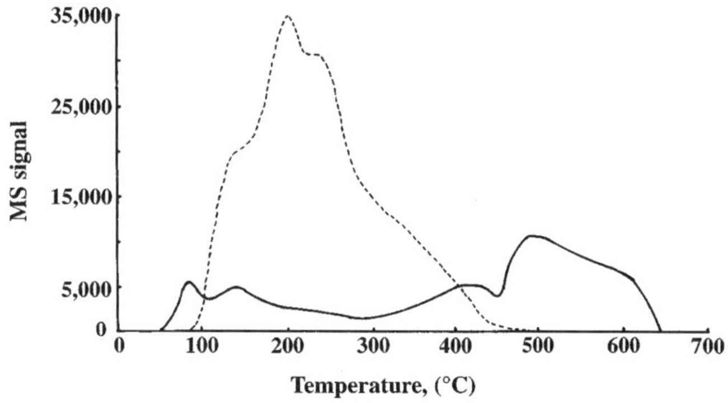

FIG. 9. Temperature-programmed desorption spectrum of CO over the reduced 13.6 wt% Ni/SiO_2 catalyst (—) CO, (- - -) CO_2.

is consistent with the thermodynamic conclusion of Claridge *et al.* (29), regarding the partial oxidation of methane, that, at low temperatures (<600°C), the CO disproportionation is the main mechanism of carbon deposition, while at high temperatures (>600°C) the CH_4 dissociation constitutes the main mechanism.

4.4. *Ni/TiO$_2$*

The CO yield is relatively small initially and decreases quickly with time over the reduced 13.6 wt% Ni/TiO_2 catalyst (Fig. 5). The initial yield is relatively small because, during reduction, TiO_x molecules migrate over the surface of Ni particles (30). This migration occurs because it leads to a decrease of the surface free energy of the system. This phenomenon was called "strong metal support interactions" (30), and, indeed, the strong interactions between TiO_x and Ni are responsible for the decrease of the free energy of the system (31). The decay of the yield with time is mostly due to carbon deposition. Indeed, carbon deposition was detected for this catalyst (Table 1).

4.5. *CO Turnover Number for the CO_2 Reforming of CH_4*

As shown in Table 1, the sequence of initial turnover numbers is 1 wt% Ni/SiO_2 > 1 wt% Ni/Al_2O_3 > the reduced mixture of NiO and Al_2O_3 (13.6 wt% Ni) > 13.6 wt% Ni/TiO_2 > 13.6 wt% Ni/SiO_2 ≈ 13.6 wt% Ni/Al_2O_3. The turnover number is much smaller for the catalysts containing a larger amount of Ni. Comparing the Ni/Al_2O_3 or Ni/SiO_2 catalysts for different Ni loadings, one can conclude that the reaction occurs more readily on the smaller crystallites, which are more likely to be present in the catalysts that contain smaller amounts of Ni. The crowding of Ni sites appears to affect negatively the CO turnover number for the CO_2 reforming of CH_4.

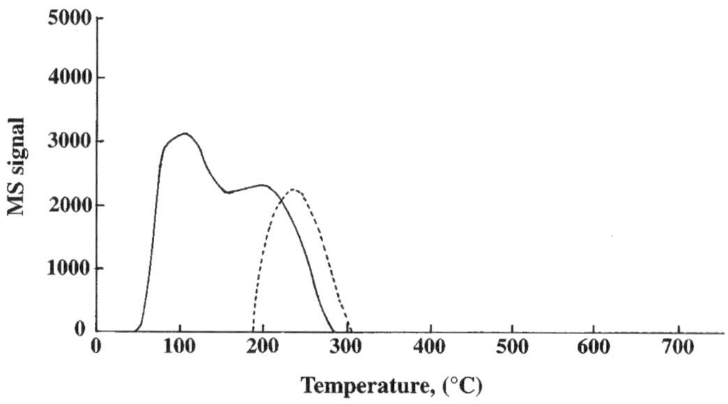

FIG. 10. Temperature-programmed desorption spectrum of CO over the reduced 13.6 wt% Ni/TiO_2 catalyst (—) CO, (- - -) CO_2.

5. CONCLUSION

The CO_2 reforming of CH_4 over Ni-supported catalysts is profoundly affected by the nature of the support. SiO_2, Al_2O_3, and TiO_2 have been used as supports and the activities and CO yields compared. For completeness, the unsupported NiO catalyst and the unsupported NiO catalyst mixed mechanically with Al_2O_3 were also investigated. While the reduced 1 wt% Ni/Al_2O_3 has high initial activity and selectivity, which decay slowly with reaction time, the reduced 13.6 wt% Ni/Al_2O_3 catalyst has high initial activity and selectivity, but the fast carbon deposition plugs the reactor in about 6 h. In contrast, 1 wt% Ni/SiO_2 has a moderate initial activity and selectivity, which decay rapidly with reaction time, while the 13.6 wt% Ni/SiO_2 has high initial activity and selectivity, which decay slowly with reaction time. The sequence of the initial CO yields of various supports (all containing 13.6 wt% Ni and prepared via the reduction of supported NiO) is: Ni/Al_2O_3 > Ni/SiO_2 > Ni/TiO_2. In most catalysts, the CO yield decreases with reaction time. However, interestingly, the activity and selectivity of the reduced NiO catalyst mixed mechanically with Al_2O_3 increase rapidly with time at the beginning and moderately later. The behavior of these catalysts is explained in terms of sintering, carbon deposition, metal–support interactions, and, in the case of NiO mixed mechanically with Al_2O_3, Ni migration to Al_2O_3.

REFERENCES

1. Hickman, D. A., and Schmidt, L. D., *Science* **259**, 343 (1993).
2. Ashcroft, A. T., Cheetham, A. K., Ford, J. S., Green, M. L. H., Grey C. P., Murrell, A. J., and Vernon, P. D. F., *Nature* **344**, 319 (1990).
3. Choudhary, V. R., Rajput, A. M., and Prabhakar, B., *J. Catal.* **139**, 326 (1993).
4. Hu, Y. H., Au, C. T., and Wan, H. L., *Chin. Sci. Bull.* **40**(4), 303 (1995).
5. Dissanayake, D., Rosynek, M. P., Kharas, K. C. C., and Lunsford, J. H., *J. Catal.* **132**, 117 (1991).
6. Dissanayake, D., Rosynek, M. P., and Lunsford, J. H., *J. Phys. Chem.* **97**, 3644 (1993).
7. Choudhary, V. R., Rajput, A. M., and Prabhakar, B., *Angew. Chem. Int. Ed. Engl.* **33**, 2104 (1994).
8. Ashcroft, A. T., Cheetham, A. K., Green, M. L. H., and Vernon, P. D. F., *Nature* **352**, 225 (1991).

9. Richardson, J. T., and Paripatyadar, S. A., *Appl. Catal.* **61**, 293 (1990).
10. Yamazaki, O., Nozaki, T., Omata, K., and Fujimoto, K., *Chem. Lett.* **10**, 1953 (1992).
11. Gadalla, A. M., and Sommer, M. E., *Chem. Eng. Sci.* **44**, 2825 (1989).
12. Gadalla, A. M., and Bower, B., *Chem. Eng. Sci.* **43**, 3049 (1988).
13. Rostrup-Nielsen, J. R., and Hansen, J-H.B., *J. Catal.* **144**, 38 (1993).
14. Kim, G. J., Cho, D. S., Kim, K. H., and Kim, J. H., *Catal. Lett.* **28**, 41 (1994).
15. Rostrup-Nielsen, J. R., *J. Catal.* **33**, 184 (1974).
16. Rostrup-Nielsen, J. R., *Stud. Surf. Sci. Catal.* **36**, 73 (1988).
17. Tang, S. B., Qiu, F. L., and Lu, S. J., *Catal. Today* **24**(3), 253 (1995).
18. Rostrup-Nielsen, J. R., *J. Catal.* **85**, 31 (1984).
19. Ruckenstein, E., and Hu, Y. H., *Appl. Catal.* **133**(1), 149 (1995).
20. Hu, Y. H., and Ruckenstein, E., *Catal. Lett.* **36**(3 & 4), 145 (1996).
21. Barret, E. P., Joyner, L. G., and Halenda, P. H., *J. Am. Chem. Soc.* **73**, 373 (1951).
22. Zhang, Z., and Verykios, X. E., *J. Chem. Soc., Chem. Commun.* **1**, 71 (1995).
23. Zdansky, E. O. F., Nilsson, A., and Martensson, N., *Surf. Sci.* **310**, L583 (1994).
24. Ruckenstein, E., and Lee, P. S., *J. Catal.* **86**, 457 (1984).
25. de Lange, J. J., and Visser, G. H., *Ingenieur* **58**, 24 (1964).
26. Mirodatos, C., Barthomeuf, D., Dalmon, J. A., and Garbowski, E. D., *Zeolites* **2**, 125 (1982).
27. Schuit, G. C. A., and van Reijen, L. L., *Adv. Catal.* **10**, 242 (1960).
28. Singley, W. J., and Carriel, J. T., *J. Am. Chem. Soc.* **75**, 778 (1953).
29. Claridge, J. B., Green, M. L. H., Tsang, S. C., York, A. P. E., Ashcroft, A. T., and Battle, P. D., *Catal. Lett.* **22**, 299 (1993).
30. Stevenson, S. A., Dumesic, J. A., Baker, R. T. K., and Ruckenstein, E., "Metal–Support Interactions in Catalysis, Sintering and Redispersion." Van Nostrand, New York, 1987.
31. Ruckenstein, E., *in* "Strong Metal–Support Interactions" (R. T. K. Baker, S. J. Tauster, and T. A. Dumesic, Eds.), p. 152. Am. Chem. Soc., Symposium Series 298, Washington, DC, 1986.

Carbon dioxide reforming of methane over nickel/alkaline earth metal oxide catalysts

Eli Ruckenstein *, Yun Hang Hu [1]

Department of Chemical Engineering, State University of New York at Buffalo, Amherst, NY 14260, USA

Received 2 June 1995; revised 27 July 1995; accepted 30 July 1995

Abstract

The CO_2 reforming of CH_4 over reduced NiO/alkaline earth metal oxide catalysts was investigated. A CO yield of 95% was obtained from a stoichiometric feed mixture of CH_4 and CO_2, at high GHSV (60 000 $cm^3\ g^{-1}\ h^{-1}$), over a reduced NiO/MgO with a weight ratio of 0.2. In addition, the catalyst had excellent stability, since the CO yield remained unchanged during 120 h. Compared to the reduced NiO/MgO catalyst, the reduced NiO/CaO, NiO/SrO and NiO/BaO catalysts had low CO yields and very low stabilities. The TPD of CO over the reduced NiO/MgO catalyst indicated a lower decomposition of CO to CO_2 than over the other catalysts investigated, hence that MgO inhibits the disproportionation reaction $2CO \rightarrow C + CO_2$ over Ni. The behavior of the reduced NiO/MgO catalyst is probably due to the formation of a NiO–MgO solution, as a result of the similar crystalline structures of NiO and MgO.

Keywords: Ni; Alkaline earth metals; Methane; Carbon dioxide; Reforming; Temperature-programmed desorption

1. Introduction

The conversion of CH_4 to the synthesis gas ($CO + H_2$), which constitutes the feed for the methanol and the Fischer–Tropsch syntheses, is very important from an industrial point of view. The dominant commercial method employed to produce synthesis gas is the steam reforming of methane [1,2],

$$CH_4 + H_2O \rightarrow CO + 3H_2 \quad \Delta H = 225.4\ \text{kJ/mol} \qquad (1)$$

This reaction has, however, several limitations: (i) a very high energy consumption, (ii) poor selectivity for CO and (iii) a high H_2/CO product ratio unsuitable for

* Corresponding author. Tel. (+1-716) 6452911/2214, fax. (+1-716) 6453822.

[1] Permanent address: Department of Chemistry, Xiamen University, Xiamen 361005, P.R. China.

This paper, which was originally published in **Applied Catalysis A**: General 133, 149–161 (1995), is republished with permission.

methanol and Fischer–Tropsch syntheses. In recent years, many researchers have attempted to convert methane to syngas via the catalytic partial oxidation of methane [3–9],

$$CH_4 + \tfrac{1}{2}O_2 \rightarrow CO + 2H_2 \quad \Delta H = -22.6\ \mathrm{kJ/mol} \tag{2}$$

Although this conversion is only mildly exothermic, a small decrease in the CO selectivity due to the total combustion of methane, which is a highly exothermic reaction, results in a large increase in the reaction temperature. In addition, a high methane conversion coupled with a high space velocity produces a large amount of heat in a small region of the catalyst. Because it is difficult to remove this heat from the reactor, particularly from a large-scale equipment, the process becomes very difficult to control [10,11].

To achieve a high CO selectivity and more suitable H_2/CO ratios, several authors proposed to reform catalytically methane with CO_2 instead of steam [12–16],

$$CH_4 + CO_2 \rightarrow 2CO + 2H_2 \quad \Delta H = 260.5\ \mathrm{kJ/mol} \tag{3}$$

This reaction has also important environmental implications, since both methane and carbon dioxide are greenhouse gases which can be converted into valuable feedstock.

Sodesawa et al. [17] studied the CO_2 reforming with a stoichiometric feed ratio, over NiO, CuO and Fe_2O_3 catalysts, reduced with H_2 at 500°C. They found that Ni supported on SiO_2 exhibited nearly the equilibrium conversion and a high selectivity to CO. The other catalysts, CuO and Fe_2O_3, deactivated rapidly, due to carbon deposition. An improvement was suggested by Topor et al. [18], who recommended the use of excess CO_2 to avoid carbon formation. Chubb [19] and Chubb et al. [20] investigated the CO_2 reforming using an excess of carbon dioxide, with CO_2/CH_4 molar feed ratios of 3:1 and 5:1, over nickel supported on alumina. They observed the powdering of the catalyst for both ratios, with a rate of disintegration smaller for the higher one. While investigations regarding the cause of deactivation were not carried out, the authors [19,20] speculated that the growth of carbon inside the pore is responsible for it. The catalyst deterioration could be alleviated by activating the catalyst with ruthenium; however, a much lower conversion was obtained.

No commercial catalyst is yet available for reaction (3), the main reason being the formation of carbon via the reaction

$$CH_4 \rightarrow C + 2H_2 \quad \Delta H = 90.5\ \mathrm{kJ/mol} \tag{4}$$

which, in the absence of steam, quickly deactivates the conventional reforming catalysts. A similar difficulty arises with the methanation catalysts, because the low H_2/CO ratio generated by reaction(3) leads to carbon formation via the reaction

$$2CO \rightarrow C + CO_2 \quad \Delta H = -170\ \mathrm{kJ/mol} \tag{5}$$

The carbon formation sequence for various metals, reported by Rostrup-Nielsen [21], is: Ni $\gg$ Pt $>$ Ru. Ashcroft et al. [12] and Rostrup-Nielsen [22] demonstrated that one can not avoid carbon formation with CO_2/CH_4 ratios 1:1 over nickel catalysts. Ashcroft et al. [12] suggested an alternative strategy for CO_2 reforming, which provided a CO yield of 89% from a stoichiometric feed of CO_2 and CH_4. The carbon deposition ('coking') was suppressed by using the platinum-group catalysts. Although the catalysts based on noble metals are less sensitive to coking than the nickel-based catalysts, it is worthwhile to develop improved stable and effective nickel-based catalysts because of the high cost and limited availability of the noble metals. In the present paper, a Ni/MgO catalyst with excellent CO yield and stability for CO_2 reforming of CH_4 is proposed. A comparison among the Ni/alkaline earth metal oxides is also made.

2. Experimental

2.1. Catalyst preparation

The NiO/MgO, NiO/CaO, NiO/SrO and NiO/BaO (with a NiO/oxide weight ratio of 20/100) were prepared by impregnating the alkaline earth metal oxides with an aqueous solution of nickel nitrate. The obtained paste was dried at room temperature in air, then decomposed and calcined at 800°C in air for 1.5 h.

2.2. Catalytic reaction

The catalytic reaction was carried out under atmospheric pressure in a flow system. A quartz tube (2 mm inside diameter) was used as reactor. The catalyst powder (weight: 0.020 g) was held on quartz wool. The molar ratio of the reactant gases, with high GHSV (60 000 $cm^3\ g^{-1}\ h^{-1}$, unless otherwise stated), was $CO_2/CH_4 = 1{:}1$. The catalyst was reduced in H_2 (20 ml/min) at 500°C for 14 h, followed by an increase in temperature to 790°C at a rate of 20°C/min, also in H_2 (20 ml/min). The analysis of reactants/products mixtures was performed with an in-situ gas chromatograph equipped with a Porapak Q column. The yield of CO is defined as the ratio in the product

$$\frac{\text{CO (mol)}}{\text{CO (mol)} + \text{CO}_2\text{ (mol)} + \text{CH}_4\text{ (mol)}} \times 100\%$$

and the yield of H_2 as

$$\text{CH}_4\text{conversion}(\%) \times \frac{\text{H}_2\text{ (mol)}}{\text{H}_2\text{ (mol)} + \text{H}_2\text{O (mol)}}$$

in the product.

2.3. *CO adsorption*

The exposed Ni metal surface area of the catalyst was measured by CO pulse adsorption in a quartz tube (6 mm inside diameter) at room temperature, assuming a stoichiometry of 1/1. The catalyst powder (weight: 0.25 g) was held on quartz wool and its reduction was carried as described above. At 790°C, the reduced catalyst was purged with He (45 ml/min) for 10 min. Ultra high purity helium was used as carrier gas, at a flow rate of 45 ml/min. The analysis of gases during the adsorption of CO was carried out with an on-line mass spectrometer (HP Quadrupole, 5971 Series Mass Selective Detector) equipped with a fast response inlet capillary system.

2.4. *Temperature-programmed desorption*

The temperature-programmed desorption (TPD) experiments were carried out at a constant heating rate (10 K/min), using ultra high purity helium as carrier gas, at a flow rate of 45 ml/min. The analysis of the desorbed gases was performed with an on-line mass spectrometer (HP Quadrupole, 5971 Series Mass Selective Detector) equipped with a fast response inlet capillary system.

2.5. *X-ray powder diffraction*

X-ray powder diffraction (XRD) was carried out on a Nicolet X-ray diffraction equipment, using the Cu Kα radiation, at 40 kV and 20 mA.

3. Results

3.1. *CO_2 reforming of CH_4 over the reduced catalysts*

We examined the stoichiometric CO_2-reforming reaction at atmospheric pressure, GHSV of 60 000 $cm^3\ g^{-1}\ h^{-1}$ and 790°C over a number of reduced NiO/alkaline earth metal oxide catalysts. It was found that the reduced NiO/MgO was the best catalyst. Indeed, this catalyst provided about 91% and 98% conversions of CH_4 and CO_2 (Fig. 1), respectively, and yields of CO and H_2 of about 95% and 91% (Fig. 2), respectively. During 120 h of reaction, the conversions of the reactants and the yields of the products remained unchanged (Fig. 1 and Fig. 2), indicating that the reduced NiO/MgO catalyst has a rather high stability. Over the reduced NiO/CaO, the initial conversions of CH_4 and CO_2 were 67% and 86%, respectively, and the initial yields of CO and H_2 77% and 65%, respectively. However, its stability was very low. During 8 h of reaction, the conversions of CH_4 and CO_2 dropped to about 11%, and the yields of CO and H_2 dropped to 11% and 2.4%, respectively. The reduced NiO/SrO catalyst had lower initial conversions

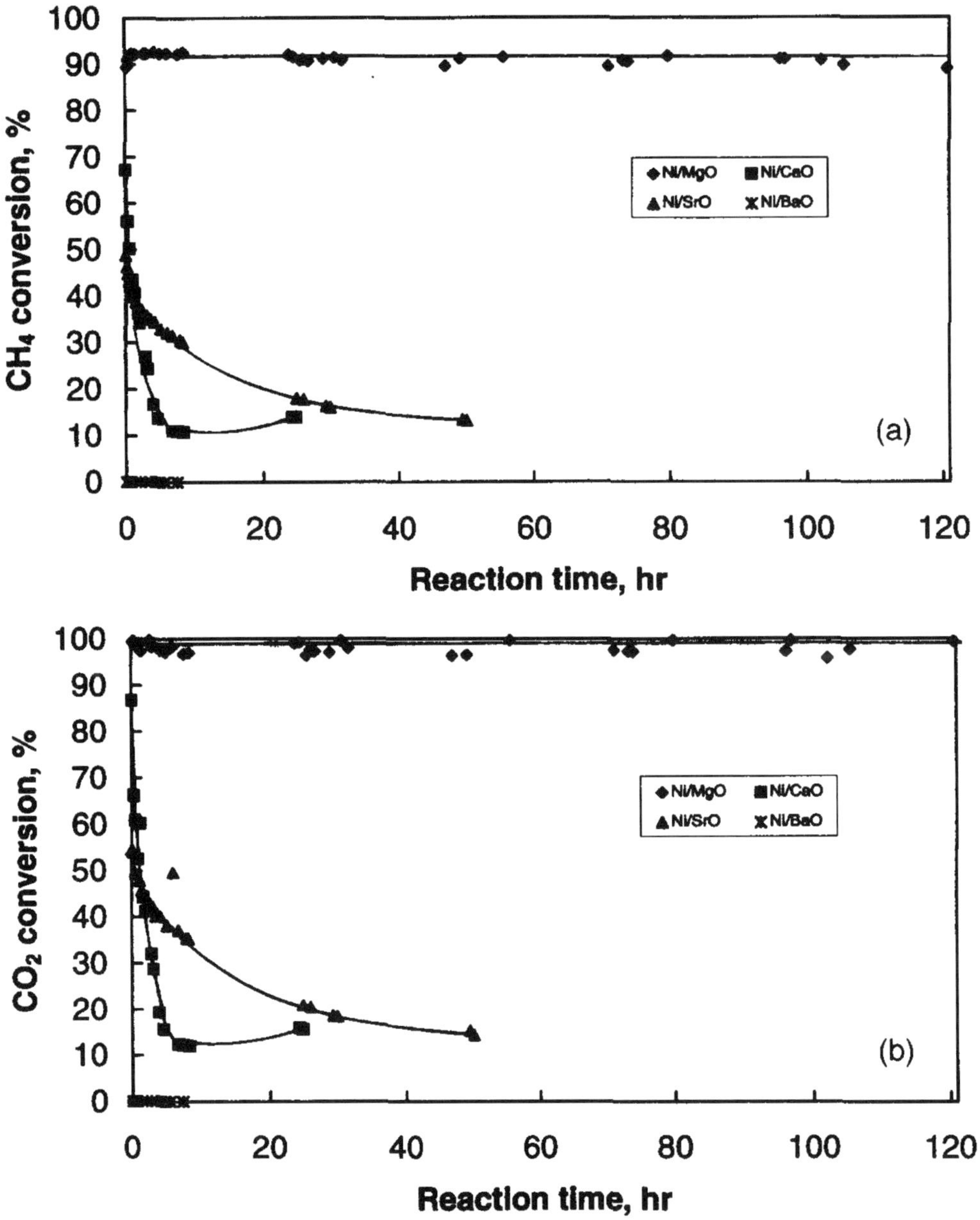

Fig. 1. CH_4 (a) and CO_2 (b) conversions for reduced NiO/alkaline earth metal oxides. Total pressure = 1 atm, $T = 790°C$, $CH_4/CO_2 = 1{:}1$, GHSV = 60 000 $cm^3\ g^{-1}\ h^{-1}$.

(CH_4: 48.8% and CO_2: 54.6%) and yields (CO: 51.7% and H_2: 36.7%). The conversions and yields decreased after 50 h of reaction time to low values. The reduced NiO/BaO catalyst had no activity.

The stabilities of the reduced NiO/MgO were compared at two space velocities (Fig. 3). We found that although the CO yield at GHSV = 253 521 $cm^3\ g^{-1}\ h^{-1}$ was about 58%, hence much smaller than the one at GHSV = 60 000 $cm^3\ g^{-1}\ h^{-1}$, the yield still remained unchanged during 105 h of reaction.

3.2. CO_2 reforming of CH_4 over unreduced catalysts

The CO_2 reforming of CH_4 was carried out previously only over unreduced catalysts [12–16]. For comparison purposes, the reaction was also carried out by

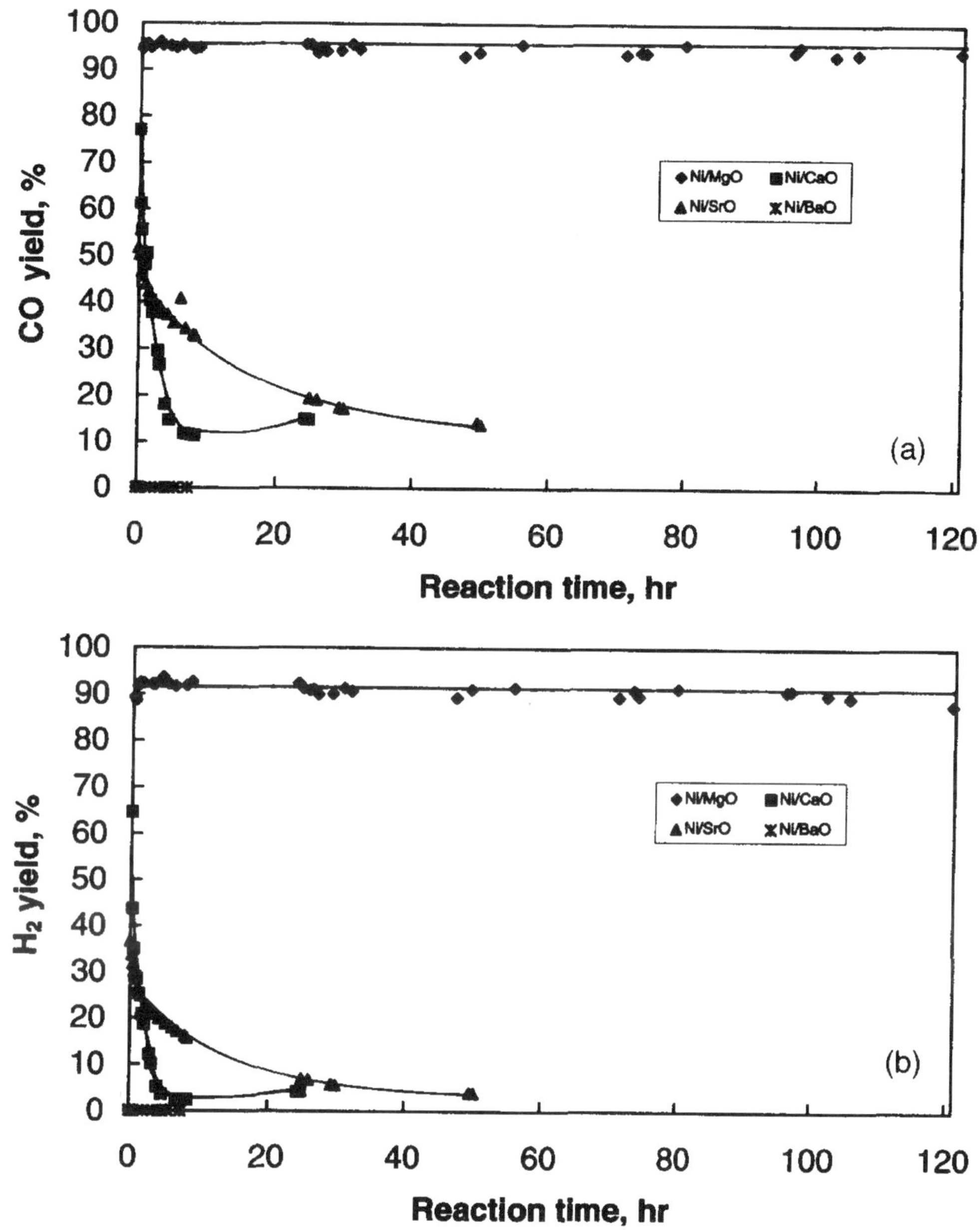

Fig. 2. CO(a) and H_2 (b) yields for reduced NiO/alkaline earth metal oxides. Total pressure = 1 atm, $T = 790°C$, $CH_4/CO_2 = 1:1$, GHSV = 60 000 $cm^3\ g^{-1}\ h^{-1}$.

us over unreduced NiO/alkaline earth metal oxide catalysts. Fig. 4 and Fig. 5 show that the unreduced NiO/MgO, NiO/SrO and NiO/BaO catalysts have very low activities. Over the unreduced NiO/CaO catalyst, the initial conversions of CH_4 and CO_2 were 51% and 62%, respectively, and the initial yields of CO and H_2 56% and 37%, respectively; the conversions of reactants and the yields of products first increase with time and then decrease.

3.3. Metal Ni surface areas

The exposed metal surface areas (Table 1) were determined via CO chemisorption. The data show that the exposed metal surface areas are small, and that their

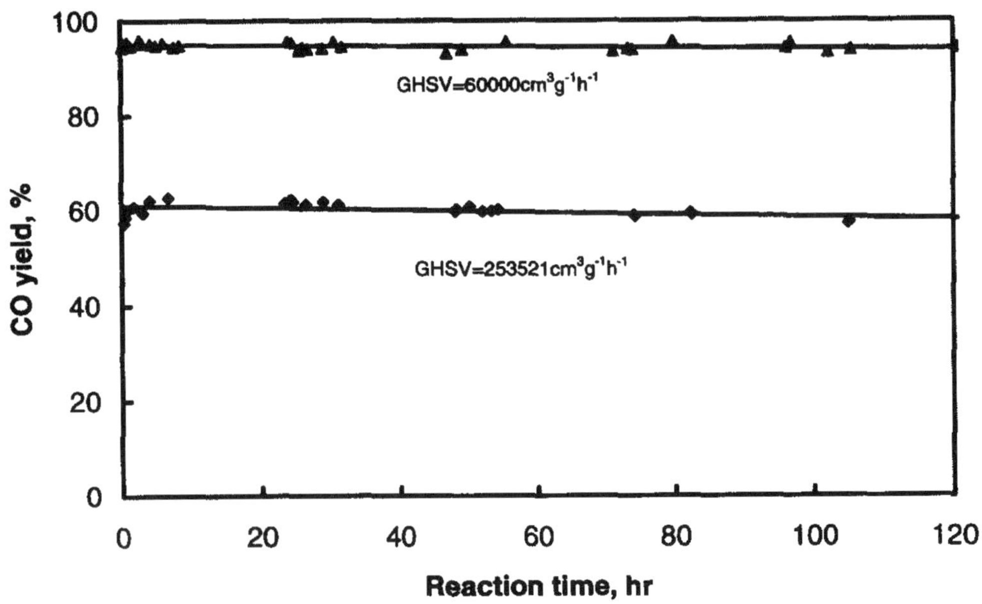

Fig. 3. Effect of space velocity of the feed gas on the CO yield for the reduced NiO/MgO catalyst. Total pressure = 1 atm, T = 790°C, CH_4/CO_2 = 1:1.

sequence for the reduced catalysts is: NiO/MgO > NiO/CaO > NiO/SrO > NiO/BaO. The metal surface areas of reduced NiO/SrO and reduced NiO/BaO were rather small because the supports sintered at the high temperature of 790°C. The advanced sintering of SrO and BaO could be observed directly.

3.4. *CO temperature-programmed desorption*

CO TPD was carried out to investigate the dissociation ability of the catalysts for CO. Fig. 6 shows that the reduced NiO/MgO, NiO/CaO and NiO/SrO catalysts have very similar TPD curves. There are two groups of overlapping peaks: at low temperatures (<260°C), and at high temperatures (>300°C). In the overlapping peaks at low temperatures, only CO is present, whereas in the overlapping peaks at high temperatures, both CO and CO_2 are detected. Compared with the reduced NiO/CaO and NiO/SrO, the NiO/MgO catalyst exhibits a much smaller disproportionation of CO at high temperatures.

4. Discussion

The reduced NiO/MgO provides very high conversions of CH_4 and CO_2 and yields of CO and H_2. Hence this catalyst activates very well both CH_4 and CO_2. In the previously investigated Ni supported on Al_2O_3 and CaO [12,22], the deactivation via carbon formation could not be avoided. In contrast, the conversion and selectivity over the reduced NiO/MgO catalyst employed by us remained unchanged for 120 h, implying that the catalyst has rather high stability. In other words, the carbon formation, resulting in deactivation, is avoided over the reduced NiO/

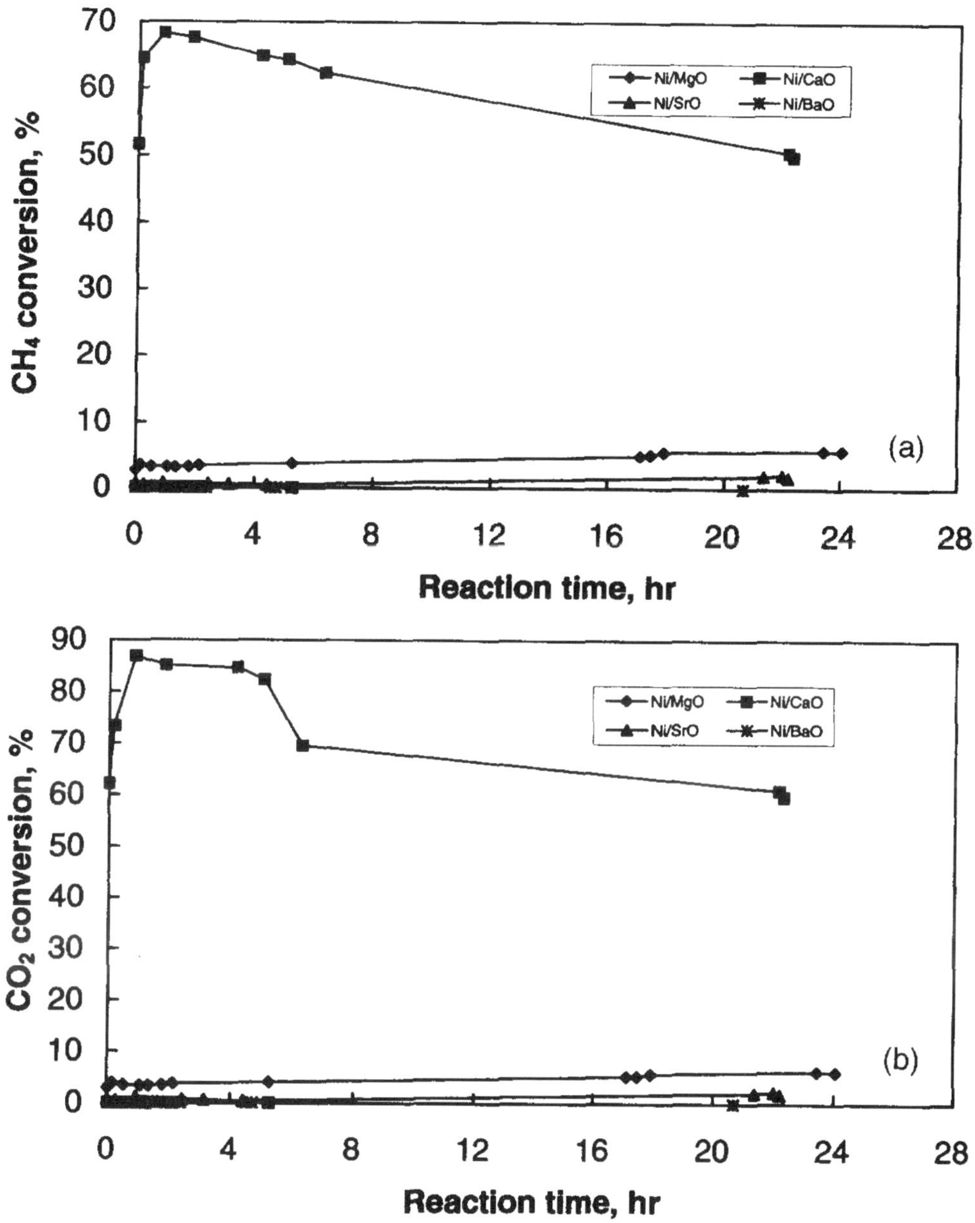

Fig. 4. CH_4 (a) and CO_2 (b) conversions for unreduced NiO/alkaline earth metal oxides. Total pressure = 1 atm, T = 790°C, CH_4/CO_2 = 1:1, GHSV = 60 000 $cm^3\ g^{-1}\ h^{-1}$.

MgO. The carbon formation is mainly a result of two reactions: CH_4 dissociation and CO disproportionation ($2CO \rightarrow C + CO_2$). The TPD results showed that the reduced NiO/MgO produces a much smaller amount of CO_2 from CO than the reduced NiO/CaO and NiO/SrO do; this implies that the high stability of the NiO/MgO is at least partially due to the inhibition of the disproportionation reaction of CO.

NiO, MgO, CaO, and SrO are face-centered cubic type oxides; however, only MgO has the lattice parameter and bond distance close to those of NiO (Table 2). For this reason, MgO and NiO can form a solid solution. This is also demonstrated by our XRD results which show that MgO and NiO/MgO have the same XRD spectra (Fig. 7). The formation of NiO/MgO solution has two important impli-

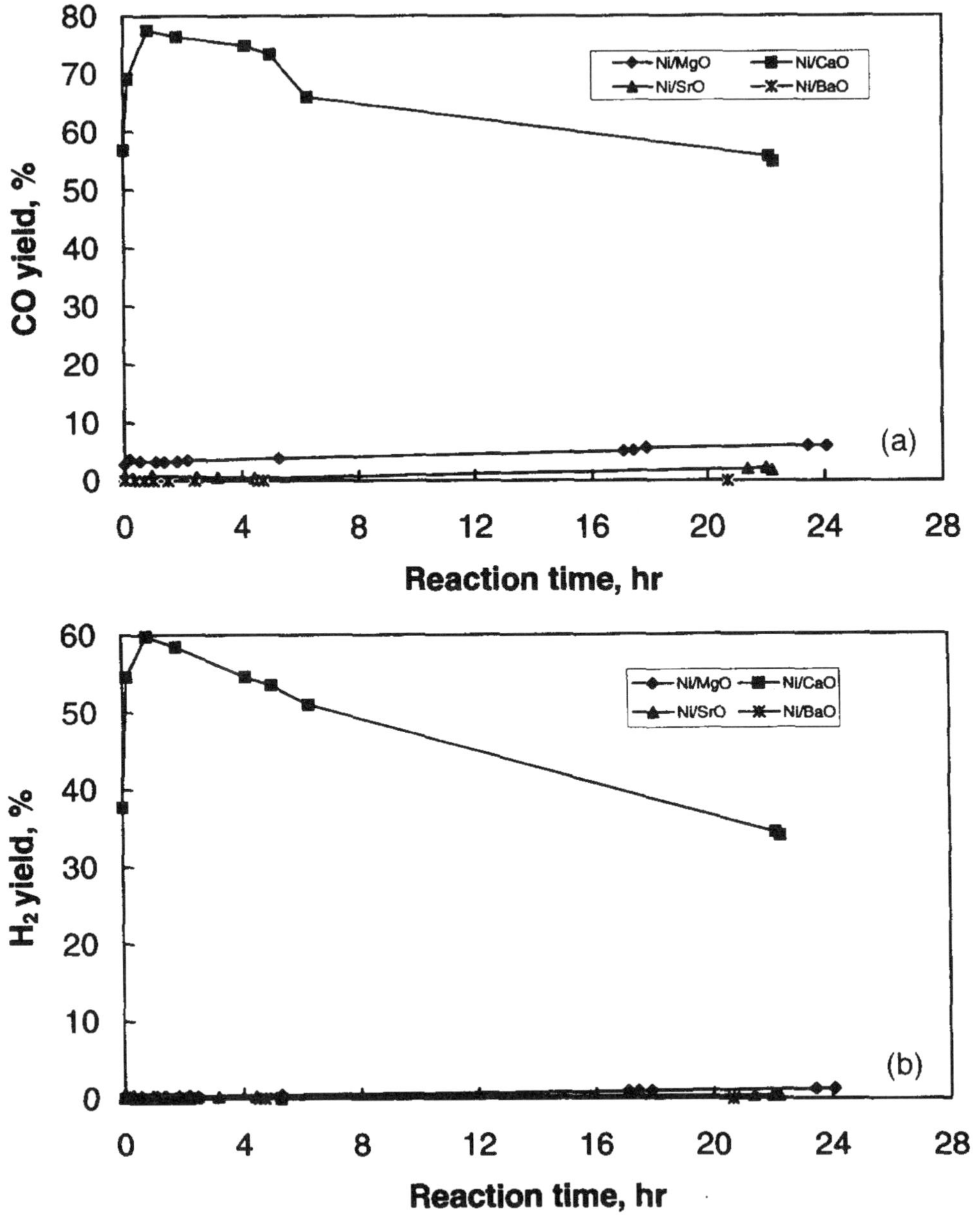

Fig. 5. CO (a) and H_2 (b) yields for unreduced NiO/alkaline earth metal oxides. Total pressure = 1 atm, $T = 790°C$, $CH_4/CO_2 = 1:1$, GHSV = 60 000 $cm^3\ g^{-1}\ h^{-1}$.

Table 1
Exposed metal surface areas

Catalyst[a]	Metal surface area (m^2/g)
NiO/MgO	0.34
NiO/CaO	0.22
NiO/SrO	≈0.02
NiO/BaO	[b]

[a] Reduced in H_2.
[b] Too small to be measured.

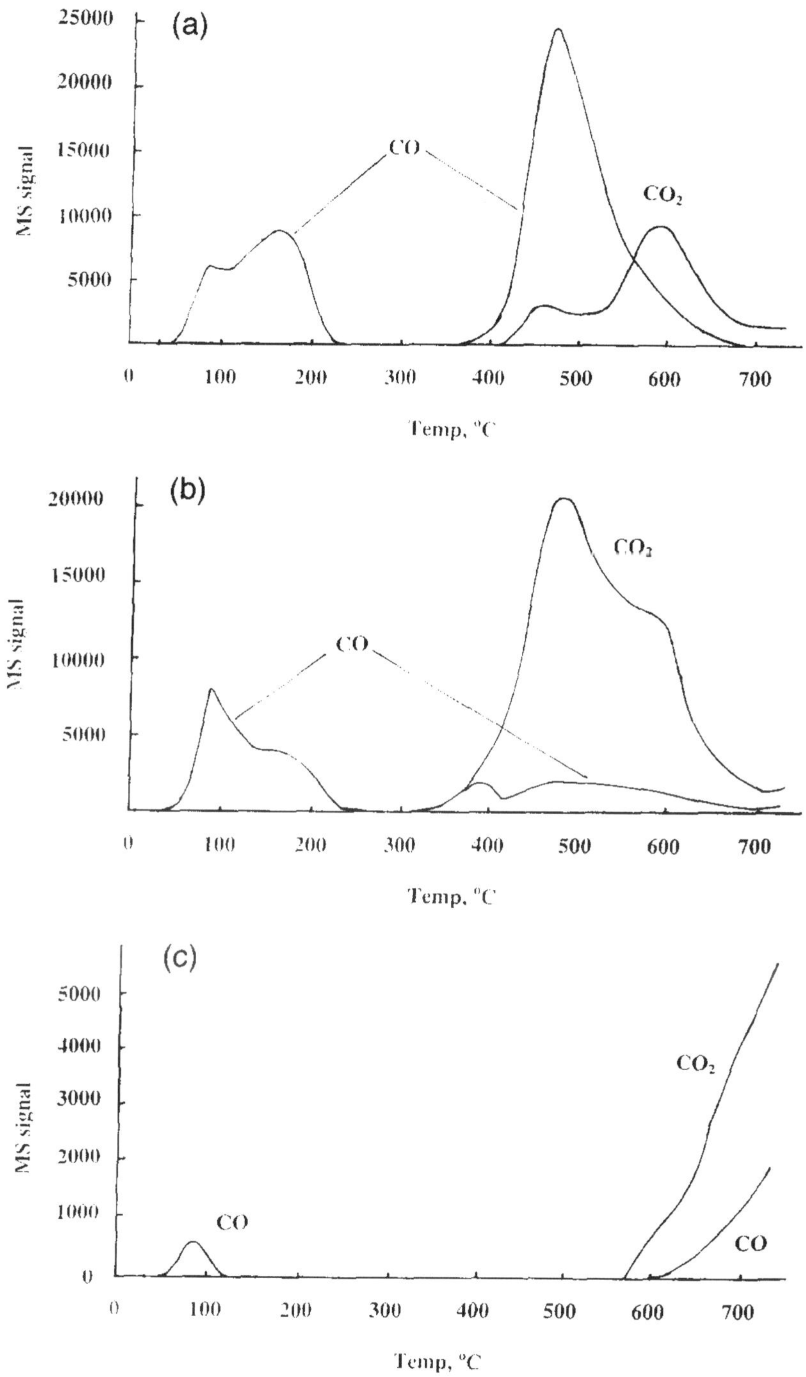

Fig. 6. Temperature-programmed desorption spectrum of CO over reduced NiO/MgO (a), NiO/CaO (b) and NiO/SrO (c).

cations: (i) Because of the dissolution of NiO in the bulk MgO, the reduction of NiO to Ni occurs with some difficulty. This provides a relatively small amount of Ni segregated as small particles on the surface of the support. As a result, little sintering occurs. The low sintering is even probably compensated by some NiO dissolved in MgO, which, being reduced by the hydrogen generated by the reaction itself, segregates at the solid-gas interface. In this manner, a self conservation of the exposed surface of Ni is probably achieved. Since NiO does not form solutions

Table 2
Crystal data for B1 type metal oxides [23]

Compound	a (Å)	$A-B$ (Å)
MgO	4.2112	2.11
CaO	4.8105	2.40
SrO	5.1396	2.57
BaO	5.5391	2.77
NiO	4.1946	2.10

with the other supports, it is more easily reduced and sintering of the generated Ni particles occurs. In addition, the other supports, particularly SrO and BaO, sinter tremendously. (ii) Because there are strong interactions between the small Ni particles and MgO, the donor ability (Lewis basicity) of Ni is decreased. The binding of the CO molecule to a transition metal atom proceeds via the donation of the lone electron pair of the carbon atom to the vacant d orbitals of the metal atom and the back donation of electrons from filled d orbitals of the metal to the vacant antibonding π^* orbitals of the CO molecule [24]. The electron donation to the antibonding orbitals of the CO molecule weakens the CO bond. However, the donor ability of the small Ni particles in the Ni/MgO catalysts, and, consequently, the ability for disproportionation of CO is reduced. When the amount of NiO employed is too low, a too small amount of Ni is segregated at the solid-gas interface during reduction. As a results, the activity of the catalyst is expected to be low. This explains the low activity observed by Yamazaki et al. [14] for 5.6 wt.-% NiO/MgO. An excess amount of Ni formed via the reduction of NiO supported on the CaO makes the relatively large Ni crystallites to behave like bulk materials. This increases the ability of CO for disproportionation. The unreduced NiO/MgO has low activity, because NiO and MgO form a solid solution and only little (if any) NiO remains exposed as particles to the chemical atmosphere. The unreduced NiO/CaO has higher activity, because it has free NiO exposed to the chemical atmosphere. NiO in NiO/SrO and NiO/BaO catalysts could not be reduced because of the sintering of SrO and BaO at the high temperature employed.

5. Conclusion

The reduced NiO/MgO is an excellent catalyst for the CO_2 reforming of CH_4. It has high conversions of the reactants and high yields of the products as well as rather high stability. The reduced NiO/CaO and NiO/SrO have lower activities and selectivities as well as very low stabilities. The formation of a NiO–MgO solution in the NiO/MgO catalyst is responsible for the very high CO yield and rather high stability.

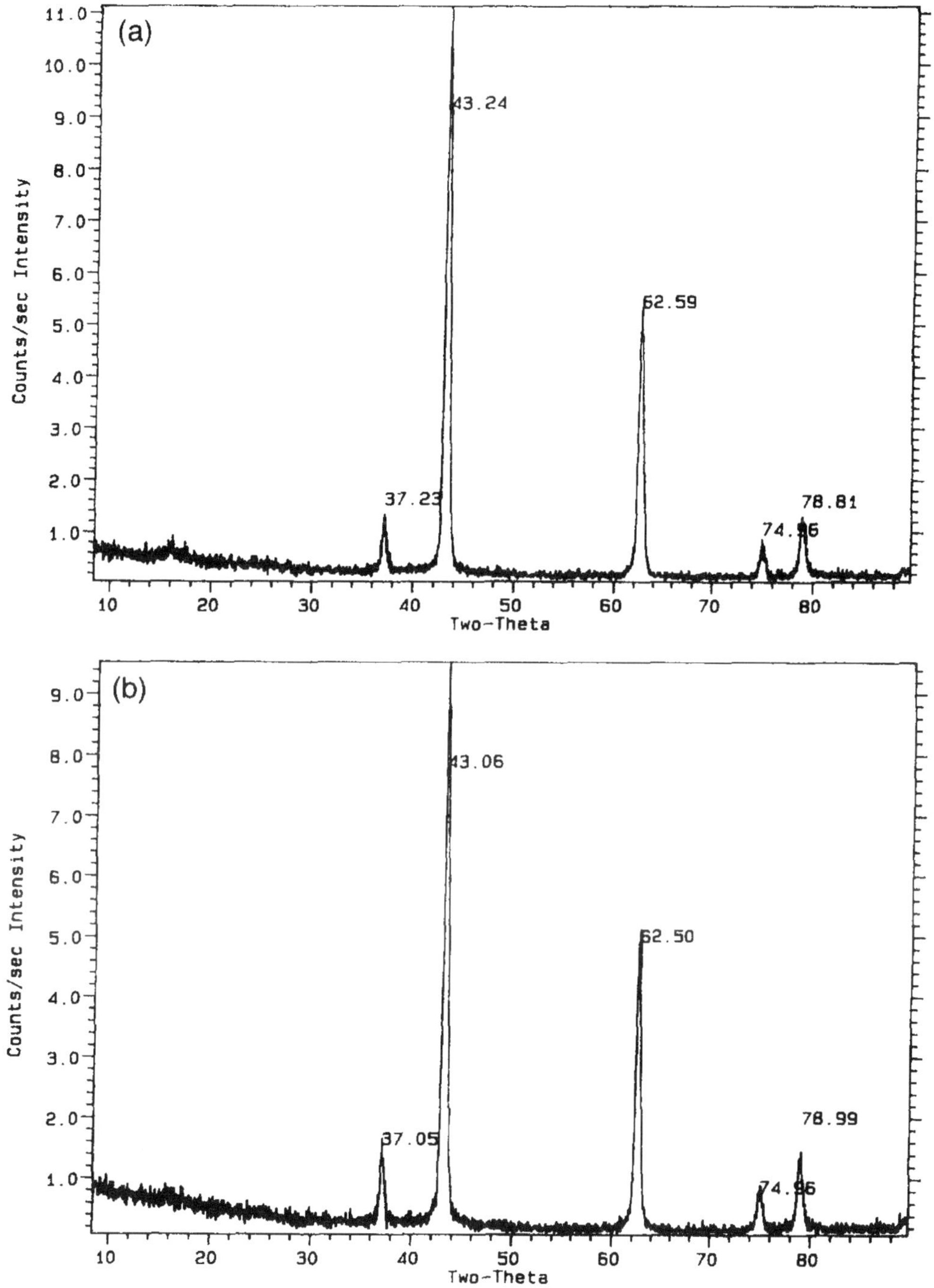

Fig. 7. XRD powder diffraction patterns of (a) calcined MgO at 790°C and (b) calcined NiO/MgO (weight ratio: 20/100) at 790°C.

References

[1] D.L. Trimm, Catal. Rev. Sci. Eng., 16 (1977) 155.

[2] J.R. Rostrup-Nielsen, in J.R. Anderson and M. Boudart (Editors), Catalysis Science and Technology, Vol. 5, Springer, New York, 1984.

[3] D.A. Hickman and L.D. Schmidt, Science, 259 (1993) 343.

[4] A.T. Ashcroft, A.K. Cheetham, J.S. Foord, M.L.H. Green, C.P. Grey, A.J. Murrell and P.D.F. Vernon, Nature (London), 344 (1990) 319.

[5] V.R. Choudhary, A.M. Rajput and B.J. Prabhakar, J. Catal., 139 (1993) 326.
[6] Y.H. Hu and E. Ruckenstein, Catal. Lett., 34 (1995) 41.
[7] Y.H. Hu, C.T. Au and H.L. Wan, Chin. Sci. Bull., 40(4) (1995) 303.
[8] C.T. Au, Y.H. Hu and H.L. Wan, Catal. Lett., 27 (1994) 199.
[9] D. Dissanayake, M.P. Rosynek, K.C.C. Kharas and J.H. Lunsford, J. Catal., 132 (1991) 117.
[10] D. Dissanayake, M.P. Rosynek and L.H. Lunsford, J. Phys. Chem., 97 (1993) 3644.
[11] V.R. Choudhary, A.M. Rajput and B.P. Prabhakar, Angew. Chem., Int. Ed. Engl., 33 (1994) 2104.
[12] A.T. Ashcroft, A.K. Cheetham, M.L.H. Green and P.D.F. Vernon, Nature (London), 352 (1991) 225.
[13] J.T. Richardson and S.A. Pripatyadar, Appl. Catal., 61 (1990) 293.
[14] O. Yamazaki, T. Nozaki, K. Omata and K. Fujimota, Chem. Lett., (1992) 1953.
[15] A.M. Gadalla and M.E. Sommer, Chem. Eng. Sci., 44 (1989) 2825.
[16] A.M. Gadalla and B. Bower, Chem. Eng. Sci., 43 (1988) 3049.
[17] T. Sodesawa, A. Dobashi and F. Nozaki, React. Kinet. Catal. Lett., 12 (1979) 107.
[18] L. Topor, L. Bejan, E. Ivana and N. Georgescu, Revue Chim. Bucharest, 30 (1979) 539.
[19] T.A. Chubb, Sol. Energy, 24 (1980) 341.
[20] T.A. Chubb, J.H. McCrary, G.E. McCrary, J.J. Nemecek and D.E. Simmons, Proc. Meet. Am. Sect. Int. Sol. Eng. Soc., 4 (1981) 166.
[21] J.R. Rostrup-Nielsen, J. Catal., 33 (1974) 184.
[22] J.R. Rostrup-Nielsen, Stud. Surf. Sci. Catal., 36 (1988) 73.
[23] I. Naray-Szabo, Inorganic Crystal Chemistry, Akademiai Kiado, Budapest, 1969, p. 237.
[24] R.R. Ford, Adv. Catal., 21 (1970) 51.

An optimum NiO content in the CO_2 reforming of CH_4 with NiO/MgO solid solution catalysts

Y.H. Hu and E. Ruckenstein [1]

Department of Chemical Engineering, State University of New York at Buffalo, Amherst, NY 14260, USA

Received 16 August 1995; accepted 10 October 1995

Reduced NiO/MgO, with a NiO content in the range 9.2–28.6 wt%, was found to be a highly effective catalyst for the CO_2 reforming of CH_4 to CO and H_2 (at 790°C, atmospheric pressure and a space velocity of 60000 $cm^3 g^{-1} h^{-1}$). For smaller or higher NiO contents, the yield was smaller, being negligible for 4.9 wt%. In contrast to the other reforming catalysts, the new catalyst has high stability, since in the optimum NiO range the CO yield remained unchanged at 95% for 120 h without any carbon deposition. The formation of a solid solution between NiO and MgO, which was demonstrated by both X-ray diffraction and temperature-programmed reduction, is most likely responsible for the high selectivity and stability in a large range of compositions of NiO/MgO.

Keywords: Ni; Mg; CH_4; CO_2 reforming

1. Introduction

Synthesis gas is an important feedstock for the methanol and the Fischer–Tropsch syntheses, reason for which the conversion of CH_4 to the synthesis gas CO/H_2 is of industrial significance. So far, three routes have been suggested to achieve such a conversion: (i) steam reforming of CH_4; (ii) catalytic partial oxidation of methane; and (iii) CO_2 reforming of CH_4.

The steam reforming of methane constitutes, for the time being, the dominant commercial method for the production of synthesis gas [1,2]. This reaction has, however, poor selectivity for CO and a high H_2/CO ratio, unsuitable for the methanol and the Fischer–Tropsch syntheses. Although the catalytic partial oxidation of CH_4 has high activity and selectivity [3–6], the process cannot be easily controlled because of the generation of hot spots in the catalyst [7]. The CO_2 reforming of CH_4 provides a high CO selectivity, more suitable H_2/CO ratios, and converts both CH_4 and CO_2, which are greenhouse gases, into valuable feedstocks [8–14]. For this reason, the latter reaction attracted increasing attention. However, no commercial catalyst is yet available for this reaction, because the generation of carbon when the conventional reforming catalysts, such as Ni/Al_2O_3, are employed results in the rapid deactivation of the catalyst. We reported in a recent paper [15] that the formation of carbon over Ni could be suppressed when a reduced NiO/MgO solid solution catalyst with a NiO/MgO weight ratio of 20/100 was used. In addition, the catalyst had high stability and yields. The scope of the present paper is to emphasize the existence of a large range of NiO/MgO compositions for which the yields have optimum, large, values.

[1] To whom correspondence should be addressed.

2. Experimental

The catalysts were prepared by impregnating MgO (or, for comparison purposes, Al_2O_3) with an aqueous solution of nickel nitrate, then drying at room temperature in air and finally calcining at 800°C in air for 1.5 h. The reduction of the catalysts was carried out in H_2 (20 ml/min) at 500°C for 14 h; this was followed by an increase in temperature to 790°C at a rate of 20°C/min also in H_2 (20 ml/min). The stoichiometric CO_2-reforming reaction was carried out at atmospheric pressure, 790°C and a space velocity of 60000 $cm^3\ g^{-1}\ h^{-1}$ over a number of prereduced NiO/MgO catalysts. Samples of 0.020 g of catalyst held on quartz wool located in a quartz tube (2 mm inside diameter) were employed. The yield of CO is defined as the ratio (in the product) CO (mol)/[CO (mol) + CO_2 (mol) + CH_4 (mol)] × 100% and the yield of H_2 as: CH_4 conversion (%) × H_2 (mol)/[H_2(mol) + H_2O (mol)] in the product. The catalysts have been examined by X-ray diffraction (XRD) both before and after reaction. Temperature-programmed reduction (TPR) experiments were also performed by heating the samples from room temperature to 800°C, at a rate of 17 K/min, in a 4% H_2/Ar gas flow rate of 47 ml/min through a quartz tube of 2 mm inside diameter.

3. Results and discussion

It was found that the reduced NiO/MgO catalysts can provide maximum yields of CO and H_2 of over 95 and 91% (fig. 1), respectively, for NiO contents between 9.2 and 28.6 wt%. The yields are higher than those previously reported for other catalysts [8–14]. The reduced catalyst had no activity for a NiO content of 4.8 wt%.

This paper, which was originally published in **Catalysis Letters**: 36, 145–149 (1996), is republished with permission.

a

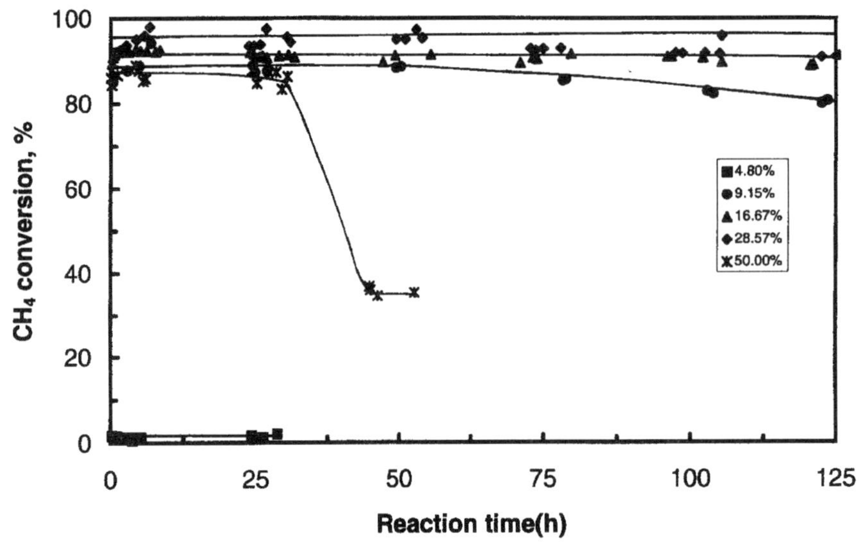

b

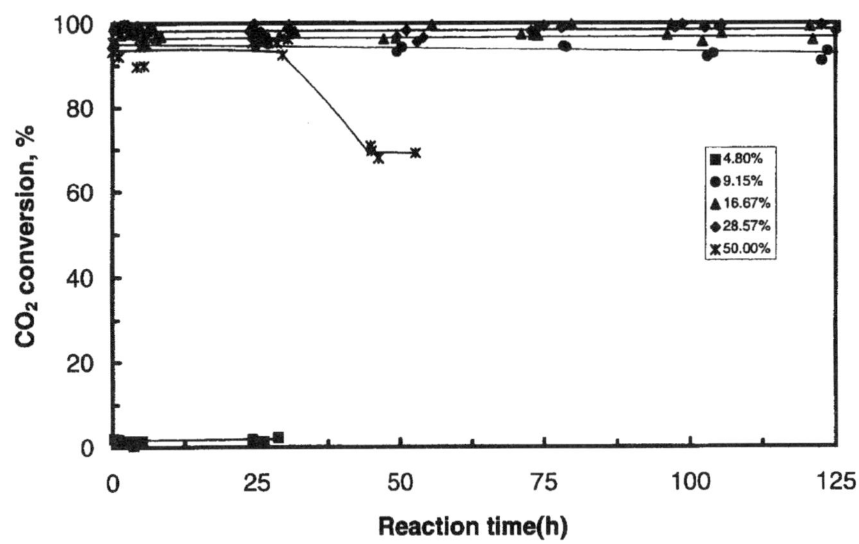

Fig. 1. (a) CH_4 conversion for reduced NiO/MgO catalysts. (b) CO_2 conversion for reduced NiO/MgO catalysts. (c) CO yield for reduced NiO/MgO catalysts. (d) H_2 yield for reduced NiO/MgO catalysts.

When the content of NiO in NiO/MgO catalysts was in the optimum range of 9.2 to 28.6 wt%, the CO yield of the catalysts hardly changed in 120 h (fig. 1) because the latter catalysts inhibit carbon formation. Indeed, while the reduced NiO/Al_2O_3 catalysts became black after a few hours, the reduced NiO/MgO catalysts maintained their initial color for NiO contents between 4.8 and 28.6 wt%. However, the catalyst with a NiO content of 50 wt% became black after about 50 h. At the higher NiO content of 50 wt%, the yield of CO decreased, from 91 to 53% after 40 h. Let us also note that, on the reduced 16.7% NiO/Al_2O_3 catalyst, carbon deposition occurred quickly, resulting in the decrease of the feed gas space velocity and the plugging of the reactor. For the unreduced NiO/MgO catalysts, the activities were also dependent on the NiO content. At 16.7 wt% NiO, the catalyst had hardly any activity, while at 28.6 wt%, the CO yield increased from 3 to 89% in about 20 h, remaining at the latter value for the remaining duration of the experiment (fig. 2).

The XRD patterns (fig. 3) reveal very similar structures for NiO/MgO and MgO, indicating that NiO and MgO form a solid solution. The XRD of the reduced 50 wt% NiO/MgO catalyst showed after reaction the presence of Ni besides the NiO/MgO solid solution phase; in the 16.7 wt% NiO/MgO catalyst, only small amounts of Ni phase could be identified after reaction.

As shown in fig. 4, there is one peak at 350°C in the TPR of NiO. In contrast, there is a small peak in the 50 wt% NiO/MgO catalyst and no obvious peak in the 16.7 wt% NiO/MgO. This indicates that NiO and MgO form indeed a solid solution.

In some investigations [8,11], it was concluded that it is impossible to avoid carbon formation for CO_2/CH_4

c

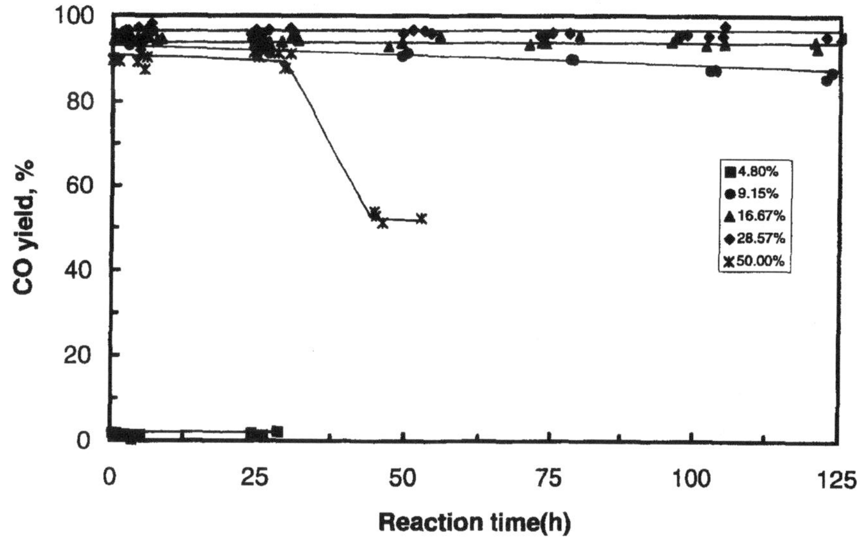

d

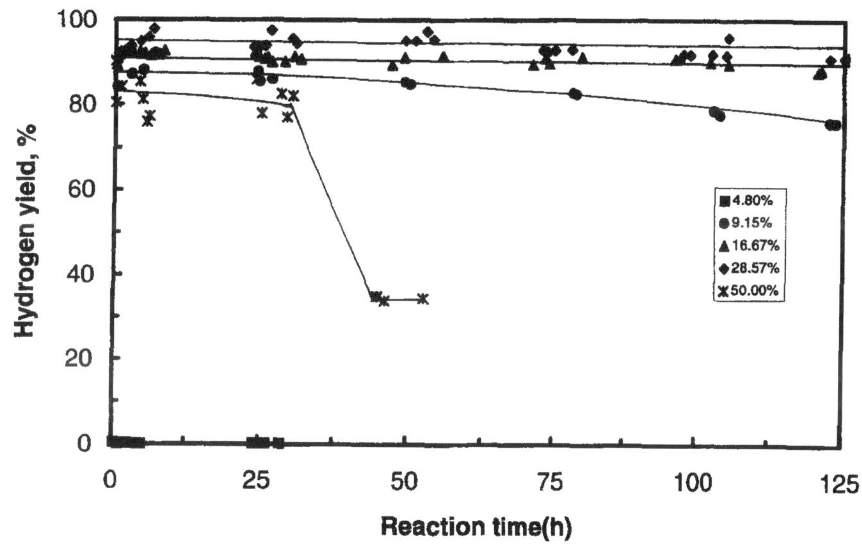

Fig. 1. (Continued.)

molar ratios of 1 : 1 on nickel based catalysts, such as Ni/Al_2O_3. The sulphur passivation, commercialized in steam reforming (the SPARG process), can inhibit to a certain extent carbon formation by blocking some of the active sites [16]; however, it reduces also the catalytic activity. Even though the noble metal catalysts can suppress efficiently the carbon deposition in the carbon dioxide reforming [8,11,17], they are expensive. For this reason, the presently proposed reduced NiO/MgO solid solution catalysts may have commercial value.

As shown in fig. 1, the reduced NiO/MgO catalysts activate very well both CH_4 and CO_2 and have rather high stability. The unexpected stability of the reduced NiO/MgO catalysts is most likely due to the almost identical crystal structures of NiO and MgO. Both NiO and MgO have a face-centered cubic structure with almost the same lattice parameters and bond distances [18]. For this reason, MgO and NiO can form solid solutions. This was demonstrated both by the XRD results, which showed that NiO/MgO forms a single phase, and by the absence of a peak in the TPR results. The formation of NiO/MgO solutions has the following consequences [15]: (i) Only a small amount of NiO is reduced to Ni which segregates as small particles over the surface of the support. Because of the small amount, little sintering occurs. This is why NiO/MgO catalysts have rather high stability in a large range of NiO contents. Since NiO does not form solutions with Al_2O_3, it is more easily reduced and the relatively large amount of Ni sinters generating relatively large particles. (ii) Because there are strong interactions between the small Ni particles which are present on the reduced NiO/MgO catalyst and MgO, the donor ability of Ni is decreased. The bonding of the CO molecule to a transition metal atom such as Ni involves the donation of the lone electron pair of the carbon atom to vacant d orbitals of the metal atom

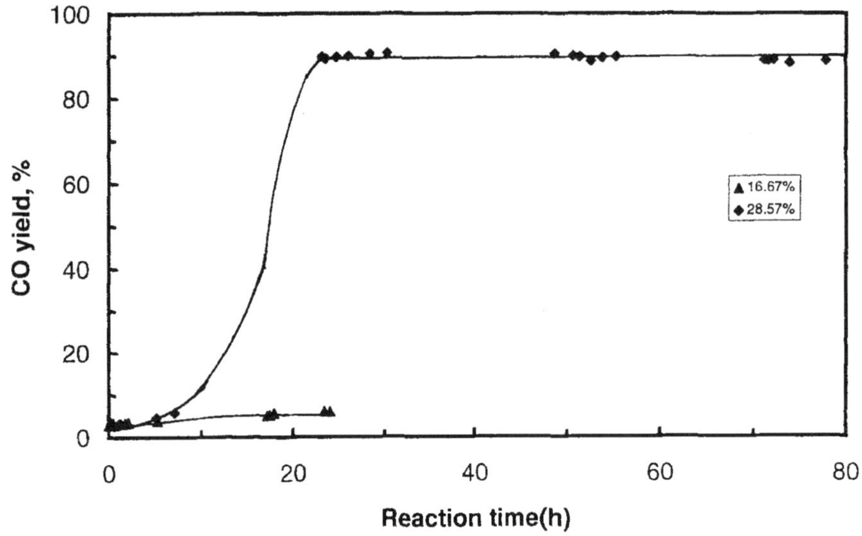

Fig. 2. CO yield for unreduced NiO/MgO catalysts.

and the back donation of electrons from the filled d orbitals of the metal to the vacant antibonding π^* orbitals of the CO molecule [19]. This weakens the CO bond, thus stimulating the CO decomposition. The donor ability of Ni and, consequently, the ability for disproportionation of CO are, however, decreased by the strong interactions between the small Ni particles and MgO. At high NiO contents (50 wt%), a larger amount of NiO is reduced. As a result, the sintering is intensified and the Ni particles become large and behave like bulk materials. Therefore, the strong interactions, which are responsible for the decrease of the donor ability, are absent and the deactivation, via the Boudouard reaction, of the catalyst is intensified. At low NiO contents (4.8 wt%), a too small amount of Ni is generated, and, hence, the activity is negligible.

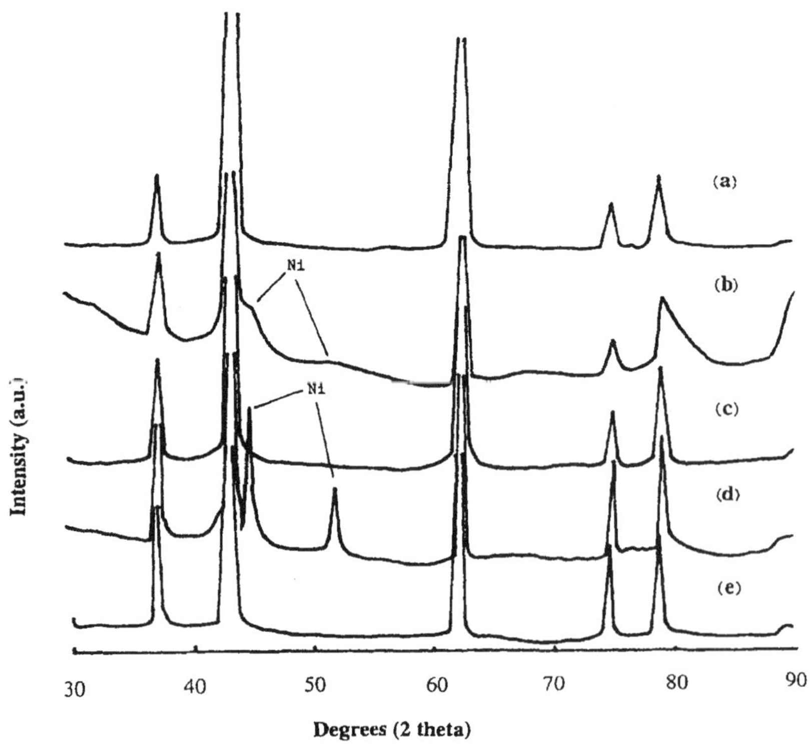

Fig. 3. X-ray powder diffraction patterns of (a) calcined MgO, (b) 16.7 wt% NiO/MgO catalyst (after reaction), (c) 16.7 wt% NiO/MgO catalyst (before reduction), (d) 50% NiO/MgO catalyst (after reaction), and (e) 50% NiO/MgO catalyst (before reduction).

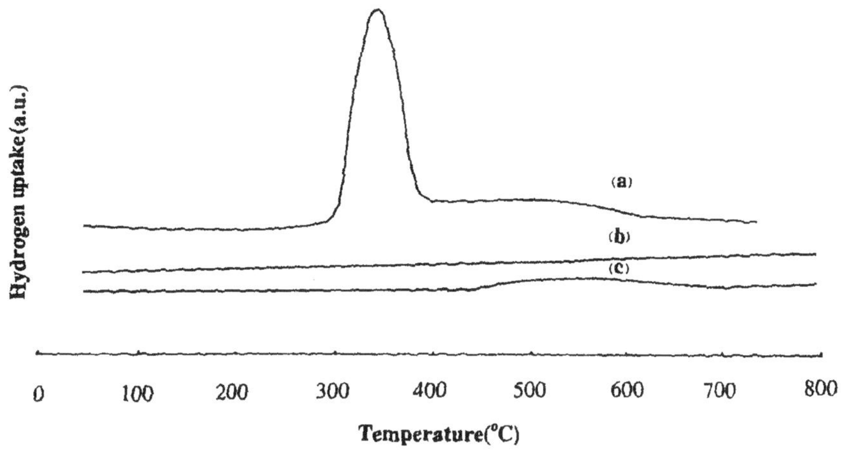

Fig. 4. TPR profiles of (a) NiO, (b) 16.7 wt% NiO/MgO catalyst and (c) 50 wt% NiO/MgO catalyst.

From the above results, one can conclude that the NiO/MgO solid solution might be an appropriate industrial catalyst for the CO_2 reforming of CH_4. In addition, solid solution catalysts may have applications to other catalytic reactions.

References

[1] D.L. Trimm, Catal. Rev. Sci. Eng. 16 (1977) 155.
[2] J.R. Rostrup-Nielsen, in: *Catalysis Science and Technology*, Vol. 5, eds. J.R. Anderson and M. Boudart (Springer, Berlin, 1984).
[3] D.A. Hickman and L.D. Schmidt, Science 259 (1993) 343.
[4] A.T. Ashcroft, A.K. Cheetham, J.S. Food, M.L.H. Green, C.P. Grey, A.J. Murrell and P.D.F. Vernon, Nature 344 (1990) 319.
[5] V.R. Choudhary, A.M. Rajput and B.J. Prabhakar, J.Catal. 139 (1993) 326.
[6] Y.H. Hu and E. Ruckenstein, Catal. Lett. 34 (1995) 41.
[7] D. Dissanayake, M.P. Rosynek and L.H. Lunsford, J. Phys. Chem. 97 (1993) 3644.
[8] A.T. Ashcroft, A.K. Cheetham, M.L.H. Green and P.D.F. Vernon, Nature 352 (1991) 225.
[9] O. Yamazaki, T. Nozaki, K. Omata and K. Fujimota, Chem. Lett. (1992) 1953.
[10] J.T. Richardson and S.A. Paripatyadar, Appl. Catal. 61 (1990) 293.
[11] J.R. Rostrup-Nielsen, Studies Surf. Sci. Catal. 36 (1988) 73.
[12] A.M. Gadalla and M.E. Sommer, Chem. Eng. Sci. 44 (1989) 2825.
[13] A.M. Gadalla and B. Bower, Chem. Eng. Sci. 43 (1988) 3049.
[14] T. Sodesawa, A. Dobashi and F. Nozaki, React. Kinet. Catal. Lett. 12 (1979) 107.
[15] E. Ruckenstein and Y.H. Hu, Appl. Catal., in press.
[16] H.C. Dibbern, P. Obsen, J.R. Rostrup-Nielsen, P.B. Tottrup and N.R. Udengaard., Hydroc. Proc. 65 (1986) 71.
[17] P.D.F. Vernon, M.L.H. Green, A.K. Cheetham and A.T. Ashcroft, Catal. Today 13 (1992) 417.
[18] I. Naray-Szabo, *Inorganic Crystal Chemistry* (Akadémiai Kiadó, Budapest, 1969) p. 237.
[19] R.R. Ford, Adv. Catal. 21 (1970) 51.

The characterization of a highly effective NiO / MgO solid solution catalyst in the CO_2 reforming of CH_4

Yun Hang Hu * and Eli Ruckenstein ‡

Department of Chemical Engineering, State University of New York at Buffalo, Amherst, NY 14260, USA

Received 22 August 1996; accepted 5 November 1996

A NiO/MgO catalyst prepared by impregnation, which reduced in H_2 had very high CO yield and stability in CO_2 reforming of methane, was investigated by XPS, XRD, BET and pulse-MS response. This catalyst was compared to that obtained by mechanical mixing of powders of the two oxides. It was found that the entire NiO formed a solid solution with MgO in the former catalyst, while only a fraction of NiO formed a solid solution with MgO in the latter one. BET revealed that, in contrast to NiO and MgO, the NiO/MgO catalyst prepared by impregnation had a high stability to sintering, because its surface area hardly changed during calcination from 1.5 to 20 h at 800°C. In the same catalyst, a surface enrichment in MgO, which was greater after than before reduction, was detected. Compared to MgO or NiO, this catalyst had a lower Mg(2p) and a higher $Ni(2p_{3/2})$ binding energy. This indicates that electron transfer from NiO to MgO took place, which, increasing the binding between the two oxides, might be responsible for the resistance of the solid solution to sintering. Because of the interactions between Ni and Mg, the clustering of Ni, which stimulates carbon deposition is inhibited. This explains the high stability of the CO yield in the CO_2 reforming of methane over the NiO/MgO catalyst prepared by impregnation. The pulse-MS response suggested that the decompositions of CO_2 to CO and O and of CH_4 to C and H are involved in the reaction mechanism of CO_2 reforming of methane over the reduced NiO–MgO solid solution catalyst.

Keywords: NiO, MgO, solid solution, XPS, XRD, pulse-MS, CO_2 reforming of methane

1. Introduction

In recent years, renewed interest in the CO_2 reforming of methane to synthesis gas

$$CH_4 + CO_2 \rightarrow 2CO + 2H_2, \quad \Delta H = 247 \text{ kJ/mol}$$

has arisen [1–7]. While this pathway provides a high CO selectivity and a more suitable H_2/CO ratio for the Fischer–Tropsch or methanol syntheses, it could not be applied industrially because of the aging of the catalyst, caused mainly by carbon deposition. Ashcroft et al. [1] and Rostrup-Nielsen [8] observed that carbon deposition cannot be avoided over nickel-supported catalysts and for a CO_2/CH_4 molar ratio of 1 : 1. Ashcroft et al. [1] suppressed the carbon deposition in CO_2 reforming by replacing Ni with platinum-group catalysts. Rostrup-Nielsen [9] noted that carbon-free deposition in steam reforming of methane can be achieved by using a sulfur-passivated nickel catalyst. However, sulfur also reduces the activity of the catalyst. Zhang et al. found that Ni/La_2O_3 has high CO yield and stability in CO_2 reforming of methane [10], and we demonstrated that its stability depends on the preparation conditions and is enhanced by a small amount of chlorine in the catalyst [11]. Recently, we found that the reduced NiO/MgO catalyst prepared by impregnation has a rather high CO yield (94%) and excellent stability, since the carbon deposition is inhibited [12–14].

Because of similar structures, NiO and MgO are completely miscible and form an ideal solid solution [15]. The value of such solid solutions for catalysis was first recognized when trying to understand how electronic interactions between cations affect the catalytic action [16,17]. A number of physico-chemical properties of the NiO/MgO solid solution [18–26], such as the influence of Ni precursor and preparation conditions on the reducibility of magnesia-supported NiO and on the diffusion of Ni^{2+} ions into the MgO lattice [23], the effects of calcination temperature and treatment atmosphere on the phase composition of NiO-MgO solid solutions [22,24], and the interactions between small gas molecules (CO, CO_2 or CH_4) and the solid solution [19 21,25,26], were investigated.

Although the NiO/MgO catalyst was widely studied, it was considered inadequate for the reforming of methane, because it had a poor stability and a low activity [27]. We found, however, that its activity depends on the kind of MgO employed and conditions used in preparation [14]. The goal of this paper is to investigate the NiO/MgO catalyst, which has high CO yield and time stability in the CO_2 reforming of methane, by the XPS, XRD, pulse-MS response and BET techniques. A possible reaction mechanism is suggested.

* Permanent address: Department of Chemistry, Xiamen University, Xiamen 361005, P.R. China.

‡ To whom correspondence should be addressed.

This paper, which was originally published in **Catalysis Letters**: 43, 71–77 (1997), is republished with permission.

2. Experimental

2.1. Catalyst preparation

The catalyst was prepared by impregnating a MgO powder (Aldrich) with an aqueous solution of nickel nitrate (Alfa Chemicals), followed by drying at room temperature in air and finally by calcination at 800°C in air for 1.5 h. The prepared catalyst contained 20 g NiO per 100 g MgO, composition which provides the optimum activity [12,13].

2.2. Catalytic reaction

The catalytic reaction was carried out under atmospheric pressure, at 790°C, in a flow system, using a vertical quartz tube (2 mm inside diameter) reactor. The catalyst powder (weight: 0.02 g) was held on quartz wool. The molar ratio of the reactant gases, with a GHSV = 253500 $cm^3\ g^{-1}\ h^{-1}$, was $CO_2/CH_4 = 1:1$. The catalyst was reduced in a flow of H_2 (20 ml/min) at 500°C for 14 h, followed by an increase in temperature to 790°C at a rate of 20°C/min, also in a flow of H_2 (20 ml/min). The analysis of the reactants/products mixtures was performed with an in situ gas chromatograph equipped with a Porapak Q column.

2.3. BET surface area and pore size distribution measurements

The surface area and the pore size distribution were determined using a Micromeritics ASAP 2000 instrument. The surface area was calculated by the BET method, while the pore size distribution curve was obtained by the BJH method [28]. The sample was degassed at 200°C in high vacuum before measurements.

2.4. X-ray powder diffraction and ESCA

X-ray powder diffraction (XRD) was carried out using a Nicolet X-ray diffraction instrument, equipped with a Cu K_α source, at 40 kV and 20 mA.

The ESCA analysis was carried out using a SSI model 100 small spot ESCA instrument, equipped with an Al K_α monochromatized X-ray source. The binding energies were referenced to the C(1s) level (285.0 eV).

2.5. Pulse reaction and dynamic response

A quartz tube (2 mm inside diameter) was used as reactor and the catalyst powder (weight, 0.02 g; bed height, 2 mm) was held on quartz wool. Pulses of 45 μl were employed and ultra-high-purity helium was used as carrier gas, with a flow rate of 45 ml/min. The analysis of gases during transients was carried out with an on-line mass spectrometer (HP Quadrupole, 5971 series mass selective detector) equipped with a fast response inlet capillary system.

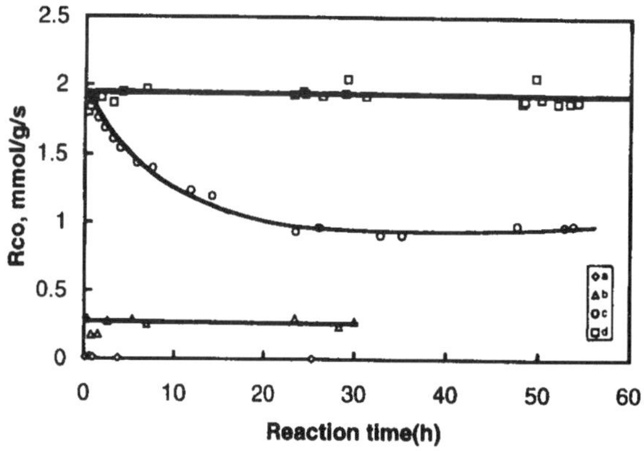

Fig. 1. Relationship between the rate of formation of CO and the reaction time over NiO/MgO catalysts: (a) catalyst obtained by mixing NiO and MgO powders (this is not followed by calcination); (b) catalyst obtained by mechanical mixing of NiO and MgO powders, followed by calcination in air at 800°C for 1.5 h; (c) catalyst obtained by mechanical mixing of NiO and MgO powders, followed by calcination in air at 800°C for 14 h; (d) catalyst prepared by impregnation, followed by calcination in air at 800°C for 1.5 h.

3. Results

3.1. The CO_2 reforming reaction of methane

As shown in fig. 1, the preparation conditions of the catalyst affect in a major way the CO_2 reforming of methane. For the reduced mechanical mixture of NiO and MgO which was not subsequently calcined, the rate of formation of CO is near zero (curve a of fig. 1). For the reduced mechanical mixture of NiO and MgO calcined for 1.5 h at 800°C, the rate of formation of CO is about 0.25 mmol $g^{-1}s^{-1}$. For the reduced mechanical mixture of NiO and MgO calcined for 14 h at 800°C, the

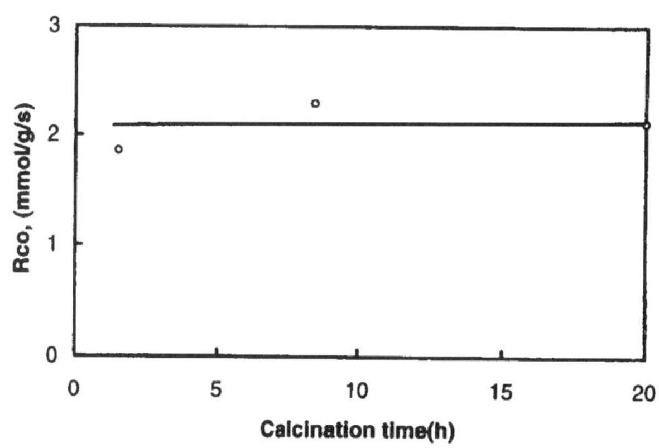

Fig. 2. Relationship between the rate of CO formation and calcination time over NiO/MgO catalysts prepared by impregnation, followed by calcination in air at 800°C.

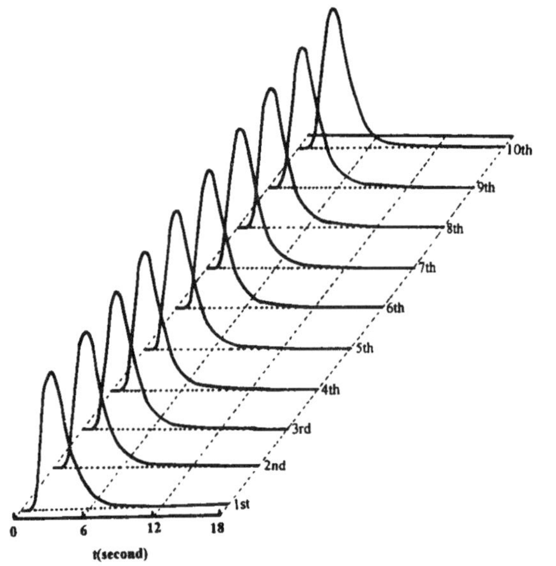

Fig. 3. Response to CH_4/CO_2 (1/1) pulses at 790°C over the unreduced NiO/MgO catalyst prepared by impregnation ((---) CO; (——) CO_2).

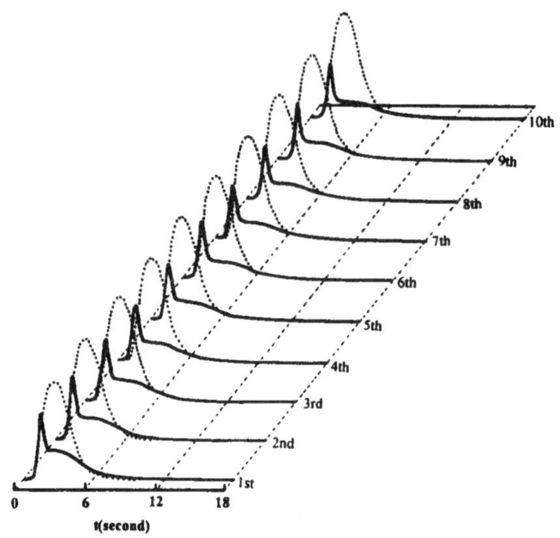

Fig. 5. Response to CH_4/CO_2 (1/1) pulses at 790°C over the NiO/MgO catalyst prepared by impregnation, reduced in H_2 and used for 2 h in the CO_2 reforming of methane ((---) CO; (——) CO_2).

rate of formation of CO is initially 1.923 mmol g^{-1} s^{-1} and then decreases with reaction time. In contrast, the reduced NiO/MgO catalyst, prepared by impregnation followed by calcination in air at 800°C for 1.5 h, has the high rate of formation of CO of 1.9 mmol g^{-1} s^{-1}, which remains unchanged with reaction time. Fig. 2 shows that for the reduced NiO/MgO prepared by impregnation, the rate of formation of CO changes little with calcination time.

3.2. *Pulse response of CO/CO_2 in CO_2 reforming of methane*

Fig. 3 shows that no CO is detected during ten CO_2/CH_4 (1/1) pulses over unreduced NiO/MgO catalyst prepared by impregnation. In contrast, over the reduced NiO/MgO catalyst, the CO concentration increases while that of CO_2 decreases with increasing number of pulses (fig. 4). For the reduced catalyst used for 2 h in the CO_2 reforming of methane, the results are similar to those over the reduced catalyst (fig. 5).

Over the reduced NiO/MgO used for 2 h in the CO_2 reforming of methane, CO and CO_2 are detected during CH_4 pulses, with concentrations which decrease with increasing number of pulses (fig. 6). Over the reduced

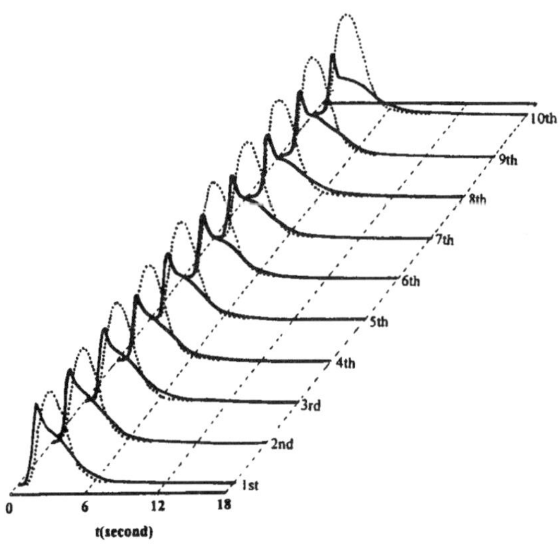

Fig. 4. Response to CH_4/CO_2 (1/1) pulses at 790°C over the reduced NiO/MgO catalyst prepared by impregnation ((---) CO; (——) CO_2).

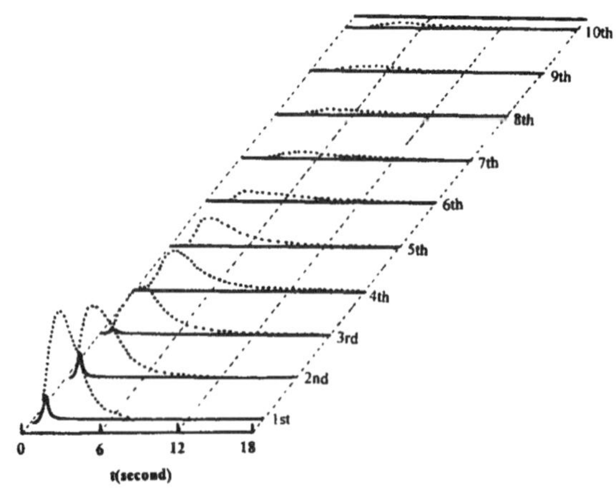

Fig. 6. Response to CH_4 pulses at 790°C over the reduced NiO/MgO catalyst prepared by impregnation, and employed for 2 h in the CO_2 reforming of methane ((---) CO; (——) CO_2).

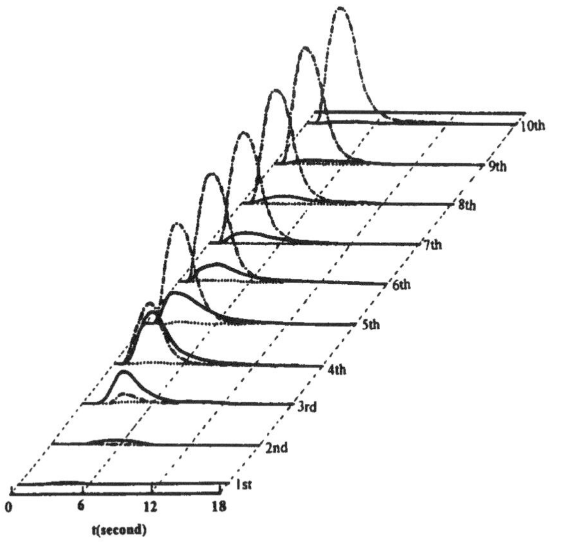

Fig. 7. Response to O_2 pulse at 790°C over the reduced NiO/MgO catalyst prepared by impregnation and employed for 2 h in the CO_2 reforming of methane ((- - -) CO; (——) CO_2; (– · –) O_2).

NiO/MgO previously used for 2 h in the CO_2 reforming of methane, a small amount of CO_2 is detected during the first O_2 pulse, the amount increases during the first four pulses and then decreases (fig. 7).

3.3. Surface area and pore size distribution

The BET measurements provide the values 54, 21, 50, 47 and 47 m^2/g for MgO without calcination, MgO calcined for 1.5 h, and for the NiO/MgO prepared by impregnation and calcined for 1.5, 8.5 or 20 h, at 800°C, respectively. This indicates that MgO sinters at 800°C, whereas NiO/MgO prepared by impregnation is stable to sintering at 800°C. Fig. 8 shows that there are two

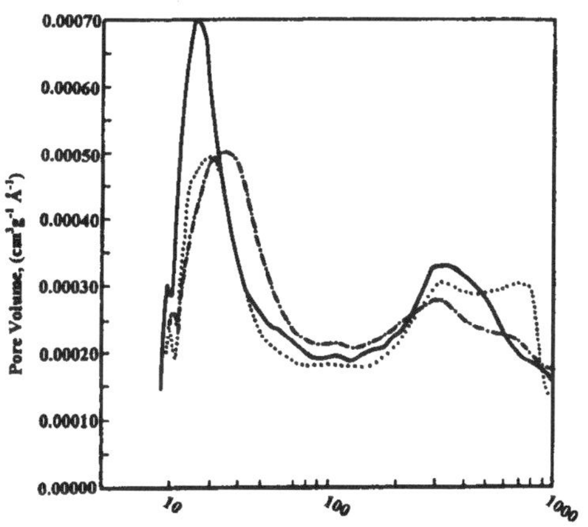

Fig. 8. Pore size distribution of NiO/MgO catalysts prepared by impregnation ((- - -) calcined at 800°C for 1.5 h; (– · –) calcined for 8.5 h; (——) calcined for 20 h).

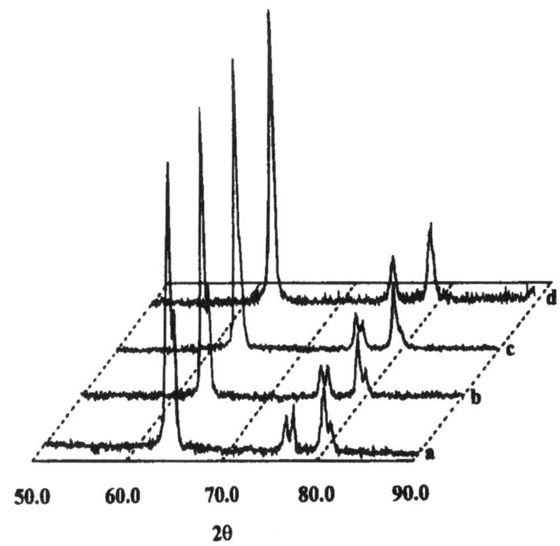

Fig. 9. XRD patterns of NiO/MgO: (a) mechanical mixture of NiO and MgO without calcination; (b) mechanical mixture of NiO and MgO calcined in air at 800°C for 1.5 h; (c) mechanical mixture of NiO and MgO calcined in air at 800°C for 14 h; (d) catalyst prepared by impregnation, followed by calcination in air at 800°C for 1.5 h.

kinds of pores, with sizes in the ranges 10–100 Å and 200–1000 Å in the NiO/MgO catalysts prepared by impregnation. The 20 Å pores are more numerous in the NiO/MgO calcined for 20 h than in those calcined for 8.5 or 1.5 h

3.4. Phase structure

The XRD patterns of mechanical mixtures of NiO and MgO and NiO/MgO prepared by impregnation are presented in fig. 9. As shown by this figure, three double peaks at 2θ of about 62.64, 75.00 and 79.00 are present for the mechanical mixtures. In contrast, the NiO/MgO catalyst prepared by impregnation and calcination has only three single peaks near the above 2θ values.

Although the XRD patterns of MgO are similar to those of NiO, there are some differences in their peak positions, by 0.6, 0.8 and 0.81° for $2\theta \approx 62.64$, 75.00 and 79.00, corresponding to the (220), (311) and (222) faces (the higher 2θ is for the NiO phase). The above three diffraction lines can be used to identify the formation of a solid solution. The NiO/MgO catalyst prepared by impregnation has three single peaks near the above 2θ values (fig. 9); this indicates that a solid solution was indeed formed. For the mechanical mixtures, the NiO peak in each of the double peaks decreases with increasing calcination time; this indicates that calcination leads to partial formation of a solid solution.

3.5. Surface composition and structure

The results of XPS analysis of three samples prepared by impregnation, but subjected to different treatments

Table 1
The binding energies of Mg(2p), Ni(2p) and O(1s)

Catalyst [a]	Binding energy (eV)		
	Mg(2p)	O(1s)	$Ni(2p_{3/2})$ [b]
sample a	48.8	530.0, 532.2	856.0
sample b	49.4	529.8, 532.0	855.4
sample c	49.3	529.5, 531.6	855.3
NiO (ref. [29])	–	529.9	854.5
MgO (ref. [30])	50.5	530.5	–

[a] Samples a, b and c are the unreduced (a), reduced with H_2 (b), and both reduced and used in the CO_2 reforming of methane (c) catalysts, prepared by impregnation and calcined at 800°C for 1.5 h.
[b] The peak at about 861.5 eV is due to the charge-transfer multielectron excitation (electron shake-up).

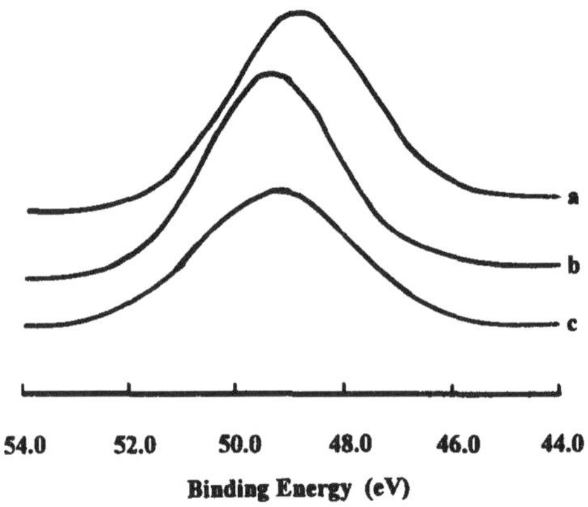

Fig. 11. XPS spectra of Mg(2p) for the NiO/MgO prepared by impregnation: (a) unreduced; (b) reduced; and (c) reduced and used for 2 h in the CO_2 reforming of methane.

(unreduced (a), reduced (b) and used in CO_2 reforming (c)), are summarized in table 1. There are two kinds of oxygen, with O(1s) binding energies of about 530 and 532 eV, respectively (fig. 10). The binding energy of Mg(2p) in sample a is 48.8 eV (fig. 11), which is lower than that in MgO (50.5 eV). The binding energy of Mg(2p) increases after reduction (sample b) and reaction (sample c). The binding energy of $Ni(2p_{3/2})$ in sample a is 856.0 eV, which is higher than 854.5 eV in NiO, and the binding energy of Mg(2p) in samples b and c are larger than in sample a. The surface compositions of the samples are listed in table 2, which shows that the unreduced catalyst (sample a) has 5.1/100, the reduced one (sample b) has 2.1/100 and sample c 2.3/100 mol ratio Ni/Mg.

4. Discussion

The XRD patterns indicate that, after calcination at 800°C for 1.5 h, the entire NiO in the NiO/MgO catalyst prepared by impregnation forms a solid solution with MgO (fig. 1). This catalyst has very high rate of CO formation and stability. Only a fraction of NiO forms a solid solution with MgO in the mechanical mixture, even when subjected to calcination at 800°C for 14 h (fig. 9). The amount of solid solution formed increases with increasing calcination time. In the CO_2 reforming of methane over the mechanical mixtures (fig. 1), the rate of CO formation is related to the calcination time; the longer the calcination time, the greater is the initial rate of formation of CO. Hence, the formation of a solid solution plays an important role in the generation of the active sites for the CO_2 reforming of methane. As pointed out by Parmaliana et al. [31], the first step in the formation of a NiO–MgO solid solution is the diffusion of NiO over the surface of MgO. This is followed by diffusion inside the MgO lattice. The longer the calcination time, the greater is the dispersion of NiO and for this reason the activity of the mechanical mixture of NiO and MgO increases with the calcination time. When the dispersion of NiO becomes sufficiently high, a large number of active sites is formed, and the initial activity is high. The activity decreases, however, with reaction time because of the clustering of the Ni atoms in particles. In contrast, for the NiO/MgO catalyst prepared by impregnation, even if the calcination time increases 13.3 times from 1.5 to 20 h, its surface area and the rate of CO formation in CO_2 reforming of methane do not change. This is because in the NiO/MgO catalyst the entire NiO is present in the solid solution, which stabilizes both oxides, and a greater calcination time does not affect the

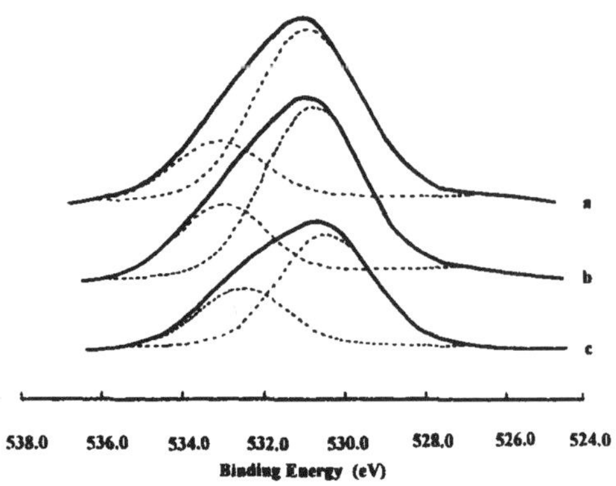

Fig. 10. XPS spectra of O(1s) for the NiO/MgO prepared by impregnation: (a) unreduced; (b) reduced; and (c) reduced and used for 2 h in the CO_2 reforming of methane.

Table 2
The Ni/Mg mol ratio of catalysts obtained by XPS

Catalyst [a]	Ni/Mg (mol. ratio)	
	total [b]	XPS
sample a	10.8/100	5.1/100
sample b	10.8/100	2.1/100
sample c	10.8/100	2.3/100

[a] The same as in table 1.
[b] Including the surface and bulk of the catalysts.

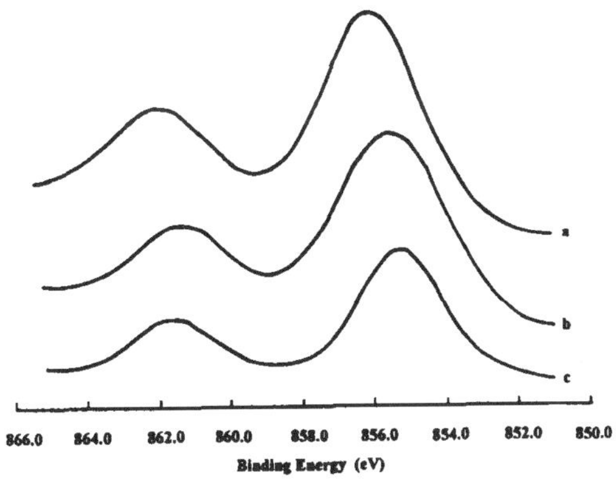

Fig. 12. XPS spectra of Ni($2p_{3/2}$) for the NiO/MgO prepared by impregnation: (a) unreduced; (b) reduced; and (c) reduced and used for 2 h in the CO_2 reforming of methane.

structure. In addition, only the NiO in the surface layer of the NiO/MgO solid solution catalyst is reduced during reduction below 1093 K [20] and the Ni sites generated remain in close contact with MgO; this hinders their sintering.

As the XPS analysis indicates, there are two kinds of oxygen with binding energies of about 530 and 532 eV in the NiO/MgO prepared by impregnation. Since the O(1s) binding energy of O^{2-} is about 530 eV [29,30], the binding energy of about 530 eV corresponds to O^{2-} in either NiO or MgO. The binding energy of 532 eV might be due either to a peroxide species (O_2^{2-}) or to a hydroxy species [32], most likely to a hydroxy species generated by the exposure of MgO to air before the measurement. Compared to MgO or NiO, the MgO/NiO solid solution prepared by impregnation has a lower Mg(2p) and a higher Ni($2p_{3/2}$) binding energy. Consequently, there is electron transfer from NiO to MgO. This increases the interactions between the two oxides and is probably responsible for the increased difficulty of reducing the NiO. Ni^0 was not detected in the reduced NiO/MgO, probably because the catalyst was reoxidized in air. However, since the Ni($2p_{3/2}$) binding energy is decreased after reduction in H_2, it is clear that at least some reduction of NiO has taken place. This strong interaction between NiO and MgO is probably responsible for the inhibition of sintering in the solid solution NiO/MgO. While NiO and MgO sinter easily, their solid solution is almost free of sintering. In addition, the strong interactions inhibit coke deposition, probably because they inhibit the formation of the large clusters needed for coke generation [9]. This explains the time stability of the CO yield.

The XPS analysis also shows that, although the total mole ratio Ni/Mg in the catalyst is 0.1, its surface mole ratio is only 0.05 for the unreduced catalyst and even lower, 0.02, for the reduced one. This indicates that the surface is enriched in MgO. The enrichment is higher in the reduced case because the interfacial free energy of a metal is higher than that of its oxide. Interestingly, the surface concentration of Ni after reaction is almost equal to that before reaction (table 2), indicating that the surface composition of the reduced NiO/MgO catalyst prepared by impregnation is stable in the CO_2 reforming of methane.

The response to pulses of CH_4/CO_2 over the catalyst prepared by impregnation reveals that the reaction occurs only on the reduced catalyst, indicating that the active site is the reduced NiO. The CO and CO_2 were detected during the response to pulses of CH_4 over the catalyst already used in the CO_2 reforming of methane. This indicates that some of the Ni active sites, were oxidized by CO_2 during reaction. This means that CO_2 decomposes to CO and O. CO and CO_2 were also detected during the response of pulses of O_2 over the catalyst already employed in the CO_2 reforming of CH_4. This indicates that C species are formed during the reforming reaction; it is, therefore, possible that the reaction includes the CH_4 decomposition to C and H species. One can, therefore, propose the following mechanism:

$$CH_4 + Ni \rightarrow CH_{4(s)}$$

$$CH_{4(s)} + Ni \rightarrow \text{---} \rightarrow C_{(s)} + 4H_{(s)}$$

$$CO_2 + Ni\ (\text{or MgO}) \rightarrow CO_{2(s)}$$

$$CO_{2(s)} + Ni \rightarrow O_{(s)} + CO_{(s)}$$

$$O_{(s)} + C_{(s)} \rightarrow CO_{(s)}$$

$$CO_{(s)} \rightarrow CO_{(g)} + Ni$$

$$H_{(s)} + H_{(s)} \rightarrow H_{2(s)} \rightarrow H_{2(g)} + Ni$$

$$H_{(s)} + O_{(s)} \rightarrow OH_{(s)}$$

$$OH_{(s)} + H_{(s)} \rightarrow H_2O_{(s)} \rightarrow H_2O_{(g)}$$

5. Conclusions

The NiO/MgO catalyst prepared by impregnation, which has very high activity and stability in the CO_2 reforming of methane, was studied by XRD, XPS, BET and pulse-MS response and compared with the catalyst prepared by mechanical mixing of the two oxides. We found that a calcination time of 1.5 h was long enough for the entire NiO to form a solid solution with MgO in the catalyst prepared by impregnation. In contrast, only a fraction of NiO formed a solid solution with MgO in mechanical mixtures of powders of NiO and MgO, even after 14 h of calcination. In the latter case, the amount of solid solution increased with calcination time. The rate of formation of CO in the CO_2 reforming of methane depends on the amount of NiO–MgO solid solution present in the catalyst; the greater the amount of solid solution, the higher is the rate of CO formation. The NiO/MgO catalyst prepared by impregnation had high stability, since the surface area and CO yield did not change either when the calcination time was increased from 1.5 to 20 h or with reaction time. The XPS analysis indicated the existence of electron transfer from NiO to MgO and of a surface enrichment in MgO. Probably the strong interactions between Ni and Mg inhibited both the sintering of NiO and MgO and the formation of the large clusters of Ni needed for coke deposition. The pulse-MS investigations suggested a possible mechanism for the CO_2 reforming of methane.

References

[1] A.T. Ashcroft, A.K. Cheetham, M.L.H. Green and P.D.F. Vernon, Nature 352 (1991) 225.
[2] J.T. Richardson and S.A. Paripatyadar, Appl. Catal. 61 (1990) 293.
[3] O. Yamazaki, T. Nozaki, K. Omata and K. Fujimota, Chem. Lett. (1992) 1953.
[4] A.M. Gadalla and M.E. Sommer, Chem. Eng. Sci. 44 (1989) 2825.
[5] A.M. Gadalla and B. Bower, Chem. Eng. Sci. 43 (1988) 3049.
[6] J.R. Rostrup-Nielsen and J.-H.B. Hansen, J. Catal. 144 (1993) 38.
[7] G.J. Kim, D.S. Cho, K.H. Kim and J.H. Kim, Catal. Lett. 28 (1994) 41.
[8] J.R. Rostrup-Nielsen, Stud. Surf. Sci. Catal. 36 (1988) 73.
[9] J.R. Rostrup-Nielsen, J. Catal. 85 (1984) 31.
[10] Z. Zhang and X.E. Verykios, J. Chem. Soc. Chem. Commun. (1995) 71.
[11] E. Ruckenstein and Y.H. Hu, J. Catal. 161 (1996) 55.
[12] E. Ruckenstein and Y.H. Hu, Appl. Catal. 133 (1995) 149.
[13] Y.H. Hu and E. Ruckenstein, Catal. Lett. 36 (1996) 145.
[14] E. Ruckenstein and Y.H. Hu, Appl. Catal., in press.
[15] H.B. Nussler and O. Kubaschewski, Z. Phys. Chem. (NF) 121 (1980) 187.
[16] E.G. Vrieland and P.W. Selwood, J. Catal. 3 (1964) 539.
[17] A. Cimino, M. Schiavello and F.S. Stone, Discussion Faraday Soc. 41 (1966) 350.
[18] A.F. Shestakov, V.A. Matishak, A.A. Kadushin and O.V. Krylov, Kinet. Catal. 20 (1979) 151.
[19] A. Zecchina, G. Spoto and S. Coluccia, J. Chem. Soc. Faraday I 80 (1984) 1875.
[20] A. Zecchina, G. Spoto and S. Coluccia, J. Chem. Soc. Faraday I 80 (1984) 1891.
[21] F. Arena, A. Parmaliana, N. Mondello, F. Frusteri and N. Giordano, Langmuir 7 (1991) 1555.
[22] A. Parmaliana, F. Arena, F. Frusteri and N. Giordano, J. Chem. Soc. Faraday Trans. 86 (1990) 2663.
[23] G.C. Bond and S.P. Sarsam, Appl. Catal. 38 (1988) 365.
[24] T. Borowiecki, Appl. Catal. 10 (1984) 273.
[25] Y.H. Hu and E. Ruckenstein, J. Catal. 163 (1996) 306.
[26] Y.H. Hu and E. Ruckenstein, Langmuir, submitted.
[27] F. Arena, B.A. Horrell, D.L. Cocke, A. Parmaliana and N. Giordano, J. Catal. 132 (1991) 58;
S. Narayanan and G. Sreekanth, J. Chem. Soc. Faraday Trans. I 85 (1989) 3785.
[28] E.P. Barret, L.G. Joyner and P.H. Halenda, J. Am. Chem. Soc. 73 (1951) 373.
[29] K.S. Kim and N. Winograd, Surf. Sci. 43 (1974) 625.
[30] J.S. Corneille, J.W. He and D.W. Goodman, Surf. Sci. 306 (1994) 269.
[31] F. Arena, F. Frusteri, A. Parmaliana, L. Plyasova and A.N. Shmakov, J. Chem. Soc. Faraday Trans. 92 (1996) 469.
[32] X.D. Peng and M.A. Barteau, Surf. Sci. 233 (1990) 283; Catal. Lett. 7 (1990) 395.

Carbon dioxide reforming of methane to synthesis gas over supported cobalt catalysts

E. Ruckenstein*, H.Y. Wang

Department of Chemical Engineering, State University of New York at Buffalo, Cliffort C. Furnas Hall, SUNY-Buffalo, Box 604200, Amherst, NY 14260-4200, USA

Received 9 March 2000; received in revised form 26 May 2000; accepted 27 May 2000

Abstract

The CO_2 reforming of CH_4 over Co supported on an alkaline earth metal oxide (MgO, CaO, SrO, or BaO) as well as on γ-Al_2O_3 and on SiO_2 was investigated. Among these supports, only MgO exhibited a high and stable activity. It provided a CO yield of 93% and a H_2 yield of 90% at the high space velocity of 60 000 ml g^{-1} h^{-1}, which remained unchanged during the period of study of 50 h. γ-Al_2O_3 provided initially a high CO yield, which, however, rapidly decayed. All the other supports exhibited low CO yields and CaO and SiO_2 also low stabilities. A solid solution of CoO and MgO was identified by XRD in the calcined MgO supported catalyst, but not in the other supports. Because the oxygens in the solid solution are shared by both Mg and Co and their interactions with Mg are strong, the solid solution is less reducible than the pure CoO and small clusters of metallic Co are generated. Being at least partially embedded in the substrate, these clusters are more stable to sintering than the usual ones; being small they do not favor coke formation. For these reasons, this catalyst exhibits high stability.

Keywords: Carbon dioxide reforming of methane; Synthesis gas; Co catalyst; Effect of support

1. Introduction

The catalytic reforming of CH_4 with CO_2 to synthesis gas (reaction 1) has received increasing attention in the past few years [1–14] because it provides a low H_2/CO ratio (≤ 1), which is suitable for the Fischer–Tropsch synthesis to higher hydrocarbons, and employs CO_2 (a greenhouse gas):

$$CH_4 + CO_2 \rightarrow 2CO + 2H_2; \quad \Delta H_{298}{}^0 = 247\,kJ/mol \qquad (1)$$

Most of the group VIII metals (e.g. Ni, Co, Ru, Rh, Pd, Pt, Ir) are more or less catalytically active towards this reaction. However, being strongly endothermic, the CO_2 reforming of CH_4 requires high temperatures (typically, 1073–1173 K) and this stimulates deactivation caused by carbon deposition and/or metal sintering. Generally, the noble-metal based catalysts exhibit less carbon deposition. However, because of their high price and limited availability, it is desirable to identify catalysts that involve the other metals of group VIII. Some supported Ni catalysts [7,8], particularly the MgO supported catalysts [6,9], have been reported to provide high activity and long life without obvious deactivation. The formation of a NiO–MgO solution in the NiO/MgO catalyst was suggested to be responsible for its high activity and stability [6,9]. As

* Corresponding author. Tel.: +1-716-645-2911/2214; fax: +1-716-645-3822.
E-mail address: feaeliru@acsu.buffalo.edu (E. Ruckenstein).

This paper, which was originally published in **Applied Catalysis A**: General 204, 257–263 (2000), is republished with permission.

Table 1
Crystal data for NaCl type metal oxides [15]

Compound	*a* (Å)	*A–B* distance (Å)
MgO	4.2112	2.11
CaO	4.8105	2.40
SrO	5.1396	2.57
BaO	5.5391	2.77
NiO	4.1940	2.10
CoO	4.2603	2.13

shown in Table 1, similar to NiO, the lattice parameter and bond distance of CoO are close to those of MgO. This suggested us to explore if MgO supported Co is also an effective catalyst for the CO_2 reforming of CH_4. Moreover, the effect of other supports such as the other alkaline earth metal oxides (CaO, SrO, and BaO), γ-Al_2O_3, and SiO_2 was also investigated.

2. Experimental

2.1. Catalyst preparation

Alkaline earth metal oxides (MgO, CaO, SrO, and BaO) as well as γ-Al_2O_3 and SiO_2 were used as supports. The supported cobalt catalysts were prepared by impregnating the supports with aqueous solutions of $Co(NO_3)_2{\cdot}6H_2O$, followed by overnight drying at 110°C and calcination in the open air of a furnace at 800°C for 8 h. The calcined catalysts are denoted $Co(O)/M_xO_y$. The catalysts reduced in H_2 are denoted Co/M_xO_y. Co loading means weight percent of Co in the completely reduced catalyst; all the catalysts had a loading of 12 wt.%.

2.2. Catalytic reaction

All the catalysts were tested under atmospheric pressure in a fixed-bed vertical quartz reactor (i.d. 3 mm), which was operated in a down flow mode with the catalyst held on a quartz wool bed. The temperatures were monitored using two thermocouples: one located in the middle of the catalyst bed and the other one immediately after the quartz wool bed to measure the temperature of the gas phase. The reduction of the catalyst was carried out by increasing the temperature from room temperature to 600°C at a rate of 20°C min^{-1} and from 600 to 900°C at a rate of 10°C min^{-1}, in a H_2 flow (20 ml min^{-1}), without holding at 900°C. After reduction, the feed gases (CH_4/CO_2=1/1) were introduced into the catalyst bed at a flow rate of 20 ml min^{-1}, which for a 20.0 mg catalyst corresponds to a space velocity (SV) of 6.0×10^4 ml h^{-1} g^{-1}. The reactants and products were analyzed with an on-line gas chromatograph equipped with a Porapak Q column. An ice-cold trap was set between the reactor exit and the GC sampling valve to remove the water formed during reaction.

In the absence of the catalyst, the blank runs with quartz wool indicated that methane conversion was below 3% at 900°C. Consequently, the contribution of the gaseous phase reaction was insignificant under the conditions employed in the present paper.

2.3. Catalyst characterization

2.3.1. BET surface area and the exposed Co surface area

The BET surface areas of the calcined catalysts, determined via nitrogen adsorption using a Micromeritics ASAP 2000 instrument, are listed in Table 2. All the samples were degassed at 200°C for at least 5 h in high vacuum before measurements.

The exposed Co metal surface area of the reduced catalyst was determined by CO pulse adsorption in a quartz tube (i.d. 4 mm) at room temperature, assuming a 1/1 stoichiometry. Fifty milligrams of calcined catalyst powder was held on a quartz wool bed in a vertical quartz tube reactor and its reduction was carried out as described in the previous section. After reduction, the reduced catalyst was purged with an ultra high purity helium flow (35 ml min^{-1}) at 900°C for 0.25 h, the sample was cooled to room temperature and CO was pulsed (10 μl per pulse) over the catalyst until no further adsorption of CO was detected. The CO left after CO adsorption was determined quantitatively with a thermal conductivity detector (TCD).

2.3.2. CH_4 decomposition

The decomposition of pure methane over the reduced catalyst was carried out in a pulse microreactor. A quartz tube (i.d. 4 mm), in which the calcined catalyst was held on a quartz wool bed, was employed as reactor. In each experiment, 50.0 mg of catalyst was used, and its reduction was carried out as

Table 2
Activities of supported Co (12 wt.% loading) catalysts for CO_2 reforming of methane at 900°C ($T_{furnace}$)[a]

Support	Surface area (m^2/g-cat.)	Co surface area (m^2/g-cat.)×100	Time after onset of reaction (h)	Temperature (°C) bed/gas	CH_4 conv. (%)	CO_2 conv. (%)	CO yield (%)	H_2 yield (%)	H_2/CO ratio
MgO	22	11	0.5	870/894	91.9	93.9	92.9	89.8	0.97
			20	875/895	91.9	93.5	92.7	89.8	0.97
CaO	5	48	0.5	889/896	38.6	54.3	46.3	31.9	0.69
			20	897/895	16.6	32.6	24.7	8.2	0.33
SrO	4	6	0.5	904/900	3.4	7.9	5.7	1.1	0.18
			20	904/900	14.8	29.0	22.0	7.2	0.32
BaO	2	–[b]	0.5	905/900	3.2	7.7	5.5	0.9	0.16
			20	903/897	11.4	23.6	17.6	4.9	0.28
γ-Al_2O_3	53	68	0.5	870/890	86.5	88.1	87.3	84.8	0.97
			20	900/895	58.8	58.7	58.8	58.9	1.00
SiO_2	451	80	0.5	896/897	24.0	36.7	30.3	17.9	0.59
			20	900/898	12.9	23.7	18.4	7.3	0.40

[a] Catalyst: 20.0 mg; flow rate: 20 ml/min (CH_4/CO_2=1/1); space velocity: 60 000 ml h^{-1} g^{-1}.
[b] Too small to be detected.

described previously. After reduction, the catalyst was purged with He (35 ml min^{-1}) at 900°C for 0.25 h and then pulses of CH_4 (250 μl) were injected in the carrier gas, which was flowing through the reactor. The products were analyzed with an on-line gas chromatograph equipped with a TCD and a Porapak Q column.

2.3.3. Temperature-programmed reduction (TPR)

During each TPR run, a high-purity 2.5% H_2/Ar mixture (35 ml/min), which was additionally purified using Hydro-Purge II and Oxy-Trap columns, was passed over a calcined sample in a vertical quartz tube reactor (i.d. 4 mm). The sample (10.0 mg) was held on a quartz wool bed and the temperature of the sample was increased from 40 to 1000°C, at a rate of 20°C min^{-1}. The water, formed during reduction, was trapped in a Hydro-Purge II column. The hydrogen consumed in TPR was monitored continuously with a thermal conductivity detector (TCD).

2.3.4. X-ray powder diffraction (XRD)

A Siemens D500 X-ray diffractometer instrument with monochromatized Cu Kα radiation was used for the X-ray determinations. The calcined catalysts were examined at a scanning rate of 1.2° min^{-1} and the resulting X-ray diffraction patterns were used to identify the major phases present.

3. Results

3.1. Carbon dioxide reforming of methane

3.1.1. Effect of support

The alkaline earth metal oxides (MgO, CaO, SrO, and BaO) as well as γ-Al_2O_3 and SiO_2 were used as supports. The stoichiometric CO_2-reforming reaction was carried out at atmospheric pressure, a space velocity of 6.0×10^4 ml g^{-1} h^{-1} and a $T_{furnace}$ of 900°C. The catalytic activities after 0.5 and 20 h on stream are listed in Table 2. The performance of the supported catalyst is strongly affected by the support. Among the supports investigated, MgO provided the highest conversions of CH_4 and CO_2 (92 and 94%, respectively) and yields to CO and H_2 (93 and 90%, respectively), and its conversions and yields after 20 h on stream remained as after 0.5 h. Over CaO, the conversions of CH_4 and CO_2 after 0.5 h on stream were about 39 and 54%, respectively, and decreased to about 17 and 33%, respectively, after 20 h. SrO provided very low conversions of CH_4 and CO_2 (3 and 8%, respectively) after 0.5 h on stream, but a slight increase in conversions was noted after 20 h on stream. A very similar behavior was observed over BaO. γ-Al_2O_3 provided after 0.5 h on stream high conversions of CH_4 and CO_2 (87 and 88%, respectively), which decreased, however, to 59% after 20 h on stream. Over SiO_2, the conversions of CH_4 and CO_2 after 0.5 h on stream were 24 and 37%, respectively, but after 20 h of reaction, they decreased to 13 and 24%, respectively. The larger conversion of CO_2 than that of CH_4 as well as the H_2/CO ratio <1 might be caused by the reverse water gas shift reaction ($CO_2+H_2=CO+H_2O$).

Table 2 also provides information about the temperatures of the catalyst bed and of the gas phase just after the quartz wool bed. Because the reforming reaction is endothermic, the temperature of the catalyst bed should be lower than that of the gas phase for sufficiently high activity reactions. Indeed, this happens over the MgO during the entire reaction period and over Al_2O_3 as long as the activity is high.

3.1.2. Effect of time on stream

The time-dependent conversions of CH_4 and CO_2 and yields to CO and H_2 of supported Co catalysts are plotted in Fig. 1, which shows that only over MgO, the conversions of CH_4 and CO_2 and yields to CO and H_2 remained unchanged during the period of study of 50 h. However, deactivation was observed over CaO, γ-Al_2O_3, and SiO_2. Over the SrO and BaO, the conversions and the yields were initially very low; they increased with time on stream to some extent, but remained very low even after 50 h on stream.

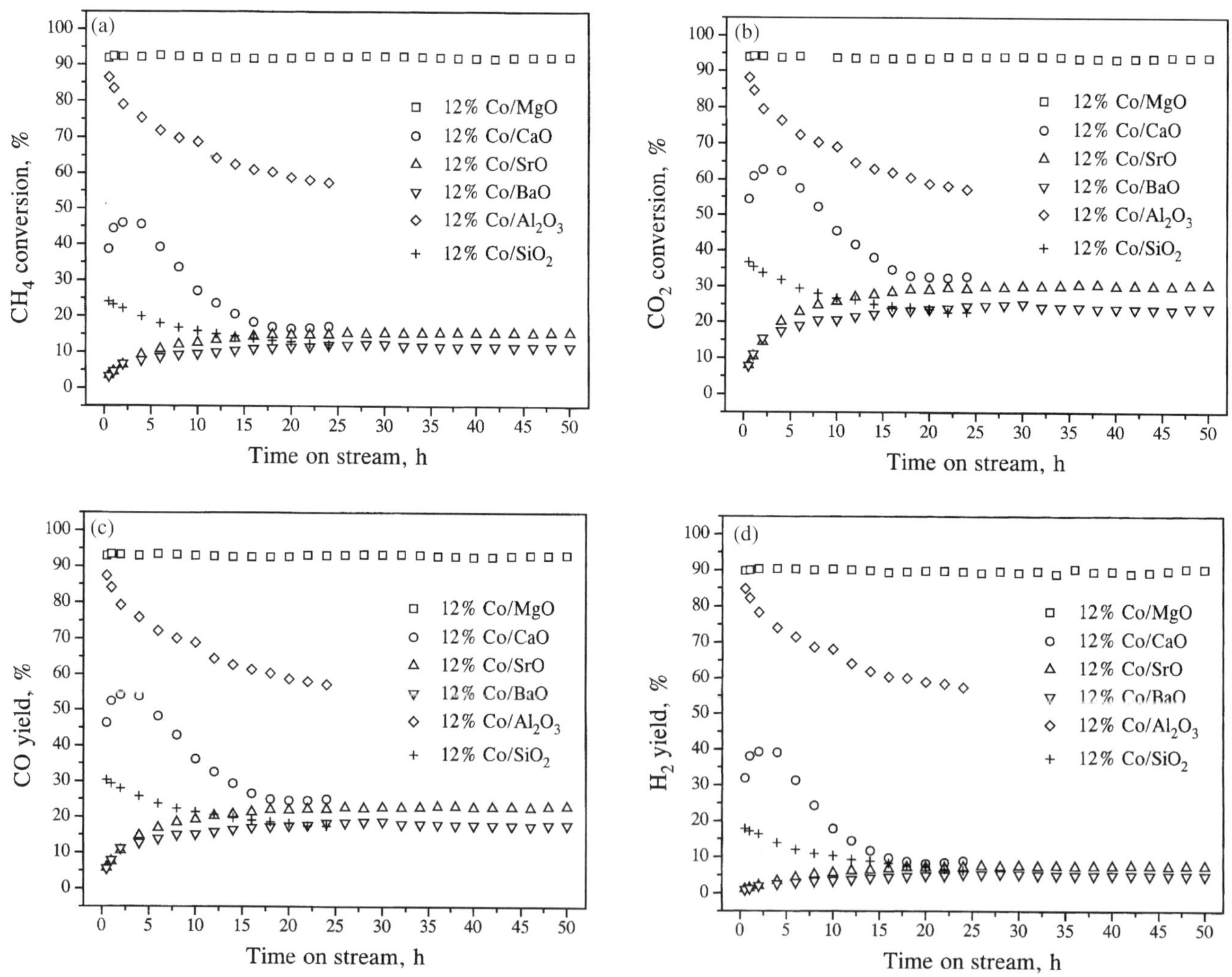

Fig. 1. Time-dependent conversions of CH_4 (a) and CO_2 (b) and yields to CO (c) and H_2 (d) over the supported Co catalysts. P=1 atm, $T_{furnace}$=900°C, CH_4/CO_2=1.0, space velocity=60 000 ml h^{-1} g^{-1}.

3.2. *Physico-chemical characterization*

3.2.1. *Exposed metallic Co surface area of the reduced catalyst*

After the calcined catalysts were reduced with H_2, the exposed metal surface areas were determined by CO chemisorption. Table 2 shows that the exposed metal surface areas were small, and their sequence was SiO_2>γ-Al_2O_3>CaO>MgO>SrO>BaO. For the last support, almost no measurable amount of adsorbed CO could be detected.

3.2.2. *Surface carbon (C̲) formed in the CH_4 decomposition*

The decomposition of CH_4 pulses over reduced catalysts was carried out at 900°C and the amount of surface carbon(C̲) generated was calculated from the carbon balance. Table 3 shows that the amount of C̲ accumulated over the MgO supported catalyst was much lower than those over SiO_2, γ-Al_2O_3 and CaO, which provided in the CO_2-reforming of CH_4 initial (after 0.5 h) CH_4 conversions higher than 20%. The amount of carbon deposited was somewhat smaller over BaO and somewhat larger over SrO than on MgO. The initial conversions of CH_4 for the reforming reaction over SrO and BaO were, however, below 5% (Table 2).

3.2.3. *Reduction characteristics of the calcined supported Co catalysts*

The TPR spectra over the calcined catalysts are presented in Figs. 2 and 3. For Co(O)/MgO, a peak started to form at about 870°C (Fig. 2a). A single peak at about 550°C was observed for Co(O)/CaO (Fig. 2b), and two peaks, one (the main) at about 480°C and the other one as a shoulder at about 695°C for Co(O)/SrO (Fig. 2c). Two discrete peaks, one at about 445°C and the other at about 830°C, were noted for Co(O)/BaO (Fig. 2d), and two peaks, one at about 460°C and the other one at about 910°C for Co(O)/γ-Al_2O_3 (Fig. 3a).

Table 3
The amount of carbon deposition formed from CH_4 decomposition over the reduced supported Co (12 wt.% loading) catalysts at 900°C

Catalyst	C̲ formed during different period (μmol)				
	1st pulse	1st 3 pulses	1st 5 pulses	1st 10 pulses	1st 20 pulses
Co/MgO	2.6	6.4	7.6	10.6	15.4
Co/CaO	9.5	17.4	20.7	26.0	32.9
Co/SrO	6.3	9.4	11.3	14.7	19.2
Co/BaO	1.7	4.3	5.8	8.8	15.0
Co/γ-Al_2O_3	6.1	17.9	21.8	26.3	32.1
Co/SiO_2	3.6	22.9	42.6	93.3	186.6

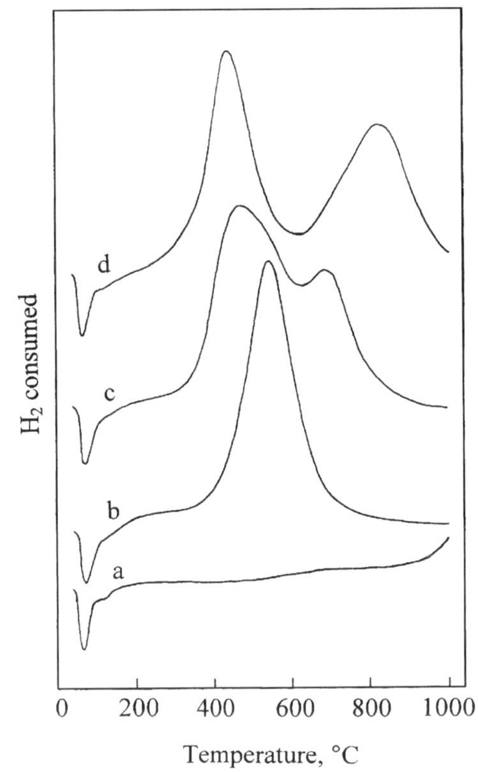

Fig. 2. TPR profiles of calcined Co (12 wt.% loading) catalysts supported on MgO (a), CaO (b), SrO (c), BaO (d).

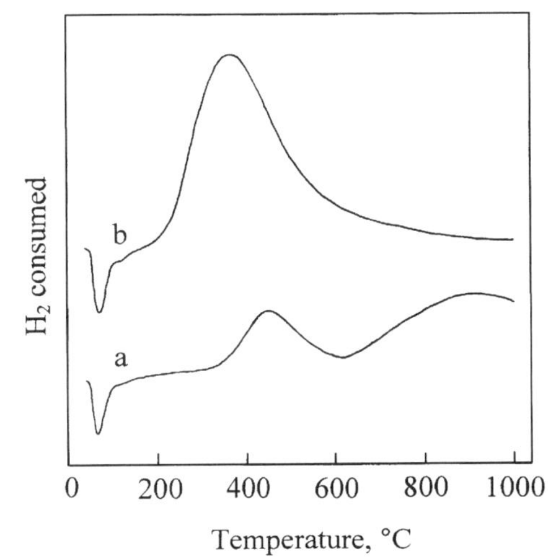

Fig. 3. TPR profiles of calcined Co (12 wt.% loading) catalysts supported on γ-Al_2O_3 (a), and SiO_2 (b).

Table 4
Data and assignments of XRD patterns of calcined supported Co (12 wt.% loading) catalysts

Catalysts[a]	*d* (Å)	Assignments
12%Co(O)/MgO	2.106, 1.490, 1.269, 1.214, 2.431	(Co, Mg)O
12%Co(O)/CaO	2.275, 2.680, 3.577	$Ca_3Co_4O_9$
12%Co(O)/SrO	2.740, 3.130, 2.045, 1.640, 1.583	$SrCoO_{2.52}$
12%Co(O)/BaO	3.394, 2.792, 2.175, 2.138, 1.706; 3.394, 2.792, 2.151, 1.706, 2.378	$BaCoO_{2.63}$[b]; $BaCoO_{2.8}$
12%Co(O)/γ-Al_2O_3	2.440, 2.016, 1.431, 2.867, 1.558	Co_3O_4, Co_2AlO_4 or $CoAl_2O_4$
12%Co(O)/SiO_2	2.435, 1.427, 1.555, 2.861, 2.028	Co_3O_4

[a] All the samples were calcined at 800°C for 8 h.
[b] Other possible matches include $BaCoO_{2.93}$ and $BaCoO_{2.70}$.

Finally, Co(O)/SiO_2 exhibited a single peak at about 370°C (Fig. 3b).

3.2.4. *Compounds present in the calcined supported Co catalysts*

The Co-containing species present in the calcined catalysts were identified by XRD. The data and the assignments of the XRD patterns for the Co(O)/M_xO_y (M=Mg, Ca, Sr, Ba, Al, and Si) catalysts are listed in Table 4. A solid solution between MgO and CoO (represented as (Co, Mg)O) was detected over the MgO supported Co catalysts, and this is in agreement with the TPR spectrum (Fig. 2a), which exhibited a high-temperature reduction peak that started to form at about 870°C. $Ca_3Co_4O_9$ was identified over Co(O)/CaO, and this is again in agreement with the TPR spectrum (Fig. 2b) which provided a single reduction peak. Only $SrCoO_{2.52}$ was detected over Co(O)/SrO, while two reduction peaks were present in the TPR spectrum (Fig. 2c). Most likely, the amount of the species responsible for the TPR shoulder was too small to be detectable by XRD. Two compounds, $BaCoO_{2.63}$ and $BaCoO_{2.8}$, were identified over Co(O)/BaO, and this is in agreement with the TPR spectrum (Fig. 2d) which indicated two discrete reduction peaks. The X-ray patterns obtained over Co(O)/γ-Al_2O_3 can be attributed to Co_3O_4, Co_2AlO_4 or $CoAl_2O_4$. Because the TPR spectrum (Fig. 3a) indicated two reduction peaks, one at about 460°C and the other at about 910°C, and Co_3O_4 can be reduced in H_2 below 500°C [16,17] and the reducibility decreases in the sequence: Co_3O_4>Co_2AlO_4>$CoAl_2O_4$, we attributed the first peak to Co_3O_4 and the second to Co_2AlO_4. Finally, over Co(O)/SiO_2, Co_3O_4 was identified, and this is in agreement with the TPR spectrum which exhibited a single peak at about 370°C (Fig. 3b).

4. Discussion

Among the supports investigated, only MgO provided a high and stable activity during the period of study of 50 h; various extents of deactivation were observed over γ-Al_2O_3, CaO, and SiO_2 and low activities over SrO and BaO (Table 2 and Fig. 1). The low activities of SrO and BaO supported catalysts were most likely due to the unavailability of accessible metallic Co sites after reduction (Table 2). This is probably caused by the sintering of the SrO and BaO and the burying of the active sites by the former. Indeed, the surface areas of the catalysts supported on CaO and SrO (and even BaO) were comparable before reduction; the exposed metal surface area after reduction was, however, much lower over SrO and negligible over BaO (Table 2). The carbon deposition and the metal sintering are the main causes for catalyst deactivation during reaction [2,12–14,18]. The latter not only decreases the number of active sites but also accelerates carbon deposition, since large metal ensembles are needed for coke formation [19]. The carbon deposition during the CH_4 decomposition was small over the MgO supported catalyst compared to most of the supported catalysts investigated (Table 3). It was somewhat smaller over BaO support and somewhat larger over SrO (Table 3), but the initial conversions of CH_4 and CO_2 were very small for the CO_2-reforming of CH_4 in the latter cases (Table 2), because of the absence of metallic active sites.

The reducibility of the catalyst is directly related to the interactions between CoO and support. The more difficult the reduction is, the stronger these interactions are. Among the supported catalysts investigated, the Co(O)/MgO is the most difficult to reduce (Figs. 2 and 3); hence the interactions in this catalyst are the strongest.

MgO, CaO, SrO, BaO, and CoO all have a NaCl type structure. However, only the MgO has a lattice parameter and bond distance close to CoO (Table 1). As a result, CoO and MgO are completely miscible and can form a solid solution (Table 4). Due to the irreducibility of MgO and the fact that the oxygen belonging to Co interacts also with Mg, it is difficult to reduce the CoO dispersed in the lattice of MgO. Indeed, Co(O)/MgO requires the highest temperature to be reduced (Figs. 2 and 3), and the exposed metal surface area after reduction is much lower over MgO supported catalyst than over CaO, γ-Al_2O_3 and SiO_2 supported catalysts (Table 2). The size of the metal clusters thus formed is expected to be small. Being generated from a solid solution, the clusters take the form of hillocks protruding from the substrate; for this reason they are more resistant to sintering than the usual crystallites. Moreover, during the catalytic process the metallic active sites are successively oxidized and reduced because of the coexistence of oxidative (CO_2 and H_2O) and reducing (CH_4 and H_2) agents. At a given moment, the clusters contain both Co^0 and CoO. The strong interactions between CoO and MgO additionally hinder the sintering of the clusters [20]. Finally, being small, the clusters do not favor coke formation. In summary, the fact that MgO and CoO can form a solid solution is responsible for the high stability of the Co/MgO catalyst.

5. Conclusion

The Co/MgO catalyst provided a high and stable activity for the CO_2-reforming of CH_4. The suppression of carbon deposition and the resistance to sintering of this catalyst are induced by the fact that CoO and MgO form a solid solution.

References

[1] A.M. Gadalla, B. Bower, Chem. Eng. Sci. 43 (11) (1988) 3049.
[2] J.T. Richardson, S.A. Paripatyadar, Appl. Catal. 61 (1990) 293.
[3] A.T. Ashcroft, A.K. Cheetham, M.L.H. Green, P.D.F. Vernon, Nature 352 (1991) 225.
[4] J.R. Rostrup-Nielsen, J.-H.B. Hansen, J. Catal. 144 (1993) 38.
[5] A. Erdöhelyi, J. Cserényi, F. Solymosi, J. Catal. 141 (1993) 287.
[6] E. Ruckenstein, Y.H. Hu, Appl. Catal. A: Gen. 133 (1995) 149.
[7] Z.L. Zhang, X.E. Verykios, J. Chem. Soc., Chem. Commun. (1995) 71.
[8] V.C.H. Kroll, H.M. Swaan, C. Mirodatos, J. Catal. 161 (1996) 409.
[9] Y.H. Hu, E. Ruckenstein, Catal. Lett. 36 (1996) 145.
[10] R.N. Bhat, W.M.H. Sachtler, Appl. Catal. A: Gen. 150 (1997) 279.
[11] H.Y. Wang, C.T. Au, Appl. Catal. A: Gen. 155 (1997) 239.
[12] S.M. Stagg, E. Romeo, C. Padro, D.E. Resasco, J. Catal. 178 (1998) 137.
[13] K. Tomishige, Y. Chen, K. Fujimoto, J. Catal. 181 (1999) 91.
[14] J.H. Bitter, K. Seshan, J.A. Lercher, J. Catal. 183 (1998) 336.
[15] I. Náray-Szabó, Inorganic Crystal Chemistry, Akadémiai Kiadó, Budapest, 1969, p. 237.
[16] K.S. Chung, F.E. Massoth, J. Catal. 64 (1980) 320.
[17] P. Arnoldy, J.A. Moulijn, J. Catal. 93 (1985) 38.
[18] Z.L. Zhang, V.A. Tsipouriari, A.M. Efstathiou, X.E. Verykios, J. Catal. 158 (1996) 51.
[19] J.R. Rostrup-Nielsen, in: J.R. Anderson, M. Bourdart (Eds.), Catalysis Science and Technology, Vol. 5, Springer, Berlin, 1984, pp. 1–118.
[20] E. Ruckenstein, in: S.A. Stevenson, J.A. Dumesic, R.T.K. Baker, E. Ruckenstein (Eds.), Metal-Support Interactions in Catalysis, Sintering, and Redispersion, Van Nostrand Reinhold, New York, 1987, pp. 141–301.

CO_2 reforming of CH_4 over Co/MgO solid solution catalysts — effect of calcination temperature and Co loading

H.Y. Wang, E. Ruckenstein*

Department of Chemical Engineering, State University of New York at Buffalo, Amherst, NY 14260, USA

Received 9 June 2000; received in revised form 4 August 2000; accepted 5 August 2000

Abstract

The effect of calcination temperature on the CO_2 reforming of CH_4 over Co/MgO of various Co loadings was investigated and the results can be summarized as follows: (i) the catalysts with Co loadings between 8 and 36 wt.% precalcined at 500 or 800°C provided high and stable activities; (ii) for Co loadings between 4 and 48 wt.%, the high calcination temperature of 900°C yielded low activities or very long activation periods; (iii) a high Co loading (48 wt.%) combined with a calcination temperature of 500 or 800°C resulted in unstable activities. Depending on the calcination temperature and Co loading, one, two or three of the following Co-containing species, Co_3O_4, $MgCo_2O_4$, and (Co, Mg)O (solid solution of CoO and MgO) were identified by combining temperature programmed reduction and X-ray diffraction (XRD) results. Their reducibility decreased in the sequence: $Co_3O_4 > MgCo_2O_4 > (Co, Mg)O$. It is suggested that the high and stable activities observed over most catalysts (except the 48 wt.% one) precalcined at 500 or 800°C as well as the low activities or the long activation periods observed over the 900°C calcined catalysts were induced by the formation of a solid solution between CoO and MgO. The activity decay observed over the 48 wt.% Co catalyst precalcined at 500 or 800°C was most likely caused by the large metallic particles, formed particularly through the reduction of Co_3O_4 and $MgCo_2O_4$. These particles sintered and stimulated coke formation.

Keywords: Carbon dioxide reforming of methane; Synthesis gas; Co catalyst; Effect of calcination temperature; Effect of Co loading

1. Introduction

Currently, the catalytic steam reforming is used for converting methane to synthesis gas (mixture of H_2 and CO). This process produces, however, a synthesis gas with a too high ratio of H_2:CO (>3) which is unsuitable for the Fischer–Tropsch synthesis of liquid hydrocarbons. The CO_2 reforming of CH_4 which produces a synthesis gas with a H_2:CO ratio close to 1, appears to be a more suitable one.

Numerous supported metal catalysts have been tested for this reaction including Ni and Co-based catalysts as well as supported noble metals (Rh, Ru, Pd, Pt, and Ir) [1–15]. The main problem encountered was the rapid deactivation caused by carbon deposition and/or sintering of the metal. In general, the noble-metal based catalysts can provide operations with lower carbon deposition. However, considering their high cost, it is desirable to employ non precious metals.

In a previous paper [16], the effect of support on the performance of Co-based catalyst was investigated and it was found that the 12 wt.% Co/MgO catalysts provided a CO yield of 93% and a H_2 yield of 90% at the high space velocity of 60,000 ml g^{-1} h^{-1}, with very high stability. It was suggested that the suppression of carbon deposition and the resistance to sintering of this catalyst were induced by the formation of

* Corresponding author.
E-mail address: feaeliru@acs.buffalo.edu (E. Ruckenstein).

This paper, which was originally published in **Applied Catalysis A**: General 209, 207-215 (2001), is republished with permission.

a solid solution between CoO and MgO, because both have a NaCl type structure and close lattice parameters and bond distances [17].

The formation of a solid solution is expected to depend on the Co loading, calcination atmosphere, calcination temperature, and calcination time. To obtain information about their effect, experiments have been carried out in which Co loading was varied between 4 and 48 wt.%, and the calcination temperature between 500 and 900°C. Temperature-programmed reduction (TPR) and X-ray diffraction (XRD) were performed to characterize the structural properties of the calcined catalysts and CO chemisorption to determine the exposed metal surface areas of the reduced catalysts. The results showed that a solid solution between CoO and MgO was indeed formed in the calcined catalysts. However, depending upon the calcination temperature and Co loading, Co_3O_4 and Co_2MgO_4 could also be generated. The reduction behavior and the catalytic performance are affected by the species formed in the catalysts and their amounts. Possible explanations for the obtained results are provided.

2. Experimental

2.1. Catalyst preparation

The MgO supported Co catalysts were prepared by impregnating MgO (32 m^2/g) with aqueous solutions of $Co(NO_3)_2 \cdot 6H_2O$, followed by overnight drying at 110°C and calcination in the open air of a furnace for 8 h at 500, 800 or 900°C. The calcined catalysts are denoted as Co(O)/MgO (500, 800 or 900°C), and the catalysts reduced in H_2 are denoted as Co/MgO (500, 800 or 900°C). The temperature inside the parenthesis indicates the calcination temperature. The Co loading means wt.% Co in the completely reduced catalyst.

2.2. Catalytic reaction

All the catalysts were tested under atmospheric pressure in a fixed-bed vertical quartz reactor (i.d. 3 mm) with the catalyst held on a quartz wool bed. The reactor was operated in a down flow mode, and the temperatures were monitored using two thermocouples: one located in the middle of the catalyst bed and the other one immediately below the quartz wool bed to measure the temperature of the gas phase. The catalyst was reduced in a H_2 flow (20 ml/min) by increasing the temperature from room temperature to 600°C at a rate of 20°C/min and then to 900°C at a rate of 10°C/min without holding at 900°C. After reduction, the feed gases (CH_4:CO_2 = 1:1) were allowed to flow through the catalyst bed at a rate of 20 ml/min. For a 20.0 mg catalyst, this corresponds to a space velocity (SV) of 60,000 ml h^{-1} g^{-1}. The reactants and products were analyzed with an on-line gas chromatograph (GC) equipped with a Porapak Q column. An ice-cold trap was located between the reactor exit and the GC sampling valve to remove the water formed during reaction.

In the absence of the catalyst, the blank runs with quartz wool and two thermocouples indicated that at 900°C, the CH_4 conversion was below 3%. Consequently, the contribution of the homogeneous reaction was insignificant under the conditions employed.

2.3. Catalyst characterization

2.3.1. CO adsorption

The exposed Co metal surface area of the reduced catalyst was determined by CO pulse adsorption in a quartz tube (i.d. 4 mm) at room temperature. Even though CO might also have been adsorbed in bridged and twinned forms besides the linear one, a 1:1 stoichiometry of CO adsorption on Co was assumed. The sample (50 mg) of calcined catalyst powder was held on a quartz wool bed in a vertical quartz tube reactor and its reduction was carried out as already described. The reduced catalyst was purged with an ultra high purity helium flow (35 ml/min) at 900°C for 0.25 h. After the sample was cooled to room temperature, CO was injected as pulses (10 μl per pulse) into the helium flow until no further adsorption of CO was detected. The CO left after CO adsorption was determined quantitatively with a thermal conductivity detector (TCD). Before use, H_2 and He were purified using Hydro-Purge II and Oxy-Trap columns.

2.3.2. Temperature-programmed reduction

During each TPR run, a high-purity 2.5% H_2/Ar mixture (35 ml/min) which was additionally purified using Hydro-Purge II and Oxy-Trap columns was passed over the calcined samples in a vertical quartz tube reactor (i.d. 4 mm). The sample (10.0 mg, unless otherwise indicated) was held on a quartz wool bed. The temperature of the sample was increased from 40 to 1000°C at a rate of 20°C/min. The water, formed during reduction was trapped in a Hydro-Purge II column. The hydrogen consumed in TPR was monitored continuously with a TCD which was calibrated using known amounts of CoO and Co_3O_4.

2.3.3. *X-ray powder diffraction*

The XRD patterns, obtained on a Siemens D500 X-ray diffractometer instrument with a Cu Kα radiation at 40 kV and 30 mA were used to identify the major phases present in the calcined catalysts.

3. Results

3.1. *Carbon dioxide reforming of methane*

The initial (after 0.5 h on-stream) catalytic activities (conversions of CH_4 and CO_2 and yields to CO and H_2) as well as the temperatures of the catalyst and the gas phase are listed in Table 1. With the exception of the 4 wt.% Co/MgO (800°C), the catalysts precalcined at 500 and 800°C provided initially high activities (CH_4 conversion >75%) and H_2:CO ratios close to 1.0. For all the catalysts precalcined at 900°C, the initial activities were low (CH_4 conversion <20%) and the H_2:CO ratio was much lower than 1.0 but both characteristics increased with increasing loading.

The time-dependent activities are plotted in Figs. 1–3 (only the conversion of CH_4 was plotted because that of CO_2 and the yields to CO and H_2 followed the same pattern with time on stream). The results can be summarized as follows: (i) at a calcination temperature of 500°C, the catalysts with Co loadings between 4 and 36 wt.% provided high and stable activities; deactivation with time on stream was, however, observed over the 48 wt.% catalyst; (ii) at a calcination temperature of 800°C, the catalysts with a Co loading between 8 and 36 wt.% yielded high and stable activities; deactivation with time on stream was, however, observed over the 48 wt.% catalyst; (iii) at a calcination temperature of 900°C, the catalysts with a Co loading ≤12 wt.% provided low activities (CH_4 conversion <15%), the 24 wt.% catalyst needed a very long activation period to reach a high steady-state activity (85%), and the 36 and

Table 1
Effect of calcination temperature and Co loading on initial (after 0.5 h) activity of Co/MgO for CO_2 reforming of methane at a $T_{furnace}$ of 900°C[a]

Co loading (wt.%)	Calcination temperature (°C)	Temperature (°C) bed/gas	CH_4 conversion (%)	CO_2 conversion (%)	CO yield (%)	H_2 yield (%)	H_2:CO ratio
4	500	887/895	85.7	87.6	86.7	84.3	0.97
	800	895/898	7.6	17.4	12.6	2.5	0.20
8	500	878/894	94.7	96.7	95.7	91.8	0.96
	800	881/891	76.0	77.8	77.0	74.6	0.97
	900	894/891	5.3	12.9	9.1	1.5	0.16
12	500	870/890	95.7	97.4	96.6	92.4	0.96
	800	870/894	91.9	93.9	92.9	89.8	0.97
	900	896/892	5.5	11.9	8.7	2.2	0.25
24	500	865/892	95.5	97.1	96.3	93.6	0.97
	800	860/894	93.0	94.9	94.0	90.9	0.97
	900	894/893	11.1	22.8	17.0	4.9	0.29
36	500	868/890	95.2	97.3	96.3	91.9	0.95
	800	857/890	94.0	95.9	95.0	91.7	0.96
	900	897/894	14.2	26.7	20.5	7.9	0.39
48	500	866/890	96.0	97.3	96.7	94.2	0.97
	800	861/891	96.1	97.2	96.6	94.4	0.98
	900	895/894	19.2	37.6	28.4	9.8	0.34

[a] Catalyst: 20.0 mg; flow rate: 20 ml/min (CH_4:CO_2 = 1:1); space velocity: 60,000 ml h^{-1} g^{-1}.

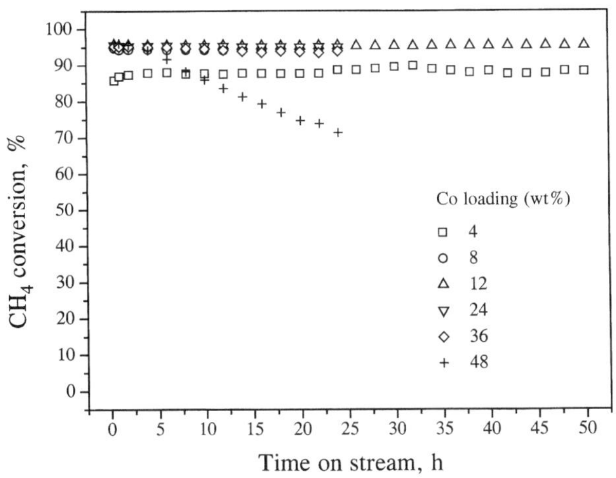

Fig. 1. Effect of Co loading on CH_4 conversion as a function of time on stream over the Co/MgO catalysts precalcined at 500°C.

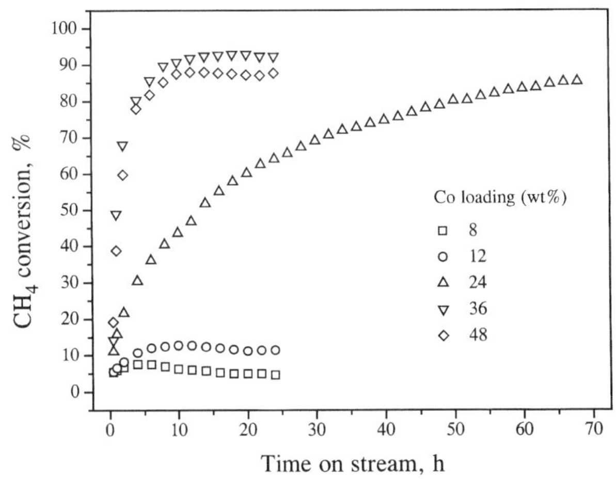

Fig. 3. Effect of Co loading on CH_4 conversion as a function of time on stream over the Co/MgO catalysts precalcined at 900°C.

48 wt.% catalysts yielded high and stable activities (93 and 87%, respectively) after a 10 h activation period.

Due to the endothermicity of the reforming reaction, the temperature of the catalyst bed is expected to be lower than that of the gas phase. Indeed, this occurred over all the catalysts (Table 1) for high conversions. At low conversions (as for the 900°C calcined catalysts), the temperature of the bed was a little higher. Because at low conversions, the conversion of CO_2 was much higher than that of CH_4, it is likely that a side reaction related to CO_2 (such as the reoxidation of the metallic sites) took place.

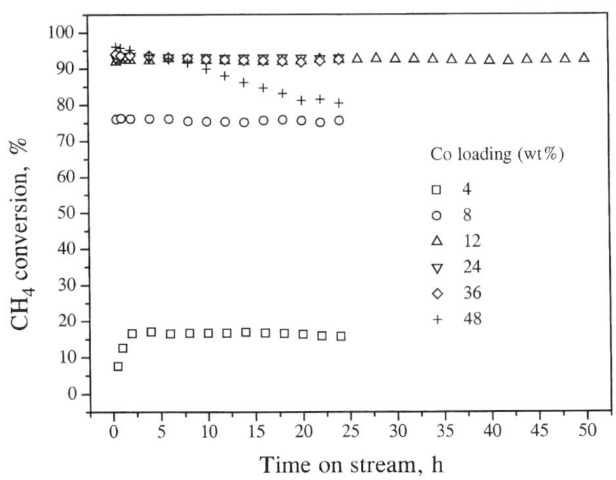

Fig. 2. Effect of Co loading on CH_4 conversion as a function of time on stream over the Co/MgO catalysts precalcined at 800°C.

3.2. Physico-chemical characterization

3.2.1. BET surface area of the calcined catalyst and the exposed metallic Co surface area of the reduced catalyst

The surface area of some calcined catalysts was determined by the BET method. The results obtained are listed in Table 2 which shows that at a given Co loading the surface area decreased with increasing calcination temperature.

After the calcined catalysts were reduced with H_2, the exposed metal surface areas were determined by CO chemisorption and the results are listed in Table 3. For the Co/MgO (500°C) catalysts, the Co surface area increased with increasing Co loading from 4 to 36 wt.%; for the 48 wt.% catalyst it was comparable to that for the 36 wt.% catalyst. For the Co/MgO (800°C) catalysts, the exposed Co surface area increased with increasing Co loading, more moderately between 8 and 36 wt.%. For the Co/MgO (900°C) catalysts, those with Co loadings between 8 and 24 wt.%

Table 2
The BET surface area of some Co(O)/MgO catalysts precalcined at different temperatures (m^2/g-catalyst)

Co loading (wt.%)	Calcination temperature (°C)		
	500	800	900
24	58	20	11
48	48	12	3

Table 3
Exposed Co surface area of the reduced Co/MgO catalysts pre-calcined at different temperatures ((m^2/g-catalyst) × 100)

Co loading (wt.%)	Calcination temperature (°C)		
	500	800	900
4	38.3	5.8	–[a]
8	73.4	10.9	8.9
12	95.3	11.2	8.7
24	155.6	12.7	8.2
36	200.6	13.2	3.0
48	198.3	30.3	2.4

[a] Not available.

possessed larger exposed metal surface areas than those with loadings of 36 and 48 wt.%. For the same Co loading, the metal surface area decreased in the sequence: Co/MgO(500°C) >> Co/MgO(800°C) > Co/MgO(900°C).

3.2.2. Reduction characteristics of the calcined supported Co catalysts

The TPR traces of the calcined supported Co catalysts, presented in Figs. 4–6 were strongly dependent on both the calcination temperature and the Co loading. A low temperature reduction peak between 260 and 300°C and a high temperature reduction peak between 660 and 690°C could be observed for most of the Co(O)/MgO (500°C) catalysts (Fig. 4). For

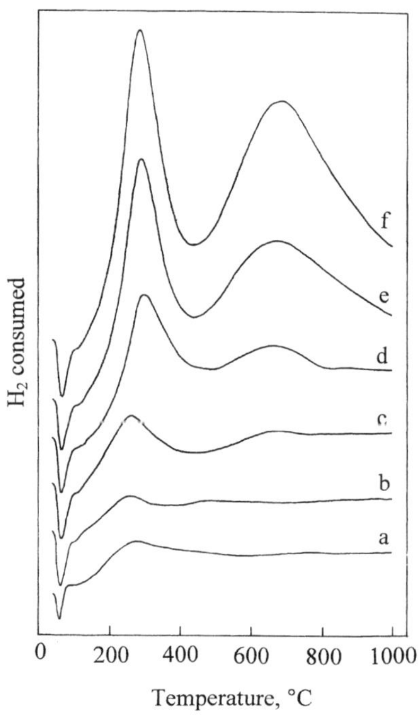

Fig. 4. The TPR profiles of the Co(O)/MgO (500°C) catalysts with Co loadings of (a) 4 wt.% (20.0 mg of catalyst), (b) 8 wt.%, (c) 12 wt.%, (d) 24 wt.%, (e) 36 wt.%, and (f) 48 wt.%.

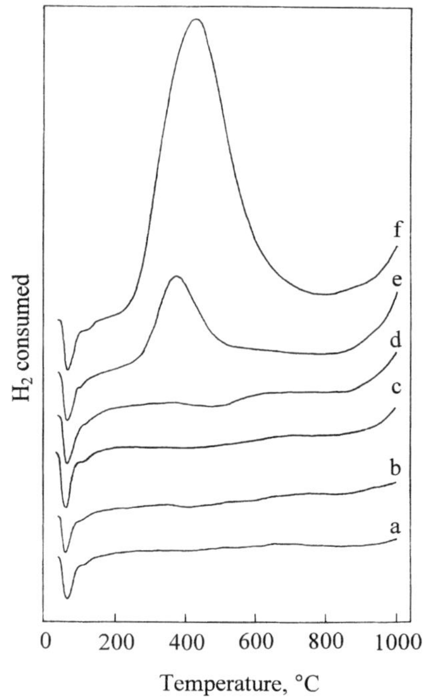

Fig. 5. The TPR profiles of the Co(O)/MgO (800°C) catalysts with Co loadings of (a) 4 wt.% (20.0 mg of catalyst), (b) 8 wt.%, (c) 12 wt.%, (d) 24 wt.%, (e) 36 wt.%, and (f) 48 wt.%.

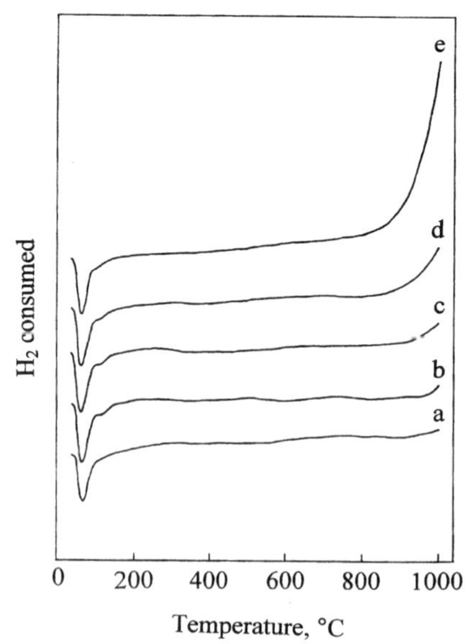

Fig. 6. The TPR profiles of the Co(O)/MgO (900°C) catalysts with Co loadings of (a) 8 wt.%, (b) 12 wt.%, (c) 24 wt.%, (d) 36 wt.%, and (e) 48 wt.%.

Table 4
Reduction degree (%) of the calcined catalysts after TPR

Co loading (wt.%)	Calcination temperature (°C)		
	500	800	900
12	22	3	–[a]
24	24	3	1
36	32	14	2
48	45	39	6

[a] Too small to be detected.

all the Co(O)/MgO (800°C) catalysts, a peak started to form at a temperature higher than 800°C; an additional low temperature reduction peak between 380 and 430°C was observed only over the catalysts with Co loadings of 36 and 48 wt.% (Fig. 5). For all the Co(O)/MgO (900°C) catalysts, a peak started to form at a temperature higher than 800°C (Fig. 6).

Table 4 lists the reduction degrees of the calcined catalysts in percentage. In all cases, the reduction degree was much less than 100%. At each calcination temperature, the reduction degree increased with increasing Co loading; for the same loading, it decreased with increasing calcination temperature.

3.2.3. Compounds present in the calcined supported Co catalysts

The Co-containing species present in the calcined catalysts were identified by XRD. The data and the assignments of the XRD patterns for some Co(O)/MgO catalysts are listed in Table 5. For the 48 wt.% Co(O)/MgO (500°C), the XRD patterns can be assigned to Co_3O_4 and/or $MgCo_2O_4$ as well as to (Co, Mg)O (a solid solution between CoO and MgO) and/or MgO. Since Co_3O_4 can be reduced in H_2 below 500°C [18,19] and as shown later a complete reduction of (Co, Mg)O requires a temperature higher than 1000°C, the peak at about 290°C can be attributed to Co_3O_4 and the one at about 690°C to $MgCo_2O_4$. Though only two reduction peaks were observed for this catalyst (Fig. 4f), its reduction degree of only 45% (Table 4) suggested that (Co, Mg)O which is much less reducible was also present besides Co_3O_4 and $MgCo_2O_4$. The absence of a TPR peak for the solid solution was most likely due to the insufficiently high upper-limit temperature during the TPR experiment. Because for the calcination temperature of 500°C and Co loadings between 12 and 48 wt.% all TPR spectra are similar (Fig. 4) and the reduction degree is much less than 100% (Table 4), one can conclude that the Co(O)/MgO (500°C) catalysts with Co loadings ≥12 wt.% contain (as did the 48 wt.% one) Co_3O_4, $MgCo_2O_4$ and (Co, Mg)O. For the 24 wt.% Co(O)/MgO (800°C), the XRD patterns (Table 5) could be attributed to (Co, Mg)O) and/or MgO. The TPR spectrum (Fig. 5d) which exhibited a reduction peak that started to form at a temperature higher than 800°C, allowed one to conclude that a solid solution was formed. Similarly, the XRD patterns (Table 5) coupled with the TPR spectrum (Fig. 5f) indicated that Co_3O_4 and (Co, Mg)O were present in the 48 wt.% Co(O)/MgO (800°C) catalyst. Among the 800°C calcined catalysts, the TPR spectra for loadings between 4 and 12 wt.% are similar to that for 24 wt.%, and the TPR spectrum for the 36 wt.% is similar to that for the 48 wt.% (Fig. 5). Consequently, Co_3O_4 and (Co, Mg)O were present in the 36 and 48 wt.% Co(O)/MgO (800°C) catalysts but only (Co, Mg)O in the Co(O)/MgO (800°C) catalysts with Co loadings ≤24 wt.%. For the 48 wt.% Co(O)/MgO

Table 5
Data and assignments of XRD patterns of calcined Co(O)/MgO catalysts

Co loading (wt.%)	Pre-treatment, (calcined for 8 h)	*d* (Å)	Assignments
24	at 800°C	2.110, 1.493, 1.272, 1.218, 2.438	(Co, Mg)O, MgO
48	at 500°C	2.456, 1.440, 2.879, 1.568, 2.038; 2.098, 1.488, 2.456	Co_3O_4, $MgCo_2O_4$; (Co, Mg)O, MgO
48	at 800°C	2.111, 1.493, 1.274, 1.219, 2.438; 2.438, 1.429, 2.860, 1.556, 2.022, 1.652	(Co, Mg)O, MgO; Co_3O_4
48	at 900°C	2.110, 1.490, 1.272, 1.218, 2.435	(Co, Mg)O, MgO

(900°C), only (Co, Mg)O was identified by coupling XRD (Table 5) with TPR (Fig. 6e) results. From the similarity of the TPR spectra, one can infer that only (Co, Mg)O was present in all Co(O)/MgO (900°C) catalysts.

4. Discussion

4.1. Effects of calcination temperature and Co loading on the species present in the calcined catalysts

Depending upon the calcination temperature and Co loading, XRD and TPR provided evidence for the existence of one, two or three of the following compounds: Co_3O_4, $MgCo_2O_4$ and (Co, Mg)O. At the calcination temperature of 900°C, only the solid solution (Co, Mg)O was present in all the catalysts. At the low calcination temperature of 500°C, CoO (formed through the decomposition of $Co(NO_3)_2$) either diffused into the MgO matrix to form a solid solution or was oxidized to Co_2O_3; the latter reacted either with CoO to form Co_3O_4 or with MgO to generate $MgCo_2O_4$. At the high calcination temperature of 800°C, Co_3O_4 and $MgCo_2O_4$ partially decomposed because they are unstable at high temperatures; a complete solid solution was formed in those with loadings ≤24 wt.%, but the Co_3O_4 phase was still present for loadings ≥36 wt.% and its amount increased steeply with increasing Co loading from 36 to 48 wt.%. At the high calcination temperature of 900°C, the Co_3O_4 phase was no longer present and (Co, Mg)O was the only Co-containing phase in all the catalysts.

Regarding the reduction characteristics of Co_3O_4, $MgCo_2O_4$ and (Co, Mg)O, Co_3O_4 can be reduced below 500°C, and $MgCo_2O_4$ below 700°C (Fig. 4). However, a temperature higher than 1000°C was required to reduce the solid solution (Co, Mg)O (Figs. 5 and 6). This happened because of the irreducibility of MgO and because both Co and Mg share the oxygen in the solid solution which thus becomes more strongly bound. Because the diffusion of CoO which generates the solid solution was deeper at higher temperatures, the reduction degree of the catalysts with the same loading decreased with increasing calcination temperature (Table 4).

4.2. Effect of calcination temperature on the exposed metal surface area of the reduced catalysts

Table 3 shows that for any Co loading the exposed metal surface area decreased with increasing calcination temperature. This is caused by the decreases in the reducibility and in the surface area of the catalyst. Regarding the former, the increase in the calcination temperature from 500 to 800°C tremendously decreased the amounts of Co_3O_4 and $MgCo_2O_4$ which are much more reducible than the solid solution. The moderate decrease from 800 to 900°C for Co loadings ≤24 wt.% occurred because in this range only a solid solution was present and CoO diffused deeper in the support at the higher temperature. The notable decrease from 800 to 900°C for Co loadings ≥36 wt.% occurred because the catalysts contained only a solid solution at the calcination temperature of 900°C but also Co_3O_4 at the 800°C one.

4.3. Effects of calcination temperature and Co loading on the activity and stability of the catalysts

The present results show that most of MgO supported Co catalysts calcined at 500 and 800°C provided high and stable activities for the CO_2 reforming of CH_4. However, the high calcination temperature of 900°C decreased tremendously the activities of the catalysts with loadings ≤12 wt.%, and generated long activation periods for those with loadings ≥24 wt.%. The high Co loading of 48 wt.% combined with a calcination temperature of 500 or 800°C resulted in an unstable activity; a stable activity was, however, reached at the high calcination temperature of 900°C.

At the calcination temperature of 500°C, most of the catalysts provided high and stable activities; significant deactivation, however, occurred over the 48 wt.% catalyst (Fig. 1). The XRD and TPR indicated that Co_3O_4 and $MgCo_2O_4$ which are more reducible than the solid solution were present in most Co(O)/MgO (500°C) catalysts and that their amounts increased notably with increasing Co loading (Fig. 4). Thus, a large number of metallic sites were generated during reduction particularly over the 48 wt.% catalyst. During reaction, the metallic sites were oxidized by CO_2 and H_2O and reduced by CH_4 and H_2. At the beginning, the oxidation is expected to dominate. The oxidized sites formed diffused during reaction into the MgO matrix in which they are soluble particularly if the CoO content of the matrix is not too high. This process is expected to be rapid because of the high reaction temperature (900°C). At steady-state which is reached after a short time (<0.5 h) by most catalysts (except the 48 wt.% one), the rate of site oxidation becomes equal to that of site reduction. The formation of the solid solution is responsible for the stability of these catalysts for reasons explained later in the paper. The rapid deactivation observed over the 48 wt.% catalyst (Fig. 1) occurred most likely because large amounts of Co_3O_4 and $MgCo_2O_4$ were present in the calcined catalyst. Being more reducible than the solid solution, they generated large metal clusters which stimulated coke formation. In addition, due to the relatively weak interactions between these clusters and the substrate, enhanced sintering could occur easily. Indeed, one can see from Table 3 that the Co surface area of the reduced 48 wt.% Co(O)/MgO (500°C) was near that of the 36 wt.% catalyst. This is an indication that larger metallic clusters were generated over the former after reduction. The sintering and the coke formation are responsible for the deactivation observed over the 48 wt.% catalyst precalcined at 500°C. The 48 wt.% Co/MgO (800°C) catalyst was unstable (Fig. 2) for the same reason (a large amount of Co_3O_4 was present in the calcined catalyst).

The Co/MgO (800°C) catalysts with a Co loading between 8 and 24 wt.% provided high and stable activities (Fig. 2). As indicated by XRD and TPR, a solid solution (Co, Mg)O was present in the calcined catalyst (Table 5 and Fig. 5). The high stability can be explained as follows. Due to the inherent difficulty of reducing a solid solution, the size of the metal clusters generated via the reduction of the solid solution is expected to be small. Being small, these clusters did not stimulate coke formation. Being formed from a solid solution, the clusters were at least partially embedded in the substrate, and thus were more resistant to sintering than the crystallites on the conventional supports (such as Al_2O_3, SiO_2, etc.).

As indicated by XRD and TPR, Co was present in all the Co(O)/MgO (900°C) catalysts solely as (Co, Mg)O (Table 5 and Fig. 6). Table 3 shows that their exposed Co surface area decreased with increasing Co loading while Table 1 reveals that the conversions of CH_4 and CO_2 after 0.5 h increased with increasing loading. This means that the activity attained after 0.5 h was not directly related to the exposed Co surface area of the fresh reduced catalyst. Most likely this occurred because the number of metallic sites was too small and most of them were oxidized by CO_2, thus, the system became catalytically less active. Only after a sufficiently large number of sites were generated by CH_4 and/or H_2 through reduction, could the catalyst become more active. For a solid solution catalyst, the generation of metallic sites is strongly affected by the calcination temperature and the Co loading. At the same calcination temperature, the lower the Co loading the smaller is its concentration near the surface and the fewer the metallic sites formed through reduction during reaction. As a result, the catalysts with higher Co loadings can be more easily activated than those with lower ones. This is the reason why the 24 wt.% Co/MgO (900°C) catalyst had a much longer activation period than the 36 and 48 wt.% catalysts (Fig. 3).

5. Summary

In summary, when the active component of the catalyst forms a solid solution with the substrate, the pretreatment of the catalyst plays a major role regarding its structural properties and the reduction behavior, and hence the catalytic performance. The variables investigated in this paper are the calcination temperature used during the preparation of the impregnated catalyst and the metal loading. The results demonstrate that the Co/MgO solid solution catalysts with loadings between 8 and 36 wt.%, precalcined at 500 and 800°C, provide high and stable activities for the CO_2 reforming of CH_4. However, a too high calcination temperature (e.g. 900°C) may cause a low activity or a long activation period, and a too high Co loading (48 wt.%) combined with a calcination temperature $\leq$800°C may result in an unstable activity.

References

[1] A.M. Gadalla, B. Bower, Chem. Eng. Sci. 43 (11) (1988) 3049.
[2] J.T. Richardson, S.A. Paripatyadar, Appl. Catal. 61 (1990) 293.
[3] A.T. Ashcroft, A.K. Cheetham, M.L.H. Green, P.D.F. Vernon, Nature 352 (1991) 225.
[4] J.R. Rostrup-Nielsen, J-H.B. Hansen, J. Catal. 144 (1993) 38.

[5] A. Erdöhelyi, J. Cserényi, F. Solymosi, J. Catal. 141 (1993) 287.
[6] E. Ruckenstein, Y.H. Hu, Appl. Catal. A: Gen. 133 (1995) 149.
[7] Z.L. Zhang, X.E. Verykios, J. Chem. Soc., Chem. Commun. 71 (1995).
[8] V.C.H. Kroll, H.M. Swaan, C. Mirodatos, J. Catal. 161 (1996) 409.
[9] Y.H. Hu, E. Ruckenstein, Catal. Lett. 36 (1996) 145.
[10] R.N. Bhat, W.M.H. Sachtler, Appl. Catal. A: Gen. 150 (1997) 279.
[11] H.Y. Wang, C.T. Au, Appl. Catal. A: Gen. 155 (1997) 239.
[12] S.M. Stagg, E. Romeo, C. Padro, D.E. Resasco, J. Catal. 178 (1998) 137.
[13] J.H. Bitter, K. Seshan, J.A. Lercher, J. Catal. 183 (1998) 336.
[14] K. Tomishige, Y. Chen, K. Fujimoto, J. Catal. 181 (1999) 91.
[15] H.Y. Wang, E. Ruckenstein, Appl. Catal. A: Gen. 204 (2000) 143.
[16] E. Ruckenstein, H.Y. Wang, Appl. Catal. A: Gen. 204 (2000) 257.
[17] I. Náray-Szabó, Inorganic Crystal Chemistry, Akadémiai Kiadó Budapest, 1969, 237 pp.
[18] K.S. Chung, F.E. Massoth, J. Catal. 64 (1980) 320.
[19] P. Arnoldy, J.A. Moulijn, J. Catal. 93 (1985) 38.

Carbon Deposition and Catalytic Deactivation during CO_2 Reforming of CH_4 over Co/γ-Al_2O_3 Catalysts

E. Ruckenstein[1] and H. Y. Wang

Department of Chemical Engineering, State University of New York at Buffalo, Amherst, New York 14260

Received May 31, 2001; revised October 22, 2001; accepted October 22, 2001; published online January 3, 2002

The reaction behavior and carbon deposition during the CO_2/CH_4 reforming reaction have been investigated over the γ-Al_2O_3-supported Co catalysts as a function of Co loading (between 2 and 20 wt%) and calcination temperature ($T_c = 500$ or 1000 C). It was found that the stability of Co/γ-Al_2O_3 catalysts was strongly dependent on the Co loading and calcination temperature. For some loadings (6 wt% for $T_c = 500$ C and 9 wt% for $T_c = 1000$ C), stable activities have been achieved. However, over the catalysts with high Co loadings (>12 wt%), notable amounts of carbon were accumulated during reforming, and deactivation was observed. Moreover, severe deactivation was also noted over the 2 wt% catalysts, both when carbon deposition occurred ($T_c = 500$ C) or was absent ($T_c = 1000$ C). In the latter case, the oxidation of the metallic sites was responsible for the deactivation. Hence, there are two different deactivation mechanisms, namely, carbon deposition and oxidation of metallic sites. The activities were stable when a balance between carbon formation and its oxidation could be achieved.

Key Words: **carbon dioxide reforming of methane; synthesis gas; Co catalyst; deactivation.**

1. INTRODUCTION

The catalytic CO_2 reforming of CH_4 (reaction 1), in which a mixture of CH_4 and CO_2, both greenhouse gases, is converted to synthesis gas, has been comprehensively studied for the past few years (1–35). The synthesis gas thus generated has a low H_2/CO ratio (1) and is therefore suitable for the Fischer–Tropsch synthesis of higher hydrocarbons and for the oxo-synthesis or synthesis of oxygenates (36, 37).

$$CH_4 + CO_2 \rightarrow 2CO + 2H_2, \quad \Delta H^0_{298} = 247 \text{ kJ/mol} \qquad [1]$$

Most of the group VIII metals are more or less catalytically active toward this reaction. Due to the inherent inertness of methane, a high temperature (typically, 800–900 C) is needed to achieve a meaningful yield. Under such severe conditions, an excess carbon deposition occurs on the surface of the catalyst, and this constitutes a major drawback. Even though the noble metal-based catalysts (such as Rh, Ru, Pd, and Pt) can provide high activity and selectivity with little or no carbon deposition (2, 5), they are unsuitable for large-scale commercial use because of their limited availability. While Ni- and Co-based catalysts are easily available, they deactivated rapidly when the conventional supports, such as Al_2O_3 and SiO_2, were employed. Exceptions were the cases in which MgO was used as support (10, 16, 34, 35). There is general agreement that the catalytic deactivation during the CO_2 reforming of methane is caused by carbon deposition and/or sintering of the metal particles (2, 3, 5, 12, 17, 24, 28, 32). Sintering accelerates the carbon deposition since large metal ensembles stimulate coke formation (38).

Numerous mechanistic studies have suggested that during the CH_4/CO_2 reforming, CH_4 is decomposed on the metallic sites to reactive carbon species (CH_x, $x = 0$–3), which are oxidized to CO by the oxygen-containing species that originate from CO_2 (2, 4, 5, 8, 13, 18, 21, 30). Consequently, the rate of carbon accumulation on the catalyst surface is determined by the relative rates of the generation of carbon species and its oxidative removal. When the former is faster than the latter, excess carbon deposition will occur. The goal of the present paper was to correlate the carbon deposition with the catalytic deactivation. For this purpose, the CO_2 reforming of CH_4 was investigated over γ-Al_2O_3-supported Co catalysts as a function of Co loading and calcination temperature (T_c) and the amount of carbon deposits generated during the reforming reaction was determined. It was found that besides the carbon deposition, the oxidation of the metallic sites by CO_2 also caused deactivation. Conditions could be identified under which the activity of the catalyst remained stable as a result of a dynamic balance between carbon formation and its oxidation.

2. EXPERIMENTAL

2.1. Catalyst Preparation

The supported cobalt catalysts were prepared by impregnating γ-Al_2O_3 with aqueous solutions of $Co(NO_3)_2$ $6H_2O$, followed by overnight drying at 110 C and calcination in the open air of a furnace at 500 or 1000 C for

[1] To whom correspondence should be addressed. Fax: (+1-716) 6453822.

This paper, which was originally published in **Journal of Catalysis**: 205, 289–293 (2002), is republished with permission.

TABLE 1

Physical Data for the γ-Al_2O_3-Supported Co Catalysts

Co loading (Wt%)	T_c (C)	BET surf. area[a] (m^2/g-cat.)	Co surf. area[b] (m^2/g-cat.) (100)
2	500	77	48
	1000	51	15
6	500	77	83
	1000	44	26
9	500	78	91
	1000	46	34
12	500	71	109
	1000	40	39
20	500	66	154
	1000	35	49

[a] For the calcined catalysts.
[b] For the reduced catalysts.

8 h. The calcined catalysts are denoted as Co(O)/γ-Al_2O_3 (500 or 1000 C). The catalysts reduced in H_2 are denoted as Co/γ-Al_2O_3 (500 or 1000 C). The temperatures inside the parentheses refer to the calcination temperature (T_c). Co loading means weight percent Co in the completely reduced catalyst.

2.2. *Catalytic Reaction*

The catalytic assay was carried out under atmospheric pressure in a fixed-bed vertical quartz reactor (i.d. 4 mm), which was operated in a down-flow mode. The calcined catalyst (20.0 mg) was held on a quartz wool bed and its reduction was carried out in a H_2 flow (30 ml/min) by increasing the temperature from room temperature to 600 C at a rate of 20 C/min and from 600 to 900 C at a rate of 10 C/min, without holding at 900 C. After reduction, the feed gases ($CH_4/CO_2 = 1/1$) were introduced into the catalyst bed at a flow rate of 20 ml/min, which for 20.0 mg of catalyst corresponds to a space velocity of 6.0 10^4 ml h ^{1} g ^{1}. The reactants and products were analyzed with an online gas chromatograph (GC) equipped with a Porapak Q column. An ice-cold trap was set between the reactor exit and the GC sampling valve to remove the water formed during reaction.

2.3. *Catalyst Characterization*

2.3.1. BET surface area and the exposed Co surface area. The BET surface area of the calcined catalysts and the exposed Co surface area of the reduced catalysts, determined as in our previous paper (34), are listed in Table 1.

2.3.2. Coke determination. The amount of carbon deposited on the catalyst surface during the CO_2/CH_4 reforming reaction was determined by oxidizing the carbon to CO with CO_2 in a pulse reactor. A quartz tube (i.d. 4 mm), in which the used catalyst was held on a quartz wool bed, was employed as reactor. The used catalyst was heated to 900 C in a flow of He (35 ml/min) to remove the possible adsorbed species (such as water and CO_2) and then pulses of CO_2 (500 μl) were injected into the carrier gas (He, 35 ml/min). The reaction product, CO, was analyzed with an online gas chromatograph equipped with a thermal conductivity detector (TCD) and a Porapak Q column. Through the CO_2 oxidation of the reduced catalysts, it was found that the CO that resulted via the oxidation of the metallic sites was, with the exception of the 2 and 6 wt% catalysts precalcined at 1000 C, much smaller than those that resulted from the deposited carbon. For the exceptions, the amount of carbon deposited was negligibly small.

2.3.3. X-ray photoelectron spectroscopy (XPS). Samples were mounted on double-sided adhesive tapes and the XPS data were obtained with a SSI Small Spot ESCA system using monochromatized Al*K*α radiation ($h\nu =$ 1.485 keV). The pressure in the ion-pumped analysis chamber was below 2 10 ^{9} Torr during data acquisition. The binding energies were referenced to the adventitious C 1s line (284.8 eV).

3. RESULTS AND DISCUSSION

As reported previously (39), in the γ-Al_2O_3-supported Co catalysts, the species generated during calcination were strongly dependent on the calcination temperature (T_c): Co_3O_4 was generated as a major phase at a T_c of 500 C, and Co_2AlO_4 and $CoAl_2O_4$ were generated at a T_c of 1000 C. As shown in Table 1, for the same Co loading, Co/γ-Al_2O_3 (500 C) had after reduction a larger exposed Co surface area than did Co/γ-Al_2O_3 (1000 C) because of the higher reducibility of Co_3O_4. For this reason, their catalytic performances in the CO_2/CH_4 reforming were expected to be different. The time-dependent activities of the Co/γ-Al_2O_3 (500 C) and Co/γ-Al_2O_3 (1000 C) catalysts are plotted in Figs. 1 and 2. Only the conversion of

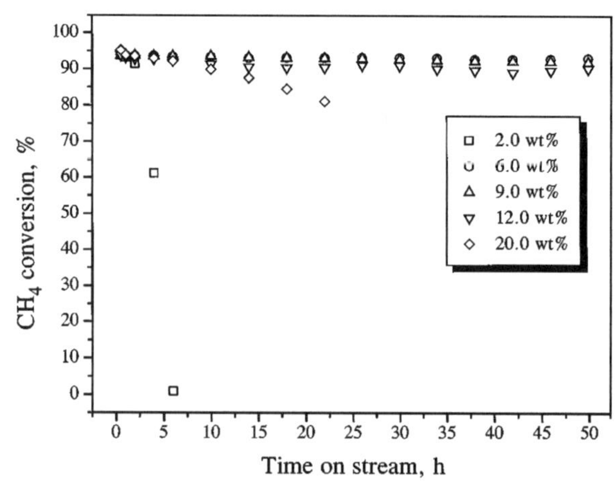

FIG. 1. Time-dependent conversion of CH_4 over the reduced Co/γ-Al_2O_3 (500 C) catalysts. $P = 1$ atm, $T = 900$ C, $CH_4/CO_2 = 1.0$, space velocity = 60 000 ml h ^{1} g ^{1}.

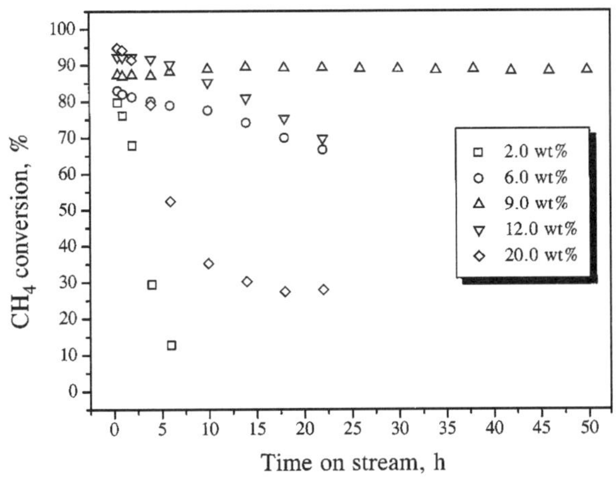

FIG. 2. Time-dependent conversion of CH_4 over the reduced Co/γ-Al_2O_3 (1000 C) catalysts. $P = 1$ atm, $T = 900$ C, $CH_4/CO_2 = 1.0$, space velocity = 60 000 ml h $^{-1}$ g $^{-1}$.

CH_4 is presented, because the conversion of CO_2 and the yields to H_2 and CO follow the same pattern. The equilibrium conversion of CH_4 at a reaction temperature of 900 C is about 96.8. As shown in Figs. 1 and 2, in most cases the initial activities of Co/γ-Al_2O_3 catalysts are close to the equilibrium value. With time on stream, some catalysts deactivated, while others maintained their initial activities. At a T_c of 500 C, severe deactivation occurred over the catalysts with low (2 wt%) and high (20 wt%) loadings, while no significant deactivation was observed during 50 h over those with loadings between 6 and 12 wt% (Fig. 1). At a T_c of 1000 C, the 9 wt% catalyst provided a stable activity during 50 h, while all the others experienced deactivation to various extents (Fig. 2). Figures 1 and 2 demonstrate that the stability of Co/γ-Al_2O_3 catalysts in the CO_2/CH_4 reforming reaction was strongly affected by the Co loading and the calcination temperature.

The amount of carbon accumulated on the catalyst surface during reaction is listed in Table 2. To correlate the carbon deposition with the catalytic deactivation, the deactivation, which is defined as the difference between the initial (after 0.5 h) and final CH_4 conversions with respect to the initial one, is also included in Table 2. Figures 3a and 3b plot the amount of carbon deposited and the deactivation as a function of Co loading for a selected reaction time. At a T_c of 500 C, both exhibited a minimum at a loading of 6 wt% (Fig. 3a), indicating that the carbon deposition, which decreases the number of exposed metallic sites, was one cause for catalytic deactivation. At a T_c of 1000 C and for Co loadings higher than 9 wt%, both increase with increasing Co loading (Fig. 3b), indicating again that the carbon

TABLE 2

Coke Deposition and Catalytic Deactivation during CO_2/CH_4 Reforming over the Reduced Co/γ-Al_2O_3 Catalysts

Co loading (wt%)	T_c (C)	Reaction time (h)	Amount of coke (mol) (10^6)	Deactivation[a] (%)
2	500	6	101.2	99.0
	1000	6	0.2	83.9
6	500	22	13.2	0.8
		50	31.1	1.1
	1000	22	0.3	19.6
9	500	22	132.5	0.8
		50	189.3	2.1
	1000	22	114.3	2.2
12	500	6	162.4	0.4
		22	258.1	3.6
		50	308.5	4.0
	1000	22	711.6	24.5
20	500	6	420.3	3.2
		22	733.9	14.7
	1000	22	1447.5	70.6

[a] Defined as the difference between the initial (after 0.5 h) and final CH_4 conversions with respect to the initial one and multiplied by 100.

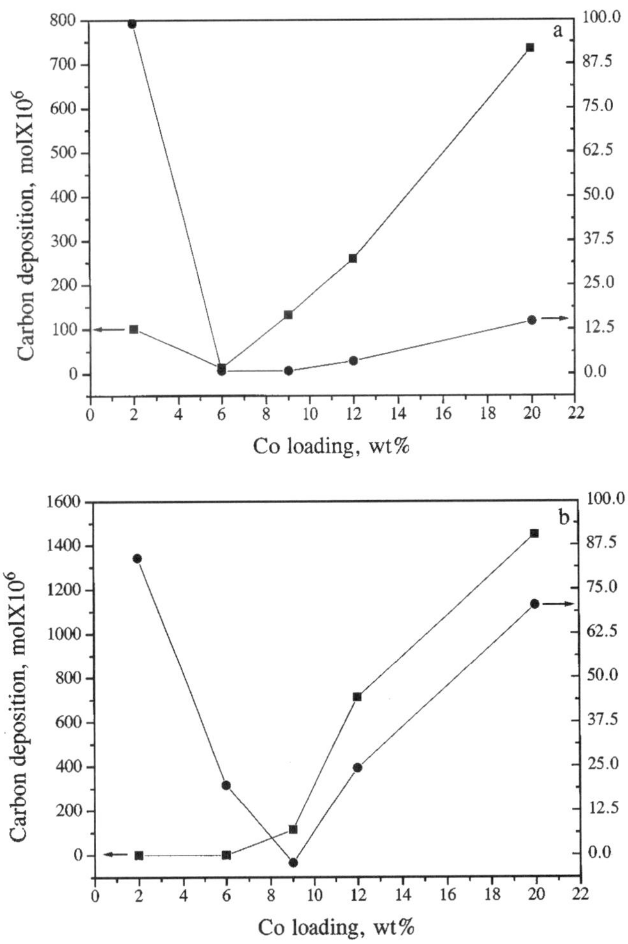

FIG. 3. Carbon deposition and deactivation as a function of Co loading over the Co/γ-Al_2O_3 (500 C) (a) and Co/γ-Al_2O_3 (1000 C) (b). (The data employed are after 22 h of reaction, with the exception of the 2 wt% catalyst which after 6 h had small CH_4 conversions.)

TABLE 3

Color of the γ-Al_2O_3-Supported Catalysts after Various Treatments

Co loading (wt%)	T_c (C)	After calcination	After reduction	After reforming reaction
2	500	Light green	Dark gray	Black
	1000	Light blue	Gray	Light blue
6	500	Black	Black	Black
	1000	Blue	Blue-gray	Blue (surface layer) Blue-gray (the rest)
9	500	Black	Black	Black
	1000	Blue	Blue-gray	Black
12	500	Black	Black	Black
	1000	Dark blue	Black	Black
20	500	Black	Black	Black
	1000	Dark blue	Black	Black

deposition was responsible for deactivation. However, for Co loadings lower than 9 wt%, severe deactivation occurred even though almost no carbon deposition could be detected (Fig. 3b). This implies that besides carbon deposition there is also another cause for catalytic deactivation. The color changes observed under various conditions provided some useful information. As shown in Table 3, the 2 wt% Co/γ-Al_2O_3 (1000 C) catalyst had after the reforming reaction the same light blue color as after calcination. The 6 wt% Co/γ-Al_2O_3 (1000 C) catalyst had after the reforming reaction a thin blue top layer, as after calcination, while the rest was blue-gray, as after reduction. Since after calcination at T_c = 1000 C, Co_2AlO_4 and $CoAl_2O_4$ were generated in the Co(O)/γ-Al_2O_3 catalysts, these color changes indicate that a large number of metallic Co sites were oxidized and restructuring of catalysts took place (to form Co_2AlO_4 and $CoAl_2O_4$) during the reforming reaction over the Co/γ-Al_2O_3 (1000 C) catalysts with Co loadings below 9 wt%. Consequently, the oxidation of the metallic sites, which decreases the total number of metallic sites, constitutes another cause for catalytic deactivation.

For the Co/γ-Al_2O_3 (500 C) catalysts, it should be noted that while the carbon deposition was much lower over the 2 wt% catalyst than over the 20 wt% one, the deactivation was much higher over the former than over the latter. To provide a reasonable explanation, the reactivity of carbon generated during the reforming reaction toward CO_2 was examined in a pulse reactor (i.d. 4 mm) at 900 C by pulsing CO_2 over the used 2, 6, 12, and 20 wt% Co/γ-Al_2O_3 (500 C) catalysts, which had been subjected for 6 h to the reforming reaction. The amount of CO formed as a function of the CO_2 pulse number is plotted in Fig. 4. For the first pulse point, the amount of CO formed was much lower over the 2 and 6 wt% catalysts than over the 12 and 20 wt% ones. Over the 2 wt% catalyst, the amount of CO formed decreased moderately during the first five pulses, and very slowly during the following 45 pulses, while over the 6 wt% catalyst, it decreased to practically zero after 12 pulses. Except for the 2 wt% catalyst, the amount of CO generated for the last pulse point was negligible compared to the total amount of CO formed. Figure 4 demonstrates that the carbon deposited on the 2 wt% catalyst was more difficult to remove with CO_2 than were those deposited over the 6, 12, and 20 wt% catalysts. This could have been caused either by the lower reactivity of the carbon deposited over the 2 wt% catalyst or by the smaller number of active sites present in this catalyst or, possibly, by both. As demonstrated by XPS (Fig. 5), the peaks of C 1s of the used 2 and 20 wt% catalysts were centered at 285.0 eV, which could be assigned to a –C–C– type deposit (26). Thus, the carbons deposited on the 2 and 20 wt% catalysts had the same

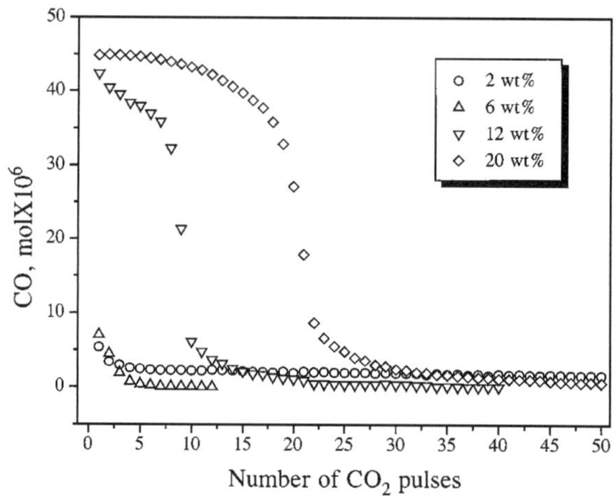

FIG. 4. Amount of CO formed as a function of CO_2 pulse number during the reaction between surface carbon and CO_2 over the Co/γ-Al_2O_3 (500 C) catalysts that were exposed to the reforming reaction for 6 h under the following conditions: $P = 1$ atm, $T = 900$ C, $CH_4/CO_2 = 1.0$, space velocity = 60 000 ml h $^{-1}$ g $^{-1}$.

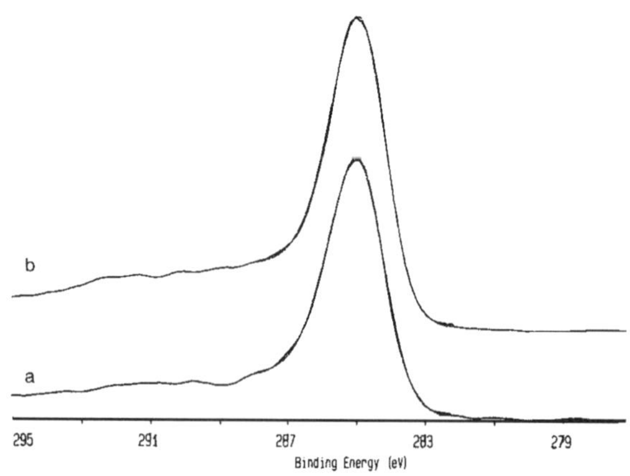

FIG. 5. C 1s XPS spectrum obtained over the 2 (a) and 20 (b) wt% Co/γ-Al_2O_3 (500 C) catalysts that were exposed to the reforming reaction for 6 h under the following conditions: $P = 1$ atm, $T = 900$ C, $CH_4/CO_2 = 1.0$, space velocity = 60 000 ml h $^{-1}$ g $^{-1}$.

nature. Consequently, the difficulty in the removal of the carbon deposited on the 2 wt% catalyst with CO_2 was due to the fewer accessible metallic Co sites available; most of them were covered by carbon. This explains why after 6 h of reaction the 2 wt% catalyst almost lost its activity for the CO_2 reforming of methane (Fig. 1) and also demonstrates that metallic Co sites are required for the reaction between the surface carbon and CO_2.

In the literature, two mechanisms have been suggested for the CO_2/CH_4 reforming. Mark and Maier (7, 15) suggested an Eley–Rideal mechanism for the CO_2 reforming, in which methane is adsorbed and decomposed over the metal (Rh) into H_2 and adsorbed carbon, and the carbon reacts directly with CO_2 from the gas phase generating CO. In an alternative mechanism (2, 4, 5, 8, 13, 18, 21, 30), methane is decomposed on the metal to yield a surface CH_x species and adsorbed hydrogen. On sorption, CO_2 dissociates to CO and adsorbed oxygen, and the oxygen reacts with the CH_x species to generate CO and hydrogen. Our results are compatible with the second mechanism because of the oxidation of the metallic sites by CO_2.

In summary, due to the coexistence of reductive (CH_4, H_2, and CO) and oxidative (CO_2 and H_2O) species in CO_2/CH_4 reforming, the atmosphere in the reactor is both reductive and oxidative. During reaction, a fraction of Co^0 is oxidized to Co–O, which is reduced again to Co^0 by carbon species, thereby generating a dynamic redox process. The reductive atmosphere stimulates the regeneration of metallic Co and the dissociative adsorption of CH_4, while the oxidative atmosphere favors the oxidation of metallic Co sites. When the former dominates (as over the 20 wt% catalysts), an excess of carbon is deposited; when the latter dominates (as over the 2 wt% Co/γ-Al_2O_3 (1000 C) catalyst), the number of metallic sites is decreased due to their oxidation and subsequent restructuring of catalysts. Both lead to severe catalytic deactivations. Consequently, the catalysts are stable when there is a balance between the generation of carbon species and its oxidative removal. By optimizing the metal loading (6 wt% for $T_c = 500$ C and 9 wt% for $T_c = 1000$ C), highly effective and stable Co/γ-Al_2O_3 catalysts can be obtained.

REFERENCES

1. Gadalla, A. M., and Bower, B., *Chem. Eng. Sci.* **43**(11), 3049 (1988).
2. Richardson, J. T., and Paripatyadar, S. A., *Appl. Catal.* **61**, 293 (1990).
3. Ashcroft, A. T., Cheetham, A. K., Green, M. L. H., and Vernon, P. D. F., *Nature* **352**, 225 (1991).
4. Solymosi, F., Kutsan, Gy., and Erdöhelyi, A., *Catal. Lett.* **11**, 149 (1991).
5. Rostrup-Nielsen, J. R., and Hansen, J-H. B., *J. Catal.* **144**, 38 (1993).
6. Erdöhelyi A., Cserényi J., and Solymosi, F., *J. Catal.* **141**, 287 (1993).
7. Mark, M. F., and Maier, W. F., *Angew. Chim. Int. Ed. Engl.* **33**(15/16), 1657 (1994).
8. Qin, D., and Lapszewicz, J., *Catal. Today* **21**, 551 (1994).
9. Nakamura, J., Aikawa, K., Sato, K., and Uchijima, T., *Catal. Lett.* **25**, 265 (1994).
10. Ruckenstein, E., and Hu, Y. H., *Appl. Catal. A* **133**, 149 (1995).
11. Zhang, Z. L., and Verykios, X. E., *J. Chem. Soc. Chem. Commun.* 71 (1995).
12. Zhang, Z. L., Tsipouriari, V. A., Efstathiou, A. M., and Verykios, X. E., *J. Catal.* **158**, 51 (1996).
13. Qin, D., Lapszewicz, J., and Jiang, X., *J. Catal.* **159**, 140 (1996).
14. Kroll, V. C. H., Swaan, H. M., and Mirodatos, C., *J. Catal.* **161**, 409 (1996).
15. Mark, M. F., and Maier, W. F., *J. Catal.* **164**, 122 (1996).
16. Hu, Y. H., and Ruckenstein, E., *Catal. Lett.* **36**, 145 (1996).
17. Zhang, Z. L., Verykios, X. E., MacDonald, S. M., and Affrossman, S., *J. Phys. Chem.* **100**, 744 (1996).
18. Hu, Y. H., and Ruckenstein, E., *Catal. Lett.* **57**, 167 (1999).
19. Bhat, R. N., and Sachtler, W. M. H., *Appl. Catal. A* **150**, 279 (1997).
20. Ruckenstein, E., and Hu, Y. H., *Appl. Catal. A* **154**, 185 (1997).
21. Wang, H. Y., and Au, C. T., *Appl. Catal. A* **155**, 239 (1997).
22. Chen, Y. G., Tomishige, K., Yokoyama, K., and Fujimoto, K., *Appl. Catal. A* **165**, 335 (1997).
23. Bitter, J. H., Seshan, K., and Lercher, J. A., *J. Catal.* **176**, 93 (1998).
24. Stagg, S. M., Romeo, E., Padro, C., and Resasco, D. E., *J. Catal.* **178**, 137 (1998).
25. Lemonidou, A. A., Goula, M. A., and Vasalos, I. A., *Catal. Today* **46**, 175 (1998).
26. Wang, S., and Lu, G. Q., *Appl. Catal. A* **169**, 271 (1998).
27. Tomishige, K., Chen, Y., and Fujimoto, K., *J. Catal.* **181**, 91 (1999).
28. Bitter, J. H., Seshan, K., and Lercher, J. A., *J. Catal.* **183**, 336 (1999).
29. Chen, Y. G., Tomishige, K., Yokoyama, K., and Fujimoto, K., *J. Catal.* **184**, 479 (1999).
30. Matsui, N., Anzai, K., Akamatsu, N., Nakagawa, K., Ikenaga, N., and Suzuki, T., *Appl. Catal. A* **179**, 247 (1999).
31. Wei, J. M., Xu, B. Q., Li, J. L., Cheng, Z. X., and Zhu, Q. M., *Appl. Catal. A* **196**, L167 (2000).
32. Kim, J. H., Sub, D. J., Park, T. J., Kim, and K. L., *Appl. Catal. A* **197**, 191 (2000).
33. Wang, H. Y., and Ruckenstein, E., *Appl. Catal. A* **204**, 143 (2000).
34. Ruckenstein, E., and Wang, H. Y., *Appl. Catal. A* **204**, 257 (2000).
35. Wang, H. Y., and Ruckenstein, E., *Appl. Catal. A* **209**, 207 (2001).
36. Trimm, D. L., *Catal. Rev.–Sci. Eng.* **16**, 155 (1977).
37. Xu, B. Q., and Sachtler, W. M. H., *J. Catal.* **180**, 198 (1998).
38. Rostrup-Nielsen, J. R., *in* "Catalysis Science and Technology" (J. R. Anderson and M. Boudart, Eds.), Vol. 5, pp. 1–118. Springer-Verlag, Berlin, 1984.
39. Wang, H. Y., and Ruckenstein, E., *Catal. Lett.* **75**, 13 (2001).

Carbon dioxide reforming of methane to synthesis gas over supported rhodium catalysts: the effect of support

H.Y. Wang, E. Ruckenstein*

Department of Chemical Engineering, State University of New York at Buffalo, Amherst, NY 14260, USA

Received 21 December 1999; received in revised form 3 March 2000; accepted 7 March 2000

Abstract

The carbon dioxide reforming of methane over the reduced supported Rh (0.5 wt.%) catalysts was investigated. Two kinds of oxides, reducible (CeO_2, Nb_2O_5, Ta_2O_5, TiO_2, and ZrO_2) and irreducible (γ-Al_2O_3, La_2O_3, MgO, SiO_2, and Y_2O_3) were used as supports. Among the irreducible metal oxides, γ-Al_2O_3, La_2O_3 and MgO provided stable catalytic activities during the period of study, and the activity increased in the sequence: $La_2O_3 < MgO \approx \gamma\text{-}Al_2O_3$. Deactivation, however, occurred over the SiO_2 and Y_2O_3 supported catalysts. Possible explanations for stability, based on the strong interactions between Rh_2O_3 and support, were proposed on the basis of TPR and XRD studies. Among the reducible oxides, Nb_2O_5 provided a low activity, ZrO_2 and CeO_2 exhibited a very long activation period, and deactivation occurred over Ta_2O_5 and TiO_2. Most of the reducible oxides provided much lower yields to CO and H_2 than the irreducible ones, and they are not suitable supports for Rh in the CO_2 reforming of methane. In summary, MgO and γ-Al_2O_3 are the most promising supports; they provided a stable high activity with a CO yield of 83–85% and a H_2 yield of 76–79% at the high space velocity of 60 000 ml/h/g.

Keywords: Carbon dioxide reforming of methane; Synthesis gas; Rh catalyst; Effect of support

1. Introduction

The carbon dioxide reforming of methane to synthesis gas (reaction 1) is an important topic of the current catalytic research. This reaction is of interest because of the low H_2/CO ratio ($\leq$1) it generates, which is suitable for the Fischer–Tropsch synthesis to higher hydrocarbons, and also because CO_2 (a greenhouse gas) is consumed in a useful manner.

$$CH_4+CO_2 \rightarrow 2CO+2H_2, \quad \Delta H^0_{298} = 247\,\text{kJ/mol} \quad (1)$$

* Corresponding author. Tel.: +1-716-645-2911/2214; fax: 1-716-645-3822.
E-mail address: feaeliru@acsu.buffalo.edu (E. Ruckenstein).

Fischer and Tropsch [1] studied the CO_2 reforming reaction of methane as early as 1928. However, this reaction attracted renewed interest recently [2–15]. Most of the Group VIII metals are more or less catalytically active towards this reaction. Because the reforming reaction is strongly endothermic, it requires high temperatures (typically, 1073–1173 K) and, under such severe conditions, deactivation of the catalyst due to carbon deposition and/or sintering of the metal often occurs. Rostrup-Nielsen and Hansen compared the catalysts based on nickel, ruthenium, rhodium, palladium, iridium and platinum, and found that ruthenium and rhodium provided high selectivities with carbon-free operations and high activities [5].

A few groups have studied the influence of the catalyst support on the catalytic performances of CO_2

This paper, which was originally published in **Applied Catalysis A**: General 204, 143–152 (2000), is republished with permission.

reforming over Rh-based catalysts. Erdöhelyi et al. [6,16] reported that the carrier has no effect on the activity of Rh catalysts. In contrast, Nakamura et al. [17] and Zhang et al. [18] observed that the initial turnover frequency (specific activity) of Rh crystallites is significantly affected by the nature of the carrier. Zhang et al. [18] also reported that the deactivation of Rh crystallites is strongly dependent on the nature of the support. Among the supports studied by Zhang et al. [18] (TiO_2, γ-Al_2O_3, MgO, SiO_2, La_2O_3, and yttria-stabilized zirconia), MgO exhibited the lowest specific activity and the highest deactivation rate.

In a previous paper [19], the effect of support on the partial oxidation of methane to synthesis gas over Rh-based catalysts was investigated; it was found that the reducible oxides supported rhodium catalysts provided, in general, much lower activities and selectivities (to H_2 and CO) than those supported on irreducible oxides. Among the irreducible oxides, MgO provided the highest catalytic activity with high product selectivity and stability. In the present paper, the effect of the support on the carbon dioxide reforming of methane over Rh-based catalysts is investigated. Again, two kinds of metal oxides, reducible and irreducible, are used as supports.

In literature, the specific activities are based on H_2 or CO chemisorption measurements conducted on the reduced catalysts. However, both the oxidation state and the morphology of the metal particles are altered during reaction, and the initial dispersion of the metal catalyst does not represent the actual dispersion under reaction conditions. In addition, as emphasized later in Section 3, due to the long activation times and to the deactivation exhibited by some catalysts, the initial activity is sometimes very different from the later ones. These factors complicate the interpretation of the specific activity results. In the present paper, the overall conversions of CH_4 and CO_2 will be used to compare the performances of the catalysts supported on various carriers.

2. Experimental

2.1. Catalyst preparation

The following supports were used: γ-Al_2O_3 (60.9 m^2/g), CeO_2 (3.3 m^2/g), Nb_2O_5 (4.9 m^2/g), Ta_2O_5 (3.8 m^2/g), ZrO_2 (3.9 m^2/g) (Alfa), La_2O_3 (2.8 m^2/g), MgO (31.7 m^2/g), SiO_2 (594.0 m^2/g), TiO_2 (9.4 m^2/g), and Y_2O_3 (9.4 m^2/g) (Aldrich). The supported rhodium catalysts were prepared by impregnating the support with an ethanol solution of $Rh(NO_3)_3{\cdot}2H_2O$ (Alfa), followed by overnight drying at 110°C and calcination in the open air of a furnace at 800°C for 4 h. The calcined catalysts are denoted $Rh(O)/M_xO_y$. The catalysts reduced in H_2 are denoted Rh/M_xO_y. Rh loading means wt.% Rh in the completely reduced catalyst.

2.2. Catalytic reaction

All the catalysts were tested under atmospheric pressure in a fixed-bed vertical quartz reactor (i.d. 3 mm), which was operated in a down flow mode. In the reactor, the catalyst was held on a quartz wool bed and the temperatures were monitored using two thermocouples: one located in the middle of the catalyst bed and the other one immediately after the quartz wool bed to measure the temperature of the gas phase. The reduction of the catalyst was carried out by increasing the temperature from room temperature to 800°C at a rate of 20°C/min in a H_2 flow (20 ml/min) without holding at 800°C. After reduction, the feed gases ($CH_4/CO_2=1/1$) were introduced into the catalyst bed at a flow rate of 20 ml/min, which for a 20.0 mg catalyst corresponds to a space velocity of 6.0×10^4 ml/h/g. The reactants and products were analyzed with an on-line gas chromatograph equipped with a Porapak Q column. An ice-cold trap was set between the reactor exit and the GC sampling valve to remove the water formed during reaction.

To achieve meaningful yields to CO and H_2, the reaction temperature was set at 800°C. In the absence of the catalyst, the blank runs with quartz wool indicated that methane conversion was below 2.0% at 800°C. Consequently, the contribution of the gaseous phase reaction can be neglected under the conditions employed in the present paper.

2.3. Catalyst characterization

2.3.1. Surface area

The surface area of the support was determined via nitrogen adsorption with a Micromeritics ASAP2000 instrument. All the samples were degassed at 200°C for at least 5 h in high vacuum before measurements.

2.3.2. Exposed Rh surface area

The exposed Rh surface area of the reduced catalyst was determined via CO adsorption at room temperature. Even though CO is also adsorbed in bridged and twinned forms besides the linear one, a 1/1 stoichiometry of CO adsorption on Rh was assumed. Calcined catalyst powder of 100 mg, held on a quartz wool bed in a vertical quartz tube reactor (i.d. 4 mm), was reduced in a H_2 flow (20 ml/min) at 650°C for 1 h. Then, at the same temperature, the reduced catalyst was purged with an ultra high purity helium flow (35 ml/min) for 1 h. After the sample was cooled to room temperature, CO was pulsed (10 μl per pulse) over the catalyst until no further adsorption of CO was detected. The CO left in CO chemisorption was determined quantitatively with a thermal conductivity detector (TCD). Before use, the hydrogen and helium were additionally purified by employing Hydro-Purge II and Oxy-Trap columns.

2.3.3. Temperature-programmed reduction (TPR)

The TPR of the calcined catalysts was conducted by heating samples (50.0 mg, unless otherwise indicated) held on a quartz wool bed in a vertical quartz tube reactor (i.d. 4 mm), from 100 to 1000°C for pure supports, and from 25 to 800°C for irreducible oxides and Ta_2O_5 supported catalysts, at a rate of 20°C/min in a flow of 2.5% H_2/Ar mixture (35 ml/min). The H_2/Ar gas mixture was purified with Hydro-Purge II and Oxy-Trap columns before use, and the hydrogen consumed in TPR was determined with a TCD.

2.3.4. X-ray powder diffraction (XRD)

The XRD was carried out with a Siemens D500 X-ray diffractometer, using the Cu Kα radiation, at 40 kV and 30 mA. The catalysts precalcined at 800°C for 4 h were examined to determine the major phases present.

3. Results

3.1. Carbon dioxide reforming of methane

3.1.1. Effect of support

Two kinds of metal oxides, reducible (CeO_2, Nb_2O_5, Ta_2O_5, TiO_2, ZrO_2) and irreducible (γ-Al_2O_3, La_2O_3, MgO, SiO_2, Y_2O_3), were used as supports. Even though all reactions were carried out at a $T_{furnace}$ of 800°C, Table 1 shows that the temperature of the catalyst bed was always much lower than that of the furnace, while the temperature of the gas phase was close to the latter. The temperature difference between the furnace and the catalyst bed was dependent on the support. In general, the temperature differences observed over the reducible oxides (except Ta_2O_5) were much smaller than those observed over the irreducible ones.

The catalytic activities obtained after 0.5 and 50 h on stream are listed in Table 1. The CO_2 conversion was always higher than that of CH_4, and the CO yield was always higher than that of H_2. A H_2/CO ratio below 1.0 was always obtained. The performance of the supported catalyst is strongly affected by the support. With the exception of Ta_2O_5, the reducible oxides provided much lower conversions of CH_4 and CO_2 and yields to CO and H_2 than the irreducible ones (Table 1). For the irreducible oxides, the conversions and yields obtained after 0.5 h on stream decreased in the sequence: $SiO_2 \approx MgO \approx \gamma\text{-}Al_2O_3 > Y_2O_3 > La_2O_3$; however, after 50 h on stream the sequence became $\gamma\text{-}Al_2O_3 \approx MgO > Y_2O_3 > SiO_2 \approx La_2O_3$. For the reducible oxides, the conversions and yields obtained after 0.5 h on stream decreased in the sequence $Ta_2O_5 \gg TiO_2 \gg ZrO_2 > Nb_2O_5 > CeO_2$, while after 50 h on stream in the sequence $ZrO_2 > Ta_2O_5 > CeO_2 \gg TiO_2 \gg Nb_2O_5$.

3.1.2. Effect of time on stream

The time-dependent activities of irreducible oxides supported Rh catalysts are plotted in Fig. 1. Over γ-Al_2O_3 and MgO, the conversions of CH_4 and CO_2 remained unchanged during the period of study, 100 h. La_2O_3 also provided a stable activity during the period of study, 50 h. However, deactivation was observed over SiO_2 and Y_2O_3. Over SiO_2, the conversions were high initially but decreased noticeably during the first 28 h, to become almost unchanged from 28 to 50 h. Compared with SiO_2, the decay of the conversion over Y_2O_3 was relatively small.

The time-dependent activities of reducible oxides supported catalysts are presented in Fig. 2. No significant deactivation occurred over Nb_2O_5, but the conversions of CH_4 and CO_2 were low. Ta_2O_5 and TiO_2 provided initially much higher conversions than the other reducible oxides, but they decayed with time on stream. Deactivation occurred mainly during the first

Table 1
Activities of Supported Rh (0.5 wt.%) catalysts for CO_2 reforming of methane at 800°C ($T_{furnace}$)[a]

Support	Rh surface area (m^2/g cat.)×100	Time after onset of reaction (h)	Temperature (°C) bed/gas	CH_4 conversion (%)	CO_2 conversion (%)	CO yield (%)	H_2 yield (%)	H_2/CO ratio
γ-Al_2O_3	59.0	0.5	742	78.5	86.5	82.5	74.6	0.90
		50	740	82.1	87.4	84.8	78.7	0.93
La_2O_3	8.5	0.5	753/792	67.8	79.3	73.6	61.3	0.83
		50	755/792	68.2	77.5	72.9	63.1	0.86
MgO	41.8	0.5	752/794	80.1	85.8	83.0	76.2	0.92
		50	750/792	80.3	85.8	83.1	76.5	0.92
SiO_2	49.8	0.5	749/782	82.3	88.2	85.3	77.6	0.91
		50	761/789	71.9	77.2	74.6	68.9	0.92
Y_2O_3	50.1	0.5	741/788	75.1	84.5	79.8	70.8	0.89
		50	748/791	73.0	82.1	77.6	67.7	0.87
CeO_2	9.9	0.5	794/797	5.4	12.0	8.7	2.0	0.23
		50	759/785	50.7	63.2	56.9	44.6	0.78
Nb_2O_5	–[b]	0.5	796/800	16.2	27.1	21.7	10.5	0.49
		50	790/794	15.3	26.6	21.0	9.4	0.44
Ta_2O_5	9.4	0.5	753/787	69.0	77.4	73.2	64.2	0.88
		50	757/788	63.4	69.7	66.6	59.7	0.90
TiO_2	10.2	0.5	780/800	50.1	65.7	57.9	42.1	0.73
		50	783/797	33.1	46.5	39.7	26.9	0.68
ZrO_2	11.1	0.5	783/794	18.5	31.9	25.2	11.7	0.46
		50	750/786	65.9	74.2	70.1	61.0	0.88

[a] Catalyst, 20.0 mg; flow rate, 20 ml/min (CH_4/CO_2=1/1); space velocity, 60 000 ml/h/g.
[b] Too small to be detected.

24 h of reaction. Over the CeO_2 and ZrO_2, the conversions were initially low, but increased with time on stream. A very long activation period was noted for both of them. The strong variation of activity with time on stream is a typical feature of the reducible oxide supports.

The deactivation characteristics of the irreducible oxides and Ta_2O_5 (Ta_2O_5 is also included because it provided an activity close to that of the irreducible oxide La_2O_3) supported Rh catalysts are summarized in Table 2. γ-Al_2O_3, MgO, and La_2O_3 provided stable activities during the period of study, while different extents of deactivation occurred over SiO_2, Y_2O_3, and Ta_2O_5 supported catalysts. In contrast to the observations of Zhang et al. [18], γ-Al_2O_3 and MgO were found to provide both high activity and high stability.

3.1.3. Effect of the amount of catalyst

As shown in Fig. 3, at a $T_{furnace}$ of 800°C, over the 0.5 wt.% Rh/γ-Al_2O_3 catalyst, the conversions of CH_4 and CO_2 and the yields to CO and H_2 increased with increasing amount of catalyst in the range investigated (2.5–20.0 mg). This behavior is different from that in the partial oxidation of methane where the methane conversion increased with increasing amount of catalyst from 1.0 to 5.0 mg, but remained almost unchanged for amounts ≥5.0 mg [20].

3.1.4. Effect of the metal particle size

The effect of particle size on the specific activity of Rh for the CO_2-reforming of CH_4 was investigated over the Rh/γ-Al_2O_3 catalyst. The particle size was changed by varying the Rh loading, and was calcu-

Table 2
Deactivation characteristics of irreducible oxides and Ta_2O_5 supported Rh (0.5 wt.%) catalysts[a]

Support	$C_{CH_4}^{50\,h}/C_{CH_4}^{0.5\,h}$	$C_{CO_2}^{50\,h}/C_{CO_2}^{0.5\,h}$	$Y_{CO}^{50\,h}/Y_{CO}^{0.5\,h}$	$Y_{H_2}^{50\,h}/Y_{H_2}^{0.5\,h}$
γ-Al_2O_3	1.04	1.01	1.03	1.05
La_2O_3	1.01	0.98	0.99	1.03
MgO	1.00	1.00	1.00	1.00
SiO_2	0.87	0.93	0.91	0.89
Y_2O_3	0.97	0.97	0.97	0.96
Ta_2O_5	0.92	0.90	0.91	0.93

[a] $C_{CH_4}^{50\,h}$ means the conversion of methane after 50 h of reaction, and $Y_{CO}^{50\,h}$ the yield to CO after 50 h of reaction.

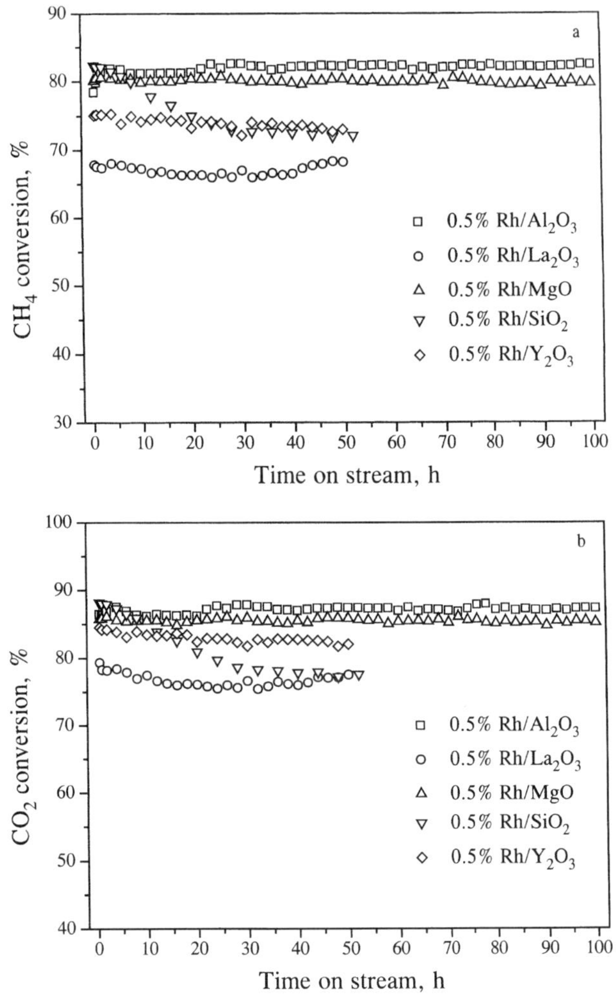

Fig. 1. Time-dependent conversions of CH_4 (a) and of CO_2 (b) over irreducible oxides supported Rh catalysts; P=1 atm, $T_{furnace}$=800°C, CH_4/CO_2=1.0, space velocity=60 000 ml/h/g.

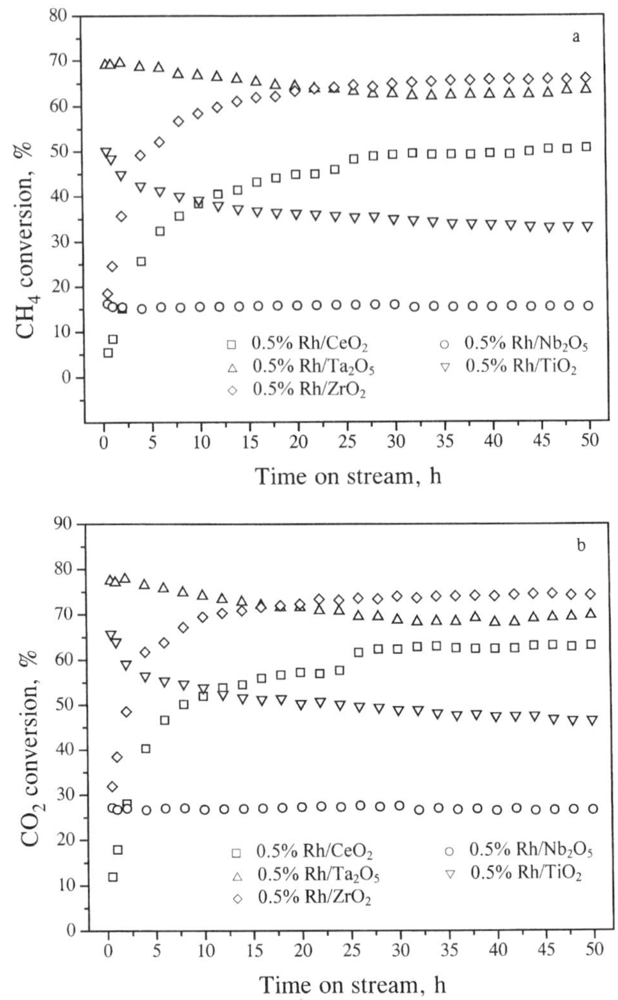

Fig. 2. Time-dependent conversions of CH_4 (a) and of CO_2 (b) over reducible oxides supported Rh catalysts; P=1 atm, $T_{furnace}$=800°C, CH_4/CO_2=1.0, space velocity=60 000 ml/h/g.

lated from the dispersion data assuming a hemisphere shape. Fig. 4 shows that the turnover frequency (TOF) depends significantly on the particle diameter of Rh in the range of 1–7 nm. This strongly indicates that the reforming reaction is structure sensitive on γ-Al_2O_3 supported Rh catalyst. A similar result was reported by Zhang et al. [18].

3.2. Physico-chemical characterization

3.2.1. Exposed metallic Rh surface area of the reduced catalyst

After the calcined catalysts were reduced with H_2, the exposed metal surface areas were determined by CO adsorption. Table 1 shows that over the irreducible oxides (except La_2O_3) the exposed Rh surface area was much higher than over the reducible ones.

3.2.2. Reducibility of reducible oxides

The reducibility of reducible oxides was studied by TPR and the results are presented in Fig. 5. A peak at about 850°C was observed for CeO_2, at about 890°C for Nb_2O_5, and a peak started to form at 800°C for TiO_2. Color changes were noted for the above three samples after the TPR experiments. No obvious reduction peaks were observed and no color changes noted for Ta_2O_5 and ZrO_2. Our TPR results indicated that the reducibility increased in the sequence: ZrO_2,

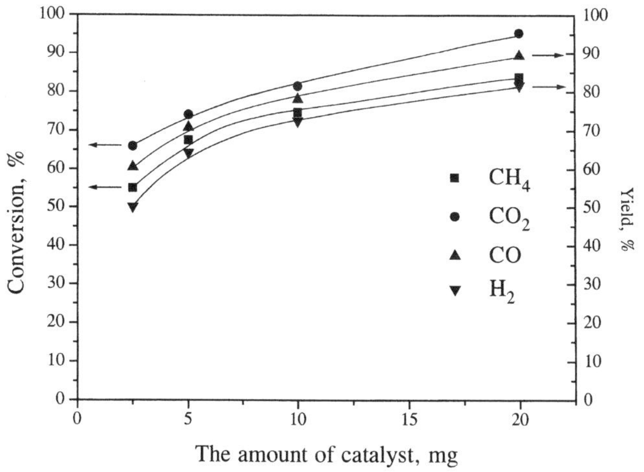

Fig. 3. Effect of the amount of catalyst on the performance of γ-Al_2O_3 supported Rh (0.5 wt.%) catalysts; $P=1$ atm, $T_{furnace}=800°C$, $CH_4/CO_2=1.0$ (20 ml/min) (data obtained after 10 h of reaction).

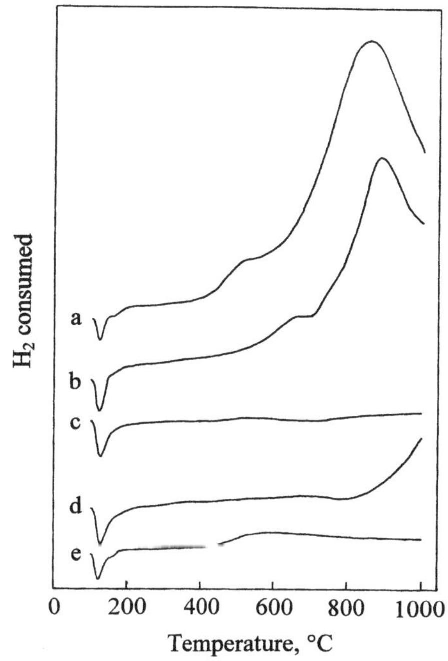

Fig. 5. TPR profiles of pure CeO_2 (a), Nb_2O_5 (b), Ta_2O_5 (c), TiO_2 (d), and ZrO_2 (e).

$Ta_2O_5 < TiO_2 < Nb_2O_5 < CeO_2$. Tauster and Fung [21] reported that the reducibility increased in the sequence $ZrO_2 < Ta_2O_5 < TiO_2 < Nb_2O_5$.

3.2.3. Reduction behavior of the irreducible oxides and Ta_2O_5 supported Rh catalysts

The results of the TPR experiments over the calcined irreducible oxides and Ta_2O_5 supported Rh catalysts are presented in Fig. 6. A single peak between 150 and 210°C was observed over the calcined γ-Al_2O_3, SiO_2, Y_2O_3, and Ta_2O_5 supported catalysts. The 0.5 wt.% Rh(O)/La_2O_3 provided two peaks, one

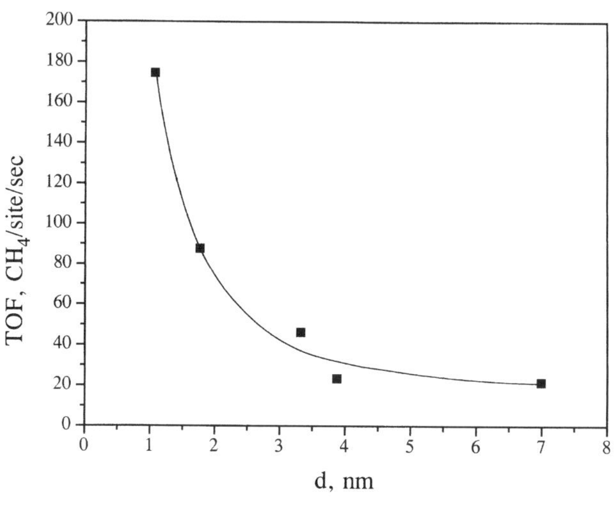

Fig. 4. Turnover frequency based on surface rhodium atoms as a function of the mean particle diameters of rhodium in the CO_2/CH_4 reaction over 0.5 wt.% Rh/γ-Al_2O_3; $P=1$ atm, $T_{furnace}=800°C$, $CH_4/CO_2=1.0$, space velocity=60 000 ml/h/g (data obtained after 10 h of reaction).

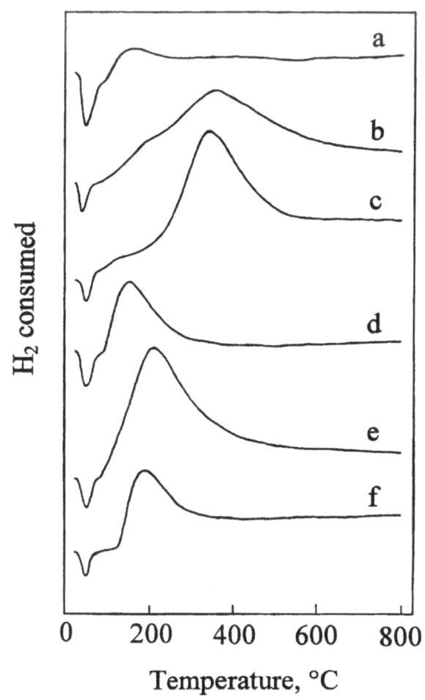

Fig. 6. TPR profiles of calcined Rh (0.5 wt.%) catalysts supported on γ-Al_2O_3 (a), La_2O_3 (b), MgO (c), SiO_2 (d), Y_2O_3 (e), and Ta_2O_5 (f) (25.0 mg was used for γ-Al_2O_3 and 50.0 for the other supports).

Table 3
Data and assignments of XRD patterns of calcined 10 wt.% Rh(O)/M_xO_y catalysts

Catalysts	Pretreatment	d (Å)	Assignments	References
10% Rh(O)/Al_2O_3	Calcined at 800°C for 4 h	1.39, 1.96	γ-Al_2O_3	[19]
		2.50, 2.68, 1.70	Rh_2O_3	
10% Rh(O)/La_2O_3	Calcined at 800°C for 4 h	2.93, 1.95, 3.34, 1.74, 1.64	La_2O_3	[19]
		2.76, 3.45	$LaRhO_3$	
		2.60, 2.79	Rh_2O_3(minor)	
10% Rh(O)/MgO	Calcined at 800°C for 4 h	2.09, 1.48, 1.21, 1.27	MgO	[19,22]
		2.53, 4.76, 2.09, 1.48, 2.42	$MgRh_2O_4$	
10% Rh(O)/SiO_2	Calcined at 800°C for 4 h	2.52, 2.69, 1.69	Rh_2O_3	[22]
10% Rh(O)/Y_2O_3	Calcined at 800°C for 4 h	3.06, 1.87, 2.65, 1.60, 4.33	Y_2O_3	[23]
		2.54, 2.71, 1.72	Rh_2O_3	
10% Rh(O)/CeO_2	Calcined at 800°C for 4 h	3.12, 1.91, 1.63, 2.70, 1.24, 1.10	CeO_2	This work
		2.61, 2.56	Rh_2O_3	
10% Rh(O)/Nb_2O_5	Calcined at 800°C for 4 h	3.93, 3.14, 2.46, 3.08, 2.42, 1.96	Nb_2O_5	This work
		2.56, 2.73, 1.72	Rh_2O_3	
10% Rh(O)/Ta_2O_5	Calcined at 800°C for 4 h	3.13, 3.85, 2.43, 3.08, 2.41, 1.65	Ta_2O_5	[23]
		2.52, 2.75, 1.70	Rh_2O_3	
10% Rh(O)/TiO_2	Calcined at 800°C for 4 h	3.52, 1.89, 1.70, 1.66, 2.38, 1.48	Ta_2O_5	This work
		2.56, 2.73, 1.71, 3.73, 1.87	Rh_2O_3	
10% Rh(O)/ZrO_2	Calcined at 800°C for 4 h	3.16, 2.84, 2.62, 1.82, 2.54, 1.85	Ta_2O_5	This work
		2.57, 2.73	Rh_2O_3	

as a shoulder at about 180°C and the other one (the main) at about 360°C. The 0.5 wt.% Rh(O)/MgO exhibited a single peak at about 350°C. The (main) peak temperature increased in the sequence SiO_2<γ-Al_2O_3<Ta_2O_5<Y_2O_3≪MgO, La_2O_3.

3.2.4. *Compounds present in the calcined supported Rh catalysts*

Because the rhodium content of 0.5 wt.% was too low for the XRD analysis, 10 wt.% Rh(O)/support samples were employed. The data and the assignments of the XRD patterns for the 10 wt.% Rh(O)/M_xO_y (M=Al, La, Mg, Si, Y, Ce, Nb, Ta, Ti, and Zr) catalysts are listed in Table 3. Only Rh_2O_3 was detected over γ-Al_2O_3, SiO_2, Y_2O_3, CeO_2, Nb_2O_5, Ta_2O_5, TiO_2, and ZrO_2 supported Rh catalysts. Over the 10 wt.% Rh(O)/La_2O_3, besides Rh_2O_3, $LaRhO_3$ was also identified, but over the 10 wt.% Rh(O)/MgO, only $MgRh_2O_4$ was detected. Results of TPR and XRD clearly indicate that the low temperature reduction peaks between 150 and 210°C are due to the reduction of Rh_2O_3 and the high temperature reduction peaks at 360°C for 0.5 wt.% Rh(O)/La_2O_3 and at 350°C for 0.5 wt.% Rh(O)/MgO are due to the reduction of $LaRhO_3$ and $MgRh_2O_4$, respectively.

4. Discussion

4.1. *Effect of support on activity*

As shown in Table 1, the activity of rhodium catalysts was strongly dependent on the support. The reducible oxides provided, in general, much lower conversions of CH_4 and CO_2 and yields to CO and H_2 than the irreducible ones. Ta_2O_5 constitutes an exception; it provided an initial activity comparable to that of La_2O_3. These patterns are very similar to those observed in the partial oxidation of methane [19]. As shown in Table 1, after the calcined reducible oxides supported catalysts were reduced with H_2, the exposed metal surface areas were, in general, much lower than those observed over the irreducible oxides. One possible explanation is that the rhodium metal particles were partially covered by islets of partially reduced oxide species. Experimental evidence for the burying by the suboxide TiO_x, formed through the reduction of the TiO_2 support, of the metal particles was brought with regard to the so-called strong metal-support interactions [24,25]. The TiO_x migrates over the exposed surface of the metal particles because this decreases the free energy of the system [26]. Indeed, as shown in

Table 4
Surface free energies of some metals [27]

Metal	σ (erg/cm^2)$\times 10^{-3}$
Rh(1 1 1)	2.53
Rh(1 0 0)	2.81
Rh(1 1 0)	2.88
Ta(poly)	2.70
Ti(poly)	1.70

Table 4, Ti has a much smaller surface free energy than Rh [27]. TiO_2 being an oxide is expected to have an even smaller surface energy than Ti. Though TiO_x may have a somewhat higher surface energy than TiO_2, the reduction generates vacancies, which enhance the surface diffusion and hence the migration of TiO_x over the exposed surface area of the metal particles. The partial coverage of rhodium particle by species generated via the reduction of the support, which occurs because the surface energy of TiO_x is smaller than that of Rh, decreases the number of active metal sites, thus resulting in a lower activity. However, as shown in Table 4, Ta has a surface free energy comparable to that of Rh. Consequently, the surface energy of Ta_2O_x may not be much smaller than that of Rh to migrate as a strong surface-active agent over the surface of the metal particles. Perhaps for this reason the Ta_2O_5 supported rhodium catalyst exhibited much higher catalytic activity than the other reducible oxides.

Based on the TPR results of the pure supports (Fig. 5) and the results reported in [21], the reducibility of the reducible oxides increases in the order $ZrO_2 < Ta_2O_5 < TiO_2 < Nb_2O_5 < CeO_2$. The initial activity of the reducible oxides decreases in the sequence $Ta_2O_5 \gg TiO_2 \gg ZrO_2 > Nb_2O_5 > CeO_2$. With the exception of ZrO_2, the activity sequence is opposite to that of reducibility. However, the sequence has changed after 50 h on stream; it became $ZrO_2 > Ta_2O_5 > CeO_2 \gg TiO_2 \gg Nb_2O_5$. As shown in Fig. 2, ZrO_2 and CeO_2 supported catalysts exhibited very long activation periods, and their activities after 50 h on stream were much higher than after 0.5 h (Table 1 and Fig. 2). The strong dependence of activity on time on stream is a typical feature of the reducible oxides supported Rh catalysts. A reconstruction of the metal particles and/or of the migrated partially reduced oxide species of the substrate might be responsible for the observed changes in activity with time on stream. Beneficial changes occurred for ZrO_2 and CeO_2 supported catalysts, and harmful ones over the Ta_2O_5 and TiO_2 supported ones. Beneficial effects might have occurred either because the cooperation between the partially reduced oxide and the metallic sites generated sites with higher activity, or because the reoxidation of the partially reduced species by the oxygen from CO_2 decreased the surface free energy over the particle and the new species migrated back to the surface of the substrate, thus exposing a larger number of sites. In summary, Nb_2O_5 provides a low activity, ZrO_2 and CeO_2 have a very long activation period during which the activity increases until a steady state is reached, and deactivation occurs over Ta_2O_5 and TiO_2. In conclusion, the reducible oxides are, in general, not suitable supports for Rh in the carbon dioxide reforming of methane.

For the irreducible oxides, the conversions and yields obtained after 0.5 h of reaction decreased in the sequence: $SiO_2 \approx MgO \approx \gamma\text{-}Al_2O_3 > Y_2O_3 > La_2O_3$. Due to deactivation that occurred over SiO_2 and Y_2O_3, the sequence became $\gamma\text{-}Al_2O_3 \approx MgO > Y_2O_3 > SiO_2 \approx La_2O_3$ after 50 h on stream. Unlike some of the reducible oxides, no irreducible oxide supported catalysts exhibited a long activation period. $\gamma\text{-}Al_2O_3$ and MgO provided the highest conversions and yields.

4.2. Stability of the irreducible oxides and Ta_2O_5 supported Rh catalysts

$\gamma\text{-}Al_2O_3$, MgO, and La_2O_3 provided stable activities during the period of study (100 h for $\gamma\text{-}Al_2O_3$ and MgO, and 50 h for La_2O_3), while different extents of deactivation were observed over SiO_2, Y_2O_3, and Ta_2O_5 (Table 2 and Figs. 1 and 2). The deactivation may be due mainly to sintering because this decreases the number of active sites and also accelerates the carbon deposition, since large metal ensembles are needed for carbon deposition.

The (main) TPR peak temperatures of La_2O_3 and MgO supported Rh catalysts are much higher than those of SiO_2, $\gamma\text{-}Al_2O_3$, Y_2O_3, Ta_2O_5 supported ones (Fig. 6), indicating that there are much stronger interactions between rhodium oxide and La_2O_3 or MgO. While no compounds between rhodium oxide and support were formed over the calcined SiO_2, $\gamma\text{-}Al_2O_3$, Ta_2O_5 and Y_2O_3 supported catalysts, $LaRhO_3$ and $MgRh_2O_4$ were formed, respectively, over the calcined

La_2O_3 and MgO supported ones (Table 3). With the exception of γ-Al_2O_3 supported catalyst, all the catalysts for which no compounds were identified after calcination exhibited decay in activity during reaction. The strong interactions between rhodium and magnesium or lanthanum oxides are most likely responsible for the high stabilities of the MgO and La_2O_3 supported catalysts. These interactions arise because at the high reaction temperature involved a fraction of Rh is reoxidized by CO_2 and the formed oxide interacts with MgO or La_2O_3. Though no compound between rhodium oxide and γ-Al_2O_3 could be identified (Fig. 6 and Table 3), the γ-Al_2O_3 supported Rh catalyst was stable (Fig. 1). Some interactions may occur between rhodium oxide and alumina, which do not lead to a chemical compound, but are sufficiently strong to ensure the stability of the catalyst. Indeed, our TPR experiments indicated that the alumina supported Rh catalyst was much more resistant to metal sintering than the silica supported one [23].

4.3. Additional features of carbon dioxide reforming of methane

The carbon dioxide reforming of methane is a strongly endothermic reaction (see reaction 1). Due to its endothermicity, the temperature of the catalyst bed was always much lower than that of the furnace (Table 1). The temperature difference between the furnace and the catalyst bed over the reducible oxides (except Ta_2O_5) supported catalysts were, in general, much smaller than those observed over the irreducible ones because the activity of the former was, in general, much lower than that of the latter and the higher the activity, the higher is the energy needed for reaction. For the same reason, an increase in T_{bed} was noted when the activity of a catalyst (e.g. Rh/SiO_2) decreased with time on stream, while a decrease in T_{bed} was observed when the activity of a catalyst (e.g. Rh/ZrO_2) increased with time on stream.

Over all the catalysts, the CO_2 conversion was higher than that of CH_4, and the CO yield was higher than that of H_2. A H_2/CO ratio below 1.0 was obtained in all cases. This might indicate that a reverse water gas shift reaction (reaction 2) takes place as a side reaction.

$$CO_2 + H_2 \rightarrow CO + H_2O \tag{2}$$

The conversions of CH_4 and CO_2 and the yields to CO and H_2 increased with increasing amount of catalyst in the range investigated (2.5–20.0 mg) in the carbon dioxide reforming of methane (Fig. 3). In contrast, in the partial oxidation of methane, the methane conversion increased with increasing amount of catalyst from 1.0 to 5.0 mg, but remained almost unchanged for amounts $\geq$5.0 mg [20]. This happens because in the partial oxidation of methane, which is exothermic, the reaction occurs essentially within a thin layer of about 5.0 mg (or even less) at the beginning of the catalyst bed and equilibrium is reached; the carbon dioxide reforming of methane, which is endothermic, needs a much longer catalyst bed to reach equilibrium.

5. Conclusions

The support affects in a major way, the activity of the supported Rh catalyst for the CO_2-reforming of CH_4. The reducible oxides are not suitable supports for this reaction because most of them provide relatively low conversions and yields. Among the irreducible oxides, SiO_2 and Y_2O_3 are not suitable because the conversions and yields decay with time on stream. MgO and γ-Al_2O_3 are the most promising supports; they provide high activities and are also stable with time.

References

[1] F. Fischer, H. Tropsch, Brennst. Chem. 3 (1928) 39.
[2] A.M. Gadalla, B. Bower, Chem. Eng. Sci. 43 (11) (1988) 3049.
[3] J.T. Richardson, S.A. Paripatyadar, Appl. Catal. 61 (1990) 293.
[4] A.T. Ashcroft, A.K. Cheetham, M.L.H. Green, P.D.F. Vernon, Nature 352 (1991) 225.
[5] J.R. Rostrup-Nielsen, J-H.B. Hansen, J. Catal. 144 (1993) 38.
[6] A. Erdöhelyi, J. Cserényi, F. Solymosi, J. Catal. 141 (1993) 287.
[7] E. Ruckenstein, Y.H. Hu, Appl. Catal. A: Gen. 133 (1995) 149.
[8] Z.L. Zhang, X.E. Verykios, J. Chem. Soc., Chem. Commun. (1995) 71.
[9] Y.H. Hu, E. Ruckenstein, J. Catal. 163 (1996) 306.
[10] R.N. Bhat, W.M.H. Sachtler, Appl. Catal. A: Gen. 150 (1997) 279.
[11] H.Y. Wang, C.T. Au, Appl. Catal. A: Gen. 155 (1997) 239.
[12] J.H. Bitter, K. Seshan, J.A. Lercher, J. Catal. 176 (1998) 93.

[13] S.M. Stagg, E. Romeo, C. Padro, D.E. Resasco, J. Catal. 178 (1998) 137.
[14] K. Tomishige, Y. Chen, K. Fujimoto, J. Catal. 181 (1999) 91.
[15] M.C.J. Bradford, M.A. Vannice, J. Catal. 183 (1999) 69.
[16] A. Erdöhelyi, J. Cserényi, E. Rapp, F. Solymosi, Appl. Catal. A: Gen. 108 (1994) 205.
[17] J. Nakamura, K. Aikawa, K. Sato, T. Uchijima, Catal. Lett. 25 (1994) 265.
[18] Z.L. Zhang, V.A. Tsipouriari, A.M. Efstathiou, X.E. Verykios, J. Catal. 158 (1996) 51.
[19] E. Ruckenstein, H.Y. Wang, J. Catal. 187 (1999) 151.
[20] E. Ruckenstein, H.Y. Wang, Appl. Catal. A: Gen. 198 (2000) 33.
[21] S.J. Tauster, S.C. Fung, J. Catal. 55 (1978) 29.
[22] H.Y. Wang, E. Ruckenstein, J. Catal. 186 (1999) 181.
[23] E. Ruckenstein, H.Y. Wang, J. Catal. 190 (2000) 32.
[24] S.A. Stevenson, G.B. Raupp, J.A. Dumesic, S.J. Tauster, R.T.K. Baker, in: S.A. Stevenson, J.A. Dumesic, R.T.K. Baker, E. Ruckenstein (Eds.), Metal-Support Interactions in Catalysis, Sintering, Redispersion, Van Nostrand Reinhold, New York, 1987, p. 34.
[25] KH.M. Minachev, E.S. Shpiro, in: KH.M. Minachev, E.S. Shpiro (Eds.), Catalyst Surface: Physical Methods of Studying, CRC Press, Boca Raton, 1990, p. 255 (G. Leib, Trans. from the Russian).
[26] E. Ruckenstein, in: S.A. Stevenson, J.A. Dumesic, R.T.K. Baker, E. Ruckenstein (Eds.), Metal-Support Interactions in Catalysis, Sintering, Redispersion, Van Nostrand Reinhold, New York, 1987, p. 141.
[27] Personal information from Dr. W.F. Egelhoff, Jr., Magnetic Material Group, NIST, Gaithersburg, MD 20899.

Transient Response Analysis via a Broadened Pulse Combined with a Step Change or an Isotopic Pulse. Application to CO_2 Reforming of Methane over NiO/SiO_2

Yun Hang Hu and Eli Ruckenstein*

Department of Chemical Engineering, State University of New York at Buffalo, Amherst, New York 14260

Received: May 22, 1997[⊗]

In this paper, a transient response analysis is developed by combining a broadened pulse with a step change or an isotopic pulse. The combined method provides information about the activation of the catalyst and the catalytic reaction mechanism. The method is applied to the CO_2 reforming of methane over NiO/SiO_2 at 800 °C. The response to the broadened CH_4/CO_2 (1:1) pulse over fresh NiO/SiO_2 exhibited an induction time during which no reaction between CH_4 and CO_2 occurred, indicating that NiO/SiO_2 had no activity. As the NiO was gradually reduced by CH_4, the reforming reaction took place, indicating that Ni^0 is the active site for this reaction. During the transient response to a step change to He in the broadened pulse of CH_4/CO_2 over a reduced NiO/SiO_2 catalyst at 800 °C, a long CO tail was observed; however, no such tail was noted in the response after a step change to He in the broadened pure CO pulse at 800 °C. This indicates that the CO desorption is rapid and that the surface reaction between C and O species constitutes the rate-determining step of the reforming reaction. The transient response to a sharp $^{18}O_2$ pulse introduced into the broadened $CH_4/C^{16}O_2$ pulse indicated that, over a reduced NiO/SiO_2 catalyst, $^{18}O_2$ replaced some $C^{16}O_2$ in the reaction with CH_4 to generate CO^{18} and that some Ni^0 was oxidized to $Ni^{18}O$. The long tail of CO^{18} indicates that carbon species reduce this NiO^{18} to Ni^0.

1. Introduction

One of the most important uses of methane is in the preparation of synthesis gas. As for now, steam reforming is the dominant commercial method for producing synthesis gas.[1,2] However, this process has a poor selectivity for CO and a too high H_2/CO product ratio, unsuitable for the methanol and the Fischer–Tropsch syntheses. Although the partial oxidation of methane can provide high activity and selectivity, and a suitable CO/H_2 ratio,[3–10] this process cannot be easily controlled due to the difficulty of removing the reaction heat from the reactor, particularly from large-scale equipment.[11,12] Therefore, in recent years, renewed interest in methane conversion to synthesis gas via the CO_2 reforming reaction has arisen.[13–22] This pathway provides a high CO selectivity and a more suitable H_2/CO ratio. Recently, Ni-based catalysts with high selectivity and activity and excellent stability have been reported.[18–22] However, there are few investigations regarding the activation process of NiO-based catalysts in the reactant atmosphere and the reaction mechanism.

Transient methods were often used in catalytic research to gain insight into the adsorption–desorption kinetics and the reaction mechanism.[23–28] They involve two kinds of techniques: the pulse technique, which provides unsteady-state information, and the step change, which provides information about the transient process from one steady state to another one. A steady-state tracing technique involving a step change in which one of the compounds is labeled with an isotope was also employed.[29] However, in the common pulse techniques,[26] the pulse width is $1-10^2$ s; it is about 1 ms in the temporal analysis of products (TAP) technique.[30] It is, therefore, difficult to make a step change in the reactant or to introduce an isotopic sharp pulse during a regular pulse to investigate the catalytic process.

In this paper, a transient response analysis is developed which combines a broadened pulse with a step change in reactants or a sharp isotopic pulse which is injected any time in the broadened pulse. Using such a method, the CO_2 reforming of methane over NiO/SiO_2 catalyst is investigated.

2. Transient Response Analysis of the Broadened Pulse Combined with a Step Change or an Isotopic Pulse

Figure 1 shows how the broadening of the pulse was achieved. A gas pulse of 2.5 mL was injected almost instantaneously, using the six-way valve (5) and the sample tube (6), into a carrier gas (He) and allowed to flow via the four-way valve (7) into the broadening tube (8). In the latter tube, which has a diameter of 40 mm and a length of 164 mm, the pulse was broadened to about 20 min. From the broadening tube, the pulse flowed more slowly through the four-way valve (7) and then through the six-way valve (9) to the reactor and finally to the MS equipment. During the broadened pulse, a step change in the reactants, using the four-way valve (7), or a sharp isotopic pulse of 3 μL, using the six-way valve (9) and the sample tube (10), was introduced at selected times of the broadened pulse. The reactant and products have been continuously detected with MS equipment (HP quadrupole, 5971 Series mass selective detector).

3. Experimental Section

3.1. Catalyst Preparation. NiO/SiO_2 (containing 13.6 wt % Ni) catalysts were prepared by impregnating SiO_2 (Aldrich) with an aqueous solution of nickel nitrate (Alfa). The obtained paste was dried at room temperature in air and then decomposed and calcined at 800 °C in air for 1.5 h. The surface area and the pore size distribution of the catalyst were determined by nitrogen adsorption, using a Micromeritics ASAP 2000 instrument. The BET surface area and the average diameter of the pores were 362 m^2/g and 52 Å, respectively.

3.2. Transient Response Analysis. The transient experiments were performed at 800 °C under atmospheric pressure with the equipment presented in Figure 1. The catalyst powder (weight, 0.02 g) was held on quartz wool in a vertical quartz

⊗ Abstract published in *Advance ACS Abstracts,* September 1, 1997.

This paper, which was originally published in **Journal Physical Chemistry B**: 101, 7563–7565 (1997), is republished with permission.

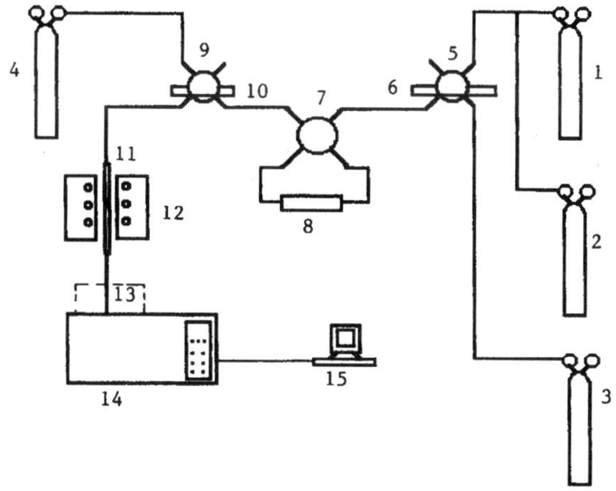

Figure 1. Transient response system: (1) CH_4/CO_2 (1:1) cylinder, (2) Ar cylinder, (3) He cylinder, (4) $^{18}O_2$ cylinder, (5) six-way valve, (6) sample tube, (7) four-way valve, (8) broadening tube, (9) six-way valve, (10) sample tube, (11) reactor, (12) electrical furnace, (13) inlet capillary system, (14) MS, (15) computer.

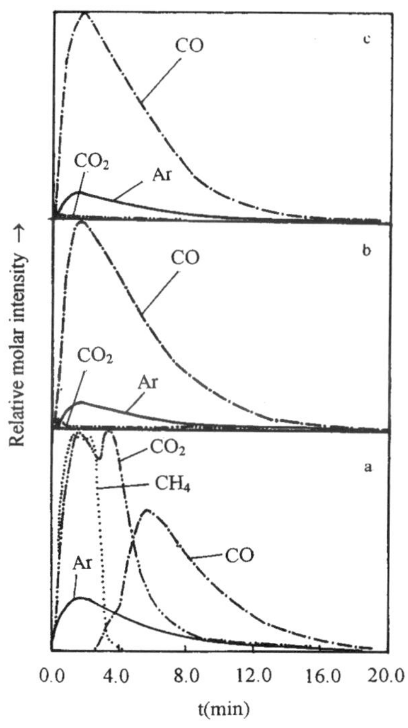

Figure 2. Responses of CH_4, CO_2, CO, and Ar as a function of time for broadened pulse of $CH_4/CO_2/Ar$ (2.5 mL) over a fresh NiO/SiO_2 catalyst at 800 °C: (a) first pulse, (b) second pulse, and (c) third pulse.

tube reactor (2 mm i.d.) in an electronically controlled furnace of large heat capacity to keep the temperature constant. Helium (60 mL/min) was used as the carrier gas. The feed gas contained CO_2, CH_4 (CO_2/CH_4 1:1), and 13.4% Ar as inner standard. The pulses were of 2.5 mL and 3 μL for the broadened reactant pulse and the short isotopic pulse, respectively.

4. Results and Discussion

4.1. Transient Response Analysis of the Broadened Pulse. Figure 2 presents the response curves when only the broadened pulse is present; it contains the responses of three succesive pulses, each introduced after the previous one was exhausted. In the response curves of the first pulse, there are two peaks

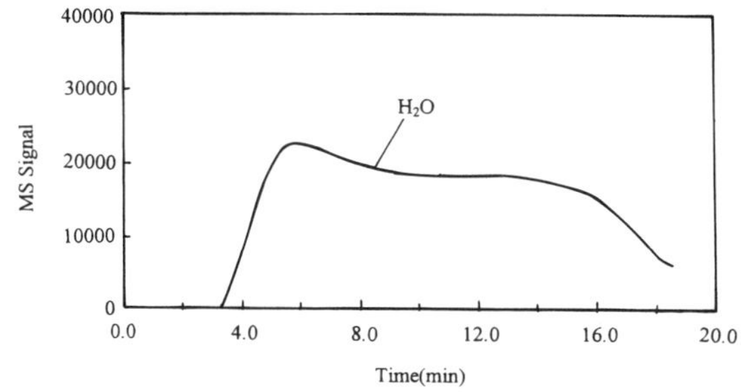

Figure 3. Response curve of H_2O as a function of time for the first broadened pulse of $CH_4/CO_2/Ar$ (2.5 mL) in Figure 2.

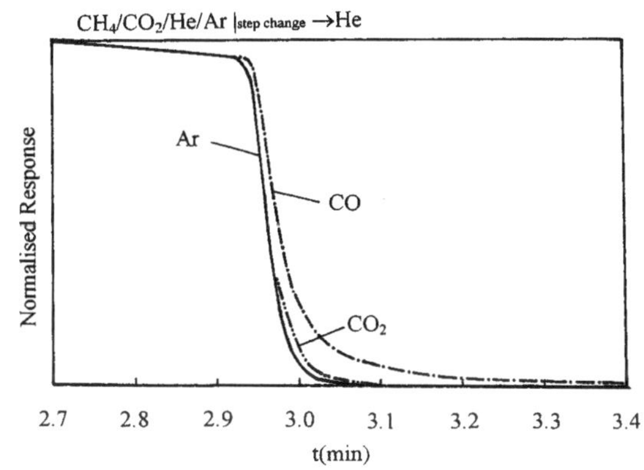

Figure 4. MS signal response curves of CO_2, CO, and Ar as a function of time for a broadened pulse of $CH_4/CO/Ar$ combined with a step change to pure carrier (He) over NiO/SiO_2 subjected previously to 10 broadened succesive exhausted pulses.

for CO_2 at 1.8 and 3.5 min, a peak for CO at about 5.8min, and a peak for CH_4 at about 1.8 min. For the second pulse, there are very small peaks for CO_2 and CH_4 and a high one for CO. For the third pulse, the response is similar to the second one. This figure shows that the CO peak increases and the CO_2 peak decreases with increasing pulse number.

Figure 2 shows that, for the first pulse, the first CO_2 peak appears at the same position as the peaks of Ar and CH_4 and that no CO is present at that position. This indicates that the first CO_2 peak can be attributed to the CO_2 unreacted in the feed gas. The second CO_2 peak appears after that of Ar (with a time difference of 1.8 min), at a time at which CH_4 is no longer present. It can, therefore, be attributed to the CH_4 reaction with NiO, which results in the formation of Ni^0 and CO_2. It is worth noting that water was detected in the response curve after the first CO_2 peak but before the second CO_2 peak (Figure 3). This confirms that NiO is reduced by CH_4 to Ni^0. Calculations based on the second CO_2 peak of Figure 2a indicate that about 44% of the NiO of the catalyst was reduced to Ni^0. The CO peak appears after the second CO_2 peak, indicating that Ni^0 constitutes the active site for the production of CO. This also explains why the response curve of CO exhibits an induction time and why CO increases and CO_2 decreases with the pulse number (Figure 2).

4.2. Transient Response Analysis of the Broadened Pulse Combined with a Step Change. The response to a step change in the broadened $CH_4/CO_2/Ar$ pulse was carried out over a catalyst previously subjected to 10 succesively exhausted CH_4/CO_2 (1:1) pulses. Figure 4 shows that, when $CH_4/CO_2/Ar/He$ is replaced via a step change with He, the CO_2 response curve almost coincides with that of Ar, whereas the CO response curve has a long tail. The long tail of CO may be due to (a) the CO desorption from the catalyst surface or (b) the CO formation

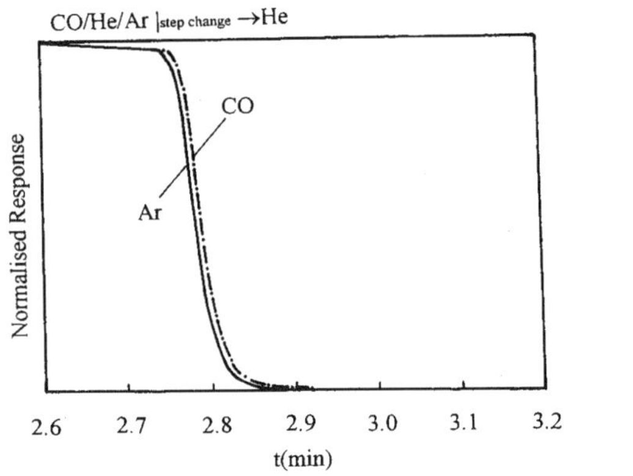

Figure 5. MS signal response curves of CO and Ar as a function of time for a broadened pulse of CO/Ar combined with a step change to pure carrier over NiO/SiO_2 subjected previously to 10 broadened succesively exhausted pulses.

by the surface reaction between the C and O species. To make a choice between the two, the response curve to a step change from CO/Ar/He to He in a broadened pulse of CO/Ar is presented in Figure 5, which shows that the CO response is near that of Ar. This eliminates the possibility (a) that the desorption of CO is responsible for the long tail; in other words, the long tail is due to the surface reaction between the C and O species, which constitutes the rate-determining step. In addition, the long CO tail also indicates that the C species cannot easily desorb. Consequently, the adsorbed species containing C cannot be CH_4, because of its small adsorption energy (about 6 kcal/mol). It is, therefore, reasonable to suggest that a dissociative adsorption of CH_4 takes place, i.e., $CH_{4(g)} + Ni \rightarrow \cdots \rightarrow CH_{x(s)} + (4 - x)H_{(s)}$.

4.3. Transient Response Analysis of the Broadened Pulse Combined with an Isotopic Sharp Pulse. The combination between the CO_2 reforming and the partial oxidation of CH_4 may become a useful process,[13] because the CO_2 reforming is endothermic, while the partial oxidation is exothermic. For this reason, a transient response of a sharp $^{18}O_2/Ar$ pulse of 3 μL into a broadened $CH_4/C^{16}O_2$ pulse of 2.5 mL over a NiO/SiO_2 catalyst, previously subjected to 10 CH_4/CO_2 succesively exhausted pulses, was carried out. Figure 6 shows that the responses of CO^{18} and $C^{16}OO^{18}$ are much longer than that of Ar. The tail of CO^{18} is about 10 min long, whereas the Ar pulse has a width of only 0.1 min. This suggest that ^{18}O forms relatively stable species on the catalyst surface by oxidizing Ni^0 to $Ni^{18}O$. Figure 6 also shows that the response of $C^{16}O$ decreases and that of $C^{16}O_2$ increases rapidly when the $^{18}O_2/Ar$ pulse is injected, and that, subsequently, they recover rapidly the original curves. However, no methane response was detected. The $C^{16}O$ decreases because $^{18}O_2$, being more reactive, replaces some of $C^{16}O_2$ in its reaction with CH_4 and generates $C^{18}O$. The $C^{16}O_2$ increases because it is replaced by $^{18}O_2$. No methane response is detected because it is completely consumed by its reaction with $C^{16}O_2$ and $^{18}O_2$.

5. Conclusion

This paper emphasizes that the transient response of a broadened pulse combined with a step change or a sharp isotopic pulse is a useful method in the study of heterogeneous processes. The following useful information was obtained regarding the CO_2 reforming of methane over NiO/SiO_2: (1)Ni^0 is the active site; (2) the reaction between C and O species on Ni^0 is the rate-determining step; and (3) the introduction of a small amount of $^{18}O_2$ in the broadened $CH_4/C^{16}O_2$ pulse replaces some $C^{16}O_2$ in the reaction with CH_4 and generates $C^{18}O$.

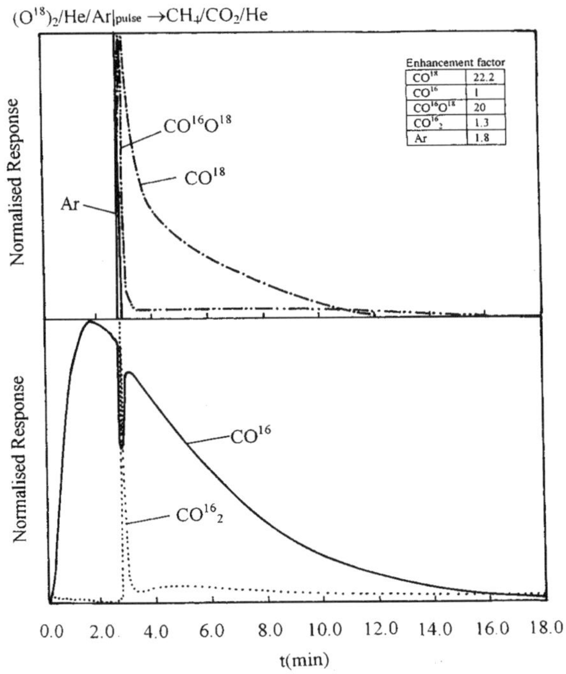

Figure 6. Normalized responses of $C^{16}O_2$, $C^{16}O$, $C^{18}O$, $C^{16}O^{18}O$ and Ar as a function of time for the broadened pulse of $CH_4/C^{16}O_2/Ar$ combined with a sharp pulse of $^{18}O_2/Ar$ over NiO/SiO_2 subjected previously to 10 broadened succesively exhausted pulses.

References and Notes

(1) Trimm, D. L. *Catal. Rev. Sci. Eng.* **1977**, *16*, 155.
(2) Rostrup-Nielsen, J. R. In *Catalysis Science and Technology*; Anderson, J. R., Boudart, M., Eds; Springer: New York, 1984; Vol. 5.
(3) Hickman, D. A.; Schmidt, L. D. *Science* **1993**, *259*, 343.
(4) Ashcroft, A. T.; Cheetham, A. K.; Foord, J. S.; Green, M. L. H.; Grey, C. P.; Murrell, A. J.; Vernon, P. D. F. *Nature* **1990**, *344*, 319.
(5) Choudhary, V. R.; Rajput, A. M.; Prabhakar, B. *J. Catal.* **1993**, *139*, 326.
(6) Hu, Y. H.; Ruckenstein, E. *J. Catal.* **1996**, *158*, 260.
(7) Hu, Y. H.; Ruckenstein, E. *Catal. Lett.* **1995**, *34*, 41.
(8) Ruckenstein, E.; Hu, Y. H. *Catal. Lett.* **1995**, *35*, 265.
(9) Dissanayake, D.; Rosynek, M. P.; Kharas, K. C. C.; Lunsford, J. H. *J. Catal.* **1991**, *132*, 117.
(10) Hu, Y. H.; Au, C. T.; Wan, H. L. *Chin. Sci. Bull.* **1995**, *40*, 303.
(11) Dissanayake, D.; Rosynek, M. P.; Lunsford, J. H. *J. Phys. Chem.* **1993**, *97*, 3644.
(12) Choudhary, V. R.; Rajput, A. M.; Prabhakar, B. *Angew. Chem., Int. Ed. Engl.* **1994**, *33*, 2104.
(13) Ashcroft, A. T.; Cheetham, A. K.; Green, M. L. H.; Vernon, P. D. F. *Nature* **1991**, *352*, 225.
(14) Richardson, J. T.; Paripatyadar, S. A. *Appl. Catal.* **1990**, *61*, 293.
(15) Yamazaki, O.; Nozaki, T.; Omata, K.; Fujimoto, K. *Chem. Lett.* **1992**, *10*, 1953.
(16) Gadalla, A. M.; Sommer, M. E. *Chem. Eng. Sci.* **1989**, *44*, 2825.
(17) Rostrup-Nielsen, J. R.; Hansen, J.-H. B. *J. Catal.* **1993**, *144*, 38.
(18) Zhang, Z.; Verykios, X. E. *J. Chem. Soc., Chem. Commun.* **1995**, *1*, 71.
(19) Hu, Y. H.; Ruckenstein, E. *J. Catal.* **1996**, *163*, 306.
(20) Ruckenstein, E.; Hu, Y. H. *Appl. Catal.* **1995**, *133*, 149.
(21) Hu, Y. H.; Ruckenstein, E. *Catal. Lett.* **1996**, *36*, 145.
(22) Hu, Y. H.; Ruckenstein, E. *Catal. Lett.* **1997**, *43*, 71.
(23) Winslow, P.; Bell, A. T. *J. Catal.* **1985**, *91*, 142.
(24) Biloen, P. *J. Mol. Catal.* **1983**, *21*, 17.
(25) Soong, Y.; Krishna, K.; Biloen, P. *J. Catal.* **1986**, *97*, 330.
(26) Greten, G.; Lafyatis, D. S.; Froment, G. F. *J. Catal.* **1995**, *154*, 151.
(27) Bennett, C. O. *Catal. Rev.-Sci. Eng.* **1976**, *13*, 121.
(28) Efstathiou, A. M.; Lacombe, S.; Mirodatos, C.; Verikios, X. E. *J. Catal.* **1994**, *148*, 639.
(29) Happel, J. *Isotopic Assessment of Heterogeneous Catalysis*; Academic Press: San Diego, 1986.
(30) Gleaves, J. T.; Ebner, J. R. U.S. Patent 4,626,421, 1986.

2

CATALYTIC CONVERSION OF METHANE TO SYNTHESIS GAS BY PARTIAL OXIDATION

INTRODUCTION

The partial oxidation of methane,

$$CH_4 + ½\, O_2 \rightarrow CO + 2H_2, \qquad \Delta H^0_{298} = -36 \text{ kJ/mol}$$

is a mildly exothermic reaction; it produces a synthesis gas with a H_2/CO ratio of around 2, suitable for methanol and Fischer–Tropsch syntheses. Compared with steam or CO_2 reforming, the partial oxidation of methane is more energy efficient, requiring little or no external energy input.

The initial study of methane partial oxidation can be traced back to 1940s, when Prettre et al. [1] reported the formation of synthesis gas through the catalytic partial oxidation of methane over a refractory supported Ni (10 wt%) catalyst at temperatures between 973 and 1173 K. Thermodynamic equilibrium corresponding to the temperature of the catalyst bed exit was achieved under all the conditions investigated. In 1970, Huszar et al. [2] investigated the effect of diffusion on the reactivity of a single grain of Ni/mullite catalyst at temperatures ranging from 1033 to 1173 K and the ignition and extinction characteristics of this catalyst. They reported that the Ni catalyst deactivated in an oxidative environment but was reactivated on reduction. In 1990, Ashcroft et al. [3] reported, for a lanthanide ruthenium oxide catalyst ($Ln_2Ru_2O_7$, where Ln = Pr, Eu, Gd, Dy, Yb, or Lu), that about 90% conversion of CH_4 and higher than 90% selectivity to CO and H_2 were achieved at a temperature of about 1048 K, atmospheric pressure, and a GHSV of 4×10^4 mL/h/mL catalyst. In 1992, Hickman and Schmidt [4] achieved high selectivity to CO and H_2 in the partial oxidation of methane with platinum monoliths. In the subsequent decade or so, various noble metal catalysts were examined for the partial oxidation of methane [5–19]. In general, noble metal catalysts provide excellent activity and selectivity as well as high stability.

In view of the limited availability of noble metals, nonnoble metal catalysts, especially Ni-based catalysts, have been widely studied since 1990. Lunsford et al. [20] reported CO selectivity close to 95% and near complete conversion of CH_4 over a 25 wt% Ni/Al_2O_3 catalyst at temperatures ranging from 723 to 1173 K. Choudhary et al. examined Ni/CaO [21], Ni/Al_2O_3 [22], NiO-rare earth oxide [23], and Co/rare earth oxide [24] catalysts, and reported both high conversions and high

Heterogeneous Catalysis: Experimental and Theoretical Studies, First Edition. Edited by Eli Ruckenstein and Haiyou Wang.

selectivities to CO and H_2. The major issue with these non-noble metal catalysts is their relatively low stability. Carbon deposition and metal sintering have been regarded as the main causes for catalyst deactivation. Efforts to improve the stability of Ni-based catalyst have included the utilization of suitable supports, such as La_2O_3 [25], $SrTiO_3$ [26], and CeO_2 [27], and the addition of suitable promoters, such as La_2O_3 [28, 29], Li_2O [29], and iron oxide [30], as well as novel catalyst preparation methods, such as the solid phase crystallization [31], sol-gel [32], and citrate [33] methods.

Our work was focused on the effect of various supports on the reactivity of supported metal catalysts and the structure-property relationships. A great deal of effort was expended to shed light on the mechanism of methane partial oxidation by using the pulse-transient analysis method and the CH_4/CD_4 isotope effect method. To sum up, Chapter 2 consists of three parts: the first is concerned with Rh-based catalysts (**2.1–2.2**); the second with Ni and Co-based catalysts (**2.3–2.4**); and the third with mechanistic studies of methane partial oxidation (**2.5**, **2.6**, **2.7**, and **2.8**).

The catalysts that have been reported to be active and selective toward synthesis gas formation were mainly supported noble metals, such as Rh, Ru, Pd, and Pt, supported nonprecious metals, such as Ni and Co, as well as some pyrochlore type oxides, such as $Ln_2Ru_2O_7$, and perovskite type oxides, such as $LaMO_3$ (with M = Ni, Ph, Co, Cr). The nature of the support was found to have a significant impact on the performance (activity and stability) of supported Rh catalysts for methane partial oxidation (**2.1**). In general, the irreducible oxides, such as La_2O_3, MgO, and so on, provided much higher activities and selectivities than the reducible ones, such as CeO_2, Nb_2O_5, and so on. Among the irreducible metal oxides, La_2O_3 and MgO provided stable catalytic activities and selectivities, whereas deactivation occurred over Y_2O_3 and SiO_2. The strong interactions between rhodium and magnesium or lanthanum oxides (especially the formation of $MgRh_2O_4$ or $LaRhO_3$) were most likely responsible for the high stability of MgO and La_2O_3 supported Rh catalysts. The stability of the Y_2O_3 supported catalyst was notably improved by conducting the partial oxidation of methane at temperatures at which the compound $YRhO_3$ could be formed (**2.2**). Compared with the noble-metal based catalysts, the Ni- and Co-based ones are particularly promising due to their low prices. Ni/MgO catalysts, obtained through the reduction of NiO–MgO solid solutions containing 9.7–35 mol% NiO, were found to provide high and stable activities for methane partial oxidation (**2.3**). Like NiO, CoO can also form an "ideal" solid solution over the entire mole fraction range with MgO. However, unlike Ni, Co has two oxidation states (Co^{2+} and Co^{3+}). Consequently, additional cobalt and magnesium species can be generated during the preparation of the catalyst. The species that form are expected to depend on the preparation conditions, such as calcination temperature and metal loading. The species present in a catalyst affect its reducibility, the size and morphology of the metal particles, and ultimately the catalytic performance provided by the catalyst. A structure–property relationship was noted over Co catalysts supported on MgO and other alkaline earth metal oxides (**2.4**).

Mechanistic studies of methane partial oxidation remain a great challenge since this exothermic, fast reaction results in an extremely high catalyst temperature. As a result, the mechanism of methane partial oxidation is still debated. Two mechanisms have been suggested for the partial oxidation of methane to synthesis gas: (i) the combustion-reforming mechanism, in which the oxidation reaction involves an initial strongly exothermic oxidation of CH_4 to CO_2 and H_2O followed by the endothermic reforming reaction of the remaining CH_4 by the CO_2 and H_2O formed in the previous step; (ii) the pyrolysis-oxidation mechanism, in which methane pyrolysis is followed by surface carbon oxidation and desorption of CO and H_2. A pulse-transient analysis method was used to study the kinetics of methane oxidation over Ni/La_2O_3 (**2.5**). The transient response curves revealed that CO_2 generation is delayed in comparison with that of CO, implying that CO is the direct product and CO_2 is only subsequently generated from CO; in other words, the reaction follows the pyrolysis mechanism. An isotopic pulse combined with GC-MS was used to study the mechanism of methane partial oxidation over a reduced NiO/SiO_2 catalyst (**2.6**). It was found that the reaction of $CH_4/CD_4/O_2$ (1/1/1) occurs via the pyrolysis mechanism, with the dissociation of methane faster than the oxidation of CH_x. Mechanistic studies over both unreduced and reduced NiO/SiO_2 catalysts (**2.7**) were conducted using a new transient method, in which, instead of sharp pulses, both broadened pulses combined with step changes and broadened pulses combined with sharp isotopic pulses were employed. It was observed that over the unreduced NiO/SiO_2 catalyst, the reaction between CH_4 and O_2 occurs via the Eley–Rideal mechanism, in which O_2 in the adsorbed state reacts with CH_4 in the gas phase. In contrast, over the reduced catalyst, the reaction takes place via the Langmuir–Hinshelwood mechanism, in which adsorbed CH_4 and O_2 species react with each other. Lastly, the CH_4/CD_4 isotope effect was determined by conducting $CH_4 + O_2$ and $CD_4 + O_2$ reactions alternately over 0.05 and 1.0 wt% Rh/γ-Al_2O_3 catalysts in a pulse microreactor at temperatures ranging from 400 to 850°C (**2.8**). The results show that the mechanism of methane partial oxidation to synthesis gas is dependent on both metal loading and temperature. It is likely that at low loadings (e.g., 0.05 wt%), the combustion-reforming mechanism is responsible for the reaction, while at high loadings (e.g., 1.0 wt%), a combination of the combustion-reforming and pyrolysis-oxidation mechanism is responsible at low temperatures (≤500°C) and the pyrolysis-oxidation mechanism becomes dominant at high temperatures (≥650°C).

REFERENCES

[1] Prettre, M., Eichner, C., and Perrin, M., *Trans. Faraday Soc.*, **43**, 335 (1946).

[2] Huszar, K., Racz, G., and Szekely, G., *Acta Chim. Acad. Sci. Hungary*, **70**, 287 (1971).

[3] Ashcroft, A.T., Cheetham, A.K., Foord, J.S., Green, M.L.H., Grey, C.P., Murrell, A.J., and Vernon, P.D.F., *Nature*, **344**, 319 (1990).

[4] Hickman, D.A., and Schmidt, L.D., *J. Catal.*, **138**, 267 (1992).

[5] Hickman, D.A., and Schmidt, L.D., *Catal. Lett.*, **17**, 223 (1993).

[6] Mallens, E.P.J., Hoebink, J.H.B.J., and Marin, G.B., *Catal. Lett.*, **33**, 291 (1995).

[7] Soick, M., Buyevskaya, O., Hohenberger, M., and Wolf, D., *Catal. Today*, **32**, 163 (1996).

[8] Nakagawa, K., Suzuki, T., Kobayashi, T., and Haruta, M., *Chem. Lett.*, 1029 (1996).

[9] Boucouvalas, Y., Zhang, Z.L., and Verykios, X.E., *Catal. Lett.*, **40**, 189 (1996).

[10] Boucouvalas, Y., Efstathiou, A.M., Zhang, Z.L., and Verykios, X.E., *Natural Gas Conversion IV. Stud. Surf. Sci. Catal.*, **107**, 435 (1997).

[11] Laa, J.C., Berger, R.J., and Marin, G.B., *Catal. Lett.*, **43**, 63 (1997).

[12] Hofstad, K.H., Hoebink, J.H.B.J., Holmen, A., and Marin, G.B., *Catal. Today*, **40**, 157 (1998).

[13] Kikuchi, E., and Chen Y., *Natural Gas Conversion V. Stud. Surf. Sci. Catal.*, **119**, 441 (1998).

[14] Elmasides, C., Ioannides, T., and Verykios, X.E., *Natural Gas Conversion V. Stud. Surf. Sci. Catal.*, **119**, 801 (1998).

[15] Yan, Q.G., Chu, W., Gao, L.Z., Yu, Z.L., and Yuan, S.Y., *Natural Gas Conversion V. Stud. Surf. Sci. Catal.*, **119**, 855 (1998).

[16] Nakagawa, K., Ikenaga, N., Suzuki, T., Kobayashi, T., and Haruta, M., *Appl. Catal. A: Gen.*, **169**, 281 (1998).

[17] Wang, H.Y., and Ruckenstein, E., *J. Catal.*, **186**, 181 (1999).

[18] Wang, H.Y., and Ruckenstein, E., *Catal. Lett.*, **59**, 121 (1999).

[19] Pantu, P., and Gavalas, *Appl. Catal. A: Gen.*, **223**, 253 (2002).

[20] Dissanayake, D., Rosynek, M.P., and Lunsford, J.H., *J. Catal.*, **132**, 117 (1991).

[21] Choudhary, V.R., Rajput, A.M., and Prabhakar, B., *Catal. Lett.*, **15**, 363 (1992).

[22] Choudhary, V.R., Rajput, A.M., and Prabhakar, B., *J. Catal.*, **139**, 326 (1993).

[23] Choudhary, V.R., Rajput, A.M., and Rane, V.H., *J. Phys. Chem.*, **96**, 8696 (1992).

[24] Choudhary, V.R., Sansare, S.D., and Mamman, A.S., *Appl. Catal. A: Gen.*, **90**, L1 (1992).

[25] Tsipouriari, V.A., Zhang, Z.L., and Verykios, X.E., *J. Catal.*, **179**, 283 (1998).

[26] Takehira, K., Shishido, T., and Kondo, M., *J. Catal.*, **207**, 307 (2002).

[27] Zhu, T.L., and Flytzani,-Stephanopoulos, M., *Appl. Catal. A: Gen.*, **208**, 403 (2001).

[28] Cao, L., Chen, Y., and Li, W., *Natural Gas Conversion IV. Stud. Surf. Sci. Catal.*, **107**, 467 (1997).

[29] Liu, S.L., Xiong, S.S.,Miao, Q., and Yang, W.S., *Natural Gas Conversion V. Stud. Surf. Sci. Catal.*, **119**, 747 (1998).

[30] Wang, J.G., Liu, C.J., Zhang, Y.P., Zhu, X.L., Zou, J.J., Yu, K.L., and Eliasson, B., *Chem. Lett.*, 1068 (2002).

[31] Shiozaki, R., Andersen, A.G., Hayakawa, T., Hamakawa, S., Suzuki, K., Shimizu, M., and Takehira, K., *Third World Congress on Oxidation Catalysts. Stud. Surf. Sci. Catal.*, **110**, 701 (1997).

[32] Zhang, Y.H., Xiong, G.X., Sheng, S.S., and Yang, W.S., *Catal. Today*, **63**, 517 (2000).

[33] Hayakawa, T., Harihara, H., Andersen, A.G., Suzuki, K., Yasuda, H., Tsunoda, T., Hamakawa, S., York, A.P.E., Yoon, Y.S., Shimizu, M., and Takehira, K., *Appl. Catal. A: Gen.*, **149**, 391 (2001).

Effect of Support on Partial Oxidation of Methane to Synthesis Gas over Supported Rhodium Catalysts

E. Ruckenstein[1] and H. Y. Wang

Department of Chemical Engineering, State University of New York at Buffalo, Amherst, New York 14260

Received March 1, 1999; revised May 24, 1999; accepted May 25, 1999

Two kinds of oxides, reducible and irreducible, were used as supports. It was found that the reducible-oxides-supported rhodium catalysts provide, in general, much lower activities and selectivities than those supported on irreducible oxides. The exposed metal surface areas of the reduced catalysts were determined, and the reaction behaviors of CH_4/O_2 (2/1) over the pure supports and the precalcined 1% Rh(O)/M_xO_y catalysts were investigated in a pulse microreactor. It is suggested that the partial coverage of rhodium sites by the reducible oxides and the combustion of methane that occurs over these oxide sites are responsible for the lower activities and selectivities observed over this kind of supported catalysts. Among the irreducible metal oxides, γ-Al_2O_3, La_2O_3, and MgO provided stable catalytic activities and selectivities during 100 h of reaction, and the activity increased in the sequence $La_2O_3 < \gamma$-$Al_2O_3 \leq$ MgO. Possible explanations for stability are proposed on the basis of TPR and XRD experiments. In addition, the nature of the active sites is examined based on pulse reaction experiments. © 1999 Academic Press

Key Words: **methane partial oxidation; syngas; rhodium-based catalyst; effect of support.**

1. INTRODUCTION

Steam reforming constitutes the dominant commercial process employed to produce synthesis gas (a mixture of CO and H_2) from methane (1, 2):

$$CH_4 + H_2O \rightarrow CO + 3H_2 \quad \Delta H^0_{298} = 206\ \text{kJ/mol}. \qquad [1]$$

However, this process requires a large energy input and provides a syngas with a too high H_2/CO ratio (>3) for the methanol and Fischer–Tropsch syntheses. For these reasons, research efforts have been made to obtain the suitable ratio of 2 via methane partial oxidation to H_2 and CO (3–28):

$$CH_4 + 1/2O_2 \rightarrow CO + 2H_2 \quad \Delta H^0_{298} = -36\ \text{kJ/mol}. \qquad [2]$$

This reaction is mildly exothermic, and the catalysts were mainly supported noble metals, such as Rh, Ru, Pd, Pt (3–10), and supported Ni catalysts (11–23), as well as some pyrochlore-type oxides, such as $Ln_2Ru_2O_7$ (24, 25), and perovskite type oxides, such as LaMO$_3$ (M = Ni, Rh, Co, Cr) (26–28).

The support is usually a surface that allows the spreading of the metal catalyst as small clusters. In some cases, the support may actually contribute to the catalytic activity. For the CO_2 reforming of methane over Rh-based catalysts, the activity sequence of the supports was $Al_2O_3 > TiO_2 > SiO_2$ (29). For both the partial oxidation of methane and the CO_2 reforming of methane over Ir-based catalysts, the activity sequence was $TiO_2 \geq ZrO_2 \geq Y_2O_3 > La_2O_3 > MgO \geq Al_2O_3 > SiO_2$ (30). For the partial oxidation of hydrocarbons over noble metal monoliths, the effect of the support was small (31).

It is well known that, after a high-temperature H_2 reduction, the hydrogen and carbon monoxide chemisorption may be suppressed over the group VIII metal crystallites supported on semiconductor oxides (e.g., CeO_2, Nb_2O_5, Ta_2O_5, TiO_2, ZrO_2), without appreciable change in their dispersion (32, 33). Because this effect, called strong metal–support interactions (SMSI), can play an important role, it is of interest to compare the performances of group VIII metal catalysts supported on reducible and irreducible oxides in the partial oxidation of methane. In the present paper, two kinds of metal oxides, reducible and irreducible, were used as supports for rhodium. It was found that the irreducible oxides provided much higher activities and selectivities than the reducible ones. Among the irreducible oxides, γ-Al_2O_3 and MgO, exhibited high activities and selectivities with high stability. A somewhat lower activity but comparable selectivities and stability were provided by the La_2O_3 support. Possible explanations based on pulse reactions, CO chemisorption, TPR, and XRD experiments are suggested.

2. EXPERIMENTAL

2.1. *Catalyst Preparation*

The following supports were used: γ-Al_2O_3, CeO_2, Nb_2O_5, Ta_2O_5, ZrO_2 (Alfa); and La_2O_3, $MgO_2 \cdot x$MgO,

[1] To whom correspondence should be addressed.

This paper, which was originally published in **Journal of Catalysis**: 187, 151–159 (1999), is republished with permission.

SiO_2, TiO_2, Y_2O_3 (Aldrich). The supported rhodium catalysts were prepared by impregnating the support with an ethanol solution of $Rh(NO_3)_3 \cdot 2H_2O$ (Alfa), followed by overnight drying at 110°C and calcination in air at 800°C for 4 h. The calcined catalysts are denoted $Rh(O)/M_xO_y$. Since $MgO_2 \cdot x MgO$ was completely converted into MgO after calcination at 800°C for 4 h, its supported rhodium catalyst is also denoted Rh(O)/MgO. The catalysts reduced in H_2 are denoted Rh/M_xO_y. Rh loading means wt% Rh in the reduced catalyst.

2.2. Activity Assay

All activity assays were conducted under atmospheric pressure in a fixed-bed vertical quartz reactor (3 mm inside diameter) located in an electronically controlled furnace, with the catalyst powder held on quartz wool. Five (5.0) mg catalyst was used in each run. The prepared catalyst was heated in an H_2 flow (20 ml/min) up to 750°C at 20°C/min, after which, maintaining the furnace temperature at 750°C, H_2 was switched to a CH_4/O_2 mixture with a molar ratio of 2.0, at a GHSV of 720000 ml g^{-1} h^{-1}. The reactants and products were analyzed with an on-line gas chromatograph equipped with Porapak Q and 5A molecular sieve columns. A thermocouple was inserted into the middle of the catalyst bed to measure the temperature of the catalyst.

Methane (99.97% purity, from Matheson) and oxygen (99.9% purity, from Cryogenic) were used without further purification. The gases were premixed before they were introduced into the reactor.

2.3. Pulse Reaction

A quartz tube (4 mm inside diameter) was used as reactor with the calcined catalyst held on quartz wool. A constant flow of helium (35 ml/min) was employed, and a reactant gas mixture was injected in the carrier gas. In each run, 50.0 mg catalyst was used and the pulse volume was 50 μl CH_4/O_2 (2/1). The reactants and products were analyzed with an on-line gas chromatograph equipped with a thermal conductivity detector (TCD) and a Porapak Q column. The carrier gas helium was purified using Hydro-Purge II and Oxy-Trap columns.

2.4. Catalyst Characterization

2.4.1. Surface area. The surface areas of the calcined catalysts were determined by nitrogen adsorption, using a Micromeritics ASAP2000 instrument. The sample was degassed at 200°C for at least 5 h in high vacuum before the measurement.

2.4.2. The exposed metal surface area. The exposed surface areas of rhodium of the reduced catalysts were determined by CO chemisorption at room temperature by assuming a 1/1 stoichiometry. The exposed surface area of rhodium was calculated by assuming a hemisphere shape for the crystallites. One hundred (100.0) mg catalyst powder held on quartz wool was reduced in an H_2 flow (20 ml/min) at 550°C for 1.5 h. Further, at the same temperature, the reduced catalyst was purged with ultrahigh purity helium (35 ml/min) for 1 h. After the temperature was decreased to room temperature, CO (10 μl per pulse) was pulsed over the catalyst until no further adsorption of CO was detected. The CO left during CO chemisorption was determined quantitatively with a thermal conductivity detector (TCD). Both the hydrogen and helium were purified with Hydro-Purge II and Oxy-Trap columns before use.

2.4.3. Temperature-programmed reduction (TPR). The TPR of the catalyst was conducted by heating the calcined catalyst from 50 to 850°C at a rate of 20°C/min in a flow of 2.5% H_2/Ar mixture (35 ml/min). The hydrogen consumed in TPR was determined with a thermal conductivity detector (TCD). Fifty (50.0) mg sample precalcined in air at 800°C for 4 h was used in each TPR run.

2.4.4. X-ray powder diffraction (XRD). X-ray powder diffraction (XRD) was carried out on a Siemens D500 X-ray diffractometer, using Cu $K\alpha$ radiation, at 40 kV and 30 mA.

3. RESULTS

3.1. Continuous Reaction

3.1.1. Effect of support on the performance of Rh catalysts. Two kinds of metal oxides, reducible (e.g., CeO_2, Nb_2O_5, Ta_2O_5, TiO_2, and ZrO_2) and irreducible (e.g., γ-Al_2O_3, La_2O_3, MgO, SiO_2, Y_2O_3), were used as supports. As shown in Table 1, the performance of the supported rhodium catalyst is strongly affected by the support. Figure 1 shows the effect of time on stream on the activity of rhodium supported on the irreducible oxides and on the reducible Ta_2O_5. (The conversions in this paper have been always lower than the equilibrium conversions (34) at the temperatures of the catalysts.) The catalytic performances of the supported rhodium catalysts can be summarized as follows:

(i) With the exception of Ta_2O_5, the reducible oxides provided much lower methane conversions and selectivities to CO and H_2 than the irreducible ones (Table 1).

(ii) Nb_2O_5 and TiO_2 not only exhibited poor activities and selectivities to CO and H_2 but also provided a syngas with low H_2/CO ratio (1.0 and 1.5, respectively). The other supports provided an H_2/CO ratio of about 2, as expected from the stoichiometry of the reaction (Table 1).

(iii) Among the irreducible metal oxides, γ-Al_2O_3, La_2O_3, and MgO provided stable activities during 100 h of reaction, and the activity increased in the sequence $La_2O_3 < \gamma\text{-}Al_2O_3 \leq MgO$; deactivation was observed over the SiO_2- and Y_2O_3-supported catalysts (Fig. 1).

TABLE 1

Activity of Supported Rh (1 wt%) Catalysts for Methane Partial Oxidation at 750°C ($T_{furnace}$)[a]

Catalysts	Surf. area (m^2/g-cat.)	Rh surf. area (m^2/g-cat. × 10^2)	Catal. temp. (°C)	CH_4 conv. (%)	CO select. (%)	H_2 select. (%)	H_2/CO (ratio)
Rh/CeO_2	3.2	2.7	796	58.1	71.6	72.1	2.0
Rh/Nb_2O_5	6.1	0.1	840	33.0	35.3	17.5	1.0
Rh/Ta_2O_5	5.0	—[b]	782	69.1	95.8	92.2	1.9
Rh/TiO_2	10.1	3.2	815	47.1	68.2	52.5	1.5
Rh/ZrO_2	4.1	2.4	770	55.4	72.7	69.4	1.9
Rh/Al_2O_3	59.3	28.1	759	81.6	94.0	93.5	2.0
Rh/La_2O_3	6.2	7.9	766	72.5	90.3	91.6	2.0
Rh/MgO	37.4	17.9	761	75.5	92.1	96.4	2.1
Rh/SiO_2	573.7	41.5	752	80.1	86.7	93.5	2.2
Rh/Y_2O_3	9.4	14.9	789	68.3	85.9	88.7	2.1

[a] Catalyst, 5.0 mg; flow rate, 60.0 ml/min ($CH_4/O_2=2.0$); space velocity, 720000 h^{-1} ml/g-catalyst; data obtained after 6 h of reaction; the oxygen conversion was always 100%.
[b] Data not obtainable because the reaction $2CO \rightarrow CO_2+\underline{C}$ occurred over this catalyst at room temperature.

(iv) Though Ta_2O_5 provided much higher methane conversion and selectivities to CO and H_2 compared to the other reducible oxides, its activity decayed with time (Fig. 1).

(v) A relatively long induction time was observed over the MgO-supported rhodium catalyst.

3.2. Pulse Reaction

3.2.1. Reactivities of CH_4/O_2 (2/1) pulses over the oxide supports. The reaction of methane and oxygen over the metal oxides was investigated at 750 and 850°C by using pulses of CH_4/O_2 (2/1), and some results are listed in Table 2. The carbon balance indicated that no carbon deposition occurred during reaction. Except at 850°C over CeO_2, for which some CO and H_2 was formed, the products of the reaction during the first pulse were exclusively CO_2 and H_2O. Except with ZrO_2 at 750°C and with CeO_2 at

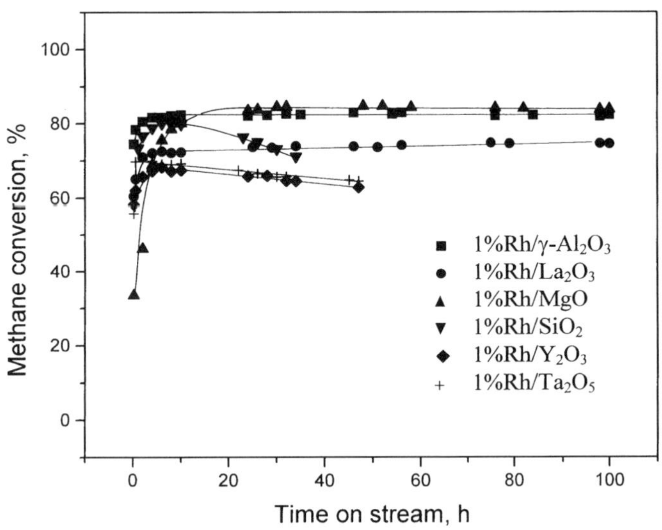

FIG. 1. Effect of time on stream on the activity of rhodium supported on the irreducible oxides and Ta_2O_5. $P=1$ atm, $T_{furnace}=750$°C, $CH_4/O_2=2.0$, GHSV = 720 000 ml g^{-1} h^{-1}.

TABLE 2

Reactivities of CH_4/O_2 (2/1) Pulses over Metal Oxides[a,b]

Supports	Surf. area[b] (m^2/g)	Temp. (°C)	CH_4 conv. (%)	O_2 conv. (%)	CO sel. (%)	CO_2 sel. (%)
CeO_2	3.3	750	21.9[c]	91.6	0.0	100.0
		850	51.2[c]	100.0	1.3	98.7
			44.6[d]	100.0	12.4	87.6
			32.7[e]	100.0	30.5	69.5
Nb_2O_5	4.9	750	6.1[c]	24.7	0.0	100.0
		850	13.6[c]	54.0	0.0	100.0
Ta_2O_5	3.8	750	—[f]	—	—	—
		850	1.0[c]	4.6	0.0	100.0
TiO_2	9.4	750	0.6[c]	2.6	0.0	100.0
		850	5.8[c]	24.3	0.0	100.0
ZrO_2	3.9	750	2.9[c]	2.8	0.0	100.0
		850	17.2[c]	62.0	0.0	100.0
γ-Al_2O_3	60.9	750	23.4[c]	88.8	0.0	100.0
		850	21.8[c]	82.1	0.0	100.0
La_2O_3	4.9	750	16.5[c]	76.1	0.0	100.0
		850	21.5[c]	86.9	0.0	100.0
MgO[g]	38.3	750	23.4[c]	93.5	0.0	100.0
		850	24.6[c]	92.3	0.0	100.0
SiO_2	594.0	750	9.0[c]	41.8	0.0	100.0
		850	16.0[c]	63.9	0.0	100.0
Y_2O_3	9.4	750	18.2[c]	88.5	0.0	100.0
		850	21.0[c]	81.4	0.0	100.0

[a] Catalysts, 50.0 mg; volume of CH_4/O_2 (2/1) pulse, 50.0 μl.
[b] After calcination at 800°C for 4 h.
[c] Data obtained for the first pulse.
[d] Data obtained for the second pulse.
[e] Data obtained for the third pulse.
[f] Almost no reaction occurred.
[g] Obtained from $MgO_2 \cdot xMgO$ by calcination at 800°C for 4 h.

850°C, the conversion of O_2 was approximately four times that of CH_4, indicating that the overall reaction proceeded according to the combustion reaction,

$$CH_4 + 2O_2 = CO_2 + 2H_2O. \quad [3]$$

Among the reducible oxides, CeO_2 is the most reducible. At 850°C, CeO_2 was partially reduced by methane during the first pulse, as indicated by a methane conversion much higher than 25%. With the increase in the extent of reduction during successive pulses, the methane conversion decreased while the CO selectivity increased (Table 2).

Compared to the other reducible oxides, Ta_2O_5 exhibited much lower conversions of methane and oxygen; almost no reaction occurred at 750°C, and only 1.0% methane was converted at 850°C.

3.2.2. Reaction of CH_4/O_2 pulses over the 1 wt% $Rh(O)/M_xO_y$. The reaction between methane and oxygen over the oxidized catalysts was investigated by using CH_4/O_2 pulses with a stoichiometric feed ratio of 2/1. The products contained CO, H_2, CO_2, and H_2O. The carbon balance indicated that no carbon was deposited on the catalyst. In Figs. 2 and 3, the reactivities of CH_4/O_2 over the precalcined rhodium supported on irreducible and reducible oxides are compared at 750°C. Over the irreducible-oxide-supported catalysts, O_2 was completely converted during each pulse and the methane conversion was equal or close to 100% (Fig. 2a). During the first pulse, no CO or H_2 was formed, CO_2 being the only carbon-containing product (Fig. 2b). Starting with the second pulse, CO was formed and its selectivity increased rapidly over all the irreducible supports except MgO. Over the MgO-supported catalyst, no CO was formed until the fourth pulse. The high methane conversion (>95%) and CO selectivity (>92%) at the steady state over all the irreducible-oxide-supported rhodium catalysts indicate that the combustion reaction provides only a minor contribution. Among the reducible oxide supports, Ta_2O_5 exhibited reactivities similar to the irreducible ones (Fig. 3); this is consistent with the results of the continuous flow reaction (Table 1). Over the CeO_2- and ZrO_2-supported Rh catalysts, CO was formed starting with the third pulse and its selectivity greatly increased at the fourth pulse (Fig. 3c). Even though during the first two pulses the methane conversion was high (equal or close to 100%), it decreased notably starting with the third pulse. Both methane conversion and CO selectivity were much lower after the eighth pulse onward than those observed over the irreducible oxides (compare Figs. 3a and 3b to Fig. 2a); this is again consistent with the results of the continuous flow reaction (Table 1). Over the Nb_2O_5- and TiO_2-supported Rh catalysts, no H_2 or CO was formed, hence no partial oxidation reaction occurred (Fig. 3c). As over the pure reducible oxides (Table 2), the conversion of O_2 was approximately four times that of methane (Figs. 3a and 3b), indicating that the overall reaction proceeded according to the combustion reaction.

The effect of temperature on the reactivities of CH_4/O_2 (2/1) was also investigated. As shown in Fig. 4, over the γ-Al_2O_3-supported Rh catalyst, at both 650 and 750°C, CO was formed during the second pulse, but the CO selectivity was higher at 750°C than at 650°C. At 850°C, CO was formed during the first pulse. The higher the temperature, the more easily it induces the partial oxidation reaction. Similar results were observed over the other irreducible-oxide-supported rhodium catalysts.

As shown in Fig. 5, over the Nb_2O_5-supported rhodium catalyst, at 650 and 750°C, no H_2 or CO was formed,

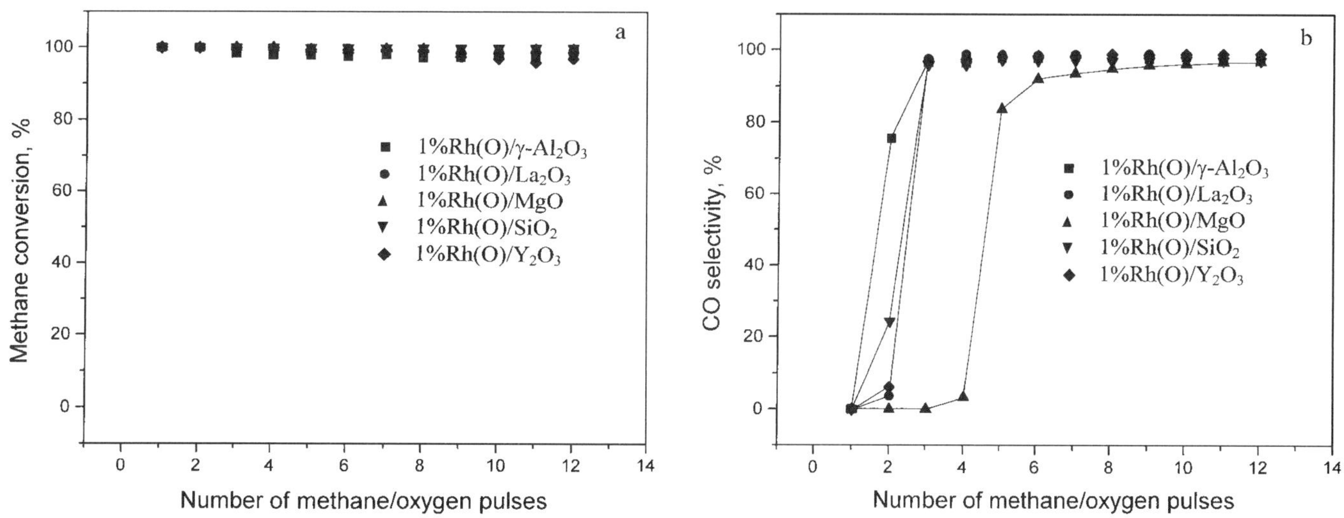

FIG. 2. Methane conversion (a) and CO selectivity (b) against the number of CH_4/O_2 (2/1) pulses over the irreducible-oxide-supported Rh catalysts at 750°C.

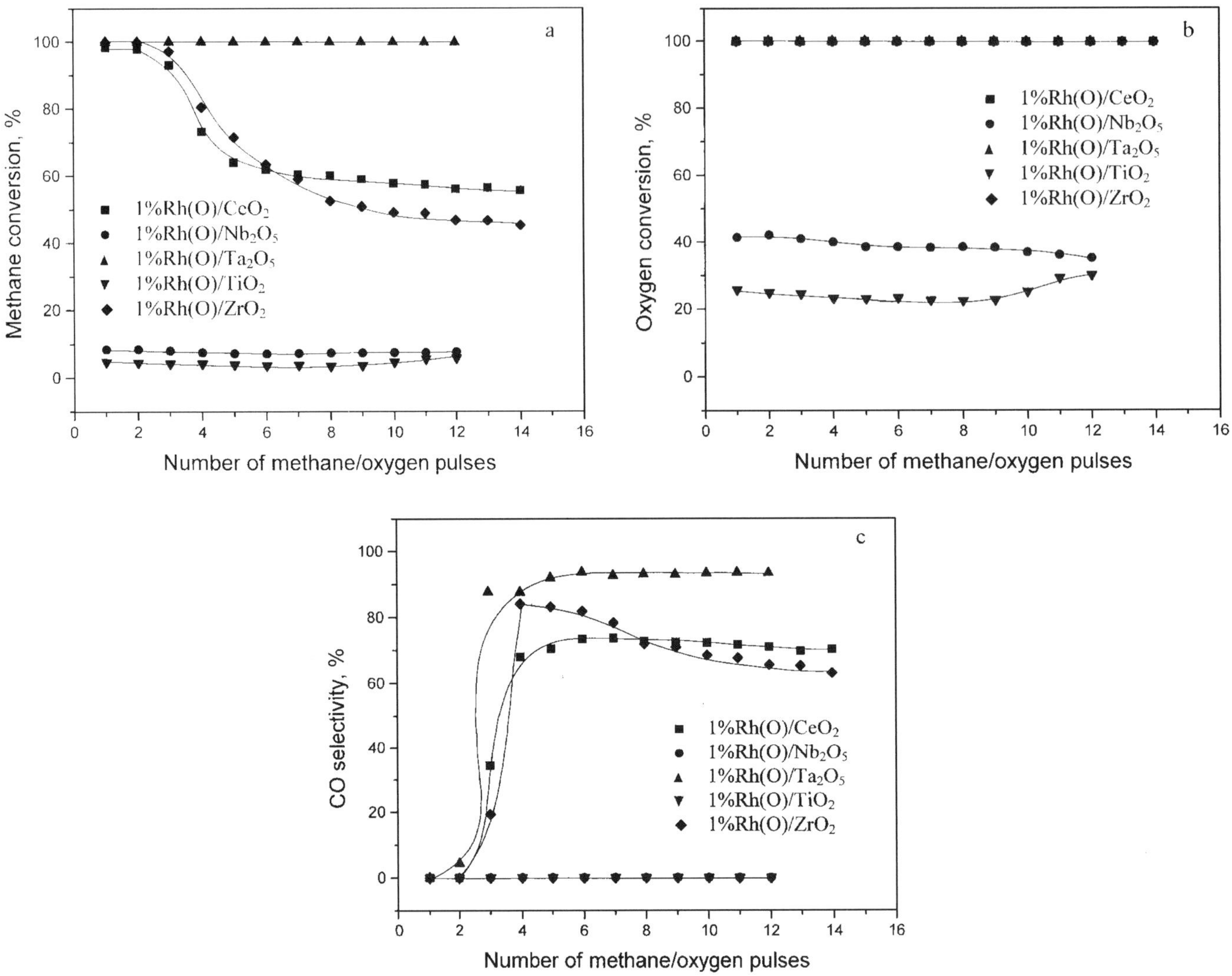

FIG. 3. Methane conversion (a), oxygen conversion (b), and CO selectivity (c) against the number of CH_4/O_2 (2/1) pulses over the reducible-oxide-supported Rh catalysts at 750°C. (Remark: ■, ▲, and ◆ are overlapping in b; and ● and ▼ in c.)

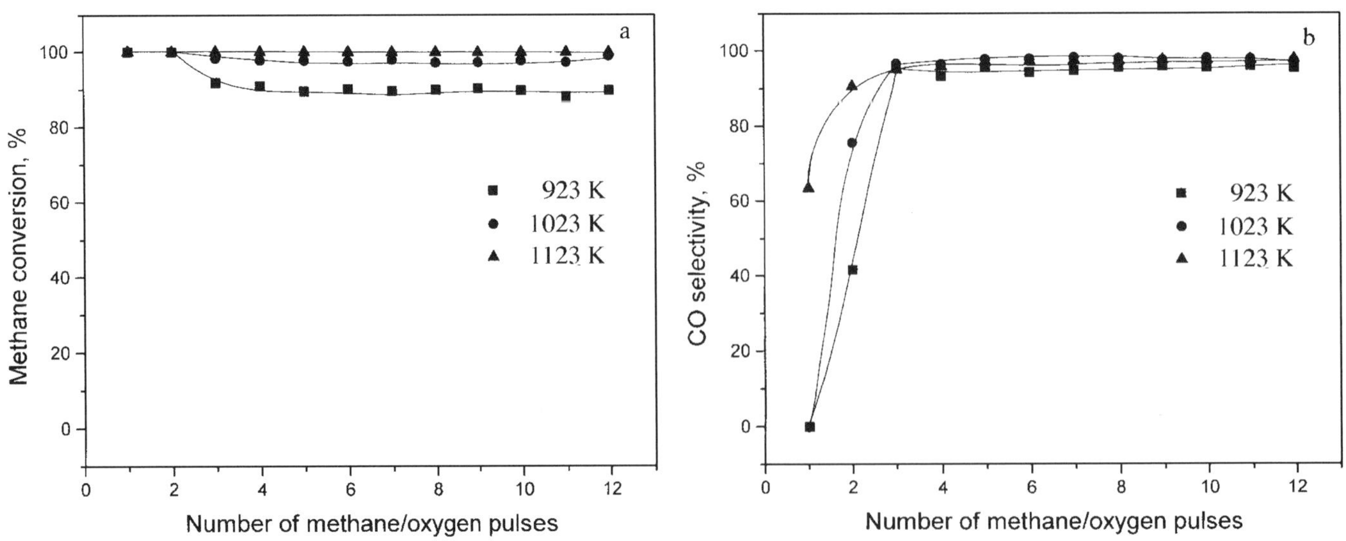

FIG. 4. Methane conversion (a) and CO selectivity (b) against the number of CH_4/O_2 (2/1) pulses over the 1% Rh(O)/γ-Al_2O_3 at 650, 750, and 850°C.

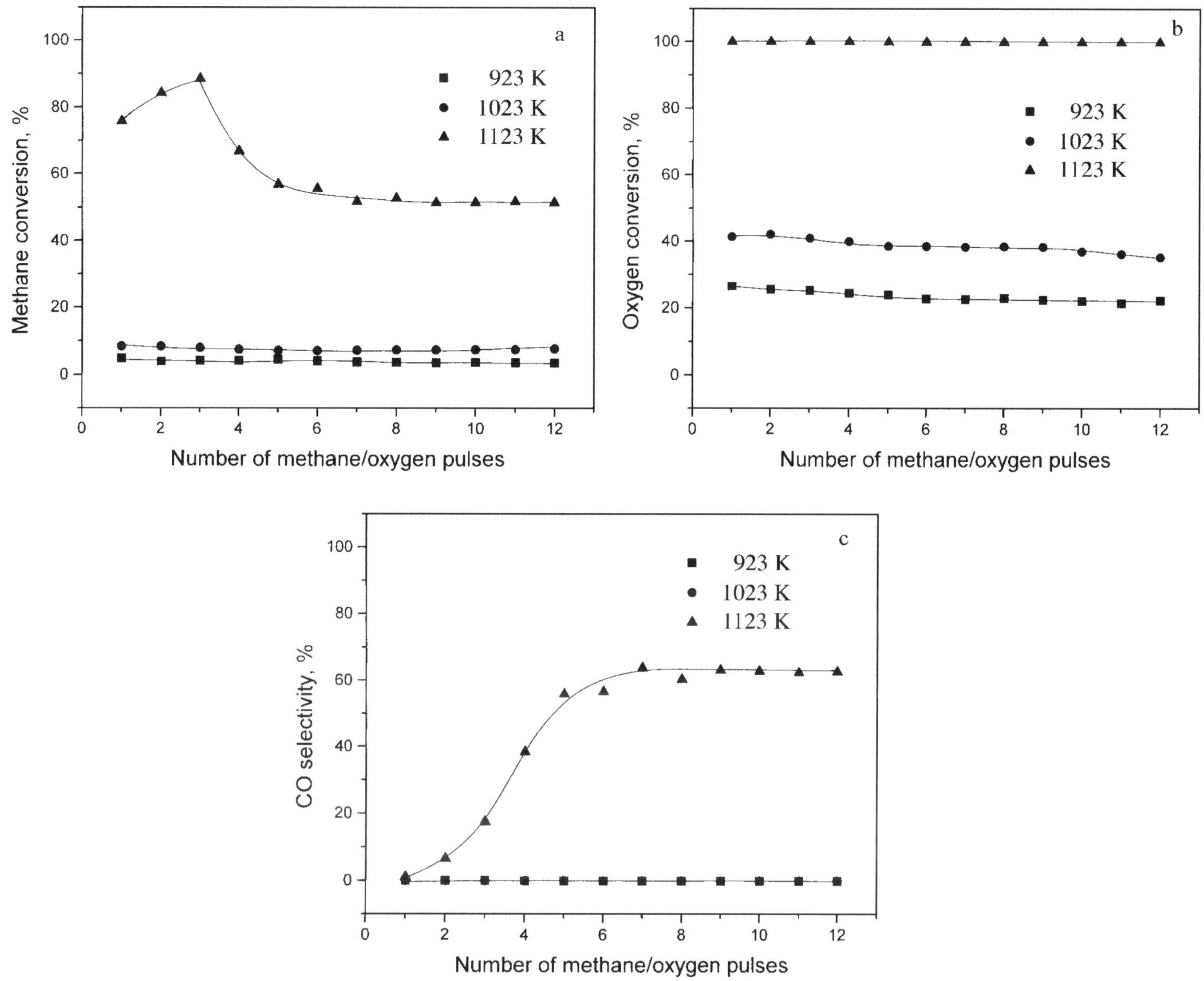

FIG. 5. Methane conversion (a), oxygen conversion (b), and CO selectivity (c) against the number of CH_4/O_2 (2/1) pulses over the 1% Rh(O)/Nb_2O_5 at 650, 750, and 850°C. (Remark: ■ and ● are overlapping in c.)

and the methane and oxygen conversions were very low. However, when the reaction temperature was increased to 850°C, the CO selectivity increased gradually with the increase in the number of pulses, indicating that the higher temperature activated the catalyst and induced the partial oxidation of methane. However, even at this high temperature, the CO selectivity remained much lower than that over the irreducible-oxide-supported rhodium catalysts. Similar results were observed over the TiO_2-supported catalyst. These results are again consistent with those of the continuous flow reaction (Table 1).

3.3. *Physico-Chemical Characterization*

After the calcined catalysts were reduced with H_2, the exposed metal surface areas were determined by CO chemisorption. As shown in Table 1, over the reducible-oxide-supported Rh catalysts, the exposed metal surface areas are much lower than those over the irreducible-oxide-supported ones.

TPR experiments were performed for all the calcined irreducible-oxide-supported catalysts and for the reducible Ta_2O_5 one. As shown in Fig. 6, the TPR peak temperature increased in the sequence

$$SiO_2 \leq \gamma\text{-}Al_2O_3 \leq Ta_2O_5 < Y_2O_3 \ll La_2O_3 < MgO.$$

XRD analysis was conducted for γ-Al_2O_3-, La_2O_3-, and MgO-supported rhodium catalysts, which provided high activities and selectivities with high stability (Fig. 1). Because the rhodium content of 1 wt% was too low for the XRD analysis, 10 wt% Rh(O)/support samples were prepared.

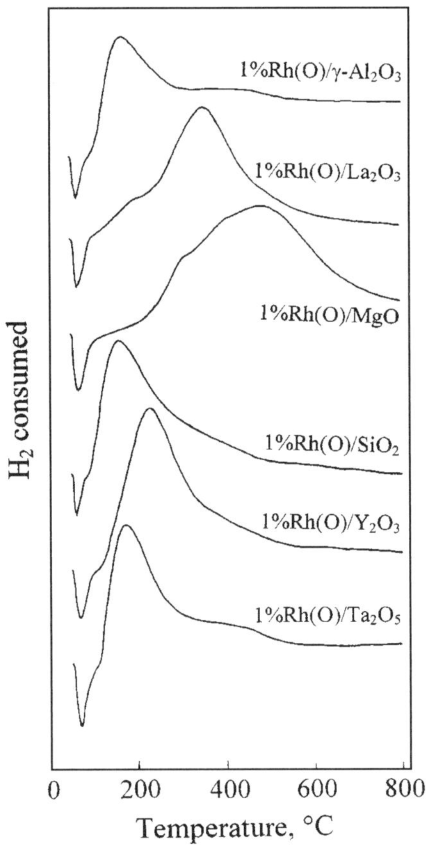

FIG. 6. TPR profiles of 1% Rh(O)/M_xO_y.

As shown in Table 3, Rh_2O_3, $LaRhO_3$ and Rh_2O_3 (minor), and $MgRh_2O_4$ were detected over the 10 wt% Rh(O)/γ-Al_2O_3, 10 wt% Rh(O)/La_2O_3, and 10 wt% Rh(O)/MgO, respectively. $LaRhO_3$ has the structure of a perovskite and $MgRh_2O_4$ the structure of a spinel.

4. DISCUSSION

4.1. Effect of Support

As shown in Table 1, the activity of rhodium strongly depends on the support. The irreducible supports provide, in general, much higher activities and selectivities to CO and H_2 than the reducible ones. As an exception, the reducible Ta_2O_5 provided activity and selectivities comparable to the irreducible support Y_2O_3. No simple, direct correlation between the catalytic activity (methane conversion) and the exposed metal surface area was found.

TABLE 3

Data and Assignments of XRD Patterns of Calcined 10% Rh(O)/M_xO_y Catalysts

Catalysts	d (Å)	Assignments
10% Rh(O)/Al_2O_3	1.39, 1.96; 2.50, 2.68, 1.70	γ-Al_2O_3; Rh_2O_3
10% Rh(O)/La_2O_3	2.93, 1.95, 3.34, 1.74, 1.64; 2.76, 3.45; 2.60, 2.79	La_2O_3; $LaRhO_3$; Rh_2O_3(minor)
10% Rh(O)/MgO	2.09, 1.48, 1.21, 1.27; 2.53, 4.76, 2.09, 1.48, 2.42	MgO; $MgRh_2O_4$

As shown in Table 1, after the calcined reducible-oxide-supported Rh catalysts were reduced with H_2, the exposed metal surface areas were much lower than those observed over the irreducible-oxide-supported ones. One possible explanation is that the rhodium metal particles were partially covered by islets of reduced oxide. Experimental evidence for the burying by the suboxide TiO_x, formed through the reduction of the TiO_2 support, of the metal particles was brought with regard to the so-called strong metal–support interactions (32, 33). The migration of TiO_x over the exposed surface of the metal particles takes place because it decreases the free energy of the system (32b). Indeed, depending upon the crystal face, Rh has a surface energy at room temperature between 2.5×10^3 and 2.9×10^3 erg/cm^2, while Ti has a value of 1.7×10^3 erg/cm^2 (35). TiO_2 is expected to have a smaller surface energy than Ti, because an oxide has a lower surface energy than the metal (32b). While TiO_x may have a somewhat higher surface energy than TiO_2, the reduction generates vacancies, which enhance the surface diffusion and hence the migration of TiO_x over the exposed surface area of the metal particles. The partial coverage of a rhodium particle by species generated via the reduction of support decreases the number of active metal sites, thus resulting in a lower activity. Even though no direct correlation between activity and the exposed metal surface area is evident (Table 1) for the reducible-oxide-supported catalysts, the Nb_2O_5 supported Rh catalyst, which has the lowest exposed metal surface area after reduction, exhibited the lowest catalytic activity.

As shown in Table 2, combustion of methane took place when a mixture of CH_4 and O_2 was pulsed over the pure (free of Rh) reducible or irreducible metal oxides. However, over the irreducible-oxide-supported catalysts, the combustion reaction played a minor role (Fig. 2), because CH_4 and O_2 can be easily activated by the metal sites and converted to CO and H_2, and the combustion which takes place over the oxide sites is much slower. With the exception of Ta_2O_5, which behaved like the irreducible oxides, for the reducible-oxide-supported catalysts, the selectivity to CO_2 was 100% over Nb_2O_5- and TiO_2-supported catalysts and important over the CeO_2- and ZrO_2-supported ones (Fig. 3). Under the continuous reaction conditions, a part of the suboxide generated via the reduction of a reducible oxide migrated over the metal surface, where it was reoxidized by the oxygen contained in the feed gas. The suboxide (or oxide) present on the surface of the metal particles decreases the number of active rhodium sites, thus resulting in a lower activation of CH_4 and O_2 and conversion to CO and H_2; on the other hand, the combustion reaction is promoted by the numerous oxide sites available. As revealed by Table 2, the pure Ta_2O_5 exhibited negligible activity for the combustion reaction of CH_4 and O_2 at 750 and 850°C compared to other reducible oxides. One possible reason might be that the ignition temperature in the presence of this oxide is higher than the temperature employed. In addition, the surface energy of Ta is 2.7×10^3 erg/cm^2, hence comparable to that of Rh and much higher than that of Ti (35). Even though Ta_2O_5 is reducible (much less than the other reducible oxides (32a)), the surface energy of Ta_2O_x is not sufficiently smaller than that of Rh to allow the former to migrate as a strong surface active agent over the surface of the metal particles. For these reasons the Ta_2O_5-supported rhodium catalyst exhibited much higher catalytic activity and selectivities to H_2 and CO than the other reducible oxides.

The sintering is very harmful for the supported metal catalysts, because the aggregation of metal crystallites decreases the number of active sites and also accelerates the carbon deposition since large metal ensembles stimulate the carbon deposition. In general, the strong interactions between metal and support are beneficial in the stabilization of the tiny metal crystallites (32b). As shown in Fig. 1, among the irreducible supports, γ-Al_2O_3-, La_2O_3-, and MgO-supported Rh catalysts provided a stable catalytic activity for the partial oxidation of methane at the high GHSV of 720000 ml g^{-1} h^{-1}. As shown in Fig. 6, the TPR peak temperatures of La_2O_3- and MgO-supported Rh catalysts are much higher than those of SiO_2-, γ-Al_2O_3-, Ta_2O_5-, and Y_2O_3-supported ones, indicating that there are much stronger interactions between rhodium oxide and La_2O_3 or MgO. In the latter two catalysts, $LaRhO_3$ and $MgRh_2O_4$ were identified, respectively, in the precursor catalysts by means of XRD (Table 3). These compounds are most likely responsible for the stability of the La_2O_3- and MgO-supported Rh catalysts. Nevertheless, while no strong interactions between rhodium oxide and γ-Al_2O_3 could be identified (Fig. 6 and Table 3), the γ-Al_2O_3-supported Rh catalyst was stable. A possible explanation is as follows: Under the reaction conditions of methane partial oxidation, the atmosphere in the reactor is both reductive and oxidative. Sintering is stimulated by the reductive atmosphere and the oxidative atmosphere stimulates redispersion (32b). When on average they compensate one another, the system can remain stable and this probably happens with the Al_2O_3-supported catalysts but not with the SiO_2- and Y_2O_3-supported catalysts.

4.2. Active Sites for the Partial Oxidation of Methane

Many researchers concluded that the metal is the active site for the partial oxidation of methane (6, 12, 26, 36–46). New evidence is given below in this paper.

As shown in Fig. 5, over the precalcined Nb_2O_5- (similar results have been obtained with TiO_2) supported Rh catalysts, only total oxidation products, CO_2 and H_2O, were formed in the pulse reaction between CH_4 and O_2 at 650 and 750°C; both methane and oxygen conversions remained constant with increasing number of pulses. The following two steps are most likely involved in this reaction:

1. The reaction between methane and surface lattice oxygen, which gives CO_2 and H_2O and reduces Rh_2O_3 to Rh^0,

$$3CH_4(g) + 8Rh^{3+} + 12[O^{2-}] \rightarrow 3CO_2(g) + 6H_2O(g) + 8Rh^0 + 12[*]. \quad [4]$$

2. Reoxidation of Rh^0 by the gaseous oxygen, which restores the catalyst to its original state,

$$2Rh^0 + 3O_2 + [*] \rightarrow 2Rh^{3+} + 3[O^{2-}], \quad [5]$$

where [*] denotes an anion vacancy.

However, when the temperature was increased to 850°C, during the first pulse the oxygen was completely converted and the methane conversion was about 75% ($\gg$25%, 25% corresponding to the complete combustion based on the oxygen present in the feed) with a CO_2 selectivity close to 100%. This means that some of the oxygen was provided by the lattice. Consequently, one can conclude that at 850°C metallic rhodium was generated according to reaction [4]. The amount of metallic rhodium increased with the number of CH_4/O_2 pulses until a steady state was reached. Since the selectivity to CO increased in parallel with the number of metal sites, it is clear that the latter is responsible for the partial oxidation of methane. As shown in Fig. 4, over γ-Al_2O_3 and other irreducible-oxide-supported Rh catalysts, metallic rhodium can be generated even at 650°C during the first pulse of CH_4/O_2, thus inducing the methane partial oxidation at a lower temperature than over the reducible oxides. Since the rate of reaction is directly related to the state of the Rh particles, the relative ease of generation of metallic rhodium for irreducible-oxide-supported catalysts implies that they have much higher activity for syngas formation compared to the reducible-oxide-supported ones. This conclusion is in agreement with the continuous flow reaction experiments (Table 1).

5. CONCLUSION

The effect of support on the partial oxidation of methane over rhodium-based catalysts was investigated. The performance of the supported Rh catalyst is strongly affected by the support. The reducible oxides are, in general, not suitable supports for rhodium for this reaction. A possible reason is that the suboxide generated via the reduction of the reducible oxides migrates onto the surface of the metal particles, decreasing the number of active rhodium sites and hence the catalytic activity and promoting the combustion reaction. Among the irreducible oxides, MgO provides the highest catalytic activity with high product selectivity and stability. The strong interactions between rhodium and magnesium oxides (especially the formation of $MgRh_2O_4$) are responsible for the high stability of MgO-supported Rh catalyst. Metallic rhodium can be generated via the reaction between methane and rhodium oxide (or rhodium-containing compound) even in the presence of oxygen and induces the partial oxidation of methane. Therefore, it is clear that the metal sites are responsible for the partial oxidation of methane.

REFERENCES

1. Rostrup-Nielsen, J. R., *Catal. Today* **18**, 305 (1993).
2. Fox, J. M., III, *Catal. Rev. Sci. Eng.* **35**, 169 (1993).
3. Hickman, D. A., and Schmidt, L. D., *J. Catal.* **138**, 267 (1992).
4. Bhattacharaya, A. K., Breach, J. A., and Chand, S., *et al., Appl. Catal. A Gen.* **80**, L1 (1992).
5. Hickman, D. A., Haupfear, E. A., and Schmidt, L. D., *Catal. Lett.* **17**, 223 (1993).
6. Hickman, D. A., and Schmidt, L. D., *Science* **259**, 343 (1993).
7. Torniainen, P. M., Chu, X., and Schmidt, L. D., *J. Catal.* **146**, 1 (1994).
8. Bharadwaj, S. S., and Schmidt, L. D., *Fuel Processing Technol.* **42**, 109 (1995).
9. Witt, P. M., and Schmidt, L. D., *J. Catal.* **163**, 465 (1996).
10. Nakagawa, K., Ikenaga, N., and Suzuki, T., *et al., Appl. Catal. A Gen.* **169**, 281 (1998).
11. Prettre, M., Eichner, C., and Perrin, M., *Trans. Faraday Soc.* **43**, 335 (1946).
12. Dissanayake, D., Rosynek, M. P., Kharas, K. C. C., and Lunsford, L. H., *J. Catal.* **132**, 117 (1991).
13. Vermeiren, W. J. M., Blomsma, E., and Jacobs, P. A., *Catal. Today* **13**, 427 (1992).
14. Choudhary, V. R., Mamman, A. S., and Sansare, S. D., *Angew. Chem. Int. Ed. Engl.* **31**, 1189 (1992).
15. Choudhary, V. R., Rajput, A. M., and Prabhakar, B., *J. Catal.* **139**, 326 (1993).
16. Choudhary, V. R., Uphade, B. S., and Mamman, A. S., *Catal. Lett.* **32**, 387 (1995).
17. Chu, Y., Li, S., and Lin, J., *et al., Appl. Catal. A Gen.* **134**, 67 (1996).
18. Choudhary, V. R., Uphade, B. S., and Mamman, A. S., *J. Catal.* **172**, 281 (1997).
19. Miao, Q., Xiong, G., and Sheng, S., *et al., Appl. Catal. A Gen.* **154**, 17 (1997).
20. Lu, Y., Liu, Y., and Shen, S., *J. Catal.* **177**, 386 (1998).
21. Tang, S., Lin, J., and Tan, K. L., *Catal. Lett.* **51**, 169 (1998).
22. Drago, R. S., Jurczyl, K., Kob, N., Bhattacharyya, A., and Masin, J., *J. Catal.* **51**, 177 (1998).
23. Chen, P., Zhang, H. B., Lin, G. D., and Tsai, K. R., *Appl. Catal. A Gen.* **166**, 343 (1998).
24. Ashcroft, A. T., Cheetham, A. K., and Foord, J. S., *et al., Nature* **344**, 319 (1990).
25. Jones, R. H., Ashcroft, A. T., and Waller, D., *et al., Catal. Lett.* **8**, 169 (1991).
26. Slagtern, A., and Olsbye, U., *Appl. Catal. A Gen.* **110**, 99 (1994).
27. Choudhary, V. R., Uphade, B. S., and Belhekar, A. A., *J. Catal.* **163**, 312 (1996).
28. Lago, R., Bini, G., Pena, M. A., and Fierro, J. L. G., *J. Catal.* **167**, 198 (1997).
29. Nakamura, J., Aikawa, K., Sato, K., and Uchijima, T., *Catal. Lett.* **25**, 265 (1994).
30. Nakagawa, J., Anzai, K., and Matsui, N., *et al., Catal. Lett.* **51**, 163 (1998).
31. Bodke, A. S., Bharadwaj, S. S., and Schmidt, L. D., *J. Catal.* **179**, 138 (1998).
32. (a) Stevenson, S. A., Raupp, G. B., Dumesic, J. A., Tauster, S. J., and Baker, R. T. K., *in* "Metal–Support Interactions in Catalysis, Sintering, and Redispersion" (S. A. Stevenson, J. A. Dumesic, R. T. K. Baker, and E. Ruckenstein, Eds.), p. 34. Van Nostrand–Reinhold, New York, 1987; (b) Ruckenstein, E., in "Metal–Support Interactions in Catalysis, Sintering, and Redispersion" (S. A. Stevenson, J. A. Dumesic, R. T. K. Baker, and E. Ruckenstein, Eds.), p. 297. Van Nostrand–Reinhold, New York, 1987.
33. Minachev, Kh. M., and Shpiro, E. S., *in* "Catalyst Surface: Physical Methods of Studying" (Kh. M. Minachev and E. S. Shapiro, Eds., translated from the Russian by G. Leib), p. 255. CRC Press, Boca Raton, FL, 1990.
34. Dissanayake, D., Rosynek, M. P., and Lunsford, J. H., *J. Phys. Chem.* **97**, 3644 (1993).
35. Personal information from Dr. W. F. Egelhoff, Jr., Magnetic Material Group, NIST, Gaithersburg, MD 20899.
36. Buyevskaya, O. V., Wolf, D., and Baerns, M., *Catal. Lett.* **29**, 249 (1994).
37. Hu, Y. H., and Ruckenstein, E., *Catal. Lett.* **34**, 41 (1995).
38. Hu, Y. H., and Ruckenstein, E., *J. Catal.* **158**, 260 (1996).
39. Au, C. T., Wang, H. Y., and Wan, H. L., *J. Catal.* **158**, 343 (1996).
40. Wang, D. Z., Dewaele, O., Groote, A. M. D., and Froment, G. F., *J. Catal.* **159**, 418 (1996).
41. Au, C. T., and Wang, H. Y., *Catal. Lett.* **41**, 159 (1996).
42. Mallens, E. P., Hoebink, J. H. B. J., and Marin, G. B., *J. Catal.* **167**, 43 (1997).
43. Au, C. T., and Wang, H. Y., *J. Catal.* **167**, 337 (1997).
44. Looij, F. van, and Geus, J. W., *J. Catal.* **168**, 154 (1997).
45. Hu, Y. H., and Ruckenstein, E., *J. Phys. Chem. A* **102**, 230 (1998).
46. Hu, Y. H., and Ruckenstein, E., *J. Phys. Chem. A* **102**, 10568 (1998).

Temperature-Programmed Reduction and XRD Studies of the Interactions in Supported Rhodium Catalysts and Their Effect on Partial Oxidation of Methane to Synthesis Gas

E. Ruckenstein[1] and H. Y. Wang

Department of Chemical Engineering, State University of New York at Buffalo, Amherst, New York 14260

Received May 21, 1999; revised October 5, 1999; accepted October 26, 1999

In the present paper, comprehensive temperature-programmed reduction and X-ray diffraction studies were conducted, with particular emphasis on the effect of calcination temperature on the interactions between rhodium oxide and support. The formation of a compound between rhodium and support oxide is strongly dependent on the nature of the support and the calcination temperature. No such compounds were formed over γ-Al_2O_3 and SiO_2 supported Rh catalysts even after a calcination at a temperature as high as 950°C. However, $LaRhO_3$, $MgRh_2O_4$, $YRhO_3$, and $RhTaO_4$ could be formed over La_2O_3, MgO, Y_2O_3, and Ta_2O_5 supported catalysts after their calcination at suitable temperatures. The ease of formation of these compounds decreased in the sequence MgO > $La_2O_3 \gg Y_2O_3 \geq Ta_2O_5$. The lower reducibilities of these compounds compared to that of Rh_2O_3 brought evidence of increased Rh–O bond strength. In other words, the formation of these compounds after calcination is an indication of strong interactions between rhodium and support. At a reaction temperature of 750°C, La_2O_3 and MgO provided stable catalytic activities and selectivities during 100 h of reaction, whereas deactivation was observed over the Y_2O_3 and Ta_2O_5 supported catalysts. However, because compounds between the latter two supports and Rh_2O_3 could be formed at higher temperatures, reactions at higher temperatures have also been carried out. It was found that the stability of the 1 wt% Rh/Ta_2O_5 could be notably improved at a reaction temperature of 900°C, and that the stability of the 1 wt% Rh/Y_2O_3 could be improved at a reaction temperature of 850°C.

Key Words: **methane partial oxidation; syngas; Rh catalyst; TPR; XRD.**

1. INTRODUCTION

The background of the present work and the relevant references were given in a previous paper (1). In that paper, the effect of the nature of the support on the performance of Rh-based catalysts in the partial oxidation of methane to synthesis gas was investigated. It was found that, with the exception of Ta_2O_5, the reducible oxide supports provided much lower methane conversions and selectivities to CO and H_2 than the irreducible ones. Among the irreducible metal oxides, γ-Al_2O_3, La_2O_3, and MgO provided stable catalytic activities and selectivities during 100 h of reaction at a furnace temperature of 750°C, while deactivation occurred over SiO_2 and Y_2O_3. The reducible Ta_2O_5 provided activity, selectivities, and deactivation comparable to those of the irreducible Y_2O_3. It was suggested that the strong interactions between rhodium and magnesium or lanthanum oxides (especially the formation of $MgRh_2O_4$ or $LaRhO_3$) were responsible for the high stability of MgO and La_2O_3 supported Rh catalysts. No such compounds were detected for Y_2O_3 and Ta_2O_5 at a calcination temperature of 800°C. An alternate sintering–redispersion mechanism was proposed to explain the high stability of γ-Al_2O_3 supported catalyst for which no compound between Al_2O_3 and Rh_2O_3 could be identified.

In the work described in present paper, comprehensive temperature-programmed reduction (TPR) and X-ray diffraction (XRD) studies were conducted, with the emphasis on the effect of calcination temperature. It was found that not only MgO and La_2O_3 but also tantalum and yttrium oxides form compounds with Rh_2O_3 after calcination at suitable high temperatures. The stabilities of the latter two supported catalysts could be notably improved when the reaction was conducted at temperatures at which the compounds $YRhO_3$ or $RhTaO_4$ could be formed.

2. EXPERIMENTAL

2.1. Catalyst Preparation

The following powder supports were used: γ-Al_2O_3 and Ta_2O_5 from Alfa, and La_2O_3, MgO (obtained from $MgO_2 \cdot x MgO$ via calcination at 800°C for 4 h), SiO_2, and Y_2O_3 from Aldrich. The supported rhodium catalysts were prepared by impregnating the support with an ethanol solution of $Rh(NO_3)_3 \cdot 2H_2O$ (Alfa), followed by overnight drying at 110°C. The samples thus obtained were subjected to calcination in air at 800°C for 4 h, unless otherwise stated. The calcined catalysts will be denoted Rh(O)/M_xO_y(800°C), where the temperature inside the

[1] To whom correspondence should be addressed.

This paper, which was originally published in **Journal of Catalysis**: 190, 32–38 (2000), is republished with permission.

parentheses indicates the calcination temperature during catalyst preparation. The catalysts reduced in H_2 will be denoted Rh/M_xO_y. Rh loading means weight percent of Rh in the reduced catalyst.

2.2. *Catalyst Testing*

All activity assays were conducted under atmospheric pressure in a fixed-bed vertical quartz reactor (3 mm inside diameter) immersed in an electronically controlled furnace. Five or ten milligrams of catalyst, which was held on quartz wool, was used in each run. A thermocouple was inserted into the middle of the catalyst bed to measure the temperature of the catalyst and another was located in the gas near the exit of the catalyst bed to measure the temperature of the gaseous phase. Before the testing was performed, the catalyst was reduced. The reduction was carried out in the reactor by increasing the temperature up to 750°C at 20°C/min in a H_2 flow (20 ml/min). The reactants and products were analyzed with an online gas chromatograph equipped with Porapak Q and 5A molecular sieve columns. An ice-cold trap was located between the reactor exit and the GC sampling valve to remove the water formed during reaction.

Methane (99.97% purity, from Matheson) and oxygen (99.9% purity, from Cryogenic) were used without further purification. The gases were premixed ($CH_4/O_2 = 2/1$) before being introduced into the reactor.

The methane conversion and product selectivities were calculated on the basis of the carbon and hydrogen balances by neglecting any coking (some coking was observed as carbon deposition on the thermocouple downstream of the catalyst bed after a long time of reaction). The effect of the nature of the support on the conversion and product selectivities over the 1 wt% Rh catalysts was reported in a previous paper (1).

2.3. *Catalyst Characterization*

The surface areas after calcination and the Rh surface areas after reduction for the 1 wt% Rh catalysts were reported previously (1). Because 1 wt% Rh loading is too low for XRD analysis, 10 wt% Rh(O)/M_xO_y (M = Al, La, Mg, Si, Y, and Ta) samples were prepared for the XRD and TPR characterizations. The TPR results for the 10 wt% Rh catalysts calcined in an air flow at 800°C are in agreement with those (1) of the 1 wt% Rh ones calcined in air at the same temperature.

2.3.1. Temperature-programmed reduction (TPR). The TPR profile of the catalyst was obtained by heating the sample, held on quartz wool in a quartz reactor (4 mm inside diameter) immersed in an electronically controlled furnace, from 50 to 850°C at a rate of 20°C/min in a flow of 2.5% H_2/Ar mixture (35 ml/min). The H_2/Ar gas mixture was purified with Hydro-Purge II and Oxy-Trap columns before use and the hydrogen consumed in TPR was determined with a thermal conductivity detector (TCD). Twelve and one-half milligrams of sample calcined in air at 800°C for 4 h was used in the first TPR run. To investigate the effect of calcination temperature on the TPR patterns of a sample, the sample used in the first TPR run was reoxidized in an air flow (30 ml/min) for 1 h at a new calcination temperature and then a new assay was conducted after the sample was cooled to 50°C, and so on. The calcination temperatures were in the range 600°C $\le T \le$ 950°C, in increasing order.

2.3.2. X-ray powder diffraction (XRD). The X-ray powder diffraction (XRD) was performed on a Siemens D500 X-ray diffractometer, using the Cu$K\alpha$ radiation, at 40 kV and 30 mA.

3. RESULTS

3.1. *Catalyst Characterization*

The effect of calcination temperature on the TPR patterns of 10 wt% Rh/M_xO_y (M = Al, La, Mg, Si, Y, and Ta) was comprehensively investigated and the results are presented in Figs. 1–6.

Two TPR peaks, one at 180–200°C and the other at 350–400°C, were observed for the La_2O_3 supported Rh catalyst (Fig. 1). At the calcination temperature of 700°C, the

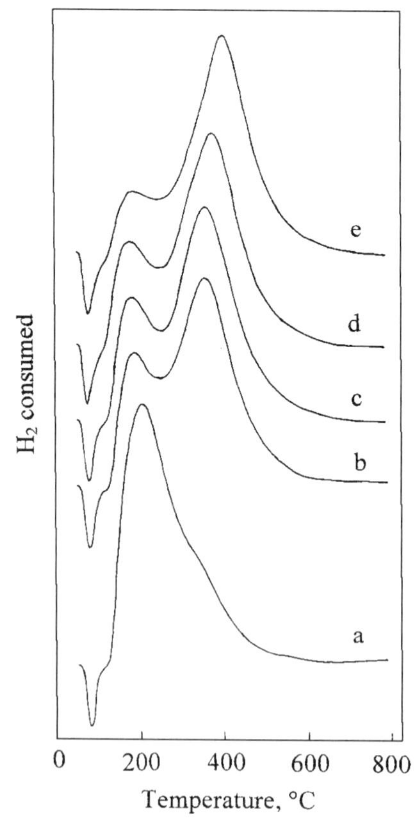

FIG. 1. TPR profiles of 10 wt% Rh(O)/La_2O_3 calcined in an air flow for 1 h at 700°C (a), 800°C (b), 850°C (c), 900°C (d), and 950°C (e).

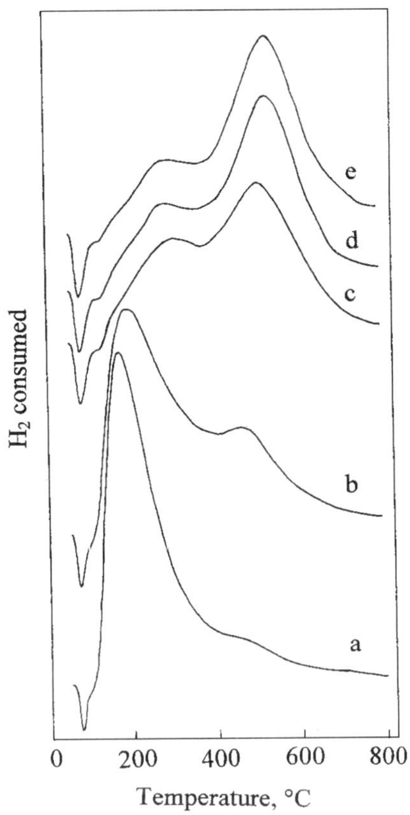

FIG. 2. TPR profiles of 10 wt% Rh(O)/MgO calcined in an air flow for 1 h at 600°C (a), 700°C (b), 800°C (c), 900°C (d), and 950°C (e).

high-temperature reduction peak (denoted HTR) appeared as a shoulder. It became, however, distinct when the calcination temperature was increased to 800°C. With the increase in the calcination temperature, the intensity of the HTR peak increased, whereas that of the low-temperature reduction peak (denoted LTR) decreased.

Similarly, two TPR peaks, at 180–300°C and at 470–530°C, were observed for the MgO supported catalyst (Fig. 2). The LTR peak shifted by about 20°C when the calcination temperature was increased from 600 to 700°C, but shifted notably from about 190°C to about 310°C when the calcination temperature was increased from 700 to 800°C. A further increase in the calcination temperature had no notable influence on the position of the LTR peak.

As shown in Fig. 3, over the Y_2O_3 supported catalyst, only one peak at about 220°C was present for a calcination temperature of 800°C. Two peaks at about 220 and 380°C, however, appeared after the catalyst was calcined at 850°C, indicating that a new compound containing Rh was formed. With the further increase of the calcination temperature to 950°C, the LTR peak almost disappeared and the HTR peak shifted to about 410°C.

As shown in Fig. 4, over the Ta_2O_5 supported catalyst, only one peak at about 180°C appeared for a calcination temperature of 800°C. A HTR peak at about 470°C appeared, however, when the calcination temperature was increased to 850°C, and at 950°C the LTR peak almost disappeared.

For the γ-Al_2O_3 supported catalyst, only one TPR peak, at about 160–170°C, was present (Fig. 5). The intensity of the TPR peak decreased to some extent after the catalyst was calcined in an air flow at 950°C, but was restored to a level comparable to that observed in the first TPR run (in which the catalyst was calcined in air at 800°C for 4 h) after it was calcined again in an air flow at 800°C for 1 h; hence, the process was reversible.

Only one TPR peak at about 160–180°C was noted for the SiO_2 supported catalyst (Fig. 6). The intensity of the peak decreased tremendously after the catalyst was calcined in an air flow at 950°C for 1 h. However, unlike the γ-Al_2O_3 supported catalyst, no notable change in intensity occurred after the catalyst was calcined again in an air flow at 800°C; hence, the process was irreversible.

The data and the assignments of the XRD patterns of the 10 wt% Rh(O)/M_xO_y (M = Al, La, Mg, Si, Y, and Ta) catalysts are listed in Table 1. Only Rh_2O_3 was detected over the 10 wt% Rh(O)/γ-Al_2O_3(800°C) and 10 wt% Rh(O)/SiO_2(800°C) catalysts, and this is in agreement with the

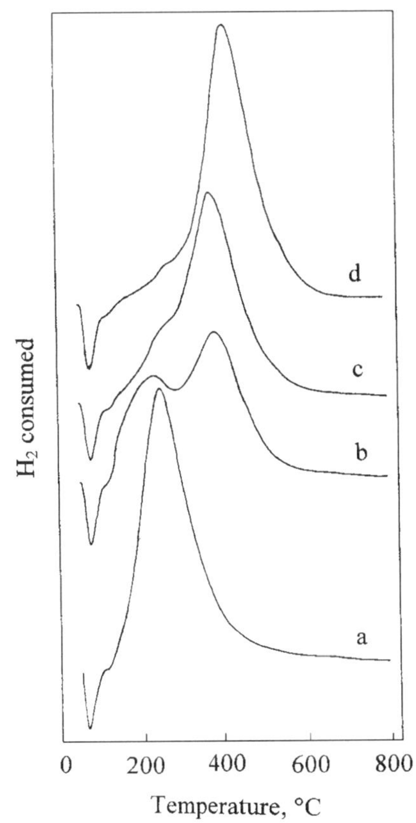

FIG. 3. TPR profiles of 10 wt% Rh(O)/Y_2O_3 calcined in an air flow for 1 h at 800°C (a), 850°C (b), 900°C (c), and 950°C (d).

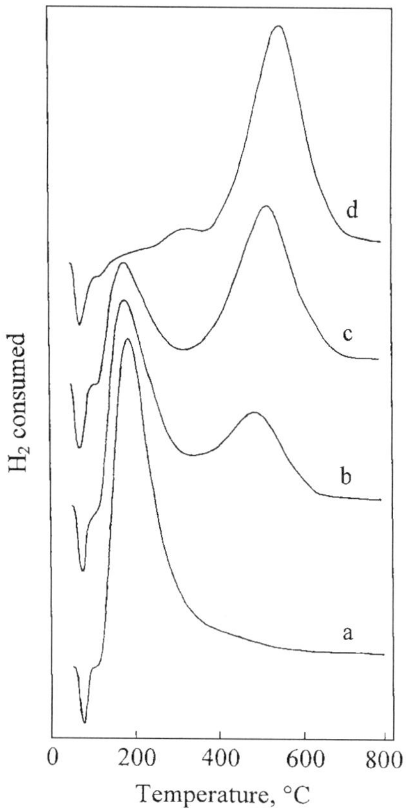

FIG. 4. TPR profiles of 10 wt% Rh(O)/Ta_2O_5 calcined in an air flow for 1 h at 800°C (a), 850°C (b), 900°C (c), and 950°C (d).

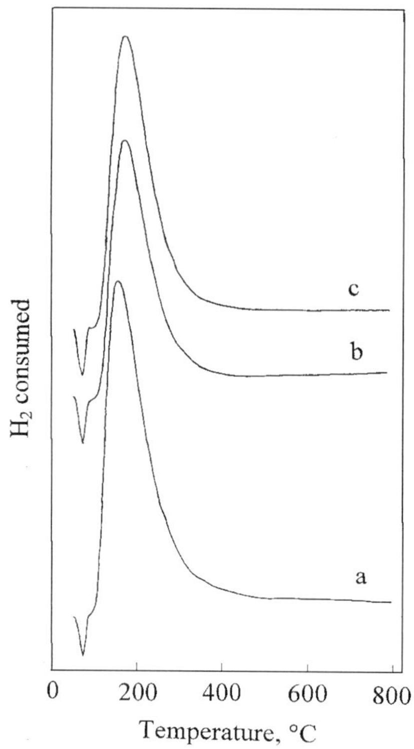

FIG. 5. TPR profiles of 10 wt% Rh(O)/γ-Al_2O_3 calcined in air at 800°C for 4 h (a), and calcined in an air flow for 1 h at 950°C (b) and at 800°C (c).

TPR results (Figs. 5 and 6). Over the 10 wt% Rh(O)/La_2O_3(800°C), besides Rh_2O_3, $LaRhO_3$ was also detected, and this is again in agreement with the TPR results which exhibited two TPR peaks (Fig. 1b). However, over the 10 wt% Rh/MgO(800°C), only the compound $MgRh_2O_4$ was detected, while two peaks were present in the TPR spectra (Fig. 2c). Over the 10 wt% Rh/Y_2O_3(800°C) and 10 wt% Rh/Ta_2O_5(800°C) catalysts, only Rh_2O_3 was identified, and this is again in agreement with the TPR spectra, which exhibited only LTR peaks (Figs. 3a and 4a). However, after the two catalysts were further calcined in air at 950°C for 2 h, only the compounds $YRhO_3$ and $RhTaO_4$ could be identified. This is again in agreement with the TPR spectra, which exhibited only HTR peaks (Figs. 3d and 4d).

3.2. Stabilities at Different Reaction Temperatures over 1 wt% Rh/Ta_2O_5 and 1 wt% Rh/Y_2O_3 Catalysts

As shown in Table 2, the stability of the 1 wt% Rh/Ta_2O_5 catalyst was notably influenced by the space velocity (SV). At SV = 720,000 ml g^{-1} h^{-1}, the activity decreased during the first 24 h of reaction, as indicated by the values of C_{10h}/C_{1h} and C_{24h}/C_{1h} (<1), but it remained unchanged during the first 24 h of reaction at a SV of 90,000 ml g^{-1} h^{-1}. In conclusion, the stability of the catalyst improved notably

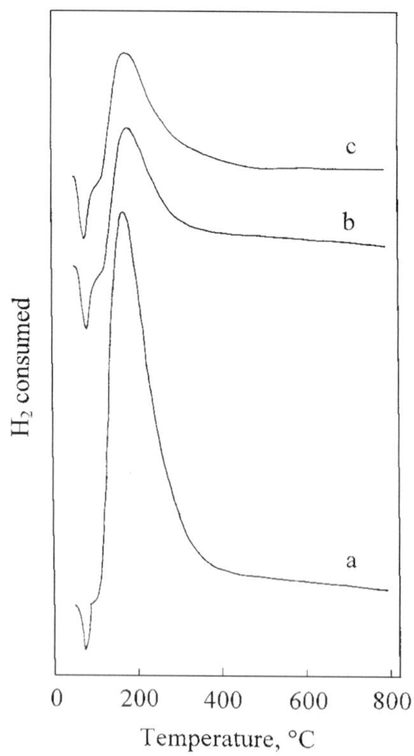

FIG. 6. TPR profiles of 10 wt% Rh(O)/SiO_2 calcined in air at 800°C for 4 h (a), and calcined in an air flow for 1 h at 950°C (b) and at 800°C (c).

TABLE 1

Data and Assignments of XRD Patterns of Calcined 10 wt% Rh(O)/M_xO_y Catalysts

Catalyst	Pretreatment	d (Å)	Assignments	Refs.
10 wt% Rh(O)/Al_2O_3	Calcined at 800°C for 4 h	1.39, 1.96;	γ-Al_2O_3	1
		2.50, 2.68, 1.70	Rh_2O_3	
10 wt% Rh(O)/La_2O_3	Calcined at 800°C for 4 h	2.93, 1.95, 3.34, 1.74, 1.64;	La_2O_3	1
		2.76, 3.45;	$LaRhO_3$	
		2.60, 2.79	Rh_2O_3(minor)	
10 wt% Rh(O)/MgO	Calcined at 800°C for 4 h	2.09, 1.48, 1.21, 1.27;	MgO	1, 2
		2.53, 4.76, 2.09, 1.48, 2.42	$MgRh_2O_4$	
10 wt% Rh(O)/SiO_2	Calcined at 800°C for 4 h	2.52, 2.69, 1.69	Rh_2O_3	2
10 wt% Rh(O)/Y_2O_3	(a) Calcined at 800°C for 4 h	3.06, 1.87, 2.65, 1.60, 4.33;	Y_2O_3	this work
		2.54, 2.71, 1.72	Rh_2O_3	
	(b) Calcined at 950°C for 2 h following (a)	3.06, 1.87, 2.65, 1.60, 4.33;	Y_2O_3	
		2.71, 2.62, 2.85, 1.89, 3.44	$YRhO_3$	
10 wt% Rh(O)/Ta_2O_5	(a) Calcined at 800°C for 4 h	3.13, 3.85, 2.43, 3.08, 2.41, 1.65, 1.94, 1.82;	Ta_2O_5	this work
		2.52, 2.75, 1.70	Rh_2O_3	
	(b) Calcined at 950°C for 2 h following (a)	3.13, 3.84, 2.44, 3.08, 2.41, 1.65, 1.94, 1.82;	Ta_2O_5	
		1.65, 2.53, 3.29	$RhTaO_4$	

with decreasing SV. Consequently, meaningful conclusions about the stability of a catalyst should be obtained by carrying out the reaction at a sufficiently high SV. We selected SVs $\geq$ 360,000 ml g^{-1} h^{-1} for the stability testing. All the methane conversions reported in Table 2 are lower than the thermodynamic equilibrium data (3) at the temperatures of the catalyst bed.

As indicated by the TPR and XRD experiments, no compounds between rhodium and yttrium or tantalum oxides were formed at a calcination temperature of 800°C; $YRhO_3$ and $RhTaO_4$ were, however, generated at a calcination temperature of 850°C, and the intensities of the HTR peaks corresponding to $YRhO_3$ and $RhTaO_4$ became stronger than those of Rh_2O_3 at calcination temperatures of 850 and 900°C, respectively. Consequently, it is of interest to compare the reaction behaviors at different reaction temperatures, since the interactions between rhodium and yttrium or tantalum oxides are strongly dependent on temperature. We carried out the partial oxidation reaction at furnace temperatures of 750 and 850°C over the 1 wt% Rh/Y_2O_3, and at 750 and 900°C over the 1 wt% Rh/Ta_2O_5. The catalyst temperature was recorded during the reaction and was found in all cases to be only a little higher than the furnace temperature. The deactivation is defined as $[(C^{\max} - C)/C^{\max}] \times 100$, where C denotes the conversion of methane and $C^{\max}$ the maximum conversion. After 48 h of reaction, the deactivations over the Y_2O_3 supported catalysts, calculated from Fig. 7, were 8.5% at 750°C and 4.2% at 850°C. Over the Ta_2O_5 supported catalysts, the

TABLE 2

Effect of the SV on the Deactivation Characteristics of 1 wt% Rh/Ta_2O_5

SV (ml g^{-1}h^{-1})	Reaction temp., furnace/bed (°C)	C_{1h}[a] (%)	C_{10h}/C_{1h}	C_{24h}/C_{1h}	C_{32h}/C_{1h}
720,000	750/828	61.1	0.92	0.88	n.a.[b]
360,000	750/813	64.9	0.94	0.92	0.89
240,000	750/798	71.6	0.98	0.95	n.a.
180,000	750/794	80.4	0.99	0.99	0.98
90,000	750/780	83.6	1.01	1.00	0.99

[a] C_{1h} means methane conversion after 1 h of reaction.
[b] n.a. means not available.

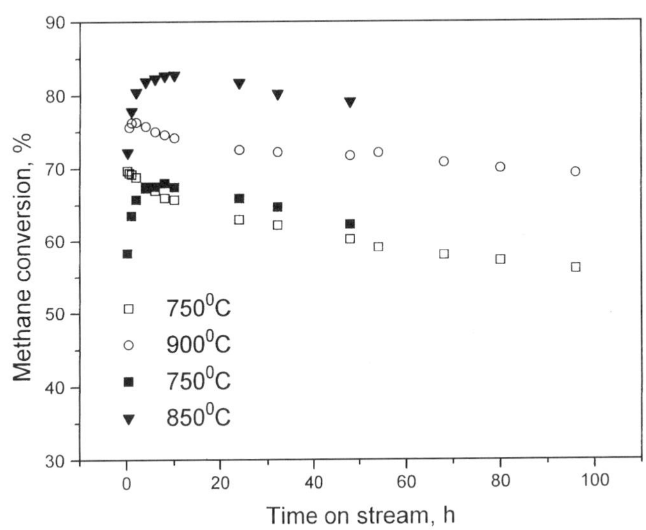

FIG. 7. Effect of time on stream on the activity of 1 wt% Rh/Y_2O_3 (solid) and 1 wt% Rh/Ta_2O_5 (open). $P = 1$ atm, $CH_4/O_2 = 2.0$, SV = 720,000 ml g^{-1} h^{-1} for the former and 360,000 ml g^{-1} h^{-1} for the latter.

deactivations after 96 h of reaction were 19.1% at 750°C and 9.2% at 900°C. One can conclude that the stability can be notably improved for the above two kinds of supported catalysts when suitable reaction temperatures are selected.

4. DISCUSSION

Over the 10 wt% Rh/γ-Al_2O_3(800°C) and 10 wt% Rh/SiO_2(800°C) catalysts, only Rh_2O_3 was detected by XRD (Table 1), and only a single TPR peak (below 200°C) was observed for each of the two catalysts (Figs. 5 and 6). Because Rh_2O_3 precursor can be easily reduced (4, 5), it is reasonable to attribute the LTR peaks in Figs. 1–4 to the reduction of Rh_2O_3. The HTR peaks can be attributed to the reduction of the compound formed between the rhodium and support oxide during the calcination process. The compounds identified by XRD (Table 1) are $LaRhO_3$, $MgRh_2O_4$, $YRhO_3$, and $RhTaO_4$, respectively, for the calcined La_2O_3, MgO, Y_2O_3, and Ta_2O_5 supported catalysts. $LaRhO_3$ and $YRhO_3$ have the structure of a perovskite, $MgRh_2O_4$ has the structure of a spinel, and $RhTaO_4$ has a tetragonal structure. With the increase in calcination temperature, no significant temperature shift could be noted for the LTR peak, except for the 10 wt% Rh(O)/MgO catalyst. In the latter case, a positive shift of about 120°C, from 190 to 310°C, was observed when the calcination temperature was increased from 700 to 800°C (Fig. 2). Because only the $MgRh_2O_4$ phase was detected by XRD (Table 1), it is likely that either the amount of Rh_2O_3 generated or the size of the Rh_2O_3 particles was too small to be detectable by XRD.

The formation of a compound between the rhodium and support oxide is strongly dependent on the nature of the support and the calcination temperature. No compound was formed over the calcined γ-Al_2O_3 and SiO_2 supported catalysts even for calcination temperatures as high as 950°C. MgO formed most easily a compound with rhodium oxide, as revealed by the distinct HTR shoulder in the TPR spectrum of the catalyst calcined in an air flow at 700°C. For La_2O_3 and MgO supported Rh catalysts and at a calcination temperature of 800°C, the amounts of $LaRhO_3$ and $MgRh_2O_4$ were greater than those of Rh_2O_3 (as indicated by the relative intensities of the HTR and LTR peaks); no compound was formed over the Y_2O_3 and Ta_2O_5 supported catalysts calcined at 800°C. However, $YRhO_3$ and $RhTaO_4$ were formed when the calcination temperature was increased to 850°C. In conclusion, the ease of formation of these compounds decreased in the sequence MgO > $La_2O_3 \gg Y_2O_3 \geq Ta_2O_5$. The lower reducibilities of these compounds ($LaRhO_3$, $MgRh_2O_4$, $YRhO_3$, and $RhTaO_4$) compared to that of Rh_2O_3 are an indication for the strong interactions between rhodium and support.

In a previous paper (1), we reported that at a reaction temperature of 750°C the Rh catalysts supported on γ-Al_2O_3, La_2O_3, and MgO and calcined at 800°C provided stable catalytic activities during 100 h of reaction. However, deactivation occurred over the SiO_2, Y_2O_3, and Ta_2O_5 supported catalysts. With the exception of γ-Al_2O_3 supported catalyst, the catalysts for which no compounds were formed after calcination exhibited a decay in activity during reaction. The strong interactions between rhodium and magnesium or lanthanum oxides were most likely responsible for the high stabilities of the MgO and La_2O_3 supported catalysts. This implies that if one could generate compounds between Rh_2O_3 and Ta_2O_5 or Y_2O_3 one could increase the stability of the corresponding supported catalysts. The TPR experiments indicated that such compounds did form if high enough calcination temperatures (>800°C) were used. We found that at a reaction temperature of 750°C the 1 wt% Rh/Ta_2O_5 catalyst calcined at 950°C and reduced by increasing the temperature to 750°C in a flow of H_2 still decayed at a rate comparable to that of the same catalyst calcined at 800°C; the deactivation after 96 h of reaction was 16.0% for the former and 19.1% for the latter. However, as shown in Fig. 7, the stability of the 1 wt% Rh/Ta_2O_5 calcined at 950°C was notably improved at a reaction temperature of 900°C. Over the 1 wt% Rh/Y_2O_3 catalyst, similar results were obtained, the stability being higher at a reaction temperature of 850°C than at 750°C.

The stability of the catalyst can be explained as follows. During reduction, metallic sites, which are responsible for the formation of the synthesis gas, are generated via the reduction of Rh_2O_3 and of the compound. Under reaction conditions, the atmosphere in the reactor is both reductive and oxidative, because of the coexistence of reductive (CH_4, CO, and H_2) and oxidative (O_2, CO_2, and H_2O) compounds. The reduced catalyst is reoxidized and a redox process occurs over the working catalyst (6–9). One could speculate that a catalyst is stable when a quasi-steady-state concentration of metallic sites is generated (the rates of reduction and oxidation become comparable). The strong interactions with the compound lessen the metal sintering and moderate the rates of the oxidation and reduction. This moderation makes the achievement of a quasi-steady state possible. Consequently, if the reaction takes place at a suitable temperature at which strong interactions between metal and support can be induced, the activity of the catalyst is expected to have a higher stability.

As shown in Fig. 6, the intensity of the TPR peak of the SiO_2 supported Rh catalyst decreased tremendously (compared to that of the TPR peak of the catalyst calcined in air at 800°C for 4 h) after the catalyst was calcined in an air flow at 950°C. This is an indication of metal sintering, because the intensity of the peak is expected to be higher for a larger exposed surface area of the metal. After a subsequent calcination in an air flow at 800°C, the intensity of the TPR peak remained as small as that after the calcination at 950°C. This indicates that the metal sintering that occurred over the SiO_2 supported Rh catalyst was irreversible. However, over the alumina supported Rh catalyst, the intensity of the TPR peak decreased to some extent (compared to that of the TPR peak of the catalyst calcined at 800°C for 4 h) after the catalyst was calcined in an air flow at 950°C, but it was almost restored to a level comparable to that observed in the first TPR run after a subsequent calcination in an air flow at 800°C. While no strong interactions could be identified on the alumina supported Rh catalyst, the latter catalyst exhibited resistance to metal sintering compared to the silica supported one, probably because of the comparable but opposite effects of the reductive and oxidative atmospheres (10). As already reported, alumina has a much higher stability than silica (1).

5. CONCLUSION

The interactions between Rh_2O_3 and various substrates were investigated by TPR and XRD. These investigations have shown that, for certain supports, compounds were generated between Rh_2O_3 and support at suitable calcination temperatures which were dependent upon the nature of the substrate. The paper demonstrates that the strong interactions resulting from the formation of compounds increase the stability of the catalysts. Rh catalysts supported on Y_2O_3 and Ta_2O_5 exhibited higher rates of deactivation at a reaction temperature of 750°C than at 850 and 900°C, respectively. The TPR and XRD experiments indicated that compounds did form at 850 and 900°C, respectively, but not at 750°C.

REFERENCES

1. Ruckenstein, E., and Wang, H. Y., *J. Catal.* **187**, 151 (1999) [and references therein].
2. Wang, H. Y., and Ruckenstein, E., *J. Catal.* **186**, 181 (1999).
3. Dissanayake, D., Rosynek, M. P., and Lunsford, J. H., *J. Phys. Chem.* **97**, 3644 (1993).
4. Fornasiero, P., Kaspar, J., and Graziani, M., *J. Catal.* **167**, 576 (1997).
5. Burch, R., and Loader, P. K., *Appl. Catal. A* **143**, 317 (1996).
6. Santos, A., Menendez, M., and Monzon, A., *et al., J. Catal.* **158**, 83 (1996).
7. Mallens, E. P., Hoebink, J. H. B. J., and Marin, G. B., *J. Catal.* **167**, 43 (1997).
8. Hu, Y. H., and Ruckenstein, E., *Ind. Eng. Chem. Res.* **37**, 2333 (1998).
9. Chen, P., Zhang, H. B., Lin, G. D., and Tsai, K. R., *Appl. Catal. A* **166**, 343 (1998).
10. Ruckenstein, E., *in* "Metal-Support Interactions in Catalysis, Sintering, and Redispersion" (S. A. Stevenson, J. A. Dumesic, R. T. K. Baker, and E. Ruckenstein, Eds.), p. 297. Van Nostrand–Reinhold, New York, 1987.

Methane partial oxidation over NiO/MgO solid solution catalysts

Eli Ruckenstein*, Yun Hang Hu[1]

Department of Chemical Engineering, State University of New York at Buffalo, Amherst, NY 14260, USA

Received 24 November 1998; received in revised form 9 February 1999; accepted 15 February 1999

Abstract

The following issues regarding the partial oxidation of methane to synthesis gas, over NiO/MgO solid solution catalysts, at a very high space velocity of the feed gas ($CH_4/O_2=2/1$, GHSV=720 000 cm^3/g h), were investigated: (i) the effect of the solid solution formation between NiO and MgO on the reaction; (ii)the effect of the composition of the catalyst; and (iii) the reaction mechanism. It was found that the mechanical mixtures of NiO and MgO have low activity and selectivity; however, their activity and selectivity became higher with increasing calcination time, due to the formation of a NiO–MgO solid solution. In contrast, the solid solution catalysts prepared via impregnation provided high activity and selectivity, as well as high stability over a large range of NiO concentrations. However, the activity was low for too low concentrations of NiO in NiO/MgO and the catalyst was no longer stable for too high concentrations of NiO. By comparing, under the same reaction conditions, the rate of partial oxidation of methane for $CH_4/O_2=2/1$ to that of the reforming reaction of a mixture $CH_4/H_2O/CO_2$ obtained from the former by complete combustion, one could conclude that the pyrolysis mechanism was dominant at 700°C and 750°C, but at 850°C, the combustion reforming mechanism also played a part.

Keywords: NiO/MgO; Solid solution; Partial oxidation of methane; Hot layer; Mechanism of partial oxidation of methane

1. Introduction

In the last 20 years, the conversion of methane to more useful chemicals became a most important field in catalysis. The conversion processes can be classified in two major classes: the direct and indirect conversion. The attempts regarding the direct conversion of CH_4 have focused on the oxidative coupling to ethylene and ethane [1–4] and on the oxygenation to methanol and formaldehyde [5–8]. However, these processes provide either low conversions of CH_4 or low selectivities. The indirect route to liquid fuel from methane is through the synthesis gas intermediate, a CO and H_2 mixture, which can be easily transformed to either methanol or hydrocarbons through the Fischer–Tropsch reaction [9–13]. Currently, the synthesis gas is typically prepared by steam reforming (reacting CH_4 with H_2O at ~800°C) [14]

$$CH_4 + H_2O \rightarrow CO + 3H_2, \quad \Delta H = 225.4\,\text{kJ/mol}. \tag{1}$$

This process provides, however, either low conversion of methane or low selectivity to CO and an

*Corresponding author. Tel.: +1-716-645-2911; fax: +1-716-645-3822; e-mail: feaeliru@acsu.buffalo.edu

[1]Current address: Strategic Research Center, Mobil Technology Company, 13777 Midway Rd., Dallas, TX75244, USA.

This paper, which was originally published in **Applied Catalysis A**: General 183, 85–92 (1999), is republished with permission.

unsuitable H_2/CO ratio for the Fischer–Tropsch and methanol syntheses. Therefore, in recent years, many researchers attempted to convert methane to the synthesis gas via the CO_2 reforming (reaction (2)) [15–21] or via the catalytic partial oxidation of methane (reaction (3)) [22–35]

$$CH_4 + CO_2 \rightarrow 2CO + 2H_2, \quad \Delta H = 260.5\,\text{kJ/mol}, \tag{2}$$

$$CH_4 + 1/2O_2 \rightarrow CO + 2H_2, \quad \Delta H = -22.6\,\text{kJ/mol}. \tag{3}$$

The catalytic partial oxidation offers the greatest potential for a fast, efficient and economical conversion of methane to synthesis gas, due to the high conversion, high selectivity, suitable H_2/CO ratio, and very short residence time. For this reason, the investigations regarding the partial oxidation of methane, its mechanism, and the characterization of the catalyst became one of the most active topics in the catalytic conversion of CH_4.

In previous work, we reported that a 9.7 mol% NiO/MgO catalyst has high activity and selectivity, as well as excellent stability in the CO_2 reforming of methane [18,20] and for the combination reaction between the partial oxidation of methane and the CO_2 reforming of methane [36]. Tang et al. [37] also noted that the above catalyst has high activity and stability for the partial oxidation of methane.

Lunsford and coworkers [35] found that a hot spot was generated on the top layer of the catalyst (NiO/Y_2O_3) during the oxidation of CH_4. We observed, during the catalytic oxidation of CH_4 over NiO/Al_2O_3, NiO/SiO_2 or NiO/MgO, the formation of a hot layer of about 2 mm thickness. This hot layer remained fixed over NiO/Al_2O_3 and NiO/MgO but moved down and up over the NiO/SiO_2 catalyst, generating oscillations in the catalyst temperature [38].

Two mechanisms have been proposed for the partial catalytic oxidation of methane to synthesis gas: (i) The combustion-reforming mechanism, in which CO_2 and H_2O are the primary products, and CO is the secondary product of their reaction with CH_4. (ii) The pyrolysis mechanism, in which CO is produced directly without the preformation of CO_2. Prettre et al. [39], who were the first to study the CH_4 partial oxidation over a nickel catalyst, concluded that the process involves two steps: an initial exothermic oxidation of CH_4 to CO_2 and H_2O, followed by the endothermic reactions $CH_4+CO_2 \rightarrow 2CO+2H_2$ and $CH_4+H_2O \rightarrow CO+3H_2$. Some recent investigations brought support to this mechanism [22,26–28]. The other mechanism also involves two steps, namely the pyrolysis $CH_4 \rightarrow C+4H$, followed by the oxidation of C to CO [17,28–33]. Recently, it was demonstrated by the authors [40] that the partial oxidation of methane follows different mechanisms over the unreduced and reduced Ni-based catalysts. Over the unreduced NiO/SiO_2 catalyst, CH_4 in the gas phase reacts with O_2 in the adsorbed state and leads to CO_2 (Eley–Rideal mechanism). In contrast, over the reduced NiO/SiO_2 catalyst, CH_4 and O_2 react in the adsorbed states via the pyrolysis mechanism, and lead to CO and H_2 (Langmuir–Hinshelwood mechanism).

In the present paper, three issues are investigated regarding the partial oxidation of methane to synthesis gas, over NiO/MgO solid solution catalysts, at a very high space velocity of the feed gas: (i) the effect of the solid solution formation between NiO and MgO; (ii) the effect of the composition of the NiO/MgO solid solution catalysts; and (iii) the reaction mechanism.

2. Experimental

2.1. Catalyst preparation

NiO/MgO catalyst was prepared by impregnating MgO (Aldrich) with an aqueous solution of nickel nitrate (Alfa). The paste thus obtained was dried at room temperature in air and then decomposed and calcined at 800°C in air for 1.5 h.

2.2. X-ray powder diffraction (XRD)

The phase analysis was carried out by X-ray powder diffraction (XRD) on a Nicolet X-ray diffractometer. The XRD patterns of MgO are very similar to those of NiO. For MgO, the peak positions corresponding to the (2 2 0), (3 1 1) and (2 2 2) faces are at 2θ=62.64, 75.00 and 79.00, respectively. For NiO, they are by 0.60°, 0.80° and 0.81°, respectively, larger. The above three diffraction lines can be used to identify the formation of a solid solution.

2.3. ESCA

The XPS analysis was carried out using a SSI Model 100 small spot ESCA instrument, equipped with an Al K_α monochromatized X-ray source. The binding energies were referenced to the C(1s) level (285.0 eV).

2.4. BET surface area

The surface area and the pore size distribution were determined using a Micromeritics ASAP 2000 instrument. The surface area was calculated by the BET method. The sample was degassed at 200°C in high vacuum before measurements.

2.5. Catalytic reaction

The catalytic reaction was conducted in a flow system, using a vertical quartz tube (3 mm inside diameter) as reactor at atmospheric pressure. The catalyst powder (0.005 g, <200 mesh) was held on quartz wool. Mixtures of $CH_4/O_2(2/1)$ at 60 ml/min were employed in the investigation. In the experiments with reduced catalysts, the reduction was carried out with H_2 (20 ml/min), by temperature-programmed heating from room temperature to 800°C, at a rate of 20°C/min. In the experiments with the "unreduced" catalysts, the system was heated in the feed gas flow (60 ml/min) to the reaction temperature at a rate of 20°C/min. The catalyst temperature, T, was measured by inserting a thermocouple into the top layer of the catalyst at a depth of 1 mm. The analysis of the reactants/products mixtures was performed with an in situ gas chromatograph equipped with both Porapak Q and 5A molecular sieve columns.

3. Results

3.1. Comparison, for reduced catalysts, between the mechanical mixture of NiO and MgO and the NiO/MgO catalyst prepared by impregnation

Table 1 shows that the initial CH_4 conversion, and the CO and H_2 selectivities are 40%, 60% and 51% over a mechanical mixture of NiO and MgO, containing 9.7 mol% NiO, calcined for 1.5 h. They become 53%, 75% and 75% over the mechanical mixture calcined for 14 h, and 87%, 96% and 100% over the NiO/MgO catalyst prepared by impregnation and containing the same amount of NiO. Compared with the NiO/MgO catalyst prepared by impregnation, the mechanical mixture of NiO and MgO has a lower activity and selectivity. However, the increase of the calcination time of the mechanical mixture results in increases of both the activity and selectivity.

The phase analysis was carried out by XRD (Fig. 1) for both the mechanical mixtures and the catalyst prepared by impregnation. As shown in Fig. 1, three double peaks near $2\theta=62.64$, 75.00 and 79.00, which correspond to the (2 2 0), (3 1 1) and (2 2 2) faces, are present for the mechanical mixtures. In contrast, the NiO/MgO catalyst prepared by impregnation and calcination has only three single peaks near the above 2θ values. This indicates that a solid solution was indeed formed. For the mechanical mixtures, the NiO peak in each of the double peaks decreases with increasing calcination time, because the calcination leads to the partial formation of a solid solution.

The binding energies of Mg(2p) and Ni(2p3/2) were determined by XPS (Table 2). The binding energy of Mg(2p) in the NiO/MgO catalyst is 48.8 eV, which is lower than that in MgO (50.5 eV). The binding energy

Table 1
Comparison for reduced catalysts of methane partial oxidation over mechanical mixtures and over NiO/MgO prepared by impregnation (both contain 9.7 mol% NiO)

NiO/MgO catalyst	CH_4 conversion (%)	Selectivity (%)	
		CO	H_2
Sample prepared by mechanical mixing and calcined at 800°C for 1.5 h	40	60	51
Sample prepared by mechanical mixing and calcined at 800°C for 14 h	53	75	75
Sample prepared by impregnation and calcined at 800°C for 1.5 h	87	96	100

Reaction conditions: $T=850$°C (catalyst temperature), GHSV=720 000 cm^3/h g, $CH_4/O_2=2/1$.

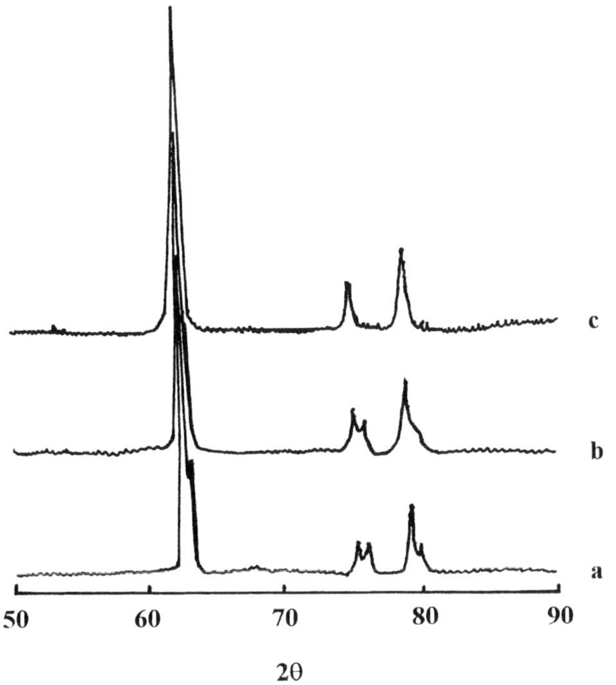

Fig. 1. XRD patterns of 9.7 mol% NiO/MgO (a) mechanical mixture of NiO and MgO calcined at 800°C for 1.5 h, (b) mechanical mixture of NiO and MgO calcined at 800°C for 14 h, (c) NiO/MgO catalyst prepared by impregnation.

of Ni(2p3/2) is 856.0 eV, which is higher than that in NiO (854.5 eV).

3.2. *The effect of NiO/MgO composition in catalysts prepared by impregnation on the reaction*

The partial oxidation of methane was carried out at 850°C (catalyst temperature) at a GHSV of 720 000 cm^3/h g, over both reduced and unreduced NiO/MgO catalysts. Fig. 2 shows that over the reduced catalysts, the initial activities are 16% and 22% over the 2.6 mol% NiO/MgO and pure NiO, respectively, but become about 86% over the 9.7–

Table 2
XPS data for the 9.7 mol% NiO/MgO catalyst prepared by impregnation

Sample	Binding energy	
	Mg(2p)	Ni(2p3/2)
NiO/MgO catalyst	48.8	856.0
NiO [44]	–	854.5
MgO [45]	50.5	–

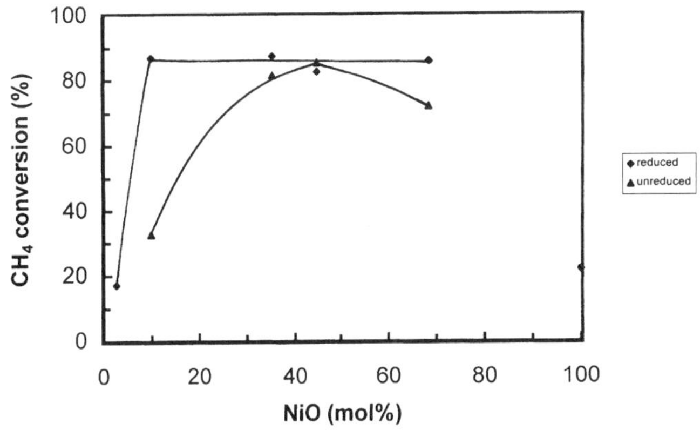

Fig. 2. CH_4 conversion vs. NiO mol% over the NiO/MgO catalyst. T=850°C, feed gas: CH_4/O_2(2/1) at 720 000 cm^3/h g.

68 mol% NiO/MgO catalysts. Fig. 3 shows that over the reduced catalysts the selectivities to CO are 20% and 5% over a 2.6 mol% NiO/MgO and a pure NiO catalyst, respectively, but about 96% over the 9.7–68 mol% NiO/MgO catalysts. The H_2 selectivities are very similar to those of CO (Fig. 4). One can conclude that when the NiO/MgO composition changes from 9.7 to 68 mol%, the initial activity and selectivity remain almost constant over the reduced catalysts.

For the unreduced catalysts, Fig. 2 shows that the initial activities are 32%, 80%, 85%, 72% over a 9.7, 35, 45 and 68 mol% NiO/MgO catalyst, respectively. Fig. 3 indicates that the CO selectivity is 32% over a 9.7 mol% NiO/MgO catalyst and about 94% over the 35–68 mol% NiO/MgO catalysts. Fig. 4 shows that over the unreduced catalysts the H_2 selectivity is 28% over the 9.7 mol% NiO/MgO catalyst but 90% over the 35–68 mol% NiO/MgO catalysts.

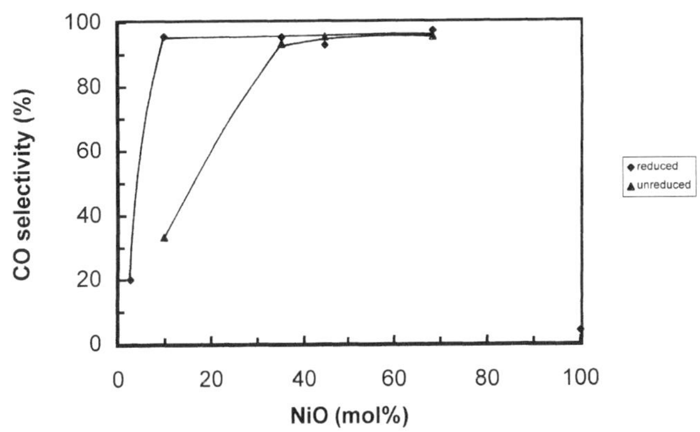

Fig. 3. CO selectivity vs. NiO mol% over the NiO/MgO catalysts. T=850°C, feed gas: CH_4/O_2(2/1) at 720 000 cm^3/h g.

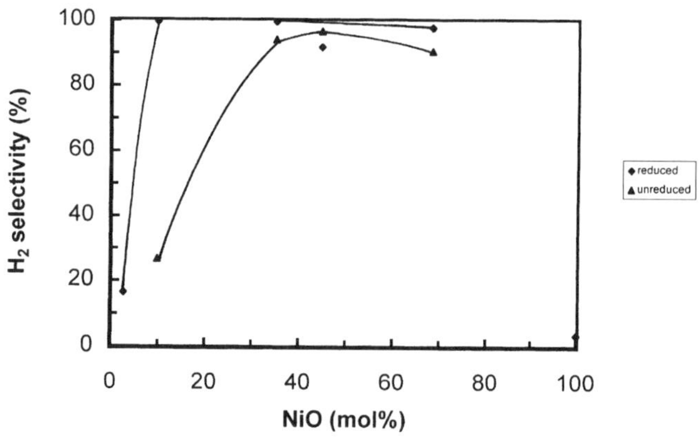

Fig. 4. H_2 selectivity vs. NiO mol% over the NiO/MgO catalysts. T=850°C, feed gas: CH_4/O_2(2/1) at 720 000 cm^3/h g.

For the reduced catalysts, Figs. 5 and 6 reveal that the CH_4 conversion and CO selectivity change with reaction time over the 68 mol% NiO/MgO catalyst, but do not change with reaction time over the 35 and 45 mol% NiO/MgO catalysts. This indicates that the catalyst with a too high content of NiO has a lower stability.

3.3. Comparison between the rates of CO_2–H_2O reforming of methane and the partial oxidation of methane

The combustion–reforming mechanism of the partial oxidation involves two kinds of reactions: combustion,

$$CH_4 + 2O_2 \rightarrow 2H_2O + CO_2$$

and reforming,

$$CH_4 + H_2O \rightarrow CO + 3H_2$$

$$CH_4 + CO_2 \rightarrow 2CO + 2H_2$$

A comparison between the rates of CO_2–H_2O reforming of methane and of the partial oxidation is made because it provides some insight into the mechanism of partial oxidation. If the partial oxidation would occur via the combustion–reforming mechanism, then its rate should have an upper bound in the rate of a reaction in which a very rapid combustion is followed by the reforming reaction. When the composition of the feed gas in the partial oxidation is CH_4/O_2=2/1, the gas composition after a complete combustion will be $CH_4/H_2O/CO_2$=3/2/1. The reforming reaction rate of such a mixture, under the same conditions as those for the partial oxidation and at the temperature of the hot layer of the latter, will be used in what follows for comparison.

Table 3 shows that for the preceding reforming reaction at 650°C, 750°C and 850°C, the CH_4 conversions are 0%, 3% and 26%, the CO selectivities 0%, 27% and 78%, and the H_2 selectivities 0%, 12% and 76%, respectively.

Fig. 7 compares the CO formation rate in the CO_2–H_2O reforming experiments to those in the partial oxidation of methane. At 850°C, the rate of CO formation due to reforming is about half that in the partial oxidation of methane, but it is 7 and 10 times lower at 750°C and 700°C, respectively.

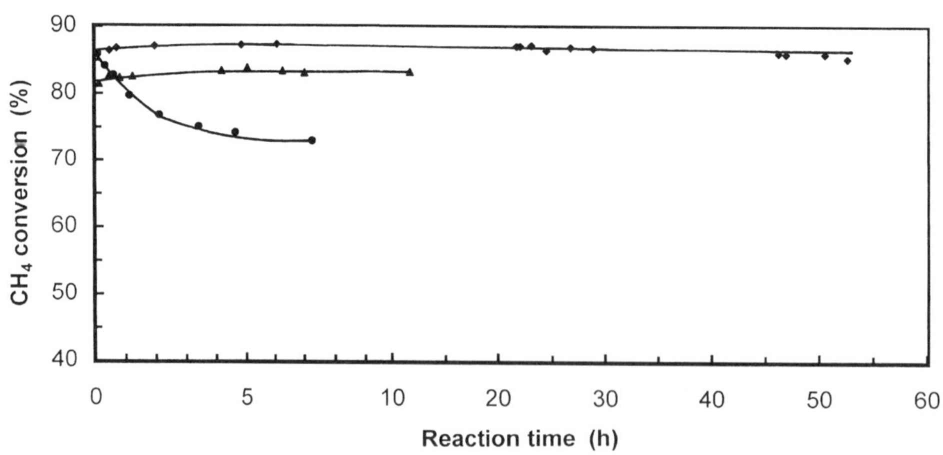

Fig. 5. CH_4 conversion vs. reaction time over reduced NiO/MgO catalysts. T=850°C, feed gas: CH_4/O_2(2/1) at 720 000 cm^3/h g. ((◆) 35 mol% NiO, (▲) 45 mol% NiO, (●) 68 mol% NiO).

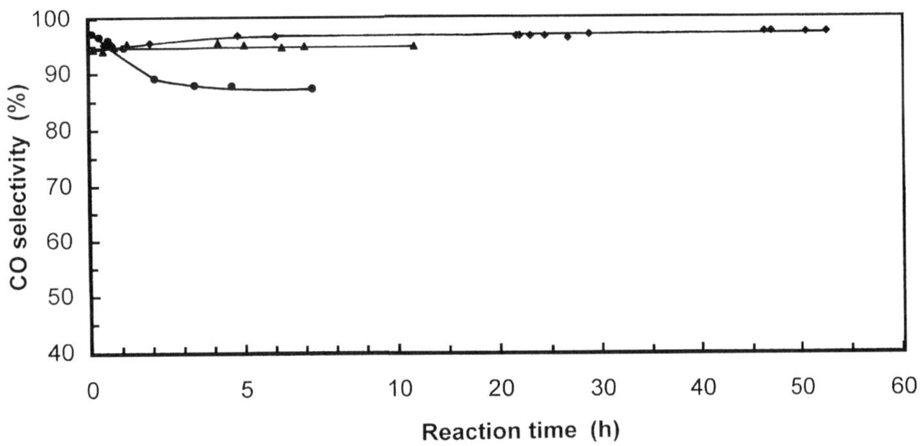

Fig. 6. CO selectivity vs. reaction time over reduced NiO/MgO catalysts. $T=850°C$, feed gas: $CH_4/O_2(2/1)$ at 720 000 cm^3/h g. ((◆) 35 mol% NiO, (▲) 45 mol% NiO, (●) 68 mol% NiO).

4. Discussion

4.1. The effect of the solid solution formation of NiO and MgO on the reaction

Because of similar structures, NiO and MgO are completely miscible and form an ideal solid solution [41]. Indeed, the XRD indicates that NiO and MgO form a solid solution in the NiO/MgO catalysts prepared by impregnation and calcined at 800°C (Fig. 1). However, only a fraction of NiO formed a solid solution with MgO in the mechanical mixture, even when the mixture was subjected to calcination at 800°C for 14 h. As pointed out by Arena et al. [42], the first step in the formation of a NiO–MgO solid solution is the diffusion of NiO over the surface of MgO. This is followed by diffusion inside the MgO lattice. Obviously, the amount of solid solution formed should increase with increasing calcination time. The activity and selectivity are higher over the NiO/MgO catalyst prepared by impregnation than over the mechanical mixture of NiO and MgO calcined at 800°C for 1.5 or 14 h (Table 1). Therefore, the larger the amount of NiO that forms a solid solution, the higher the conversion and selectivity, indicating that the formation of a solid solution plays an important role in the generation of the active sites for the partial oxidation of methane. The XPS analysis indicated that compared to MgO or NiO, in the MgO/NiO solid solution Mg(2p) has a lower and Ni(2p3/2) a higher binding energy (Table 2). This means that electron transfer from NiO to MgO takes place, which, by increasing the interactions between the two oxides, slows down the reduction of the NiO during the preparation of the reduced catalyst. This leads to Ni

Table 3
The reaction of $CH_4/H_2O/CO_2(3/2/1)$ over reduced NiO/MgO (9.7 mol%), GHSV=720 000 cm^3/h g

T (°C)	CH_4 conversion (%)	CO selectivity (%)	H_2 selectivity (%)
850	26	78	76
750	3	27	12
650	0	0	0

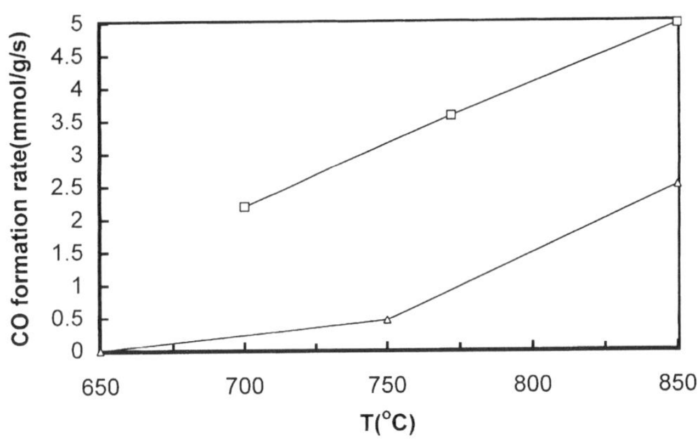

Fig. 7. Comparison of CO formation rate between the CO_2–H_2O reforming of methane (△) and the partial oxidation of methane (□), feed gas: $CH_4/O_2(2/1)$ at 720 000 cm^3/h g over reduced 9.7 mol% NiO/MgO.

atoms segregated over the surface of the catalyst, resulting in a high dispersion of Ni° [42], which is responsible for the high activity of the catalyst. In addition, (i) the segregated Ni atoms interact strongly with those remaining in the substrate and this inhibits their sintering, and (ii) the high dispersion of Ni inhibits the coke formation which requires large clusters of Ni.

4.2. Effect of the composition of the NiO/MgO catalysts

For NiO contents in the range 9.7–35 mol%, the NiO/MgO catalysts prepared by impregnation provide high conversions and selectivities, as well as high stabilities. For too small NiO contents, the conversion and selectivity are low, and for high contents the stability is low. This can be explained as follows: The XRD patterns indicate that, after calcination at 800°C for 1.5 h, the NiO in the NiO/MgO catalyst prepared by impregnation forms a solid solution with MgO (Fig. 1 and [43]). As noted above, the electron transfer from NiO to MgO induces strong interactions between NiO and MgO, which inhibit the reduction of NiO. At low concentrations of NiO (2.6 mol%), only few NiO can be reduced, and the small number of active sites, is responsible for the low activity. As the amount of NiO increases, the number of Ni° sites increases both because of the higher concentration of NiO and of the weakening of the interactions with MgO. This is why the 9.7–35 mol% NiO/MgO catalysts have high activity. However, when the content of NiO becomes too high, numerous Ni° atoms are generated during reduction, which sinter particularly during reaction to form large particles. This decreases the number of active sites and also stimulates coke formation. This explains the low stability of the 68 mol% NiO/MgO catalyst.

The almost constant initial activity in the range from 9.7–68 mol% NiO/MgO might be a result of two factors which compensate one another. On one hand, the larger the NiO content, the easier its reduction and the larger the number of active sites. On the other hand, as shown in Fig. 8, the surface area of the catalyst decreases as the NiO content increases from 9.7 to 68 mol%, and the decrease in the surface area decreases the number of active sites exposed to the chemical atmosphere. Another possible explanation

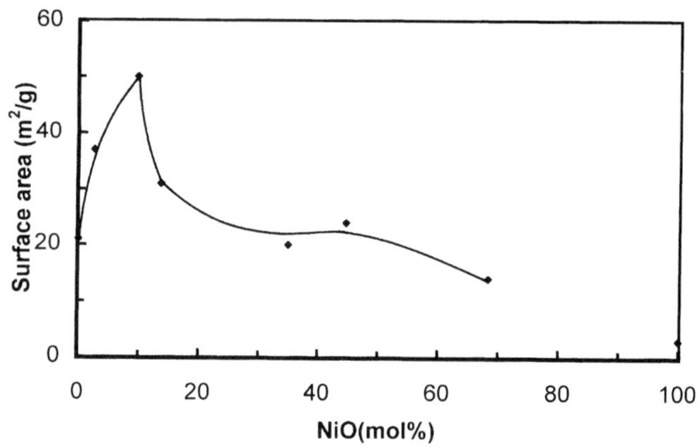

Fig. 8. BET surface area vs. NiO mol% in the NiO/MgO catalysts.

for the constant initial activity is that the reaction is diffusion limited or even that the catalyst bed which has a height of about 3 mm is sufficiently long for thermodynamic equilibrium to be achieved at its exit. Indeed, the activity and the selectivity determined by us are near the equilibrium values [35] at a temperature of about 770°C, which might be the temperature at the bed exit (this temperature was not determined by us).

It is of interest to note that the "unreduced" catalysts containing 35–68 mol% NiO/MgO provide high activity and selectivity. This probably occurs because in that range of NiO concentrations a sufficiently large number of Ni° active sites is generated via its reduction by the flowing gas.

4.3. Reaction mechanism

The reforming experiments described in a preceding section show that the CO formation rate is much smaller in the reforming reaction than in the partial oxidation of methane. At 750°C and 700°C, the CO formation rates are about 7 and 10 times larger in the partial oxidation reaction (Fig. 7). This indicates that the combustion–reforming mechanism plays a negligible role at 700°C and 750°C. However, at 850°C, the CO formation rate in the partial oxidation reaction is only twice as high as in the reforming reaction. Consequently, at 700°C and 750°C, the pyrolysis mechanism is responsible for the partial oxidation reaction, while at 850°C both the pyrolysis and the combustion–reforming mechanism play a part.

5. Conclusion

The following conclusions can be drawn regarding the partial oxidation of methane to the synthesis gas over NiO/MgO solid solution catalysts:

1. The NiO/MgO solid solution catalyst has high activity and selectivity, whereas the corresponding mechanical mixtures have low activity and selectivity. Nevertheless, the activity and the selectivity increase with the calcination time of the mechanical mixtures.
2. In the composition range of 9.7–35 mol% NiO, the NiO/MgO solid solution catalysts have high activities and selectivities, as well as high stabilities. For too low NiO concentrations in the solid solution, the activities and selectivities are low, while for too large concentrations, the stability is low.
3. At 850°C, the reaction takes place via both the pyrolysis and the combustion–reforming mechanisms, whereas at the lower temperatures of 700°C and 750°C, it takes place mainly via the pyrolysis mechanism.

References

[1] G.E. Keller, M.M. Bhasin, J. Catal. 73 (1982) 9.
[2] J.S. Lee, S.T. Oyama, Catal. Rev.-Sci. Eng. 30(2) (1988) 249.
[3] Y. Amenomiya, V.I. Birss, M. Goledzinowski, J. Galuszka, A.R. Sanger, Catal. Rev.-Sci. Eng. 33(3) (1990) 163.
[4] J.H. Lunsford, Stud. Surf. Sci. Catal. 81 (1994) 1.
[5] R. Pitchai, K. Klier, Catal. Rev.-Sci. Eng. 28(1) (1986) 13.
[6] N.R. Foster, Appl. Catal. 19 (1985) 1.
[7] T.J. Hall, J.S.J. Hargreaves, G.J. Hutchings, R.W. Joyner, S.H. Taylor, Fuel Process. Technol. 42 (1995) 151.
[8] R.A. Periana, D.J. Taube, S. Gamble, H. Taube, T. Satoh, H. Fujii, Science 280 (1998) 560.
[9] R.G. Herman, K. Klier, G.W. Simmons, B.P. Finn, J.B. Bulko, J. Catal. 56 (1979) 407.
[10] H.H. Kung, Catal. Rev.-Sci. Eng. 22(2) (1980) 235.
[11] H. Pichler, Adv. Catal. 4 (1952) 271.
[12] M.A. Vannice, Catal. Rev.-Sci. Eng. 14 (1976) 153.
[13] J.C.J. Bart, R.P.A. Sneeden, Catal. Today 2 (1987) 1.
[14] D.L. Trimm, Catal. Rev.-Sci. Eng. 16 (1977) 155.
[15] A.M. Gadalla, B. Bower, Chem. Eng. Sci. 43 (1988) 3049.
[16] A.T. Ashcroft, A.K. Cheetham, M.L.H. Green, P.D.F. Vernon, Nature 352 (1991) 225.
[17] J.R. Rostrup-Nielsenand, J.-H.B. Hansen, J. Catal. 144 (1993) 38.
[18] E. Ruckenstein, Y.H. Hu, Appl. Catal. A 133 (1995) 149.
[19] Z. Zhang, X.E. Verykios, J. Chem. Soc., Chem. Commun. (1) (1995) 71.
[20] Y.H. Hu, E. Ruckenstein, Catal. Lett. 43 (1997) 71.
[21] R.N. Bhat, W.M.H. Sachtler, Appl. Catal. A 150 (1997) 279.
[22] D. Dissanayake, M.P. Rosynek, J.H. Lunsford, J. Catal. 132 (1991) 117.
[23] D.A. Hickman, L.D. Schmidt, J. Catal. 138 (1992) 267.
[24] F. van Looij, J.W. Geus, J. Catal. 168 (1997) 154.
[25] M. Alibrando, H.S. Hahm, E.E. Wolf, Catal. Lett. 49 (1997) 1.
[26] Y. Boucouvalas, Z.L. Zhang, X.E. Verykios, Catal. Lett. 27 (1994) 131.
[27] H. Papp, P. Schuler, Q. Zhuang, Catal. Lett. 3 (1996) 299.
[28] Y. Boucouvalas, Z.L. Zhang, X.E. Verykios, Catal. Lett. 40 (1996) 189.
[29] V.R. Choudhary, A.M. Rajput, B.J. Prabhaker, Catal. Lett. 15 (1992) 363.
[30] D.A. Hickman, L.D. Schmidt, Science 259 (1993) 343.
[31] Y.H. Hu, E. Ruckenstein, J. Catal. 158 (1996) 260.
[32] C.T. Au, M.S. Liao, C.F. Ng, Chem. Phys. Lett. 267 (1997) 44.
[33] O.V. Buyeskaya, K. Walter, D. Wolf, M. Baerns, Catal. Lett. 38 (1996) 81.
[34] E.P.J. Mallens, J.H.B.J. Hoebink, G.B. Marin, J. Catal. 167 (1997).
[35] D. Dissanayake, M.P. Rosynek, J.H. Lunsford, J. Phys. Chem. 97 (1993) 3644.
[36] E. Ruckenstein, Y.H. Hu, Ind. Eng. Chem. Res. 37(5) (1998) 1744.
[37] S. Tang, J. Lin, K.L. Tan, Catal. Lett. 51 (1998) 169.
[38] Y.H. Hu, E. Ruckenstein, Ind. Eng. Chem. Res. 37(6) (1998) 2333.
[39] M. Prettre, C. Eichner, M. Perrin, Trans. Faraday Soc. 42 (1946) 335.
[40] Y.H. Hu, E. Ruckenstein, J. Phy. Chem. B 102 (1998) 230.
[41] H.B. Nussler, O. Kubaschewski, Z. Phys. Chem. (N.F.) 121 (1980) 187.
[42] F. Arena, F. Fruster, A. Parmaliana, L. Plyasova, A.N. Shmakov, J. Chem. Soc., Faraday Trans. 92 (1996) 469.
[43] Y.H. Hu, E. Ruckenstein, J. Catal. 163 (1996) 306.
[44] K.S. Kim, N. Winograd, Surf. Sci. 43 (1974) 625.
[45] J.S. Corneille, J.W. He, D.W. Goodman, Surf. Sci. 306 (1994) 269.

Partial Oxidation of Methane to Synthesis Gas over Alkaline Earth Metal Oxide Supported Cobalt Catalysts

H. Y. Wang and E. Ruckenstein[1]

Department of Chemical Engineering, State University of New York at Buffalo, Amherst, New York 14260

Received September 15, 2000; accepted January 29, 2001; published online March 27, 2001

The partial oxidation of methane to synthesis gas was investigated over the alkaline earth metal oxide supported Co catalysts. Only MgO has proved to be a suitable support. The reaction behavior of the MgO-supported Co catalysts was significantly influenced by the calcination temperature (T_c) and Co loading. The 24 wt% Co/MgO catalyst precalcined at 800°C provided a high and stable activity; those precalcined at higher temperatures provided a lower stable activity and those precalcined at lower temperatures a lower stability. Because of the two oxidative states of Co, the structural characteristics of the Co catalysts are strongly affected by the nature of the support, calcination temperature, and Co loading. In the 24 wt% MgO-supported catalysts, only a solid solution of CoO and MgO was identified for $T_c \geq 800°C$, while one or two of the following compounds, Co_3O_4 and $MgCo_2O_4$, were additionally detected for $T_c < 800°C$. Correlations among the species present in the catalyst, its reducibility, and its catalytic performance were established.

Key Words: **partial oxidation of methane; synthesis gas; Co catalyst; effect of support; effect of calcination temperature; effect of Co loading.**

1. INTRODUCTION

The partial oxidation of methane, which was first investigated by Prettre *et al.* as early as 1946 (1), has gained renewed interest during the past years (2–22). The catalysts that have been reported to be active and selective toward the synthesis gas formation were mainly supported noble metals, such as Rh, Ru, Pd, Pt (2–8), supported Ni (9–17), and Co (16–18) catalysts, as well as some pyrochlore type oxides, such as $Ln_2Ru_2O_7$ (19, 20), and perovskite type oxides, such as $LaMO_3$ (M = Ni, Rh, Co, Cr) (21, 22).

Compared to the noble-metal based catalysts, the Ni- and Co-based catalysts are particularly promising due to their low prices. In recent papers, Ni/MgO catalysts, obtained through the reduction of NiO–MgO solid solutions, were reported to provide high and stable activities for the methane conversion to synthesis gas via CO_2 reforming (23), partial oxidation (12, 15), and their combination (24). Like NiO, CoO can also form an "ideal" solid solution over the entire mole fraction range with MgO (25, 26). However, unlike Ni, Co has two oxidative states (Co^{2+} and Co^{3+}). Consequently, additional species can be generated between the cobalt and magnesium oxides during the preparation of the catalyst, which are expected to depend on the preparation conditions, such as calcination temperature and metal loading. The species present in a catalyst affect its reducibility and the size and morphology of the metal particles thus generated and hence the catalytic pattern of the catalyst. Of course, during reaction other species may form as well in the catalyst. It is therefore of interest to identify the species formed in the catalyst and their role in the catalytic process. For this purpose, the reaction behavior of the Co/MgO catalysts in the partial oxidation of methane was investigated as a function of calcination temperature and Co loading. For comparison purposes, the Co catalysts supported on the other alkaline earth metal oxides were also included in the investigation.

2. EXPERIMENTAL

2.1. *Catalyst Preparation*

The catalysts were prepared by impregnating the supports (MgO, CaO, SrO, and BaO) with aqueous solutions of $Co(NO_3)_2 \cdot 6H_2O$, followed by overnight drying at 110°C. The samples were then calcined in the open air of a furnace for 8 h at 800°C, when CaO, SrO, and BaO were used as supports, and at various temperatures between 500 and 1000°C, when MgO was employed. The calcined catalysts are denoted as Co(O)/MO (M = Mg, Ca, Sr, or Ba) and the catalysts reduced in H_2 as Co/MO. Co loading means wt% Co in the completely reduced catalyst.

2.2. *Catalytic Reaction*

All the catalysts were tested under atmospheric pressure and at a $T_{furnace}$ of 850°C in a fixed-bed vertical quartz reactor (i.d. 3 mm), which was operated in a downflow mode with the catalyst held on a quartz wool bed. The reduction

[1] To whom correspondence should be addressed. Fax: +1-716-645-3822. E-mail: feaeliru@acsu.buffalo.edu.

This paper, which was originally published in **Journal of Catalysis**: 199, 309–317 (2001), is republished with permission.

of the catalyst was carried out by increasing the temperature from room temperature to 600°C at a rate of 20°C/min and from 600 to 850°C at a rate of 10°C/min, in a H_2 flow (20 ml/min), without holding at 850°C. After reduction, the feed gases ($CH_4/O_2 = 2/1$) were introduced into the catalyst bed at a flow rate of 60 ml/min. The reactants and products were analyzed with an on-line gas chromatograph equipped with a Porapak Q column. An ice-cold trap was located between the reactor exit and the GC sampling valve to remove the water formed during reaction.

2.3. *Catalyst Characterization*

2.3.1. BET surface area and the exposed Co surface area. The BET surface areas of the calcined catalysts, determined via nitrogen adsorption using a Micromeritics ASAP2000 instrument, are listed in Table 1.

The exposed Co metal surface area of the reduced catalyst was determined by CO pulse adsorption in a quartz tube (i.d. 4 mm) at room temperature, assuming a 1/1 stoichiometry. Fifty milligrams of the calcined catalyst powder was charged and its reduction was carried out as described in the previous section. After reduction, the catalyst was purged with an ultra high purity carrier gas (He, 35 ml/min) at 850°C for 0.25 h, and then the sample was cooled to room temperature. CO pulses (10 μl per pulse) were then injected in the carrier gas and passed over the reduced catalyst. The CO uptake in each pulse was monitored using a thermal conductivity detector (TCD).

2.3.2. CH_4 decomposition. The decomposition of pure methane over the reduced catalyst was carried out in a quartz tube (i.d. 4 mm) reactor containing 25.0 mg of calcined catalyst. After reduction, which was carried out as described previously, the catalyst was purged with a carrier gas (He, 35 ml/min) at 850°C for 0.25 h and then pulses of CH_4 (250 μl) were injected in the carrier gas. The products were analyzed with an on-line gas chromatograph equipped with a TCD and a Porapak Q column.

2.3.3. Temperature-programmed reduction (TPR). During each TPR run, a high-purity 2.5% H_2/Ar mixture (35 ml/min) was passed over a calcined sample in a vertical quartz tube reactor (i.d. 4 mm). The sample (10.0 or 20.0 mg) was held on a quartz wool bed and the temperature of the sample was increased from 40 to 1000°C at a rate of 20°C/min. The water, formed during reduction, was trapped in a Hydro-Purge II column. The hydrogen consumed in TPR was monitored continuously with a TCD, which was calibrated using known amounts of CoO and Co_3O_4.

2.3.4. X-ray powder diffraction (XRD). XRD, performed with a Siemens D500 X-ray diffractometer using filtered Cu$K\alpha$ radiation, was used to identify the major phases of the calcined catalysts.

TABLE 1

Physical Data for the Supported Co Catalysts

Support	Co loading (wt%)	Calcination temp. (°C)	Surface area[a] (m^2/g cat.)	Co surface area[b] (m^2/g cat.) × 100
MgO	6	800	26	7
MgO	12	800	22	7
MgO	24	500	58	201
MgO	24	700	28	52
MgO	24	800	20	9
MgO	24	900	11	5
MgO	24	1000	5	—[c]
MgO	36	800	14	8
MgO	48	800	12	25
CaO	24	800	5	79
SrO	24	800	4	8
BaO	24	800	1	—[c]

[a] For the calcined catalysts.
[b] For the reduced catalysts.
[c] Too small to be detected.

3. RESULTS

3.1. *Partial Oxidation of Methane to Synthesis Gas*

3.1.1. Comparison of the catalysts supported on different alkaline earth metal oxides. The time-dependent conversions of CH_4 and O_2 and selectivities to CO and H_2 of the 24 wt% Co catalysts supported on MgO, CaO, SrO, and BaO precalcined at 800°C are plotted in Fig. 1. Over the Co/MgO, after an activation period of about 2 h, the conversion of CH_4 and selectivities to CO and H_2 remained high and unchanged during the period of study of 120 h and a H_2/CO ratio of about 2 was obtained. Over CaO-, SrO-, and BaO-supported catalysts the initial (after 8 min) conversions of CH_4 and selectivities to CO and H_2 were much lower than those over MgO and, in addition, notable deactivation was observed.

3.1.2. Effect of calcination temperature on the MgO-supported Co catalysts. The time-dependent activities of the MgO-supported catalysts precalcined at temperatures between 500 and 1000°C are plotted in Fig. 2. Only the conversion of CH_4 is presented because that of O_2 was always 100%, and only the selectivity to CO is plotted because that to H_2 followed the same pattern. The results can be summarized as follows: (i) the catalysts precalcined at temperatures ≥800°C exhibited stable activities and selectivities, which decreased with increasing calcination temperature; (ii) some deactivation with time on stream occurred over the catalysts precalcined at 500 and 700°C; (iii) the initial conversion of methane and CO selectivity decreased with increasing calcination temperature from 500 to 1000°C; (iv) activation periods were observed, particularly for the catalysts calcined at temperatures ≥800°C.

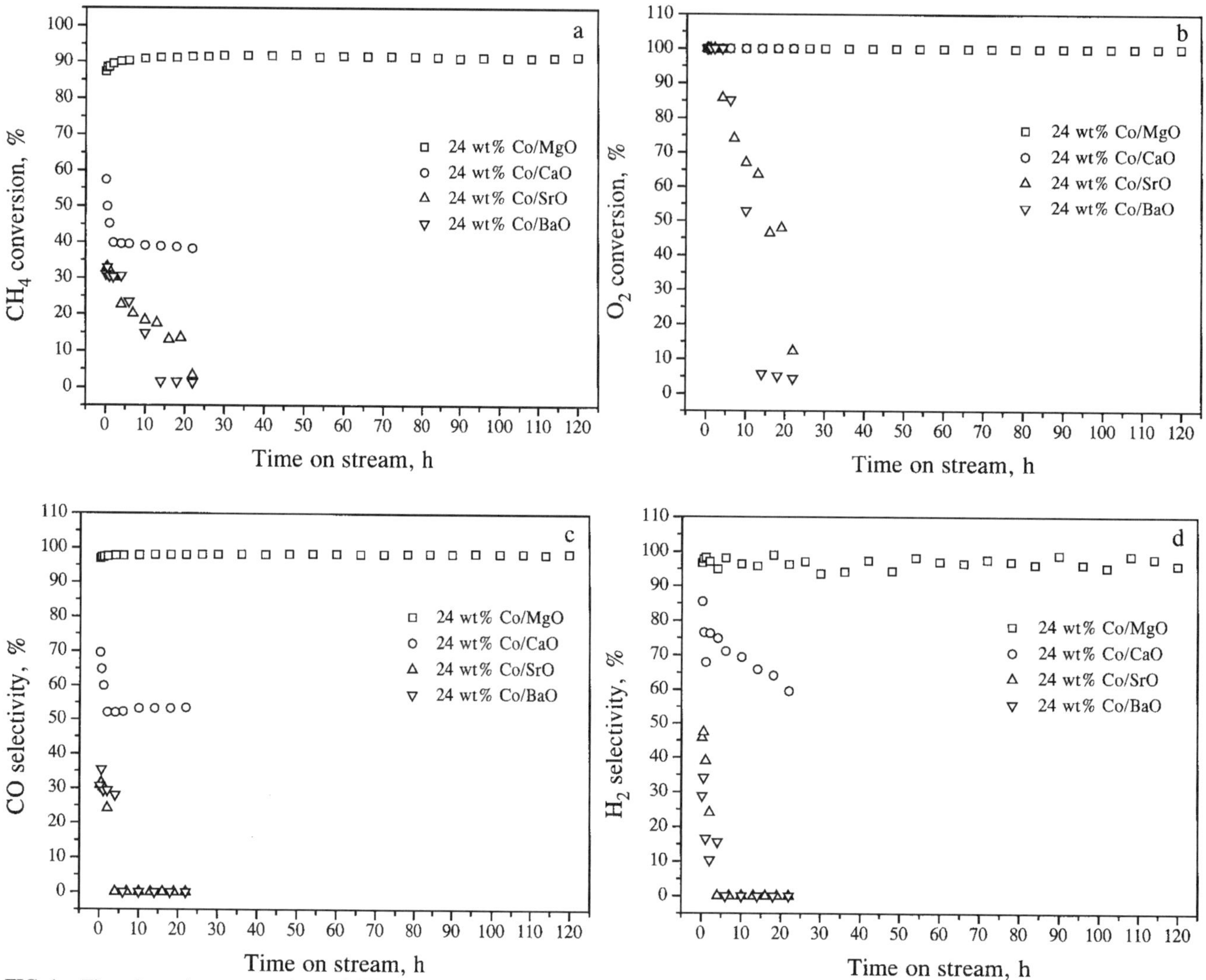

FIG. 1. Time-dependent conversions of CH_4 (a) and O_2 (b) and selectivities to CO (c) and H_2 (d) over the alkaline metal oxide-supported Co catalysts (precalcined at 800°C). P = 1 atm, $T_{furnace}$ = 850°C, CH_4/O_2 = 2.0, space velocity = 720,000 ml h^{-1} g^{-1}.

3.1.3. Effect of Co loading on the MgO-supported Co catalysts. The time-dependent activities of the MgO-supported catalysts with loadings between 6 and 48 wt% (precalcined at 800°C) are plotted in Fig. 3, which shows: (i) over the 6 wt% catalyst, the conversion of CH_4 was below 20%, and no CO and H_2 were formed; (ii) over those with loadings ≥12 wt%, the conversion of CH_4 was above 80%, and CO and H_2 were formed with high selectivities (>90%); (iii) with the increase in Co loading from 12 to 24 wt%, the conversion of CH_4 and the selectivity to CO increased somewhat, but remained almost unchanged with the further increase from 24 to 48 wt%.

3.1.4. Effect of the amount of catalyst. As shown in Fig. 4, over the 24 wt% Co/MgO catalyst (precalcined at 800°C), the conversion of CH_4 and the selectivity to CO increased with increasing amount of catalyst from 1.5 to 5.0 mg, but remained almost unchanged when the amount was increased from 5.0 to 15.0 mg.

3.2. Physicochemical Characterization

3.2.1. The exposed metallic Co surface area of the reduced catalyst. As shown in Table 1, for the 800°C calcined Co catalysts supported on different alkaline earth metal oxides, the exposed metal surface area after reduction decreased in the sequence CaO ≫ MgO ≈ SrO ≫ BaO. For the last support, the Co surface area was too small to be detected. For the Co/MgO catalysts, Table 1 shows that for the same loading (24 wt%), the exposed Co surface area decreased with increasing calcination temperature from 500 to 1000°C; for the same calcination temperature (800°C), it was almost the same for loadings between 6 and 36 wt%, but increased notably with increasing Co loading from 36 to 48 wt%.

3.2.2. Surface carbon (C̱) formed in the CH_4 decomposition. The decomposition of CH_4 pulses over the reduced catalysts was carried out at 850°C and the amount of surface carbon (C̱) generated was calculated from the carbon

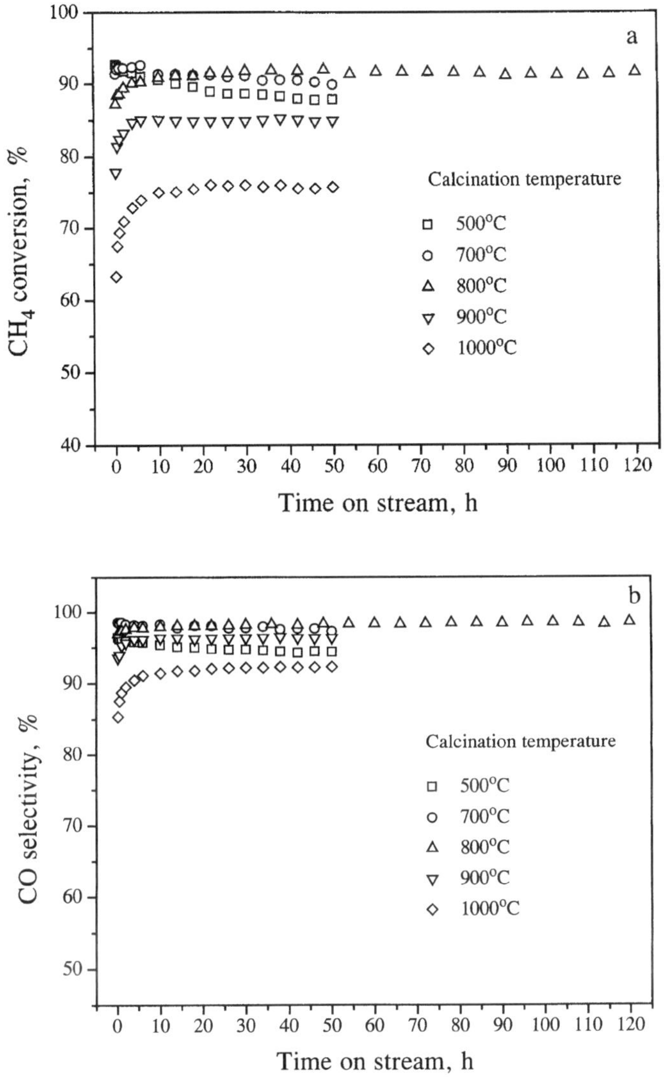

FIG. 2. Time-dependent conversion of CH_4 (a) and selectivity to CO (b) over the 24 wt% Co/MgO catalysts precalcined at various temperatures. $P = 1$ atm, $T_{furnace} = 850°C$, $CH_4/O_2 = 2.0$, space velocity = 720,000 ml h^{-1} g^{-1}.

balance. Table 2 shows that, for the 800°C calcined Co catalysts supported on different alkaline earth metal oxides, the amount of C̲ accumulated during the same period over MgO was comparable to that over BaO, but much lower than those over CaO and SrO. The low carbon deposition over BaO was due to the low exposed metal surface area. Table 2 also shows that, for the 24 wt% Co/MgO catalysts, the amount of C̲ formed decreased monotonically with increasing calcination temperature from 500 to 1000°C.

3.2.3. Reduction characteristics of the calcined supported Co catalysts. The TPR spectra of the 800°C calcined 24 wt% Co catalysts supported on different alkaline earth metal oxides are plotted in Fig. 5. For Co(O)/MgO, a peak started to form at a temperature higher than 800°C (Fig. 5a). A single peak at about 540°C was observed for Co(O)/CaO (Fig. 5b), and two peaks, one (the main) at about 505°C and the other one as a shoulder at about 710°C for Co(O)/SrO (Fig. 5c). Two peaks, one (the main) at about 455°C and the other one as a shoulder at about 790°C, were noted for Co(O)/BaO (Fig. 5d). The reduction extent over the MgO supported catalyst after TPR was estimated to be below 5%, while nearly 100% reduction was reached over the other alkaline earth metal oxide supported ones.

The TPR profiles of the 24 wt% Co(O)/MgO catalysts calcined at temperatures between 500 to 1000°C are presented in Fig. 6. The 500°C calcined 24 wt% Co(O)/MgO exhibited two reduction peaks, one at about 320°C and the other at about 695°C. The reduction extent of the sample

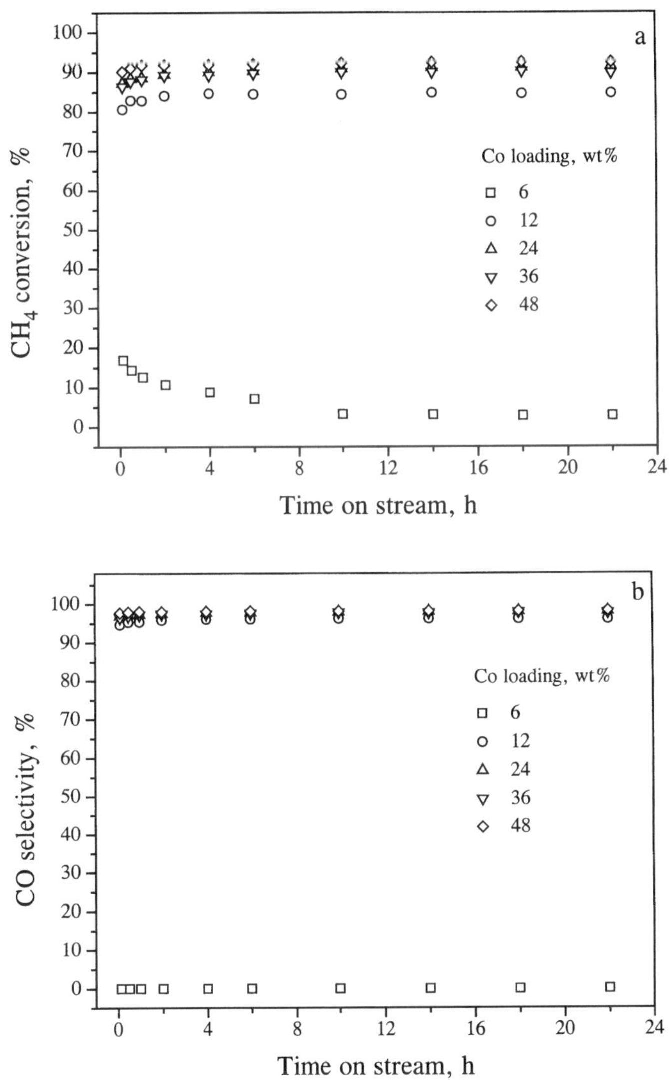

FIG. 3. Time-dependent conversion of CH_4 (a) and selectivity to CO (b) over the Co/MgO catalysts (precalcined at 800°C) with different Co loadings. $P = 1$ atm, $T_{furnace} = 850°C$, $CH_4/O_2 = 2.0$, space velocity = 720,000 ml h^{-1} g^{-1}.

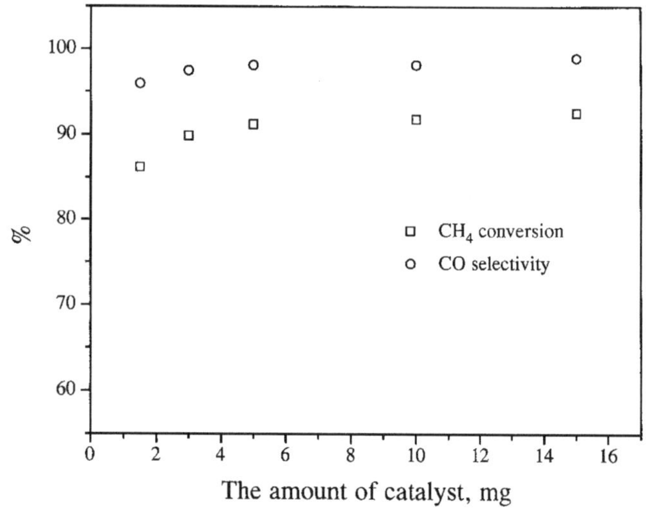

FIG. 4. Effect of the amount of catalyst on the CH_4 conversion and CO selectivity of the 24 wt% Co/MgO (precalcined at 800°C). Data obtained after 22 h of reaction; reaction conditions: P = 1 atm, $T_{furnace}$ = 850°C, CH_4/O_2 = 2.0 (60 ml/min).

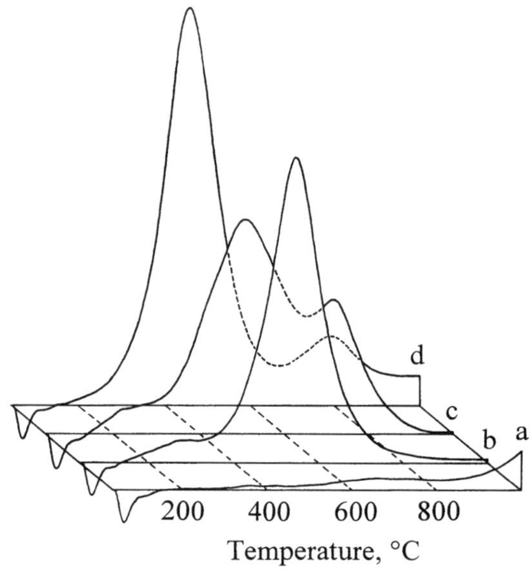

FIG. 5. TPR profiles of the 800°C calcined 24 wt% Co catalysts supported on MgO (a), CaO (b), SrO (c), and BaO (d).

was below 30%. At the calcination temperature of 700°C, in addition to a wide single peak at 430°C, a peak started to form at a temperature higher than 800°C. For calcination temperatures ≥800°C only a single peak started to form at temperatures higher than 800°C.

For all the 800°C calcined Co(O)/MgO catalysts with Co loadings between 6 and 24 wt%, a peak started to form at a temperature higher than 800°C; an additional low-temperature reduction peak between 380 and 430°C was observed over the catalysts with Co loadings of 36 and 48 wt%.

3.2.4. Compounds present in the calcined supported Co catalysts. The Co-containing species present in the calcined catalysts were identified by combining the XRD and TPR results. The data and the assignments of the XRD patterns for the calcined Co catalysts are listed in Table 3. For the 800°C calcined 24 wt% Co(O)/MgO, the XRD patterns could be attributed to (Co, Mg)O (solid solution of CoO and MgO) and/or MgO (Table 3). The reduction peak that started to form at a temperature higher than 800°C (Fig. 5a) allowed us to conclude that a solid solution was formed. $Ca_3Co_4O_9$ was identified over Co(O)/CaO, in agreement with the TPR spectrum (Fig. 5b) which provided a single reduction peak. Only $SrCoO_{2.52}$ was detected over Co(O)/SrO, while two reduction peaks were present in the TPR spectrum (Fig. 5c). Most likely, the amount of the species corresponding to the TPR shoulder was too small to be detectable by XRD. Thus, the major phase present in the Co(O)/SrO was $SrCoO_{2.52}$. Similarly, the major phase present in the Co(O)/BaO was found to be $BaCoO_{2.8}$.

The XRD results for the 500°C calcined 24 wt% Co(O)/MgO catalyst can be assigned to Co_3O_4 and/or $MgCo_2O_4$, as well as to (Co, Mg)O and/or MgO (Table 3).

TABLE 2

Amount of Carbon Deposited during CH_4 Decomposition over the Reduced Supported Co (24 wt% Loading) Catalysts at 850°C

		C formed during different periods (μmol)				
Support	Calcination temp. (°C)	First pulse	First 3 pulses	First 5 pulses	First 10 pulses	First 20 pulses
MgO	500	8.2	16.8	19.1	23.5	29.4
MgO	700	6.5	9.4	11.0	14.1	18.4
MgO	800	2.2	5.1	6.6	9.7	14.9
MgO	900	1.5	3.3	4.0	5.8	9.0
MgO	1000	1.2	2.8	3.6	4.9	6.9
CaO	800	9.0	22.7	28.2	36.1	45.7
SrO	800	6.6	10.1	12.7	17.1	22.8
BaO	800	2.2	4.3	6.0	9.6	16.3

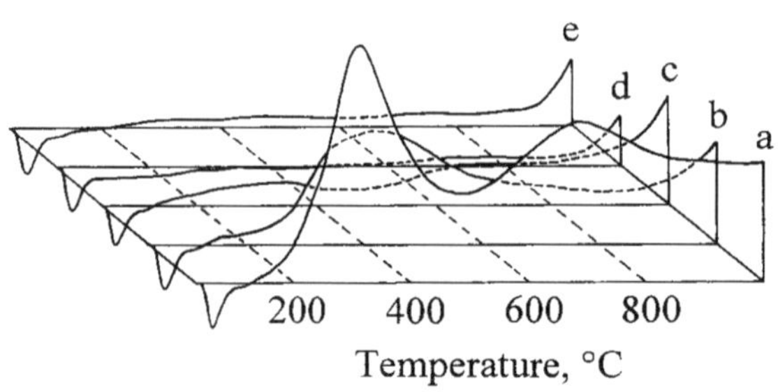

FIG. 6. TPR profiles of the 24 wt% Co(O)/MgO catalysts calcined at (a) 500, (b) 700, (c) 800, (d) 900, and (e) 1000°C.

TABLE 3

Data and Assignments of XRD Patterns of the Calcined Supported Co Catalysts

Support	Co loading (wt%)	Calcination temp. (°C)	d (Å)	Assignments	Co-containing species present[a]
MgO	24	500	2.438, 1.440, 2.869, 1.560, 2.038; 2.106, 1.490, 1.216,1.270, 2.438	Co_3O_4 and/or $MgCo_2O_4$; (Co, Mg)O and or MgO	Co_3O_4 and $MgCo_2O_4$; (Co, Mg)O
MgO	24	700	2.436, 1.430, 2.859, 1.557, 2.020; 2.108, 1.491, 1.218, 1.271, 2.436	Co_3O_4 and/or $MgCo_2O_4$; (Co, Mg)O and/or MgO	Co_3O_4; (Co, Mg)O
MgO	24	800	2.110, 1.493, 1.272, 1.218, 2.438	(Co, Mg)O and/or MgO	(Co, Mg)O
MgO	48	800	2.111, 1.493, 1.274, 1.219, 2.438; 2.438, 1.429, 2.860, 1.556, 2.022, 1.652	(Co, Mg)O and/or MgO; Co_3O_4 and/or $MgCo_2O_4$	(Co, Mg)O; Co_3O_4
CaO	24	800	2.686, 2.273, 3.587, 2.406, 2.083, 1.878	$Ca_3Co_4O_9$	$Ca_3Co_4O_9$
SrO	24	800	2.740, 3.105, 2.051, 1.642, 1.582, 1.372	$SrCoO_{2.52}$	$SrCoO_{2.52}$
BaO	24	800	3.401, 2.806, 2.164, 1.711, 2.379, 1.454	$BaCoO_{2.8}$	$BaCoO_{2.8}$

[a]Concluded by combining TPR and XRD results.

Since Co_3O_4 can be reduced in H_2 below 500°C (27, 28) and a complete reduction of (Co, Mg)O requires a temperature higher than 1000°C (Fig. 5a), the peak at about 320°C can be attributed to Co_3O_4 and the one at about 695°C to $MgCo_2O_4$. Though only two reduction peaks were observed for this catalyst (Fig. 6a), its reduction extent below 30% suggested that (Co, Mg)O was also present. Its reduction peak was not observed most likely because the upper-limit temperature during the TPR experiment was not high enough. Similarly, the XRD patterns (Table 3) coupled with the TPR spectra (Figs. 6b and 6c) indicated that Co_3O_4 and (Co, Mg)O were present in the 700°C calcined catalyst and solely (Co, Mg)O in the 800°C calcined catalyst. Based on the similarity of the TPR spectra among the catalysts calcined at temperatures ≥800°C, one can conclude that only (Co, Mg)O was present in the 900 and 1000°C calcined catalysts.

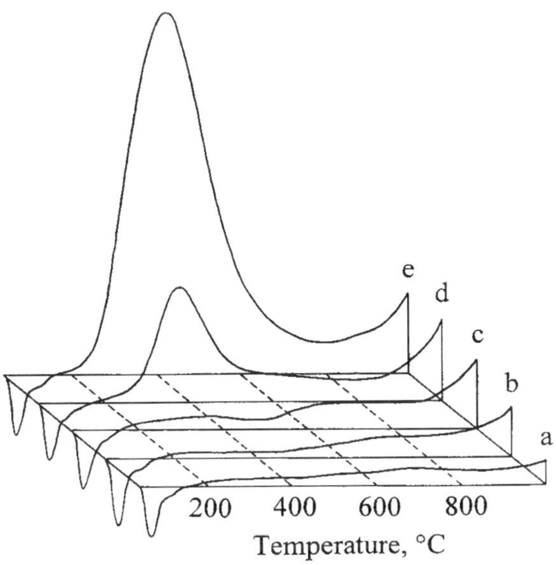

FIG. 7. TPR profiles of the 800°C calcined Co(O)/MgO catalysts with Co loadings of 6 wt% (a), 12 wt% (b), 24 wt% (c), 36 wt% (d), and 48 wt% (e).

The XRD patterns (Table 3) coupled with the TPR spectrum (Fig. 7e) indicated that Co_3O_4 and (Co, Mg)O were present in the 48 wt% Co(O)/MgO(800°C) catalyst. Among the 800°C calcined catalysts with different Co loadings, the TPR spectra for loadings of 4 and 12 wt% were similar to that for 24 wt%, and the TPR spectrum for the 36 wt% was similar to that for the 48 wt% (Fig. 7). Consequently, one can conclude that Co_3O_4 and (Co, Mg)O were present in the 36 and 48 wt% Co(O)/MgO(800°C) catalysts, but only (Co, Mg)O in the Co(O)/MgO(800°C) catalysts with Co loadings ≤24 wt%.

4. DISCUSSION

4.1. Influence of Support, Calcination Temperature, and Co Loading on the Species Present in the Co Catalysts

First, it should be noted that for the same calcination conditions and the same Co loading the species formed in Co catalysts are dependent on the nature of the support. MgO, CaO, SrO, BaO, and CoO all have a NaCl-type structure. However, only MgO has a lattice parameter and bond distance close to CoO (Table 4). As a result, a solid solution of CoO and MgO was formed in the MgO-supported catalyst

TABLE 4

Crystal Data for NaCl Type Metal Oxides (29)

Compound	a (Å)	A–B distance (Å)
MgO	4.2112	2.11
CaO	4.8105	2.40
SrO	5.1396	2.57
BaO	5.5391	2.77
CoO	4.2603	2.13

after calcination at 800°C, while a compound, in which Co was present as Co^{n+} ($n \geq 3$), was present as major phase in the other alkaline earth metal oxide supports (Table 3).

Second, one should emphasize that the species present in the MgO supported catalysts are strongly affected by the calcination temperature and Co loading. Because of the two oxidative states of Co, other species can be generated between cobalt and magnesium oxides besides the solid solution. Indeed, depending upon the calcination temperature and Co loading, XRD (Table 3) and TPR (Figs. 6 and 7) brought evidence for the existence of one, two, or three of the following compounds: Co_3O_4, $MgCo_2O_4$, and (Co, Mg)O. For the 24 wt% Co catalysts, Co_3O_4, Co_2MgO_4, and (Co, Mg)O were present in the 500°C calcined catalyst, Co_3O_4 and (Co, Mg)O in the 700°C calcined one, and solely (Co, Mg)O in the catalysts calcined at temperatures ≥800°C. For the 800°C calcined Co(O)/MgO catalysts, Co_3O_4 and (Co, Mg)O were present in the 36 and 48 wt% catalysts, but only (Co, Mg)O in the catalysts with Co loadings ≤24 wt%.

4.2. *Influence of the Species Present on the Reducibility of Co Catalysts*

The reducibility of a catalyst is directly related to the compounds formed during calcination. Due to the irreducibility of MgO and because the oxygen belonging to Co interacts also with Mg, it is difficult to reduce the CoO dispersed in the lattice of MgO. Indeed, the 800°C calcined 24 wt% Co(O)/MgO required the highest reduction temperature (Fig. 5) and its extent of reduction after the TPR was below 5%. For this reason, the exposed Co surface area over the Co/MgO was much lower than that over the Co/CaO (Table 1). Most likely, the low exposed Co surface area over the Co/SrO and Co/BaO (Table 1) was caused by the sintering of the metallic crystallites and their burying induced by the sintering of SrO and BaO.

The reducibility of the three Co-containing species identified in the MgO-supported catalysts decreased in the sequence $Co_3O_4 > Co_2MgO_4 >$ (Co, Mg)O. Consequently, under the conditions that favor the formation of a solid solution the percentage of cobalt that can be reduced to metal is expected to decrease. Indeed, the exposed metal surface area of the 24 wt% Co/MgO catalysts decreased monotonically as the calcination temperature was increased from 500 to 1000°C (Table 1).

4.3. *The Partial Oxidation of Methane to Synthesis Gas*

4.3.1. Effect of support on the reaction behavior of Co catalysts. Among the four alkaline earth metal oxides investigated, only MgO provided a high and stable activity during the period of study of 120 h; various extents of deactivation were observed over CaO, SrO, and BaO (Fig. 1). The carbon deposition and the metal sintering are the main causes for catalyst deactivation in the partial oxidation of methane. Sintering decreases not only the number of active sites, but also accelerates the carbon deposition, since large metal ensembles stimulate coke formation (30).

The size and morphology of metal clusters in the reduced catalysts is strongly affected by its reducibility. The higher the reducibility, the larger the size of metal particles is expected to be. When only a solid solution is present, hence for an MgO supported catalyst precalcined at $T_c \geq 800$°C, the size of the metal clusters is expected to be small. Parmaliana *et al.* brought transmission electron microscopy (TEM) evidence for a quasi-uniform distribution of small Ni particles ($d = 25$ Å) originated through the reduction of Ni^{2+} ions dispersed in the MgO matrix of a NiO–MgO solid solution (31). Being small, the clusters do not favor coke formation. Indeed, Table 2 indicates that the carbon deposition during the decomposition of pure CH_4 was much lower over the Co/MgO than over the Co/CaO and Co/SrO (while the carbon deposition was almost equally low over the Co/BaO, no metallic sites were present in this catalyst). Being extracted from the substrate, the generated metallic clusters remain partially embedded into the MgO lattice. Parmaliana *et al.* brought TEM evidence that such embedding is present in the reduced NiO–MgO solid solution systems (31). Being partially embedded into the MgO lattice, they are more resistant to sintering than the crystallites supported on the conventional supports (32). In conclusion, the high stability of the Co/MgO catalyst is related to the formation of a solid solution of CoO and MgO during calcination.

4.3.2. Effect of calcination temperature on the reaction behavior of Co/MgO catalysts. The calcination temperature has a significant influence on the stability of the Co/MgO catalysts. The 24 wt% Co catalysts precalcined at temperatures ≥800°C exhibited stable activities during the period of study, while some deactivation was observed over the catalysts precalcined at 500 and 700°C (Fig. 2). The XRD and TPR indicated that only the solid solution was present in the catalysts calcined at ≥800°C and that Co_3O_4 and $MgCo_2O_4$ were additionally present in the 500°C and Co_3O_4 in the 700°C calcined ones (Table 3). Being more reducible than the solid solution, they generated large metal clusters that stimulated carbon deposition. Indeed, as shown in Table 2, the carbon deposition during the CH_4 decomposition was much greater over the 500 and 700°C calcined catalysts than over those calcined at temperatures ≥800°C. In addition, due to the relatively weak interactions between these clusters and support, enhanced sintering occurred. Carbon deposition and sintering were responsible for deactivation.

On the other hand, with increasing calcination temperature from 500 to 1000°C, the initial conversion of methane in partial oxidation of methane decreased (Fig. 2). This reflects the trend of the exposed Co surface area (Table 1) and implies that the metallic sites are the active sites for the synthesis gas formation. The concentration of CoO near the surface decreased with increasing calcination temperature, because the higher the calcination temperature the deeper is the penetration of CoO into the MgO lattice. For this reason, fewer metallic sites were present on the surface of the catalysts precalcined at higher temperatures and the initial activity decreased with increasing calcination temperature. In addition, an activation period can be noted over the catalysts precalcined at temperatures $\geq$800°C (Fig. 2a), indicating that additional metallic sites were generated during reaction via the reduction of the solid solution by CH_4 and/or H_2. Consequently, the solid solution catalyst acts like a "reservoir" which can provide additional active metallic sites during reaction.

4.3.3. Effect of Co loading on the reaction behavior of Co/MgO catalysts. The reaction behavior of Co/MgO catalysts is affected by Co loading. As shown in Fig. 3, over the 6 wt% catalyst, the conversion of CH_4 was below 20% and only total oxidation of CH_4 occurred. However, the exposed Co surface area of this catalyst after reduction was not negligible (Table 1). This suggests that the initially available metallic sites were oxidized. When the Co loading was increased to 12 wt%, partial oxidation took place. With the increase in Co loading from 12 to 24 wt%, the conversion of CH_4 and the selectivity to CO increased somewhat, but remained almost the same with a further increase from 24 to 48 wt%. The almost equal methane conversions might have occurred either because the reaction was diffusion limited, or because the number of metallic sites (including those generated during reaction) became sufficiently large for thermodynamic equilibrium to be achieved. The results obtained regarding the effect of the amount of catalyst provide useful information in this direction. Indeed, Fig. 4 shows that, for the 1.5 mg of catalyst, the methane conversion was about 86%, with a CO selectivity of about 96%. As the amount of catalyst was increased, the CH_4 conversion and CO selectivity increased, but remained almost unchanged when the amount of catalyst became larger than 5.0 mg. Consequently, the reactions occurred essentially within a thin layer of about 5.0 mg (or even less) at the beginning of the catalyst bed and equilibrium was reached.

5. SUMMARY

The structural characteristics of the Co catalysts are strongly affected by the nature of the support, calcination temperature, and Co loading, and so are their reducibility and catalytic performance. Among the alkaline earth metal oxides (MgO, CaO, SrO, and BaO)-supported Co catalysts (precalcined at 800°C), only Co/MgO has proved to be a highly efficient and stable catalyst. It provided a CH_4 conversion of 91% and a CO selectivity of 98% at the high space velocity of 720,000 ml g^{-1} h^{-1}, without any deactivation for the duration of the experiment (120 h). A solid solution of CoO and MgO was identified in the 800°C calcined MgO-supported catalyst, which is responsible for its superior activity. Due to its low reducibility, the metal crystallites formed during the reduction of the solid solution were small and hence little carbon deposition occurred. Originating from a solid solution, the crystallites were partially embedded into the substrate and thus were resistant to sintering. The reaction behavior of Co/MgO catalysts was strongly affected by the calcination temperature and Co loading. Deactivation was noted for the low calcination temperatures of 500 and 700°C. Besides the solid solution, Co_3O_4 and $MgCo_2O_4$ were identified in the 500°C calcined catalyst and Co_3O_4 in the 700°C calcined one, compounds which can be more easily reduced than the solid solution. The relatively large metal particles, formed during the reduction of the more reducible Co-containing species (Co_3O_4 and $MgCo_2O_4$), stimulated carbon deposition. For calcination temperatures $\geq$800°C, the activity decreased with increasing calcination temperature. Since only the solid solution was present in the catalysts calcined at $\geq$800°C, the higher the calcination temperature the deeper was the penetration of CoO into the MgO lattice. For this reason, fewer metallic sites were present on the surface of the catalysts precalcined at higher temperatures.

REFERENCES

1. Prettre, M., Eichner, C., and Perrin, M., *Trans. Faraday Soc.* **43**, 335 (1946).
2. Hickman, D. A., and Schmidt, L. D., *J. Catal.* **138**, 267 (1992).
3. Bhattacharaya, A. K., Breach, J. A., Chand, S., *et al., Appl. Catal. A: Gen.* **80**, L1 (1992).
4. Hickman, D. A., and Schmidt, L. D., *Science* **259**, 343 (1993).
5. Torniainen, P. M., Chu, X., and Schmidt, L. D., *J. Catal.* **146**, 1 (1994).
6. Nakagawa, K., Ikenaga, N., Suzuki, T., *et al., Appl. Catal. A: Gen.* **169**, 281 (1998).
7. Fathi, M., Hofstad, K. H., Sperle, T., *et al., Catal. Today* **42**, 205 (1998).
8. Ruckenstein, E., and Wang, H. Y., *J. Catal.* **187**, 151 (1999).
9. Dissanayake, D., Rosynek, M. P., Kharas, K. C. C., and Lunsford, L. H., *J. Catal.* **132**, 117 (1991).
10. Vermeiren, W. J. M., Blomsma, E., and Jacobs, P. A., *Catal. Today* **13**, 427 (1992).
11. Choudhary, V. R., Rajput, A. M., and Prabhakar, B., *J. Catal.* **139**, 326 (1993).
12. Tang, S., Lin, J., and Tan, K. L., *Catal. Lett.* **51**, 169 (1998).
13. Tsipouriari, V. A., Zhang, Z., and Verykios, X. E., *J. Catal.* **179**, 283 (1998).
14. Provendier, H., Petit, C., Estournès, C., *et al., Appl. Catal. A: Gen.* **180**, 163 (1999).
15. Ruckenstein, E., and Hu, Y. H., *Appl. Catal. A: Gen.* **183**, 85 (1999).
16. Slagtern, Å., Swaan, H., Olsbye, U., *et al., Catal. Today* **46**, 107 (1998).
17. Santos, A., Menéndez, M., Monzón, A., *et al., J. Catal.* **158**, 83 (1996).
18. Sokolovskii, V. D., Coville, N. J., Parmaliana, A., *et al., Catal. Today*, **42**, 191 (1998).
19. Ashcroft, A. T., Cheetham, A. K., Foord, J. S., *et al., Nature* **344**, 319 (1990).
20. Jones, R. H., Ashcroft, A. T., Waller, D., *et al., Catal. Lett.* **8**, 169 (1991).
21. Slagtern, A., and Olsbye, U., *Appl. Catal. A: Gen.* **110**, 99 (1994).
22. Lago, R., Bini, G., Pena, M. A., and Fierro, J. L. G., *J. Catal.* **167**, 198 (1997).
23. Ruckenstein, E., and Hu, Y. H., *Appl. Catal. A: Gen.* **133**, 149 (1995).
24. Ruckenstein, E., and Hu, Y. H., *Ind. Eng. Chem. Res.* **37**, 1744 (1998).

25. Hagan, A. P., Lofthouse, M. G., Stone, F. S., and Trevethan, M. A., *in* "Preparation of Catalyst II" (B. Delmon, P. Grange, P. A. Jacobs, and G. Poncelet, Eds.), p. 417. Elsevier, Amsterdam, 1979.
26. Gazzoli, D., Occhiuzzi, M., Cimino, A., Cordischi, D., Minelli, G., and Pinzari, F., *J. Chem. Soc. Faraday Trans.* **92**, 4567 (1996).
27. Chung, K. S., and Massoth, F. E., *J. Catal.* **64**, 320 (1980).
28. Arnoldy, P., and Moulijn, J. A., *J. Catal.* **93**, 38 (1985).
29. Náray-Szabó, I., "Inorganic Crystal Chemistry." Akadémiai Kiadó, Budapest, 1969.
30. Rostrup-Nielsen, J. R., *in* "Catalysis Science and Technology" (J. R. Anderson and M. Boudart, Eds.), Vol. 5, pp. 1–118. Springer, Berlin, 1984.
31. Parmaliana, A., Arena, F., Frusteri, F., *et al., J. Catal.* **141**, 34 (1993).
32. Highfield, J. G., Bossi, A., and Stone, F. S., *in* "Preparation of Catalyst III" (G. Poncelet, P. Grange, and P. A. Jacobs, Eds.), p. 181. Elsevier, Amsterdam, 1983.

Transient Kinetic Studies of Partial Oxidation of CH_4

Y. H. Hu[1] and E. Ruckenstein[2]

Department of Chemical Engineering, State University of New York at Buffalo, Amherst, New York 14260

Received April 14, 1995; revised August 22, 1995; accepted August 29, 1995

The kinetics of methane oxidation over Ni/La_2O_3 at low temperatures (450–700°C), atmospheric pressure, and a residence time of 3.8 ms was studied by the pulse-transient analysis method. The transient response curves revealed that CO_2 generation is delayed in comparison to that of CO, hence that CO is the primary product and CO_2 is only subsequently generated from CO. The generation of C and O species over the surface of the catalyst during CH_4/O_2 pulses and the desorption of CO were much faster than the generation of CO; hence the reaction between the surface carbon species and the surface O species constitutes the rate-controlling step. High CH_4 conversions and almost 100% CO selectivities were obtained for small pulses of 45 μl of CH_4/O_2 (2/1) over freshly reduced catalysts, with He as the carrier gas; over the unreduced catalyst the CH_4 conversion and CO selectivity were low. The experimental data also indicated that the CH_4 conversion and CO selectivity decreased when the surface of the reduced catalyst was oxidized. The results demonstrated that Ni atoms, obtained by the reduction of the catalyst, constitute the active sites in CH_4 partial oxidation. One concludes that the CH_4 conversion depends on its dissociation, whereas the CO selectivity is largely dependent on the strength of O species binding to the catalyst. Over the reduced catalyst, the O species oxidize with difficulty, because of their strong binding to Ni atoms, CO to CO_2, whereas over the unreduced catalyst (which contains Ni oxide), the oxygen species easily oxidize, because they are weakly adsorbed, CO to CO_2.

1. INTRODUCTION

The conversion of CH_4 to interesting compounds is extremely meaningful. In the past decades (1, 2), attempts have been made to achieve this via the oxidative coupling of CH_4 to ethane and ethylene. Their usefulness was, however, limited by the low conversion of CH_4 and/or the low selectivity for hydrocarbons. A conventional route for CH_4 conversion is through the syngas intermediate (CO/H_2), which can be easily transformed to either methanol, over the Cu/Zn/Al catalysts, or hydrocarbons, via the Fischer–Tropsch reaction, over Fe catalysts (3, 4). The dominant commercial method employed to produce the synthesis gas is, however, the steam reforming of hydrocarbons, which produces from methane a H_2/CO ratio of 3/1 (5, 6).

$$CH_4 + H_2O \rightarrow CO + 3H_2, \quad \Delta H = +225.4 \text{ kJ/mol.} \quad [1]$$

This reaction is endothermic, and typical reaction conditions for a Ni/Al_2O_3 catalyst are 15 to 30 atm and 850 to 900°C.

In recent years, there has been renewed interest in the catalytic oxidation of CH_4 to syngas (7–20), via the reaction

$$CH_4 + \tfrac{1}{2}O_2 \rightarrow CO + 2H_2, \quad \Delta H = -22.6 \text{ kJ/mol.} \quad [2]$$

This reaction is mildly exothermic and provides a molar ratio H_2/CO of 2/1, which can be directly used as feed for the Fischer–Tropsch reaction or methanol synthesis. Compared to the steam-reforming reaction, the direct oxidation reaction is more energy efficient, is more selective, and can produce a better H_2/CO ratio. Very recently, steady state procedures, with high selectivities for CO and H_2 and excellent methane conversions, over a series of catalysts have been reported (12–17).

The thermodynamic calculation of Ashcroft *et al.* (7) has shown that formation of CO is more favorable than that of CO_2 for $T > 693$°C, for a CH_4/O_2 ratio of 2/1, whereas Choudhary *et al.* (13) reported that at high space velocities and temperatures below 700°C, methane conversion as well as CO and H_2 selectivities exceeded the thermodynamic predictions. However, Lunsford and co-workers (9), measuring the temperature with an optical pyrometer, pointed out that the observations of Choudhary's group were due to the presence of hot spots with temperatures 300°C higher than those measured by the latter researchers. Although some good catalysts were found (12–17), the reaction mechanism and the active sites responsible for the reaction were not identified.

In 1946, Prettre *et al.* (21), studying CH_4 partial oxidation over a nickel catalyst, concluded that the overall direct oxidation involves an initially strongly exothermic oxida-

[1] Permanent address: Department of Chemistry, Xiamen University, Xiamen 361005, People's Republic of China.

[2] To whom correspondence should be addressed.

This paper, which was originally published in **Journal of Catalysis**: 158, 260–266 (1996), is republished with permission.

tion of CH_4 to CO_2 and H_2O, followed by the endothermic reactions [1] and

$$CH_4 + CO_2 = 2CO + 2H_2, \quad \Delta H_{1000\,K} = 260.5 \text{ kJ/mol.} \tag{3}$$

The above opinion was also supported by some recent work (7, 9, 10). Hickman and Schmidt (8, 11, 18) and Choudhary *et al.* (12–14) suggested, however, that the oxidation of methane to synthesis gas follows first the pyrolysis mechanism $CH_4 \rightarrow C + 4H$, after which H_2 desorbs and C is oxidized to CO without the preformation of CO_2.

Regarding the active sites, Dissanayake *et al.* (10) found three regions in the used Ni/Al_2O_3 catalyst and concluded that CH_4 is oxidized to CO_2 and H_2O in the regions containing $NiAl_2O_4$ and $NiO + Al_2O_3$, and the steam and/or CO_2 reforming of CH_4 to CO and H_2 occur in the region containing Ni/Al_2O_3.

In the present paper, we used pulses and the transient analysis of products to investigate the CH_4/O_2 reaction over Ni/La_2O_3 catalysts. The results obtained allowed us to identify the active sites and to examine the reaction mechanism.

2. EXPERIMENTAL

2.1. Catalyst Preparation

The Ni/La_2O_3 catalyst containing 20 wt% nickel was prepared by impregnating La_2O_3 powder (Aldrich) with a nickel nitrate (Alfa) solution. The paste generated was dried at 110°C in air, decomposed at 500°C, and calcined at 800°C under O_2 (20 ml/min) for 4 h in a ceramic tube (6 mm inside diameter). The catalyst was reduced *in situ* at 600°C with H_2 (20 ml/min) for 2 h in the reactor described below before it was used for the pulse reaction and transient response. The Ni atoms (mmol/g catalyst) exposed to the chemical atmosphere by the reduced catalyst were determined by the pulse chemisorption of CO at room temperature, assuming 1/1 stoichiometry, using a pulse of 45 μl, ultra-high-purity helium as carrier gas and a mass spectrometer (HP Quadrupole, 5971 Series Mass Selective Detector) as detector.

2.2. Pulse Reaction and Dynamic Response

A quartz tube (2-mm inside diameter) was used as reactor. The catalyst powder (weight, 0.02 g; bed height, 2 mm) was held on quartz wool. The reactant, CH_4/O_2 (2/1), contained 10% argon as an internal standard, and the pulse volume was 45 μl. Ultra-high-purity helium was used as carrier gas, with a flow rate of 100 ml/min (GHSV, 300,000 $cm^3\,g^{-1}\,h^{-1}$). The analysis of gases during transients was carried out with an on-line mass spectrometer (HP Quadrupole, 5971 Series Mass Selective Detector) equipped with a fast response inlet capillary system. The calibration of the mass spectrometer was carried out with prepared mixtures of known composition. The cracking coefficients of methane, O_2, CO, and CO_2 have been determined and used to calculate their concentrations. The CO selectivity is defined as $CO/(CO + CO_2) \times 100\%$.

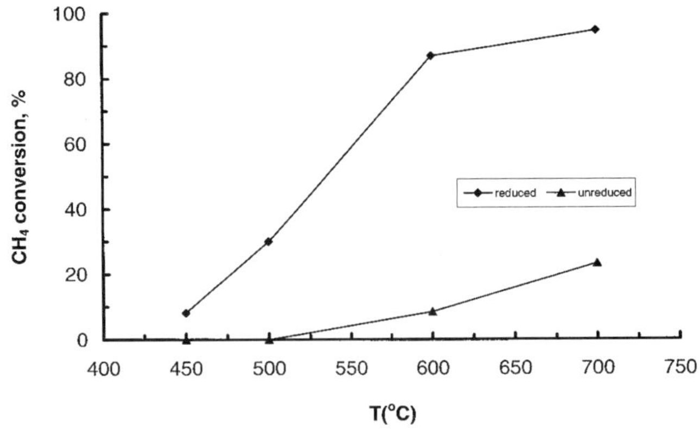

FIG. 1. Temperature dependence of CH_4 conversion for CH_4/O_2 pulses over 20% Ni/La_2O_3 catalyst.

3. RESULTS

3.1. CH_4 Conversion and CO Selectivity under Unsteady State Conditions

Figures 1 and 2 show that for CH_4/O_2 pulses over freshly reduced NiO/La_2O_3 catalysts, the CH_4 conversion decreases from 95 to 8% when the temperature changes from 700 to 450°C; the CO selectivities are, however, almost constant and nearly 100%. The unreduced catalyst, which contains Ni oxide, has much lower CH_4 conversion and CO selectivity than the freshly reduced catalyst. One can

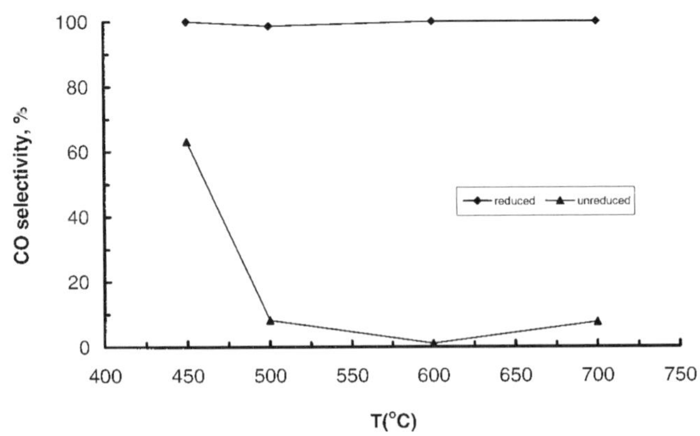

FIG. 2. Temperature dependence of the CO selectivity for CH_4/O_2 pulses over 20% Ni/La_2O_3 catalyst.

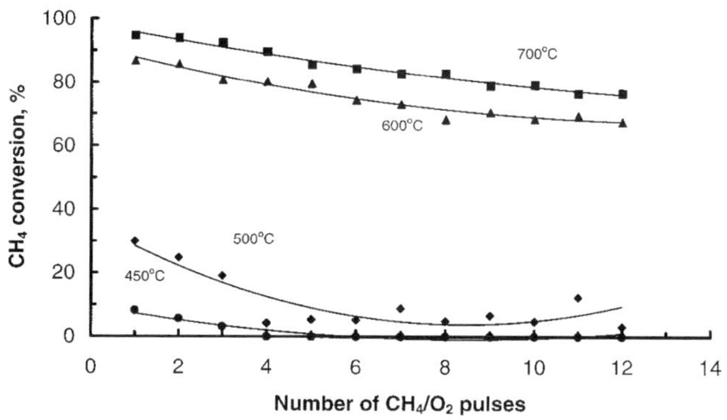

FIG. 3. Relationship between Ch_4 conversion and the number of CH_4/O_2 pulses over the reduced 20% Ni/La_2O_3 catalyst.

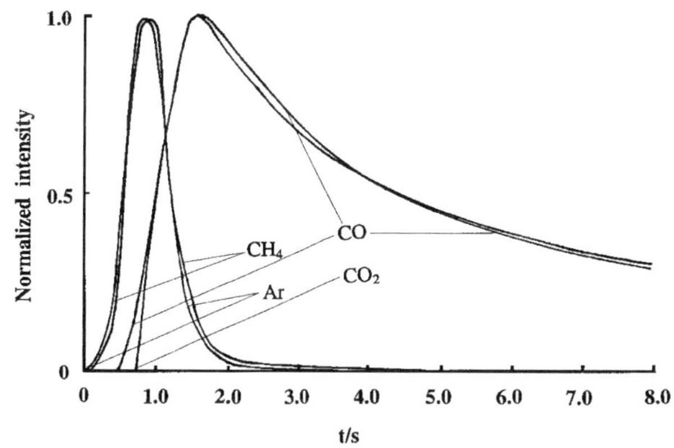

FIG. 5. Transient responses of the products in a CH_4/O_2 pulse at 500°C over the reduced 20% Ni/La_2O_3 catalyst.

see from Figs. 3 and 4 that, over an initially reduced catalyst, the CH_4 conversion and CO selectivity decrease as the number of CH_4/O_2 pulses increases, implying that the reduced catalyst is oxidized gradually with an increasing number of CH_4/O_2 pulses.

3.2. Transient MS Analysis of CH_4/O_2 Pulses

Figure 5 shows the dynamic responses of CH_4, CO, and CO_2 for a CH_4/O_2 (2/1) pulse at 500°C over a freshly reduced Ni/La_2O_3 catalyst. One can see that the initial time at which CO is detected is shorter than that for CO_2 by about 0.2 s. The experiment was repeated several times and was reproducible with an absolute deviation from the average of 28%. This indicates that either the CO generation occurs earlier or the CO desorption is faster than that of CO_2. However, the dynamic curves for pure pulses of CO and CO_2 (Fig. 6) show that the CO and CO_2 have very similar adsorption and desorption response curves. Consequently, the delay of CO_2 generation compared to that of CO is not due to desorption, but to the earlier generation of CO. Figures 5 and 6 reveal that the CO and CO_2 response curves for a CH_4/O_2 pulse have much longer times for the peaks and much longer tails than the CH_4 response curve for a CH_4/O_2 pulse (Fig. 5) and the CO response curve for the pure CO pulse (Fig. 6). This indicates that the CH_4 adsorption and desorption and the CO and CO_2 desorptions are fast compared to CO and CO_2 generation. Figures 5, 7, and 8 show that the times for the peak and tail of CO response curves for CH_4/O_2 pulses become shorter as the temperature increases. This happens because the increase in temperature increases the rate of reaction.

3.3. The Transient Analysis of Surface Species

From the dynamic response curves of CH_4/O_2 pulses, one can analyze the dynamic behavior of surface species. In the oxidation of CH_4 to syngas, the surface species

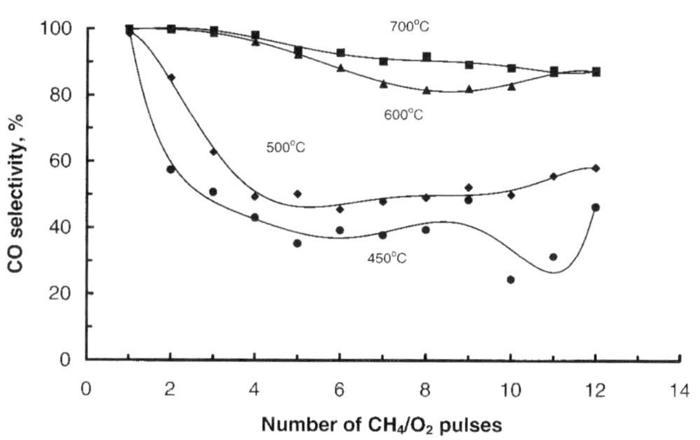

FIG. 4. Relationship between CO selectivity and the number of CH_4/O_2 pulses over the reduced 20% Ni/La_2O_3 catalyst.

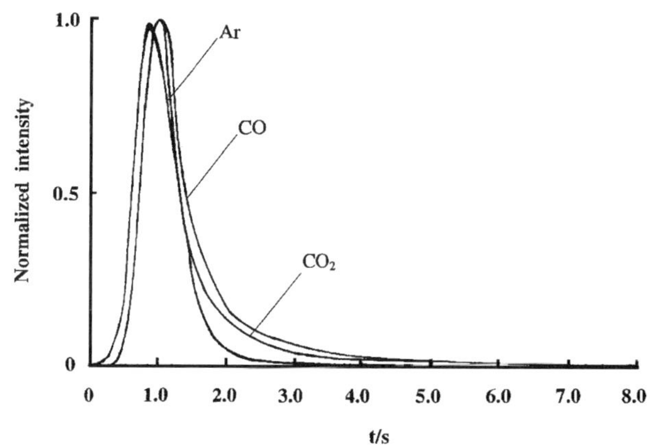

FIG. 6. Transient responses of CO and CO_2 in the pure pulses of CO and CO_2 at 500°C over the reduced 20% Ni/La_2O_3 catalyst.

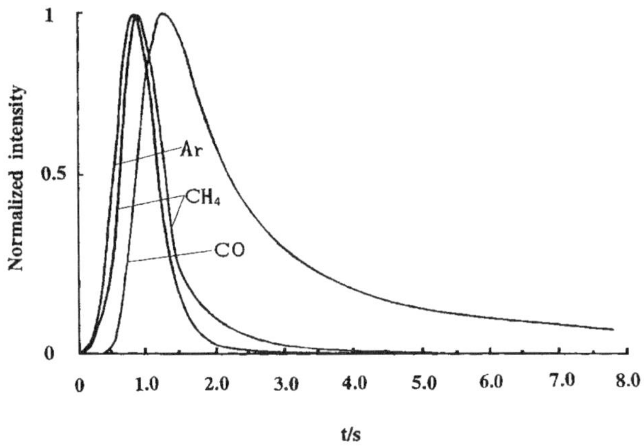

FIG. 7. Transient responses of the products in a CH_4/O_2 pulse at 600°C over the reduced 20% Ni/La_2O_3 catalyst.

TABLE 1

Activation Energies of Elementary Reactions for Methane Decomposition over Ni(111) (23)

Reaction step	Activation energy (kcal/mol)
$CH_{4(s)} \rightarrow CH_{3(s)} + H_{(s)}$	14
$CH_{3(s)} \rightarrow CH_{2(s)} + H_{(s)}$	24
$CH_{2(s)} \rightarrow CH_{(s)} + H_{(s)}$	23
$CH_{(s)} \rightarrow C_{(s)} + H_{(s)}$	5
$C_{(s)} + O_s \rightarrow CO_{(s)}$	35

containing C or O can be calculated using the expressions

$$M_C(t) = M_{CH_4}(f) - M_{CH_4}(t) - M_{CO}(t) - M_{CO_2}(t)$$

$$M_O(t) = 2M_{O_2}(f) - 2M_{O_2}(t) - M_{CO}(t) - 2M_{CO_2}(t) - M_{H_2O}(t)$$

$$\theta_C(t) = \frac{M_C(t)}{M_S}, \quad \theta_O(t) = \frac{M_O(t)}{M_S},$$

where $M_C(t)$ is the surface C species (moles) at time t; $M_O(t)$ is the surface O species (moles) at time t; $M_i(t)$ is the exit amount (moles) in the gas phase at time t of compound i, where i = CO, CO_2, CH_4, O_2, or H_2O; $M_j(f)$ is the entrance amount (moles) of compound j, where j = CH_4 or O_2; M_S is the number of Ni atoms (moles) on the catalyst exposed to the chemical atmosphere; θ_C is the fraction covered by C species; and θ_O is the fraction of covered by O species.

Since the desorption of CO and CO_2 are rapid in comparison with their generation and water was not detected, it is likely that the surface is mostly occupied by C, CH_x, and O. Since the activation energies for the decompositions $CH_x \rightarrow CH_{x-1}$ + H of the CH_x adsorbed species are small compared to that for the oxidation of the C adsorbed species (see Table 1), one can consider that θ_C consists primarily of C species. Since water was not detected and CO and CO_2 desorb fast, θ_O consists primarily of O species. The dynamic curves of C species and O species on the catalyst surface (Figs. 9–11) were calculated using the above expressions and Figs. 5, 7, and 8. Figure 9 shows that at 500°C, O species have a coverage greater than unity. This indicates that the O species are adsorbed not only on the top layer, but also in the sublayer of the catalyst. At 500°C, the C coverage is much smaller than that of O species. When the reaction was terminated, an O coverage of 1.2 still remained. At 600°C (Fig. 10), both the C and O coverages were greater than unity for a period of time; this indicates that the C species can also adsorb in the sublayer. When the reaction was terminated, there was still an O coverage of 0.25. The results for 700°C are pre-

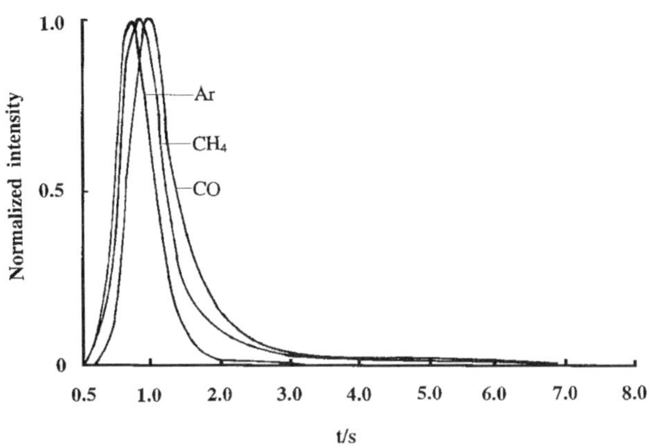

FIG. 8. Transient responses of the products in a CH_4/O_2 pulse at 700°C over the reduced 20% Ni/La_2O_3 catalyst.

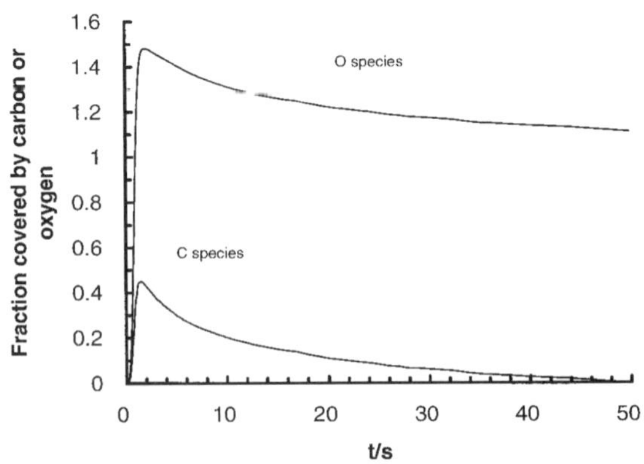

FIG. 9. Transient responses of surface and O species obtained in a pulse of CH_4/O_2 at 500°C over the reduced 20% Ni/La_2O_3 catalyst.

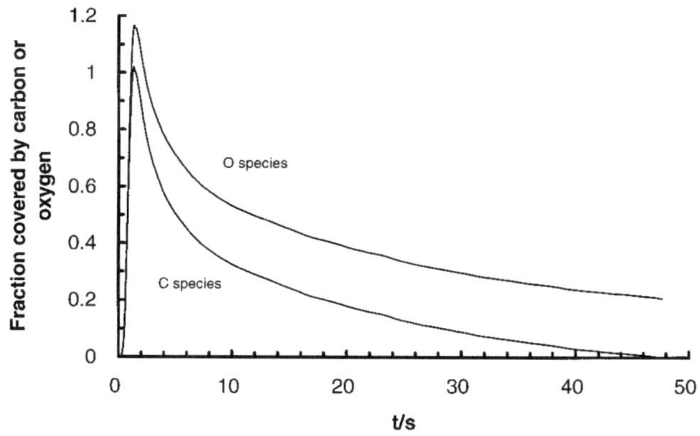

FIG. 10. Transient responses of surface C and O species obtained in a pulse of CH_4/O_2 at 600°C over the reduced 20% Ni/La_2O_3 catalyst.

TABLE 2

Thermodynamic Equilibrium Data for Methane Oxidation (9)[a]

Temperature (°C)	CH_4 conversion (%)	CO selectivity (%)
300	26	0.2
400	29	3
500	38	20
600	55	59
700	78	88
800	92	97
900	97	99

[a] Initial CH_4/O_2 = 2/1, at a pressure of 1 atm.

sented in Fig. 11, which shows a coverage of 0.1 for O when the reaction was terminated.

4. DISCUSSION

4.1. Active Sites for CH_4 Partial Oxidation

The freshly reduced catalyst has a much higher CH_4 conversion than the unreduced catalyst (Fig. 1). This demonstrates that the reduced catalyst activates CH_4 much more easily than Ni oxide does. This observation is consistent with the ESCA results obtained for CH_4 over Ni crystal surfaces (22). The dynamic curves of C species also demonstrate that surface C species are easily formed over the freshly reduced catalyst even at the low temperature of 500°C. The dynamic curves of O species reveal that surface O species are also easily generated over a freshly reduced catalyst, and that, after a pulse of CH_4/O_2 has reacted completely, O species are still present on the catalyst, hence that the reduced catalyst was partially oxidized. This explains why the CH_4 conversion decreases as the number of CH_4/O_2 pulses increases. Over the freshly reduced catayst, the CO selectivity was near 100%, whereas over the unreduced catalyst, the CO selectivity was very low even at 700°C. The CO selectivity decreases over the freshly reduced catalyst as the number of CH_4/O_2 pulses increases. These results indicate that the reduced catalyst favors the CO but not the CO_2 generation. Over the reduced catalyst, O species are easily formed; they are, however, strongly bound to the metal. Over the unreduced catalyst, the surface O species have strong oxidation ability (hence they easily oxidize $CO_{(S)}$ to $CO_{2(S)}$ because they are weakly bound to Ni oxide. The low activity of the unreduced catalyst occurs because the metal oxide does not easily dissociate CH_4 to surface C species. One can, therefore, conclude that in the oxidation reaction of CH_4, the activity is dependent on CH_4 dissociation and the CO selectivity is dependent on the O binding ability of the catalyst. It is also reasonable to consider that the Ni atoms of the reduced catalyst constitute the active sites, at least at low temperature (<700°C).

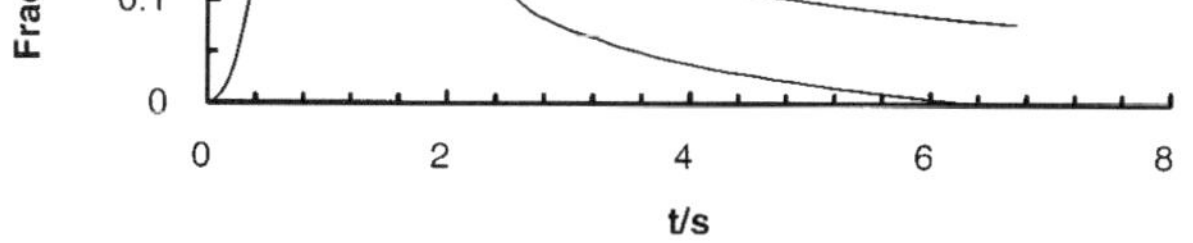

FIG. 11. Transient responses of surface C and O species obtained in a pulse of CH_4/O_2 at 700°C over the reduced 20% Ni/La_2O_3 catalyst.

In the present work, small pulses of 45 μl of $CH_4/O_2/$Ar and very short residence times (3.8 ms) were employed, in order to minimize the oxidation of the reduced catalyst during reaction. When, however, 900-μl pulses of CH_4/O_2 were employed over a freshly reduced catalyst (24), we found that at 600°C the CH_4 conversion and the CO selectivity decreased to 21 and 5%, respectively. This observation further confirms that the CO selectivity is mainly dependent on the state of the surface Ni present in the catalyst.

The near 100% CO selectivity and the high CH_4 conversion exceed the values predicted by thermodynamic equilibrium (Table 2) at 450–700°C, indicating that the reaction was controlled under the present conditions by kinetics.

4.2. Mechanism of CH_4 Oxidation

The dynamic curves of Fig. 5 show that CO_2 starts to be generated with a delay of about 0.2 s compared to CO.

This demonstrates that CO is the direct product and that CO_2 is formed subsequently from CO. This proves that the partial oxidation reaction of CH_4 follows the pyrolysis mechanism of CH_4 (8, 11, 18). A similar observations was made recently, after our paper was submitted, by Mallens *et al.* for the partial oxidation of methane over platinum (25). These authors used the high-vacuum technique TAP (temporal analysis of products) and concluded that "in the presence of both methane and dioxygen at a stoichiometric feed ratio the dominant pathways are the direct formation of CO and H_2 followed by their consecutive oxidation."

The dynamic curves of Figs. 9–11 show that surface C or O species are formed easily over the freshly reduced catalyst. The transient response curves (Figs. 5 and 6) indicate that the desorption of CO is much faster than its generation. One can, therefore, conclude that the formation of $CO_{(s)}$ from $C_{(s)}$ and $O_{(s)}$ is the rate-controlling step in the oxidation of CH_4 to CO/H_2. A possible mechanism is as follows:

$$O_{2(g)} + \text{site} \rightarrow 2O_{(s)} \qquad [1]$$
$$CH_{4(g)} + \text{site} \rightarrow \bullet\bullet\bullet \rightarrow C_{(s)} + 4H_{(s)} \qquad [2]$$
$$C_{(s)} + O_{(s)} \rightarrow CO_{(s)}\ (\text{RDS}) \qquad [3]$$
$$CO_{(s)} \rightarrow CO_{(g)} \qquad [4]$$
$$CO_{(s)} + O_{(s)} \rightarrow CO_{2(s)} \qquad [5]$$
$$CO2_{(s)} \rightarrow CO_{2(g)} \qquad [6]$$
$$H_{(s)} + H_{(s)} \rightarrow H_{2(s)} \qquad [7]$$
$$O_{(s)} + 2H_{(s)} \rightarrow H_2O_{(s)}. \qquad [8]$$

4.3. *Kinetics of CH_4 Oxidation*

The very high flow rate of the carrier gas and the very short bed length of the catalyst make the residence time 2/1000 of the pulse width. Hence, the catalytic reactor can be considered a differential reactor. According to the above mechanism, the oxidation of surface C (step 3) is the rate-limiting step. Since the CO_2 selectivity is very low (<2%), the CO generation rate can be taken as equal to

$$\frac{dP_{CO}}{dt} = k_3\theta_C\theta_O, \qquad [\text{E-1}]$$

where P_{CO} is the partial pressure of CO at the exit, θ_C and θ_O are the coverages by C and O species, respectively, and K_3 is the rate constant of step 3. Equation [E-1] can be rewritten as

$$\ln\left(\frac{dP_{CO}}{dt}\right) = \ln(k_3) + \ln(\theta_C\theta_O). \qquad [\text{E-2}]$$

From the dynamic curves of Figs. 9–11, the values of $\ln(dP_{CO}/dt)$ and $\ln(\theta_C\theta_O)$ were obtained and plotted in Fig. 12. One can see that, as predicted by Eq. [E-2], the curves are linear. The linear equations representing the data are given in Table. 3. The slopes are near 1, indicating that the experimental data are consistent with the kinetic equation [E-2]. Employing the Arrhenius equation, an activation energy of 30.5 kcal/mol was obtained from the data. This activation energy is consistent with the theoretical value of 35 kcal/mol obtained for the C(s) + O(s) reaction over Ni(111) (23). This result additionally supports the observation that the oxidation of surface C species constitutes the rate-limiting step.

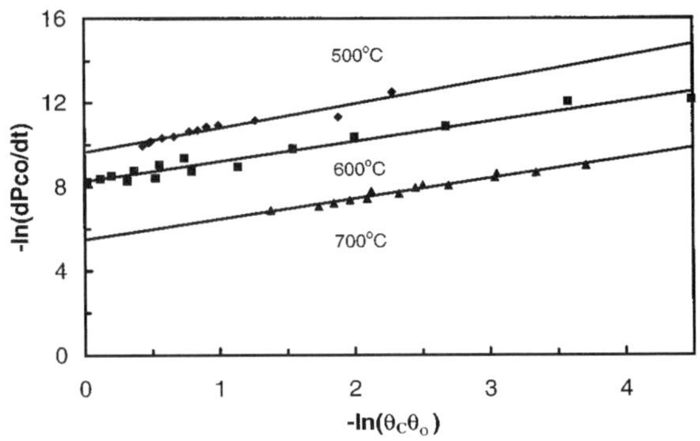

FIG. 12. Relationship between $-\ln(dP_{CO}/dt)$ and $-\ln(\theta_C\theta_O)$ over the reduced 20% Ni/La_2O_3 catalyst.

5. CONCLUSION

From the above dynamic analysis and kinetic calculations one can conclude that CO is the direct product and that CO_2 is formed subsequently from CO, i.e., that the reaction follows the pyrolysis mechanism. the oxidation of surface C formed via the CH_4 dissociation constitutes the rate-limiting step. The high CH_4 conversion over the reduced catalyst is due to the dissociation of CH_4, and the very high selectivity for CO is due to the weak oxidation ability of O species which are bound strongly to the cata-

TABLE 3

The Relationship between $\ln(dP_{CO}/dt$ and $\ln(\theta_C\theta_O)$

Temperature (°C)	Equation
500	$\ln\left(\frac{dP_{CO}}{dt}\right) = -9.64 + 1.14\ln(\theta_C\theta_O)$
600	$\ln\left(\frac{dP_{CO}}{dt}\right) = -8.20 + 0.97\ln(\theta_C\theta_O)$
700	$\ln\left(\frac{dP_{CO}}{dt}\right) = -5.50 + 0.97\ln(\theta_C\theta_O)$

lyst. Over the unreduced catalyst, which contains Ni oxide, the low dissociation of CH_4 and the strong oxidation ability of O species, which are weakly adsorbed by Ni oxide, result in low activities and CO selectivities.

REFERENCES

1. Lunsford, J. H., *Catal. Today* **6,** 235 (1990).
2. Anderson, J. R., *Appl. Catal.* **47,** 177 (1989).
3. Killer, G. E., *Adv. Catal.* **31,** 243 (1982).
4. Hindermann, J. P., Hutchings, G. J., and Kiennemann, A., *Catal. Rev.-Sci. Eng.* **35,** 1 (1993).
5. Trimm, D. L., *Catal. Rev.-Sci. Eng* **155,** 16 (1977).
6. Rostrup-Nielsen, R., *in* "Catalysis: Science and Technology" (J. R. Anderson and M. Boudart, Eds.), Vol. 5, Springer-Verlag, New York, 1984.
7. Ashcroft, A. T., Cheetham, A. K., Ford, J. S., Green, M. L. H., Grey, C. P., Murrell, A. J., and Vernon, P. D. F., *Nature* **344,** 319 (1990).
8. Hickman, D. A., and Schmidt, L. D., *Science* **259,** 343 (1993).
9. Dissanayake, D., Rosynek, M. P., and Lunsford, J. H., *J. Phys. Chem.* **97,** 3644 (1993).
10. Dissanayake, D., Rosynek, M. P., Kharas, K. C. C., and Lunsford, J. H., *J. Catal.* **132,** 117 (1991).
11. Hickman, D. A., and Schmidt, L. D., *J. Catal.* **138,** 267 (1992).
12. Choudhary, V. R., Rajput, A. M., and Prabhaker, B. J., *Catal. Lett.* **15,** 363 (1992).
13. Choudhary, V. R., Rajput, A. M., and Prabhakar, B. J., *J. Catal.* **139,** 326 (1993).
14. Choudhary, V. R., Rane, V. H., and Rajput, A. M., *Catal. Lett.* **22,** 289 (1993).
15. Matsumura, Y., and Moffat, J. B., *Catal. Lett.* **24,** 59 (1994).
16. Poirier, M. G., and Trudel, J., *Catal. Lett.* **21,** 99 (1993).
17. Hayakawa, T., Andersen, A. G., Shimizu, M., Suzuki, K., and Takehira, K., *Catal. Lett.* **22,** 307 (1993).
18. Hickman, D. A., and Schmidt, L. D., *AIChE J.* **39,** 1164 (1993).
19. Au, C. T., Hu, Y. H., and Wan, H. L., *Catal. Lett.* **27,** 199 (1994).
20. Choudhary, V. R., Rajput, A. M., and Rane, V. H., *Catal. Lett.* **22,** 289 (1993).
21. Prette, M., Eichner, C., and Perrin, M., *Trans. Faraday Soc.* **43,** 335 (1964).
22. Campbell, R. A., Lenz, J. P., and Goodman, D. W., *Catal. Lett.* **17,** 39 (1993).
23. Shustorovich, E., *Adv. Catal.* **37,** 101 (1990).
24. Hu, H. Y., and Ruckenstein, E., *Catal. Lett.* **34,** 41 (1995).
25. Mallens, E. P. J., Hoebink, J. H. B. J., and Marin, G. B., *Catal. Lett.* **33,** 291 (1995).

Isotopic GCMS Study of the Mechanism of Methane Partial Oxidation To Synthesis Gas

Yun Hang Hu† **and Eli Ruckenstein***

Department of Chemical Engineering, State University of New York at Buffalo, Amherst, New York 14260

Received: September 17, 1998

An isotopic pulse is combined with a gas chromatography−mass spectrum to investigate the mechanism of the partial oxidation of methane. For a pulse reaction of $CH_4/CD_4/O_2$ (1/1/1) at 600 °C over a reduced NiO/SiO_2 catalyst, besides CH_4 and CD_4, large amounts of CHD_3, CH_2D_2, and CH_3D were detected in the exit gas. This indicates that CH, CH_2, and CH_3 species are formed during the reaction, i.e., that CH_4 is activated via its dissociation before its oxidation. The amount of methane involved in the isotopic exchange reaction was larger than that converted to CO and CO_2. Consequently, the dissociation of methane is not rate determining. Over an unreduced NiO/SiO_2 catalyst, CH_4 and CD_4, without any CH_xD_y ($x + y = 4$), were detected in the exit gas, indicating that methane reacts directly with oxygen without its predissociation. For the reaction of CH_4/CD_4(1/1) with the lattice oxygen of the catalyst, CH_4 and CD_4, without any CH_xD_y ($x + y = 4$), were detected in the exit gas, indicating that methane reacts directly with the lattice oxygen without predissociation. Moreover, the $CH_4/^{16}O_2$ reaction over an unreduced NiO^{18} catalyst revealed that the lattice oxygen of the catalyst participates in the reaction of CH_4/O_2.

1. Introduction

Although methane is the most abundant of alkanes, it is also the least reactive. Therefore, its selective conversion to more useful chemical compounds is an important and challenging process. In the last 20 years, the attempts of direct conversion of CH_4 have focused on the oxidative coupling to ethylene and ethane[1,2] and on the oxygenation to methanol and formaldehyde.[3,4] Unfortunately, at the high temperatures (>700 °C) needed for high conversion, the formation of CO_2 is a highly favorable reaction ($\Delta G < -800$ kJ/mol). Therefore, renewed interest in methane transformation to synthesis gas,[5−15] which can be used to produce a variety of chemicals by the Fischer−Tropsch and methanol syntheses,[16,17] has arisen. Currently, the synthesis gas is commercially prepared by steam reforming. The process provides, however, low conversion or selectivity and too large a H_2/CO ratio for the above syntheses.

The catalytic partial oxidation offers the greatest potential for a fast, efficient, and economical conversion of methane to synthesis gas, due to the high conversion of methane, high selectivity, suitable H_2/CO ratio, and very short residence time. The investigations regarding the partial oxidation of methane, its mechanism, and the characterization of the catalyst have become one of the most active topics in catalysis.[11,18−32]

Two mechanisms have been suggested for the partial catalytic oxidation of methane to synthesis gas: (i) the combustion−reforming mechanism, in which CO_2 and H_2O are the primary products, and CO is a result of their reaction with CH_4; (ii) the pyrolysis mechanism, in which CO is produced via the pyrolysis products of CH_4 without the preformation of CO_2. Prettre et al.,[32] who were the first to study the CH_4 partial oxidation over a nickel catalyst, concluded that the overall oxidation involves two steps, an initial exothermic oxidation of CH_4 to CO_2 and H_2O, followed by the endothermic reactions $CH_4 + CO_2 \rightarrow 2CO + 2H_2$ and $CH_4 + H_2O \rightarrow CO + 3H_2$, i.e., the combustion−reforming mechanism. Some recent investigations brought support to this mechanism.[6,7,21−23] The other mechanism involves the pyrolysis $CH_4 \rightarrow CH_x + (4 - x)H$, followed by the oxidation of C species to CO.[11,23−29] Recently,[31] we demonstrated that the partial oxidation of methane follows different mechanisms over unreduced and reduced Ni-based catalysts. Over the unreduced NiO/SiO_2 catalyst, the reaction occurs via the Eley−Rideal mechanism; that is CH_4 in the gas phase reacts with O_2 in the adsorbed state and leads to CO_2. In contrast, over the reduced NiO/SiO_2 catalyst, the reaction takes place via a Langmuir−Hinshelwood mechanism; that is CH_4 and O_2 react in the adsorbed states via the pyrolysis mechanism and lead to CO and H_2. Nevertheless, no direct evidence appears to be available regarding whether methane is dissociated before its oxidation.

In this paper, we combined an isotope pulse with a gas chromatography−mass spectrum (GC−MS) to gather (i) evidence regarding whether methane is dissociated before its oxidation and (ii) information about the rate-determining step during the reaction between CH_4 and O_2 over unreduced and reduced NiO/SiO_2 catalysts.

2. Experimental Section

2.1. Catalyst Preparation. The NiO/SiO_2 (containing 13.6 wt % Ni) catalyst was prepared by impregnating SiO_2 (Aldrich) with an aqueous solution of nickel nitrate (Alfa). The paste thus obtained was dried at room temperature in air and then decomposed and calcined at 800 °C in air for 1.5 h. The BET surface area and average diameter of the pores, which were determined via nitrogen adsorption by using a Micromeritics ASAP 2000 instrument, were 362 m^2/g and 52 Å, respectively.

The $Ni^{18}O/SiO_2$ catalyst was prepared by reducing the $Ni^{16}O/SiO_2$ at 600 °C for 45 min with H_2 and reoxidizing with pulses of $^{18}O_2$ until no O_2 was consumed.

* To whom correspondence should be addressed.
† Current address: Strategic Research Center, Mobil Technology Corporation, Dallas, TX.

This paper, which was originally published in **Journal Physical Chemistry A**: 102, 10568–10571 (1998), is republished with permission.

2.2. GC–MS Analysis of a Pulse Reaction. The experiments were performed mostly at 600 °C under a pressure of 20 psi. The catalyst powder (weight, 0.02 g) was held on quartz wool in a vertical quartz tube reactor of 2 mm inside diameter, in an electronically controlled furnace of large heat capacity in order to keep the temperature constant. The reduced catalyst was prepared by the in situ temperature-programmed reduction (H_2 flow rate = 20 mL/min) of the NiO/SiO_2 from room temperature to 600 °C; further, the catalyst was kept for 30 min in flowing H_2 at 600 °C. Helium (28 mL/min) was used as the carrier gas. The gas pulse contained CH_4, CD_4, and O_2, and the ratio $CH_4/CD_4/O_2$ = 1/1/1 was employed in the experiments.

The experimental method can be simply described as follows. A gas pulse containing 0.3 mL methane (CH_4/CD_4 = 1/1) and 0.15 mL O_2 was injected through a 10-port valve into a carrier gas (He), which was allowed to flow through the reactor. The reactants and products have been detected with a GC–MS instrument (HP 5890 Series II gas chromatograph and HP 5971 Series quadrupole mass selective detector) equipped with a Porapak Q column.

3. Analysis of MS Data

In the mass spectra of mixtures containing CD_4, CD_3H, CD_2H_2, CDH_3, and CH_4, some molecular ion peaks overlap with fragment ion peaks. Therefore, to obtain the intensity of each molecular ion of the mixture sample, one must subtract all possible fragment ion contributions from each molecular ion peak.

CD_4 and CD_3H molecular ion intensities can be calculated directly from the peaks of mass 20 and 19, because there are no fragment ions of D-isotopic methanes that contribute to the two peaks. (It should be noted that the contribution of the ^{13}C-methane impurity to the MS peaks must be subtracted.) The CD_2H_2 molecular ion intensity can be obtained by subtracting the contribution of the CD_3 fragment ion (resulting from the cleavage of CD_4^+ or CD_3H^+) from the peak of mass 18. The CDH_3 molecular ion intensity can be obtained by subtracting the contribution of the CD_2H fragment ion (resulting from the cleavage of $CD_2H_2^+$ or CD_3H^+) from the peak of mass 17. The CH_4 molecular ion intensity can be obtained by subtracting the contribution of the CD_2 and CH_2D fragment ions (resulting from the cleavage of $CD_2H_2^+$ or CD_3H^+) from the peak of mass 16. To perform the calculations, one must carry out experiments with standard samples for calibration.

4. Results

4.1. Reaction of CH_4/CD_4 (1/1) with O_2 over NiO/SiO_2 Catalysts. The reaction between methane and O_2 was carried out using pulses of $CH_4/CD_4/O_2$ (1/1/1) at 600 °C. Figure 1 shows that 12% methane and 40% oxygen were converted over unreduced and 28% methane and 100% oxygen were converted over reduced catalysts, that CO_2 without CO was formed over the unreduced NiO/SiO_2 catalyst, and that 41% CO_2 and 59% CO were produced over the reduced NiO/SiO_2 catalyst.

To obtain information about the intermediate species, the isotopic composition after reaction was determined (Figure 2). As shown in Figure 2, only CH_4 and CD_4 without CH_xD_y ($x + y = 4$, x and $y > 0$) were present in the product obtained over the unreduced catalyst. In contrast, besides CH_4 and CD_4, CH_xD_y were detected in the product obtained over the reduced catalyst (3% CH_4, 31% CH_3D, 38% CH_2D_2, 20% CHD_3, and 8% CD_4).

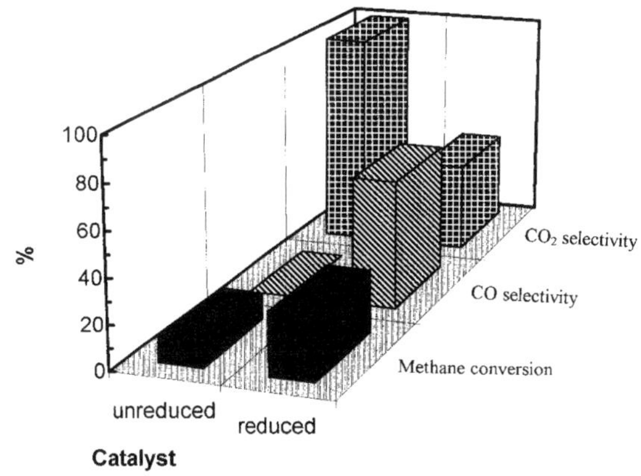

Figure 1. Methane conversion and CO and CO_2 selectivities for the reaction of a $CH_4/CD_4/O_2$ (1/1/1) pulse over unreduced and reduced NiO/SiO_2 catalysts at 600 °C. (Oxygen conversion was 40% over the unreduced catalyst and 100% over the reduced one.)

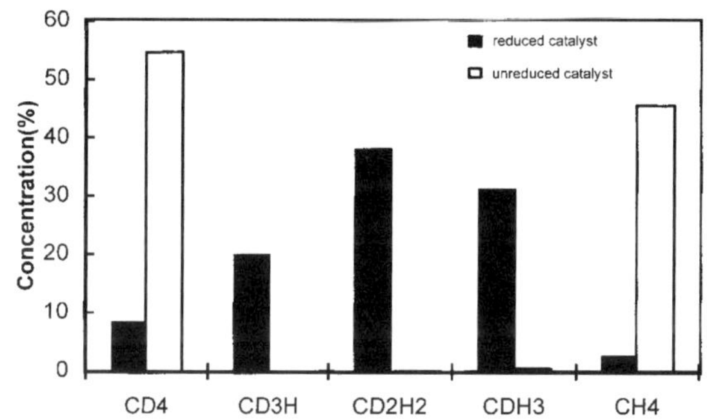

Figure 2. Isotopic species distribution in the methane converted to CO and CO_2 after the $CH_4/CD_4/O_2$ (1/1/1) pulse reaction over unreduced and reduced NiO/SiO_2 catalysts at 600 °C.

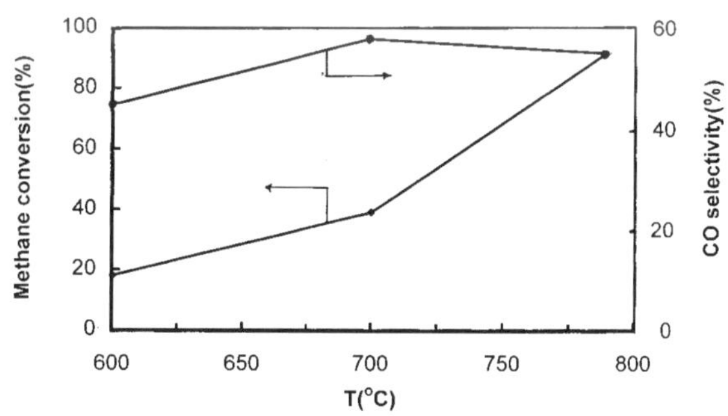

Figure 3. Methane conversion and CO selectivity for the reaction of the CH_4/CD_4 (1/1) pulse over the unreduced NiO/SiO_2 catalyst.

While 28% of the methane (CH_4 and CD_4) in the feed gas was converted to CO and CO_2, a larger amount (65%) of methane participated in the isotopic exchange reaction over the reduced catalyst. This indicates that the isotopic exchange reaction is faster than the partial oxidation over the reduced catalyst. In contrast, over the unreduced catalyst, no isotopic exchange reaction occurred (Figure 2).

4.2. The Reaction between Methane and Lattice Oxygen over NiO/SiO_2 Catalyst. The reaction between the lattice oxygen of NiO/SiO_2 and methane free of oxygen was carried out by using pulses of CH_4/CD_4 (1/1). Figure 3 shows that the methane conversion was 18% at 600 °C, 39% at 700 °C, and 92% at 790 °C and that the selectivity for CO was 45% at 600 °C, 58% at 700 °C, and 55% at 790 °C. Figure 4 reveals that,

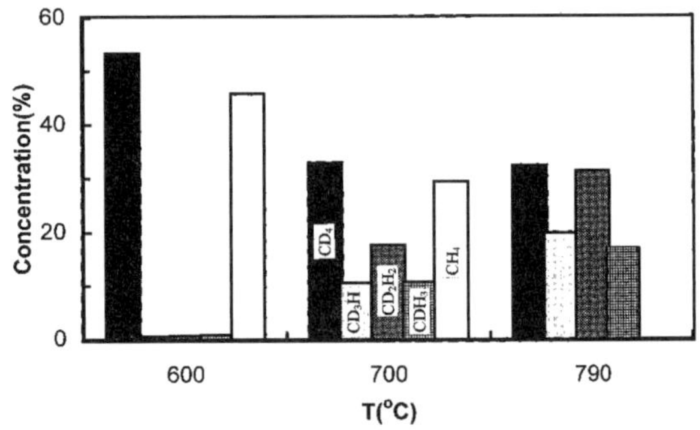

Figure 4. Isotopic species distribution in the methane converted to CO and CO_2 after the reaction of the CH_4/CD_4 (1/1) pulse over the unreduced NiO/SiO_2 catalyst.

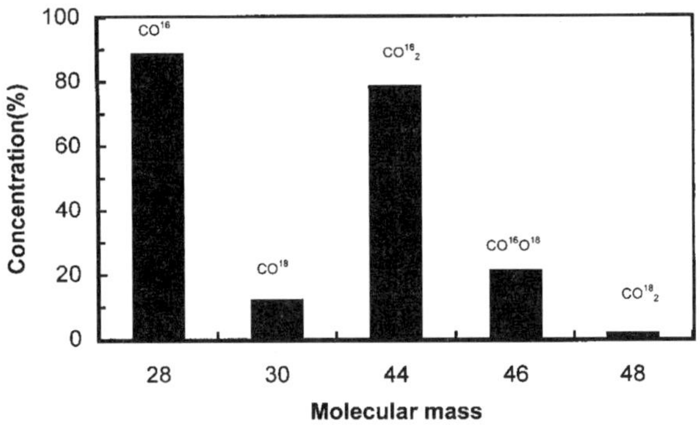

Figure 5. ^{18}O Isotopic species distribution of CO and CO_2 for the $CH_4/^{16}O_2$ (1/1/1) pulse reaction over unreduced $Ni^{18}O/SiO_2$ catalysts at 600 °C.

at 600 °C, only CH_4 and CD_4 free of any CH_xD_y were present in the exit gas. However, above 700 °C, CH_xD_y were also detected.

4.3. Reaction between Methane and $^{16}O_2$ over the Isotopic $Ni^{18}O/SiO_2$ Catalyst. The partial oxidation of methane was carried out by introducing pulses of $CH_4/^{16}O_2$ (2/1) over the unreduced $Ni^{18}O/SiO_2$ catalyst. As shown in Figure 5, besides $C^{16}O$ and $C^{16}O_2$, $C^{18}O$, $C^{16}O^{18}O$, and $C^{18}O_2$ were detected in the product, namely $C^{18}O_2/C^{16}O_2 = 0.018$, $C^{16}O^{18}O/C^{16}O_2 = 0.265$, and $C^{18}O/C^{16}O = 0.13$. This indicates that the lattice oxygen participates in the reaction over the unreduced catalyst, even when the reactant mixture contains oxygen.

5. Discussion

Even though many authors suggested that the dissociation of methane constitutes the first step in the methane oxidation, no direct evidence was yet brought in this direction. Our experiments revealed that, over the reduced catalyst, large amounts of CHD_3, CH_2D_2, and CH_3D were formed in the pulse reaction of methane with oxygen ($CH_4/CD_4/O_2 = 1/1/1$) (Figure 2). This indicates that CH, CH_2, and CH_3 species are formed on the catalyst and demonstrates that the reaction follows the pyrolysis mechanism. Moreover, over the reduced catalyst, the amount of methane involved in the exchange between CH_4 and CD_4 is larger than that involved in the conversion to CO and CO_2. Consequently, the exchange between CH_4 and CD_4 is faster than the conversion of methane to CO and CO_2, and the methane cleavage cannot be rate determining as suggested in ref 33. In other words, the rate-determining step is the reaction of the CH_x species with oxygen. Therefore, the reaction mechanism over the reduced catalyst can be expressed as follows:

$$CH_4 + Ni \rightarrow CH_{x(s)} + (4-x)H_{(s)} \tag{1}$$

$$O_{2(g)} + Ni \rightarrow 2O_{(s)} \tag{2}$$

$$CH_{x(s)} + (1+x)O_{(s)} \rightarrow CO_{(s)} + xHO_{(s)} \tag{3}$$

$$CO_{(s)} \rightarrow CO_{(g)} \tag{4}$$

$$CO_{(s)} + O_{(s)} \rightarrow CO_{2(s)} \rightarrow CO_{2(g)} \tag{5}$$

$$2H_{(s)} \rightarrow H_{2(s)} \rightarrow H_{2(g)} \tag{6}$$

$$HO_{(s)} + Ni \rightarrow O_{(s)} + H_{(s)} \tag{7}$$

$$OH_{(s)} + H_{(s)} \rightarrow H_2O_{(s)} \rightarrow H_2O_{(g)} \tag{8}$$

$$O_{(s)} + 2H_{(s)} \rightarrow H_2O_{(s)} \rightarrow H_2O_{(g)} \tag{9}$$

For the unreduced catalyst, only CD_4 and CH_4, free of any CD_xH_y, were detected. This means that no isotopic exchange reaction occurs over the unreduced catalyst. In other words, methane reacts directly with oxygen without its predissociation. The reaction mechanisms are different over the reduced and unreduced catalysts because only in the former case there are Ni^0 sites, which activate the methane. Methane is activated by Ni^0 present on the reduced Ni-based catalyst as a result of the $\sigma-\pi$ coordination between the σ bonds of C−H in CH_4 and the d orbitals of Ni. This decreases the C−H binding energy via the electron donation from the σ bond of C−H to a vacant d orbital of Ni and the electron back-donation from the filled d orbital of Ni to the vacant antibonding σ^* orbital of C−H. Over the unreduced NiO/SiO_2 catalyst, methane is not activated before reacting with oxygen.[31]

In the reaction at 600 °C between the lattice oxygen and CH_4/CD_4 (1/1) free of O_2, a negligible amount of CH_xD_y was detected. This indicates that no cleavage of methane takes place before its oxidation. At higher temperatures (>600 °C), the Ni^0 formed via the reduction of NiO by CH_4, activates the CH_4 molecules, resulting in the formation of CH_xD_y (Figure 4).

Over the unreduced $Ni^{18}O/SiO_2$, $C^{16}O^{18}O$ and $C^{18}O$ were detected during the reaction of methane with $^{16}O_2$ at 600 °C. This means that the lattice oxygen participates in the reaction even in the presence of gas-phase oxygen. This further confirms our previous results.[31]

Over the unreduced catalyst, the methane conversion for its reaction with the lattice oxygen (18%) is higher than for its reaction with gas oxygen (12%). This means that the reaction with the lattice oxygen is more facile than that with the gas oxygen. According to the above results, it is reasonable to suggest that CH_4 is oxidized substantially by the oxygen of the lattice, which is replenished by the oxygen of the gas atmosphere. This mechanism can be expressed as follows:

$$CH_4 + 2NiO \rightarrow CO_2 + 2H_2 + 2Ni \tag{10}$$

$$2Ni + O_{2(g)} \rightarrow 2NiO \tag{11}$$

Because the Ni^0 oxidation is fast, the reaction between the lattice oxygen and methane may constitute the rate-determining step.

The above mechanisms, suggested on the basis of isotopic pulses, describe the reactions over the initial states of the reduced and unreduced catalysts. Under steady-state conditions, some NiO will be reduced by the CH_4 of the feed gas and some Ni^0 will be reoxidized by the O_2 of the feed gas. Consequently, under steady-state conditions, there are both Ni^0 and NiO sites on the catalyst surface. Because the activity of Ni^0 is much higher than that of NiO, Ni^0 constitutes the main active site of the steady-state catalyst and the pulse reaction over the reduced catalyst can represent the steady-state reaction.

6. Conclusion

In the present paper, an isotopic pulse method is combined with GC−MS to obtain information about the mechanism of the partial oxidation of methane. The reaction of $CH_4/CD_4/O_2$ (1/1/1) over the reduced NiO/SiO_2 catalyst occurs via the pyrolysis mechanism, with the cleavage of methane faster than the oxidation of CH_x. This indicates that the dissociation of methane is not rate determining.

Over the unreduced NiO/SiO_2 catalyst, no isotopic exchange reaction of methane occurs. This indicates that there is no dissociation of methane before its oxidation.

For the reaction of CH_4/CD_4 (1/1) with the lattice oxygen of the NiO/SiO_2 catalyst at 600 °C, the mechanism is as for the reaction of methane with the gas-phase oxygen over the unreduced catalyst. Further, the reaction of CH_4 with $^{16}O_2$ over the $Ni^{18}O/SiO_2$ catalyst indicates that the lattice oxygen participates in the reaction even in the presence of the gas-phase oxygen. As a possible mechanism, we suggest that methane reacts substantially with the lattice oxygen, while O_2 of the gaseous phase reoxidizes the reduced NiO.

References and Notes

(1) Lee, J. S.; Oyama, S. T. *Catal. Rev. Sci. Eng.* **1988**, *30* (2), 249.

(2) Lunsford, J. H. *Catal. Today* **1990**, *6*, 235.

(3) Pitchai, R.; Klier, K. *Catal. Rev. Sci. Eng.* **1986**, *28* (1), 13.

(4) Foster, N. R. *Appl. Catal.* **1985**, *19*, 1.

(5) Gadalla, A. M.; Bower, B. *Chem. Eng. Sci.* **1988**, *43*, 3049.

(6) Ashcroft, A. T.; Cheetham, A. K.; Foord, J. S.; Green, M. L. H.; Grey, C. P.; Murrelland, A. J.; Vernon, P. D. F. *Nature* **1990**, *344*, 319.

(7) Dissanayake, D.; Rosynek, M. P.; Lunsford, J. H. *J. Catal.* **1991**, *132*, 117.

(8) Ashcroft, A. T.; Cheetham, A. K .; Green, M. L. H.; Vernon, P. D. F. *Nature* **1991**, *352*, 225.

(9) Hickman, D. A.; Schmidt, L. D. *J. Catal.* **1992**, *138*, 267.

(10) Rostrup-Nielsen, J. R.; Hansen, J.-H. B. *J. Catal.* **1993**, *144*, 38.

(11) Hu, Y. H.; Ruckenstein, E. *Catal. Lett.* **1995**, *34*, 41.

(12) Hu, Y. H.; Au, C. T.; Wan, H. L. *Chin. Sci. Bull.* **1995**, *40*, 303.

(13) Ruckenstein, E.; Hu, Y. H. *Appl. Catal., A* **1995**, *133*, 149.

(14) Ruckenstein, E.; Hu, Y. H. *Ind. Eng. Chem. Res.* **1998**, *37* (5), 1744.

(15) Zhang, Z.; Verykios, X. E.. *J. Chem. Soc., Chem. Commun.* **1995**, 71.

(16) Bart, J. C. J.; Sneeden, R. P. A. *Catal. Today 2* **1987**, 1.

(17) Kung, H. H. *Catal. Rev. Sci. Eng.* **1980**, *22* (2), 235.

(18) vanLooij, F.; Geus, J. W. *J. Catal.* **1997**, *168*, 154.

(19) Alibrando, M.; Hahm, H. S.; Wolf, E. E. *Catal. Lett.* **1997**, *49*, 1.

(20) Hu, Y. H.; Ruckenstein, E. *Ind. Eng. Chem. Res.* **1998**, *37* (6), 2333.

(21) Boucouvalas, Y.; Zhang, Z. L.; Verykios, X. E. *Catal. Lett.* **1994**, *27*, 131.

(22) Papp, H.; Schuler, P.; Zhuang, Q. *Catal. Lett.* **1996**, *3*, 299.

(23) Boucouvalas, Y.; Zhang, Z. L.; Verykios, X. E. *Catal. Lett.* **1996**, *40*, 189.

(24) Choudhary, V. R.; Rajput, A. M.; Prabhakar, B. J. *Catal. Lett.* **1992**, *15*, 363.

(25) Hickman, D. A.; Schmidt, L. D. *Science* **1993**, *259*, 343.

(26) Hu, Y. H.; Ruckenstein, E. *J. Catal.* **1996**, *158*, 260.

(27) Au, C. T.; Liao, M. S.; Ng, C. F. *Chem. Phys. Lett.* **1997**, *267*, 44.

(28) Buyevskaya, O. V.; Walter, K.; Wolf, D.; Baerns, M. *Catal. Lett.* **1996**, *38*, 81.

(29) Au, C. T.; Wang, H. Y. *J. Catal.* **1997**, *167*, 337.

(30) Mallens, E. P. J.; Hoebink, J. H. B. J.; Marin, G. B. *J. Catal.* **1997**, *167*, 43.

(31) Hu, Y. H.; Ruckenstein, E. *J. Phys. Chem. B* **1998**, *102*, 230.

(32) Prettre, M.; Eichner, C.; Perrin, M. *Trans. Faraday Soc.* **1946**, *42*, 335.

(33) Lapszewicz, J. A.; Jiang, X. Z. *Prepr. Pap.—Am. Chem. Soc., Div. Pet. Chem.* **1992**, *37*, 252.

Broadened Pulse-Step Change—Isotopic Sharp Pulse Analysis of the Mechanism of Methane Partial Oxidation to Synthesis Gas

Yun Hang Hu and Eli Ruckenstein*

Department of Chemical Engineering, State University of New York at Buffalo, Amherst, New York 14260

Received: August 28, 1997; In Final Form: October 22, 1997[⊗]

The mechanism of the partial oxidation of CH_4 was studied by a transient response analysis of a broadened pulse combined with either a step change or a sharp isotopic pulse. Over the unreduced NiO/SiO_2 catalyst, the reaction between CH_4 and O_2 occurs via the Eley–Rideal mechanism, CH_4 in the gas phase (or weakly adsorbed) reacting with O_2 in the adsorbed state. In contrast, over the reduced NiO/SiO_2 catalyst, the reaction takes place by a Langmuir–Hinshelwood mechanism, CH_4 and O_2 reacting in the adsorbed states. In addition, isotopic pulses of $^{18}O_2$ revealed that over the reduced catalyst lattice oxygen is formed and reduced by the carbon species; hence, on the reduced catalyst a dynamic redox process occurs.

1. Introduction

Two mechanisms have been suggested for the partial catalytic oxidation of methane to synthesis gas. In one, CO_2 and H_2O are the primary products, and CO is a result of their reaction with CH_4. In the other, CO is produced directly without the preformation of CO_2. Prettre et al.,[1] who were the first to study the CH_4 partial oxidation over a nickel catalyst, concluded that the overall oxidation involves two stages: an initial exothermic oxidation of CH_4 to CO_2 and H_2O followed by the endothermic reaction $CH_4 + CO_2 \rightarrow 2CO + 2H_2$. Some recent investigations brought support to this mechanism.[2–6] The other mechanism involves the pyrolysis $CH_4 \rightarrow C + 4H$ followed by the oxidation of C to CO.[6–12]

Regarding the active catalytic phase for the reaction, some of the researchers considered that the metal is the active phase.[7–11,13,14] However, Dissanayake et al.[3] identified three constituents in the used Ni/Al_2O_3 catalyst and suggested that CH_4 is oxidized to CO_2 and H_2O over the regions containing $NiAl_2O_4$ and/or $NiO + Al_2O_3$ and that the steam and/or CO_2 reforming of CH_4 to CO and H_2 occur over the regions containing Ni/Al_2O_3. Van Looij and Geus[15] suggested that, in the parallel combustion of methane to CO_2 and reforming of methane, the combustion proceeds partly over NiO and more completely over partially oxidized nickel and that the reforming of methane to synthesis gas proceeds over metallic nickel sites.

Recently,[16] a new transient method, in which, instead of sharp pulses, both broadened pulses combined with step changes and broadened pulses combined with sharp isotopic pulses are employed, was developed. This combination of pulses provides richer dynamic information than the individual pulses. In the present paper it will be shown that this method allows us to obtain some information about the mechanism of partial oxidation of CH_4 over both the unreduced and reduced NiO/SiO_2 catalysts. Different mechanisms have been thus identified: (i) the CH_4 of the gas phase (or in a weakly adsorbed state) reacts with the O_2 adsorbed on the catalyst over the unreduced catalyst; (ii)a surface reaction between adsorbed CH_4 and O_2 species takes place over the reduced catalyst.

* To whom correspondence should be addressed.

⊗ Abstract published in *Advance ACS Abstracts*, December 1, 1997.

2. Experimental Section

2.1. Catalyst Preparation. NiO/SiO_2 (containing 13.6 wt % Ni) catalysts were prepared by impregnating SiO_2 (Aldrich) with an aqueous solution of nickel nitrate (Alfa). The paste thus obtained was dried at room temperature in air and then decomposed and calcined at 800 °C in air for 1.5 h. The BET surface area and average diameter of the pores were 362 m^2/g and 52 Å, respectively.

2.2. Transient Response Analysis of a Broadened Pulse Combined with a Step Change or with a Sharp Isotopic Pulse. The transient analysis experiments were performed at 800 °C under atmospheric pressure. The catalyst powder (weight: 0.02 g) was held on quartz wool in a vertical quartz tube reactor of 2 mm inside diameter in an electronically controlled furnace of large heat capacity in order to keep the temperature constant. Helium (60 mL/min) was used as the carrier gas. The feed gas contained CH_4 and O_2 (CH_4/O_2 = 2:1) and the isotopes $^{13}CH_4$ (99%) and $O^{18}{}_2$ (99%) (Isotec Inc) were employed in the experiments.

The method, which was proposed in our previous paper,[16] can be simply described as follows. A gas pulse of 2.5 mL is injected almost instantaneously into a carrier gas (He) and allowed to flow into a large tube (40 mm diameter and 164 mm length) in which the pulse is broadened to a width of about 20 min. From the broadening tube, the pulse flows more slowly to the reactor and finally to the MS equipment. During the broadened pulse, a step change from the feed gas to He, or a sharp isotopic pulse of 3 μL, is introduced at selected times in the broadened pulse. The reactants and products have been continuously detected with a MS equipment (HP Quadrupole, 5971 Series mass-selective detector) equipped with a fast response inlet capillary system.

3. Results

3.1. Transient Response to a Broadened CH_4/CO_2 Pulse over NiO/SiO_2. Over the unreduced catalyst, the response curves to a broadened pulse (Figure 1) show that no CO is present and only a relatively small amount of CO_2 is formed. Most of CH_4 and O_2 remain unreacted. After NiO/SiO_2 was reduced with six pulses of H_2 (2.5 mL), the response (Figure

This paper, which was originally published in **Journal Physical Chemistry B**: 102, 230-233 (1998), is republished with permission.

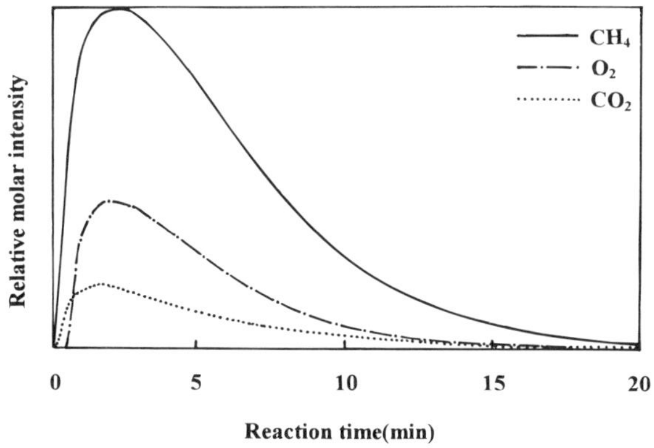

Figure 1. Responses to a broadened pulse of CH_4/O_2 (2.5 mL) over an unreduced NiO/SiO_2 catalyst at 800 °C.

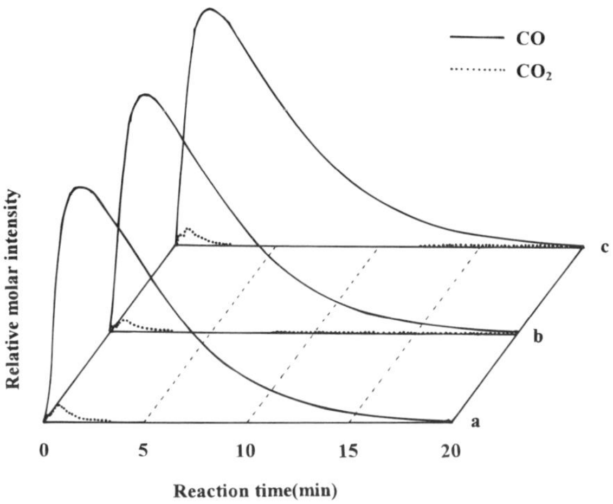

Figure 2. Responses to a broadened pulse of CH_4/O_2 (2.5 mL) over a reduced NiO/SiO_2 catalyst at 800 °C: (a) first pulse; (b) second pulse; (c) third pulse.

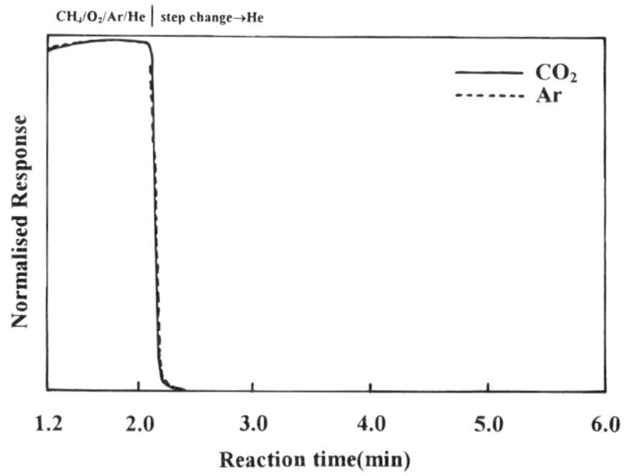

Figure 3. Normalized responses to a broadened pulse of $CH_4/O_2/Ar$ combined with a step change to pure carrier (He) over an unreduced NiO/SiO_2 catalyst at 800 °C.

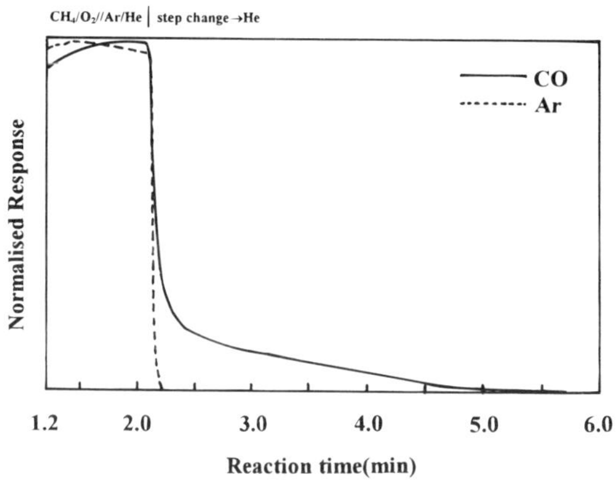

Figure 4. Responses to a broadened pulse of $CH_4/CO/Ar$ combined with a step change to He over a reduced NiO/SiO_2 catalyst at 800 °C (no CO_2 was detected after the step change).

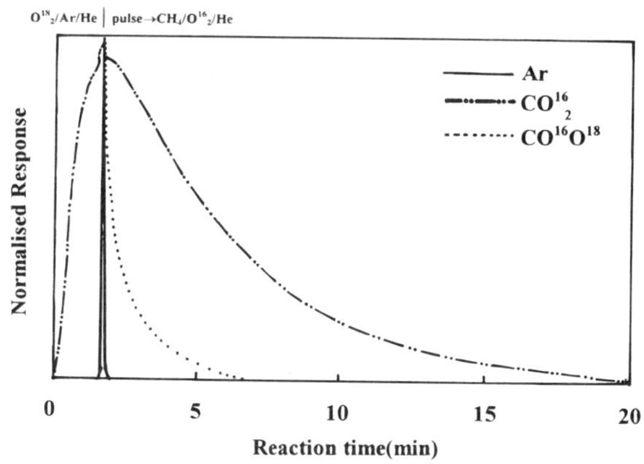

Figure 5. Normalized responses to a broadened pulse of $CH_4/^{16}O_2$ combined with a sharp pulse of $^{18}O_2/Ar$ over an unreduced NiO/SiO_2 catalyst at 800 °C.

2) to the broadened pulse of CH_4/O_2 shows the presence of CO and of a small amount of CO_2 and is free of CH_4, O_2, and water. Three successive pulses have been carried out. Since they are very similar, it is clear that the pulses do not change appreciably the surface of the catalyst.

3.2. Transient Response to a Broadened Pulse Combined with a Step Change. The response to a step change to He during the broadened $CH_4/O_2/Ar/He$ (13.4% Ar as internal standard) pulse shows that, over the unreduced catalyst (Figure 3), the CO_2, O_2, and CH_4 curves coincide with that of Ar; hence, the product CO_2 does not exhibit a delay tail. In contrast, over the reduced catalyst (Figure 4), the tail of the CO response is 3.3 min longer than that of Ar.

3.3. Transient Response Analysis of a Broadened Pulse Combined with an Isotopic Sharp Pulse. Sharp $^{18}O_2/Ar$ (13.4% Ar) or $^{13}CH_4$ (5% Ar) pulses were introduced during broadened $^{12}CH_4/^{16}O_2$ pulses (Figures 5−7). Figure 5 shows that, over the unreduced catalyst, the response curve of $C^{16}O^{18}O$ to a sharp pulse of $^{18}O_2$ has a much longer tail, 5 min longer, than Ar, which has a width of only 0.1 min, and that no $C^{18}O_2$ is present. Figure 6 shows that, over the reduced catalyst, the response curves of $C^{18}O$ and $C^{16}O^{18}O$ to a sharp pulse of $^{18}O_2/Ar$ have much longer tails than Ar, being 1.8 min longer. Figure 7 shows that the response of ^{13}CO to a $^{13}CH_4$ sharp pulse, over the reduced catalyst, has a somewhat longer tail than Ar. In addition, one can see that the tail of the $C^{18}O$ response to an $^{18}O_2$ sharp pulse is much longer than the tail of the ^{13}CO response to a $^{13}CH_4$ sharp pulse (compare Figures 6 and 7).

4. Discussion

Over the unreduced NiO/SiO_2 catalyst, the response curves of CO_2, O_2, and CH_4 to a step change from the feed gas to He coincide with that of Ar (Figure 3). Because in the absence of

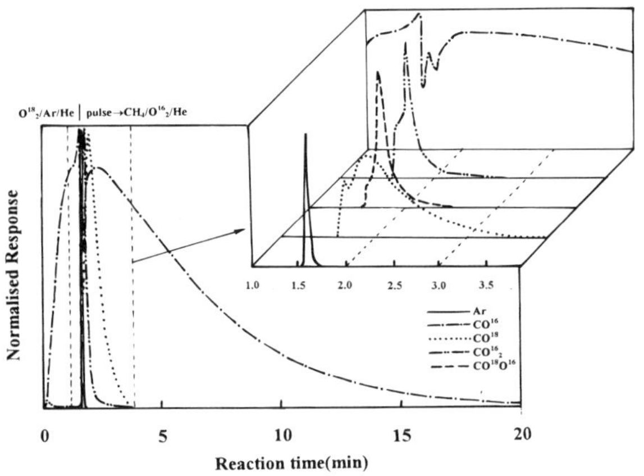

Figure 6. Normalized responses to a broadened pulse of $CH_4/C^{16}O_2/Ar$ combined with a sharp pulse of $^{18}O_2/Ar$ over a reduced NiO/SiO_2 catalyst at 800 °C.

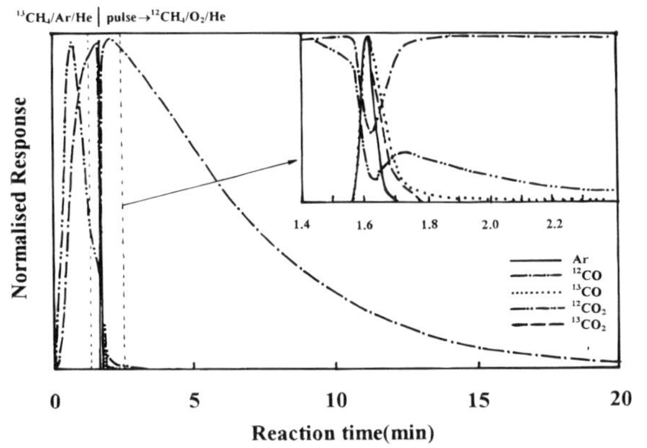

Figure 7. Normalized responses to a broadened pulse of $^{12}CH_4/O_2$ combined with a sharp pulse of $^{13}CH_4/Ar$ over a reduced NiO/SiO_2 catalyst at 800 °C.

the catalyst the conversion is extremely small, the above observations indicate that the formation of CO_2 over the unreduced catalyst can occur (a) between weakly adsorbed species, (b) between one weakly adsorbed species and another species in the gas phase, or (c) between one strongly adsorbed and another weakly adsorbed or gas-phase species with rapid desorption of the product. However, the response of $C^{18}O^{16}O$ over the unreduced catalyst to a sharp pulse of $^{18}O_2/Ar$ during a broadened pulse of $CH_4/^{16}O_2$ has a longer tail than Ar (Figure 5). This reveals that a strongly adsorbed oxygen or lattice oxygen species participates in the reaction. Therefore, over the unreduced catalyst, the reaction occurs between the gas-phase or weakly adsorbed CH_4 and the strongly adsorbed or lattice oxygen (Eley−Rideal mechanism). Because during the sharp pulse of $^{18}O_2/Ar$, $C^{16}O^{18}O$ (and not $C^{18}O_2$) is formed, it is clear that $^{18}O_2$ first dissociates to two ^{18}O before reacting with CH_4. The fact that $C^{16}O^{18}O$ has a long tail indicates that at least a part of ^{18}O is stabilized by its exchange with the lattice O. The temperature programmed reaction experiments between CH_4 and NiO [17] demonstrated that the lattice oxygen can be reduced by CH_4.

Over the reduced catalyst, the CO response to a step change to He, during the $CH_4/O_2/Ar$-broadened pulse, has a longer tail than that of Ar (Figure 4). The delay can be a result of (a) CO desorption and (b) CO formation by the surface reaction between the C and O species. Previously,[16] we have shown that the CO response to a step change from CO/Ar/He to He during a broadened pulse of CO/Ar is near that of Ar. This indicates that the delay is not due to the desorption but to the surface reaction between the C and the O species. Consequently, the reaction over the reduced NiO/SiO_2 catalyst takes place via the Langmuir−Hinshelwood mechanism, in which the adsorbed CH_4 and O_2 species are involved. This is further confirmed by the responses of $^{13}C^{16}O$ or $^{12}C^{18}O$ to sharp pulses of $^{13}CH_4/Ar$ (Figure 7) or $^{18}O_2/Ar$ (Figure 6) during $^{12}CH_4/^{16}O_2$-broadened pulses, which exhibit longer tails than that of Ar.

The long tail in the response of CO to a step change to He, during the broadened pulse of $CH_4/O_2/Ar/He$ (Figure 4), indicates the presence of C species on the reduced catalyst. In contrast, there are no C species over the unreduced catalyst (Figure 3). Therefore, one can conclude that Ni^0 is responsible for the activation of CH_4. This occurs because of the $\sigma-\pi$ coordination between the σ bonds of C−H in CH_4 and the d orbitals of Ni. This reduces the C−H binding energy via the electron donation from the σ bond of C−H to a vacant d orbital of Ni and the back-donation from the filled d orbital of Ni to the vacant antibonding σ^* orbital of C−H. Indeed, the experiments of Ceyer[18] demonstrated that the CH bond is dissociated over Ni.

Over the reduced catalyst, the tail of the $^{12}C^{18}O$ response to a sharp pulse of $^{18}O_2/Ar$ (Figure 6) is much longer than that of the $^{13}C^{16}O$ response to a sharp pulse of $^{13}CH_4$ (Figure 7), both during the broadened pulse of $^{12}CH_4/^{16}O_2$. This indicates that the C species react more easily than the O species. Because the long tail of $C^{18}O$ to the sharp pulse indicates that the ^{18}O species are relatively stable (Figure 6), it is reasonable to consider that at least some of the ^{18}O species are lattice oxygen atoms. Hence, a slow dynamic redox process consisting of lattice oxygen formation and its reduction by C species is at least partly responsible for the CO formation. This explains why after a pulse the catalyst does not remain oxidized (Figure 2). However, Figure 4 also reveals that initially, after the step change, the responses of Ar and CO coincide. This may be caused by another O species present on the surface of the catalyst, which reacts more rapidly with the C species.

5. Conclusion

The reaction mechanism between CH_4 and O_2 over the unreduced NiO/SiO_2 catalyst differs from that over the reduced catalyst. The former takes place via the Eley−Rideal mechanism in which O_2 in the adsorbed state reacts with the CH_4 in the gas phase, while the latter takes place via the Langmuir−Hinshelwood mechanism in which adsorbed CH_4 and O_2 species react with each other.

References and Notes

(1) Prettre, M.; Eicher, C.; Perrin, M. *Trans. Faraday Soc.* **1964,** *43*, 335.

(2) Ashcroft, A. T.; Cheetham, A. K.; Foord J. S.; Green, M. L. H. *Nature* **1990,** *344*, 319.

(3) Dissanayake, D.; Rosynek, M. P.; Lunsford, J. H. *J. Catal.* **1991,** *132*, 117.

(4) Boucouvalas, Y.; Zhang, Z. L.; Verykios, X. E. *Catal. Lett.* **1994,** *27*, 131.

(5) Papp, H.; Schuler, P.; Zhuang, Q. *Catal. Lett.* **1996,** *3*, 299.

(6) Boucouvalas, Y.; Zhang, Z. L.; Verykios, X. E. *Catal. Lett.* **1996,** *40*, 189.

(7) Choudhary, V. R.; Rajput, A. M.; Prabhaker, B. J. *Catal. Lett.* **1992,** *15*, 363.

(8) Hickman, D. A.; Schmidt, L. D. *Science* **1993,** *259*, 343.

(9) Hu, Y. H.; Ruckenstein, E. *Catal. Lett.* **1995,** *34*, 41.

(10) Hu, Y. H.; Ruckenstein, E. *J. Catal.* **1996,** *158*, 260.

(11) Buyerskaya, O. V.; Walter, K.; Wolf, D.; Baerns, M. *Catal. Lett.* **1996,** *38*, 81.
(12) Mallens, E. P. J.; Hoebink, J. H. B. J.; Marin, G. B. *J. Catal.,* **1997,** *167*, 43.
(13) Wang, D. Z.; Dewaele, O.; Degroote, A. M.; Froment, G. F. *J. Catal.* **1996,** *159*, 418.
(14) Chu, Y. L.; Li, S. B.; Lin, J. Z.; Gu, J. F.; Yang, Y. L. *Appl. Catal. A* **1996,** *134*, 67.
(15) Van Looij, F.; Geus, J. W. *J. Catal.* **1997,** *168*, 154.
(16) Hu, Y. H.; Ruckenstein, E. *J. Phys. Chem. B* **1997,** *101*, 7563.
(17) Hu, Y. H.; Ruckenstein, E. *Langmuir* **1997,** *13*, 2055.
(18) Ceyer, S. T. *Science* **1990,** *249*, 133.

CH_4/CD_4 Isotope Effect and the Mechanism of Partial Oxidation of Methane to Synthesis Gas over $Rh/\gamma\text{-}Al_2O_3$ Catalyst

H. Y. Wang and E. Ruckenstein*

Department of Chemical Engineering, State University of New York at Buffalo, Amherst, New York 14260

Received: July 13, 1999; In Final Form: October 29, 1999

Partial oxidation pulse experiments with CH_4/O_2 (2/1) and with CD_4/O_2 (2/1) have been carried out over the 0.05 and 1.0 wt % $Rh/\gamma\text{-}Al_2O_3$ catalysts to determine the isotope effect of the reaction. On the basis of the results obtained for the isotope effect as a function of temperature and Rh loading, one could conclude that the mechanism of the partial oxidation of methane to synthesis gas is dependent on both metal loading and temperature. It is likely that at low loadings (e.g., 0.05 wt %), the combustion–reforming mechanism is responsible for the reaction, while at high loadings (e.g., 1.0 wt %) a combination of the combustion–reforming and pyrolysis–oxidation mechanisms is responsible at low temperatures (≤500 °C) and the pyrolysis–oxidation mechanism becomes dominant at high temperatures (≥650 °C).

1. Introduction

Currently, synthesis gas is produced industrially by the steam reforming of methane:[1,2]

$$CH_4 + H_2O \rightarrow CO + 3H_2 \qquad \Delta H^0_{298} = 206 \text{ kJ/mol} \qquad (1)$$

However, this process is strongly endothermic and provides a synthesis gas with too high a H_2/CO ratio for the Fischer–Tropsch and methanol syntheses. Recently, there has been renewed interest in preparing synthesis gas via the partial oxidation of methane because of the mild exothermicity of the reaction and the suitable H_2/CO ratio obtained:

$$CH_4 + {}^1/_2O_2 \rightarrow CO + 2H_2 \qquad \Delta H^0_{298} = -36 \text{ kJ/mol} \qquad (2)$$

The research regarding this reaction, its mechanism, and the characterization of the catalysts became one of the most active topics in catalysis.[3-21] However, the mechanism of partial oxidation of methane to synthesis gas is still debated. Two main points of view have been suggested. Prettre et al.[22] proposed a mechanism (the combustion–reforming mechanism) in which the oxidation reaction involved an initial strongly exothermic oxidation of CH_4 to CO_2 and H_2O followed by the endothermic reforming reactions of the remaining CH_4 by the H_2O and CO_2 formed in the previous step. In contrast, Schmidt et al.[5,7,23] suggested that the synthesis gas is formed directly, without the intermediary involvement of CO_2. In their pyrolysis–oxidation mechanism, the methane pyrolysis, which constituted the first step, was followed by the surface carbon oxidation and the desorption of CO and hydrogen. Numerous researchers[4,6,18,19,24-27] supported the combustion–reforming mechanism; there are, however, quite a few authors[11-15,17,28] who favored the pyrolysis–oxidation mechanism. The main concern of the debate is whether CO_2 or CO is the primary product of the reaction.

To clarify the mechanism, experiments regarding the deuterium isotope effect in the partial oxidation of methane were carried out by performing $CH_4 + O_2$ and $CD_4 + O_2$ reactions alternately in a pulse microreactor at temperatures ranging from 400 to 850 °C. The isotope effect was determined over the 0.05 and 1.0 wt % $Rh/\gamma\text{-}Al_2O_3$ catalysts. The goal of the paper is to present the results obtained and to extract from them information about the reaction mechanism.

2. Experimental Section

2.1. Catalyst Preparation. The supported rhodium catalysts were prepared by impregnating $\gamma\text{-}Al_2O_3$ (Alfa, 60.9 m^2/g) with an ethanol solution of $Rh(NO_3)_3 \cdot 2H_2O$ (Alfa); this was followed by overnight drying at 110 °C and calcination in air at 800 °C for 4 h. Rh loading means wt % Rh in the reduced catalyst.

2.2. Pulse Reaction. The reaction was carried out in a pulse reaction apparatus. A vertical quartz tube (3 mm inside diameter) was used as reactor, which was operated in the down flow mode. Helium (35 mL/min), which was purified using Hydro-Purge II and Oxy-Trap columns, was employed as the carrier gas. In the pulse experiments, the reactant gas (250 μL) was injected via a six-way sampling valve into the flowing carrier gas. The reactants and products were analyzed with an on-line gas chromatograph equipped with a thermal conductivity detector (TCD) and a Porapak Q column.

2.3. CH_4/CD_4 Isotope Effects. Five milligrams of catalyst, which was supported on quartz wool, was employed in each experiment. The catalyst was first reduced in a flow of H_2 (20 mL/min) at 600 °C for 0.5 h and then purged with a He (35 mL/min) flow for 10 min at the same temperature. After pretreatment, CH_4/O_2 (2/1) pulses were injected into the He flow to perform the $CH_4 + O_2$ reaction at a selected temperature, and subsequently, CD_4/O_2 (2/1) pulses were injected to perform the $CD_4 + O_2$ reaction at the same temperature. The conversions of methane and oxygen and yields of CO and CO_2 in the two kinds of reactions were obtained by averaging over 10 successive pulses. The time interval between two pulses of the same or different kind was 6 min. The results obtained were employed to calculate the isotope effects as the ratios of the conversion of methane, the yields of CO and CO_2 for the $CH_4 + O_2$ reaction to those for the $CD_4 + O_2$ reaction.

* To whom correspondence should be addressed.

This paper, which was originally published in **Journal Physical Chemistry B**: 103, 11327-11331 (1999), is republished with permission.

TABLE 1: Effect of Temperature on the Reactivities of CH_4/O_2 (2/1) and CD_4/O_2 (2/1) Pulses and Deuterium Isotope Effect over the 0.05 wt % Rh/γ-Al_2O_3 Catalyst[a]

temp (°C)	$CH_4 + O_2$						$CD_4 + O_2$						deuterium isotope effect		
	C_{CH_4}	C_{O_2}	S_{CO}	S_{CO_2}	Y_{CO}	Y_{CO_2}	C_{CD_4}	C_{O_2}	S_{CO}	S_{CO_2}	Y_{CO}	Y_{CO_2}	for methane conversion	for CO formation	for CO_2 formation
450	5.52	22.83	0.00	100.00	0.00	5.52	3.80	15.87	0.00	100.00	0.00	3.80	1.45		1.45
525	31.80	100.00	26.70	73.30	8.49	23.31	9.17	36.38	0.00	100.00	0.00	9.17	3.47		2.54
550	35.04	100.00	37.38	62.62	13.10	21.94	10.60	40.90	0.00	100.00	0.00	10.60	3.31		2.07
600	39.93	100.00	46.92	53.08	18.74	21.19	35.17	100.00	36.47	63.53	12.83	22.34	1.14	1.46	0.95
650	60.91	100.00	73.04	26.96	44.49	16.42	52.17	100.00	64.06	35.94	33.42	18.75	1.17	1.33	0.88
750	80.38	100.00	90.77	9.23	72.96	7.42	74.13	100.00	89.53	10.47	66.37	7.76	1.08	1.10	0.96
850	92.56	100.00	97.38	2.62	90.13	2.43	87.96	100.00	95.73	4.27	84.21	3.75	1.05	1.07	0.65

[a] *C* means conversion, *S* selectivity, and *Y* yield; all in percentage units.

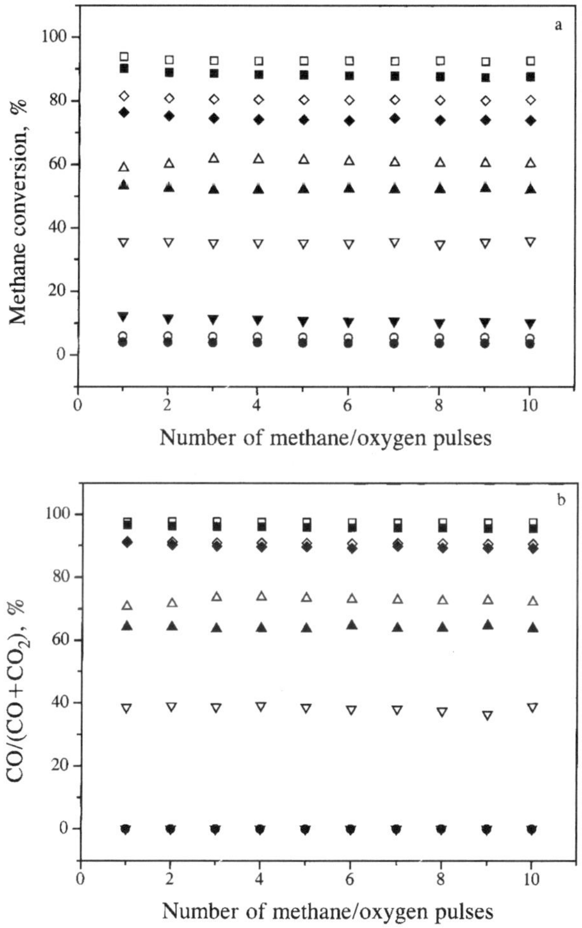

Figure 1. Methane conversions (a) and CO selectivities (b) in the $CH_4 + O_2$ (open symbols) and $CD_4 + O_2$ (solid symbols) reactions over the 0.05 wt % Rh/γ-Al_2O_3 catalyst at 450 (circle), 550 (down triangle), 650 (up triangle), 750 (diamond), and 850 °C (square).

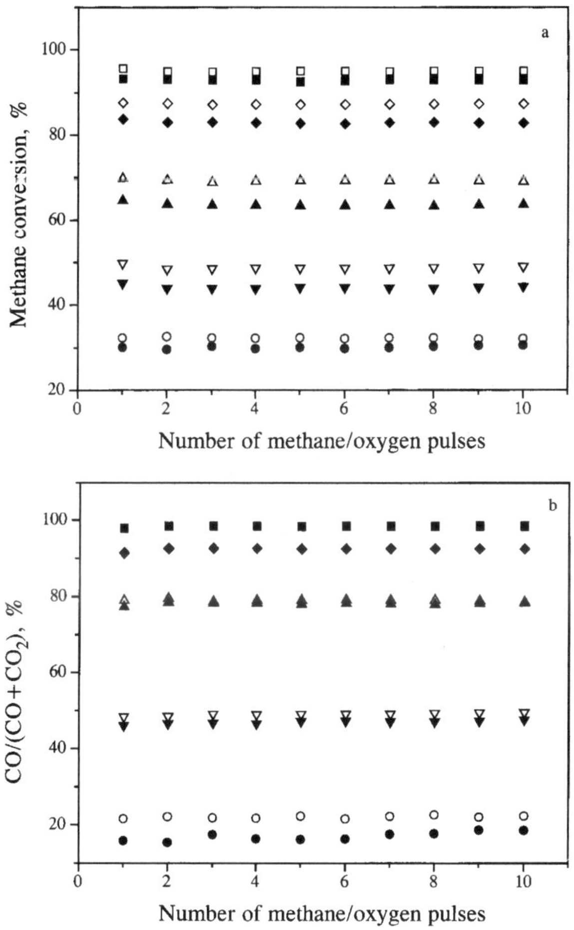

Figure 2. Methane conversions (a) and CO selectivities (b) in the $CH_4 + O_2$ (open symbols) and $CD_4 + O_2$ (solid symbols) reactions over the 1.0 wt % Rh/γ-Al_2O_3 catalyst at 450 (circle), 550 (down triangle), 650 (up triangle), 750 (diamond), and 850 °C (square).

3. Results

3.1. CH_4/CD_4 Isotope Effects. The products formed during the $CH_4(CD_4) + O_2$ reaction contained CO, $H_2(D_2)$, CO_2, and H_2O (D_2O). The carbon balance indicated that no carbon was deposited on the catalyst. The CH_4/CD_4 isotope effect was determined by performing the $CH_4 + O_2$ and $CD_4 + O_2$ reactions alternately at temperatures ranging from 400 to 850 °C. The results obtained over the 0.05 wt % Rh/γ-Al_2O_3 catalyst are presented in Figure 1 and Table 1, and those over the 1.0 wt % Rh/γ-Al_2O_3 catalyst are presented in Figure 2 and Table 2.

Over the 0.05 wt % catalyst, at 450 °C, only CO_2 was formed as a carbon-containing product in both $CH_4 + O_2$ and $CD_4 + O_2$ reactions and the conversion of oxygen was approximately 4 times that of methane (Table 1). This indicates that only complete oxidation of methane occurred at this temperature. Table 1 also shows that the rate of CO_2 formation at 450 °C for $CH_4 + O_2$ reaction was 1.45 times that for $CD_4 + O_2$. At 550 °C, CO was formed as well and O_2 was completely converted in the $CH_4 + O_2$ reaction but not in the $CD_4 + O_2$ one (Figure 1 and Table 1). In the latter reaction, the conversion of oxygen was still about 4 times that of methane, indicating the complete oxidation of CD_4. As shown in Table 1, at 550

TABLE 2: Effect of Temperature on the Reactivities of CH_4/O_2 (2/1) and CD_4/O_2 (2/1) Pulses and Deuterium Isotope Effect over the 1.0 wt % Rh/γ-Al_2O_3 Catalyst[a]

temp	$CH_4 + O_2$						$CD_4 + O_2$						deuterium isotope effect		
(°C)	C_{CH_4}	C_{O_2}	S_{CO}	S_{CO_2}	Y_{CO}	Y_{CO_2}	C_{CD_4}	C_{O_2}	S_{CO}	S_{CO_2}	Y_{CO}	Y_{CO_2}	for methane conversion	for CO formation	for CO_2 formation
400	13.00	57.48	0.00	100.00	0.00	13.00	8.43	39.06	0.00	100.00	0.00	8.43	1.54		1.54
450	32.31	100.00	21.97	78.03	7.10	25.21	30.02	100.00	16.59	83.41	4.98	25.04	1.08	1.43	1.01
500	39.68	100.00	29.37	70.63	11.36	28.31	36.44	100.00	23.70	76.30	8.64	27.78	1.09	1.31	1.02
550	48.72	100.00	49.16	50.84	23.95	24.77	44.04	100.00	47.07	52.93	20.73	23.31	1.11	1.16	1.06
650	69.18	100.00	78.98	21.02	54.63	14.54	63.44	100.00	78.21	21.79	49.62	13.82	1.09	1.10	1.05
750	87.20	100.00	92.62	7.38	80.77	6.44	82.86	100.00	92.55	7.45	76.69	6.17	1.05	1.05	1.04
850	94.94	100.00	98.40	1.60	93.42	1.52	92.85	100.00	98.48	1.52	91.44	1.41	1.02	1.02	1.08

[a] *C* means conversion, *S* selectivity, and *Y* yield; all in percentage units.

°C the methane conversion and CO_2 yield for the reaction involving CH_4 were 3.3 and 2.07 times, respectively, those involving CD_4. At 525 °C, the results were similar to those at 550 °C (Table 1). For temperatures of ≥600 °C, O_2 was completely consumed and CO was formed besides CO_2 in both cases. Both the methane conversion and the CO selectivity for the $CH_4 + O_2$ reaction were higher than for the $CD_4 + O_2$ one (Table 1 and Figure 1). With increasing reaction temperature, the difference between the CO selectivities in the two reactions became increasingly smaller; at temperatures of ≥750 °C, they were very close to each other (Figure 1). Regarding the isotope effect at temperatures of ≥600 °C, that for the CO formation decreased monotonically from 1.46 to 1.07 with an increase in the reaction temperature from 600 to 850 °C; that for the CO_2 formation was always below 1.00, and that for the methane conversion exhibited a maximum at 650 °C (Table 1).

Over the 1.0 wt % Rh/γ-Al_2O_3 catalyst, at 400 °C, only complete oxidation of methane occurred for both $CH_4 + O_2$ and $CD_4 + O_2$ reactions, as over the 0.05 wt % catalyst at 450 °C. As shown in Table 2, at 400 °C the rate of CO_2 formation for the $CH_4 + O_2$ reaction was 1.54 times that for the $CD_4 + O_2$ reaction. At temperatures of ≥450 °C, O_2 was completely consumed and CO was formed besides CO_2 in both cases. The methane conversion for the $CH_4 + O_2$ reaction was always higher than for the $CD_4 + O_2$ reaction at temperatures between 450 and 850 °C (Table 2 and Figure 2). One can observe that the difference between the CO selectivities in the two reactions decreased with increasing temperature, becoming almost zero at temperatures of ≥650 °C (Figure 2b and Table 2). Regarding the isotope effect at temperatures of ≥450 °C, that for the CO formation decreased monotonically from 1.43 to 1.02 with increasing reaction temperature from 450 to 850 °C; that for the CO_2 formation was always greater than 1.00 and higher at temperatures of ≥550 °C than at temperatures of ≤500 °C, and that for the methane conversion exhibited a maximum at 550 °C (Table 2).

In summary, at temperatures of ≥600 °C, the isotope effects for the conversion of methane and for the CO formation were much larger over the 0.05 wt % catalyst than over the 1.0 wt % one. In addition, the deuterium isotope effect for the CO_2 formation reaction was lower than 1.00 over the 0.05 wt % catalyst but higher than 1.00 over the 1.0 wt % catalyst.

3.2. Interactions of CH_4 Pulses with the 0.05 and 1.0 wt % Rh/γ-Al_2O_3 Catalysts. The interactions of methane pulses with the reduced 0.05 and 1.0 wt % Rh/γ-Al_2O_3 catalysts were investigated at temperatures ranging from 550 to 850 °C. The amount of surface carbon formed in the reaction was calculated from the carbon balance. Figure 3 presents the changes of methane conversion and surface carbon (C̲) yield over the two catalysts as a function of the number of CH_4 pulses. Over the 0.05 wt % catalyst, only CO was detected by GC as a gaseous carbon-containing product. Since no oxygen was present in the feed gas, the residual oxygen of the catalyst was responsible for the formation of CO. The carbon balance indicated that carbon deposition occurred to some extent at 850 °C but not at 550 and 700 °C (Figure 3b). As shown in Figure 3b, over the 0.05 wt % Rh catalyst and 850 °C, the C̲ yield increased during the first three pulses, passed through a maximum close to 11%, and then decreased. In contrast, over the 1.0 wt % Rh catalyst, both the methane conversion and the C̲ yield were much higher during the first six pulses than over the 0.05 wt % catalyst (Figures 3). Over the 1.0 wt % catalyst, the C̲ yield decreased

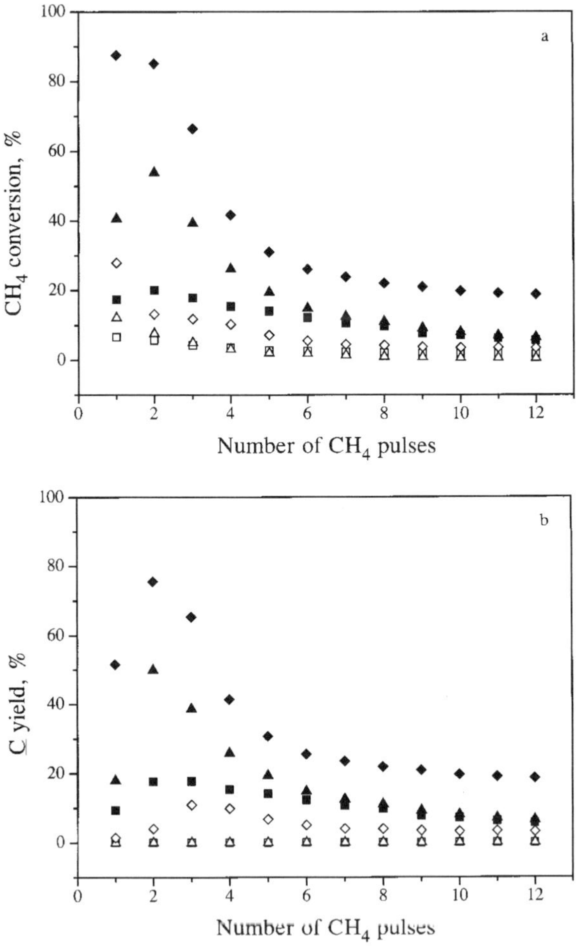

Figure 3. Methane conversions (a) and C̲ yields (b) in the reaction of pure methane with the 0.05 (open symbols) and 1.0 (solid symbols) wt % Rh/γ-Al_2O_3 catalysts at 550 (square), 700 (up triangle), and 850 °C (diamond). Catalyst: 50.0 mg. CH_4 pulse: 250 μL.

gradually at 550 °C and more rapidly at 700 and 850 °C after passing through a maximum. The maxima of the C yields at 550, 700, and 850 °C were 17.8, 49.8, and 75.6%, respectively. In summary, the catalyst with a 1.0 wt % Rh loading had a much stronger ability for methane dissociation than the one with 0.05 wt % loading. In addition, the rate of carbon deposition during the first six pulses increased greatly with increasing temperature over the 1.0 wt % Rh catalyst. It is clear that the coverage of the metallic rhodium sites by carbon species is responsible for the decay of methane conversion (Figure 3a).

4. Discussion

A complete conversion of oxygen and a rapid rise of the methane conversion could be regarded as a sign of ignition of the reaction. At 450 °C over the 0.05 wt % Rh catalyst and at 400 °C over the 1.0 wt % one, both the methane and oxygen conversions were relatively low (the reactions were not ignited) and only complete oxidation products, i.e., CO_2 and H_2O were formed. As shown in Tables 1 and 2, under the above conditions, the CO_2 formation reaction exhibited a noticeable deuterium isotope effect. In general, the breaking of a bond with H(D) during the rate-determining step of a reaction should provide a normal (>1) deuterium isotope effect.[29] Consequently, the existence of the deuterium isotope effect indicates that the rate-determining step of the reaction involves the breakage of a C−H(D) bond.

Over the 0.05 wt % catalyst, at 525 and 550 °C, Table 1 shows that the $CH_4 + O_2$ reaction was ignited while the $CD_4 + O_2$ reaction was not. Obviously, this occurs because, owing to its lower vibration frequency, CD_4 is less reactive than CH_4. At temperatures of ≥600 °C, both the $CH_4 + O_2$ and $CD_4 + O_2$ reactions were ignited over the 0.05 wt % catalyst; a normal deuterium isotope effect was observed for the CO formation and the conversion of methane, but the isotope effect was below unity for the CO_2 formation. This means that the rate-determining step in the CO formation included the cleavage of a C−H(D) bond.

In summary, over the 0.05 wt % catalyst, only CO_2 was formed at 450 °C (below the ignition temperature) as a carbon-containing product from both CH_4 and CD_4. Between 450 and 600 °C, some CO was generated in the reactions involving CH_4 but not in those involving CD_4. As already mentioned, the former reactions but not the latter were ignited. Consequently, CO was formed only after ignition. The absence of CO below the ignition temperature is most likely due to the oxidative atmosphere caused by the incomplete consumption of O_2. The isotope effect for the CO_2 formation was much higher between 450 and 600 °C than at 450 °C; this probably occurred because of the different temperature dependencies of the reaction rates of CH_4 and CD_4 with O_2. The ignition temperature was between 450 and 525 °C for CH_4 and between 550 and 600 °C for CD_4. When the reaction was carried out close to but above the ignition temperature, e.g., 525 °C for $CH_4 + O_2$ reaction and 600 °C for $CD_4 + O_2$, the CO_2 yield was below but close to 25% in both cases. These results suggest that the combustion of methane occurred with the formation of CO_2 and H_2O and that CO was formed via the reforming reactions. At temperatures greater than 600 °C, methane combustion constituted the first step with complete consumption of O_2 in both cases. Because the reaction is oxygen-limited, only 25% of methane (CH_4 or CD_4) was converted and equal amounts of CO_2 were formed in the first step in both cases. In the second step, CO was generated through the reforming of methane by the CO_2 and H_2O formed in the first step. Because of the normal deuterium isotope effect, more CO was formed and more CO_2 consumed in the reaction involving CH_4 than CD_4. As a result, more CO_2 remained in the $CD_4 + O_2$ reaction and the isotope effect for CO_2 formation became less than 1.00 (Table 1). The combustion of methane is much faster than the CO_2 and H_2O reforming of methane. Therefore, the reaction rate for CO formation was controlled by the cleavage of a C−H(D) bond that occurred during the reforming process. The yield of CO became dominant at temperatures greater than 650 °C because the rates of the reforming reactions increase rapidly with temperature.

The 1.0 wt % Rh catalyst had a lower ignition temperature than the 0.05 wt % one. Over the former catalyst, at temperatures of ≥450 °C, both reactions were ignited. Compared to the 0.05 wt % catalyst, a smaller normal deuterium isotope effect for the CO formation was observed over the 1.0 wt % catalyst. While the isotope effect for the CO_2 formation was always below 1.00 over the 0.05 wt % catalyst, it was always above 1.00 over the 1.0 wt % catalyst and higher at temperatures of ≥550 °C than at temperatures of ≤500 °C. We suggest that the reaction involves the two mechanisms (combustion−reforming and pyrolysis−oxidation) occurring simultaneously at low temperatures (≤500 °C) and that the pyrolysis−oxidation mechanism is dominant at high temperatures (≥650 °C). Indeed, at low temperatures (≤500 °C), the isotope effects for the CO and CO_2 formation reactions were very different and an isotope effect above unity was observed for the CO_2 formation (Table 2). In the pyrolysis−oxidation mechanism, the CO and CO_2 originate from the same carbon-containing species, CH_x (x = 0−3), and consequently, comparable normal isotope effects (>1.00) are expected for CO and CO_2. Since the two isotope effects are very different at temperatures below 500 °C, it is likely that the pyrolysis−oxidation mechanism is not (or is not alone) responsible for the reaction. Similarly, the combustion−reforming mechanism cannot alone explain the experimental results because this mechanism would lead to an isotope effect smaller than unity for CO_2. However, the pyrolysis−oxidation mechanism can provide a deuterium isotope effect for CO_2 greater than unity and the combustion−reforming can provide different values for the isotope effects for CO and CO_2. Consequently, a combination of the two mechanisms is compatible with the experimental results. In addition, Table 2 shows that, particularly at 500 °C, the yield of CO_2 is greater than 25%. This implies that an additional pathway for the formation of CO_2 (e.g., CO oxidation) was present besides combustion. Therefore, it is likely that both mechanisms played a role at low temperatures. At high temperatures (≥650 °C), the isotope effects for CO and CO_2 were similar to each other and the CO selectivities were almost the same in both reactions (Table 2 and Figure 2b); consequently, the experimental results are compatible with the pyrolysis−oxidation mechanism. A similar result for the 0.5 wt % Rh/SiO_2 catalyst at 700 °C was also reported.[17] An additional argument for this mechanism is as follows. In the pyrolysis−oxidation mechanism, the CO and CO_2 selectivities are mainly determined by the following two competitive reactions: (i) the desorption of CO_s to CO_g, (ii) the oxidation of CO_s to $CO_{2,s}$ and desorption of $CO_{2,s}$. The rates of CO_s desorption to CO_g and of its further oxidation to $CO_{2,s}$ and desorption are free of any deuterium isotope effect. For this reason, the same product selectivity should occur in both $CH_4 + O_2$ and $CD_4 + O_2$ reactions. Figure 2b shows that at high temperatures (≥650 °C), the CO selectivities are almost the same in both cases. This strengthens the conclusion that the pyrolysis−oxidation mechanism is mainly responsible for the partial oxidation reaction over the 1.0 wt % catalyst at high temperatures. At low temperatures, the combustion−reforming mechanism also plays a role, and for that reason, the selectivities are no longer equal (Figure 2b and Table 2).

The experiments carried out with CH_4 pulses also provide some insight into the mechanism of the partial oxidation of methane. Indeed, in the reaction of pure methane with the reduced 0.05 and 1.0 wt % Rh catalysts, a noticeable amount of carbon was deposited at temperatures between 550 and 850 °C over the latter catalyst but not over the former catalyst. Of course, the generation of surface carbon from pure methane over the metallic rhodium sites follows a direct pyrolysis mechanism: $CH_{4,s} \rightarrow CH_{x,s} \rightarrow C_s$. Because no carbon was deposited over the 0.05 wt % catalyst, it is unlikely that the pyrolysis−oxidation mechanism played a major role over that catalyst. In contrast, carbon depositions occurred over the 1.0 wt % Rh catalyst, and this indicates that the pyrolysis−oxidation mechanism might have played a role in that case. Because at temperatures below 550 °C only relatively small amounts were deposited upon the 1.0 wt % catalyst, while at higher temperatures the amounts deposited were much higher, the role of the latter mechanism is expected to be greater at higher than at lower temperatures. These conclusions are in agreement with those revealed by the isotope experiments.

5. Conclusion

Pulse isotope effect experiments for the partial oxidation of CH_4 (CD_4) to synthesis gas over Rh/γ-Al_2O_3 catalysts as well as CH_4 pulse experiments over the same catalysts revealed that the mechanism of oxidation is dependent on the loading of the catalyst and the reaction temperature. It is likely that at low loadings (0.05 wt % Rh) a combustion−reforming mechanism is responsible for the reaction, while at high loadings (1.0 wt % Rh) a combination of the combustion−reforming and pyrolysis−oxidation mechanisms is responsible at low temperatures and the pyrolysis−oxidation constitutes the main mechanism at high temperatures.

References and Notes

(1) Trimm, D. L. *Catal. Rev. Sci. Eng.* **1977**, *16*, 155.
(2) Rostrup-Nielsen, J. R. *Catal. Today* **1993**, *18*, 305.
(3) Ashcroft, A. T.; Cheetham, A. K.; Foord, J. S.; et al. *Nature* **1990**, *344*, 319.
(4) Dissanayake, D.; Rosynek, M. P.; Kharas, K. C. C.; Lunsford, L. H. *J. Catal.* **1991**, *132*, 117.
(5) Hickman, D. A.; Schmidt, L. D. *J. Catal.* **1992**, *138*, 267.
(6) Dissanayake, D.; Rosynek, M. P.; Lunsford, L. H. *J. Phys. Chem.* **1993**, *97*, 3644.
(7) Hickman, D. A.; Schmidt, L. D. *Science* **1993**, *259*, 343.
(8) Torniainen, P. M.; Chu, X.; Schmidt, L. D. *J. Catal.* **1994**, *146*, 1.
(9) Buyevskaya, O. V.; Wolf, D.; Baerns, M. *Catal. Lett.* **1994**, *29*, 249.
(10) Slagtern, A.; Olsbye, U.; *Appl. Catal. A* **1994**, *110*, 99.
(11) Hu, Y. H.; Ruckenstein, E. *Catal. Lett.* **1995**, *34*, 41.
(12) Hu, Y. H.; Ruckenstein, E. *J. Catal.* **1996**, *158*, 260.
(13) Au, C. T.; Wang, H. Y.; Wan, H. L. *J. Catal.* **1996**, *158*, 343.
(14) Choudhary, V. R.; Uphade, B. S.; Belhekar, A. A. *J. Catal.* **1996**, *163*, 312.
(15) Mallens, E. P. J.; Hoebink, J. H. B. J.; Marin, G. B. *J. Catal.* **1997**, *167*, 43.
(16) Lago, R.; Bini, G.; Pena, M. A.; J. Fierro, L. G. *J. Catal.* **1997**, *167*, 198.
(17) Au, C. T.; Wang, H. Y. *J. Catal.* **1997**, *167*, 337.
(18) Looij, F. van; Geus, J. W. *J. Catal.* **1997**, *168*, 154.
(19) Tsipouriari, V. A.; Zhang, Z.; Verykios, X. E. *J. Catal.* **1998**, *179*, 283.
(20) Tsipouriari, V. A.; Verykios, X. E. *J. Catal.* **1998**, *179*, 292.
(21) Wang, H. Y.; Ruckenstein, E. *J. Catal.* **1999**, *186*, 181.
(22) Prettre, M.; Eichner, C.; Perrin, M. *Trans. Faraday Soc.* **1946**, *42*, 335.
(23) Hickman, D. A.; Hauofear, E. A.; Schmidt, L. D. *Catal. Lett.* **1993**, *17*, 223.
(24) Vernon, P. D. F.; Green, M. L. H.; Cheetham, A. K.; Ashcroft, A. T. *Catal. Lett.* **1990**, *6*, 181.
(25) Chang, Y. F.; Heinemann, H. *Catal. Lett.* **1993**, *21*, 215.
(26) Matsumura, Y.; Moffat, J. B. *Catal. Lett.* **1994**, *24*, 59.
(27) Nakagawa, K.; Anzai, K.; Matsui, N.; et al. *Catal. Lett.* **1998**, *51*, 163.
(28) Wang, D.; Dewaele, O.; Groote, A. M. De; Froment, G. F. *J. Catal.* **1996**, *159*, 418.
(29) Westheimer, F. H. *Chem. Rev.* **1961**, *61*, 265.

3

CATALYTIC COMBUSTION OF CLEAN AS WELL AS NITROGEN-BOUND FUELS OVER TRANSITIONAL METAL OXIDES

INTRODUCTION

With accelerating depletion of fossil fuel reserves and stricter regulations on pollutant (e.g., NO_x) emission to the atmosphere, there has been increased interest in combustion techniques that reduce pollutant emission levels from combustors while simultaneously increasing combustion efficiency. Soot, CO, nitrogen oxides, and unburned hydrocarbons are often found besides CO_2 and H_2O during the combustion of hydrocarbon fuel. The formation of CO, soot, and unburnt hydrocarbons is presumably caused by the incomplete oxidation of hydrocarbons. Their emission can generally be reduced by adopting higher primary zone temperatures and longer residence times, as well as by premixing and vaporizing the fuel. The nitrogen oxides (NO_x) emitted by combustion sources are predominantly in the form of nitric oxide (NO) with a residual, usually lower than 5%, in the form of nitrogen dioxide (NO_2). These oxides are formed either through the oxidation of atmospheric N_2 at high temperatures or through the oxidation of nitrogen-containing compounds present in the fuel. The relative contributions from these two sources are determined by the combustor design, nitrogen content in the fuel, and combustion conditions.

The following three mechanisms for the formation of NO_x are generally accepted: (i) the Zeldovich mechanism (thermal NO_x) [1], which becomes dominant under lean fuel conditions at high temperatures (>1500°C), (ii) the prompt NO mechanism [2], which is favored under fuel-rich conditions and moderate conditions, and (iii) the fuel-N oxidation mechanism [3], which is favored under fuel-lean conditions. For nitrogen-free fuel, NO_x forms primarily via the Zeldovich mechanism, which is strongly temperature dependent. If the combustor is operated under sufficiently fuel-lean conditions, which is unfavorable for the prompt NO mechanism, the resulting temperatures will be sufficiently low to minimize the formation of thermal NO_x. These temperatures are adequate for many practical purposes, even though the fuel-lean conditions are not suitable for the conventional combustors because the flame is not stable. They are, however, feasible for catalytic combustors, where the heterogeneous reactions combined with the thermal inertia of the bed stabilizes the combustion process. As a result, a reduction of one or two orders of magnitude in thermal and prompt NO_x emissions can be achieved using catalytic combustion. In the case of nitrogen-bound fuels, the fuel-N compounds are more readily oxidized to NO_x than the atmospheric nitrogen [4]. For this reason, the conventional combustors, in fuel-lean operations, tend to preferentially oxidize the fuel-N to NO_x, generating high emission levels of NO_x. Nevertheless, the use of a catalytic combustor can, in principle, alleviate this problem, if the catalyst can selectively oxidize the fuel, while converting the fuel-N to

Heterogeneous Catalysis: Experimental and Theoretical Studies, First Edition. Edited by Eli Ruckenstein and Haiyou Wang.

N_2.Therefore, the catalytic combustor can serve to reduce the emission of these undesired NO_x species as a result of temperature control and judicious selection of catalyst.

The principle of catalytic combustion is straightforward. Mixtures of fuel and air are passed over a catalyst kept at a temperature sufficiently high to promote complete oxidation. When reaction takes place, oxidized products, such as CO_2 and H_2O, are formed from organic fuels, and energy is released. The catalyst system is usually composed of an active component (catalyst) and a supporting component (substrate and washcoat). The choice of both components has a critical effect on the performance of the catalytic combustor. Key properties of an optimal catalyst include (i) low light-off (ignition) temperature, (ii) high catalytic activity to maintain complete oxidation of hydrocarbon fuels at the lowest levels of air preheat, (iii) low selectivity to NO_x for nitrogen-bound fuels, (iv) low pressure drop, excellent thermal shocking resistance, and high working temperature, as well as (v) long-term stability at elevated operation temperatures (1000–1500°C).

In general, noble metal catalysts, metal oxide catalysts, and a mixture of the two are effective as combustion catalysts. The use of noble metals other than Pt and Pd seems to be limited because of their high volatility, facile oxidation, and limited availability [5]. It has been reported that Pt and Pd are among the most active catalysts for several fuels [6, 7]. The high activities of these metals are presumably due to their ability to activate the H_2, O_2, C–H, and O–H bonds. Compared with base metal oxides, the noble metal catalysts are characterized by higher specific activities for hydrocarbon oxidation, greater resistance to the loss of low temperature activity, and better resistance to deactivation by sulfur at temperatures below 500°C [8]. Nevertheless, issues with Pt and Pd catalysts do exist because the pure metals sinter rapidly at temperatures ranging from 500 to 900°C, which results in a decrease of active surface and ultimately, of catalytic activity.

The catalytic activity of base metal oxides for the oxidation of CO and hydrocarbons by oxygen has been extensively studied [9–12]. Those with metal ions that can assume more than one valence state and can participate in redox reaction often provide high oxidation activities. Transition metal oxides have been found to be the most active ones [5], and the catalytic activities of these oxides are determined by the d-shell electron configuration of the metal [13]. Maxima in activity are generally observed for 3, 6, and 8 d-electrons, while minima for 0, 5, and 10 d-electrons. For the fourth period of transition metal oxides, Dixon et al. found peaks in activity for Cr_2O_3 and Co_3O_4 [14]. In addition, it is worthwhile to note that mixtures of oxides generally provide higher activity and stability than the single oxides [9, 15]. Compared with noble metal catalysts, metal oxide catalysts usually exhibit lower catalytic activity and higher ignition temperatures. The addition of trace amounts of noble metal, such as Pt, was found to significantly improve the performance of certain base metal oxide catalysts [16, 17].

Chapter 3 is devoted to the catalytic combustion of clean as well as nitrogen-bound fuels over transitional metal oxides. In brief, this chapter consists of three parts: the first is concerned with the catalytic oxidation of clean fuel over transitional metal oxides (**3.1**); the second with the oxidation of fuel bound nitrogen in a transitional metal oxide catalytic combustor (**3.2**); and the third is a review of catalytic combustion (**3.3**).

The development of catalytic combustors has two primary objectives: to attain levels of NO_x emissions substantially lower than those obtained with conventional ones, and to carry out stable combustion at low concentrations of fuel in air. The choice of a suitable catalyst is pivotal for the catalytic combustor to perform efficiently. In this regard, the oxides of chromium, magnesium, manganese, cobalt, and iron in single or in binary mixtures deposited on alumina washcoated ceramic substrates were evaluated in an experimental catalytic combustor (**3.1**). The most active catalyst among them was found to be Cr_2O_3–Co_3O_4, which provided complete conversion of very lean fuel mixtures, equivalence ratio 0.196, at low values of the inlet temperature, 575 K. The oxidation of a nitrogen-bound fuel over a binary transition metal oxide catalyst (5.63 wt% Cr_2O_3, 1.87 wt% Co_3O_4, balance alumina), which was demonstrated to be suitable for lean fuel propane oxidation (**3.1**), was investigated (**3.2**). The results indicate that, using this catalyst, remarkably low fuel-N conversions (10–20%) to NO_x are attainable even under fuel-lean operation if T_{in} (the inlet temperature) is sufficiently high.

In the review article entitled "Catalytic Combustions" (**3.3**), attempts were made to summarize the significant developments in various aspects of catalytic combustion. The article covers the following main topics: (i) principles of catalytic combustion, (ii) NO_x formation mechanisms and its control by catalytic combustion, (iii) catalyst systems, (iv) kinetics of fuel oxidation, and (v) modeling of catalytic combustion.

REFERENCES

[1] Zeldovich, Ya. B., *Acta Physicochem. USSR*, **21**, 557 (1946).

[2] Fenimore, C.P., *Thirteenth International Symposium on Combustion*, The Combustion Institute, 1971, p.373.

[3] Song, Y.H., Blair, D.W., Siminski, V.J., and Bartok, W., *Eighteenth International Symposium on Combustion*, University of Waterloo, 1980.

[4] Appleton, J.P., and Heywood, J.B., *Fourteenth International Symposium on Combustion*, The Combustion Institute, 1973, p.777.

[5] Kesselring, J.P., Krill, W.V., Atkins, H.L., Kendall, R.M., Kelley, R.M., and Kelley, J.T., EPA-600/7-79-181, August, 1979.

[6] Dixon, J.K., and Longfield, F.E., *Catalysis*, , **7** (Emmett, P.H., ed.), Reinhold, New York, 1960, p. 183.

[7] Bond, G.C., *Catalysis by Metals*, Academic, New York, 1962, p. 464.

[8] Daniel W.A., and Wentworth, J.T., Paper 486B, March Meeting of the Society of Automotive Engineers, 1962.

[9] Dadyburjor, D.B., Jewur, S.S., and Ruckenstein, E., *Catal. Rev. – Sci. Eng.*, **19** (2), 293 (1979).

[10] Morook, Y., and Ozaki, A., *J. Catal.*, **5**, 116 (1966).

[11] Yu Yao, Y.F., *J. Catal.*, **39**, 104 (1975).

[12] Margolis, L.Y., *Adv. Catal. Related Subjects*, **14**, 429 (1963).

[13] Dowden, D.A., and Wells, D., Actes Int. Congr. Catal., 2e, Paris, 1961, p. 1499.

[14] Dixon, G.M., Nicholls, D., and Steiner, H., *Proceedings of the Third International Congress on Catalysis*, Vol. **2**, Amsterdam, 1964, p. 815.

[15] Dwyer, F.G., *Catal. Rev.*, **6** (2), 261 (1972).

[16] Evans, R.C., *An Introduction to Crystal Chemistry*, Cambridge University Press, London, 1952.

[17] Johnson, D.W., Gallagher, P.K., Wertheim, G.K., and Vogel, E.M., *J. Catal.*, **48**, 87 (1977).

Catalytic Combustion of Propane Using Transitional Metal Oxides

RAVI PRASAD, LAWRENCE A. KENNEDY and ELI RUCKENSTEIN *Faculty of Engineering and Applied Sciences, State University of New York at Buffalo, Amherst, New York 14260*

(*Received July 30, 1979; in final form January 22, 1980*)

Abstract—The oxides of chromium, magnesium, manganese, cobalt and iron in single or in binary mixtures were used for the catalytic combustion of propane/air in an adiabatic tubular reactor. Each of the catalysts was deposited on an alumina coated honeycomb ceramic substrate. The effect of inlet temperature and equivalence ratio on the combustion efficiency, temperature and emission levels were investigated. Among the catalysts examined, the most suitable was the binary mixture of cobalt oxide and chromium oxide. Complete combustion started at an equivalence ratio of 0.196 and an inlet temperature of 580°K. The residence time was 19 ms and the NO_x emission index for these conditions was 0.03 g/kg fuel. Chromium oxide alone gives a maximum of 89 percent conversion at inlet temperatures slightly higher than those used with cobalt oxide–chromium oxide catalysts.

INTRODUCTION

The use of catalytic combustors has two primary objectives: to attain levels of pollutant emissions substantially below those possible with conventional ones, and/or to carry out stable combustion for low concentrations of fuel in air.

The most important pollutants from today's combustion equipment are carbon monoxide, particulate matter, unburnt hydrocarbons, sulfur-oxides and nitrogen-oxides. The formation of the first three is due to incomplete oxidation of fuels. Their emission can usually be decreased by several techniques such as higher primary zone temperatures, longer residence times, premixing and prevaporization of fuel. The emission of sulfur-oxides is due to the sulfur content of the fuel being oxidized, and for liquid or gaseous fuels these normally do not constitute a major pollutant. NO_x forms through the oxidation of the nitrogen contained in the fuel or in the air. Since the activation energy for the latter reaction is very high (~79 Kcal/gm mole) (Nelson, 1976; Glassman, 1977), NO_x forms significantly only at high temperatures and its concentration in the burnt gas can be maintained within acceptable limits if the temperatures are held below 1500°C. This suggests that significant formation of NO_x can be avoided by carrying out complete combustion at lower temperatures. Although sufficiently low emission temperatures are acceptable for many practical purposes, the equivalence ratios to which they correspond are too lean for the conventional flame to be stable. Because of the large thermal inertia associated with catalytic combustion, the lower limit of flame stability is decreased. In diffusion flames, the NO_x emission levels are typically 70–100 ppm and only using special procedures can they be reduced to levels as low as 30 ppm (Kesselring and Brown, 1976). Similar levels are obtained from premixed flames.

The oxidation process in the conventional flame is a homogeneous reaction taking place in the bulk gas phase. The objective of introducing a catalyst in the combustor is to carry out heterogeneous oxidation on the surface of the catalyst. By the choice of a suitable catalyst, one can ensure that the activation energy for the heterogeneous catalytic reaction is much lower than that for the purely homogeneous case. In a typical catalytic combustor burning light hydrocarbon fuels (*e.g.*, propane), the activation energy for the homogeneous oxidation reaction is approximately 25–50 Kcal/gm-mole (Hottel *et al.*, 1965; Fenn and Calcote, 1965; Kydd and Foss, 1965), while that for the heterogeneous catalytic oxidation reaction is approximately 11–15 Kcal/gm-mole (Hawthorne, 1975; Rader and Weller, 1974). Consequently, appreciable heterogeneous oxidation rates can be achieved for temperatures and fuel concentrations much lower than those required for the homogeneous reactions to proceed. In a

This paper, which was originally published in **Combustion Science and Technology**: 22, 271–280 (1980), is republished with permission.

catalytic combustor therefore, the reaction in the initial part of the bed is primarily catalytic and the rate is controlled by the rate of the surface reaction. The catalytic reaction increases the gas and surface temperature and therefore the surface reaction rate constant, which increases exponentially with temperature, soon becomes so large that the rate of transport of reactants to the surface becomes the controlling step. Thereafter, the overall heterogeneous reaction rate is controlled by the mass transfer rate to the surface. The energy release rate in the mass transfer controlled regime is typically orders of magnitude smaller than those obtainable in the conventional flames (Glassman, 1977). However, at sufficiently high temperatures, homogeneous reactions are initiated in addition to the heterogeneous reactions in the catalytic bed. The conversion then rapidly goes to completion and energy release rates comparable to conventional flames are achieved. The low inlet temperatures coupled with the lean equivalence ratios that are used, results in relatively low exit temperatures where NO_x formation is minimized.

The choice of the catalyst itself is based on three main considerations: (a) it should promote the total oxidation of hydrocarbons; (b) it should not favor the formation of NO_x; and (c) it should have good thermal stability.

During the time the present research project was being conducted at SUNY/Buffalo, other studies were being performed elsewhere on catalytic combustors. Experimental work done by Pfefferle *et al.* (1975), Pfefferle (1977) and Blazowski *et al.* (1975) demonstrated the low levels of NO_x emission attainable using catalytic combustors. Anderson *et al.* (1975) and DeCorso *et al.* (1976) studied the use of catalytic combustion for gas turbines, and their results confirmed the low NO_x emission levels which could be achieved. The above-mentioned studies and most others in this field have used the noble metal catalysts (typically Pt–Pd). The objective of the research herein described is to study the potential of several transition metal oxides as catalysts for catalytic combustors and to examine the effect of inlet temperature and equivalence ratio on the combustor performance.

CHOICE OF CATALYST

The transition metal oxides and noble metals are materials that are generally effective as combustion catalysts. Dowden and Wells (1961) have shown that the catalytic activity of transition metal oxides is determined by their d-shell electron configuration. Maxima in activity are usually observed for three, six and eight d-electrons while minima for zero, five and ten d-electrons. Dixon *et al.* (1974) using the fourth period transition metal oxides as catalysts, show that peaks in activity occur for Cr_2O_3 and Co_3O_4.

In a recent paper Blazowski *et al.* (1975) indicates that Co_3O_4, Cr_2O_3, CuO, NiO, MnO_2 and V_2O_5 are potential catalysts for hydrocarbon oxidation. This is in agreement with the above discussion. Research work done elsewhere, have however indicated that mixtures of oxides often exhibit greater stability and activity than a single oxide (Dwyer, 1972).

From the literature, it would appear that solely from the viewpoint of activity, the oxides of cobalt, chromium, or manganese would be the most promising among the metal oxides with platinum, palladium and silver the most promising among the metals. Metals in general show activity at lower temperature than the oxides, but it should be noted that the activities are about the same at temperatures above 670°K (Blazowski, 1975). In the research study undertaken, it was decided to use transition metal oxides for the following reasons:

a) At the temperatures present in the combustor, their activity would not be much lower than that of platinum catalysts.

b) Many of the transition metal oxides of interest are capable of withstanding very high temperatures (Cr_2O_3 and MgO are refractory materials) and high temperature stability is an extremely important criterion for combustion catalysts.

c) There was a possibility of cost reduction by use of transition metal oxides as compared to noble metals.

From Table I, it is seen that mixtures of oxides such as $MnO_2+Cr_2O_3$, $CoO+Cr_2O_3$, $MgO+Cr_2O_3$, $MnO_2+Fe_2O_3$, $NiO+Cr_2O_3$, $CuO+Cr_2O_3$, $FeO+Cr_2O_3$ and $CuO+Fe_2O_3$ are suitable candidates from the points of view of stability and activity. The maximum temperature capabilities of these different oxides were also evaluated from literature. Using the above discussions as a base, five of the above mixtures of oxides were selected as catalysts for purposes of experimentation. Since most of the above oxide combinations contain Cr_2O_3 and

TABLE I

Comparative activity of some metals, oxides and binary oxide compounds

Initial oxides	Activity	Stability	Initial oxides	Activity	Stability
$ZnO+Cr_2O_3$	+	+	$CoO+Fe_2O_3$	++	+
$ZnO+Cr_2O_3$	++	+	$CuO+Fe_2O_3$	+++	+
$NiO+Cr_2O_3$	++	+	$ZnO+Fe_2O_3$	+	—
$NiO+Cr_2O_3$	+++	+	$MnO_2+Fe_2O_3$	+++	+
$MnO_2+Cr_2O_3$	++	+	$HgO+Fe_2O_3$	+	—
$MnO_2+Cr_2O_3$	+++	+	Ag_2O+MnO_2	++	+
$CoO+Cr_2O_3$	++	+	$CoO+MnO_2$	++	+
$CoO+Cr_2O_3$	+++	+	$PbO+MnO_2$	+	—
$HgO+Cr_2O_3$	++	+	$BaO+TiO_2$	Does not work	
$CuO+Cr_2O_3$	++	+	$CaO+TiO_2$	Does not work	
$CuO+Cr_2O_3$	+++	+	Cr_2O_3	++	—
$FeO+Cr_2O_3$	++	+	Co_2O_3	++	—
$FeO+Cr_2O_3$	+++	+	CuO[a]	+	—
$MgO+Cr_2O_3$	+++	—	MnO_2	+++	—
$CuO+Al_2O_3$	++	+	NiO	++	—
$FeO+Al_2O_3$	++	+	Al_2O_3	Does not work	
$NiO+Al_2O_3$	++	+	CaO	Does not work	
$MgO+Al_2O_3$	Does not work		MgO	Does not work	
$ZnO+Al_2O_3$	Does not work		Pt	+++	+
$CaO+Al_2O_3$	Does not work		Pd	+++	+

[a] Reduced at high temperatures to the inactive form Cu_2O (Ref. Dwyer, 1972).

since it also corresponds to a maximum in activity, the sixth catalyst contained Cr_2O_3 alone. The compositions of the various catalysts tried and their melting or decomposition temperatures are given in Table II. It should be noted that the temperatures in Table II are for single oxides and improved temperature stability is obtained for the mixed oxides.

TABLE II

Maximum temperature capability of some oxides

Oxide	Maximum temperature, °C
Co_3O_4	900–950 (decomposes)
CoO_4	1795 (melting point)
Cr_2O_3	2266 (melting point)
Fe_2O_3	1565 (melting point)
FeO	1369 (melting point)
Fe_3O_4	1594 (melting point)
MgO	2852 (melting point)
MnO_2	535 (decomposes)
Mn_3O_4	1564 (melting point)
MnO	1785 (melting point)

CATALYST PREPARATION

The catalysts used were transition metal oxides deposited on alumina washcoated ceramic substrates. The supports used in the experiments were extruded monolithic ceramic substrates provided by Corning (material code CELCOR 9475 (EX-20)) having a mean cell density of 300 sq cell/sq inch. The monoliths were chosen as substrates because they are characterized by excellent thermal shock resistance, high strength, light weight, chemical inertness and reasonably high melting temperature. Moreover, they give rise to significantly lower pressure drop than other kinds of catalyst support systems. However, the BET surface area for these substrates are low, and it was found necessary to increase it by washcoating with high surface area alumina using a technique provided by Corning. The washcoating procedure consisted of dipping the monolith in a slurry of high surface area alumina under a partial vacuum followed by centrifugation, drying and calcination. Two washcoats were applied and resulted in the following: mean percent coating of alumina = 11.29 percent and approximate BET surface area (after 2 hr of calcination at 875°K) = 21.8 m^2/gm of coated substrate.

TABLE III

Mean composition[a] of catalyst	Range of inlet temperature	Range of equivalence ratios	Inlet velocity m/sec	Maximum conversion	Inlet temperature for maximum conversion	Minimum equivalence ratio for maximum conversion	NO_x Emission at maximum conversion g/kg fuel
3.94% Cr_2O_3, 1.31% MgO	370–455 °C	0.07–0.59	2.6	53%	417 °C	0.442	0.05
4.01% Cr_2O_3, 4.01% MgO	400–455 °C	0.07–0.59	2.6	100%	417 °C	0.589	0.09
2.81% Cr_2O_3, 0.94% Co_3O_4	300–455 °C	0.07–0.59	2.6	60.2%	310 °C	0.368	0.05
4.27% Cr_2O_3, 1.42% Co_3O_4	300–345 °C	0.07–0.66	2.6	100%	310 °C	0.196	0.03
2.26% Cr_2O_3, 0.76% MnO_2	300–370 °C	0.07–0.51	2.6	100%	310 °C	0.316	0.08
4.79% Cr_2O_3, 1.61% MnO_2	300–345 °C	0.07–0.51	2.6	67.7%	310 °C	0.196	0.03
1.85% Cr_2O_3, 0.62% Fe_2O_3	290–400 °C	0.07–0.51	2.6	80.8%	297 °C	0.268	0.05
5.07% Cr_2O_3	315–455 °C	0.7–0.27	2.6	81.9%	400 °C	0.125	0.03 <

[a] Balance is alumina and substrate.

The substrate was dried for 3 hr in an oven at 200°C, weighed and impregnated with a solution of the nitrate or nitrates of the respective oxides. Next, it is dried in a hot air stream (425°K) and calcinated at 975°K for 4 hr to decompose the nitrates and deposit the oxides on the surface. Subsequently, it was weighed and knowing the initial weight and the concentration of the coating solution, the composition of the catalyst was computed. The composition of various catalysts are given in Table III.

EXPERIMENTAL FACILITY

A schematic sketch of the experimental facility is shown in Figure 1. The facility consists of a 2-in diameter tubular stainless steel test section in which the catalyst coated honeycomb substrate was mounted and the oxidation of fuel took place. The inlet air to the test section was provided by a compressor-preheater assembly capable of supplying air over a wide range of pressures and temperatures. Propane was injected into the air stream in the mixing chamber preceding the test section. This chamber had a set of four large mesh wire screens near the inlet, designed to increase the turbulence in the flow and a set of five fine mesh screens near the exit to prevent flashback from the test section.

The air flow rate was measured using a standard orifice and the propane flow was controlled with needle valves and monitored using a set of two rotameters. Temperature and pressure probes were present at various points in the experimental set-up. At the exit of the test section, a gas sampling probe was mounted which could radially sample the exhaust. It was connected by a gas sampling line to the gas analysis instrumentation.

Measurement of oxygen, nitrogen, carbon monoxide and propane concentrations were done using a Varian 1420 dual column gas chromatograph. The columns contained either a 13X molecular sieve or Porapak Q. A Varian 3700 flame ionization chromatograph was used to monitor the unburnt hydrocarbons. The NO_x was monitored by a Thermal Electron Model 44 chemiluminescent analyzer. In addition a Dragerwerk bulb-type

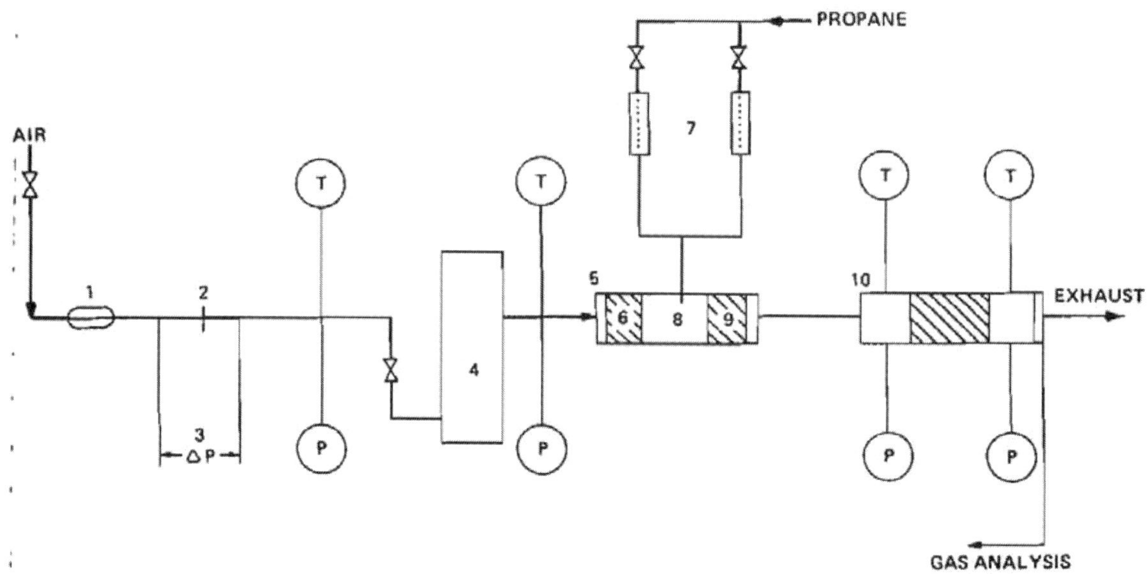

FIGURE 1 Schematic of experimental facility. (1) pressure controller, (2) orifice meter, (3) manometer bank, (4), air preheater, (5) mixing chamber, (6) mixing screens, (7) rotameters, (8) fuel injector, (9) flashback screens, (10) test section, (11) catalyst bed. *T*—Temperature probe, *P*—Pressure probe.

analyzer was available for analysis of the above gases and NO_x. Mass spectrometric analysis was also carried out to detect the trace species formed and also to verify the analysis done by the other devices.

The experimental set-up was designed to study the oxidation of lean, premixed, preheated propane air mixtures. Propane was chosen as fuel because it is representative of typical gaseous hydrocarbon fuels. The apparatus was built to operate over the following ranges of operating parameters: Equivalence ratio from 0.07 to 0.5, inlet velocity from 1 to 20 m/sec, inlet temperature from 315 to 730°K, and exit temperature from 975 to 1505°K.

QUALITATIVE ASSESSMENT OF CATALYTIC COMBUSTION OPERATION

The operation of catalytic combustors can in general be classified into three broad regimes. For very low values of inlet temperature, the rate of the surface reactions are negligible and consequently little or no oxidation takes place in the bed. As the inlet temperature is moderately increased, the surface reaction rate increases rapidly and some oxidation occurs in the bed. Here, the rate of the overall process is controlled by the rate of the reaction taking place on the catalyst surface. Under such conditions, the combustion efficiency is strongly dependent on the inlet temperatures. Typically efficiences of 0–35 percent are observed under such operations, and this constitutes the regime of surface kinetically-controlled operation. The temperatures that develop in this case are too low for homogeneous gas phase reactions to be of significance.

At somewhat higher values of inlet temperature, the combustor operation enters the regime of mass transfer control. In such a case, the reaction in the initial section of the bed is surface kinetically-controlled as before. However, oxidation taking place in this zone causes the gas and surface temperatures to increase rapidly. The surface reaction rate constant increases exponentially with increasing surface temperature and therefore the concentration of fuel is practically zero at the surface. Hence, the rate of transport of fuel to the surface is the limiting step. Thereafter, the overall oxidation rate in the bed is controlled by the transport of reactants to the surface. Thus, in this regime of operation, the reaction in the inlet section of the bed is kinetically controlled while the rest is controlled by the rate of mass transfer to the surface. Since mass transfer rates are not strongly temperature-dependent, the combustion efficiency becomes relatively independent of inlet temperature. Typically, 40–70 percent efficiencies are observed under such operation for various catalysts. As before, the gas temperatures in the bed do not become sufficiently high for homogeneous reactions to proceed at reasonable rates.

If the inlet temperatures are increased sufficiently, the gas temperatures in the latter part of the bed becomes high enough for homogeneous gas phase reactions to begin and thereafter the reaction rapidly proceeds to completion. Combustion efficiencies of nearly 100 percent occur in this regime of operation and these efficiencies are relatively independent of inlet temperature, once homogeneous reactions have begun.

At a fixed residence time, the temperature T_1 at which transition occurs from the surface kinetically-controlled region to the mass transfer-controlled region and the temperature T_2 at which transition occurs from the mass transfer-controlled region to the region of homogeneous reactions, decrease with increasing equivalence ratio. They also depend on the type of fuel which is oxidized and the properties of the catalyst.

Most conventional catalytic combustors, such as the ones used for the turbines driven by the tail gases in nitric acid plants and the automobile exhaust converters, operate under mass transfer control conditions. Because of this, complete conversions are not obtained in these applications and the energy release rates are not comparable with conventional combustors. For the combustor to be efficient and yield comparable energy release rates, operation in the third regime (*i.e.*, with homogeneous reactions) is necessary. Consequently, for the catalytic combustor to be successful, the value of T_2 should be sufficiently low and should be readily attainable for a wide range of equivalence ratios. Moreover, long catalyst life is also an important consideration.

EXPERIMENTAL RESULTS

Behavior of Various Catalysts

A summary of the important parameters studied and the results obtained for the combustor operation with each catalyst is given in Table III. The observations made for each of the catalysts are discussed below.

Cr_2O_3: MgO

This mixture exhibited a high light off temperature of the order of 690°K. Results for two different

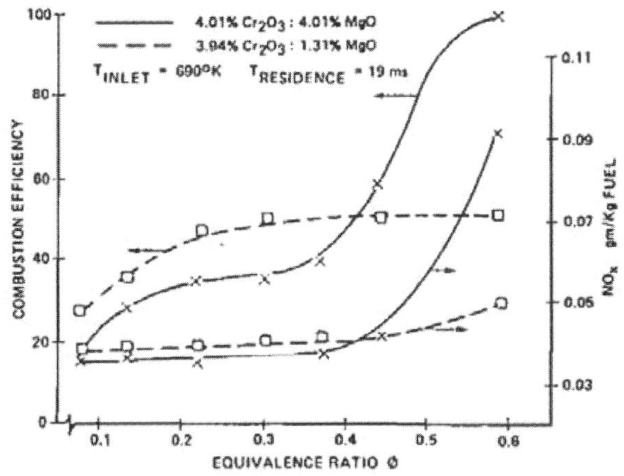

FIGURE 2 Combustion efficiency and NO_x emission index as a function of equivalence ratio for the Cr_2O_3–MgO catalyst.

surface concentrations of this catalytic mixture are presented on Figure 2 for an inlet temperature of 690°K.

For a catalyst containing 3.94 percent Cr_2O_3 and 1.31 percent MgO, the percent conversion of fuel rises from 28 percent at a low equivalence ratio of $\phi = 0.075$ to approximately 50 percent for $\phi = 0.25$. For $\phi = 0.25$ the process becomes diffusion limited and the combustion efficiency never rises above 51 percent. The NO_x emission index remains at approximately 0.04 gm/Kg fuel over the range $0.074 < \phi < 0.45$ and never exceeds 0.055 gm/Kg fuel.

Increasing the concentration of MgO in the catalyst mixture, 4.01 percent Cr_2O_3 and 4.01 percent MgO, the percent conversion of the fuel rises from 16 percent at the extremely low equivalence ratio of $\phi = 0.075$ to 100 percent at $\phi = 0.589$. The process is diffusion limited for $0.2 < \phi < 0.35$ with a combustion efficiency of approximately 35 percent. For equivalence ratios greater than 0.35, the combustion efficiency increases with ϕ attaining 100 percent at $\phi = 0.589$. The associated level of NO_x is less than 0.03 gm/Kg fuel for $\phi < 0.45$ and reached a maximum value of 0.09 gm/Kg fuel at 100 percent combustion efficiency.

Increasing the concentration of MgO in the catalyst mixture resulted in an increase in the combustion efficiency only for $\phi > 0.42$. Below this value of ϕ, surprisingly the combustion efficiency was somewhat less than achieved for the smaller concentration.

Cr_2O_3: Co_3O_4

Two surface concentrations of Cr_2O_3:Co_3O_4 were tested: 2.81 percent Cr_2O_3 and 0.94 percent Co_3O_4; 4.27 percent Cr_2O_3 and 1.42 percent Co_3O_4. Both these catalytic mixtures have a lower light off temperature than the Cr_2O_3 and MgO mixtures, typically of the order of 560°K. For the lower surface concentration mixture, the combustion efficiency increased with increasing equivalence ratio for inlet temperatures greater than 560°K. Figure 3 illustrates results for $T = 580$°K wherein the combustion efficiency starts from 14 percent at $\phi = 0.074$ and rises only to about 60 percent for $\phi > 0.4$. A possible explanation for this relatively poor conversion is that this catalyst combustor had an insufficient number of active sites to accommodate the reactive molecules; hence the catalyst concentration was increased to 4.27 percent Cr_2O_3 and 1.42 percent Co_3O_4. Figure 3 illustrates the improved performance. Starting from a combustion efficiency of 53 percent at $\phi = 0.067$, the combustion efficiency rapidly increased with increasing ϕ, attaining complete combustion at $\phi = 0.196$ with an exit temperature of 1105°C. The NO_x emission index at this point was 0.03 gm/Kg fuel. Further increase of the equivalence ratio showed a tendency for the combustion efficiency to drop to 90 percent with a resulting rise of the NO_x emission index to 0.11 gm/Kg fuel. This tendency again indicates insufficient active sites and may be overcome by further increase in the concentration of the catalyst or lengthening the bed.

The combustion was stable in all runs and deactivation was insignificant. The tests on this catalyst were terminated by meltdown of the substrate when the equivalent ratio reached 0.660 at an inlet velocity of 2.0 m/sec. The adiabatic flame temperature corresponding to these conditions exceeded the melting temperature of the substrate and hence meltdown should have been expected.

Cr_2O_3: MnO_2

Two levels of surface concentrations of Cr_2O_3: MnO_2 were examined, 4.79 percent Cr_2O_3 and 1.61 percent MnO_2 and 2.26 percent Cr_2O_3 and 0.76 percent MnO_2. Both mixtures exhibited good activity at inlet temperatures as low as 530°K but deactivated easily at high temperatures probably due to the volatility of MnO_2.

Figure 4 illustrates the significant deactivation which was observed for Cr_2O_3 and MnO_2. On the first run at an inlet temperature of 580°K, conversion increased rapidly from about 42 percent between $0.067 < \phi < 0.125$ to 100 percent at $\phi = 0.316$. Shutting the combustor off and subsequently making a second run showed a drastic reduction in combustion efficiency. No fuel conversion occurred below an equivalence ratio of 0.268 and subsequently only a 60–70 percent combustion efficiency was obtained for $\phi = 0.511$.

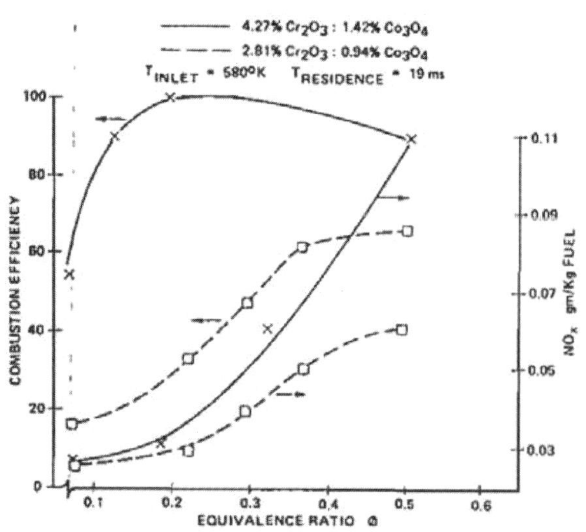

FIGURE 3 Combustion efficiency and NO_x emission index as a function of equivalence ratio for the Cr_2O_3–Co_3O_4 catalyst.

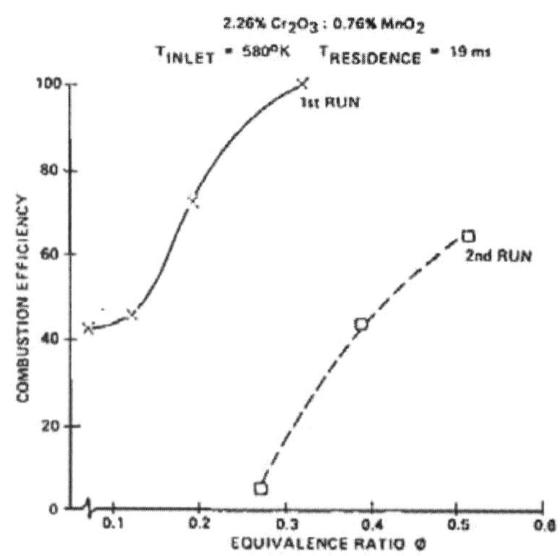

FIGURE 4 Decrease in combustion efficiency between two experimental runs.

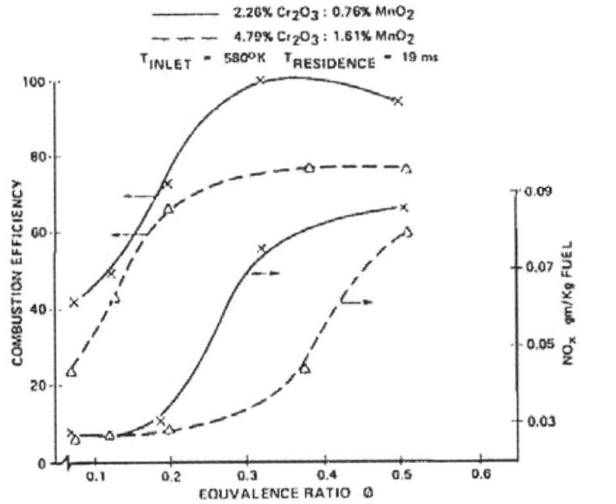

FIGURE 5 Combustion efficiency and NO_x emission index as a function of equivalence ratio for the Cr_2O_3–MnO_2 catalyst.

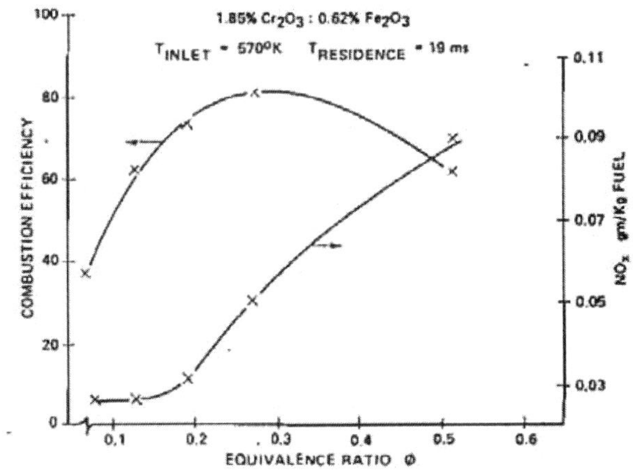

FIGURE 6 Combustion efficiency and NO_x emission index as a function of equivalence ratio for the Cr_2O_3–Fe_2O_3 catalyst.

Representative data for the conversion as a function of equivalence ratio are given for the two concentrations of the surface catalyst in Figure 5. Increasing the concentrations of Cr_2O_3 and MnO_2 did not result in improved performance. For the catalyst with the higher surface concentrations the process never passed beyond the mass transfer limited regime. In addition, its resistance to thermal degradation showed no improvement over the lower concentrations.

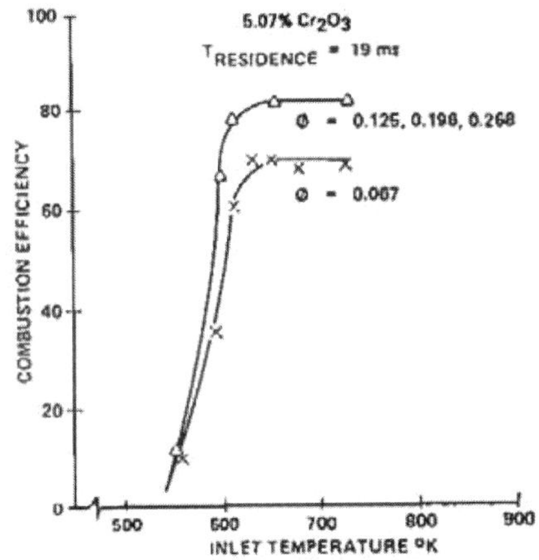

FIGURE 7 Combustion efficiency as a function of inlet temperature and various equivalence ratios for the Cr_2O_3 catalyst.

Cr_2O_3: Fe_2O_3

A mixture of 1.85 percent Cr_2O_3 and 0.62 percent Fe_2O_3 was examined and was found to yield good activity at low temperatures. The combustion efficiency was found to increase with increasing equivalence ratio up to a value of $\phi \approx 0.3$, but decreases for $\phi > 0.3$. This is thought to be a result of insufficient active sites and could be overcome with higher concentrations of the catalyst or extending the length of the bed. Figure 6 gives representative results of these results for an inlet temperature of 570°K.

Cr_2O_3

The single catalyst, Cr_2O_3, was examined for only one value of surface concentration (5.07 percent) which yielded a maximum combustion efficiency of 85 percent. However, this catalyst was cycled through many runs and proved to be very stable. Further work will examine if the combustion efficiency can be raised with increased surface concentration of this material. Figure 7 illustrates the combustion efficiency as a function of inlet temperature for Cr_2O_3.

Lastly it should be noted that all the catalysts tried were used in the combustor for a period of 10–15 hr of continuous operation. Catalyst Cr_2O_3 was used for over 40 hr of continuous operation and hence showed good stability for high temperature operation.

EMISSION OF POLLUTANTS

In the cases in which complete combustion was achieved, the emission levels of NO_x, CO, UHC were very low. The carbon monoxide and unburnt propane concentration could not be monitored on the gas chromatographs and NO_x emission index was below 0.11 gm/Kg fuel.

In cases where the combustion was incomplete (because of operation under kinetic or mass transfer control), the chief pollutant was unburnt propane. Low levels of CO emission were observed in some cases where the exit temperatures were high and conversion incomplete. NO_x emission levels were always very low and found to be strongly temperature dependent as would be expected. In no case did the concentration of NO_x exceed 0.11 gm/Kg fuel.

CATALYST MELTDOWN

As mentioned earlier, meltdown was experienced in a few of the cases studied. Such meltdowns were unexpected because the exit gas temperatures were significantly below the substrate melting point. To explore possible reasons for this phenomena, a mathematical model of our combustor was developed and numerically solved (Kennedy and Prasad, 1979). This model employed global rate expressions for the homogeneous reactions and included the enthalpy of reaction released in the heterogeneous reactions. The results showed that when the surface reaction is mass transfer controlled, the temperature at some points along the catalyst surface could exceed the gas temperature and could approach or exceed a temperature corresponding to the adiabatic flame temperature of the fuel/air mixture. For example with the catalyst Cr_2O_3–Co_3O_4, meltdown occurred at an equivalence ratio of 0.66 and an inlet temperature of 565°K. The measured gas temperature at the exit plane was 1336°K. However the corresponding adiabatic flame temperature for these conditions is approximately 2000°K. Thus under mass transfer controlled conditions, points along the catalyst surface could approach or exceed 2000°K which is above the melting temperature of the substrate.

CONCLUSION

The performance of several catalysts, consisting of binary mixtures of transition metal oxides deposited on alumina washcoated ceramic substrates, were evaluated in an experimental catalytic combustor. The most active catalyst among them was found to be Cr_2O_3–Co_3O_4 which gave complete conversion of very lean fuel mixtures, equivalence ratio 0.196, at low values of inlet temperature, 575°K. This catalyst was also found suitable for operation over a wide range of equivalence ratios and inlet temperatures. The NO_x emission index using this catalyst never exceeded 0.11 gm/Kg fuel—this being an order of magnitude smaller than that present in most conventional combustion systems. Catalytic deactivation was insignificant even after operating over several cycles.

Another suitable candidate was Cr_2O_3. Although this catalyst only yielded a maximum conversion of approximately 85 percent, it did not deactivate or melt even after prolonged operation at high temperatures. The NO_x emission was also very low being consistently below 0.09 gm/Kg fuel. It is believed that complete conversion can be achieved using this catalyst if a higher surface area substrate or larger surface catalyst concentrations than those employed in our experiments are used.

REFERENCES

Anderson, T. N., Tacina, R. R., and Mroz, T. S. (1975). Performance of a catalytic reactor at simulated gas turbine operating conditions. NASA TM X-71747.

Blazowski, W. S., and Walsh, D. E. (1975). Catalytic combustion: an important consideration for future applications. *Combustion Science and Technology*, **10**, 233.

De Corso, S. M., Mumford, S., Carrubba, R., and Heck, R. (1976). Catalysts for gas turbine combustors—experimental test results. ASME Paper No. 76-GT-4.

Dixon, G. M., Nicholls, D., and Steiner, H. (1964). *Proc. Third Intern. Congr. Catalysis*, **2**, 815, Amsterdam.

Dowden, D. A., and Wells, D. (1961). *Aetes 2e Congr. Intern. Catalyse*, **2**, 1489, Technip, Paris.

Dwyer, F. G. (1972). Catalyst for control of automotive emissions. *Catalysis Review*, **6**(2), 261.

Fenn, J. B., and Calcote, H. G. (1953). Activation energies in high temperature combustion. Laminar combustion and detonation waves. *Fourth Symposium (International) on Combustion*, The Combustion Institute, Pittsburgh, p. 231.

Glassman, I. (1977). *Combustion*. Academic Press, New York.

Hawthorne, R. D. (1974). Afterburner catalysts—effects of heat and mass transfer between gas and catalyst surface. AIChE Symposium Series, *Recent Advances in Air Pollution Control*, 428–438.

Hottel, H. C., Williams, G. C., Nerheim, N. M., and Schneider, G. R. (1965). Kinetic studies in stirred reactors: combustion of carbon monoxide and propane. *Tenth Symposium (International) on Combustion*, The Combustion Institute, Pittsburgh, p. 111.

Kennedy, L. A., and Prasad, R. (1979). Heat and mass transfer in a catalytic combustor. ASME/AIChE 18th National Heat Transfer Conference, San Diego, California, paper 79-HT-57.

Kesselring, J. P., and Brown, R. A. (1976). Catalytic oxidation of fuels for NO_x control from area sources. EPA-600/2-76-037.

Kydd, P. H., and Foss, W. L. (1965). Combustion of fuel-lean mixtures in adiabatic well stirred reactors. *Tenth Symposium (International) on Combustion*, The Combustion Institute, Pittsburgh, p. 101.

Nelson, H. F. (1976). Nitric oxide formation in combustion. *AIAA Journal*, **14**(9), 1177.

Pfefferle, W. C., Carrubba, R. V., Heck, R. M., and Roberts, G. W. (1975). Catathermal combustion: a new process for low-emissions fuel conversion. ASME Publication 75-WA/FU-1.

Pfefferle, W. C. (1977). The catalytic combustor—a look back, a look forward. Paper No. WSS, CI 77-31, presented at the 1977 Fall Meeting, Western States Section, Combustion Institute.

Rader, C. G., and Weller, S. W. (1974). Ignition on catalytic wires: kinetic parameter determination by the heated-wire technique. *AIChE Journal*, **20**(3), 515.

Oxidation of Fuel Bound Nitrogen in a Transitional Metal Oxide Catalytic Combustor

RAVI PRASAD, LAWRENCE A. KENNEDY and ELI RUCKENSTEIN *Faculty of Engineering and Applied Science, State University of New York at Buffalo, Buffalo, NY 14260*

(Received June 2, 1981; in final form July 22, 1981)

Abstract—Experimental results are reported concerning the oxidation of nitrogen containing fuels in a binary transition metal oxide catalytic combustor for both fuel-lean and fuel-rich operations.

The catalyst consisted of chromium and cobalt oxides deposited on high surface area alumina pellets (5.63 percent Cr_2O_3, 1.87 percent Co_3O_4, balance alumina). The fuel used was propane doped with 1 percent by wt. ammonia and the oxidizer was a mixture of O_2/Ar to eliminate thermal NO_x formation. The experiments were conducted over a wide range of inlet conditions and for every set of these, the exit concentrations of the dominant fixed nitrogen species (NO_x, NH_3 and HCN) as well as other major species (like O_2, CO_2, CO, H_2O and C_3H_8) were measured.

Under fuel-lean operation the sole product of fuel-N conversion was NO_x, while in fuel-rich operation HCN and unconverted NH_3 were also detected. The conversion to HCN and the amount of unconverted NH_3 increase, and the conversion to NO_x decreases, as the operation becomes more fuel rich. The concentration of HCN increases, while those of NH_3 and NO_x decrease as the inlet temperature (T_{in}) increases. The sum of the fixed nitrogen species shows a minimum at an equivalence ratio (Φ_{in}) of about 1.43.

The most significant result was that at sufficiently high T_{in} (600–640°C) the total emission levels of fixed nitrogen compounds (NO_x, NH_3 and HCN combined) can be reduced to very low levels (5–20 percent of the fuel-N present at the combustor inlet) in both fuel-lean as well as fuel-rich regions. Most of the fuel-N is converted to molecular nitrogen. This was also found to be true for lean fuel combustion when air was the oxidant. By comparison, fuel-lean operation with noble metal catalysts, typically yields 50–90 percent fuel-N conversion to NO_x.

1 INTRODUCTION

The reduction of nitrogen oxides (NO_x) emissions from combustion sources is currently an area of major interest. Catalytic combustion is a novel technique which appears to have potential for the reduction of NO_x emissions from the combustion of clean as well as nitrogen bound fuels. In this technique, an active surface (catalyst) is used to promote the combustion process.

For nitrogen free fuels, NO_x forms primarily via the Zeldovich mechanism, which is strongly temperature dependent. If the combustor is operated under sufficiently fuel-lean conditions, then the temperatures that result would be low enough to minimize the formation of thermal NO_x. These temperatures are adequate for many practical purposes, even though the fuel-lean conditions they correspond to are not suitable in conventional combustors because the flame is not stable. However, they are feasible in catalytic combustors, where the heterogeneous reactions combined with the thermal inertia of the bed stabilizes the combustion process. The type of catalyst used does not significantly affect the formation of thermal NO_x and a reduction of one to two orders of magnitude in thermal NO_x emission has been achieved using catalytic combustion (Pfefferle *et al.*, 1975; Blazowski *et al.*, 1975; DeCorso *et al.*, 1975; Anderson *et al.*, 1975; Prasad *et al.*, 1980).

In the case of nitrogen bound fuels, the fuel-N compounds are more readily oxidized to NO_x than the atmospheric nitrogen is (Appleton and Heywood, 1973). As a result, conventional combustors, in fuel-lean operation, tend to preferentially oxidize the fuel-N to NO_x, yielding high emission levels of NO_x. The use of a catalytic combustor can, in principle, alleviate this problem, if the catalyst could selectively oxidize the fuel, while converting the fuel-N to N_2. However, the catalytic combustion results reported in literature (Chu *et al.*, 1979; Pogson *et al.*, 1979; Folsom *et al.*, 1979; Folsom *et*

This paper, which was originally published in **Combustion Science and Technology**: 27, 45–54 (1980), is republished with permission.

al., 1980; Matthews *et al.*, 1977) show a high degree of conversion of fuel-N to NO_x in fuel-lean operation, implying that the catalysts used in those experiments did not possess this type of selectivity. Only by operating in the fuel-rich region, where the amount of oxygen is insufficient for complete combustion, was it possible to significantly reduce the emission level of the total fixed nitrogen (TFN) compounds (such as NO_x, NH_3, HCN) (Chu *et al.*, 1979; Folsom *et al.*, 1979; Folsom *et al.*, 1980; Matthews and Sawyer, 1977).

Almost all the studies on fuel-N oxidation in catalytic combustors have been done with noble metal catalysts. Since Pt is the commercially used catalyst for ammonia (which is representative of fuel bound nitrogen compounds) oxidation to NO, it is expected that, in fuel-lean combustion, a high degree of fuel-N conversion to NO_x would occur using this catalyst. This agrees with the results reported in literature (Chu *et al.*, 1979; Matthews *et al.*, 1977; Folsom *et al.*, 1980) and indicates additionally that, in catalytic combustion, the fuel-N conversion mechanism is dominated by the surface reactions. Consequently, the nature of the catalyst used would be expected to have a strong effect on the conversion of fuel-N. This is substantiated by the results of Folsom *et al.* (1980), who showed that significant reduction in fuel-N conversion, in the fuel-rich region, could be achieved by using a NiO–Pt catalyst as opposed to a Pt catalyst alone.

The objective of the experiments reported in this paper was to study the oxidation of a nitrogen bound fuel over a binary transition metal oxide catalyst. The catalyst used was one that had been found in an earlier study (Prasad *et al.*, 1980) to be suitable for lean fuel propane oxidation. The nitrogen bound fuel was simulated using propane doped with 1 percent by wt. ammonia. The oxidizer used was a mixture of O_2/Ar which eliminated the formation of thermal NO_x from air. The concentration of oxygen in this mixture was chosen so that, under stoichiometric conditions, the adiabatic flame temperature would remain below approximately 1225°K, for all inlet temperatures (T_{in}) of interest. This ensured that catalyst deactivation would be negligible during the experiment.

The equivalence ratio (Φ_{in}), inlet temperature (T_{in}) and inlet velocity (V_{in}) have been varied systematically over a wide range and their effects on the emission levels of NO_x, NH_3 and HCN have been studied. The results indicate that, using this catalyst, remarkably low fuel-N conversions (10–20

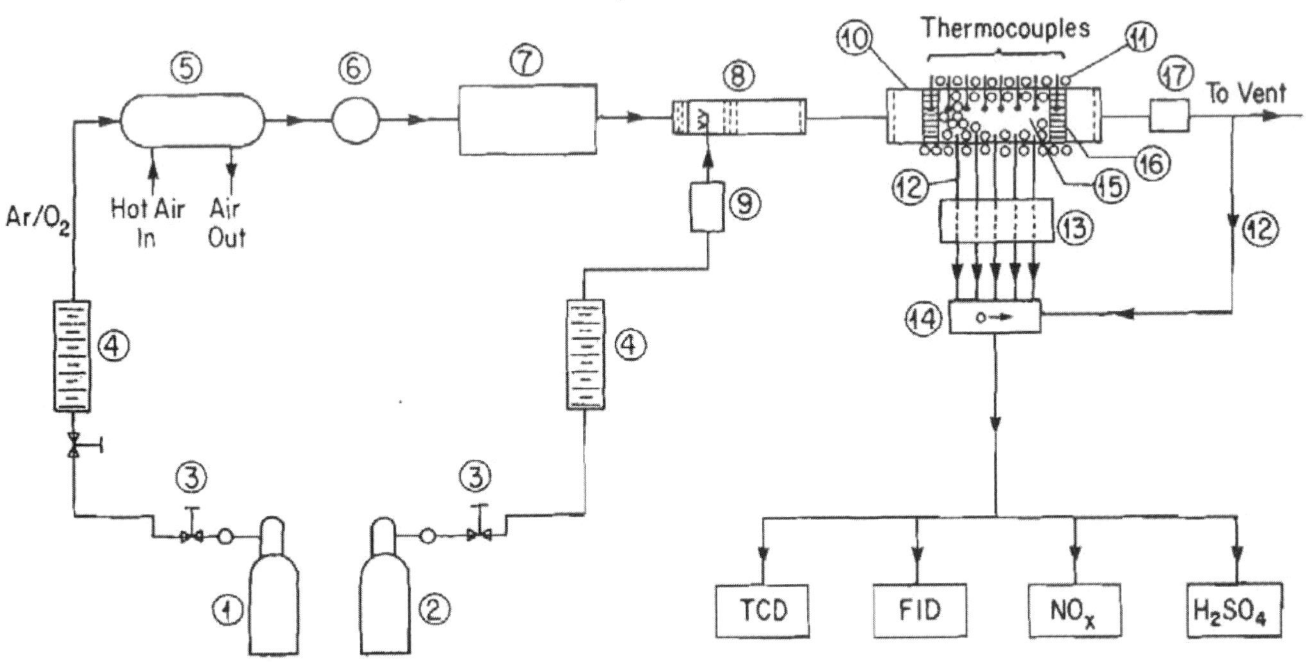

FIGURE 1 Schematic sketch of experimental facility: (1) Ar/O_2 gas cylinder; (2) C_3H_8/NH_3 gas cylinder; (3) pressure regulators; (4) flow meters; (5) heat exchanger; (6) upstream location of fuel injector; (7) furnace; (8) fuel injector—downstream location; (9) fuel preheater; (10) combustor; (11) heating elements; (12) gas sampling lines; (13) water bath; (14) gas sample line selector; (15) pellet bed; (16) blank monolith; (17) water jacketed probe. TCD—thermal conductivity detector; FID—flame ionization detector; NO_x—chemiluminescent NO_x analyzer; H_2SO_4—wet chemistry for NH_3.

percent) are attainable even under fuel-lean operations if T_{in} is sufficiently high.

2 EXPERIMENTAL DETAILS

A schematic sketch of the experimental facility is shown in Figure 1. The test section consists of a stainless steel tube, 7.6 cm long and 1.65 cm in diameter (i.d.), in which the catalyst bed is mounted and the oxidation of the fuel takes place. Pieces of monolith without catalyst, 1.5 cm long each, were placed before and after the catalyst bed to make the flow more uniform and to reduce heat losses from the combustor in the axial direction. In addition, to minimize radial heat losses, the test section was heavily insulated and heated from the outside using electrical guard heaters. The power supply to these guard heaters could be regulated to vary the heating rate.

The catalyst used consisted of binary transition metal oxides deposited on high surface area alumina pellets (5.63 per cent wt. Cr_2O_3, 1.87 percent Co_3O_4, balance alumina). The fuel used was propane doped with 1 percent by weight ammonia. This simulated other nitrogen bound fuels as well, because the fuel-N conversion is only weakly dependent on the type of fuel-N species (Chu *et al.*, 1979; Matthews *et al.*, 1977). The oxidizer used was a mixture of 2.4 percent O_2 and 97.6 percent Ar, to eliminate the formation of thermal NO_x. The oxygen concentration was chosen so that the adiabatic flame temperature, for all inlet conditions, would remain below approximately 1225°K. This ensured that catalyst deactivation would be negligible during the experiment.

The O_2/Ar mixture entering the test section was heated in two stages. In the first stage, consisting of a stainless steel heat exchanger, the above mixture is heated to approximately 300°C, using a hot-air stream. The second stage consisted of a three-zone Lindberg electric furnace whose temperature could be accurately controlled. Using this furnace the temperature of the O_2/Ar mixture could be further increased to approximately 700°C. The ammonia-doped fuel was injected into the gas stream using a multi-port opposed jet injector. The injector could be located either upstream or downstream of the Lindberg furnace and the propane could be preheated, if necessary, up to 800°C. Flow rates of O_2/Ar and C_3H_8/NH_3 were monitored and controlled using pre-calibrated Linde flowmeters.

Temperature measurements were made using Alumel–Chromel thermocouples located at the inlet, exit and at six points along the centerline within the bed. Five stainless steel gas sampling probes were present in the bed to measure the axial gas concentrations therein. In addition, there was another gas sampling probe located before the bed to measure the gas composition at the inlet of the combustor. Due to spatial limitations, these probes could be water cooled only outside the combustor. The exit gas analysis was carried out using a water-jacketed stainless steel probe which could be moved axially.

Measurements of oxygen, nitrogen, carbon monoxide, ammonia and carbon dioxide concentrations were carried out using a Varian 3700 dual column gas chromatograph. Low concentrations of ammonia were detected by passing known volumes of gas samples through 0.05 N H_2SO_4 solution which was subsequently analyzed. The concentrations of unburnt hydrocarbons and HCN were determined using a Varian 3700 flame ionization chromatograph. The NO_x concentration was measured with a Thermal Electron Model 44 Chemiluminescent Analyzer. No attempt has been made to determine the precise fate of the total fuel-N introduced. Instead, following other researchers (Chu *et al.*, 1979; Folsom *et al.*, 1980), only the dominant fixed nitrogen species (such as NH_3, HCN and NO_x) have been measured.

The range of inlet conditions studied was $0.5 < \Phi_{in} \leqslant 1.9$, $410 \leqslant T_{in} \leqslant 650$°C, $1.96 \leqslant V_{in} \leqslant 4.07$ m/sec.

Two types of experiments have been carried out. In the first type, T_{in} and V_{in} were fixed, while Φ_{in} was varied in steps from fuel-lean to fuel-rich conditions. Thereafter T_{in} or V_{in} have been changed and the experiments repeated. In the second type, Φ_{in} and V_{in} were fixed, while T_{in} was varied from a low (410–430°C) to a high (620–650°C) value. Subsequently Φ_{in} or V_{in} have been altered and the experiments were repeated. Note that due to the possible existence of multiple steady states (Prasad *et al.*, 1981), all these experiments were conducted in a heating up mode.

For every set of inlet conditions selected, the axial temperature profile was measured and gas samples from the inlet and exit of the combustor were analyzed to determine the concentrations of N_2, O_2, CO, CO_2, NH_3, NO_x, HCN and unburnt hydrocarbons. Attempts were also made to monitor the axial gas concentration profiles in the bed. In most cases, however, these measurements could not be carried out because the open ends of the gas sampling probes, inside the bed, were blocked by catalyst pellets drawn to them by the suction. Consequently, the results reported herein are based on exit measurements of gas composition.

3 RESULTS AND DISCUSSION

3.1 *Effect of Injector Location and Fuel Preheating*

Four different combinations of injector location and fuel preheating were tried. These were: (1) injector upstream of furnace and no preheating of fuel; (2) injector upstream of the furnace and fuel preheating; (3) injector downstream of the furnace and fuel preheating; (4) injector downstream of the furnace and no preheating of fuel.

Placing the injector upstream of the furnace resulted in significant oxidation of both fuel and fuel-N inside the furnace. Preheating the fuel to the same temperature as the exit gas temperature from the furnace, yielded very uniform temperature profiles at the inlet of the combustor, but introduced two difficulties. First, the preheating of fuel increased the probability of flashback and consequently, in many cases where preheating was used, a flame was found at the injector ports. The second difficulty was that, at high degrees of preheating, the propane was found to partially dissociate on the stainless steel walls of the fuel line. This led to carbon deposits on the walls that grew with time and finally choked the fuel line.

In view of these results the configuration used in most of the experiments was the one with no preheating of fuel and the injector downstream of the furnace. This was also found, in general, to yield the lowest emission levels of Total Fixed Nitrogen (TFN), while the configuration with the injector upstream of the furnace and preheating of fuel had the highest emission levels of TFN. It should be mentioned that the configuration chosen has very significant effects on both the NO_x emission levels as well as the dependence of the NO_x emission level on inlet parameters. Thus in Figure 2 it is seen that for $V_{in}=2.93$ m/sec, and $\Phi_{in}=0.72$ the fuel-N conversion is very different with the injector upstream and preheating of fuel, as opposed to the injector downstream and no preheating of fuel. In particular, at high T_{in} the former configuration shows increasing fuel-N conversion to NO_x, while the latter configuration shows decreasing fuel-N conversion to NO_x.

3.2 *Effect of Φ_{in} on Fuel-N Conversion*

Figures 3–5 show the conversion, at the combustor exit, of fuel-N to various fixed nitrogen species as a function of percent Theoretical Oxidizer (TO)

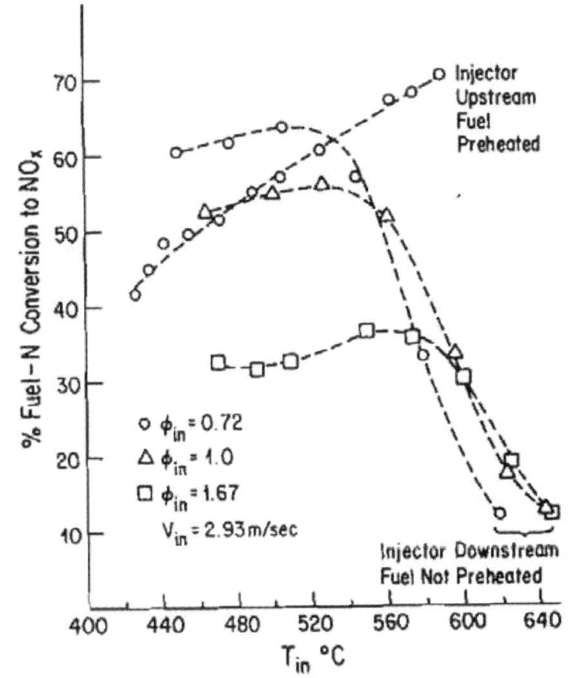

FIGURE 2 Effect of T_{in} on the fuel-N conversion to NO_x for various equivalence ratios. $V_{in}=2.93$ m/sec.

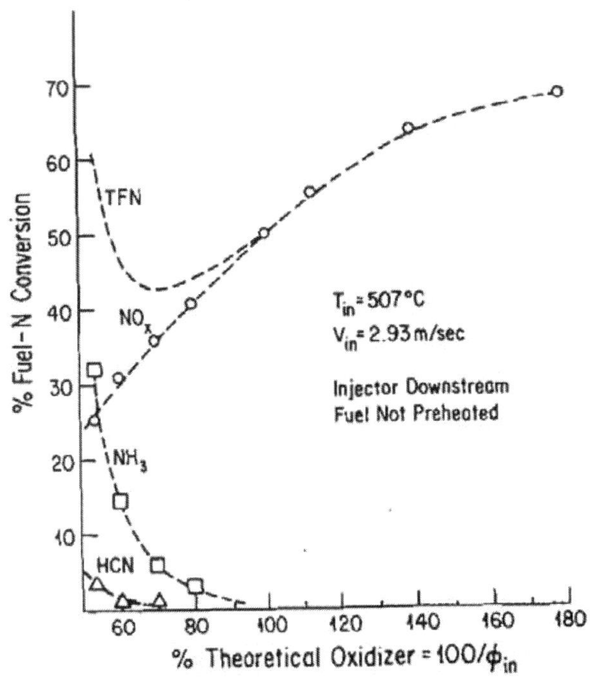

FIGURE 3 Effect of percent theoretical oxidizer on the fuel-N conversion to various fixed nitrogen species. $T_{in}=507$°C, $V_{in}=2.93$ m/sec.

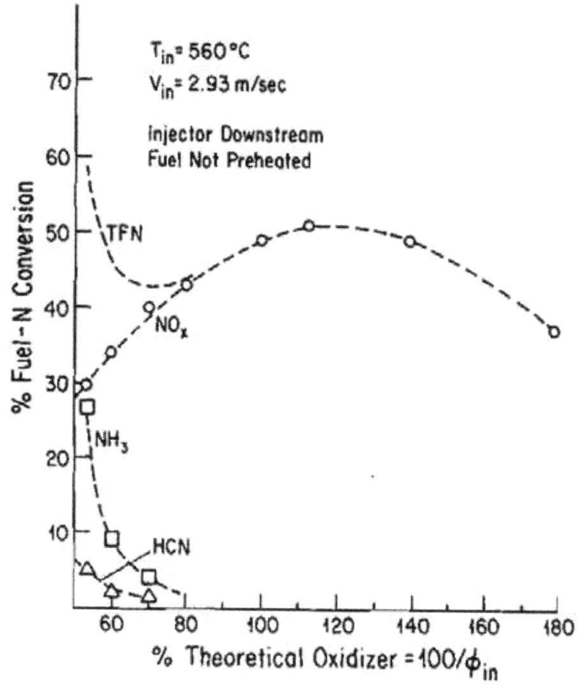

FIGURE 4 Effect of percent theoretical oxidizer on the fuel-N conversion to various fixed nitrogen species. T_{in}=560°C, V_{in}=2.93 m/sec.

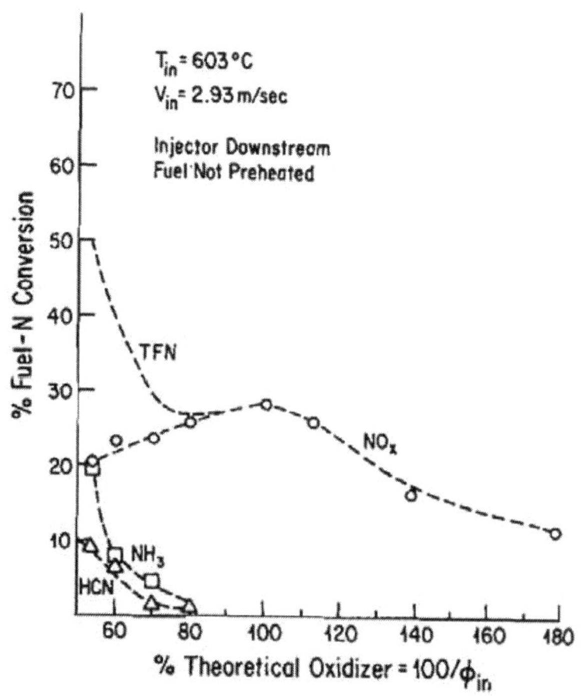

FIGURE 5 Effect of percent theoretical oxidizer on the fuel-N conversion to various fixed nitrogen species. T_{in}=603°C, V_{in}=2.93 m/sec.

(which is in the present case the equivalent of the Theoretical Air) for three different T_{in}. Since percent Theoretical Oxidizer = $100/\Phi_{in}$, values of TO greater than 100 denote fuel-lean operations, while values less than 100 correspond to fuel-rich operations. All three figures show that under fuel-lean operations, the sole product of fuel-N oxidation is NO_x. Only in the fuel-rich regions do HCN and unconverted NH_3 appear. This is consistent with the results reported for noble-metal based catalytic combustors (Chu *et al.*, 1979; Folsom *et al.*, 1980).

Under fuel-lean operations, the conversion of fuel-N to NO_x depends strongly on the inlet temperature and tends to decrease with increasing T_{in}. Thus for TO = 170 percent, percent fuel-N conversion to NO_x is 68 percent at T_{in} = 507°C, 40 percent at T_{in} = 560°C and only 12 percent at T_{in} = 603°C. In addition, the fuel-N conversion to NO_x exhibits a maximum with respect to TO. As T_{in} decreases the maximum displaces to larger values of TO, being at TO = 100 for T_{in} = 603°C and TO = 120 for T_{in} = 560°C. For T_{in} = 507°C the maximum occurs at more fuel-lean conditions than those plotted in Figure 3. The peak of fuel-N conversion to NO_x decreases as T_{in} increases, being 51 percent at T_{in} = 560°C, but only 28 percent at T_{in} = 603°C.

It should be noted that using mixed oxide catalysts, even in lean fuel operations and sufficiently high values of T_{in} (*e.g.* 603°C, Figure 5), the fuel-N conversion to NO_x is very low (12–28 percent). The major nitrogenous species produced is N_2. This represents a significant improvement over the results reported for the noble metal catalysts. On these catalysts, typically, 50–90 percent fuel-N conversion to NO_x is obtained under lean fuel conditions.

For TO < 100 (fuel-rich), the fuel-N conversion to NO_x decreases monotonically with decreasing TO (increasing Φ_{in}). Close to stoichiometric conditions (80–100 percent TO), NO_x is the dominant fixed nitrogen species and the conversion to HCN and NH_3 are small. However, as the operation becomes increasingly fuel rich (TO < 70 percent), the concentration of NH_3 and HCN rise sharply and they soon become major fixed nitrogen species. This is to be expected because under fuel-rich conditions, the formation of NO_x is inhibited by the lack of available oxygen. Similar results have also been found with noble-metal based catalytic combustors as well as in conventional combustors (Chu *et al.*, 1979; Song *et al.*, 1980). In those combustors, however, somewhat lower levels of NO_x emission were observed in the fuel-rich region. This is believed to be because the adiabatic flame temperatures used in them were significantly higher, leading to more effective reduction of the NO_x by the hydrocarbon species or CO present. Indeed it may be seen from Figure 2 that at high T_{in} the NO_x production is very low with this catalyst in the fuel-rich region.

The total of the three fixed nitrogen species measured (NO_x, HCN, NH_3), denoted by Total Fixed Nitrogen (TFN), exhibits a local minimum under fuel-rich conditions. This occurs because, in the fuel-rich region, there is a tradeoff between decreasing fuel-N conversion to NO_x and increasing fuel-N conversion to HCN and NH_3. The minimum occurs at about TO = 70 and its magnitude diminishes at higher T_{in}. At $T_{in} = 603°C$ the minimum is barely evident.

3.3 *Effect of T_{in} on Fuel-N Conversion*

O_2/Ar mixture as oxidizer Figures 6 and 7 show the fuel-N conversion to various fixed nitrogen species as a function of T_{in} for $\Phi_{in} = 1.67$ and two different values of V_{in}. It is seen that the conversion to NO_x first increases slightly and then drops sharply with increasing T_{in}. The decrease in fuel-N conversion to NO_x begins at about 580°C and the decline is more rapid at lower V_{in}. The concentration of NH_3 decreases monotonically, while that of HCN increases monotonically with increasing T_{in}. In the range of T_{in} used, the TFN decreases continuously with increasing T_{in}. However, it appears likely that if T_{in} is increased beyond some critical value, the conversion of fuel-N to TFN will start either to increase or to level off, due to the rising conversion to HCN. This should represent the value of T_{in} to be used for the catalytic combustor operation, in the fuel-rich region, to minimize the emission levels of TFN.

The effect of T_{in} on the fuel-N conversion to NO_x is shown in Figures 2, 8, 9 and 10 for various Φ_{in} and V_{in}. It is seen that, at low T_{in} (below 500°C), the fuel-N conversion to NO_x increases with increasing T_{in}, reaching a maximum in the range of 500–560°C. The maximum fuel-N conversion depends strongly on Φ_{in}, being higher in the fuel-lean region and decreasing to much lower values in the fuel-rich region (Figure 2). Thus the peak fuel-N conversion is 60–70 percent for $\Phi_{in} = 0.72$, but only 30–38 percent for $\Phi_{in} = 1.67$ (Figures 8, 10). Further increases in T_{in} cause the fuel-N conversion to drop sharply and reach very low values (5–20 percent) if T_{in} is in the range of 600–640°C. At such

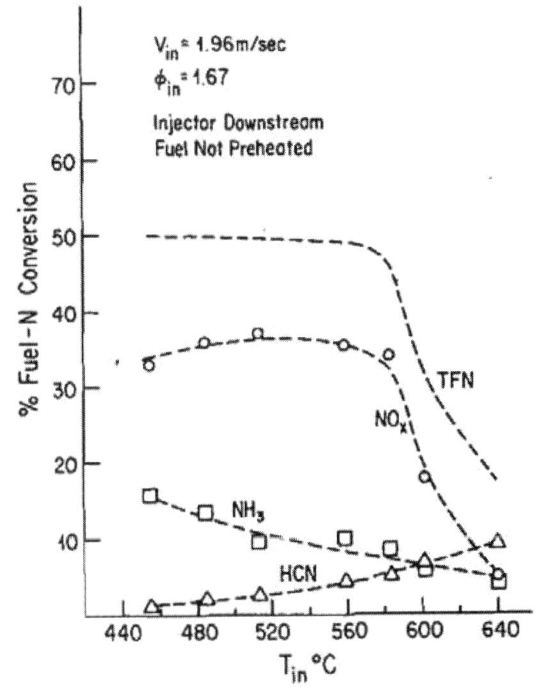

FIGURE 6 Effect of T_{in} on the fuel-N conversion to various fixed nitrogen species. $\Phi_{in} = 1.67$, $V_{in} = 1.96$ m/sec.

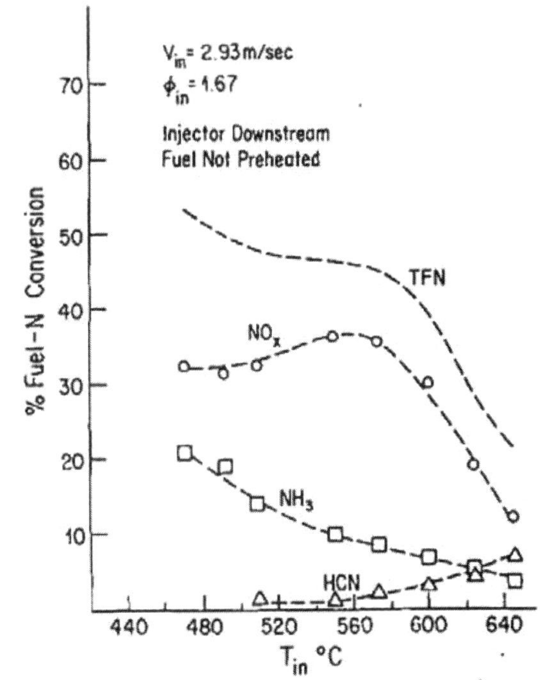

FIGURE 7 Effect of T_{in} on the fuel-N conversion to various fixed nitrogen species. $\Phi_{in} = 1.67$, $V_{in} = 2.93$ m/sec.

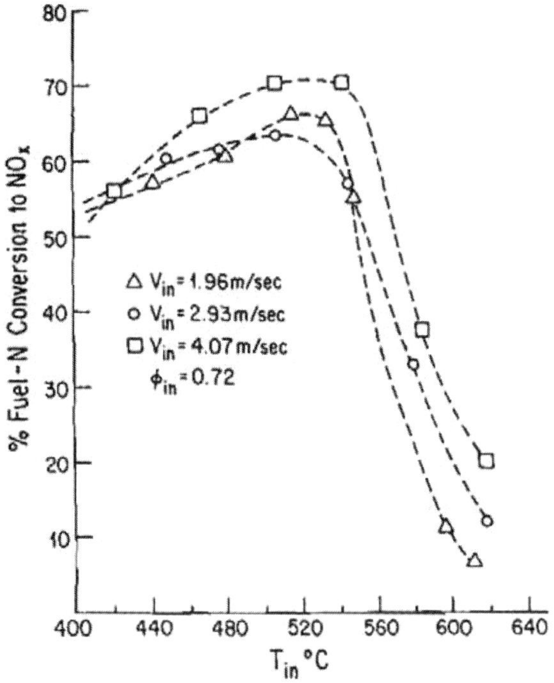

FIGURE 8 Effect of T_{in} on the fuel-N conversion to NO_x for various V_{in}. $\Phi_{in}=0.72$.

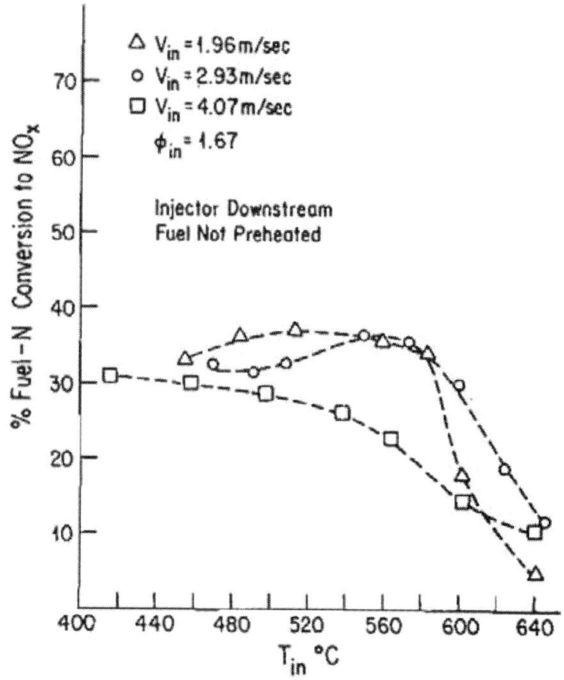

FIGURE 10 Effect of T_{in} on the fuel-N conversion to NO_x for various V_{in}. $\Phi_{in}=1.67$.

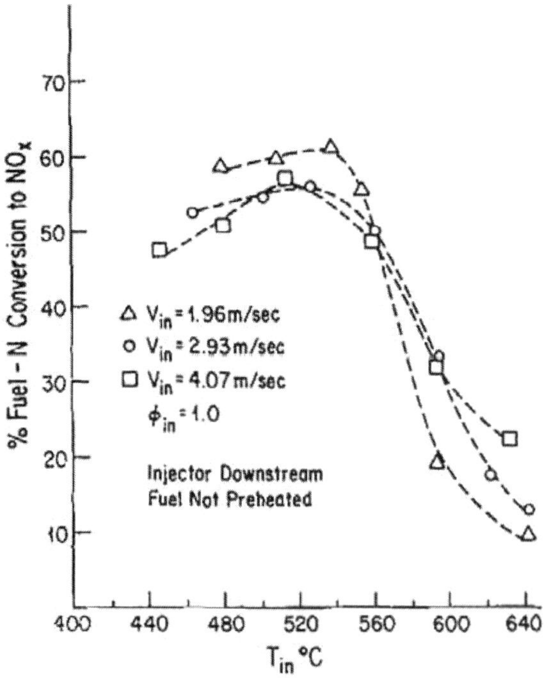

FIGURE 9 Effect of T_{in} on the fuel-N conversion to NO_x for various V_{in}. $\Phi_{in}=1.0$.

high T_{in} most of the fuel-N is converted to N_2. The decline in fuel-N conversion is very rapid under fuel-lean conditions and less rapid as the operation becomes fuel rich. Decreasing V_{in} also causes the fuel-N conversion to decline more rapidly at higher T_{in}.

The above observations on the variation of fuel-N conversion to NO_x with temperature in the fuel-lean and fuel-rich regions may be explained in the following manner:

Fuel rich region At low T_{in} the NO_x formation reactions are inhibited due to the lack of sufficient oxygen. This results in relatively low levels of fuel-N conversion to NO_x, but significant emission of unconverted fuel-N might occur (Figures 6 and 7). Hence at low T_{in}, the reactions leading to the formation of molecular nitrogen do not appear to be dominant. As T_{in} increases two things happen: (1) the reaction leading to the formation of N_2 and HCN from fuel-N become more significant, leading to a decrease in the emission of unconverted fuel-N (Figures 6 and 7). (2) The NO_x formed is more effectively reduced by CO and hydrocarbon species. Due to both these mechanisms, fuel-N conversion to NO_x and TFN drops with increasing T_{in}. This explanation is consistent with the fact that, at low T_{in}, the fuel-N conversion to NO_x is higher than that reported for noble-metal based catalytic combustors where much higher combustor temperatures were used.

Fuel lean region At low T_{in} the NO_x formation reactions are the dominant reactions in fuel-N conversion and the reactions leading to N_2 formation do not play a major role. Increasing T_{in} shifts the concentration and temperature profiles towards the entrance of the bed. Consequently, the reactions that at lower T_{in} were significant only towards the end of the bed, now become significant further upstream in the bed. Thus, the reactions leading to the oxidation of fuel-N now occur earlier in the bed. If this is true, then one would expect to see high conversions of fuel-N to NO_x early in the bed for those cases where the exit conversion to NO_x is small. This indeed was found in the few cases where concentration measurements could be made at a few points inside the bed. The sharp decline in fuel-N conversion at high T_{in} implies, however, that the NO_x formed in the initial part of the bed is reduced in the latter part of the bed, possibly by CO or hydrocarbon species. These reduction reactions become more rapid at higher temperatures and also occur over a greater portion of the bed length. In addition it is possible that, at high T_{in}, the reaction leading to N_2 formation becomes more significant. The combined effects cause a sharp decrease in fuel-N conversion at high T_{in}.

It is important to note that, for this catalyst, at sufficiently high temperatures (T_{in}=600–640°C) the fuel-N conversion to TFN is very low (5–20 percent). By contrast, with Pt catalysts, typically 50–90 percent fuel-N conversion to NO_x may be expected under fuel-lean operation. This observation represents the most significant contribution of this paper. The fact that the results obtained using this catalyst are significantly different from those obtained using noble-metal catalysts, or with conventional combustors, indicates that the fuel-N conversion reactions are primarily taking place on the catalyst surface.

Air as Oxidizer

All the experiments reported in this paper hitherto have used an O_2/Ar mixture for the oxidizer. The question arises as to whether the trends observed would be valid if air is used as the oxidizer. To check this further, the combustor was operated under fuel-lean conditions using air as the oxidizer and the ammonia-doped propane as fuel.

Figure 11 shows the fuel-N conversion to NO_x as a function of T_{in} for Φ_{in}=0.155, using air as the oxidizer. The conversion of fuel-N to NO_x is seen to first increase with T_{in} reaching a maximum of about 73 percent at T_{in}=550°C. For T_{in} higher than this value, the conversion to NO_x falls rapidly being only 30 percent at T_{in}=650°C. Hence the behavior is similar to that obtained using O_2/Ar mixture as the oxidizer. Note that the fuel-N conversion to NO_x shown in the figure includes the NO_x formed due to thermal fixation. However, for such fuel-lean conditions, this catalyst had, in an earlier study (Prasad *et al.*, 1980), produced only <5 ppm of thermal NO_x. Therefore the error introduced is not expected to be large.

In the fuel-rich region, since little thermal NO_x forms from the atmospheric nitrogen, one would not expect to see significant differences if air is used instead of O_2/Ar as the oxidizer.

3.4 *Effect of* V_{in} *on Fuel-N Conversion*

The effect of V_{in} on the conversion of fuel-N to NO_x is plotted in Figures 8 to 10. They show that for low T_{in}, in fuel-rich as well as fuel-lean operations,

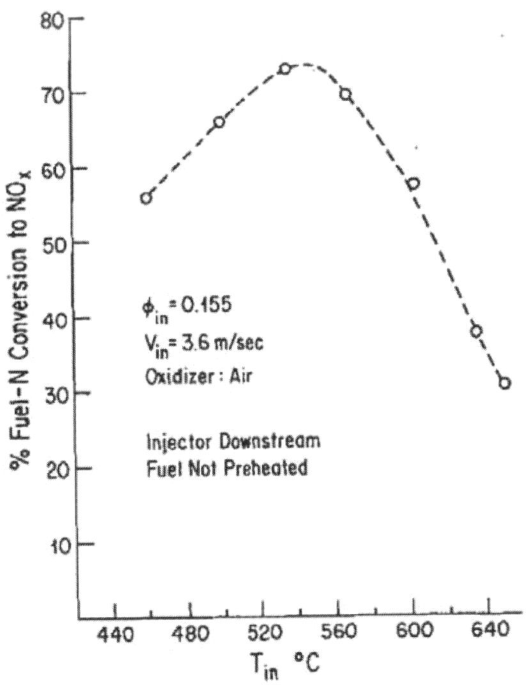

FIGURE 11 Effect of T_{in} on fuel-N conversion to NO_x. Φ_{in}=0.155, V_{in}=3.6 m/sec. Oxidizer: air.

the conversion to NO_x increases as V_{in} decreases. This is consistent with the earlier explanation that the NO_x formation reactions are dominant at low T_{in}. Consequently, increasing V_{in} (decreasing the residence time) should lead to lower NO_x emissions, and also to increasing amounts of NH_3. This was indeed observed experimentally, since at $V_{in}=4.07$ m/sec, for low T_{in}, NH_3 was detected at the combustor exit, even for fuel-lean operations.

At higher T_{in}, the fuel-N conversion to NO_x increases as V_{in} increases. This again is consistent with the earlier explanation because, at high T_{in}, the reaction leading to the formation of N_2 and reduction of NO_x become more significant. Therefore with increasing residence time (decreasing V_{in}), one would expect to obtain lower fuel-N conversion to NO_x.

The foregoing arguments suggest that the effect of V_{in} is primarily through its effects on the residence time. Therefore at low T_{in}, as V_{in} increases, one would expect to see increased NH_3 and decreased HCN emissions, while, at high T_{in}, as V_{in} decreases, one would expect to see decreased NH_3 and increased HCN emissions. These have again been observed experimentally.

4 CONCLUSION

In this paper we have studied the effect of inlet parameters (Φ_{in}, V_{in} and T_{in}) on the oxidation of fuel bound nitrogen in a mixed transition, metal oxides based, catalytic combustor. The nitrogen bound fuel was simulated by propane doped with 1 percent (by weight) ammonia. The oxidizer was a mixture of O_2/Ar which eliminated the formation of (thermal) NO_x from N_2 contained in air. In addition, the concentration of O_2 was selected so that the adiabatic flame temperature under stoichiometric conditions would be low enough to prevent catalyst deactivation.

The most significant result found was that by operating the combustor at sufficiently high T_{in}, the fuel-N conversion to NO_x could be reduced to very low values (5–20 percent), even under lean fuel conditions. By comparison, noble-metal catalysts typically produce 50–90 percent conversion of fuel-N to NO_x under lean fuel conditions.

At low values of T_{in}, significant fuel-N conversion (~70 percent) to NO_x can occur under lean fuel operation. This suggested that at high T_{in}, the NO_x formed in the initial part of the bed is subsequently destroyed, by secondary reactions, later in the bed. The measurements appear to support this explanation.

Under lean fuel operation the sole product of fuel-N conversion is NO_x, while under fuel-rich conditions the concentrations of NH_3 and HCN become significant. The conversion to NH_3 and HCN increases and to NO_x decreases as the operation becomes increasingly fuel rich. Increasing T_{in} causes the conversion to HCN to increase, while decreasing the concentration of NH_3.

The effect of V_{in} occurs primarily through the residence time. At low T_{in}, increasing V_{in} decreases the conversion to NO_x and HCN and increases that to NH_3. At high T_{in}, decreasing V_{in} decreases NH_3 and NO_x and increases HCN.

Lastly it should be mentioned that the results presented here indicate that the type of catalyst used has a significant effect on the oxidation of fuel-bound nitrogen. It appears possible to selectively oxidize the fuel while converting the fuel-N primarily to N_2. Further research, however, has to be done in the development of catalysts that may have better thermal stability and possess this type of selectivity.

ACKNOWLEDGEMENTS

This work was partially supported by DOE grant No. DE-FG22-80PC30220 and by an unrestricted Ford Motor Co. grant to the Combustion Laboratory. Portions of this work were completed while one of the authors (LAK) was supported as a Goebel Visiting Professor at the University of Michigan.

REFERENCES

Appleton, J. P., and Heywood, J. B. (1973). The effect of imperfect fuel–air mixing in a burner on NO formation from nitrogen in the air and fuel. *Fourteenth Symposium (International) on Combustion*, pp. 777–786, The Combustion Institute.

Anderson, D. N., Tacina, R. R., and Mroz, T. S. (1975). Performance of a catalytic reactor at simulated gas turbine operating conditions, NASA TM X-71747, NASA Lewis Research Center.

Blazowski, W. S., and Walsh, D. E. (1975). Catalytic combustion: An important consideration for future applications. *Combustion Science and Technology* **10**, 233–244.

Chu, E. K., and Kesselring, J. P. (1979). Fuel-NO_x Control in Catalytic Combustion. Environmental Protection Agency Publication EPA-600/7-79-038, pp. 281–330.

DeCorso, S. M., Mumford, S., Carrubba, R., and Heck, R. (1976). Catalysts for gas turbine combustors—Experimental test results. ASME Paper No. 76-GT-4.

Folsom, B. A., Courtney, C. W., and Heap, M. P. (1979). Environmental aspects of low Btu gas-fired catalytic combustion. Environmental Protection Agency Publication. EPA-600/7/79-038, pp. 345–384.

Folsom, B. A., Clark, W. D., Courtney, C. W., Heap, M. P., and Seeker, W. R. (1980). Fuel nitrogen conversion—The impact of catalyst type. Environmental Protection Agency Publication EPA-600/9-80-035, pp. 432–455.

Matthews, R. D., and Sawyer, R. F. (1977). Fuel nitrogen conversion and catalytic combustion. Paper No. 77-40, Fall Meeting, Western States Section/Combustion Institute.

Pfefferle, W. C., Carrubba, R. V., Heck, R. M., and Roberts, G. W. (1975). Catathermal combustion: A new process for low-emissions fuel conversion. ASME Paper No. 75-WA/FU-1.

Pogson, J., and Monsour, M. N. (1979). Catalytic combustion of No. 6 fuel oil. Environmental Protection Agency Publication EPA-600/7-79-038, pp. 111–138.

Prasad, R., Kennedy, L. A., and Ruckenstein, E. (1980). Catalytic combustion of propane using transitional metal oxides. *Combustion Science and Technology*, **22**, 271–280.

Prasad, R., Tsai, H. L., Kennedy, L. A., and Ruckenstein, E. (1981). Occurrence of multiple steady states in the catalytic combustion of propane. *Combustion Science and Technology*, **26**, 51–64.

Song, Y. H., Blair, D. W., Siminski, V. J., and Bartok, W. (1980). Conversion of fixed nitrogen to N_2 in rich combustion. 18th International Symposium on Combustion, August 17–22.

Catalytic Combustion

RAVI PRASAD
Linde Division
Union Carbide Corporation
Tonawanda, New York

LAWRENCE A. KENNEDY† AND ELI RUCKENSTEIN*
Faculty of Engineering & Applied Science
State University of New York at Buffalo
Amherst, New York 14260

*To whom correspondence should be addressed.
†Presently associated with the Department of Mechanical Engineering, the Ohio State University, Columbus, Ohio 43210.

This paper, which was originally published in **Catalysis Reviews – Science and Engineering**: 26(1), 1–58 (1984), is republished with permission.

I INTRODUCTION

Increased awareness of the problems caused by air pollution and the decrease of fossil fuel reserves have resulted in great interest in combustion techniques that reduce pollutant emission levels from combustors while simultaneously increasing combustion efficiency. The development of catalytic combustors has two primary objectives: to attain levels of NO_x emissions substantially below those possible with conventional ones, and/or to carry out stable combustion for low concentrations of fuel in air.

Significant research has been ongoing in this field since the early 1970s. Initial interest was primarily focused on the reduction of thermal NO_x from gas turbine (lean burn) combustors. The pioneering works of Pfefferle et al. [1], Blazowski et al. [2], Anderson et al. [3], and Wampler et al. [4] demonstrated that thermal NO_x reduction of one to two orders of magnitude is possible with lean burn catalytic combustors using noble metal catalysts.

In recent years, the use of catalytic combustion for a variety of applications—gas turbines and aircraft afterburners [5-14], systems with heat extraction [15-17], and small scale (domestic) heating [18-19]—has been investigated. At the same time, significant progress has been made in the development of improved catalyst systems such as the novel graded cell catalysts [20], oxide base catalysts [21-24], and active monoliths [23, 24]. In addition, the catalytic combustor has been successfully operated with a variety of liquid fuels such as #2 distillate oil, JP-4, and #6 Fuel Oil [1, 2, 6, 9, 16, 25, 26]. The use of liquid fuels has spurred the development of several injector systems suitable for use in catalytic combustors [27-32, 121].

More recently, much interest has been directed toward the combustion of alternate fuels such as heavy distillates, residual fuels, low Btu gas, and synthetic fuels using catalytic combustors [25, 27, 28, 32-36]. Since these fuels generally contain significant amounts of fuel-bound nitrogen, there have been numerous studies that have investigated the oxidation of fuel-N in catalytic combustors [22, 25, 37-40]. In general, under lean fuel conditions, catalytic combustors using noble metal catalysts produce very high levels of fuel-N conversion to NO_x (50-90%) although this can be reduced to much lower levels by operating in the fuel rich region ($\phi_{in} \sim 1.4$). (ϕ_{in} represents the ratio of the fuel concentration present to that required for all the oxygen in the air to be consumed. In the combustion literature this ratio is referred to as the equivalence ratio.) Somewhat improved performance in the fuel-rich region has been obtained by using a NiO-Pt catalyst [22]. However, of greater interest is the finding [40] that transition metal oxide based catalytic combustors yield very low levels of fuel-N conversion (10-20%), even in lean fuel operation ($\phi \sim 0.4$).

A number of papers aimed at studying the fundamental operating characteristics of catalytic combustors as well as the detailed physicochemical phenomena involved therein have appeared in the literature. These include works on the heterogeneous kinetics [41-44], the effect of inlet parameters and bed geometry [23, 45-48], the existence of multiple steady states together with ignition instabilities [49, 50], and the relative roles of surface and gas-phase reactions [51-53]. In addition, a number of mathematical models which successfully describe various aspects of catalytic combustion have been developed. The majority of these are steady state, one- and two-dimensional models [54-60]; two are transient, one-dimensional models [61, 62].

From the point of view of practical application, catalytic combustion has progressed to a stage where prototype system studies are being pursued. Perhaps the most ambitious is a joint Westinghouse-Englehard program in which full-scale, full-pressure tests have been successfully conducted on catalytic burners for combustion turbines [31]. The work being pursued at the Acurex Corporation on the use of catalytic combustors for boiler applications [63, 64], as well as the NASA program for the development of catalytic combustors for automotive gas turbines [24], appears to be promising.

In the present paper an attempt is made to summarize significant developments in various aspects of catalytic combustion in recent years. The material is presented on a topical basis; tables, figures, and/or data from the various works are presented where such are deemed necessary for clarification.

II. PRINCIPLES OF CATALYTIC COMBUSTION

The objective of introducing a catalyst in the combustor is to carry out heterogeneous oxidation on the surface of the catalyst.

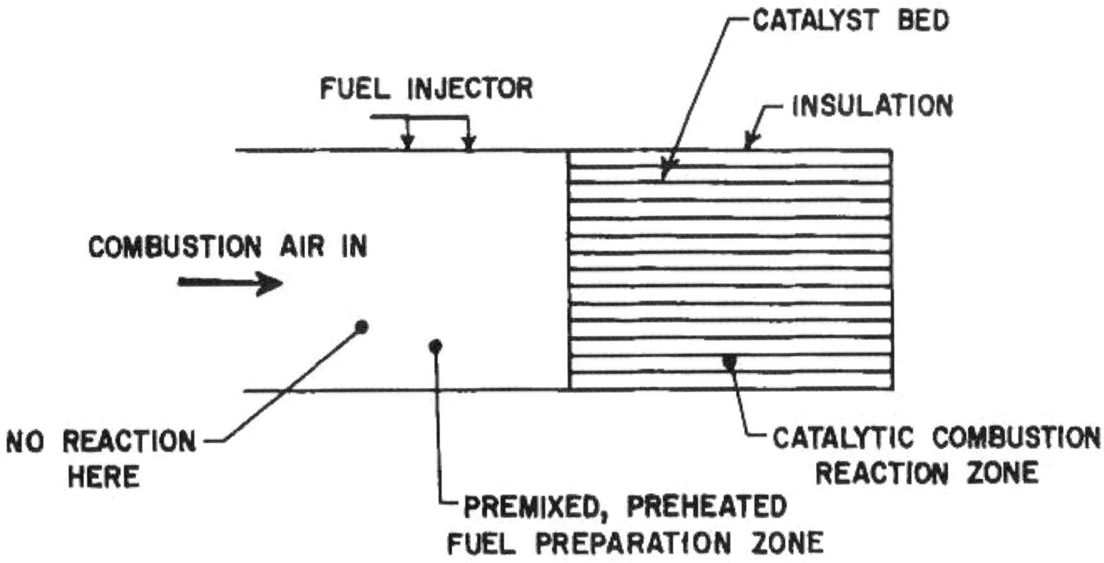

FIG. 1. A schematic sketch of the catalytic combustor.

By choice of a suitable catalyst, one can ensure that the activation energy for the heterogeneous reaction is much lower than that for the purely homogeneous case. In a typical catalytic combustor burning light hydrocarbon fuels (e.g., propane), the activation energies for the homogeneous oxidation reactions are approximately 25-50 kcal/g·mol, while those for the heterogeneous oxidation reactions are approximately 10-20 kcal/g·mol (refer to Tables 3 and 5). Consequently, appreciable heterogeneous oxidation rates can be achieved for temperatures and fuel concentrations much lower than those required for the homogeneous reactions to proceed. Because of this, and also due to the large thermal inertia of the bed, catalytic combustors can sustain stable combustion at fuel concentrations much below those possible in conventional combustors.

Functionally, the catalytic combustor (Fig. 1) consists of a catalyst bed through which a premixed and preheated fuel/air mixture is passed. In the first part of the bed, where the catalyst temperature is low, the reaction is "kinetically controlled," and the rate of the overall process is controlled by the intrinsic surface kinetics. This region is designated by A in Fig. 2. In the kinetically controlled region, the reaction rate increases exponentially with temperature. For more active catalysts or those operated at higher temperatures or fuel concentrations, the intrinsic surface reaction rate can be so high that the overall process is

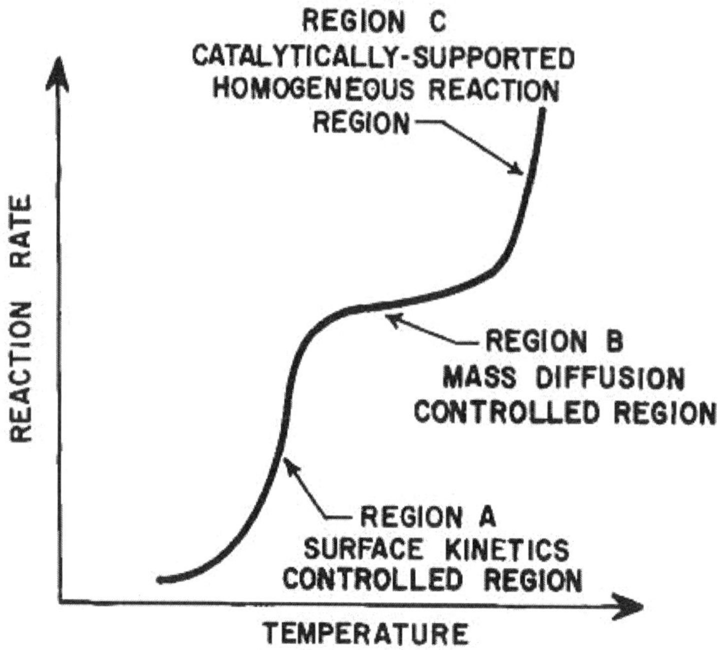

FIG. 2. Overall reaction rate vs temperature for a typical reactor used in catalytic combustion. Regions A, B, and C denote regimes of heterogeneous kinetics control, mass transfer control, and homogeneous kinetics control, respectively.

limited by the rate of species transport to the surface. The reaction is then "mass transfer controlled," and in this region (Region B, Fig. 2) the observed reaction rate is quite insensitive to changes in the temperature because the mass transfer coefficient is a rather weak function of temperature. It is, therefore, impossible to distinguish between catalysts of different activity under mass transfer controlled operation because the heterogeneous kinetics play a secondary role in determining the reaction rate. As the exothermic surface reaction proceeds further, the axial temperature increases, and finally, at some point in the bed, the bulk gas temperature becomes so high that the gas phase reactions occur simultaneously with the catalytic reactions and the combustion goes to completion. This is designated by Region C in Fig. 2.

Thus, in a catalytic combustor, the combustion process is initiated and stabilized* by the catalytic substrate. The surface reaction,

*The reaction front in ordinary flames becomes unstable at low fuel concentration resulting in its extinguishment.

coupled with the large thermal inertia of the bed, enables catalytic combustors to carry out stable, highly efficient (99.9%) combustion over a wide range of equivalence ratios. The ability of catalytic combustors to carry out complete combustion at low fuel concentrations (and hence low combustion temperatures) enables them to operate at temperatures where thermal NO_x formation can be minimized. Thermal NO_x reduction of up to two orders of magnitude below those possible in conventional combustors has been achieved. This is discussed in greater detail in the following section. In addition, because the fuel and air are premixed, and because of the plug flow inlet conditions often used, catalytic combustors provide very uniform temperature and velocity profiles at the exit. Due to the high combustion efficiencies, soot formation is absent. The pressure drop across the combustor is also small because most catalytic combustors use monolithic honeycomb substrates as catalyst support.

III. NO_x FORMATION AND CONTROL

A. Formation Mechanisms

The nitrogen oxides emitted by combustion sources are predominantly in the form of nitric oxide (NO) with the residual, usually less than 5%, in the form of nitrogen dioxide (NO_2). These oxides are formed either by the oxidation of atmospheric nitrogen at high temperatures or by the oxidation of nitrogen compounds in the fuel. The relative contributions from these two sources depend on the combustor design, nitrogen content of the fuel, and operating conditions. The kinetics of NO_x formation from both atmospheric and fuel nitrogen are reasonably well understood for homogeneous reactions, such as those present in flames. However, virtually no information is available regarding the kinetics of oxidation or reduction of nitrogeneous species on the surface of catalysts used in catalytic combustion. Hence, the kinetics and mechanisms discussed here are for homogeneous reactions only. Nevertheless, it is important to note that recent research [22, 40] indicates that surface catalytic reactions could significantly influence the oxidation of fuel-bound nitrogen.

It is generally accepted that the formation of NO_x occurs by three mechanisms: the Zeldovich mechanism (thermal NO_x) [98], Prompt NO mechanism [99], and the fuel-N oxidation mechanism. Only the first two occur in the case of clean (nitrogen-free) fuels while the third, which occurs for nitrogen-bound fuels, consists of the first two mechanisms and a number of other reactions as well. We discuss them separately here.

1. Thermal NO_X and Prompt NO Formation

For clean fuels, under fuel lean conditions, the Zeldovich mechanism is the predominant method of NO_X formation. This reaction is strongly temperature-dependent and the NO_X formation rate is significant only at temperatures greater than ∿1500°C, thereafter doubling for every increase in flame temperature of about 40°C [123]. Hence, the name "thermal NO_X."

The extended Zeldovich mechanism which models the oxidation of N_2 by O_2 is given by the following steps [100]:

$$N_2 + O \underset{K_{-1}}{\overset{K_1}{\rightleftharpoons}} NO + N \tag{1}$$

$$N + O_2 \underset{K_{-2}}{\overset{K_2}{\rightleftharpoons}} NO + O \tag{2}$$

$$N + OH \underset{K_{-3}}{\overset{K_3}{\rightleftharpoons}} NO + H \tag{3}$$

The kinetic expressions for the various steps are listed in Table 1 (Thermal NO Path and N_2 Production Path). The rate of the above reactions is basically controlled by the forward step of Reaction (1) due to its high activation energy. Because of this, "thermal NO_X" formation is highly temperature dependent.

The Zeldovich mechanism gives the following expression for the rate of NO production:

$$\frac{d[NO]}{dt} = K_1[O][N_2] - K_{-1}[N][NO] + K_2[N][O_2] - K_{-2}[O][NO] + K_3[N][OH] - K_{-3}[H][NO] \tag{4}$$

The steady-state approximation for the concentration of N is generally used because this species is present in very small amounts compared to the remaining species in the range of temperatures of interest in ordinary combustion. This yields

$$[N]_{SS} = \frac{[O]\{K_1[N_2] + K_{-2}[NO]\} + K_{-3}[H][NO]}{K_{-1}[NO] + K_2[O_2] + K_3[OH]} \tag{5}$$

TABLE 1 [112]

Reactions Involving Fixed Nitrogen during Combustion[a]

	Reaction	Rate coefficient ($cm^3/mol \cdot s$)
	Thermal NO Path	
1	$N_2 + O \rightarrow NO + N$	$7.6 \times 10^{13} \exp(-75{,}500/RT)$
2	$N + O_2 \rightarrow NO + O$	$6.4 \times 10^{9} T \exp(-6{,}250/RT)$
3	$N + OH \rightarrow NO + H$	3×10^{13}
	Fuel NO Path	
4	Fuel N $\rightsquigarrow$ HCN	Fast
5	$HCN + OH \rightarrow CN + H_2O$	$2 \times 10^{11} T^{0.634} \exp(-4{,}600/RT)$
6	$HCN + H \rightarrow CN + H_2$	$2 \times 10^{11} T^{0.69} \exp(-18{,}400/RT)$
7	$HCN + O \rightarrow CN + OH$	$1.4 \times 10^{11} T^{0.68} \exp(-16{,}900/RT)$
8	$CN + OH \rightarrow NCO + H$	5.6×10^{13}
9	$HCN + OH \rightarrow HNCO + H$	2×10^{11}
10	$NCO + H \rightarrow NH + CO$	2×10^{13}
11	$HNCO + H \rightarrow NH_2 + CO$	1×10^{13}
12	$NH_2 + H \rightarrow NH + H_2$	$1.4 \times 10^{11} T^{0.67} \exp(-4{,}300/RT)$
13	$NH_2 + O \rightarrow NH + OH$	$9.2 \times 10^{11} T^{0.5}$
14	$NH_2 + OH \rightarrow NH + H_2O$	$3 \times 10^{10} T^{0.68} \exp(-1{,}300/RT)$
15	$NH + OH \rightarrow N + H_2O$	$1.6 \times 10^{12} T^{0.56} \exp(-1{,}500/RT)$
16	$NH + O \rightarrow N + OH$	$8.4 \times 10^{12} T^{0.7} \exp(-100/RT)$
17	$NH + H \rightarrow N + H_2$	$1 \times 10^{12} T^{0.68} \exp(-1{,}900/RT)$
18	$NH + O \rightarrow NO + H$	$5 \times 10^{11} T^{0.5} \exp(-5{,}000/RT)$
19	$NH + OH \rightarrow NO + H_2$	$1.6 \times 10^{12} T^{0.56} \exp(-1{,}500/RT)$
	Prompt NO Path	
20	$N_2 + CH \rightarrow HCN + N$	$4.0 \times 10^{11} \exp(-13{,}600/RT)$
21	$N_2 + C_2 \rightarrow 2CN$	Unavailable
22	$N_2 + C \rightarrow CN + N$	Unavailable

(continued)

TABLE 1 [112] (continued)

	Reaction	Rate coefficient ($cm^3/mol \cdot s$)
	NO Conversion Path	
23	$CH + NO \rightarrow N + HCO$	1×10^{14}
24	$CH + NO \rightarrow HCN + O$	1×10^{13}
25	$CH_2 + NO \rightarrow N + CH_2O$	$1.6 \times 10^{12} \exp(-700/RT)$
26	$CH_2 + NO \rightarrow HCN + OH$	Unavailable
	N_2 Production Path	
27	$NH + NO \rightarrow N_2 + OH$	2.4×10^{12}
28	$NO + N \rightarrow N_2 + O$	$3.1 \times 10^{13} \exp(-300/RT)$
29	$NH + NH \rightarrow N_2 + H_2$	$3.6 \times 10^{11} T^{0.55} \exp(-1,900/RT)$
30	$NH + N \rightarrow N_2 + H$	$6.3 \times 10^{11} T^{0.5}$

[a] R is in cal/mol and °K, T is in °K, and mol refers to g-mol.

Combining Eq. (5) with Eq. (4), one obtains the following expression for the rate of formation of nitric oxide:

$$\frac{d[NO]}{dt} = \frac{2K_1[N_2][O]}{1 + A} - \frac{2A[NO]}{1 + A}\{K_{-2}[O] + K_{-3}[H]\} \tag{6}$$

in which

$$A \equiv K_{-1}[NO]/\{K_2[O_2] + K_3[OH]\} \tag{7}$$

Equations (6) and (7) show that the rate of formation of NO is influenced by the concentrations of N_2, O_2, O, H, OH, and NO. Because [NO] is small, A is usually much less than unity. Consequently,

$$d[NO]/dt = 2K_1[N_2][O] \tag{8}$$

This equation shows that the rate of NO formation is directly proportional to the concentration of atomic oxygen. This simplified rate expression has been used widely to predict NO concentration, especially when NO is far from equilibrium.

In the past it has been assumed that the rate of NO formation is slow compared to the rate of combustion. Therefore, the onset of NO production was considered to occur in the secondary combustion zone after the combustion products have reached their equilibrium values. It was also customary to consider that the concentration of [O] corresponds to its value in equilibrium with [O_2]. However, the agreement with experiment was not satisfactory, even in isothermal conditions such as in reflected shocks and stirred reactors.

Fenimore [99] was the first to report the existence of NO in the combustion region of premixed flames. He found concentrations as large as the equilibrium values of NO. This "Prompt NO" implied production rates that greatly exceeded those given by the Zeldovich mechanism with equilibrium atomic oxygen concentration. Fenimore [99], along with Iverach et al. [101], attributed the high formation rate of NO to the attack of carbon or hydrocarbon radicals on nitrogen molecules because "Prompt NO" did not appear for fuel-lean hydrocarbon flames or in H_2 or CO flames. The maximum "Prompt NO" occurred at equivalence ratios of about 1.4.

There is considerable controversy and confusion concerning the formation of "Prompt NO." Bowman [102] states, on the basis of methane/air shock-tube (constant temperature) data, that the extended Zeldovich mechanism gives the correct rates of formation in both the primary and secondary zones of flames if the [O] and [OH] are evaluated using nonequilibrium chemistry. His analysis, using the Zeldovich mechanism along with detailed methane oxidation kinetics, adequately predicts NO formation in lean to moderately rich methane-air mixtures behind reflected shock waves. Because reflected shocks produce a constant temperature test volume, his NO chemistry proceeds isothermally. In addition, the shock temperatures (2900 K) are high and favor the Zeldovich mechanism. Sarofim and Pohl [103] postulate that the "Prompt NO" can be explained by the Zeldovich mechanism if the nonequilibrium concentrations of O, OH, and H are considered and that hydrocarbon species play no role in this reaction. Although their computations match experimental results reasonably well, it does not explain why "Prompt NO" is only found in rich hydrocarbon flames. Studies on the oxidation of H_2, CO, and C_3H_8 in a jet-stirred reactor show that the experimental NO data for H_2 and CO oxidation is predicted correctly by the Zeldovich mechanism using nonequilibrium O concentration [104]. However, the theory predicts values that are 4 to 10 times lower than experimental results for propane flames.

At present there appears to be a general consensus that the extended Zeldovich mechanisms, even with nonequilibrium O, OH, and H concentrations, cannot adequately account for NO formation in rich hydrocarbon flames. The following reactions, the kinetics for which are given in Table 6, are thought to be the reason for "Prompt NO:"

$$N_2 + CH \rightarrow HCN + N \tag{9}$$

$$N_2 + C_2 \rightarrow 2CN \tag{10}$$

$$N_2 + C \rightarrow CN + N \tag{11}$$

A portion of the cyanides formed is subsequently oxidized to NO. Only very limited quantitative information is available on the formation of HCN and CN by the above reactions.

2. Fuel Nitrogen Oxidation Mechanism

The conversion of nitrogen in the fuel to NO_x appears to be insensitive to the form in which the nitrogen was originally bound to the fuel [105-107]. In hydrocarbon flames, gas-phase nitrogen compounds are very rapidly converted to HCN as a common intermediate in the primary reaction zone [108-110]. The HCN is subsequently converted to amine species (NH_i). The amines are either oxidized to NO or converted to N_2 by reactions with NO or other NH_i species.

The five major paths available for fixed nitrogen* formation and conversion are shown in Fig. 3. The rate expressions corresponding to the various elementary reaction steps are listed in Table 1. The "Thermal NO" and "Prompt NO" mechanisms are the same as those mentioned earlier and will not be discussed here.

Fuel NO. The conversion of fuel nitrogen to HCN is extremely rapid and the details of this reaction are not known. However, this lack of information is not a limitation for establishing the rate-controlling processes in fuel nitrogen conversion, since all available evidence points to the role of HCN as the "effective" initial gas-phase nitrogen species in fuel-rich combustion.

Haynes et al. [111] postulate that the chain carrier free radicals (O, OH, H) abstract a hydrogen atom from HCN to produce CN radicals (Reactions 5-7, Table 1). HCN can also react with OH

---

*Fixed nitrogen refers to the total sum of the compounds NH_i and NO_x.

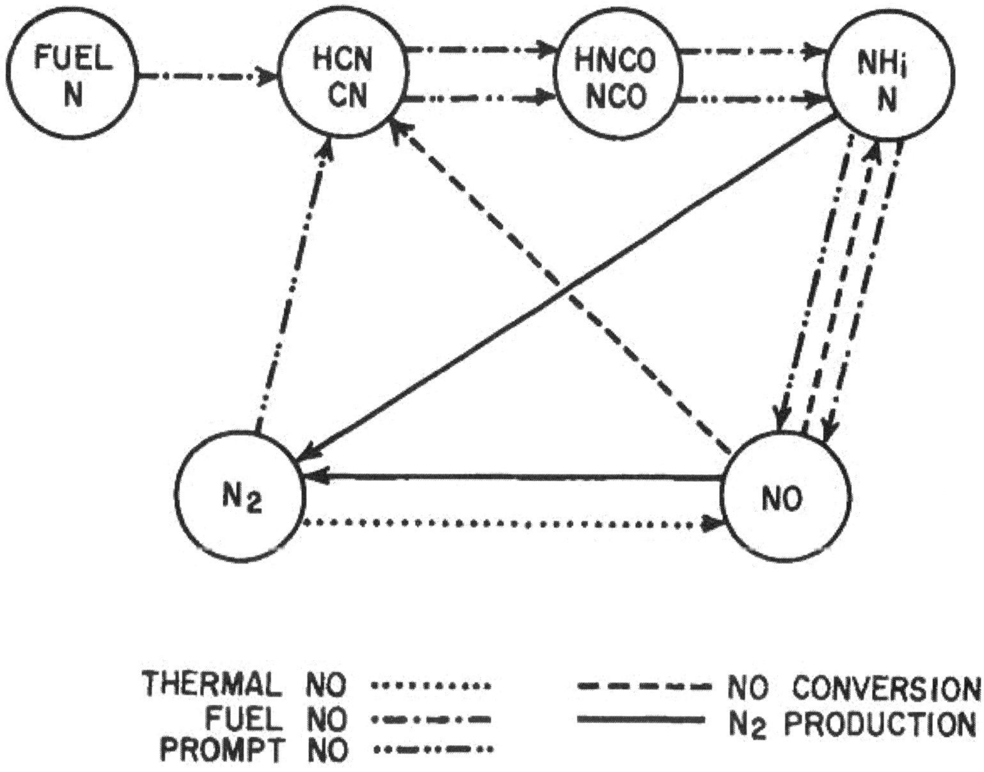

FIG. 3. Major paths for fixed nitrogen formation and conversion [112].

radicals to a lesser extent to form the intermediate HNCO (Reaction 9, Table 1). The CN radicals react with OH to form NCO as an intermediate (Reaction 8, Table 1).

Both NCO and HNCO are unstable at high temperatures (and hence have never been observed in flames). They react rapidly with H atoms to form amine species, NH_i (Reactions 10 and 11, Table 1). The NH_2 and NH radicals can both react with the chain carrier radicals (OH, H, O) to produce NH radicals (Reactions 12-14, Table 1) and N atoms (Reactions 15-17, Table 1). "Fuel NO" is then formed via the reactions of NH or N with OH or O (Reactions 3, 18, and 19, Table 1).

NO Conversion. The conversion of NO into other fixed nitrogen species is the result of its reaction with hydrocarbon radicals. Two competing reaction paths are involved: one is the reduction of NO to N atoms via reactions with CH and CH_2 radicals (Reactions 23 and 25, Table 1); the other involves the conversion of NO into HCN (Reactions 24 and 26). Thus, some of the NO produced through different reaction paths can be converted and cycled back into the "Fuel NO" reaction path.

N_2 Production. Molecular N_2 is generally believed to be produced via the reaction of NO with NH_i species or N atoms (Reactions 27 and 28, Table 1). In addition, N_2 can also be produced through the reaction of two amine species, including atomic N (Reactions 29 and 30, Table 1).

Although the above set of reactions obviously does not include all possible reactions in fuel nitrogen chemistry, it is believed to include the more significant ones. Thus, one can divide the elementary processes occurring into two types:

(a) Bimolecular reactions of a fixed nitrogen species (NO, CN, NH_i) with free radicals
(b) Bimolecular reactions involving two fixed nitrogen species (e.g., Reactions 27-30, Table 1)

Only the second set of reactions yields molecular nitrogen. Experimental evidence suggests [112] that near the flame front bimolecular reactions of two fixed nitrogen species (Type b) control the rate of conversion of fixed nitrogen to molecular N_2, while at longer reaction times the rate-controlling step appears to be the reaction of HCN with radical chain carriers (Type a).

B. NO_x Control by Catalytic Combustion

1. Clean Fuels

It has been stated earlier that for clean fuels, under lean fuel conditions, the Zeldovich mechanism has the predominent role in NO_x formation. The reactions in this mechanism are strongly temperature dependent; hence, thermal NO_x forms significantly only at high temperatures (>1500°C). This suggests that significant formation of NO_x can be avoided by carrying out complete combustion at lower temperatures. At this point, two questions arise:

(a) Are temperatures low enough to minimize thermal NO_x formation adequate for practical applications?
(b) Is it possible to design combustors that have high combustion efficiencies but whose peak temperatures do not exceed the temperature requirement for minimum thermal NO_x formation?

The first question may be answered in the affirmative because temperatures below 1500°C are adequate for several applications like gas turbines, boilers, superheaters, and domestic heaters. Indeed, certain practical applications are limited by material constraints (e.g., gas turbine blading, heat exchanger integrity)

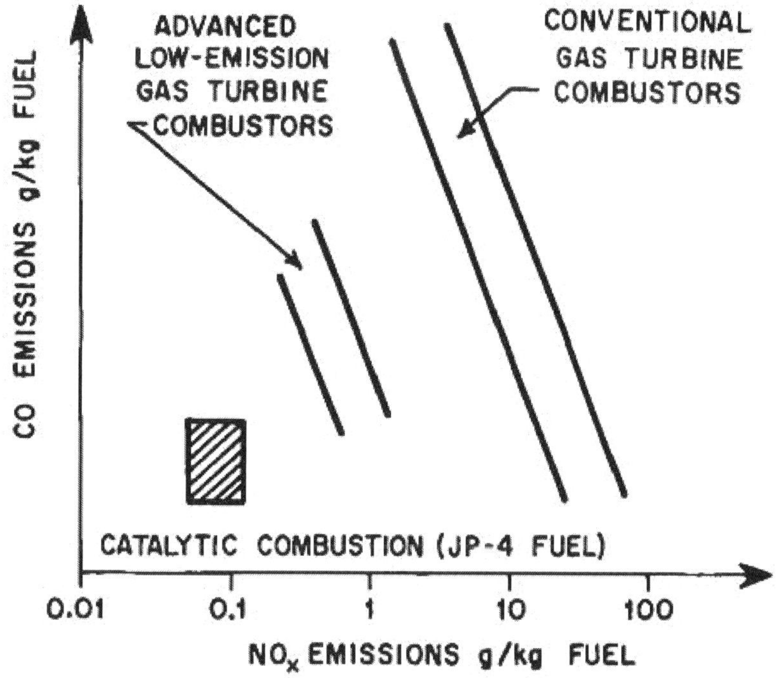

FIG. 4. CO and NO_x emissions of different combustion systems [1].

and the lower temperature is an advantage. Conventional combustors can be designed to reduce thermal NO_x by operating with a lean, premixed, prevaporized primary zone [124]. Unfortunately, such a combustor suffers from poor combustion stability. Reducing the combustion temperature, in general, also leads to increased emission of CO (and hydrocarbons). The reason for this is that the rate of the major reaction responsible for the oxidation of CO to CO_2 [125],

$$CO + OH \rightarrow CO_2 + H \tag{12}$$

decreases as the temperature decreases. Thus, conventional combustors typically have a tradeoff between CO and NO_x emissions from the combustor. This is shown in Fig. 4 which covers a wide range of conventional gas turbine combustion systems, primarily aircraft systems, as well as "advanced low emission combustors" of the premixed, prevaporized design.

The primary reason that catalytic combustors can significantly reduce thermal NO_x formation is due to their capability to carry out stable, highly efficient combustion of very lean fuel mixtures (ϕ = 0.1-0.4) in a manner such that the peak temperature in the combustor is below the temperature at which appreciable amounts of NO_x form. This is possible because catalytic combustors typically operate with well-premixed, preheated fuel-air mixtures whose oxidation is initiated on the surface of the catalyst. Since the catalyst is usually supported on a monolith comprised of a large number of small, uniform, parallel channels, the temperature of the gas is approximately constant across the cross section and increases downstream in a uniform manner. The temperature of the gas mixture increases due to the surface reactions until it becomes high enough for homogeneous reactions to come into play and dominate; at this point complete fuel conversion is obtained. The surface reactions along with the thermal inertia of the bed stabilize and promote the combustion process. This eliminates the problems related to flame stability and enables catalytic combustors to conduct stable, highly efficient combustion of lean mixtures which yield temperatures low enough to minimize thermal NO_x. Thermal NO_x reductions of one to two orders of magnitude have been achieved via catalytic combustion [1-4, 7, 9, 10, 12, 21]. The formation of Prompt NO is possible and does occur in conventional combustors in which fuel-rich pockets are present in the combustion zone. Since catalytic combustors typically use well-premixed, prevaporized, lean fuel-air mixtures, Prompt NO does not form.

It is important to note that if combustor exit temperatures greater than 1500°C are required, then thermal NO_x would also form in catalytic combustors. A reduction in NO_x from that obtained in conventional combustors may still be expected due to the more uniform radial temperature profiles in catalytic combustors which reduces the occurrence of high local temperatures and the absence of fuel-rich pockets. Both are consequences of the uniform fuel distribution at the inlet. However, at temperatures greater than 1600-1800°C, catalytic combustors tend to give rise to severe problems related to substrate degradation and catalyst deactivation.

2. Nitrogen Bound Fuels

It was discussed earlier that the oxidation of fuel-bound nitrogen leads to the formation of a number of fixed nitrogen species. The major fixed nitrogen species are NO, NH_i, and HCN, and their sum total is referred to as Total Fixed Nitrogen (TFN). In general, it is of interest to minimize the TFN from combustion systems.

For nitrogen-bound fuels, the fuel-N compounds are more readily oxidized to NO_x than the atmospheric nitrogen [113].

As a result, conventional combustors in fuel lean operation tend to preferentially oxidize the fuel-N to NO_x, yielding high emission levels of NO_x. The use of a catalytic combustor can, in principle, alleviate this problem if the catalyst selectively oxidizes the fuel while converting the fuel-N to N_2. However, the catalytic combustion results reported in the literature [22, 25, 37-39] show a high degree of conversion (40-90%) of fuel-N to NO_x in lean fuel operations. This implies that the catalysts used in those experiments did not possess this type of selectivity.

The results reported in literature indicate that under lean fuel operation, NO_x is the predominant product of fuel-N oxidation. As the operation becomes fuel rich, the NO_x formation drops off sharply and the NH_i and HCN concentrations start to rise. This results in a minimum in TFN in the fuel-rich region (ϕ = 1.1-1.5, typically). By operating at the minimum, the fuel-N conversion to TFN can be reduced to 15-35%. Hence, for these catalysts (noble metal), staged catalytic combustion (first stage fuel-rich and the second fuel-lean) appears to be promising for fuel-NO_x control. However, the main problem in the operation of a fuel-rich stage is that the bed temperatures are very high and severe catalyst degradation could occur.

The details of the oxidation of fuel-N on the surface of combustion catalysts are not well understood. Chu and Kesselring [38] report that their data tend to agree with the reaction mechanism proposed for homogeneous reactions; i.e.,

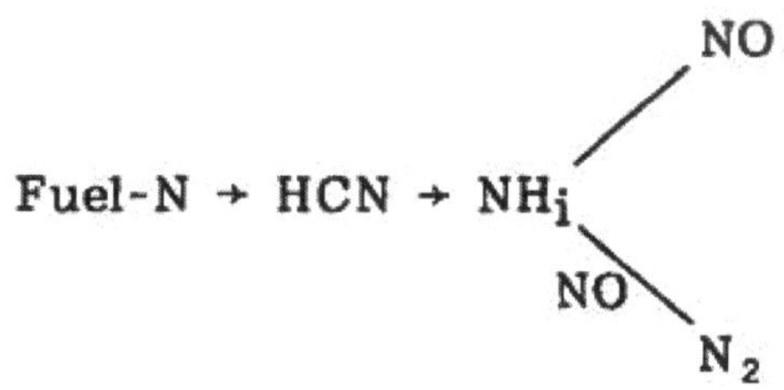

They further postulate that the reaction takes place on the catalyst surface.

Almost all the studies of the fuel-N oxidation in catalytic combustors have been done with noble metal catalysts. Since Pt is the commercially used catalyst for ammonia (which is representative of fuel nitrogen compounds) oxidation to NO, it is expected that in fuel-lean combustion a high degree of fuel-N conversion to NO_x would occur using this catalyst. This is indeed what has been observed [22, 25, 37-39] and appears to corroborate the postulate that fuel-N oxidation is dominated by surface reactions. Therefore, the nature of the catalyst used would be expected to have a strong effect on the conversion of fuel-N. This is substantiated by the results of Folsom et al. [22] and Prasad et al.

[40]. The former showed that significant reduction in fuel-N conversion in the fuel-rich region could be achieved by using a NiO-Pt catalyst as opposed to a Pt catalyst alone. The latter have shown that by using a binary transitional metal oxide catalyst, it is possible to reduce fuel-N conversion to NO_x to extremely low levels (5-20%) in both fuel-lean as well as fuel-rich regions. Thus, the use of oxide-based catalysts appears to be very promising for the control of both thermal as well as fuel-NO_x.

The effect of various parameters on fuel-N oxidation has been reported in the literature. The formation of fuel-NO_x is insensitive to the form in which the nitrogen was originally bound to the fuel [37, 38, 105-107]. Chu and Kesselring [38] state that in lean-fuel operation the fuel-N conversion to NO_x decreases as the nitrogen content of the fuel increases. However, Matthews and Sawyer [37] report exactly the opposite trend. This is believed to be a consequence of the very low velocities and large radial heat losses present in the latter experiments. The bed temperature may or may not influence fuel-NO_x formation depending on whether the surface reaction is in the kinetically or diffusion-controlled regime. The mass throughput is also reported to have a strong effect on fuel-N conversion for some catalysts [38]. Finally, the presence of H_2S is believed to enhance NO formation under fuel-rich conditions and inhibit it under stoichiometric conditions [38].

IV. CATALYST SYSTEMS

The catalyst system is in general composed of two components—the active material (catalyst) and the support system (substrate and washcoat). Choice of both components has a significant influence on the development of a successful catalytic combustor. To obtain optimal performance from the combustor, the catalyst system should possess the following properties:

1. The catalyst should be capable of igniting a fuel/air mixture at the lowest possible temperature (low light off (ignition) temperature).
2. The catalyst activity should be sufficiently high to maintain complete combustion at the lowest levels of air preheat and the highest values of mass throughput.
3. The support should have a large geometric surface area, low pressure drop, excellent thermal shock resistance, and high working temperature.

4. The support (in general, a washcoated substrate) should be capable of maintaining high surface area under high temperature combustion conditions (∿1100°C).
5. The catalyst system should maintain high activity after prolonged use at elevated temperature (1000-1500°C).

A. Choice of the Catalyst

High temperature catalysts for hydrocarbon or CO oxidation are in general comprised of noble metals or base metal oxides. These are discussed separately below.

1. Noble Metal Catalysts

A general sequence of the activity of noble metals for oxidation is Ru, Rh, Pd, Os, Ir and Pt. However, for catalytic combustion applications, the use of noble metals other than Pt and Pd appears to be limited due to their high volatility, ease of oxidation, and limited supply [46]. Considerable data are available in the literature concerning the use of the noble metals Pt and Pd as oxidation catalysts [65-67], and it has been found that these metals are among the most active catalysts for several fuels [68, 69]. The high activity of these metals is related to their ability to activate H_2, O_2, C–H, and O–H bonds. For the oxidation of CO, olefins, and methane, the specific activity of Pd is larger than that of Pt while for aromatic compounds they are similar. For the oxidation of paraffinic hydrocarbons, C_3 and greater, Pt is more active than Pd.

Compared to base metal oxides, noble metals have certain advantages, including higher specific activity for hydrocarbon oxidation, greater resistance to the loss of low temperature activity, and less deactivation by sulfur at temperatures below 500°C [70]. In addition, Pt and Pd are readily prepared in a highly dispersed form on a number of support materials. For these reasons, in addition to their high activity, only small amounts (0.1-0.5 wt%) are typically necessary for good combustion catalysts.

However, problems in the operation of Pt and Pd catalysts do exist because the pure metals sinter rapidly at temperatures of 500-900°C. Sintering results in a reduction in active surface and hence in catalyst activity. In addition, it has been found that in an oxidizing atmosphere, the noble metals can disperse as oxides on the Al_2O_3 surface at temperatures below the decomposition temperatures of the oxides (PtO_2, 585°C; PdO, 790°C), or even at higher temperatures if the complex between the noble metal oxide and the support surface possesses sufficient stability [71a, 71b].

This interaction between PdO and a γ-Al_2O_3 surface is sufficient to give considerable thermal stability to Pd-Al_2O_3 catalysts in an oxidizing atmosphere [72]. PtO_2, however, interacts less strongly with a γ-Al_2O_3 surface so that at temperatures greater than 600°C, in an oxidizing atmosphere, the Pt dispersion decreases and large metallic crystallites form. The formation of PdO as well as of large crystallites of Pt results in much reduced activity.

In spite of these problems, the development of stable noble metal catalyst capable of long-term operation at temperatures of 1250°C appears realizable. Kesselring et al. [46] have conducted extensive studies on the development of catalysts for combustion applications. They suggest that the loss of Pt surface area at high temperatures may not affect low temperature lightoff characteristics to any significant extent. Thus, a high activity catalyst may not be required for catalytic combustor applications. Moreover, although sintering is severe at 1250°C, poisoning is a less severe problem because sulfur and lead compounds decompose or volatilize at high temperatures.

Most of the research done to date in catalytic combustion has been with the use of noble metal catalysts. However, in many of the cases, due to the proprietary nature of the catalysts used, it is not clear as to whether the catalyst contains active substances (e.g., oxides) besides noble metals. Again, the activity of the final catalyst is affected greatly by factors like the presence of promoters, the nature of the dispersion on the surface, the techniques employed in obtaining a deposit, and other details of the manufacturing process. Unfortunately, very little information is available regarding these details in the open literature. It is reported [46] that hydrogen sulfide (H_2S) fixation of Pt catalysts promotes retention of platinum surface area and appears beneficial in extending platinum activity with time.

2. Metal Oxide Catalysts

The catalytic activity of base metal oxides for the oxidation of CO and hydrocarbons by oxygen has been studied by a number of research groups [73-76]. In general, high oxidation activity requires metal ions that can assume more than one valence state and can participate in reduction-oxidation (redox) reactions. The oxides of transition metals have been found to be the most active of the simple oxides [46], and the catalytic activities of these oxides are determined by the d-shell electron configuration of the metal [77]. Maxima in activity are usually observed for three, six, and eight d-electrons while minima are seen for zero, five, and ten d-electrons. For the fourth period transition metal oxides, peaks in activity occur for Cr_2O_3 and Co_3O_4 [78]. Interestingly, mixtures of oxides often exhibit greater stability and activity than the single oxides [21, 73, 79]. This is of importance in catalytic combustion applications.

Among the oxides, only those that are refractory and can maintain their activity at high temperatures have potential combustion applications. In general, for CO and hydrocarbon oxidation, Co_3O_4 is the most active catalyst with NiO, Cr_2O_3, and CuO being promising candidates. Choice of the catalyst is also constrained by the necessity that the support and catalyst be compatible. This means that there must be mutual insolubility (or low solubility) or that the melting points of the materials must be higher than the maximum operating temperature by a factor of at least 1.5. Thus, alumina, with a melting point of about 2030°C , would be expected to react readily with most catalytic materials at temperatures above 1400–1500° C [46]. In such cases the reaction product itself may be a suitable catalyst (e.g., cobalt aluminate); however, the interaction can alter the strength and thermal shock capabilities of the support. Recrystallization which decreases the catalyst activity, becomes appreciable at about one-half the melting point of the material in degrees Kelvin. The effect of recrystallization may be readily seen in Fig. 5, which shows a NiO-Pt catalyst before and after aging at 1423°C for 1000 h.

With the proper choice of materials, highly active stable catalysts of mixed oxides are believed possible. Indeed, a binary transition metal oxide catalyst comprised of Cr_2O_3-Co_3O_4 supported on alumina washcoated ceramic (Cordierite) support has been successfully demonstrated for the catalytic combustion of lean propane-air mixtures [21]. This catalyst has been tested for up to 400 h operation [49] at maximum temperatures of up to 1100°C without significant deactivation. Catalysts comprised of La_2O_3-Cr_2O_3 and NiO alone also appear to be very promising at even higher temperatures [24]. Also, it may be possible to fabricate honeycomb structures directly from catalytic oxide composition. Perovskites, for example, have been proposed as catalyst supports [80]. The low surface areas of these materials may not be a problem because any catalyst will rapidly sinter at combustor conditions, resulting in the loss of high surface area. It is usually better to start with a low surface area (presintered) catalyst since this would result in better retention of strength and thermal shock properties as well as less change in catalyst activity during operation [24, 49].

Metal oxide catalysts typically have lower catalytic activity and higher lightoff temperatures than noble metal catalysts.

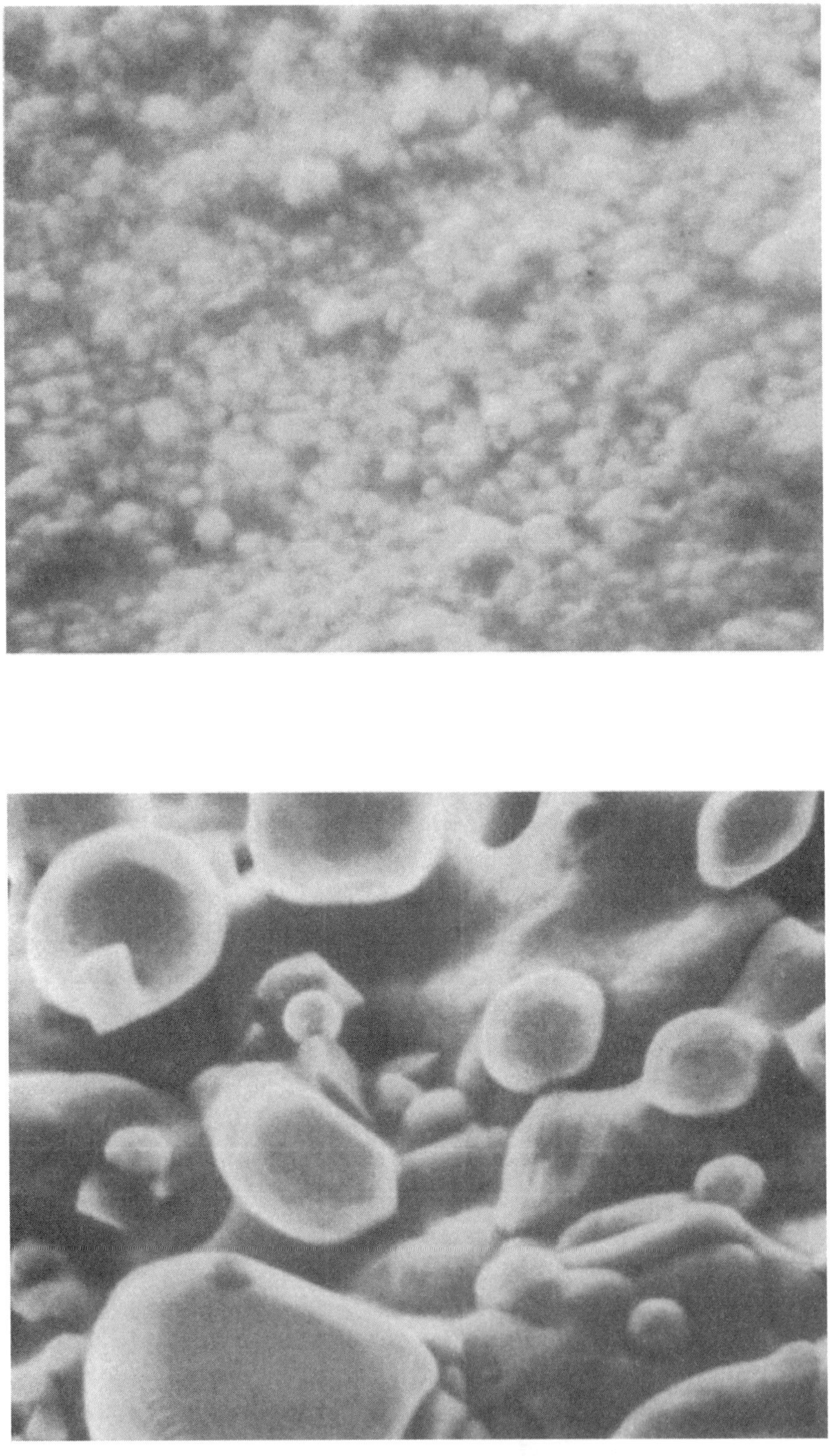

FIG. 5. SEM photograph of the surface of a NiO-Pt catalyst before (upper) and after (lower) 1000 h aging at 1423°C [24]. Catalyst type: 5-10% wt NiO, 1.5% wt Pt on Corning MCB-12.

However, under typical catalytic combustor operating conditions, their activities are comparable to those of the noble metals [2]. One should note that, for the most part, the surface reactions in a catalytic combustor are diffusion limited and under such operation, if the rate is sufficiently high, the catalytic activity plays a secondary role in determining the reaction rate.

3. Metal Oxide-Noble Metal Catalysts and Active Monoliths

The use of traces of noble metals may greatly enhance the performance of certain base metal oxide catalysts. One example of such behavior involves the use of platinum with mixed oxides of Perovskite structure [82]. Traces of platinum have been shown to render certain base metal Perovskite catalysts insensitive to sulfur poisoning [81]. Such catalysts can be expected to operate at a much higher temperature without loss of platinum than would be the case for a straight-platinum catalyst. In this case the platinum is chemically bound in the Perovskite structure and the vapor pressure of platinum oxide is accordingly reduced.

Catalysts comprised of NiO-Pt and La_2O_3-Cr_2O_3-Pt have been fabricated and, from the limited experimental data, they appear to have good potential [24]. Oxide catalysts with some platinum deposited on the upstream end of the substrate to promote lightoff have also been successfully tested [22]. It is suspected that most of the noble metal catalysts used currently in catalytic combustion research (usually supplied by catalyst manufacturers) have substantial amounts of base metal oxides incorporated in them. Perhaps the most promising development in catalysts for combustion application is the concept of active monoliths. In this type of catalyst an active material or mixed metal oxide is incorporated into the structure of the monolith when the monolith is formed. This type of catalyst solves many of the problems inherent in noble metal and oxide-based catalyst systems. It depends on base metal oxides; thus the problems of noble metal agglomeration and volatilization are eliminated. Moreover, by incorporating the active oxide directly into the monolith, interactions between catalyst and substrate are controlled during the formulation steps and further interaction does not occur during operation. Much of the pioneering work in the development of active monoliths has been done by the ACUREX Corporation [23, 24]. Improvements in manufacturing techniques and material formulations are required, however, to increase the operational life of these systems.

B. Choice of the Support

The support serves three important functions in the catalyst system:

1. It increases the surface area of the metal or metal oxide by providing a matrix that enables their dispersion as very small particles.
2. It can reduce sintering of the active catalyst material and improve its thermal stability.
3. It could enhance the activity of the catalyst material.

Catalyst supports typically consist of two parts—the substrate and the washcoat. These are discussed separately.

1. Substrates

The substrates are usually fabricated from low surface area ceramics or special metal alloys (e.g., Kanthal). Of the two, ceramics have been far more widely used in the past. Table 2 lists some of the properties of high temperature ceramics.

It is seen that the most common high temperature ceramic family is alumina. For strengthening purposes, the alumina is alloyed with silica and/or chromium. Aluminas are relatively inexpensive, reasonably resistant to thermal shock and can operate at high temperatures (>1480°C). Zirconia ceramics can be used at temperatures up to 2210°C, the highest temperature among ceramics. Zirconia is also extremely inert to most metals, even at high temperature; hence it is a good candidate as catalyst support. The chief advantages of the other ceramic supports are listed in Table 2. The substrate that has been most widely used in the past for catalytic combustion applications is cordierite. Presently, it is being replaced by other materials (such as zirconia) which have higher temperature capability, better inertness and greater structural integrity in operation.

Metal substrates are now being made to overcome the thermal shock and material stability problems sometimes encountered in the use of ceramics [84]. Johnson Matthey Corporation has developed a metal monolith for use with platinum catalysts. These metal monoliths are composed of ferroalloy steels consisting of up to 20% chromium, 0.5-12% aluminum, 0.1-3% yttrium, and the balance iron. Kentucky Metals is also fabricating metal monoliths from an alloy called Kanthal A-1 which consists of 5.5% aluminum, 22% chromium, 0.5% cobalt, and the balance iron. One of the

TABLE 2

Ceramic Properties [83]

Ceramic family	Material cost	Thermal shock resistance	Thermal strength	Thermal conductivity	Other
Alumina	Low	Fair	Good	Low	Most common high-temperature ceramic
Beryllia	High	Excellent	Good	High	Highly toxic
Zirconia	Moderate-high	Fair-good	Good	Low	Can be used at temperatures above 2478 K
Lithium-aluminum-silicate (Beta Spodumene)	Moderate	Good	Good	Low	Not resistant to sulfur, sodium
Magnesium-aluminum-silicate (cordierite)	Moderate	Good	Good	Low	More corrosion-resistant than LAS
Aluminum-silicate (aluminous keatite or mullite)	Moderate	Fair-good	Good	Low	Good corrosion resistance
Silicon carbide	Low-high	Excellent	Excellent	Low	Does not self-bond easily
Silicon nitride	Low-high	Excellent	Excellent	Low	Does not self-bond easily
Boron carbide	High	Fair	Poor	Low	High hardness, low density

FIG. 6. Corning high temperature graded cell monolithic catalyst support.

problems with metal substrates is that it is often difficult to obtain a stable coat of catalyst (especially oxides) on them.

The substrate can be made in a variety of geometric shapes such as monolithic honeycombs, pellets, tubes (by coating the inside of a conventional firetube boiler with catalyst), fiber pads, wire mesh (alloy), and corrugated sheets. From a practical viewpoint, the monolithic honeycomb (Fig. 6) is the most technologically advanced substrate for catalytic combustion applications. These are composed of small parallel channels, available in a variety of shapes and hydraulic diameters. Table 3 presents some of the significant characteristics as well as the manufacturers of several commercially available monoliths. It has been shown [20] that significantly improved combustor performance (in terms of maximum mass throughput) can be achieved by placing the monoliths in the graded cell configuration. This configuration employs the largest cells in the front of the bed, intermediate cells in the center, and the smallest cells at the downstream end of the bed, in order to achieve a tradeoff between a low pressure drop, high combustion efficiency, and mass throughput. Besides, the graded cell combustor has a high, relatively uniform radial temperature profile [23]. This helps to minimize possible poisoning effects during prolonged, high temperature operation. An example of a graded cell catalyst system is shown in Fig. 6.

2. Washcoat

The low surface area of the monolith structure can be increased by the application of a thin coat of metal oxide material, such as Al_2O_3. This washcoat strongly adheres to the ceramic support and provides a high surface area. The most commonly used washcoat material is γ-Al_2O_3. At temperatures above 900°C, the high surface area γ-Al_2O_3 undergoes a phase change to relatively low surface area α-Al_2O_3 with concomitant sintering. This phase change causes a reduction of the surface area from 300 to 5 m^2/g of washcoat. This sintering thus results in pore closure and a burying of active catalytic sites in the alumina washcoat. The use of presintered Al_2O_3 washcoats, Al_2O_3 washcoats stabilized with CeO_2 or Cs_2O, or more thermally resistant washcoats such as ZrO_2 would reduce the problems associated with sintering. These processes reduce the surface area, but the necessity of maintaining a high surface area, which is required for low temperature catalysis may not be required for high temperature catalytic combustion catalysts. Indeed, it is believed that in these catalysts, one of the major functions of the washcoat is to enable the deposition of a uniform and stable catalyst coat.

V. KINETICS OF FUEL OXIDATION

The chemical kinetics of importance in catalytic combustors involve both heterogeneous as well as homogeneous reactions. Typically, in the initial part of the bed, where temperatures are low, the surface reactions are dominant, while in the latter part of the bed where temperatures are high, the gas-phase reactions are more important. In reality they are related processes; i.e., the reactions on the surface produce radicals which could affect the homogeneous reactions and vice versa. Experimental observations, in fact, suggest that such hetero-homogeneous oxidation mechanisms do exist [49, 76, 85]. However, from a practical point of view, the determination of these hetero-homogeneous reaction mechanisms presents a most formidable task—especially in the light of the fact that even the homogeneous reaction mechanisms for most fuels are not known. Consequently, for the sake of simplicity, one often considers them as two separate and distinct sets of reactions and studies them separately.

TABLE 3

Monolithic Support Data [46]

Manufacturer	Product	Ceramic	Temperature limit, K (°F)	Thermal shock resistance
American Lava Corporation (3M Company)	Thermacomb 843	Lithia-alumina-silica	1367 (2000)	Excellent
	Thermacomb LTE	Cordierite	1478 (2200)	Excellent
	Thermacomb 795	Cordierite	1478 (2200)	Excellent
	Thermacomb 784	Zircon-mullite	1756 (2700)	Good
	Thermacomb MD-3	Mullite	1700 (2600)	Good
	Thermacomb 614	Dense 96% alumina	1811 (2800)	Fair
Corning Glass Works	Celcor 9475	Cordierite	1478 (2200)	Excellent
		Mullite-alumina-titanate	1923 (3000)	Good
		Zirconia-spinel	1978 (3100)	Fair

E. I. du Pont de Nemours & Company	Torvex	Alumina	1773 (2732)	Fair
		Mullite	1623 (2462)	Fair
General Refractories Company	Versagrid	Cordierite	1672 (2550)	Good
		Mullite	1922 (3000)	Fair
W. R. Grace & Company	Poramic	Cordierite	1478 (2200)	Excellent
Johnson Matthey Corporation	Fecralloy steels		1573 (2371)	Excellent
Norton Company	Spectramic Honeycomb RX 387	Silicon carbide	1922 (3000)	Good
	Spectramic Honeycomb RX 384	Silicon nitride	1811 (2800)	Good
Kentucky Metals	Kanthal (metal)		1678 (2560)	Excellent

A. Homogeneous Reactions

From a fundamental point of view, the correct method for modeling the homogeneous reactions is via the reaction mechanism approach. In this approach the overall reaction is broken down into a number of elementary reactions which detail the molecular steps in the conversion of reactants to products. Quite often, in such cases, the rate of one or more of these elementary reactions controls the rate of the overall process and, in such cases, a somewhat simplified picture would result. However, in practice, the use of the reaction mechanism approach is limited because of the following reasons:

1. For most practical fuels, the reaction mechanisms are not known. Only for simpler fuels like H_2, CO, and C_2H_4 are they reasonably well understood.
2. Even when the reaction mechanism is available, the kinetic data for some intermediate steps are often unavailable or poorly understood.
3. They are, in general, complicated and not suited for most practical design purposes.

The alternate approach is to assume, for practical purposes, that the overall reaction can be adequately expressed by one or more global rate expressions. The global rate expressions are generally of the form

$$r_i = k_i [C_i]^{n_{1i}} [C_2]^{n_{2i}} \ldots [C_j]^{n_{ji}}$$

where k_i = $A_i e^{-E_i/RT}$ or $(A_i'/T)e^{-E_i/RT}$
r_i = reaction rate of species i
C_j = concentration of species j
n_{ji} = reaction order with respect to species j
A_i, A_i' = preexponential factor for the reaction
E_i = activation energy for the reaction

The term $A_i e^{-E_i RT}$ represents the Arrhenius temperature dependence of the rate constant.

The use of such rate equations is often limited because of two important reasons. First, it may not be possible at all to adequately represent the reaction by a simple global-type rate

TABLE 4

Homogeneous Kinetics[a]

Fuel	Chemical equation	Rate expression	Refs.
C_3H_8	$C_3H_8 + \frac{7}{2}O_2 \xrightarrow{K_1} 3CO + 4H_2O$	$K_1 = 1.8 \times 10^{12} T^{-1/2} \exp\left(-\frac{28,000}{RT}\right) m^3 \cdot {}^\circ K^{-1/2}/kg \cdot s$	41
	$CO + \frac{1}{2}O_2 \xrightarrow{K_2} CO_2$	$K_2 = 1.5 \times 10^{12} \exp\left(-\frac{24,000}{RT}\right) m^3/kg \cdot s$	86
		$r_i = K_i \rho_A \rho_B$, where ρ_A, ρ_B = concentration of reactants, kg/m^3	
C_3H_8	$C_3H_8 + \frac{1}{2}O_2 \xrightarrow{r_1} \frac{3}{2}C_2H_4 + H_2O$	$r_1 = 1.4 \times 10^{11} \exp\left(-\frac{40,608}{RT}\right)[O_2]^{1.04} \times$	87
	$C_2H_4 + 2O_2 \xrightarrow{r_2} 2CO + 2H_2O$	$\left\{\frac{[C_3H_8]}{[C_3H_8]_{in}}\right\}^{0.25} mol/cm^3 \cdot s$	
	$CO + \frac{1}{2}O_2 \xrightarrow{r_3} CO_2$	$r_2 = 1.4 \times 10^{13} \exp\left(-\frac{60,184}{RT}\right)[O_2]^{1.18} \times$	87
		$\left\{\frac{[C_2H_4]}{[C_3H_8]_{in}}\right\}^{0.96} mol/cm^3 \cdot s$	

(continued)

TABLE 4 (continued)

Fuel	Chemical equation	Rate expression	Refs.
		$r_3 = 3.98 \times 10^{14} \exp(-\frac{40,000}{RT})[H_2O]^{0.5} \times [CO][O_2]^{0.25}$ mol/cm^3·s	87
		$r_1 = 1.4 \times 10^{8} \exp(-\frac{40,608}{RT})[O_2]^{1.04} \times \left\{\frac{[C_3H_8]}{[C_3H_8]_{in}}\right\}^{0.25}$ mol/cm^3·s	59
		$r_2 = 4.8 \times 10^{14} \exp(-\frac{70,000}{RT})[O_2]^{1.18} \times \left\{\frac{[C_2H_4]}{[C_3H_8]_{in}}\right\}^{0.96}$ mol/cm^3·s	59
		$r_3 = 1.0 \times 10^{10} \exp(-\frac{30,000}{RT})[H_2O]^{0.5} \times [CO][O_2]^{0.25}$ mol/cm^3·s	59

C_3H_8	$C_3H_8 + \frac{7}{2}O_2 \xrightarrow{r_1} 3CO + 4H_2O$	$r_1 = 1.5 \times 10^5 p^{0.3}[C_3H_8]^{0.5}[O_2]T \times$	61
	$CO + \frac{1}{2}O_2 \xrightarrow{r_2} CO_2$	$\exp\left(-\frac{24,000}{RT}\right)$ mol/cm^3·s	
		where p = dimensionless pressure = P/inlet pressure	
		$r_2 = 0.71 \times 10^{14}[CO][O_2]^{0.25}[H_2O]^{0.5} \times$	61
		$\exp\left(-\frac{40,000}{RT}\right)$ mol/cm^3·s	
CH_4	$CH_4 + 2O_2 \xrightarrow{r_1} CO_2 + 2H_2O$	$r_1 = 1.3 \times 10^8 \exp\left(-\frac{48,400}{RT}\right)[C_3H_8]^{-0.3} \times$	88
		$[O_2]^{1.3}$ mol/cm^3·s	
C_2H_6	$C_2H_6 + 3\frac{1}{2}O_2 \xrightarrow{r_1} 2CO_2 + 3H_2O$	$r_1 = 1.1 \times 10^{12} \exp\left(-\frac{30,000}{RT}\right)[C_2H_6]^{0.1} \times$	88
		$[O_2]^{1.65}$ mol/cm^3·s	
C_3H_8	$C_3H_8 + 5O_2 \xrightarrow{r_1} 3CO_2 + 4H_2O$	$r_1 = 8.6 \times 10^{11} \exp\left(-\frac{30,000}{RT}\right)[C_3H_8]^{0.1} \times$	88
		$[O_2]^{1.65}$ mol/cm^3·s	

(continued)

TABLE 4 (continued)

Fuel	Chemical equation	Rate expression	Refs.
C_8H_{18}	$C_8H_{18} + 12\frac{1}{2}O_2 \xrightarrow{r_1} 8CO_2 + 9H_2O$	$r_1 = 4.6 \times 10^{11} \exp(-\frac{30,000}{RT})[C_8H_{18}]^{0.25} \times [O_2]^{1.5}$ mol/cm³·s	88
$C_{10}H_{22}$	$C_{10}H_{22} + 15\frac{1}{2}O_2 \xrightarrow{r_1} 10CO_2 + 11H_2O$	$r_1 = 3.8 \times 10^{11} \exp(-\frac{30,000}{RT}) \times [C_{10}H_{22}]^{0.25} [O_2]^{1.5}$ mol/cm³·s	88
CH_3OH	$CH_3OH + 1\frac{1}{2}O_2 \xrightarrow{r_1} CO_2 + 2H_2O$	$r_1 = 3.2 \times 10^{12} \exp(-\frac{30,000}{RT}) \times [CH_3OH]^{0.25} [O_2]^{1.5}$ mol/cm³·s	88
C_3H_8	$C_3H_8 + 5O_2 \xrightarrow{r_1} 4H_2O + yCO + (3 - y)CO_2 + \frac{y}{2}O_2$	$r_1 = 2.9 \times 10^{10} \exp(-\frac{15,000}{RT}) f_{O_2}^{0.35} \times f_{CO} f_{H_2O}^{0.4} (P/RT)^{1.75}$ mol/cm³·s	89

		where P = pressure in atm f = mole fraction of species T = temperature in °K	
CO	$CO + \frac{1}{2}O_2 \xrightarrow{r_1} CO_2$	$r_1 = 12 \times 10^{10} \exp(-\frac{16,000}{RT}) f_{O_2}^{0.3} f_{CO} f_{H_2O}^{0.5} (P/RT)^{1.8}$ mol/cm^3·s where P = pressure in atm f = mole fraction of species T = temperature in °K	89
CO	$CO + \frac{1}{2}O_2 \xrightarrow{r_1} CO_2$	$r_1 = 4.2 \times 10^6 \exp(-\frac{14,090}{T})$ kg·mol/m^3·s	58

[a] Activation energies are in calories per gram mole, temperatures T are in °K, R is in calories per gram mole °K, and mol refers to g mol.

expression. This could be due to important intermediate reactions, the effect of walls, the presence of chain reactions, etc. Second, one often finds that the global rate expressions are good approximations only in situations similar to the one in which the original data were measured. In spite of these drawbacks, because of their great simplicity, numerous attempts have been made at determining global rate expressions for various hydrocarbon oxidation processes. Most of the practical combustor analyses, in fact, utilize such global models.

Due to the above-mentioned reasons, only global rate expressions, which might be of use to researchers in this field for design or modeling purposes, are reported here. Table 4 lists some of the rate expressions reported in the literature for the oxidation of CO and some hydrocarbons. For each case the references are cited, and it is recommended that they be consulted to determine appropriate ranges of application before these expressions are used. In addition, the activation energies for several high temperature oxidation reactions are given in Refs. 90 and 91. Readers who are interested in the reaction mechanism approach may find details of the reaction mechanisms for CH_4, C_2H_6, and $H_2-CO-CO_2-NO$ system in Refs. 92-94.

B. Heterogeneous Reactions

The details of the reaction mechanisms for the oxidation of hydrocarbons on catalysts are not well known for most catalysts. Indeed, the mechanisms would be expected to differ from catalyst to catalyst depending on the intermediates formed and the types of active sites involved. Hence, as in the case of homogeneous reactions, the approach here will be to detail some of the global reaction mechanisms and activation energies reported in the literature for certain catalysts and fuels of interest.

Information in the literature [5, 95] indicates that the rate of the oxidation of hydrocarbons on platinum catalysts is typically first order in fuel concentration and independent of oxygen concentration (provided that oxygen is in excess). In general, only carbon dioxide and water are present in the products. For oxidation of streams containing several fuel species, it is usually assumed that each species is oxidized independently at its own rate. Hawthorn [5] has reported a few estimates of chemical rate parameters for aircraft afterburner catalysts. These were usually Pt or Pt-Pd catalysts. These estimates, given in Table 5, were based on normal design specifications, published ignition and operating temperatures, and performance data reported in literature. It may be noted that the activation energies for the oxidation

TABLE 5 [5]

Estimated Chemical Rate Parameters for Afterburner Catalysts[a]

$$k_{chem,i} = A_i \exp[(\frac{E_a}{R})(\frac{1}{T_{ref}} - \frac{1}{T_s})] \text{ cm/s}$$

Component	A_i (cm/s)	E_a (kcal/g·mol)	E_a/R (°R)
Alumina-based catalysts:			
H_2	8.0	8	7,250
CO	0.45	11	10,000
Solvent hydrocarbons	0.1-0.2	11	10,000
CH_4	0.005	15	13,600
Metal-based catalysts:			
Toluene/xylene	0.2-0.3	15	13,000
MEK	0.03	15	13,000
n-Heptane	0.005	15	13,000

[a] T_{ref} = reference temperature = 1010°R
T_s = surface temperature, °R
$k_{chem,i}$ = chemical rate constant for oxidation of species i (first-order reaction assumed)

of hydrocarbons by these catalysts are typically in the range of 10-15 kcal/g·mol. This is considerably lower than the activation energies for the homogeneous reactions (Table 4) which lie in the 25-40 kcal/g·mol range.

The kinetics for the oxidation of CO and hydrocarbons by oxygen over various transition metal oxides is available in the literature [42, 75]. In general, one finds that for the pure oxides, the rate of oxidation is not greatly influenced by the oxygen partial pressure, but is inhibited by water vapor and exhibits a first- to half-order dependence on the partial pressure of CO or hydrocarbon. The inhibition by water is due to the strong chemisorption of water on surfaces. The stronger dependence of the rate on the pressure of CO or hydrocarbons than on the pressure of oxygen would appear to indicate that the reduction of the active surface is slower than its reoxidation. The apparent activation energies for CO and C_2H_4 oxidation range from 10-25 kcal/g·mol. The competitive oxidation of CO in the presence of hydrocarbons has been measured in a few cases [96]. CO can inhibit the oxidation of paraffin hydrocarbons, and propylene can inhibit the oxidation of CO.

Table 6 lists the global reaction rates reported in the literature for some catalysts and fuels of interest. Note that most of the rate expressions are first order in fuel concentration, and this is consistent with the above discussion. Again the reader is cautioned to check the references cited before using the rate expressions to ensure that they are only applied under valid conditions. Moreover, the rate expressions are only applicable to the specific catalysts mentioned. For any new catalyst developed, the rate expressions would have to be determined experimentally. In the absence of any rate data, it is best to assume that the surface reactions are completely mass transfer limited. This is a reasonable approximation over much of the bed length for most catalytic combustors of interest.

VI. MODELING OF CATALYTIC COMBUSTORS

Mathematical models applicable to catalytic combustion analysis in general fall into two categories: steady-state models and transient models. By far, the majority of the models available are steady-state ones, and these are discussed first.

A. Steady-State Models

In the modeling of catalytic combustors, it is important to recognize the importance of both heterogeneous and homogeneous reactions. At low temperatures, the homogeneous reaction rates are small and the heterogeneous reactions tend to dominate, while at very high temperatures the homogeneous reactions become more important. At temperatures of interest in most catalytic combustors, however, the contributions of both types of reaction are significant.

Among the early models that considered both homogeneous and heterogeneous reactions was that of Khitrin and Solovyeva [115] who used a simple isothermal plug flow model to investigate the role of surface reactions in the combustion of CO inside platinum and copper tubes. Krongelb and Strandberg [116] used a finite difference technique to solve a two-dimensional model incorporating a second-order homogeneous reaction and a first-order wall reaction in an isothermal tubular reactor. Walker [117] obtained the asymptotic solution to a two-dimensional model of both first-order homogeneous and heterogenous reactions in an isothermal reactor, allowing for both axial and radial diffusion. Solomon and Hudson [118] established criteria for an isothermal reactor which specified conditions under which the concentration profiles can be described adequately by considering either the homogeneous or the heterogeneous reaction alone. For other conditions, both types of reactions had to be considered for adequate treatment of the problem. Models for consecutive and concurrent homogeneous and surface reactions in packed beds have been developed by Holton and Trimm [119].

The models discussed so far were developed for isothermal reactors with both homogeneous and heterogeneous reactions; they were not developed specifically for catalytic combustors. Votruba et al. [114] published an analytical study on heat and mass transfer in monolithic honeycomb catalysts of the type used in catalytic combustors. The objective of their model was to predict multiple steady states in catalytic combustors. Axial heat conduction in the substrate was included while heat and mass transfer normal to the channel walls was treated using transfer coefficients, enabling the equations to be reduced to one-dimensional form. It was shown that two steady states may exist in some regions of operation, and that all-metal monolith supports are more prone to multiplicity of steady states than the ceramic ones. Unfortunately, their model does not consider the homogeneous reactions in the bed, making its application to real catalytic combustors difficult.

Young and Finlayson [55] developed one-, two-, and three-dimensional models for monolithic catalytic converters with various channel cross sections. As in Votruba's model, gas-phase reactions in the bed are ignored. Diffusion of energy and species normal to the flow axis is considered as well as axial conduction in the wall. They found that three-dimensional effects associated with peripheral temperature and concentration variations did occur in noncircular cross sections; however, their effect on overall bed operation was small. Axial heat conduction in the substrate was shown to be important in some situations, its effect being to reduce the temperature gradient at the point where the reaction lights off. Their results also indicated that lumped parameter models which hold Nusselt and Sherwood numbers constant over the bed length do not provide an adequate approach.

Cerkanowicz et al. [54] developed a one-dimensional model for a catalytic combustor incorporating one-step homogeneous and heterogeneous reactions with heat and mass transfer handled using transport coefficients. Gas velocity was allowed to vary axially, and heat conduction in the bed was included. Computer solutions were obtained for axial profiles of conversion, gas and

TABLE 6

Heterogeneous Kinetics[a]

Catalyst	Fuel	Chemical equation	Rate expression	Refs.
Cr_2O_3-Co_3O_4 (3:1 wt) 5% wt on Cordierite	C_3H_8	$C_3H_8 + 5O_2 \xrightarrow{r_1} 3CO_2 + 4H_2O$	$r_1 = 1.148 \times 10^6 \exp(-\frac{20,270}{RT})[C_3H_8]$ g·mol/cm²·s	42
		$C_3H_8 + 3\frac{1}{2}O_2 \xrightarrow{r_2} 3CO + 4H_2O$	$r_2 = 2.69 \times 10^4 \exp(-\frac{15,840}{RT})[C_3H_8]$ g·mol/cm²·s	
		$CO + \frac{1}{2}O_2 \xrightarrow{r_3} CO_2$	$r_3 = 1.53 \times 10^3 \exp(-\frac{11,550}{RT})[CO]$ g·mol/cm²·s	
0.1% wt Pt on Cordierite	C_3H_8	$C_3H_8 + \frac{7}{2}O_2 \xrightarrow{r_1} 3CO + 4H_2O$	$r_1 = 0.5 \times 10^6 \exp(-\frac{9,000}{RT})\rho_{C_3H_8}\rho_{O_2}$ kg/m²·s	41
		$CO + \frac{1}{2}O_2 \xrightarrow{r_2} CO_2$	$r_2 = 0.5 \times 10^6 \exp(-\frac{15,000}{RT})F\rho_{CO}\rho_{O_2}$ kg/m²·s	

			where ρ = density, kg/m^3	
			$F = \frac{\rho_{CO}}{\rho_{C_3H_8}}(1 - \exp(-\frac{\rho_{C_3H_8}}{\rho_{CO}}))$	
Pt wire	C_4H_{10}	$C_4H_{10} + 6\frac{1}{2}O_2 \xrightarrow{r_1} 4CO_2 + 5H_2O$	$r_1 = 7.9 \times 10^7 \exp(-\frac{20,100}{RT})[C_4H_{10}]$ g·mol/cm^2·s	97
Pt/γ-Al_2O_3/ Cordierite	C_3H_8	$C_3H_8 + 5O_2 \xrightarrow{r_1} 3CO_2 + 4H_2O$	$r_1 = 1.09 \times 10^9 \exp(-\frac{17,600}{RT})[C_3H_8]$ g·mol/cm^2·s	59
		$C_2H_4 + 3O_2 \xrightarrow{r_2} 2CO_2 + 2H_2O$	$r_2 = 10^{7.04} \exp(-\frac{12,000}{RT})[C_2H_4]$ g·mol/cm^2·s	
		$CO + \frac{1}{2}O_2 \xrightarrow{r_3} CO_2$	$r_3 = 3.83 \times 10^3 \exp(-\frac{24,900}{RT})[CO]$ g·mol/cm^2·s	

(continued)

TABLE 6 (continued)

Catalyst	Fuel	Chemical equation	Rate expression	Refs.
Pt on Cordierite	C_3H_8	$C_3H_8 + 5O_2 \xrightarrow{r_1} 3CO_2 + 5H_2O$	$r_1 = 2.5 \times 10^3 \exp(-\frac{10,000}{RT})[C_3H_8]$ g·mol/cm^2·s	61
	CO	$CO + \frac{1}{2}O_2 \xrightarrow{r_2} CO_2$	$r_2 = 10^5 \exp(-\frac{17,800}{RT})[CO]$ g·mol/cm^2·s	
Supported Pt	CO	$CO + \frac{1}{2}O_2 \xrightarrow{r_1} CO_2$	$r_1 = \frac{K_1[CO][O_2]}{(1 + K_2[CO])^2}$	58
			$K_1 = 4.14 \times 10^8 \exp(-\frac{12,600}{T})$ (kg·mol total)^2/kg·mol·O_2·m^2·s	
			$K_2 = 65.5 \exp(961/T)$ kg·mol total/kg·mol·CO	

[a] All activation energies are in calories per gram mole and temperatures T are in °K.

substrate temperature, velocity, and pressure. Model predictions were reported to be in good agreement with experimental catalytic combustion data. Catalytic light-off, hysteresis, and the contribution of homogeneous reactions are also discussed.

Heck [120], using a finite difference procedure, compared one- and two-dimensional catalytic converter model predictions. He concluded that one-dimensional models that hold the heat transfer coefficient constant along the length are not adequate. However, he showed that if the transfer coefficients are allowed to be functions of distance and wall temperature, characteristic of developing boundary layers, predictions comparable to the two-dimensional results could be achieved at a fraction of the cost.

Harrison and Ernst [58] have developed a two-dimensional model for the catalytic combustion of CO in an adiabatic tubular reactor under laminar flow conditions. The homogeneous case, the heterogeneous case, and the heterogeneous-homogeneous case were examined. It was shown that the behavior of the reactor approaches that of the heterogeneous model for temperatures less than 650 K, while at temperatures exceeding 1150 K the reactor behavior approaches that of the homogeneous model. Between 650 and 1150 K, both reactions were found to be significant.

Kennedy and Prasad [60] developed a one-dimensional model for the catalytic combustor assuming that the surface reactions are mass transfer limited. Although this assumption is not valid near the entrance of the combustor (where the temperatures are low), it is, in general, valid over most of the bed length. The system was assumed to be adiabatic, and the transport of energy and species to the surface are modeled using transfer coefficients. The effects of several dimensionless parameters on the steady-state combustor performance are discussed.

Perhaps the most comprehensive model developed to date for the catalytic combustor was that by Kelley et al. [57]. The model is one-dimensional but the transfer coefficients are computed as functions of distance, wall temperature, initial gas temperature, and inlet flow conditions. In addition to incorporating homogeneous and heterogeneous reactions, the model also includes bed heat conduction, radiative heat transfer, and axial diffusion. Their predictions show that the homogeneous reactions in catalytic combustors are "flame-like" in nature and that beyond a certain mass flow rate they are extinguished, leading to an abrupt increase in hydrocarbon emissions (breakthrough). This represents the upper mass throughput for low unburned hydrocarbon emissions. Increasing the mass throughput to levels much above the breakthrough point was found to cause the front end of the bed to become cool. Subsequently, small increases in mass throughput cause the cool region to spread downstream, extinguishing the wall reactions in the entire bed. This condition, called "blowout," represents the maximum mass throughput for hot bed operation. The effect of several design/operating parameters on breakthrough and blowout are discussed. Based on the model predictions, a catalytic combustor design (the graded cell concept) was suggested which utilizes a segmented bed which has large diameter channels in the first segment and very small diameter channels in the last segment. The large channels at the front end were designed to permit bed operation at large throughputs while the small channels at the back prevent breakthrough. Subsequent experimental testing has shown this concept to be very successful, leading to significant performance improvement of the catalyst system.

Recently, Bracco et al. [59] have developed a two-dimensional, axisymmetrical, compressible fluid flow model for the oxidation of propane in a catalytic combustor. Both the surface and gas phase reactions were modeled using three-step mechanisms. Conduction in the substrate was considered although radiation was neglected. Their computations produced axial and radial concentration profiles of propane, CO, and other species as well as temperature profiles for wall and gas under a variety of operating conditions. In general, the trends were correct although the agreement with experimental data was sometimes poor. They also show the relative contribution of gas-phase versus catalytic reaction to fuel conversion and indicate that most of the fuel oxidation occurs on the surface. Their results are illustrative for the importance of two-dimensional effects within catalyst channels.

B. Transient Models

Transient models for catalytic combustors are scarce in the literature. Tien [61] has developed a quasi-steady-state, one-dimensional model in which the substrate energy equation is assumed to be transient while the gas-phase energy and species equations are the steady-state ones. This assumption is based on the fact that the characteristic time for substrate heatup is much greater than the characteristic times of other important phenomena (such as mass transfer, reaction, and gas residence time) in a typical combustor. The assumption considerably simplified computation and enabled much larger time steps to be used in the numerical calculations. The surface as well as gas-phase reactions are modeled using two-step semiglobal mechanisms. The model correctly predicts a number of trends such as the effect of inlet temperature, substrate thickness, channel diameter, and bed length on the transient response time. However, because of the quasi-steady-state assumption, it does not accurately predict the propagation of a step change in inlet temperature (T_{in}) or fuel concentration (ϕ_{in}) through the combustor. More importantly, this model does not consider axial heat conduction and radiation in the bed or heat losses from the combustor. Axial conduction plays an important role in causing the zone of complete conversion to propagate upstream as steady state is achieved. In addition, axial conduction, radiation, and heat losses have important effects on the breakthrough and blowout in such systems.

Prasad, Kennedy, and Ruckenstein [127] have developed a one-dimensional transient model incorporating axial conduction and heat losses but neglecting radiation. The model uses three-step semiglobal mechanisms for both homogeneous and heterogeneous reactions with the latter being that determined experimentally by the authors [42]. For small times the model is fully transient and solves time-dependent equations for solid temperature, gas temperature, and gas species concentration. This model correctly predicted many important features, especially those relevant to small times after a change in inlet conditions. For example, it shows how a step change in ϕ_{in} is advected through the combustor and how a pulse in solid temperature follows much more slowly. However, because of the small characteristic gas residence times in the combustor, it was necessary to use time steps of the order of 1 ms to solve this problem, making the solution computationally expensive. To alleviate this difficulty, for times much larger than that required for a disturbance at inlet to be advected downstream, a quasi-steady-state model was used in which only the substrate energy equation was time dependent while the others were steady-state ones. The temperature and concentration profiles computed by the fully transient model were used as input to the quasi-steady-state model. The results from the model have provided insight into the influence of heat loss, axial conduction, and several design variables on the operation of the combustor during startup/shutdown as well as in steady state. In addition, it illustrates the upstream propagation of the combustion front as steady state is approached. One of the interesting findings using this model was that the length required for achieving complete conversion in steady-state operation is shorter than that required in transient operation. The implication of this is that catalytic combustor designs based on steady-state analysis would be of insufficient length to start the combustor up from cold conditions.

VII. PROTOTYPE SYSTEM DEVELOPMENTS

In the last decade considerable progress has been made in the development and testing of catalytic combustion concepts and components relevant to practical applications. Some of these are discussed here.

The use of liquid fuels and low Btu gas has been investigated by a number of researchers. Blazowski et al. [2] have successfully burnt JP-4 fuel over a wide range of inlet conditions. Pfefferle et al. [1] report the successful use of #2 diesel, JP-4, as well as synthetic low Btu gas in their experiments. DeCorso et al. [9] present results obtained using #2 distillate oil and low Btu synthetic coal gas in an experimental catalytic combustor for gas turbine applications. In their tests, significant improvement in emissions were obtained in comparison to a conventional combustor tested even though the catalytic combustor operated at considerably higher exit temperatures. Kesselring et al. [23] have made detailed studies on a number of different types of catalysts using diesel fuel. Krill et al. [16] have used #2 fuel oil in three prototype catalyst systems. Osgerby et al. [13] have used Jet A fuel in their simulation of the use of a catalytic reactor for jet engine emissions clean-up at idle mode conditions.

More recently there has been much interest in the combustion of alternate fuels, such as heavy distillates, residual fuels, and synthetic fuels using catalytic combustors. Pogson et al. [25] have demonstrated the use of #6 fuel oil in catalytic combustors without any apparent formation of thermal NO_X. Rosjford [33] has studied the use of #6 oil and blends of residual fuel (#6 oil or shale residual oil) and #2 oil in catalytic combustors. His results showed that no level of #6 oil was conducive to satisfactory performance unless complete vaporization was achieved. Shale oil blends (which has a lower final distillation point) gave satisfactory performance even when the fuel was incompletely vaporized. Dodds [32] has conducted experiments using #6 residual oil in a catalytic combustor employing a palladium catalyst. Although he obtained good combustion efficiencies at sufficiently high inlet temperatures, nearly 100% conversion of fuel-bound nitrogen to NO_X was observed. Chu [28] discusses criteria applicable for the design of catalytic combustors for use with heavy fuel oils. Lee et al. [34] have successfully demonstrated the use of low and medium Btu gas fuels (from coal gasification) in catalytic combustors. Tong et al. [35] have examined the use of simulated low Btu gas, coal-derived liquids, and #6 fuel in catalytic combustors. Their results showed that various alternative fuels can be efficiently burned in catalytic reactors; however, problems which may be unique to the combination of fuel, application, and operating condition still exist. Some of the more significant problems were stated to be materials and systems related (e.g., catalyst deactivation, soot formation, ash deposition, catalyst plugging). Chu et al. [36] have made detailed studies on the combustion of H-Coal and SRC II coal-derived liquids in catalytic combustors. Their results showed that coal-derived liquids could be burned efficiently catalytically, but the catalysts were prone to deactivation or poisoning.

The use of liquid fuels in catalytic combustors has necessitated the development of several fuel injector and ignitor concepts. Perhaps the most popular design is the multiple venturi fuel injector developed by Tacina [121]. Variations of this design have been fabricated and tested by a number of investigators [30, 33]. Anderson et al. [30] have investigated the use of atomizing spray nozzle injectors to improve mixing and vaporization of liquid fuels. In that paper they have also discussed two combustor ignitor concepts—the opposed jet ignitor and the aft-end torch ignitor. Pillsbury et al. [31] use a ring of eight spray nozzles to spray fuel radially into the mixing section. In their design a preburner was used to raise the inlet temperature sufficiently to achieve catalyst light off. The degree of heating provided by the preburner was maximum at ignition (startup) and small during steady-state operation.

Results relevant to the durability of catalyst systems and performance at elevated pressures have also appeared in the literature. Osgerby et al. [26] have conducted a 1000-h aging study on a proprietary catalyst at pressures as high as 5 atm using #2 fuel oil. The catalyst maintained its physical integrity throughout the 1000-h, 5-atm life test at operating temperatures in the vicinity of 1530 K and withstood the thermal shock of numerous startups and shutdowns. However, some irreversible catalyst deactivation was observed after the tests at elevated pressures. Carl et al. [122] have conducted catalyst combustion tests at pressures up to 15 atm and report very low emissions of CO, NO_x, and UHC. However, they have not investigated the durability of the catalyst element at such high pressures.

At present, catalytic combustion appears to be close to the practical demonstration stage in a number of applications. One of the most ambitious programs is the joint Westinghouse-Englehard project [31] in which full-scale, full-pressure tests are being conducted on catalytic combustors designed for applications in stationary combustion turbines. The feasibility of the concept has been demonstrated in steady-state operation although further work is required concerning the startup and transient operation, performance, and durability. The NASA-ACUREX program for the development of catalytic combustors for advanced automotive gas turbine engines [24] is also promising. The development of catalytic gas turbine, watertube boiler, and firetube boiler combustors at ACUREX [16] has good potential for thermal NO_x reduction with clean fuels.

VIII. FUTURE AREAS OF RESEARCH

The development of active catalyst systems capable of long-life operation at high temperatures, without structural and/or performance degradation, still remains the most important problem in catalytic combustion. The catalysts most commonly used for combustion applications are the noble metals (typically Pt or Pd) supported on Cordierite, mullite, or alumina substrates. However, these catalysts tend to sinter at elevated temperatures, leading to deactivation. Mixed oxides catalysts exhibit somewhat higher ignition temperatures and lower activities than Pt catalysts, but show promise for use at higher temperatures. The optimum mixed oxide catalyst designed for high combustion efficiency and low NO_x formation is fuel dependent, but little work has been done in this direction.

Both with noble metals and mixed oxides catalysts the structural integrity of the supported catalyst is a serious problem during prolonged high temperature operations. Attempts to overcome this difficulty by using active monoliths that incorporate the catalyst material have only met with limited success, because only a small part of the catalyst is exposed to the reactants. A more porous monolith may improve the performance due to a greater exposed active area.

The use of nitrogen-bound fuels in catalytic combustion is an area of interest, primarily because several synthetic fuels contain bound nitrogen. With such fuels, noble metal catalysts typically produce high levels of NO_x under fuel-lean conditions and low levels of NO_x only in the fuel-rich region. Thus to burn nitrogen-bound fuels in noble metal catalytic combustion and produce low levels of NO_x, the development of a two-stage catalytic combustor with a first-stage fuel rich and a second one fuel lean is worthwhile. Metal oxide catalysts which have been found to produce low levels of NO_x even under fuel lean conditions, have been tested only for a few nitrogen-bound fuels. Moreover, further development is required in improving their durability at high temperatures.

There is scope for the development of other components necessary for efficient catalytic combustion. For example, it is important that the fuel distribution at the inlet of the combustor be very uniform to prevent the occurrence of hot spots and of excessive thermal stresses in the monolithic substrate. The operation of the combustor is sensitive to the velocity, temperature, and fuel concentration of the gas at the inlet. Consequently, systems for closely controlling the flow rates of air and fuel as well as the preheater for the incoming air are required.

Heat extraction from the combustor is another area that is of practical importance but has not been adequately studied. The heat removal could be carried out in a number of ways, such as the introduction of water tubes directly in the catalyst bed, having a heat exchanger downstream of the bed, or by using concentric cooling jackets around the combustor. The temperature could be controlled by directly injecting water into the bed.

Operating procedures for starting up and shutting down the combustor during practical applications, as well as optimal ways to vary the inlet conditions to the combustor to match changing power requirements, are important areas that need further study.

Successful modeling and scale up of catalytic combustion requires the understanding of heterogeneous kinetics, homogeneous kinetics, and the possible role played by surface formed species in initiating or promoting homogeneous reactions. However, very little information is available in the literature on the heterogeneous kinetics and the role of hetero-homogeneous mechanisms. Furthermore, although multiple steady states have been observed in the operation of catalytic combustors, virtually no model is as yet available for predicting them satisfactorily. Such a model would be important in predicting the stability of operation of catalytic combustors.

REFERENCES

[1] W. C. Pfefferle, R. V. Carrubba, R. M. Heck, and G. W. Roberts, Catathermal Combustion: A New Process for Low-Emissions Fuel Conversion, ASME Paper #75-WA/FU-1.

[2] W. S. Blazowski and D. E. Walsh, "Catalytic Combustion: An Important Consideration for Future Application," Combust. Sci. Technol., 10, 233 (1975).

[3] D. N. Anderson, R. R. Tacina, and T. S. Mroz, Performance of a Catalytic Reactor at Simulated Gas Turbine Operating Conditions, NASA TM X-71747, June 1975.

[4] F. B. Wampler, D. W. Clark, and F. A. Gaines, Catalytic Combustion of C_3H_8 on Pt Coated Mondith, Paper No. 74-36, Western States Section, The Combustion Institute, October, 1974.
[5] R. D. Hawthorn, "Afterburner Catalysts—Effects of Heat and Mass Transfer between Gas and Catalyst Surface," AIChE Symp. Ser., Recent Adv. Air Pollut. Control, 70(137), 428 (1974).
[6] T. J. Rosjford, Catalytic Combustors for Gas Turbine Engines, AIAA Paper 76-46, Washington, D.C., January 1976.
[7] W. S. Blazowski and G. E. Bresowar, Preliminary Study of the Catalytic Combustor Concept, As Applied to Aircraft Gas Turbines, AFAPL-TR-74-32, 1974.
[8] V. J. Siminski, Development of a Catalytic Combustor for Aircraft Gas Turbine Engines, AFAPL-TR-76-80, 1976.
[9] S. M. DeCorso, S. Mumford, R. Carrubba, and R. Heck, Catalysts for Gas Turbine Combustors—Experimental Test Results, ASME Paper No. 76-GT-4, 1976.
[10] W. V. Krill et al., Catalytic Combustion for Gas Turbine Applications, ASME Paper No. 79-GT-188, 1979.
[11] D. N. Anderson, Preliminary Results from Screening Tests of Commercial Catalysts with Potential Use in Gas Turbine Combustors: Part II: Combustion Test Rig Evaluation, NASA TM X-73412, 1976.
[12] D. N. Anderson, Emissions and Performance of Catalysts for Gas Turbine Catalytic Combustors, NASA TM, X-73543, 1977.
[13] I. T. Osgerby, G. L. Gleason, and E. J. Mulraz, Combustion Catalyst Studies for Simulated Aircraft Idle Mode Operation, Presented at 4th Workshop on Catalytic Combustion, Cincinnati, Ohio, May 1980.
[14] B. A. Olson, H. C. Lee, I. T. Osgerby, R. M. Heck, and H. Hess, Durability Testing at 5 Atmospheres of Advanced Catalysts and Catalyst Supports for Gas Turbine Engine Combustors, NASA, CR-159839, 1980.
[15] W. V. Krill et al., Catalytic Combustion for System Applications, ASME Paper No. 79-HT-54, 1979.
[16] W. V. Krill, J. P. Kesselring, and S. J. Anderson, Prototype Catalytic Systems, Presented at 4th Workshop on Catalytic Combustion, Cincinnati, Ohio, May 1980.
[17] D. Bulzan, Experimental Evaluation of Catalytic Combustion with Heat Removal at Near Stoichiometric Conditions, Presented at 4th Workshop on Catalytic Combustion, Cincinnati, Ohio, May 1980.

[18] R. E. Thompson, D. W. Pershing, and E. E. Berkau, Catalytic Combustion, a Pollution-Free Means of Energy Conversion?, EPA-65012-73-018, August 1973.

[19] M. R. Dongworth and A. Melvin, "Diffusive Catalytic Combustion," Sixteenth Symposium (International) on Combustion, The Combustion Institute, 1977, p. 255.

[20] J. P. Kesselring, W. V. Krill, and R. M. Kendall, Design Criteria for Stationary Source Catalytic Combustors, Presented at Western States Section/The Combustion Institute Fall Meeting on Catalytic and Fluidized Bed Combustion, Paper No. 77-32, Stanford, California, October 17-18, 1977.

[21] R. Prasad, L. A. Kennedy, and E. Ruckenstein, "Catalytic Combustion of Propane Using Transitional Metal Oxides," Combust. Sci. Technol., 22, 271 (1980).

[22] B. A. Folsom, W. D. Clark, C. W. Courtney, M. P. Heap, and W. R. Seeker, Fuel-Nitrogen Conversion—The Impact of Catalyst Type, Presented at 4th Workshop on Catalytic Combustion, Cincinnati, Ohio, May 1980.

[23] J. P. Kesselring, W. V. Krill, and M. J. Angwin, Development of Improved Catalyst Systems, Presented at 4th Workshop on Catalytic Combustion, Cincinnati, Ohio, May 1980.

[24] H. Tong, G. C. Snow, E. K. Chu, R. L. S. Chang, M. J. Angwin, and S. L. Pessagno, Development of a High-Temperature Durable Catalyst for Turbine Engines, NASA CR-165396, September 1981.

[25] J. T. Pogson and M. N. Mansour, #6 Fuel Oil Catalytic Combustion, Presented at Third Workshop on Catalytic Combustion, Asheville, North Carolina, October 3-4, 1978.

[26] I. T. Osgerby, B. A. Olson, and H. C. Lee, CATCOM Catalyst 5 Atm 1000 Hour Aging Study Using #2 Fuel Oil, Presented at 4th Catalytic Combustion Workshop, Cincinnati, Ohio, May 1980.

[27] T. J. Rosjford, Catalytic Combustion of Heavy Partial-Vaporized Fuels, Presented at 4th Workshop on Catalytic Combustion, Cincinnati, Ohio, May 1980.

[28] E. Chu, Catalytic Reactor Design for Combustion of Partially Vaporized No. 6 Oil, Acurex Project 7543, Report, 1979.

[29] R. Tacina, Experimental Evaluation of Premixing-Prevaporizing Fuel Injection Concepts for a Gas Turbine Catalytic Combustor, NASA TM-73755, 1977.

[30] S. J. Anderson, W. V. Krill, and J. P. Kesselring, Fuel Injector, Ignition, and Temperature Measurement Techniques for Catalytic Combustors, Presented at 4th Workshop on Catalytic Combustion, Cincinnati, Ohio, May 1980.

[31] P. W. Pillsbury, P. E. Scheihing, and J. L. Toof, Full Scale, Full-Pressure Tests of Catalytic Burners for Combustion Turbines, Presented at Fifth Workshop on Catalytic Combustion, San Antonio, Texas, September 15-16, 1981.

[32] W. J. Dodds, Demonstration of Catalytic Combustion with Residual Fuel, Presented at Fifth Workshop on Catalytic Combustion, San Antonio, Texas, September 15-16, 1981.

[33] T. J. Rosjford, Catalytic Combustion with Incompletely Vaporized Residual Fuel, NASA CR-165161, March 1981.

[34] H. C. Lee and I. T. Osgerby, Catalytic Combustion Characteristics of Low and Medium BTU Gas Fuels, Presented at 4th Catalytic Combustion Workshop, Cincinnati, Ohio, May 1980.

[35] H. Tong, E. K. Chu, and G. C. Snow, Catalytic Combustion of Alternative Fuels, Presented at 4th Workshop on Catalytic Combustion, Cincinnati, Ohio, May 1980.

[36] E. K. Chu, G. C. Snow, and H. Tong, Catalytic Combustion of Coal Derived Liquids, EPRI AP-1666, January 1981.

[37] R. D. Matthews and R. F. Sawyer, Fuel Nitrogen Conversion and Catalytic Combustion, Paper No. 77-40, Fall Meeting, Western States Section/Combustion Institute, 1977.

[38] E. K. Chu and J. P. Kesselring, Fuel NO_x Control in Catalytic Combustion, Presented at Third Workshop on Catalytic Combustion, EPA-600/7-79-038, 1979.

[39] B. A. Folsom, C. W. Courtney, and M. P. Heap, The Effects of LBG Composition and Combustor Characteristics on Fuel NO_x Formation, ASME Paper 79-GT-185.

[40a] R. Prasad, L. A. Kennedy, and E. Ruckenstein, "Oxidation of Fuel Bound Nitrogen in a Transition Metal Oxide Catalytic Combustor," Combust. Sci. Technol., 27(1&2), 45 (1981).

[40b] L. A. Kennedy, R. Prasad, and E. Ruckenstein, Minimizing Fuel NO_x Using a Transitional Metal Oxide Catalytic Combustor, Presented at 5th Workshop on Catalytic Combustion, San Antonio, Texas, September 1981.

[41] P. J. Marteney and A. S. Kesten, Kinetics of Surface Reactions in Catalytic Combustion, Presented at Eighteenth Symposium (International) on Combustion, The Combustion Institute, Waterloo, Canada, 1981.

[42] R. Prasad, L. A. Kennedy, and E. Ruckenstein, "Kinetics of Catalytic Combustion of Propane on Transitional Metal Oxides," Combust. Sci. Technol., 27(5&6), 171 (1982).

[43] C. Bruno, Catalytic Combustion of CO/Air on Platinum, Presented at Fifth Workshop on Catalytic Combustion, San Antonio, Texas, September 15-16, 1981.

[44] G. Snow, Mechanism and Kinetics in Catalytic Combustion, Presented at Fifth Workshop on Catalytic Combustion, San Antonio, Texas, September 15-16, 1981.

[45] R. Prasad, H. L. Tsai, L. A. Kennedy, and E. Ruckenstein, "Effect of Inlet Parameters and Bed Length on the Operating Characteristics of a Catalytic Combustor," Combust. Sci. Technol., 25(3&4), 71 (1981).

[46] J. P. Kesselring, W. V. Krill, H. L. Atkins, R. M. Kendall, R. M. Kelley, and J. T. Kelley, Design Criteria for Stationary Source Catalytic Combustion Systems, EPA-600/7-79-181, August 1979.

[47] D. N. Anderson, The Effect of Catalyst Length and Downstream Reaction Distance on Catalytic Combustor Performance, Presented at Fourth Workshop on Catalytic Combustion, Cincinnati, Ohio, May 1980.

[48] I. T. Osgerby, B. A. Olson, H. Lew, and A. Cohen, CATCOM Catalyst Axial Temperature Profile Measurements, Presented at Fourth Workshop on Catalytic Combustion, Cincinnati, Ohio, May 1980.

[49] R. Prasad, H. L. Tsai, L. A. Kennedy, and E. Ruckenstein, Occurrence of Multiple Steady States in the Catalytic Combustion of Propane, Combust. Sci. Technol., 26(1&2), 51 (1980).

[50] R. Masel, Ignition Instabilities in Catalytic Combustion, Presented at Fifth Workshop on Catalytic Combustion, San Antonio, Texas, September 15-16, 1981.

[51] R. Robben, R. Schefer, V. Agrawal, and I. Namer, Catalyzed Combustion in a Flat Plate Boundary Layer 1. Experimental Measurements and Comparison with Numerical Calculations, Presented at Fall Meeting of the Western States Section of the Combustion Institute, Stanford, California, October 17-18, 1977.

[52] R. W. Schefer, The Role of Surface Reactions in Catalytic Combustion, Presented at Fifth Workshop on Catalytic Combustion, San Antonio, Texas, September 15-16, 1981.

[53] P. Marteney, The Effect of Gases and Surfaces in Catalytic Combustion, Presented at Fifth Workshop on Catalytic Combustion, San Antonio, Texas, September 15-16, 1981.

[54] A. E. Cerkanowicz, R. B. Cole, and J. A. Stevens, "Catalytic Combustion Modelling: Comparison with Experimental Data," J. Eng. Power, 99 (Ser. A, No. 4), 593 (1977).

[55] L. C. Young and B. A. Finlayson, "Mathematical Models of the Monolithic Catalytic Converter: Part II, Application to Automobile Exhaust," AIChE J., 22(2), 343 (1976).

[56] R. Schefer and F. Robben, Catalyzed Combustion in a Flat Plate Boundary Layer II. Numerical Calculations, Presented at Fall Meeting of the Combustion Institute, Stanford, California, October 17-18, 1977.

[57] J. T. Kelley, R. M. Kendall, E. Chu, and J. P. Kesselring, Development and Application of the PROF-HET Catalytic Combustor Code, Presented at Fall Meeting of the Combustion Institute, Stanford, California, October 17-18, 1977.

[58] K. Harrison and W. R. Ernst, "Catalytic Combustion in Cylindrical Channels: A Homogeneous-Heterogeneous Model," Combust. Sci. Technol., 19, 31 (1978).

[59] F. V. Bracco, C. Bruno, Y. Yaw, and P. M. Walsh, Computed and Measured Emissions from the Catalytic Combustion of Propane/Air Mixtures, Presented at Fourth Workshop on Catalytic Combustion, Cincinnati, Ohio, May 14-15, 1980.

[60] L. A. Kennedy and R. Prasad, Heat and Mass Transfer in a Catalytic Combustor, ASME Paper #79-47-57.

[61] J. S. Tien, Modelling of Transient Operation of Catalytic Combustor, Presented at Fourth Workshop on Catalytic Combustion, Cincinnati, Ohio, May 14-15, 1980.

[62] R. Prasad, "Transient Model of the Catalytic Combustor," PhD Thesis, State University of New York, Buffalo, 1982.

[63] M. A. Friedman, A Catalytic Combustion Water Tube Boiler System, Presented at Fifth Workshop on Catalytic Combustion, San Antonio, Texas, September 15-16, 1981.

[64] E. B. Merrick, Development of a Low-NO_x Combustion for Gas-Fired Firetube Boilers, Presented at Fifth Workshop on Catalytic Combustion, San Antonio, Texas, September 15-16, 1981.

[65] G. J. K. Acres, "Platinum Catalyst for the Control of Air Pollution," Platinum Met. Rev., 14, 2 (1970).

[66] A. Schwartz, L. L. Holbrook, and H. Wise, "Catalytic Oxidation Studies with Platinum and Palladium," J. Catal., 21, 199 (1971).

[67] T. Engel and G. Ertl, "Reactive and Non-Reactive Scattering of CO and O_2 from Pd(111)," J. Chem. Phys., 69, 1267 (1978).

[68] J. K. Dixon and F. E. Longfield, "Hydrocarbon Oxidation in Catalysis-Oxidation, Hydration, Dehydration and Cracking Catalysts," in Catalysis, Volume 7 (P. H. Emmett, ed.), Reinhold, New York, 1960, p. 183.

[69] G. C. Bond, Catalysis by Metals, Academic, New York, 1962, p. 464.

[70] W. A. Daniel and J. T. Wentworth, Paper 486B, presented at March Meeting of The Society of Automotive Engineers, 1962.

[71a] H. H. Lee and E. Ruckenstein, "Catalyst Sintering and Reactor Design," Catal. Rev.–Sci. Eng., 25, 475 (1983).
[71b] E. Ruckenstein and D. B. Dadyburjor, "Sintering and Redispersion of Supported Metal Catalysts," Rev. Chem. Eng., 1, 251 (1983).
[72] R. L. Klimisch, J. C. Summers, and J. C. Schlatter, Catalysts for the Control of Automotive Pollutants, Adv. Chem. Ser., 1975, p. 143.
[73] D. B. Dadyburjor, S. S. Jewur, and E. Ruckenstein, "Selective Oxidation of Hydrocarbon on Composite Oxides," Catal. Rev.–Sci. Eng., 19(2), 293 (1979).
[74] Y. Morooka and A. Ozaki, "Regularities in Catalytic Properties of Metal Oxide in Propylene Oxidation," J. Catal., 5, 116 (1966).
[75] Y. F. Yu Yao, Ibid., 39, 104 (1975).
[76] L. Y. Margolis, "Catalytic Oxidation of Hydrocarbons," Adv. Catal. Related Subjects, 14, 429 (1963).
[77] D. A. Dowden and D. Wells, "A Crystal Field Interpretation of Some Activity Patterns," Actes Int. Congr. Catal., 2e, Paris, p. 1499 (1961).
[78] G. M. Dixon, D. Nicholls, and H. Steiner, "Activity Pattern in the Disproportionation and Dehydrogenation of Cyclohexane to Cyclohexene and C_6H_6 over the Oxides of the First Series of Transition Metals," in Proceedings of the Third International Congress, Catalysis, Vol. 2, Amsterdam, 1964, p. 815.
[79] F. G. Dwyer, "Catalyst for Control of Automotive Emissions," Catal. Rev., 6(2), 261 (1972).
[80] E. L. McCann, "Perovskite Oxides as Carriers for Transition Metal Oxides," German Patent Offen 2,615,352.
[81] D. W. Johnson et al., "Nature and Effects of Platinum in Perovskite Catalysts," J. Catal., 48, 87 (1977).
[82] R. C. Evans, An Introduction to Crystal Chemistry, Cambridge University Press, London, 1952.
[83] J. C. Bittence, "Sorting out Ceramics," Mach. Des., 49(28), 1977.
[84] A. S. Pratt and J. A. Cairns, "Noble Metal Catalysts on Metallic Substrates," Platinum Met. Rev., 21(3), (1977).
[85] J. A. Caton, Heterogeneous Catalysis of Lean Ethylene/Air Mixtures by Platinum Coated Wire Screens, ASME Paper #76-WA/GT-2, 1976.
[86] D. Burgess and M. Hertzberg, "The Flammability Limits of Lean Fuel-Air Mixtures," ISA Trans., 14, 129 (1975).

[87] C. Bruno, P. M. Wash, D. A. Santavieca, N. Sinha, Y. Yaw, and F. V. Bracco, "Catalytic Combustion of Propane/Air Mixtures on Platinum," Combust. Sci. Technol., In Press.

[88] C. K. Westbrook and F. L. Dryer, Simplified Reaction Mechanisms for the Oxidation of Hydrocarbon Fuels in Flames, Presented at Spring Technical Meeting, Central States Section, The Combustion Institute, Warren, Michigan, 1981.

[89] H. C. Hottel and G. C. Williams, "Kinetic Studies in Stirred Reactors: Combustion of Carbon Monoxide and Propane," in Tenth Symposium (International) on Combustion, The Combustion Institute, 1965, p. 111.

[90] J. B. Fenn and H. F. Calcote, "Activation Energies in High Temperature Combustion," in Fourth Symposium (International) on Combustion, Cambridge, Massachusetts, 1952, p. 231.

[91] P. H. Kydd and W. I. Foss, "Combustion of Fuel Lean Mixtures in Adiabatic, Well Stirred Reactors," in Tenth Symposium (International) on Combustion, 1965, p. 101.

[92] C. K. Westbrook, "An Analytical Study of the Shock Tube Ignition of Mixtures of Methane and Ethane," Combust. Sci. Technol., 20, 5 (1979).

[93] C. J. Jachinowski, High Temperature Chemical Kinetic Study of the H_2-CO-CO_2-NO Reaction System, NASA TN D-7897, May 1975.

[94] C. T. Bowman, "An Experimental and Analytical Investigation of the High Temperature Oxidation Mechanism of Hydrocarbon Fuels," Combust. Sci. Technol., 2, 161 (1970).

[95] Y. Moro-Oka, Y. Morihana, and A. Ozaki, "Regularity in the Catalytic Property of Metal Oxides in Hydrocarbon Oxidation," J. Catal., 7, 2 (1967).

[96] Y. F. Yu Yao, Ibid., 28, 139 (1973).

[97] C. G. Rader and S. W. Weller, "Ignition on Catalytic Wires: Kinetic Parameter Determination by the Heated Wire Technique," AIChE J., 20(3) (1974).

[98] Ya. B. Zeldovich, "The Oxidation of Nitrogen in Combustion and Explosions," Acta Physicochem. USSR, 21, 557 (1946).

[99] C. P. Fenimore, "Formation of Nitric Oxide in Premixed Hydrocarbon Flames," in Thirteenth Symposium (International) on Combustion, The Combustion Institute, 1971, p. 373.

[100] A. A. Westenberg, "Kinetics of NO and CO in Lean Premixed Hydrocarbon-Air Flames," Combust. Sci. Technol., 4, 59 (1971).

[101] D. Iverach, K. S. Basden, and N. Y. Kirov, "Formation of Nitric Oxide in Fuel-Lean and Fuel-Rich Flames," in Thirteenth Symposium (International) on Combustion, The Combustion Institute, 1973, p. 767.

[102] C. T. Bowman, "Kinetics of Nitric Oxide Formation in Combustion Processes," in Fourteenth Symposium (International) on Combustion, The Combustion Institute, 1973, p. 729.

[103] A. F. Sarofim and J. H. Pohl, "Kinetics of Nitric Oxide Formation in Premixed Laminar Flames," in Fourteenth Symposium (International) on Combustion, The Combustion Institute, 1973, p. 739.

[104] V. S. Engleman, W. Bartak, J. P. Longwell, and R. B. Edleman, "Experimental and Theoretical Studies on NO_x Formation in a Jet-Stirred Combustion," in Fourteenth Symposium (International) on Combustion, The Combustion Institute, 1973, p. 755.

[105] G. B. Martin and E. E. Berkau, Preliminary Evaluation of Flue Gas Recirculation as a Control Method for Thermal and Fuel Related Nitric Oxide Emissions, Presented at Combustion Institute (Western States Section), Fall Meeting, Berkeley, California, 1971, Paper 71-41.

[106] D. W. Turner, R. L. Andrews, and C. W. Siegmund, "Influences of Combustion Modification and Fuel Nitrogen Content on Nitrogen Oxide Emissions from Fuel Oil Combustion," Combustion, 44, 21 (1972).

[107] W. Bartok, A. R. Crawford, and A. Skopp, "Control of NO_2 Emissions from Stationary Sources," Chem. Eng. Prog., 67, 64 (1971).

[108] G. G. DeSoete, "Overall Reaction Rates of NO and N_2 from Fuel Nitrogen," in Fifteenth International Symposium on Combustion, The Combustion Institute, 1975, p. 1093.

[109] C. P. Fenimore, "The Ratio NO_2/NO in Fuel-Lean Flames," Combust. Flame, 25, 85 (1975).

[110] B. S. Haynes, "Reactions of Ammonia and Nitric Oxide in the Burnt Gases of Fuel-Rich Hydrocarbon-Air Flames," Ibid., 28, 81 (1977).

[111] B. S. Haynes, D. Iverach, and N. Y. Kirov, "Behavior of Nitrogen Species in Fuel Rich Hydrocarbon Flames," in Fifteenth International Symposium on Combustion, The Combustion Institute, 1975, p. 1103.

[112] Y. H. Song, D. W. Blair, V. J. Siminski, and W. Bartok, "Conversion of Fixed Nitrogen to N_2 in Rich Combustion," in Eighteenth International Symposium on Combustion, University of Waterloo, 1980.

[113] J. P. Appleton and J. B. Heywood, "The Effects of Imperfect Fuel-Air Mixing in a Burner on NO Formation in the Air and in the Fuel," in Fourteenth Symposium (International) on Combustion, The Combustion Institute, 1973, p. 777.

[114] J. Votruba, J. Sinkule, V. Hlavacek, and J. Skriranek, "Heat and Mass Transfer in Monolithic Honeycomb Catalysts—I," Chem. Eng. Sci., 30, 117 (1975).

[115] L. N. Khitrin and L. S. Solovyeva, "Homogeneous-Heterogeneous Combustion of Carbon Monoxide in Narrow Tubes (Channels)," in Seventh Symposium (International) on Combustion, Reinhold, New York, 1958, Paper No. 74, p. 532.

[116] S. Krongelb and M. W. D. Strandberg, "Use of Paramagnetic-Resonance Technique in the Study of Atomic Oxygen Recombination," J. Chem. Phys., 31, 1196 (1959).

[117] R. E. Walker, "Chemical Reaction and Diffusion in a Catalytic Tubular Reactor," Phys. Fluids, 4(10), 1211 (1961).

[118] R. L. Solomon and J. L. Hudson, "Heterogeneous and Homogeneous Reactions in a Tubular Reactor," AIChE J., 13(3), 545 (1967).

[119] R. D. Holton and D. L. Trimm, "Mathematical Modelling of Heterogeneous-Homogeneous Reactions," Chem. Eng. Sci., 31, 549 (1976).

[120] R. H. Heck, J. Wei, and J. R. Katzer, "Mathematical Modeling of Monolithic Catalysts," AIChE J., 22(3), 477 (1976).

[121] R. R. Tacina, Experimental Evaluation of Fuel Preparation Systems for an Automotive Gas Turbine Catalytic Combustor, DDE/NASA/1011-78/23, 1977.

[122] D. E. Carl, M. C. Krepinevich, and I. T. Osgerby, High Pressure Test Results, Presented at 4th Workshop on Catalytic Combustion, Cincinnati, Ohio, May 1980.

[123] A. F. Sarofim and R. C. Flagan, "NO_x Control for Stationary Combustion Sources," Prog. Energy Combust. Sci., 2, 1 (1976).

[124] D. N. Anderson, Effect of Premixing on Nitric Oxide Formation, Presented at the AIChE National Meeting, 75th, Detroit, Michigan, June 3-6, 1973.

[125] R. M. Fristrom and A. A. Westenberg, Flame Structure, McGraw-Hill, New York, 1965.

[126] F. J. Verkamp, A. J. Verdouw, and J. G. Tomlinson, Impact of Emission Regulations on Future Gas Turbine Engine Combustors, AIAA Paper No. 73-1277, 1973.

[127] R. Prasad, L. A. Kennedy, and E. Ruckenstein, "A Model for the Transient Behavior of Catalytic Combustors," Combust. Sci. Technol., 66, 1 (1982).

4

ZEOLITES AND THEIR APPLICATIONS AS CATALYSTS AND/OR CATALYST SUPPORTS

INTRODUCTION

High-silica-content zeolites have been widely used for many industrial chemical processes due to their hydrothermal and acidic stability, high resistance to coking, and high activity and selectivity for specific types of reaction [1]. The β-zeolite, which has a SiO_2/Al_2O_3 ratio in the range from 5 to 100 [2], belongs to this class. It has a three-dimensional 12-ring pore system, and its structure is a highly faulted intergrowth of two distinct, but closely related structures: polymorph A (tetragonal) and polymorph B (monoclinic) [3, 4]. Accordingly, the β-zeolite is a combination of the polymorph A structure, which has straight channels of 0.73×0.60 nm and tortuous channels of 0.56×0.56 nm, and polymorph B, which has straight channels of 0.73×0.68 nm and tortuous channels of 0.55×0.55 nm. Because of its pore opening, β-zeolite can accommodate molecules with diameters larger than those that can be accommodated by ZSM-5, and can provide shape selectivity for intermediate-size molecules to a higher degree than X and Y faujasites.

Since its initial synthesis in 1967, β-zeolite has been used as a catalyst or a catalyst support. As a catalyst, Lapierre et al. [5–7] reported that β-zeolite has beneficial effects in the catalytic hydrodewaxing of petroleum oils by lowering their pour point. Corma et al. [8, 9] compared the activities of H-β-zeolite and HY zeolite for various reactions. They found that β-zeolite is less active and produces less gasoline and more coke than HY zeolite in the cracking of gas-oil [8], but is slightly more active in the cracking of n-heptane [9]. Wang et al. [10] evaluated the reactivity of β-zeolite in the disproportionation and transalkylation of toluene and trimethylbenzene and found that it provides excellent catalytic stability and transalkylation selectivity. As a support, β-zeolite was used to support Pt. The β-zeolite supported Pt catalyst has been employed for various reactions. Martens et al. [11] examined the catalytic activity of Pt/H-β-zeolite for the isomerization and hydrocracking of long chain paraffins and found that metal-supported β-zeolite behaves as an ideal bifunctional catalyst that exhibits high activity for primary cracking and selectivity for isomerization. Ratnasamy et al. [12] examined the reactions of aromatic hydrocarbons, such as the methylation and disproportionation of toluene and the isomerization of *m*-xylenes over Pt/β-zeolite, and found that due to its large pores, it does not exhibit any product shape selectivity for the

Heterogeneous Catalysis: Experimental and Theoretical Studies, First Edition. Edited by Eli Ruckenstein and Haiyou Wang.

isomerization of *m*-xylene to *o*- and *p*-xylenes; however, in contrast to the 10-membered ring pores zeolite such as ZSM-5, which has smaller pores, it favors the disproportionation reaction.

Amorphous aluminosilicates have been widely employed in acid-catalyzed reactions [13–17]. The activity of the aluminosilicates is dependent upon the total amount of Brønsted and Lewis acid sites, and the acid strength distribution. It is generally accepted that the generation of these acid sites is due to the isomorphous substitution of Si^{4+} by Al^{3+} in the silicate lattice [18–20], which produces an exchangeable cation. This aluminum ion can coordinate to oxygen tetrahedrally, or octahedrally, or a combination of the two. The relative amounts of these coordination states are determined by both the alumina content in the sample and the pretreatment temperature. There are two types of Brønsted acid sites present: (i) bridged hydroxyl groups and (ii) water molecules coordinated to a trigonal aluminum atom (the trigonal aluminum itself becoming a Lewis acid site). Borovkov et al. [21] reported that the hydroxyl groups of the amorphous aluminosilicates are quite similar to those present in zeolites, but their concentration is much lower. Since only the tetrahedrally held aluminum results in the generation of a bridged hydroxyl group, it is clear that the earlier-mentioned sites are responsible for the catalytic activity. In other words, the alumina content and heat treatment conditions determine the relative amounts of tetrahedral versus octahedral aluminum sites and ultimately the catalytic activity.

Furthermore, amorphous silica-alumina can be used as diluent in the preparation of composite zeolite catalysts. Composite catalytic particles have been used extensively in several industrial processes. The catalytic cracking of gas oil over X or Y faujasites embedded in an amorphous silica-alumina matrix [22] and the conversion of methanol to gasoline over ZSM-5 composite catalysts [23] are two classic examples of the applicability of composite zeolites. The large-scale use of composite zeolites is a result of their enhanced activity, selectivity, and stability. As suggested by Ruckenstein [24], the increased activity of a composite catalytic particle composed of microspheres of active porous catalyst uniformly imbedded in a more porous substrate can be attributed to the advantageous diffusional characteristics that the substrate provides. Since the diffusion coefficient is greater in the more porous material but the volume available for reaction decreases with increasing dilution, the overall reaction rate will exhibit a maximum as a function of the volume fraction of the active catalyst. The performance of various composite zeolite catalysts was extensively examined both theoretically and experimentally. Ramkrishna and Amundson [25] studied the transport in composite materials (multiphase media) by using the theory of self-adjoint operators. Varghese and Wolf [26] examined the performance of a composite catalytic plate under pore-mouth poisoning and concluded that composite particles exhibit longer lifetimes than conventional catalysts. Lee and Ruckenstein [27] studied the selectivity of composite pellets for parallel and series types of reactions for various distributions of the active phase. Dean and Dadyburjor [28] investigated experimentally the activity and coke deposition on composite cracking catalysts during cracking of n-hexadecane and found that the composite cracking catalysts exhibit a beneficial behavior in terms of coke deactivation.

Chapter 4 is devoted to various zeolites used as catalysts and/or catalyst supports. It consists of two parts: the first is concerned with the use of β-zeolite supported Pt as a reforming catalyst (**4.1**, **4.2**, **4.3**, and **4.4**), whereas the second is concerned with silica-alumina catalyst (**4.5**) and silica-alumina containing composite zeolite catalysts (**4.6**, **4.7**, and **4.8**).

The application of β-zeolite supported Pt as a bifunctional catalyst in various reforming reactions was explored. A comparison between Pt/β-zeolite and the conventional Pt/γ-Al_2O_3 in the reforming of n-hexane, methylcyclopentane, and cyclohexane under low severity conditions was carried out (**4.1**). It was found that because of its acidity, the β-zeolite leads to (i) the skeletal isomerization of the olefins generated by the dehydrogenation of n-paraffins on Pt sites and (ii) the isomerization of the cycloolefinic intermediates generated from methylcyclopentane on Pt sites. The latter isomerization leads to the enlargement of the C_5-cyclic ring of methylcyclopentane to the C_6-cyclic ring, which, via dehydrogenation on Pt sites, leads to benzene. The performance of composite powders consisting of Pt/β-zeolite and Pt/γ-Al_2O_3 was then investigated in the reforming of naphthenes (**4.2**). Because of its relatively high Pt dispersion, Pt/γ-Al_2O_3 is responsible for the dehydrogenation reactions leading to methylcycloolefins. Pt/β-zeolite, which has a much higher Brønsted acidity than γ-Al_2O_3, is responsible for the bimolecular alkylation reactions leading to aromatics. The transalkylation reactions of aromatics, and the isomerization reactions of methylcycloolefins to C_6 cycloolefins then lead, via dehydrogenation, to benzene. For these reasons, the composite powders consisting of Pt/β-zeolite and Pt/γ-Al_2O_3 exhibit advantageous behaviors in the reforming of naphthenes. The dependence of the effectiveness factor of composite cylindrical pellets made of Pt supported on both β-zeolite and γ-Al_2O_3 was determined as a function of zeolite content and temperature for methylcyclopentane and methylcyclohexane (**4.3**). The balancing of the acidity provided by Pt/β-zeolite with the dehydrogenation capabilities of Pt/γ-Al_2O_3

leads to a superior reforming catalyst that upgrades the octane number of naphthenes. The performance of the individual platinum-loaded L, β, and USY zeolites and of the composites of Pt/BaKL with either Pt/β or Pt/USY zeolite was investigated for reforming reactions (**4.4**). The reactions of various mixtures of n-hexane, methylcyclopentane, or methylcyclohexane over the composites resulted in increased selectivities to C_{7+} aromatics, which was attributed to the synergism of the components of the composites. It was shown that the balancing of a basic zeolite (L zeolite) with an acidic zeolite (β or USY zeolite) via composites can provide a selective route to C_8 and C_9 aromatics for industrial feeds with large contents of C_{6+} paraffins.

Synthetic amorphous silica-alumina can be used as catalysts or catalyst supports. Examples of processes that use silica-alumina catalysts are the production of pyridine from crotonaldehyde, formaldehyde, steam, air and ammonia and the cracking of hydrocarbons. As mentioned above, the activity of aluminosilicates in acid-catalyzed reactions presumably depends on the total amount of Brønsted and Lewis acid sites, as well as on the acid strength distribution. Nevertheless, the performance of silica-alumina catalysts is affected by numerous additional factors. Silica-aluminas with 15 wt% Al_2O_3 and subjected to identical heat treatments were synthesized in aqueous and alcohol solvents, and different activities and selectivities were observed in the methanol conversion to hydrocarbons (**4.5**).

In contrast to the amorphous silica-alumina, zeolites are crystalline, porous aluminosilicates. Both the zeolite and the silica-alumina can be independently used as catalysts. Composite catalysts that contain both zeolite and amorphous silica-alumina exhibit advantages over either the zeolite alone or the amorphous silica-alumina catalyst alone. A composite catalyst was prepared by adding crystallites of HAlZSM-5 ($SiO_2/Al_2O_3 = 35$) during the cogelation procedure used to prepare amorphous silica-alumina catalyst. The addition of a small amount (5–10 wt%) of HAlZSM-5 to the amorphous silica-alumina improved the resistance of the catalyst to deactivation by coking during methanol conversion to hydrocarbons. It also decreased slightly the selectivity to lower molecular-weight alkenes and increased that to aromatics and higher molecular weight aliphatics (**4.6**). The same composite catalyst was used for the oligomerization of ethene and propene (**4.7**). It was found that the silica-alumina catalysts containing 5 and 10 wt% ZSM-5 exhibited much greater activities than the pure silica-alumina, whereas the selectivity of the reaction remained comparable to that of silica-alumina and not of HAlZSM-5. Composites of ZSM-5 and amorphous silica-alumina were prepared under conditions at which a ZSM-5 phase can be crystallized *in situ* from an alumina-rich gel at low TPA^+. The activity and selectivity of the resulting composite catalysts for the cracking of n-hexane were compared with those of pure ZSM-5 and to mechanical mixtures prepared by dispersing the preformed ZSM-5 in an alumina-silica gel (**4.8**). It was found that the *in situ* crystallized composite catalysts with 10–40 wt% HAlZSM-5 exhibited significantly higher activities. Since both types of catalysts, namely those crystallized *in situ* and those prepared mechanically, had comparable numbers of strong acid sites, the better connectivity between the two phases was suggested to be responsible for the greater activity of the *in situ* crystallized composite catalysts.

REFERENCES

[1] Scherzer, J., *ACS Synp.*, Vol. **248**, p. 157, ACS, Washington, DC, 1984.

[2] Wadlinger, R.L., Kerr, G.T., and Rosinski, E.J., U.S. Patent 3,308,064 (1967).

[3] Newsam, J.M., Treacy, M.M.J., Koetsier, W.T., and DeGruyter, C.B., *Proc. R. Soc. London Ser. A*, **420**, 375 (1988).

[4] Treacy, M.M.J., and Wadlinger, R.L., *Nature*, **332**, 249 (1988).

[5] Lapierre, R.B., and Patridge, R.D., *Eur. Pat. Appl.*, 94827 (1983).

[6] Lapierre, R.B., Patridge, R.D., Chen, N.Y., and Wong, S.S., European Patent Appl., 95303 (1983).

[7] Lapierre, R.B., Randell, D.P., Chen, N.Y., and Wong, S.S., U.S. Patent, 4,501,926 (1985).

[8] Corma, A., Fornes, V., Melo, P., and Perez-Pariente, J., *Prepr.-Am. Chem. Soc. Div. Pet. Chem.*, 632 (1987).

[9] Corma, A., Fornes, V., Monton, J.B., and Orchilles, A.V., *J. Catal.*, **107**, 288 (1987).

[10] Wang, I., Tsai, T.C., and Huang, S.T., *Ind. Eng. Chem. Res.*, **29**, 2005 (1990).

[11] Martens, J.A., Perez-Pariente, J., and Jacobs, P.A., in "Chemical Reactions in Organic and Inorganic Constrained Systems" (E. Selton, ed.), Kluwer, Dordrecht, 1986, NATO aSI Ser. C., Vol. **165**, p. 115.

[12] Ratnasamy, P., Bhat, R.N., Pokhriyal, S.K., Hegde, S.G., and Kumar, R., *J. Catal.*, **119**, 65 (1989).

[13] Tanabe, K., *Solid Acids and Bases*, Kodansha, Tokyo, 1970.

[14] Tamele, M.W., *Discuss. Faraday Soc.*, **8**, 270 (1950).

[15] Ward, J.W., and Hansford, R.C., *J. Catal.*, **13**, 154 (1969).

[16] Mills, G.A., Boedecker, E.R., and Oblad, A.G., *J. Am. Chem. Soc.*, **72**, 1554 (1950).

[17] Holm, V.C.F., Bailey, G.C., and Clark, A., *J. Phys. Chem.*, **63**, 129 (1959).

[18] Thomas, C.L., Ind. Eng. Chem., 41, 2564 (1949).

[19] Leonard, A.J., Ratnasamy, P., Declerck, F.D., and Fripiat, J.J., *Faraday Soc.*, **52**, 98 (1971).

[20] Miesserov, K.G., *J. Catal.*, **13**, 169 (1969).

[21] Borovkov, V. Yu, Alexeev, A.A., and Kazansky, V.B., *J. Catal.*, **80**, 462 (1983).

[22] Eastwood, S.C., Drew, R.D., and Gartzell, F.D., *Oil Gas J.*, 152 (1962).

[23] Meisel, S.L., McCullough, J.P., Lechthaler, C.H., and Weisz, P.B., *Chem. Tech.*, **6**, 86–89.

[24] Ruckenstein, E., *A.I.Ch.E. J.*, **16**, 151 (1970).

[25] Ramkrishna, D., and Amundson, N.R., Chem. Engng Sci., 29, 1457 (1974).

[26] Varghese, P., and Wolf, E.E., *A.I.Ch.E. J.*, **26**, 55 (1980).

[27] Lee, S.H., and Ruckenstein, E., *Chem. Engng Commun.*, **46**, 43 (1986).

[28] Dean, S.W., and Dadyburjor, D.B., *Ind. Engng Chem. Res.*, **28**, 271 (1989).

Comparison between Zeolite β and γ-Al_2O_3 Supported Pt for Reforming Reactions

PANAGIOTIS G. SMIRNIOTIS AND ELI RUCKENSTEIN[1]

Department of Chemical Engineering, State University of New York at Buffalo, Buffalo, New York 14260

Received July 7, 1992; revised November 12, 1992

The performance of zeolite β supported platinum as a bifunctional catalyst for the reforming of *n*-hexane, methylcyclopentane, and cyclohexane, typical representatives of reforming feeds, has been examined under low severity conditions. A comparison under the same conditions with γ-Al_2O_3 supported Pt was carried out. Both catalysts were impregnated with 0.5 wt% Pt. It is found that β-zeolite exhibits skeletal isomerization reactions to a higher extent than alumina, producing molecules with much higher octane number than the feed molecules. Bimolecular alkylation reactions that take place in the zeolite pores have as a result the formation of products with a carbon number larger than that of the feed molecule. The enlargement of the cyclic ring of methylcyclopentane leads at suitable temperatures to aromatics, a reaction of great importance for reforming. This behavior is a result of the relatively large acidity of the zeolite and of its pore structure that favors the proximity among different molecules increasing the probability for reaction. In contrast, Pt/Al_2O_3 stimulates dehydrogenation–hydrogenation and to a smaller extent skeletal isomerization. The β-zeolite is deactivated more easily than alumina due to the coke build up on the acidic sites. The characterization of the catalysts has been carried out using XRD, NH_3-TPD, SEM, FTIR, BET, TEM, and H_2-Chemisorption.

INTRODUCTION

It is well known that high-silica-content zeolites are attractive catalysts for many industrial processes because of their hydrothermal and acidic stability, high resistance to coking, and high activity and selectivity for specific types of reactions (*1*). The β-zeolite, which was first synthesized by Wadlinger *et al.* (*2*), belongs to this class since it can possess a SiO_2/Al_2O_3 ratio in the range of 5 to 100 (*3*). Its structural characteristics have been investigated recently by Newsam *et al.* (*4*) and Treacy and Newsam (*5*). They found that zeolite β possesses a three-dimensional 12-ring pore system and that its structure is a highly faulted intergrowth of two distinct, but closely related structures: polymorph A (tetragonal) and polymorph B (monoclinic). Thus, the β-zeolite is a combination of the polymorph A structure, which has straight channels of 0.73×0.60 nm and tortuous channels of 0.56×0.56 nm, and polymorph B, which has straight channels of 0.73×0.68 nm and tortuous channels of 0.55×0.55 nm. Because of its pore opening, zeolite β can accommodate molecules with diameters larger than those which can be accommodated by ZSM-5, and can provide shape selectivity for intermediate size molecules to a higher extent than X and Y faujasites. Very recently a second monoclinic polymorph C of zeolite β has been reported by Higgins *et al.*(*6*).

The crystallization rate of the β-zeolite and its crystal size depend strongly on the total alkali content and the relative proportion of Na^+ and K^+ cations (*7*). Studies of the crystallization process of zeolite β have been carried out by Perez-Pariente *et al.* (*8, 9*). The zeolite crystallinity is altered significantly by the water to alumina and the template to alumina molar ratios. It was also

[1] To whom correspondence should be addressed.

This paper, which was originally published in **Journal of Catalysis**: 140, 526-542 (1993), is republished with permission.

found that a very crystalline zeolite can be obtained after two weeks at 160°C by using tetraethylammonium hydroxide (TEAOH) as template, if the H_2O/Al_2O_3 molar ratio is kept below 600 (*10*). For high crystallinity, the molar ratio $TEAOH/Al_2O_3$ has to be above 2.

There are only a few papers in the literature on the catalytic behavior of zeolite β, since it has been relatively disregarded after its initial synthesis in 1967. Zeolite β has beneficial effects in the catalytic hydrodewaxing of petroleum oils by lowering their pour point (*11–13*). The catalytic activity of Pt/H-β zeolite has been examined by Martens *et al.* (*14*) for the isomerization and hydrocracking of long chain paraffins. They found that metal-supported β-zeolite behaves as an ideal bifunctional catalyst that exhibits high activity for primary cracking and selectivity for isomerization. The activity of H-β zeolite for cracking of gas-oil has been compared to that of HY zeolite (*15*). It was found that zeolite β is less active and produces less gasoline and more coke than HY zeolite. For the cracking of *n*-heptane it was observed that zeolite β exhibits a slightly higher activity than HY zeolite and lower hydrogen transfer capability (*16*). The reactions of aromatic hydrocarbons, such as the methylation and disproportionation of toluene and the isomerization of *m*-xylene over Pt/β zeolite, have been also investigated (*17*). It was found that because of its relatively large pores, it does not provide any product shape selectivity for the isomerization of *m*-xylene to *o*- and *p*-xylenes, but in contrast to the 10-membered ring pores zeolites such as ZSM-5 which has smaller pores, it favors the disproportionation reactions. On the other hand it suppresses the formation of the larger trimethylbenzene molecules. The disproportionation and transalkylation of toluene and trimethylbenzene over β-zeolite have been also examined (*18*). It was found that β-zeolite exhibits excellent catalytic stability and transalkylation selectivity.

In the present work the performance of β-zeolite and γ-Al_2O_3, both impregnated with Pt, for the reforming reaction of *n*-hexane, methylcyclopentane, and cyclohexane is examined. All reactants have six carbon atoms, and constitute typical representatives of reforming feeds. However, as it was noted by Weisz (*19*), Ciapetta and Wallace (*20*), and Ramage *et al.* (*21*), they follow different mechanisms of reaction. Low severity operating conditions (relatively low temperatures and atmospheric total pressure) have been chosen in order to minimize possible hydrocracking (*22*).

The zeolite β stimulates the skeletal isomerization of *n*-hexane to a higher extent than alumina, producing even dibranched hexanes. This is a result of the increased ability of β-zeolite for isomerization over its acidic sites of the olefinic intermediates which are generated on the metal. Its ability for production of aromatics by dehydrocyclization is also higher than that of Al_2O_3, probably because of the geometrical constraint imposed by its pore structure. A small degree of hydrocracking of *n*-hexane was found to occur in the zeolite. Zeolite β transforms methylcyclopentane to dibranched molecules with a cyclic ring of six carbon atoms and dibranched aromatic molecules to a much greater extent than Al_2O_3. These products, which have higher octane number than the feed molecule, are a result of: (1) the stimulation of bimolecular alkylation reactions between cycloolefinic intermediates generated by the dehydrogenation of the feed molecules over the Pt sites, with carbonium ions produced by the ring opening of other feed molecules over the acidic sites of the zeolite, and (2) the disproportionation of the alkylated intermediate molecules formed. The reactions of cyclohexane show that Al_2O_3 has a higher ability for dehydrogenation-hydrogenation than β-zeolite, probably because of the higher Pt dispersion. Instead, the zeolite possesses a significantly higher ability for the isomerization of cyclohexane. The rate of deactivation of β-zeolite is higher than that of alumina because of some coke build up on the acidic sites.

EXPERIMENTAL

Synthesis and Preparation of the Catalyst

Zeolite β has been synthesized hydrothermally from an aluminosilicate gel with the nominal composition 3TEAOH–30 SiO_2–Al_2O_3–0.9Na_2O–250H_2O containing tetraethylammonium hydroxide (TEAOH) as template. The reagents used were: colloidal silica (LUDOX HS-40, Du Pont) containing 40 wt% SiO_2 and 0.41 wt% Na_2O; sodium aluminate (Fisher Scientific); tetraethylammonium hydroxide (Aldrich 40 wt% solution in water); sodium chloride (Fisher Scientific); and distilled water. Individual solutions of the above chemicals were mixed with vigorous stirring for 45 min in the appropriate proportions. The pH of the resultant gel was 13. The gel was loaded in a Teflon lined stainless-steel autoclave (Autoclave Engineers) and kept at 160 ± 1°C under autogeneous pressure for 10 days without stirring. At the end of the crystallization period the autoclave was quenched in cold water and the zeolite was separated by filtering the content of the autoclave and washing the precipitate thoroughly with distilled water. After drying the resulting zeolite overnight at 140°C it was found that a high yield, 97 wt% of the solid material originally used for the gel preparation, has been achieved.

The protonated form of the zeolite has been prepared as follows: TEAOH-β samples were heated at 500 ± 2°C for 5 h by increasing the temperature at a rate of 10°C/min in an ultra high purity O_2 flow (5 ml/min) in order to burn the template occluded in the pores of the zeolite. At this calcination temperature, the structure of the zeolite is preserved unchanged, as found by Perez-Pariente *et al.* (*23*) and Liu *et al.* (*24*). The zeolite was exchanged with 2M ammonium sulfate (Fisher) for 4 h at 90 ± 1°C. The remaining sodium after the cation exchange was 0.021 wt% while the molar ratio of SiO_2/Al_2O_3 was 23. A second calcination at 450°C for 1 h has followed. High purity γ-Al_2O_3 (PURALOX SB a-230) was kindly donated to us by Condea Chemie GmbH. The sample had a surface area of 228 m^2/g, and a Na_2O content below 0.005 wt%. The preparation of the platinum-loaded catalysts has been carried out by impregnation with H_2PtCl_6 (Aldrich 8 wt% solution in water). The Pt loading was 0.5 wt% for both catalysts (estimated on a dry basis). After impregnation the catalysts were dried overnight at 130°C.

X-ray Diffraction and Infrared Spectroscopy

The X-ray diffraction spectra of zeolite β were obtained with a Nicolet powder X-ray diffractometer equipped with a CuK_α source. For the quantification of its crystallinity we followed the procedure described by Yarlagada *et al.* (*25*). Samples made with various proportions of high crystallinity β-zeolite having a molar ratio of SiO_2/Al_2O_3 = 25 (PQ Corporation, CP 811 BL-25), were mixed thoroughly with MgO (Baker analyzed 99.6 wt%). The ratio of the area under the two more intense peaks of the reference β-zeolite ($2\theta \simeq 7.8°$ and $2\theta \simeq 22.6°$) to the area under the main peak of MgO vs the weight percentage of β-zeolite was employed as the calibration curve. Calcined samples of the prepared β-zeolite mixed with MgO in various proportions were compared with the calibration curve to determine the extent of their crystallinity.

For the infrared investigation of the zeolite, a Fourier transform-infrared spectrometer (Mattson Alpha Centauri) has been used. Transmittance spectra have been obtained in the range of 4000–400 cm^{-1} with a resolution of 4 cm^{-1}. Prior to the actual run, thin wafers of the zeolite (1 wt%) in KBr have been prepared in a dry atmosphere under a pressure of 56 atm.

NH_3-TPD Experiments

For the determination of the acidity of the catalysts, NH_3 temperature-programmed desorption has been employed. A sample of 0.2 g H-β zeolite or γ-Al_2O_3 was heated for 1 h at 400°C in high purity helium in order to remove possible impurities and moisture from the sample. The bed was heated at a rate of 10° C/min. After the bed was cooled down, pulses of anhydrous ammonia (Matheson) were introduced into the bed until complete saturation ($T = 25°C$) of the catalyst was achieved. After the adsorption step, the bed was flushed with helium for 1 h at $T = 25°C$. The temperature-programmed desorption took place in the range of 25 to 700°C with a heating rate of 10°C/min and a carrier gas flow of He of 20 ± 0.5 cm^3/min. The detection of the desorbed ammonia has been made with a tungsten-rhenium thermal conductivity detector of a Perkin–Elmer (Sigma 2000) gas chromatograph. The dead volume from the bed to the detector has been kept minimal in order to increase the accuracy. A mixture of liquid nitrogen and heptanol has been used as a trap after the bed in order to collect any possible moisture removed from the zeolite pores or produced by the hydroxyl groups of alumina at high temperatures.

H_2-Chemisorption

Hydrogen chemisorption has been employed for the determination of the metal dispersion. After calcination and reduction, the catalysts were loaded in a pyrex-glass sample holder located in a furnace of a vacuum glass system equipped with a pressure transducer. A reduction in flowing H_2 at 200°C for 1 h has followed in order to eliminate the possible partial oxidation of the catalysts during their transfer to the glass holder. The samples were evacuated at 10^{-5} Torr and 400°C for 1 h and then cooled down to room temperature ($T = 24°C$) under the same vacuum. High purity H_2 was introduced at pressures between 10 to 400 Torr for the determination of the total amount (irreversible and reversible) of H_2 adsorbed. After this step, the samples were evacuated at 10^{-5} Torr and room temperature for the elimination of the weakly bound H_2, and a second adsorption curve at room temperature over the latter samples regarding the weakly bound hydrogen was determined. The amount of strongly chemisorbed H_2 was obtained after subtracting the two isotherms and extrapolating the results to zero pressure. The metal dispersion was evaluated assuming that H/Pt is 1 : 1.

Catalytic Runs

The catalytic experiments were conducted in a plug flow differential reactor operating at 1 atm pressure. Prior to the reaction, the catalysts were activated by oxidation in situ in the reactor by flowing 5 ± 0.5 ml/min ultra high purity oxygen (99.999%) at 400°C for 2 h. In all the experiments the reactor was heated at a rate of 10°C/min. After this step, a purging of the catalyst with high purity He (99.995%) for 15 min, followed by its reduction in flowing 5 ± 0.5 ml/min ultra high purity H_2 (99.999%) at 400°C for 2 h, brought the catalysts to their final state for reaction. Hydrogen, at a predetermined flow rate, was bubbled through a bubbler containing the appropriate hydrocarbon to be fed into the reactor, at 24°C. The composition of the feed stream was found to be independent of the flow rate through the bubbler, indicating that the stream is saturated with the feed molecules. The reactants were *n*-hexane (Aldrich 99 + wt%), methylcyclopentane (Aldrich 98 wt%) and cyclohexane (anhydrous Aldrich 99 + wt%). With these compounds the H_2/Oil molar ratios were 4.4, 4.8, and 7.4, respectively.

In all the catalytic runs, 100 ± 2 mg (on a dry basis) of catalyst were loaded on a glass wool plug located at the middle of a vertical stainless steel tube reactor $\frac{1}{8}$ in. in diameter. High-resolution gas chromatography has been used for the on-line analysis of the effluent stream. The analysis was carried out with a gas chromatograph (Hewlett-Packard 5890 Series II) associated with a Mass Spectrometer (Hewlett-Packard 5971). The chromatograph was loaded with a Poraplot capillary column (ID 0.53 mm, length 28 m with a film thickness of 20 μm), a molecular sieve packed column (5A) and a high performance capillary column (HP-1, ID 0.2 mm, length 12 m, with a film thickness of 0.33 μm).

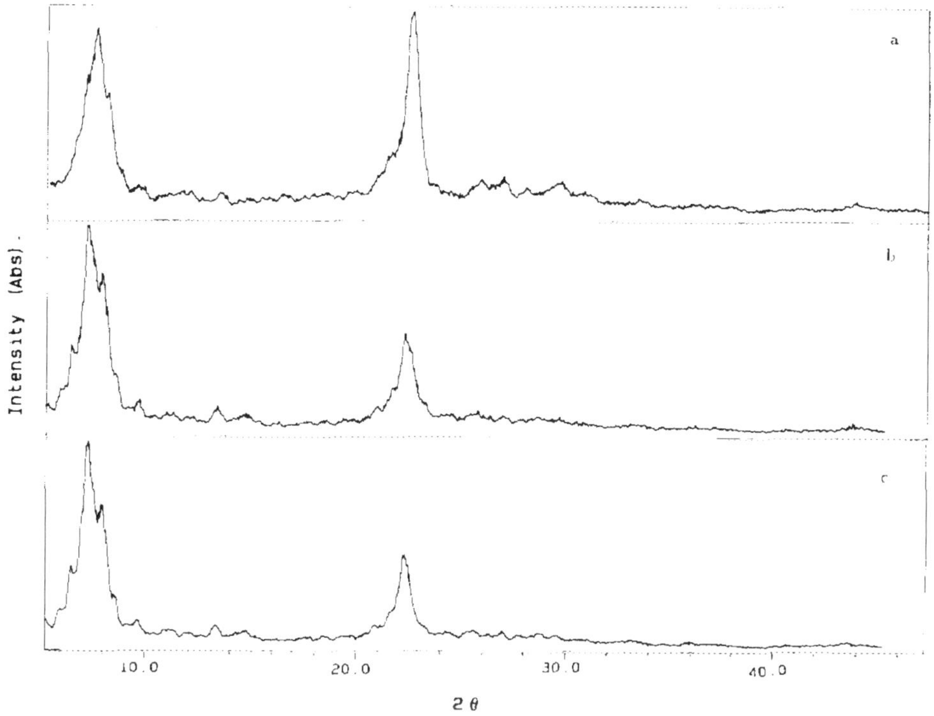

FIG. 1. X-ray diffraction pattern of β-zeolite (a) as produced, (b) calcined at 550°C in O_2 for 5 h, and (c) after four calcinations at 550°C in O_2 for 5 h each.

RESULTS AND DISCUSSION

X-ray

The X-ray diffraction pattern of the β-zeolite matched closely with that published earlier (*26*). It was found that the synthesized zeolite is highly crystalline, what is expected because of the relatively high SiO_2/Al_2O_3 ratio in the initial gel (*9*). A comparison of the X-ray pattern of the prepared β zeolite with those computed as a function of fault probability by Newsam *et al.* (*4*) indicated the existence of almost equal amounts of polymorphs A and B. The decrease of the height of the peak at $2\theta \simeq 22.6°$ after calcination, shown in Fig. 1, is probably due to microstructural defects that are introduced (*15*). Additional four successive calcinations of the zeolite at 550°C for 5 h each, led to negligible decreases in the height of this peak. The morphology of the zeolite crystals was examined with a scanning electron microscope (Hitachi S-800). They have spherical shape with an average diameter of about 1 μm.

BET Measurements

The effect of calcination time and temperature on the surface area of zeolite β is given in Table 1. The measurements have been carried out with a surface area, pore volume analyzer (Model 2100D Micromeritics), after degassing the samples overnight at

TABLE 1

Surface Area of Zeolite β as a Function of Calcination Conditions

T_C(°C)	t_C(h)[a]	S_g(m^2/g)
Uncalcined[b]	—	79.3
500	5	400.3
500	12	398.6
600	5	387.5
600	12	277.0

[a] T_C = calcination temperature, t_C = calcination time.

[b] For the uncalcined zeolite the degassing was carried out at 200°C to avoid the contamination of the BET apparatus with the unburned template.

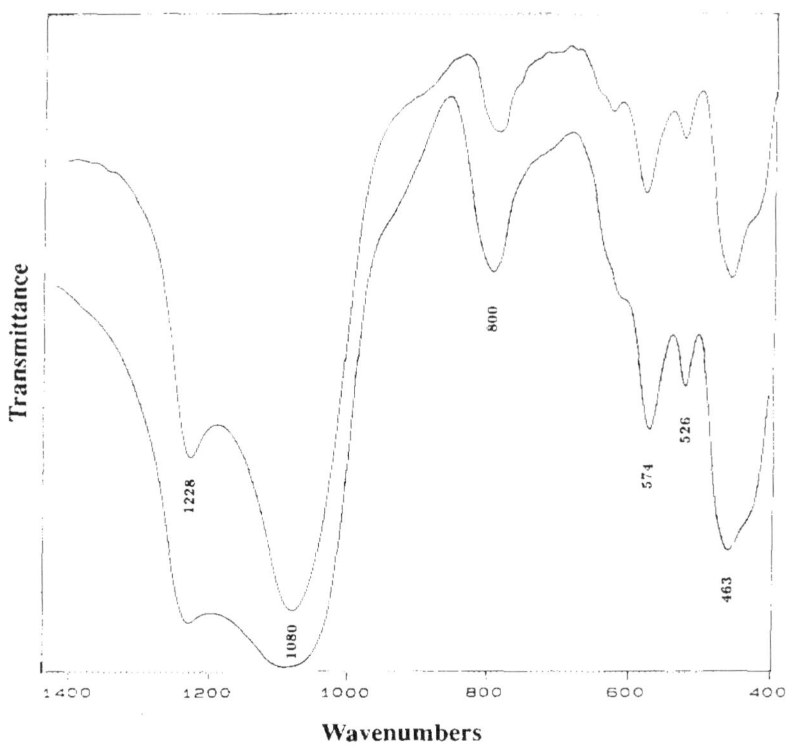

FIG. 2. Mid-Infrared spectra of β-zeolite (the top curve is for calcination at T_C = 600°C for 5 h, the bottom curve is for calcination at T_C = 500°C for 5 h).

350°C. One can note a relatively large decrease of the surface area after long calcination at high temperatures (above 600°C). The small value of S_g for the uncalcined sample is probably due to the unburned template.

Infrared Spectroscopy

The identification of the prepared powder as the β-zeolite is also supported by its mid-infrared spectrum given in Fig. 2, where the characteristic peak of β at 526 cm^{-1} appears (*9, 26*). The complete burning of the template during calcination has been achieved since no peaks were found which can be assigned to the ethyl groups of the template (3100–2800 cm^{-1} for stretching vibrations and 1500–1300 cm^{-1} for bending vibrations). By examining the region of the asymmetric stretching vibrations, which is between 1250 and 950 cm^{-1} (*27*), one can observe a decrease of the peak height at 1228 cm^{-1} with increasing calcination temperature. The relative decrease of the height of this peak indicates a crystallographic change in the structure of the zeolite at high temperatures (above 600°C). The peak at 1080 cm^{-1} is insensitive to the heat treatment, like all the peaks which are due to internal tetrahedra (*27*). In the range of symmetric stretching vibrations (950–650 cm^{-1}), there is a main peak at 800 cm^{-1} which can be assigned to the external linkages of AlO_4 tetrahedra vibrations. Its height decreases with increasing calcination temperature, indicating a crystallographic change. The peaks at 574 cm^{-1} and 526 cm^{-1} (the latter is specific to zeolite β only) are in the range of the double ring vibrations of tetrahedra, while the one at 463 cm^{-1} can be assigned to TO_4 bending vibrations of the internal tetrahedra. The additional peak at 526 cm^{-1}exhibited by the β-zeolite results probably from the double four ring (D4R) vibrations of β, since its framework contains a large fraction of 4-membered rings (*4, 6*). On the other hand,

TABLE 2

Platinum Dispersion D (=H/Pt) of Pt/Zeolite β and Pt/γ-Al_2O_3

	T_O(°C)	T_R(°C)[a]	D = H/Pt
γ-Alumina	400	400	0.60
β-Zeolite	300	300	0.23
β-Zeolite	400	400	0.06

Note. For both catalysts the metal loading is 0.5 wt%. The duration of oxidation and reduction was 2 h each.

[a] T_O = oxidation temperature, T_R = reduction temperature.

calculations (*28*) show that the infrared vibrations for D4R occur around 630, 539, and 527 cm^{-1}, while for D6R, around 605, 584, and 510 cm^{-1}. Pentasil zeolites, such as the ZSM-5 do not possess such peaks since its framework consists exclusively of five-membered rings (*29, 30*).

NH_3-TPD

The NH_3-TPD spectrum of β-zeolite matched well with those presented in the earlier work of Hedge *et al.* (*31*). An intense low temperature peak appears due to the weakly bounded ammonia on silanol groups. However, the acidity of β-zeolite is correlated with another less intense peak that appears in the region of 400°C. This peak was found to be very broad and is shifted to lower temperatures compared to that of ZSM-5 (500–550°C), indicating the stronger acidity (*32*) of the latter catalyst. The spectrum of γ-Al_2O_3 shows a less intense peak than that of zeolite β. Alumina possesses (*33*) Lewis acidity which is produced after the water removal during heating at high temperatures (*34*), and a much lower Brønsted acidity.

H_2-Chemisorption

As results from Table 2, the dispersion of Pt is higher on γ-Al_2O_3 than on the zeolite. Transmission electron micrographs (JEOL-100 CX operating at 100 KV) barely show Pt particles located on the surface of γ-Al_2O_3 (Fig. 3a), indicating that a large fraction of Pt is well dispersed. Under the present oxidation and reduction conditions (400°C for 2 h), the metal dispersion over the zeolite β is significantly lower than that on Al_2O_3. Figure 3b shows a large number of Pt particles (around 50 Å) located on the external surface of the zeolite crystals. The oxidation and reduction of the zeolite under less severe conditions, namely 300°C for 2 h (not given), showed fewer Pt particles located on the zeolite crystal surface. Of course, the metal dispersion is higher in this case. A similar behavior was observed earlier (*35*) with Pt over NaY zeolite oxidized and reduced at higher temperatures. Originally, during the calcination step, Pt^{2+} ions are transferred to the sodalite cages. During the reduction step, the Pt^0 atoms formed migrate from the sodalite cages, through the zeolite channels, to the external surface of the zeolite. As a result, the availability of Pt inside the zeolite pores is reduced. The chloride concentrations in both catalysts under different oxidation-reduction conditions are given in Table 3. The catalysts were dissolved in HNO_3 and the Cl^- concentration was measured potentiometrically with a selective ion chloride probe.

Catalytic Behavior

n-Hexane. The product distribution of γ-Al_2O_3 given in Table 4 shows negligible hydrocracking of *n*-hexane, due to its small Brønsted acidity. As a result, the production of C_1, C_2 molecules is very low. The (moderate) acidity of β-zeolite is expected to favor mainly the skeletal isomerization of the olefinic intermediates formed on Pt sites. This is confirmed by the product distribution given in Table 5. The probabilities for primary and secondary cracking are small, otherwise the ratios C_1/C_5 and C_2/C_4 would be even larger than unity (*36, 37*). This is supported by the fact that at elevated temperatures, up to 400°C, only small amounts of C_1, C_2 hydrocarbons are formed. The generation of propane, which increases with in-

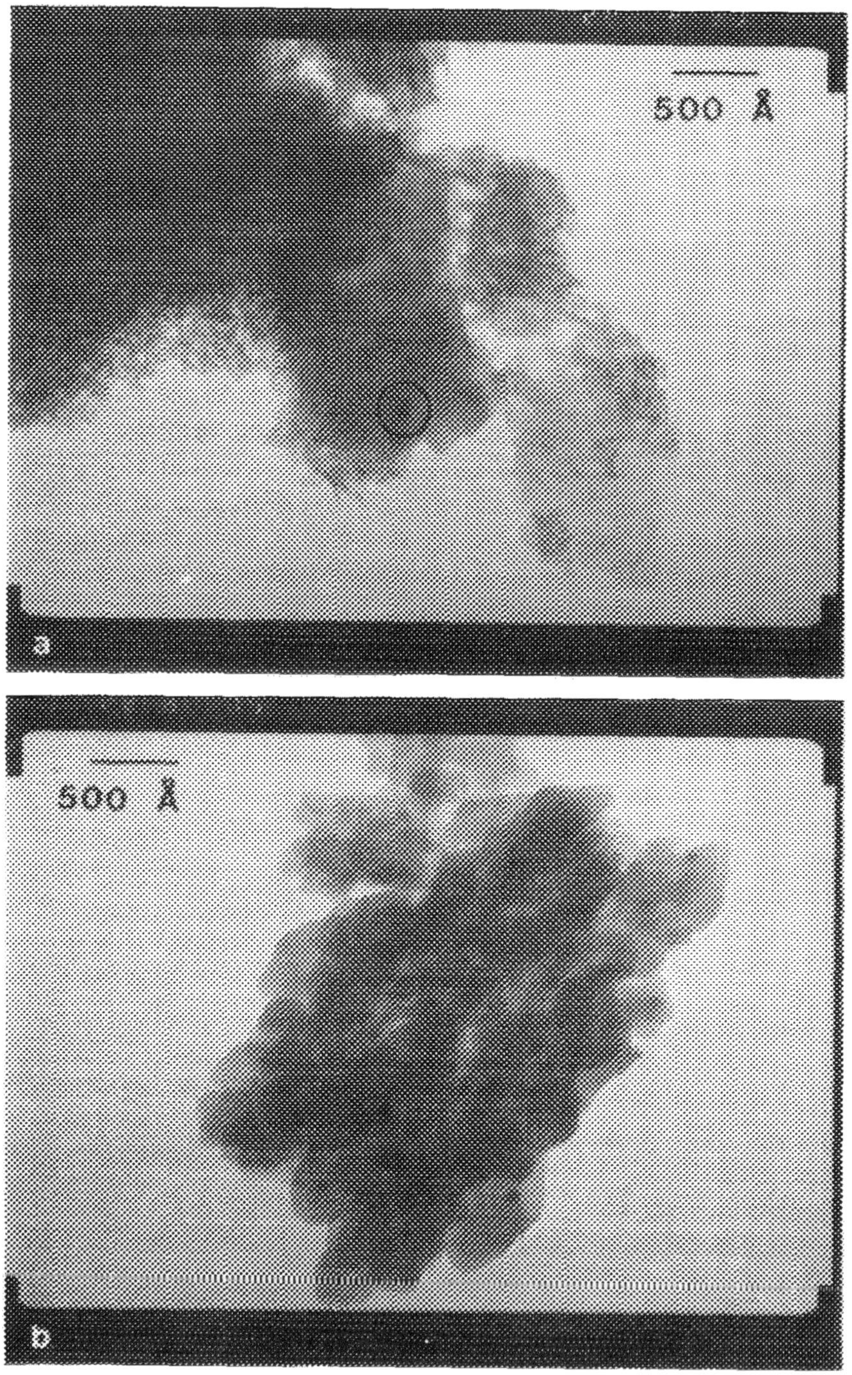

FIG. 3. (a) Transmission electron micrograph of Pt/γ-Al_2O_3 calcined at 400°C for 2 h and reduced at 400°C for 2 h and (b) transmission electron micrograph of Pt/β zeolite calcined at 400°C for 2 h and reduced at 400 C for 2 h.

creasing temperature, can be partially attributed to a multistep hydrocracking (*19*) consisting of the dehydrogenation of hexane to 3-hexene, followed by the "center cracking" of hexene to C_3 olefins on the acid sites, and finally by the hydrogenation of the latter molecules over Pt sites to propane. The production of linear and branched C_4, C_5 hydro-

TABLE 3

Chloride Concentration of β-Zeolite and γ-Al_2O_3 under Different Oxidation Reduction Conditions

	T_O(°C)	T_R(°C)[a]	Cl (mg/g cat.)
β-Zeolite	As impregnated		3.53
β-Zeolite	Calculated[a]		2.59
β-Zeolite	300	300	2.75
β-Zeolite	400	400	1.63
β-Zeolite	550	550	0.82
γ-Alumina	As impregnated		3.44
γ-Alumina	400	400	5.07

[a] The calculation is based on the fact that both catalysts were impregnated with H_2PtCl_6 for a metal loading of 0.5 wt%.

carbons cannot be a result of the direct cracking of *n*-hexane via the β-scission of a C_6 cation, since as already mentioned, C_1 and C_2 are present in small amounts compared to C_4 and C_5 hydrocarbons. Their generation probably takes place via the alkylation reactions of C_6 molecules with olefinic molecules having less than six carbon atoms. This bimolecular reaction is stimulated by the appropriate size of the β-zeolite pores. A subsequent cracking of the $C_{>6}$ cations thus produced, can explain the C_3–C_5 hydrocarbon yields. Olefins have not been detected as final products in the present experiments probably because of their hydrogenation over Pt. A similar behavior has been observed earlier by Poutsma and Schaffer (*38*) with *n*-hexane over REY and $RE(NH_4)Y$ (rare earth faujasit). They suggest that the origin of C_4 and C_5 products from a C_6 precursor is not a result of direct cracking, but a result of the alkylation of an olefinic intermediate that generates a molecule with more than six carbon atoms, followed by cracking. The branched C_4 and C_5 hydrocarbons which are produced in relatively large amounts by zeolite β (Table 4), have significantly larger octane numbers than *n*-hexane (*39, 40*).

The zeolite promotes skeletal isomerization to a much higher extent than γ-Al_2O_3 (compare Table 4 with column 2 of Table 5). It generates even dibranched C_6 molecules, while γ-Al_2O_3 does not possess such a capability. Alumina produces from *n*-hexane mainly 2-methyl and 3-methylpentane, by the isomerization mechanism of linear paraffins known to occur over bifunctional nonzeolitic catalysts (*19, 20*). According to this mechanism, the dehydrogenation of the paraffin over the metal (Pt) results in an olefinic intermediate which is further isomerized on the acidic sites to produce a monobranched intermediate. In the final step, the hydrogenation of the branched intermediate over Pt generates a monobranched paraffin. The isomerization step of the olefin is the rate-controlling step, since independent experiments with both *n*-pentane and 1-pentene have shown the same isomerization rate (*41*). The same mechanism can be applied to the formation of monobranched isomers of *n*-hexane over the Pt loaded β zeolite.

TABLE 4

Product Distribution from *n*-Hexane (WHSV = 1.95 h^{-1}), Methylcyclopentane (WHSV = 1.76 h^{-1}), and Cyclohexane (WHSV = 1.22 h^{-1}) over Pt-Supported γ-Al_2O_3 after 1 h on Stream at 300°C

Products	mole%
n-Hexane WHSV = 1.95 h^{-1}	
Propane	0.1
Butane	0.1
n-Pentane	0.2
2-Methylpentane	8.4
3-Methylpentane	6.2
n-Hexane	85.0
Methylcyclopentane WHSV = 1.76 h^{-1}	
Propane	0.1
n-Pentane	0.6
2-Methylpentane	4.2
3-Methylpentane	2.0
Methylcyclopentane	87.0
Benzene	5.9
Toluene	0.2
Cyclohexane WHSV = 1.22 h^{-1}	
Propane	0.7
Methylcyclopentane	0.1
Benzene	92.8
Cyclohexane	2.2
Toluene	4.2

TABLE 5

Product Distribution from n-Hexane against WHSV over Pt-Supported Zeolite β after 1 h on Stream at $T = 300°C$

Products (mole%)	WHSV (h^{-1})					
	0.47	1.95	3.55	6.51	8.97	11.71
Methane	0.4	0.3	0.3	0.8	0.4	0.4
Ethane	0.4	0.1	0.5	0.8	0.3	0.2
Propane	12.0	10.3	9.6	11.2	3.5	2.4
i-Butane	23.0	10.6	9.3	5.1	5.6	4.5
n-Butane	6.4	2.7	3.4	2.6	1.6	1.0
i-Pentane	14.3	7.0	6.6	3.6	3.2	2.3
n-Pentane	6.1	4.8	3.6	2.3	1.3	1.0
2,2-Dimethylbutane	5.1	5.0	5.3	1.2	2.7	2.6
2,3-Dimethylbutane	2.3	4.3	4.1	3.5	4.2	4.6
2-Methylpentane	12.3	20.1	22.5	26.2	27.0	27.5
3-Methylpentane	9.9	19.6	20.6	18.6	20.9	22.3
n-Hexane	7.7	15.2	14.2	23.7	28.8	31.2

The suitable size of the zeolite pores can in addition stimulate the transformation of a part of the monobranched olefinic intermediates into dibranched ones by bond shift of carbon atoms. The hydrogenation over Pt of the dibranched olefinic intermediate produces dibranched paraffins. This behavior is probably a combined result of the acidity (both Brønsted and Lewis (*16*)) of the zeolite and of its pore structure, which enables different carbon atoms of the same molecule to be brought closer. As suggested by Martens *et al.* (*42*), high isomerization yields can be achieved if the dehydrogenation-hydrogenation function and the Brønsted acidity of the bifunctional catalyst are in balance. The isomerization yields of a well-balanced bifunctional zeolite depends strongly on the carbon number of the *n*-paraffin and on the pore size of the zeolite. The lower isomerization of Al_2O_3 is due to its lower acidity while no dibranching occurs probably because of its larger pores.

To explain the production of dibranched hexane molecules in Y zeolite, Bolton and Lanewala (*37*) proposed the existence of a bimolecular transition state intermediate generated through the close adsorption of two hexane molecules on the catalyst surface. The final product distribution of the hexane isomers is a result of the disproportionation reactions of different transition states of the two *n*-hexane molecules. These bimolecular transition states cannot be completely excluded for zeolite β, but they occur to a much smaller extent than with Y zeolite and probably only in the straight channels of β which are comparable in size with those of Y.

The ratio of the products 2-methylpentane/3-methylpentane (2-MP/3-MP) is higher than unity for both Al_2O_3 and β zeolite catalysts under all conditions (*37*). This indicates that the equilibrium between the methylpentanes is shifted towards 2-methylpentane. A cyclic reaction mechanism which involves a methylcyclopentane intermediate was also suggested to explain the isomerization of *n*-hexane (*43*). This mechanism does not seem to occur in either of the two catalysts employed in the present study, since the C_6 isomers generated from *n*-hexane and methylcyclopentane are different (compare Tables 5 and 6).

The effect of the temperature on the conversion of *n*-hexane over β zeolite and γ-Al_2O_3 is presented in Figs. 4, 5, respectively. Zeolite β has a much higher activity than Al_2O_3. The activity of zeolite β increases with temperature and reaches a pla-

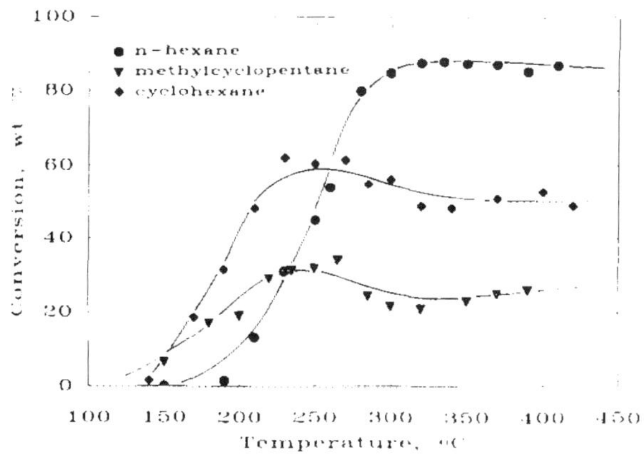

FIG. 4. Conversion vs temperature for the three feeds over β-zeolite (after 1 h on stream, WHSV = 1.95 h^{-1} for n-hexane, WHSV = 1.76 h^{-1} for methylcyclopentane, and WHSV = 1.22 h^{-1} for cyclohexane).

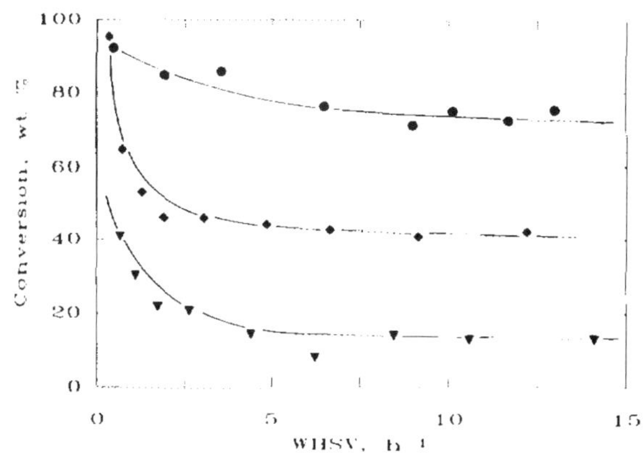

FIG. 6. Conversion vs WHSV for the three feeds over β-zeolite (after 1h on stream at T = 300°C, (●) for n-hexane, (▼) for methylcyclopentane, and (◆) for cyclohexane).

teau. Small amounts of toluene and xylenes are produced at high temperatures. The increase of the temperature decreases the yield of the C_6 isomers and increases those of the C_1–C_5 alkanes. Under the same operating conditions, alumina did not produce any aromatics. The pores of the zeolite promote to a higher extent than those of Al_2O_3 branching and some cyclization probably because of their higher acidity and because their size can stimulate the bending around of the linear C_6 molecules.

The effect of the WHSV on the conversion of both catalysts is presented in Figs. 6 and 7. The conversion over Al_2O_3 is relatively low and falls to very small values with increasing WHSV. The product distribution of alumina as a function of the residence time remains relatively unchanged. The conversion over the zeolite is relatively large at small values of WHSV and decreases moderately for high values of WHSV. For the zeolite, the decrease of the residence time reduces the concentrations of C_4, C_5 alkanes but increases those of the C_6 isomers. This indicates that the direct isomerization of the C_6 olefinic intermediate to methylpentanes is faster than its bimolec-

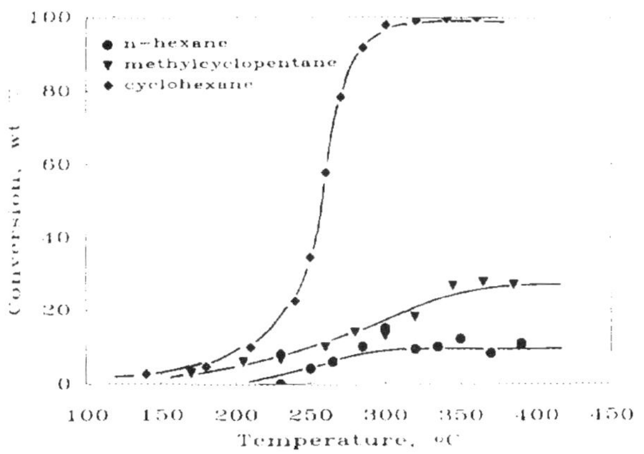

FIG. 5. Conversion vs temperature for the three feeds over γ-Al_2O_3 (after 1 h on stream, WHSV = 1.95 h^{-1} for n-hexane, WHSV = 1.76 h^{-1} for methylcyclopentane, and WHSV = 1.22 h^{-1} for cyclohexane).

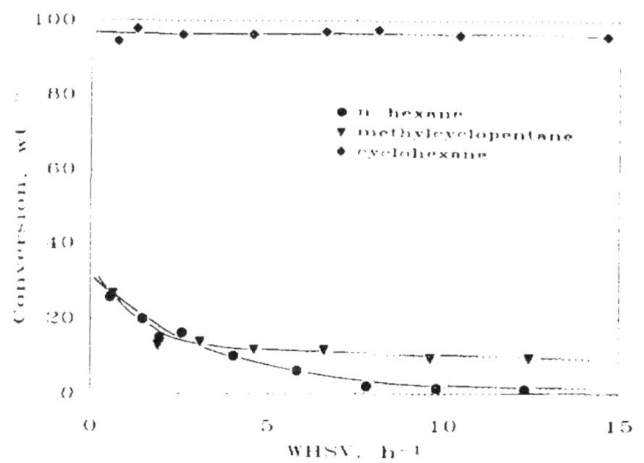

FIG. 7. Conversion vs WHSV for the three feeds over γ-Al_2O_3 (after 1 h on stream at T= 300°C).

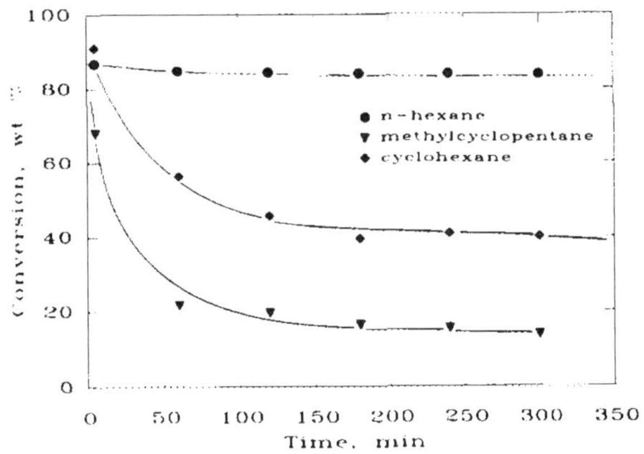

FIG. 8. Conversion vs temperature for the three feeds over β-zeolite (after 1 h on stream at $T = 300°C$, WHSV = 1.95 h^{-1} for n-hexane, WHSV = 1.76 h^{-1} for methylcyclopentane, and WHSV = 1.22 h^{-1} for cyclohexane).

ular reaction with smaller molecules or with itself which lead to larger molecules whose rupturing generates C_3–C_5 alkanes.

The effect of the deactivation of both catalysts as a function of time on stream at 300°C is presented in Figs. 8 and 9 for the zeolite and the Al_2O_3, respectively. Low deactivation is observed for both catalysts, probably because of the small amounts of aromatic or cyclic intermediates, mainly methylcyclopentane (*44*), formed which are precursors for coke formation. At higher temperatures the activity decreases faster, particularly for the zeolite because of coke formation.

Methylcyclopentane. The production distributions for both catalysts under the same operating conditions are given in Tables 4 and 6. As suggested earlier (*19, 40*), the reforming of alkylcyclopentanes on Al_2O_3 obeys the following mechanism: (a) dehydrogenation of the methylpentane to methylpentene over Pt; (b) some hydrogenolysis over the metal sites resulting in small amounts of small molecular weight hydrocarbons, and mainly isomerization of methylcyclopentene to cyclohexene over the acidic sites via a carbenium ion mechanism; and finally (c) dehydrogenation of cyclohexene to benzene over Pt. The above mechanism explains the generation of aromatics and hydrocracked molecules. The opening of the cyclopentane ring can also occur and generates hexane isomers. This mechanism is indeed followed by Al_2O_3 in the present experiments (Table 4). The absence of cyclohexane indicates that alumina favors the dehydrogenation over the Pt sites of the intermediate cyclohexene molecules generated over its acidic sites.

The above reaction scheme cannot explain completely the product distribution of the zeolite (Table 6) which contains relatively large yields of C_4, C_5, and of aromatic hydrocarbons with larger number of carbon atoms than the feed molecule. The very small ratios of C_1/C_5 and C_2/C_4 exclude hydrocracking and hydrogenolysis as main reactions. The C_6 alkanes are produced by the opening of the cyclic ring of methylcyclopentane. If each of the C–C bonds of methylcyclopentane had equal probabilities for rupturing, the product distribution among the hexane isomers would be n-Hex. : 2-MP : 3-MP = 2 : 2 : 1. However, deviations from the above distribution exist for both catalysts. n-Hexane is not generated by any of the two catalysts, indicating that the probability of opening of the tertiary–secondary C–C bonds is much smaller than the probability of opening of the secondary–secondary C–C bonds (*43*). In addi-

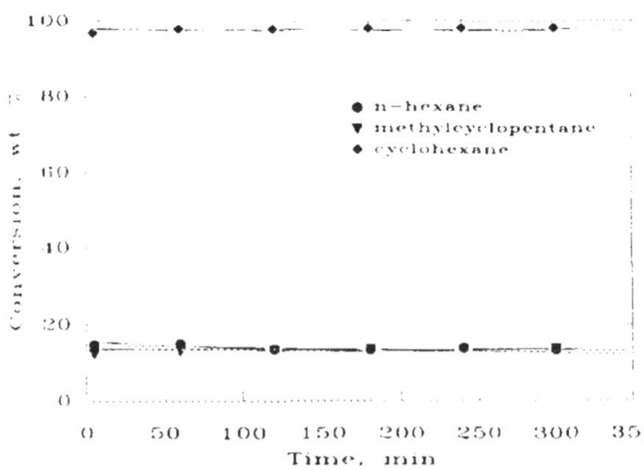

FIG. 9. Conversion vs temperature for the three feeds over γ-Al_2O_3 (after 1 h on stream at $T = 300°C$, WHSV = 1.95 h^{-1} for n-hexane, WHSV = 1.76 h^{-1} for methylcyclopentane, and WHSV = 1.22 h^{-1} for cyclohexane).

TABLE 6

Product Distribution from Methylcyclopentane against WHSV over Pt-Supported Zeolite β after 1 h on Stream at T = 300°C

Products (mole%)	WHSV (h^{-1})				
	0.66	1.76	4.41	6.22	10.57
Ethane	0.2	0.1	—	—	—
Propane	2.3	0.9	0.7	0.7	0.5
i-Butane	6.4	3.5	2.4	2.4	2.1
n-Butane	1.6	0.7	0.5	0.5	0.5
i-Pentane	3.9	2.1	1.3	1.4	1.1
n-Pentane	0.4	0.2	0.1	0.2	0.3
Cyclopentane	1.2	0.4	0.4	0.3	0.4
2-Methylpentane	2.4	1.0	1.0	1.1	0.9
Methylcyclopentane	59.2	78.4	85.7	85.9	87.1
Benzene	3.0	1.6	1.2	0.9	1.9
Cyclohexane	5.8	4.4	2.8	3.1	1.5
1,3-Dimethylcyclopentane	1.7	1.3	0.6	0.6	0.6
Methylcyclohexane	2.6	2.4	1.1	1.0	1.0
Toluene	2.9	0.8	0.8	0.7	0.8
1,4-Dimethylcyclohexane	0.9	1.1	0.3	0.1	0.3
Ethylbenzene	0.5	0.2	0.2	—	—
o-Xylene	0.6	0.1	—	—	—
m-Xylene	1.7	0.4	0.5	0.4	0.4
p-Xylene	0.8	0.2	0.2	0.2	0.2
1,2,3-Trimethylbenzene	1.0	0.1	—	—	—
1-Ethyl-3-methylbenzene	0.5	0.1	0.2	—	—
1-Ethyl-4-methylbenzene	0.4	0.1	—	—	—

tion, for the zeolite the ratio of 2-MP : 3-MP deviates significantly from 2 : 1. Other researchers (*45*) noted a similar deviation. They suggested that the hydrogenolytic splitting of methylcyclopentane on Pt sites proceeds predominantly on the C–C bond of the ring in the β-position to the carbon atom bearing the methyl group. The low dispersion of Pt in the zeolite is probably responsible for the low dehydrogenation of cyclohexene (produced over the acidic sites) to benzene.

It is worth noting that the zeolite produces dibranched aromatic and cyclic molecules with six carbon atoms to a much greater extent than Al_2O_3, particularly at small WHSV. This increases significantly the octane number of the feed molecule, fact which is very important since alkylcyclopentanes constitute a large fraction of the napthenic feed stocks used in reforming industry. As mentioned earlier (*19, 20*), a catalyst that has the ability to increase the carbon number of the C_5 cyclic ring, is extremely valuable. A possible reaction mechanism that explains this behavior of zeolite β is the following: In the first step, the feed molecules are dehydrogenated over the metal to generate C_6-cycloolefins. Bimolecular alkylation reactions between these olefinic intermediates and carbonium ions formed by the ring opening of other feed molecules, produce $C_{>6}$ unstable intermediates. In the final step, the latter intermediates undergo disproportionation reactions which produce C_4, C_5 alkanes and C_8, C_7 cyclic and aromatic hydrocarbons. This mechanism is supported by the fact that approximately equal molar amounts of C_5 and C_7 and of C_4 and C_8 are formed at any

WHSV. Other researchers (*46, 47*) have observed a similar behavior of Pt/Ca-Y faujasite. This bimolecular alkylation reaction is perhaps favored by the suitable size of the pores of β zeolite which provides just enough space for the two molecules to be in close contact, thus increasing the probability for reaction. Such type of reaction is less likely to occur in the large pores of the less active Al_2O_3. The above scheme is favored by high residence times, indicating that the bimolecular alkylation reaction is slow and constitutes the rate-controlling step. This ability of the zeolitic structure to favor bimolecular reactions compared with other acid catalysts was also emphasized by Bolton *et al.* (*48*).

The effect of the temperature on the activity of both catalysts is presented in Figs. 4 and 5. For the zeolite the conversion passes through a maximum, because of the increased rate of deactivation above a certain temperature. Al_2O_3 behaves differently. As already suggested (*49*), hydrogen-deficient hydrocarbon residues, formed by extensive dehydrogenation over the metal sites, cover the metallic surface and have as a result the reduction of the hydrogenation–dehydrogenation capability of the catalyst. In addition, coke can be generated on the acidic sites via the polymerization of the precursors transferred from the metal to the acid sites. This results in maxima (at 350 and 270°C) for the selectivities of aromatics and napthenes, respectively. This is expected to occur to a greater extent in the zeolite because of its higher acidity and more facile blocking of its smaller pores.

The increase of WHSV reduces the reforming ability of both catalysts. This occurs mainly because the isomerization over the acid sites and the bimolecular alkylation reactions are relatively slow. The higher decrease in the zeolite activity can be additionally attributed to its higher rate of coke build up. The behavior of both catalysts as a function of time on stream is presented in Figs. 8 and 9. Alumina exhibits negligible deactivation in contrast to zeolite. The higher acidity of the zeolite favors deactivation, because of the formation of unsaturated polycyclic compounds over its more numerous and stronger acid sites. Higher hydrogen pressures are required to minimize the coke precursors, especially if the temperature is high. Re can be used as a second metal to improve the performance of the catalyst (*20*).

Cyclohexane. From the product distribution of γ-Al_2O_3 (Table 3), one can see that alumina generates almost exclusively benzene above a certain temperature. This reaction involves the dehydrogenation of the feed molecule to a cycloolefin C_6 intermediate, followed by its further dehydrogenation to benzene. This behavior is a combined result of the relatively high Pt dispersion on Al_2O_3 and its low Brønsted acidity. It is worth noting that this reaction has been used for the characterization of the dehydrogenation ability of the reforming catalysts.

Zeolite β has the ability to produce aromatic and cyclic molecules with a larger number of carbon atoms than the feed molecule. The bimolecular alkylation reactions described before followed by disproportionation, result in C_4, C_5 isomers and branched cyclic or aromatic hydrocarbons with a larger number of atoms than the feed molecule. Other researchers (*50*) have noted the occurrence of bimolecular reactions between alkyl carbonium ions and benzene over REX and REY zeolites.

The isomerization of cyclohexene to methylcyclopentene and finally to methylcyclopentane takes place to a great extent since the yield of the latter molecule is relatively large.

The effect of the temperature on the conversion of cyclohexane over the two catalysts is presented in Figs. 4 and 5. For the zeolite, a weak maximum can be observed because deactivation becomes significant after a certain temperature. For the present feed molecule there is a monotonic increase of aromatics and of C_2–C_6 alkanes with temperature while the opposite is true for methylcyclopentane. This indicates that the

TABLE 7

Product Distribution from Cyclohexane against WHSV over Pt-Supported Zeolite β after 1 h on Stream at T = 300°C

Products (mole%)	WHSV (h^{-1})				
	0.34	1.22	3.05	6.66	9.15
Ethane	0.4	0.2	—	—	—
Propane	4.7	1.1	0.6	0.4	0.3
i-Butane	11.8	2.4	2.0	1.1	1.0
n-Butane	3.3	0.8	0.5	0.3	0.3
i-Pentane	6.2	1.2	1.0	0.5	0.5
n-Pentane	1.2	0.4	0.2	0.2	0.2
Cyclopentane	2.6	0.6	0.4	0.3	0.2
2-Methylpentane	2.2	0.6	0.5	0.2	0.2
3-Methylpentane	0.7	—	—	—	—
Methylcyclopentane	22.2	21.1	19.2	19.8	18.1
Benzene	17.8	18.6	18.0	18.5	18.8
Cyclohexane	4.6	46.8	54.1	57.4	59.2
1,3-Dimethylcyclopentane	1.5	0.6	0.6	0.3	0.1
Methylcyclohexane	1.9	1.0	1.2	0.3	0.1
Toluene	7.8	1.8	0.7	0.7	0.6
1,4-Dimethylcyclohexane	2.0	0.5	0.2	—	—
Ethylbenzene	0.7	0.6	0.1	0.1	0.1
o-Xylene	1.2	0.2	0.1	—	—
m-Xylene	3.5	0.8	0.4	0.1	0.2
p-Xylene	1.8	0.4	0.2	—	0.1
1,2,3-Trimethylbenzene	1.0	0.2	—	—	—
1-Ethyl-3-methylbenzene	0.4	0.1	0.1	—	—
1-Ethyl-4-methylbenzene	0.5	—	—	—	—

bimolecular alkylation of cyclohexene followed by the disproportionation of the produced intermediate is favored by an increase of temperature to a larger extent than the simple isomerization of the cyclohexene over the acidic sites. For Al_2O_3, an abrupt increase in conversion takes place at temperatures higher than 250°C mainly because the dehydrogenation of cyclohexane to benzene is accelerated by increasing the temperature.

Regarding the effect of the residence time, one can see (Fig. 7) that the conversion remains almost unchanged for alumina as a function of the WHSV. This is because the production of benzene, which is the main product and takes place on Pt sites, is rapid and does not involve the migration of olefinic intermediates to the acidic sites for isomerization, which is a relatively slow process. For Al_2O_3, which has a lower acidity than the zeolite, the possibility for isomerization is reduced and the change of residence time has an insignificant effect since the dehydrogenation step is fast. This is not however, the case with zeolite β (Fig. 6). The slow isomerization reaction that takes place over its acidic sites and the bimolecular alkylation reactions are affected by the residence time and have as a result the reduction of catalyst activity as WHSV increases. As with the previous feed molecule, the molar amounts of C_5 and C_7 and of C_4 and C_8 are approximately equal (Table 7), validating the existence of the bimolecular alkylation mechanism.

The conversion as a function of time on stream for the zeolite (Fig. 8) shows a decrease which is due to coke formation. However, the activity is higher than when methylcyclopentane is used as a feed, indicating that the coke build up is higher for feeds with C_5 rings than with those having C_6 rings. This was also noted by Parera *et al.* (*44*). Because of its lower acidity, the time on stream behavior of Al_2O_3 (Fig. 9) shows negligible deactivation, indicating that coke build up on the metal crystallites is small under the present operating conditions since in the reaction of cyclohexane only the metal sites of the catalyst are involved.

CONCLUSIONS

Zeolite β promotes skeletal isomerization reactions that produce mono- and di-branched hexane isomers from *n*-hexane. This behavior can be partially attributed to its acidity which is responsible for the isomerization of olefinic intermediates. On the other hand, its pores have the suitable size to stimulate the C–C bond shift of carbonium ions leading to dibranched isomers, and the bending of paraffinic molecules leading to chain closure.

For methylcyclopentane the acidity and the pore structure of β-zeolite facilitate isomerization and bimolecular alkylation reactions. As a result, molecules with carbon numbers in excess of the parent hydrocarbon molecule are generated. Thus enlargements of the cyclic ring of methylcyclopentane leading to dibranched C_6 cyclic and aromatic molecules occur. This capability of zeolite β is of key importance for reforming since it can increase significantly the octane number of the feed stock. Similarly, zeolite β can enrich the cyclohexane with additional carbon atoms. The deactivation of the catalyst under the present operating conditions is probably due to the coke build up on its acid sites.

In contrast to the zeolite β, alumina does not promote either isomerization or enlargement of the feed molecules under the present conditions. This is due to the lower acidity and to the much larger pores it possesses. A combination of metal with both zeolite β and γ-Al_2O_3 may result in a bifunctional catalyst of value for reforming.

ACKNOWLEDGMENTS

The authors are indebted to professor C. R. F. Lund and to Mr. M. C. Hausladen and Mr. S. S. Peri for their valuable help in the H_2-Chemisorption experiments.

REFERENCES

1. Scherzer, J., "Catalytic Materials: Relation between Structure and Reactivity," ACS Symp., Vol. 248, p. 157. ACS, Washington, DC, 1984.
2. Wadlinger, R. L., Kerr, G. T., and Rosinski, E. J., U.S. Patent 3,308,069 (1967).
3. Wadlinger, R. L., Kerr, G. T., and Rosinski, E. J., U.S. Patent 3,308,064 (1975).
4. Newsam, J. M., Treacy, M. M. J., Koetsier, W. T., and DeGruyter, C. B., *Proc. R. Soc. London Ser. A* **420,** 375 (1988).
5. Treacy, M. M. J., and Newsam, J. M., *Nature* **332,** 249 (1988).
6. Higgins, J. B., LaPierre, R. B., Schlenker, J. L., Rohrman, A. C., Wood, J. D., Kerr, G. T., and Rohrbangh, W. J., *Zeolites* **8,** 446 (1988).
7. Camblor, M. A., and Perez-Pariente, J., *Zeolites* **11,** 202 (1991).
8. Perez-Pariente, J., Martens, J. A., and Jacobs, P. A., *Zeolites* **8,** 46 (1988).
9. Perez-Pariente, J., Martens, J. A., and Jacobs, P. A., *Appl. Catal.* **31,** 35 (1987).
10. Leu, L. J., Hou, Y. L., Kang, B. C., Li, C., Wu, S. T., and Wu, J. C., *Appl. Catal.* **69,** 49 (1991).
11. Lapierre, R. B., and Partridge, R. D., Eur. Pat. Appl. 94827 (1983).
12. Lapierre, R. B., Partridge, R. D., Chen, N. Y., and Wong, S. S., Eur. Pat. Appl. 95303 (1983).
13. Lapierre, R. B., Randell, D. P., Chen, N. Y., and Wong, S. S., U.S. Patent 4,501,926 (1985).
14. Martens, J. A., Perez-Pariente, J., and Jacobs, P. A., *in* "Chemical Reactions in Organic and Inorganic Constrained Systems" (E. Selton, Ed.) Kluwer, Dordrecht, 1986. NATO ASI Ser. C., Vol. 165, p. 115.
15. Corma, A., Fornes, V., Melo, P., and Perez-Pariente, J., *Prepr.-Am. Chem. Soc. Div. Pet. Chem.*, 632 (1987).
16. Corma, A., Fornes, V., Monton, J. B., and Orchilles, A. V., *J. Catal.* **107,** 288 (1987).
17. Ratnasamy, P. Bhat, R. N., Pokhriyal, S. K., Hegde, S. G., and Kumar, R., *J. Catal.* **119,** 65 (1989).
18. Wang, I., Tsai, T. C., and Huang, S. T., *Ind. Eng. Chem. Res.* **29,** 2005 (1990).
19. Weisz, P. B., *Adv. Catal.* **13,** 137 (1962).
20. Ciapetta, F. G., and Wallace, D. N., *Catal. Rev.* **5,** 67 (1971).
21. Ramage, M. P., Graziani, K. R., Schipper, P. H.,

Krambeck, F. J., and Choi, B. C., *Adv. Chem. Eng.* **13,** 193 (1987).

22. Satterfield, C. N., "Heterogeneous Catalysis in Industrial Practice," McGraw–Hill, New York, 1991.
23. Perez-Pariente, J., Sanz, J., Fornes, V., and Corma, A., *J. Catal.* **124,** 217 (1990).
24. Liu, S. B., Wu, J. F., Ma, L. J., Tsai, T. C., Wang, I., *J. Catal.* **132,** 432 (1991).
25. Yarlagadda, P., Lund, C. R. F., and Ruckenstein, E., *J. Catal.* **133,** 28 (1992).
26. Jacobs, P. A., and Martens, J. A., *in* "Synthesis of High-Silica Aluminosilicate Zeolites" (B. Delmon and J. T. Yates, Eds.), Vol. 33. Elsevier, Amsterdam, 1986.
27. Flanigen, E. M., *in* "Zeolite Chemistry and Catalysis" (J. A. Rabo, Ed.), ACS Monograph, ACS, Vol. 171, p. 80. Washington, DC, 1976.
28. Flanigen, E. M., Khatani, H., and Szymanski, H. A., *Adv. Chem. Ser.* **101,** 201 (1971).
29. Szostak, R., "Molecular Sieves: principles of Synthesis and identification," Van Nostrand–Reinhold Catalysis Series, New York, 1989.
30. Jacobs, P. A., Beyer, H. K., and Valyon, J., *Zeolites* **1,** 161 (1981).
31. Hegde, S. G., Kumar, R., Bhat, R. N., and Ratnasamy, P., *Zeolites* **9,** 231 (1989).
32. Babu, G. P., Hegde, S. G., Kulkarni, S. B., and Ratnasamy, P., *J. Catal.* **81,** 471 (1983).
33. Peri, J. B., *J. Phys. Chem.* **69,** 231 (1965).
34. Hindin, S. G., Weller, S. W., *J. Phys. Chem.* **60,** 1501 (1956).
35. Moretti, G., and Sachtler, W. M. H., *J. Catal.* **116,** 350 (1989).
36. Matsumoto, H., Saito, Y., and Yoneda, Y., *J. Catal.* **19,** 101 (1970).
37. Bolton, A. P., and Lanewala, M. A., *J. Catal.* **18,** 1 (1970).
38. Poutsma, M. L., and Schaffer, S. R., *in* "Zeolite Chemistry and Catalysis" (J. A. Rabo, Ed.), ACS Monograph, ACS, Vol. 171, p. 512. Washington, DC, 1976.
39. Scherzer, J., "Octane-Enhancing Zeolites FCC Catalysts." Dekker, New York, 1990.
40. Ciapetta, F. G., Dobres, R. M., and Baker, R. W., "Catalysis" (P. H. Emmett, Ed.), Vol. 6, p. 495. Reinhold, New York, 1958.
41. Sinfelt, J. H., Hurwitz, H., and Rohrer, J. C., *J. Phys. Chem.* **64,** 892 (1960).
42. Martens, J. A., Tielen, M., and Jacobs, P. A., *in* "Zeolites as Catalysts Sorbents and Detergent Builders" (H. G. Karge and J. Weitkamp, Eds.), Vol. 46, p. 49. Elsevier, Amsterdam, 1989.
43. Gault, F. G., *Adv. Catal.* **30,** 1 (1981).
44. Parera, J. M., Verderone, R. J., and Querini, C. A., *in* "Catalyst Deactivation" (B. Delmon and G. F. Froment Eds.), Vol. 34, p. 135. Elsevier, Amsterdam, 1987.
45. Kochloefl, K., and Bazant, V., *J. Catal.* **10,** 140 (1968).
46. Schulz, H., Weitkamp, J., and Eberth, H., *in* "Proceedings, 5th International Congress on Catalysis, Palm Beach, 1972" (J. W. Hightower, Ed.), Vol. 2, p. 1229. North-Holland, Amsterdam, 1973.
47. Weitkamp, J., Jacobs, P. A., Ernst, S., *in* "Structure and Reactivity of Modified Zeolites" (P. A. Jacobs, Ed.), Vol. 18, p. 279. Elsevier, Amsterdam, 1984.
48. Bolton, A. P., Lanewala, M. A., and Pickert, P. E., *J. Org. Chem.* **33,** 1513 (1968).
49. Sinfelt, J. H., and Rohrer, J. C., *J. Phys. Chem.* **65,** 978 (1961).
50. Venuto, P. B., Hamilton, L. A., Landis, P. S., *J. Catal.* **5,** 484 (1966).

Synergism of Pt/γ-Al_2O_3 and Pt/β Zeolite in the Reforming of Naphthenes

Panagiotis G. Smirniotis and Eli Ruckenstein*

Department of Chemical Engineering, State University of New York at Buffalo, Buffalo, New York 14260

Composite catalysts of Pt supported on both zeolite β and γ-Al_2O_3 are employed for the reforming of methylcyclopentane and methylcyclohexane. Due to its relatively high Pt dispersion, Pt/γ-Al_2O_3 is responsible for dehydrogenation reactions leading to methylcycloolefins. Pt/β zeolite, which has a much higher Brønsted acidity than γ-Al_2O_3, is responsible for the bimolecular alkylation reactions leading to aromatics, the transalkylation reactions of aromatics, and the isomerization reactions of methylcycloolefins to C_6 cycloolefins which further via dehydrogenation lead to benzene. The residence time plays an important role because the dehydroisomerization reactions are much slower than the bimolecular alkylation reactions. At a high residence time and with methylcyclopentane as feed, the composites generate a synergistic effect which leads to a maximum in the benzene yield at about 40 wt % Pt/β zeolite. With methylcyclohexane as feed, high aromatic (mainly toluene) yields are obtained for zeolite contents below 30 wt %. Bimolecular alkylation reactions, which are stimulated by the zeolite pore structure and the high Brønsted acidity of zeolite β, also occur and lead to aromatics with a larger number of carbon atoms than the feed naphthene and to a better utilization of $C_{<6}$ olefins and paraffins generated by cracking. Beneficial effects can be achieved by balancing the acidic with the dehydrogenation functions of a reforming catalyst through the incorporation of Pt/β zeolite into Pt/γ-Al_2O_3.

1. Introduction

Over the past decades reforming has been used extensively for the octane enhancement of naphthenic feedstocks. Reforming is commonly conducted at temperatures in the vicinity of 500 °C and at pressures of 10–30 atm. High hydrogen pressures are needed to reduce the rate of coke formation by keeping the catalyst surface free of hydrogen-deficient hydrocarbon residues (Sinfelt and Rohrer, 1961). The traditional reforming catalysts are bifunctional in nature and consist of a metal dispersed on a relatively acidic support, γ-Al_2O_3. Mills et al. (1953) emphasized the existence of the two separate catalytic functions of the bifunctional catalysts. The metal, mainly Pt, acts as a dehydrogenation–hydrogenation agent, while the acidic support is responsible for the cracking and the isomerization reactions of the olefinic intermediates formed over the metal sites (Ciapetta and Wallace, 1971). Today's reforming technology employs bimetallic catalysts such as Pt–Re or Pt–Ir (Kluksdahl, 1968; Sinfelt, 1981) supported on Al_2O_3. Weisz and Swegler (1957) examined the roles of the acidic and metal functions of a bifunctional catalyst by employing composites of Pt/SiO_2 and SiO_2–Al_2O_3 in the isomerization of saturated hydrocarbons. The main advantages of the bimetallic combinations in comparison with the conventional Pt reforming catalyst consist of the increased stability even at low reactor pressures and the high selectivities for $C_{\geq 5}$ reformates.

There are only a few papers in literature that examine the reforming capabilities of zeolites. Recently, Hughes et al. (1986) demonstrated the advantageous behavior in the aromatization of n-paraffins of Ba-exchanged L zeolite impregnated with Pt in comparison with the conventional bifunctional catalysts. This catalyst (AROMAX) is basic and exhibits high aromatization selectivities for C_6 and C_7 n-paraffins (Tamm et al., 1988). This behavior of the Pt supported L zeolite in the aromatization of n-hexane to benzene was attributed to its unique channel geometry which enhances the 1–6 cyclization of n-hexane (Derouane and Vanderveken, 1988). Zeolites with large pores, such as Y faujasite, β, ZSM-3, and ZSM-20, impregnated with Pt have been used as bifunctional catalysts for the isomerization of long-chain paraffins (Martens et al., 1989). It was found that high isomerization yields can be achieved if the dehydrogenation–hydrogenation activity and the Brønsted acidity of the catalyst are in balance.

Zeolite β has been synthesized by Mobil (Wadlinger et al., 1967), and its structure has been investigated by Newsam et al. (1988) and Treacy and Newsam (1988). It possesses a three-dimensional 12-ring pore configuration consisting of two distinct but closely related structures: polymorph A and polymorph B. The pore opening of zeolite β being between those of ZSM-5 and of X or Y faujasites can provide shape selectivities for intermediate size molecules like those employed in reforming. It was also found that zeolite β has beneficial effects in the hydrodewaxing of petroleum oils (Lapierre et al., 1983; 1985). Zeolite β loaded with Pt has been employed for hydrocracking and isomerization of long-chain paraffins (Martens et al., 1986). It was also found that H/β zeolite is less active than HY zeolite in the cracking of gas-oil, producing less gasoline and slightly more coke than the latter (Corma et al., 1987). Zeolite β favors the disproportionation and transalkylation reactions among aromatics (Wang et al., 1990). The performance of Pt/β zeolite in the reforming of n-hexane, methylcyclopentane, and cyclohexane, typical representatives of reforming feeds, was examined recently under low-severity conditions (Smirniotis and Ruckenstein, 1993), and a comparison with the conventional Pt/γ-Al_2O_3 catalyst was carried out. It was found that Pt/β zeolite stimulates to a greater extent the skeletal isomerization reactions of paraffins than Pt/γ-Al_2O_3. It was also observed that the C_5-cyclic ring of methylcyclopentane is enlarged over Pt/β zeolite to C_6-rings; this is followed by the generation of benzene. In addition, the pore structure of zeolite β stimulates bimolecular alkylation reactions of the feed naphthenes which lead to their enlargement via the disproportionation of the $C_{>6}$ olefinic alkylated intermediates generated. Regarding the Pt/γ-Al_2O_3, it was found that it has a higher

* Author to whom correspondence should be addressed.

This paper, which was originally published in **Industrial & Engineering Chemistry Research**: 33, 493–503 (2000), is republished with permission.

dehydrogenation capability than Pt/β zeolite because Pt is better dispersed over the former than the latter substrate.

In the present work we examine the performance of composites containing Pt supported on both zeolite β and γ-Al_2O_3 in the reforming reaction of methylcyclopentane and methylcyclohexane. The above two species constitute 20–50 wt % of the naphthenic fraction of the feeds used in catalytic reforming (Ciapetta and Wallace, 1971; Ramage et al., 1987). The performance of the composites is examined as a function of the zeolite content, of the Pt loading, and of the weight hourly space velocity (WHSV). Most of the measurements have been carried out at 5 min on stream in order to obtain selectivities free of deactivation.

We found that, with methylcyclopentane as feed, there is an optimum amount of zeolite β which maximizes the yield of aromatics (mainly benzene) generated. This behavior can be attributed to the balancing of the dehydrogenation capability of the composites, determined particularly by the Pt/Al_2O_3 matrix, which enhances the production of cycloolefinic intermediates, with their acidic function, determined mainly by the zeolite β, which is responsible for the isomerization of the latter molecules. For methylcyclohexane large yields of aromatics (mainly toluene) are obtained with composites containing below 30 wt % Pt/β zeolite. For larger zeolite contents, the yield of toluene decreases, while those of benzene, $C_{>7}$ aromatics, and C_1-C_6 paraffinic and olefinic hydrocarbons increase. With both feed naphthenes it was observed that a higher zeolite content enhances the bimolecular alkylation reactions followed by disproportionation (Schulz et al., 1973; Weitkamp et al., 1984; Smirniotis and Ruckenstein, 1993) which lead to aromatics with different number of carbon atoms than the feed naphthene.

2. Experimental Section

2.1. Catalyst Preparation. Details regarding the preparation and characterization of zeolite β with XRD, SEM, FTIR, BET, TEM, H_2-chemisorption, and NH_3-temperature programmed desorption (TPD) have been presented in our earlier work (Smirniotis and Ruckenstein, 1993). The zeolite β was prepared hydrothermally from an aluminosilicate gel with a nominal composition of 30 SiO_2:Al_2O_3:0.9Na_2O:3TEAOH:250H_2O using tetraethylammonium hydroxide (TEAOH) as template. The gel was kept at 160 ± 1 °C and autogenous pressure in a Teflon-lined stainless-steel autoclave for 10 days. At the end of the crystallization period, the synthesized zeolite β had a SiO_2/Al_2O_3 molar ratio of 23 and consisted of spherical microcrystals of about 1 μm in size. The protonated form of the zeolite β was prepared by calcination of the templated zeolite at 510 °C in high-purity O_2 (99.999%) for 5 h, followed by cation exchange with 2 M ammonium sulfate (Fisher) for 4 h at 90 °C under reflux, and finally calcination of the ammonium form of the zeolite at 450 °C for 1 h in O_2. High-purity γ-Al_2O_3 with a surface area S_g = 230 m^2/g was employed (Condea Chemie GmbH). Wet impregnation with H_2PtCl_6 has been employed for both catalysts to obtain a 0.5 wt % Pt loading. Composite powders of Pt/β zeolite and Pt/γ-Al_2O_3 have been prepared in various proportions after grinding, sieving each of the powders with a 150 mesh size sieve (particle size below 100 μm), and drying them at 200 °C for 2 h. A thorough mixing for 2 h of the composite powders with a shaker mixer was used to produce homogeneous composites. The weight fraction of zeolite β in the composites is denoted by ε.

2.2. TEM and XRD. The effect of the heat treatment on the size of the Pt particles over both catalyst was examined with transmission electron microscopy (TEM) (JEOL 100-CX II). A drop of the suspension of each catalyst in water was evaporated on a copper grid. The specimens thus obtained were subjected to observation at magnifications ranging from 8000× to 200000× at 80 kV.

The identification of the zeolite prepared as zeolite β as well as its crystallinity were examined with X-ray diffraction (XRD). The X-ray diffraction patterns of β zeolite after impregnation were recorded in order to determine any possible structural changes due to the successive oxidation–reduction cycles. The patterns were obtained with a Nicolet powder X-ray diffractometer instrument equipped with a Cu Kα source.

2.3. Cl^- Determination. The Cl^- concentrations of the catalysts after various heat treatments were measured potentiometrically with a selective ion chloride probe (Orion Research EA 940). Prior to analysis the catalysts were dissolved in HNO_3 and the suspensions thus obtained were kept under agitation for 2 days.

2.4. Acidic Strength of the Catalysts. For the determination of the acid strength of β and γ-Al_2O_3, the butylamine titration method (Benesi, 1956) has been employed. This method determines the distribution of acid strength (both Brønsted and Lewis acid sites) of a solid catalyst on the basis of the color change of the basic indicators used to that of their conjugate acid form, which indicates that the H_0 values of the surface are equal to or lower than the pK_a values of the indicator employed. Lower values of H_0 correspond to greater acid strength. Experimental details of the method are given in the earlier work of Tanabe (1970). In the acidity measurements, benzene (Aldrich, 99+ %) was chosen as the nonpolar solvent. Each indicator was dissolved in benzene to obtain a 1.0 wt % solution.

2.5. Catalytic Tests. The catalytic experiments were performed in a vertical quartz tubular reactor 40 cm long with an internal diameter of 1.9 cm. The reactor was heated in a temperature programmable controlled furnace. In all the experiments 140 ± 2 mg (on dry basis) of catalyst was used. The total pressure was atmospheric, and the reaction temperature was varied between 440 and 500 °C. The activation of the catalysts was carried out in situ. All the fresh samples were subjected to an initial heating at 550 °C for 1 h in ultra-high-purity oxygen (99.999%), followed by purging of the catalyst with high-purity He (99.995%) for 15 min. As the final step, a reduction at 550 °C in ultra-high-purity H_2 (99.999%) for 1 h was carried out. Before any catalytic experiment the catalysts were regenerated at 550 °C for 1 h in ultra-high-purity O_2, purged in He for 15 min, and reduced at 500 °C for 1 h. The changes in the bed temperature were carried out at a rate of 15 °C/min. Most experiments were carried out at 5 min on stream, in order to obtain the initial product distribution free of deactivation. Experiments at larger times on stream have been also performed.

Ultra-high-purity H_2 was passed through a metallic bubbler containing the feed hydrocarbon. The reactants were methylcyclopentane (Aldrich 98 wt %) and methylcyclohexane (Aldrich 99 wt %). It was found that complete saturation of the feed stream was achieved regardless of the flow rate through the bubbler. The composition of the feed stream (the temperature of the bubbler was 24 °C in all the experiments) was 17.10 and 5.66 mol % for methylcyclopentane and methylcyclohexane, respectively. For the on-line analysis of the products, high-resolution gas chromatography has been employed. A gas chromatograph (GC) (Hewlett-Packard 5890 series II) equipped with a mass spectrometer (Hewlett-Packard

5971 series) and a PC unit for data manipulations have been used. The GC was equipped with a molecular sieve packed column (5A), a high-performance capillary column (HP-1 cross-linked methyl silicone, 0.2-mm i.d., 12-m length, and film thickness of 0.33 μm), and a fused-silica PoraPLOT wide bore capillary column (0.53-mm i.d., 28-m length, and film thickness of 20 μm). The products were sampled to the GC through a heating line kept at 120 °C in order to avoid the condensation of the high molecular weight hydrocarbons.

3. Results and Discussion

Reproducibility after each oxidation–reduction cycle has been observed, since successive experiments under identical conditions gave the same conversions. This is probably due to the initial severe reduction of the catalyst that reduces from the beginning the Pt dispersion and the Cl^- amount to a level from which the repeated regeneration–reduction cycles have no significant effect on the catalysts. Sushumna and Ruckenstein (1987) have observed that an initial oxidation and reduction of Pt/Al_2O_3 under severe conditions, followed by oxidation–reduction cycles under less severe conditions, had as a result the alternation in the size of the metal particles. It was shown that in an oxidizing atmosphere (regeneration at 500 °C) the size of the metal particles decreases because of the formation of a thin film as an atmosphere around each Pt particle. Subsequent heating in a H_2 atmosphere at 500 °C resulted in the regaining by the metal particles of the size they had prior to the oxygen treatment, because the thin film coalesces with the particle. This behavior has repeated for many cycles.

The Pt dispersions on zeolite β and γ-Al_2O_3 prior to the first catalytic run, measured by H_2-chemisorption were 0.06 and 0.46, respectively. The low Pt dispersion on the β zeolite is a result of the oxidation–reduction conditions employed in the present study. It was found earlier (Gallezot et al., 1975; Moretti and Sachtlèr, 1989) that the calcination of Pt/NaY at $T_C \geq 550$ °C leads to the complete migration of Pt^{2+} ions into the sodalite cages. Reduction under severe conditions leads to the formation of Pt^0 atoms which migrate through the channels and cages to the outside of the crystals more rapidly than the platinum ions (Jaeger et al., 1984). The mobility of Pt^0 is higher than that of Pt^{2+} because the latter counterions are subjected to electrostatic attraction by the oppositely charged zeolite lattice. Hence, large metal crystals are formed during reduction on the external surface of the zeolite which result in low Pt dispersions. Preparation of the Pt/zeolite by ion exchange will probably result in a better dispersion (Alvarez et al., 1989).

3.1. TEM. No Pt particles were observed in the micrographs of Al_2O_3 samples even after five successive oxidation–reduction cycles. This indicates that the metal remains relatively well dispersed over the support after each reduction. A different behavior was found for β zeolite with a 0.5 wt % Pt loading. After a severe oxidation–reduction at 550 °C, Pt particles of about 50 Å are formed (Figure 1a). An oxidation–reduction cycle (the oxidation was carried out at 550 °C for 1 h followed by a He purging for 15 min, and finally by a reduction at 500 °C for 1 h) of the Pt/β zeolite led to a slight enlargement of the metal particles (Figure 1b). The TEM observations are in agreement with the lower Pt dispersion on β zeolite than on γ-Al_2O_3 measured by H_2-chemisorption. Four additional oxidation–reduction cycles have not generated any detectable changes in the particle size. This indicates that,

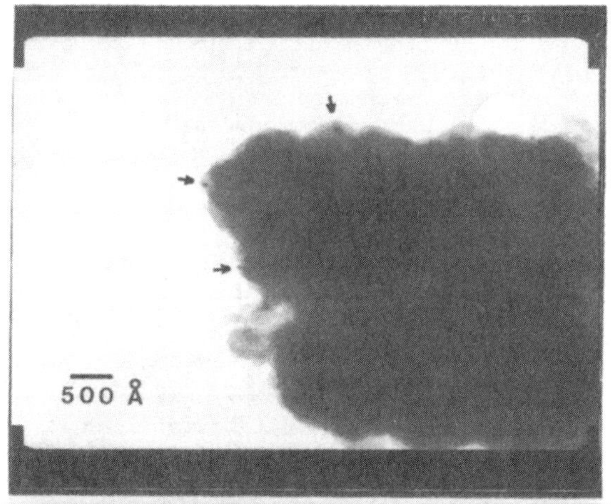

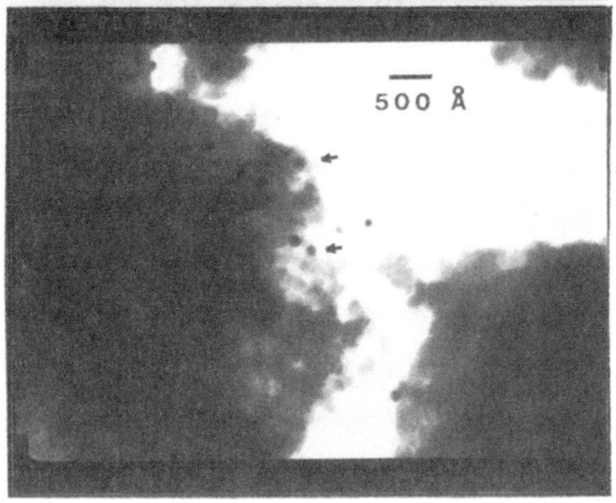

Figure 1. Transmission electron micrographs of (a, top) Pt/β zeolite after oxidation at 550 °C for 1 h, purging with He for 15 min, and reduction at 550 °C for 1 h; (b, bottom) Pt/β zeolite oxidized at 550 °C for 1 h and reduced at 500 °C for 1 h after the above oxidation–reduction cycle. The sample was loaded with 0.5 wt % Pt.

after an initial severe oxidation and reduction of the zeolite, successive oxidation–reduction cycles do not change the Pt condition to a measurable extent. Probably, an alternation in the metal particle size during oxidation and reduction takes place, and therefore no net changes occur.

3.2. XRD. The thermal stability of zeolite β was examined earlier (Liu et al., 1991; Perez-Pariente et al., 1990), and it was found that calcination up to 560 °C does not cause any structural changes. The X-ray diffraction patterns of zeolite β (Figure 2) show that there are no structural alternations during the successive oxidation–reduction cycles performed in the present study. Hence, zeolite β can be used as a reforming catalyst since its crystallinity can be preserved unchanged under regeneration cycles at 550 °C.

3.3. Cl^- Concentration. In the experiments of the present study no extra chloride was added to the catalysts, a step that could increase their acidity (Tanaka and Ogasawara, 1970). Therefore the only source of chloride in Al_2O_3 comes from the H_2PtCl_6 used in the impregnation step. The Cl^- concentrations in both catalysts are given in Table 1 as a function of the heat treatment conditions. One can observe a lower rate of decrease of the Cl^- concentration in γ-Al_2O_3 than in β zeolite as a function of the number of oxidation–reduction cycles. Even after five successive oxidation–reduction cycles the Cl^- concentration in Al_2O_3 remains only slightly below that employed in

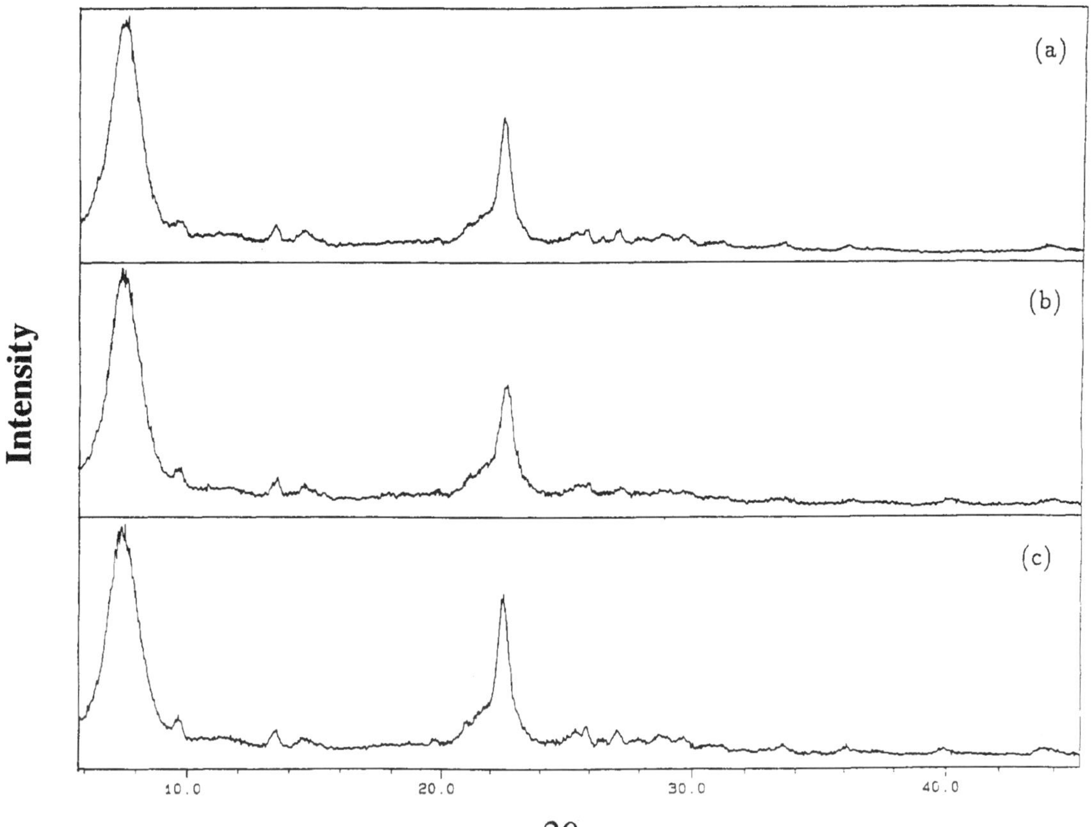

Figure 2. X-ray diffraction patterns of β zeolite (a) calcined after preparation at 500 °C in O_2 for 5 h for the burning of the template; (b) oxidized at 550 °C and reduced at 500 °C for 1 h each (after an initial oxidation and reduction at 550 °C for 1 h each); (c) oxidized at 550 °C and reduced at 500 °C for 1 h each five times (after an initial oxidation and reduction at 550 °C for 1 h each).

Table 1. Chloride Concentration of Pt/β Zeolite and Pt/γ-Al_2O_3 After Successive Oxidation and Reduction Cycles[a]

	T_O[b] (°C)	T_R[b] (°C)	Cl^- (mg/g of cat.)
β-zeolite as impregnated			3.53
γ-Al_2O_3 as impregnated			3.44
β-zeolite after 1 cycle	550	500	0.68
γ-Al_2O_3 after 1 cycle	550	500	2.63
β-zeolite after 5 cycles	550	500	0.23
γ-Al_2O_3 after 5 cycles	550	500	2.58
HCl/γ-Al_2O_3 as impregnated[c]			145.00
HCl/γ-Al_2O_3 after 1 cycle	550	500	4.94

[a] The oxidation and the reduction lasted 1 h each; a He purging of the catalyst for 15 min has been employed between them. [b] T_O = oxidation temperature; T_R = reduction temperature. [c] HCl was added during the Pt impregnation. (The amount of Cl was about 35 times higher than that introduced with the chloroplatinic acid.)

industry (Sinfelt and Rohrer, 1961; Satterfield, 1992), which is typically in the range of 0.3–1.0 wt %.

In a few experiments, HCl was added to γ-Al_2O_3 during impregnation, in order to increase its Cl^- concentration. The amount of Cl^- thus incorporated was about 35 times higher than that due to H_2PtCl_6. However, as expected, the chloride content of the catalyst decreased tremendously during the first cycle (Table 1). A similar behavior has been observed earlier (Castro et al., 1981) for chloride-rich (3 wt %) alumina. Other researchers (Barbier et al., 1992) have observed an abrupt decrease in Cl^- concentrations for samples rich in chloride. It was found that a large amount of hydrochloric acid was desorbed during reduction from alumina impregnated with H_2PtCl_6 at high Pt loading (6 wt %). In contrast, for a 0.6 wt % Pt on alumina, the chloride content remained almost unchanged. The decreased amount of chloride in zeolite β after heat treatment does not affect its isomerization capability since β is already highly acidic.

Table 2. Acid Strength of the Catalysts Employed[a]

	$H_0 = pK_a$						
catalyst	6.8	2.0	0.8	−3.0	−10.1	−12.0	−13.8
β	+	+	+	+	+	−	−
γ-Al_2O_3	+	+	+	−	−	−	−

[a] +, Change of the indicator color from the basic to the acidic form; −, no color change of the indicator from the basic to the acidic form.

3.4. Catalyst Acidity. As shown in Table 2, the β zeolite suspension changed its color for indicators possessing a $pK_a > -12.0$. Hence, the β zeolite possesses under the present preparation conditions acidic strengths H_0 larger than a value between −10.1 and −12.0. As suggested by Tanabe et al. (1989) the value of $H_0 = -12.0$ is an upper limit for superacidity and hence catalysts with $H_0 \leq -12.0$ can be considered as superacids. Infrared investigations with pyridine adsorbed on acidic zeolites (Borade and Clearfield, 1992) have shown that β zeolite possesses a higher Brønsted than Lewis acidity. It was found that the Brønsted to Lewis (B/L) acid sites ratio was about 3 for a β zeolite sample with a SiO_2/Al_2O_3 ratio similar to that employed in the present study. For alumina, the suspension containing the indicator changed its color for indicators with $pK_a > -3.0$, indicating that the acidic strength of γ-Al_2O_3 is much lower than that of β zeolite.

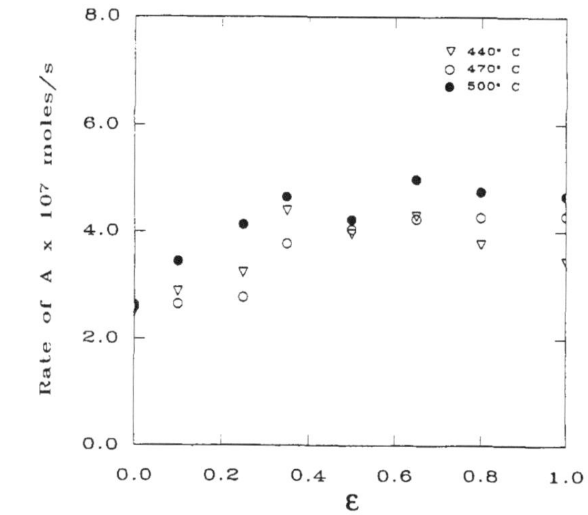

Figure 3. Rate of reaction of methylcyclopentane as a function of zeolite content (5 min on stream, WHSV = 1.26 h^{-1}, Pt loading 0.5 wt %).

Earlier investigations (Peri, 1965) have shown that alumina possesses mainly Lewis acid sites.

3.5. Catalytic Results with Composites. The most important reactions occurring during catalytic reforming include dehydroisomerization of naphthenes to aromatics, dehydrocyclization of *n*-paraffins to naphthenes and aromatics, isomerization of *n*-paraffins to isoparaffins, and hydrocracking and hydrogenolysis of alkanes and cycloalkanes to lower molecular weight alkanes (Weisz and Prater, 1957; Ciapetta and Wallace, 1971). In the present work the performance of naphthenes is examined. Methylcyclopentane was selected for investigation because it possesses a cyclic ring with five carbon atoms whose enlargement to the C_6 cyclic ring, which is a precursor for aromatics, is of high value in reforming (Weisz, 1962). Methylcyclohexane was chosen because it has already a cyclic ring with six carbon atoms, and its simple dehydrogenation can lead to aromatics.

It has been shown earlier (Mills et al., 1953; Keulemans and Voge, 1959) that the reforming of methylcyclopentanes over the bifunctional catalyst Pt/γ-Al_2O_3 occurs in the following sequence: (1) dehydrogenation of the feed molecule over the metal sites to methylcyclopentenes and migration of the latter molecules to the acidic sites of the catalyst; (2) formation of carbenium ions from the methylcyclopentenes over the Brønsted acidic sites, followed by their isomerization–enlargement to a C_6 cycloolefin; (3) desorption of the latter molecule from the acidic site and further dehydrogenation of the latter molecule over the metallic sites to benzene. The above pathway is summarized in the following reaction scheme:

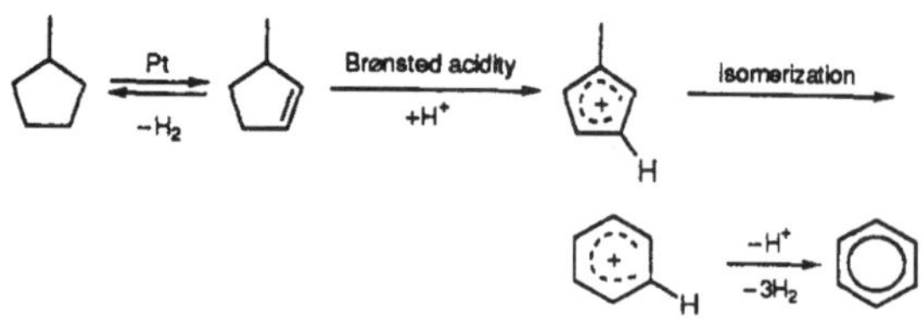

(I)

The slow step for the generation of aromatics was found to be the isomerization of the methylcyclopentenes over the acidic sites of the catalyst; the dehydrogenation steps and the migration of the olefinic intermediates to the acidic sites are much faster (Keulemans and Voge, 1959).

In the presence of the Brønsted acidity of the zeolites, bimolecular alkylation reactions which result in the generation of $C_{>6}$ aromatics occur. These reactions take place between methylcyclopentenes (formed via dehydrogenation of methylcyclopentane over the Pt sites) with carbonium ions formed by the protonation of light paraffins generated by hydrocracking and by the ring opening of methylcyclopentane over the acidic sites. The generated $C_{>6}$ intermediates are disproportionated, producing a $C_{>6}$ aromatic and a lighter paraffin than that involved originally in the alkylation reaction. The above pathway is summarized for the reaction of methylcyclopentane with a C_4 paraffin in scheme II.

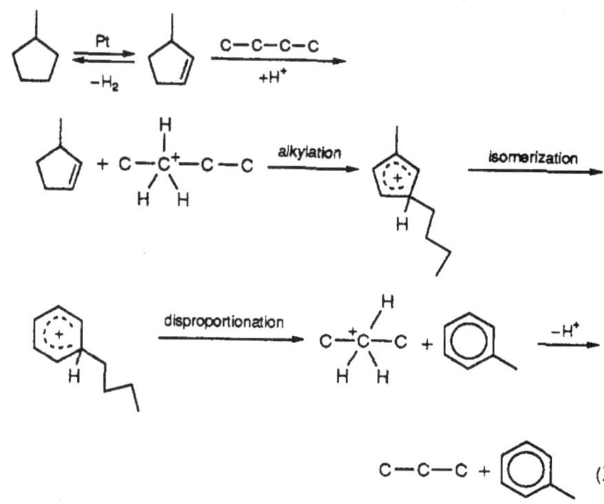

(II)

This type of reaction is stimulated by the size of the β zeolite pores which enhances the effective bimolecular collisions because the pores of the zeolite β are large enough to allow the motion of several molecules but also sufficiently small to generate a favorable steric constraint. Such bimolecular alkylations do not occur in Al_2O_3, because its "open" structure does not enhance bimolecular collisions. Hence, by combining the two Pt supported catalysts, one can better balance the higher dehydrogenation capabilities of Pt/γ-Al_2O_3 (due to the higher metal dispersion) with the isomerization and bimolecular alkylation capabilities of the Pt/β zeolite.

The activity of the composites as a function of the zeolite content ϵ is presented in Figure 3 for methylcyclopentane as feed. All the data have been recorded at 5 min on stream in order to minimize the effect of coke formation on the product selectivities. The rate of transformation increases with increasing ϵ up to about $\epsilon = 0.35$. A further increase in the zeolite content does not essentially change the activity of the composites. This behavior arises because the Pt/Al_2O_3 component stimulates mainly dehydrogenation reactions, while the zeolite component stimulates the ring opening and some cracking as primary steps. The increase in activity with increasing ϵ for composites poor in zeolite is due to the enhancement of (a) the isomerization of the methylcyclopentene intermediates generated over the Pt sites to C_6 cycloolefins and (b) the bimolecular alkylation reactions. Both reactions shift the dehydrogenation reaction of the feed naphthene (which is rapid and can be considered at equilibrium) to the right, having as a result the increase of the conversion of the feed. For values of ϵ larger than about 0.35, the overall dehydrogenation capability of the composites, which provides cycloolefinic intermediates for isomerization and bimolecular alkylation reactions, decreases. The higher rates of ring opening and cracking counterbalance, however, the decrease in the dehydrogenation rate. As a result, the overall rate of transformation of the feed molecules remains constant.

As mentioned earlier, with methylcyclopentane the main reaction which occurs over these bifunctional catalysts is the isomerization over the acidic sites of the methylcycloolefinic intermediates formed over the Pt sites. The catalytic performance of a Pt/γ-Al_2O_3 enriched in chloride, which can enhance its isomerization capabilities, has been examined for the reaction of methylcyclopentane at 470 °C. It was observed that the level of conversion was the same as that of the sample not enriched in Cl^-. This is as expected, since as shown in Table 1 the Cl^- concentration reaches a low value after the original severe oxidation–reduction cycle.

Methylcyclohexane has an entirely different behavior (Figure 4) than methylcyclopentane; the activity remains essentially unchanged with the variation of the zeolite content. Composites rich in γ-Al_2O_3 favor the direct dehydrogenation of methylcyclohexane to toluene because of the relatively high Pt dispersion over these catalysts. The increase of the Brønsted acidity of the composites results in the partial transalkylation of the generated toluene to benzene and C_8 aromatics. In addition, bimolecular alkylation reactions between toluene and carbenium ions generated from olefins produced by cracking via their protonation over the Brønsted acid sites result in the generation of $C_{>7}$ alkylbenzenium ions (Weitkamp, 1985). The latter intermediates are disproportionated to a $C_{>7}$ aromatic and a lighter olefin than that involved originally which can participate in other alkylation reactions. This pathway can be summarized as follows:

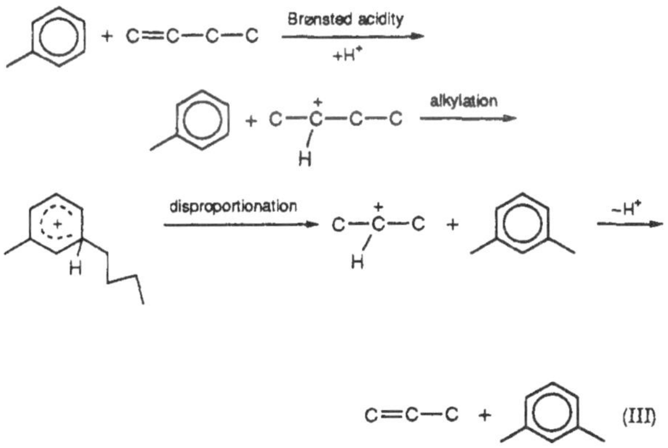

(III)

A typical product distribution for various composites with methylcyclopentane as feed is given in Table 3.

The selectivities for benzene, xylenes, and C_3, C_4, and C_5 hydrocarbons increase while that for toluene decreases with increasing zeolite content (Table 3). The activity of the composites is expected to decrease with increasing ϵ, since the Pt dispersion of the composites decreases, and hence the rate of methylcyclohexane dehydrogenation to toluene decreases. However, the increase in the toluene transalkylation and bimolecular alkylation reactions with increasing ϵ shifts the methylcyclohexane dehydrogenation to toluene. This explains why the activity remains almost unchanged with varying ϵ.

The yields of the products lumped in two groups, aromatics and C_1–C_6 paraffinic and olefinic hydrocarbons, are plotted in Figure 5 as a function of the zeolite content

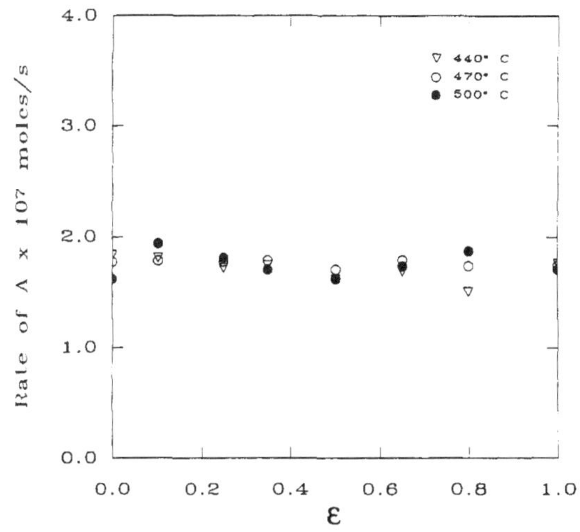

Figure 4. Rate of reaction of methylcyclohexane as a function of zeolite content (5 min on stream, WHSV = 0.49 h^{-1}, Pt loading 0.5 wt %).

Table 3. Product Distribution and Conversion with Methylcyclopentane (WHSV = 1.26 h^{-1}) and Methylcyclohexane (WHSV = 0.49 h^{-1}) at 470 °C for Various Amounts of Zeolite β Incorporated into the Composite (5 min on Stream, Pt Loading 0.5 wt %)

	methylcyclopentane			methylcyclohexane		
products (mol %)	$\epsilon = 0$	$\epsilon = 0.35$	$\epsilon = 0.80$	$\epsilon = 0$	$\epsilon = 0.35$	$\epsilon = 0.80$
methane	1.1	0.3	0.8	0.5	0.9	1.4
ethene		0.6	0.9		0.5	2.2
ethane	0.9	1.3	2.7		0.7	1.6
propene	0.7	1.1	2.6		1.1	1.0
propane	3.0	12.4	18.0	1.2	1.9	2.3
isobutane	1.8	4.0	8.1		2.6	3.3
1-butene	0.9		1.6			
n-butane	2.2	2.7	6.4	0.1	2.7	3.9
2-methyl-1-propene	0.4	0.1	0.7		2.1	0.9
isopentane	3.1	2.2	2.6		0.3	0.4
n-pentane	4.6	0.9	1.4	0.1	0.2	0.7
cyclopentane	3.3	0.1	0.4		0.3	0.7
methylcyclopentenes	10.5	2.5	0.5			
cyclohexene	25.4	2.8	1.7			
benzene	32.2	50.9	30.1	1.2	8.9	17.4
toluene	6.7	13.5	15.7	96.8	70.9	51.5
ethylbenzene	0.4	0.7	1.4			0.3
o-xylene	0.4	1.4	0.9		0.9	2.6
m-xylene	0.6	0.5	1.6		1.7	3.0
p-xylene	0.9	1.7	1.7	0.2	3.5	6.2
conversion (mol %)	54.0	78.7	80.2	86.8	87.4	90.8

for methylcyclopentane. The yield Y_i of species i is defined as

$$Y_i = XS_i$$

where X is the conversion of the feed molecule and S_i is the selectivity of species i defined as the ratio of the moles of species i generated to the moles of all the products. A broad maximum at about ϵ = 40 wt % occurs for the yield of aromatics, while the yields of C_1–C_6 paraffinic and olefinic hydrocarbons and $C_{>6}$ aromatics increase monotonically with ϵ. The maximum in aromatics is mainly due to benzene (Figure 6). As mentioned before, benzene is generated via the dehydrogenation of methylcyclopentane over Pt sites to methylcyclopentenes and the subsequent isomerization–enlargement of the latter molecules over the Brønsted acidic sites of the catalyst followed by dehydrogenation on the metal sites (reaction scheme I) of the generated C_6 cycloolefin. The maximum in the benzene yield is a result of the combination of the complementary

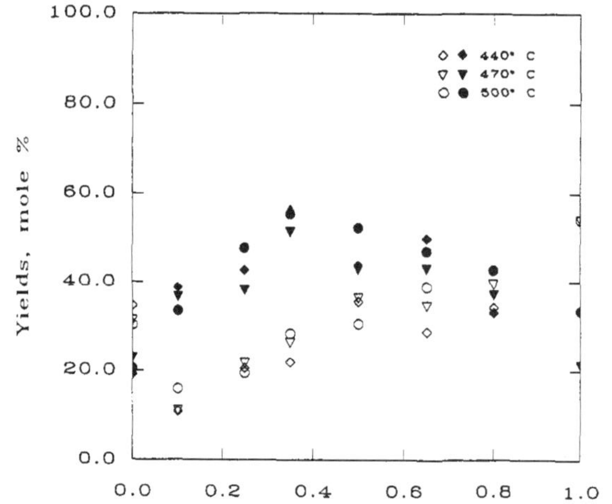

Figure 5. Yields of lumped products (filled symbols stand for aromatics, open symbols stand for paraffinic and olefinic C_1–C_6 hydrocarbons) as a function of zeolite content for methylcyclopentane (5 min on stream, WHSV = 1.26 h^{-1}, Pt loading 0.5 wt %). The generated methylcyclopentenes and cyclohexenes are included in the estimation of the C_6 hydrocarbon selectivity.

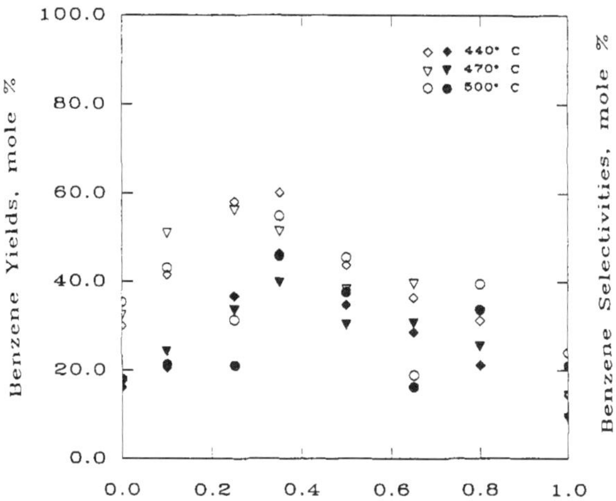

Figure 6. Benzene yields (filled symbols) and selectivities (open symbols) as a function of zeolite content for methylcyclopentane (5 min on stream, WHSV = 1.26 h^{-1}, Pt loading 0.5 wt %).

features of Pt/γ-Al_2O_3 and the Pt/β zeolite. The alumina enhances the dehydrogenation of the feed naphthene to methylcyclopentenes, but since it lacks Brønsted acidity, it cannot stimulate the isomerization–enlargement of the latter olefins to C_6 cycloolefins which are precursors to aromatics. On the other hand, the involvement of zeolite β enhances the isomerization reactions of methylcyclopentenes because of its Brønsted acidity. Hence, composites containing both catalysts can maximize the rates of aromatization of alkylcyclopentanes.

Bimolecular alkylation reactions (reaction schemes II and III) lead to aromatics with a larger number of carbon atoms than methylcyclopentane. These reactions are responsible for the increase in the molar yields of C_3 and C_4 hydrocarbons and of $C_{>6}$ aromatics with increasing content of zeolite and are stimulated by the suitable pore size of zeolite β which enhances bimolecular collisions. Because the dispersion of Pt is much higher over Al_2O_3 than over the zeolite β, experiments with a composite containing equal amounts of Pt/γ-Al_2O_3 and β zeolite free of Pt were carried out at 470 °C. It was found that the

Table 4. Conversion and Selectivities for Methylcyclopentane over Composites Containing Either Pt/β or β at 470 °C (WHSV = 1.26 h^{-1}, 5 min on Stream, Pt Loading 0.5 wt %)

		selectivities		
ε	conversion	C_1–C_6 paraffins and olefins	C_6 cycloolefins	$C_{\geq 6}$ aromatics
Pt/β				
0.0	54.0	33.1	25.4	41.5
0.5	79.1	44.3	1.6	54.1
1.0	87.5	61.8	0.2	38.0
β				
0.0	54.0	33.1	25.4	41.5
0.5	72.9	37.4	9.5	53.1
1.0	82.1	58.7	6.2	35.1

aromatic yields and the conversion were smaller than those obtained when Pt/β zeolite was employed (Table 4). This indicates that the presence of Pt in zeolite β contributes to some extent to the performance of the composites despite its low dispersion. The presence of platinum in the β zeolite enhances, in spite of its low dispersion, the intimacy of the two catalytic functions. Weisz (1962) developed a general criterion for polystep reaction systems which determines the conditions under which there is cooperation between the acidic and the dehydrogenation functions. Such a cooperation exists when the following inequality is satisfied:

$$\frac{dN}{dt}\frac{1}{B_{eq}}\frac{R^2}{D} < 1$$

where R is the radius of the zeolite particles, B_{eq} is the concentration of methylcyclopentenes, dN/dt is the rate of generation of methylcyclopentenes, and D is the diffusion coefficient of the latter molecules in the medium which provides the acidic function (β zeolite). If one uses for the diffusivity of the methylcycloolefins in the β zeolite crystals the value of 10^{-7} cm^2/s (Palekar and Rajadhyaksha, 1986), for the radius of the zeolite crystals the value of 0.5 μm, and for dN/dt the value of 4 × 10^{-7} mol/s (Figure 3), the left side of the above relation at 470 °C becomes 0.23 indicating that cooperation between the two functions of the composites is achieved. If relatively large zeolite crystals or very active zeolite samples are employed, the intimacy criterion for cooperation between the two functions can be violated. Hence, the use of Pt in the β zeolite, despite its low dispersion, can enhance the dehydrogenation capability of the composite.

An entirely different behavior was found for methylcyclohexane. A high yield of aromatics, mainly toluene, was produced by Pt/γ-Al_2O_3 and by composites containing less than about 30 wt % zeolite β. The increase in the zeolite content decreases the total aromatic yield (Figure 7). The main product of the reaction of methylcyclohexane over any of the present composites is toluene, which is produced by the direct dehydrogenation of the feed naphthene. The effect of the zeolite content on the toluene yield and selectivity is presented in Figure 8 for various temperatures. Higher yields of toluene are obtained at lower temperatures probably because of lower cracking rates. Pt/Al_2O_3 provides a higher toluene yield than any of the present composites, because of its higher Pt

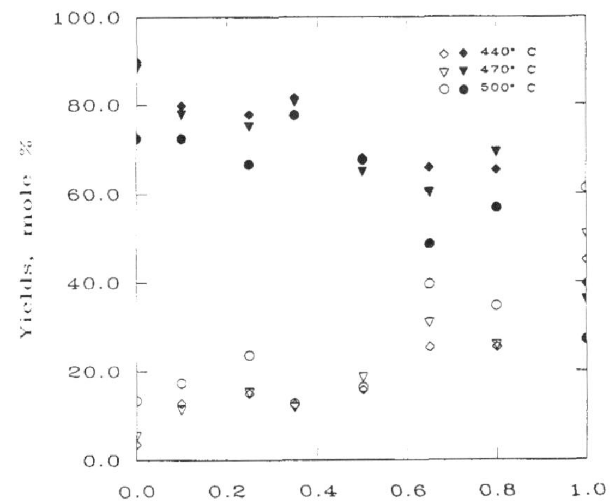

Figure 7. Yields of lumped products (filled symbols stand for aromatics, open symbols stand for C_1–C_6 paraffinic and olefinic hydrocarbons) as a function of zeolite content for methylcyclohexane (5 min on stream, WHSV = 0.49 h^{-1}, Pt loading 0.5 wt %).

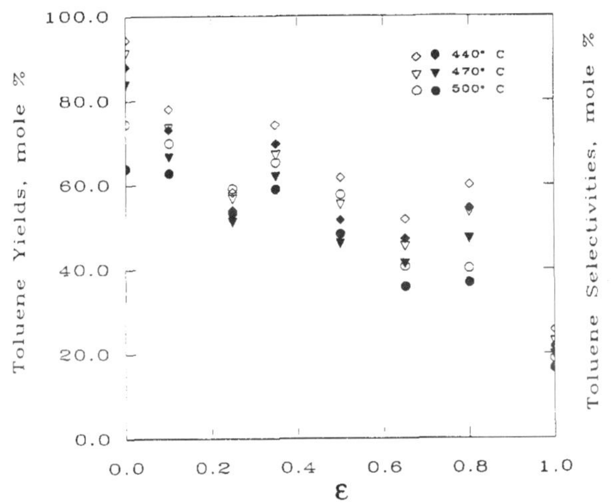

Figure 8. Toluene yields (filled symbols) and selectivities (open symbols) as a function of zeolite content for methylcyclohexane (5 min on stream, WHSV = 0.49 h^{-1}, Pt loading 0.5 wt %).

dispersion. The presence of zeolite decreases the toluene yields since the dehydrogenation capability of the composites decreases. However, the yields of benzene and xylenes increase with the content of zeolite as a result of the toluene transalkylation reaction and bimolecular alkylation reactions. As shown in Table 3, Pt/γ-Al_2O_3 produces only a small amount of benzene and no xylenes. This indicates that the latter catalyst does not stimulate alkylation reactions, because of its lower Brønsted acidity and its larger pores which do not enhance the bimolecular collisions.

3.6. Effect of Pt Loading. The effect of Pt loading is presented in Table 5 for both naphthenes, for a composite containing equal metal content of γ-Al_2O_3 and β zeolite. The increase of the loading plays a beneficial role in the aromatization of methylcyclopentane. The conversion increases monotonically with Pt loading, more significantly below 0.5 wt % Pt. Increased Pt loadings enhance the dehydrogenation capability of the composite, generating larger amounts of methylcycloolefins which can be further isomerized–enlarged to cyclohexenes. Indeed, the yield of aromatics (mainly benzene) increases with Pt loading

Table 5. Conversion and Aromatic Yields for Methylcyclopentane (WHSV = 1.26 h^{-1}) and Methylcyclohexane (WHSV = 0.49 h^{-1}) at 470 °C for Various Pt Loadings (5 min on Stream, ϵ = 0.5)[a]

Methylcyclopentane			
Pt (wt %)	conversion	benzene yield	aromatic yield
0.1	65.9	19.6	21.2
0.5	79.1	30.3	42.8
1.0	87.8	36.8	46.6
Methylcyclohexane			
Pt (wt %)	conversion	toluene yield	aromatic yield
0.1	81.2	23.4	31.3
0.5	83.4	46.1	64.9
1.0	92.5	43.7	61.3

[a] The metal loading is the same for both γ-Al_2O_3 and zeolite β.

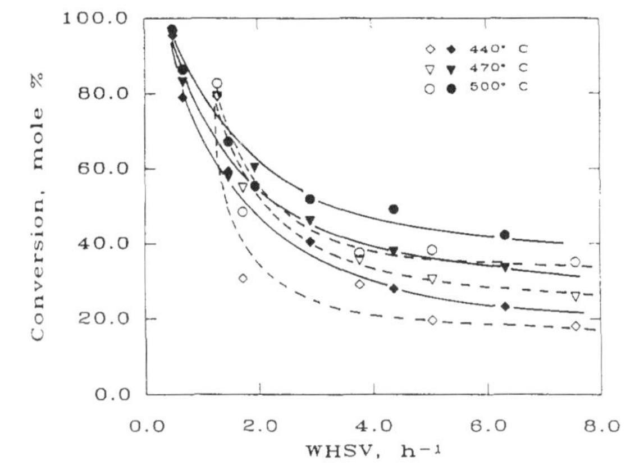

Figure 9. Conversion of methylcyclopentane (open symbols) and methylcyclohexane (filled symbols) as a function of WHSV for a composite of ϵ = 0.5 at 470 °C and 5 min on stream.

(Table 5). The increases in conversion, and in benzene and aromatic yields, are attenuated by increasing the Pt loading.

For methylcyclohexane as feed, higher conversions are observed than for methylcyclopentane. The main product is toluene, which results from the direct dehydrogenation of the feed molecule. The increase in Pt loading from 0.1 to 0.5 wt % increases significantly the toluene and aromatic yields, due to the higher rates of dehydrogenation. Aromatics with a carbon atom number different than C_7 are also produced as a result of toluene transalkylation and of bimolecular alkylation reactions stimulated by the β zeolite. The increase of the Pt loading above 0.5 wt % does not enhance the aromatization performance of the composites, probably because of increased sintering. The yield of aromatics decreases slightly at any temperature for composites with 1.0 wt % Pt in comparison with 0.5 wt % Pt (Table 5), despite the fact that the conversion is maximum at the former metal loading.

3.7. Effect of WHSV. The effect of WHSV on conversion is presented in Figure 9. Relatively high conversions are achieved at small flow rates for both feed molecules; for methylcyclohexane the conversion is almost complete at any temperature. An increase in WHSV results in an abrupt reduction in conversion, while at large values of WHSV the conversion keeps decreasing. For methylcyclopentane, the benzene selectivity (Table 6) is the highest at small values of WHSV, but decreases significantly as WHSV increases. The selectivities for other aromatics, which are produced by bimolecular alkylation reactions, do not decrease in such a significant manner. Moreover, at any value of WHSV except the

Table 6. Aromatic Product Distribution (mol %) for Methylcyclopentane and Methylcyclohexane as a Function of WHSV at Various Temperatures (5 min on stream, ϵ = 0.5, Pt Loading 0.5 wt %)

Methylcyclopentane

	440 °C		470 °C		500 °C	
WHSV (h^{-1})	benzene	$C_{>6}$ aromatics	benzene	$C_{>6}$ aromatics	benzene	$C_{>6}$ aromatics
1.26	34.7	9.0	30.3	14.5	37.7	14.5
1.71	1.8	7.5	5.5	12.7	8.2	12.2
3.78	2.1	4.4	3.6	11.8	3.8	11.3
5.04	1.1	4.6	2.4	10.2	2.9	12.1
7.55	0.9	4.2	1.6	6.6	2.7	11.1
10.07	1.2	2.7	2.8	3.4	2.3	7.3

Methylcyclohexane

	440 °C		470 °C		500 °C	
WHSV (h^{-1})	toluene	C_6, $C_{>7}$ aromatics	toluene	C_6, $C_{>7}$ aromatics	toluene	C_6, $C_{>7}$ aromatics
0.49	70.5	14.8	46.3	20.1	59.4	25.2
0.66	38.4	12.1	46.5	13.3	22.7	21.4
1.46	35.9	9.7	23.9	10.4	21.5	13.2
1.94	31.4	9.2	23.1	8.9	20.8	10.0
2.92	21.7	8.1	22.7	7.9	15.6	5.4

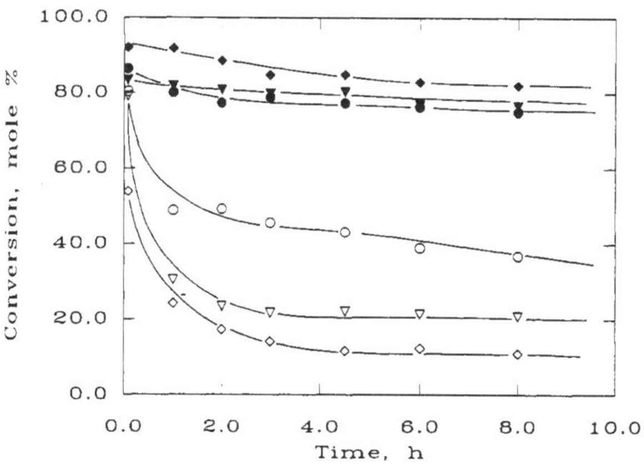

Figure 10. Conversion as a function of time on stream of methylcyclopentane over (◇) γ-Al_2O_3, (▽) composite with ϵ = 0.5, (○) β zeolite, and of methylcyclohexane over (◆) γ-Al_2O_3, (▼) composite with ϵ = 0.5, (●) β zeolite as a function of time on stream at 470 °C (Pt loading 0.5 wt %, WHSV = 1.26 h^{-1} for methylcyclopentane and WHSV = 0.49 h^{-1} for methylcyclohexane).

smallest one (Table 6), the selectivity for $C_{>6}$ aromatics is higher than that for benzene. The fact that the reduction of the residence time in the reactor decreases more significantly the benzene generation than that of $C_{>6}$ aromatics indicates that the isomerization reactions of methylcyclopentenes over the acidic sites to cyclohexene and further to benzene are much slower than the bimolecular alkylation reactions. The transalkylation of the $C_{\geq 7}$ aromatics over the Brønsted sites may have a small contribution to the formation of benzene.

For methylcyclohexane, the generation of toluene is larger than that of any other aromatics at any flow rate (Table 6). Hence, the dehydrogenation of the feed naphthene dominates the bimolecular alkylation reactions. In Table 6 it is shown that both types of reactions have comparable rates of decrease as a function of the residence time through the bed.

3.8. Time on Stream Behavior. The effect of deactivation on the activity of Pt/γ-Al_2O_3, Pt/β zeolite, and a composite containing equal weights of the above catalysts is given in Figure 10 for both naphthenes at 470 °C. For methylcyclopentane, one can observe that the conversion decreases significantly with time on stream due to the high deactivation rate. The conversion reaches a constant value after 4 h on stream, indicating that there are no further changes in the catalyst condition. Other researchers (Beltramini et al., 1983; Parera et al., 1987) have emphasized that naphthenes with a C_5 ring (cyclopentane, methylcyclopentane) are the most effective coke precursors among hydrocarbons with the same number of carbon atoms. High hydrogen pressures can decrease significantly the rate of coke formation. The role of hydrogen is to keep the Pt and the acidic sites of the catalyst free of hydrogen-deficient hydrocarbon residues which act as poisons by decreasing the activities of both functions of the catalyst (Myers et al., 1961). The inclusion of Pt/β zeolite in the composite leads to a higher initial activity and better time on stream behavior than that of Pt/γ-Al_2O_3. All the data presented in the present paper, with the exception of Figure 10, are after 5 min on stream in order to minimize the extent at which coke deactivation alters the intrinsic properties of the composites.

For methylcyclohexane, the conversion is higher than for methylcyclopentane for small time on stream, because its dehydrogenation to toluene, which constitutes the main reaction, is more rapid than the dehydroisomerization of methylcyclopentane. A slight decrease in conversion occurs at short times on stream, while for times on stream greater than 4 h the conversion remains practically unchanged. The initial rate of decrease of conversion with time on stream is significantly smaller for methylcyclohexane than for methylcyclopentane. The reaction of methylcyclohexane, which involves mainly the Pt sites, exhibits a small deactivation. In contrast, the reaction of methylcyclopentane, which involves in addition the isomerization of methylcycloolefins over the acidic sites, is expected to be more severely altered by the coke deposition which takes place with preference on the acidic sites.

4. Conclusions

It is demonstrated that composite catalysts of Pt/zeolite β and Pt/γ-Al_2O_3 possess beneficial effects in the reforming of naphthenes. The advantageous behavior of the composites can be attributed to the synergism between Pt/β-Al_2O_3 and Pt/β zeolite due to the complementary features that each of them provides. Alumina, because of its higher Pt dispersion, favors dehydrogenation reactions which generate methylcyclopentene intermediates, while zeolite β favors, due to its Brønsted acidity, the bimolecular alkylation reactions, the transalkylation reaction, and the isomerization–enlargement reactions of methylcyclopentenes to C_6 cycloolefins. Since the latter reactions are slow, they occur at sufficiently large residence times. At large residence times, the yield of aromatization (mainly benzene) of methylcyclopentane passes through a maximum at about 40 wt % β zeolite. With methylcyclohexane, a large amount of toluene is generated for a β zeolite content below 30 wt %. A decrease in the toluene yield associated with increasing yields of benzene and xylenes was observed with further increases in zeolite content, due to the transalkylation of the former molecule. The increase of the zeolite content in the composites results in an increase of bimolecular alkylation reactions for both feeds.

It was found that the dehydroisomerization reaction of methylcyclopentane is slower than the dehydrogenation and bimolecular alkylation reactions; the latter two reactions are comparably fast. Coke is formed mainly over the acidic sites of the catalyst.

Nomenclature

B_{eq} = equilibrium concentration of methylcyclopentenes
D = diffusion coefficient of methylcyclopentenes in the acidic function (β zeolite)
H_0 = Hammett acidity function
dN/dt = rate of generation of methylcyclopentenes
R = radius of the zeolite particles
S_i = selectivity of i lumped species
X = conversion of the feed hydrocarbon
Y_i = yield of i lumped species

Greek Symbols

ϵ = weight fraction of zeolite β in the composites

Literature Cited

Alvarez, F.; Montes, A.; Perot, G.; Guisnet, M. Hydroisomerization and Hydrocracking of Alkanes. Methylcyclohexane transformation on PtUSHY catalysts. In *Zeolites: Facts, Figures, Future*; Studies in Surface Science and Catalysis 49; Elsevier: Amsterdam, 1989; pp 1367–1376.

Barbier, J.; Bahloul, D.; Marecot, P. Reduction of Pt/Al_2O_3 catalysts: Effect of Hydrogen of Water and of Hydrochloric Acid Vapor on the Accessibility of Platinum. *J. Catal.* **1992**, *137*, 377–384.

Benesi, H. A. Acidity of Catalyst Surfaces. I) Acid Strength from Indicators. *J. Am. Chem. Soc.* **1956**, *78*, 5490–5494.

Beltramini, J. M.; Martinelli, E. E.; Churin, E. J.; Figoli, N. S.; Parera, J. M. Pt/Al_2O_3-Cl in pure hydrocarbon reforming. *Appl. Catal.* **1983**, *7*, 43–55.

Borade, R. B.; and Clearfield, A. Characterization of acid sites in beta and ZSM-20 zeolites. *J. Phys. Chem.* **1992**, *96*, 6729–6737.

Castro, A. A.; Scelza, O. A.; Benvenuto, E. R.; Baronetti, G. T.; Parera, J. M. Regulation of the Chloride Content on Pt/Al_2O_3 Catalyst. *J. Catal.* **1981**, *69*, 222–226.

Ciapetta, F. G.; Wallace, D. N. Catalytic Naphtha Reforming. *Catal. Rev.—Sci. Eng.* **1971**, *5*, 67–158.

Corma, A.; Fornes, V.; Monton, J. B.; Orchilles, A. V. Catalytic Activity of Large-pore High Si/Al Zeolites: Cracking of Heptane on H-beta and Dealuminated HY zeolite. *J. Catal.* **1987**, *107*, 288–295.

Derouane, E. G.; Vanderveken, D. J. Structural Recognition and Preorganization in Zeolite Catalysis: Direct Aromatization of n-Hexane on Zeolite L-based catalysts. *Appl. Catal.* **1988**, *45*, L15–L22.

Gallezot, P.; Alarcon-Diaz, A.; Dalmon, J.-A.; Imelik, B. Location and Dispersion of Pt in PtY Zeolites. *J. Catal.* **1975**, *39*, 334–349.

Hughes, T. R.; Buss, W. C.; Tamm, P. W.; Jacobson, R. L. Aromatization of Hydrocarbons over Platinum Alkaline Earth Zeolites. In *New Developments in Zeolite Science and Technology*; Studies in Surface Science and Catalysis 28; Kodansha-Elsevier: Tokyo, 1986; pp 725–732.

Jaeger, N. I.; Ryder, P.; Schulz-Ekloff, G. Concepts of Reduction and Dispersion of Metals in Zeolites. In *Structure and Reactivity of Modified Zeolites*; Studies in Surface Science and Catalysis 18; Elsevier: Amsterdam, 1984; pp 299–311.

Keulemans, A. I. M.; Voge, H. H. Reactivities of naphthenes over a platinum reforming catalyst by a gas chromatographic technique. *J. Phys. Chem.* **1959**, *63*, 476–480.

Kouksdahl, H. E. Reforming a sulfur-free naphtha with a Pt-Re catalyst. U.S. Patent 3,415,737, Dec 10, 1968.

Lapierre, R. B.; Partridge, R. D.; Chen, N. Y.; Wong, S. S. Catalytic Dewaxing Process. Eur. Pat. Appl. 95,303, May 17, 1983.

Lapierre, R. B.; Randell, D. P.; Chen, N. Y.; Wong, S. S. Catalytic dewaxing process with zeolite beta. U.S. Patent 4,501,926, Feb 26, 1985.

Liu, S. B.; Wu, J. F.; Ma, L. J.; Tsai, T. C.; Wang, I. On the thermal stability of zeolite beta. *J. Catal.* **1991**, *132*, 432–439.

Martens, J. A.; Perez-Pariente, J.; Jacobs, P. A. Selectivity induced by the void structure of zeolite beta and ferrierite in hydroconversion reaction of n-decane. In *Chemical Reaction in Organic and Inorganic Constrained Systems*; Reidel Publishing Co.: Dordrecht, 1986; Vol. 165, pp 115–129.

Martens, J. A.; Tielen, M.; Jacobs, P. A. Relation between Paraffin Isomerization capability and Pore Architecture of large-pore bifunctional zeolites. In *Zeolites as Catalysts, Sorbents and Detergent Builders*; Studies in Surface Science and Catalysis 46; Elsevier: Amsterdam, 1989; pp 49–60.

Mills, G. A.; Heinemann, H.; Milliken, T. H.; Oblad, A. G. Catalytic Mechanism: Houdriforming Reactions. *Ind. Eng. Chem.* **1953**, *45* (1), 134–137.

Moretti, G.; Sachtler, W. M. H. Geometric Causes of the Methylcyclopentane, Ring Opening Selectivity over Pt/NaY Catalysts. *J. Catal.* **1989**, *116*, 350–360.

Myers, L. G.; Lang, W. H.; Weisz, P. B. Aging of platinum reforming catalysts. *Ind. Eng. Chem.* **1961**, *53* (4), 299–302.

Newsam, J. M.; Treacy, M. M. J.; Koetsier, W. T.; DeGruyter, C. B. Structural characterization of zeolite beta. *Proc. R. Soc. London A* **1988**, *420*, 375–405.

Palekar, M. G.; Rajadhyaksha, R. A. Sorption accompanied by chemical reaction on zeolites. *Catal. Rev.—Sci. Eng.* **1986**, *28*, 371–429.

Parera, J. M.; Verderone, R. J.; Querini, C. A. Coking on Bifunctional Catalysts. In *Catalyst Deactivation*; Studies in Surface Science and Catalysis 34; Elsevier: Amsterdam, 1987; pp 135–145.

Peri, J. B. Infrared study of adsorption of ammonia on dry γ-Al_2O_3. *J. Phys. Chem.* **1965**, *69*, 231–239.

Ramage, M. P.; Graziani, K. R.; Schipper, P. H.; Krambeck, F. J.; Choi, B. C. KINPTR (MOBIL's Kinetic Reforming Model): A review of Mobil's Industrial Process. *Adv. Chem. Eng.* **1987**, *13*, 193–266.

Satterfield, C. N. In *Heterogeneous Catalysts in Industrial Practice*; McGraw-Hill: New York, 1992.

Schulz, H.; Weitkamp, J.; Eberth, H. New Disproportionation Reaction of Alicyclic Hydrocarbons on Bifunctional Catalysts in the Presence of Hydrogen. In *Procedings of the Fifth International Congress on Catalysis*; North-Holland Publishing Co.: Amsterdam, 1973; Vol. 2, pp 1229–1239.

Sinfelt, J. H. Catalytic Reforming of Hydrocarbons. In *Catalysis, Science and Technology*; Springer-Verlag: New York, 1981; Vol. 1, pp 257–300.

Sinfelt, J. H.; Rohrer, J. C. Kinetics of the Catalytic Isomerization-Dehydroisomerization of Methylcyclopentane. *J. Phys. Chem.* **1961**, *65*, 978–981.

Smirniotis, P. G.; Ruckenstein, E. Comparison between zeolite β and γ-Al_2O_3 supported Pt for Reforming Reactions. *J. Catal.* **1993**, *140*, 526–542.

Sushumna, I.; Ruckenstein, E. Redistribution of Pt/Alumina via Film Formation. *J. Catal.* **1987**, *108*, 77–96.

Tamm, P. W.; Mohr, D. H.; Wilson, C. R. Octane enhancement by selective reforming of light paraffins. In *Catalysis 1987*; Studies in Surface Science and Catalysis 38; Elsevier: Amsterdam, 1988; pp 335–353.

Tanabe, K. In *Solids Acids and Bases*; Kodansha: Tokyo, and Academic Press: New York, 1970.

Tanabe, K.; Misono, M.; Ono, Y.; Hattori, H. In *New Solids Acids and Bases: Their Catalytic Properties*; Studies in Surface Science and Catalysis 51; Elsevier: Amsterdam, 1989.

Tanaka, M.; Ogasawara, S. Infrared Studies of the Adsorption and the Catalysis of Hydrogen Chloride on Alumina and on Silica. *J. Catal.* **1970**, *16*, 157–163.

Treacy, M. M. J.; Newsam, J. M. Two new three-dimensional twelve-ring zeolite frameworks of which zeolite beta is a disordered intergrowth. *Nature* **1988**, *332*, 249–251.

Wadlinger, R. L.; Kerr, G. T.; Rosinski, E. J. Catalytic Composition of a Crystalline Zeolite. U.S. Patent 3,308,069, March 7, 1967.

Wang, I.; Tsai, T. C.; Huang, S. T. Disproportionation of toluene and of trimethylbenzene and their transalkylation over zeolite beta. *Ind. Eng. Chem. Res.* **1990**, *29*, 2005–2012.
Weisz, P. B. Polyfunctional Heterogeneous Catalysis. *Adv. Catal.* **1962**, *13*, 137–190.
Weisz, P. B.; Prater, C. B. Basic activity properties for Pt-type reforming catalysts. *Adv. Catal.* **1957**, *9*, 575–586.
Weisz, P. B.; Swegler, E. W. *Science* **1957**, *126*, 31–32.
Weitkamp, J. Alkylation of hydrocarbon with zeolite catalysts-commercial applications and mechanistic aspects. In *Proceedings International Symposium on Zeolite Catalysis*, Acta Phys. Chem. Hungarica, Siofok, Hungary; 1985; pp 271–290.
Weitkamp, J.; Jacobs, P. A.; Ernst, S. Shape selective Isomerization and Hydrocracking of Naphthenes over Pt/HZSM-5 zeolite. In *Structure and Reactivity of Modified Zeolites*; Studies in Surface Science and Catalysis 18; Elsevier: Amsterdam, 1984; pp 279–290.

Received for review September 2, 1993
Revised manuscript received December 2, 1993
Accepted December 13, 1993[⊗]

[⊗] Abstract published in *Advance ACS Abstracts,* February 15, 1994.

Effect of the platinum/aluminum sites ratio on the performance of platinum/β zeolite in the reforming of alkylcyclopentanes

Panagiotis G. Smirniotis[1], Eli Ruckenstein*

Department of Chemical Engineering, State University of New York at Buffalo, Buffalo, New York 14260, USA

Received 28 March 1994; revised 27 May 1994; accepted 29 May 1994

Abstract

The reforming of methylcyclopentane or ethylcyclopentane over Pt/β zeolite catalysts of various SiO_2/Al_2O_3 molar ratios has been examined at various H_2 pressures. The balancing of the Brønsted acidity with the Pt loading at an appropriate pressure maximizes the dehydroisomerization of the naphthenes. A maximum aromatic yield occurs for both naphthenes for a catalyst with a Pt (determined by H_2 chemisorption)/Al atoms ratio of about 3×10^{-2} (SiO_2/Al_2O_3 molar ratio of 130 and a Pt loading of 0.5 wt.-%) and at a pressure of about 800 kPa. The control of the Brønsted acidity via dealumination minimizes coke formation and hydrocracking and results in the preservation of the catalyst activity for many hours-on-stream. The coke was characterized using ^{1}H-NMR, FTIR, and TGA.

Keywords: Alkylcyclopentanes; Coking; Dealumination; Platinum/β zeolite; Reforming

1. Introduction

Zeolites have been successfully employed in the reforming of hydrocarbons. The acidic zeolites supported Pt enhance significantly the extent of bimolecular alkylation reactions that lead to aromatization with the enlargement of the feed hydrocarbon [1–3]. They also favor the isomerization enlargement of methylcyclopentane [4]. The reforming of paraffinic feedstocks has been also carried out over the basic L zeolite supported Pt [5,6]; this catalyst transforms very selectively the C_6, C_7 paraffinic hydrocarbons in aromatics. The two functions of

* Corresponding author. Tel. (+1-716)6452911 ext. 2214 or 2203, fax. (+1-716)6453822.

[1] Present address: Chemical Engineering Dept., University of Cincinnati, Cincinnati OH 45221-0171, USA.

This paper, which was originally published in **Applied Catalysis A**: General 117, 75–96 (1994), is republished with permission.

the metal-supported zeolite catalysts, in combination with the structure of the zeolite have also other favorable effects. For instance, Parton et al. [7] noted that the generation of C_{10} dibranced alkanes is maximized over composites of Pt/HZSM-22 and partially protonated Pt/Y faujasite. A maximum in the dehydroisomerization of methylcyclopentane has been observed over various composites of Pt/γ-Al_2O_3 with Pt/β zeolite [4].

The disadvantages of acidic zeolites are: (i) they lose their activity because of excessive coke formation and (ii) the relatively high extent of hydrocracking of the hydrocarbon feed. In order to improve their behaviour, their acidity should be controlled. Similarly, an excess of Pt generates excessive hydrogenolysis of the feed hydrocarbons. In the present study we examine the performance of Pt/β zeolite with different amounts of acidity and Pt loadings in the reforming of methylcyclopentane and ethylcyclopentane at various pressures. The acidity was changed by varying via dealumination the SiO_2/Al_2O_3 molar ratio. We found that a very pronounced maximum in the dehydroisomerization of both alkylcyclopentanes occurs for a catalyst with a SiO_2/Al_2O_3 molar ratio of about 130 and 0.5 wt.-% Pt and a pressure of about 800 kPa. The above catalyst had a good time-on-stream behavior, since the coke build-up remained at a low level. As a result, the activity was preserved in time and the aromatic selectivity remained high. Some preliminary results have been reported in a short communication [8].

2. Experimental

2.1. Synthesis and preparation of the catalysts

The steps for the synthesis of zeolite β employed in the present study were described in our earlier work [3]. The nominal composition of the initial aluminosilicate gel was 30 SiO_2 : Al_2O_3 : 0.9 Na_2O_3 : 3 TEAOH : 250 H_2O with tetraethylammonium hydroxide (TEAOH) as template. The prepared zeolite β had a high crystallinity, as determined by XRD, and a SiO_2/Al_2O_3 molar ratio of 28, as determined by atomic absorption spectroscopy (AAS). The obtained zeolite was calcined at 510 ± 2°C for 5 hours in high purity O_2 (99.999%), in order to burn the template occluded in the zeolite pores. The calcination conditions were selected to ensure that the zeolite crystallinity was preserved [9]. Samples with various SiO_2/Al_2O_3 molar ratios were prepared by the dealumination of the synthesized zeolite β with HCl [10]. For dealumination, 2 g of catalyst were introduced in a 1-liter solution containing a suitable amount of HCl acid (see Table 1). The SiO_2/Al_2O_3 molar ratios of the dealuminated zeolites and their preparation conditions are listed in Table 1. The crystallinity of the dealuminated zeolites was determined with a Nicolet powder X-ray diffractometer equipped with a Cu Kα source. MgO (Baker analyzed 99.6 wt.-%) was added (20 wt.-%) to each sample followed by grinding and sieving (150 mesh size sieve). The ratios of the area under the peak of the

Table 1
Catalysts employed, crystallinity based on the nondealuminated sample (CR), activation conditions, Pt loading (wt.-%), Pt dispersion $D=(H/Pt, \%)$, Pt amount dispersed (wt.-% of the catalyst), number of Al atoms (n_{Al}), and n_{Pt}/n_{Al} ratio

Both the oxidation and the reduction steps lasted 1 h each. After oxidation and before reduction the catalyst was purged with He for 15 min

Sample	SiO_2/Al_2O_3 (molar ratio)	CR	T_O[c]	T_R	Pt	D	Pt dispersed	n_{Al}/g cat.	n_{Pt}/n_{Al}[d]
B1 zeolite β	28	100	550	550	0.1	22	0.022	6.76×10^{20}	1.00×10^{-3}
B2 zeolite β	28	100	550	550	0.5	6	0.030	6.76×10^{20}	1.36×10^{-3}
B3[a] zeolite β	130	91	300	300	0.5	29	0.145	1.52×10^{20}	2.90×10^{-2}
B4[b] zeolite β	270	89	300	300	1.0	21	0.210	7.36×10^{19}	8.78×10^{-2}
Si SiO_2	–	–	300	300	1.0	25	0.250	–	∞

[a]The dealumination was carried out with 2 *M* HCl for 4 h (2 g catalyst in 1 liter solution).
[b]The dealumination was carried out 3.5 *M* HCl for 4 h (2 g catalyst in 1 liter solution).
[c]T_O = oxidation temperature in °C, T_R = reduction temperature in °C.
[d]The n_{Pt} was determined by H_2 chemisorption.

XRD pattern of β zeolite at $2\theta \simeq 22.6°$ and the area under the peak of MgO ($2\theta \simeq 42°$) for the dealuminated and non-dealuminated zeolites were compared. The protonated forms of the various zeolites were prepared by cation exchange for 4 hours at 95 ± 1°C, under reflux, of 2 g of catalyst in 1 liter of 4 *M* NH_4Cl (Fisher). The zeolites were transformed to their acidic form by an additional calcination step at 450°C for 1 hour in O_2.

Wet impregnation with H_2PtCl_6 (Aldrich 8 wt.-% sol.) was employed. After impregnation, the catalysts were dried overnight at 130°C. This was followed by grinding and sieving of the catalysts. The catalysts employed in the present study and their Pt loadings are listed in Table 1. The SiO_2 support was purchased from Aldrich (Aldrich, grade 951). The catalysts were prepared to have a gradual increase of the n_{Pt}/n_{Al} ratio from B1 through Si catalysts, where n_{Pt} and n_{Al} stand for the number of Pt atoms detected by H_2 chemisorption and the number of Al atoms, both per gram of catalyst, respectively.

2.1.1. *H_2 Chemisorption*

Hydrogen chemisorption at room temperature was employed for the determination of Pt dispersion. After the calcination and reduction of the catalyst (the conditions are given in Table 1), 1 ± 0.01 g (on a dry basis) of catalyst was loaded in a Pyrex glass U-shaped sample holder. The holder was attached to a vacuum glass system equipped with a pressure transducer. After an initial reduction step at 200°C for 1 hour to eliminate any possible oxidation of the catalyst, the sample was evacuated at 400°C for 1 h under vacuum (10^{-5} Torr). Further, the sample was cooled to the ambient temperature and a first isotherm was determined for a pressure range between 10 to 400 Torr. After evacuating the system at room temperature to 10^{-5} Torr, a second adsorption isotherm was determined for the above pressure range. The amount of strongly chemisorbed H_2 was estimated by subtracting the two isotherms, and the metal dispersion evaluated by assuming dissociative adsorption of H_2 on the Pt atoms.

2.1.2. Acidity and acidic strength distribution

The acidity and the acid strength distribution of all the catalysts were determined by the butylamine titration method using appropriate indicators [11]. Experimental details are provided by Tanabe [12]. The acid strength of the catalyst, both Brønsted and Lewis, was determined on the basis of the color change of the basic indicators to their conjugate acidic forms. From the pK_a values of the indicators employed, the H_0 values of the surface catalyst were obtained. The acidity of the catalyst at a given H_0 is usually expressed as mmole of butylamine per unit weight of the solid, and can be obtained by measuring the amount of butylamine reacted with the solid acid. In all the titrations, benzene (Aldrich, 99+%) was used as the nonpolar solvent.

2.1.3. Coke analysis

2.1.3.1. Thermogravimetric analysis (TGA). The coke deposited on the catalysts was determined with a thermogravimetric balance (CAHN 2000) by measuring the weight change of the catalyst during its oxidation. The deactivated catalyst was evacuated in high vacuum for 1 hour and further heated in a He atmosphere at 400°C, also for 1 hour for cleaning. The catalyst was cooled to the ambient temperature, followed by a final heating in high purity O_2 in the range of 25–750°C for coke burning. To eliminate any fluid transport effect, an additional heating in O_2 of the regenerated catalyst under identical conditions was also performed. The actual weight loss of the catalyst was obtained by the substraction of the two weight profiles. The temperature changes were carried out at a rate of 10°C/min.

2.1.3.2. ^{1}H-NMR and infrared spectroscopy (FTIR) studies of the soluble coke. For the NMR and the infrared investigation of the soluble coke, the deactivated catalyst was dissolved in 40 wt.-% hydrofluoric acid solution (Aldrich) overnight [13]. The soluble coke was removed from the acidic solution by extraction with CH_2Cl_2. The nonsoluble part of the coke (high-molecular-weight polycyclic saturated compounds) remained as small black particles. For the ^{1}H-NMR studies, the CH_2Cl_2 was evaporated in a stream of air and the soluble coke was dissolved in chloroform-*d* (Aldrich, 100.0 atom-% D). The ^{1}H-NMR analysis was performed with a Varian (VXR-400 S) spectrometer.

For the infrared investigations of the soluble coke, a Fourier transform-infrared spectrometer (Mattson Alpha Centauri) was employed. Transmittance spectra were obtained in the range of 4000–100 cm^{-1}, with a 4 cm^{-1} resolution. The samples employed in these analyses were those used for the NMR studies. For this reason, prior to the actual run, the spectrum of pure chloroform-*d* was also recorded. The

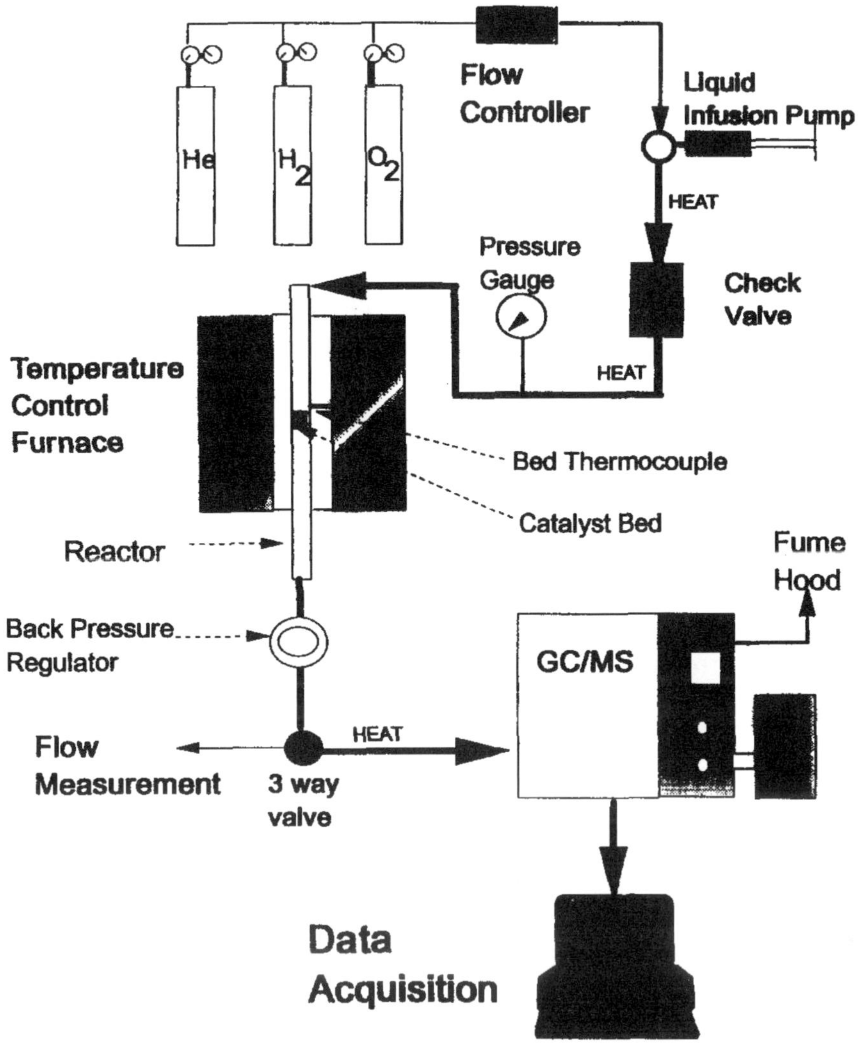

Fig. 1. Schematic diagram of the high-pressure reactor employed.

spectrum of the sample was obtained by substracting the chloroform-*d* spectrum from the spectrum of the sample dissolved in chloroform-*d*.

Elemental analysis (performed by Quantitative Technologies Inc., NJ) was employed for the determination of the H/C atomic ratio of the total coke.

2.1.4. Catalytic tests

A vertical plug-flow stainless steel tube of 0.60 cm I.D. and 40 cm length was employed in the catalytic experiments. The reaction temperature was varied between 400 and 450°C, and the total pressure between atmospheric and 1820 kPa. The reactor was heated to the reaction temperature in a programmable temperature controlled furnace, and the pressure regulation was achieved with a back-pressure regulator. In all experiments, 100 ± 2 mg (on a dry basis) of fresh catalyst was loaded on a glass wool plug located in the middle of the reactor. The catalysts were activated in situ by oxidation in ultra-high purity oxygen (99.999%, 5 ± 0.5 cm^3/min) for 1 hour, followed by the purging of the catalyst with high purity He (99.995%) for 15 min, and finally by the reduction with ultra-high purity H_2

(99.999%, 5 ± 0.5 cm^3/min) for 1 hour. The oxidation and reduction temperatures were varied to alter the Pt dispersion and are listed in Table 1. The temperature changes were carried out at a rate of 10°C/min. The reactants were methylcyclopentane (Aldrich, 99+%) and ethylcyclopentane (DWR Research Co., 99.7%). The feed hydrocarbons were introduced using an infusion liquid pump through a septum injection port into the heating line carrying H_2 to the reactor at the desired flow rate. The flow of H_2 was controlled by a flow controller. For the on-line analysis of the products, high resolution gas chromatography was employed. The analysis was carried out with a gas chromatograph (HP 5890 series II) associated with a mass spectrometer (HP 5971 series). The GC was loaded with a fused silica PoraPLOT wide bore capillary column (0.53 mm I.D., 28 m length with a film thickness of 20 μm), a molecular sieve packed column (5A) and a high performance capillary column (HP-1 cross-linked methyl silicone, 0.2 mm I.D., 12 m length with a film thickness of 0.33 μm). A PC was attached to the GC/MS unit for storage of the data. The reactor effluent line, including the back-pressure regulator, was heated at 150°C in order to avoid any possible condensation of the products. The experimental set up employed in the present catalytic experiments is presented in Fig. 1.

3. Results and discussion

3.1. H_2 chemisorption

The Pt dispersion determined by H_2 chemisorption and the amount of metal dispersed over the catalysts employed are listed in Table 1. The metal dispersion is relatively low in all the samples. This happens because the Pt^{2+} ions formed during the oxidation step are reduced to Pt^0 atoms in the hydrogen atmosphere [14] which, being no longer attracted electrostatically by the zeolite lattice, migrate through the pores to the particle surface. It appears that the platinum atoms have a relatively high mobility in the β zeolite, even when the reduction temperature is as low as 300°C. This decreases the fraction of Pt located in the zeolite pores, and is in agreement with the transmission electron micrographs of Pt/β zeolite (not given), in which metal clusters larger than 50 Å are present at the edges of the zeolite particles. However, the amount of metal dispersed increases from sample B1 to Si (Table 1), satisfying our requirement for an increased number of Pt atoms dispersed when the total acidity decreases.

3.2. Acidity and acid strength distribution

The acidity and the acid strength distribution of the Pt free H/β zeolites employed were determined by n-butylamine titration. As shown in Table 2, the nondealuminated Pt/β zeolite (SiO_2/Al_2O_3 = 28) changed the colors for all the indicators with

Table 2
Acid amount in mmole of n-butylamine/g cat. (first column) or in molecules of n-butylamine/g cat. (second column) and strength distribution of H/β zeolite with various SiO_2/Al_2O_3 molar ratios

SiO_2/Al_2O_3	28		130		270	
Al atoms/g cat.	6.76×10^{20}		1.52×10^{20}		7.39×10^{19}	
H_0	Acid amount (mmole of n-butylamine/ g cat)	Acid amount (molecules of n-butylamine/ g cat) $\times10^{-20}$	Acid amount (mmole of n-butylamine/ g cat)	Acid amount (molecules of n-butylamine/ g cat) $\times10^{-20}$	Acid amount (mmole of n-butylamine/ g cat)	Acid amount (molecules of n-butylamine/ g cat) $\times10^{-20}$
6.8	4.08	24.57	3.00	18.07	2.80	16.86
2.0	1.12	6.74	0.60	3.61	0.68	4.09
0.8	0.60	3.61	0.44	2.65	0.36	2.17
−3.0	0.52	3.13	0.40	2.41	0.36	2.17
−10.1	0.08	0.48	0.00	0.00	0.00	0.00
−12.0	0.00	0.00	0.00	0.00	0.00	0.00

$pK_a \geq -10.1$. Hence, this catalyst has the smallest acidic strength H_0 between −10.1 and −12.0. The latter value (−12.0) of H_0 is considered as the upper limit for superacidity [15]; catalysts that have H_0 values below −12.0 are considered superacids. Hence, the catalysts employed in the present study are not superacids. Reforming catalysts which are superacids are not desirable because they generate extensive hydrocracking of the feed hydrocarbons.

The two dealuminated samples B3 and B4 (Table 1) possess a lower acid strength than the nondealuminated samples B1 and B2 since they no longer change color with the indicator with $pK_a = -10.1$. The B3 and the B4 samples change color for the indicator with $pK_a = -3.0$. The acidity of the catalyst at a certain acid strength is measured by n-butylamine titration of the benzene suspension containing the indicator after its color changes from the base to the acid-form. The results expressed in mmoles of n-butylamine per gram of catalyst are presented for all the samples in Table 2. Small differences exist in the acidities of the two dealuminated samples. SiO_2 did not change the color of any of the indicators used because of its negligible acidity.

The above results provide information about the total acidity of the catalysts. Infrared investigations with pyridine adsorbed on β zeolite with SiO_2/Al_2O_3 molar ratios of about 30 [16] showed that the Brønsted-to-Lewis acid site ratios (B/L) was about 3. One expects the Brønsted-to-Lewis ratio to decrease with dealumination, since the Brønsted acidity is related to the aluminum framework content, whereas the Lewis acidity is related to the extraframework aluminum [17]. The performance of the catalysts can be controlled by varying the extent of dealumination; this alters the amount of Brønsted acidity, thus avoiding hydrocracking. However, enough Brønsted acid sites must be kept for isomerization to take place.

No extra chloride was added to the catalysts to increase their acidity, as is usually done with Al_2O_3-based reforming catalysts [18], because they have sufficient acidity. The only source of chloride is that introduced with H_2PtCl_6 during the impregnation step. As shown earlier [19], the level of chloride over Pt/β zeolite decreases significantly after the activation step. However, the decrease of the Cl^- does not affect the isomerization capability, since all the samples possess enough acidic sites. Hence, compared with the traditionally employed Pt/Al_2O_3 catalyst [18], the Pt/β zeolite with controlled acidity has the advantage that it can operate for long times without the periodic introduction of Cl^- for acidity preservation.

3.3. Catalytic behavior

3.3.1. Methylcyclopentane

As demonstrated by Keulemans and Voge [20], the reforming of methylcyclopentane over the bifunctional Pt/Al_2O_3 catalyst consists of the following steps: (a) the dehydrogenation of methylcyclopentane over the noble metal to cycloalkenes and migration of the latter molecules to the acidic sites of the catalyst; (b) the formation of carbenium ions over the Brønsted acid sites of the catalyst, followed by the isomerization enlargement of the latter intermediates to C6 cyclic rings; (c) the desorption from the acid sites and further dehydrogenation of the latter molecules to benzene. The above pathway for the aromatization of methylcyclopentane is presented in Scheme 1.

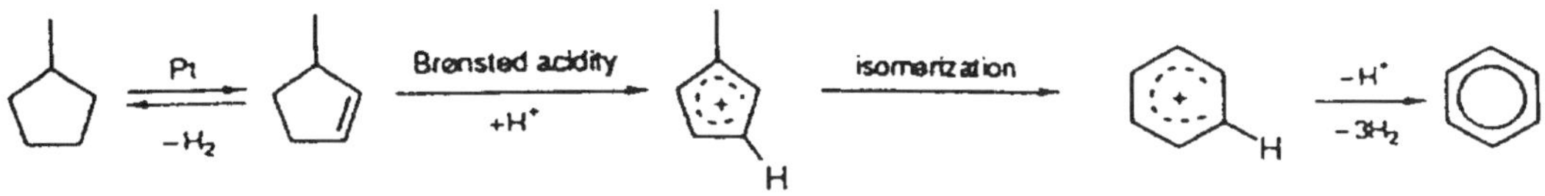

(I)

As previously noted, the rate controlling step is the isomerization step, while the dehydrogenation steps can be considered to occur at equilibrium [21]. The benzene generation from methylcyclopentane can be considered as a measure of the dehydroisomerization capability of a bifunctional catalyst.

This type of reaction, which is of great importance in the upgrading of the feed naphthene [22], involves the cooperation of the two functions of the catalyst. Enough Pt should exist to provide sites for the formation of cycloalkenes, but large amounts of noble metal should be avoided to minimize hydrogenolysis. Similarly, the acidity, plays an important role. A suitable amount of Brønsted acid sites which can provide isomerization enlargement of the C_5 cyclic ring to a C_6 one, is absolutely necessary. However, an excess of acidity stimulates coke formation and hydrocracking.

For the catalysts employed in the present study the optimum balancing of the two functions is achieved by the B3 catalyst (SiO_2/Al_2O_3 molar ratio = 130, 0.5 wt.-% Pt), which maximizes the dehydroisomerization of methylcyclopentane to benzene. As shown in Fig. 2, where the benzene yield is plotted against the n_{Pt}/n_{Al} ratio, the yield possesses a very steep maximum for this catalyst. The yield of species i is defined as $Y_i = X\,S_i$, where X is the conversion and S_i is the selectivity of any product or any lumped species i.

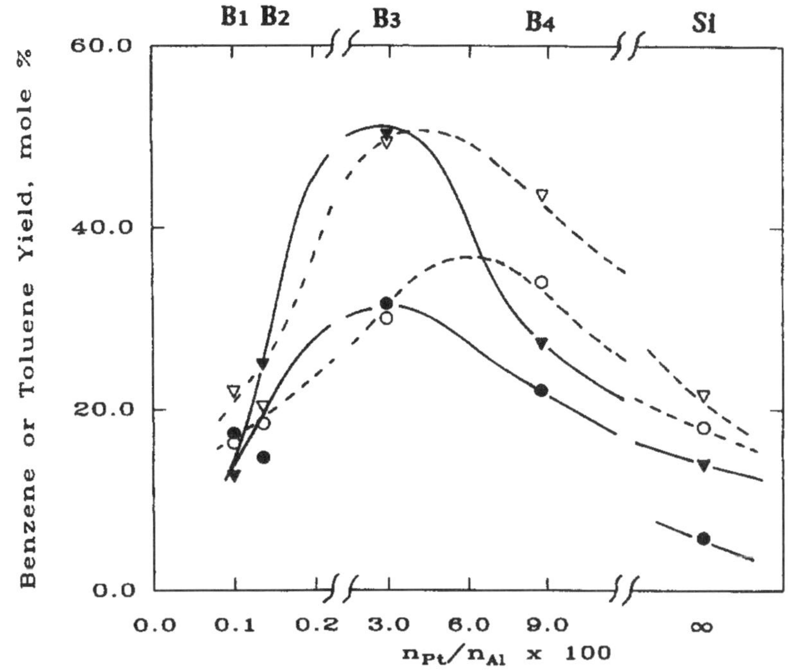

Fig. 2. Benzene yield (filled symbols) for methylcyclopentane and toluene yield (open symbols) for ethylcyclopentane versus the metallic and the acidic functions of the catalyst (pressure 790 kPa, 2 hours-on-stream, WHSV = 2.0 h^{-1}, H_2/oil = 12.0 (molar ratio)). (○, ●) 400°C, (▽, ▼) 450°C.

The product distribution for methylcyclopentane over all the catalysts is given in Table 3. The table shows that the maximum of the benzene yield for the B3 catalyst is associated with a decreased C_1–C_6 paraffinic and olefinic selectivity compared to those for B1 and B2. The higher Brønsted acidity of the B1 and B2 samples is responsible for this behavior. The $C_{\geq 7}$ aromatic selectivity of the B1 and B2 samples is slightly higher than that of the B3 catalyst. The alkylation reactions between the benzene generated and alkenes [2] or between methylcyclopentane and alkanes [19,23] which are enhanced by the higher Brønsted acidity of the non-dealuminated B1 and B2 catalysts explain this behavior. Two features of the β zeolite are responsible for the bimolecular alkylation reactions: (i) The zeolite pore structure, with pore sizes which are neither too large, as in nonzeolitic supports, nor too small, as in the 8 or 10 ring pore zeolites, that increase the proximity among the hydrocarbon molecules, and (ii) the Brønsted acidity. The sample B4, which possesses a higher SiO_2/Al_2O_3 molar ratio than the B3 catalyst, provides lower aromatization. In contrast, the C_1–C_6 paraffinic and olefinic selectivity of the B4 catalyst is higher than that of B3, due to the increased hydrogenolysis caused by the Pt loading, which for the B4 sample is 1.0 wt.-%. It appears that 0.5 wt.-% Pt provides sufficient dehydrogenation capability, despite the relatively low Pt dispersion, while keeping the hydrogenolysis of the naphthene feed at a low level.

SiO_2 supported Pt (catalyst Si) with no zeolite content and a relatively high Pt loading was also employed. The C_1–C_5 alkanes and alkenes and $C_{\geq 6}$ aromatic selectivities are low due to the absence of acidity. In contrast, the selectivities for the C_6 isomers are high due to the ring opening of the naphthene feed over the Pt sites. The equilibrium among the isomers is shifted towards the methylpentanes,

Table 3
Product distribution and conversion over B1, B2, B3, B4 and Si catalysts at $T=450°C$ and 790 kPa total pressure (2 h-on-stream, WHSV = 2.0 h^{-1}, H_2/oil = 12.0 (molar ratio) for both naphthenes)

Products (mol-%)	Methylcyclopentane					Ethylcyclopentane				
	B1	B2	B3	B4	Si	B1	B2	B3	B4	Si
Methane	–	–	0.1	2.1	–	3.5	1.6	0.7	1.3	1.2
Ethane	0.8	1.5	1.3	3.9	0.4	2.8	8.1	1.4	2.2	1.8
Propane	21.9	17.0	16.4	23.9	2.2	21.5	29.6	11.4	13.9	2.1
i-Butane	10.7	12.3	–	5.5	1.0	13.9	13.2	10.0	10.1	1.1
2-Butene	2.4	0.3	–	–	–	–	–	–	–	–
n-Butane	10.0	16.2	9.9	8.3	2.0	13.8	15.4	8.1	9.5	2.7
i-Pentane	7.1	8.5	6.6	5.3	2.7	4.5	1.8	2.0	2.3	1.4
n-Pentane	4.8	5.3	4.1	3.5	4.2	3.0	1.1	1.2	1.5	2.6
2-Pentene	2.9	1.0	0.7	–	0.8	–	–	–	–	0.5
2-Methylpentane	6.8	1.5	1.1	6.1	28.1	–	–	–	0.5	2.2
3-Methylpentane	2.9	0.9	0.7	4.0	24.0	–	–	–	0.3	2.0
n-Hexane	–	–	–	2.2	17.8	–	–	0.1	0.8	1.9
Benzene	16.6	25.8	52.4	28.0	14.8	5.0	4.4	9.4	10.9	2.1
Cyclohexane	1.3	–	–	–	–	–	–	–	–	–
2-Methylhexane	–	–	–	–	–	–	–	–	–	8.8
3-Methylhexane	–	–	–	–	–	–	–	–	–	26.0
Heptane	–	–	–	–	–	–	–	–	–	20.0
Toluene	6.9	7.9	3.8	5.8	0.9	21.9	12.5	50.4	43.5	21.5
Ethylbenzene	1.3	0.4	0.4	0.2	–	–	1.9	0.6	0.7	–
o-Xylene	1.0	0.2	0.3	0.2	–	2.1	2.7	0.9	0.3	–
m-Xylene	0.4	0.3	0.5	0.3	–	1.8	1.4	0.4	0.2	–
p-Xylene	2.0	1.0	0.9	0.9	–	5.9	6.1	2.7	2.1	–
Conversion (mol-%)	74.5	93.1	97.2	96.9	93.0	100.0	100.0	100.0	100.0	100.00

more than as expected from the statistical distribution (n-hexane:2-MP:3-MP=2:2:1) among the isomers. A similar behavior was noted earlier by Gault [24] during methylcyclopentane hydrogenolysis at relatively low Pt dispersions. The molar ratio of 3-methylpentane to n-hexane is larger than unity, an indication that selective hydrogenolysis to methylpentanes occurs [24]. Gault suggested that hydrogenolysis occurs for Pt particles larger than 25 Å. Our TEM investigations demonstrated that for the present Pt/SiO_2 catalyst the metal particle is larger than the above limiting size.

3.3.2. Ethylcyclopentane

The reforming of ethylcyclopentane follows the reaction pathway described for methylcyclopentane. For ethylcyclopentane, the C_7 cycloolefinic intermediates generated over the Pt sites are isomerized enlarged over the Brønsted acid sites to a C_6 cyclic ring intermediate, which finally leads to toluene after its dehydrogenation over the Pt sites. From Table 3 one can observe that toluene is the main aromatic molecule generated. A pronounced maximum of the toluene yield is observed in Fig. 2 for ethylcyclopentane between the B3 and B4 catalyst. As for methylcyclopentane this maximum is a result of the balancing of the Brønsted acidity with the amount of Pt dispersed.

For catalyst B2, the selectivities for other aromatics than toluene are higher than over the Bl, B3 and B4 catalysts. This is due to the bimolecular alkylation reactions and to the partial transalkylation of toluene to xylenes and benzene, reactions which are enhanced by the higher Brønsted acidity and suitable Pt loading of B2. For the range of temperatures employed in the present study the toluene yield passes through a broad maximum located between the B3 and B4 catalysts. In contrast, the previous naphthene had a maximum dehydroisomerization for the B3 catalysts. Hence, the optimum balancing of the two catalytic functions is affected in some measure by the number of carbons of the naphthene feed. A similar behavior has been observed by Martens et al. [25] in the isomerization of long-chain alkanes over bifunctional Pt/zeolites; they found that an optimum loading of the noble metal exists which depends on the chain length of the alkane. As shown in Table 3 the $C_{\leq 6}$ paraffinic and olefinic selectivities for both naphthene feeds over B4 are higher than those over B3 despite the fact that the SiO_2/Al_2O_3 molar ratio of the former catalyst is higher. This occurs, because at relatively high metal loadings the hydrogenolysis becomes important.

The conversion of ethylcyclopentane is higher than that of methylcyclopentane under identical conditions (Table 3). This is in accordance with the classical bifunctional mechanism, and is a result of the increased reactivity of the larger cycloalkenes over the acidic sites [20].

The increase of the reaction temperature favors the dehydroisomerization of both naphthenes, leading to higher benzene and toluene yields, respectively (Fig. 2).

3.4. *Kinetics of alkylcyclopentanes conversion*

As already mentioned, the rate controlling step for the reaction of alkylcyclopentanes is the isomerization of alkylcycloalkenes over the Brønsted acidic sites. Hence, the reaction rate of the naphthene is given by the following expression:

$$r = k_s \, \theta_{A_{ol}} \tag{1}$$

where $\theta_{A_{ol}}$ is the fraction of the acidic sites covered by the feed hydrocarbon and k_s is the surface rate constant. Assuming a Langmuir equilibrium isotherm for the adsorption–desorption of the olefinic intermediates on the Brønsted sites, the coverage of the acidic sites $\theta_{A_{ol}}$ is given by

$$\theta_{A_{ol}} = \frac{K \, C_{A_{ol}}}{1 + K \, C_{A_{ol}}} \tag{2}$$

where $C_{A_{ol}}$ is the molar concentration of alkylcyclopentenes in the gas phase. Since the dehydrogenation of the naphthene feeds is rapid, it can be considered to take place at equilibrium [21]. Consequently, one can connect $A_{A_{ol}}$ to C_A, the concentration of the hydrocarbon feed, through the relation

Table 4
Values of the parameters of Eq. (4) for methylcyclopentane and ethylcyclopentane over B1 and B3 catalysts at 450°C and 790 kPa (1 h-on-stream)

Catalysts	$K_s \times 10^5$ (moles/g cat. s)	K_A (cm^3/mole)	Goodness of fit (%)
Methylcyclopentane			
B1	8.25	6383.8	99.7
B3	4.41	11466	99.9
Ethylcyclopentane			
B1	4.88	77286.8	98.3
B3	1.06	39843.3	99.0

$$K_e = \frac{C_{A_{ol}} C_{H2}}{C_A} \tag{3}$$

From Eqs. (1)–(3) one can obtain the following reaction rate of alkylcyclopentanes in terms of the gas concentration C_A:

$$r = \frac{k_s K_A C_A}{1 + K_A C_A} \tag{4}$$

where K_A is the apparent adsorption constant given by the relation:

$$K_A = \frac{K\, K_e}{C_{H_2}} \tag{5}$$

Typical examples of the fitting of the experimental data to Eq. (4) were carried out using a least square analysis. Since the H_2 partial pressure is much higher than those of the feeds, it was considered constant. The parameters of Eq. (4) for both naphthenes over the B1 and B3 catalysts are listed in Table 4, which shows that the goodness of fit is above 99%. One can conclude that our experiments are compatible with a mechanism in which the isomerization enlargement constitutes the rate controlling step.

3.5. Coke formation

Table 5 shows that the catalysts have different coking capabilities for the two feeds, the amount of coke deposited being higher for methylcyclopentane than for ethylcyclopentane. Parera et al. [26] already noted that under identical reforming conditions the coke deposited on the catalysts was higher with cyclopentane than with methylcyclopentane. The reason for this behavior, is the reduction of the polymerization rate of the feed naphthenes (via Diels–Alder reactions) leading to coke, with increasing size of the alkyl group. Larger alkyl substituents provide higher steric hindrance and therefore the alkylcarbenium intermediates located on the acidic sites cannot react (polymerize) with the gas phase cycloalkenes to

Table 5
Coke deposited (g/100 g cat.) over the catalysts at 450°C and 790 kPa total pressure (WHSV = 2.0 h^{-1}, for both naphthenes H_2/oil = 12.0 (molar ratio))

Duration (h)	Methylcyclopentane					Ethylcyclopentane				
	B1	B2	B3	B4	Si	B1	B2	B3	B4	Si
3	10.8	8.1	0.5	0.8	1.0	2.6	1.0	0.2	0.5	1.1
72	13.4	nd[a]	5.6	nd[a]	1.3	5.5	nd[a]	0.6	nd[a]	1.2

[a] nd stands for not determined.

generate coke (polyaromatic and saturated rings). The presence of the alkyl group has also a negative effect on the dehydrogenation of the feed naphthene. Parera et al. [26] have indeed shown that the extent of dehydrogenation of cyclopentane is higher than that of methylcyclopentane under identical conditions.

3.6. Analysis of the soluble coke by ^{1}H-NMR, FTIR and GC/MS

^{1}H-NMR indicates that after 72 hours-on-stream the aromaticity of the soluble coke, when methylcyclopentane is used as feed, decreases with increasing dealumination of the β zeolite, since (Fig. 3) the percentage of aromatic protons (H_{AR}, 7 ppm $\leq \delta \leq$ 9 ppm) detected decreases. In contrast, the percentage of protons that belong to saturated cycles (H_{AL}, 1.5 ppm $\leq \delta \leq$ 1.7 ppm) increases. This indicates that for the B1 catalyst (the sample with the highest acidity) the aromatic coke is preferably deposited on the acidic sites; for the Si catalyst the coke is located on the Pt sites. This is in agreement with the H/C ratios of the coked samples (Table 6) determined by elemental analysis. The H/C ratio of the catalysts, which corresponds to the total coke (soluble and nonsoluble), increases with increasing SiO_2/Al_2O_3 molar ratio and Pt loading. Higher H/C values are observed for methylcyclopentane than ethylcyclopentane indicating that the coke formed with

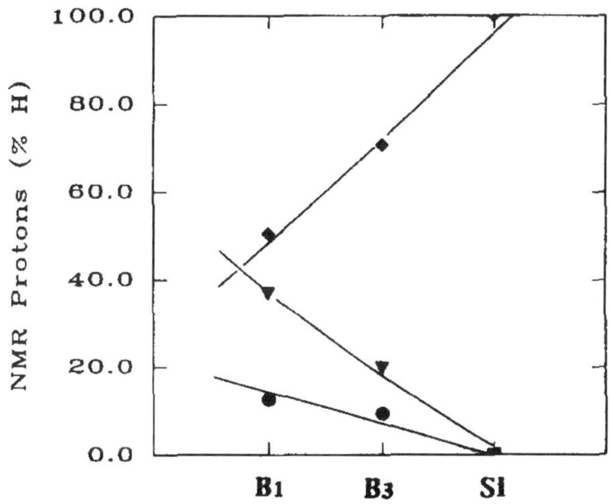

Fig. 3. Change of the percentages of aromatic protons (H_{AR}, ●), of alkylaromatic protons (H_{AA}, ▼) and of aliphatic protons (H_{AL}, ◆) as a function of the SiO_2/Al_2O_3 molar ratio with methylcyclopentane at 450°C after 72 hours-on-stream (790 kPa).

Table 6
Hydrogen-to-carbon atomic ratio (H/C) of coked samples for methylcyclopentane and ethylcyclopentane over B1, B3 and Si catalysts under various operating conditions at 450°C determined by elemental analysis (QTI Labs., NJ)

H/C values for the Si catalyst are unrealistic and are due to excessive moisture adsorbed on the samples

Catalyst	Time-on-stream (h)	Pressure (kPa)	H/C
Methylcyclopentane			
B1	36	101	4.17
B1	36	1270	0.99
B1	3	790	1.67
B1	72	790	1.51
B3	72	790	3.56
Si	72	790	51.9
Ethylcyclopentane			
B1	72	790	0.80
B3	72	790	0.90
Si	72	790	43.80

the latter feed is more aromatic than with the former. The present behavior agrees with the observations of Barbier [27], who found that the coke deposited on Pt is less dehydrogenated than that deposited on chlorided Pt/Al_2O_3. The percentage of alkylaromatic protons (H_{AA}, 2 ppm $\leq \delta \leq$ 2.5 ppm) decreases from sample B1 through Si.

For methylcyclopentane, the H_{AA}/H_{AR} ratio is about 2 for the B1 and B3 catalysts, indicating that the soluble coke is appreciably branched. The presence of dibranched and tribranched naphthalenes and of monobranched anthracene in the soluble coke was confirmed by GC/MS investigations.

The above observation is in agreement with the infrared studies (Fig. 4), since the peaks at 2900 and 3000 cm^{-1}, which correspond to alkane and alkene groups [28], respectively, are lower for the B1 catalyst than for the Si sample. This indicates that the hydrogen content of the coke of the SiO_2 sample is higher than that of the B1 catalyst. In addition, the peaks in the range of 1250 to 1000 cm^{-1}, which correspond to aromatics are more intense for the B1 catalyst than for the Si sample.

3.7. *Effect of pressure*

The effect of hydrogen pressure on the aromatic yield for methylcyclopentane over the B1 and the B3 catalysts is presented in Fig. 5, for various times-on-stream. For both catalysts there is a maximum of the aromatic yield at some intermediary pressure. For relatively low pressures, the aromatic yield is small because of the excessive coke deposition on the catalyst surface. At moderate pressures, the increase of the H_2 pressure increases the methylcyclopentane aromatization, because the rate of saturation of deficient hydrocarbon residues (coke precursors)

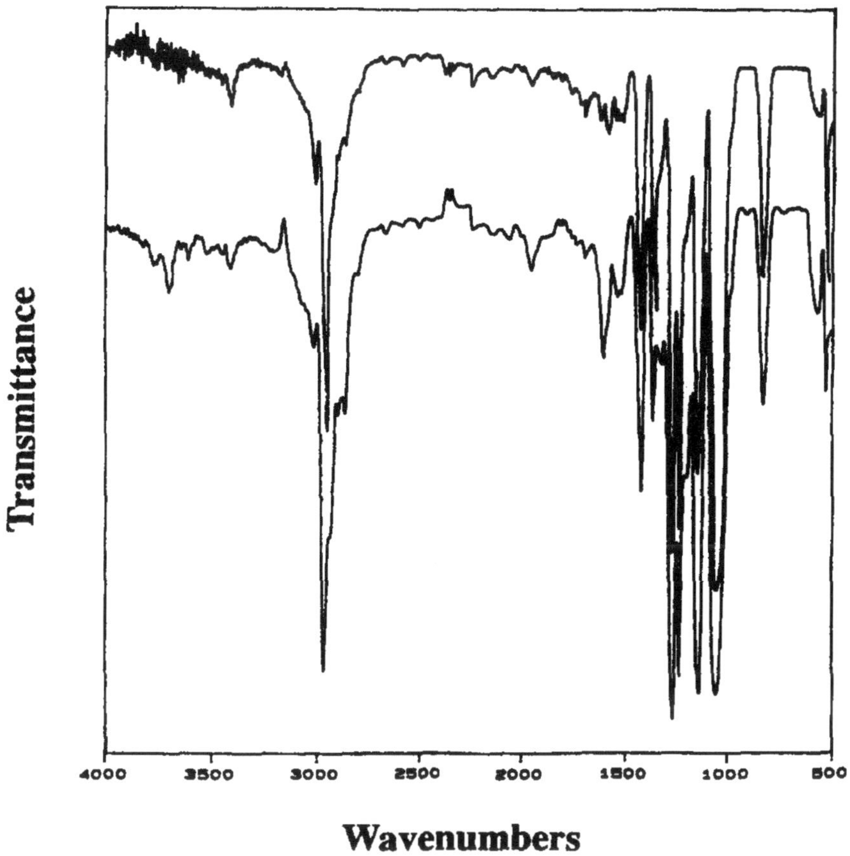

Fig. 4. Infrared spectra of the soluble coke over the B1 catalyst (top) and the Si catalyst (bottom) with methylcyclopentane as feed at 450°C (790 kPa total pressure, 72 hours-on-stream, WHSV = 2.0 h^{-1}, H_2/oil = 12.0 (molar ratio)).

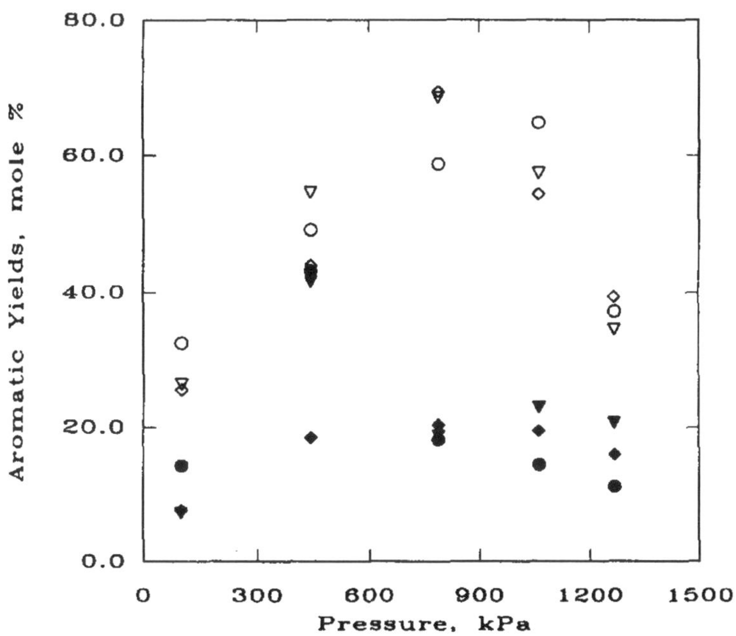

Fig. 5. Aromatic yield as a function of pressure over B1 (filled symbols) and over B3 (open symbols) for methylcyclopentane for various times-on-stream at 450°C (WHSV = 2.0 h^{-1}). (●, ○) 2 h, (▼, ▽) 10 h, (◆, ◇) 36 h.

Table 7
Product distribution and conversion over B1 and B3 catalysts with methylcyclopentane at $T=450°C$ and various pressures (10 h-on-stream, $WHSV=2.0\ h^{-1}$)

Product	B1 (pressures (kPa))			B3 (pressures (kPa))		
	101	790	1270	101	790	1270
Methane	1.5	0.5	4.1	0.5	0.2	4.4
Ethane	0.9	1.7	2.7	0.8	0.4	4.1
Propene	2.6	–	0.3	–	–	–
Propane	5.1	19.8	41.1	1.9	5.9	28.3
i-Butane	3.3	13.7	9.5	1.0	3.4	4.2
n-Butane	1.2	11.4	13.3	1.1	3.1	6.3
i-Pentane	1.2	4.9	4.3	2.6	3.0	5.8
n-Pentane	0.5	3.9	2.8	1.7	1.9	3.8
2-Methylpentane	–	–	–	0.3	2.3	3.7
3-Methylpentane	–	6.1	–	6.3	1.3	2.4
Methylcyclopentenes	39.0	–	–	18.1	–	–
Benzene	31.9	20.3	9.8	53.6	69.4	25.9
Cyclohexane	3.8	5.8	–	1.2	–	–
Toluene	5.0	7.1	6.6	4.5	5.6	2.5
Ethylbenzene	0.5	1.0	1.4	0.7	0.6	0.6
o-Xylene	0.7	0.8	1.2	1.3	0.7	3.6
m-Xylene	0.5	0.5	0.7	0.4	0.2	0.4
p-Xylene	3.1	2.5	2.2	3.4	2.1	4.3
Conversion (mol-%)	17.8	39.4	99.3	41.2	98.1	95.0

located on both kind of sites increases, thus preserving their activity [29]. A maximum of the aromatic yield occurs for both B1 and B3 catalysts at about 400 and 800 kPa, respectively. At even higher hydrogen pressures, the aromatic yield decreases because the dehydrogenation of methylcyclopentane is hindered. Hydrocracking over the Brønsted sites and hydrogenolysis over the Pt sites (both kind of sites are free of coke because of the high hydrogen pressures employed) of the feed naphthene constitute the main reactions. Indeed, as shown in Table 7, at high pressures the selectivities for aromatics decrease while those for the C_1–C_6 alkanes and alkenes increase for both catalysts with pressure.

The maximum in the aromatic generation is a result of the maximization of the methylcyclopentane dehydroisomerization. The hydrogen pressure regulates the extent of methylcyclopentane dehydrogenation equilibrium, as well as the coke deposition which alters the availability of both kind of sites of the catalyst. The aromatic selectivity over the B3 catalyst is higher than that over the B1 at any operating pressure because of the better balance of the two functions of the catalyst.

The increase of the pressure results in a decrease of the H/C atomic ratio of the coke (Table 6), indicating that the coke becomes more aromatic. A similar trend has been observed by Barbier [27] over Pt/Al_2O_3. This observation combined with the fact that the coke over the Pt sites is more hydrogenated (compare samples B1 and Si with both naphthenes after 72 hours-on-stream (Table 6)), indicates that

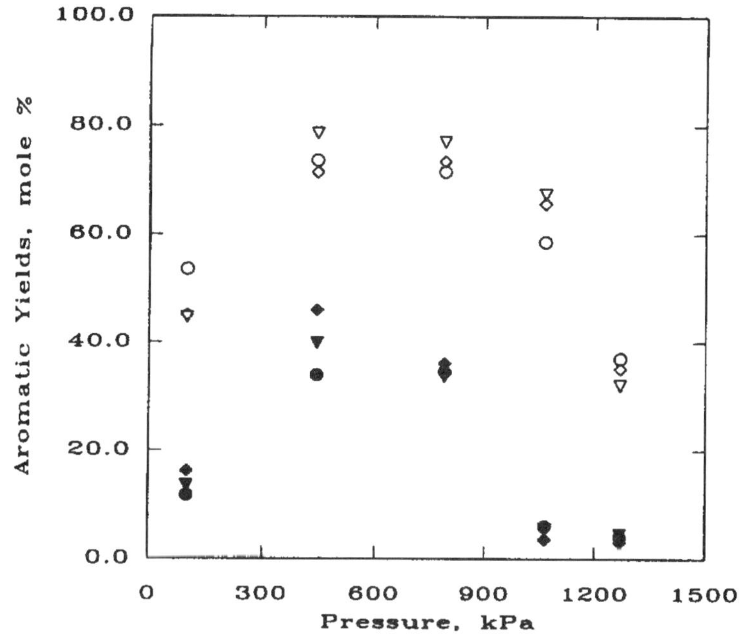

Fig. 6. Aromatic yield as a function of total pressure over B1 (filled symbols) and over B3 (open symbols) for ethylcyclopentane for various times-on-stream at 450°C (WHSV = 2.0 h^{-1}). (●, ○) 2 h, (▼, ▽) 10 h, (◆, ◇) 36 h.

with increasing pressure there is a shift in the location of the coke to the acidic sites.

The effect of pressure on the aromatic yield for ethylcyclopentane is presented in Fig. 6. A behavior similar to that of the previous feed is observed with maxima for aromatization located at about 400 and 800 kPa, respectively, for the catalysts B1 and B3. The aromatic yield for ethylcyclopentane is higher than that for methylcyclopentane, because methylcyclopentane is a better coke precursor (see also Table 8). A detailed product distribution for ethylcyclopentane at various pressures is given in Table 9.

The amount of coke deposited on catalysts B1 and B3 after 36 hours-on-stream as a function of pressure is listed in Table 8. For the nondealuminated catalyst, the amount of coke deposited is high even at the highest pressure. In contrast, for the dealuminated catalyst, there is a steep decrease of coke weight with increasing H_2

Table 8
Amount of coke (g/100 g cat.) deposited on B1 and B3 catalysts with methylcyclopentane and ethylcyclopentane at 450°C at various H_2 pressures (for both naphthenes WHSV = 2.0 h^{-1}, 36 h-on-stream)

Total pressure (kPa)	Methylcyclopentane		Ethylcyclopentane	
	B1	B3	B1	B3
101	18.07	12.50	9.82	4.22
445	14.82	5.96	5.67	1.52
790	11.85	2.73	2.96	0.35
1063	9.12	1.09	0.55	0.27
1270	7.33	0.34	0.17	0.19

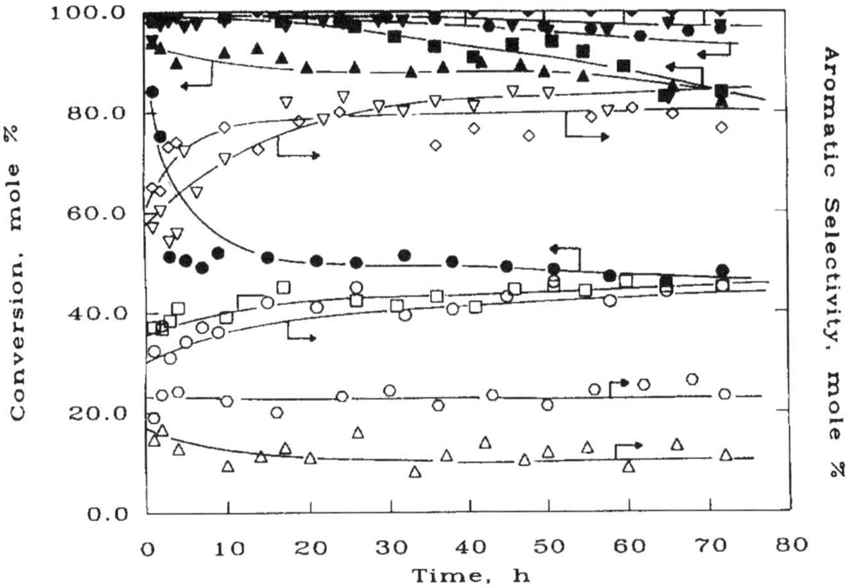

Fig. 7. Conversion over B1 (●), B3 (▼) and Si (▲) for methylcyclopentane and over B1 (■), B3 (◆) and Si (⬣) for ethylcyclopentane, and aromatic selectivity over B1 (○), B2 (▽) and (△) for methylcyclopentane and over B1 (□), B3 (◇) and Si (⬡) for ethylcyclopentane as a function of time-on-stream at 450°C and 790 kPa total pressure (WHSV = 2.0 h^{-1}, H_2/oil molar ratio = 12.0).

pressure. It is worthwhile noting that the operation at atmospheric pressure even with the B3 catalyst results in a large coke deposition, indicating the important role of increased hydrogen pressures in avoiding the aging of the catalysts by coking. The significant difference between the coking of B1 and B3 catalysts especially at high pressures, indicates that coke is deposited preferably on the Brønsted sites, since the amount of coke deposited on the B3 catalyst, which has a smaller number of Brønsted sites, is smaller.

3.8. Time-on-stream behavior

The effect of the time-on-stream on the conversion and aromatic selectivity for methylcyclopentane at 790 kPa is presented in Fig. 7. The conversion over the nondealuminated Pt/β zeolite (B1 sample) decreases abruptly within the first minutes-on-stream due to coking. For times on-stream larger than 4 hours, the conversion remains essentially unchanged at about 50%. The aromatic selectivity remains unchanged as a function of time-on-stream and is relatively low, due to the high rate of hydrocracking favored by the high acidity and the high H_2 pressure. In contrast to the B1 catalyst, over the dealuminated B3 catalyst the conversion remains unchanged even after 3 days-on-stream. The aromatic selectivity is high, approximately twice that over the B1 catalyst, and increases with time, reaching a plateau after about 15 hours. This exceptional performance is a result of the moderate amount of Pt and moderate acidity of the B3 catalyst (0.5 wt.-% Pt and acidity corresponding to $SiO_2/Al_2O_3 = 130$) which minimize hydrogenolysis and hydrocracking. The suitable pressure keeps the catalyst surface almost free of coke. The small amount of coke deposited on the Brønsted sites of B3 during the initial stages

Table 9
Product distribution and conversion over B1 and B3 catalysts with ethylcyclopentane at $T = 450°C$ and various pressures (10 h-on-stream, $WHSV = 2.0\ h^{-1}$)

Product	B1 (pressures (kPa))			B3 (pressures (kPa))		
	101	790	1270	101	790	1270
Methane	0.3	1.5	0.4	0.8	0.5	0.4
Ethane	0.3	4.1	2.8	0.9	1.2	–
Propene	4.4	–	–	–	–	–
Propane	13.3	25.8	44.6	14.8	7.0	23.6
i-Butane	14.7	14.0	22.1	18.1	6.0	26.7
n-Butane	8.9	16.0	19.6	7.6	4.9	15.1
i-Pentane	10.6	3.0	2.2	3.6	1.0	0.1
n-Pentane	6.2	2.2	2.0	1.5	0.7	0.3
2-Methylpentane	–	–	0.2	0.2	–	–
3-Methylpentane	–	–	–	–	–	–
Benzene	2.7	4.6	0.3	2.6	6.8	2.1
Cyclohexane	–	–	–	–	–	–
Toluene	28.1	18.1	4.1	44.4	55.3	31.6
Ethylbenzene	0.7	0.6	0.1	0.3	1.1	–
o-Xylene	0.7	2.2	0.1	0.4	2.3	0.1
m-Xylene	1.7	2.3	0.3	1.4	2.9	0.1
p-Xylene	7.2	5.4	0.2	3.4	10.1	–
Conversion (mol-%)	33.2	100.0	100.0	85.1	100.0	100.0

of reaction of methylcyclopentane appears to enhance the aromatic selectivity, probably because of the selective poisoning of the strong cracking sites. The conversion over Pt/SiO_2 remains practically unchanged at a relatively high level, but the aromatic selectivity is low.

With ethylcyclopentane (Fig. 7), the catalysts exhibit, with regard to selectivity, behaviors similar to those of the previous naphthene. The activities of B3 (dealuminated β zeolite) and SiO_2 remain at high levels as a function of time-on-stream. In contrast to methylcyclopentane, with ethylcyclopentane the activity of the B1 catalyst starts decreasing only after about 30 hours-on-stream. Of course, the conversion of ethylcyclopentane over the B1 catalyst after 3 days-on-stream is higher than that of methylcyclopentane, since the former naphthene generates less coke.

As shown in Table 5, the amount of coke deposited at various times over the catalysts is less for ethylcyclopentane than for methylcyclopentane, the difference being more pronounced for the most acidic catalyst (B1). After 3 hours-on-stream, the B3 catalyst had for both feed naphthenes a significantly lower amount of coke than any other catalyst. The catalysts B4 and Si, which are less acidic than B3 but have higher Pt loadings, have somewhat higher coke depositions, indicating that coke is formed over the Pt sites as well. For methylcyclopentane, the minimum coke deposit is displaced from B3 at 3 hours to Si catalyst at 72 hours, while for ethylcyclopentane, for which the amount of coke deposition is lower, the minimum remains over the entire duration at B3.

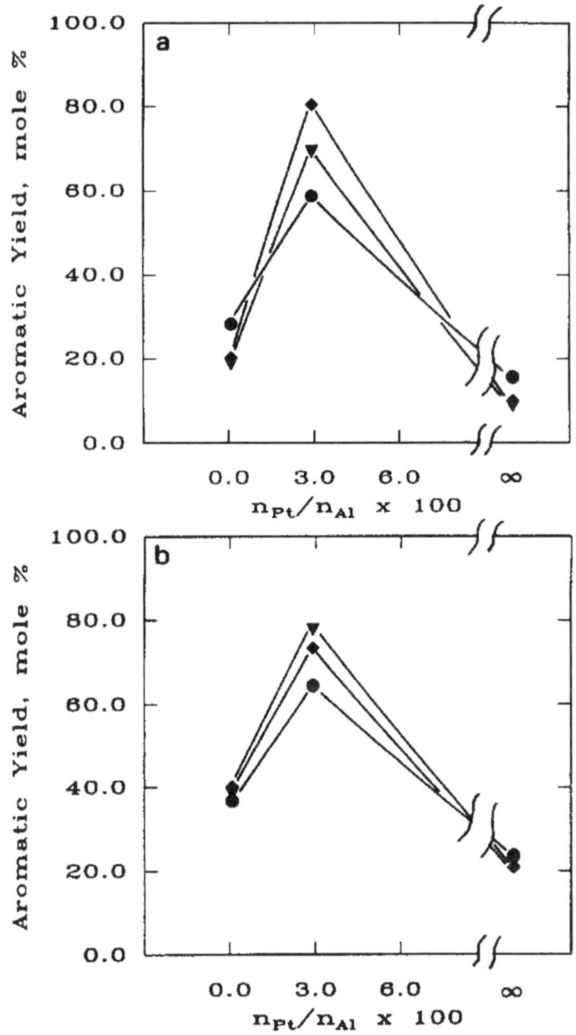

Fig. 8. (a) Aromatic yields over B1, B3 and Si at 480°C and 790 kPa total pressure with methylcyclopentane at various times-on-stream (WHSV = 2.0 h^{-1}, H_2/oil = 12.0 (molar ratio)). (b) Aromatic yields over B1, B3 and Si at 480°C and 790 kPa total pressure with ethylcyclopentane at various times-on-stream (WHSV = 2.0 h^{-1}, H_2/oil = 12.0 (molar ratio)). (●) 2 h, (▼) 10 h, (◆) 36 h.

Table 10
Product selectivities and conversion for methylcyclopentane and ethylcyclopentane at 450°C for low levels of conversion under 790 kPa total pressure 3 h-on-stream, WHSV = 2.0 h^{-1}, H_2/oil = 12.0 (molar ratio))

Catalysts	Conversion	Benzene selectivity	C_1–C_6 paraffinic and olefinic selectivity
Methylcyclopentane			
B1	32.4	28.6	44.7
B3	35.1	68.7	82.4
Si	29.8	23.4	72.8
Ethylcyclopentane			
B1	36.7	30.3	57.3
B3	41.3	59.2	24.9
Si	37.9	27.1	65.1

We have already emphasized the existence of an optimum n_{Pt}/n_{Al} ratio for a catalyst close to B3 at 790 kPa and 2 hours-on-stream, which leads to the maximization of the dehydroisomerization of alkylcyclopentanes. For longer times-on-stream, the optimum yield remains at B3 but with increased values [Figs. 8a and b].

3.9. Effect of the conversion level

A few experiments have been carried out at low and almost constant conversions with both naphthenes. As shown in Table 10, the selectivities for benzene and toluene, with methylcyclopentane and ethylcyclopentane, respectively, pass through maxima for the B3 catalyst, regardless of the conversion.

4. Conclusions

Balancing the Brønsted acidity of the β zeolite with the Pt loading maximizes the dehydroisomerization capability of the catalyst. As a result, the aromatic yield possesses a maximum for a catalyst with a n_{Pt}/n_{Al} ratio of about 3.0×10^{-2} (SiO_2/Al_2O_3 molar ratio of 130 and 0.5 wt.-% Pt). The H_2 pressure plays an important role in the performance of the catalysts, since it regulates the extent of coke build-up and hydrocracking. The increase in pressure results in a relative reduction of the coke deposited on the Pt sites compared to that deposited on the acidic sites. It was found that the size of the alkyl substituent of alkylcyclopentanes affects the coke deposition. At a pressure of about 790 kPa, the above catalyst exhibited excellent time-on-stream behavior, high aromatic selectivity and minimal coke deposition even after many hours-on-stream. The dealuminated β zeolite may be a promising catalyst for the reforming of naphthenes under industrial conditions.

References

[1] H. Schulz, J. Weitkamp and H. Eberth, in Proc. of the 5th Congress on Catalysis, North-Holland, Amsterdam, 1973, Vol. 2, p. 1229.
[2] J. Weitkamp, in Proc. of the Int. Symp. on Zeolite Catalysis, Siofok, Hungary, 1985, Acta Phys. Chem., 31 (1985) 271.
[3] P.G. Smirniotis and E. Ruckenstein, J. Catal., 140 (1993) 526.
[4] P G. Smirniotis and E. Ruckenstein, Chem. Eng. Sci., 48 (1993) 3263.
[5] T R. Hughes, W.C. Buss, P.W. Tamm and R.L. Jacobson, in Y. Murakami, A. Iijima and J.W. Ward, (Editors), New Developments in Zeolite Science and Technology (Studies in Surface Science and Catalysis, Vol. 28), Kodansha–Elsevier, Tokyo–Amsterdam, 1986, p. 725.
[6] P.W. Tamm, D.H. Mohr and C.R. Wilson, in J.W. Ward (Editor), Catalysis 1987 (Studies in Surface Science and Catalysis, Vol. 38), Elsevier, Amsterdam, 1988, p. 335.
[7] R. Parton, L. Uytterhoeven, J.A. Martens, P.A. Jacobs and G.F. Froment, Appl. Catal., 76 (1991) 131.
[8] P.G. Smirniotis, and E. Ruckenstein, Catal. Lett., 25 (1994) 351.
[9] S.B. Liu, J.F. Wu, L.J. Ma, T.C. Tsai and I. Wang, J. Catal., 133 (1991) 432.

[10] R.B. Lapierre, R.D. Partridge, N.Y. Chen and S.S. Wong, US Patent 4 501 926 (Feb. 26, 1985).
[11] H.A. Benesi, J. Am. Chem. Soc., 78 (1956) 5490.
[12] K. Tanabe, in Solid Acids and Bases, Kodansha–Academic Press, Tokyo–New York, 1970.
[13] M. Guisnet and P. Magnoux, Appl. Catal., 54 (1989) 1.
[14] P. Gallezot, A. Alarcon-Diaz, J.-A. Dalmon, A.I. Renouprez and B. Imelik, J. Catal., 79 (1975) 334.
[15] K. Tanabe, M. Misono, Y. Ono and H. Hattori, New Solids Acids and Bases: Their Catalytic Properties (Studies in Surface Science and Catalysis, Vol. 51), Elsevier, Amsterdam, 1989, p. 206.
[16] R.B. Borade and A. Clearfield, J. Phys. Chem., 96 (1992) 6729.
[17] M. Maache, A. Janin, J.C. Lavalley, J.F. Joly and E. Benazzi, Zeolites, 13 (1993) 419.
[18] J.H. Sinfelt, in J.R. Anderson and M. Boudart (Editors) Catalysis, Science and Technology, Vol. 1, Spring-Verlag, New York, 1981, p. 257.
[19] P.G. Smirniotis and E. Ruckenstein, Ind. Eng. Chem. Res., 33 (1994) 493.
[20] A.I.M. Keulemans and H.H. Voge, J. Phys. Chem., 63 (1959) 476.
[21] B.C. Gates, J.R. Katzer and G.C. Schuit, Chemistry of Catalytic Processes, McGraw-Hill, New York, 1979.
[22] P.B. Weisz, Adv. Catal., 13 (1962) 137.
[23] J. Weitkamp, P.A. Jacobs and S. Ernst, in P.A. Jacobs, N.I. Jaeger, P. Jiru, V.B. Kazansky and G. Schulz-Ekloff (Editors), Structure and Reactivity of Modified Zeolites (Studies in Surface Science and Catalysis, Vol. 18), Elsevier, Amsterdam, 1984, p. 279.
[24] F.G. Gault, Adv. Catal., 30 (1981) 1.
[25] J.A. Martens, M. Tielen, and P.A. Jacobs, in H.G. Karge and J. Weitkamp (Editors), Zeolites as Catalysts, Sorbents and Detergent Builders (Studies in Surface Science and Catalysis, Vol. 46) Elsevier, Amsterdam, 1989, p. 49.
[26] J.M. Parera, R.J. Verdelone and C.A. Querini, in B. Delmon and G.F. Froment. (Editors), Catalyst Deactivation (Studies in Surface Science and Catalysis, Vol. 31) Elsevier, Amsterdam, 1987, p. 135.
[27] J. Barbier, in B. Delmon and G.F. Froment (Editors), Catalyst Deactivation 1987 (Studies in Surface Science and Catalysis, Vol. 51), Elsevier, Amsterdam, 1987, p. 1.
[28] J. Colthup, J. Opt. Soc. Am., 40 (1950) 397.
[29] J.H. Sinfelt, and J.C. Rohrer, J. Phys. Chem., 65 (1961) 978.

Increased aromatization in the reforming of mixtures of n-hexane, methylcyclopentane and methylcyclohexane over composites of Pt/BaKL zeolite with Pt/β or Pt/USY zeolites

Panagiotis G. Smirniotis, Eli Ruckenstein *

Department of Chemical Engineering, State University of New York at Buffalo, Buffalo, NY 14260, USA

Received 9 May 1994; revised 19 October 1994; accepted 19 October 1994

Abstract

The performance of the individual L, β and USY zeolites supported platinum and of composites of Pt/BaKL with either Pt/β or Pt/USY zeolite is examined for the reforming reactions of n-hexane, methylcyclopentane, or methylcyclohexane and their mixtures. The reactions of different mixtures of the above three hydrocarbons over the composites result in increased selectivities for $C_{\geqslant 7}$ aromatics compared to those calculated as molar averages by multiplying the selectivities corresponding to the individually reacting feed hydrocarbons with their molar fractions in the feed mixture. This behavior is attributed to the synergism caused by the components of the composites in combination with the components of the feed mixture. The balancing of the L zeolite with an acidic zeolite via composites can provide a selective route to C_8, C_9 aromatics for industrial feeds with large contents of $C_{\geqslant 6}$ paraffins.

Keywords: Acidic zeolites; Aromatization; Basic zeolites; Beta zeolite; Composite basic and acidic zeolites; Hydrocarbon mixtures; Reforming; USY zeolite; Zeolites

1. Introduction

Over the last decades catalytic reforming has been one of the most important petroleum processes. The conventional catalysts employed in reforming are platinum, or a combination of platinum with rhenium or iridium, supported on Al_2O_3 [1,2], which as noted by Mills et al. [3] are bifunctional catalysts. Later, the cooperation between the metallic and the acidic functions of bifunctional catalysts

* Corresponding author. Tel. (+ 1-716) 6452911, fax. (+ 1-716) 6453822.

This paper, which was originally published in **Applied Catalysis A**: General 123, 59-88 (1995), is republished with permission.

was demonstrated in the isomerization of saturated hydrocarbons [4] by employing a catalyst containing Pt/SiO_2 and SiO_2–Al_2O_3.

Metal supported zeolites have been also considered for reforming reactions. NaY and HNaY supported platinum or PtRe have been tested in the reforming of n-heptane [5] while twelve ring pore zeolites, such as ZSM-3, ZSM-20, β, Y or USY supported platinum, have been employed in the reforming of long-chain n-paraffins [6]. With any of the above zeolites and feed paraffins a maximum isomerization yield can be attained by the proper balancing of the two catalyst functions. The investigation of the reforming of naphthenes over zeolites [7–10] has shown that they enhance the bimolecular alkylation reactions, which lead to aromatics with carbon numbers different from that of the feed naphthene, and the dehydroisomerization of methylcyclopentane to benzene. Recently, we examined the reforming of n-hexane, methylcyclopentane or cyclohexane over Pt/β zeolite under low severity conditions [11] and the results compared with those obtained over Pt/γ-Al_2O_3. In experiments carried out between 200 and 400°C it was found that the zeolite β stimulates to a much higher extent than γ-Al_2O_3 the skeletal isomerization reactions of n-hexane and the production of benzene by the enlargement of the C_5-cyclic ring of methylcyclopentane to a C_6-cyclic ring. In contrast, Pt/γ-Al_2O_3 enhances to a greater extent than Pt/β zeolite the direct dehydrogenation of cyclohexane to benzene. The performance of composite pellets and powders of Pt/γ-Al_2O_3 and Pt/β zeolite in the reforming of methylcyclopentane or methylcyclohexane [12] has also been examined. It was demonstrated that a suitable balancing of the metal function with the Brønsted acidity results in the maximization of methylcyclopentane dehydroisomerization to benzene.

Bernard [13] reported that Pt/KL zeolite is very selective in the n-hexane aromatization at 733° K and atmospheric pressure. Besoukhanova et al. [14] observed that the reforming activity of L zeolite in the n-hexane dehydrocyclization increases with its basicity. Chevron researchers noted [15–17] that Pt/BaKL exhibits a beneficial behavior in the selective conversion of light naphthas to aromatics.

This study was carried out to examine the reforming reactions over composites containing a highly selective basic dehydrocyclization zeolite (L zeolite) and a twelve ring pore acidic zeolite (such as β or USY). The latter two acidic zeolites have been selected because our preliminary experiments have shown that among the β, ZSM-5, USY and Y zeolites, the β zeolite and USY provide the most beneficial behavior in reforming. The reforming of n-hexane, methylcyclopentane or methylcyclohexane and their mixtures over the above composites was investigated.

The main observation is that the selectivity for $C_{\geqslant 7}$ aromatics, when mixtures of the above three hydrocarbons react over the composites, is higher than that calculated as a molar average by multiplying the selectivities corresponding to the individually reacting hydrocarbons with their molar fractions in the feed stream. As explained later in detail, this behavior is a result of complex cooperations among the components of the catalyst and of the feed mixtures.

2. Experimental

2.1. Catalyst preparation

The zeolite β has been synthesized from an aluminosilicate gel with a nominal molar composition of 30 SiO_2:Al_2O_3:0.9 Na_2O:3 TEAOH:250 H_2O, containing tetraethylammonium hydroxide (TEAOH) as template. The gel was prepared by mixing individual solutions of: colloidal silica (Ludox HS-40, Du Pont) containing 40 wt.-% SiO_2 and 0.41 wt.-% Na_2O; tetraethylammonium hydroxide (Aldrich 35 wt.-% sol. in water); sodium aluminate (Fisher Scientific); and sodium chloride (Fisher Scientific). After vigorous stirring for 45 min, the resultant gel was loaded in a Teflon lined stainless-steel autoclave and kept at 160°C without stirring under autogenous pressure. After a crystallization period of ten days, the generated zeolite β was filtered and dried overnight at 140°C. X-ray diffraction (XRD) investigations have shown that the prepared zeolite β was highly crystalline. The SiO_2/Al_2O_3 molar ratio of the prepared β zeolite, determined by atomic absorption spectroscopy (AAS) was 28.

The KL zeolite was supplied by UOP (Lot No 13923-02) and had a SiO_2/Al_2O_3 molar ratio of 6.0, Na_2O = 0.16 wt.-%, K_2O = 16.01 wt.-% and a surface area $S_g = 353$ m^2/g. USY was provided by W.R. Grace (SMR 6-2811-0992) and had the composition: SiO_2 = 71 wt.-%, Al_2O_3 = 24.28 wt.-% and Na_2O = 4.47 wt.-%. Its surface area was $S_g = 712$ m^2/g and its unit cell size = 24.58 Å.

The protonated form of β zeolite was obtained as follows: The TEAOH-β zeolite was heated at 510°C for 5 h in ultra high purity oxygen (99.999%). Under these calcination conditions the burning of the template employed was surely achieved [18], and the crystallinity of the zeolite preserved [19]. This was followed by the grinding and sieving (150 mesh size sieve; particle size below 100 μm) of the zeolite β. Zeolites β and USY were cation exchanged under reflux with 4 *M* NH_4Cl (Aldrich) at 95°C for 3 h. In all cation exchanges, 2 g of zeolite was used in 1 l solution. The remaining Na content of the zeolite in the form of Na_2O was 0.17 and 0.34 wt.-% for zeolite β and USY, respectively. The zeolites were transformed to their acidic forms by calcination at 450°C for 1 h in oxygen. The KL zeolite was exchanged with 0.3 *M* $Ba(NO_3)_2$ for 3 h at 95°C under reflux. A calcination in air at 600°C for 16 h followed. The elemental analysis (AAS) has shown that the amount of K_2O was reduced to 6.47 wt.-% from 16.01 wt.-%, while that of BaO became 12.61 wt.-%.

Wet impregnation with H_2PtCl_6 (Aldrich 8 wt.-% solution in water) was employed for both acidic zeolites. For the preparation of Pt/BaKL, wet impregnation with an aqueous solution of platinum(II) tetraammine nitrate (Alfa No 88960) was employed. Chloroplatinic acid has not been employed in the case of L zeolite, since it is a source of Cl^- which generates acidity in the catalyst. After impregnation, all catalysts were dried overnight at 130°C. Composite catalysts with various weight fractions of acidic zeolite were prepared after grinding, sieving the zeolites with a 150 mesh size sieve, and drying them at 200°C for 2 h. A thorough mixing for 2 h with a shaker mixer (Turbula) of the composite powders followed. The weight fraction of the acidic zeolite in the composites is denoted by ϵ; ϵ equal to zero corresponds to Pt/BaKL zeolite while and equal to unity corresponds to the acidic zeolite. The homogeneity of the composites was tested by XRD.

2.2. *Transmission electron microscopy (TEM)*

The effect of the heat treatment on the platinum particles size during the activation step of the catalyst was examined with a transmission electron microscope (JEOL 100-CX II). A small amount of each sample was thoroughly ground and dispersed ultrasonically in acetone for 1 h. A drop of the suspension was loaded and then evaporated on a carbon covered copper grid. The samples were examined at magnifications ranging from 20,000 to 200,000× with an accelerating voltage of 80 KV. The mean size of the platinum crystallites (d_s) was estimated with the following expression.

$$d_s = \frac{\sum n_i d_i^3}{\sum n_i d_i^2}$$

where n_i is the number of counted particles in the range $d_i \pm 5$ Å. Several micrographs (about 10 for each zeolite) have been used to obtain the values of n_i.

2.3. *Hydrogen chemisorption*

The platinum dispersion of the catalyst has been determined by hydrogen chemisorption at room temperature. After the oxidation and reduction of the catalyst in various conditions, 1 g of dry catalyst was loaded in a pyrex-glass U shape sample holder. The holder was attached to a vacuum glass system equipped with a pressure transducer. To remove the effect of the possible partial oxidation of the catalyst during its transferring to the holder, a reduction at 200°C for 1 h was first carried out. For cleaning, the samples were heated at 400°C for 1 h under vacuum (10^{-5} Torr, 1 Torr = 133.3 Pa). Under the same vacuum the temperature of the samples was brought to the ambient temperature. High purity hydrogen was employed at pressures between 10 to 400 Torr for the determination of the total amount of hydrogen, both weakly and strongly adsorbed. After this step, the samples were evacuated at 10^{-5} Torr and ambient temperature for the removal of the weakly adsorbed hydrogen. A second adsorption isotherm was determined at ambient temperature in the same pressure range as the first one. The difference between the two adsorption isotherms after extrapolation to zero pressure provides the amount of strongly chemisorbed hydrogen. The platinum dispersion was estimated assuming dissociative adsorption of hydrogen on the metal. The amount of platinum exposed is determined mainly by the metal present inside the L zeolite channels as small particles or atomically dispersed. The remaining of the platinum loading is located outside the zeolite crystals as large crystallites.

2.4. Cl^- determination

In order to account for a possible contribution of the Cl^- introduced during impregnation, the Cl^- concentration in the acidic zeolites was measured after different heat treatments. The determination was carried out potentiometrically with a selective ion chloride probe (Orion Research EA 940). Before analysis, both Pt/β and Pt/USY zeolites were dissolved in HNO_3 with vigorous stirring for two days.

2.5. Catalytic tests

The catalytic experiments were performed in a vertical plug flow stainless steel reactor of 0.60 cm I.D. and 40 cm long. The reactor was heated in a temperature-programmable controlled furnace. Most of the catalytic experiments were carried out at 5 min on stream, in order to obtain the initial product selectivity free of deactivation, but also at various times-on-stream. In all runs, 100 ± 2 mg of fresh catalyst was loaded on glass wool located in the middle of the reactor. Prior to reaction, the catalysts were activated in situ at 300°C by flowing high purity oxygen (99.999%) for 1 h at 5 ± 0.5 ml/min. A purging of the catalyst with a flow-rate of about 100 ml/min of high purity helium (99.995%) for 15 min followed. The reduction of the catalyst was carried out at 500°C with a high purity hydrogen (99.999%) flow-rate of 5 ± 0.5 ml/min for 1 h. All changes in the bed temperature were carried out at a rate of 15°C/min. The experiments were carried out at 480°C and either atmospheric or 100 psig total pressure. The reactants employed in the present study were n-hexane (Aldrich, 99+ wt.-%), methylcyclopentane (Aldrich, 98 wt.-%) and methylcyclohexane (Aldrich, 99 wt.-%). All feeds were injected into the reactor with a programmable liquid infusion pump through a heated line at the appropriate flow-rate.

High resolution gas chromatography (GC) was employed for the on-line analysis of the products. A gas chromatograph (Hewlett-Packard 5890 series II) equipped with a mass spectrometer (Hewlett-Packard 5971 series) was used. A PC was attached to the GC/MS unit for data manipulation. The GC was equipped in series with a fused silica PoraPLOT wide bore capillary column (0.53 mm I.D., 28 m length with a film thickness of 20 μm), a molecular sieve packed column (5A), and a high performance capillary column (HP-1 cross-linked Methyl Silicone, 0.2 mm I.D., 12 m length with a film thickness of 0.33 μm). The reactor effluent stream was sampled to the GC through a heating line kept at 140°C in order to avoid the condensation of the products. The dead volume of the heating line was included in the calculations of the time-on-stream.

The comparison among the catalysts was made at comparable conversions and, therefore, the weight hourly space velocity (WHSV) has to vary with the fraction of the acidic zeolite in the composite.

3. Results and discussion

3.1. TEM and hydrogen chemisorption

It was suggested [14] that there are four types of platinum particles in the KL zeolite: large (100–600 Å) particles outside the zeolite channels, crystallites of 10–25 Å size inside the channels, small metallic cylinders outside the channels, and very small particles in the cavities. A representative TEM micrograph of Pt/BaKL is presented in Fig. 1a where one can identify platinum particles on the external zeolite surface with an average diameter of $d_s = 57$ Å. The number density of the platinum crystallites in L zeolite is lower than in the acidic zeolites. The oxidation and reduction under more severe conditions, namely 550°C for 1 h each, resulted in larger platinum particles. The platinum dispersion in BaKL zeolite determined by hydrogen chemisorption is given in Table 1. The values determined are somewhat lower than those reported in the literature [20], which are in the range of 0.9 to 1.0, probably because of the higher reduction temperature employed in the present study. Another reason for the lower platinum dispersions observed is probably the preparation method employed (impregnation), in contrast to the cation exchange method employed in the above references.

Larger platinum crystallite sizes were observed on the external surface of the zeolite particles in the acidic zeolites (Fig. 1b). The average particle diameter was 77 and 81 Å for zeolite β and USY, respectively. The platinum particle size distribution presented in Fig. 2 shows that the distribution in the acidic zeolites is broader and shifted to larger sizes than in the L zeolite. These observations are in agreement with the hydrogen chemisorption results listed in Table 1, which show that the platinum dispersions in the acidic zeolites are lower than in the L zeolite. More severe oxidation and reduction conditions have as a result the generation of larger platinum particles and a further decrease in dispersion (see Table 1). It is worthwhile to note that the metal dispersion in L zeolite is higher than that in the acidic zeolites even under severe heat treatment. This behavior can be attributed to the different platinum compounds employed in the impregnation of the basic and acidic zeolites. Since tetraammine Pt^{2+} cations were used in the preparation of Pt/BaKL, The Pt^{2+} cations were ion exchanged with K^+ and/or Ba^{2+} thus giving highly dispersed Pt^{2+} ions. Upon reduction, good platinum dispersion was therefore achieved, though, as indicated by TEM data, an appreciable amount of relatively large platinum particles was present on the external surface of the L zeolite. A different mechanism for the platinum particles formation occurred in the acidic zeolites, where H_2PtCl_6 was employed for impregnation. In this case, $PtCl_6^{2-}$ anions

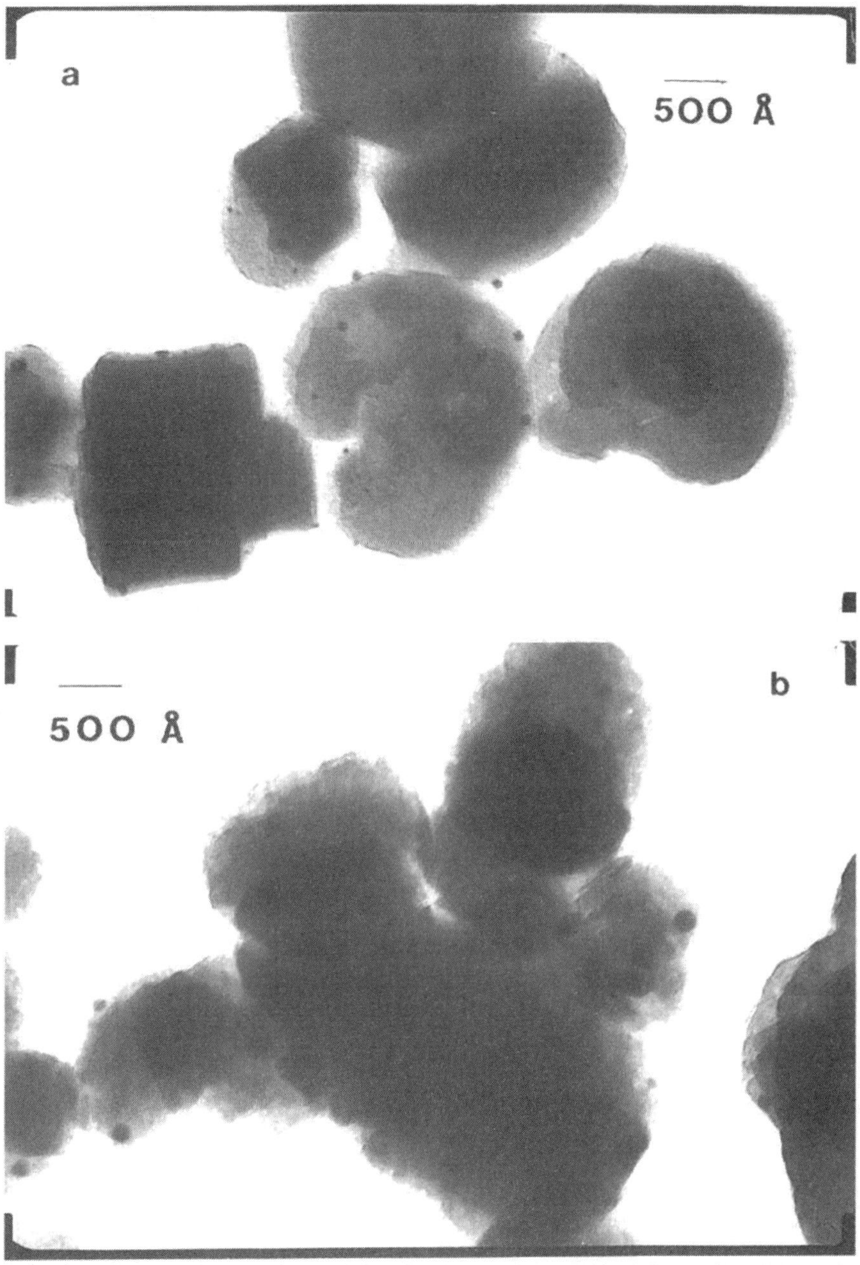

Fig. 1. Transmission electron micrographs of (a) Pt/BaKL zeolite and (b) Pt/β zeolite. All zeolites were loaded with 0.5 wt.-% Pt, were oxidized at 300°C for 1 h and reduced at 500°C for 1 h.

cannot be ion exchanged with the protons and/or the Na^+ cations, and platinum deposition is due to pore filling and formation of salt platinum particles when the water is removed. These particles decompose to relatively large platinum oxide or oxychloride particles upon calcination in air and generate large platinum particles upon reduction.

Preparation of the catalysts by ion exchange would probably result in a higher platinum dispersion into the zeolite pores providing more metallic sites close to the acidic sites. Indeed, as shown earlier [9] the loading of platinum in HY faujasite by ion exchange resulted in higher dispersions, namely about 90%, under similar activation conditions.

Table 1
Platinum dispersion D ($=$H/Pt) of Pt/BaKL, Pt/β and Pt/USY zeolites

Catalysts	T_0 (°C)	T_R (°C)[a]	D=H/Pt
Pt/BaKL	300	500	0.67
Pt/β	300	500	0.15
Pt/USY	300	500	0.14
Pt/BaKL	550	550	0.60
Pt/β	550	550	0.06
Pt/USY	550	550	0.04

For all catalysts the Pt loading was 0.5 wt.-%. The oxidation and the reduction steps lasted 1 h each.
[a] T_0 = oxidation temperature, T_R = reduction temperature.

3.2. Cl^- concentration

It is well known that the presence of chloride in alumina based reforming catalysts increases their acidity and consequently their isomerization capabilities [21]. The Cl^- concentrations of both acidic zeolites under different heat treatments are given in Table 2. The amount of Cl^- introduced via H_2PtCl_6 during impregnation is between 0.35 and 0.42 wt.-%, while the limits for bifunctional reforming catalysts employed in industry based on alumina are between 0.3 and 1.0 wt.-% [22]. The heat treatment of the catalyst decreases the content of chloride, the relative decrease being larger for the β zeolite. We found that the relative Cl^- concentration decrease is higher for both acidic zeolites than for γ-Al_2O_3. The isomerization capability of the acidic catalysts employed here is not diminished by the decrease in the Cl^- content during catalyst preparation, since, in contrast to the traditionally used alumina, both zeolite β and USY have relatively high Brønsted acidities.

3.3. Catalytic behavior

The feed molecules employed in the present study, n-hexane, methylcyclopentane and methylcyclohexane are typical representatives of the hydrocarbons present in virgin or cracked naphthas. Each of these hydrocarbons exhibits a different reforming behavior. The n-hexane or more generally the C_6–C_8 paraffins, which constitute a relatively large fraction of the light naphthas [23], can lead to aromatics via dehydrocyclization. For methylcyclopentane the enlargement of the C_5 ring to a C_6 ring, which is a precursor for aromatics, is the crucial step for its reforming. Methylcyclohexane represents the naphthenes that already possess a C_6 cyclic ring and therefore their direct dehydrogenation leads to aromatics.

Most of the catalytic measurements were carried out at 5 min on stream in order to obtain initial product selectivities free of deactivation. However, measurements at longer times-on-stream have been also performed. The experiments have been carried out at constant conversion, namely for about 50 or 10% conversions. Of course, the residence time had to be varied with the composition of the composites.

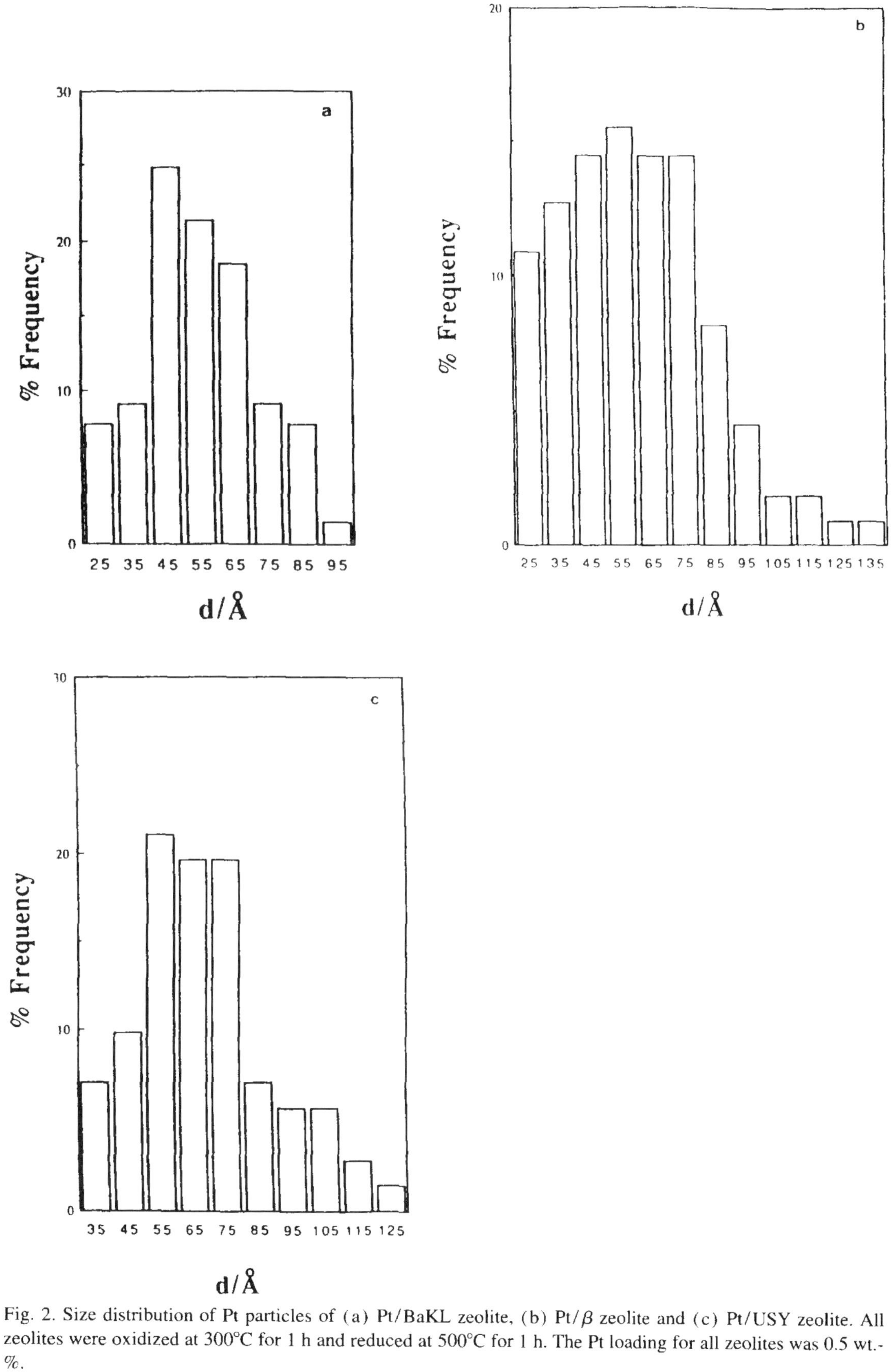

Fig. 2. Size distribution of Pt particles of (a) Pt/BaKL zeolite, (b) Pt/β zeolite and (c) Pt/USY zeolite. All zeolites were oxidized at 300°C for 1 h and reduced at 500°C for 1 h. The Pt loading for all zeolites was 0.5 wt.-%.

Table 2
Chloride concentration in Pt/β zeolite and Pt/USY under different activation heat treatments (Pt loading 0.5 wt.-%)

Catalysts		T_O (°C)[a]	T_R (°C)	Cl^- (mg/g cat.)
	calculated[b]			5.40
β-zeolite	as impregnated			3.53
USY-zeolite	as impregnated			4.19
β-zeolite		300	500	1.08
USY-zeolite		300	500	1.92
β-zeolite		550	550	0.95
USY-zeolite		550	550	1.87

The oxidation and the reduction steps lasted 1 h each, while a helium purging of the catalysts for 15 min has been employed between them.
[a] T_O = oxidation temperature, T_R = reduction temperature.
[b] The calculated Cl^- concentration was obtained for 0.5 wt.-% Pt loading with H_2PtCl_6.

Experiments have been also performed (but are not included in the paper) at constant residence time for the composites. In the latter cases, the conversion increased with the fraction of acidic zeolite employed. The trends of the selectivities have been the same in both cases. For this reason we include in the paper only the results at constant conversion. The conversion X and the selectivities S_i of the lumped species i are defined as:

$$X = \left(1 - \frac{\text{moles of the reactant hydrocarbons unconverted}}{\text{moles of the reactant hydrocarbons fed into the reactor}}\right) \times 100$$

$$S_i = \left(\frac{\text{moles of } i \text{ species produced}}{\text{moles of all products}}\right) \times 100$$

n-Hexane

The product distribution for n-hexane over Pt/β, Pt/USY, Pt/BaKL and their composites containing equal weights of acidic and L zeolite is presented in Table 3. Hydrogenolysis and cracking take place to a small extent over the acidic zeolites and composites rich in acidic zeolites, since the C_1/C_5 and C_2/C_4 molar ratios are much smaller than unity. One can note that propane is formed with the highest selectivity among the $C_{<6}$ hydrocarbons. This can be attributed to the 'center cracking' over the acidic sites via a carbenium ion mechanism of C_6 olefins generated over the platinum sites [24]. C_6 isomers of n-hexane are not detected, indicating that, under the present operating conditions, the skeletal isomerization over the acidic sites of the C_6 olefinic intermediates (produced over the platinum sites) does not take place to a measurable extent. Different results were obtained at the lower temperatures of 200–400°C [11], where over Pt/β zeolite the propane selectivity was smaller while those of C_4, C_5 and C_6 isomers higher. The selectivity for C_1–C_6 paraffins and olefins over Pt/β zeolite is slightly higher than that over Pt/USY. In general, one can note that both acidic zeolites behave in a similar way,

Table 3
Product distribution and conversion with n-hexane as feed ($T=480°C$, 5 min on stream, atmospheric pressure, 0.5 wt.-% Pt loading, H_2/Oil = 4.2 mol/mol)

Products (mol-%)	Pt/BaKL	Pt/β-Pt/L	Pt/USY-Pt/L	Pt/β	Pt/USY
WHSV (h^{-1})	2.0	2.0	2.6	2.6	2.6
Methane	0.4	1.1	0.4	1.4	1.3
Ethane	0.8	7.5	8.2	5.9	4.0
Propene	–	6.8	5.6	4.9	4.6
Propane	2.5	40.8	39.9	44.6	52.6
i-Butane	0.8	9.1	11.1	10.2	3.3
2-Butene	–	1.0	0.8	0.8	0.3
n-Butane	3.7	7.9	9.1	9.7	4.2
i-Pentane	1.9	–	2.1	0.2	–
n-Pentane	7.9	3.3	2.9	0.5	0.4
2-Methylpentane	23.0	–	0.2	–	–
3-Methylpentane	16.0	–	0.4	–	–
Benzene	37.8	12.4	10.1	11.5	16.3
Toluene	1.9	5.1	4.4	4.2	7.4
o-Xylene	0.4	0.5	0.4	0.7	0.3
m-Xylene	0.4	1.2	1.1	1.1	1.4
p-Xylene	0.8	2.3	1.9	3.1	2.2
1-Ethyl-4-methylbenzene	0.2	0.3	0.8	0.7	0.9
1,3,5-Trimethylbenzene	0.6	0.6	0.5	0.6	1.1
Conversion (mol-%)	48.1	51.4	49.1	44.7	45.1

The composites Pt/β-Pt/L and Pt/USY-Pt/L contain equal weights of L zeolite with either β or USY zeolite. The values of WHSV have been selected in order to compare the catalysts at comparable conversions.

probably because they possess similar intracrystalline steric constraints, since their spaciousness indexes (SI = i-butane/n-butane) are very similar [25].

From the aromatic product distribution over the acidic zeolites (Table 3), one can note a relatively large amount of $C_{>6}$ aromatics. Benzene is produced via cyclization, over the acidic sites, of the olefinic intermediates formed over the platinum sites by the dehydrogenation of n-hexane [26]. The formation of $C_{>6}$ aromatics occurs in the following two ways: (i) The alkylation of C_6 cycloolefinic intermediates (formed by the dehydrogenation and cyclization of the hexane molecule) with carbonium ions (formed over the Brønsted sites from $C_{<6}$ paraffins generated by cracking). These reactions lead to $C_{>6}$ intermediates, which via disproportionation form $C_{>6}$ aromatics and lighter paraffins. (ii) The electrophilic attack on the benzene π-electrons by the carbenium ions, produced from $C_{<6}$ olefins over the Brønsted sites, which generate alkylbenzenium ions. The disproportionation of the latter unstable molecules results in $C_{>6}$ aromatics and lighter olefins [27].

As expected, because of its basic nature, zeolite L exhibits a lower C_1–C_6 paraffinic and olefinic selectivity than the acidic zeolites. In contrast to the acidic zeolites, the selectivity for C_6 isomers over L zeolite is much higher, since the metal function of the Pt/BaKL zeolite stimulates the methyl shifts [28] and hence enhances the skeletal rearrangements of n-hexane. Benzene is the main aromatic compound generated over the L zeolite, with much higher selectivity than over the acidic zeolites. The generation of benzene from n-hexane over the monofunctional Pt/BaKL catalyst occurs via a mechanism [29,30] that differs from that over bifunctional catalysts. These authors consider that the physical interactions between n-hexane and the active environment generated by platinum in the super cages of L zeolite cause the 1–6 ring closure. This cyclization is promoted by the migration, during calcination at high temperatures, of Ba^{2+} cations present in the open positions of the channels to the locked exchanged sites in the cages [31]. The modeling of the various steps of the n-hexane cyclization over Pt/BaKL by Derouane and Vanderveken [29] has considered that the migration of Ba^{2+} cations favors: (i) the diffusion of n-hexane through the open zeolite L channels which are free of bulky cations and Brønsted acid sites, and (ii) the orientation for desorption (through the locked cations in the cages) of the generated benzene. The selectivity for $C_{>6}$ aromatics is very small, because, in contrast to the acidic zeolites, (i) zeolite BaKL cannot generate the light paraffins and olefins needed for the alkylation reactions of the C_6 ring, which can lead via disproportionation to the enrichment in carbon atoms, and (ii) it lacks the Brønsted acidity needed for alkylation reactions. Lately, Iglesia and Baumgartner [32] suggested that the one-dimensional channels of zeolite L inhibit the bimolecular transition state intermediates involved in the polymerization–coking reactions. They also found that highly dealuminated HY zeolites (Si/Al $>$ 100) provide initially slightly higher n-heptane dehydrocyclization turnover rates than the L zeolite, but are less effective due to coke formation. Pt/SiO_2 is an effective catalyst for n-heptane aromatization as well, but it provides very small resistance to coking due to its 'open' structure. It seems that the unique properties of L zeolite in the reforming of paraffins are a result of the transport restrictions that its structure imposes.

The dependence of the $C_{\geqslant 6}$ aromatic and C_1–C_6 hydrocarbon selectivities on the weight fraction ϵ of the acidic component of the composite is presented in Fig. 3. One can observe a monotonic decrease of the $C_{\geqslant 6}$ aromatic selectivity and a monotonic increase of the C_1–C_6 paraffinic and olefinic selectivity with increasing ϵ. However, while the total aromatic selectivity decreases, the $C_{\geqslant 7}$ selectivity increases due to bimolecular alkylation reactions. The product selectivities vary in a similar way for both types of composites.

Methylcyclopentane

Zeolite L provides relatively high benzene and low paraffinic and olefinic selectivities in comparison with the acidic zeolites (Table 4). In contrast, the $C_{\geqslant 7}$ aromatic selectivities over L zeolite are lower than over the acidic ones. The L zeolite and the dealuminated Y faujasite have similar structures; channels of about 7.2 Å connecting widened channel cavities or supercages of 12.4 Å, respectively [28]. However, only the acidic zeolites enhance to a significant extent the bimolecular alkylation reactions, because the Brønsted acidity is involved in the for-

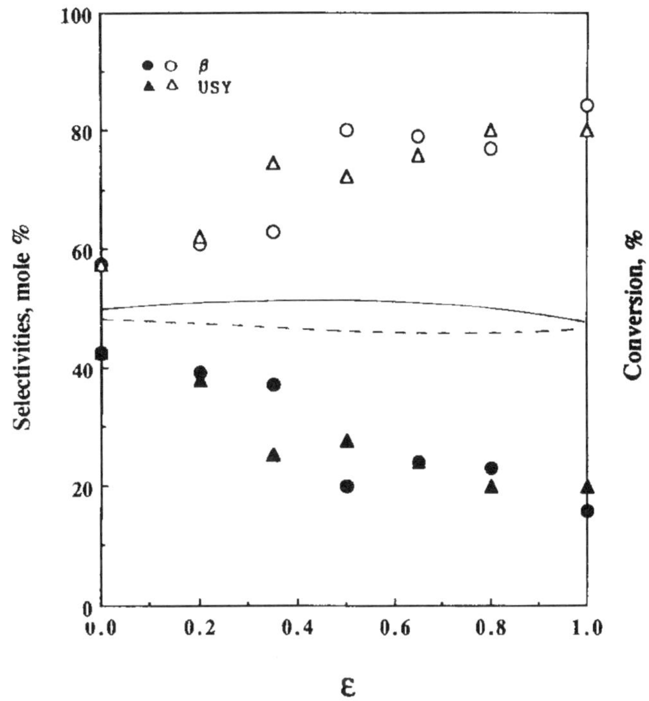

Fig. 3. Selectivities for aromatics (filled symbols) and C_1–C_6 paraffins and olefins (open symbols) at 480°C for various composites of Pt/β zeolite with Pt/BaKL or Pt/USY with Pt/BaKL for n-hexane as feed (5 min on stream, 0.5 wt.-% Pt loading, atmospheric total pressure, H_2/Oil = 4.2 mol/mol, WHSV range = 2.0–2.6 h^{-1}). The continuous line corresponds to the conversion over the composites containing β zeolite while the dashed line corresponds to the conversion over the composites containing USY. In order to compare the catalysts at comparable conversions, the WHSV had to be varied. A similar comment is valid in all the figures in which a range of values of WHSV is given.

mation of carbenium or carbonium ion intermediates. The suitable pore sizes of the 12 ring pore acidic zeolites contribute to increased collisions between the latter ions and aromatics, thus stimulating the bimolecular alkylation reactions. It was shown earlier [11] that the 'open' structure of the acidic γ-Al_2O_3 does not lead to bimolecular alkylation reactions. Hence, the combination of acidity and zeolite pore structure is responsible for this beneficial behavior. As noted earlier [17], methylcyclopentane cannot form a six-membered ring directly over the L zeolite. It must first undergo a ring opening to n-hexane, which is transformed to benzene via the cyclization mechanism described before. The high benzene selectivity over the L zeolite (Table 4) involves probably the above mechanism. The first column of Table 4 shows unexpectedly that, over the basic L zeolite, small amounts of $C_{\geqslant 7}$ aromatics are generated, even though their formation must involve acidic sites. As will be discussed more fully in the section regarding the methylcyclohexane, protons are formed via the reduction of Pt^{2+} by H_2; they become associated to the O^{2-} of the zeolite matrix, thus generating some acidic sites.

The benzene selectivity decreases with increasing amount of acidic zeolite in the composites, and the C_1–C_6 paraffinic and olefinic selectivities increase moderately (Table 4). This occurs because the hydrocracking and the bimolecular alkylation increase with increasing Brønsted acidity. The selectivity for $C_{\geqslant 7}$ aromatics increases with increasing amount of acidic zeolite because of the bimolecular

Table 4
Product distribution and conversion with methylcyclopentane ($T = 480°C$, 5 min on stream, atmospheric pressure, 0.5 wt.-% Pt loading, H_2/Oil = 4.4 mol/mol)

Products (mol-%)	Pt/BaKL	Pt/β-Pt/L	Pt/USY-Pt/L	Pt/β	Pt/USY
WHSV (h^{-1})	1.8	2.2	1.8	3.0	3.0
Methane	0.9	0.2	1.3	0.7	0.4
Ethane	0.7	0.4	0.8	2.5	2.2
Propene	0.2	0.3	0.5	0.2	–
Propane	2.4	10.1	1.6	13.4	6.9
i-Butane	0.4	5.7	0.5	8.9	7.3
2-Butene	0.4	1.3	0.5	0.2	0.9
n-Butane	1.1	1.7	0.8	3.2	1.9
i-Pentane	1.1	2.0	0.8	0.3	1.1
n-Pentane	3.1	1.3	1.0	0.5	1.4
2-Pentene	0.7	1.7	0.2	–	–
n-Hexane	1.8	–	–	–	–
Methylcyclopentenes	26.9	4.9	18.3	1.4	5.0
Benzene	49.7	36.1	46.9	23.2	34.7
Cyclohexane	1.1	0.3	1.0	–	–
Toluene	4.3	16.1	7.0	22.6	21.9
Ethylbenzene	0.4	1.4	1.0	2.4	1.7
o-Xylene	0.9	2.1	2.3	2.9	2.0
m-Xylene	1.1	2.7	2.0	9.2	2.4
p-Xylene	1.6	7.1	4.9	6.7	7.1
1-Ethyl-4-methylbenzene	–	3.6	1.5	0.9	2.3
1-Ethyl-3-methylbenzene	–	1.4	1.0	0.7	2.1
1,3,5-Trimethylbenzene	–	–	2.6	0.2	1.9
Conversion (mol-%)	45.0	49.1	53.1	51.4	57.0

The composites Pt/β-Pt/L and Pt/USY-Pt/L contain equal weights of L zeolite with either β or USY zeolite. The values of WHSV have been selected in order to compare the catalysts at comparable conversions.

alkylation reactions followed by disproportionation. The fact that the C_1/C_5 and C_2/C_4 ratios are very small for all zeolites indicates that hydrocracking and hydrogenolysis occur to low extent. Table 4 also shows that no C_6 alkanes isomers are generated as final products over the acidic zeolites, indicating that the carbocations generated over the acidic sites via the ring opening of the feed molecules or the protonation of paraffins are employed either in bimolecular alkylation reactions or are further cracked. A small amount of n-hexane is detected over zeolite L. Regarding the methylcyclopentenes generated, one can observe that zeolite L provides the highest selectivity. Since zeolite L is essentially basic it cannot favor the isomerization of the methylcyclopentenes to C_6 cyclic olefins. As expected, the acidic zeolites provide low selectivities for methylcyclopentenes because they do favor the enlargement to a C_6 ring.

The $C_{\geqslant 6}$ aromatic selectivity and C_1–C_6 paraffinic and olefinic selectivity over composites of Pt/β zeolite or Pt/USY with Pt/BaKL are presented in Fig. 4. One can note an increase of the total aromatic selectivity with increasing ϵ, for $\epsilon \leqslant 0.5$, followed by a plateau and possibly by a decrease after $\epsilon \approx 0.8$. The selectivity for

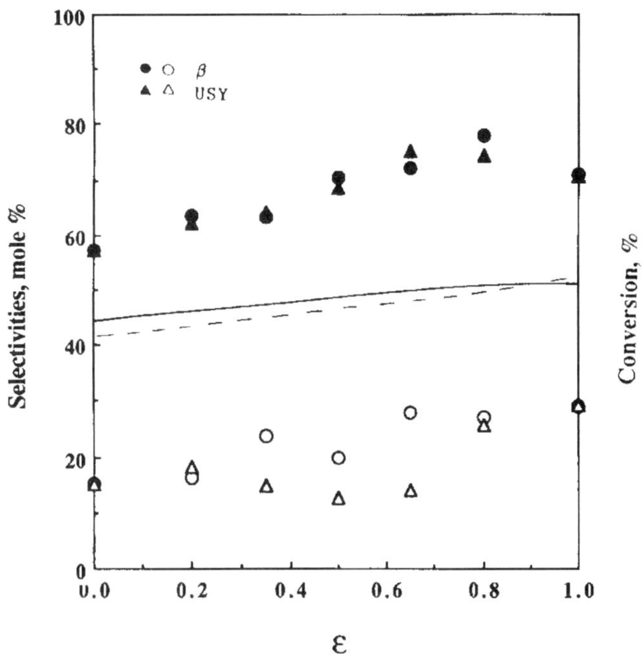

Fig. 4. Selectivities for aromatics (filled symbols) and C_1–C_6 paraffins and olefins (open symbols) at 480°C for various composites of Pt/β zeolite with Pt/BaKL or Pt/USY with Pt/BaKL for methylcyclopentane as feed (5 min on stream, 0.5 wt.-% Pt loading, atmospheric total pressure, H_2/Oil = 4.4 mol/mol, WHSV range = 1.8–3.0 h^{-1}). The continuous line corresponds to the conversion over the composites containing β zeolite while the dashed line corresponds to the conversion over the composites containing USY.

C_1–C_6 paraffins and olefins is in the vicinity of 20% and increases slightly with increasing ϵ. The aromatic selectivities and yields for β and USY zeolites vary in a similar way.

Methylcyclohexane

The product distribution of methylcyclohexane over Pt/BaKL, presented in Table 5, shows that this feed molecule undergoes mainly dehydrogenation to toluene. As noted earlier [33], the dehydrogenation of naphthenes is the fastest among the reforming reactions and, therefore, the methylcyclohexane molecules are predominantly dehydrogenated to toluene. The relatively high platinum dispersion in L zeolite is responsible for the above reaction. The selectivity of C_1–C_6 paraffins and olefins is at a very low level, indicating that hydrogenolysis occurs only to a low extent. It is surprising that some benzene (2.2%) and xylenes are also obtained, because the transalkylation of toluene, which can generate equal molar amounts of benzene and xylenes, as well as the alkylation reactions of toluene with lower paraffins or olefins, which are probably responsible for the excess of xylenes, require acidic sites. However, as previously suggested [34,35], for NaY zeolite loaded with platinum or palladium, the reduction of Pt^{2+} or Pd^{2+} cations by hydrogen can lead to the formation of protons which become associated to the O^{2-} of the matrix. These acidic sites were responsible for the enlargement capability [35] of Pd/NaY zeolite in the aromatization of methylcyclopentane to benzene and are also probably responsible for the formation of small amounts of benzene and xylenes in the present case.

Table 5
Product distribution and conversion with methylcyclohexane as feed ($T = 480°C$, atmospheric pressure, 5 min on stream, 0.5 wt.-% Pt loading, H_2/Oil = 4.4 mol/mol)

Products (mol-%)	Pt/BaKL	Pt/β-Pt/L	Pt/USY-Pt/L	Pt/β	Pt/USY
WHSV (h^{-1})	2.3	7.7	7.7	15.4	15.4
Methane	–	0.3	0.5	0.9	0.3
Ethane	–	1.3	0.4	0.3	1.9
Propane	0.3	3.0	1.2	1.6	5.5
i-Butane	0.2	2.2	0.5	1.9	2.5
n-Butane	0.1	0.8	2.4	2.5	5.2
i-Pentane	0.4	0.9	2.9	3.0	1.6
n-Pentane	0.3	0.8	1.7	4.5	1.9
Methylcyclopentane	2.6	–	–	–	–
Benzene	2.2	13.5	11.1	20.3	22.5
Dimethylcyclopentane	11. 1	–	–	–	–
Toluene	78.0	55.3	65.0	44.9	38.1
Ethylbenzene	0.3	0.3	2.9	2.0	1.8
o-Xylene	0.7	2.6	2.3	2.7	2.5
m-Xylene	1.3	2.1	1.0	1.7	1.9
p-Xylene	2.4	9.1	4.9	9.4	8.5
1-ethyl-4-Methylbenzene	–	4.4	1.0	1.5	1.5
1,3,5-Trimethylbenzene	–	3.2	1.9	2.4	4.3
Conversion (mol-%)	50.2	51.4	53.2	48.1	50.9

The composites Pt/β-Pt/L and Pt/USY-Pt/L contain equal weights of L zeolite with either β or USY zeolite. The values of WHSV have been selected in order to compare the catalysts at comparable conversions.

The product distributions of methylcyclohexane over the acidic zeolites differs from that over the L zeolite. The total selectivity for aromatics remains approximately that over the L zeolite, but that for toluene is reduced to half, because of its partial transformation to benzene and $C_{\geq 8}$ aromatics. The toluene generated by direct dehydrogenation of the feed naphthene is partially transalkylated to equal molar amounts of benzene and C_8 aromatics. Additional alkylation–dealkylation reactions between C_7 and C_8 aromatics [36,37] can explain the excess of benzene with respect to xylenes and the formation of C_9 aromatics. The above reactions are caused by the Brønsted acidity of the USY and β zeolite. The C_1–C_6 paraffinic and olefinic selectivity is at a low level even for the individual β or USY zeolite, indicating that cracking reactions of methylcyclohexane occur only to a small extent. The selectivity for toluene decreases over the composites compared to the selectivity over the L zeolite due to the partial transalkylation stimulated by the acidic zeolites. Instead, the selectivities for benzene and $C_{\geq 8}$ aromatics increase.

The selectivity for C_1–C_6 paraffins and olefins and for total aromatics over the composites of L zeolite with β or USY zeolites are presented in Fig. 5. One can note that the selectivities for both lumped species remain almost unchanged with ϵ. A small decrease of the total aromatic selectivity and a slight increase of the C_1–C_6 paraffinic and olefinic selectivity is observed for composites with $\epsilon \geq 0.5$ due to their increased acidity. Both acidic zeolites behave in a similar way.

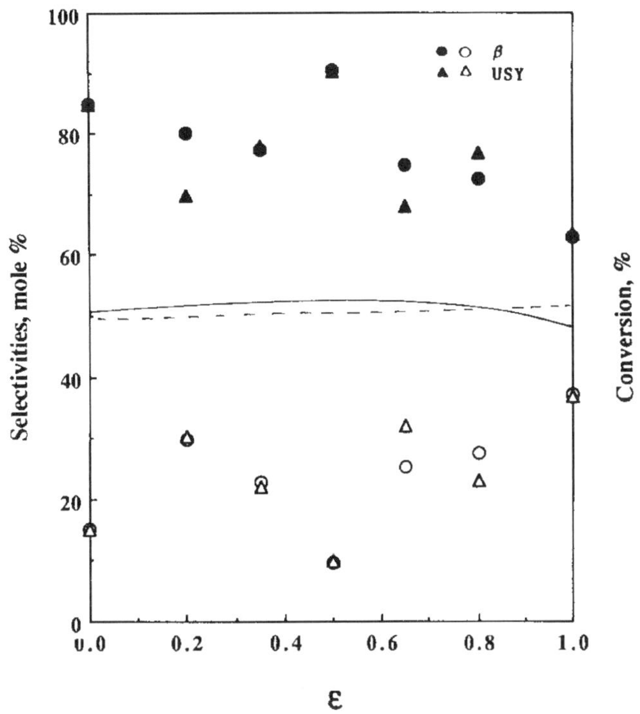

Fig. 5. Selectivities for aromatics (filled symbols) and C_1–C_6 paraffins and olefins (open symbols) at 480°C for various composites of Pt/β zeolite with Pt/BaKL or Pt/USY with Pt/BaKL for Methylcyclohexane as feed (5 min on stream, 0.5 wt.-% Pt loading, atmospheric total pressure, H_2/Oil = 4.4 mol/mol, WHSV range = 2.3–15.4 h^{-1}). The continuous line corresponds to the conversion over the composites containing β zeolite while the dashed line corresponds to the conversion over the composites containing USY.

3.4. Catalytic behavior of the composites with hydrocarbon mixtures

The performance of the composites has been examined with different mixtures of n-hexane, methylcyclopentane and methylcyclohexane in order to investigate any possible interactions among the different feed molecules. Three mixtures have been employed corresponding to real feeds with different features: (a) mixture M1 with molar ratios of n-hex:MCP:MCH = 60:20:20; (b) mixture M2 with equal molar percentages of n-hex:MCP:MCH = 1:1:1 and (c) mixture M3 which has molar ratios of n-hex:MCP:MCH = 20:40:40.

Mixture M1

This feed mixture of hydrocarbons corresponds to a typical paraffinic industrial feed stock. The selectivities of the lumped products (filled symbols) over Pt/β-Pt/L and Pt/USY-Pt/L composites at 480°C are presented in Figs. 6 and 7, respectively. The same quantities, calculated as molar averages by multiplying those corresponding to the individually reacting hydrocarbons with the molar fractions of the components in the feed mixture, are also included in the figures (open symbols). By comparing Figs. 6 and 7 one cannot find any significant difference between β and USY zeolites. The $C_{\geq 6}$ aromatic selectivities over the Pt/β-Pt/L and Pt/USY-Pt/L composites are significantly higher than those calculated, and the C_1–C_6 paraffinic and olefinic selectivities are much lower. Details regarding the

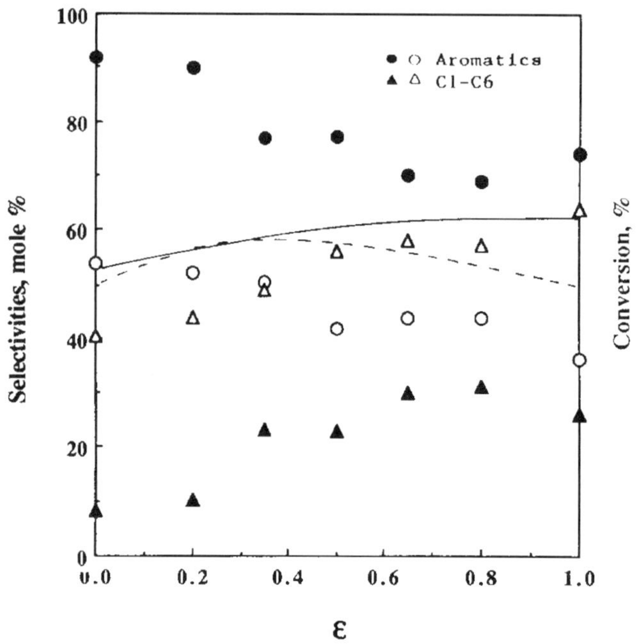

Fig. 6. Selectivities for C_1–C_6 paraffins and olefins (▲) and Aromatics (●) for M1 (filled symbols) over the Pt/β-Pt/L zeolite composites at 480°C. The open symbols stand for the selectivities calculated from the individual hydrocarbon selectivities by weighting them with their molar fractions in the feed (H_2/Oil = 4.2 mol/mol, 5 min on stream, atmospheric pressure, 0.5 wt.-% Pt, WHSV range = 2.0–2.6 h^{-1}). The continuous line corresponds to the conversion measured, while the dashed line corresponds to the conversion calculated as the sum of the individual conversions multiplied with the molar fractions of the components in the feed.

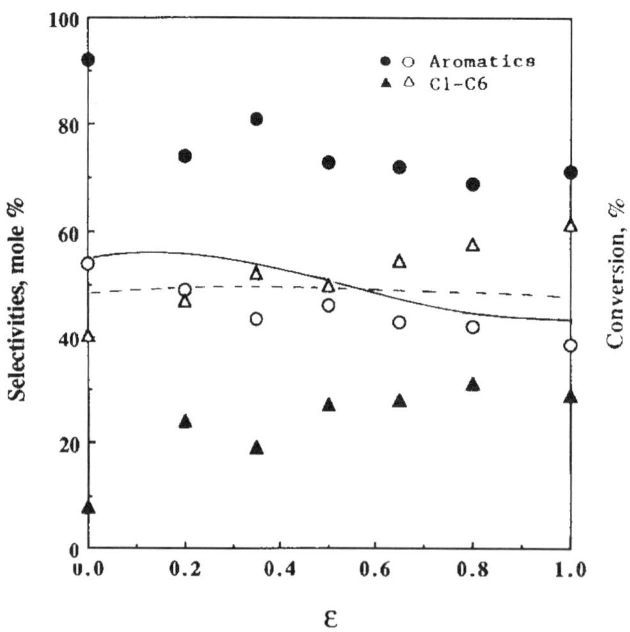

Fig. 7. Selectivities for C_1–C_6 paraffins and olefins (▲) and Aromatics (●) for M1 (filled symbols) over the Pt/USY-Pt/L zeolite composites at 480°C. The open symbols stand for the selectivities calculated from the individual hydrocarbon selectivities by weighting them with their molar fractions in the feed (H_2/Oil = 4.2 mol/mol, 5 min on stream, atmospheric pressure, 0.5 wt.-% Pt, WHSV range = 2.0–2.6 h^{-1}). The continuous line corresponds to the conversion measured, while the dashed line corresponds to the conversion calculated as the sum of the individual conversions multiplied with the molar fractions of the components in the feed.

Table 6
Product selectivities over Pt/β-Pt/L and Pt/USY-Pt/L composites for Mixture M1 ($T = 480°C$, 5 min on stream, atmospheric pressure, 0.5 wt.-% Pt loading, H_2/Oil = 4.4 mol/mol)

Weight fraction	X	WHSV (h^{-1})	C_1–C_6 selectivity	Benzene selectivity	Toluene selectivity	$C_{>7}$ aromatic selectivity	Aromatic selectivity
Pt/β-Pt/L composites							
$\epsilon = 0.00$	52.7	2.0	8.1	44.4	43.1	4.4	91.9
$\epsilon = 0.20$	44.5	2.0	10.1	32.9	50.6	6.4	89.9
$\epsilon = 0.35$	66.9	2.0	23.1	13.6	50.8	12.5	76.9
$\epsilon = 0.50$	51.9	2.0	22.9	14.9	51.6	10.6	77.1
$\epsilon = 0.65$	55.3	2.6	29.9	15.7	48.7	5.7	70.1
$\epsilon = 0.80$	67.5	2.6	31.1	19.9	44.3	4.6	68.9
$\epsilon = 1.00$	60.8	2.6	26.2	23.4	42.4	8.0	73.8
Pt/USY-Pt/L composites							
$\epsilon = 0.00$	52.7	2.0	8.1	44.4	43.1	4.4	91.9
$\epsilon = 0.20$	63.9	2.0	24.0	20.7	42.9	10.3	73.9
$\epsilon = 0.35$	59.3	2.0	19.1	15.9	47.7	17.2	80.8
$\epsilon = 0.50$	44.0	2.6	27.3	9.4	48.2	15.1	72.7
$\epsilon = 0.65$	45.1	2.6	28.1	10.1	50.9	10.9	71.9
$\epsilon = 0.80$	44.2	2.6	31.2	11.9	47.9	9.0	68.8
$\epsilon = 1.00$	51.6	2.6	28.9	10.6	48.1	12.3	71.0

The values of WHSV have been selected in order to compare the catalysts at comparable conversions.

benzene, the toluene, and the $C_{\geqslant 8}$ aromatic selectivities as a function of ϵ are given in Table 6.

These observations regarding the product selectivities when the mixture M1 is employed are a result of complex cooperative reactions involving the components of the feed mixture and those of the composite. A possible pathway of these reactions is summarized in Fig. 8 and discussed below. Individual β or USY zeolites transform n-hexane predominantly to C_1–C_6 paraffins and olefins, methylcyclopentane to benzene and $C_{\geqslant 7}$ aromatics, and methylcyclohexane to aromatics. In contrast, zeolite L transforms n-hexane and methylcyclopentane to benzene, and methylcyclohexane to toluene. Because of the Brønsted acidity of the composites, the benzene molecules are alkylated with olefins formed mainly from n-hexane over the acidic zeolite, and the resulting intermediates generate by disproportionation $C_{\geqslant 7}$ aromatics and lighter olefins. Additional $C_{>6}$ aromatics are generated by the alkylation of methylcyclopentane with light paraffins formed mainly from n-hexane over the acidic zeolite. These alkylation reactions result in the enlargement of aromatics and therefore, in the utilization of the paraffinic fraction of the mixture (n-hexane) in the generation of higher aromatics. In the presence of a mixture of hydrocarbons, the consumption of benzene in the alkylation reactions shifts the cyclization of n-hexane over L zeolite to benzene, thus moderating the cracking of n-hexane.

The above cooperative reactions are responsible for the increased $C_{>6}$ aromatic selectivity and decreased benzene and C_1–C_6 paraffinic and olefinic selectivities. Consequently, a suitable amount of L zeolite in the composite optimizes the selec-

Table 7
Product selectivities over Pt/β-Pt/L and Pt/USY-Pt/L composites for Mixture M2 ($T = 480°C$, 5 min on stream, atmospheric pressure, 0.5 wt.-% Pt loading, H_2/Oil = 4.4 mol/mol)

Weight fraction	X	WHSV (h^{-1})	C_1–C_6 selectivity	Benzene selectivity	Toluene selectivity	$C_{>7}$ aromatic selectivity	Aromatic selectivity
Pt/β-Pt/L composites							
$\epsilon = 0.00$	41.6	2.2	8.5	37.8	53.1	2.2	91.5
$\epsilon = 0.20$	41.7	2.2	22.9	21.8	47.4	7.9	77.1
$\epsilon = 0.35$	50.2	2.2	19.3	15.6	46.9	17.1	79.7
$\epsilon = 0.50$	62.5	2.2	20.6	14.6	46.0	18.8	79.4
$\epsilon = 0.65$	60.7	2.2	23.1	15.9	43.1	17.9	76.9
$\epsilon = 0.80$	61.1	2.2	20.1	16.8	42.0	19.3	78.1
$\epsilon = 1.00$	69.4	2.2	19.1	17.5	42.8	20.6	80.9
Pt/USY-Pt/L composites							
$\epsilon = 0.00$	41.6	2.2	8.5	37.8	53.1	2.2	91.5
$\epsilon = 0.20$	46.2	2.2	15.9	12.9	65.9	5.2	84.0
$\epsilon = 0.35$	48.1	2.2	20.2	13.2	60.8	5.8	79.8
$\epsilon = 0.50$	51.4	2.2	25.1	11.8	55.2	7.8	74.9
$\epsilon = 0.65$	56.2	2.2	27.9	11.6	49.2	11.4	72.2
$\epsilon = 0.80$	58.0	2.2	27.5	13.3	42.9	16.2	72.4
$\epsilon = 1.00$	51.9	2.2	25.6	16.6	40.0	17.7	74.3

The values of WHSV have been selected in order to compare the catalysts at comparable conversions.

Table 8
Product selectivities over Pt/β-Pt/L and Pt/USY-Pt/L composites for Mixture M3 ($T = 480°C$, 5 min on stream, atmospheric pressure, 0.5 wt.-% Pt loading, H_2/Oil = 4.4 mol/mol)

Weight fraction	X	WHSV (h^{-1})	C_1–C_6 selectivity	Benzene selectivity	Toluene selectivity	$C_{>7}$ aromatic selectivity	Aromatic selectivity
Pt/β-Pt/L composites							
$\epsilon = 0.00$	47.8	7.7	9.2	39.0	47.6	4.2	90.8
$\epsilon = 0.20$	45.0	7.7	18.5	24.1	46.1	11.3	81.5
$\epsilon = 0.35$	43.2	7.7	14.2	8.5	48.8	22.1	85.8
$\epsilon = 0.50$	57.4	7.7	20.0	7.8	37.9	34.3	80.0
$\epsilon = 0.65$	51.1	15.4	29.2	11.2	36.1	23.5	70.8
$\epsilon = 0.80$	47.2	15.4	29.0	24.9	35.7	10.4	71.0
$\epsilon = 1.00$	49.2	15.4	28.1	25.3	35.1	11.4	71.8
Pt/USY-Pt/L composites							
$\epsilon = 0.00$	47.8	7.7	9.2	39.0	47.6	4.2	90.8
$\epsilon = 0.20$	47.3	7.7	23.2	19.6	49.2	7.9	76.7
$\epsilon = 0.35$	46.0	7.7	23.0	18.5	48.3	10.1	76.9
$\epsilon = 0.50$	51.3	7.7	25.1	9.2	54.1	11.6	74.9
$\epsilon = 0.65$	50.1	15.4	30.3	10.1	45.6	14.0	69.7
$\epsilon = 0.80$	49.7	15.4	25.4	13.2	42.1	19.3	74.6
$\epsilon = 1.00$	60.6	15.4	22.2	24.5	40.0	13.3	77.8

The values of WHSV have been selected in order to compare the catalysts at comparable conversions.

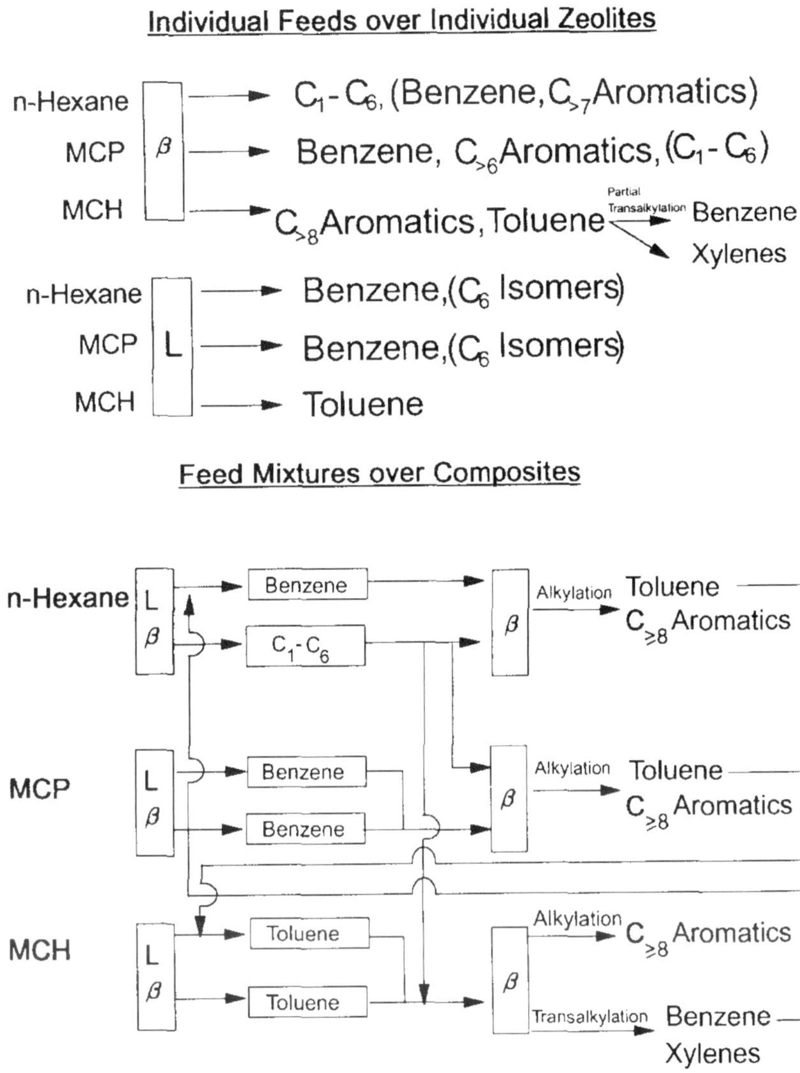

Fig. 8. Top: Products from the hydrocarbon feeds employed over the acidic (represented by β zeolite) and the basic zeolites. Bottom: pathways of the cooperative reactions occurring between the hydrocarbons of the mixtures catalyzed by the components of the composites. Note: The selectivities for the hydrocarbons in the parentheses are much lower than those of the other hydrocarbons.

tivity for $C_{>6}$ aromatics. The aromatic selectivities with M1 over the Pt/β zeolite, Pt/USY zeolite and their composites with the L zeolite are higher than those calculated as the molar averages by multiplying the selectivities corresponding to the individually reacting feed hydrocarbons with their molar fractions in the feed. This is due to the alkylation reactions which utilize the light paraffins and olefins generated mainly from hexane which occur to a greater extent when the mixture M1 is employed than any of its components individually.

Mixture M2

The performance of the composites with equimolar amounts of the three hydrocarbons is presented in Figs. 9 and 10. One can observe that the aromatic selectivity is higher than that calculated as the molar average by multiplying the selectivities of the individually reacting feed hydrocarbons with their molar fractions in the

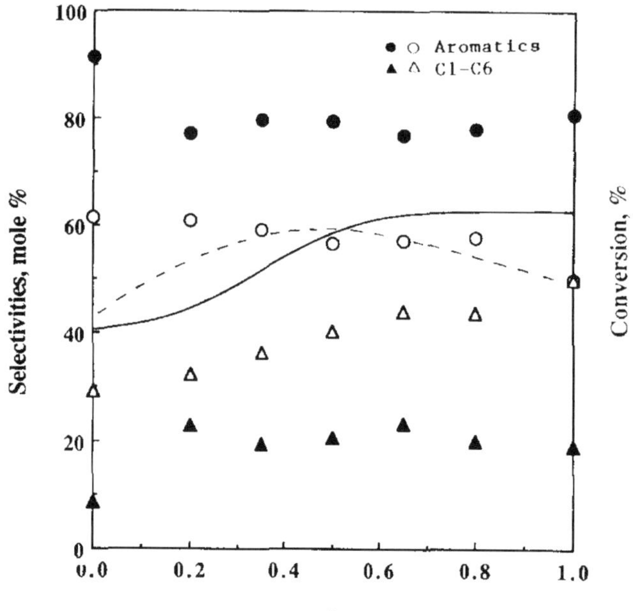

Fig. 9. Selectivities for C_1–C_6 paraffins and olefins (▲) and aromatics (●) for M2 (filled symbols) over the Pt/β-Pt/L zeolite composites at 480°C. The open symbols stand for the selectivities calculated from the individual hydrocarbon selectivities by weighting them with their molar fractions in the feed (H_2/Oil = 4.4 mol/mol, 5 min on stream, atmospheric pressure, 0.5 wt.-% Pt, WHSV = 2.2 h^{-1}). The continuous line corresponds to the conversion measured, while the dashed line corresponds to the conversion calculated as the sum of the individual conversions multiplied with the molar fractions of the components in the feed.

feed. The C_1–C_6 paraffinic and olefinic selectivity is smaller than that of the M1 mixture since the molar percentage of n-hexane whose cracking generates light paraffins and olefins is in the present feed mixture lower than in M1. On the other hand, the higher concentrations of naphthenes in the feed mixture enhance the alkylation reactions.

Mixture M3

The performance of the composites with the most naphthenic (M3) mixture is given in Figs. 11 and 12 for Pt/β-Pt/L and Pt/USY-Pt/L zeolite composites. With both types of composites the product selectivities are similar to those of M2.

3.5. Time-on-stream behavior and effect of the hydrogen pressure

At atmospheric pressure, the conversion of the catalyst decreased significantly within the first 10 h on stream and then remained unchanged for three days (Fig. 13a). Zeolite L provided the highest final activity in comparison with those of the acidic zeolites, indicating that coke is formed preferentially over the acidic sites rather than over the platinum clusters. Indeed, after reaction the Pt/BaKL zeolite had a grey color, in contrast to Pt/β and Pt/USY zeolite which were black. Under identical conditions the rate of deactivation of USY was higher than that of β zeolite. The more numerous acidic sites of USY is a possible reason for this behavior. The generation of bulky coke deposits in the supercages of the USY

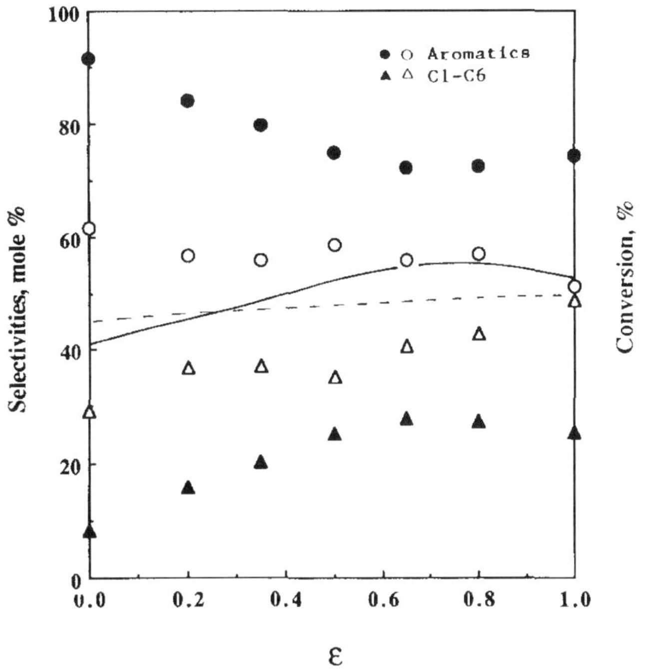

Fig. 10. Selectivities for C_1–C_6 paraffins and olefins (▲) and aromatics (●) for M2 (filled symbols) over the Pt/USY-Pt/L zeolite composites at 480°C. The open symbols stand for the selectivities calculated from the individual hydrocarbon selectivities by weighting them with their molar fractions in the feed (H_2/Oil = 4.4 mol/mol, 5 min on stream, atmospheric pressure, 0.5 wt.-% Pt, WHSV = 2.2 h^{-1}). The continuous line corresponds to the conversion measured, while the dashed line corresponds to the conversion calculated as the sum of the individual conversions multiplied with the molar fractions of the components in the feed.

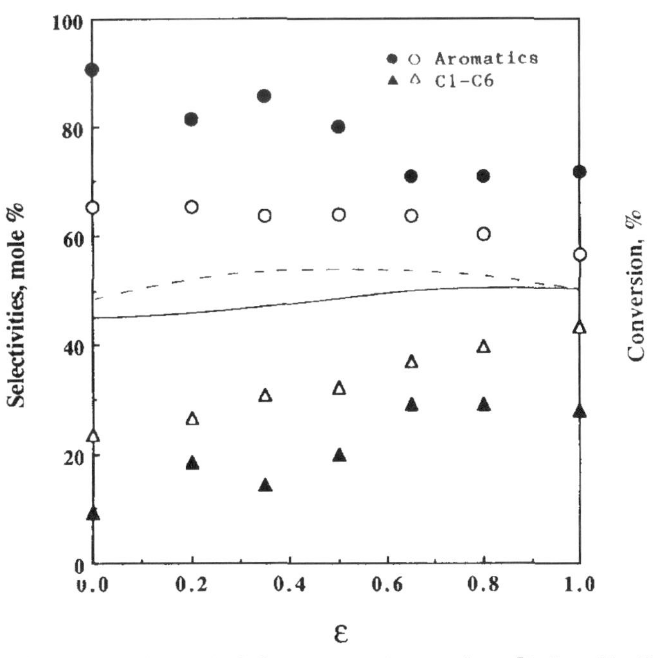

Fig. 11. Selectivities for C_1–C_6 paraffins and olefins (▲) and aromatics (●) for M3 (filled symbols) over the Pt/β-Pt/L zeolite composites at 480°C. The open symbols stand for the selectivities calculated from the individual hydrocarbon selectivities by weighting them with their molar fractions in the feed (H_2/Oil = 4.4 mol/mol, 5 min on stream, atmospheric pressure, 0.5 wt.-% Pt, WHSV range = 7.7–15.4 h^{-1}). The continuous line corresponds to the conversion measured, while the dashed line corresponds to the conversion calculated as the sum of the individual conversions multiplied with the molar fractions of the components in the feed.

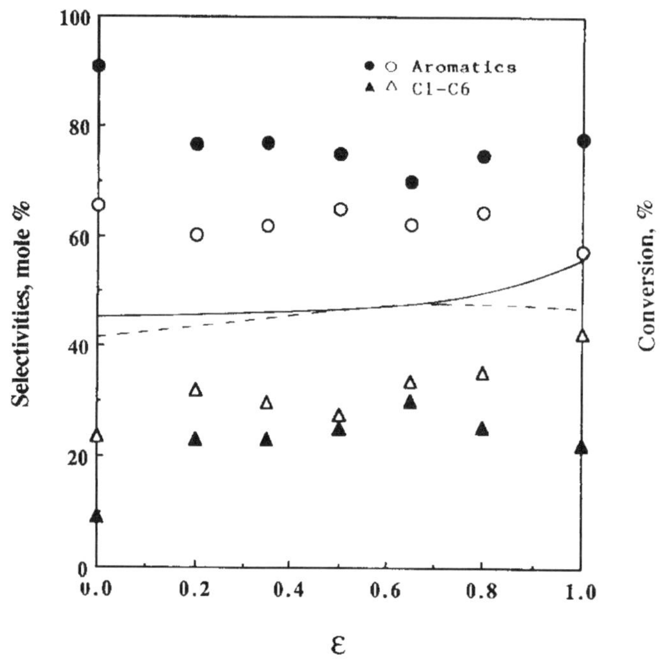

Fig. 12. Selectivities for C_1–C_6 paraffins and olefins (▲) and aromatics (●) for M3 (filled symbols) over the Pt/USY-Pt/L zeolite composites at 480°C. The open symbols stand for the selectivities calculated from the individual hydrocarbon selectivities by weighting them with their molar fractions in the feed (H_2/Oil = 4.4 mol/mol, 5 min on stream, atmospheric pressure, 0.5 wt.-% Pt, WHSV range = 7.7–15.4 h^{-1}). The continuous line corresponds to the conversion measured, while the dashed line corresponds to the conversion calculated as the sum of the individual conversions multiplied with the molar fractions of the components in the feed.

zeolite results in the blocking of its pores, thus preventing the access of the reactants to the active sites [38,39]. The selectivity for aromatics at atmospheric pressure over the individual Pt/L, Pt/β and Pt/USY zeolites is presented in Fig. 13b. For all zeolites the aromatic selectivity decreases during the first five hours on stream and then remains relatively unchanged with time. The aromatic selectivity of USY is slightly higher than that of the β zeolite. This behavior is a result of the higher acidity of USY, which by leading to a larger amount of coke that covers some acidic sites decreases the cracking capability of the catalyst and increases the aromatic selectivity. Consequently, the aromatic selectivity of USY is larger than that of the β zeolite. The aromatic selectivity of the L zeolite is comparable to that of the USY.

Over the acidic zeolites, the conversion was higher at a hydrogen pressure of 100 psig (Table 9, 1 psig = 6.895 kPa) than at the atmospheric pressure, due to higher rates of hydrocracking. The activity over Pt/USY at 100 psig decreased slightly after a few hours on stream, while that over Pt/β zeolite remained almost constant. It seems that even at elevated hydrogen pressures some bulky carbonaceous compounds are still formed into the supercages of the USY zeolite. The preservation of activity is a result of the fact that elevated hydrogen pressures saturate the olefinic intermediates (monocyclic and dicyclic diolefins) located on the acidic and platinum sites, thus keeping them clean [40]. As shown in Fig. 13b, the total aromatic selectivities over Pt/β and Pt/USY zeolites at 100 psig pressure are relatively low during the initial stages, but become larger after 10 h on stream.

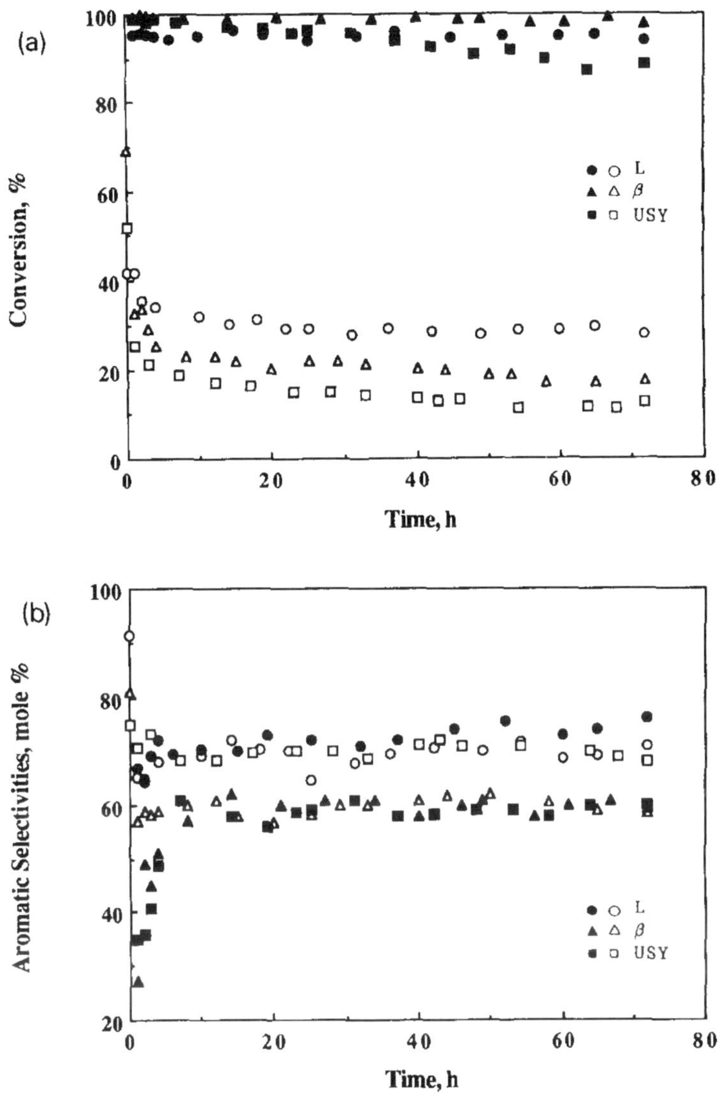

Fig. 13. (a) Conversion of M2 over Pt/L, Pt/β and Pt/USY zeolites as a function of time-on-stream at 480°C under atmospheric pressure (open symbols, WHSV = 2.0 h^{-1}, H_2/Oil = 4.4 mol/mol, 0.5 wt.-% Pt) and under 100 psig pressure (filled symbols, WHSV = 2.0 h^{-1}, H_2/Oil = 12.0 mole/mole, 0.5 wt.-% Pt). (b) Aromatic selectivity of M2 over Pt/L, Pt/β and Pt/USY zeolites as a function of time-on-stream at 480°C under atmospheric pressure (open symbols, WHSV = 2.0 h^{-1}, H_2/Oil = 4.4 mole/mole, 0.5 wt.-% Pt) and under 100 psig pressure (filled symbols, WHSV = 2.0 h^{-1}, H_2/Oil = 12.0 mol/mol, 0.5 wt.-% Pt).

This behavior is a result of the selective coking of some strong acidic sites, which are potential cracking sites. One can conclude that acidity regulation, in order to reduce overcracking, and a suitable increase of the hydrogen pressure in order to reduce coke build-up and simultaneously minimize hydrocracking, can result in a catalyst with constant activity with time-on-stream and high aromatization yield.

The conversion over Pt/L (Fig. 13a) is significantly higher at 100 psig than at atmospheric pressure. The aromatic selectivities remain relatively unchanged as a function of time-on-stream. The absence of acidity is responsible for this behavior. One can observe that the aromatic selectivities over Pt/L zeolite are slightly higher at 100 psig than at the atmospheric pressure at large times on stream.

Table 9
Product selectivities and conversion with Mixture M2 over Pt/L, Pt/β and Pt/USY zeolites at 480°C after 2 h on stream at atmospheric and at 100 psig pressure (0.5 wt.-% Pt loading, WHSV = 2.0 h^{-1})

Products (mol-%)	Atmospheric pressure			100 psig		
	Pt/BaKL	Pt/β	Pt/USY	Pt/BaKL	Pt/β	Pt/USY
Methane	0.2	0.2	0.3	0.6	4.4	3.1
Ethane	0.4	1.4	0.8	1.0	1.0	2.1
Propene	1.7	2.7	1.6	–	–	–
Propane	3.4	10.1	2.8	3.9	23.9	21.1
i-Butane	3.5	4.7	2.5	8.8	7.8	6.6
2-Butene	–	2.7	–	–	0.6	–
n-Butane	5.6	4.5	3.5	4.1	8.1	6.5
i-Pentane	1.0	3.5	0.9	2.7	3.2	2.8
n-Pentane	1.7	5.1	4.1	3.1	1.8	1.7
Methylpentanes	–	–	–	10.6	–	1.0
Methylcyclopentenes	15.7	–	–	–	–	–
Benzene	21.2	10.0	10.9	35.4	16.2	16.3
Cyclohexane	1.1	–	0.3	–	–	–
Dimethylcyclopentanes	1.6	6.8	9.9	–	–	–
Toluene	42.3	34.4	50.0	29.7	28.3	28.2
Ethylbenzene	–	0.4	0.3	–	0.7	0.4
o-Xylene	–	2.1	2.6	–	0.3	2.3
m-Xylene	–	2.5	2.0	–	0.4	0.3
p-Xylene	0.8	5.0	4.3	0.1	3.1	6.1
C_9 aromatics	–	3.8	3.2	–	0.2	2.2
Conversion (mol-%)	35.7	33.8	21.3	95.9	99.9	99.0

3.6. *Effect of platinum*

The effect of platinum loading in the individual zeolites with individual feed molecules as feeds is presented in Tables 10–12 at comparable levels of conversion. For n-hexane (Table 10), one can observe a monotonic increase in benzene and $C_{\geqslant 7}$ aromatics with the metal loading. This indicates that platinum is absolutely necessary for the enhancement of the catalyst activity. The increase in platinum loading results in the reduction of the C_1--C_6 paraffinic and olefinic selectivity and in the increase of the total aromatic selectivity over all zeolites. With 0.1 wt.-% Pt, L zeolite produces mainly the C_6 isomers of n-hexane, while the acidic zeolites produce mainly propane. Increased platinum loading enhances the cyclization of n-hexane to benzene over L zeolite. Platinum stimulates the aromatization of n-hexane over the acidic zeolites as well, since their total aromatic selectivities, especially for 1.0 wt.-% Pt, are significantly larger than those at smaller loadings. USY exhibits a slightly better performance than β zeolite regarding the generation of benzene at any metal loading.

The benzene and $C_{\geqslant 7}$ aromatic selectivities with methylcyclopentane (Table 11) over the L zeolite increase monotonically with the platinum loading. For any platinum loading the aromatic selectivity is smaller than those over the acidic

Table 10
Effect of the platinum loading on the activity and product selectivities over individual zeolites with n-hexane as feed ($T = 480°C$, 5 min on stream, atmospheric pressure, H_2/Oil = 4.2 mol/mol, WHSV = 2.0 h^{-1})

Catalysts	Conversion	C_1–C_6 selectivity	Benzene selectivity	Aromatic selectivity
0.1 wt.-% Pt				
Pt/BaKL	46.1	60.9	28.7	39.1
Pt/β	45.8	81.5	5.1	18.5
Pt/USY	51.7	82.9	8.9	17.1
0.5 wt.-% Pt				
Pt/BaKL	48.1	57.6	37.8	42.4
Pt/β	44.7	78.1	11.5	21.9
Pt/USY	45.1	70.4	16.3	29.6
1.0 wt.-% Pt				
Pt/BaKL	50.3	22.1	53.4	77.7
Pt/β	51.7	42.7	18.1	57.3
Pt/USY	49.4	44.1	21.7	55.8

zeolites. Platinum is again absolutely necessary for the reaction to occur, since BaKL without platinum, under the same operating conditions, provides a conversion below 10%. This indicates that the ring opening of methylcyclopentane at the tertiary–secondary C–C bonds, which leads to hexane and further to benzene, occurs only to a very small extent over BaKL free of platinum. The increase of the platinum loading in the L zeolite decreases the C_1–C_6 paraffinic and olefinic selectivity. Both USY and β zeolite behave in a similar way. The total aromatic selectivity increases with the metal loading, while the selectivity for C_1–C_6 paraffins and olefins decreases. This indicates that apart of the acidic function of the zeolite which enhances the isomerization (enlargement) of the C_5 cyclic ring of methylcyclopentane to a C_6 ring, the increase of the metal loading also plays a beneficial role.

Table 11
Effect of the platinum loading on the activity and product selectivities over individual zeolites with methylcyclopentane as feed ($T = 480°C$, 5 min on stream, atmospheric pressure, H_2/Oil = 4.4 mol/mol, WHSV = 1.8 h^{-1})

Catalysts	Conversion	C_1–C_6 selectivity	C_6 cycloolefins	Benzene selectivity	Aromatic selectivity
0.1 wt.-% Pt					
Pt/BaKL	41.1	26.1	26.8	35.0	47.1
Pt/β	49.5	41.0	–	6.9	59.0
Pt/USY	45.3	33.4	–	13.0	66.5
0.5 wt.-% Pt					
Pt/BaKL	45.0	12.9	26.9	49.7	58.0
Pt/β	51.4	31.2	–	23.2	68.8
Pt/USY	57.0	23.9	–	34.7	76.1
1.0 wt.-% Pt					
Pt/BaKL	55.8	7.2	13.6	39.3	79.2
Pt/β	50.1	10.8	–	29.4	89.0
Pt/USY	48.7	2.4	–	44.9	97.6

Table 12
Effect of the platinum loading on the activity and product selectivities over individual zeolites with methylcyclohexane as feed ($T=480°C$, 5 min on stream, atmospheric pressure, H_2/Oil = 4.4 mol/mol, WHSV = 2.3 h^{-1} for L zeolite, WHSV = 15.4 h^{-1} for β and USY zeolites)

Catalysts	Conversion	C_1–C_6 selectivity	Toluene selectivity	Aromatic selectivity
0.1 wt.-% Pt				
Pt/BaKL	51.2	27.9	69.6	72.1
Pt/β	48.1	48.6	37.1	51.3
Pt/USY	54.1	57.1	39.2	42.8
0.5 wt.-% Pt				
Pt/BaKL	50.2	3.6	78.0	82.7
Pt/β	48.1	35.4	44.9	64.6
Pt/USY	50.9	41.4	38.1	58.6
1.0 wt.-% Pt				
Pt/BaKL	52.3	3.9	86.6	96.1
Pt/β	49.1	7.2	44.8	92.8
Pt/USY	52.3	1.4	42.4	85.7

This occurs because the increased metal loading results in a larger generation of cycloolefins from methylcyclopentane. It appears that, in the range of platinum loadings employed, the rate of hydrogenolysis of the feed hydrocarbon remains almost constant.

The C_1–C_6 paraffinic and olefinic selectivity decreases and the total aromatic selectivity increases for methylcyclohexane (Table 12), over any of the zeolites, with increasing platinum loading.

3.7. Effect of conversion level

A few experiments have been carried out at a low level of conversion. Pt/β-Pt/L and Pt/USY-Pt/L composites with and $\epsilon=0.5$ have been employed for the reforming of mixture M2. Experiments with individual hydrocarbons have been also performed over the above composites. From Table 13 one can observe that at a level of conversion of about 10% the experimentally observed aromatic selectiv-

Table 13
Experimental and calculated aromatic and C_1–C_6 hydrocarbon selectivities for M2 at 480°C and atmospheric pressure at low conversion over Pt/β-Pt/L and Pt/USY-Pt/L composites (5 min on stream, 0.5 wt.-% Pt loading, H_2/Oil = 4.4 mol/mol, $\epsilon=0.5$, WHSV = 72 h^{-1} for the composite containing β zeolite, WHSV = 53 h^{-1} for the composite containing USY)

	Pt/β-Pt/L			Pt/USY-Pt/L		
	Conversion	C_1–C_6 selectivity	Aromatic selectivity	Conversion	C_1–C_6 selectivity	Aromatic selectivity
Experimental	9.6	18.5	81.5	13.2	26.7	73.3
Calculated	9.2	38.7	61.2	11.1	34.5	65.6

ities are higher than those calculated as the molar averages by multiplying those corresponding to the individual hydrocarbons with their molar fractions in the feed. This indicates that the cooperations observed at higher conversions also occurs at low levels of conversion.

4. Conclusions

From the present study of the reforming of n-hexane, methylcyclopentane or methylcyclohexane and their mixtures over Pt/β-Pt/L zeolite and Pt/USY-Pt/L zeolite composites, the following conclusions can be drawn:

(1) For n-hexane as feed, Pt/β-Pt/L zeolite and Pt/USY-Pt/L zeolite composites increase the C_1–C_6 paraffinic and olefinic selectivity and the selectivity for $C_{>6}$ aromatics when the acidic fraction of the composite increases. With methylcyclopentane as feed the increase of the acidic content of the composite increases the aromatization, while the C_1–C_6 paraffinic and olefinic selectivity remains low for any composite. For methylcyclohexane as feed, the increase of the acidic component of the composite increases the transalkylation of toluene, which is generated over the metal sites, to benzene and C_8 aromatics.

(2) Reaction with mixtures of the above hydrocarbons over the composites results in increased $C_{>6}$ aromatic selectivity and decreased benzene selectivity in comparison with those calculated as molar averages by multiplying the selectivities corresponding to the individually reacting feed hydrocarbons components with their molar fractions in the feed. Pathways for the complex cooperative reactions responsible for this behavior are suggested.

(3) The increase of hydrogen pressure leads to the preservation of activity as a function of time for all the catalysts employed. If, in addition to operating at elevated hydrogen pressures, the catalyst acidity is controlled in order to avoid excessive hydrocracking, a beneficial performance for the $C_{>6}$ aromatic selectivity can be achieved.

(4) Platinum is absolutely necessary for the enhancement of the activity of any of the zeolites. The increase of the platinum loading increases the aromatic selectivity for all the individual hydrocarbons. Under the present operating conditions the extent of hydrogenolysis is low.

(5) The aromatization selectivity of the composite catalysts with a mixture of hydrocarbons is higher than that calculated as the molar average by multiplying those corresponding to the individual components with their molar fractions in the feed.

Acknowledgements

The comments of the Referees have been very helpful.

References

[1] H.E. Kluksdahl, US Patent 3 415 737 (1968).
[2] J.H. Sinfelt, in J.R. Anderson and M. Boudart (Editors), Catalysis, Science and Technology, Spring, New York, Vol. 1., 1981, p. 257.
[3] G.A. Mills, H. Heinemann, T.H. Milliken and A.G. Oblad, Ind. Eng. Chem., 45 (1953) 134.
[4] P.B. Weisz and E.W. Swegler, Science (Washington), 126 (1957) 31.
[5] C. Dossi, C.M. Tsang and W.M.H. Sachtler, Energy Fuels, 3 (1989) 468.
[6] J.A. Martens, M. Tielen and P.A. Jacobs, Stud. Surf. Sci. Catal., 46 (1985) 49.
[7] H. Schulz, J. Weitkamp and H. Eberth, in J.W. Hightower (Editor), Proceedings of the 5th Congr. Catal., North Holland, Amsterdam, Vol. 2, 1973, p. 1229.
[8] J. Weitkamp, P.A. Jacobs and E. Ernst, Stud. Surf. Sci. Catal., 18 (1984) 279.
[9] Z. Alvarez, A. Montes, G. Perot and M. Guisnet, Stud. Surf. Sci. Catal., 49 (1989) 1367.
[10] P.G. Smirniotis and E. Ruckenstein, Catal. Lett., 17 (1993) 341.
[11] P.G. Smirniotis and E. Ruckenstein, J. Catal., 140 (1993) 526.
[12] P.G. Smirniotis and E. Ruckenstein, Chem. Eng. Sci., 48 (1993) 3263.
[13] J.R. Bernard, in L.V.C. Rees (Editor), Proceedings 5th Int. Conf. Zeolites, Heyden, London, 1980, p. 686.
[14] C. Besoukhanova, J. Guidot, D. Barthomeuf, M. Breysse and J.P. Bernard, J. Chem. Soc., Faraday Trans. 1, 77 (1981) 1595.
[15] T.R. Hughes, W.S. Buss, P.W. Tamm and R.J. Jacobson, Stud. Surf. Sci. Catal., 28 (1986) 725.
[16] D.W. Law, P.W. Tamm and C.M. Detz, Energy Progress, 7 (4) (1987) 215.
[17] P.W. Tamm, D.H. Mohr and C.R. Wilson, Stud. Surf. Sci. Catal., 38 (1988) 335.
[18] A. Corma, V. Fornes, P. Melo and J. Perez-Pariente, Prepr. Am. Chem. Soc. Div. Pet. Chem., (1987) 632.
[19] S.B. Liu, J.F. Wu, L.J. Ma, T.C. Tsai and I. Wang, J. Catal., 132 (1991) 432.
[20] D.J. Ostgard, L. Kustov, K.R. Poeppelmeier and W.M.H. Sachtler, J. Catal., 133 (1992) 342.
[21] M. Tanaka and S. Ogasawara, J. Catal., 16 (1970) 157.
[22] C.N. Satterfield, in Heterogeneous Catalysts in Industrial Practice, McGraw-Hill, New York, 1992.
[23] F.G. Ciapetta and D.N. Wallace, Catal. Rev.-Sci. Eng., 5 (1971) 67.
[24] C.G. Meyers and G.W. Jr. Munns, Ind. Eng. Chem., 50 (1958) 1727.
[25] J. Weitkamp, S. Ernst and R. Kumar, Appl. Catal., 27 (1986) 207.
[26] M.L. Poutsma, in J.A. Rabo (Editor), Zeolite Chemistry and Catalysis, ACS Monograph, Washington, D.C., Vol. 171, 1976, p. 518.
[27] J. Weitkamp, in Proceedings Int. Symp. Zeolite Catalysis, Acta Physica et Chemica, Siofok, Hungary, 1985, p. 271, Vol. 31.
[28] C. Corolleur, S. Corolleur and F.G. Gault, J. Catal., 24 (1972) 385.
[29] E.G. Derouane and D.J. Vanderveken, Appl. Catal., 45 (1988) 15.
[30] S.J. Tauster and J.J. Steger, J. Catal., 125 (1990) 387.
[31] P.A. Newell and L.V.L. Rees, Zeolites, 3 (1982) 22.
[32] E. Iglesia and J.E. Baumgartner, in R. von Ballmoos, J.B. Higgins and M.M.J. Treacy (Editors), Proceedings of the 9th International Zeolite Conference, Vol. II, Butterworth-Heinemann, 1993, p. 421.
[33] B.C. Gates, J.R. Katzer and G.C.A. Schuit, Chemistry of Catalytic Processes, McGraw-Hill, New York, 1979, p. 189.
[34] M. Tzou, B.K. Teo and W.M.H. Sachtler, J. Catal., 113 (1988) 220.
[35] X. Bai and W.M. Sachtler, J. Catal., 129 (1991) 181.
[36] J. Das, Y.S. Bhat and A.B. Halgeri, Ind. Eng. Chem. Res., 32 (1993) 2525.
[37] P.G. Smirniotis and E. Ruckenstein, Ind. Eng. Chem. Res., submitted for publication.
[38] R.A. Dalla-Beta and M. Boudart, in J.W. Hightower (Editor), Proc. 5th Inter. Congr. Catal., North Holland, Amsterdam, Vol. II, 1973, p. 1329.
[39] R. Parton, L. Uytterhoven, J.A. Martens and P.A. Jacobs, Appl. Catal., 76 (1987) 131.
[40] P. Magnoux, P. Cartraud, S. Mignard and M. Guisnet, J. Catal., 106 (1987) 242.

Different Activities and Selectivities of Silica–Alumina Catalysts Synthesized in Aqueous and Alcohol Solvents

P. YARLAGADDA, C. R. F. LUND, AND E. RUCKENSTEIN[1]

Department of Chemical Engineering, State University of New York at Buffalo, Amherst, New York 14260

Received January 30, 1990; revised April 16, 1990

Silica–aluminas having 15 wt% Al_2O_3 and subjected to identical heat treatment were synthesized starting from (i) sodium silicate and aluminum nitrate in water (ASA) and (ii) tetraethyl orthosilicate and aluminum isopropoxide in ethanol solvent (SA-SG). Both catalysts possess strong Brønsted acid sites which are capable of oligomerizing propene or ethene. With methanol as a feed, only ASA had a high hydrocarbon yield, while SA-SG produced mostly the dehydration product, dimethyl ether. While SA-SG possesses strong acidic sites, it is suggested that these sites become ineffective because of the strong chemisorption of the water resulting from dehydration. The weaker acidic sites also present on the SA-SG catalyst are probably responsible for the formation of the dehydration product.

INTRODUCTION

Numerous investigations (*1–5*) on amorphous aluminosilicates have shown that their activity in acid-catalyzed reactions depends to a great extent on the amount of alumina present in the catalyst and to a lesser extent on other factors such as the method of preparation and the temperature at which the samples were heat treated prior to reaction. The nature of the acid sites present on the amorphous silica–aluminas, whether of Brønsted or Lewis type or both, has been the subject of innumerable studies (*6–9*), but still there is no consensus regarding the origin of these sites. However, it is widely accepted that the generation of these acid centers is due to the isomorphous substitution of Si^{4+} by Al^{3+} in the silica lattice (*10–12*) which generates an exchangeable cation. This aluminum ion can coordinate to oxygen either tetrahedrally or octahedrally or a combination of both. The relative amounts of these coordination states depend both on the alumina content in the sample and on the temperature at which it was heat treated. It has been shown (Table 1 of Ref. *13*) that the $Al^{IV}/(Al^{IV} + Al^{VI})$ ratio is independent of the alumina content when it is less than ca. 30 wt%; above this amount, the fraction of Al^{IV} decreases. It has also been observed that when silica–alumina is heated at 700°C for 24 h, all aluminum present in the sample becomes located in tetrahedrally coordinated sites, over the entire range of alumina content.

The activity of the aluminosilicates in acid-catalyzed reactions depends on the total amount of Brønsted and Lewis acid sites, as well as on the acid strength distribution. Two types of Brønsted acid sites are known: (i) bridged hydroxyl groups and (ii) water molecules coordinated to a trigonal aluminum atom (the trigonal aluminum itself being a Lewis acid site). Recent studies (*14*) have shown that the hydroxyl groups of the amorphous aluminosilicates are quite similar in their acidic and spectral features to those present in zeolites, but their concentration is much lower. Since only the tetrahedrally held aluminum leads to the formation of a bridged hydroxyl group, it is obvious that the above sites are responsible for the catalytic activity. Thus, one can gen-

[1] To whom correspondence should be addressed.

This paper, which was originally published in **Journal of Catalysis**: 125, 421–426 (1990), is republished with permission.

erally conclude that the $Al_2O_3/(SiO_2 + Al_2O_3)$ ratio and the conditions of heat treatment determine the tetrahedral versus octahedral aluminum sites and hence the catalytic activity. Recently we observed significant differences in the catalytic behavior of two types of silica–aluminas with a similar alumina content but synthesized from gels formed in water and ethanol solvents, and the results are presented herein.

EXPERIMENTAL

Silica–alumina samples were prepared by cogelation starting with different sources of silica and alumina and in an aqueous or ethanol medium.

(*i*). *Silica–alumina synthesized in an aqueous medium* (*ASA*). To a 500-ml batch of 0.15 *M* aqueous sodium silicate (Fisher Scientific) solution whose pH was initially adjusted to 4 by the addition of 0.2 *M* hydrochloric acid, 250 ml of 0.06 *M* aqueous aluminum nitrate (Baker) was added. The pH of the above mixture was adjusted to about 6.5 by dropwise addition of 0.2 *M* NH_4OH solution under vigorous stirring. The resulting gel was allowed to age for 24 h at room temperature under slow stirring and then was filtered. The resultant cake was thoroughly washed with distilled water, dried at 100°C overnight, and calcined at 600°C for 1 h in air. Later, the sample was treated with 1 *M* aqueous $(NH_4)_2SO_4$ solution at room temperature for 3 h. Finally, this NH_4^+ exchanged sample was calcined at 550°C for 1 h. Chemical analysis carried out by Galbraith Laboratories showed that the above sample, designated as ASA, contained 15.5 wt% Al_2O_3, and 0.07 wt% Na. Before the ammonium exchange, the Na level was 0.82 wt%. The BET surface area (N_2 as adsorbent) and the pore volume were 410 $m^2\ g^{-1}$ and 0.69 $cm^3\ g^{-1}$, respectively. X-ray powder diffraction showed the sample to be amorphous.

(*ii*). *Silica–alumina synthesized in an ethanol-rich medium* (*SA-SG*). The pH of a solution containing 30 ml of tetraethyl orthosilicate (Aldrich) dissolved in 200 ml of ethanol was adjusted to pH $\leqq 1$ by adding nearly 8 ml of nitric acid. To the above solution, 5 g of powdered aluminum isopropoxide (Aldrich) was added, and the whole mixture was stirred until all the aluminum isopropoxide had dissolved. Subsequently, the pH of the solution was adjusted to nearly 8.5 by adding 20 ml of 14% aqueous ammonia solution, under stirring. At this stage, a thick gel formed, to which, in order to be able to filter, an additional amount of 200 ml ethanol was added and the whole mass was kept under agitation overnight at room temperature. The mixture was filtered and the cake was washed with ethanol. After the cake was dried overnight at 70°C in vacuum, the sample was subjected to NH_4^+ exchange and to heat treatment under the conditions described earlier. The sample, designated as SA-SG, was analyzed and found to contain 15 wt% Al_2O_3 and a sodium content less than 0.03 wt%. The BET surface of the sample was 490 $m^2\ g^{-1}$ and the XRD pattern showed it to be amorphous.

The infrared spectra of the silica–alumina samples were recorded with a Perkin–Elmer 727B instrument. A fine powder of the sample was thoroughly mixed with an equivalent amount of KBr (Baker) and the mixture was pressed under a pressure of 180 $kg\ cm^{-2}$ to obtain a thin plate. The pressed sample was heated at 400°C overnight in air and then transferred into a quartz cell with CaF_2 windows. Before the IR spectrum was recorded, the sample was further heated at 150°C inside the quartz cell and continuously purged with helium. The IR spectrum was recorded at room temperature with a grating placed in the path of the reference beam.

Temperature-programmed desorption (TPD) spectra of the silica–alumina samples were obtained using ammonia as the probe base molecule. Details of the TPD setup and the procedure adopted for recording the spectra are similar to those reported previously (*15*, *16*).

Methanol to hydrocarbon conversion and olefin (ethene and propene) oligomerization experiments were conducted in a fixed-bed microreactor, in the manner described elsewhere (*15, 17*). In order to clarify some mechanistic features, additional experiments were conducted with ethanol or isopropanol as feed. The silica–alumina samples, after having been loaded into the reactor, were maintained at the reaction temperature for 4 h in helium before the commencement of the reaction.

RESULTS AND DISCUSSION

The activity and selectivity data obtained during the methanol conversion and olefin oligomerization reactions are given in Table 1. Silica–alumina catalysts deactivate rapidly (over ca. 10–20 h) and severely under methanol conversion conditions. The results in Table 1 were obtained after 2 h on stream, during which the catalysts deactivated by less than 5% of their fresh activity.

For the ethene or propene oligomerization reactions, both the ASA and the SA-SG catalysts led to comparable conversions. With regard to the selectivity, the following observations can be made: When ethene was the feed and SA-SG the catalyst, the selectivity was higher for the C_3, C_4 olefins; for the ASA catalyst, the selectivity was higher for the C_4 alkanes. More pronounced differences in selectivity were observed with propene as reactant; a larger fraction of C_8 and C_9 hydrocarbons were formed via polymerization, followed by cracking to aromatics, on the SA-SG catalyst.

A striking difference in the behavior of the two silica–alumina samples was, however, observed during the methanol conversion reaction. Table 1 shows that the conversion of methanol to hydrocarbons (excluding dimethyl ether) is around 94 mol% for the ASA catalyst and only 2.5% for the SA-SG catalyst, in spite of the fact that both catalysts have almost the same alumina content and have been heat treated in the same way. Furthermore, no hydrocarbons higher than C_4 were detected during the reaction over the SA-SG catalyst, while a wide spectrum of hydrocarbons has resulted over the ASA catalysts.

TABLE 1

Activity and Selectivity Data Obtained during Catalytic Reactions over the Silica–Alumina Samples

	ASA	SA-SG
a. CH_3OH conversion reaction (375 ± 5°C; 1.0 ± 0.1 h^{-1} WHSV; 1 atm total pressure)		
Conversion (mol%)	94.2	2.5
Selectivity (wt%)		
CH_4	2.4	24.2
Total C_2–C_4 olefins	63.9	66.1
Total C_2–C_4 alkanes	17.1	9.7
C_{5+} nonaromatics	11.6	—
Aromatics	5.0	—
b. C_3H_6 oligomerization reaction (365 ± 5°C; 3.4 ± 0.1 h^{-1} WHSV; 1 atm total pressure)		
Conversion (mol%)	8.4	8.1
Selectivity (wt%)		
CH_4	0.2	0.1
C_2H_4	0.4	0.5
C_2H_6	0.6	0.8
C_3H_8	22.4	19.3
$(i + n)$ C_4H_{10}	4.7	6.2
C_4 olefins	51.6	48.1
$C_5 + C_6$	5.2	2.9
C_7	3.6	2.6
C_8	6.1	10.3
C_9	3.3	7.8
C_{9+}	1.9	1.4
	100.0	100.0
Total aromatics	1.7	3.1
c. C_2H_4 oligomerization reaction (410 ± 5°C; 3.7 ± 0.1 h^{-1} WHSV; 1 atm total pressure)		
Conversion (mol%)	4.1	5.3
Selectivity (wt%)		
CH_4	0.1	0.2
C_2H_6	9.2	8.0
C_3H_6	15.0	20.9
C_3H_8	5.4	2.5
$(i + n)$ C_4H_{10}	30.5	11.3
C_4 olefins	39.8	57.1

This difference might be attributed to the deactivation of the catalyst due to coking, but, as already noted, the activities of the catalysts after 2 h on stream (the only ones

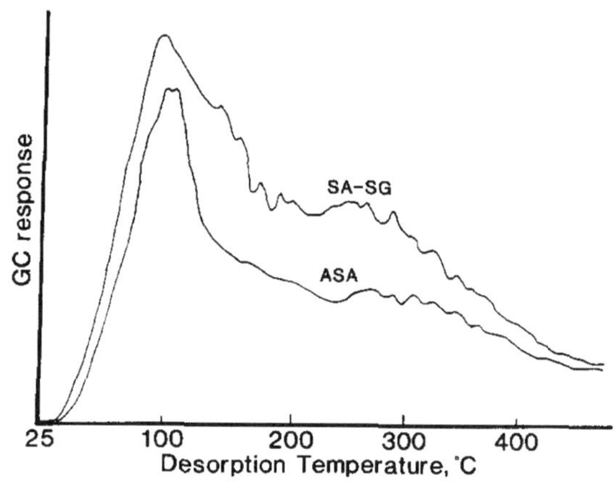

FIG. 1. Temperature-programmed desorption spectra of adsorbed ammonia for 0.23 g of SA-SG and 0.20 g of ASA.

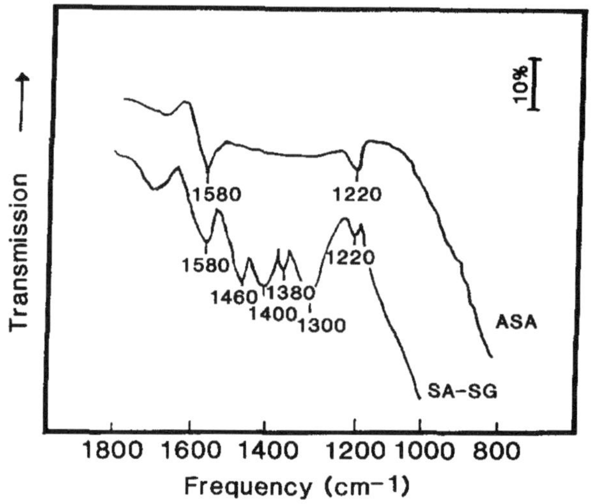

FIG. 2. IR spectra of silica–alumina samples.

listed in Table 1) had decreased by less than 5% of their initial activity. Therefore, the level of deactivation and/or coking does not appear to be responsible for the difference in behavior of the two catalysts.

Since the reactions occurring in the present study are acid catalyzed (*18–20*), it is likely that the difference in the behavior of the two catalysts can be related to their acidic characteristics. From the ammonia TPD spectra shown in Fig. 1, it is apparent that both the ASA and the SA-SG catalysts possess strong acid sites (sites from which ammonia desorbs above 200°C) which are heterogeneous in nature. The number of strong acid sites as well as the total number of acidic sites is greater on the SA-SG than on the ASA catalyst surface (using the spectral area as a measure of the acid sites after accounting for the sample weight).

Additional evidence for the presence of both strong and weak acid sites comes from the infrared spectra of the catalysts in the hydroxyl frequency range. The IR spectra of both catalysts in the frequency region of 1800–800 cm^{-1} are shown in Fig. 2. Both samples exhibit absorption peaks at 1580 cm^{-1}, which can be ascribed to water molecules bound to octahedral aluminum ions (*21*). Such sites constitute weak acid sites. Interestingly, the SA-SG catalyst has additional peaks at 1460, 1400, 1380, and 1300 cm^{-1} which are not present in the ASA catalyst. The absorption peak at 1460 cm^{-1} is due to the water molecules coordinated either on an Al^{IV}–Al^{VI} or an Al^{VI}–Al^{VI} ion pair, which constitute relatively weak acid sites. The 1380 cm^{-1} peak is due to the tetrahedrally coordinated aluminum ions (*21*) which lead to strong acid sites. The peaks at 1400 and 1300 cm^{-1} have not yet been identified. However, it is known (*21*) that the peaks in the lower frequency region represent aluminum ions with coordination states smaller than 6 and, hence, with varying capabilities to coordinate water molecules leading to strong acidic sites.

Given the above TPD and IR features which indicate a greater number of strong acid sites for the SA-SG catalyst, the SA-SG catalyst would be expected to be more active than the ASA catalyst. However, experiment shows that: (1) SA-SG and ASA catalysts are oligomerizing ethene and propene in comparable conversions, and (2) the ASA catalyst produces higher hydrocarbons from methanol while the SA-SG catalyst produces only dimethyl ether. The fact that SA-SG catalyst produces only dimethyl ether and not higher hydrocarbons, may be due either to a geometrical arrangement of the sites such that they do not allow the initial C–C bond formation or, more likely, to their preference to chemisorb water molecules instead of organic moieties. Another possibility can be related to the difference in Na levels, since sodium is known to poison acid sites. However, it is the catalyst (ASA) with the greater amount of sodium (i.e., more sites poisoned) which is more active in the methanol conversion to hydrocarbons.

The methanol conversion process involves the formation of dimethyl ether by dehydration, primary C–C bond formation, and aromatization accompanied by hydrogen transfer (*22*). According to Parera and Figoli (*23*), the ether formation over amorphous silica–alumina involves a Brønsted acid site and occurs by the following steps:

$$CH_3OH + H^+ \rightarrow CH_3OH_2^+$$

$$CH_3OH_2^+ + CH_3OH \rightarrow (CH_3)_2O + H_2O + H^+$$

The formation of the initial C–C bond from dimethyl ether on the amorphous silica–alumina can be explained either by the rake mechanism (*25*) or by the alkoxonium route (*25*). In the former scheme, the chain growth occurs by carbene insertion into the surface alkoxy species, which are later transformed into olefins via a carbenium ion intermediate. The latter route involves the formation of a dimethyl oxonium ion by the interaction of the dimethyl ether with a Brønsted acid site which then reacts with another molecule of dimethyl ether to form trimethyl oxonium ion which, finally, via a Stevens-type intramolecular arrangement results in a methyl ethyl oxonium ion. Thus, the above mechanisms require at least one Brønsted acid site (alkoxonium) and preferably two adjacent acid sites in the rake mechanism for the C–C bond formation. In the case of SA-SG such sites or the necessary topological arrangement of such sites may be absent. Another possibility is that such sites, even if present, prefer to interact strongly with the water molecules released during the dehydration step. In support of the latter possibility is the fact that the IR spectrum in Fig. 2 of the SA-SG catalyst shows multiple peaks at frequencies below 1580 cm^{-1}; as already mentioned, these peaks are due to hydroxyl groups attached to Al cations of coordination less than 6, which are strong acid sites. Consequently, water and/or oxygenate molecules may bind strongly to these sites, thus partially poisoning the catalyst. The reactions which do proceed, such as the transformation of methanol to dimethyl ether, take place on the weak acid sites.

In order to confirm the difference in behavior of the two catalysts, we conducted experiments with other oxygenates (ethanol or isopropanol) as feed. Unlike methanol, upon dehydration both of these oxygenates lead directly to their corresponding olefins, namely, ethene and propene. (In contrast to methanol, the first C–C bond formation does not need in these cases two neighboring sites.) The resulting olefins could undergo further oligomerization reactions upon the acid sites. Interestingly, with the SA-SG catalyst, ethene formed from ethanol and propene from isopropanol with selectivities greater than 96 wt%. With both oxygenates only small amounts of other hydrocarbons were detected. In contrast, with the ASA catalyst, a wide range of hydrocarbon products were obtained from both ethanol and isopropanol. These results suggest that in the case of SA-SG, the acidic sites which are active in the oligomerization reactions are indeed easily poisoned with the water molecules formed during the dehydration reaction, the latter being strongly chemisorbed on these sites. The weaker acid sites which are also present on the SA-SG catalyst are responsible for the formation of the dehydration product.

Since the SiO_2/Al_2O_3 ratio and the heat treatment procedure are almost the same for both silica–aluminas, the environment during gel formation itself is likely to affect the generation of sites of specific coordination. In the case of zeolites, it has been shown (*26–28*) that the solvent in which the gel is formed plays an important role in determining the structure.

CONCLUSIONS

The present work has shown that depending on the type of solvent in which the silica–alumina gels are prepared, the catalytic activity for oxygenate conversion to hydrocarbons can vary markedly while the activity for olefin oligomerization remains nearly constant. Thus, on the catalyst prepared in alcohol, the Brønsted acid sites responsible for oligomerization appear to be ineffective in the presence of water molecules released during the dehydration step. As a result, a low hydrocarbon yield is obtained. In contrast with the catalyst whose gel was prepared with water as a solvent, the hydrocarbon yield is large.

ACKNOWLEDGMENTS

The experimental assistance of Mr. M. Schiek is highly appreciated. Occidental Chemical Corporation's financial support to C.L. through the NSF PYI program (CBT-8857100) is also appreciated and acknowledged.

REFERENCES

1. Tanabe, K., "Solid Acids and Bases." Kodansha, Tokyo, 1970.
2. Tamele, M. W., *Discuss. Faraday Soc.* **8**, 270 (1950).
3. Ward, J. W., and Hansford, R. C., *J. Catal.* **13**, 154 (1969).
4. Mills, G. A., Boedecker, E. R., and Oblad, A. G., *J. Amer. Chem. Soc.* **72**, 1554 (1950).
5. Holm, V. C. F., Bailey, G. C., and Clark, A., *J. Phys. Chem.* **63**, 129 (1959).
6. Hirschler, A. E., *J. Catal.* **6**, 1 (1966).
7. Hansford, R. C., "Advances in Catalysis," Vol. 4, p. 17. Academic Press, San Diego, 1952.
8. Milliken, Jr., T. H., Mills, G. H., and Oblad, A. G., *Discuss. Faraday Soc.* **8**, 279 (1950).
9. Hall, W. K., Lutinski, F. E., and Gerberich, H. R., *J. Catal.* **3**, 512 (1964).
10. Thomas, C. L., *Ind. Eng. Chem.* **41**, 2564 (1949).
11. Leonard, A. J., Ratnasamy, P., Declerck, F. D., and Fripiat, J. J., *Discuss. Faraday Soc.* **52**, 98 (1971).
12. Miesserov, K. G., *J. Catal.* **13**, 169 (1969).
13. Leonard, A., Suzuki, S., Fripiat, J. J., and DeKimpe, C., *J. Phys. Chem.* **68**(9), 2608 (1964).
14. Borovkov, V. Yu, Alexeev, A. A., and Kazansky, V. B., *J. Catal.* **80**, 462 (1983).
15. Yarlagadda, P., Lund, C. R. F., and Ruckenstein, E., *Appl. Catal.* **54**, 139 (1989).
16. Nelson, H., Lussier, R. J., and Still, M. E., *Appl. Catal.* **7**, 113 (1983).
17. Yarlagadda, P., Lund, C. R. F., and Ruckenstein, E., *Appl. Catal.*, in press.
18. Shephard, F. E., Rooney, J. J., and Kemball, C., *J. Catal.* **1**, 379 (1962).
19. Poutsma, M. L., *in* "Zeolite Chemistry and Catalysis" (J. A. Rabo, Ed.), ACS Monograph 171, p. 483. Amer. Chem. Soc. Washington, DC, 1976.
20. Chang, C. D., "Hydrocarbons From Methanol." Dekker, New York, 1983.
21. Vlaev, L., Damyanov, D., and Mohamed, M. M., *Colloids Surf.* **36**, 427 (1989).
22. Chang, C. D., and Silvestri, A. J., *J. Catal.* **47**, 249 (1977).
23. Parera, J. M., and Figoli, N. S., *J. Catal.* **14**, 303 (1969).
24. Cormerias, F. X., Perot, G., Chevalier, F., and Guisnet, M., *J. Chem. Res. Synop.* 362 (1980).
25. van den Berg, J. P., Wolthiuzen, J. P., and van Hooff, J. H. C., "Proceedings Fifth Conf. Zeolites, Naples, Italy, 1980," p. 649.
26. Bibby, D. M., and Dale, M. P., *Nature (London)* **317**, 157 (1985).
27. van Erp, W. A., Kouwenhoven, H. W., and Nanne, J. M., *Zeolites* **7**, 286 (1987).
28. Sand, L. B., "Molecular Sieves" (R. M. Barrier, Ed.), p. 71. Soc. Chem. Ind., London, 1968.

Conversion of Methanol to Hydrocarbons over Silica–Alumina

Selective Formation of Lower Olefins

P. YARLAGADDA, C.R.F. LUND and E. RUCKENSTEIN*

Department of Chemical Engineering, State University of New York at Buffalo, Amherst, New York 14260 (U.S.A.)

(Received 23 January 1989, revised manuscript received 8 May 1989)

ABSTRACT

The catalytic conversion of methanol to hydrocarbons, particularly to the lower olefins, has been studied in a fixed-bed flow reactor over amorphous silica–alumina and over silica–alumina containing small amounts of zeolite. The amorphous silica–alumina catalysts were prepared by the cogelation method such that the alumina content varied from 1.5 to 30 wt.-%. The composite catalysts were prepared by adding crystallites of HAlZSM-5 ($SiO_2/Al_2O_3=35$) during the cogelation procedure. In each of these composite catalysts the silica–alumina matrix contained 15 wt.-% alumina whereas the ratio of the HAlZSM-5 to the silica–alumina matrix varied from 4 to 19 wt.-%. The conversion of methanol to hydrocarbons other than dimethyl ether at 375°C, 1 atm and 1 h^{-1} space velocity, over the synthesized amorphous silica–alumina catalysts increased from 22.5 to 96.5% as the alumina content increased from 5 to 20 wt.-%. The selectivity to C_2–C_4 olefins was essentially constant (62 to 65%) over this range of SiO_2/Al_2O_3 ratios. These amorphous silica–alumina catalysts deactivated significantly over a period of 15 h. When small amounts (5 to 10 wt.-%) of HAlZSM-5 zeolite were added to the silica–alumina matrix having 15 wt.-% Al_2O_3, a more stable and active catalyst for the conversion of methanol to lower olefins was obtained.

INTRODUCTION

Selective conversion of methanol to lower olefins is a process of economic interest [1]. Although there is some controversy regarding the mechanism of formation of the initial hydrocarbon or hydrocarbons, it is generally accepted that lower olefins are intermediates in the conversion of methanol to higher hydrocarbons over ZSM-5 class zeolite catalysts [2]. The reaction pathway can be represented as follows:

$$[2CH_3OH \rightleftharpoons (CH_3)_2O + H_2O] \rightarrow C_2\text{–}C_4 \text{ olefins} \rightarrow (\text{paraffins} + \text{aromatics})$$

The extensive literature concerning the ZSM-5 zeolite-mediated conversion of methanol to higher hydrocarbons as well as that to lower olefins has been re-

This paper, which was originally published in **Applied Catalysis**: 54, 139-157 (1989), is republished with permission.

viewed by Chang [3,4]. The formation of mixtures of aromatic and paraffinic hydrocarbons from methanol involves a number of reactions including olefin formation, alkylation, oligomerization, aromatization, cracking and disproportionation, and these reactions are coupled [5]. Given a reaction sequence like that above, at least two approaches can be used to obtain a higher selectivity to C_2–C_4 olefins. One approach would simply be to operate at shorter residence times; as expected, it has been observed that at low methanol conversions, the lower olefins are formed in abundance, while at high conversions aromatic and C_{5+} aliphatic hydrocarbons [6] are favored. The second approach would be to develop a catalyst which can selectively convert methanol, at high conversions, to lower olefins.

The HAlZSM-5 catalyst is considered to be the most active and stable catalyst for the conversion of methanol to hydrocarbons [7]. The two dimensional structure and channel openings of 5.4 Å×5.6 Å [8] are responsible for impeding the formation of coke precursors, such as fused-ring aromatics. However, pure ZSM-5 catalyst is not very selective towards the formation of lower olefins, the latter being further converted to aromatics and paraffins. This is due to the high intermolecular hydrogen transfer among the reacting species. Thus with this catalyst the approach has been to alter the process conditions (temperature, pressure and space velocity) so that most of the reaction stops after olefin formation. High selectivities for the lower (C_2–C_4) olefins have been achieved by increasing the reaction temperature [9], by decreasing the contact time [10] or by reducing the oxygenate partial pressure [11]. However, these methods suffer from a number of drawbacks such as (i) low conversion of methanol which introduces the need for recycling, (ii) rapid deactivation and (iii) deep cracking at high reaction temperatures.

The other approach to enhance selectivity to lower olefins also has been pursued by making various modifications to the ZSM-5 catalyst. These include: (i) the treatment of ZSM-5 with compounds of phosphorus [12], magnesium [13], or silicon [14]; (ii) the use of ZSM-5 catalysts with high SiO_2/Al_2O_3 ratios [15], and, (iii) the framework substitution of Si^{4+} by Fe^{3+} [16]. Unfortunately, such modifications sometimes significantly decrease the catalytic activity of the zeolite. For this reason, the reaction temperatures must be increased considerably to achieve higher methanol conversions. This in turn increases the coke formation and reduces the catalyst life.

Zeolites with narrower pores, such as ZSM-34 [17], ZK5 [18] and clinoptilolite [19] have also been examined for selective formation of lower olefins from methanol. Although the selectivity to C_2–C_4 olefins is high, the activity decreases rapidly due to coke deposition.

The majority of studies on methanol conversion reactions have involved the use of pure zeolite crystallites, either in the form of powder, or in the form of small compressed particles. Binders such as silica [20] and alumina [21] have been used in some instances. As silica or alumina possess little acidity [22], their direct contribution to the formation of hydrocarbons is negligible. However, they may contribute to the overall activity of a composite catalyst by their effect on the diffusion of the species involved. Both the activity and selectivity of such a catalyst can be optimized by diluting the zeolite in an inactive or active matrix of relatively high porosity [23,24]. In the present investigation the effect of a small addition of zeolite to a silica–alumina catalyst has been studied. In the studies cited above the zeolite is the primary active component. In the present study, silica–alumina which is known to be a solid acid capable of initiating hydrocarbon transformation reactions [25–32] is also a major active component. Hence both components of the catalyst are active. The aspects of this complex catalyst system which have been examined are

(i) the influence of Si/Al ratio in the pure silica–alumina component on the activity and selectivity,

(ii) the time-on-stream behavior of pure silica–alumina catalysts both in terms of activity and selectivity, and

(iii) the performance of composite catalysts containing varying amounts of HAlZSM-5 zeolite dispersed in a silica–alumina matrix.

EXPERIMENTAL

Preparation of silica–alumina catalysts

Numerous samples have been prepared under a variety of synthesis conditions, i.e. by changing the pH for cogelation between 3 and 9, by mixing preformed silica and alumina gels or by addition of an aluminum salt solution to preformed silica gel. The corresponding calcined samples have been catalytically screened for methanol conversion activity and selectivity. Most of the samples were able to dehydrate methanol to dimethyl ether and small amounts of methane. Only those samples which have been prepared by the cogelation procedure described below have been found to be active in the formation of hydrocarbons higher than C_2.

Silica–alumina catalysts were prepared by cogelation with an alumina content that varied from 1.5 to 30 wt.-%. The pH of a 500 ml batch of 0.25 *M* aqueous sodium silicate (Fisher Scientific) was adjusted to 4 by adding 0.2 *M* hydrochloric acid. To the resultant solution, 250 ml of 0.08 *M* aqueous aluminum nitrate (Baker) was added. By slowly adding 0.2 *M* NH_4OH the pH of the mixture was adjusted to about 6.5. At this stage, gelation took place. The entire mass was agitated for 24 h and then filtered. The resultant cake was thoroughly washed and dried at 100°C overnight. The dried mass was finely ground and then calcined at 600°C for 1 h in air. The calcined material was subsequently subjected to NH_4^+ exchange, by treating it with 1 *M* aqueous $(NH_4)_2SO_4$ solution at room temperature for 3 h. A final calcination at 550°C for 1 h followed.

Preparation of composite zeolite catalyst

The ZSM-5 zeolite was synthesized by the method of Chen et al. [33] with a SiO_2/Al_2O_3 molar ratio of 35. The zeolite was converted to its protonated form by NH_4^+ exchange followed by calcination at 550°C for 12 h. X-ray diffraction (XRD) showed that the synthesized zeolite was highly crystalline and the *d*-values matched well with those reported for ZSM-5 zeolite [33].

To prepare the composite SiO_2–Al_2O_3/zeolite catalysts, a modification of the previously described cogellation procedure was used. A solution containing sodium silicate and aluminum nitrate was prepared such that the resultant solid would contain 15 wt.-% Al_2O_3. The pH was adjusted to 6.5–7.0 using 0.2 *M* NH_4OH. Just before the onset of gelation, a measured quantity of HAlZSM-5 zeolite powder was added to the solution. Vigorous stirring was maintained for 24 h, after which the slurry was filtered. The resulting cake was thoroughly washed, dried and calcined, and the calcined material was ion exchanged with NH_4^+. Finally a second calcination was performed under conditions similar to those mentioned for the silica–alumina catalysts.

The chemical compositions, surface areas and information regarding crystallinity are presented in Table 1 for the catalysts synthesized in the present study. The XRD patterns for the three composite catalysts are generally weak and broadened. This is not an indication that the zeolite component has lost its crystallinity. Fig. 1 shows the XRD patterns for the pure HAlZSM-5, a physical mixture of 5% of the pure HAlZSM-5 in SiO_2 and two of the composite catalysts prepared by cogellation. The weakness and broadening of the peaks is clearly present in the physical mixture. Given the high thermal and hydro-

TABLE 1

Properties of catalysts used in the present study

Designation	Nominal composition		BET surface area (m^2g^{-1})	Pore volume (cm^3g^{-1})	XRD Data[b]
	Al_2O_3 (wt.-%) in silica-alumina	Amount of zeolite (wt.-%)			
ASA-1	1.5	–	264	0.59	Amorphous
ASA-2	5.0	–	261	0.62	Amorphous
ASA-3	10.0	–	290	0.66	Amorphous
ASA-4	15.0	–	410	0.69	Amorphous
ASA-5	20.0	–	304	0.61	Amorphous
ASA-6	30.0	–	174	0.52	Amorphous
ZSA-1[a]	15.0	5.0	400	0.68	Line broadening
ZSA-2[a]	15.0	10.0	425	0.72	Weak peaks
ZSA-3[a]	15.0	20.0	420	0.72	Weak peaks

[a]HAlZSM-5 ($SiO_2/Al_2O_3=35$) zeolite was used.
[b]Line broadening at $2\theta=23°$; weak peaks at $2\theta=23.34°, 9.39°, 8.66°$.

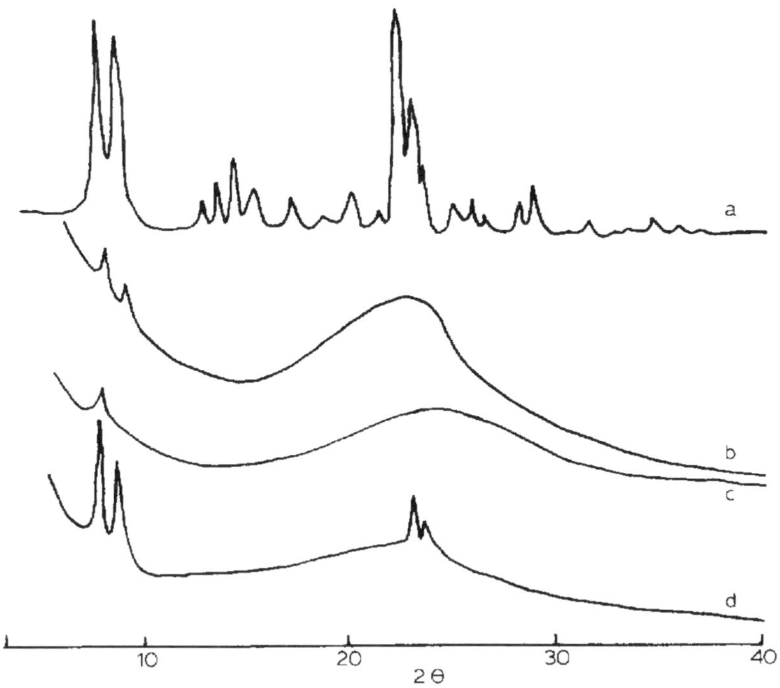

Fig. 1. Comparison of the X-ray diffraction patterns observed for (a) pure HAlZSM-5, (b) a physical mixture of 5% HAlZSM-5 with silica, (c) the 5 wt.-% composite catalyst (ZSA-1) prepared by cogellation and (d) the 10 wt.-% composite catalyst (ZSA-2) prepared by cogellation.

thermal stability of ZSM-5 zeolites [34] a loss of crystallinity is not to be expected, and the strong similarity between the physical mixture and the composite catalysts supports this expectation.

Temperature-programmed desorption (TPD)

The TPD spectra of the catalyst samples were obtained using the conventional method of temperature-programmed desorption of a chemisorbed base [35]. The TPD set up was similar to that reported in the literature [36]. Anhydrous ammonia was used as the probe base molecule. The desorbed ammonia molecules were detected by a Pt–Re thermal conductivity detector. A typical TPD experiment is described below. A shallow packed bed of the catalyst (0.2 g) was heated in a helium flow at 500°C for 2 h to remove moisture and subsequently cooled down to the adsorption temperature (25°C). Ammonia was then pulsed over the sample in aliquots of known concentration until saturation, as detected by its breakthrough curve. Before starting desorption, the sample tube was flushed with helium for 2 h. The TPD run was carried out from 25°C to 600°C, with a heating rate of 10°C min^{-1}. Helium (Linde, High Purity Grade) was used as the carrier gas at a flow-rate of 30 sccm (1 sccm = 1 cm^3 (STP) min^{-1}).

Catalytic studies

The catalytic conversion of methanol to hydrocarbons was carried out in a fixed-bed (6 mm I.D.) stainless steel microreactor. In all the experiments, 0.5 g of catalyst (60/80 mesh) was held between two small quartz wool plugs located at the mid-point of the reactor tube. Prior to introducing the methanol feed, each catalyst sample was heated in flowing oxygen at 500 °C for 3 h, followed by complete purging with helium. Experiments were carried out at 1 atm total pressure, 375 ± 5 °C, WHSV (g of methanol/g catalyst.h) of $1.0 \pm 0.1\ h^{-1}$, and with a feed of 30% methanol in helium. The gases emerging from the reactor were analyzed for (i) C_1–C_4 hydrocarbons on a VZ-10 column (6 feet × 1/8 inch) using a flame ionization detector, (ii) C_{5+} hydrocarbons on a 10% SE-30 Chrom W-HP column (6 feet × 1/8 inch) again using a flame ionization detector, and (iii) CH_3OH, $(CH_3)_2O$, CO, and H_2O on a Porapak-T column (6 feet × 1/8 inch) using a thermal conductivity detector. The methanol had a purity of 99.9% (certified ACS grade from Fisher Scientific). All the isomers of the C_2–C_4 fraction were well resolved and identified on the VZ-10 column. Within the C_{5+} fraction, the aromatic hydrocarbons up to C_{12} have been identified by comparing their retention times with those of the pure compounds, but the aliphatics have been lumped together in the presentation of the data.

Throughout the present work, the percentage conversion has been defined as

$$x = \frac{n^{o}_{CH_3OH} - (n_{CH_3OH} + 2n_{DME})}{n^{o}_{CH_3OH}} \cdot 100\%$$

where x is the conversion, $n^{o}_{CH_3OH}$ is the moles of the methanol fed to the reactor per unit time, n_{CH_3OH} is the moles of methanol leaving the reactor per unit time, and n_{DME} is the moles of dimethyl ether leaving the reactor per unit time. That is, dimethyl ether is treated as equivalent to unreacted methanol, not as a reaction product. Selectivities are reported here as the weight of the desired species leaving the reactor divided by the total weight of aliphatic and aromatic hydrocarbons leaving the reactor. That is, the selectivities reported do not include carbon monoxide or carbon dioxide. The instantaneous carbon balance was generally observed to close to within 3–5%.

RESULTS AND DISCUSSION

Conversion of methanol to hydrocarbons over amorphous silica–alumina catalysts

The methanol conversion and hydrocarbon product distribution obtained over a series of silica–alumina catalysts (ASA-2 to ASA-6) are given in Table 2. The data in Table 2 were measured after 60 min on-stream. When a series

TABLE 2

Methanol conversion and hydrocarbon product distribution over amorphous silica–alumina catalysts

$375 \pm 5\,^\circ C$; $1.0 \pm 0.1\ h^{-1}$ WHSV; 1 atm total pressure; time-on-stream = 60 min.

Catalyst	ASA1[a]	ASA2	ASA3	ASA4	ASA5	ASA6	HAlZSM-5 ($SiO_2/Al_2O_3 = 35$)
Hydrocarbon distribution (wt.-%)							
CH_4		16.9	15.3	2.5	2.7	5.9	1.3
C_2H_6		0.1	0.1	0.1	0.1	0.4	0.4
C_2H_4		13.2	17.8	15.0	19.1	17.0	1.7
C_3H_8		1.1	1.3	1.1	1.5	5.5	13.1
C_3H_6		28.5	25.5	25.2	26.6	21.4	3.2
i-C_4H_{10}		13.4	15.3	14.9	14.7	11.2	22.3
n-C_4H_{10}		0.6	0.7	0.7	0.6	2.1	2.9
1-C_4H_8		2.1	5.2	7.5	3.6	1.0	1.3
i-C_4H_8		6.1	5.8	5.5	3.8	3.6	1.9
(*cis* + *trans*) 2-C_4H_8		10.3	2.4	2.8	6.0	5.7	1.1
C_4H_6		5.4	7.3	8.1	3.6	5.8	2.8
C_{5+} non-aromatics		2.3	2.8	11.5	7.8	8.3	14.6
Aromatics		–	0.5	5.1	9.9	12.1	43.4
Total C_2–C_4 olefins		65.6	64.0	64.1	62.7	54.5	12.0
Conversion		22.5	62.9	95.8	96.5	83.1	100.0
C_2–C_4 olefin/paraffin ratio		4.3	3.7	3.8	3.7	2.8	0.3

[a]Main product dimethyl ether, some methane and no C_{2+} hydrocarbons.

of similar catalysts are compared at equal feed and operating conditions, changes in observed selectivity may be due to catalyst properties or due to conversion–selectivity relationships. Additional complication arises from the unsteady behavior of these catalysts. By comparing at a fixed time-on-stream, different catalysts are compared at unequal degrees of deactivation. Ideally the catalysts should be evaluated under identical conditions of conversion, degree of deactivation, etc. Here two comparisons will be used: first a comparison as represented in Table 2, and additionally, a comparison of time-on-stream behavior.

The results obtained, under similar reaction conditions, over a binderless HZSM-5 zeolite catalyst ($SiO_2/Al_2O_3 = 35$) are also included in Table 2 for comparison. As expected, with the HZSM-5 zeolite, methanol was found to be totally converted and the C_{5+} hydrocarbon selectivity was 60 wt.-%, with aromatics accounting for 72 wt.-% of this fraction. Over the amorphous silica–alumina catalysts, the methanol conversion was found to increase from approximately 22% to around 96% as the alumina content increased from 5% (ASA-2) to 20% (ASA-5). On the silica–alumina catalyst which had 1.5% Al_2O_3 (ASA-1), methanol essentially dehydrated to give dimethyl ether together with a small amount of methane, but no higher hydrocarbons. At still higher aluminum content the conversion decreased slightly, but other catalyst properties also changed, as discussed later.

Among the hydrocarbon products formed over the active silica–alumina catalysts (ASA-2 through ASA-5), the lower olefins constituted a major fraction. Interestingly, the selectivity for the C_2–C_4 olefins remained essentially constant at ca. 64%, irrespective of the level of methanol conversion which ranged from 22.5 to 96.5%. However, with the highest alumina containing catalyst (ASA-6), both the methanol conversion and the selectivity for the lower olefins were found to decrease. Examination of the hydrocarbon products (Table 2) shows that the methane selectivity markedly decreases with an increase in the alumina content (>10 wt.-%) of the silica–alumina catalyst, while, the C_{5+} hydrocarbon fraction including aromatics increased. The reaction mechanisms suggested by Ono and Mori [37] and Dejaifve et al. [38] involve the formation of methoxy groups on the surface Brønsted acid sites of the catalyst. These methoxy groups can act either as hydrogen transfer agents, yielding methane, or as a source of methyl carbonium ions which can then enter into the hydrocarbon growth pathway. At lower conversions there are fewer olefins, and hence less possibility for the hydrocarbon growth pathway. Thus the methane pathway is favored at lower conversion, and less favored as conversion increases.

The time-on-stream behavior of the two most active silica–alumina catalysts (ASA-4 and ASA-5) was studied over a period of about 15 h and the results are presented in Tables 3 and 4, respectively. The methanol conversion over both catalysts decreased with the time-on-stream, the decrease being more significant for the ASA-5 catalyst. On the ASA-4 catalyst, the selectivity for the C_2–C_4 olefins at the beginning (fifteen minutes after switching to the methanol feed) was 61.7 wt.-% and it increased to 69.1 wt.-% after 1.5 h on-stream. During the subsequent part of the experiment, the selectivity for the lower olefins remained unchanged. However, during this reaction period, the selectivity towards methane increased nearly four-fold from 2.7 to 9.5 wt.-%. The C_{5+} non-aromatic fraction decreased from 15.3 to 5.4 wt.-% during the first 1.5 h and remained essentially the same later. On the other hand, the aromatic hydrocarbons increased from 0.8 to about 4 wt.-% during the first 1.5 h and then remained constant. As seen from Table 4, the performance of the silica–alumina catalyst ASA-5 was similar to that of ASA-4. The methanol conversion decreased from an initial value of 96.7 to 47.3% over 15 h, while the selectivity for the C_2–C_4 olefins remained at about 60 wt.-%. As compared to the ASA-4 catalyst, the selectivity for methane over the ASA-5 silica–alumina catalyst was high (15–25 wt.-% as compared to 2.7–9.5 wt.-%). The increase in the selectivity to methane as a function of time-on-stream has been observed in other studies [29,39] and is believed to be related to the coking of the cat-

TABLE 3

Variation in activity and hydrocarbon product selectivity with time-on-stream over amorphous (ASA-4) silica-alumina catalyst

Time-on-stream (h)	Hydrocarbon product distribution, (wt.-%)					
	0.25	1.5	4.0	7.0	10.0	14.0
CH_4	0.5	2.7	2.9	4.4	9.1	9.5
C_2H_6	0.2	0.1	0.2	0.2	0.3	0.3
C_2H_4	19.3	16.1	17.5	16.5	16.2	16.0
C_3H_8	2.4	1.9	2.0	1.9	1.6	1.6
C_3H_6	25.4	27.2	27.0	27.0	27.5	27.8
i-C_4H_{10}	16.3	16.0	16.8	14.8	12.9	13.0
n-C_4H_{10}	0.8	0.7	0.8	0.8	0.6	0.6
1-C_4H_8	5.5	8.1	6.1	6.3	5.9	5.8
i-C_4H_8	3.1	5.9	5.4	7.1	6.7	6.5
(*cis*+*trans*) 2-C_4H_8	2.6	3.0	3.7	3.2	3.2	2.8
C_4H_6	7.8	8.8	9.0	8.7	7.5	7.5
C_{5+} non-aromatics	15.3	5.4	4.7	4.8	4.1	4.4
Aromatics	0.8	4.1	3.9	4.3	4.4	4.2
Total C_2–C_4 olefins	61.7	69.1	68.7	68.8	67.0	66.4
Conversion	96.2	93.1	92.8	83.1	62.5	58.9

TABLE 4

Variation in activity and selectivity during the methanol to hydrocarbons reaction over amorphous (ASA-5) silica-alumina catalyst

Time-on-stream (h)	Hydrocarbon product distribution (wt.-%)					
	0.20	2.3	4.0	6.0	12.0	15.0
CH_4	3.0	15.8	14.4	24.5	21.4	19.4
C_2H_6	0.2	0.1	0.1	0.4	0.2	0.2
C_2H_4	21.3	19.0	18.8	15.6	15.9	17.6
C_3H_8	1.7	2.0	2.3	1.5	2.2	2.1
C_3H_6	29.6	25.6	24.9	23.0	21.0	23.6
i-C_4H_{10}	16.4	14.7	15.9	11.4	14.0	14.5
n-C_4H_{10}	0.6	0.8	0.8	0.6	0.8	0.9
1-C_4H_8	4.0	5.6	5.9	4.0	4.7	5.3
i-C_4H_8	4.2	4.0	4.1	5.1	3.6	3.9
(*cis*+*trans*) 2-C_4H_8	6.7	2.5	2.6	1.4	2.2	2.3
C_4H_6	4.1	4.1	4.2	6.2	7.7	5.6
C_{5+} non-aromatics	4.9	2.0	2.1	2.5	2.6	0.9
Aromatics	3.3	3.8	3.9	3.8	4.1	3.7
Total C_2–C_4 olefins	69.9	60.8	60.5	55.3	55.1	58.3
Conversion	96.7	92.3	82.1	73.9	48.6	47.3

alyst. As coking proceeds, i.e., with increasing time-on-stream, the conversion decreases, and as previously noted decreased conversion favors the methane pathway. Furthermore, the ASA-5 silica–alumina catalyst deactivated more rapidly than the ASA-4 catalyst (Fig. 2), despite the fact that the initial level of conversion was similar.

The results presented in Table 2, indicate that the methanol conversion (activity) rather than the selectivity is strongly dependent upon the SiO_2/Al_2O_3 ratio of the amorphous silica–alumina catalysts. A number of studies, summarized in ref. 22, have shown that the acid strength and also the amount of Brønsted acidity of the amorphous silica–aluminas increase with their alumina content up to about 30 wt.-% and later decrease. Furthermore, it has been observed that the activity of the ZSM-5 zeolite catalyst for methanol conversion is related to the strength and number of Brønsted acid sites [40]. It appears that the activity of amorphous silica–alumina catalysts for methanol conversion also increases with their acidity.

The temperature-programmed desorption of ammonia from these catalysts has been used as a qualitative measure of the acid site density and acid strength distribution. Desorption spectra are shown in Fig. 3 for each catalyst in the series which exhibited hydrocarbon production. All the catalysts showed a sharp peak at approximately 100°C, and there is no apparent correlation between the intensity of this peak and the activity of the catalysts. Thus it appears that this peak represents acid sites which are too weak to catalyze methanol con-

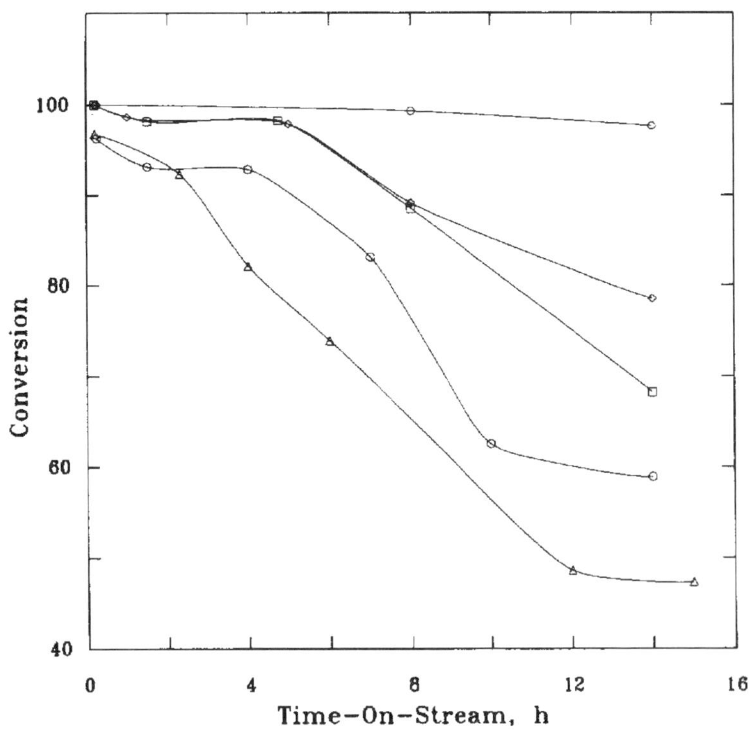

Fig. 2. Decrease in methanol conversion as a function of time-on-stream. (○) ASA-4, (△) ASA-5, (◇) ZSA-1, (○) ZSA-2, (□) ZSA-3.

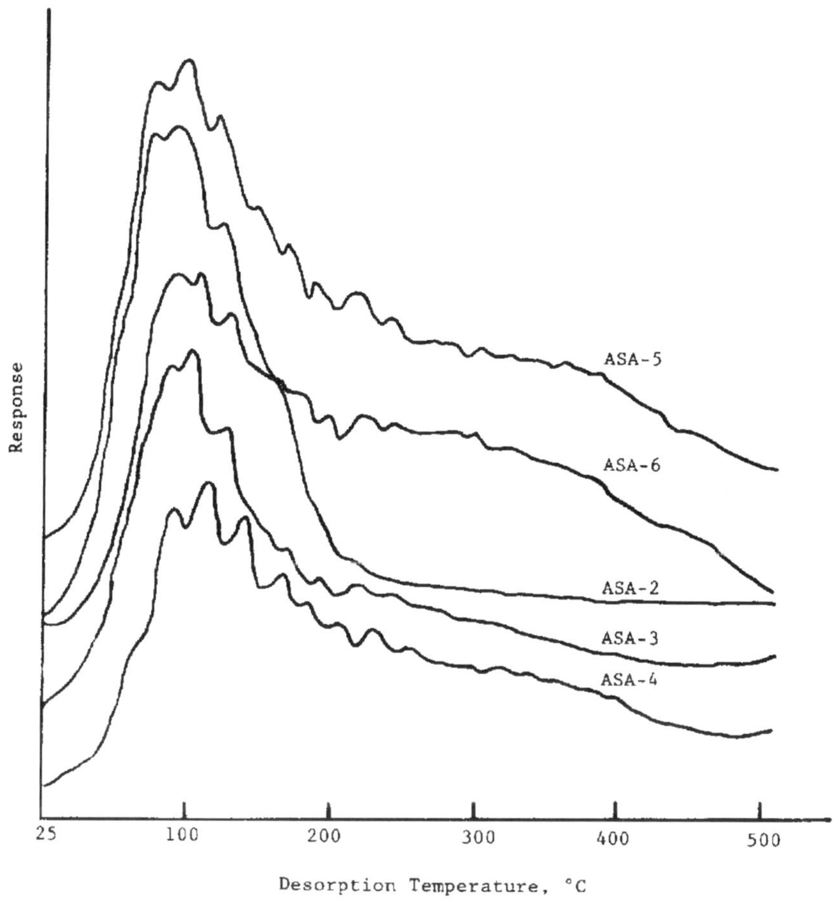

Fig. 3. Ammonia temperature-programmed desorption spectra from the amorphous silica–aluminas. Ammonia adsorbed at 25 °C and 1 atm; ammonia desorbed into flowing helium (30 sccm) by heating 0.2 g sample at a rate of 10 °C min^{-1}.

version under reaction conditions. In most of the spectra there is also a weak shoulder at about 200 °C which tails out to approximately 400 °C. This shoulder is absent in the 5% alumina sample, grows steadily through the 20 wt.-% alumina sample and then declines slightly in the 30 wt.-% sample. This correlates very well with the trend for activity (conversion). In the case of the 30 wt.-% alumina sample there is also a significant decrease in the specific surface area. Thus the lower activity of this sample is not necessarily due to a decrease in acid site density, it may be due to a loss in total sites.

It is interesting to note that the conversion–selectivity relationship observed with the present silica–alumina catalysts differs from that reported for ZSM-5 zeolite catalysts. Chang et al. (Fig. 1 in ref. 9) noted that an increase in the conversion of methanol (due to an increase in the ZSM-5 catalysts' Brønsted acidity) causes the following selectivity changes: (i) a continuous decrease in C_2–C_4 olefins, (ii) an increase in aromatics and C_1–C_4 paraffins and (iii) a decrease in the C_{5+} nonaromatics fraction after passing through a maximum. However, with the amorphous silica–alumina catalyst the decrease in the selectivity for C_2–C_4 olefins was small and the increase in aromatics was modest. A plausible explanation for this behavior may lie in the interplay of acidity and diffusion in the macro and mesopores of the silica–alumina catalysts and in the absence of a confining channel network. In the micropores of zeolite catalyst, the olefins which are formed initially undergo a number of reactions (outlined in ref. 41) to form aromatics and paraffins. This is in part because the small size of the zeolite channels precludes these olefins from diffusing away from the active acid sites. Consequently, relatively few low molecular weight olefins survive and leave the zeolite crystal to enter the bulk gas stream. In the pores of an amorphous silica–alumina the resistance for these products to diffuse away from the active acid sites will be smaller. As a result, the extent of the secondary reactions (olefins$\rightleftharpoons$naphthenes; olefins + naphthenes$\rightarrow$ aromatics + paraffins) in the series reaction scheme will be smaller. This is even more likely when the olefins react primarily with reactive intermediates formed between methanol and the acid sites, and not directly with the acid sites themselves. At the entrance of the reactor the bulk gas phase olefin concentration is low and diffusion away from the active sites is favored. As a result little secondary reactions will occur. As the reaction proceeds through the bed the driving force for diffusion of olefins away from the sites decreases, but less unreacted methanol/dimethyl ether is available for secondary reactions with the olefins. In the zeolite, however, the close confines of the channels preclude

TABLE 5

Methanol conversion and hydrocarbon product distribution over composite zeolite catalysts

$375 \pm 5\,^{\circ}C$, $1.0 \pm 0.1\ h^{-1}$ WHSV; 1 atm total pressure, time-on-stream = 60 min.

Catalyst	ZSA-1	ZSA-2	ZSA-3
Hydrocarbon distribution (wt.-%)			
CH_4	2.2	2.8	3.3
C_2H_6	0.1	0.1	Trace
C_2H_4	10.3	7.8	7.8
C_3H_8	1.6	2.1	2.3
C_3H_6	28.2	24.5	12.8
i-C_4H_{10}	11.8	11.3	14.9
n-C_4H_{10}	1.0	0.6	0.9
1-C_4H_8	5.2	5.0	4.0
i-C_4H_8	4.3	4.7	2.6
(*cis* + *trans*) 2-C_4H_8	2.4	2.5	2.1
C_4H_6	8.8	6.5	9.7
C_{5+} non-aromatics	16.8	21.7	23.7
Aromatics	8.3	10.4	15.9
Total C_2–C_4 olefins	59.2	51.0	39.0
Conversion	98.6	100.0	100.0
C_2–C_4 olefin/paraffin ratio	4.1	3.6	2.2

TABLE 6

Performance of composite zeolite catalyst (ZSA-1) during the methanol to hydrocarbon conversion reaction

Time-on-stream (h)	Hydrocarbon product distribution (wt.-%)				
	0.20	1.0	5.0	8.0	14.0
CH_4	1.8	2.2	2.2	1.8	3.9
C_2H_6	0.2	0.1	0.1	0.2	0.1
C_2H_4	10.9	10.3	9.8	16.0	12.4
C_3H_8	2.3	1.6	1.8	1.7	1.1
C_3H_6	28.1	28.2	27.7	29.6	32.0
i-C_4H_{10}	12.2	11.8	13.7	9.6	9.6
n-C_4H_{10}	1.0	1.0	1.5	0.6	0.5
1-C_4H_8	5.2	5.2	5.2	5.6	5.6
i-C_4H_8	3.9	4.3	4.6	4.9	5.6
(*cis*+*trans*)2-C_4H_8	2.3	2.4	2.8	2.8	2.8
C_4H_6	6.8	8.8	8.0	7.2	6.4
C_{5+} non-aromatics	17.2	16.8	10.2	7.9	7.4
Aromatics	7.1	8.3	12.4	12.1	12.6
Total C_2-C_4 olefins	57.2	59.2	58.1	66.1	64.8
Conversion	100.0	98.6	97.8	89.1	78.5

the olefins from diffusing away from the sites even at the front of the reactor. Hence a higher extent of secondary reactions occurs throughout. Other shape selectivity effects such as a cut-off in the maximum carbon number or a high *p*-xylene selectivity are not observed with the amorphous catalysts [32,42]. Nevertheless, with the amorphous silica–alumina catalysts synthesized in the present study, a high selectivity for C_2–C_4 olefins was obtained and this may be due to the combination of their acidic properties and their pore structure. There remains the problem of severe deactivation, however.

Methanol conversion reaction over composite zeolite catalysts

The activities and hydrocarbon product distributions obtained over the composite zeolite catalysts (ZSA-1 to ZSA-3) are presented in Table 5. In all these composite catalysts, the amount of alumina in the silica–alumina matrix was maintained at 15 wt.-% and the zeolite (HAlZSM-5, $SiO_2/Al_2O_3=35$) content was varied from 5 wt.-% (ZSA-1) to 20 wt.-% (ZSA-3). All the three composite zeolite catalysts were found to be highly active for the conversion of methanol (>98%). This is not surprising as the matrix alone exhibits a 96% conversion and HAlZSM-5 exhibits 100% (Table 2). However, the hydrocarbon product distributions (Tables 5 and 2) show that the composite zeolite catalysts exhibit a higher selectivity to C_2–C_4 olefins (39 to 59 wt.-%) compared to pure

TABLE 7

Variation in methanol conversion and hydrocarbon product selectivity over composite zeolite (ZSA-2) catalyst

Time-on-stream (h)	Hydrocarbon product distribution (wt.-%)		
	0.25	8.0	14.0
CH_4	3.3	2.0	2.2
C_2H_6	0.1	0.1	0.1
C_2H_4	8.1	7.1	7.9
C_3H_8	2.2	1.3	0.9
C_3H_6	23.6	26.1	25.3
i-C_4H_{10}	13.8	7.1	6.1
n-C_4H_{10}	0.7	0.5	0.1
1-C_4H_8	4.4	7.2	5.9
i-C_4H_8	4.6	6.2	5.7
(*cis*+*trans*) 2-C_4H_8	2.7	2.9	4.2
C_4H_6	10.2	5.3	8.1
C_{5+} non-aromatics	17.6	22.0	17.9
Aromatics	8.7	12.2	15.6
Total C_2-C_4 olefins	53.6	54.8	57.1
Conversion	100.0	99.3	97.6

TABLE 8

Methanol conversion and hydrocarbon product selectivity data over composite zeolite (ZSA-3) catalyst

Time-on-stream (h)	Hydrocarbon product distribution (wt.-%)				
	0.20	1.5	4.75	8.0	14.0
CH_4	3.3	3.2	4.0	6.1	8.4
C_2H_6	Trace	0.1	0.1	0.1	0.1
C_2H_4	7.3	7.9	5.7	10.1	10.8
C_3H_8	2.4	2.1	1.5	1.0	1.1
C_3H_6	7.4	15.5	15.9	26.1	27.0
i-C_4H_{10}	15.2	10.3	7.9	4.4	3.9
n-C_4H_{10}	0.9	0.8	0.5	0.4	0.4
1-C_4H_8	3.3	5.6	6.4	5.4	5.1
i-C_4H_8	Trace	3.4	7.2	4.2	4.2
(*cis*+*trans*) 2-C_4H_8	1.4	2.7	6.2	4.6	4.1
C_4H_6	10.3	8.4	6.7	3.9	4.1
C_{5+} non-aromatics	32.5	23.9	22.1	19.2	13.6
Aromatics	16.0	16.1	15.8	14.5	17.2
Total C_2-C_4 olefins	29.7	43.5	48.1	54.3	55.3
Conversion	100.0	98.1	98.2	88.5	68.2

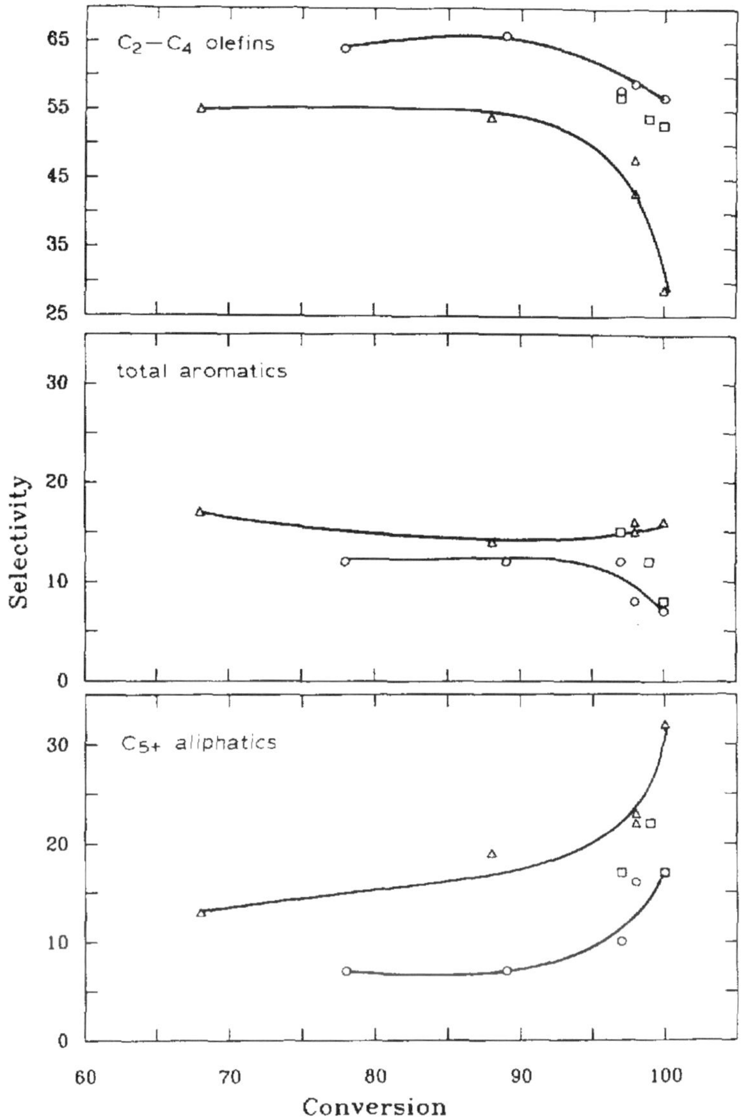

Fig. 4. Hydrocarbon selectivity as a function of conversion for the composite catalysts. (○) ZSA-1, (□) ZSA-2, (△) ZSA-3.

HAlZSM-5 (12 wt.-%). As the amount of zeolite in the composite catalyst increased, the selectivity for the lower olefins decreased and the C_{5+} fraction including the aromatic hydrocarbons increased. The methanol conversion remained essentially the same. Methanol conversion and hydrocarbon product distribution were measured for all catalysts as a function of time-on-stream, and the results are summarized in Tables 6 through 8. Analysis of duplicate preparations of samples demonstrated the reproducibility of the data. As seen from Fig. 3, the composite zeolite catalyst ZSA-2 (10 wt.-% zeolite) showed better activity maintenance than catalyst ZSA-1 (5 wt.-% zeolite) and catalyst ZSA-3 (20 wt.-% zeolite).

Unlike the amorphous silica–alumina catalysts (Tables 3 and 4), with the composite zeolite catalysts (Tables 6 to 8) as the methanol conversion de-

TABLE 9

Methanol conversion and hydrocarbon selectivity as a function of space velocity over ZSA-1 catalyst $373 \pm 5\,°C$, 1 atm total pressure, time-on-stream = 1 h.

	WHSV h^{-1}		
	0.33	1.0	2.1
Hydrocarbon distribution (wt.-%)			
CH_4	3.7	2.2	5.0
C_2H_6	0.2	0.1	0.2
C_2H_4	8.3	10.3	9.4
C_3H_8	2.6	1.6	1.5
C_3H_6	24.7	28.2	36.8
i-C_4H_{10}	8.3	11.8	5.6
n-C_4H_{10}	1.0	1.0	0.7
1-C_4H_8	5.1	5.2	7.2
i-C_4H_8	4.9	4.3	7.0
(*cis* + *trans*) 2-C_4H_8	1.9	2.4	3.4
C_4H_6	6.1	7.8	4.5
C_{5+} non-aromatics	25.1	16.8	14.2
Aromatics	8.1	8.3	4.5
Total C_2–C_4 olefins	50.4	59.2	68.3
Conversion	100	100	ca. 65

creased with time-on-stream, the selectivity to the lower olefins increased. For example on ZSA-1 catalyst as the methanol conversion decreased from 100% to 78.5% over a period of 14 h, the C_2–C_4 olefin selectivity increased from 57.2 to 64.8 wt.-% as shown in Fig. 4. Taken alone, this might be regarded as a residence time effect, however during the same run period, the selectivity for aromatics also increased while that for the C_{5+} non-aromatic hydrocarbons decreased. Similar trends were observed on the other (ZSA-2, ZSA-3) composite zeolite catalysts. At higher zeolite loadings (>5 wt.-%) of the composite catalyst, the selectivity to C_{5+} hydrocarbons including aromatics was found to increase while that for the lower olefin fraction decreased.

In order to verify the effect of the residence time on hydrocarbon selectivity, the residence time for the catalyst ZSA-1 was increased three-fold. The selectivities for various hydrocarbon fractions changed, however, very little (Table 9). In contrast, comparing the catalysts ZSA-1 and ZSA-3, which have 5 and 20% zeolite, respectively, the lower olefin selectivity is much higher in the former than in the later case. Hence, for the residence times employed, the selectivity is affected much more by the composition of the catalyst than by the residence time.

From the results presented in Tables 6 to 8, it is interesting to note that at intermediate zeolite loadings (ca. 10 wt.-%), the high initial activity of the

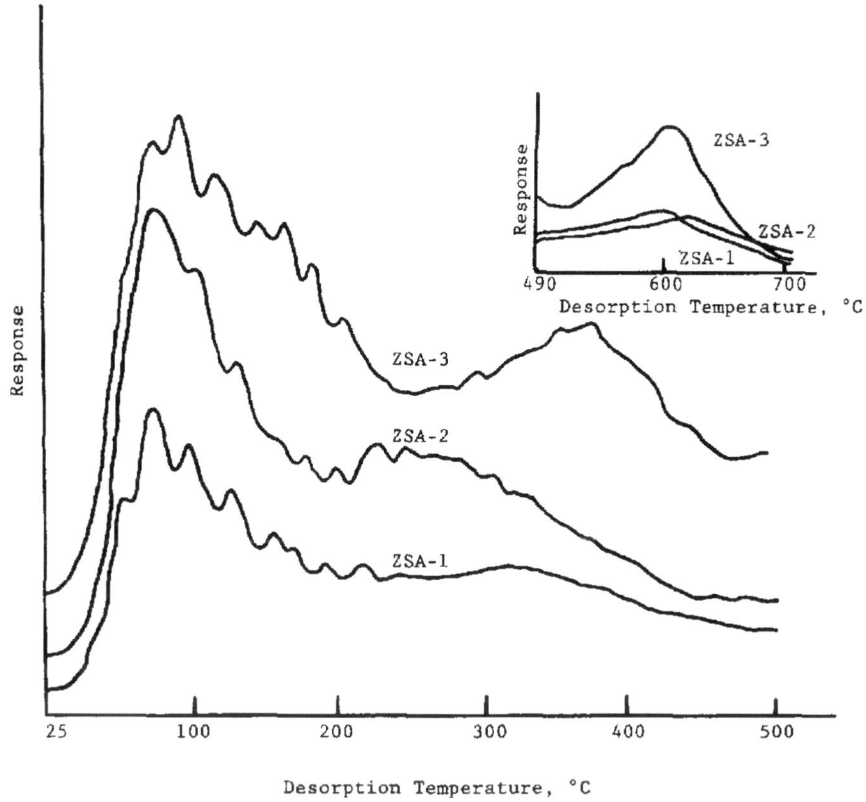

Fig. 5. Ammonia TPD from composite catalysts under the conditions used in Fig. 3.

composite zeolite catalyst (ZSA-2) was maintained for a longer time-on-stream. Furthermore, the yield of C_2–C_4 olefins at the intermediate zeolite loadings (even after longer runs) was higher than that observed with either amorphous silica–alumina (ASA-4) or with the other composite zeolite catalysts (ZSA-1 and ZSA-3). It is interesting to observe in Fig. 5 that catalyst ZSA-2 exhibits a pronounced peak in its ammonia TPD spectrum in the range from 200 to 400°C. This feature is not nearly so pronounced for the other catalysts.

More importantly, the addition of even small amounts of HAlZSM-5 resulted in much greater activity maintenance than was observed with the amorphous silica–aluminas. One possible reason for this may be that a significant fraction of the coke precursors which are formed over the silica–alumina component may enter the zeolite component. In this way these coke precursors may be directed along reaction pathways which favor the production of C_{5+} hydrocarbons instead of coke. This is consistent with the increase in the C_{5+} fractions observed.

CONCLUSIONS

The present investigation has shown that greater than 60% selectivity to lower (C_2–C_4) olefins can be achieved at high methanol conversions by using amorphous silica–alumina catalysts. However, these catalysts deactivate severely with time-on-stream. Low rates of deactivation can be achieved by incorporating small amounts of HAlZSM-5 with the silica–alumina to form a composite catalyst. The amount of zeolite added to form the composite catalyst affects both the life of the catalyst and the selectivity to C_2–C_4 olefins.

REFERENCES

1 N.Y. Chen and T.F. Degnan, Chem. Eng. Prog., (February 1988) 39.
2 C.D. Chang and A.J. Silvestri, J. Catal., 47 (1977) 249.
3 C.D. Chang, Catal. Rev.-Sci. Eng., 25 (1983) 1.
4 C.D. Chang, Catal. Rev.-Sci. Eng., 26 (1984) 323.
5 W.W. Kaeding and S.A. Butter, J. Catal., 61 (1980) 155.
6 C.T.-W. Chu and C.D. Chang, J. Catal., 86 (1984) 297.
7 S.L. Meisel, J.P. McCullough, C.H. Lechthaler and P.B. Weisz, ChemTech, 6 (1976) 86.
8 D.H. Olson, G.T. Kokotailo, S.L. Lawton and W.M. Meier, J. Phys. Chem., 85 (1981) 2238.
9 C.D. Chang, C.T.-W. Chu and R.F. Socha, J. Catal., 86 (1984) 289.
10 C.D. Chang, W.H. Lang and A.J. Silvestri, U.S. Patent 4 052 479.
11 C.D. Chang, W.H. Lang and R.L. Smith, J. Catal., 56 (1979) 169.
12 W.W. Kaeding and S.A. Butter, J. Catal., 61 (1980) 155.
13 G. Chen and J. Liang, China-Japan-U.S. Symposium, Heterogeneous Catalysis, Dalian, China, August 31-September 2, 1982, Paper A01C.
14 P.G. Rodewald, U.S. Patent 4 145 315.
15 M.M. Wu and W.W. Kaeding, J. Catal., 88 (1984) 478.
16 T. Inui, H. Matsuda, O. Yamase, H. Nagata, K. Fukuda, T. Ukawa and A. Miyamoto, J. Catal., 98 (1986) 491.
17 F.N. Givens, C.J. Plank and E.J. Rosinski, U.S. Patent 4 074 095 and 4 079 096.
18 C.D. Chang, W.H. Lang and A.J. Silvestri, U.S. Patent 4 062 905.
19 G.J. Hutchings, T. Themistocleous and R.G. Copperthwaite, Appl. Catal., 43 (1988) 133.
20 U. Hammon, M. Kotter and L. Riekert, Appl. Catal., 37 (1988) 155.
21 D. Shihabi, W.E. Garwood, P. Chu, J.N. Miale, R.M. Lago, C.T.-W. Chu and C.D. Chang, J. Catal., 93 (1985) 471.
22 K. Tanabe, Solid Acids and Bases, Kodansha, Tokyo, 1970.
23 E. Ruckenstein, AIChE J., 16 (1970) 151.
24 P.J. Nandapurkar and E. Ruckenstein, Chem. Eng. Sci., 39 (1984) 371.
25 H. Adkins and P.D. Perkins, J. Phys. Chem., 32 (1928) 221.
26 K.V. Topchieva and A.V. Ballod, Dokl. Akad. Nauk SSSR, 75 (1950) 247.
27 E. Gorin and M.H. Gorin, U.S. Patent 2 456 584.
28 F.X. Cormerais, Y. Chen, M. Kern, N.S. Gnop, G. Perot and M. Guisnet, J. Chem. Res., (1981) 290.
29 J.W. Evans, D.L. Trimm and M.S. Wainwright, Ind. Eng. Chem. Prod. Res. Dev., 22 (1983) 242.
30 R.L. Espinoza, C.M. Stander and W.G.B. Mandersloot, Appl. Catal., 6 (1983) 11.
31 U.A. Sedran, R.A. Comelli and N.S. Figoli, Appl. Catal., 11 (1984) 227.
32 U.A. Sedran and N.S. Figoli, Appl. Catal., 19 (1985) 317.
33 N.Y. Chen, J.N. Miale and W.J. Reagan, U.S. Patent 4 112 056.
34 P.S. Yarlagadda, H. Yaoliang and N.N. Bakhshi, Ind. Eng. Chem., Prod. Res. Dev., 25(2) (1986) 251.

35 Y. Amenomiya and R.J. Cvetanovic, J. Phys. Chem., 67 (1963) 144.
36 H.C. Nelson, R.J. Lussier and M.E. Still, Appl. Catal., 7 (1983) 113.
37 Y. Ono and T. Mori, J. Chem. Soc. Faraday Trans., 77 (1981) 2209.
38 P. Dejaifve, J. Vedrine, V. Bolis and E.G. Derouane, J. Catal., 63 (1980) 331.
39 B.E. Langner, Appl. Catal., 2 (1982) 289.
40 H. Itoh, C.V. Hidalgo, T. Hattori, M. Niwa and Y. Murakami, J. Catal., 85 (1984) 521.
41 W.O. Haag, R.M. Lago and P.G. Rodewald, J. Mol. Catal., 17 (1982) 161.
42 J. Gordon Robinson and D.I. Barnes, Appl. Catal., 16 (1985) 255.

Oligomerization of Ethene and Propene over Composite Zeolite Catalysts

P. YARLAGADDA, C.R.F. LUND and E. RUCKENSTEIN*

Department of Chemical Engineering, State University of New York at Buffalo, Amherst, NY 14260 (U.S.A.)

(Received 15 November 1989, revised manuscript received 1 February 1990)

ABSTRACT

The conversion of ethene and propene over a series of composite zeolite catalysts has been studied in a fixed-bed reactor. The composite catalysts were synthesized by incorporating 1 to 10 wt.-% of HAlZSM-5 (molar ratio $SiO_2/Al_2O_3=35$) zeolite in a silica–alumina matrix. The silica–alumina matrix contained 15 wt.-% Al_2O_3 and was prepared by the cogelation method. Pure ethene and propene were used as the feed gases. Experiments were conducted over a temperature range of 200 to 600°C at a space velocity of 3.4 ± 0.1 $g_{alkene}/g_{catalyst}/h$ and atmospheric pressure. In addition, a few experiments were carried out at the space velocities of 34 and 68 h^{-1}. For all the catalysts, ethene and propene conversion increases with increasing reaction temperature. In the case of propene, the conversion increases rapidly over a narrow temperature range, for a ZSM-5 zeolite content of 10 wt.-%. However, a moderate increase with temperature was observed when ethene was the reactant and the composite catalyst had a zeolite loading of less than 10 wt.-%. With both reactants, silica–alumina catalysts containing 5 and 10 wt.-% ZSM-5 exhibited activities much greater than the pure silica–alumina. The selectivity of the reaction remained comparable to that of silica–alumina however, not of HAlZSM-5. The concomitant formation of aromatic hydrocarbons and lower alkanes (C_1–C_4), indicates that hydrogen transfer reactions occur on all catalysts. This transfer reaction is sensitive to the zeolite content of the composite. Both the amorphous silica–alumina and the pure HAlZSM-5 zeolite exhibit high aromatization activity, because of hydrogen transfer, whereas, at intermediate zeolite loading (5 wt.-%), the hydrogen transfer activity is lower, resulting in higher alkenes. For both the 5% and 10% zeolite composites, the coke content expressed as a percentage of the converted propene was less than 1.5 wt.-%, a value close to that for the pure HAlZSM-5 zeolite. In contrast, the amorphous silica–alumina as well as the 1.5 wt.-% of zeolite composite resulted in a larger coke fraction. Furthermore, the temperature-programmed oxidation profiles indicate that the coke could be oxidized more easily in the case of 5% zeolite composite than the 10% zeolite composite or the pure zeolite.

Keywords: zeolites, silica–alumina, ethene oligomerization, propene oligomerization, selectivity (alkenes), acid sites, coke formation, catalyst preparation (cogelation).

INTRODUCTION

The selectivity of the methanol conversion reaction using amorphous silica–alumina catalysts recently has been observed [1] to favor lower-molecular-

This paper, which was originally published in **Applied Catalysis**: 62, 125–139 (1990), is republished with permission.

weight alkenes over aromatics and higher molecular weight aliphatics. It was further observed that the addition of a small amount of HAlZSM-5 to the amorphous silica–alumina improved the resistance of the catalyst to deactivation by coking. Adding the HAlZSM-5 to the amorphous silica–alumina decreased slightly the selectivity to lower-molecular-weight alkenes and increased that to aromatic and higher molecular weight aliphatics. The observed changes in the selectivity were not a simple average of the behaviour which would be observed with pure amorphous silica–alumina and pure HAlZSM-5. The reaction on silica-alumina might be represented as: methanol→alkenes (a small amount of alkenes converts to aromatics and alkanes); while that on pure ZSM-5 might be represented as: methanol→alkenes→aromatics+alkanes (a small amount of alkenes remains).

In the present study, the selectivity during the conversion of alkenes over the same catalyst has been studied in an attempt to provide insight into how the diffusional properties of a composite catalyst interplay with the acid properties of the catalyst in determining overall selectivity. The findings are also relevant to the previous study of methanol conversion over these catalysts.

Solid acids such as amorphous silica–aluminas and zeolites are known to catalyze the polymerization of lower alkenes to a broad range of hydrocarbons [2–4]. The overall transformation of alkenes to higher hydrocarbons involves oligomerization, cyclization, isomerization, aromatization, hydrogen transfer and coke formation [5]. The relative rates of these processes are affected by the local environmental variables (temperature, pressure, and composition). The variation of these local environmental variables throughout the reactor is affected both by operational parameters (space velocity, etc.) and by the pore structure and active site distribution of the catalyst. By using a composite catalyst it is possible to alter the active site distribution relative to the pore structure.

Alkene oligomerization has been carried out using a phosphoric-acid-impregnated silica clay catalyst [6]. Medium pore zeolites, such as HAlZSM-5, now are finding increasing application in the conversion of lower alkenes to hydrocarbons in the gasoline and diesel range [7–9]. ^{13}C-NMR [10] and TPD-TGA [11,12] have been employed in an attempt to understand the events occurring in the pores of HAlZSM-5 zeolite during the alkene oligomerization. The results indicate that at low temperatures (<300 K) linear oligomers form in the intracrystalline zeolites pores. As the reaction temperature progressively increases (<500 K), the pore mouth becomes blocked by the product oligomers, and the reaction proceeds predominantly on the external acidic sites, i.e., in the intercrystalline pores. As a result, a large fraction of branched oligomers is formed, without shape selectivity. Under practical reaction conditions of still higher temperatures (>500 K), the oligomerized products begin to crack, thereby opening the pore mouths. Consequently, the reaction rates are found to significantly increase, with some improvement in shape selectivity. Furthermore, the rates of aromatization and hydrogen transfer also become significant at higher temperatures [13].

Previous studies with amorphous silica–aluminas [14,15] have shown that the oligomerization of propene starts at temperatures as low as 330 K. It has been observed that Brønsted acidity is essential for the initiation of the oligomerization reaction. In contrast, Lewis acidity promotes aromatization, hydrogen transfer and coke formation reactions [14]. As the hydrogen transfer reaction is bimolecular, its rate is expected to be higher in zeolites than in the amorphous silica–aluminas because the pores are much larger in the latter. Chen and Haag [16] have reported, however, that medium pore zeolites, such as ZSM-5, exhibit a hydrogen transfer activity which is lower than that of amorphous silica–alumina. They ascribe this to the size constraint imposed upon the transition state complex by the narrow pores of the ZSM-5 type zeolites.

From the short review presented above, it is obvious that much information has been gathered regarding the chemistry of the conversion of lower alkenes using either pure ZSM-5 zeolites or pure amorphous silica–alumina. Such information is not available however for composite catalysts comprised of both components. Thus, the effect of reaction temperature upon the conversion of ethene and propene over a series of composite catalysts is reported in this paper. This is also relevant to the formation of gasoline range hydrocarbons [17,18] from methanol, since the lower alkenes are intermediates in this process. Furthermore, from the temperature-programmed oxidation profiles, the nature of coke deposits on the composite catalysts during the propene conversion reaction is also discussed.

EXPERIMENTAL

Preparation of catalysts

Amorphous silica–alumina with an alumina content of 15 wt.-% was prepared by the cogelation method described previously [1]. In short, a solution of sodium silicate of pH 4 was mixed with a solution of aluminium nitrate with vigorous agitation. The final pH was adjusted to pH 6.5, the solution agitated for 24 h, and filtered. The solids were washed thoroughly with distilled water, dried at 373 K and calcined at 873 K for 1 h in air. Subsequently, the above calcined mass was subjected to ammonium ion exchange, by treating it with 1 *M* aqueous $(NH_4)_2SO_4$ solution at room temperature for 3 h, then again was water washed and finally calcined at 823 K for 1 h. Chemical analysis (Galbraith Laboratories) confirmed the nominal 15% alumina content.

The ZSM-5 zeolite was synthesized by the method of Chen et al. [19]. In order to decompose the tetrapropyl ammonium cation, the washed and dried wet cake was calcined at 823 K. The SiO_2/Al_2O_3 molar ratio of the synthesized zeolite was 35. The sodium form of the zeolite was converted to its protonated form by NH_4^+ exchange and finally calcined at 823 K for 12 h in air.

The composite zeolite catalysts were prepared by the addition of a predetermined amount of HAlZSM-5 zeolite powder to a solution, containing sodium silicate and aluminium nitrate whose pH was adjusted to 6.5–7.0, as soon as a turbidity due to the onset of gelation appeared in the solution. The slurry was agitated for 24 h, followed by filtration, washing, drying, calcination and NH_4^+ exchange under conditions similar to those mentioned for the amorphous silica–alumina catalyst. The zeolite content in the finished catalyst samples was varied from 1.5 to 10 wt.-%, whereas the Al_2O_3 content in the amorphous silica–alumina component was maintained at about 15 wt.-%. All the composite zeolite catalysts were subjected to a final heat treatment at 823 K for 1 h in air.

Reaction studies

The alkene oligomerization experiments were conducted in a fixed-bed stainless steel microreactor (6 mm I.D.) described previously [1]. Ethene (Linde, CP Grade, minimum purity 99.5%) and propene (Linde, CP Grade, minimum purity 99.0%) without any dilution, were used as the feed gases. In all the runs 0.5 g of catalyst, (60/80 mesh) held between two small quartz wool plugs located at the mid-point of the reactor tube, was used. Each catalyst sample was pretreated in flowing oxygen at 500°C and 1 atm (101.325 kPa) for 3 h, followed by complete purging with helium. Experiments were carried out at 1 atm total pressure, WHSV (weight hourly space velocity = g of alkene/g catalyst/h) of 3.4 ± 0.1 h^{-1} and at reaction temperatures between 200 and 600°C. For each set of reaction conditions, the experiments have been carried out with a fresh catalyst. Furthermore, the data have been collected before major deactivation (which occurs over 10–20 h) has set in. The gases leaving the reactor were cooled to room temperature, and the liquid product was collected in a glass trap immersed in an ice-salt bath. The condensed gases were analyzed by on-line gas chromatography (GC) for (i) C_1–C_4 hydrocarbons on a VZ-10 column (6 ft. × 1/8 in.) and (ii) C_{5+} hydrocarbons on a 10% SE-30 Chrom W-HP column (6 ft. × 1/8 in.) using a flame ionization detector. The liquid products which were accumulated over a period of experimental time were weighed and later analyzed on the 10% SE-30 column. The C_1–C_4 hydrocarbons, except for the *cis* and *trans* isomers of butene were well resolved and identified on the VZ-10 column. Hydrocarbons containing five or more carbon atoms have been grouped in terms of their respective carbon number. The different carbon numbers were well-resolved in the GC, but within any one carbon number the isomers were not always well-resolved. For this purpose, the following standards (Alltech Associates) were injected, and their retention times have been used in determining the weight fractions: C6 : 1–C12 : 1 alkenes (stock no. 5398); C12:1–C18:1 alkenes (stock # 5400); C_6–C_8 aromatics (stock no. 5404) and C_9 aromatics (stock no. 6406).

The conversion and the product distribution were calculated as follows:

$$\text{conversion} = \left(1 - \frac{\text{Number of moles of ethene or propene leaving the reactor}}{\text{Number of moles of ethene or propene entering the reactor}}\right) \times 100\%$$

$$\text{selectivity} = \frac{\text{weight of the desired hydrocarbon in the exist stream}}{\text{total weight of the product stream excluding the alkene reactant}} \times 100\%$$

Each catalytic experiment has been carried out thrice with catalysts freshly prepared. The conversions differed by less than 5%. The data presented in the tables represent arithmetic averages of the three runs.

Temperature programmed oxidation of coke deposits

Coke deposited on the catalysts was characterized by temperature-programmed oxidation (TPO). Prior to starting the TPO, each of the spent catalysts was heated in flowing helium from 25 to 325 °C at 10 °C/min and held at 325 °C for 3 h. The latter temperature was about 25 °C lower than the propene conversion reaction temperature. This heat treatment removed any adsorbed air, water and occluded free organics from the catalyst. The treated samples then were subjected to TPO from 325 °C to 700 °C at 10 °C/min in a stream of oxygen. The carbon dioxide formed was determined by a gas chromatograph using a thermal conductivity detector. In the TPO experiments, about 100 mg of the spent catalyst was used.

RESULTS AND DISCUSSION

Reaction of propene

Fig. 1 shows the conversion of propene measured at 1 atm and WHSV $= 3.4 \pm 0.1\ h^{-1}$ as a function of reaction temperature. There are clearly two distinct behavior patterns revealed in the figure. One behavior pattern is characterized by a sigmoid shape of the plot, which is the behavior pure ZSM-5 exhibits. This behavior for pure ZSM-5 zeolite has been reported previously for propene oligomerization [20,21] and also for the conversion of methanol to hydrocarbons [22]. The sudden increase in the conversion has been attributed to an autocatalytic reaction [20]. This is more likely to occur for a meth-

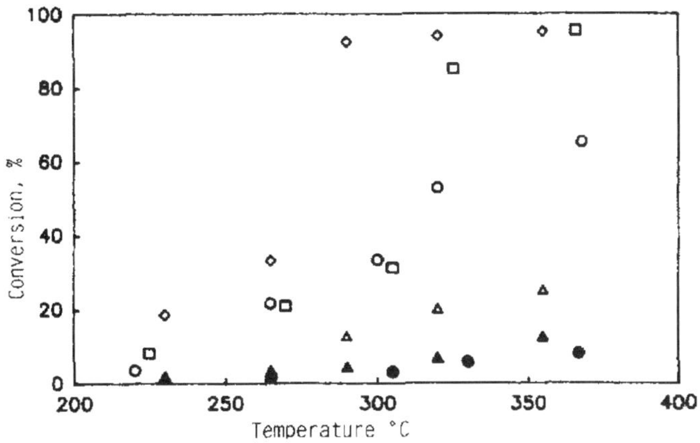

Fig. 1. Propene conversion as a function of reaction temperature at 1 atm, $3.4 \pm 0.1\ h^{-1}$ WHSV. (◇) HAlZSM-5, (□) SA+10.0% HAlZSM-5, (○) SA+5.0% HAlZSM-5, (△) SA+2.5% HAlZSM-5, (▲) SA+1.5% HAlZSM-5, (●) SiO_2-15% Al_2O_3 (SA).

anol feed because in such a case the reactant species methylate the primary hydrocarbon products leading to a chain reaction. However, in the case of an alkene feed, where the methylating species are absent, the sudden increase in reaction rate may be due to the unplugging of the zeolite pore mouths as discussed in the introduction. The second behavior pattern seen in Fig. 1 shows a gradual increase in conversion with temperature. Pure silica–alumina is representative of this behavior. The absence of a sharp transition to higher conversion may be a consequence of either of the following two factors. First, the contribution to the overall conversion made by the zeolite channels may be insignificant for these catalysts since there is little zeolite present. Then, even if the channels become unplugged, there is no sharp rise in conversion. The second possibility is that the silica–alumina alters the selectivity of the initial propene reacted, and this selectivity change does not lead to high conversions (i.e. chain reactions). It is not surprising that the composite samples with higher zeolite content exhibit the first behavior pattern and the samples with a low zeolite content exhibit the second behavior pattern.

Table 1 presents selectivity data for each catalyst at a number of temperatures which span the range of conversions presented in Fig. 1. The data in Table 1 for temperatures below that where the conversion suddenly increases shows propane to be the major product for all catalysts containing any amount of silica–alumina. The second and third most abundant products for all these catalysts are C_4 alkenes and isobutane, respectively. Pure ZSM-5 on the other hand, exhibits markedly different behavior: C_4 alkenes are the major products, with propane selectivity second but an order of magnitude smaller. In summary, at lower temperatures, all the composites employed in the present study perform similar to pure silica–alumina with respect to reaction selectivity.

The behavior at higher temperatures is of some interest, however, because it is at higher temperatures that the two different activity profiles are mani-

TABLE 1

C_1-C_4 hydrocarbon selectivities as a function of reaction temperature with propene as reactant

C_3H_6 WHSV $= 3.4 \pm 0.1$ h^{-1}; 1 atm total pressure; time-on-stream at each temperature = 2 h. Even though at high temperatures liquid (C_{5+}) forms, the selectivity is defined here only with respect to the C_1-C_4 fraction.

SiO_2-15% Al_2O_3 (SA)			
Temperature (°C)	221	275	357
% Conversion	2.8	3.5	6.7
CH_4	trace	0.25	0.30
C_2H_6	1.26	0.81	0.75
C_2H_4	0.29	0.36	0.45
C_3H_8	51.76	46.02	28.03
i-C_4H_{10}	13.02	10.25	5.66
n-C_4H_{10}	trace	0.10	0.30
C_4 alkenes	33.67	42.21	64.51
SA + 1.5% HAlZSM-5			
Temperature (°C)	270	330	385
% Conversion	1.9	6.1	9.4
CH_4	0.20	0.25	0.26
C_2H_6	1.84	1.62	1.45
C_2H_4	0.59	0.81	1.02
C_3H_8	47.25	42.68	24.19
i-C_4H_{10}	11.96	7.26	5.77
n-C_4H_{10}	trace	0.25	0.46
C_4 alkenes	38.16	47.13	66.85
SA + 2.5% HAlZSM-5			
Temperate (°C)	285	325	386
% Conversion	10.3	20.8	28.6
CH_4	0.18	0.14	0.05
C_2H_6	2.26	1.28	0.56
C_2H_4	trace	0.15	0.29
C_3H_8	50.71	41.36	10.00
i-C_4H_{10}	11.69	7.18	3.38
n-C_4H_{10}	trace	trace	trace
C_4 alkenes	35.16	49.89	85.72
SA + 5.0% HAlZSM-5			
Temperature (°C)	214	310	350
% Conversion	2.9	45.1	65.3
CH_4	trace	0.04	0.06
C_2H_6	3.59	0.34	0.31
C_2H_4	0.68	0.35	0.75
C_3H_8	58.84	6.78	5.19
i-C_4H_{10}	5.09	2.09	2.06
n-C_4H_{10}	trace	0.26	0.41
C_4 alkenes	31.80	90.14	91.22

TABLE 1 (*continued*)

SA + 10.0% HAlZSM-5			
Temperature (°C)	195	263	366
% Conversion	3.8	20.8	96.1
CH_4	0.13	trace	0.26
C_2H_6	2.79	1.51	0.50
C_2H_4	trace	0.46	0.72
C_3H_8	48.60	20.05	15.37
i-C_4H_{10}	11.84	3.79	34.35
n-C_4H_{10}	trace	1.18	10.34
C_4 alkenes	36.64	73.01	38.46
HAlZSM-5 (SiO_2/Al_2O_3=35)			
Temperature (°C)	255	320	370
% Conversion	30.1	94.1	95.8
CH_4	0.02	0.51	0.55
C_2H_6	0.36	2.58	4.16
C_2H_4	0.82	1.54	1.05
C_3H_8	8.74	26.99	41.37
i-C_4H_{10}	2.18	17.81	32.18
n-C_4H_{10}	1.38	4.00	10.24
C_4 alkenes	86.50	46.57	10.45

TABLE 2

Product selectivity obtained during the propene conversion reaction

355±5°C; 3.4±0.1 h^{-1} WHSV; 1 atm total pressure; time-on-stream=3 h. The selectivity is defined here with respect to all the products (gas and liquid excluding coke).

Catalyst	SiO_2-15% Al_2O_3 (SA)	SA+1.5% HAlZSM-5	SA+2.5% HAlZSM-5	SA+5.0% HAlZSM-5	SA+10.0% HAlZSM-5	HAlZSM-5 (SiO_2/Al_2O_3=35)
Selectivity (wt.-%)						
CH_4	0.3	0.3	trace	trace	0.4	0.3
C_2-C_4 alkanes	21.0	17.8	11.9	8.3	16.8	18.6
C_2-C_4 alkenes	62.7	65.2	70.4	73.6	51.6	4.5
C_{5+} aliphatics	13.9	14.7	16.2	16.7	28.5	65.7
Aromatics	2.1	2.0	1.5	1.4	2.7	10.9

fested (Fig. 1). Table 2 presents the selectivities for all the components. In the lower hydrocarbon fraction, there is a distinct difference between pure ZSM-5 and all the composite catalysts. The pure ZSM-5 yields alkanes (C_3H_8, i-C_4H_{10} and n-C_4H_{10}) as the major reaction products, whereas C_4 alkenes predominate for all the composite catalysts. This may be taken as an indication that hydrogen transfer occurs much more readily on the pure ZSM-5. Within the liquid hydrocarbons (C_{5+}), the pure ZSM-5 exhibits the highest aromatics selectivity. Since hydrogen transfer (leading to alkanes) constitutes an integral part of the aromatization process, the observations are consistent. In summary, at high temperature, just as at low temperature, all the composites studied exhibit a selectivity similar to pure silica–alumina. It is interesting to note that the pure silica–alumina gives a conversion of only 6.7%, whereas the 10% ZSM-5 catalyst gives a conversion of 96% (Fig. 1), yet their selectivities are comparable to each other and very different from those on the pure ZSM-5.

It might be possible that if the 5%, 10% and 100% ZSM-5 samples were compared at equal space velocities based upon only the ZSM-5 present, then their selectivities would be comparable. Experiments were therefore conducted at the constant reaction temperature of 355 °C, but varying the space velocity between 3 and 70 h^{-1}. Figs. 2 and 3 compare the selectivities to aliphatics and to C_4 alkenes, respectively, on this basis. The figures clearly show that the differences between the catalysts are not simply due to residence time effects. Furthermore, all three catalysts exhibit high conversions, so the differences

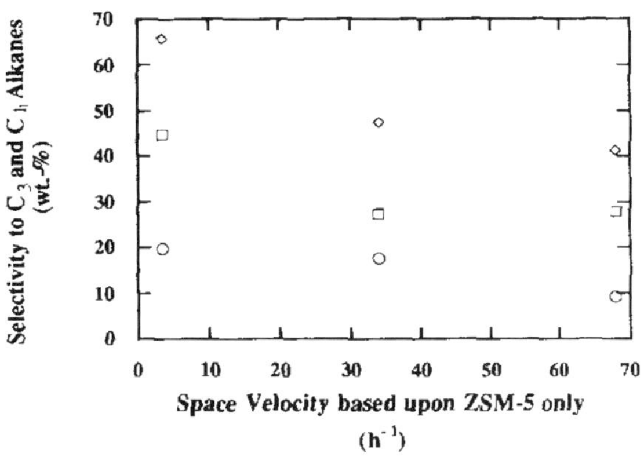

Fig. 2. Dependence of alkane selectivity upon space velocity based upon the zeolite only. (○) 5% ZSM-5, (□) 10% ZSM-5, (◇) 100% ZSM-5.

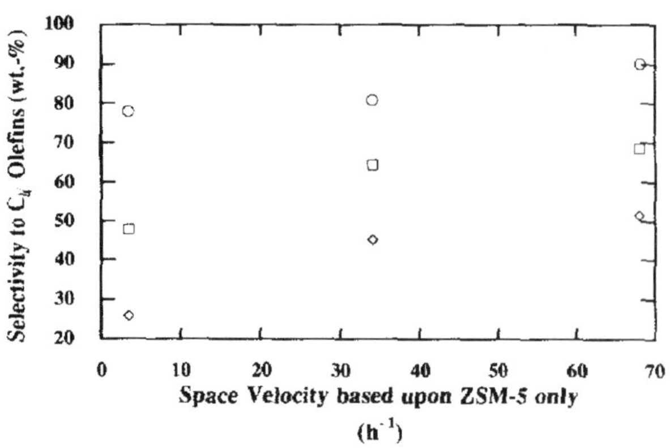

Fig. 3. Dependence of the butene selectivity upon space velocity based upon the zeolite only. Symbols as in Fig. 2.

can not be caused by the extent of conversion. Thus, it appears that the interplay between the two pore structures and the distribution of acid sites between them can significantly affect catalyst performance.

The coke content of the catalysts after use in propene conversion, for 3 h, at a reaction temperature of 355 °C, and a space velocity of 3.4 h^{-1}, is given in Table 3. The amount of coke is the largest for the amorphous silica–alumina catalyst and the lowest for the pure HAlZSM-5 zeolite, with a gradual decrease with increasing zeolite content in the composite zeolites. In the amorphous silica–alumina catalyst, the coke accounted for nearly 15 wt.-% of the converted propene while for the composite zeolites with a zeolite content of 5 and 10% as well as for the pure zeolite, the coke content based on the converted propene was less than 1.5 wt.-%. The HAlZSM-5 zeolite is known to have a greater resistance to coke formation, a characteristic which has been attributed to its pore dimensions which sterically hinder the formation of coke precursors [23]. In spite of this, the ZSM-5 zeolite, as well as the other acid catalysts, eventually deactivates due to the accumulation of coke deposits both in the pores [24,25] and on the external surface [26]. A recent study [27] with 1-hexene as reactant and pure zeolite as catalyst has shown that at reaction temperatures less than 320 °C, the residue buildup was mainly within the pores, whereas at higher temperatures (>320 °C), this residue migrated to the external zeolite surface and finally converted to a nonvolatile polymeric material. On the other hand, the amorphous silica–alumina with a wide pore size distribution is expected to deactivate rapidly due to coke buildup. Our results on the amorphous silica–alumina and the pure zeolite are in agreement with the earlier observations. It is, however, striking to observe that for zeolite contents of only 5% to 10%, the coke buildup is similar to that of the pure zeolite.

The temperature-programmed oxidation (TPO) profiles of the coked samples are presented in Fig. 4. Except for the 5% zeolite composite, the peaks are

TABLE 3

Coke content[a] of the catalysts after use in propene conversion at the reaction conditions given in Table 2.

Catalyst	Coke content (mg/g catalyst)	Coke expressed as a percentage of the converted propene
SiO_2-15% AlO_3(SA)	102.6	15.1
SA+1.5% HAlZSM-5	98.5	11.8
SA+2.5% HAlZSM-5	91.3	3.6
SA+5% HAlZSM-5	85.4	1.3
SA+10% HAlZSM-5	76.1	0.9
HAlZSM	70.2	0.8

[a]Determined in a thermogravimetric analyzer by burning of coke in a flow of air at 500 °C till a constant weight of the sample is obtained.

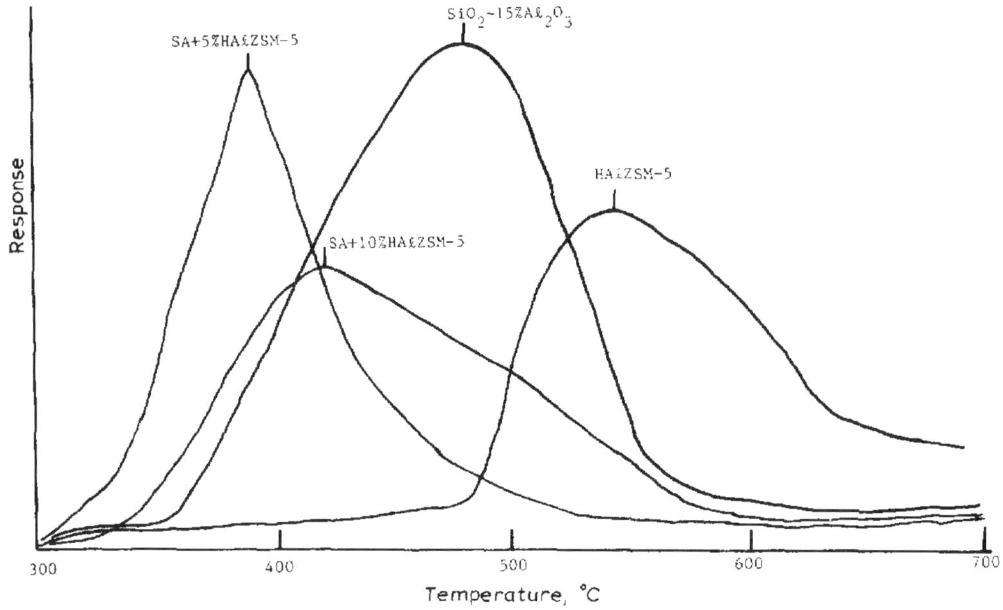

Fig. 4. Temperature programmed oxidation profiles of the catalysts used in the propene conversion at reaction conditions mentioned in Table 2.

broad, reflecting the broad nature of the deposited coke. The TPO profile shifted to higher temperatures in the order: 5% zeolite composite < 10% zeolite composite < amorphous silica–alumina < pure HAlZSM-5 zeolite. Thus the peak temperature of the oxidation profile for the pure zeolite is around 520 °C, while for the 5% zeolite composite it is close to 400 °C. High oxidation temperatures are required either if the coke is highly aromatic in nature and/or if the coke is predominantly concentrated in the zeolite pores. The ease of coke oxidation for the 5 and 10% zeolite composites suggests that the coke is mainly located external to the zeolite. One may note that the coke formed is not a linear combination of the coke formed on pure silica–alumina and pure zeolite. A synergistic effect between the two components of the composite exists, as was already noted in our previous paper [1].

The reaction of ethene

From Fig. 5, one can see that unlike propene, the ethene conversion over both the amorphous silica–alumina and the composite catalysts (zeolite content > 5 wt.-% and < 10%) increased monotonically with temperature. In contrast, the increase in ethene conversion was fairly large on the pure HAlZSM-5 over a relatively small temperature range. This latter observation is in agreement with the results of Garwood on pure zeolite [2]. It is likely that for the above composition zeolite catalysts, no pore plugging has set in and, consequently, no diffusional hindrance occurs. However, in pure zeolites, blockage does occur as a result of oligomer formation. In contrast to the case of propene,

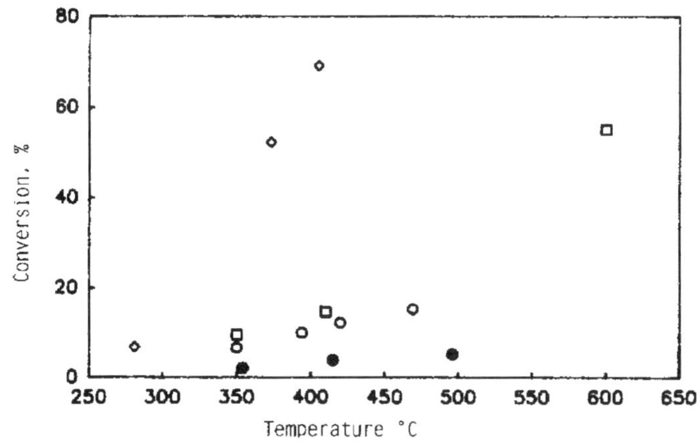

Fig. 5. Ethene conversion as a function of reaction temperature at 1 atm, $3.7 \pm 0.1\ h^{-1}$ WHSV. Symbols as in Fig. 1.

in the case of ethene conversion over composite catalysts such a blockage does not occur. It is likely that different types of oligomer chains form from ethene and propene, and those that are generated in the former case more easily undergo scission over silica–alumina.

As seen from Table 4, for both the 5% and 10% zeolite composite catalyst, the C_4 alkene selectivity increases with increasing reaction temperature when the temperature is relatively low and after passing through a maximum decreases appreciably at relatively high temperatures. For the amorphous silica–alumina, the C_4 alkene selectivity underwent a small increase throughout the entire temperature range, whereas in the case of the pure HAlZSM-5 zeolite it decreases sharply with a rise in temperature. Among the alkanes and for all the catalysts employed, isobutane was the major component followed by propane and by a relatively small amount of ethane. The 5% zeolite composite produced mostly C_3 and C_4 alkenes with the alkanes accounting for a small amount of the C_1–C_4 hydrocarbon fraction. Thus, with ethene, as with propene, the 5% zeolite composite exhibits a low C_2–C_4 alkane selectivity. This again indicates that the hydrogen transfer process occurs to a smaller extent over the composite catalysts than over pure ZSM-5.

By true oligomerization, only $C_6^=$, $C_9^=$, $C_{12}^=$, etc. compounds are expected to form from propene, and $C_4^=$, $C_6^=$, $C_8^=$, etc. compounds from ethene. In reality, oligomerization is coupled with cracking, resulting in the formation of intermediate carbon number products such as $C_5^=$, $C_7^=$, etc. Further, dehydrocyclization followed by hydrogen transfer to the alkene also takes place, producing aromatics and alkanes. As noted by Bessell and Seddon [13], at low reaction temperatures, alkanes can be formed not necessarily by hydrogen transfer to the alkene but rather by methyl and hydride transfer from a coke precussor to the monomer. From the present results (Tables 1 and 4), it is evident that to a varying extent, dependent on temperature and zeolite content, all the above reactions occur on the catalysts employed.

TABLE 4

Ethene conversion and C_1–C_4 hydrocarbon selectivities as a function of reaction temperature

C_2H_4 WHSV = 3.7 ± 0.1 h^{-1}; 1 atm total pressure; time-on-stream at each temperature = 2 h. In the cases presented the amount of liquid (C_{5+}) was negligible.

SiO_2-15% Al_2O_3 (SA)				
Temperature (°C)	354	415	496	
% Conversion	2.1	3.9	5.2	
CH_4	0.06	0.14	0.20	
C_2H_6	8.24	9.11	8.25	
C_3H_8	2.49	5.24	5.38	
C_3H_6	14.90	15.06	15.27	
i-C_4H_{10}	32.75	29.58	28.15	
n-C_4H_{10}	2.96	1.91	1.86	
C_4 alkenes	38.87	38.96	41.89	
SA + 5.0% HAlZSM-5				
Temperature (°C)	350	394	420	469
% Conversion	6.7	10.0	12.2	15.3
CH_4	0.02	0.14	0.17	0.58
C_2H_6	2.62	2.74	3.22	6.12
C_3H_8	1.19	0.72	0.78	1.48
C_3H_6	26.39	32.54	26.11	45.12
i-C_4H_{10}	16.58	5.08	4.95	3.81
n-C_4H_{10}	0.88	0.61	0.78	0.79
C_4 alkenes	52.32	58.17	64.04	42.10
SA + 10.0% HAlZSM-5				
Temperature (°C)	350	410	600	
% Conversion	9.5	14.6	55.1	
CH_4	trace	0.30	8.16	
C_2H_6	2.08	1.83	14.48	
C_3H_8	0.37	1.47	10.24	
C_3H_6	33.95	19.56	20.17	
i-C_4H_{10}	3.12	10.58	11.77	
n-C_4H_{10}	0.38	2.14	3.65	
C_4 alkenes	60.10	64.12	31.53	
HAlZSM-5 (SiO_2/Al_2O_3 = 35)				
Temperature (°C)	281	373	405	
% Conversion	6.8	52.3	69.2	
CH_4	trace	0.09	0.31	
C_2H_6	1.68	2.18	3.27	
C_3H_8	9.85	16.54	21.33	
C_3H_6	14.73	18.37	19.16	
i-C_4H_{10}	5.41	13.45	23.10	
n-C_4H_{10}	6.24	7.06	8.25	
C_4 alkenes	62.09	42.31	24.58	

Studies on amorphous silica–aluminas [14,28] have shown that the Lewis acid sites are primarily active in the aromatization and coke forming reactions. Hence, it is expected that the rate of hydrogen transfer will be parallel to the Lewis acidity of the catalysts. However, from the ammonia temperature-programmed desorption spectra (Figs. 3 and 5 of ref. 1) of the catalyst used in the present study, no relationship between the hydrogen transfer activity and acidity could be observed.

CONCLUSIONS

It has been observed that the zeolite content of the composite catalyst influences the activity and to a lesser extent the selectivity during the alkene conversion process. Thus, with propene and for a 10% zeolite composite, the conversion increases rapidly over a narrow temperature range, a behavior akin to that of the pure ZSM-5 zeolite, yet the selectivity of the reaction remains similar to that for silica–alumina. With ethene no such feature regarding the effect of temperature on conversion is noticed. The conversion has a gradual increase with temperature, which is, however, more rapid than for silica–alumina. Nevertheless, the selectivity remains close to that of silica–alumina.

As shown by the product analysis, during both ethene and propene conversion over composite zeolites a variety of acid catalyzed reactions occur. A comparison of hydrocarbon selectivities in the C_1–C_4 and C_{5+} ranges (Table 2) shows that, particularly, the hydrogen transfer process leading to the formation of lower alkanes and aromatics depends upon the zeolite content. Thus at 5% zeolite loading, the selectivities for aromatics and for the lower alkanes are low, while that for alkenes (C_4) is high.

During propene conversion, the amount of coke and its nature for composite catalysts are found to vary between those of amorphous silica–alumina, and pure zeolite. As expected, the amount of converted propene appearing as coke is the highest for amorphous silica–alumina and the smallest for the pure zeolite. Interestingly, for 5% and 10% zeolite composites, the coke content is very close to that of the pure ZSM-5 zeolite. The TPO profiles of the spent catalysts shows that the oxidation peak of the 5% zeolite composite is about 100°C lower than that of the pure zeolite, while the corresponding peaks for the 10% composite and the amorphous silica–alumina are located between the above two extremes.

ACKNOWLEDGEMENTS

Partial support of this research by the NSF (Grant No. CBT 8857100) and by Occidental Chemical Company are gratefully acknowledged.

REFERENCES

1 P. Yarlagadda, C.R.F. Lund and E. Ruckenstein, Appl. Catal., 54 (1989) 139.
2 W.E. Garwood, in G.D. Stucky and F.G. Dwyer (Editors), Intrazeolite Chemistry, ACS Symposium Series 218, American Chemical Society, Washington, D.C., (1983) p. 383.
3 K.G. Wilshier, P. Smart, R. Western, T. Mole and T. Behrsing, Appl. Catal., 31 (1987) 339.
4 J.C.Q. Fletcher, M. Kojima and C.T. O'Connor, Appl. Catal., 28 (1986) 181.
5 J.C. Vedrine, P. Dejaifve, E.D. Garbowski, and E.G. Derouane, in B. Imelik, C. Naccache, Y. Ben Taarit, J.C. Vedrine, G. Coudurier and H. Praliaud (Editors), Catalysis by Zeolites, (Studies in Surface Science and Catalysis, Vol. 5), Elsevier, Amsterdam, 1980, p. 29.
6 V.N. Ipatieff and G. Egloff, Oil Gas J. 33(52) (1934) 31.
7 S.A. Tabak, F.J. Krambeck and W.E. Garwood, AIChE J., 32(9) (1986) 1526.
8 S.A. Tabak, U.S. Patent 4 254 295 (1981).
9 N.Y. Chen and T.Y. Yan, Ind. Chem. Process Des. Dev., 25 (1) (1986) 151.
10 J.P. van den Berg, J.P. Wolthuizen, A.D.H. Clague, G.R. Hays, R. Huis and J.H.C. van Hooff, J. Catal., 80 (1983) 130.
11 J.P. van den Berg, J.P. Wolthuizen and J.H.C. van Hooff, J. Catal., 80 (1983) 139.
12 T.J. Gricus Kofke and R.J. Gorte, J. Catal., 115 (1989) 223.
13 S. Bessell and D. Seddon, J. Catal., 105 (1987) 270.
14 F.E. Shephard, J.J. Rooney and C. Kemball, J. Catal., 1 (1962) 379.
15 F.H. Gayer, Ind. Eng. Chem., 25 (1933) 1122.
16 N.Y. Chen and W.O. Haag, in Z. Paal and P.G. Menon (Editors), Hydrogen Effects in Catalysis, Marcel Dekker, New York, 1988, pp. 695–722.
17 C.D. Chang, Catal. Rev. Sci. Eng., 25 (1983) 1.
18 C.D. Chang and A.J. Silvestri, J. Catal., 47(2) (1977) 249.
19 N.Y. Chen, J.N. Miale and W.J. Reagan, U.S. Patent 4 112 056 (1978).
20 Y. Ono and T. Mori, J. Chem. Soc., Faraday Trans. 1, 77 (1981) 2209.
21 R.L. Espinoza, Appl. Catal., 26 (1986) 203.
22 N.Y. Chen and W.J. Reagan, J. Catal., 59 (1981) 123.
23 L.D. Rollman, J. Catal., 47 (1977) 113.
24 D.M. Bibby, N.B. Milestone, J.E. Patterson and L.P. Aldridge, J. Catal., 97, (1986) 493.
25 B.E. Langner, Appl. Catal., 3 (1982) 289.
26 P. Dejaifve, A. Auroux, P.C. Gravelle and J.C. Vedrine, J. Catal., 70 (1981) 123.
27 J.R. Anderson, Y.F. Chang and R.J. Western, J. Catal., 118 (1989) 466.
28 J.J. Rooney and R.C. Pink, Trans. Faraday Soc., 58 (1962) 1632.

The Synthesis and Cracking Behavior of Zeolite-Amorphous Silica–Alumina Composites Prepared Using Gels with High Alumina and Low Organic Template Content

PRASAD YARLAGADDA, CARL R. F. LUND, AND ELI RUCKENSTEIN[1]

Department of Chemical Engineering, State University of New York at Buffalo, Amherst, New York 14260

Received March 25, 1991; revised July 16, 1991

Composites of ZSM-5 and amorphous silica–alumina have been prepared from starting gels with SiO_2/Al_2O_3 molar ratios of 10 and below and tetrapropyl ammonium $(TPA^+)/Al_2O_3$ molar ratios below 0.5. The syntheses have been followed by periodic removal of aliquots from the gel with subsequent analysis. At SiO_2/Al_2O_3 = 10 and TPA^+/Al_2O_3 = 0.44 mordenite formed. When the SiO_2/Al_2O_3 ratio was reduced to 8.3, but the ratio TPA^+/Al_2O_3 was kept at 0.45, crystallization progressed from the initial formation of P_c to a mixture of P_c plus ZSM-5 to finally only ZSM-5 at the end of the run. At the lower SiO_2/Al_2O_3 ratio of 6 the P_c phase persisted, in a mixture with ZSM-5, until the end of the run. At the lowest SiO_2/Al_2O_3 ratio employed, 3.5, only P_c was detected for a TPA^+/Al_2O_3 ratio of 0.4. In all cases an amorphous silica–alumina coexisted with the zeolites. Samples removed periodically from gels having initial molar ratios of SiO_2/Al_2O_3 = 8.3 and TPA^+/Al_2O_3 = 0.45 were calcined and subsequently converted to the H^+ form of ZSM-5 and evaluated in the cracking of *n*-hexane. These catalysts, which contained between 10 and 42 wt% HAlZSM-5 were more active than either pure HAlZSM-5 or mechanical mixtures prepared by the addition of preformed ZSM-5 powder (from 10 to 42 wt%) to an amorphous silica–alumina gel. All catalysts examined did show comparable selectivity to C_6–C_{10} aromatics when compared at equal conversions. Both types of catalysts, namely those crystallized *in situ* and those prepared mechanically, had comparable numbers of strong acid sites (relative to the amount of HAlZSM-5) as measured by the temperature-programmed desorption of ammonia. The better connectivity between the two phases is suggested to be responsible for the greater activity of the composites in which the ZSM-5 crystallizes *in situ*, compared to the mechanical mixtures.

INTRODUCTION

Jacobs *et al.* (*1*) have observed that by interrupting the synthesis of ZSM-5 during its preparation from a gel of composition 125 TPAOH–Na_2O–Al_2O_3–200 SiO_2–5500 H_2O, a solid that is amorphous to X-rays but has IR framework vibrations and shape selectivity characteristics of crystalline ZSM-5 is obtained. They suggested that this solid consisted of submicrocrystalline ZSM-5 nuclei with sizes well below the threshold limit of X-ray detection dispersed in an amorphous matrix. The catalytic activity of this material in the hydrocracking of *n*-decane, while much smaller than the activity of pure ZSM-5, was due mainly to the zeolite fraction. The amorphous matrix of the catalyst was predominantly silica and therefore is not expected to have catalytic activity since it has no acidic sites. However, it could still affect the activity and/or selectivity of the composite via diffusion (*2, 3*). Such an effect is due to the higher rate of diffusion in the larger pores of the matrix than that in the pores (5–6 Å) of the zeolite.

More complex behavior is expected to occur when the matrix also possesses active sites. In order to prepare such a catalyst in a manner analogous to that of Jacobs *et al.* (*1*), a high alumina content is necessary in the starting gel. Then a fraction of the alu-

[1] To whom correspondence should be addressed.

This paper, which was originally published in **Journal of Catalysis**: 133, 28–41 (1992), is republished with permission.

mina can be incorporated in the zeolitic framework, and the remaining part, along with the unconverted silica, can form an amorphous aluminosilicate matrix that contains acidic sites. The first objective of the present study was to synthesize a composite catalyst that contains a highly siliceous zeolite, such as ZSM-5, and an alumina-rich amorphous silica–alumina matrix. It has been found that there are conditions under which a ZSM-5 phase can be crystallized *in situ* from an alumina-rich gel at low TPA^+ content. The second objective was to compare the activity and selectivity of the composite catalyst to those of pure ZSM-5 and to corresponding samples prepared by dispersing preformed ZSM-5 in an alumina–silica gel.

As noted in numerous papers (*4–6*), the formation of the ZSM-5 phase depends on the SiO_2/Al_2O_3, TPA^+/SiO_2, and $(Na^+ + TPA^+)/Na^+$ ratios, as well as the duration and temperature of synthesis. Synthesis in the absence of organic additives (*7, 8*) has revealed that for SiO_2/Al_2O_3 molar ratios <30 in the gel, the mordenite phase forms predominantly. Only in the very narrow range of molar ratios between 40 and 45 is the ZSM-5 phase formed, albeit at a molar ratio Na_2O/Al_2O_3 of 4.5. However, the presence of organic additives such as tetrapropyl ammonium cations (TPA^+) permits the formation of the ZSM-5 phase from starting gels with a very wide range of SiO_2/Al_2O_3 molar ratios (20–100) (*9, 10*). It has also been noted that an optimum TPA^+/SiO_2 ratio must be used (*11*) in order to obtain highly crystalline ZSM-5.

Another important variable that determines the type of zeolite as well as its yield is the synthesis temperature. Temperatures greater than 100°C decrease the induction time (*11*), a trend normally encountered in the synthesis of many zeolites. Both Barrer (*12*) and Breck (*13*) have presented a detailed discussion regarding the low-temperature zeolite formation. Starting from a gel mixture containing the following four components, $Na_2O–Al_2O_3–SiO_2–H_2O$, and using temperatures in the range of 60–100°C, the Zeolites A, P, X, Y, R, and S can be synthesized. The inclusion of a quaternary alkyl ammonium compound (TPA^+, TMA^+) in the gel leads to additional types of zeolites including the more siliceous ones, like ZSM-5 and ZSM-11.

In spite of the wealth of information available regarding various aspects of the synthesis of the pentasil family of high-silica zeolites starting from gels rich in silica and in organic template, no information is available for alumina-rich gels of low organic content. The present investigation is focused on synthesis using starting gels having (i) SiO_2/Al_2O_3 molar ratios of 10 or less; (ii) $TPA/Al_2O_3 < 0.5$, and (iii) a synthesis temperature of less than 100°C. The first condition was chosen in order to have 10–15 wt% Al_2O_3 in the amorphous phase and thus to generate acidic sites. The second condition ensures that only a part of the silica–alumina from the gel converts to the ZSM-5 phase. The progress of crystallization and the identification and quantification of the phases formed have been carried out using X-ray powder diffraction, supplemented with FT–IR investigations.

The catalytic activity and selectivity of the above type of composite catalysts in their protonated form were assessed using *n*-hexane cracking as a probe reaction. For comparison purposes, the *n*-hexane cracking was also carried out over catalysts prepared by dispersing preformed ZSM-5 in a silica–alumina gel, and also over pure ZSM-5 zeolite. Furthermore, the acidic characteristics of the composite catalysts of both types have been determined using the temperature-programmed desorption of adsorbed ammonia molecules.

EXPERIMENTAL

Gel Preparation and Synthesis

Technical sodium silicate aqueous solution (40–42° Bé, Fisher Scientific) containing 28.5 wt% SiO_2 and 7.7 wt% Na_2O was used as the silica source, and purified sodium aluminate (Fisher Scientific) containing 43.5 wt% Al_2O_3 and 30.0 wt% Na_2O was used as the alumina source. The other chemicals employed were tetrapropyl ammonium bromide (Aldrich), sodium chloride (Fisher Scientific), and sulfuric acid (Fisher Scientific). A typical gel was obtained as follows: individual solutions containing the above chemicals in such quantities so as to obtain the required molar compositions were first prepared. These solutions were mixed under stirring for at least 15 min in a blender after which the pH was adjusted to 12 ± 0.5 followed by an additional 15 min of stirring. Subsequently, the gel was transferred to a 1-liter three-neck clean glass flask provided with a stirrer, water-cooled condenser, and a sampling port. The gel was heated under reflux at 95°C and stirred continuously. Samples from the slurry were withdrawn periodically and were immediately cooled to room temperature by the addition of distilled water followed by filtration, washing, and drying at 100 ± 5°C overnight. In some cases, when needed for sample characterization, additional heat treatment was conducted at 550 ± 10°C for 24 h in a muffle furnace in air.

X-Ray Diffraction and Infrared Studies

The various phases have been identified by X-ray diffraction with a STOE powder X-ray diffractometer that possesses a curved, position-sensitive detector and a Cu$K\alpha$ source of radiation. In order to quantify the amounts of the various phases, samples were finely mixed with 20 wt% MgO (99.6% purity, Baker analyzed) whose major reflection peak served as standard. The relative amounts of each of the phases present were evaluated as the ratio of the area under the selected reflection peak(s) ($2\theta = 23.2$ for ZSM-5; 28.2 for P_c; the arithmetic average corresponding to 9.8 and 26 for mordenite) to the area under the reflection peak $2\theta = 43.1$ of MgO after correcting each of the areas for the background. (For mordenite, the arithmetic average of the areas corresponding to the two intense peaks at $2\theta = 9.8$ and 26 was used for the calculation of the crystallinity to improve the quantitative determinations.)

For the determination of the absolute crystallinity of the samples, a calibration plot was constructed by using intimate physical mixtures containing 0–80 wt% pure ZSM-5 (Mobil, MZ-44), 80–0 wt% amorphous silica–alumina, and 20 wt% MgO.

The absorption infrared spectra were obtained with a Mattson Alpha Centauri Fourier transform-infrared spectrophotometer equipped with a data station in the mid-infrared (1400–400 cm^{-1}) range and having a resolution of 0.5 cm^{-1}. The thin wafers, used for recording the IR spectra, were prepared by employing the KBr pellet technique (1% wt sample in KBr).

Catalyst Preparation

The samples were prepared for use as catalysts by calcining in air at 550°C for 24 h, followed by ammonium ion exchange with 1 *M* $(NH_4)_2SO_4$ solution under reflux for 24 h, drying, and a final calcination at 550°C for 4 h.

Temperature-Programmed Desorption (TPD)

Temperature-programmed desorption (TPD) spectra of some of the catalyst samples using ammonia as the probe base molecule were recorded. The TPD setup and the experimental protocol were similar to those employed in our earlier work (*15*).

Catalyst Activity Measurements

The catalytic behavior of the composite catalysts prepared by *in situ* crystallization of ZSM-5, hereafter denoted as CA, was studied using *n*-hexane cracking as a probe reaction. The reaction was conducted in a plug-flow microreactor at atmospheric pressure. Purified helium at a predetermined flow rate was bubbled through an *n*-hexane saturator (at 20°C) before being passed into the reactor. In all the catalytic runs, 0.15 g of catalyst powder (particle size <150 μm) was used. The catalysts were pretreated by heating at 500°C under helium flow for 1 h before starting the reaction. The exit gas stream was analyzed for C_1 to C_{20} hydrocarbons on a 10% SE-30 column equipped with a flame ionization detector. The reaction temperature was varied from 380 to 490°C, the space time from 0.2 to 1.5 (g catalyst)h/(g hexane), and the feed gas contained 5 mol% n-C_6H_{14} in helium. From the GC results, the n-hexane fractional conversion (x) and the C_6–C_{10} aromatic selectivity were calculated as follows:

$$\text{Conversion } (x) = \left(1 - \frac{\text{mol of } n\text{-}C_6H_{14} \text{ in the exit stream}}{\text{mol of } n\text{-}C_6H_{14} \text{ in the feed stream}}\right)$$

$$\text{Selectivity} = \left(\frac{\text{weight of } C_6\text{–}C_{10} \text{ aromatic fraction}}{\text{total weight of all hydrocarbons excluding exit } n\text{-hexane}}\right) \times 100.$$

For each of the catalysts tested, the rate constant (k) was obtained from the slope of the first-order plot of $-\ln(1 - x)$ against space time at a given reaction temperature. The activation energy E_a was calculated from an Arrhenius plot of ln k vs $1/T$.

The cracking of n-hexane was also studied using pure ZSM-5 and catalysts (hereafter denoted as CB) obtained by dispersing pure ZSM-5 particles in an amorphous silica–alumina gel. The pure zeolite was synthesized following the general procedure of Chen *et al.* but using different SiO_2/Al_2O_3 molar ratios than those given in their examples (*14*), while the CB type catalysts were made as described in our earlier work (*15*). Briefly, the procedure for preparing the CB composites involved the addition of a definite amount of preformed ZSM-5 zeolite powder to a solution containing sodium silicate and aluminum nitrate as soon as a turbidity due to the onset of gelation appeared in the solution. The pH was adjusted to 6.5–7.0. The zeolite content was varied from 10 to 45 wt%, whereas the Al_2O_3 content in the amorphous silica–alumina was kept at about 15 wt%.

RESULTS AND DISCUSSION

A. Preparation of Composites via in situ Crystallization

The molar compositions of the gels employed in the present study are summarized in Table 1. The crystallization temperature was maintained at 95°C during all these syntheses. The phases identified were mordenite, P_c and ZSM-5; they are listed in Table 1 in the order in which they were detected. The SiO_2/Al_2O_3 ratios of the starting gels and of the final composite powders for some samples are given in Table 2.

The SiO_2/Al_2O_3 molar ratio of the starting gel appears to be the main factor that determines the sequence in which the zeolite phases form as well as their yields. Thus, at a molar ratio of 10, and for both levels of TPA^+ template used (TPA^+/Al_2O_3 = 0.14 and 0.44), mordenite was the only crystalline phase detected (in addition to an amorphous phase); there was an induction period of 9–12 days before any crystalline material was detected. However, for comparable synthesis times, the crystallinity increased from 25 to 54 wt% when the TPA^+/Al_2O_3 molar ratio increased from 0.14 to 0.44. The induction period remained essentially unchanged.

For a starting SiO_2/Al_2O_3 molar ratio of 8.3, the zeolite P_c (also denoted B_1 in the literature) formed exclusively in the early stages of synthesis. Its amount increased with time, attained a maximum, and subsequently decreased to zero after 15 days of synthesis at 95°C. As the amount of P_c decreased, a concurrent increase in the amount of ZSM-5 was observed. As shown in Fig. 1, the maximum ZSM-5 XRD crystallinity was 42 wt%. The XRD patterns as a function of time for this batch are presented in Fig. 2. It should be noted that while the TPA^+/Al_2O_3 ratios were comparable for batches 2 and 3, the zeolite phases produced were completely different (mordenite for

TABLE 1

Gel Molar Composition

Batch no.	Composition (mol)					Zeolite phases (with increasing synthesis time)	Product composition of the final sample (wt%)	Duration of run (days)
	TPA Br	Na_2O	Al_2O_3	SiO_2	H_2O			
1	0.14	4.6	1	10.0	460	Mordenite (MOR)	25% MOR + 75% amorphous phase	15
2	0.44	4.8	1	10.0	440	Mordenite (MOR)	54% MOR + 46% amorphous phase	15
3	0.45	4.4	1	8.3	455	P_c, P_c + ZSM-5, ZSM-5	42% ZSM-5 + 58% amorphous phase	15
4	0.43	3.5	1	6.0	422	P_c, P_c + ZSM-5	26% P_c + 37% ZSM-5 + 37% amorphous phase	16
5	0.40	3.5	1	3.0	450	P_c	60% P_c + 40% amorphous phase	16
6	—	4.4	1	8.3	455	P_c	64% P_c + 36% amorphous phase	15
7[a]	0.45	4.4	1	8.3	455	P_c, P_c + ZSM-5	14% P_c + 32% ZSM-5 + 54% amorphous phase	17

Note. All the batches contained ~2 mol of NaCl per mol of Al_2O_3 in the starting gel.
[a] TPA Br solution was added to the synthesis mixture after 3 days.

batch 2 (SiO_2/Al_2O_3 = 10), and the P_c phase followed by P_c + ZSM-5 and finally by ZSM-5 for batch 3 (SiO_2/Al_2O_3 = 8.3)).

For a SiO_2/Al_2O_3 molar ratio of 6 (batch 4), P_c was the initial zeolite phase observed. It transformed in time to the ZSM-5 phase. In contrast to batch 3 (SiO_2/Al_2O_3 = 8.3), P_c remained present even after 16 days of synthesis but the ZSM-5/P_c ratio increased with time. Crystallization curves are presented in Fig. 3.

At a SiO_2/Al_2O_3 molar ratio of 3.5 (batch 5), P_c was the only zeolitic phase observed during the entire crystallization process of 16 days. As seen from Fig. 4, the concentration of P_c passes through a maximum at around 5 days. The decrease in the P_c phase with the synthesis time is expected to be followed by the appearance of a different

TABLE 2

Silica/Alumina Ratios of the Starting Mixtures and of the Final Composite Powders[a]

Batch no.	SiO_2/Al_2O_3 ratio (mol/mol)	
	Starting gel	Final sample
2	10.0	9.4
3	8.3	12.9
4	6.0	11.8
5	3.0	4.3
6	8.3	6.0

[a] Analyzed by the wet analytical method using atomic absorption spectroscopy.

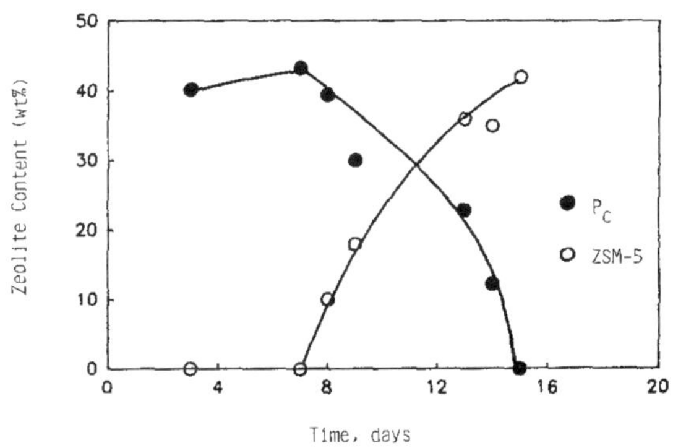

FIG. 1. Change with time of the zeolite content during the synthesis of the composite catalyst in batch 3.

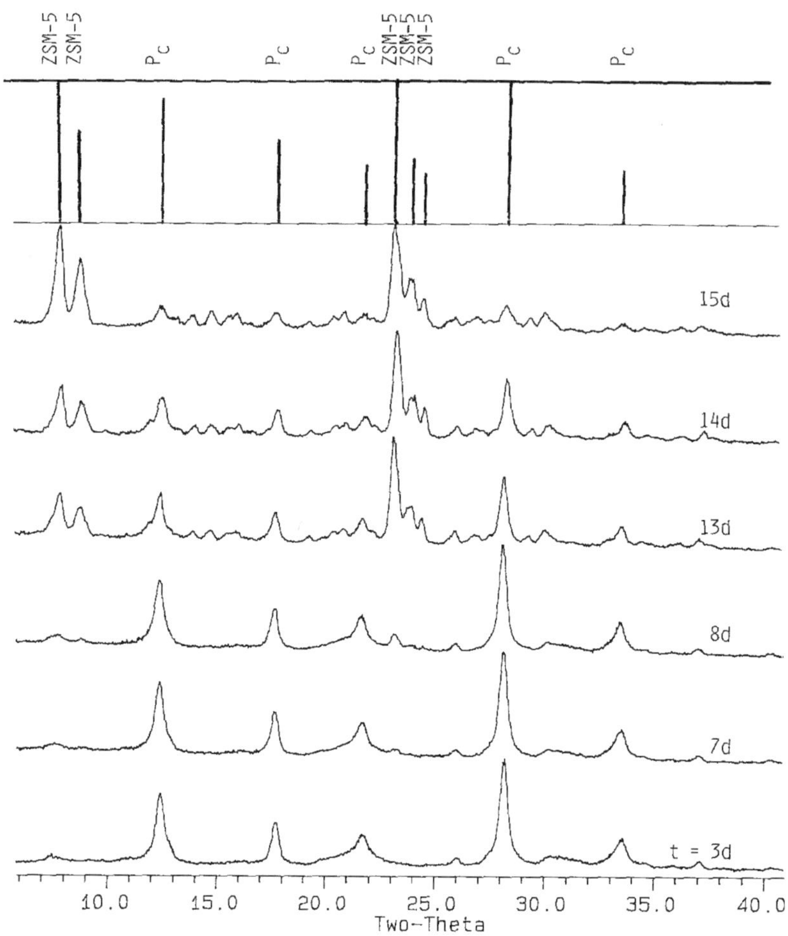

FIG. 2. XRD patterns of the solid samples withdrawn at different time intervals during the synthesis of the composite catalyst in batch 3.

and more stable zeolite phase rather than its conversion to an amorphous phase. However, the XRD patterns have not revealed any new phase. This might be due to the small size of the crystallites formed. As discussed later, IR can detect such clusters. Consequently, in order to confirm such a possibility, some of the samples were examined by IR. The IR spectra of the samples corresponding to 5 (maximum P_c phase) and 10 days (less P_c phase) of synthesis were similar and neither indicated the presence of five-membered ring structures, which are known to be precursors for mordenite and ZSM-5. Hence, it appears that the P_c phase transforms into an amorphous phase with time.

The synthesis conducted without any TPA cations in the gel (batch 6), for a SiO_2/Al_2O_3 molar ratio of 8.3, resulted in the P_c phase as the only crystalline phase even after 15 days. The crystallization curve is also presented in Fig. 4. This experiment clearly indicates that the presence of the organic cation is absolutely necessary for the formation of the ZSM-5 phase, but is not necessary for the formation of the P_c phase. One may note the presence of a maximum in P_c at 10 days. Beyond 10 days the crystallinity decreases because the P_c phase transforms into an amorphous phase.

Recent studies of Katovic *et al.* (*16*) showed that a gel with composition 4.2 Na_2O–Al_2O_3–3.6SiO_2–320H_2O (a composition comparable to that of batch 5 of Table 1) led initially (at a temperature of 80°C) to

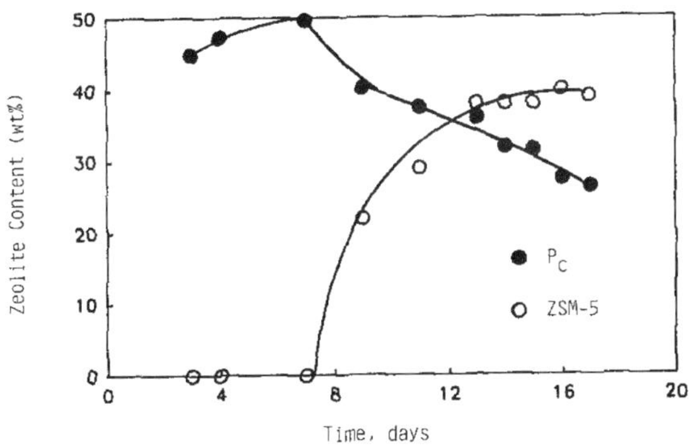

FIG. 3. Change with time of the zeolite during the synthesis of the composite catalyst in batch 4.

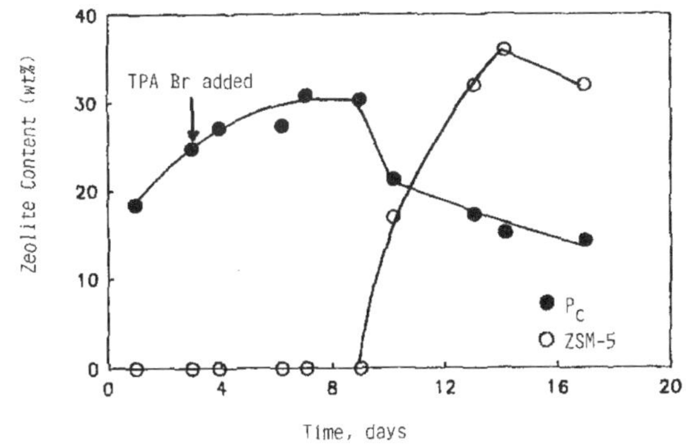

FIG. 5. Change with time of the amount of crystallization during the synthesis of the composite catalyst in batch 7.

the tetragonal form (P_t) of Zeolite P, which, however, completely transformed in time to its cubic form (P_c). In their study, the P_t phase formed after 6 h and completely transformed to the P_c phase after an additional 5 h. The P_t phase was not detected in the present experiments perhaps because the sampling was performed only once per day.

The data from batches 3, 4, and 5 show that the maximum concentration of the P_c phase was reached after 5–6 days of synthesis in the presence of the TPA cation template. However, in its absence (batch 6, Fig. 4), the maximum shifted from 5 to about 10 days. Thus the presence of the TPA^+ species in the gel enhances the rate of formation of the P_c phase apart from inducing the transformation of P_c to the ZSM-5 phase.

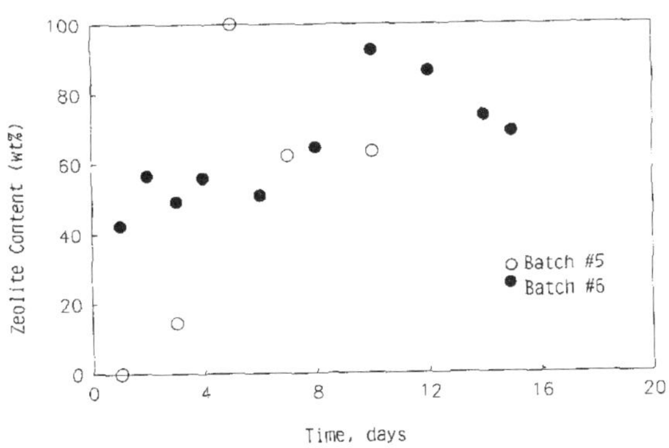

FIG. 4. Change with time of the P_c zeolite content during the synthesis of the composite catalyst in batches 5 and 6.

It is unlikely that the tetrapropyl ammonium cation with a diameter of 8–9 Å (*13*) can be accommodated in the pores of Zeolite P, which has a free aperture size of 3–4 Å (*13*), even though a P-type zeolite that contains tetramethyl ammonium cations has been reported (*18*). Furthermore, the absence of IR stretching frequencies characteristic of the tetrapropyl groups in samples that contained only P_c (e.g., batch 3 before the transformation of P_c to ZSM-5) indicates that TPA cations were not present in composite catalysts comprised of amorphous alumino-silicate and the crystalline P_c phase. On the other hand, a number of studies have indicated that TPA cations constitute templates for the formation of the ZSM-5 framework (*19–21*).

In order to verify whether the TPA cations are necessary from the beginning of the crystallization process for ZSM-5 to be generated, an experiment was conducted during which the organic cation was added after 3 days (batch 7) time at which the P_c phase has been already formed in significant amount. The molar composition of this batch was identical to that of batch 3. A comparison of the crystallization curves of batch 7 (Fig. 5) and of batch 3 (Fig. 1) shows that (i) the time required to reach the maximum concentration of P_c is increased and (ii) the maximum yield of the ZSM-5 phase reflected in the crystallinity of the system

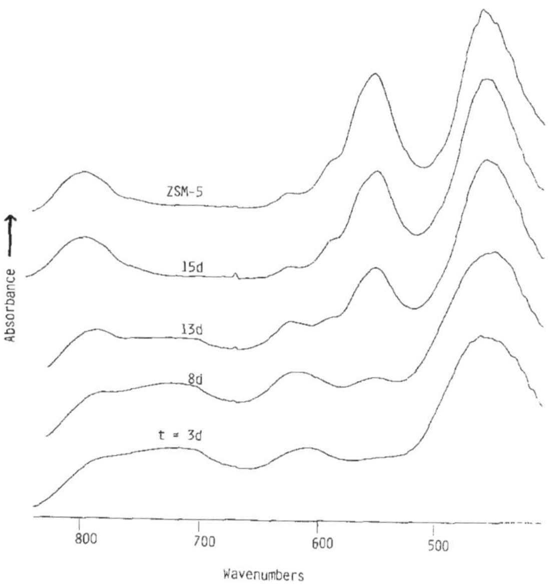

FIG. 6. IR spectra of the composite catalysts, heat treated at 550°C, withdrawn at different times from batch 3.

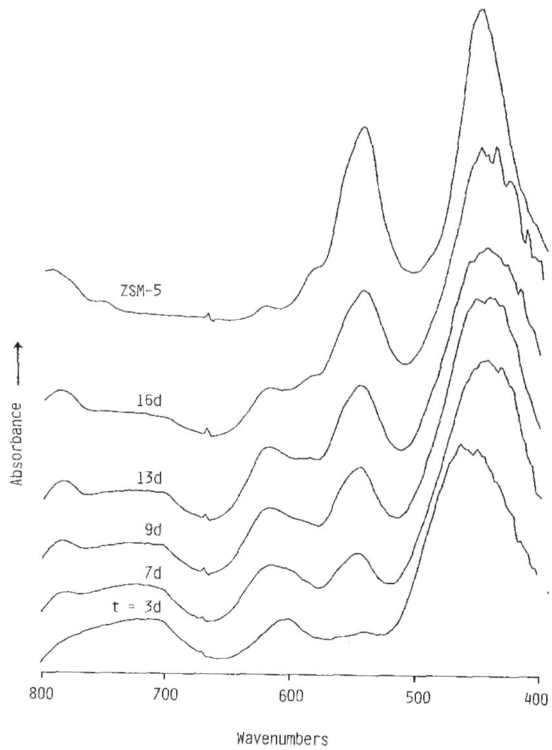

FIG. 7. Infrared spectra of the composite catalysts, heat treated at 550°C, withdrawn at different times from batch 4.

is slightly decreased (within the limits of experimental error), when the addition of the TPA cations is delayed until the end of the third day. On this basis, one can conclude that P_c and ZSM-5 phases are in competition only in the presence of TPA. Due to a small induction period coupled with a high rate of crystallization, the P_c phase forms more rapidly and consequently is more easily detected by XRD. In contrast, the ZSM-5 precursors do not generate nuclei for a long time and are not detectable by XRD (1) but may be detected by IR. For this purpose, the samples of batches 3 and 4 that contain TPA^+ were examined by IR. The IR spectra (800 cm^{-1}–400 cm^{-1}) of these samples together with that of pure ZSM-5 (Mobil, MZ-44) are presented in Figs. 6 (batch 3) and 7 (batch 4), respectively. In Fig. 6, a peak at 550 cm^{-1}, characteristic of the five-membered rings (*22–24*), which are known to be precursors for some zeolitic structures, such as mordenite and pentasil, appears after 8 days with a weak intensity. Its intensity increases, however, with time. The 450 cm^{-1} peak, which is present in all the samples, can be assigned to the Al–O or Si–O linkage (*26*). Furthermore, a shoulder at 1222–1225 cm^{-1} (not shown in the figures) was observed only for the samples for which a well-developed peak at 550 cm^{-1} was also identified. The peak at 1225 cm^{-1}, which is due to the external asymmetric (structure sensitive) stretching vibration, is characteristic of the ZSM-5 (25). Similar trends in IR were also observed for batch 4 (Fig. 7). Every structure (precursors, nuclei, small crystallites undetectable by XRD, and crystallites detectable by XRD) that contains five-membered rings can be detected by IR spectra. The optical density ratio at 550 and 450 cm^{-1} can be taken as a measure of such rings (this ratio is 0.75 for 100% ZSM-5). The results presented in Table 3 compare IR and XRD measurements. It is obvious that the IR values are higher because they account for every five-membered ring structure.

Although at this stage one cannot explain

TABLE 3

IR Detectable Structures and XRD Crystallinity

	IR detectable structures[a]	XRD crystallinity
Pure ZSM-5 (mobile, MZ-44)	100%	100%
Batch 3		
t = 3d	—	—
7d	—	—
13d	44%	35%
15d	73%	42%
Batch 4		
t = 3d	—	—
4d	—	—
7d	19%	—
9d	27%	18%
11d	37%	29%
17d	48%	38%

[a] With respect to pure ZSM-5.

the mechanism by which the P_c phase is transformed into the ZSM-5 phase, a possible scheme is outlined in Table 4. This scheme suggests that (Al, Si)O_4 tetrahedra formed in the gel can form either five-membered rings with an attached tetrahedron (5 − 1) or single four-membered rings (S4R). At the highest SiO_2/Al_2O_3 ratio studied (*10*), the latter do not compete favorably with the former and mordenite results. At the lowest SiO_2/Al_2O_3 ratio or for any SiO_2/Al_2O_3 ratio but in the absence of TPA^+, S4Rs predominate and P_c results. At intermediate SiO_2/Al_2O_3 ratio and in the presence of TPA^+ there is a competition between the S4R and 5 − 1 structures and while the former yield P_c more rapidly, the process is reversible so that in time, ZSM-5 is the final product. Additional evidence in favor of this scheme is provided by the results of chemical analysis presented below.

TABLE 4

Scheme for the Formation of Different Zeolitic Phases

(i) amorphous gel → (Al, Si)O_4 tetrahedra → 5 − 1 or 4 + 6 → mordenite (batches 1 and 2)

(ii) amorphous gel → (Al, Si)O_4 tetrahedra → S4R → P_c (batches 5 and 6)

(iii) amorphous gel → (Al, Si)O_4 tetrahedra ⇌ S4R ⇌ P_c
↘ 5 − 1 → ZSM-5 (↖ from S4R)

(batches 3, 4, 7)

5 − 1: five ring with an attached tetrahedra
4: four-membered ring
6: six-membered ring
S4R: Single four-membered ring

The chemical analysis (Table 2) shows that the SiO_2/Al_2O_3 molar ratio of the final composite sample is very close to that of the starting gel for batch 2. In batches 5 and 6 the ratios are comparable, but not in as close agreement as for batch 2. For batches 3 and 4, in which the ZSM-5 phase appeared, the final composite samples have SiO_2/Al_2O_3 molar ratios significantly higher than those of the starting gels. In batch 2, mordenite was the only crystalline phase present, the remaining material being the amorphous phase. P_c was the only crystalline phase in both batches 5 and 6. In contrast to mordenite or P_c, the ZSM-5 zeolite (batches 3 and 4) has a very low aluminum content. As P_c is converted into ZSM-5, Al species from the P_c phase must return to the solution, to permit the enrichment in silica needed for the formation of ZSM-5. This Al is apparently not incorporated into the amorphous phase. These results, in conjunction with the observation that the P_c phase is free of the TPA cation, suggest that the P_c phase transforms into ZSM-5 by first "redissolving" followed by a template-stimulated nucleation crystallization.

The ideal unit cell composition of mordenite is $Na_8[(AlO_2)_8(SiO_2)_{40}]\,24H_2O$. In reality, it can have a silica/alumina molar ratio between 8.3 and 10 (*27*). In batches 1 and 2 (see Table 1), the starting gels have a SiO_2/Al_2O_3 molar ratio of 10 and this appears to be an appropriate composition for the formation of mordenite. The induction period in these batches was long (9–12 days), but this is due mainly to the low temperature used. Elsewhere (*10*) it has been noted that mordenite is the initial phase that is formed, followed closely by ZSM-5; they transform later into the analcime phase. In a different study (*28*), the ZSM-5 phase was found to form initially, followed by its transformation to mordenite; after a long time, both phases transform into α-quartz. The gel compositions and the synthesis temperatures employed in Refs. (*10, 28*) were different from each other and also from those employed in the present study. In the present work, only the mordenite phase was detected in batches 1 and 2 after 16 days.

Zeolite P has the following ideal unit cell composition: $Na_6[(AlO_2)_6\ (SiO_2)_{10}]\ 15\ H_2O$. In real synthesized samples the silica/alumina ratio varies between 2.2 and 5 (*18*). For this reason, the P phase is expected to form easily in batches 3–6. Because Zeolite P has more Al atoms per unit cell (6 out of 16 tetrahedral atoms), the relatively high repulsive forces between the charged aluminum atoms make the lattice metastable. In contrast, the Zeolite ZSM-5 has at most 8 Al atoms in a total of 96 tetrahedral atoms per unit cell (*29*), and for this reason it is more stable than the P_c phase. Hence, the formation of ZSM-5 phase at the expense of the P_c phase observed in batches 3 and 4 appears to be in qualitative agreement with the stabilities of different zeolites.

B. Activity and Selectivity in the Conversion of n-hexane

Three CA composite zeolite samples were withdrawn from the gel mixture during the synthesis of batch 3 (Table 1) at the end of 8, 9.5, and 15 days of synthesis and were used as catalysts in the *n*-hexane conversion reaction. From XRD analysis, the above samples were found to contain 10, 20, and 42 wt% ZSM-5 (Fig. 1), respectively. Prior to calcination, the first two of the above three samples contained, in addition to ZSM-5, 38 and 32 wt% P_c. However, during the calcination step (550°C for 24 h), which was necessary to burn off the organic cation, the P_c phase collapsed into an amorphous phase, while there was no measurable change in the amount of ZSM-5. The final composite zeolite, after calcination, NH_4^+ exchange and a final calcination, had only ZSM-5 as the crystalline phase apart from the amorphous silica–alumina. Details regarding all the catalysts used in the reaction studies are given in Table 5.

Table 6 summarizes the kinetic parameters and the rate constants for *n*-hexane cracking over various catalysts. As an example, a typical plot of $-\ln(1-x)$ vs space time is provided in Fig. 8a, and a typical Arrhenius plot, is shown in Fig. 8b. For all the CA and the CB composite catalysts, the *n*-hexane cracking rate was found to be first-order over the temperature range of 370–490°C. It should be noted that the conversion data were collected during the initial time period in which the deactivation due to coking was low. For the amorphous silica–alumina sample, the deactivation was severe, and hence the activity at 723 K was determined by extrapolating to zero time the conversion versus time plot.

The C_6–C_{10} aromatic selectivity, which reflects the extent of secondary reactions that occur mainly in the ZSM-5 pores, was plotted against the *n*-hexane conversion for various reaction temperatures (Fig. 9). The plots include data from all the catalysts investigated here.

Figure 9 shows that there is some correlation between the selectivity and activity with a scattering that is larger at 380°C and smaller at higher temperatures. A comparison between the *n*-hexane cracking rate constants (last column of Table 6) shows that (i) the 10 wt% CA composite zeolite has an activity comparable to that of the pure zeolite, (ii) among the CA composite zeolites, the rate constant increases nonlinearly with increasing ZSM-5 content, (iii) the CB mechanical mixture composites have activities significantly lower than those of their CA composite counterparts or the pure zeolite, and (iv) among the CB composites the activity increases nonlinearly with increasing ZSM-5 content. The activation energies E_a over a temperature range of about 120 K were mostly between 10 and 13 Kcal/mol and this is approximately the value for the

TABLE 5

Catalysts Used in *n*-Hexane Conversion Reaction

Sample designation	Nominal chemical composition			Average ZSM-5 crystallite size (from SEM) (μm)	Method of preparation
	Amorphous phase	Zeolite phase			
	Alumina content (wt%) (by chemical analysis of total sample)	Zeolite content (wt%) (from XRD)	SiO_2/Al_2O_3 ratio (mol/mol) (from EDX)		
(a) Composite zeolite samples (CA)					Present study (from batch 3; Table 1 and Fig. 1)
10% Composite	N.D.	10	20	2	
20% Composite	N.D.	20	22	3	
42% Composite	15	42	22	3	
(b) Mechanical mixtures (CB)					As in Ref. (*15*)
10% Mechanical	15	10	30	5	
20% Mechanical	15	20	30	5	
42% Mechanical	15	42	30	5	
(c) Pure zeolite sample	—	100	30	5	As in Ref. (*14*) for different SiO_2/Al_2O_3 ratios

Note. nd, not determined.

TABLE 6

Kinetic Parameters and Rate Constants for *n*-Hexane Conversion Reaction over Various Catalysts Tested[a]

Catalyst	Kinetic parameters		k at 723 K g h^{-1} (g catalyst)$^{-1}$ (atm)$^{-1}$
	$\ln k_0$ g h^{-1} (g catalyst)$^{-1}$ (atm)$^{-1}$	E_a (kcal/mol)	
(a) CA composites			
10% Zeolite	7.3564	10.1	1.39
20% Zeolite	11.5130	15.7	1.79
42% Zeolite	9.8688	12.6	3.00
(b) CB composites			
10% Zeolite	5.6794	11.6	0.09
20% Zeolite	6.4145	10.0	0.58
42% Zeolite	8.2628	11.8	1.00
(c) Pure zeolite	8.7748	12.2	1.33
(d) Amorphous silica–alumina			0.03[b]

[a] Assuming first-order kinetics $r = k P_{n\text{-hexane}}$, where $k = k_0 e^{-E_a/RT}$ and $P_{n\text{-hexane}}$ is the partial pressure of *n*-hexane.

[b] Initial rate constant obtained from conversion versus time-on-stream plot.

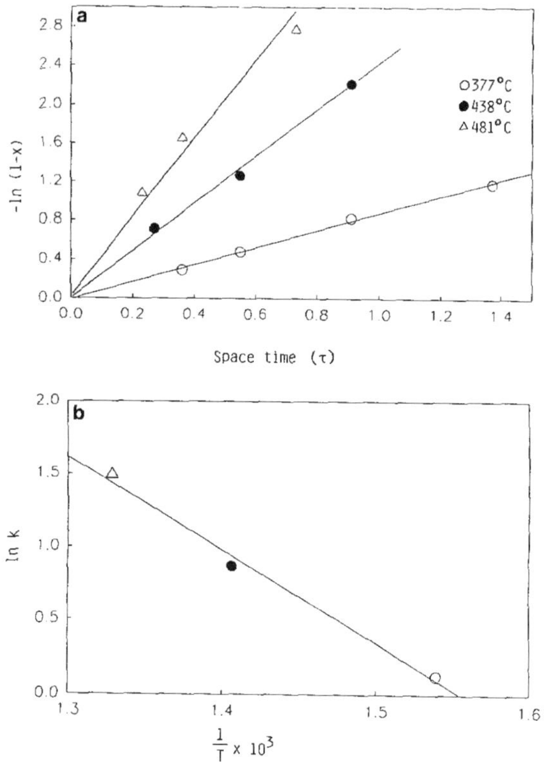

FIG. 8. (a) Conversion plot used to determine rate constants for 42 wt% CA composite. (b) Arrhenius plot for 42 wt% CA composite for *n*-hexane cracking reaction.

The observed differences in *n*-hexane cracking activity could arise from either differences in crystallite size and/or the different SiO_2/Al_2O_3 ratios (or acid site density) of the ZSM-5 zeolites present in the composites. SEM indicates that for the pure ZSM-5, the average crystallite size is about 5 μm diameter. This pure ZSM-5 was also used in the preparation of the CB-type composites. For the CA composites, the size of the zeolite particles was in the range of 1–3 μm diameter. For such crystallite sizes, the effectiveness factor for *n*-hexane (diffusion coefficient 3×10^{-4} $cm^2\ s^{-1}$) cracking reaction is essentially unity (*31*). Hence, the measured rate constants for the reaction given in Table 6 are not believed to be influ-

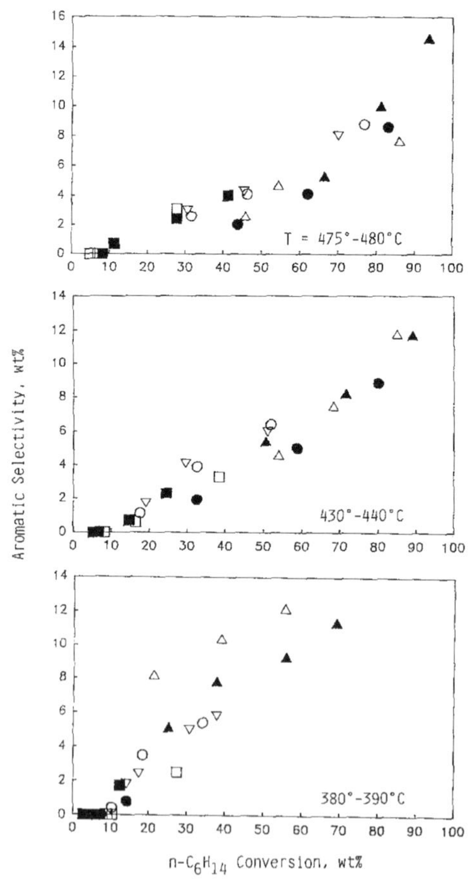

FIG. 9. Variation of aromatic selectivity with *n*-hexane conversion at different reaction temperatures. ○ Pure ZSM-5; ● 10% CA composite; △ 20% CA composite; ▲ 42% CA composite; ■ 10% CB composite; □ 20% CB composite; ▽ 42% CB composite.

pure ZSM-5. This behavior, in conjunction with the observation that the pure silica–alumina sample has a low activity, indicates that the zeolite component of the composite is the locus of the *n*-hexane cracking. Furthermore, from Fig. 9, it is clear that at a given reaction temperature, the C_6–C_{10} aromatic selectivity depends on the extent of *n*-hexane conversion rather than on the nature of the catalyst samples (method of preparation or zeolite content). The hydrogen transfer reaction leading to the formation of aromatics is sensitive to the aluminum content of the ZSM-5 framework (*30*) and also to the pore opening (*30*). As long as these values remain comparable or in a range where their influence is negligible, the aromatic selectivity is expected to depend only on the conversion level. The data presented in Fig. 9 confirm to a large extent such a conclusion.

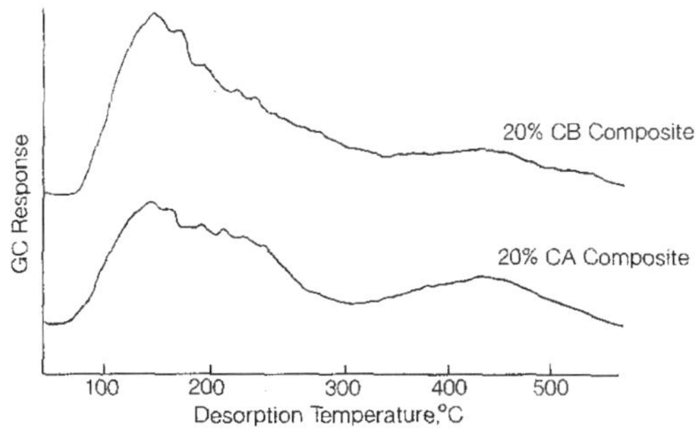

FIG. 10. Ammonia TPD spectra of 20% CA composite and 20% CB composite samples. Conditions: Weight of sample = 0.1 g; NH_3 adsorption temperature = 20°C; heating rate 10°C/min; TC detector attn × 4.

enced by the diffusion in the ZSM-5 particles.

As the *n*-hexane cracking reaction is acid catalyzed and occurs either by monomolecular protolytic cracking (*32*) and /or by carbenium ion formation followed by β-scission (*33*), the presence of strong acid sites is necessary for the primary reaction. The low activity of the amorphous SiO_2/Al_2O_3 sample indicates that the sites associated with the tetrahedral Al cations on the amorphous silica–alumina surface are less active in causing the primary reaction. Thus it becomes necessary to assess the intrazeolitic sites that have a high intrinsic site activity (*34*) both in the CA and CB samples. The SiO_2/Al_2O_3 ratio of the zeolitic component has been probed via EDX in a SEM. Table 5 shows that the SiO_2/Al_2O_3 molar ratios, measured by EDX, of individual ZSM-5 crystallites present in the CA samples vary from 20 to 22; in the CB samples the molar ratio is about 30. The uncertainty in these values is expected to be large.

Thus the EDX data suggest that the number of Al sites in the ZSM-5 component of the CA catalysts is greater than in the CB catalysts or pure ZSM-5. In other words, the number of acid sites is likely to be greater in the former than in latter catalysts.

The TPD of ammonia was used to probe the difference in acidities. The TPD spectra in Fig. 10 compare the 20% CA catalyst to the 20% CB catalyst, and Fig. 11 similarly compares the 42% catalysts. There is no discernable difference, for 42% zeolite content, in the number or distribution of strong acid sites (desorption above 350°C). For 20% zeolite the CA composite does show a somewhat higher acid concentration above 350°C than the corresponding CB sample. There are subtle differences between 100 and 350°C, where the CB catalysts have a broader distribution and a greater population. This behavior does not explain why the CB catalysts are less active, however.

Perhaps the activity is not solely dependent on the acidity and another cause of a physical nature is responsible for the higher activity of the CA catalysts. A possible explanation is the better connectivity between the two phases that is achieved when the ZSM-5 catalysts are generated via *in situ* crystallization.

CONCLUSIONS

The present investigation shows that ZSM-5-containing composites can be prepared from gels rich in alumina and low in TPA^+ templates in a narrow range of SiO_2/Al_2O_3 molar ratios in the starting gels. A SiO_2/Al_2O_3 molar ratio of 10 led to a composite containing mordenite as the crystalline phase while a molar ratio of 3 led to P_c zeolite. For molar ratios of SiO_2/Al_2O_3 between 8.3 and 6, the phase sequence is amorphous gel → P_c → ZSM-5. When the molar ratio of SiO_2/Al_2O_3 is 8.3, the amount

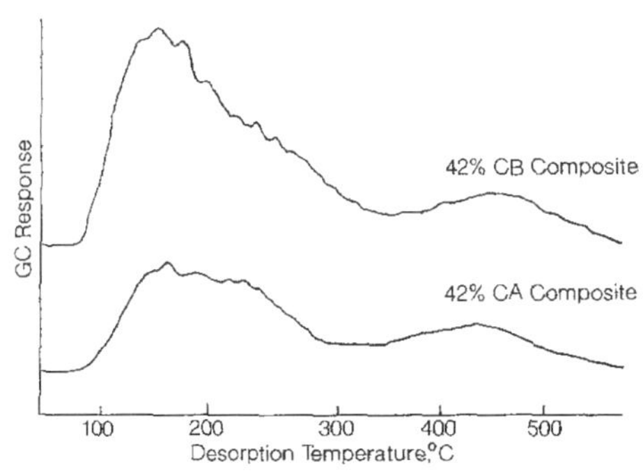

FIG. 11. Ammonia TPD spectra of 42% CA composite and 42% CB composite samples. (Conditions the same as those in Fig. 10.)

of P_c passes through a maximum, and subsequently decreases until, after 15 days, the P_c phase is no longer detected.

From the gels based on initial molar composition of SiO_2/Al_2O_3 = 8.3 and TPA^+/Al_2O_3 = 0.45, samples have been obtained after crystallization *in situ* for 8, 9.5, and 15 days of synthesis. These samples have been calcined at 550°C for 24 h, ammonium ion exchanged, and again calcined at 550°C for 4 h. XRD showed that they have 10, 20, and 42 wt% ZSM-5 zeolite, respectively, the balance being amorphous silica–alumina. For the *n*-hexane cracking reaction, the above samples exhibited activities significantly higher than those of the corresponding (same zeolite content) composite samples prepared by dispersing preformed ZSM-5 zeolite particles in an aluminosilicate gel. A 10 wt% composite prepared by the *in situ* crystallization method exhibited an activity comparable to that of the ZSM-5 zeolite. Although the amounts of strong acid sites as probed via NH_3 desorption were comparable in both types of composites, the *in situ* crystallized composites probably have a better connectivity between the two phases than the catalysts based upon preformed ZSM-5. This may then be the cause for the differences in activity.

REFERENCES

1. Jacobs, P. A., Derouane, E. G., and Weitkamp, J., *J. Chem. Soc. Chem. Commun.* 591 (1981).
2. Ruckenstein, E., *AIChE J.* **16,** 151 (1970).
3. Nandapurkar, P. J., and Ruckenstein, E., *Chem. Eng. Sci.* **39,** 371 (1984).
4. Derouane, E. G., and Gabelica, Z., *J. Solid State Chem.* **64,** 296 (1986).
5. Jacobs, P. A., and Martens, J. A., "Synthesis of High-Silica Aluminosilicate Zeolites." Elsevier, The Netherlands, 1986.
6. Szostak, R., "Principles of Synthesis and Identification." Van Nostrand–Reinhold Catalysis Series, Van Nostrand–Reinhold, New York, 1989.
7. Narita, E., Sato, K., Yatabe, N., and Okabe, T., *Ind. Eng. Chem. Prod. Res. Dev.* **24,** 507 (1985).
8. Shiralkar, V. P., and Clearfield, A., *Zeolites* **9,** 363 (1989).
9. Argauer, R. J., and Landolt, G. R., US Patent 3,702,886 (1972).
10. Erdem, A., and Sand, L. B., *J. Catal.* **60,** 241 (1979).
11. Chao, K-J., Tasi, T. C., and Chen, M. S., *J. Chem. Soc. Faraday Trans. 1* **77,** 547 (1981).
12. Barrer, R. M., "Hydrothermal Chemistry of Zeolites." Academic Press, San Diego, 1982.
13. Breck, D. W., "Zeolite Molecular Sieves." Robert E. Krieger Publishing, Malabar, FL, 1984.
14. Chen, N. Y., Miale, J. N., and Reagan, W. J., US Patent 4,112,056.
15. Yarlagadda, P., Lund, C. R. F., and Ruckenstein, E., *Appl. Catal.* **54,** 139 (1989).
16. Katovic, A., Subotic, B., Smit, T., and Despotovic, L. A., *Zeolites* **9,** 45 (1989).
17. Lok, B. M., Cannan, T. R., and Messina, C. A., *Zeolites* **3,** (1983).
18. Baerlocher, C. H., and Meier, W. M., *Helv. Chim. Acta* **53,** 1285 (1970).
19. Scholle, K. F. M. G. J., Veeman, W. S., Frenken, P., and van der Velden, G. P. M., *Appl. Catal.* **17,** 233 (1985).
20. Dutta, P. K., and Puri, M., *J. Phys. Chem.* **91,** 4329 (1987).
21. Gabelica, Z., Nagy, J. B., Bodart, P., and Nastro, A., *Thermochim. Acta* **93,** 749 (1985).
22. Shukla, D. B., and Pandya, V. P., *J. Chem. Technol. Biotechnol.* **44,** 147 (1989).
23. Jacobs, P. A., Beyer, H. K., and Valyon, J., *Zeolites* **1,** 161 (1981).
24. Coudurier, G., Naccache, C., and Vedrine, J. C., *J. Chem. Soc., Chem. Commun.*, 1413 (1982).
25. Jansen, J. C., van der Gaag, F. J., and Bekkum, H., *Zeolites* **4,** 369 (1982).
26. Flanigen, E. M., Khatami, H., and Szymanski, H. A., *in* "Advances in Chem. Series 101," p. 201, 1971.
27. Bajpai, P. K., *Zeolites* **6,** 2 (1986) [and references therein].
28. Dai, F. Y., Suzuki, M., Takahasi, H., and Saito, Y., *in* "New Developments in Zeolite Science and Technology" (Murakami, Iijima, and Ward, Eds.), p. 233, Elsevier, Amsterdam, 1986.
29. van der Gaag, F. J., Jansen, J. C., and van Bekkum, H., *Appl. Catal.* **17,** 261 (1985).
30. Wielers, A. F. H., vaarKamp, M., and Post, M. F. M., *J. Catal.* **127,** 51 (1991).
31. Haag, W. O., Lago, R. M., and Weisz, P. B., *Discuss. Faraday Chem. Soc.* **72,** 317 (1981).
32. Haag, W. O., and Dessau, R. M., *in* "Proceedings, 8th International Congress on Catalysis, Berlin, 1984," Vol. 2, p. 305. Dechema, Frankfurt-am-Main, 1984.
33. Greensfelder, B. S., Voge, H. H., and Good, G. M., *Ind. Eng. Chem.* **41,** 2573 (1949).
34. Lombardo, E. A., and Hall, W. K., *J. Catal.* **112,** 565 (1988).

5

SYNTHESIS OF MESOPOROUS V–Mg–O OXIDES AND THEIR APPLICATIONS AS CATALYSTS

INTRODUCTION

Since the pioneering synthesis of the M41S family of silicates by Mobil researchers in the early 1990s [1, 2], the self-assembly of ordered mesoporous materials, templated by long-chain surfactants, have been extensively studied both for their synthesis and their applications. Because of their unique textural properties, that is, large pores, narrow pore size distributions, and high specific surface areas and thermal stabilities, the mesoporous materials found applications in numerous domains of science and technology [3–8]. The synthesis of mesoporous materials is mainly based on the sol-gel technology. The controlled hydrolysis of the inorganic precursor species and their subsequent condensation, templated by supramolecular assembly of surfactant micelles, constitute the key steps. Depending on synthesis conditions, including the nature of the template employed, the molar ratios of inorganic sources to template, temperature, pH, and reaction time, numerous mesophases have been synthesized [9–18]. While silica has been widely used to synthesize mesostructured silicates, nonsilica oxides have been used as sources for the preparation of mesoporous materials. Among various nonsilica-based mesoporous materials, those containing vanadium are especially important due to the unique catalytic applications of vanadium-based oxides in the selective oxidative dehydrogenation of hydrocarbons. In this regard, various mesostructured vanadium oxides, including a hexagonal mesostructured vanadium oxide [19], a layered vanadium oxide [20], a mixture of lamellar and hexagonal mesophases [21], and vanadium oxide nanotubes [22], were reported. Yagi et al. [23] examined the transformations of mesoporous vanadium oxide template by hexadecyltrimethylammonium surfactant, which occur during low temperature calcination. They observed that the lamellar mesostructure was transformed to other mesophases during calcination at 443 K and that the mesostructures were stable only below 523 K. Consequently, it is desirable to synthesize vanadium-containing mesostructures that are stable at the catalytic reaction temperatures of interest.

The oxidative dehydrogenation of light alkanes (e.g., ethane and propane) into corresponding alkenes (e.g., ethene and propene) has received extensive attention. Both ethene and propene are raw materials for a wide variety of products. Compared with the commercial production of ethene (propene as well) through direct dehydrogenation reaction [24, 25], the oxidative dehydrogenation process is potentially advantageous because this reaction is exothermic, requiring much less energy input. Nevertheless, the oxidative dehydrogenation reaction has relatively low selectivity due to the occurrence of side reactions, such as cracking and

Heterogeneous Catalysis: Experimental and Theoretical Studies, First Edition. Edited by Eli Ruckenstein and Haiyou Wang.

combustion. At moderate temperatures, these processes are mainly heterogeneous and could be controlled by the judicious selection of a catalyst. The development of such a catalyst with a sufficiently high activity and selectivity is thus highly desirable. V–Mg–O catalyst systems appear to be among the most promising ones [26–31]. However, most V–Mg–O catalyst prepared using conventional methods provided low activities at temperatures below 500°C [29–31].

Chapter 5 is therefore devoted to the synthesis of V–Mg oxides and their applications in catalysis, including the oxidative dehydrogenation of alkanes to alkenes. In short, this Chapter consists of two parts: the first is concerned with the synthesis of mesoporous V–Mg–O oxides (**5.1**, **5.2**); and the second with the performance of V–Mg–O oxides as catalysts for the oxidative dehydrogenation of alkanes to alkenes (**5.3**, **5.4**).

Mesoporous materials are generally obtained as aggregates and loose agglomerates of small particles. However, for numerous applications, including as catalysts, chemical sensors, and optical devices, specific morphologies may be required. In this regard, nanofibrous mesoporous V–Mg–O was synthesized using V_2O_5 and $MgCl_2$ as raw materials and the surfactant CTAB (cetyltrimethylammonium bromide) or the mixture of surfactants CTAB and HDA (hexadecylamine) as template (**5.1**). Further, the effects of a number of factors including the nature of the surfactant, the pH, and the vanadium/magnesium molar ratio in the formation of mesoporous V–Mg–O phases were investigated (**5.2**). The anionic surfactants sodium dodecyl sulfate (SDS) and sodium dodecylbenzene sulfonate (SDBS) generated amorphous structures, while the cationic surfactant cetyltrimethylammonium bromide (CTAB) generated a lamellar mesoporous structure with various degrees of crystallinity. The formation of lamellar mesostructured V–Mg–O was promoted by high pH values, by binary mixtures of anionic and cationic surfactants as templates, and by equimolar vanadium and magnesium contents.

The oxidative dehydrogenation of light alkanes such as ethane and propane has been suggested to replace the direct dehydrogenation route to reduce the energy consumption. The reaction becomes exothermic and can occur at relatively low temperatures. Nevertheless, it is necessary to remove the heat of reaction to avoid over oxidation to CO_2, thus providing a high selectivity to alkenes. The latter issue could be solved by finding proper catalysts. In this regard, for oxidative dehydrogenation of ethane (**5.3**) and of propane (**5.4**), the V–Mg oxide catalysts prepared from their mesoporous precursors were more active and selective than those prepared via the conventional solid reaction.

REFERENCES

[1] Kresge, C.T., Leonowicz, M.E., Roth, W.J., Vartuli, J.C., and Beck, J.S., *Nature*, **359**, 710 (1992).

[2] Beck, J.S., Vartuli, J.C., Roth, W.J., Leonowicz, M.E., Kresge, C.T., Schmitt, K.D., Chu, C.T.W., Olson, D.H., Sheppard, E.W., McCullen, S.B., Higgins, J.B., and Schlenker, J.L., *J. Am. Chem. Soc.*, **114**, 10834 (1992).

[3] Behrens, P., *Adv. Mater.*, **5**, 127 (1993).

[4] Sayari, A., *Chem. Mater.*, **8**, 1840 (1996).

[5] Corma, A., *Chem. Rev.*, **97**, 2373 (1997).

[6] Brunel, D., *Microporous Mesoporous Mater.*, **27**, 329 (1999).

[7] Ying, J.Y., Mehnert, C.P., and Wong, M.D., *Angew. Chem. Int. Ed.*, **38**, 56 (1999).

[8] Selvam, P., Bhatia, S.K., and Sonwane, C.G., *Ind. Eng. Chem. Res.*, **40**, 3237 (2001).

[9] Raman, N.K., Anderson, M.T., and Brinker, C.J., *Chem. Mater.*, **8**, 1682 (1996).

[10] Huo, Q., Margolese, D.I., Ciesla, U., Demuth, D.G., Feng, P., Gier, T.E., Sieger, P., Firouzi, A., Chmelka, B.F., Schuth, F., and Stucky, G.D., *Chem. Mater.*, **6**, 1176 (1996).

[11] Sakamoto, Y., Kaneda, M., Terasaki, O., zhao, D.Y., Kim, L.M., Stucky, G.D., Shin, H.J., and Ryoo, R., *Nature*, **408**, 449 (2000).

[12] Huo, Q.,Leon, R., Petroff, P.M., and Stucky, G.D., *Science*, **268**, 1324 (1995).

[13] Zhao, D.Y., Huo, Q., Feng, J.L., Chmelka, B.F., and Stucky, G.D., *J. Am. Chem. Soc.*, **120**, 6024 (1998).

[14] Zhao, D.Y., Huo, Q., Feng, J.L., Kim, J., Han, Y, and Stucky, G.D., *Chem. Mater.*, **11**, 2668 (1999).

[15] Tanev, P.T., and Pinnavaia, T.J., *Science*, **267**, 865 (1995).

[16] Bagshav, S.A., Prouzet, E., and Pinnavaia, T.J., *Science*, **269**, 1242 (1995).

[17] Kim, S.S., Zhang, W., Pinnavaia, T.J., *Science*, **282**, 1302 (1998).

[18] Ryoo, R., Kim, J.M., Shin, C.H., and Lee, J.Y., *Stud. Surf. Sci. Catal.*, **105A**, 45 (1996).

[19] Luca, V., and Hook, J.M., *Chem. Mater.*, **9**, 2731 (1997).

[20] Janauer, G.G., Dobley, A., Guo, J.D., Zavalij, P.Y., and Whittungham, M.S., *Chem. Mater.*, **8**, 2096 (1996).

[21] Liu, P., Moudrakovski, I.L., Liu, J., and Sayari, A., *Chem. Mater.*, **9**, 2513 (1997).

[22] Spahr, M.E., Bitterli, P., Nesper, R., Muller, M., Krumeich, F., and Nissen, H.U., *Angew. Chem. Int. Ed.*, **37** (9), 1263 (1998).

[23] Yagi, Y., Zhou, H., Miyayama, M., Kudo, T., and Honma, I., *Langmuir*, **17**, 1328 (2001).

[24] Albright, L.F., Crynes, B.L., and Corcoran, W.H., (eds), *Pyrolysis: Theory and Industrial Practice* (Academic Press, London, 1983).

[25] Resasco, D.E., and Haller, G.L., *Catalysis (London)*, **11**, 379 (1994).

[26] Kung, H.H., *Adv. Catal.*, **40**, 1 (1994).

[27] Lemonidou, A.A., Tjatjopoulos, G.J., and Vasalos, I.A., *Catal. Today*, **45**, 65 (1998).

[28] Banaras, M.A., *Catal. Today*, **51**, 319 (1999).

[29] Pantazidis, A., Aurox, A., Herrmann, J.M., and Mirodatos, C., *Catal. Today*, **32**, 81 (1996).

[30] Martra, G., Arena, F., Coluccia, S., Frusteri, F., and Parmaliana, A., *Catal. Today*, **63**, 197 (2000).

[31] Pak, C., Bell, A.T., and Tilley, T.D., *J. Catal.*, **206**, 49 (2002).

Synthesis of Mesoporous V–Mg–O Nanofibers

Eli Ruckenstein* and Zi-Sheng Chao

Department of Chemical Engineering, State University of New York at Buffalo, Amherst, New York 14260

Received September 24, 2001; *Revised Manuscript Received October 18, 2001*

ABSTRACT

A mesostructured V–Mg–O was synthesized for catalytic purposes using vanadium(V) oxide and magnesium chloride as sources and various surfactants or surfactant mixtures as templates. Among the surfactants employed, it was found that cetyltrimethylammonium bromide (CTAB) and the mixture of CTAB and hexadecylamine (HDA) templated fibrous lamellar-like mesostructures. The mixture of CTAB + HDA was more efficient in templating the fibrous mesoporous V–Mg–O than the individual surfactants.

Introduction. Because of their promising characteristics, such as large and tailorable pore sizes, narrow pore size distributions, long-range ordering of the pore packing, high specific surface areas, variable framework compositions and, in most cases, high thermal stability, the mesoporous materials have been utilized or could be potentially used for selective adsorption[1–3] and catalysis[4–7] and as functional materials.[8–12] So far, the synthesis of mesoporous materials was focused on (i) the enlargement of the pores and the stabilization of the mesostructure; (ii) the modification of the mesoporous silicate by incorporating heteroatoms and the preparation of mesoporous materials from other elements; and (iii) the morphology control. The emphasis was, however, mostly on the first two features. Generally, the mesoporous materials were obtained as aggregates and loose agglomerates of small particles. However, for numerous applications in catalysis, chemical sensors, or as optical devices, specific morphologies may be required. In the present letter, we report about the synthesis of nanofibrous mesoporous V–Mg–O, whose formation we noted during the preparation of a catalyst for the oxidation of alkanes.

Experimental Section. *Syntheses.* All reagents were purchased from Aldrich and used without further purification. A certain amount of vanadium(V) oxide was first dissolved in a 0.1 M dilute sodium hydroxide solution with stirring and heating at the boiling temperature of the solution. Then, an aqueous solution containing magnesium chloride and a surfactant (dodecylbenzenesulfonic sodium salt (SDBS), cetyltrimethylammonium bromide (CTAB), hexadecylamine (HDA), or triblock copolymer poly(ethylene oxide)-poly-(propylene oxide)-poly (ethylene oxide) (PEO–PPO–PEO)) was added with intense stirring to the vanadium-containing solution. The pH had to be adjusted to a value in the range 3~4, using a dilute hydrogen chloride solution. The final mixture had the molar composition of $V_2O_5/MgCl_2/0.25$

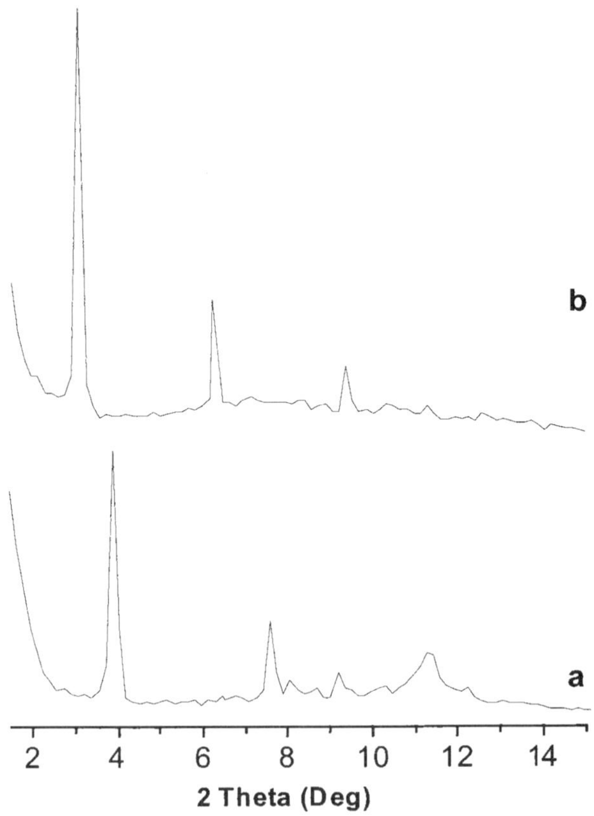

Figure 1. XRD spectra of mesostructured V–Mg–O synthesized with (a) CTAB and (b) CTAB + HDA as surfactant.

* Corresponding author. Tel: 1-716-645-2911 ext. 2214. Fax: 1-716-645-3822. E-Mail: feaeliru@acsu.buffalo.edu (E. Ruckenstein).

This paper, which was originally published in **Nano Letters**: 1(12), 739–742 (2001), is republished with permission.

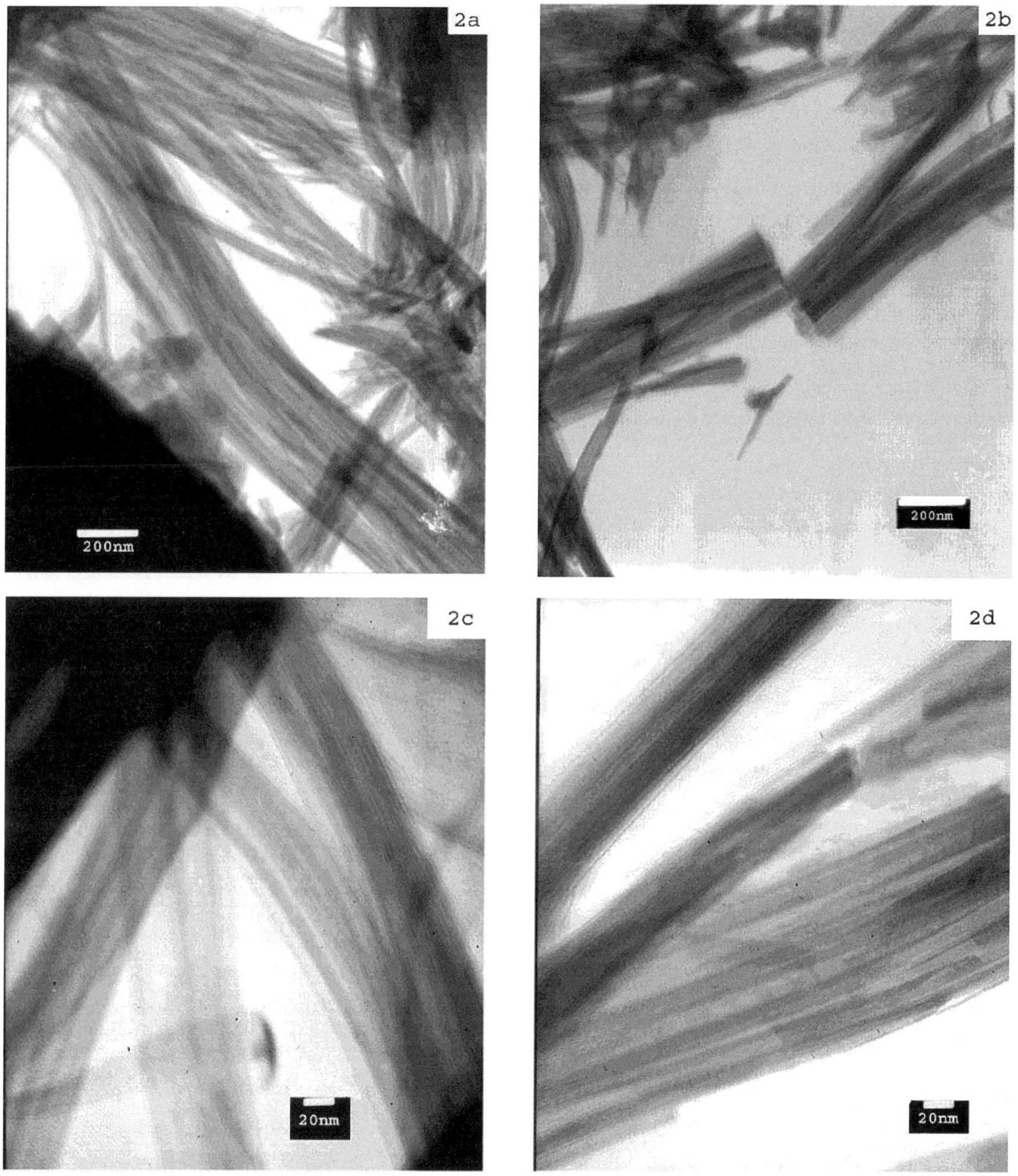

Figure 2. TEM micrographs of mesostructured V−Mg−O synthesized with (a) and (c) CTAB; and (b) and (d) CTAB + HDA as surfactant. ((a) and (b) are with lower and (c) and (d) with higher resolutions.)

surfactant/295 H_2O. The gel obtained was stirred at room temperature for about 24 h and then kept for aging for 2 days. The solid product was recovered by filtration, washed with distilled water, dried at 373 K for 12 h, and further used for the X-ray diffraction and transmission electron microscopy characterization.

Characterization. X-ray diffraction (XRD) spectra were obtained on a Siemens D500 diffractometer using Cu Kα radiation of wavelength 1.5406 Å. The diffraction data were recorded for 2 θ angles between 1 and 15°, with a resolution of 0.02°.

Transmission electron microscopy (TEM) was carried out on a JEM-2010 electron microscope equipped with a tungsten gun operating at an accelerating voltage of 200 kV. The samples were ground in methanol and supported on a holey carbon film located on a Cu grid.

Nitrogen adsorption−desorption isotherms were determined at 77 K with a Micromeritics ASAP 2010 gas sorption and porosimetry apparatus. Before the measurements, the synthesized samples were pretreated in a flow of nitrogen from room temperature to 623 K at a heating rate of 10 K/min and kept at 623 K for 8 h to remove the surfactant template. Then, the sample was transferred to an ASAP 2010 apparatus and outgassed at 623 K for 4 h under a vacuum of 1×10^{-3} Torr. The specific surface area was determined by using the BET method and the pore volumes were estimated at a relative pressure $P/P_0 > 0.99$, assuming full surface saturation with nitrogen. The pore size distribution was obtained from the N_2 desorption branch by using the BJH (Barrett−Joyner−Halenda) method.

The energy dispersive spectroscopy (EDS) investigation was carried out with specimens treated as for the N_2 adsorption experiments, using a Hitachi S-4000 field emission SEM, operating at 20 KeV. The spectra were obtained using a PGT IMIX X-ray microanalysis equipment and were collected for 100 seconds at a 44.9° takeoff angle.

Results and Discussion. In the preparation of the mesoporous V−Mg−O, various surfactants, which are popular in the synthesis of the mesoporous silicate, such as SDBS, CTAB, HDA, or PEO−PPO−PEO, have been tested. It was found that SDBS, HDA, and PEO−PPO−PEO were inactive in templating a mesoporous V−Mg−O phase, as evidenced by the XRD spectra, which exhibited no diffraction peaks for $2\theta < 15°$. However, a mesoporous structure could be obtained when the cationic surfactant CTAB was employed. We have also examined the following equimolar mixtures of surfactants, SDBS + HDA, SDBS + CTAB, and CTAB + HDA, and found that only the mixture CTAB + HDA was effective in the formation of a mesoporous V−Mg−O phase. Figure 1 presents the XRD spectra of two of the synthesized samples. One can see that for the sample prepared with CTAB, there are three main peaks at 2θ = 3.848°, 7.548°, and 11.299° (Figure 1 a), corresponding to d spacings of 2.294 (d_{001}), 1.170 (d_{002}), and 0.782 (d_{003}) nm, respectively. Similarly, three main peaks can be observed at 2θ = 3.082, 6.180, and 9.281 (Figure 1 b) for the sample prepared with the mixture CTAB + HDA, corresponding to d spacings of 2.864 (d_{001}), 1.429 (d_{002}), and 0.952 (d_{003}) nm, respectively. Both samples have a lamellar-like mesostructure, due to the equidistance of the three peaks in Figures 1a and 1b. There are also some weak peaks, for instance the weak peak at 2θ = 9.156° in Figure 1a, which is probably caused by a nonindexed compound between V, Mg, and O. The intensities of the main peaks in Figure 1b are much stronger than those in Figure 1a. The above results indicate that the sample prepared with CTAB + HDA has a more regular structure than that prepared with CTAB alone.

The structure and morphology of the samples were also examined by TEM and electron diffraction, and the micrographs and the diffraction patterns are presented in Figures 2 and 3, respectively. One can see from the TEM micrographs that the samples prepared with CTAB (Figure 2a) and with CTAB + HDA (Figure 2b) have the morphology of fibers with an average diameter of about 50 nm and an average length longer than 1 μm. These fibers have a lamellar-like structure, being composed of numerous fine fibers with average diameters of about 0.6 nm and separation distances of 1.7 and 2.2 nm for the samples prepared with CTAB (Figure 2c) and CTAB + HDA (Figure 2d), respectively. The selected area electron diffraction patterns (Figure 3a and 3b) exhibit sharp reflections with layered symmetry extended to the fourth order, indicating a long-range order

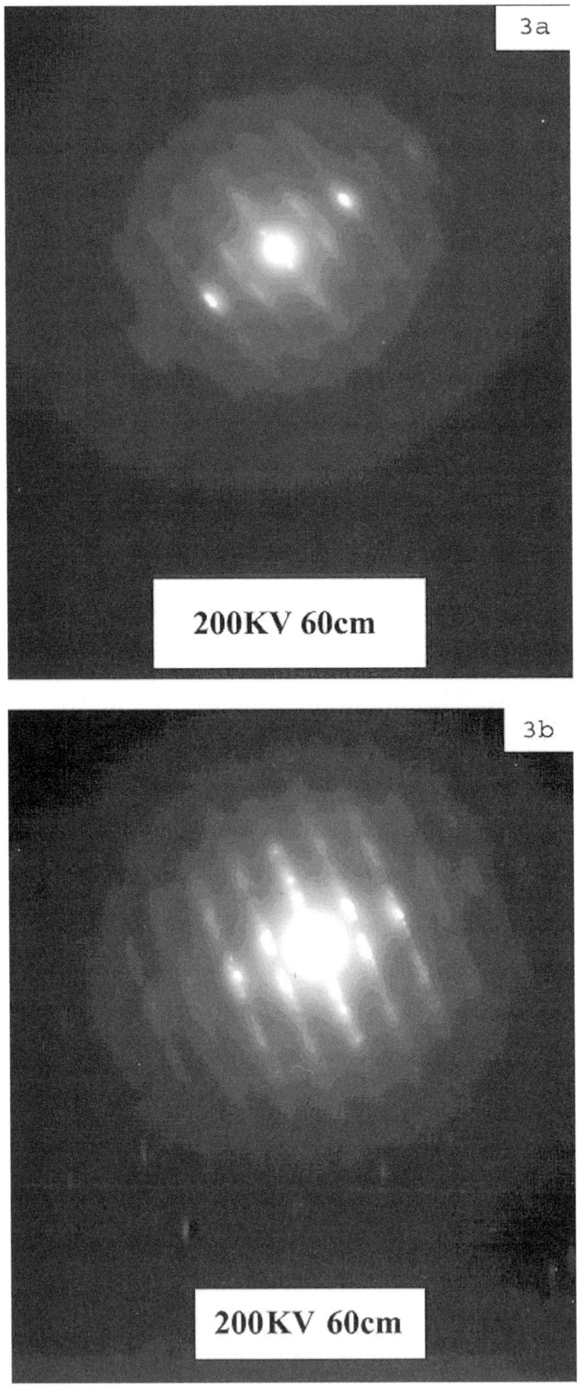

Figure 3. Electron diffraction patterns of mesostructured V−Mg−O synthesized with (a) CTAB and (b) CTAB + HDA as surfactant.

in the fine structure of the mesophase. The arrangement of the diffraction spots is more ordered for the sample prepared with CTAB + HDA (Figure 3b) than that prepared with CTAB alone (Figure 3a), indicating a more ordered structure in the former case. There are two sets of diffraction spots in both cases, with one set normal to the other one. One can calculate that the *d* spacings corresponding to these diffraction spots are about 0.675 and 0.185 nm, respectively.

The samples heat-treated as mentioned above were also examined via nitrogen adsorption−desorption. The isotherms of both samples prepared with CTAB and CTAB + HDA were of type IV, according to the IUPAC classification, and exhibited the characteristic hysteresis loop of mesoporous materials. The specific surface areas, pore volumes, and the average pore sizes are, respectively, 237.67 m^2/g, 0.45 cm^3/g, and 7.56 nm for the sample prepared with CTAB + HDA and 177.96 m^2/g, 0.30 cm^3/g, and 6.74 nm for those prepared with CTAB alone. The pore size distribution was much narrower for the sample prepared with CTAB than with CTAB + HDA.

The surface V/Mg mole ratios of the samples, which were heat-treated as for the nitrogen adsorption−desorption experiments, were determined by EDS. The results show that there is an enrichment in vanadium on the surfaces of the samples. The vanadium/magnesium ratio is somewhat larger on the surface of the sample prepared with CTAB (V/Mg = 1: 0.15) than with CTAB + HDA (V/Mg = 1: 0.22). The enrichment in vanadium is caused by the much lower surface free energy of V_2O_5 than MgO (V_2O_5 has a surface free energy of 90 $ergs/cm^2$ at 963 K and MgO a surface free energy of 1000 $ergs/cm^2$ at $T = 298$ K[13]).

Conclusion. Mesostructured V−Mg−O samples were prepared using V_2O_5 and $MgCl_2$ as sources and CTAB or the mixture CTAB + HDA as templates. The synthesized samples exhibited a fibrous morphology (50 nm in diameter) consisting of numerous fine fibers (separated by an average distance of about 2 nm). The mixture of surfactant CTAB + HDA templated a mesoporous V−Mg−O with a larger pore size and higher regularity than CTAB alone.

References

(1) Sonwane, C. G.; Bhatia, S. K.; Calos, N. *Ind. Eng. Chem. Res.* **1998**, *37*, 2271.
(2) Morishige, K.; Shikimi, M. *J. Chem. Phys.* **1998**, *108*, 7821.
(3) Floquet, N.; Coulomb, J. P.; Giorgio, S.; Grillet, Y; Llewellyn, P. L. *Stud. Surf. Sci. Catal.* **1998**, *117*, 583.
(4) Tanev, P. T.; Chibwe, M.; Pinnavaia, T. J. *Nature* **1994**, *368*, 321.
(5) Carrazza, J.; Gonzalez, F.; Adrian, R.; Djaouadi, D.; Moore, J. G.; Shahriari, D. Y.; Landry, C. C.; Lujano, J. In *Proceedings of the 12th International Zeolite Conference*, Baltimore, MD, 1998.
(6) Zhang, S.; Fujji, Y.; Yamashita, H.; Koyano, K.; Tatsumi, T.; Anpo, M. *Chem. Lett.* **1997**, *7*, 659.
(7) Wu, C. G.; Bein, T. *Chem. Mater.* **1994**, *6*, 1109.
(8) Agger, J. R.; Anderson, M. W.; Pemble, M. E.; Terasaki, O.; Nozue, Y. *J. Phys. Chem. B* **1998**, *102*, 3345.
(9) Carter, C. *Molecular Electronic Devices*; Marcel Dekker: New York, 1982.
(10) Aviram, A. *J. Am. Chem. Soc.* **1988**, *110*, 5687.
(11) Mann, S.; Archibald, D. D.; Didymus, T. D.; Heywood, B. R.; Meldrum, F. C.; Reeves, N. J. *Science* **1993**, *261*, 1286.
(12) Sayari, A. *Chem. Mater.* **1996**, *8*, 1840.
(13) Overbury, S. H.; Bertrand, P. A.; Somorjai, G. A. *Chem. Rev.* **1995**, *75*, 547.

NL015634Y

Effect of the Nature of the Templating Surfactant on the Synthesis and Structure of Mesoporous V−Mg−O

Zi-Sheng Chao and Eli Ruckenstein*

Department of Chemical Engineering, State University of New York at Buffalo, Amherst, New York 14260

Received September 4, 2001. In Final Form: November 16, 2001

The mesostructured V−Mg−O was synthesized at room temperature using vanadium acetylacetonate and magnesium chloride as source materials. The effects of a number of factors, such as the nature of the surfactant, the pH, and the vanadium/magnesium molar ratio in the formation of mesoporous V−Mg−O phases were investigated for a fixed molar concentration of surfactant. The synthesized samples were examined by X-ray diffraction, Brunauer−Emmett−Teller analysis, transmission electron microscopy, and energy dispersive spectroscopy. The anionic surfactants sodium dodecyl sulfate (SDS) and sodium dodecylbenzene sulfonate (SDBS) generated amorphous structures, while the cationic surfactant cetyltrimethylammonium bromide (CTAB) generated lamellar mesoporous structures with various degrees of crystallinity. Equimolar mixtures of an anionic surfactant, SDS, and a cationic surfactant (dodecyltrimethylammonium bromide (DTAB), myristyltrimethylammonium bromide (MTAB), cetylpyridium bromide (CPB), and cetyltrimethylammonium bromide (CTAB)), particularly the mixture SDS + CTAB, generated ordered lamellar structures with high crystallinity. It was found that the ordered arrangement ("crystallinity") of the pores was strongly dependent on the vanadium/magnesium molar ratios in the initial batches; too large or too small ratios were unfavorable to the crystallinity and generated distorted hexagonal (wormlike) mesostructures. The optimum vanadium/magnesium molar ratio regarding the crystallinity was found to be in the range 1:1 to 1:1.5, and the mesoporous phase had a lamellar structure. For both the single and the mixtures of anionic and cationic surfactants, the higher the pH, the more favorable was the formation of a lamellar-structured mesoporous V−Mg−O phase.

1. Introduction

Since the first synthesis of MCM-41, the most famous member of the M41S family of silicate mesoporous materials, reported by Mobil in 1992,[1,2] the ordered mesoporous structures have attracted a great deal of attention. These materials have very high specific surface areas, up to 1500 $m^2\ g^{-1}$, large pore sizes, between 2 and 10 nm, a narrow pore-size distribution, and high thermal stability. Depending on the synthesis conditions and particularly on the surfactant/SiO_2 ratio, various phases could be obtained, such as the hexagonal MCM-41,[2] the cubic MCM-48,[3] the lamellar MCM-50,[4] and the wormlike MSU-X[5] phase. Several synthesis pathways have been developed, involving various kinds of interactions between the templating surfactant molecules and the inorganic precursors. Huo[6] and co-workers found that the mesoporous materials could be synthesized using routes involving the electrostatic interactions between the surfactant and the inorganic precursor. Tanev[7] and Bagshaw[8] and their co-workers prepared mesoporous silica using neutral routes involving the hydrogen bonding between a noncharged amine or poly(ethylene oxide) and neutral inorganic species.

The mesoporous materials are attractive candidates as catalyst,[9−12] catalyst supports,[13−17] adsorbents,[18−30] and

* Corresponding author: tel, (+1-716) 6452911/2214; fax, (+1-716) 6453822; e-mail, feaeliru@acsu.buffalo.edu (E. Ruckenstein).

(1) Kresge, C. T.; Leonowicz, M. E.; Roth, W. J.; Vartuli, J. C.; Beck, J. S. *Nature* **1992**, *359*, 710.
(2) Beck, J. S.; Vartuli, J. C.; Roth, W. J.; Leonowicz, M. E.; Kresge, C. T.; Schmitt, K. D.; Chu, C. T. W.; Olson, D. H.; Sheppard, E. W.; McCullen, S. B.; Higgeng, J. B.; Schlenker, J. L. *J. Am. Chem. Soc.* **1992**, *114*, 10834.
(3) Monnier, A.; Schuth, F. et al., *Science* **1993**, *261*, 1299.
(4) Ayyappan, S.; Rao, C. N. R. *Chem. Commun.* **1997** 575.
(5) Herrier, G.; Blin, J. L.; Su, B. L. *Langmuir* **2001**, *17*, 4422.
(6) Huo, Q.; Margolese, D.; et al. *Nature* **1994**, *368*, 317.
(7) Tanev, P. T.; Pinnavaia, T. J. *Science* **1995**, *267*, 865.
(8) Bagshaw, S. A.; Prouzet, E. *Science* **1995**, *269*, 1242.
(9) Tanev, P. T.; Chibwe, M.; Pinnavaia, T. J. *Nature* **1994**, *368*, 321.
(10) Carrazza, J.; Gonzalez, F.; Adrian, R.; Djaouadi, D.; Moore, J. G.; Shahriari, D. Y.; Landry, C. C.; Lujano, J. In Proceedings of 12th International Zeolite Conference, Baltimore, 1998.
(11) Zhang, S.; Fujji, Y.; Yamashita, H.; Koyano, K.; Tatsumi, T.; Anpo, M. *Chem. Lett.* **1997**, *7*, 659.
(12) Wu, C. G.; Bein, T. *Chem. Mater.* **1994**, *6*, 1109.
(13) Cauvel, A.; Brunel, D.; DiRenzo, F.; Fajula, F. *Organic Coatings, Proceedings of the 53rd International Meeting of Physical Chemistry*; American Institute of Physics: New York, 1996; Vol. 354, p 477.
(14) Brunel, D.; Cauvel, A.; Fajula, F.; DiRenzo, F. *Stud. Surf. Sci. Catal.* **1995**, *97*, 173.
(15) Mashmeyer, T.; Rey, F.; Sankar, G.; Thomas, J. M. *Nature* **1995**, *378*, 159.
(16) Corma, A. *Chem. Rev.* **1997**, *97*, 237.
(17) Bowes, C. L.; Malek, A.; Ozin, G. A. *Chem. Vap. Dep.* **1996**, *2*, 97.
(18) Llewellyn, P. L.; Ciesla, U.; Decher, H.; Stadler, R.; Schuth, F.; Unger, K. K. *Stud. Surf. Sci. Catal.* **1994**, *84*, 2013.
(19) Branton, P. J.; Hall, P. G.; Sing, K. S. W.; Reichert, H.; Schuth, F.; Unger, K. K. *J. Chem. Soc., Faraday Trans.* **1994**, *90*, 2965.
(20) Rathousky, J.; Franks, A.; Schulz-Ekloff, G. *J. Chem. Soc., Faraday Trans.* **1995**, *91*, 937.
(21) Schmidt, R.; Stocker, M.; Hansen, E.; Akporiaye, D.; Ellestad, O. H. *Microporous Mater.* **1995**, *3*, 443.
(22) Ravikovitch, P. I.; Wei, D.; Chueh, W. T.; Haller, G. L.; Neimark, A. V. *J. Phys. Chem. B* **1997**, *101*, 3671.
(23) Kruk, M.; Jaroniec, M.; Sayari, A. *J. Phys. Chem.* **1997**, *B101*, 583.
(24) Maddox, M. W.; Olivier, J. P.; Gubbins, K. E. *Langmuir* **1997**, *13*, 1737.
(25) Ravikovitch, P. I.; Haller, G. L.; Neimark, A. V. *Adv. Colloid Interface Sci.* **1998**, *203*, 76.
(26) Ravikovitch, P. I.; Haller, G. L.; Neimark, A. V. *Stud. Surf. Sci. Catal.* 1998, *117*, 77.
(27) Neimark, A. V.; Ravikovitch, P. I.; Grun, M.; Schuth, F.; Unger, K. K. *J. Colloid Interface Sci.* **2000**, *129*, 597.
(28) Sonwane, C. G.; Bhatia, S. K.; Calos, N. *Ind. Eng. Chem. Res.* **1998**, *37*, 2271.
(29) Morishige, K.; Shikimi, M. *J. Chem. Phys.* **1998**, *108*, 7821.
(30) Floquet, N.; Coulomb, J. P.; Giorgio, S.; Grillet, Y.; Llewellyn, P. L.; *Stud. Surf. Sci. Catal.* **1998**, *117*, 583.

This paper, which was originally published in **Langmuir**: 18, 734–743 (2002), is republished with permission.

host materials for nanocomposites.[31–36] The mesoporous silica was modified through the incorporation of heteroatoms, such as Ti,[37] Zr,[38–44] Pd,[45] Mn,[46] and V,[47–50] and reviews were published by Tuel[51] and Ying.[52] Pure mesoporous materials other than silica, such as those for the oxides of Pb,[53] Sb,[54] Ti,[55] Nb,[56] Ta,[57] Al,[58] Mn,[59] Ce,[60] Ha,[61] Zr,[62] V,[63] and W,[64] and composite mesoporous materials, such as Al−B−O,[65] Al−P−O,[66] V−P−O,[67] Zr−P−O,[68] and Al−P−V−O,[69] were also prepared. In the present paper, we are concerned with the preparation of vanadium−magnesium oxide mesoporous materials, because the non-mesostructured mixture of the two oxides was previously used as a very active catalyst for the oxidative dehydrogenation (ODH) of light alkanes (C_2−C_4) to the corresponding alkenes. The ODH reaction is of industrial importance, since it offers a most useful alternative for the production of valuable unsaturated hydrocarbons from light alkanes. We have synthesized the mesostructured V−Mg−O for its potential application to the ODH reaction.

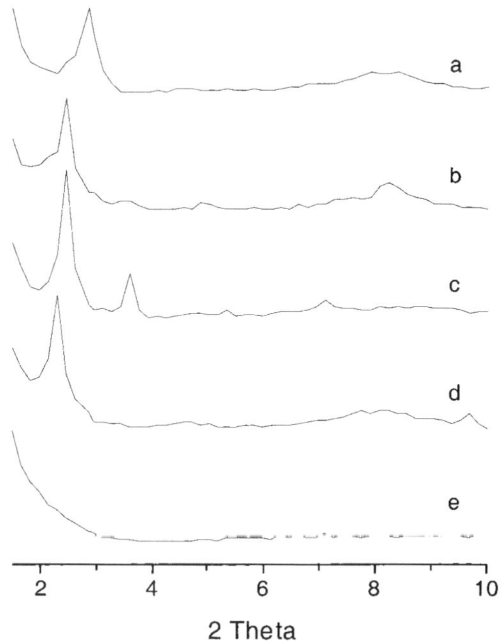

Figure 1. XRD patterns of the as-synthesized mesostructured V−Mg−O prepared, at a pH = 10, using CTAB as surfactant and a V/Mg molar ratio of (a) 4:1, (b) 1.5:1, (c) 1:1, (d) 1:1.5, and (e) 1:4 in the initial batches.

2. Experimental Section

2.1. Synthesis of Mesostructured V−Mg−O. All reagents were purchased from Aldrich and used without further purification. Vanadium(III) acetylacetonate ($V(acac)_3$) and magnesium chloride were used as sources of V and Mg, respectively. A number of cationic surfactants, such as benzyltrimethylammonium bromide (BTAB), dodecyltrimethyammonium bromide (DTAB), myristyltrimethylammonium bromide (MTAB), cetylpyridium bromide (CPB), and cetyltrimethylammonium bromide (CTAB), and anionic surfactants, such as sodium dodecyl sulfate (SDS) and sodium dodecylbenzene sulfonate (SDBS), were employed either alone or as anionic/cationic mixtures. The surfactant together with $MgCl_2$ was dissolved with stirring in a dilute hydrogen chloride solution, possessing a pH lower than 2. Then, $V(acac)_3$ was dispersed at room temperature into distilled water with vigorous stirring for about 2 h, and the dispersion formed was added to the above surfactant and $MgCl_2$-containing solution, again with vigorous stirring and at room temperature. To reach a selected pH between 2 and 10, a dilute sodium hydroxide solution was employed. The final mixture was stirred at room temperature for about 24 h and then was allowed to age for 2 days. The product was recovered by filtration, washed with distilled water, and dried at 373 K for 12 h.

2.2. Characterization Techniques. The X-ray diffraction (XRD) patterns of the as-synthesized samples were obtained with a SIEMENS D500 diffractometer using the Cu Kα radiation of wavelength 1.5406 Å. The diffraction data were recorded for 2θ angles between 1.5° and 10° with a resolution of 0.02°.

The transmission electron microscopy (TEM) investigation was performed with a JEM-2010 electron microscope equipped with a tungsten gun operating at a voltage of 200 kV. The samples were ground in methanol and supported on a holey carbon film located on a Cu grid, before the TEM measurements.

The nitrogen adsorption−desorption isotherms were determined at 77 K using a Micromeritics ASAP 2010 gas sorption

(31) Wu, C. G.; Bein, T. *Chem. Mater.* **1994**, *6*, 1109.
(32) Agger, J. R.; Anderson, M. W.; Pemble, M. E.; Terasaki, O.; Nozue, Y. *J. Phys. Chem. B* **1998**, *102*, 3345.
(33) Carter, C. *Molecular Electronic Devices*; M. Dekker: New York, 1982.
(34) Aviram, A. *J. Am. Chem. Soc.* **1988**, *110*, 5687.
(35) Mann, S.; Archibald, D. D.; Didymus, T. D.; Hoywood, B. R.; Meldrum, F. C.; Reeves, N. J. *Science* **1993**, *261*, 1286.
(36) Sayari, A. *Chem. Mater.* **1996**, *8*, 1840.
(37) Bagshaw, S. A.; DiRenzo, F.; Fajula, F. *Chem. Commun.* **1996**, 2209.
(38) Wang, X. X.; Lefebvre, F.; Patarin, J.; Basset, J. M. *Microporous Mesoporous Mater.* **2001**, *42*, 269.
(39) Gontier, S.; Tuel, A. *Appl. Catal., A* **1996**, *143*, 125.
(40) Tuel, A.; Gontier, S.; Teissier, R. *J. Chem. Soc., Chem. Commun.* **1996**, 651.
(41) Jones, D. J.; Jimenez, J. J.; Lopz, A. J.; Torres, P. M.; Pastor, P. O.; Castellon, E. R.; Roziere, J. *J. Chem. Soc., Chem. Commun.* **1974**, 431.
(42) Luan, Z.; Kevan, L. *J. Phys. Chem. B* **1997**, *101*, 2020.
(43) Biz, S.; Oceli, M. L. *Catal. Rev. Sci. Eng.* **1998**, *40*, 327.
(44) Beck, J. S.; Vartuli, J. C.; Roth, W. J.; Leonowicz, M. E.; Kresge, C. T.; Schmitt, K. D.; Chu, C. T. W.; Olson, D. H.; Sheppard, E. W.; McCullen, S. B.; Higgins, J. B.; Schlenker, J. L. *J. Am. Chem. Soc.* **1992**, *114*, 10834.
(45) Mehnert, C. P.; Ying, J. Y. *Chem. Commun.* **1997**, 2215.
(46) Zhang, J.; Goldfarb, D. *J. Am. Chem. Soc.* **2000**, *122*, 7034.
(47) Baltes, M.; Cassiers, K.; Van Der Vort, P.; Weckhuysen, B. M.; Schoonheydt, R. A.; Vansant, E. F. *J. Catal.* **2001**, *197* 160.
(48) Daai, L. X.; Tabata, K.; Suzuki, E. *J. Mater. Sci. Lett.* **2000**, *19*, 2071.
(49) Daai, L. X.; Tabata, K.; Suzuki, E.; Tatsumi, T. *Chem. Mater.* **2001**, *13*, 208.
(50) Morey, M. S.; Davidson, A.; Stucky, G. D. *J. Porous Mater.* **1998**, *5*, 195.
(51) Tuel, A. *Microporous Mesoporous Mater.* **1999**, *27*, 151.
(52) Ying, Y. J.; Mehnert, C. P.; Wong, M. S. *Angew Chem., Int. Ed.* **1999**, *38*, 56.
(53) Huo, Q.; Margolese, D. I.; Ciesla, U.; Demuth, D. G.; Feng, P.; Gier, T. E.; Sieger, P.; Firouzi, A.; Chmelka, B. F.; Schuth, F.; Stucky, G. D.; *Chem. Mater.* **1994**, *6*, 1176.
(54) Huo, Q.; Margolese, D. I.; Ciesla, U.; Feng, P.; Gier, T. E.; Sieger, P.; Leon, R.; Petroff, P. M.; Schuth, F.; Stucky, G. D. *Nature* **1994**, *368*, 317.
(55) Antonelli, D. M. *Microporous Mesoporous Mater.* **1999**, *30*, 315.
(56) Sun, T.; Ying, J. Y. *Nature* **1997**, *389*, 704.
(57) Antonelli, D. M.; Ying, J. Y. *Chem. Mater.* **1996**, *8*, 874.
(58) Holland, B. T.; Isbester, P. K.; Blanford, C. F.; Munson, E. J.; Stein, A. *J. Am. Chem. Soc.* **1997**, *119*, 6796.
(59) Tian, Z. R.; Tong, W.; Wang, J. Y.; SN. G. Duan, Krishnan, V. V.; Suib, S. L. *Science* **1997**, *276*, 926.
(60) Terribile, D.; Trovarelli, A.; De Leitenburg, C.; Dolcetti, G.; Llorca, J.; *Chem. Mater.* **1997**, *9*, 2676.
(61) Liu, P.; Liu, L.; Sayari, A. *Chem. Commun* **1997**, 577.
(62) Michael, S. W.; Ying, J. K. *Chem. Mater.* **1998**, *10*, 2067.
(63) Luca, V.; Hook, J. M. *Chem. Mater.* **1997**, *9*, 2731.
(64) Huo, Q.; Margolese, D. I.; Ciesla, U.; Demuth, D. G.; Feng, P.; Gier, T. E.; Sieger, P.; Firouzi, A.; Chmelka, B. F.; Schuth, F.; Stucky, G. D. *Chem. Mater.* **1996**, *6*, 1840.
(65) Ayyappan, S.; Rao, C. N. R. *Chem. Commun.* **1997**, 575.
(66) Kimura, T.; Sugahara, Y.; Kuroda, K. *Chem. Commun.* **1998**, 559.
(67) Mizuno, N.; Hatayama, H.; Uchida, S.; Taguchi, A. *Chem. Mater.* **2001**, *13*, 179.
(68) Ciesla, U.; Schacht, S.; Stucky, G. D.; Unger, K. K.; Schuth, F.; *Angew. Chem., Int. Ed. Engl.* **1996**, *35*, 541.
(69) Gougeon, R. D.; Bodart, P. R.; Harris, R. K.; Kolonia, D. M.; Petrakis, D. E.; Pomonis, P. J. *Phys. Chem. Chem. Phys.* **2000**, *2*, 5286.

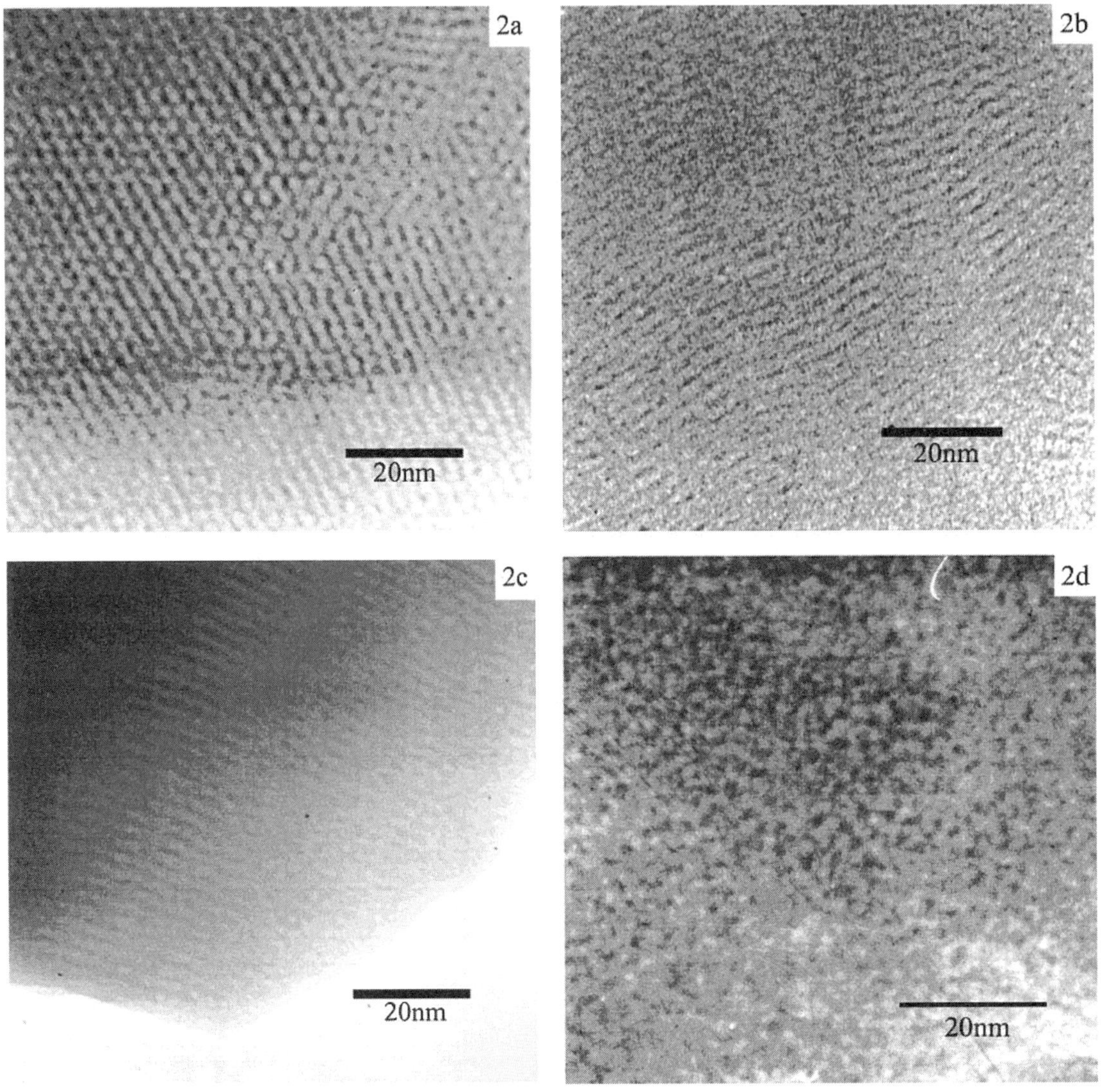

Figure 2. TEM micrographs of the as-synthesized mesostructured V−Mg−O prepared, at a pH = 10, using CTAB as surfactant and a V/Mg molar ratio of (a) 4:1, (b) 1.5:1, (c) 1:1, and (d) 1:1.5 in the initial batches.

and porosimetry apparatus. Before the measurements, the synthesized samples were pretreated in a flow of nitrogen from room temperature to 623 K at a heating rate of 10 K/min and kept at 623 K for 4 h in order to remove the surfactant template. Then, the samples were transferred to the ASAP 2010 apparatus and outgassed at 623 K for another 4 h under a vacuum of 1×10^{-3} Torr. The specific surface area was determined by the BET method, and the pore volume was estimated at a relative pressure of $P/P_0 > 0.99$, assuming full surface saturation with nitrogen. The pore size distribution was obtained from the N_2 desorption branch by the BJH (Barrett−Joyner−Halenda) method.

The energy dispersive spectroscopy (EDS) investigation was carried out using a Hitachi S-4000 field emission scanning electron microscope operating at 20 keV. The spectra were obtained using PGT IMIX X-ray microanalysis equipment.

3. Results and Discussion

3.1. Effect of Composition. The mesostructured V−Mg−O was synthesized at room temperature at a pH of 10 and molar composition of *a* $V(acac)_3$/*b* $MgCl_2$/0.25 CTAB/150 H_2O, where $a + b = 2$. Figure 1 presents the X-ray diffraction patterns of the samples synthesized with various $V(acac)_3/MgCl_2$ ratios. The XRD patterns were scanned starting from $2\theta = 1.5°$, because of the strong background scattering below 1.5°. One can see that, for $V(acac)_3/MgCl_2$ ratios larger than 1:1.5, all the samples (Figure 1a−d) exhibited the characteristic diffraction patterns of a mesostructured phase, i.e., a strong peak at a low 2θ and several weaker peaks at higher values. For $V(acac)_3/MgCl_2$ ratios lower than 1:1.5, almost no diffraction peaks could be observed (Figure 1e), but only a baseline rise in the low 2θ range, suggesting that the sample was either amorphous or its strong diffraction peak was below 1.5° because of its very large *d* spacing. For a $V(acac)_3/MgCl_2$ ratio of 1:1, the synthesized sample exhibited a set of diffraction peaks at $2\theta = 2.45°$, 3.58°, 5.34°, and 7.10° (Figure 1c). Because the distance between the first two diffraction peaks is smaller than the other distances between successive peaks, the synthesized sample is expected to have a slightly distorted lamellar structure, with an interlayer distance of about 36.1 Å. At higher $V(acac)_3/MgCl_2$ ratios, such as 1.5:1, most of the diffraction peaks of the lamellar structure are weakened, and other diffraction peaks at $2\theta = 2.18°$ (a relatively strong one) and weak ones at 2.93°, 3.42°, and 4.87° (Figure

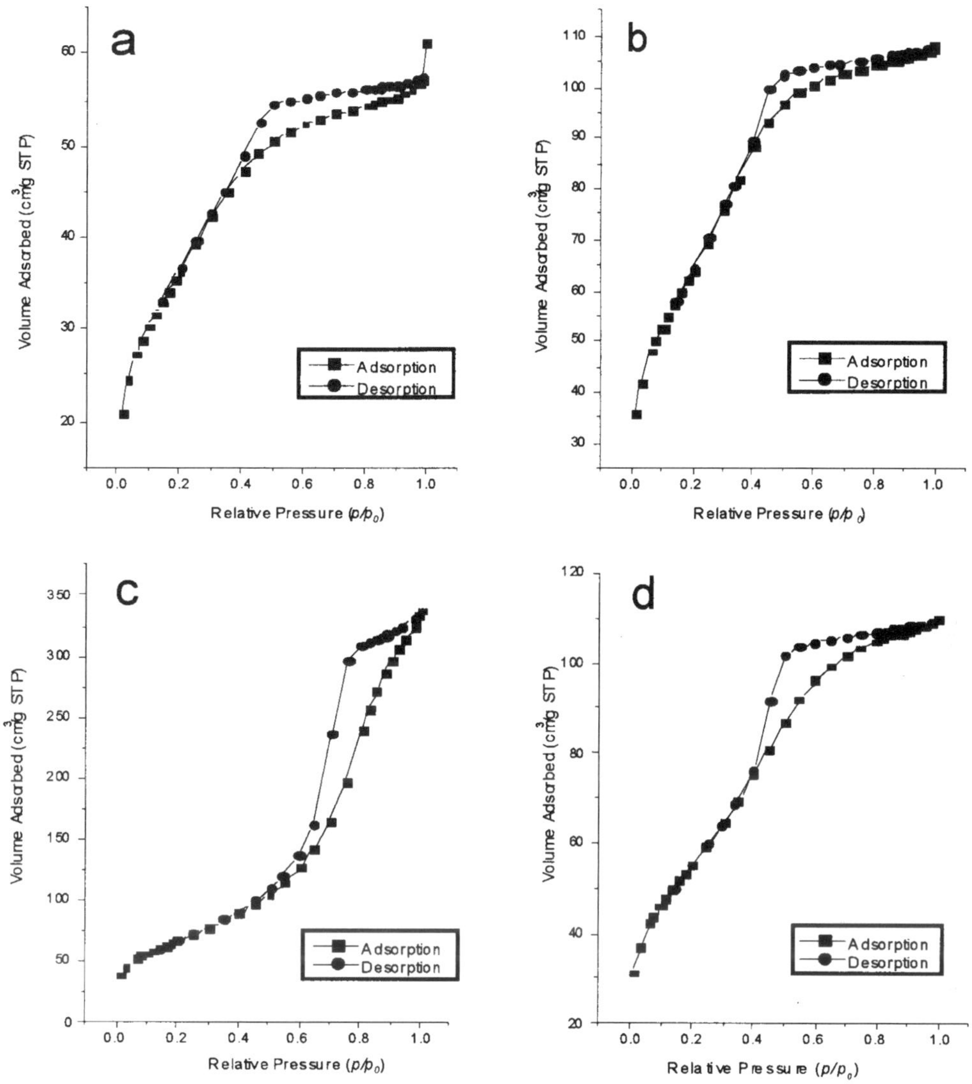

Figure 3. N_2 adsorption–desorption isotherms of mesostructured V–Mg–O synthesized, at a pH = 10, using CTAB as surfactant and a V/Mg molar ratio of (a) 4:1, (b) 1.5:1, (c) 1:1, and (d) 1:1.5, respectively, in the initial batches.

1b) appear, indicating that the synthesized sample consisted of mixed distorted hexagonal and lamellar phases. At much higher $V(acac)_3/MgCl_2$ ratios, such as 4:1, the diffraction peaks of the lamellar structure are absent and only a strong diffraction peak at $2\theta = 2.850°$ (Figure 1a), indicating a distorted hexagonal structure, is present. When the $V(acac)_3/MgCl_2$ ratio is lower than 1:1, such as 1:1.5 (Figure 1d), a diffraction pattern similar to that for $V(acac)_3/MgCl_2 = 4:1$ is observed. For the samples prepared with $V(acac)_3/MgCl_2$ ratios from 1:1.5 to 4:1 (except 1:1), there are several broad weak diffraction peaks between $2\theta = 7.5°$ and 9°. One can speculate that the peaks at high 2θ angles are caused by some nonindexed compounds between vanadium and magnesium. Table 1 lists the d spacings, the repeat distances of the mesoporous arrays, and the phase assignments of the synthesized mesostructured V–Mg–Os.

The mesostructures of the above samples were also investigated by TEM, and the micrographs obtained are presented in Figure 2; the pore sizes and the wall thicknesses determined from them are listed in Table 2. One can see that the sample prepared using a $V(acac)_3/MgCl_2$ molar ratio of 1:1 has a lamellar structure (Figure 2c), with an interlayer distance (the gray part) of about 2.40 nm and a wall thickness (the black part) of about 1.22 nm. When the $V(acac)_3/MgCl_2$ ratio increases from 1:1 to 1.5:1, the sample acquires a distorted lamellar structure in some areas and a wormlike structure in other

Table 1. XRD Characterization of the As-synthesized Mesostructured V–Mg–O Prepared from Batches with Various $V(acac)_3$:$MgCl_2$ Molar Ratios, Using CTAB as Surfactant and a pH of 10

V:Mg (mol)	d spacing[a] (nm)	repeat distance[b] (nm)	phase assignments
4:1	3.0974	3.5766	hexagonal
1.5:1	3.5782	3.5782	lamellar plus hexagonal
1:1	3.6104	3.6104	lamellar
1:1.5	3.8297	3.8297	wormlike
1:4			amorphous

[a] For the lamellar structure, the d spacing refers to d_{001}, while for the hexagonal structure the d spacing refers to d_{100}. [b] For the lamellar structure, the repeat distance equals the d spacing, while for the hexagonal structure the repeat distance is $2d/\sqrt{3}$.

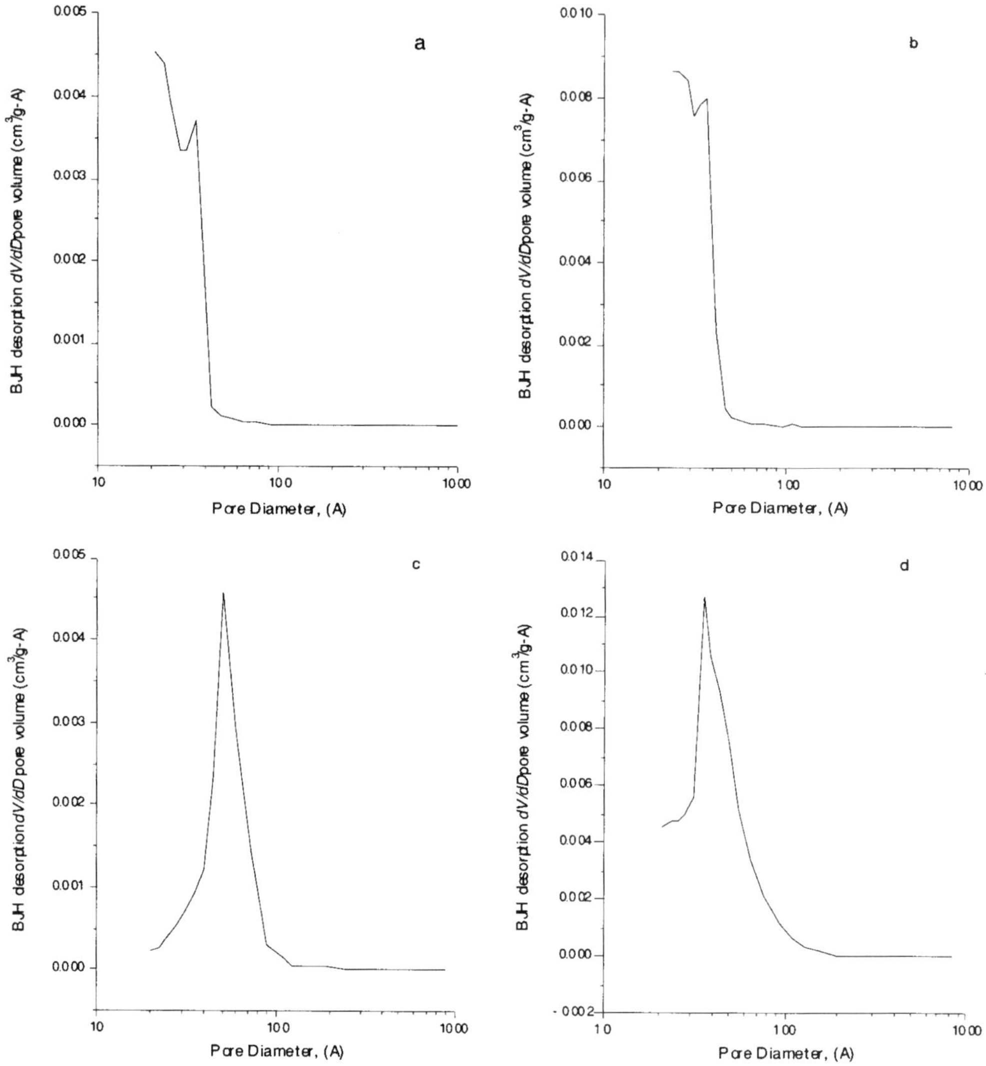

Figure 4. Pore size distribution of mesostructured V−Mg−O synthesized, at a pH = 10, using CTAB as surfactant and a V/Mg molar ratio of (a) 4:1, (b) 1.5:1, (c) 1:1, and (d) 1:1.5, respectively, in the initial batches.

Table 2. Effect of V/Mg Molar Ratio in the Initial Batches on the Textural Properties of the Mesostructured V−Mg−O Synthesized Using CTAB as Surfactant and a pH of 10

V:Mg (mol)	specific surface area (m^2/g)	pore vol (cm^3/g)	pore size (nm) N_2 adsoption	pore size (nm) TEM	wall thickness (nm)
4:1	129.53	0.085	2.63	2.34	1.20
1.5:1	235.90	0.165	2.80	2.35	1.20
1:1	238.28	0.173	2.91	2.40	1.22
1:1.5	196.55	0.167	3.38	>2.0	>1.0

areas (Figure 2b). The interlayer distance and wall thickness are 2.35 and 1.20 nm, respectively. When the $V(acac)_3/MgCl_2$ molar ratio is further increased to 4:1, the TEM picture clearly shows that the sample acquires a deformed hexagonal structure (Figure 2a), with a pore size and wall thickness of 2.34 and 1.20 nm, respectively. When the $V(acac)_3/MgCl_2$ molar ratios are lower than 1:1.5, the structure of the sample becomes wormlike (Figure 2d), and the pore size and wall thickness can hardly be determined accurately from the TEM micrographs, because the pores have very different shapes and their arrangement is disordered. However, one can evaluate that the sizes of the pores and the wall thicknesses are greater than 2.0 and 1.0 nm, respectively.

Figure 3 presents the N_2 adsorption−desorption isotherms determined at the liquid nitrogen temperature. All of them are of type IV, according to the IUPAC classification, and exhibit the characteristic hysteresis loop of mesoporous materials. The starting point of the hysteresis loops increases from $p/p_0 = 0.25$ to $p/p_0 = 0.5$ as the $V(acac)_3/MgCl_2$ ratio decreases from 4:1 to 1:1. Similarly, the sharpness of the hysteresis loop decreases with increasing ratio, indicating a decrease of the average pore size and a broadening of the pore size distribution. This can be also clearly seen in the pore size distribution curve (Figure 4). The textural properties of the samples are listed in Table 2, which shows that with increasing $V(acac)_3/MgCl_2$ molar ratio from 1:1 to 4:1, the specific surface area, the pore volume, and the average pore size decrease. However, for the 1:1.5 molar ratio, the specific surface area and the pore volume are somewhat lower than those obtained for the ratio 1:1.

From Tables 1 and 2 and Figures 1−4, one can see that the XRD, TEM, and N_2-adsorption results are consistent

Table 3. V/Mg Ratios in the Batches and on the Surface of the Synthesized Samples[a]

	V/Mg molar ratio				
in the batches	4:1	1.5:1	1:1	1:1.5	1:4
on the surface of the samples	*b*	*b*	1:0.0225	1:0.258	1:0.608

[a] The samples were prepared at pH = 10 using CTAB as template. [b] Mg could not be detected.

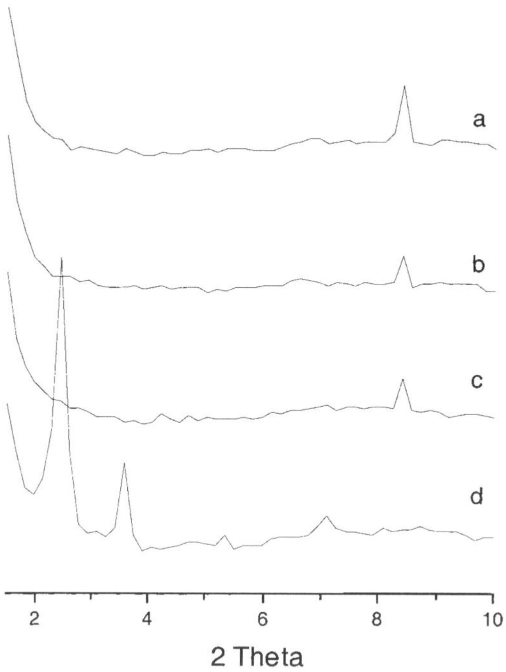

Figure 5. XRD patterns of the as-synthesized mesostructured V–Mg–O prepared with CTAB as surfactant, a 1:1 molar ratio of V/Mg in the initial batches and at a pH of (a) 2, (b) 4, (c) 8, and (d) 10.

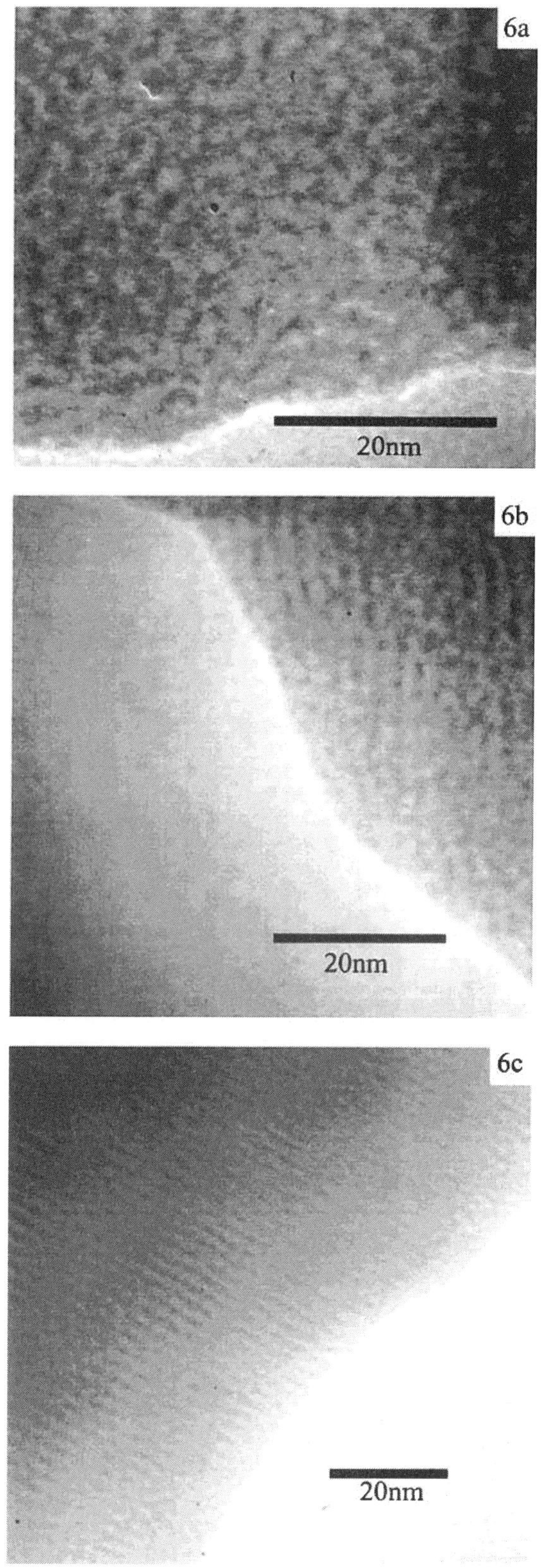

Figure 6. TEM micrographs of the as-synthesized mesostructured V–Mg–O prepared with CTAB as surfactant, a 1:1 molar ratio of V/Mg in the initial batches and at a pH of (a) 4, (b) 8, and (c) 10.

with each other. Indeed, the structures of the samples provided by the XRD patterns are also reproduced by the TEM micrographs. The repeat distance of the pore array, which for TEM is the sum of the pore size and wall thickness, has nearly the same value and the same trend with increasing $V(acac)_3/MgCl_2$ molar ratio as that provided by XRD. There is also the same trend for the average pore size, which increases with increasing $V(acac)_3/MgCl_2$ molar ratio for both the N_2 adsorption and TEM. One should also notice that the wall thickness remains almost unchanged with increasing $V(acac)_3/MgCl_2$ molar ratio. Lower ratios than 1:1.5 and ratios that are too large (4:1) are unfavorable to the formation of mesoporous structures. Indeed, for the ratios of 4:1 and 1:1.5, distorted hexagonal and wormlike structures are, respectively, present (panels a and d of Figure 2). As a conclusion, the $V(acac)_3/MgCl_2$ molar ratio of 1:1 appears to generate the optimum lamellar crystallinity for mesostructured V–Mg–O.

Table 3 presents the results obtained from the EDS spectra of the samples prepared using various $V(acac)_3/MgCl_2$ molar ratios in the initial batches. When the ratio was higher than 1:1, no magnesium could be detected in EDS spectra. For the molar ratio of $V(acac)_3/MgCl_2 = 1:1$, only a small signal of magnesium was present; the signal of magnesium increased, however, rapidly with decreasing $V(acac)_3/MgCl_2$ molar ratio. Because the surface free energy of V_2O_5 is much smaller than that of MgO (MgO has a surface free energy of 1000 ergs/cm^2 at $T = 298$ K

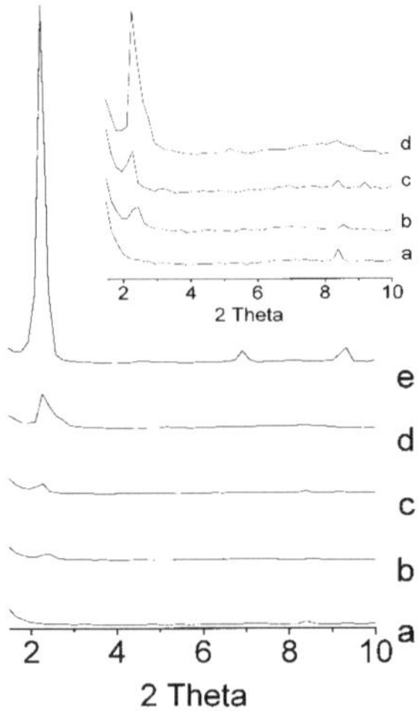

Figure 7. XRD patterns of the as-synthesized mesostructured V−Mg−O prepared using a 1:1 molar ratio of V/Mg in the initial batches and (a) SDS as surfactant at a pH = 2 and (b−e) a mixture of CTAB and SDS as surfactant at pH = 2, 7, 9, and 10, respectively. In the inset, the vertical scale is finer.

Table 4. Textural Properties of Mesostructured V−Mg−O Synthesized at Various pH Values Using CTAB as Surfactant and a V/Mg Molar Ratio of 1:1 in the Initial Batches

pH	specific surface area (m^2/g)	pore vol (cm^3/g)	pore size (nm) N_2 adsorption	pore size (nm) TEM	wall thickness (nm)
2	31.58	0.018	2.10		
4	85.06	0.047	2.24	2.05	1.00
8	133.89	0.089	2.88	2.30	1.10
10	238.28	0.173	2.91	2.40	1.25

Table 5. XRD Characterization of the As-Synthesized Mesostructured V−Mg−O Prepared at Various pH Values, Using Mixed Surfactants CTAB + SDS and a V/Mg Molar Ratio of 1:1 in the Initial Batches

pH	*d* spacing (nm)	phase assignments
2	3.633	wormlike
7	3.852	wormlike + distorted lamellar
9	3.876	lamellar
10	3.885	lamellar

and V_2O_5 a surface free energy of 90 ergs/cm^2 at 963 K[70]), a surface enrichment in vanadium is expected to occur. Indeed, for a ratio of 1:1 in the initial batch, the bulk ratio V/Mg of the specimen, determined by the ICP method by Galbraith Laboratories Inc., was 3.08, which is much smaller than the surface ratio.

3.2. Effect of pH. The effect of pH on the synthesis of mesostructured V−Mg−O was investigated at room temperature for the molar composition of 1 V(acac)$_3$/1 $MgCl_2$/0.25 CTAB/150 H_2O. Figure 5 presents the XRD patterns of the synthesized samples. For the samples prepared in an acidic environment, e.g., pH = 2 and 4 (Figure 5a, b), the characteristic XRD peak for mesoporous materials in the low 2θ region is reduced to a shoulder in the background, indicating a poor crystallinity. When the sample was prepared in a weak alkaline environment (pH = 8 (Figure 5c)), the characteristic XRD peak for mesoporous materials in the low 2θ range was absent, indicating an even poorer order in the structure than in acidic environments. It is possible that the characteristic

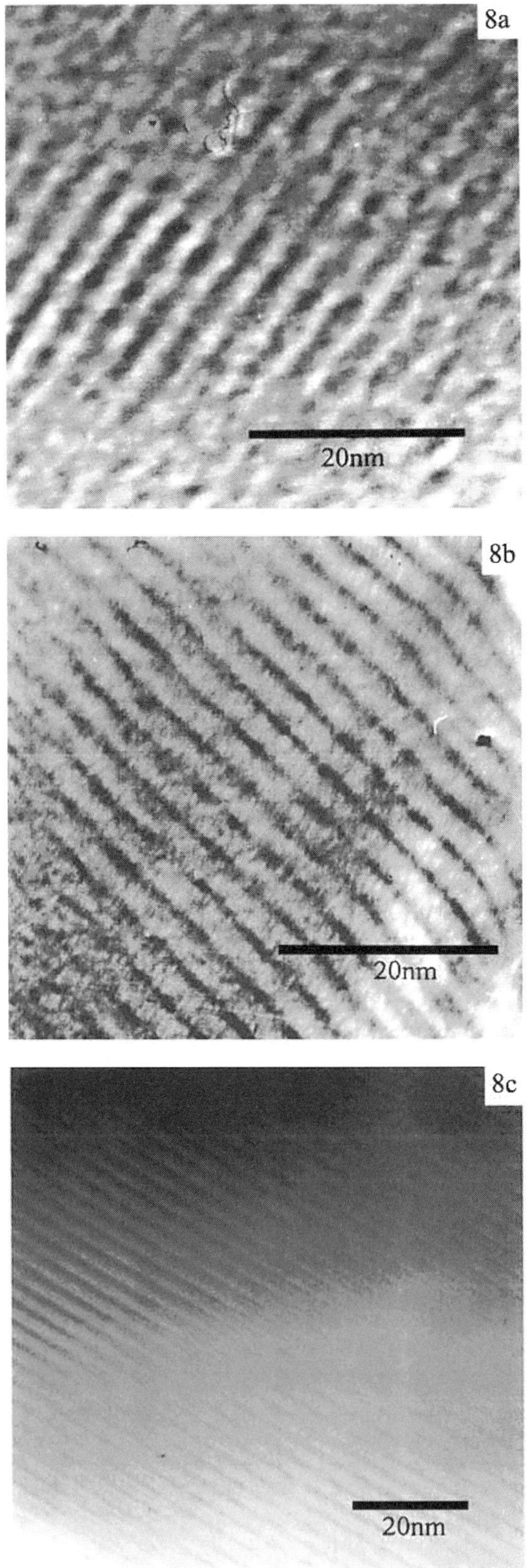

Figure 8. TEM micrographs of the as-synthesized mesostructured V−Mg−O prepared using a mixture of CTAB and SDS as template, a 1:1 molar ratio of V/Mg in the initial batches, and a pH of (a) 7, (b) 9, and (c) 10.

(70) Overburg, S. H.; Bertrand, P. A.; Somorjai, G. A. *Chem. Rev.* **1975**, *75*, 547.

Table 6. Textural Properties of Mesostructured V–Mg–O Synthesized at Various pH Values Using CTAB + SDS as Surfactant and a V/Mg Molar Ratio of 1:1 in the Initial Batches[a]

pH	specific surface area (m^2/g)	pore vol (cm^3/g)	pore size (nm) N_2 adsoption	pore size (nm) TEM	wall thickness (nm)
2	80.02	0.065	3.24		
7	94.5	0.079	3.33	2.56	1.25
9	219.57	0.192	3.55	2.95	1.28
10	278.9	0.256	3.68	3.10	1.35

[a] Differences in pore sizes determined by adsorption and TEM occur because in the former case the specimens were also thermally treated after synthesis, while in the latter they were not.

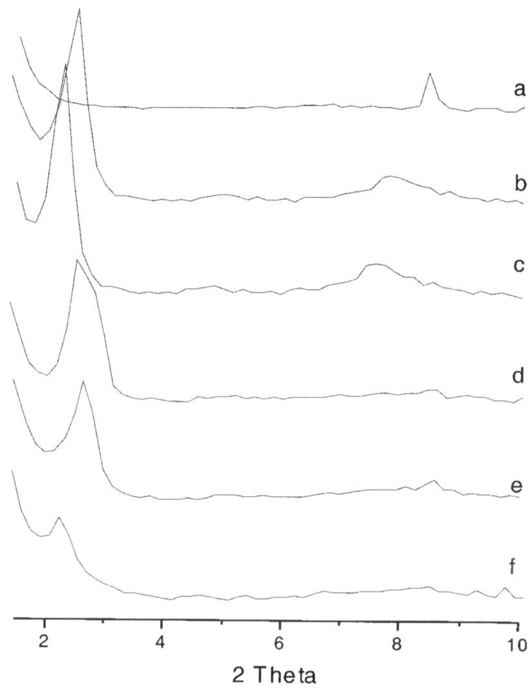

Figure 9. XRD patterns of the as-synthesized mesostructured V–Mg–O prepared using a 1:1 molar ratio of V/Mg in the initial batches at a pH = 9 and a mixture of an anionic and a cationic surfactant: (a–e) the anionic surfactant is SDS and the cationic surfactants are BTAB, CPB, CTAB, MTAB, and DTAB; (f) the anionic and cationic surfactants are SDBS and CTAB, respectively.

peak was displaced to a $2\theta < 1.5°$, which cannot be detected by the normal XRD technique employed by us but by a more specific technique, such as the small-angle X-ray scattering (SAXS). However, when the sample was prepared in a strong alkaline environment, e.g., pH = 10 (Figure 5d), the characteristic XRD peaks of the mesoporous materials are present, indicating a lamellar structure with $d_{001} = 3.61$ nm. A small peak is also present at the higher end of the 1.5–1 0° range (Figure 5a–c), which can be probably ascribed to a nonindexed compound of V, Mg, and O, just as in section 3.1. The TEM micrographs (Figure 6) show that the sample prepared at pH = 4 has a wormlike structure (Figure 6a); at pH = 8, the structure is still wormlike (Figure 6b); for the sample prepared at pH = 10, an ordered lamellar structure can be observed (Figure 6c).

The results obtained via N_2 adsorption at the liquid nitrogen temperature are listed in Table 4, which shows that as the pH increases from 4 to 10, the specific surface area, pore volume, and average pore size increase.

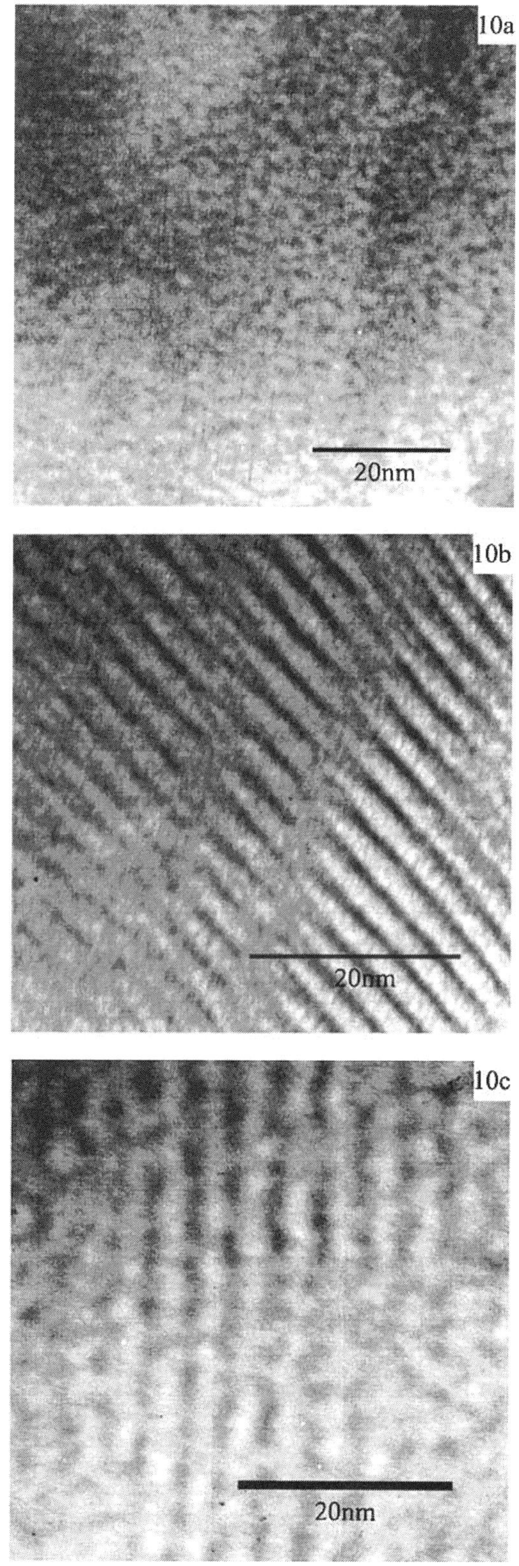

Figure 10. TEM micrographs of the as-synthesized mesostructured V–Mg–O prepared using a 1:1 molar ratio of V/Mg in the initial batches at a pH = 9 and a mixture of an anionic and a cationic surfactant: the anionic surfactant is SDS and the cationic surfactants are (a) BTAB; (b) CTAB; (c) MTAB.

Table 7. Textural Properties of Mesostructured V−Mg−O Synthesized Using Binary Mixed Surfactants at a pH of 9 and a 1:1 V/Mg Molar Ratio

surfactants	structure	specific surface area (m^2/g)	pore vol (cm^3/g)	pore size (nm)[a] N_2 adsoption	pore size (nm)[a] TEM	wall thickness (nm)
BTAB + SDS	wormlike	15.66	0.015	3.89		
CPB + SDS	wormlike	38.68	0.024	2.51		
CTAB + SDS	lamellar	219.57	0.192	3.55	2.95	1.28
MTAB + SDS	lamellar	172.18	0.138	3.27	2.53	1.26
DTAB + SDS	wormlike	54.72	0.044	3.22		
CTAB + SDBS	wormlike	77.31	0.065	3.32		

[a] Differences in pore sizes determined by adsorption and TEM occur because in the former case the specimens were also thermally treated after synthesis, while in the latter they were not.

3.3. Effect of the Nature of the Surfactant. Mesostructured V−Mg−O materials were also prepared at room temperature, using a single anionic surfactant or a mixture of cationic and anionic surfactants. The batches had a general molar composition of 1 $V(acac)_3$:1 $MgCl_2$:0.25 surfactant:150 H_2O, in which either a single surfactant or an equimolar mixture of anionic and cationic surfactants was present. Figure 7 presents the XRD patterns, and Table 5 summarizes the results. When an anionic surfactant, SDS or SDBS, was employed, no mesoporous phase was formed. One can see from Figure 7a that for SDS and a pH of 2, only the baseline rises in the low 2θ range; the small peaks in the high 2θ range are probably due to a nonindexed compound between V and Mg. When the synthesis involving SDS was carried out at a higher pH, e.g., pH = 9, no solid was produced. Similar results were also obtained when SDBS was employed as surfactant. However, when a mixture of anionic and cationic surfactants, e.g., SDS + CTAB, was used, a mesostructured V−Mg−O was generated. The curves b, c, and d of Figure 7 show that the "crystallinity" of the samples increases with increasing pH, as for a single cationic surfactant. At a pH of 10 (Figure 7e), the strength of the XRD peak in the low 2θ range is by an order of magnitude higher for SDS + CTAB than for CTAB alone. Table 5 lists the *d* values and the phase assignments for the samples prepared using a binary mixture of surfactants. With increasing pH, the structure of the sample was transformed gradually from wormlike to distorted lamellar and, finally, to an ordered lamellar structure. Information about the samples was also obtained via TEM (Figure 8) and N_2 adsorption. The results are listed in Table 6, which shows that the higher the pH, the greater the specific surface area, pore volume, and pore size. Compared with the mesostructured V−Mg−O prepared using a single surfactant (Table 4), those prepared with binary mixed surfactants (Table 6) have larger pore volumes and pore sizes.

A series of quaternary ammonium bromides, BTAB, DTAB, MTAB, CPB, and CTAB, were used together with SDS or SDBS as binary surfactant mixtures in the synthesis of mesostructured V−Mg−O. Figure 9 presents the XRD patterns of the samples. The figure shows that for the SDS−BTAB mixture (Figure 9a), no characteristic diffraction peaks of the mesophase are present (the weak peaks in the high 2θ region are probably due to a nonindexed compound between V, Mg, and O). However, when the quaternary ammonium surfactant has a longer alkyl, the diffraction patterns of the specimens exhibit the peaks characteristic to a mesophase (Figure 9b−e). The longer the alkyl chain, the stronger are the characteristics of a lamellar structure and the crystallinity of the mesostructure. One can also notice that when CTAB is coupled with SDBS (Figure 9f), the crystallinity of the mesostructured V−Mg−O is much lower than for the mixture of CTAB and SDS (Figure 9c).

The structure of the samples was also examined by TEM and N_2 adsorption−desorption. The TEM micrographs are presented in Figure 10, and the textural properties of the samples are listed in Table 7. One can see that when the binary BTAB + SDS is used as template, the structure of the sample is wormlike (Figure 10a). Similar structures have been observed with CPB + SDS, DTAB + SDS and CTAB + SDBS. With CTAB + SDS (Figure 10b) or MTAB + SDS (Figure 10c), the structures of the samples are lamellar. The interlayer distance is 2.95 and 2.53 nm for CTAB + SDS and MATB + SDS, respectively.

The above results indicate that the mixed surfactants composed of an anionic surfactant (SDS) and a cationic one (with a long alkyl chain), such as CTAB, CPB, and MTAB, can template an ordered mesostructured V−Mg−O phase. Among them, SDS + CTAB can be considered the optimum template. This conclusion is also confirmed by the N_2 adsorption−desorption results, listed in Table 7. One can see that the samples prepared with CTAB + SDS, CPB + SDS, or MTAB + SDS as templates have higher specific surface areas and larger pore sizes than those prepared with the other mixed surfactants. As the length of the alkyl chain in alkyltrimethylammonium surfactant increases, the specific surface area and pore size increase.

The structure of the mesoporous materials is expected to depend on the structure of the surfactant solution. While numerous phase diagrams are available, only few involve two surfactants—an anionic plus a cationic one. One of them[71] considers a mixture of cetyltrimethylammonium tosylate (CTAT) and sodium dodecylbenzene sulfonate (SDBS) and reveals that there are relatively large domains in which lamellar structures are formed. No phase diagram is available for the time being for our optimum system (CTAB + SDS) and for the conditions employed in our experiments. Because of the lamellar structure of the mesoporous material obtained, it is plausible to consider that the lamellar structure of the template is responsible for the lamellar mesoporous material.

Conclusion

Mesostructured V−Mg−Os was synthesized using $V(acac)_3$ and $MgCl_2$ as source materials. Depending upon conditions, the synthesized mesostructured V−Mg−O acquired a lamellar, hexagonal, or wormlike structure. The formation of lamellar mesostructured V−Mg−O was promoted by higher pH values, by binary mixtures of anionic and cationic surfactants as templates, and by equimolar vanadium and magnesium contents. The cat-

(71) Kaler, E. W.; Murthy, A. K.; Rodriguez, B. E.; Zasadzinski, J. A. N. *Science* **1989**, *245*, 1371.

ionic surfactant, CTAB, at high pH values, and particularly the equimolar mixture of CTAB and SDS, provided the most ordered lamellar structure. It should be emphasized that the above mixture templated at pH = 10 a structure with a degree of crystallinity by an order of magnitude higher than that templated by CTAB alone.

LA011391D

Noncatalytic and catalytic conversion of ethane over V–Mg oxide catalysts prepared via solid reaction or mesoporous precursors

Zi-Sheng Chao and Eli Ruckenstein *

Department of Chemical Engineering, State University of New York at Buffalo, Amherst, NY 14260, USA

Received 5 May 2003; revised 7 November 2003; accepted 12 November 2003

Abstract

This paper deals with both noncatalytic and catalytic conversions of ethane. The effects of reactor configuration (empty tube, tube containing inert material or containing catalyst, and inert material), reaction temperature, reactant composition, and flow rate, as well as catalyst composition and structure, were systematically investigated. Two groups of V–Mg oxides, namely Meso-VMg (originated from mesostructured V–Mg oxide) and Mix-VMg (prepared via a solid reaction between vanadia and magnesia), were employed as catalysts. High conversions and selectivities were obtained at high temperatures during the ethane thermolysis to ethene, accompanied, however, by high carbon depositions, especially in the presence of a catalyst. The contribution of homogeneous reactions to the ODH of ethane was important at high, but less important at low temperatures particularly when the fraction of reactor occupied by inert silica granules was large. For the Mix-VMg catalysts, the ethane conversion and the ethene yield increased but the selectivity to ethene decreased with decreasing V/Mg atomic ratio. The Meso-VMg catalysts exhibited in most cases higher yields and selectivities to ethene than the Mix-VMg ones. In addition, a higher fraction of ethane converted to oxygenates (mainly formaldehyde) as liquid products was obtained over the Meso-VMg catalysts. Magnesium vanadates were identified in the Mix-VMg catalysts; they may represent the active phase for this series of catalysts. A V_2O_3 phase, which may contain highly dispersed magnesia, was identified and suggested to be responsible together with the large surface area for the high performance of the Meso-VMg catalysts.

Keywords: Ethane; Ethene; Oxidative dehydrogenation; Oxygenates; Catalyst; Mesoporous; Vanadium; Magnesium

1. Introduction

Activation of light alkanes is of high economic importance, although it is well known that these saturated hydrocarbon molecules are weakly reactive due to the absence of lone electron pairs, empty orbitals, and polarity in their C–H bonds. Production of olefins and oxygenates represents a most useful way for effective utilization of light alkanes. Ethene is one of the most basic feedstocks in chemical industry and its demand is steadily increasing. The main commercial routes for production of ethene are steam thermal cracking and FCC (fluid catalytic cracking) processes [1]. They involve, however, the drawbacks of high energy input required by the highly endothermic reactions, high operation costs due to coke deposition on catalyst and reactor, and generation of low molecular weight alkanes. Furthermore, the reserves of raw materials for these processes are becoming increasingly limited. Consequently, alternative processes with higher efficiency, which utilize more abundant and economic sources for ethene production, are becoming increasingly necessary. Ethane is the second major component of natural gas and is also abundant in refinery gas. Production of ethene via oxidative dehydrogenation (ODH) of ethane has received increasing attention, owing to its potential advantages, such as exothermic reaction heat and less coke deposition. Furthermore, the ODH can be carried out at relatively low temperatures in the presence of properly selected catalysts. Until now, numerous catalysts were employed for the ODH of ethane, such as composite oxides between alkaline earths and rare earths [2–4], halogen (particularly F and Cl) and/or alkali ion-promoted oxides [5–8], as well as some transition metal (Mo, V, Bi, etc.) oxide-based catalysts [9,10]. As an alternative to the heterogeneous route, a hetero-homogeneous process for the ODH of ethane at temperatures higher than 900 °C, the so-called autothermal oxidative dehydrogenation, was also employed [11]. The

* Corresponding author.
E-mail address: feaeliru@buffalo.edu (E. Ruckenstein).

This paper, which was originally published in **Journal of Catalysis**: 222, 17-31 (2004), is republished with permission.

contribution of the homogeneous reaction to the ODH of ethane was reported in Refs. [12–17].

Oxygenates, such as alcohols, aldehydes, and carboxylic acids are very important chemicals in the hydrocarbon processes and organic synthesis. They have attracted attention as fuel alternatives or as additives to gasoline, due to their lower environmental pollution and higher octane numbers. The current methods for oxygenate production require multiple stages, expensive catalysts, large residence times, and careful temperature control [18,19]. Thus, the direct production of oxygenates from light alkanes is of significant industrial interest. In contrast to the large number of studies regarding the production of ethene from ethane via the ODH process, few reports on the conversion of ethane to oxygenates could be found in literature [10,20–25].

Mesoporous silicates, such as MCM 41 [26], are expected to be useful in selective adsorption and catalysis [26–32], because they facilitate diffusion. However, silica possesses low catalytic activities for most reactions. To enhance the catalytic effect, active components must be introduced into the framework of the mesosilicates. Recently, Zhang et al. [33] and Solsona et al. [34] studied the ODH of ethane and found that the activity increased with increasing vanadium content in MCM-41 support. An ethane conversion of up to 18% with ethene selectivity of up to 50% at 550 °C and an ethene yield of up to 10.8% with a selectivity of up to 38.3% at 600 °C were obtained by Zhang et al. [33] and Solsona et al. [34], respectively.

In our laboratory, mesostructured V–Mg–O were synthesized [35,36] and the catalytic conversions of light alkanes over these materials were investigated. Some preliminary results have been published [37]. In the present paper, the catalytic oxidative conversion of ethane was carried out over Meso-VMg catalysts (originated from mesostructured V–Mg oxides) and Mix-VMg (prepared via a solid reaction between vanadia and magnesia) catalysts. Furthermore, the thermolysis of ethane to ethene and the noncatalytic conversion of ethane were also examined.

2. Experimental

2.1. Catalyst preparation

All reagents were purchased from Aldrich and used without further purification. Two groups of Meso-VMg catalysts were prepared using either V_2O_5 or vanadium (III) acetylacetonate ($V(acac)_3$) as vanadium source and magnesium chloride ($MgCl_2 \cdot 6H_2O$) as magnesium source. Surfactants such as cetyltrimethylammonium bromide (CTAB), sodium dodecylbenzene sulfonate (SDBS), benzyltrimethylammonium bromide (BTAB), and hexadecylamine (HDA) were used as templates in the synthesis of the mesoporous materials. First, $MgCl_2 \cdot 6H_2O$ and the template were dissolved into an aqueous solution of hydrochloric acid. Second, $V(acac)_3$ was dispersed homogeneously into distilled water with vigorous stirring, or V_2O_5 was dissolved into an aqueous solution of sodium hydroxide. Further, the solution containing vanadium was added slowly into that containing magnesium, with vigorous stirring at room temperature. The pH of the mixture was adjusted to 4.0 or 10.0 when the mesostructured V–Mg oxides was prepared using V_2O_5 or $V(acac)_3$ as vanadium source, respectively. The final mixtures had the molar composition a $V(acac)_3$:b $MgCl_2$:0.25 surfactant: 295 H_2O, or V_2O_5:$MgCl_2$:0.25 surfactant:295 H_2O, where a/b was varied between 4:1 and 1:4 and $a+b=2$. After stirring at room temperature for 24 h, the mixture was allowed to age statically at room temperature for 2 days. The solid formed was recovered by filtration, washed with distilled water, and dried at 100 °C for 12 h. To remove the surfactant, the prepared specimens were heated in a flow of argon from room temperature up to 750 °C at a rate of 10 °C min^{-1} and kept at that temperature for 4 h. Meso-V was prepared using the same procedure, except that no magnesium was introduced and the pH of the mixture was adjusted to 7.0. The as-obtained materials were used to examine their textural properties and employed as catalysts.

The Mix-VMg catalysts were prepared via a solid-state reaction. Powders of V_2O_5 and $Mg(NO_3)_2 \cdot 6H_2O$ were mixed together and ground thoroughly in a mortar, and the mixture obtained was calcined at 750 °C for 2 h after it was heated from room temperature at a rate of 4 °C min^{-1}.

2.2. Catalyst characterization

X-ray diffraction (XRD) was performed on a Siemens D500 diffractometer with a Cu-K_α radiation of 1.5406 Å. N_2 adsorption measurements were performed on a Micromeritics ASAP 2010 gas sorption and porosimetry instrument. The specific surface area was determined by the BET method and the pore volume was calculated at a relative pressure of $P/P_0 > 0.99$. The pore-size distribution was calculated using the N_2 desorption branch by the BJH (Barrett–Joyner–Halenda) method. The chemical compositions of the Meso-VMg catalysts were determined by atomic absorption spectrophotometry, using a Perkin–Elmer Model AAS 3030 spectrophotometer. Magnesium was determined using an acetylene/air flame at a wavelength of 285.2 nm and vanadium using an acetylene/nitrous oxide flame at a wavelength of 318.4 nm. The X-ray photoelectron spectroscopy (XPS) was carried out on a VG-ESCA lab-210 spectrometer at a vacuum of 10^{-10} Torr, using Mg-K_α radiation as excitation source. The C_{1s} line with the binding energy of 284.8 eV was used as reference.

Table 1 provides details regarding the preparation of the Meso-VMg and Mix-VMg catalysts.

2.3. Catalytic performance evaluation

The catalytic performance was determined at atmospheric pressure in a tubular fixed-bed quartz microreactor (internal diameter = 5 mm, operation length = 30 cm). Unless otherwise mentioned, the reactor was packed as follows: the

Table 1
Preparation of Meso-VMg and Mix-VMg catalysts

Specimens	V and Mg sources	Surfactant	Method	V/Mg atomic ratio	
				Initial batches	Prepared specimens
Meso-VMg-1	$V(acac)_3 + MgCl_2$	CTAB	Sol-gel	4.00	48.23
Meso-VMg-2	$V(acac)_3 + MgCl_2$	CTAB	Sol-gel	1.50	35.29
Meso-VMg-3	$V(acac)_3 + MgCl_2$	CTAB	Sol-gel	0.67	10.65
Meso-VMg-4	$V(acac)_3 + MgCl_2$	CTAB	Sol-gel	0.25	1.39
Meso-VMg-5	$V(acac)_3 + MgCl_2$	CTAB + SDBS[a]	Sol-gel	1.00	3.19
Meso-VMg-6	$V_2O_5 + MgCl_2$	BTAB + HDA[b]	Sol-gel	1.00	21.59
Meso-V	V_2O_5	CTAB + HDA[c]	Sol-gel	–	–
Mix-VMg-1	$V_2O_5 + Mg(NO_3)_2$	None	Solid reaction	4.00	4.00
Mix-VMg-2	$V_2O_5 + Mg(NO_3)_2$	None	Solid reaction	2.00	2.00
Mix-VMg-3	$V_2O_5 + Mg(NO_3)_2$	None	Solid reaction	1.00	1.00
Mix-VMg-4	$V_2O_5 + Mg(NO_3)_2$	None	Solid reaction	0.50	0.50
Mix-VMg-5	$V_2O_5 + Mg(NO_3)_2$	None	Solid reaction	0.25	0.25

[a] Molar ratio CTAB:SDBS = 1:1.
[b] Molar ratio BTAB:HDA = 1:1.
[c] Molar ratio CTAB:HDA = 1:1.

middle of the reactor was plugged with quartz wool and a catalyst (about 0.25 g) was located over it. The space of the reactor above the catalyst bed was filled with quartz granules. The reactor was placed into a tubular furnace with the catalyst bed located in the constant temperature zone. In addition, four other reactor configurations free of catalyst were employed in the study of the noncatalytic conversion of ethane, namely, ET (the empty tube), FQ (the reactor filled with quartz granules up to a height of 1 cm), HFQ (the upper half of the reactor was filled with quartz granules and the rest of the reactor was empty), and FFQ (the entire reactor was filled with quartz granules). Two thermocouples were employed to monitor and control the temperature. One of them was embedded in the furnace, and the other one was located in the center of the catalyst bed and tightly contacted the external surface of the quartz reactor. The temperatures measured by these two thermocouples were almost the same. The compositions of reactant mixtures (N_2, O_2, and C_2H_6) and gaseous effluents from reactor were determined on-line using a SRI 8610C gas chromatograph equipped with both a thermal conductivity detector (TCD) and a flame ionization detector (FID). The components were separated at programmed temperatures and pressures on a Hayesep DB (30 ft × 1/8 in.) packed column and then allowed to flow into TCD and FID detectors. The liquid effluent from the reactor was captured in a cooled trap and analyzed using another SRI 8610C gas chromatograph. A Porapak Q (12 ft × 1/8 in.) packed column was employed for the separation of the liquid components. For the gaseous product, the conversion of ethane and oxygen, the selectivity to a carbon-containing product, and the yield to ethene are defined as follows:

$$\text{Ethane conversion } X_{C_2H_6}\ (\%) = \frac{\sum \nu_i C_{p,i}}{2C_u + \sum \nu_i C_{p,i}} \times 100,$$

$$\text{Oxygen conversion } X_{O_2}\ (\%) = \frac{\sum \nu_i' O_{p,i}}{2O_u + \sum \nu_i' O_{p,i}} \times 100,$$

$$\text{Selectivity to a product: } S_i\ (\%) = \frac{\nu_i C_{p,i}}{\sum \nu_i C_{p,i}} \times 100,$$

$$\text{Yield of ethene: } Y_{C_2H_4}\ (\%) = \frac{X_{C_2H_6} \times S_{C_2H_4}}{100},$$

where $C_{p,i}$ and $O_{p,i}$ are the moles of gaseous carbon-containing products and those of gaseous oxygen-containing products, C_u and O_u the moles of unconverted ethane and oxygen in the product stream, and ν_i and ν_i' the number of carbon atoms and of oxygen atoms in the product i, respectively.

The fraction of ethane converted to nongaseous species (carbon deposits and liquid products) was calculated, according to the equation

$$X_{\text{Car\&Oxy}} = \frac{C_{in} - C_{out}}{C_{in}} \times 100\%,$$

where C_{in} and C_{out} are the moles of carbon in the reactant mixture and those in the gaseous effluent.

The fraction of ethane converted to oxygenates, X_{oxy}, and the selectivity, S_j, to component j in the oxygenates were calculated using the expressions

$$X_{oxy}\ (\%) = \frac{\sum \nu_j N_j}{2t_r R_f \chi_{C_2H_6}},$$

$$S_j\ (\%) = \frac{\nu_j N_j}{\sum \nu_j N_j} \times 100,$$

where N_j is the number of moles of component j in the oxygenates produced during reaction time t_r, $\chi_{C_2H_6}$ is the mole fraction of ethane in the feed gas, t_r is the reaction time in hours, R_f is the flow rate of the feed gas in mol h^{-1}, and ν_j is the number of carbon atoms in component j.

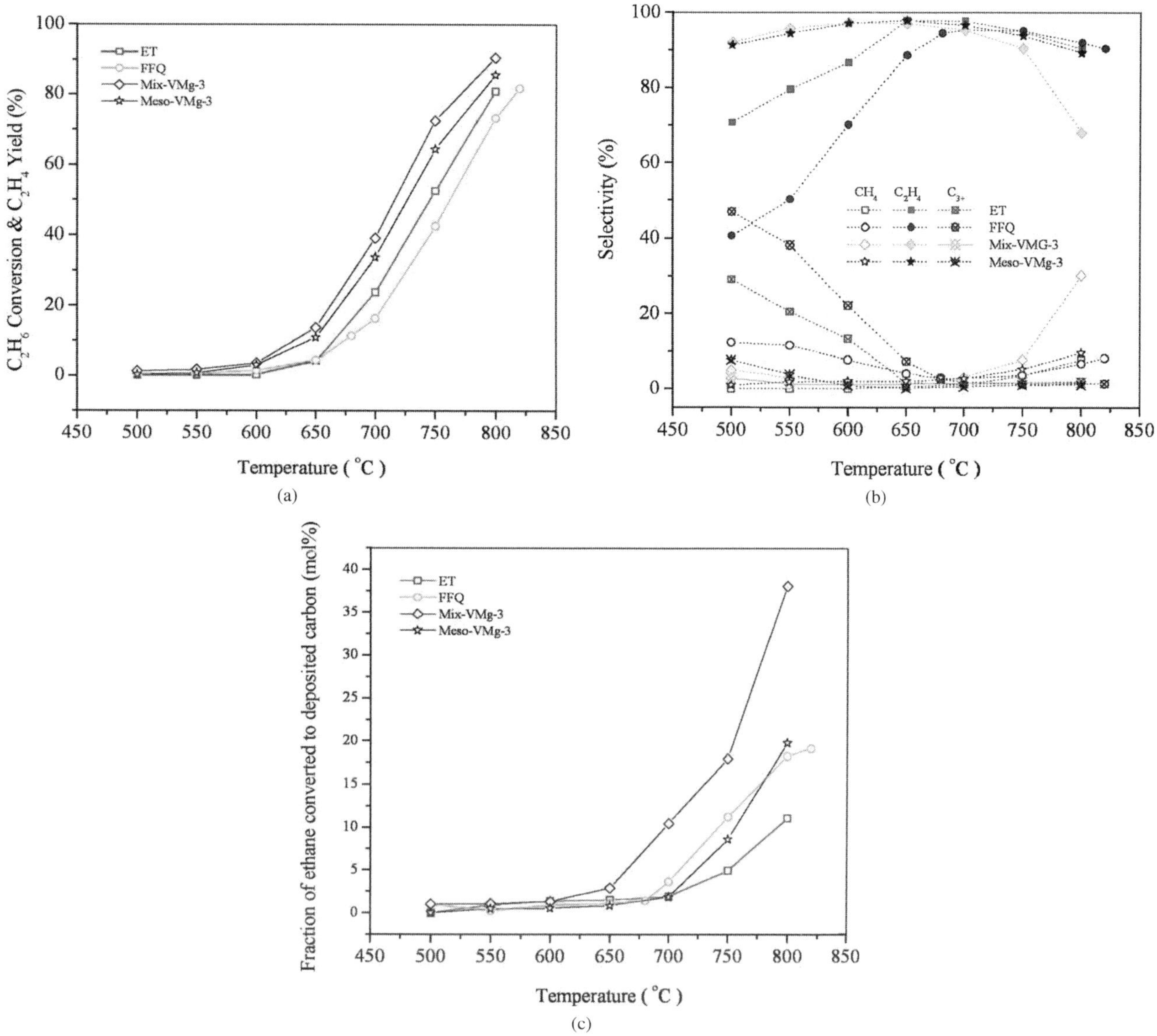

Fig. 1. Reactivities vs temperature for noncatalytic and catalytic thermolysis of ethane. (a) Conversion of ethane; (b) selectivities to various gaseous products; (c) fraction of ethane converted to deposited carbon. (The reactions were carried out using a feed gas with a mole composition of N_2:O_2:C_2H_6 = 93:0:7 and a flow rate of 2.43 ml s^{-1}.)

3. Results and discussion

3.1. *Thermolysis of ethane*

Fig. 1a provides the conversions of C_2H_6 in the absence of O_2 for the ET and FFQ reactor configurations as well as over the Mix-VMg-3 and Meso-V Mg-3 catalysts. It shows that the conversions of ethane were very small at low temperatures (⩽ 550 °C) and increased rapidly with increasing temperature. The conversion of ethane followed in most cases the sequence FFQ < ET < Meso-VMg-3 < Mix-VMg-3. The selectivities to gaseous products are presented in Fig. 1b, which shows that the selectivity to ethene passed through a maximum at about 700 °C. The selectivities to ethene followed the sequence FFQ < ET < Mix-VMg-3 ≈ Meso-VMg-3 for temperatures < 700 °C, but became close to each other at higher temperatures. An exception was observed for the Mix-VMg-3, over which the selectivity to ethene was much lower.

It must be, however, pointed out that although high conversions of ethane and selectivities to ethene could be obtained over catalysts, quartz granules, and even from the empty reactor, carbon deposition was equally high, particularly at high temperatures and over the catalysts. Fig. 1c provides the fraction of ethane converted to deposited carbon as a function of temperature, which increases in the sequence ET < Meso-VMg-3 ~ FFQ < Mix-VMg-3.

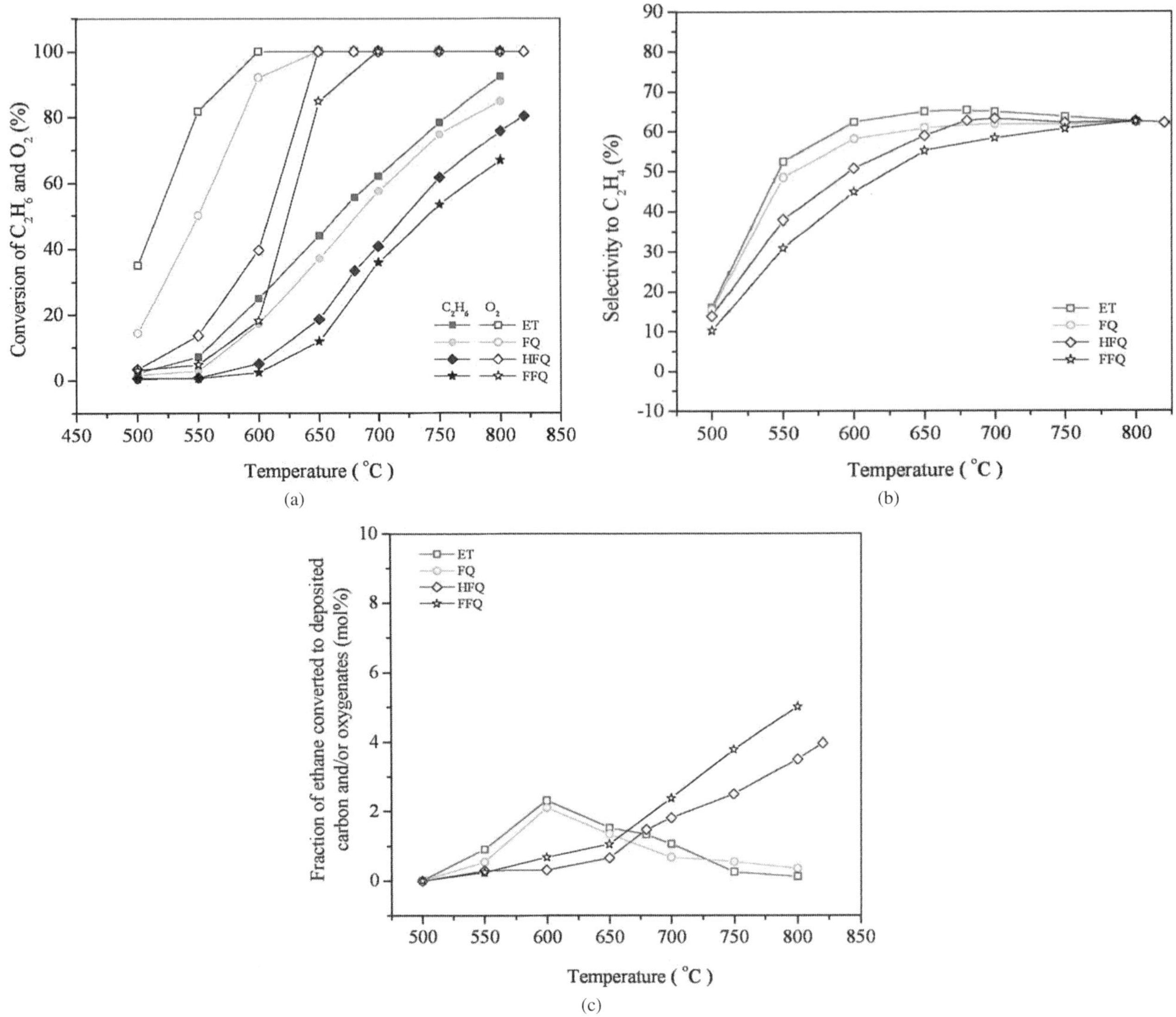

Fig. 2. Reactivities vs temperature for noncatalytic ODH of ethane in various reactor configurations. (a) Conversions of ethane and oxygen; (b) selectivity to ethene; (c) fraction of ethane converted to deposited carbon and/or oxygenates. (The reactions were carried out using a feed gas with a mole composition of N_2:O_2:C_2H_6 = 90:4:6 and a flow rate of 2.43 ml s^{-1}.)

3.2. Noncatalytic oxidative dehydrogenation of ethane

The noncatalytic oxidative conversion of ethane was investigated for four reactor configurations, i.e., ET, FQ, HFQ, and FFQ. The conversions of ethane and oxygen are presented in Fig. 2a, which shows that they increase in the sequence FFQ < HFQ < FQ < ET. The conversions of ethane and oxygen increased with increasing temperature. The conversion of oxygen increased fast and finally attained 100% at temperatures higher than 650 °C. The selectivity to C_2H_4 is presented in Fig. 2b. In the entire temperature range, the main by-products to ethene were CO_x. With increasing temperature, the selectivity to ethene attained its highest value at about 700 °C, after which it remained constant.

These results indicate that the noncatalytic reactivity of ethane is affected by the reactor configuration. The larger the space of the reactor occupied by quartz granules, the smaller the conversion of ethane, due to the recombination on the surface of the granules of the homogeneously formed radicals. When the reactor was completely filled with quartz granules, almost no homogeneous reaction took place at temperatures < 650 °C. However, at temperatures $\geqslant 650$ °C, the contribution of the homogeneous reactions to the conversion was important even for the FFQ reactor configuration.

Fig. 2c provides the fraction of ethane converted to nongaseous products, i.e., carbon deposits and oxygenates, as a function of temperature. One can see that the fraction of ethane converted to nongaseous products in the HFQ and FFQ configurations is much larger at high temperatures but smaller at low temperatures than those for the ET and FQ configurations.

Table 2
Fraction of ethane converted to oxygenates, X_{oxy}, and distribution of components in oxygenates for noncatalytic ODH of ethane[a]

	Temperature (°C)	Composition of feed gas (mol%)			X_{oxy} (%)	Selectivities (%)								
						Alcohol			Aldehyde		Ketone		Acid	
		N_2	O_2	C_2H_6		C_1	C_2	C_{3+}–OH[b]	C_1	C_2	C_3	C_4	C_1	C_2
ET	600	90.21	3.89	5.90	2.31	8.67	11.70	0.26	64.62	12.61	0.03	0.18	0.07	1.87
FQ	600	90.36	3.83	5.81	2.04	4.41	6.62	0.24	80.17	6.41	0.07	0.05	0.01	2.03
HFQ	550	88.74	4.47	6.79	0.29	8.35	11.60	0.06	64.72	12.14	0.08	0.00	0.21	2.83
HFQ	650	88.74	4.47	6.79	0.64	4.50	13.76	0.12	73.07	6.54	0.00	0.08	0.00	1.93
FFQ	550	89.53	4.17	6.30	0.24	20.33	3.78	0.11	45.20	29.57	0.03	0.06	0.08	0.84

[a] Flow rate = 2.43 ml s^{-1}.
[b] C_{3+}–OH: 1-propanol, 2-propanol, 1-butanol, and 2-butanol.

The fraction of ethane converted to oxygenates and the distribution of their components are listed in Table 2, which shows that the main component of oxygenates was formaldehyde. In addition, small amounts of methanol, ethanol, and acetaldehyde, as well as trace amounts of ketones and acids, were present.

3.3. *Catalytic oxidative dehydrogenation of ethane*

3.3.1. *Mix-V–Mg oxide catalysts prepared via solid reaction*

Fig. 3a presents the conversions of ethane and oxygen as a function of temperature over the pure V_2O_5 and MgO as well as Mix-VMg catalysts with various V/Mg ratios. The conversion of ethane increased with increasing temperature and changed with catalyst composition in the sequence V_2O_5 < MgO < Mix-VMg-1 < Mix-VMg-2 < Mix-VMg-3 < Mix-VMg-4 < Mix-VMg-5. They reveal a synergistic effect between V_2O_5 and MgO and that the activity increased with increasing magnesium content of the catalyst. The differences between the conversions of ethane among various catalysts were relatively large in the middle range of temperatures, but smaller in the lower and higher ranges. The conversion of oxygen increased rapidly with increasing temperature, particularly for catalysts with atomic ratios V/Mg < 0.5 (the Mix-VMg-5 and Mix-VMg-4 catalysts), and attained 100% within a small temperature range. The same sequence as for the conversion of ethane was also present for the conversion of oxygen over the catalysts. The catalysts with higher magnesium content were more active regarding oxygen activation. In the entire range of temperature, the main by-products to ethene were found to be CO_x. The effect of temperature on the selectivity to ethene is presented in Fig. 3b, which shows that at low temperatures (< 650 °C), the selectivity to ethene was high over the pure V_2O_5 and low over the pure MgO, and intermediary over the Mix-VMg catalysts. With increasing temperature, the selectivity to ethene decreased rapidly over the pure V_2O_5 and increased over the pure MgO and the Mix-VMg catalysts. The selectivity to ethene followed the sequence MgO < Mix-VMgO-5 < Mix-VMgO-4 < Mix-VMgO-3 < Mix-VMgO-2 < Mix-VMgO-1 < V_2O_5, revealing an increase with decreasing magnesium content. At high temperatures, the change of selectivity to ethene with temperature as well as the differences of ethene selectivities among the pure V_2O_5, the pure MgO, and the Mix-VMg catalysts became very small.

The fraction of ethane converted to carbon deposits and oxygenates is provided by Fig. 3c, which shows that it increases with increasing temperature.

The fraction of ethane converted to oxygenates and their component distributions are listed in Table 3. The higher the reaction temperature, the larger the fraction of ethane converted to oxygenates. Formaldehyde was found to be the main component in oxygenates; the selectivities to other components were very small. The fraction of ethane converted to oxygenates increased in the sequence MgO < V_2O_5 < Mix-VMg catalysts. Among the Mix-VMg catalysts, the fraction of ethane converted to oxygenates exhibited a maximum with increasing V/Mg atomic ratio.

3.3.2. *V–Mg oxide catalysts prepared from mesoporous materials*

Fig. 4a presents the conversions of ethane and oxygen as a function of temperature over the Meso-V and Meso-VMg catalysts. One can see that the conversion of ethane increased with increasing temperature and followed the sequence Meso-V < Meso-VMg-1 < Meso-VMg-4 < Meso-VMg-6 < Meso-VMg-5 < Meso-VMg-2 < Meso-VMg-3. Over the Meso-VMg-2, Meso-VMg-3, Meso-VMg-5, and Meso-VMg-6 catalysts, the conversion of oxygen increased rapidly with increasing temperature and attained 100% at temperatures higher than 650 °C. Over the Meso-VMg-1, Meso-VMg-4, and Meso-V, the 100% oxygen conversion was reached only after a slow increase with increasing temperature. The selectivity to ethene is presented in Fig. 4b. The main by-products were CO_x, and the selectivities to CO_2 and C_{3+} decreased and those to CO and CH_4 increased, with increasing temperature. The selectivity to ethene over the Meso-V was higher than over the Meso-VMg catalysts at very low temperatures but smaller than for some of the latter at high temperatures. The selectivity to ethene over most of the Meso-VMg catalysts first increased, passed through a maximum, and then decreased slightly, with increasing temperature. Except for the low (500 °C) and high (800 °C) temperatures, the selectivity to ethene followed the sequence

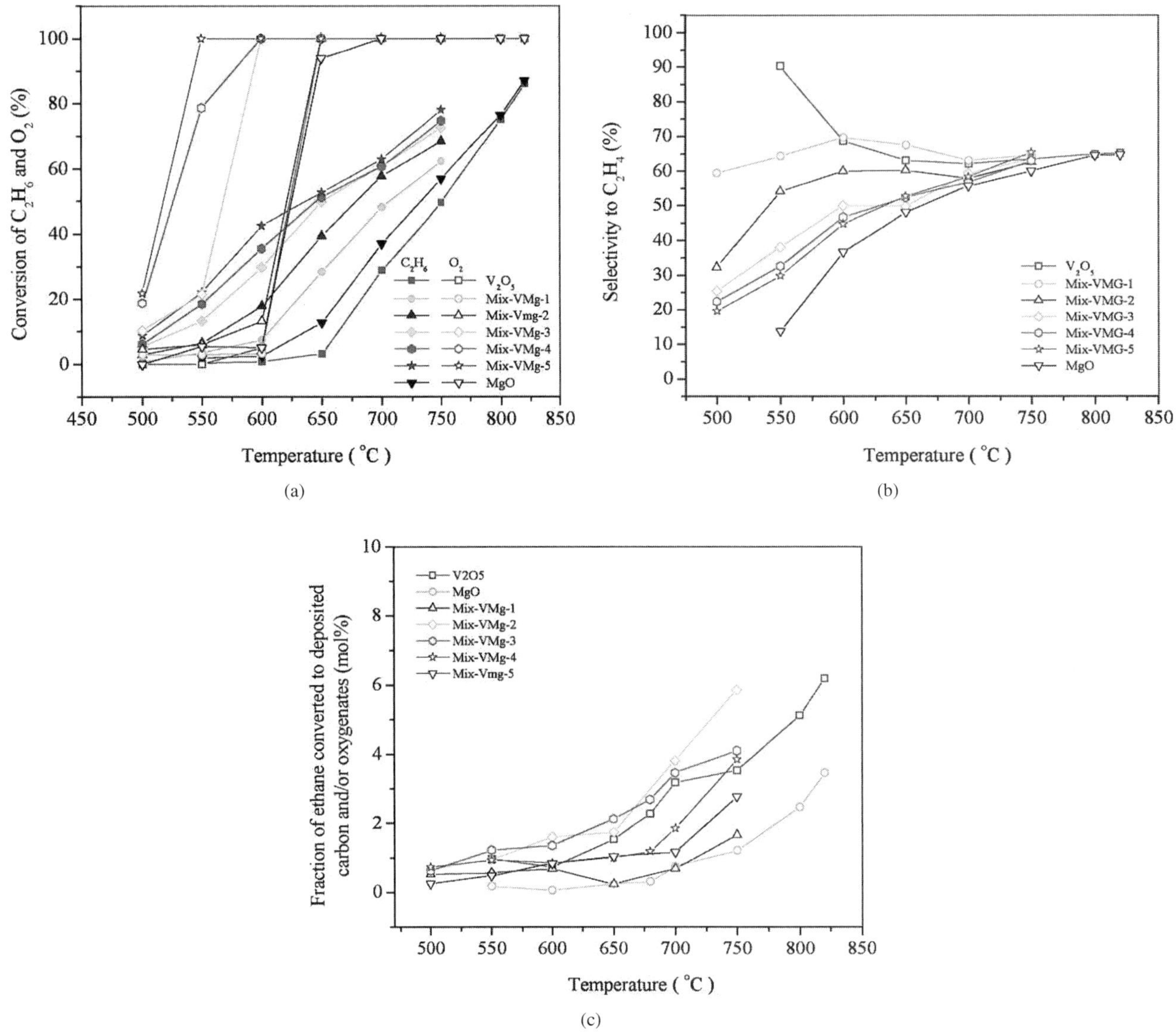

Fig. 3. Reactivities vs temperature for ODH of ethane over the Mix-VMg catalysts. (a) Conversions of ethane and oxygen; (b) selectivity to ethene; (c) fraction of ethane converted to deposited carbon and/or oxygenates. (The reactions were carried out using feed gas with a mole composition of $N_2:O_2:C_2H_6 = 90:4:6$ and a space velocity of 35,027 ml g^{-1} h^{-1}.)

Table 3
Fraction of ethane converted to oxygenates, X_{oxy}, and distribution of components in oxygenates over Mix-VMg catalysts[a]

Catalysts	Temperature (°C)	Composition of feed gas (mol%)			X_{oxy} (%)	Selectivities (%)								
						Alcohol			Aldehyde		Ketone		Acid	
		N_2	O_2	C_2H_6		C_1	C_2	C_{3+}–OH[b]	C_1	C_2	C_3	C_4	C_1	C_2
Mix-VMg-1	600	91.19	3.47	5.24	0.68	0.74	3.57	0.07	94.25	1.07	0.00	0.05	0.00	0.25
Mix-VMg-3	600	89.04	4.32	6.64	1.21	5.45	0.77	0.00	85.29	7.92	0.16	0.00	0.00	0.41
Mix-VMg-5	550	90.33	3.85	5.81	0.49	0.21	4.18	0.00	94.32	0.31	0.00	0.00	0.00	0.98
Mix-VMg-5	600	90.33	3.85	5.81	0.75	3.62	0.00	0.00	90.95	5.27	0.00	0.00	0.00	0.15
Mix-VMg-5	650	90.33	3.85	5.81	0.95	0.40	26.64	0.50	70.97	0.58	0.00	0.55	0.00	0.36
MgO	600	90.19	3.91	5.90	0.05	2.11	0.61	0.10	93.61	3.07	0.05	0.11	0.00	0.34
V_2O_5	600	89.53	3.99	6.03	0.73	0.50	1.64	0.07	96.88	0.73	0.00	0.08	0.00	0.10

[a] Space velocity = 35,027 ml g^{-1} h^{-1}.
[b] C_{3+}–OH: 1-propanol, 2-propanol, 1-butanol, and 2-butanol.

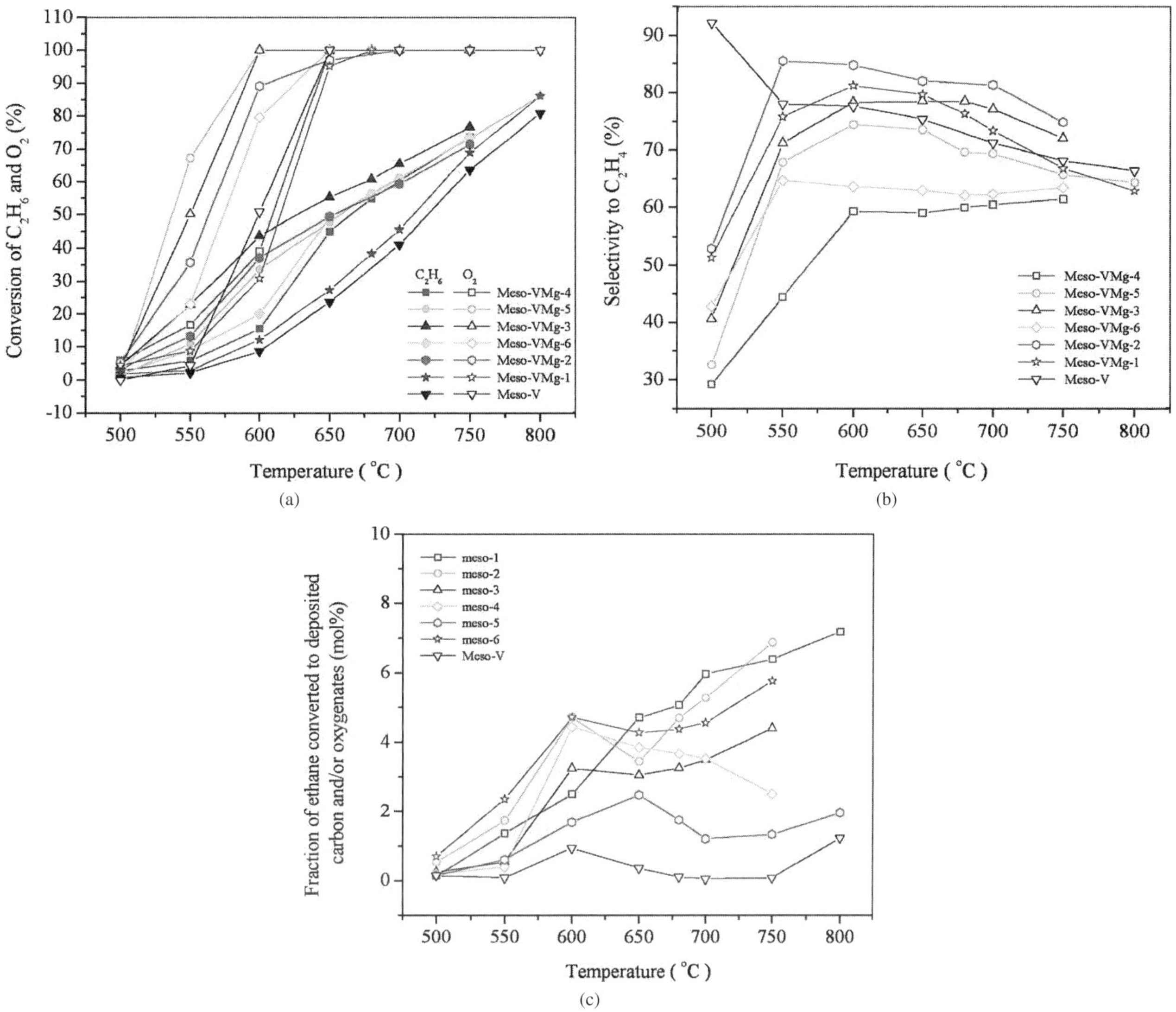

Fig. 4. Reactivities vs temperature for ODH of ethane over the Meso-VMg catalysts. (a) Conversions of ethane and oxygen; (b) selectivity to ethene; (c) fraction of ethane converted to deposited carbon and/or oxygenates. (The reactions were carried out using a feed gas with a mole composition of N_2:O_2:C_2H_6 = 90:4:6 and a space velocity of 35,027 ml g^{-1} h^{-1}.)

Meso-VMg-4 < Meso-VMg-6 < Meso-VMg-5 < Meso-V < Meso-VMg-3 < Meso-VMg-1 < Meso-VMg-2.

The effect of C_2H_6/O_2 mole ratio in the feed gas on the ODH of ethane over the Meso-VMg-2 catalyst is examined in Table 4. One can see that with increasing C_2H_6/O_2 mole ratio, the conversion of ethane, the selectivity to CO_x and the yield to ethene decreased, but the selectivity to ethene and CH_4 increased. The selectivity to C_{3+} changed only slightly. The conversion of oxygen at 550 °C increased with increasing C_2H_6/O_2 mole ratio, and a 100% oxygen conversion was reached for all C_2H_6/O_2 mole ratios at 650 °C.

The effect of the space velocity on the oxidative dehydrogenation of ethane over the Meso-VMg-2 catalyst is examined in Table 5, which shows that with increasing space velocity, the selectivity to ethene increased but the conversion of ethane decreased.

The fraction of ethane converted to deposited carbon and liquid products is presented in Fig. 4c.

The fraction of ethane converted to oxygenates and the component distribution of oxygenates over the Meso-VMg catalysts are listed in Table 6. The main component in oxygenates was formaldehyde and the selectivities to other components were relatively small. The fraction of ethane converted to oxygenates was larger over the Meso-VMg catalysts than over the Meso-V. With increasing temperature, the fraction exhibited a maximum both over the Meso-VMg catalysts and over the Meso-V. Comparing Table 6 with Tables 2 and 3, one can conclude that the fraction of ethane converted to oxygenates was larger over the Meso-VMg catalysts than over the Mix-VMg ones and the noncatalytic ODH of ethane.

Table 4
Effect of molar composition of feed gas on ethane ODH over Meso-VMg-2 catalyst[a]

Feed composition			Conversion (%)		Selectivity (%)						Yield to
N_2	O_2	C_2H_6	C_2H_6	O_2	CO	CO_2	CH_4	C_2H_4	C_3H_6	C_3H_8	C_2H_4 (%)
					550 °C						
87.10	8.14	4.76	35.55	18.63	49.98	3.93	3.64	42.02	0.02	0.41	14.94
87.62	7.24	5.14	32.21	42.38	47.89	2.8	3.95	44.01	0.29	1.06	14.18
87.46	6.72	5.83	29.43	42.48	47.96	1.24	3.16	46.84	0.14	0.66	13.79
87.79	6.02	6.19	28.44	58.66	47.18	1.17	2.63	47.12	0.02	1.88	13.40
88.48	4.99	6.53	26.88	64.75	46.63	1.13	3.42	48.41	0.12	0.29	13.01
87.59	4.94	7.47	22.96	77.18	46.72	0.96	3.08	48.88	0.18	0.18	11.22
88.18	4.34	7.48	21.09	82.61	39.76	0.63	5.34	52.13	0.62	1.52	10.99
90.30	3.12	6.57	20.2	82.84	38.78	0.41	5.60	53.18	0.29	1.74	10.74
90.20	3.01	6.79	19.04	85.09	35.26	0.39	6.00	54.56	0.92	2.87	10.39
90.09	2.76	7.15	17.88	86.06	36.09	0.31	6.28	56.69	0.27	0.36	10.14
90.51	2.03	7.46	13.25	87.92	32.64	0.25	6.89	58.68	0.55	0.99	7.78
90.30	1.37	8.34	9.53	100	30.56	0	6.87	65.51	0.55	0.51	6.24
91.11	1.16	7.73	8.13	100	20.44	0	8.13	69.31	0.94	1.18	5.63
92.31	0.00	7.69	0.46	–	0	0	8.37	90.63	0.51	0.49	0.42
					650 °C						
87.90	7.51	4.59	81.57	100	36.29	16.18	7.75	37.87	1.27	0.65	30.89
89.62	4.97	5.41	67.18	100	31.59	12.56	9.03	44.45	1.52	0.85	29.86
87.48	5.35	7.17	55.54	100	31.50	8.00	8.08	49.07	2.13	1.22	27.25
89.49	4.03	6.48	45.63	100	35.06	8.10	6.83	57.71	1.55	0.75	26.33
90.30	3.12	6.57	45.00	100	24.17	6.42	7.87	58.2	1.86	1.47	26.19
87.90	2.07	10.03	25.03	100	13.89	0	6.79	77.28	2.09	0.05	19.34
93.25	0.00	6.75	3.43	–	0	0	3.23	95.69	1.08	0	3.28

[a] Space velocity = 23,071 $ml\,g^{-1}\,h^{-1}$.

Table 5
Effect of space velocity on ethane ODH over Meso-VMg-2 catalyst[a]

Space velocity	Conversion (%)		Selectivity (%)						Yield to
($ml\,g^{-1}\,h^{-1}$)	C_2H_6	O_2	CO	CO_2	CH_4	C_2H_4	C_3H_6	C_3H_8	C_2H_4 (%)
35027	49.44	96.97	5.27	1.62	6.85	82.01	2.32	1.93	40.55
21262	56.35	100	32.21	3.41	6.94	54.61	1.93	0.9	30.77
14854	57.18	100	34.01	6.56	7.37	49.43	1.84	0.79	28.26
12322	57.21	100	43.5	8.19	7.43	38.28	1.82	0.78	21.90
8888	63.3	100	50.55	9.03	8.5	29.37	1.8	0.75	18.59
3866	82.18	100	58.86	11.01	9.31	19.02	1.62	0.18	15.63
1355	92.59	100	66.14	12.56	9.89	10.94	0.47	0	10.13

[a] Mole composition of feed gas: N_2:O_2:C_2H_6 = 89.1:4.1:6.9; reaction temperature = 650 °C.

Table 6
Fraction of ethane converted to oxygenates, X_{oxy}, and distribution of components in oxygenates over Meso-VMg catalysts[a]

Catalysts	Temperature	Composition of			X_{oxy}	Selectivities (%)								
	(°C)	feed gas (mol%)			(%)	Alcohol			Aldehyde		Ketone		Acid	
		N_2	O_2	C_2H_6		C_1	C_2	C_{3+}–OH[b]	C_1	C_2	C_3	C_4	C_1	C_2
Meso-VMg-1	650	88.79	4.46	6.75	4.58	0.75	4.95	0.00	92.36	1.09	0.10	0.00	0.00	0.75
Meso-VMg-2	550	89.09	4.32	6.59	1.73	0.71	14.71	0.00	83.24	1.03	0.04	0.19	0.00	0.08
Meso-VMg-2	600	89.09	4.32	6.59	4.72	0.07	1.51	0.00	98.25	0.11	0.06	0.00	0.00	0.00
Meso-VMg-2	650	89.09	4.32	6.59	3.40	0.46	7.64	0.00	91.04	0.66	0.00	0.20	0.00	0.00
Meso-VMg-3	600	88.74	4.50	6.76	3.21	0.95	1.06	0.00	96.18	1.38	0.43	0.00	0.00	0.00
Meso-VMg-4	600	88.79	4.46	6.75	4.22	6.58	10.03	1.80	67.03	9.58	1.79	0.00	0.78	2.40
Meso-VMg-5	650	89.06	4.37	6.57	2.42	1.80	1.31	0.00	93.57	2.62	0.00	0.00	0.00	0.70
Meso-VMg-6	600	89.09	4.36	6.55	4.71	2.85	5.10	0.00	86.60	4.14	0.00	0.00	0.00	1.32
Meso-V	550	89.15	4.34	6.52	0.08	0.26	3.31	0.59	95.12	0.38	0.00	0.00	0.26	0.09
Meso-V	600	89.15	4.34	6.52	0.94	2.71	1.91	0.01	91.07	3.94	0.34	0.00	0.01	0.01
Meso-V	650	89.15	4.34	6.52	0.26	1.10	2.63	0.22	93.83	1.60	0.00	0.39	0.10	0.14

[a] Space velocity = 35,027 $ml\,g^{-1}\,h^{-1}$.
[b] C_{3+}–OH: 1-propanol, 2-propanol, 1-butanol, and 2-butanol.

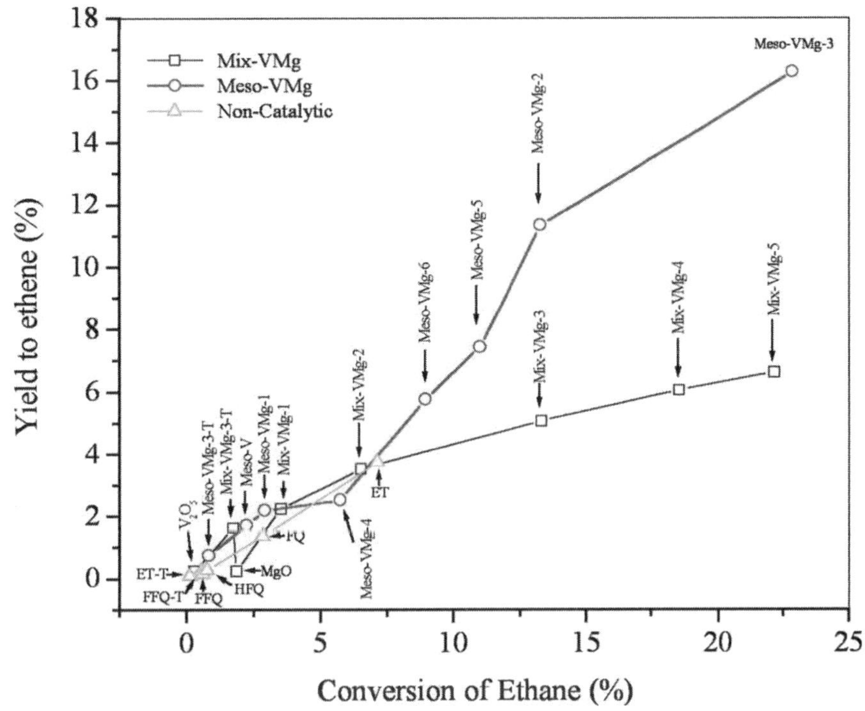

(a)

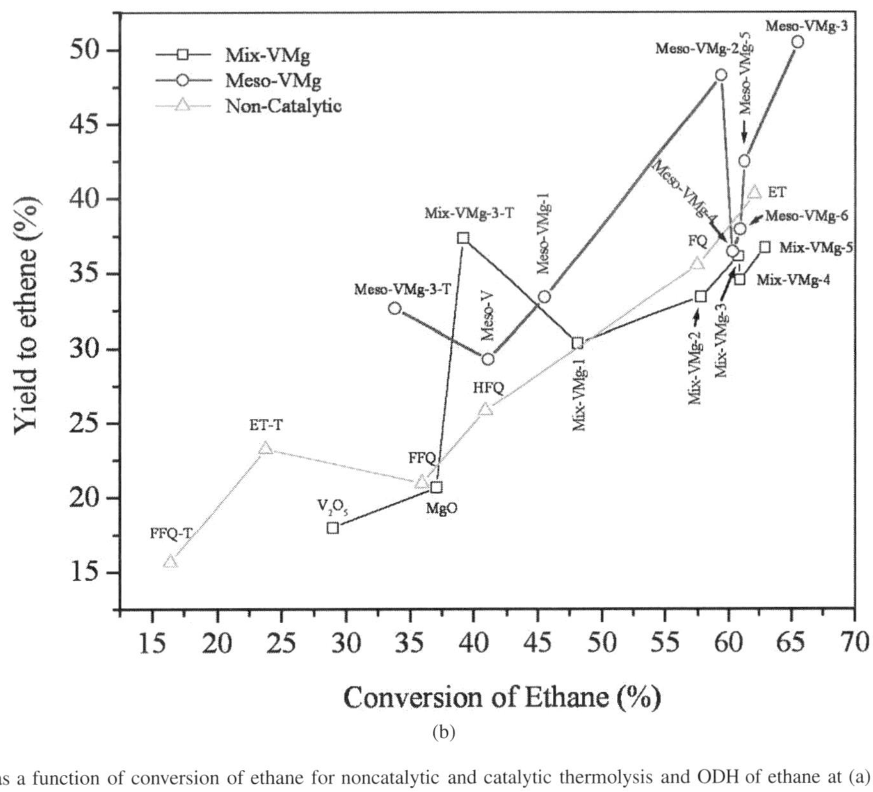

(b)

Fig. 5. Yield of ethene as a function of conversion of ethane for noncatalytic and catalytic thermolysis and ODH of ethane at (a) 550 and (b) 700 °C. (The thermolysis cases are marked by -T.)

3.3.3. Comparison between the noncatalytic and catalytic conversion of ethane

Figs. 5a and 5b present the yield of ethene against the conversion of ethane for the noncatalytic and catalytic thermolysis and ODH of ethane at 550 and 700 °C, respectively. At low temperatures, the differences between the yields of ethene for the noncatalytic and catalytic thermolysis and ODH of ethane are small when the conversions of ethane are low. The highest yield of ethene for the noncatalytic ODH occurs for a conversion of about 7%; however, the yield to ethene for the catalytic ODH can be much higher, being larger for the Meso-VMg catalysts than the Mix-VMg ones and the highest for the Meso-VMg-3 catalyst. At higher temperatures, thermolysis provides much larger yield to ethene; but the yields to ethene are even higher for the ODH processes. Furthermore, the differences between the yields to ethene for the ODH over the Mix-VMg catalysts and the noncatalytic ODH are in most cases small; they are, however, surpassed by the ODH over the Meso-VMg catalysts.

Figs. 5a and 5b reveal that at low temperatures, the contribution of thermolysis and noncatalytic ODH of ethane to the production of ethene are small, compared to those of the catalytic ODH processes. However, at high temperatures, the contributions of both thermolysis and noncatalytic ODH of ethane are important. The Meso-VMg catalysts provide the highest yield to ethene.

3.4. *Structures and textural properties of the catalysts*

Fig. 6 presents the XRD patterns of the as-synthesized mesoporous V–Mg–O. When $V(acac)_3$ was used as vanadium source and for an atomic ratio $V/Mg > 0.25$ in the initial batches the synthesized mesoporous materials had a wormlike or distorted hexagonal structure, characterized by a strong peak at $2\theta = 2$–$3°$ (Fig. 6, a–c, and e). For V/Mg atomic ratios $\leqslant 0.25$, the XRD patterns exhibited only a weak peak on the raised baseline in the 2θ range of 2–3° (Fig. 6, d), suggesting a very low degree of crystallinity of the mesostructure or perhaps a larger *d* spacing for a diffraction line located at a very small 2θ which could not be detected by our equipment. When V_2O_5 was used as vanadium source, the synthesized mesoporous materials had a lamellar structure, characterized by equidistanced peaks in the XRD patterns (Fig. 6, f and g). When the specimens were thermally treated at high temperatures, the mesostructures, as expected, were destroyed; however, some crystalline phases could be identified. Fig. 7 presents the XRD patterns of the specimens thermally treated at 750 °C. For atomic ratios $V/Mg > 3.2$ in the catalyst (Fig. 7, c–g), neither MgO nor V_2O_5 phases could be identified, but solely the V_2O_3 phase. For atomic ratios $V/Mg \leqslant 3.2$ (Fig. 7, b), $Mg_{1.5}VO_4$ and MgO phases were also detected, besides the V_2O_3 phase. This suggests that for atomic ratios $V/Mg > 3.2$, magnesium was highly dispersed in the V_2O_3 phase.

Table 7 summarizes the *d* spacings of the as-synthesized mesoporous V–Mg oxides and their textural characteristics after the thermal treatment at 750 °C. One can see that the *d* spacings of the as-synthesized specimens were larger for those based on $V(acac)_3$ than on V_2O_5. After the thermal

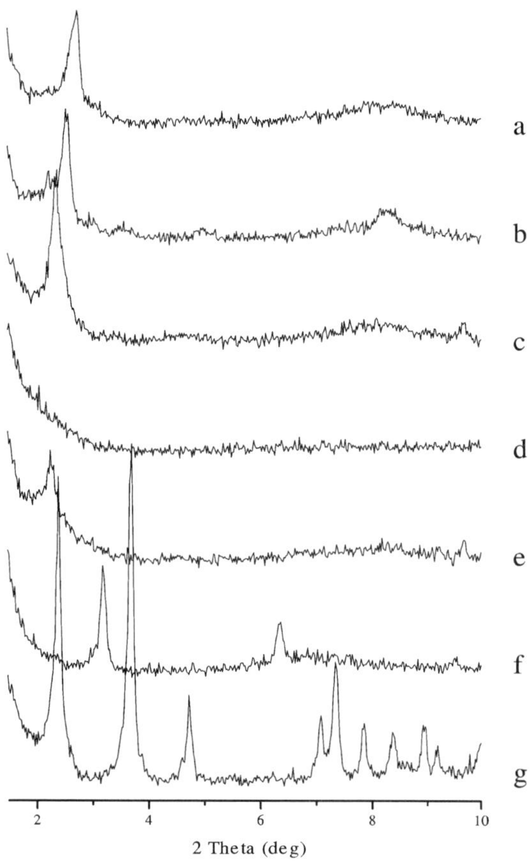

Fig. 6. XRD patterns of as-synthesized specimens. (a) Meso-VMg-1; (b) Meso-VMg-2; (c) Meso-VMg-3; (d) Meso-VMg-4; (e) Meso-VMg-5; (f) Meso-VMg-6; (g) Meso-V. (For their synthesis conditions see Table 1.)

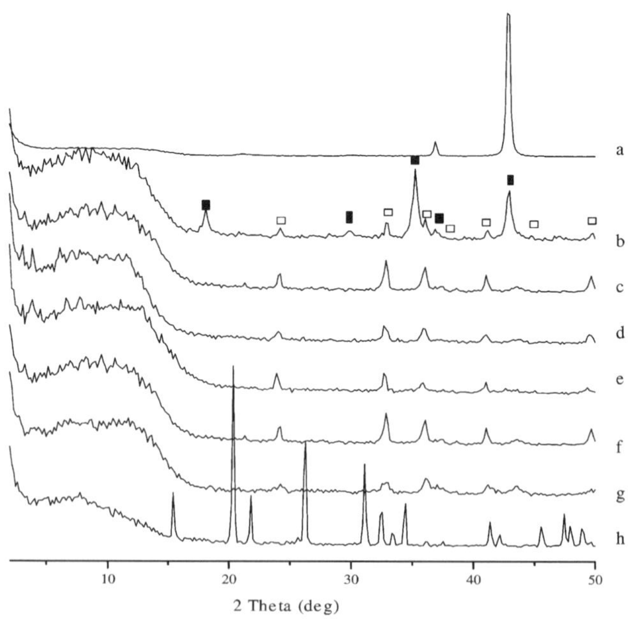

Fig. 7. XRD patterns of the Meso-VMg catalysts prepared via the calcination of mesoporous precursors at 750 °C. (a) MgO; (b) Meso-VMg-5; (c) Meso-VMg-3; (d) Meso-VMg-6; (e) Meso-VMg-2; (f) Meso-VMg-1; (g) Meso-V; (h) V_2O_5. (For their synthesis conditions see Table 1.) The intensities of the peaks for MgO were reduced by a factor of 10 to be able to present several spectra in a single figure. (Symbols: (□) V_2O_3 and (■) $Mg_{1.5}VO_4$.)

Table 7
Textural properties of the Meso-VMg catalysts

Specimens	As-prepared	Thermally treated at 750 °C		
	d spacings (nm)	Specific surface area ($m^2 g^{-1}$)	Pore volume ($cm^3 g^{-1}$)	Average pore size (nm)
Meso-VMg-1	3.10	151.0	0.17	4.51
Meso-VMg-2	3.58	155.1	0.18	4.66
Meso-VMg-3	3.83	159.1	0.16	4.05
Meso-VMg-4	–	N.A.[a]	N.A.	N.A.
Meso-VMg-5	4.17	158.5	0.16	3.99
Meso-VMg-6	2.85	96.82	0.174	7.17
Meso-V	2.86	148.6	0.281	7.56

[a] N.A.: not available.

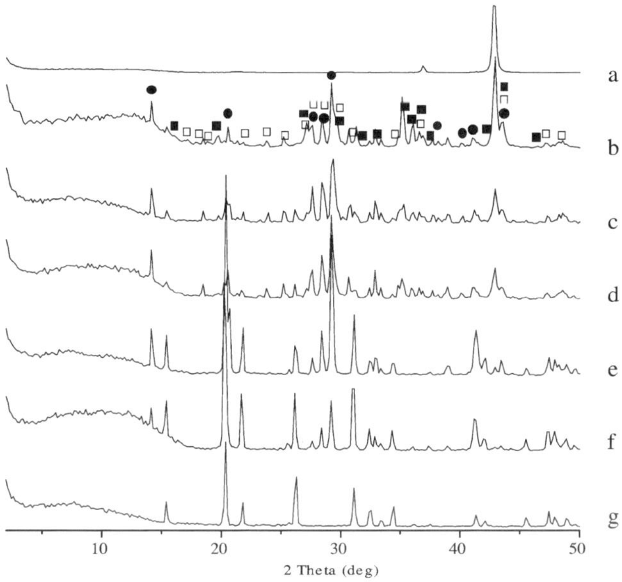

Fig. 8. XRD patterns of the Mix-VMg catalysts prepared via solid reaction between vanadia and magnesia. (a) MgO; (b) Mix-VMg-5; (c) Mix-VMg-4; (d) Mix-VMg-3; (e) Mix-VMg-2; (f) Mix-VMg-1; (g) V_2O_5. (For their synthesis conditions see Table 1.) The intensities of the peaks for MgO were reduced by a factor of 10 to be able to present several spectra in a single figure. (Symbols: (●) *meta*-MgV_2O_6; (□) *pyro*-$Mg_2V_2O_7$; (■) *ortho*-$Mg_3(VO_4)_2$.)

treatment, the BET specific surface areas were larger but the average pore sizes smaller for the specimens based on $V(acac)_3$ than on V_2O_5. The pore-size distributions of these specimens were found to be relatively narrow.

Fig. 8 presents the XRD patterns of Mix-VMg, MgO, and V_2O_5 catalysts. In the Mix-VMg catalysts with atomic ratios V/Mg > 1, two phases, V_2O_5 (major) and *meta*-MgV_2O_6 (minor), were present. With the decrease of the atomic ratio of V/Mg, the intensity of the diffraction peak specific to the V_2O_5 phase decreased, but that for the *meta*-MgV_2O_6 increased (Fig. 8, e–g). In the Mix-VMg catalysts with atomic ratios V/Mg $\leqslant$ 1, no V_2O_5 phase but the *meta*-MgV_2O_6 and MgO as well as other phases could be identified (Fig. 8, a–d). For example, a *pyro*-$Mg_2V_2O_7$ phase was identified in the Mix-VMg catalysts with V/Mg atomic ratios of 0.25–1 (Fig. 8, c and d) and an *ortho*-$Mg_3(VO_4)_2$ phase in those with V/Mg atomic ratios $\leqslant$ 0.25 (Fig. 8, b).

Table 8
Textural properties of the Mix-VMg catalysts

Catalysts	BET ($m^2 g^{-1}$)	Pore volume ($cm^3 g^{-1}$)	Average pore size (nm)
V_2O_5	4.75	0.01	8.28
Mix-VMg-1	1.02	0.0028	10.83
Mix-VMg-2	2.19	0.0071	12.65
Mix-VMg-3	6.36	0.017	10.63
Mix-VMg-4	6.54	0.017	10.61
Mix-VMg-5	7.16	0.023	12.78
MgO	14.10	0.043	12.20

Table 8 summarizes the textural properties of the Mix-VMg catalysts. One can see that the surface areas and pore volumes of pure V_2O_5 and MgO and mixed oxide catalysts are very small. The mixed oxides catalysts with atomic ratios V/Mg > 1.0 had lower specific surface areas and pore volumes than the pure V_2O_5. The specific surface areas and the pore volumes of the catalysts increased with decreasing V/Mg atomic ratio and for V/Mg $\leqslant$ 1.0 became larger than those of pure V_2O_5. In all the cases, the specific surface areas and the pore volumes of the mixed oxide catalysts were smaller than those of pure MgO. The pore-size distributions were broad. The porosity of these materials is probably due mainly to the interparticle pores.

3.5. *Activation of ethane over V–Mg–O catalysts*

The vanadium-containing catalysts, especially the V–Mg–O ones, are usually considered among the most active and selective catalysts in the ODH of light alkanes to the corresponding alkenes [38–60]. The V–Mg–O catalysts were prepared via either wet impregnation or solid-phase reaction between magnesium and vanadium oxides. Although attempts were made to investigate the nature of the active phases in these catalysts, no final conclusion was yet reached. For instance, the active and selective phase was attributed to $Mg_3(VO_4)_2$ by Chaar et al. [38,39] and Michalkos et al. [40], or to $Mg_2V_2O_7$ by Sam et al. [41] and Dula et al. [54]. Solsona et al. [42] and other researchers [43,44, 56,59,60] proposed, however, that the coexistence of magnesium vanadates and MgO phases is responsible for the activity of the catalysts. Pak et al. [55] found that the most active V–Mg–O catalysts were those that contain very small $Mg_3(VO_4)_2$ domains on the surface of MgO rather than crystallites of bulk $Mg_3(VO_4)_2$. Many investigations were also concerned with supported vanadium oxide catalysts, and it was found that factors, such as the oxidation state of vanadium, the dispersion of vanadium species, and the acid–base properties of the supports, affected the ODH of hydrocarbons. For examples, Anderson and Kung [61,62] found that the selectivities to $C_4^{=}$ for the butane ODH over the alumina-supported vanadium oxide catalyst increased rapidly as the catalyst was progressively reduced, while

Rybarczyk et al. [63] proposed V^{5+} to be more selective but less active than V^{4+}. Haber [64] and many other researchers [65,66] proposed that the interaction of a hydrocarbon molecule with a bridging oxygen in vanadium oxide clusters is more favorable than with a terminal one (V=O). Cortez and Banares [67] proposed, however, that the active site for the ODH of propane over alumina-supported vanadium oxide catalyst is a single VO_4 site rather than a V–O–V bond.

In the present paper, we found that the conversion of ethane increased and the selectivity to ethene decreased with increasing magnesium content. The conversion of ethane over the Mix-VMg catalysts was much larger than over pure vanadia or magnesia and the selectivities to ethene were much higher than over pure magnesia but smaller than over pure vanadia. Furthermore, the specific surface areas of the catalysts were smaller than that of MgO (see Table 8). It is clear that the active phases should have been generated during the catalyst preparation. The XRD results revealed that *meta*-MgV_2O_6 and V_2O_5 phases were present in the Mix-VMg catalysts with low Mg content, whereas *pyro*-$Mg_2V_2O_7$, *ortho*-$Mg_3(VO_4)_2$, and MgO phases were present in the catalysts with high Mg content. For low magnesium contents, V_2O_5 may have acted as support, over which magnesium was dispersed and the *meta*-MgV_2O_6 phase was formed. The V_2O_5 phase and/or the coexisting V_2O_5 and *meta*-MgV_2O_6 phases were responsible for the high selectivity. In contrast, for high magnesium contents, MgO may have played the role of support over which vanadium was dispersed and the *pyro*-$Mg_2V_2O_7$ and *ortho*-$Mg_3(VO_4)_2$ phases were formed. Thus, the *pyro*-$Mg_2V_2O_7$ and *ortho*-$Mg_3(VO_4)_2$ might have been the active phases responsible for the ODH of ethane. Because of the higher oxidation state of vanadium than magnesium in the vanadates, the electrons of oxygen in the V–O–Mg bond tended to be pulled toward the vanadium end, the oxygen thus becoming deficient in electrons. The as-formed oxygen species are more active in the activation of the hydrocarbon than the lattice oxygen of the bulk vanadia or magnesia. It is also possible that O_2 molecules were adsorbed on vanadium sites rich in electrons, generating active species, such as O_2^{2-} or O_2^-. However, when too many such active species became present over the catalyst, the overoxidation of the reactant and/or intermediates took place, thus reducing the selectivity to ethene.

In the Meso-VMg catalysts, only the V_2O_3 phase was identified in the high range of V/Mg atomic ratios, and the $Mg_{1.5}VO_4$ and MgO phases could be additionally detected in the catalysts with V/Mg ratios smaller than 3.2. The XPS examination of the Meso-VMg-5 and Meso-VMg-3 indicated that the surface V/Mg atomic ratios were 5.12 and 7.05, whereas in bulk they were 3.19 and 10.65, respectively (see Table 1). Consequently, the surface of the Meso-VMg-3 catalyst was enriched in magnesium. Because of the high dispersion over the large surface area of V_2O_3 (see Table 7), magnesium could not be detected by XRD. The V/Mg atomic ratio on the surface of the Meso-VMg-5 was larger than in bulk, perhaps because of a poor dispersion of magnesium, which formed separate MgO and $Mg_{1.5}VO_4$ phases. The catalytic performance of the Meso-VMg catalysts revealed that for the ODH of ethane, the conversion of ethane followed the sequence Meso-V < Meso-VMg-1 < Meso-VMg-4 < Meso-VMg-6 < Meso-VMg-5 < Meso-VMg-2 < Meso-VMg-3, while the selectivity to ethene followed the sequence Meso-VMg-4 < Meso-VMg-6 < Meso-VMg-5 < Meso-V < Meso-VMg-3 < Meso-VMg-1 < Meso-VMg-2. In most cases higher conversions of ethane were obtained over the catalyst with a lower V/Mg ratio and higher selectivities to C_2H_4 over those with a higher V/Mg ratio. The presence of exceptions suggests that the reactivity of the catalysts is determined not only by the V/Mg ratio but also by some structural properties.

It is not yet possible to explain why the performances of the Meso-VMg catalysts are superior to those of the Mix-VMg ones. The more facile diffusion of the species and the larger surface areas in the former catalysts are some of the reasons. Probably the organization of the V and Mg species by the template plays the main role and Mg species moderate the redox capability of V_2O_3, thus increasing the activity and selectivity of the catalyst.

While the Meso-VMg catalysts possess much larger specific surface areas than the Mix-VMg ones, one should not attribute simply the higher catalytic performance of the Meso-VMg catalysts to the differences in specific surface areas between these two kinds of catalysts. In fact, it is well known that higher specific surface areas are usually favorable to deep oxidations in the catalytic conversion of hydrocarbons. It is possible that the high BET specific surface area of the Meso-VMg catalysts allowed a higher dispersion of magnesium over the surface of vanadium species and, as a result, they provided high selectivities.

The mechanism of the activation of light hydrocarbons was not yet completely clarified, though many possible reaction schemes have been suggested. It is believed that the first step in the activation of light alkanes is the initiation of alkyl radicals via a homogeneous or heterogeneous pathway. The overall process involves a complex series of parallel and/or sequential steps, which can be catalytic surface reactions or noncatalytic ones in the gas phase, dependent on many factors, such as reaction temperature, reactant composition, and characteristics of the reactor in noncatalytic processes and those of the catalyst in the catalytic ones.

In the present paper, the differences of ethane activities are relatively small at high temperatures but large at low and moderate temperatures. This indicates that the mechanism of ethane ODH over V–Mg-based catalysts is strongly dependent on temperature. Heterogeneous processes are mainly involved at low temperatures, while both heterogeneous and homogeneous ones are involved at high temperatures. Because oxygenates were identified during the ODH of ethane, a series of alkyl peroxides and/or alkoxides species are believed to occur as intermediates in the ODH of ethane, which have been described in Refs. [68–70].

4. Conclusion

The thermolysis of ethane provided a high conversion of ethane and selectivity to ethene at high temperatures, which were, however, associated with large carbon depositions. The noncatalytic ODH of ethane also provided high activity and selectivity at high temperatures, but much lower at low temperatures, particularly when a large fraction of the reactor was occupied by inert silica granules. The Meso-VMg catalysts were selective and in most case more active than the Mix-VMg ones. For both Mix-VMg and Meso-VMg catalysts, higher magnesium contents provided in most cases higher ethane conversions but lower selectivities to ethene. A higher fraction of ethane was converted to oxygenates over the Meso-VMg catalysts than over the Mix-VMg ones. The higher performance for ODH of ethane the Meso-VMg catalysts than the Mix-VMg ones may be due to the V_2O_3 phase containing highly dispersed magnesium species and possessing large specific surface areas in the former cases. The Mg species probably moderate the redox capability of V_2O_3, thus controlling the activations of ethane and oxygen. The activation mechanism of ethane over these catalysts is dependent on temperature. Heterogeneous processes occur at low temperatures, whereas heterogeneous–homogeneous ones account for the behavior of the catalysts at high temperatures.

References

[1] L. Kniel, O. Winter, K. Stork, Ethylene: Keystone to the Petrochemical Industry, Dekker, New York, 1980.
[2] O.J. Velle, A. Anderson, K.J. Jens, Catal. Today 6 (1990) 567.
[3] V.R. Choudhary, B.S. Uphade, S.A.R. Mulla, Angew. Chem. Int. Ed. Engl. 34 (1995) 665.
[4] S.A.R. Mulla, O.V. Buyevskaya, M. Baerns, Appl. Catal. A 226 (2002) 73.
[5] L. Ji, J. Liu, Chem. Commun. (1996) 1203.
[6] S.J. Conway, D.J. Wang, J.H. Lunsford, Appl. Catal. A 79 (1991) L1.
[7] X.P. Zhou, Z.S. Chao, H.L. Wan, K.R. Tsai, Appl. Catal. A 133 (1995) 263.
[8] Z.S. Chao, X.P. Zhou, H.L. Wan, K.R. Tsai, China patent, ZL 94 1 17229 5, B01J 27/12.
[9] W. Ueda, S.W. Lin, I. Tohmoto, Catal. Lett. 44 (1997) 241.
[10] E.M. Thorsteinson, T.P. Wilson, F.G. Young, P.H. Kasai, J. Catal. 52 (1978) 116.
[11] F. Donsi, R. Pirone, G. Russo, J. Catal. 209 (2002) 51.
[12] R. Burch, S.C. Tsang, Appl. Catal. 65 (1990) 259.
[13] R. Burch, E.M. Crabb, Appl. Catal. A 97 (1993) 49.
[14] P. Ciambelli, L. Lisi, R. Pirone, G. Ruoppolo, G. Russo, Catal. Today 61 (2000) 317.
[15] S.A.R. Mulla, O.V. Buyevskaya, M. Baerns, J. Catal. 197 (2001) 43.
[16] S.A.R. Mulla, O.V. Buyevskaya, M. Baerns, Appl. Catal. A 226 (2002) 73.
[17] K. Otsuka, Catal. Today 13 (1992) 667.
[18] C.N. Satterfield, Heterogeneous Catalysis in Industrial Practice, McGraw-Hill, New York, 1991.
[19] M.V. Twigg, Catalyst Handbook, Manson Publishing, London, 1996.
[20] M. Roy, M. Gubelmann-Bonneau, H. Ponceblanc, J.C. Volta, Catal. Lett. 42 (1996) 93.
[21] W. Ueda, K. Oshihara, Appl. Catal. A 200 (2000) 135.
[22] L. Tessier, E. Bordes, M.G. Bonneau, Catal. Today 24 (1995) 335.
[23] Y. Wang, K. Otsuka, J. Catal. 171 (1997) 106.
[24] Y. Uragami, K. Otsuka, J. Chem. Soc., Faraday Trans. 88 (1992) 3605.
[25] N.U. Zhanpeison, K. Otsuka, React. Kinet. Catal. Lett. 52 (1994) 27.
[26] J.S. Beck, J.C. Vartuli, W.J. Roth, M.E. Leonowicz, C.T. Kresge, K.D. Schmitt, C.T.W. Chu, D.H. Olson, E.W. Sheppard, S.B. McCullen, J.B. Higgins, J.L. Schlenker, J. Am. Chem. Soc. 114 (1992) 10834.
[27] P. Behrens, Adv. Mater. 5 (1993) 127.
[28] A. Sayari, Chem. Mater. 8 (1996) 1840.
[29] A. Corma, Chem. Rev. 97 (1997) 2373.
[30] D. Brunel, Micropor. Mesopor. Mater. 27 (1999) 329.
[31] J.Y. Ying, C.P. Mehnert, M.D. Wong, Angew. Chem. Int. Ed. 38 (1999) 56.
[32] P. Selvam, S.K. Bhatia, C.G. Sonwane, Ind. Eng. Chem. Res. 40 (2001) 3237.
[33] Q. Zhang, Y. Wang, Y. Ohishi, T. Shishido, K. Takehira, J. Catal. 202 (2001) 308.
[34] B. Solsona, T. Blasco, J.M. Lopez Nieto, M.L. Pena, F. Rey, A. Vidal-Moya, J. Catal. 203 (2001) 443.
[35] Z.S. Chao, E. Ruckenstein, Langmuir 18 (2002) 734.
[36] Z.S. Chao, E. Ruckenstein, Chem. Mater. 14 (2002) 4611.
[37] Z.S. Chao, E. Ruckenstein, Catal. Lett. 88 (2003) 147.
[38] M.A. Chaar, D. Patel, H.H. Kung, J. Catal. 109 (1988) 463.
[39] M.A. Chaar, D. Patel, H.H. Kung, J. Catal. 105 (1987) 483.
[40] P.M. Michalkos, M.N. Kung, I. Jahan, H.H. Kung, J. Catal. 140 (1993) 226.
[41] D.S.H. Sam, V. Soenen, J.C. Volta, J. Catal. 123 (1990) 417.
[42] B. Solsona, A. Dejoz, M.I. Vazquez, F. Marquez, J.M.L. Nieto, Appl. Catal. A 208 (2001) 99.
[43] A. Corma, J.M.L. Nieto, N. Paredes, J. Catal. 144 (1993) 425.
[44] X. Gao, P. Ruiz, Q. Xin, X. Guo, B. Delmon, J. Catal. 148 (1994) 56.
[45] S.R.G.C. Peres, J.P. Bernard, M. Ruwet, P. Ruiz, B. Delmon, J. Catal. 158 (1996) 452.
[46] A. Pantazidis, A. Auroux, J.M. Herrmann, C. Mirodatos, Catal. Today 32 (1996) 81.
[47] D.L. Stern, J.N. Michaels, L. Decaul, R.K. Grasselli, Appl. Catal. A 153 (1997) 21.
[48] H.H. Kung, M.C. Kung, Appl. Catal. A 157 (1997) 105.
[49] J.M.L. Nieto, A. Dejoz, M.I. Vasquez, W.O. Leary, J. Cunningham, Catal. Today 40 (1998) 215.
[50] J.M.L. Nieto, J. Soler, P. Concepcion, J. Herguido, M. Menedez, J. Santamaria, J. Catal. 185 (1999) 324.
[51] R.X. Valenzuela, V.C. Corberan, Top. Catal. 11 (2000) 153.
[52] A. Pantazidis, S.A. Bucholz, H.W. Zanthoff, Y. Schuurman, C. Mirodatos, Catal. Today 40 (1998) 207.
[53] M. Fathi, R. Lodeng, E.S. Nilsen, B. Silberova, A. Holmen, Catal. Today 64 (2001) 113.
[54] R. Dula, K. Wcislo, J. Stoch, B. Grzybowska, E.M. Serwicka, F. Kooli, K. Bahranowski, A. Gawel, Appl. Catal. A 230 (2002) 281.
[55] C. Pak, A.T. Bell, T.D. Tilley, J. Catal. 206 (2002) 49.
[56] A.A. Lemonidou, G.J. Tjatjopoulos, I.A. Vasalos, Catal. Today 45 (1998) 65.
[57] A.A. Lemonidou, A.E. Stambouli, Appl. Catal. A 171 (1998) 325.
[58] A.A. Lemonidou, Appl. Catal. A 216 (2001) 277.
[59] T. Blasco, J.M.L. Nieto, Appl. Catal. A 157 (1997) 117.
[60] H.H. Kung, Adv. Catal. 40 (1994) 1.

[61] P.J. Anderson, H.H. Kung, J. Phys. Chem. 96 (1992) 3114.
[62] P.J. Anderson, H.H. Kung, in: L. Guczi, F. Solymosi, P. Teteny (Eds.), New Frontiers in Catalysis, Elsevier, Amsterdam, 1993, p. 205.
[63] P. Rybarczyk, H. Berndt, J. Radmik, M.M. Pohl, O.V. Buyevskaya, M. Baerns, A. Bruckner, J. Catal. 202 (2001) 45.
[64] J. Haber, in: E.G. Derouane, J. Haber, F. Lemos, F.R. Riberiro, M. Guisnet (Eds.), Catalytic Activation and Functionalisation of Light Alkanes: Advances and Challenges, NATO ASI Series, Kluwer Academic, Dordrecht, 1998.
[65] T.E. Wachs, J.M. Jehng, G. Deo, B.M. Weckhuysen, V.V. Guliants, J.B. Benziger, Catal. Today 32 (1996) 47.
[66] G. Deo, I.E. Wachs, J. Catal. 146 (1994) 323.
[67] G.G. Cortez, M.A. Banares, J. Catal. 209 (2002) 197.
[68] F. Cavani, F. Trifiro, Catal. Today 51 (1999) 561.
[69] K.J. Laidler, Chemical Kinetics, McGram-Hill, New York, 1950, p. 333.
[70] A.F. Wagner, I.R. Slagle, D. Sarzynski, D. Gutman, J. Phys. Chem. (1990) 1853.

V–Mg–O prepared via a mesoporous pathway: a low-temperature catalyst for the oxidative dehydrogenation of propane to propene

Zi-Sheng Chao and Eli Ruckenstein*

Department of Chemical Engineering, State University of New York at Buffalo, Amherst, NY 14260, USA

Received 14 January 2004; accepted 3 March 2004

The oxidative dehydrogenation of propane was investigated in a fixed-bed tubular flow reactor at 350 and 450 °C over V–Mg–O catalysts prepared via a solid reaction (the mix-VMg catalysts) or from mesoporous precursors (the meso-VMg catalysts). Significantly higher selectivities and yields to propene were obtained over the meso-VMg catalysts than over the mix-VMg ones. The cooperation between a much higher BET specific surface area and a novel phase, a highly dispersed-Mg-in-V_2O_3, is suggested as responsible for the higher ODP performance of the meso-VMg catalysts than of the mix-VMg ones.

KEY WORDS: mesoporous; vanadium; magnesium; catalyst; oxidative dehydrogenation; propane; propene.

1. Introduction

The oxidative dehydrogenation of propane (ODP) to propene has received extensive attention, because of the increased demand for propene. Compared to the commercial production of propene via non-oxidative dehydrogenation [1–3], the ODP process is potentially advantageous, because the reaction is exothermic and employs a cheaper and more abundant raw material. The ODP reaction has a relatively low selectivity, because propene is more reactive than propane and is prone to side reactions, such as cracking and combustion. At moderate temperatures, these processes are mainly heterogeneous and could be controlled by the appropriate choice of a catalyst. The development of such a catalyst with a sufficiently high activity and selectivity at low temperatures is thus highly desirable. Many catalysts have been reported for the ODP reaction and the V–Mg–O system appears to be among the most promising ones [4–23]. Most of the results obtained with the latter catalysts were at temperatures higher than 500 °C, and only few were concerned with 500 °C or lower temperatures [21–23]. Very small conversions and yields were obtained at low temperatures. The present paper deals with the ODP reaction at 350 and 450 °C over V–Mg–O catalysts of various V/Mg atomic ratios, prepared via a solid reaction (mix-VMg) or from mesoporous precursors (meso-VMg). The results have clearly demonstrated that the catalysts based on mesoporous V–Mg–O are more active and selective than those prepared through a solid reaction. The yields provided by the latter catalysts (mix-VMg) are comparable to those reached previously [21–23].

* To whom correspondence should be addressed.
E-mail: feaeliru@acsu.buffalo.edu

2. Experimental

Two groups of catalysts, namely mix-VMg and meso-VMg, were prepared and the information regarding their preparation and characterization is summarized in tables 1 and 2. The mix VMg catalysts were prepared via a solid reaction between V_2O_5 and $MgCl_2$ by mixing their powders and heating in air at 750 °C for 2 h after they were heated at a rate of 4 °C/min from room temperature. The mesoporous V–Mg–O were synthesized at room temperature by dispersing vanadium acetylacetonate ($V(acac)_3$) into distilled water, and dissolving $MgCl_2$ and the template (cetyltrimethylammonium bromide (CTAB) or an equimolar mixture of CTAB and sodium dodecylbenzene sulfonate (SDBS)) into a HCl aqueous solution, and then mixing the two systems. The mixture was continuously stirred for 24 h and then allowed to age statically for 2 days. The pH was adjusted to 10 during aging. A mesoporous vanadium oxide was also prepared from V_2O_5 and an equimolar mixture of CTAB and hexadecylamine (HDA) using a similar procedure, except that V_2O_5 was dissolved into a NaOH aqueous solution, and the pH was adjusted to 7 during aging. The solid specimens were recovered by filtering, washing and drying and used for characterization and preparation of the catalysts. The meso-VMg catalysts were prepared by calcining the mesoporous V–Mg–O precursors in a flow of argon at 750 °C for 4 h after they were heated at a rate of 10 °C/min from room temperature.

The phases present in the specimens were identified by X-ray diffraction (XRD) on a SIEMENS D500 diffractometer with a CuK_α radiation of 1.5406 Å. The diffraction data were recorded for 2θ angles ranging between 1 and 50°, with a resolution of 0.02°. The textural properties of the catalysts were determined by N_2 adsorption using a Micromeritics ASAP 2010 Gas

This paper, which was originally published in **Catalysis Letters**: 94(3–4), 217–221 (2004), is republished with permission.

Table 1
Preparation conditions, phases and properties of the mesoporous V–Mg oxides and the corresponding meso-VMg catalysts

Catalysts	Mole composition in initial batch	V/Mg atomic ratio		S_{BET}[a] (m^2/g)	PV[b] (cm^3/g)	APS[c] (nm)	Phases identified by XRD	
		AAS	XPS				As-synthesized	Calcination at 750 °C
meso-VMg-1	$1.6V(acac)_3 : 0.4MgCl_2 : 0.25CTAB : 295H_2O$	48.23	6.82	151.0	0.17	4.51	Hexagonal	V_2O_3
meso-VMg-2	$1.2V(acac)_3 : 0.8MgCl_2 : 0.25CTAB : 295H_2O$	35.29	0.82	155.1	0.18	4.66	Hexagonal	V_2O_3
meso-VMg-3	$0.6V(acac)_3 : 1.2MgCl_2 : 0.25CTAB : 295H_2O$	10.65	7.06	159.1	0.16	4.05	Hexagonal	V_2O_3
meso-VMg-4	$0.4V(acac)_3 : 1.6MgCl_2 : 0.25CTAB : 295H_2O$	1.39	2.09	159.8	0.19	4.77	Hexagonal	V_2O_3, $Mg_{1.5}VO_4$, MgO
meso-VMg-5	$V(acac)_3 : MgCl_2 : (0.125CTAB+0.125SDBS) : 295H_2O$	3.19	5.12	158.5	0.16	3.99	Hexagonal	V_2O_3, $Mg_{1.5}VO_4$, MgO
meso-V	$V_2O_5 : (0.125CTAB+0.125HDA) : 295H_2O$	–	–	148.6	0.28	7.56	Lamellar	V_2O_3

[a] S_{BET}: BET specific surface area.
[b] PV: pore volume.
[c] APS: average pore size.

Sorption and Porosimetry instrument. The specific surface area was determined by the BET method, the pore volume was calculated at a relative pressure of $P/P_0 > 0.99$, and the pore size distribution was calculated by the BJH method. The bulk V/Mg atomic ratios of the catalysts were determined by Atomic Absorption Spectrophotometry (AAS) using a Perkin-Elmer model AAS 3030 spectrophotometer. The surface V/Mg atomic ratios were determined by X-ray photoelectron spectroscopy (XPS) with a VG-ESCA lab-210 spectrometer, using MgK_α radiation as the excitation source.

The results of the characterization are listed in table 1 for the meso-VMg catalysts and in table 2 for the mix-VMg catalysts.

The catalytic tests for the ODP reaction were carried out at atmospheric pressure in a fixed-bed quartz tubular flow reactor (internal diameter = 5 mm and operation length = 30 cm). A small amount (0.2 g) of catalyst was loaded in the middle of the reactor, and above the catalyst bed quartz granules were introduced to quench the possible homogeneous reactions. A feed mixture with a mole composition of 90% propane, 4% oxygen and 6% nitrogen was continuously passed through the reactor, which was heated at 350 or 450 °C. The compositions of the feed and the product mixtures were analyzed on-line using a SRI 8610C gas chromatograph equipped with a thermal conductivity detector (TCD) and a flame ionization detector (FID). A Hayesep DB (30′ × 1/8″) packed column was employed to separate the mixtures.

3. Results and discussion

The ODP reactivities at 350 and 450 °C over V_2O_5, MgO and the mix-VMg catalysts are listed in table 3. No conversions in an empty reactor and in a reactor filled with quartz granules were detected. Consequently, any conversion of propane in the presence of a catalyst was the result of the catalytic process. The pure V_2O_5 exhibited a somewhat higher ODP performance than the pure MgO, with a higher selectivity to propene and a slightly higher conversion of propane. Compared to the pure V_2O_5, the mix-VMg catalysts provided much higher conversions of propane, but lower selectivities to propene. For all mix-VMg catalysts, the elevation of the reaction temperature from 350 to 450 °C increased both the conversion of propane and the selectivity to propene, and the increase of the flow rate improved the selectivity to propene but reduced the conversion of propane. With the decrease of V/Mg atomic ratios in the mix-VMg catalysts, the conversion of propane increased but the selectivity to propene passed through a maximum. The highest selectivity to propene was provided by the mix-VMg-3 catalyst. The mix-VMg catalysts provided yields comparable to those obtained previously with similar V–Mg–O catalysts [21–23].

Table 2
Properties of mix-VMg catalysts and the phases they contained

Catalyst	V/Mg atomic ratio	S_{BET}[a] (m^2/g)	PV[b] (cm^3/g)	APS[c] (nm)	Phase identified by XRD[d]
V_2O_5	∞	4.75	0.010	8.28	V_2O_5
mix-VMg-1	4.0	1.02	0.0028	10.83	V_2O_5 and *m*
mix-VMg-2	2.0	2.19	0.0071	12.65	V_2O_5 and *m*
mix-VMg-3	1.0	3.36	0.0017	10.63	MgO, *m* and *p*
mix-VMg-4	0.5	6.54	0.0017	10.61	MgO, *m* and *p*
mix-VMg-5	0.25	7.16	0.023	12.78	MgO, *m*, *p* and *o*
MgO	0	14.10	0.043	12.20	MgO

[a–c]See table 1.
[d]*m*, *p* and *o* refer to *m*-MgV_2O_6, *p*-$Mg_2V_2O_7$, *o*-$Mg_3(VO_4)_2$, respectively.

Table 3
ODP reactivities over mix-VMg catalysts

Catalysts	F[a] (mL/s)	X[b] (%)	Selectivity (%)						Y[c] (%)
			CO_x	CH_4	C_2H_4	C_2H_6	C_3H_6	C_{4+}	
350 °C									
V_2O_5	5.63	0.05	0.00	0.00	0.00	54.63	45.37	0.00	0.02
	2.37	0.44	51.00	1.00	0.00	2.77	41.21	4.02	0.18
MgO	5.63	0.00	0.00	0.00	0.00	0.00	0.00	0.00	0.00
	2.37	0.00	0.00	0.00	0.00	0.00	0.00	0.00	0.00
mix-VMg-1	2.37	3.81	81.77	0.00	0.19	6.96	5.43	5.65	0.21
mix-VMg-2	2.37	4.68	76.37	0.00	0.31	5.49	15.48	2.35	0.72
mix-VMg-3	2.37	4.80	73.90	0.00	0.56	5.42	18.55	1.57	0.89
mix-VMg-4	2.37	6.85	81.40	0.00	0.73	4.84	11.48	1.55	0.79
mix-VMg-5	2.37	7.03	86.88	0.00	0.98	3.61	6.99	1.54	0.49
mix-VMg-5	5.63	1.56	49.98	0.25	0.50	6.42	33.44	9.41	0.52
450 °C									0.00
V_2O_5	5.63	0.33	0.00	0.75	6.20	37.73	50.32	5.00	0.17
	2.37	2.01	53.88	0.29	0.71	2.21	39.52	3.39	0.79
MgO	5.63	0.06	14.01	1.78	14.28	28.91	35.44	5.58	0.02
	2.37	0.15	24.00	2.29	16.57	12.53	30.75	13.87	0.05
mix-VMg-1	2.37	4.88	77.19	0.00	0.31	5.21	12.51	4.78	0.61
mix-VMg-2	2.37	5.72	61.88	0.00	1.10	4.71	30.00	2.31	1.72
mix-VMg-3	2.37	6.58	49.33	0.00	1.50	3.73	43.89	1.55	2.89
mix-VMg-4	2.37	7.59	73.50	0.31	1.91	3.44	19.55	1.29	1.48
mix-VMg-5	2.37	8.34	76.74	0.57	2.31	2.34	17.31	0.73	1.44
mix-VMg-5	5.63	4.11	38	0.26	7.59	4.59	44.81	4.75	1.84

[a]*F*: flow rate.
[b]*X*: conversion of propane.
[c]*Y*: yield of propene.

The ODP reactivities at 350 and 450 °C over the meso-VMg catalysts are listed in table 4. It was found that the meso-V provided a higher conversion of propane but a lower selectivity to propene than V_2O_5. At the temperatures of 350 and 450 °C and a flow rate of 5.63 mL/s, propane conversions of 1.08% and 1.31% and propene selectivities of 7.62% and 14.46%, respectively, were attained over meso-V. However, over the meso-VMg catalysts, the ODP performance was significantly improved. At the temperatures of 350 and 450 °C, propane conversions of 7.73% and 9.96% and propene selectivities of 89.22% and 72.13%, respectively, were reached over the meso-VMg-5 catalyst. When the temperature was increased from 350 to 450 °C, the conversion of propane increased for all V/Mg bulk atomic ratios. However, the selectivity to propene increased with increasing temperature for the catalysts with bulk atomic ratios of V/Mg > 10 (meso-VMg-1, meso-VMg-2 and meso-VMg-3), but decreased for the other catalysts (meso-VMg-4 and meso-VMg-5). With increasing V/Mg bulk atomic ratio, the conversion of propane decreased. High selectivities to propene were achieved for most V/Mg bulk ratios. The highest selectivities were obtained over the catalysts with smaller V/Mg surface ratios. The increase in the flow rate increased the selectivity to propene but reduced the propane conversion.

Comparing tables 3 and 4, one can conclude that the meso-VMg catalysts provided much higher ODP

Table 4
ODP reactivities over meso-VMg catalysts

Catalysts	F^a (mL/s)	X^b (%)	Selectivity (%)						Y^c (%)
			CO_x	CH_4	C_2H_4	C_2H_6	C_3H_6	C_{4+}	
350 °C									
meso-V	5.63	1.08	88.06	0.03	0.07	2.32	7.62	1.89	0.08
meso-V	2.37	2.55	88.67	0.05	0.10	3.58	5.49	2.11	0.14
meso-VMg-1	2.37	3.87	79.13	0.05	0.25	1.01	19.36	0.20	0.75
meso-VMg-2	2.37	8.48	0	0	2.12	17.78	75.07	5.03	6.36
meso-VMg-3	2.37	7.37	0	0	1.66	30.49	55.96	11.89	4.12
meso-VMg-4	2.37	7.84	0	0	1.67	20.06	72.83	5.44	5.71
meso-VMg-5	2.37	8.11	0	0	0	35.03	51.61	13.36	4.19
meso-VMg-5	5.63	7.73	0	0	0	5.85	89.22	4.94	6.90
450 °C									
meso-V	5.63	1.31	83.68	0.06	0.36	0.69	14.46	0.76	0.19
meso-V	2.37	7.08	86.94	0.13	0.41	1.36	10.87	0.29	0.77
meso-VMg-1	2.37	8.61	69.15	0.10	0.38	0.82	29.37	0.18	2.53
meso-VMg-2	2.37	10.60	15.10	0.10	0.96	2.33	80.75	0.75	8.56
meso-VMg-3	2.37	11.50	0	0	2.83	15.49	80.61	1.06	9.27
meso-VMg-4	2.37	11.13	41.28	0	0.80	1.69	55.84	0.39	6.21
meso-VMg-5	2.37	10.75	0	0.68	1.51	34.82	45.84	17.15	4.93
meso-VMg-5	5.63	9.96	22.04	0	0.43	2.12	72.13	3.28	7.18

[a] *F*: flow rate.
[b] *X*: conversion of propane.
[c] *Y*: yield of propene.

performances, particularly for the selectivity to propene, than the mix-VMg catalysts. The difference between the reactivities of the meso-VMg and mix-VMg catalysts is probably caused by their different phase structures and textural properties. As one can see from tables 1 and 2, all the meso-VMg catalysts possessed much larger specific surface areas and pore volumes than the mix-VMg catalysts. However, this does not mean that the higher performance of the meso-VMg catalysts is solely caused by the increased specific surface area and pore volumes. In fact, as already noted, the meso-V catalyst (BET area = 148.6 m^2/g and pore volume = 0.28 cm^3/g) did not perform better than any of the mix-VMg catalysts (BET area ≤ 7.16 m^2/g and pore volume ≤ 0.023 cm^3/g). This hints that the phase structure of the catalyst constitutes an equally important factor that affects the ODP performance. From table 2, one can see that depending on the V/Mg atomic ratio, the mix-VMg catalysts contained either a V_2O_5 or a MgO phase together with one or more magnesium vanadate phases. In the meso-VMg catalysts, only the V_2O_3 phase could be identified by XRD for bulk atomic ratios of V/Mg > 10, and V_2O_3 and MgO together with an additional magnesium vanadate, $Mg_{1.5}VO_4$, for the catalysts with V/Mg < 10. The fact that none of Mg-containing phases could be detected by XRD for bulk atomic ratios of V/Mg > 10 may mean that magnesium was highly dispersed into or onto the V_2O_3 phase. XPS investigations have provided proof in this direction. As shown in table 1, the surface atomic ratios of V/Mg were smaller for the meso-VMg-1, meso-VMg-2 and meso-VMg-3 (for which the bulk atomic ratios of V/Mg > 10), but larger for the meso-VMg-4 and meso-VMg-5 (for which the bulk atomic ratios of V/Mg < 10) than the corresponding bulk atomic ratios. This indicates that there is an enrichment in Mg on the surface of the meso-VMg catalysts for bulk atomic ratios of V/Mg > 10, but not for the meso-VMg catalysts with bulk atomic ratios of V/Mg < 10. Thus, Mg must be highly dispersed onto the V_2O_3 phase in the former meso-VMg catalysts; otherwise, the Mg-containing phases would have been detected by XRD in the meso-VMg catalysts with bulk atomic ratios of V/Mg > 10. Consequently, the high specific surface area associated with the meso-VMg catalysts may provide a medium in which Mg species could become highly dispersed onto the V_2O_3 phase. In this manner, an active phase, which can provide active sites, for example, some active lattice oxygen species or active vanadium centers, may be formed. In addition, the high pore volume of the meso-VMg catalysts can facilitate the diffusion and thus improve the selectivity of the ODP reaction. These conclusions are similar to those reached by us regarding the oxidative dehydrogenation of ethane [24,25]. Consequently, it is expected the meso-VMg catalysts to be high-performance ODP catalysts.

4. Conclusion

The Meso-VMg catalysts, prepared by calcining mesoporous V–Mg–O precursors, were more selective and active for the ODP reaction at the low temperatures of 350 and 450 °C employed in the present paper than the mix-VMg ones, prepared via a solid reaction between vanadia and magnesia. The high specific surface areas of the meso-VMg catalysts provided a medium in which Mg species could be highly dispersed over or into the V_2O_3 phase. The highly dispersed-Mg-in-V_2O_3 might have generated active sites, such as some lattice oxygen species or vanadium centers that enhanced the ODP reaction. In addition, the high porosity promoted the diffusion of the species involved in the reaction. These characteristics of the meso-VMg catalysts make them promising candidates as high-performance ODP catalysts.

References

[1] L.F. Albright, B.L. Crynes and W.H. Corcoran, (eds), *Pyrolysis: Theory and Industrial Practice* (Academic Press, London, 1983).
[2] D.E. Resasco and G.L. Haller, Catalysis (London), 11 (1994) 379.
[3] P.B. Venuto and E.T. Habib, *Fluid Catalytic Cracking with Zeolite Catalysts* (Marcel Dekker, New York, 1979).
[4] M.A. Chaar, D. Patel, M.C. Kung and H.H. Kung, J. Catal. 109 (1988) 463.
[5] M. Fathi, R. Lodeng, E.S. Nilsen, B. Silberova and A. Holmen, Catal. Today 64 (2001) 113.
[6] H.H. Kung, and M.C. Kung, Appl. Catal. A: Gen. 157 (1997) 105.
[7] H.H. Kung, Adv. Catal. 40 (1994) 1.
[8] A.A. Lemonidou, G.J. Tjatjopoulos and I.A. Vasalos, Catal. Today 45 (1998) 65.
[9] M.A. Chaar, D. Patel, M.C. Kung and H.H. Kung, J. Catal. 105 (1987) 483.
[10] A. Corma, J.M Lopez Nieto and N. Paredes, J. Catal. 144 (1993) 425.
[11] X. Gao, P. Ruiz, Q.Xin, X. Guo and B. Delmon, J. Catal. 148 (1994) 56.
[12] A. Pantazidis and C. Mirodatos, ACS Symposium Ser. 638 (1996) 207.
[13] D. Siew Hew Sam, V. Soemen and J.C. Volta, J. Catal. 123 (1990) 417.
[14] M.A. Banaras, Catal. Today 51 (1999) 319.
[15] P.M.Michalakos, M.C. Kung, I. Jahan, and H.H. Kung, J. Catal. 140 (1993) 226.
[16] A. Pantazidis, A. Aurox, J.M. Herrmann and C. Mirodatos, Catal. Today. 32 (1996) 81.
[17] A. Burrows, C. J.Kiely, J. Perregaard, P.E. Hojlund-Nielsen, G. Vorbeck, J.J. Calvino and C. Lopez-Cartes, Catal. Lett., 57 (1999) 121.
[18] B. Solsona, A. Dejoz, M.I. Vazquez, F. Marquez and J.M. Lopez Nieto, Appl. Catal. A: Gen. 208 (2001) 99.
[19] M.A. Chaar, D. Patel and H.H. Kung, J. Catal., 109 (1988) 463.
[20] A. Pantazidis, A. Burrows and C.J. Kiely, C. Mirodatos, J. Catal. 177 (1998) 325.
[21] K. Bahranowski, G. Bueno, V.C. Cortes, F. Kooli, E.M. Serwicka, R.X. Valenzuela and K. Wcislo, App. Catal. A: General, 185 (1999) 65.
[22] G. Martra, F. Arena, S. Coluccia, F. Frusteri and A. Parmaliana, Catal. Today 63 (2000) 197.
[23] C. Pak, A.T. Bell and T.D. Tilley, J. Catal. 206 (2002) 49.
[24] Z.S. Chao and E. Ruckenstein, Catal. Lett. 88 (2003) 147.
[25] Z.S. Chao and E. Ruckenstein, J. Catal. 222 (2004) 17.

6

SYNTHESIS OF POLYMER-SUPPORTED CATALYSTS AND POLYMER-COATED SILICA SUPPORTS AND THEIR APPLICATIONS IN CATALYSIS

INTRODUCTION

The use of phase transfer catalysts (PTCs) to ameliorate the activity and selectivity of organometallic catalysts for organic synthesis, such as carbonylation, vinylation, oxidation, and reduction, represents a new trend in the field of catalysis [1]. A PTC is a catalyst that facilitates the migration of a reactant from one phase into another phase where reaction occurs. Most of the previous investigations regarding organic reactions mediated by organometallic compounds under PTC conditions were carried out using homogeneous catalysts [2, 3]. As a result, the use of very expensive organometallic compounds (usually they contain noble metal ions or sophisticated ligands) created the problem of catalyst recovery. To this end, numerous strategies have been attempted, such as tethering to inorganic and organic supports [4–7], use of liquid/liquid biphasic extractions [8], and solvent-free transformations [9], with the immobilization of the catalysts on inorganic or organic supports representing the bulk of the literature reports. Nevertheless, supported catalysts often provide lower activities (diffusion effects) and selectivities (site heterogeneity) when compared with their nonsupported equivalents. It is therefore highly desirable if one can not only recover supported metal catalysts but also achieve the desired catalytic activity and selectivity. In this regard, it is believed that synergy between the support and the catalytic moiety and easy tunability of the support are keys to achieve this goal [7]. Catalyst supports based on organic polymers ranging from insoluble resins to highly soluble oligomers are perhaps among the most versatile supports in the literature. In particular, soluble polymer supports are advantageous because of the development of controlled and living polymerization methods that are tolerant to a wide variety of functional groups, including controlled radical polymerizations [10–12] and ring-opening metathesis polymerization [13–15]. These methods allow for tuning the density and structure of the catalyst sites along the polymer chain and the easy incorporation of catalysts into copolymer structures, thereby enabling the development of structure–property relationships between a catalyst and its polymer support and ultimately the improvement of catalytic activity of the metal complexes.

Chapter 6 is accordingly devoted to the synthesis of polymer-supported catalysts and their applications. It consists of two parts: the first is concerned with polymer-supported PTCs (**6.1**, **6.2**), whereas the second with polymer-coated silica supports (**6.3**, **6.4**, and **6.5**).

For processes where organic reactions are mediated by organometallic compounds under PTC conditions,

Heterogeneous Catalysis: Experimental and Theoretical Studies, First Edition. Edited by Eli Ruckenstein and Haiyou Wang.

heterogenization by attaching both catalysts to the same polymer support provides a reaction environment that can allow a more effective cooperation between two kinds of active sites. In this regard, three heterogeneous pathways were proposed: (1) palladium sulfonated phenyl phosphine complexes are bound to a polymer matrix containing pendant benzyl tributyl ammonium groups; (2) palladium phosphate complexes are bound to a polymer matrix containing pendant benzyl methyl diethanol ammonium groups; and (3) palladium phosphine oxide complexes together with benzyl tributyl ammonium or phosphonium groups are supported on the same polymer substrate (**6.1**). The quaternary onium group and palladium phosphorus complex are bound to the same polymer matrix, thereby generating bifunctional polymer-supported catalysts. These three types of polymer-supported bifunctional catalysts were employed in the vinylation reaction of iodobenzene with methylacrylate. The yield of vinylation was much larger than in homogeneous systems because of the cooperation between palladium phosphine oxide complexes and quaternary onium groups. Polymer-supported PTCs containing various amounts of hydrophilic quaternary onium groups bound to the surface of a hydrophobic cross-linked polystyrene were also prepared. In such catalysts, the dual-surface characteristics due to the hydrophilic-bound cations and hydrophobic polystyrene can be varied by controlling the partial polymerization time of styrene/divinylbenzene and the amount of vinylbenzyl chloride (**6.2**). The effect of the accessible hydrophilicity of the supported catalyst on the activity and selectivity of the alkylation of isopropylidene malonate with ethyl bromoacetate was investigated.

One of techniques of heterogenization of the homogeneous catalysts is through the use of water-soluble complexes in water–oil biphase catalytic systems. In this regard, polymer chains of quarternary oniums were deposited on the surface of hydrophilic silica by the radical polymerization of water-soluble quarternary onium monomers present in a water medium which is adsorbed by the porous silica (**6.3**). The modified (quarternaized) silica, which contains a small amount of water, is a better support than pristine silica since it provides cation anchors for negatively charged catalytic species. It was used as substrate for the immobilization of monosulfonated triphenylphosphine and of its rhodium carbonyl complex. The immobilized catalytic system containing a small amount of water was active in the hydroformylation of styrene and ethyl 10-undecenoate in cyclohexane. A higher reaction rate was achieved over a supported liquid polymer phase catalyst in which (i) poly (hydroxyethyl methacrylate) cross-linked with N, N′-methylenebisacrylamide was deposited on a porous silica support denoted as P(HEMA-MBAM)/SiO_2, and (ii) a liquid polymer phase catalyst consisting of the oligometric ligand α, ω-bis(diphenylphosphino)-poly(ethyl glycol) and its Rh(I) carbonyl complex was coated on a the P(HEMA-MBAM) network. This provided a higher compatibility for a high molecular mass olefin with the liquid phase polymer (**6.4**). Further improvement for high regioselectivity could be achieved by using a new immobilization method that consists of (i) depositing an organic layer, either poly (hydroxyethyl acrylate crosslinked with N, N′-methylenebisacrylamide) or [3-(siloxysilyl) propyl] ethylenediamine, over the surface of silica, and (ii) coating this modified silica with suitable phosphine ligands and their Rh complexes (**6.5**).

REFERENCES

[1] Gore, E.S., *Platinum Met. Rev.*, **34**, 2 (1990).

[2] Alper, H., *Pure Appl. Chem.*, 60, 35 (1988).

[3] Miura, M., Hashimoto, H., Itoh, K., and Nomura, M., *J. Chem. Soc. Perkin. Trans.*, 1, 2207 (1990).

[4] Takizawa, S., Patil, M.L., Marubayashi, K., and Sasai, H., *Tetrahedron*, **63**, 6512 (2007).

[5] Baleizão, C., and Garcia, H., *Chem. Rev.*, **106**, 3987 (2006).

[6] Barbaro, P., *Chem.–Eur. J.*, **12**, 5666 (2006).

[7] Notestein, J.M., and Katz, A., *Chem.–Eur. J.*, **12**, 3954 (2006).

[8] Bergbreiter, D.E., and Sung, S.D., *Adv. Synth. Catal.*, **348**, 1352 (2006).

[9] Walsh, P.J.; Li, H., and de Parrodi, C.A., *Chem. Rev.*, **107**, 2503 (2007).

[10] Kamigaito, M., Ando, T., and Sawamoto, M.,*Chem. Rev.*, **101**, 3689 (2001).

[11] Hawker, C.J., *Acc. Chem. Res.*, **30**, 373 (1997).

[12] Fischer, H., *Chem. Rev.*, **101**, 3581 (2001).

[13] *Handbook of Metathesis*; Wiley-VCH, Weinheim, Germany, 2003; Vol. 3.

[14] Buchmeiser, M.R., *Chem. Rev.*, **100**, 1565 (2000).

[15] Trnka, T.M., and Grubbs, R.H., *Acc. Chem. Res.*, **34**, 18 (2001).

PREPARATION OF POLYMER-SUPPORTED CATALYSTS CONTAINING PHOSPHORUS PALLADIUM COMPLEXES AND QUATERNARY ONIUM GROUPS, AND THEIR APPLICATION TO THE VINYLATION REACTION

L. HONG and E. RUCKENSTEIN *

Department of Chemical Engineering, State University of New York at Buffalo, Buffalo, NY 14260 (USA)

(Received July 11, 1991; accepted in revised form October 13, 1991)

Micron-sized crosslinked polystyrene [poly(ST-DVB)] particles prepared by the concentrated emulsion polymerization method were coated with nonuniform layers of the following functionalized polymers: (I) poly(3- and 4-vinylbenzyl chloride) [poly(VBC)]; (II) a copolymer of VBC and 4-chlorosytrene [poly(VBC)-(CST)] and; (III) a composite of poly(VBC) and poly(3,4-dichloro-1-butene) [poly(VBC)-poly(DCB)]. A fine powder composed of small aggregates of such particles was obtained by grinding. The polymers thus prepared were converted to three types of reactive substrates containing both organophosphorus ligands and quaternary onium groups: (1) (m-sulfonate phenyl) diphenyl phosphine and poly(vinylbenzyl tributyl ammonium chloride) bound to poly(ST-DVB); (2) diphenyl ethylene phosphite and benzyl methyl diethanol ammonium groups bound to poly(ST-DVB); and (3) phosphine oxide and quaternary benzyl tributyl ammonium or phosphonium groups bound to poly(ST-DVB). The polymer-supported catalysts containing phosphorus palladium complexes and quaternary onium groups were then employed in the vinylation reaction of iodobenzene with methyl acrylate. These supported bicatalysts displayed a cooperative effect that increased the yield of vinylation by a factor greater than two when compared to its homogeneous counterpart. The relative locations of palladium complexes and quaternary onium groups are found to influence the yield.

Keywords: concentrated emulsion polymerization; quaternization; bifunctionalized substrate; vinylation; porous macroparticles

INTRODUCTION

The use of phase transfer catalysis (PTC) to ameliorate the activity and selectivity of

* To whom correspondence should be addressed.

This paper, which was originally published in **Reactive Polymers**: 16, 181–197 (1991/1992), is republished with permission.

organometallic catalysts for organic synthesis, such as carbonylation, vinylation, oxidation and reduction, represents a new trend in the field of catalysis [1]. Most of the previous investigations [2] regarding organic reactions mediated by organometallic compounds under PTC conditions were carried out using homogeneous catalysts. Heterogenization of these processes by attaching both catalysts to the same polymer support provides a reaction environment that can allow a more effective cooperation between the two kinds of active sites [3]. In this paper, we propose three heterogeneous pathways: (1) palladium sulfonated phenyl phosphine complexes are bound to a polymer matrix containing pendant benzyl tributyl ammonium groups; (2) palladium phosphite complexes are bound to a polymer matrix containing pendant benzyl methyl diethanol ammonium groups; and (3) palladium phosphine oxide complexes together with benzyl tributyl ammonium or phosphonium groups are supported on the same polymer substrate.

The polymeric substrate was prepared in two steps: Firstly, agglomerates of particles of crosslinked polystyrene coated with (I) poly(3- and 4-vinylbenzyl chloride), (II) poly-(3- and 4-vinylbenzyl chloride + 4-chlorostyrene) or (III) poly(3- and 4-vinylbenzyl chloride + 3,4-dichloro-1-butene) were prepared by the concentrated emulsion polymerization method described below. These three polymer substrates are denoted in what follows as: I, II and III. Secondly, benzyl tributyl ammonium chloride, benzyl tributyl phosphonium chloride or benzyl methyl diethanol ammonium chloride were bound to the polymer substrate *via* a quaternization reaction. In substrate I, the polymerized VBC units were thus converted to anchored benzyl tributyl ammonium groups or anchored benzyl methyl diethanol ammonium groups, generating two quaternized substrates denoted in what follows as Ia and Ia′, respectively. A fraction of the positively charged anchored onium groups of substrate Ia was further combined with the negatively charged (*m*-sulfonated phenyl) biphenyl phosphine to form polymer-bound ion pairs, which, as shown later, can serve as binding sites for palladium complexes. One can thus prepare a bifunctionalized substrate containing anchored benzyl tributyl ammonium groups as well as sulfonated phenyl diphenyl phosphine groups, which is denoted as Ia(sphP). A fraction of the ethanols of the anchored quaternary ammonium groups of substrate Ia′ was connected to diphenol phosphorus groups *via* the reaction with diphenol phosphorus chloride. The pendant benzyl methyl bis(diphenyl ethylene phosphite) ammonium groups thus obtained can serve as coordination ligands for palladium as well as the phase transfer catalyst. The bifunctionalized substrate containing anchored benzyl methyl diethanol ammonium groups as well as diphenyl ethylene phosphite groups is denoted Ia′($P(OR)_3$). In substrates II and III a fraction of the VBC units was converted to pendant benzyl tributyl ammonium or phosphonium groups. The partially quaternized polymers, which are denoted IIa, IIp, IIIa and IIIp (a refers to ammonium and p to phosphonium), were then treated with diphenyl phosphine lithium. The phosphines thus generated react with the unquaternized chlorides present in the polymer matrix to form aromatic phosphonium species; this reaction is probably catalyzed by the pendant quaternary onium groups of the polymer. Further, their deprotonation by the excess Ph_2PLi converts these species into ylids. Finally, the oxidation of these ylids generates pendant phosphine oxides. The bifunctionalized polymers thus prepared contain both phosphine oxides and quaternary onium groups, and are denoted IIa($P(O)R_3$), IIp ($P(O)R_3$), IIIa($P(O)R_3$) and IIIp($P(O)R_3$) where $P(O)R_3$ refers to phosphine oxide. Compared to the known polymer-supported systems, the reactive polymers prepared by us have: (1) functional groups attached to a polymer layer which coats non-reactive highly crosslinked polystyrene particles of micron size, and (2) trivalent or pentavalent organophosphorus ligands as well as quaternary onium groups anchored to the same polymer matrix. Polymer-supported bifunctional catalysts that are useful for some reactions are thus obtained. Because the reactions take place in the coating layer and the reactants do not have to penetrate deep into the particles, diffusion is not too restrictive.

In this paper, the above listed bifunctionalized polymers are employed as catalytic substrates for the vinylation reaction. In previous papers regarding the vinylation of organic halides (Heck-type reaction) [4], the homogeneous catalytic system containing $Pd(OAc)_2$, NBu_4Cl, $NaHCO_3$ and DMF, proposed by Jeffery [5], has been found to have high activity and stereoselectivity. The heterogenization of this homogeneous catalyst by means of the functionalized polymer substrates listed above constitutes the main goal of the present paper. Three kinds of polymer-supported catalysts were synthesized and employed in the vinylation of iodobenzene: (1) palladium (*m*-sulfonated phenyl) biphenyl phosphine complex bound to the functionalized polymer Ia(sphP); (2) palladium diphenyl ethylene phosphite complexes bound to the functionalized polymer Ia′($P(OR)_3$); and (3) palladium phosphine oxide complexes bound to the functionalized polymers IIa($P(O)R_3$), IIp($P(O)R_3$), IIIa ($P(O)R_3$) and IIIp($P(O)R_3$).

The present paper suggests pathways for the preparation of polymer-bound bicatalysts, and presents results regarding the effects of onium groups, phosphorus ligand of palladium complexes and of relative locations of Pd and onium-sites in the polymer matrix on the vinylation reaction. The results indicate that there are numerous potential advantages of this type of catalysts.

EXPERIMENT

Materials

Vinylbenzyl chloride (VBC) (*meta* 40% and *para* 60% , Kodak), 3,4-dichloro-1-butene (DCB) (98%, Aldrich), 4-chlorostyrene (CST) (97%, Aldrich), styrene (99%), divinylbenzene (DVB) (tech. 55%, Aldrich) were distilled under reduced pressure. Commercial tributylphosphine (TBP) (95%), tributylamine (TBA) (99%), chlorodiphenylphosphine (tech.), phosphorus trichloride (99.9%), phenol (99 + %) and iodobenzene (99%) were purchased from Aldrich and used without further purification. The solvents were commercially available. Lithium ribbon (99.9%), palladium(II) acetate (98%), potassium carbonate (anhydrous), sodium dodecyl sulfonate (SDS) were purchased from Aldrich and used as received.

Instruments

EDS experiments were carried out with a PGT/IMIX instrument, ESCA with a Physical Electronics PHI 5100 equipment, and FT-IR with a Mattson Alpha Centauri FT-IR spectrophotometer. Elemental analyses were performed by Quantitative Technologies, Inc., NJ. HPLC was performed with a column containing Adsorbosphere C8,5μ (Alltech) using an ISCO/2350-HPLC pump and an ISCO/V4 absorbance detector.

SYNTHESES

Typical procedure for the preparation of three types of porous functionalized polystyrene substrates (8.3 mole% of DVB as crosslinking agent)

Porous polymers composed of micron-sized particles having a core of crosslinked polystyrene [poly(ST-DVB)] nonuniformly covered by functionalized layers of poly-(VBC), poly(VBC-CST) or poly(VBC)-poly-(DCB) have been prepared by the concentrated emulsion polymerization method as follows: 29 ml mixture of styrene and DVB (11/1 molar ratio) containing AIBN (0.04 g) as initiator was added dropwise at room temperature to an aqueous solution of SDS (0.92 g SDS dissolved in 8 ml of water) under stirring. The gel-like emulsion thus obtained in the flask was immersed in a water bath (40–45°C) for 6 h for mild polymerization. Either VBC (2.67 g, 17.5 mmol), or a mixture of VBC (2.00 g, 13.2 mmol) and CST (1.83 g, 13.3 mmol), or a mixture of VBC (2.00 g, 13.2 mmol) and DCB (1.76 g, 13.2 mmol) was subsequently added to the partially polymerized concentrated emulsion under vigorous stirring. Then, this gel-like emulsion was packed into a tube by gently tapping. Complete polymerization was carried out by heating the tube at 45°C in a water bath for 48 h. The polymers thus obtained were ground in fine powders, washed in a 200 ml mixture of methanol and acetone (v/v = 4/1) at 50–55°C under stirring for 2 h, and washed again in hot methanol under stirring for 40 min. After filtration, the polymers were dried overnight in vacuum at 50°C. Three different functionalized polymer substrates were prepared: poly(VBC)/poly(ST-DVB); poly(VBC-CST)/ poly(ST-DVB) and poly(VBC)-poly(DCB)/poly(ST-DVB). The "/" sign signifies covering.

Typical procedure for the syntheses of polymer-bound benzyl tributyl ammonium or phosphonium groups

3.0 g (2.73 mmol of Cl) of poly(VBC-CST)/poly(ST-DVB) powder or 3.0 g (1.89 mmol of Cl) of poly(VBC)-poly(DCB)/ poly-(ST-DVB) powder and 1.2 equiv. of either a tertiary amine or phosphine (TBA or TBP) were introduced together with 20 ml DMF in a flask, and the system was heated under stirring and a nitrogen atmosphere at 75–80°C for 60 h. Then, a mixture of 10 ml THF and 40 ml of petroleum ether (30–60°C) was added to the flask. The precipitated polymer was separated by filtration, washed with ethyl ether (15 ml) three times and dried in vacuum for 12 h at 50°C. Four kinds of partially quaternized polymers were thus prepared involving either quaternary benzyl tributyl ammonium or phosphonium groups bound to either poly(VBC-CST)/ poly(ST-DVB) or poly(VBC)-poly(DCB)/ poly(ST-DVB).

Typical procedure for anchoring phosphine oxide groups to the partially quaternized polymer matrixes IIa, IIp, IIIa and IIIp

3.5 g of chlorodiphenylphosphine (15 mmol) was injected with a syringe into a suspension of 0.22 g lithium ribbon (32 mmol) in 15 ml anhydrous THF under stirring and nitrogen flow. With the dissolution of Li, a red color appeared. After 1 h, the red solution was carefully removed with a syringe and introduced under a nitrogen atmosphere in a flask containing 3.0 g of the quaternized polymer powder and 10 ml anhydrous THF. The heterogeneous mixture thus obtained was then stirred under reflux and nitrogen flow at 60–65°C. The red solution gradually changed its color to orange in about 6 h. After another 6 h of reaction under THF reflux, the mixture was poured into a saturated aqueous solution of ammonium chloride and stirred for 45 min at room temperature. The polymer recovered by filtration was washed in water several times. Finally the polymer was introduced into 30 ml THF, washed by stirring for 0.5 h and precipitated by the addition of 40 ml of light petroleum ether. After filtration, the functionalized polymer was dried in vacuum at 50°C. The FT-IR spectra showed the characteristic absorption bands of P=O at about 1182 and 1121 cm^{-1}. The ESCA spectrum indicated a binding energy of 132.2 eV for the pentavalent phosphine oxide species.

Preparation of the polymer-supported quaternary benzyl tributyl ammonium chloride (Ia)

The synthesis of the polymer-bound quaternary benzyl tributyl ammonium chloride proceeded as follows: 1.0 g of poly(VBC)/poly(ST-DVB) (0.87 mmol of Cl/g polymer) powder was suspended under stirring in a solution of TBA (1.85 g, 10 mmol) in 20 ml of DMF and heated at 75–80°C for 48 h. The quaternized polymer was precipitated by the addition of 40 ml ethyl ether, and purified by introducing it twice in THF under stirring at room temperature. The elemental analysis indicated 0.55 mmol of N/g polymer.

Binding of (m-sulfonate phenyl) diphenyl phosphine to the polymer-supported quaternary benzyl tributyl ammonium chloride [Ia(sphP)]

Sodium (*m*-sulfonate phenyl) diphenyl phosphine (*m*-sphP) was synthesized using the method indicated in Ref. [6]. 0.1 g of *m*-sphP was dissolved in a 20 ml mixture of water and ethanol (v/v = 1/1). Then 1.0 g of quaternized polymer Ia was added and the system was stirred for 12 h at room temperature. After filtration, the polymer was washed three times with water and dried in vacuum at 50°C. The elemental analysis indicated 0.11 mmol of P/g polymer.

Preparation of polymer-bound benzyl methyl diethanol ammonium groups (Ia')

2.75 g of poly(VBC)/poly(ST-DVB) (2.4 mmol of Cl) powder and 2.0 g of methyl diethanol amine (16.8 mmol) were introduced together with 20 ml DMF in a flask, and the system was heated under stirring at 75–80°C for 48 h. Then 40 ml of ethyl ether was added to the flask. The precipitated polymer was separated by filtration, washed again with ethyl ether (15 ml) three times and dried in vacuum overnight at 50°C. 3.03 g of the quaternized polymer were thus obtained. The elemental analysis indicated 0.78 mmol of N/g polymer.

Procedure for anchoring ethylene diphenyl phosphite to the quaternized polymer Ia'

A solution of 1.13 g phenol (12 mmol) in 10 ml CCl_4 was added dropwise under stirring and nitrogen atmosphere to a 30 ml anhydrous solution of 0.82 g of phosphorus trichloride (6 mmol) in CCl_4. The resulting mixture was heated under reflux for 2 h. Then, 3.0 g of quaternized polymer powder Ia' was added to the solution. The stirred suspension was heated under reflux and a nitrogen atmosphere for 8 h. 20 ml light petroleum ether was then added in order to precipitate the polymer. The polymer separated by filtration was washed twice under stirring with a mixture of dichloromethane and ethyl ether (v/v = 2/1). Finally, the polymer was washed three times with water and dried overnight in vacuum at 50°C. The elemental analysis indicated a content of P of 0.57 mmol/g polymer.

Preparation of polymer-bound palladium organophosphorous complexes

1.0 g of functionalized polymer powder was added under stirring at room temperature to a solution of 0.06 g (0.25 mmol) palladium acetate in 10 ml benzene. The red color of $Pd(OAc)_2$ solution in benzene faded and the polymer became yellow within 10–15 min. The polymer-supported Pd complex was precipitated by the addition of 15–20 ml of light petroleum ether. The yellow polymers were subjected to extraction in a Soxhlet for 2 h with THF to remove the uncombined ($Pd(OAc)_2$. If the extraction process lasts too long, some decomposition of palladium compound to palladium metal occurs.

Polymer-bound organophosphorus palladium complexes and quaternary onium groups catalyzed vinylation of iodobenzene with methyl acrylate

The polymer-supported bicatalyst IIp ($P(O)R_3$)-Pd complex (0.45 g, 0.10 mmol Pd plus 0.09 mmol of benzyl tributyl phosphonium) and potassium carbonate (1.0 g, 7.2 mmol) were introduced together with 10 ml of DMF in a 50 ml flask. Methyl acrylate (0.71 g, 8.2 mmol) and iodobenzene (1.32 g, 0.65 mmol) were injected under stirring with syringes. The system was placed in a dark surrounding because iodobenzene is photosensitive, and the reaction was conducted under stirring and a nitrogen atmosphere in an oil bath at 50°C for 20 h. The polymer catalyst was separated through filtration. The crude product of *trans*-methyl cinnamate was precipitated by pouring the filtrate into 30 ml of cold water. Further purification can be carried out using the column chromatography separation process as described by Miura *et al.* [2b].

The formation of *trans*-methyl cinnamate was quantitatively determined by HPLC. The conditions were: column: 200 mm × 3 mm; packing: Adsorbosphere C8, 5μ (Alltech); mobile phase: 70 vol% acetonitrile and 30 vol% water; flow rate: 0.4 ml/min; detector: UV 254 nm. *Trans*-methyl cinnamate (99%, Aldrich) was used for the calibration curve.

RESULTS AND DISCUSSIONS

Preparation of the polymer substrates

A general method for the preparation of functionalized polystyrene substrates as reactive polymers is *via* the copolymerization of the functional monomers with styrene (ST) and divinylbenzene (DVB) [8]. The main disadvantage of this method is due to the fact that a large fraction of the functional groups are buried inside the polymer matrix, especially when the employed functional monomers are more reactive than styrene. The functional groups located inside the matrix are not easily available to chemical reaction due to diffusional limitations. The optimal design of a reactive substrate should avoid as much as possible diffusional limitations for reactants and products. In this respect, the concentrated emulsion polymerization method [9] provides a convenient pathway for the preparation of heterogeneous particles, in which the polymerized functional monomers form a layer that coats crosslinked polystyrene particles. These small micron-sized polymer particles agglomerate to generate porous macroparticles. A concentrated emulsion has a large volume fraction of the dispersed phase (0.74–0.99) and the appearance of a gel. The continuous phase has a low volume fraction. When the latter is sufficiently small, it has the form of a network of thin liquid films that separate the polyhedral drops of the dispersed phase. The continuous phase contains a surfactant whose adsorption on the surface of the dispersed drops ensures the stability of the system. In the present case, a concentrated emulsion of styrene (containing divinylbenzene) in water was first prepared and subjected to partial polymerization. Then, the functional monomer was added under stirring and the system was completely polymerized. The particles thus prepared have a core of crosslinked polystyrene covered with a nonuniform layer of polymerized functional units. This method is expected to increase the efficiency of the bound functional groups because it generates a reactive surrounding with relatively low diffusional limitations. In addition, it reduces the amount of functional monomers that achieve the same concentration of functional

TABLE 1

Preparation of poly(ST-DVB) supported functionalized polymers *via* the concentrated emulsion polymerization

Substrate	ST-DVB (g)	Functional monomers (mmol)			Content of Cl in polymers [a] (mmol of Cl/g)
		VBC	CST	DCB	
I	21.8–4.48	14.5			0.87
II	21.8–4.48	13.2	13.2		0.91
III	21.8–4.48	13.2		13.2	0.63

[a] Elemental analysis data.

groups in the outer layer of polymer particles as the conventional method.

In the present experiments, 3- and 4-vinylbenzyl chloride (VBC), 4-chlorostyrene (CST) and 3,4-dichloro-1-butene (DCB) were used as functional monomers, to which catalytic functional groups can be further attached. Three functionalized substrates of polystyrene have been prepared *via* the concentrated emulsion polymerization method (Table 1). They are (I) poly(VBC)/poly(ST-DVB), (II) poly(VBC-CST)/poly(ST-DVB), and (III) poly(VBC)-poly(DCB)/poly(ST-DVB). The SEM pictures (Fig. 1a and 1b) show that coarse layers have grown nonuniformly on the surface of poly(ST-DVB) particles. This coarse layer is poly(VBC-CST) in Fig. 1a and poly(VBC-DCB) in Fig. 1b.

VBC being a versatile compound, various organic groups, such as quaternary onium,

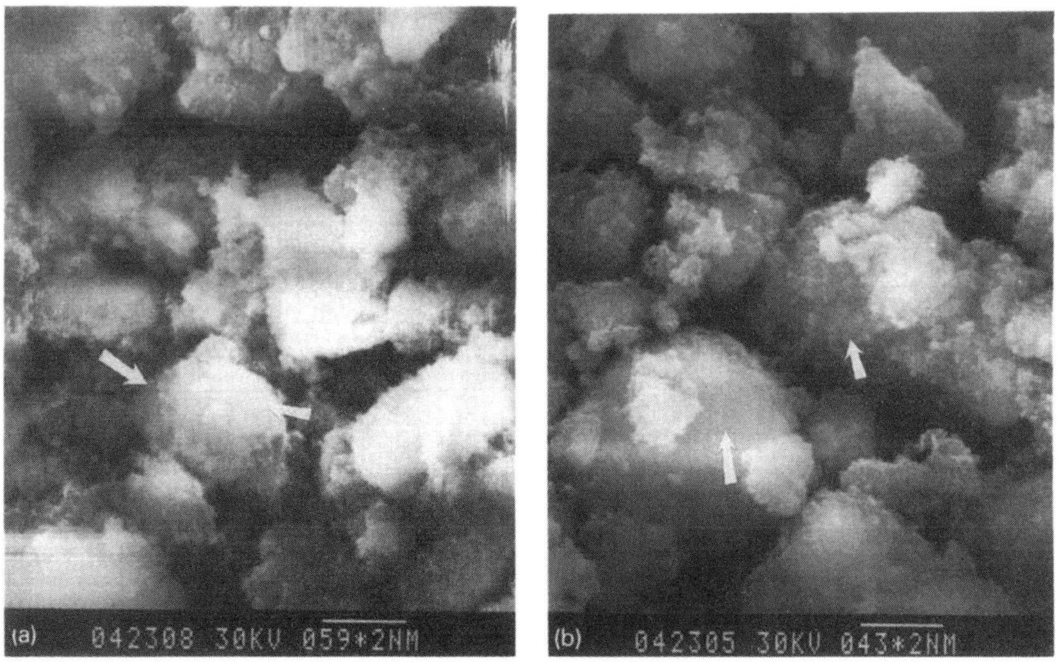

Fig. 1. SEM micrograph: (a) The macroporous morphology of poly(VBC-CST)/poly(ST-DVB) substrate, in which a nonuniform layer of poly(VBC-CST) copolymer coats micron-sized crosslinked polystyrene particles; (b) the macroporous morphology of poly(VBC)-poly(DCB)/poly(ST-DVB) substrate, in which a nonuniform layer of poly(VBC)-poly(DCB) polymer composite coats micron-sized cross-linked polystyrene particles. The arrows indicate the cross-linked polystyrene particles and the nonuniform layer.

TABLE 2

Relative $1/r_1$ sequence for styrene radicals [11]

M_2	$1/r_1$
Vinylbenzyl chloride (VBC)	1.38
4-Chlorostyrene (CST)	1.35
3,4-Dichloro-1-butene (DCB)	0.03 [a] 0.46 [b]

[a] $1/r_1$ of allylchloride; [b] $1/r_1$ of methyl allylchloride. The $1/r_1$ of DCB is estimated between these two values.

ester, tiol ester, *etc.*, can substitute the chlorine atom of the chloromethyl group. In addition, VBC can form copolymers with other vinyl monomers [10]. Therefore, in the present paper either 4-chlorostyrene or 3,4-dichloro-1-butene was used together with VBC. Since in the subsequent functionalization one reactive group is attached to polymerized VBC and another to polymerized CST or DCB units, the degree of copolymerization between VBC and CST or DCB is expected to influence the cooperation among the attached reactive groups. Table 2 provides the copolymerization ability (r value) of three functional monomers with styrene. From these data it is clear that VBC copolymerizes easily with CST, but that VBC and DCB tend to homopolymerize (Scheme 1).

Conversion of phosphines to phosphine oxides via the ylid intermediate of the functionalized polymer matrix

The homogeneous catalytic system [5] employed in the Heck-type vinylation, which involves palladium diacetate and a quaternary onium group, was selected for heterogenization by bonding both to a polymer substrate. As noted in the previous section, three kinds of substrates denoted as I, II and III have been prepared by the concentrated emulsion polymerization method. Using substrate I, poly(VBC)/poly(ST-DVB), two kinds of functionalized polymer substrates (Ia(sphP) and Ia′($P(OR)_3$), containing phosphorus coordination anchors as well as pendant quaternary onium groups, have been prepared by employing the general organic reactions presented in Scheme 2.

Substrate II, poly(VBC-CST)/poly(ST-DVB), and III, poly(VBC)-poly(DCB)/

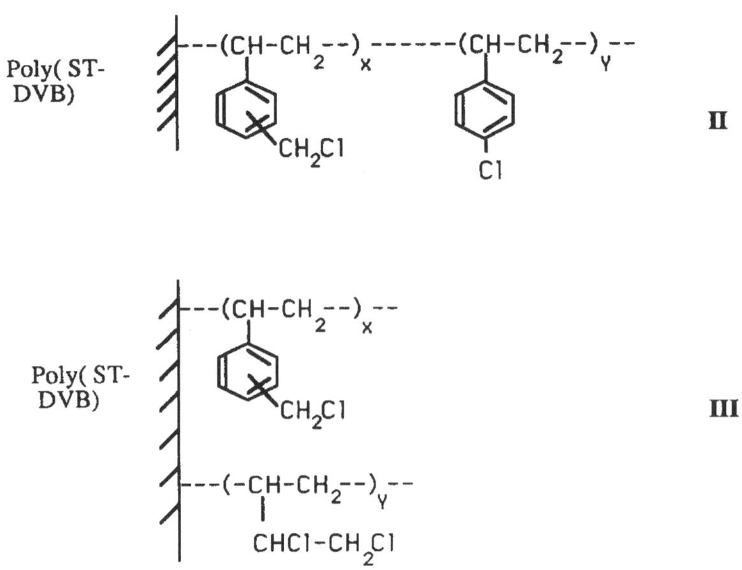

Scheme 1. The poly(VBC-CST)/poly(ST-DVB) II and poly(VBC)-poly(DCB)/poly(ST-DVB) III.

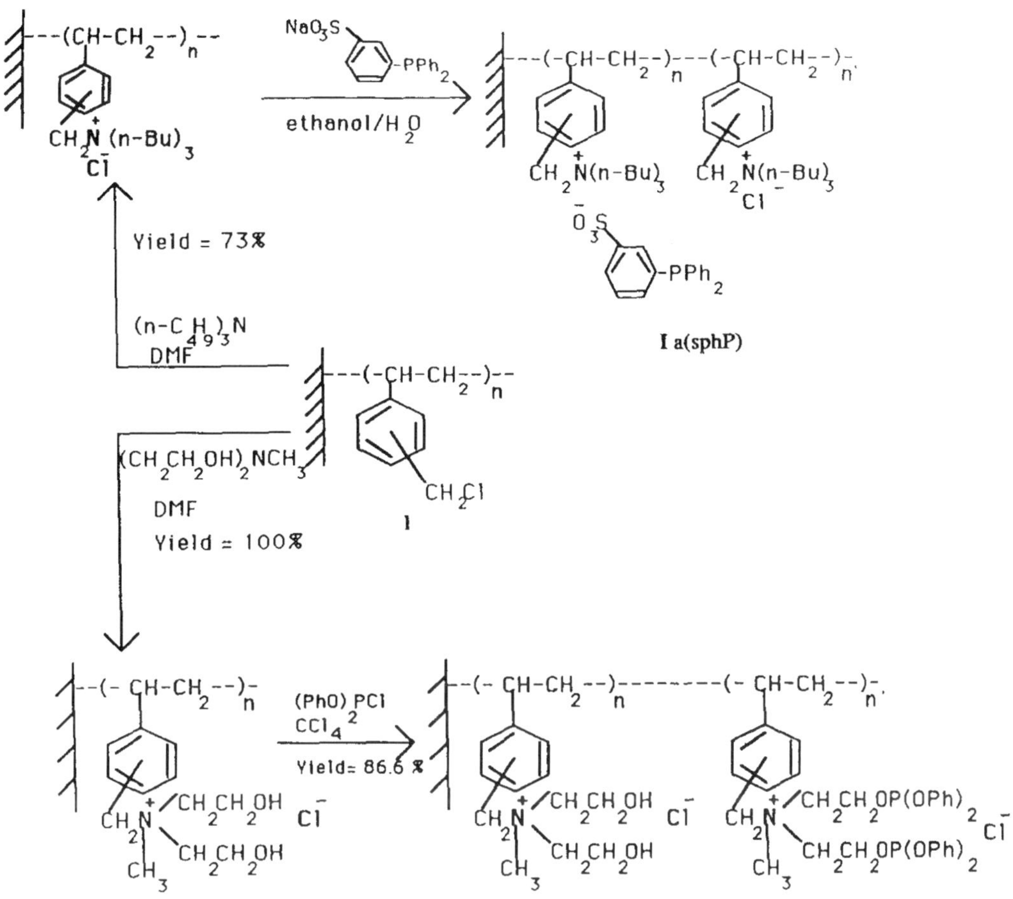

I a' [P(OR)$_3$]

Scheme 2. Preparation of the functionalized substrates Ia(sphP) and Ia'[P(OR)$_3$].

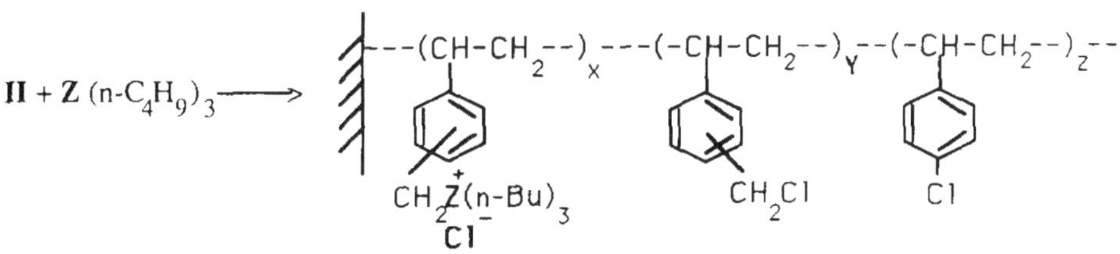

IIa (**Z= N**), **II**p (**Z= P**)

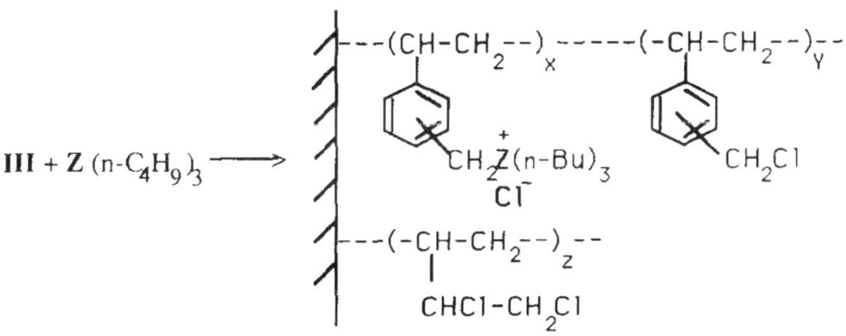

IIIa (**Z** = **N**), **III**p (**Z= P**)

Scheme 3. Quaternization is assumed to take place mainly at polymerized VBC units.

TABLE 3

The elemental analysis of the quaternized polymers

Substrate	N (mmol/ g polymer)	P (mmol/ g polymer)	Quaternization yield [a] (mole %)
Ia	0.55		73
Ia′	0.78		100
IIa	0.17		21.8
IIp		0.20	26
IIIa	0.11		19.5
IIIp		0.15	26.8

[a] Calculation based on the content of chlorine in polymers I, II and III.

poly(ST-DVB), were quaternized with a tertiary amine or phosphine. A fraction of the pendant organic chloride units was thus converted to quaternary onium groups. Most of these quaternized units can be assumed to be pendant benzyl tributyl onium (shown in Scheme 3), because methylenechloride of polymerized VBC is the most active group among the three kinds of pendant organic chlorides for nucleophilic attack by TBA or TBP.

The elemental analysis and the quaternization yields are listed in Table 3. The resulting partially quaternized substrates IIa, IIp, IIIa and IIIp were then phosphinated with Ph_2PLi under reflux. Finally, a saturated aqueous solution of NH_4Cl was introduced in the system in order to decompose the unreacted Ph_2PLi. ESCA and FT-IR spectra indicated that the phosphination of the remaining pendant benzyl chloride, phenyl chloride and ethylene dichloride groups of the partially quaternized polymers leads, unexpectedly, to the pentavalent phosphine oxides and not to the trivalent phosphine. As is well known [12], the tertiary phosphine ligands suppress the catalytic activity of palladium for the vinylation reaction due to their strong coordination ability. However, the phosphine oxide ligands coordinate to the metal through the oxygen atoms rather than phosphorus atoms. As a result, their coordination ability is weaker than that of the tertiary phosphines. This is the reason for which we employed the pathway just described to functionalize the polymers II and III prepared by the concentrated emulsion polymerization procedure.

The mechanism of the conversion of pendant organic phosphine, such as benzyl diphenyl phosphine to its counterpart phosphine oxide is not known. The following mechanism is suggested on the basis of previous research [13] (Scheme 4): Step 1. The pendant (benzyl tributyl ammonium or phosphonium groups mediate a nucleophilic substitution between the formed phosphines and the unquaternized chloride groups leading to aromatic phosphoniums. *Step 2.* The aromatic phosphoniums are deprotonated by the excess diphenylphosphorus lithium, generating ylid species. *Step 3.* An oxidation reaction takes place in the aqueous solution of NH_4Cl in the presence of air, which proceeds *via* the cleavage of ylid species to phosphine oxide and an aldehyde. The latter compound is oxidized in air to a carboxylic acid. The carboxylic group was identified from the FT-IR absorption bands at 1732 ($\nu_{C=O}$) and 2368–2342 cm^{-1} (ν_{-OH}).

The functionalized polymers were characterized by ESCA, FT-IR and elemental analysis. In the ESCA spectra (Fig. 2), a strong peak at 132.2–132.4 eV (BE) can be observed, which can be assigned to the pentavalent aromatic phosphine oxide. Relatively weak tetravalent aromatic phosphonium peaks (BE = ~ 133.3 eV and ~ 131 eV) as well as trivalent phosphine peak (BE = 130 eV) were also identified. In the FT-IR spectrum, in addition to the absorption bands of the carboxyl group, those of P=O at about 1180 and 1120 cm^{-1} were also observed. This observation was also confirmed by the appearance of an oxygen peak in the ESCA spectra.

Characterization of polymer-bound palladium phosphine oxide complexes

The FT-IR spectra [14] of the prepared polymer-supported quaternary onium groups and phosphine oxide carrying coordinated palladium acetate complexes have three types of characteristic absorption bands: (1) the $\nu_{C=O}$ of the acetate ligand at 1605 cm^{-1}; (2) the $\nu_{P=O}$ of the coordinated phosphine oxide which is shifted, by about 30 cm^{-1}, to a lower frequency compared to the uncoordinated one; and, (3) the $\nu_{Pd \leftarrow O=P}$ of phosphine oxide complexes at about 520–420 cm^{-1}. For example, the FT-IR spectra (Fig. 3) of the prepared functionalized polymer IIa($P(O)R_3$) and of its palladium containing counterpart catalyst presents the following three types of absorption bands: the bands of coordinated CH_3COO–: $\nu_{C=O} = 1605$ cm^{-1}; the band of coordinated phosphine oxides: $\nu_{P=O} = 1116$ cm^{-1}, which can be considered as the shifted band of the uncoordinated phosphine oxide at 1147 cm^{-1}; and the bands $\nu_{Pd-O} = 522$ cm^{-1}, 420 cm^{-1}. ($\nu = 540$ cm^{-1} is the absorption band of polystyrene.)

Polymer-supported palladium phosphorus complexes and quaternary onium groups catalyzed vinylation of iodobenzene with methylacrylate:

The palladium-catalyzed coupling reaction of aryl halide with alkenes constitutes a useful method for the preparation of styrenic type derivatives

$$ArX + H{-}\overset{|}{C}{=}\overset{|}{C}{-} \rightarrow Ar{-}\overset{|}{C}{=}\overset{|}{C}{-}$$

Previous reports [4] have shown that phase transfer catalysts can remarkably improve the yield and stereospecificity of the reaction and allow the reaction to proceed in milder conditions. As already mentioned, three types of heterogeneous polymer-supported catalysts are employed in the present paper (Table 4): (1) Ia(sphP)–Pd complex (palladium bis(*m*-sulfonated phenyl) diphenyl phosphine diacetate complex); (2) Ia′($P(OR)_3$)–Pd complex (palladium diphenyl ethylene phosphite diacetate complex); (3) [IIa($P(O)R_3$), IIp($P(O)R_3$), IIIa($P(O)R_3$) and IIIp($P(O)R_3$)]–Pd complex (palladium phosphine oxides diacetate complex) have been employed in the vinylation of iodobenzene with methyl acrylate. The yields of the obtained *trans*-methyl cinnamate are listed in Table 5. The catalytic efficiency (yield of vinylation product per mmol of palladium) follows the sequence Ia(sphP)–Pd complex > [IIa($P(O)R_3$), IIp($P(O)R_3$), IIIa($P(O)R_3$) and IIIp($P(O)R_3$)]–Pd complex > Ia′($P(OR)_3$)–Pd complex. The structure of the phosphorus ligand, the concentration and the structure of the quaternary onium groups influence the yield of the reaction. The ligand (*m*-sulfonate phenyl) diphenyl phosphine is the most effective among the three kinds of phosphorus ligands employed. A low concentration of pendant quaternary benzyl tributyl onium groups in the [IIa($P(O)R_3$), IIp($P(O)R_3$), IIIa($P(O)R_3$) and IIIp($P(O)R_3$)]–Pd has a negative effect on the yield of vinylation. The phosphite ligand coupled with the pendant quaternary benzyl methyl diethanol ammonium group [catalyst Ia′($P(OR)_3$)–Pd complex] displays a lower catalytic efficiency than the other bifunctional polymer supported catalysts. This is possibly due to a weaker hydrophobicity of the pendant diethanol ammonium ion compared to benzyl tributyl ammonium or phosphonium ions and a stronger coordination ability of the phosphite ligands for Pd.

The catalytic systems containing [IIa($P(O)R_3$), IIp($P(O)R_3$), IIIa($P(O)R_3$) and IIIp($P(O)R_3$)]–Pd complexes, in which the amount of pendant quaternary onium groups is 0.005–0.009 equiv., of Pd complex 0.016 equiv. and of iodobenzene 1.0 equiv., displayed a yield of 44–56% for vinylation of iodobenzene with methylacrylate. The homo-

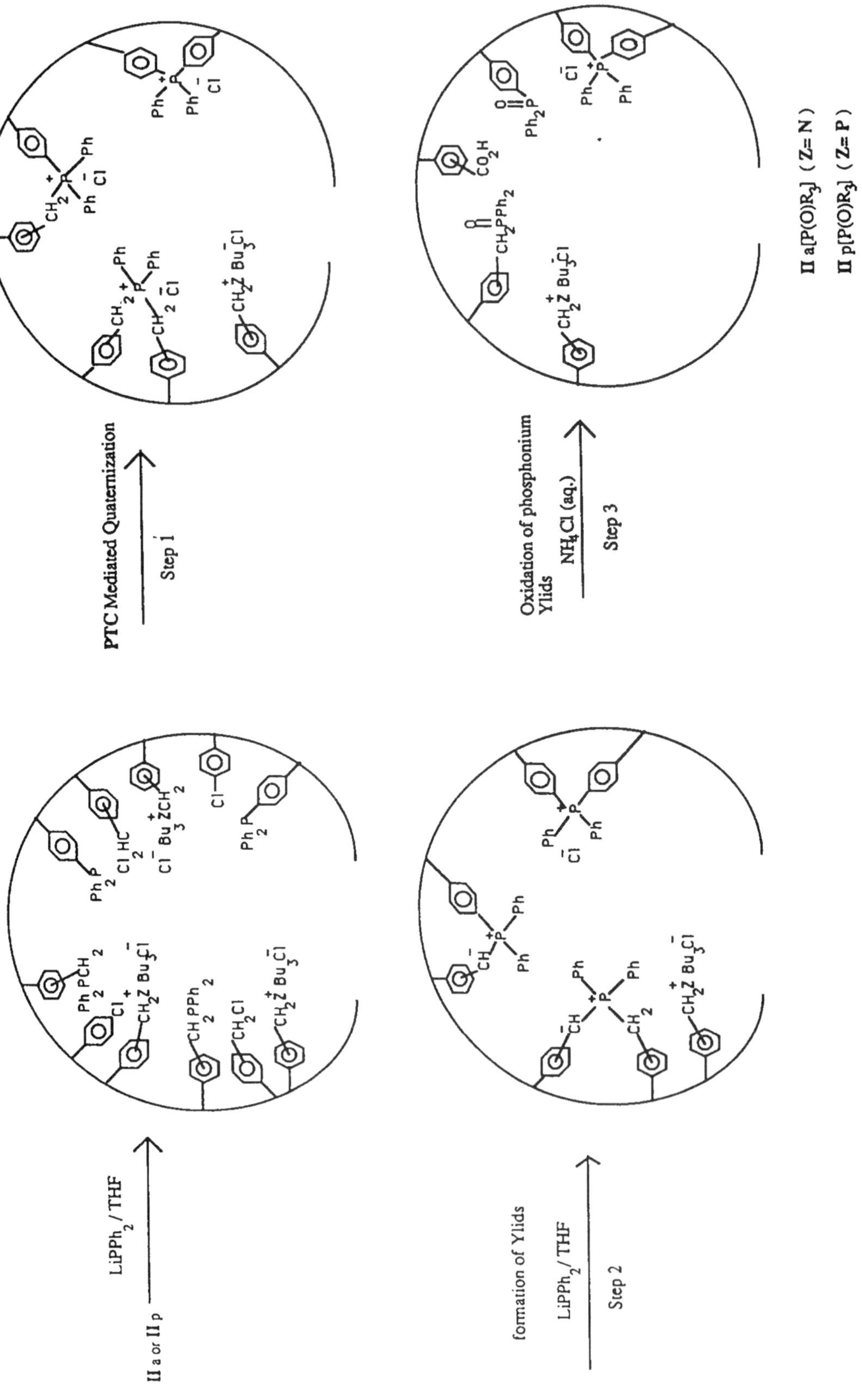

Scheme 4. The supposed mechanism of the transformation of phosphine to phosphine oxide *via* ylid intermediates.

geneous catalytic experiments carried out by Jeffery [5] with 0.1 equiv. NBu_4Cl, 0.01–0.02 equiv. $Pd(OAc)_2$ and 1.0 equiv. iodobenzene provided a yield of 20%. The homogeneous catalytic reaction (run 7, Table 5), in which 0.1 equiv. phosphonium salt $BzlPBu_3Cl$, 0.33 equiv. $(ph_3P)_2Pd(OAc)_2$ complex and 2.7 equiv. iodobenzene were used, provides a yield of 24%. Because, as confirmed by previous reports [2b,5] the yield depends upon the concentration of the phase transfer agent, the above comparison suggests that the polymer-bound bicatalyst can have great potential for further improvements through changes in the relative content of the two catalysts. The higher yield of the supported catalyst is probably due to a better cooperation between the two catalytic species, since each of them has the other one in its microenvironment. In homogeneous systems,

Fig. 3. FT-IR spectra: (a) phosphine oxide and ammonium groups attached to the substrate $IIa(P(O)R_3)$; (b) polymer-supported $IIa(P(O)R_3)$–Pd complex.

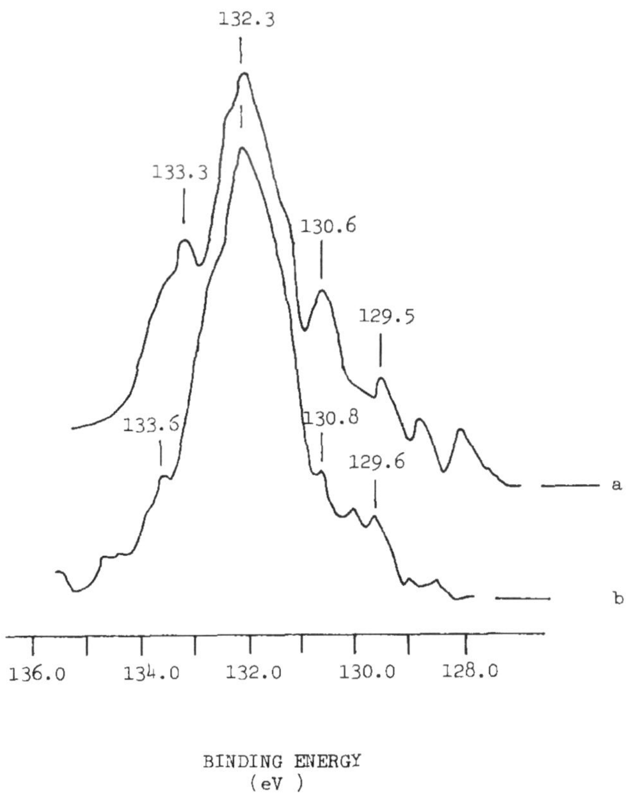

Fig. 2. ESCA spectra: (a) phosphine oxide and phosphonium groups attached to the substrate $IIp(P(O)R_3)$; (b) phosphine oxide and phosphonium groups attached to the substrate $IIIp(P(O)R_3)$.

the random motion of the two catalytic species reduces the probability of their cooperation in a common catalytic act.

In order to compare the behavior of polymer-bound bicatalysts systems with those in which only one catalyst is bound while the other is homogeneous, three systems were investigated (runs 8 to 10 in Table 5). Comparing runs 7 and 8, it is clear that the heterogeneous phase transfer catalyst plus the homogeneous Pd-complex catalyst of run 8 are more efficient than the corresponding homogeneous system (run 7). Comparing runs 6 and 8, one can note that the yields are the same; however, the amount of phase transfer catalyst is much larger in run 8. This suggests that, in comparable conditions, the yield is most likely to be higher in the heterogeneous–heterogeneous case than in heterogeneous–homogeneous case. We are inclined to believe that the high yield in run 8 is due to the adsorption of Pd-compound on the surface of the polymer and that the catalytic process is in reality a heterogeneous–heterogeneous one. One may also note by comparing runs 9 and 10, that an excess of phosphine ligand decreases tremendously the yield. This is because the phosphine ligand coordinates with palladium, thus poisoning the catalyst.

From Table 5 (runs 3–6), one can observe that the efficiency sequence of the phase transfer catalysts is phosphonium > ammonium, hence that a phase transfer catalyst with higher lipophilicity leads to a higher yield of vinylation. The catalyst IIIp(P(O)R_3)–Pd complex was used in three successive experiments (Table 5). The yield slowly decreased because of the loss of palladium caused by the relatively weak coordination ability of the phosphine oxide.

The influence of spatially separated sites of catalysts on the reaction rate

We have already noted that phosphine oxide and quaternary onium ion can be bound on the polymer matrix using either a copolymer, poly(VBC-CST), or a polymer composite poly(VBC)-poly(DCB) that coats cross linked polystyrene particles. Most of the quaternary onium ions are attached to the polymerized VBC units, while most of the phosphine oxide groups are bound to polymerized CST and DCB units. As concern the spatial location of these two kinds of functional groups, the catalysts [IIa(P(O)R_3) and IIp(P(O)R_3]–Pd complexes, based on the copolymer poly(VBC–CST), are different from the catalysts [IIIa(P(O)R_3) and IIIp (P(O)R_3]–Pd complexes obtained using the polymer-composite poly(VBC)-poly(DCB). An interesting observation was made by comparing the activities of IIp(P(O)R_3)–Pd and IIIp(P(O)R_3)–Pd complexes (Fig. 4). The former catalyst had a higher activity than the latter during the first 4 h of reaction, but thereafter the reactivity of catalyst IIIp(P(O)-R_3)–Pd complex rose more rapidly than that of IIp(P(O)R_3)–Pd complex. This behavior is

TABLE 4

Elemental analysis of the polymer-bound catalysts

Catalyst	BTBA [a] (mmol/g polymer)	BTBP [b] (mmol/g polymer)	Phosphorus (mmol of P/g polymer)	Palladium (mmol/g polymer)
Ia(sphP)–Pd	0.55		0.1	2.8×10^{-2}
Ia'(P(OR)$_3$)–Pd	1.1		0.6	0.26
IIa(P(O)R_3)–Pd	0.17		1.2	0.25
IIp(P(O)R_3)–Pd		0.20	0.77	0.21
IIIa(P(O)R_3)–Pd	0.11		0.70	0.20
IIIp(P(O)R_3)–Pd		0.15	0.97	0.23

[a] BTBA: polymer-bound benzyl tributyl ammonium chloride.
[b] BTBP: polymer-bound benzyl tributyl phosphonium chloride.

TABLE 5

Vinylation of iodobenzene catalyzed by the polymer-bound catalysts

Run	Catalyst	PTC		Pd (mmol)	Yield (mole%) of *trans*-methyl cinnamate [e]		
		BTBA (mmol)	BTBP (mmol)		1st Time	2nd Time	3rd Time
1	Ia(sphP)–Pd	0.22		1.1×10^{-2}	36		
2	Ia′($P(OR)_3$)–Pd	0.5		0.12	30		
3	IIa($P(O)R_3$)–Pd	7.6×10^{-2}		0.11	40		
4	IIp($P(O)R_3$)–Pd		9.0×10^{-2}	0.10	53		
5	IIIa($P(O)R_3$)–Pd	5.0×10^{-2}		0.09	49		
6	IIIp($P(O)R_3$)–Pd		6.8×10^{-2}	0.10	56	49.5	43.3
7	$(Ph_3P)_2Pd(OAc)_2$–$BzlPBu_3Cl$ [a]		0.24	0.08	24		
	Het. / Hom.						
8	P–$BzlPBu_3Cl/(Ph_3P)_2Pd(OAc)_2$ [b]		0.24	0.08	57		
9	P–$BzlPBu_3Cl/(Ph_3P)_2Pd(OAc)_2 + Ph_3P$ [c]		0.41	0.06	–		
10	I(PR_3)–Pd/$BzlPBu_3Cl$ [d]		0.18	0.05	2		

[a] The complex $(Ph_3P)_2Pd(OAc)_2$ was synthesized according to the method of Stephenson *et al.* [7] and the PTC catalyst $BzlPBu_3Cl$ was prepared using the general method.

[b] Polymer-bound PTC catalyst P-$BzlPBu_3Cl$ was prepared *via* the quaternization of poly(VBC) (0.6 mole% crosslinking agent) and contains 2.42 mmol of P/g polymer.

[c] The mole ratio of $(ph_3P)_2Pd(OAc)_2/Ph_3P = 1/5$.

[d] I(PR_3) was prepared *via* the reaction of substrate I with Ph_2PLi, containing 0.77 mmol of P/g polymer. After coordination with $Pd(OAc)_2$, I(PR_3)–Pd contains 0.13 mmol of Pd/g polymer.

[e] After each run, the catalyst IIIp($P(O)R_3$)–Pd was filtered out, washed with water several times and subsequently with ether and dried in vacuum. The yield was calculated using as a basis the molar amount of iodobenzene employed.

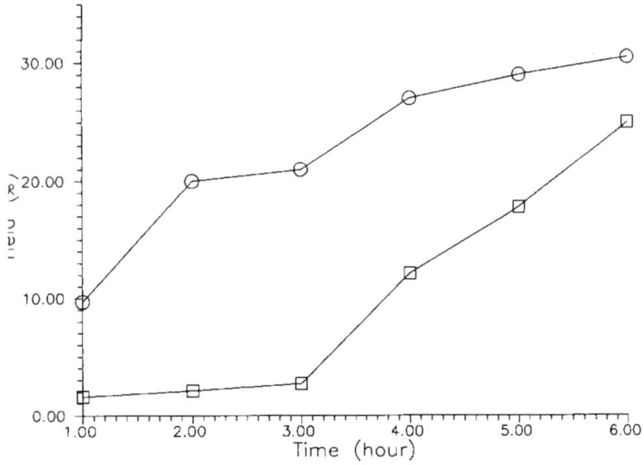

Fig. 4. The yield of methyl cinnamate in mole% for different reaction times in systems in which either IIp(P(O)R_3)–Pd (o) or IIIp(P(O)R_3)–Pd complexes (□) are employed. Reaction conditions: (o) pendant Pd complexes 5.3×10^{-2} mmol, pendant benzyl tributyl phosphonium 4.8×10^{-2} mmol; (□) pendant Pd complexes 5.3×10^{-2} mmol, pendant benzyl tributyl phosphonium 3.6×10^{-2} mmol. C_6H_5I (6.5 mmol), CH_2=$CHCO_2CH_3$ (8.2 mmol), and K_2CO_3 (1.0 g) are the same for both systems.

related to the arrangement of the bound catalytic sites of phosphine oxide palladium complexes and benzyl tributyl phosphonium chloride which can be regarded as alternately anchored on the polymer chain. Hence the cooperative effect between the two types of active species appears from the beginning of the reaction in this catalyst. In contrast, in catalyst IIIp(P(O)R_3)–Pd complex the bound phosphine oxide palladium complex and benzyl tributyl phosphonium group could possibly be attached to different polymer chains [original poly(VBC) and poly(DCB)]. Some time is therefore needed for these two active species to better cooperate as a result of the mobility of the chains stimulated by swelling.

CONCLUSIONS

The quaternary onium group and palladium phosphorus complex are bound to the same polymer matrix thus generating bifunctional polymer-supported catalysts. This polymer matrix is located as a nonuniform layer on highly crosslinked polystyrene micron-sized particles. Three types of catalysts based on the substrates [poly(VBC)/poly (ST-DVB), poly(VBC-CST)/poly(ST-DVB), or poly(VBC)-poly(DCB)/ poly(ST-DVB)] are prepared *via* multiple functionalization steps. These substrates are prepared using the concentrated emulsion polymerization method. They are composed of micron-sized crosslinked polystyrene particles covered nonuniformly by a polymerized layer containing pendant organic chloride groups. The fine powder prepared by grinding contains small porous aggregates of such particles. Among the functionalization methods a new synthetic pathway for the binding of both quaternary onium groups and phosphine oxide ligands to the polymer matrix is suggested. The three types of polymer-supported catalysts thus prepared contain: (1) palladium sulfonated phenyl phosphine complexes and quaternary ammonium groups; (2) palladium diphenyl ethylene phosphite complexes and quaternary ammonium groups; and (3) palladium phosphine oxide complexes and quaternary onium groups.

These three types of polymer-supported bifunctional catalysts (Table 4) are employed in the vinylation reaction of iodobenzene with methylacrylate. The yield of vinylation is much larger than in homogeneous systems. The cooperation between the palladium phosphine oxide complexes and the quaternary onium groups which are permanently near to one another is responsible for the higher yield. When the two catalytic centers are connected to the same polymer chain alternately [IIp(P(O)R_3)–Pd complex], the resulting catalyst has a higher activity during the initial time of the reaction than the catalyst in which the two catalytic centers are bound to different polymer chains [IIIp (P(O)R_3)–Pd complex]. However, for longer times, because of swelling and because of the rearrangements of the polymer chains, the above inequality may become inverted. Among the three supported catalysts, the one containing palladium sulfonated phenyl phosphine complexes coupled with quaternary onium groups provides the highest catalytic efficiency.

REFERENCES

1 E.S. Gore, Platinum group metals in phase transfer catalysis, Platinum Met. Rev., 34 (1990) 2.

2 (a) H. Alper, Metal catalyzed carbonylation and oxidation–reduction reactions, Pure Appl. Chem., 60 (1988) 35.
(b) M. Miura, H. Hashimoto, K. Itoh and M. Nomura, Palladium-catalysed desulphonylative vinylation of arenesulphonyl chlorides under solid–liquid phase-transfer conditions, J. Chem. Soc. Perkin. Trans. 1, (1990) 2207.
(c) H. des Abbayes, J.C. Clement, P. Laurent, G. Tanguy and N. Thilmont, Scope and mechanism of the phase-transfer carbonylation of organic halides catalyzed by penta-carbonyliron to give ketones and carboxylic acids, Organometallics, 7 (1988) 2293.

3 J.R. Blanton and J.M. Sally, Reducing alkyl halides using a polymer-bound crown ether/tin hydride cocatalyst, J. Org. Chem., 56 (1991) 490.

4 R.F. Heck, Org. React., 27 (1982) 345; Palladium-catalyzed vinylation of organic halides, Org. React., 27 (1982) 345; Palladium Reagents in Organic Synthesis, Academic Press, London, 1985.

5 T. Jeffery, Palladium-catalyzed vinylation of organic halides under solid–liquid phase transfer conditions, J. Chem. Soc., Chem. Commun., (1984) 1287.

6 S. Ahrland, J. Chatt, N.R. David and A.A. Williams, The relative affinities of coordinating atoms for silver ion, Part II. Nitrogen, phosphorus and arsenic, J. Chem. Soc., (1958) 276.

7 T.A. Stephenson, S.M. Morehouse, A.R. Powell, J.P. Heffer and G.J. Wilkinson, Carboxylates of palladium, platinum and rhodium and their adducts, J. Chem. Soc., (1965) 3632.

8 A. Guyot, Some problems in the physical and chemical characterization of functionalized supports, React. Polym., 10 (1989) 113.

9 E. Ruckenstein and K.J. Kim, Polymerization in gel-like emulsions, J. Appl. Polym. Sci., 36 (1988) 907.

10 R.H. Yocum and E.B. Nyquist, Functional Monomers – Their Preparation, Polymerization and Application", Marcel Dekker, New York, 1973, p. 251.

11 L.J. Young, Copolymerizations involving styrene or substituted styrenes as principal components, in: G.E. Ham (Ed.), Copolymerization, Wiley-Interscience, New York, 1964, Chap. 8.

12 T. Jeffery, Pd-catalysed vinylation of vinylic halides under phase transfer conditions, Tetrahedron Lett., 26 (1985) 2667.

13 A.W. Johnson, Ylid Chemistry, A series of Monographs, Vol. 7, Academic Press, New York, 1966, p. 95.

14 K. Nakamoto, Infrared and Raman Spectra of Inorganic and Coordination Compounds, 3rd Edn. Wiley, New York, 1978.

Hydrophilic Recognition by Polymer-Supported Phase Transfer Catalysts and Its Effect on Reaction Activity and Selectivity

E. RUCKENSTEIN[1] AND L. HONG

Department of Chemical Engineering, State University of New York at Buffalo, Buffalo, New York 14260

Received November 12, 1991; revised February 19, 1992

Cross-linked polystyrene latexes with surface-bound quaternary ammonium groups have been prepared by a two-step method. First the functional monomer vinylbenzyl chloride (VBC) was introduced into a partially polymerized concentrated emulsion of styrene and divinylbenzene (ST/DVB) in water and complete polymerization was carried out. Second the poly(ST/DVB) particles with surface-bound benzyl chloride groups were quaternized. The dual-surface characteristics due to the hydrophilic bound cations and hydrophobic polystyrene can be varied by controlling the partial polymerization time of ST/DVB and the amount of VBC introduced. The phase transfer catalysts thus obtained were employed in the alkylation reaction of isopropylidene malonate (Meldrum acid). The ratio of mono- to dialkylation products was observed to increase with decreasing accessible hydrophilicity of the supported catalysts. When the partial polymerization time is short, a part of VBC copolymerizes with styrene and another part homopolymerizes and is grafted to the surface of the particles. When the partial polymerization time is long, only homopolymers of VBC, which are grafted to the particles, form. The ammonium groups to which the benzyl chloride units are converted via quaternization are more accessible, for steric reasons, to the Meldrum acid and to the monosubstituted Meldrum acid when they are part of the grafted chains than when they are part of the copolymer. As a result, the activity for the conversion of the Meldrum acid increases as the partial polymerization time increases. However, the selectivity for the monosubstituted Meldrum acid increases with decreasing partial polymerization time of the concentrated emulsion, because the fraction of hydrophilic groups that are accessible decreases and the monosubstituted compound has more difficulty, for steric reasons, binding to the onium cations than the unsubstituted compound.

INTRODUCTION

We report here the preparation of polymer-supported phase transfer catalysts that contain various amounts of hydrophilic quaternary onium groups bound to the surface of hydrophobic cross-linked polystyrene particles. The dependence of the activity and selectivity of these catalysts on their degree of accessible hydrophilicity was investigated for the alkylation of isopropylidene malonate (Meldrum acid) with ethyl bromoacetate. The formation of ion pairs between the polymer-bound onium ions and the reactive (unsubstituted or monosubstituted) isopropylidene malonate anions and the transfer of the electrophilic molecules of ethyl bromoacetate from the organic solvent (chloroform) to the reactive ion pairs bound to the polymer represent the two processes involved in this nucleophilic substitution reaction. Both processes require hydrophilic and hydrophobic interactions, which provide the basis for reactant recognition by the polymer-supported catalyst. This reaction has been extensively investigated, but only under homogeneous conditions (*1*).

The strategy of using concentrated emulsions (*2*) has been employed for the preparation of the polymer-supported phase transfer catalysts. An oil in water (o/w) concentrated emulsion can be easily prepared by dispersing a hydrophobic liquid into a surfactant-containing hydrophilic liquid by mechanical stirring. The concentrated emulsion is stabilized by the adsorption of the

[1] To whom correspondence should be addressed.

This paper, which was originally published in **Journal of Catalysis**: 136, 378–391 (1992), is republished with permission.

surfactant on the surface of the dispersed droplets. The volume fraction of the dispersed phase is larger than 0.74, which represents the volume fraction of the most compact arrangement of spheres of equal radii and can be as large as 0.99. The concentrated emulsions have the appearance and behavior of gels and for this reason are easy to handle (*3*). When the dispersed liquid is a hydrophobic monomer that contains a suitable initiator, and the continuous phase is water, polymer latexes of micrometer size are obtained after polymerization. Functional groups can be bound to the surface of the insoluble supporting polymer particles by introducing hydrophobic functional monomer(s) into partially polymerized concentrated emulsions of a hydrophobic monomer (which will be called supporting monomer) in water (in the present paper, the mole ratio of functional monomer to the supporting monomer is in the range 7 to 19%) and polymerizing again. The composition and the structure of the functionalized surface layer of the polymer particles can be controlled by varying the time of partial polymerization. Thus, if the functional monomer is introduced into a concentrated emulsion that has experienced a short polymerization time, a part of the functional monomer molecules will copolymerize during subsequent polymerization with the supporting monomer, and another part will homopolymerize and be grafted to the polymer particles. On the other hand, if the functional monomer is introduced into a concentrated emulsion that has experienced a long polymerization time, most of the functional monomer molecules will homopolymerize and be grafted to the surface of the particles. In the present paper, styrene (ST) and divinylbenzene (DVB) were used as supporting monomers, and vinylbenzyl chloride (VBC) was used as the functional monomer. The surface-pendant benzyl chloride units were chemically transformed to quaternary ammonium cations. Two extreme types of catalysts were obtained. In one, copolymers of quaternary ammonium and phenyl groups as well as grafted polyquaternary ammonium chains form an active layer on the particles (type 1). In the other, only grafted polyquaternary ammonium chains are present on the surface of poly(ST/DVB) (type 2).

The alkylation of the Meldrum acid with ethyl bromoacetate was chosen to examine the effect of the accessible hydrophilicity of the supported catalyst on the activity and selectivity of this nucleophilic reaction. This choice was made because: (1) the reaction can generate both mono- and disubstituted Meldrum acid (Scheme 2), as a result of the nucleophilic attack of unsubstituted and monosubstituted anions on the alkyl halide; (2) intuition suggests that the type 2 catalyst, which has surface-pendant polycation chains, can bind the unsubstituted and monosubstituted anions more easily than the type 1 catalyst does, for steric reasons. Therefore, the conversion of the Meldrum acid and the selectivity for the disubstituted compound are expected to be higher for the type 2 catalyst.

If the reaction is catalyzed homogeneously by a quaternary onium salt or a lightly cross-linked polyquaternary onium salt, only the disubstituted compound is obtained, because, in contrast to the heterogeneous polymer catalysts with dual surface characteristics, there are no steric constraints for the two kinds of anions to approach the quaternary onium cations. The presence of styrene segments adjacent to cation segments in the type 1 catalyst generates steric constraints that are greater for the monosubstituted than for the unsubstituted anion. In the type 2 catalyst, the surroundings of the surface-pendant ammonium chains have much less steric constraints than in the type 1 catalyst. The hydrophobic steric constraints due to the polystyrene segments are responsible for the decreased recognition of the un- and monosubstituted anions by the type 1 catalyst.

EXPERIMENTAL

Main materials. Styrene (Aldrich, 99%), divinylbenzene (Aldrich, tech., 55%), butyl acrylate (Aldrich, 99 + %), and vinylbenzyl chloride (Kodak, *p* and *m*-isomeric mixture) were distilled under vacuum before use. Tributyl amine (Aldrich, 99%), tributyl phosphine (Aldrich, 92%), benzyl chloride (Aldrich, 0.25% propylene oxide), isopropylidene malonate (Aldrich, 98%), ethyl bromoacetate (Aldrich, 98%), sodium dodecyl sulfate (SDS) (Aldrich, 99%), and various solvents were used without further purification.

Instruments used. ^{1}H NMR spectra of the alkylation products were recorded on a VXR-400 spectrometer (400 MHz). Energy dispersive spectroscopy (EDS) surface analysis was performed on a PG/IMIX equipment (15 Kev). The morphologies of the polymer samples were investigated with scanning electron microscopy (SEM) (Amray 100A). Infrared absorption spectra of benzyl chloride groups bound on the polymer particles in different surroundings were obtained with a Mattson Alpha Centauri FT–IR instrument by using a KBr grate. The content of monoalkyl isopropylidene malonate was determined by HPLC (ISCO-2350 pump and ISCO – V_4 detector) using a column (4.6 mm × 25 cm) filled with Adsorbosphere C_8 of 5 μm. Elemental analyses for Cl, N, and P were carried out by Quantitative Technologies, Inc. (Bound Brook, NJ).

General Methods

Preparation of the cross-linked polystyrene particles with surface-attached benzyl chloride by the concentrated emulsion polymerization method. In a 100-ml round-bottom flask, a solution of SDS (0.30 g) in 4 ml of water was first prepared. Then, 12 ml of ST/DVB mixture (mol ratio = 10:1) containing 15 mg of initiator azobisisobutyronitrile (AIBN) was added dropwise to the aqueous solution, with mechanical stirring (about 500 rpm) using a Teflon blade, over a period of about 20 min at room temperature. The resultant gel-like concentrated emulsion (the volume fraction of the dispersed ST/DVB phase was 0.75) was polymerized at 40°C, for 4, 8, or 16 h. After the partially polymerized concentrated emulsions were cooled to room temperature, VBC (7.87 to 19.67 mmol) was added slowly with vigorous stirring (about 700 rpm). The emulsions charged with VBC were packed tightly by shaking, sealed in glass tubes, and polymerized for 48 h at 40–45°C. The polymers were ground and washed by stirring in refluxing methanol (30–40 ml) 3 times (1 h each time). After vacuum drying, the functionalized polymer powders (agglomerates of micrometer-size latexes) were analyzed for chlorine.

Preparation of the poly(ST/DVB)-bound benzyl tributyl ammonium chloride (P-BTAC). Tributyl amine was introduced into a suspension of benzyl chloride bound to cross-linked polystyrene (2.5 g) in 15 ml of DMF (5.6 eq tertiary amine for 1.0 eq bound chlorine). The mixture was stirred under a N_2 atmosphere, at 70–75°C for 72 h. After cooling, the mixture containing the quaternized polymer powder was introduced and stirred into a 60-ml mixture of ethyl ether and petroleum ether (v/v = 1/4). The filtered polymer powder was subsequently washed three times by stirring in the above mixture (30 min each time) and dried under vacuum at 50°C.

Preparation of poly(VBC)-supported benzyl tributyl ammonium chloride (P'-BTAC). Poly(VBC) cross-linked with 1.3 mol% DVB was first prepared starting from a concentrated emulsion of VBC (containing DVB) in water (volume fraction of the dispersed phase was 0.8). A total of 3.0 g of poly(VBC) powder (6.48 mmol of Cl/g polymer) and 5.56 g of tributyl amine was introduced in 30 ml of ethyl alcohol and held at 80°C for 48 h with stirring. After reaction, the quaternized polymer powder was purified twice by stirring in acetone under reflux. The obtained poly(VBC)-supported ammonium chloride contains 2.4 mmol of N/g polymer.

Synthesis of benzyl tributyl phosphonium chloride. Benzyl tributyl phosphonium chloride was synthesized by the reaction of benzyl chloride with tributylphosphine and then recrystallized from ether and chloroform according to the method of Ref. (*4*).

Alkylation reaction in the triphase reaction system. The alkylation reaction of propylidene malonate with ethyl bromoacetate was carried out under triphase (poly(ST/DVB)-supported BTAC catalyst/inorganic base/reagent and solvent) conditions. In a typical run, propylidene malonate (0.77 g, 5 mmol), a powder of potassium carbonate (0.69 g, 5 mmol), and a powder of polymer-supported catalyst (containing 0.38 mmol of BTAC) were introduced into a flask (25 ml). Then, 10 ml of chloroform containing 2 μl of water and subsequently ethyl bromoacetate (1.24 g, 7.4 mmol) were added. The stirred slurry was held at 50–55°C under N_2 for 4 h. After reaction, about 20 ml of petroleum ether was added to precipitate the polymer catalyst. The filtered polymer was washed twice with a 20-ml mixture of ethyl ether and petroleum ether (v/v = 1) in a Buchner filter. The filtered liquid was collected and the solvent removed by evaporation. A total of 80 ml of petroleum ether was added, and the solution was cooled in a refrigerator overnight to allow crystallization of the products. The petroleum ether was gently drained out, and the remaining colorless solid washed by stirring for 30 min in 20 ml of water to remove the unreacted Meldrum acid. After vacuum drying, the isolated product containing the mono- and disubstituted Meldrum acid was subjected to thin layer chromatography (TLC) analysis (10 ml $CHCl_3$ and 4 ml CH_3CN as eluent, silica plate, Aldrich) and HPLC analysis (only the unsubstituted and the monosubstituted Meldrum acid has UV absorption at 248 nm).

RESULTS AND DISCUSSION

Grafting Hydrophobic Functional Polymer Chains to the Surface of Cross-Linked Polystyrene Latexes—A Functionalized Polymerization Pathway

One of the characteristics of the concentrated emulsion polymerization (CEP) method is that hydrophobic functional groups can be attached either to the surface of cross-linked polystyrene latexes (*5a*) or of a macroporous network (*5b*). These two morphologies are obtained depending on whether the concentrated emulsion is prepared by dispersing ST/DVB into water or by dispersing water into ST/DVB. If the desired hydrophobic functional monomer(s) (here VBC) is introduced into a partially polymerized concentrated emulsion of ST/DVB in water, then during subsequent polymerization, the functional monomer(s) copolymerizes with styrene close to the interface between oil and water, and the rest homopolymerizes and is grafted to the surface of the particles. Two variables, the time of partial polymerization of the supporting monomers ST/DVB and the amount of functional monomer VBC, which affect the extent of copolymerization and homopolymerization, were varied in the present experiments (Table 1).

The samples with benzyl chloride groups attached to poly(ST/DVB) obtained for different partial polymerization times (denoted in Table 1 as SDC-1 to SDC-4) contain almost equal amounts of benzyl chloride groups. However, the SEM pictures show that their surface morphologies are different (Figs. 1a–1d). Sample SDC-1 (Fig. 1a) was prepared by polymerizing a concentrated emulsion of a mixture of ST/DVB and VBC in water without a partial polymerization of ST/DVB; in this case, the surface of the polymer particles is smooth. In contrast, samples SDC-2, SDC-3, and SDC-4 were prepared by introducing VBC in concentrated emulsions of ST/DVB in water that had already experienced 4, 8, and 16 h of partial polymerization, respectively. The SEM pictures (Figs. 1b to 1d) show that with increasing partial polymerization time the surface morphology of the polymer particles changes from a coarse layer (Fig. 1b) to a number of nodules attached to a relatively smooth surface (Figs. 1c and 1d). The surface compositions of samples SDC-1 to SDC-4 have been determined by EDS (15 keV, with a detection depth of about 100 Å) (Fig. 2). The average concentration of the

TABLE 1

Preparation of Poly(ST-DVB) Particles with Surface Copolymerized VBC and Surface-Grafted Poly(VBC) Chains by the Partially Polymerized CEP Method[a]

ST-DVB (mmol)	VBC (mmol)	Partial polymerization time (h)[b]	VBC (mmol)/g monomers	mmol of Cl/g polymer
SDC-1 87.9–14.3	7.87	0[c]	0.72	0.73
SDC-2 87.9–14.3	7.87	4	0.72	0.69
SDC-3 8.79–14.3	7.87	8	0.72	0.75
SDC-4 8.79–14.3	7.87	16	0.72	0.71
SDC-5 87.9–14.3	11.8	16	0.92	0.97
SDC-6 87.9–14.3	15.73	16	1.18	1.32
SDC-7 87.9–14.3	19.67	16	1.41	1.52

[a] The detailed procedure has been described under Experimental.

[b] The partial polymerization time of the concentrated emulsion of ST-DVB in water before VBC was introduced.

[c] In this experiment, the monomers VBC and ST-DVB were dispersed in water simultaneously to form concentrated emulsions.

bound benzyl chloride groups within the surface layer of the functionalized particles (their diameters are below 10 μm) increases with increasing partial polymerization time. On the basis of SEM and EDS results, three kinds of surface compositions can be inferred. (1) In sample SDC-1, a mixture of ST/DVB and VBC was copolymerized; since VBC and DVB are more reactive than ST (6), much of benzyl chloride groups are buried inside the polymer particles, and the surface is rich in polystyrene. (2) In sample SDC-2, the coarse layer can be considered to contain the copolymer and chains of poly(VBC) grafted to the particles. (3) In sample SDC-4, the surface-attached nodules can be considered as poly(VBC) chains grafted to the particles, because when VBC is introduced into the concentrated emulsion, the primary poly(ST/DVB) network near the interface has already been formed. The VBC molecules will therefore homopolymerize and be grafted to the surface of the dispersed droplets. The surface morphology and composition of sample SDC-3 are similar to those of SDS-4. Samples SDC-5, SDC-6, and SDC-7 were prepared by increasing the amount of VBC introduced into concentrated emulsions that experienced the same partial polymerization time (16 h). Hence, their surface morphology and compositions are like those of SDC-4.

The FT–IR spectra (Fig. 3) of samples SDC-1 to SDC-4 show that the characteristic absorption band (7) of poly(VBC) at 607 cm^{-1} were observed only in samples SDC-3 and SDC-4, the absorption of the latter being greater than that of the former. Because there is no IR absorption band around 600 cm^{-1} for the 3 or 4-alkyl benzyl chloride compound (7), this absorption band can be assigned exclusively to poly(VBC). This result of FT–IR is in agreement with the above inference.

The polymer-supported PTC catalysts were prepared via the quaternization of the polymer supports SDC-i (i = 2 − 7) to obtain the series of P-BTAC catalysts that includes six types of catalysts (denoted SDC-

FIG. 1. Scanning electron micrographs of the benzyl chloride group-bound poly(ST/DVB) particles (SDC-i series, i = 1–4; Table 1). (a) SDC-1 sample; (b) SDC-2 sample; (c) SDC-3 sample; (d) SDC-4 sample.

a
071607 25KV 079*2NM
b
090507 30KV 073*2NM

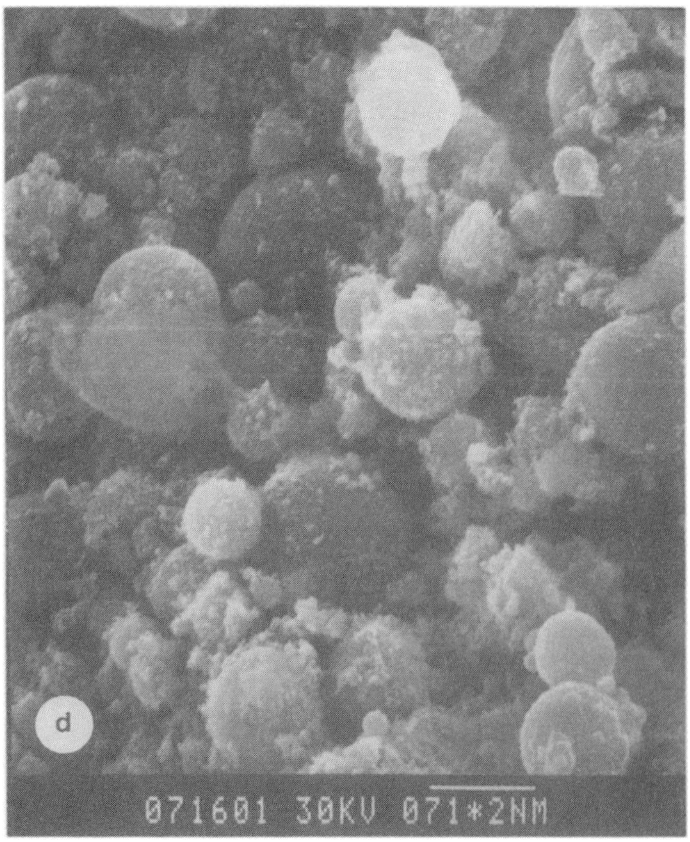

FIG. 1—*Continued*

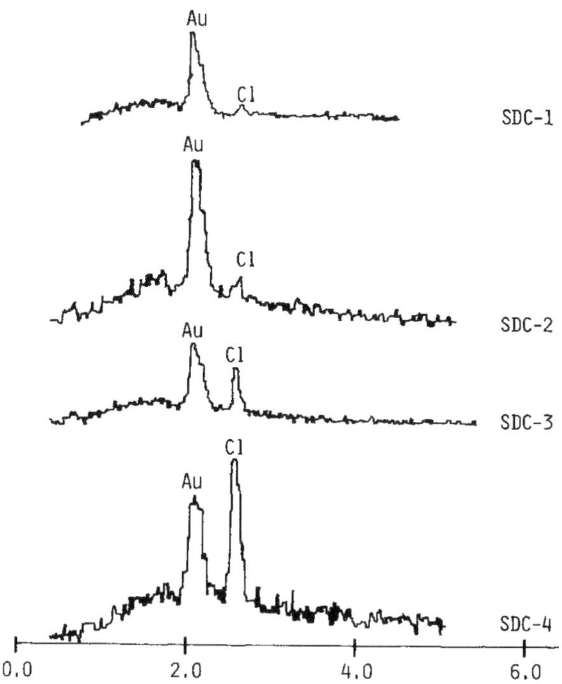

FIG. 2. Surface Analysis of the benzyl chloride group-bound poly(ST/DVB) particles (SDC-*i* series) by EDS; samples were coated with gold before analysis.

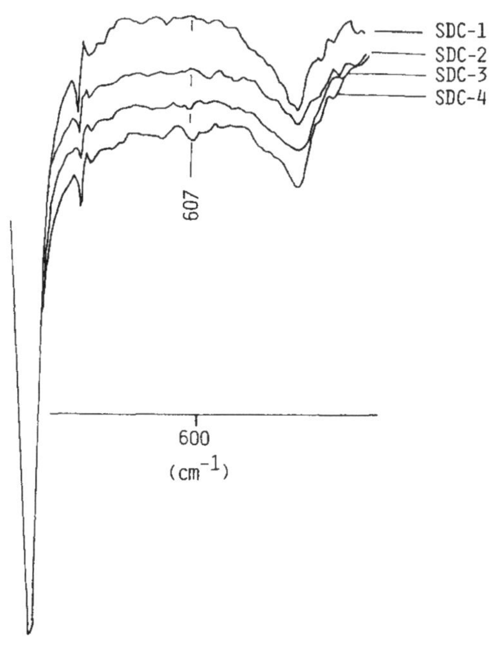

FIG. 3. FT–IR spectra of the benzyl chloride group-bound poly(ST/DVB) samples (SDC-*i* series).

TABLE 2

Quaternization of Benzyl Chloride Groups Grafted on the Surface of Poly(ST-DVB) Particles

Polymer-supported phase transfer catalysts P-BTAC	Content of pendant quaternary ammonium salt mmol of N/g polymer	Yield of quaternization (mol%)
SDC-2N	0.53	85
SDC-3N	0.59	88
SDC-4N	0.55	86
SDC-5N	0.74	88
SDC-6N	0.95	87
SDC-7N	1.09	90

2N to SDC-7N in Table 2). From the nearly equally high quaternization yields of SDC-2, SDC-3, and SDC-4 one can conclude that most benzyl chloride groups are present near the surface of cross-linked polystyrene particles and hence can react with the tributyl amine molecules. The poor ability of the quaternization medium (DMF) to swell the cross-linked polysytrene is in agreement with this conclusion.

The Effect of the Accessible Hydrophilicity on the Activity and Selectivity of the Alkylation of Isopropylidene Malonate

Three kinds of quaternary onium salts (depicted in Scheme 1) have been employed as PTC catalysts for the alkylation of isopropylidene malonate with ethyl bromoacetate (Scheme 2); see Table 3. They are: (1) the homogeneous catalyst benzyl tributyl phosphonium chloride (HOM-C), (2) poly(VBC)-supported poly(benzyl tributyl ammonium chloride) (P′-BTAC), which was prepared by the quaternization of poly(VBC) particles cross-linked with 1.3 mol% DVB (the molar yield of quaternization is 66%, other details have been described under experimental), and (3) the poly(ST/DVB)-supported benzyl tributyl ammonium chloride (P-BTAC) series. The P′-BTAC contains 2.4 mmol of quaternary ammonium per gram

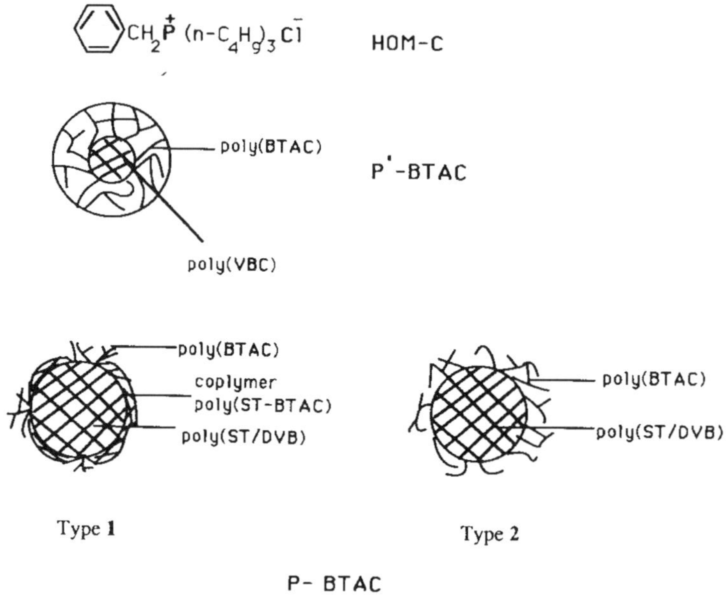

SCHEME 1. Three kinds of phase transfer catalysts employed for the alkylation of isopropylidene malonate.

of polymer; hence the bound benzyltributyl ammonium chloride is present as a thick hydrophilic layer on a poly(VBC) core. In the P-BTAC series, hydrophilic quaternary onium groups are bound to the hydrophobic support, and therefore there are both hydrophilic and hydrophobic regions on the surfaces of the polymer particles.

In the present experiments, the activity for alkylation of isopropylidene malonate is enhanced by: (1) using a highly polar and reactive compound, ethyl bromoacetate; and (2) adding a trace of water (about 1–2 μl, since too much water hydrolyzes the Meldrum acid) into the reaction medium (chloroform is the reaction medium and potassium carbonate the base). The trace amount of water provides carbonate anions for abstracting protons from the Meldrum acid and its monosubstituted compound.

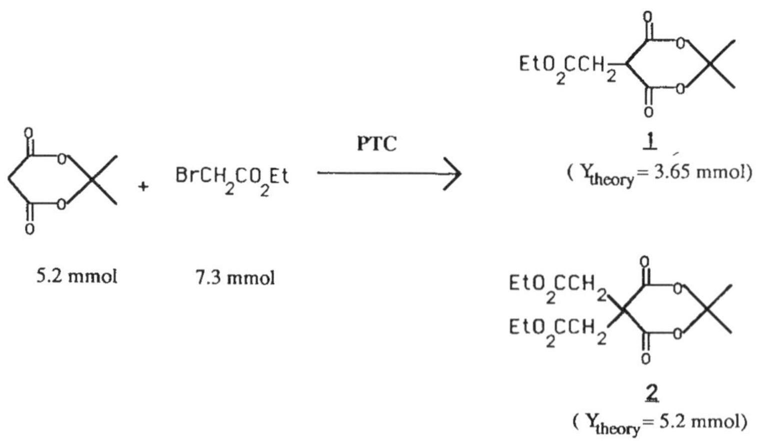

SCHEME 2. Alkylation of isopropylidene malonate with ethyl bromoacetate under phase transfer conditions.

TABLE 3

Alkylation of Isopropylidene Malonate Catalyzed by Three Kinds of Quaternary Onium Salts (PTC Catalysts)[a]

Quaternary onium (PTC catalyst)	mmol of quaternary onium salt			
	0.62	0.38	0.19	0.095
	Alkylation yields (mol%)[b]			
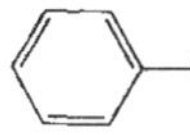—$CH_2P^+(n\text{-}C_4H_9)_3Cl^-$	66	61	44	34
P′-BTAC	61	48	43	38
P-BTAC SDC-3N	37(7)[c]	40(8)	41(8)	—

[a] In a typical run, isopropylidene malonate (5.2 mmol), ethyl bromoacetate (7.3 mmol), K_2CO_3 (5 mmol), and quaternary onium salts (the amounts are given in millimoles) in 10 ml chloroform were stirred under N_2 at 55–60°C for 4 h.

[b] Alkylation yields for the bisalkyl-substituted product **2** as indicated in Scheme 1.

[c] Yields in parentheses for the monoalkyl-substituted product **1** as indicated in Scheme 1.

The results of the alkylation experiments show that when the HOM-C and the P′-BTAC catalysts are used, the disubstituted product **2** (Fig. 4a) is selectively generated, although small amounts of catalyst were employed. However, when the catalyst series P-BATC was employed, some monosubstituted products **1** were also obtained (Fig. 4b). The catalyst SDC-2N had the lowest yields of mono- and disubstituted products but the highest ratio of the two. For the other catalysts in the series, SDC-3N to SDC-7N, the alkylation yield increases but the mono-/di- ratio decreases (Fig. 5). This sequence is due to the increased accessibility of the pendant benzyl tributyl ammonium groups on the P-BTAC particles when the time of partial polymerization or the concentration of VBC increases. The strong tendency to form the disubstituted product **2** when the HOM-C and P′-BTAC catalysts are employed arises from the accessibility of hydrophilic catalytic onium cations that are involved in the ion-pair formation with the anions of both unsubstituted and monosubstituted Meldrum acid (Scheme 3). However, in the P-BTAC series, the copolymer bound cations are surrounded by phenyl groups. The proportion of copolymer bound cations is greater when the support SDC-*i* is prepared using a short partial polymerization time. The hydrophobic steric constraints around the quaternary onium groups, such as in the case of SDC-2N, decrease the capacity of the polymer catalyst for complexing the anions or adsorbing the polar halide molecules. As a result, the activity for the conversion of the Meldrum acid decreases as the partial polymerization time decreases. The decrease of the ratio of mono-/disubstituted products with increasing partial polymerization time can be explained as follows: the monoalkylated Meldrum acid bears three electron withdrawing groups and has a chair structure (*1a*), whereas the unsubstituted Meldrum acid bears two electron withdrawing groups and has a boat structure. The former is thus more susceptible to steric constraints than the latter and the steric constraints are greater for shorter partial polymerization times.

A kinetic study regarding the formation of the monosubstituted product **1** and its conversion to the disubstituted product **2** was also performed, by employing SDC-2N, SDC-4N, and P′-BTAC as catalysts (Fig. 6). In the reaction system involving P′-BTAC,

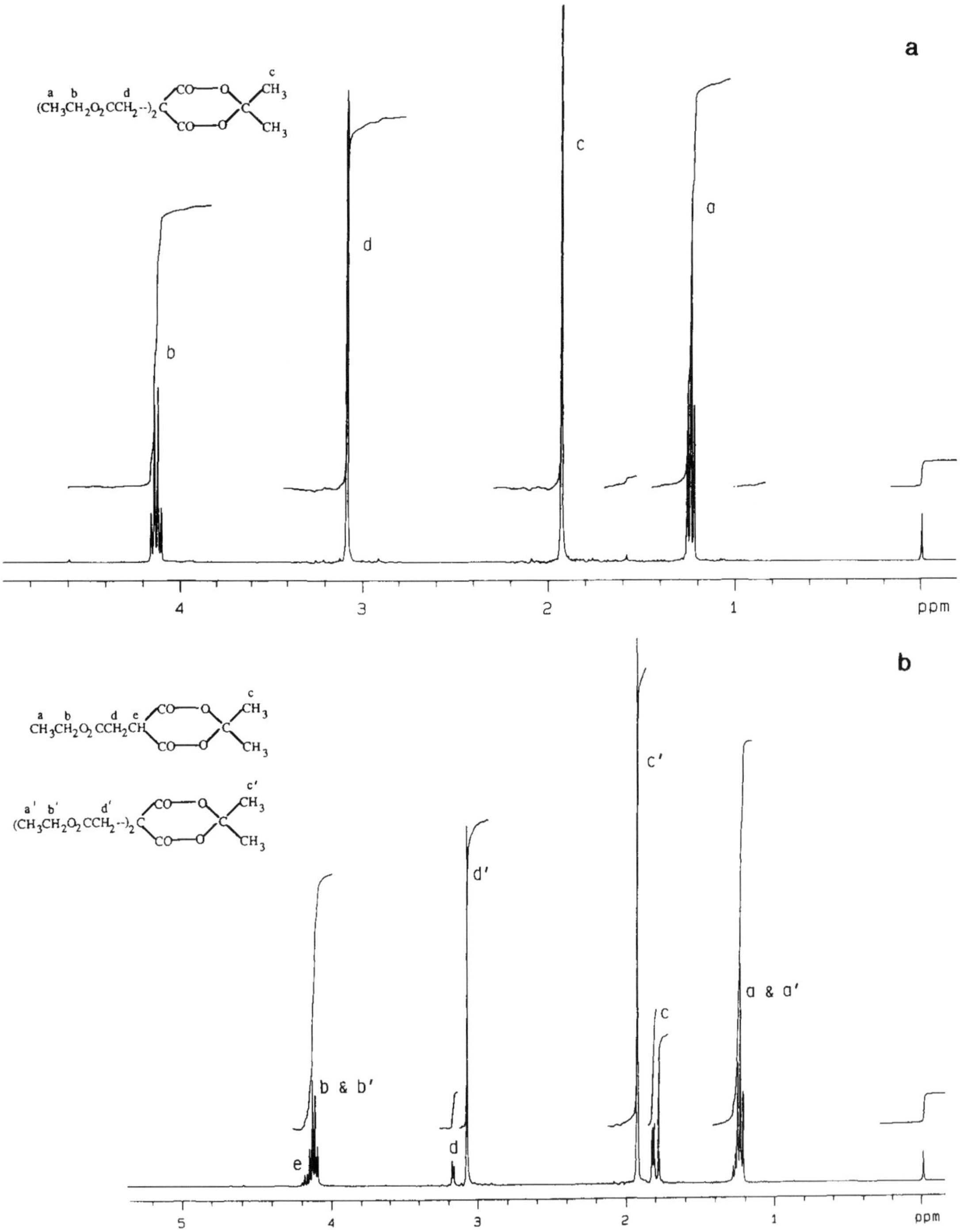

FIG. 4. H[1] NMR of the substituted Meldrum acid. (a) The disubstituted Meldrum acid, 5-di(ethoxy-carbonyl methylene) isopropylidene Malonate **2** obtained by using the catalyst P′-BTAC; (b) the mixture of disubstituted Meldrum acid and monosubstituted Meldrum acid, 5-(ethoxy-carbonyl methylene) isopropylidene malonate **1**, obtained by using the catalyst SDC-6N.

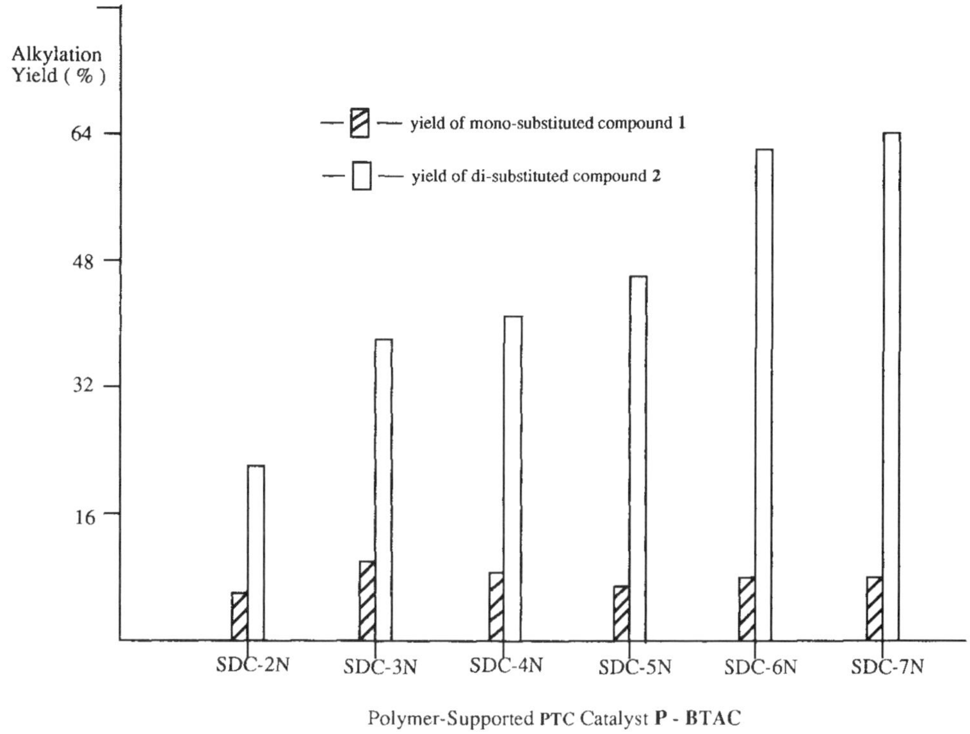

FIG. 5. Comparison of the ratios of mono-/disubstituted Meldrum acid in the products obtained by using catalysts SDC-*i*N (i = 1–7). The reaction conditions are specified under Experimental.

product **1** is more rapidly consumed than in the reaction systems involving SDC-4N or SDC-2N. This observation is in agreement with the high accessibility of the onium cations in the swollen layer of the P′-BTAC particles. In contrast, due to their dual surface properties, catalysts SDC-4N and SDC-2N show weaker tendencies to transform the Meldrum acid to products **1** and **2**. Although in Fig. 6 SDC-4N exhibits a higher yield for product **1** than SDC-2N, one can see from Fig. 5 that SDC-2N has a higher selectivity for product **1** than SDC-4N. This occurs, for steric reasons, since the former has less hydrophilic accessibility of onium cation than the latter. Because of this hydrophilic accessibility, SDC-4N leads to a higher yield of product **1** than SDC-2N. In summary, although SDC-2N has the highest selectivity for the monosubstituted product, it provides the lowest yields for the monosubstituted and disubstituted products.

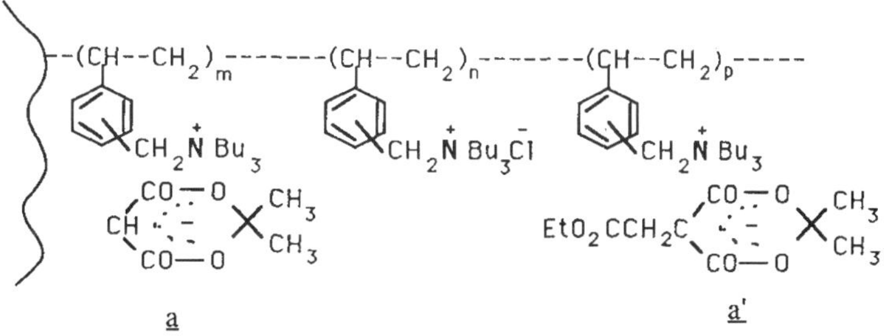

SCHEME 3. Ion pairs between the pendant quaternary onium cations and the anions of unsubstituted a and mono-substituted a′ Meldrum acid.

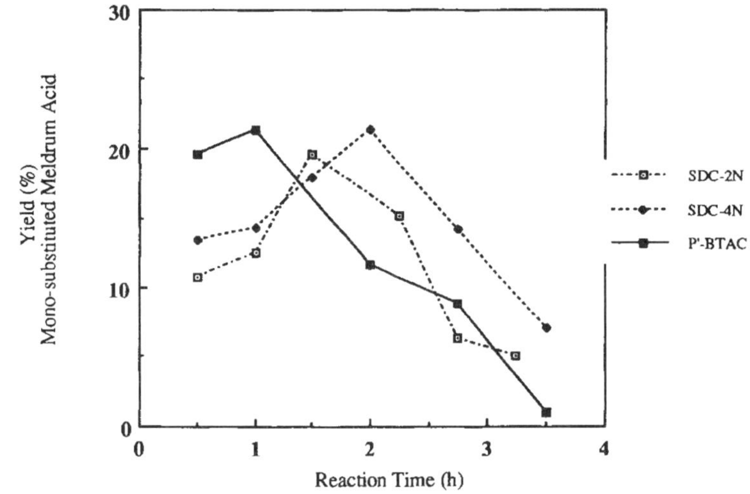

FIG. 6. Comparison of the activities of P'-BTAC, SDC-4N, and SDC-2N for the formation and conversion of the monosubstituted Meldrum acid. The measurements were carried out for the reaction conditions as specified under Experimental.

Hydrophobic and Hydrophilic Cooperations of Polymer-Supported Catalysts in Michael Addition

The SDC-7N and the HOM-C catalysts were also used for the nucleophilic addition (Michael addition) of Meldrum acid to butyl acrylate (Scheme 4). The polymer catalyst has a higher activity than the homogeneous catalyst probably because the hydrophobic butyl acrylate molecules are adsorbed on the hydrophobic region of the particle surface from where they can approach the hydrophilic quaternary onium groups that bind the Meldrum acid anions as ion pairs. From this experimental observation, one can conclude that if both hydrophilic and hydrophobic regions coexist on the surface of latexes, the dual functions can cooperate to generate a phase transfer pathway from the hydrophobic to the hydrophilic region.

SUMMARY

Benzyl chloride groups can be bound to the surface of poly(ST/DVB) latexes by introducing VBC into a partially polymerized emulsion of ST/DVB in water and polymer-

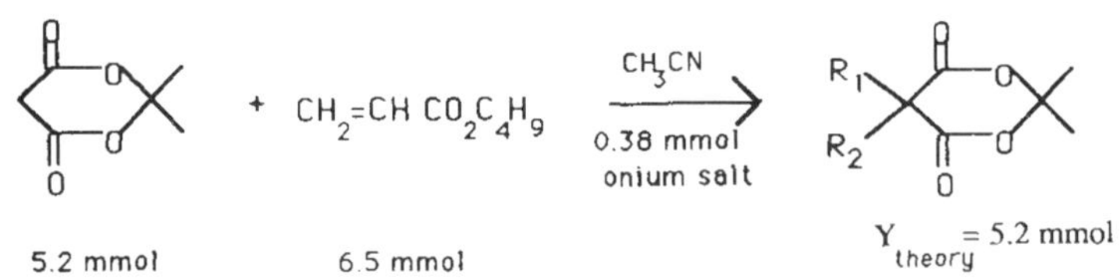

		Yield of the Alkylation (mole %)		
R_1	R_2	$C_6H_5CH_2\overset{+}{P}(n\text{-}C_4H_9)_3\overline{Cl}$	SDC-7N	H^1-NMR ($CDCl_3$/TMS) δ (ppm)
H	$CH_2CH_2CO_2C_4H_9$	7	31	0.96 (t, 3H); 1.35 (m,2H); 1.58 (m,2H); 4.08 (t,2H); 2.62 (t,2H); 2.38 (q,2H); 3.98 (t, 1H); 1.80 (t, 6H)

SCHEME 4. Michael addition of isopropylidene malonate to an electrophilic double bond.

izing again. The surface morphology and the surface composition of the prepared particles can be changed by varying the partial polymerization time and the amount of VBC introduced. If the partial polymerization time is short, VBC molecules copolymerize with styrene molecules on the surface and also homopolymerize and are grafted to the surface of the particles; if the time is long, VBC molecules homopolymerize and are grafted to the surface.

A polymer-supported phase transfer catalyst is obtained by transforming the surface bound benzyl chloride groups to benzyl tributyl ammonium chloride (BTAC) groups. Two extreme types of reactive environments can be obtained on the surface of the particles: (1), copolymer with BTAC and phenyl groups, and some polymer chains of BTAC grafted to the poly(ST/DVB) particles when the partial polymerization time is short; (2) chains of BTAC grafted to the poly(ST/DVB) particles when the time of partial polymerization is long. In the alkylation reaction of isopropylidene malonate with ethyl bromoacetate, the activity of type 1 catalyst is smaller because of the hydrophobic steric constraints for the anions of both unsubstituted and monosubstituted isopropylidene malonate due to the copolymerized fraction of BTAC. In addition, the ratio of monosubstituted to disubstituted product is increased, because the copolymerized fraction of BTAC has less recognition for the anion of monosubstituted isopropylidene malonate. In contrast, the type 2 catalytic environment promotes the activity of BTAC groups and generates the disubstituted product almost exclusively due to the accessibility of the grafted poly(BTAC) chains to both anions. The existence of separated hydrophilic and hydrophobic regions on the surface of type 2 catalyst stimulates the adsorption of both hydrophilic and hydrophobic reagents on the surface of the particles. The activity of the heterogeneous catalyst is much greater than that of the homogeneous catalyst when the Michael addition of the isopropylidene malonate to butyl acrylate is employed as a probe reaction.

REFERENCES

1. (a) McNab, H., *Chem. Soc. Rev.* **7,** 345 (1978); (b) Fieser, L. F., and Fieser, M., "Reagents for Organic Synthesis," p. 526. Wiley, New York, 1967; (c) Patil, S. V., Mane, R. B., and Salunkhe, M. M., *Chem. Ind.* **7,** 24 (1991).
2. Ruckenstein, E., and Kim, K-J., *J. Appl. Polym. Sci.* **36,** 907 (1988).
3. Ruckenstein, E., and Park, J-S., *J. Polym. Sci. Polym. Lett.* **26,** 529 (1988).
4. (a) Henderson, W. A., and Buckler, S. A., *J. Am. Chem. Soc.* **82,** 5794 (1960); (b) Starks, C. M., *J. Am. Chem. Soc.* **93,** 195 (1971).
5. (a) Hong, L., and Ruckenstein, E., *React. Polym.* **16,** 181 (1991/1992); (b) Ruckenstein, E., and Hong, L., *Chem. Mater.,* **4,** 122 (1992).
6. Guyot, A., *React. Polym.* **10,** 113 (1989); Guyot, A., and Bartholin, M., *Prog. Polym. Sci.* **8,** 277 (1982).
7. Pouchert, C. J., "The Aldrich Library of FT–IR Spectra," Vol. 2, p. 1208 and Vol. 1, pp. 985 and 992.

Onium Chains–Silica as Catalytic Support: Application to Hydroformylation

L. HONG AND E. RUCKENSTEIN[1]

Department of Chemical Engineering, State University of New York at Buffalo, Box 60, Buffalo, New York 14260-4200

Received January 29, 1993; revised April 13, 1993

Polymer chains of quaternary oniums are deposited on the surface of hydrophilic silica by the radical polymerization of water-soluble quaternary onium monomers present in a water medium which imbibes the porous silica. The modified (quaternized) silica surface which contains a small amount of water is a more suitable support than silica, since it provides anchors for the catalytic species and sites for the adsorption of reactant molecules. It is used as substrate for the immobilization of monosulfonated triphenylphosphine (DPM) and of its rhodium carbonyl complex $Cl(CO)Rh(DPM)_2$, through an ion-exchange process. The immobilized catalytic system containing a small amount of water was employed in the hydroformylation of styrene and ethyl 10-undecenoate. The presence of ethanol in the reaction medium is necessary to achieve a high yield of aldehydes. For styrene, the ratio of the isomer to the normal aldehyde is greater than 6, while for ethyl 10-undecenoate, it is $\frac{1}{8}$. Structural factors that affect the regioselectivities of the aldehydes are suggested as possible explanations for the above ratios. The performance of the catalyst has remained almost unchanged during several reuses.

INTRODUCTION

The use of very expensive complexes as homogeneous catalysts (usually they contain noble metal ions or chiral ligands) generated the problem of catalyst recovery. A number of techniques of heterogenization of the homogeneous catalysts have been therefore developed (*1*). Among them the use of water-soluble complexes in water–oil biphase catalytic systems represents an important development (*2*), which offers the following advantages over other immobilization procedures: (1) one can bind water-soluble groups such as carboxyl, sulfonate, or onium salts to various ligands (*3*); (2) the catalyst and the reaction products can be separated by phase-separation; (3) the reactivities and selectivities can be controlled by modifying the structure of the water–oil interface with surfactants, polymers, or polymerized surfactants (*4*); and (4) the catalyst often exhibits higher stability under high pressure of carbon monooxide in carbonylation reactions (*5*). The sulfonated phenylphosphines (*6*) have been frequently used as water-soluble ligands, and the transition-metal complexes thus obtained have been used to catalyze the hydroformylation (*6*), the alkylation (*7*), the hydrogenation (*8*), the ring-opening metathesis polymerization (*9*) reactions, and the photochemical reduction of water (*10*).

The hydroformylation reaction (oxo reaction) of an olefin with syn-gas (H_2/CO) to obtain an aldehyde with one additional carbon, is of industrial importance in the synthesis of a number of fine organic compounds (*11*). The technology of using sulfonated phosphine–rhodium carbonyl catalysts for the hydroformylation of propene, developed by Rhône–Poulenc and Ruhrchemie (*12*), has solved in a satisfactory manner the problem of reusing this extremely expensive catalyst which is highly active and selective under mild reaction

[1] To whom correspondence should be addressed.

This paper, which was originally published in **Journal of Catalysis**: 143, 52–63 (1993), is republished with permission.

conditions. Although this development has potential significance in the oxo reaction of higher molecular weight olefins, including olefins with functional groups, in water–oil biphase systems, the accessibility of the nonvolatile olefin molecules, that are usually dissolved in an organic solvent, to the water-soluble catalyst is always very low. For improving this shortcoming, alkyl substituents have been bound to the sulfonated phenylphosphine ligands to increase their surface activity (*12*), or a small amount of a polar organic solvent, such as ethyl acetate, ethanol, or acetone, was added to the two-phase system to generate a border layer with a suitable polarity able to mediate the contact between the catalyst and the reactant. These methods increase, however, the risk of losing rhodium species to the organic phase. One of the best solutions seems to be offered by the combination of high-performance ultrafiltration membranes of aromatic copolyamides with the catalyst; it resulted in a highly efficient recovery of rhodium complexes (*13*), but represents a more complicated technology.

Recently, an interesting innovation based on the adsorption of the water-soluble complex $HRh(CO)(TPPTS)_2$ {where TPPTS is triphenylphosphine trisulfonate $P(m\text{-}C_6H_4SO_3Na)_3$} and of a small amount of water on a high-surface-area hydrophilic support (silica or porous glass) was suggested by Davis (*14*). In this procedure, the hydrophilicities of the ligand TPPTS and of the support generate sufficiently large interaction energies to ensure the stability of the immobilization. The water content of the catalyst greatly influences its performance, the optimum water-content being 8.5 wt%. When the water content was increased to 45 wt%, the yield of the aldehyde fell to less than 10% of its optimum value (*15*). This probably happens because as the water content increases the immobilized catalyst becomes increasingly segregated from the hydrophobic olefin molecules. It is, however, not easy to uniformly coat a high-surface-area support with small amounts of water, particularly in large-scale processes.

In this paper, we suggest another method for the immobilization of the water-soluble sulfonated phenylphosphine rhodium–carbonyl complexes to a silica support. The sulfonated triphenylphosphine $Ph_2P(m\text{-}C_6H_4SO_3Na)$ (denoted DPM) is attached to a silica surface on which poly(vinylbenzyl triethylammonium chloride) {P(VEAC)} was previously deposited. The quaternization of the silica surface was carried out by immersing the porous silica into an aqueous solution of VEAC containing the initiator $K_2S_2O_8$, followed by the evaporation of most of the water. The solid containing about 60 wt% water thus obtained was introduced into a high boiling hydrocarbon medium for polymerization. The catalytic species were bound to the quaternized surface by immersing the functionalized silica powder in an aqueous solution containing the rhodium complex $Cl(CO)Rh(DPM)_2$ and the ligand DPM.

The presence of the quaternary onium groups on the water containing surface increases the compatibility between the water and oil phases, thus facilitating the contact between the olefin molecules and the catalytic sites. The performance of the present catalyst was found to be enhanced when ethanol was introduced into the system. For example, when ethyl 10-undecenoate was hydroformylated, the yield of the aldehydes increased about five-fold as the volume fraction of ethanol was increased from 1.2 to 11% in the mixture of ethanol and cyclohexane. This occurs probably because the polymer chains containing pendant ion-pairs are swellable by ethanol molecules, and this increases the flexibility and hence the accessibility of the catalytic polymeric chains.

In the present paper, both styrene and ethyl 10-undecenoate were used for hydroformylation at 85°C and 750 psi (CO/H_2 = 1); the molar ratio of the phosphine ligand to rhodium was 10, and of the olefins to the rhodium complex about 4000 for styrene and 2400 for ethyl 10-undecenoate. We observed that when the hydroformylation of styrene was carried out for 15 h, 56% of styrene was converted to aldehydes, with a ratio of 2-phenylpropanal to 3-phenylpropanal greater than 6. The latter feature is important in the synthesis of 2-(*p*-isobutylphenyl) propanal from *p*-isobutylstyrene; the subsequent oxidation of this aldehyde generates ibuprofen, an over-the-counter analgesic (*16*). The aliphatic compound ethyl ester of 10-undecylenoic acid, which is the thermolysis product of castor oil, was also employed for the oxo reaction. The linear product 11-ethoxylcarbonyl undecanal can be used for the synthesis of the musk perfumes and nylon-12. We found that the ratio of linear (n) to branched (i) aldehyde can be as high as 8, with 40% conversion to aldehydes when the ethanol volume fraction in the reaction system was 11%. The catalyst was reused in three successive runs for both reaction systems, the yields of the aldehydes and the n/i ratios remaining almost the same.

EXPERIMENTAL

Reagents

p- and *m*-Vinylbenzyl chloride and styrene (Aldrich, 99%) were passed through an inhibitor removal column before use. Ethyl 10-undecenoate (Aldrich 98%), triphenylphosphine (Aldrich, 99%), fuming sulfuric acid (Aldrich, free SO_3 content 27–33%), triethylamine (Aldrich, 99%), rhodium chloride hydrate (Johnson Matthey 99.9%), syngas (Linde, $CO/H_2 = 1$), formaldehyde (Aldrich, 37 wt% aq.), sodium hydroxide (Aldrich, 99.99%), cyclohexane (Aldrich, HPLC), and other solvents were used without further purification. Silica catalyst support (Aldrich, grade 951) and silica gel (Aldrich, TLC high-purity grade, without binder) were used as support and for the chromatographic separation, respectively.

Instruments

A 150-ml high-pressure reactor (Parr 4841) was employed for hydroformylation. Energy dispersive spectroscopy surface analysis (EDS) was performed on a PGT/IMIX field emission electron microscopy equipment. The surface functionalized silica particles were investigated with a scanning electron microscope (SEM) (Hitachi S-800) and the specific surface area was measured with a physical adsorption analyzer (Accusorb 2100E, Micromeritics). The infrared absorption spectra of the supported catalysts and of the oxo reaction products were obtained with a Mattson Alpha Centauri FT-IR instrument using a KBr cell. ^{1}H-NMR spectra of the hydroformylation products were recorded on a VXR-400 spectrometer using d-chloroform as solvent.

Preparation of the Monomer and Ligand

The *m*-, *p*-vinylbenzyl triethyl ammonium chloride (VEAC) and phosphonium chloride (VEPC) were synthesized according to reported methods (*17*). The monosulfonated triphenylphosphine (DPM) was synthesized employing the procedure of Ahrland *et al.* (*18*).

Polymerization on the Surface of Silica Particles

A powder of 2.0 g silica was introduced with magnetic stirring into an aqueous solution containing 0.6 g of VEAC, 8 mg of $K_2S_2O_8$ and 10 ml water, placed in a 50-ml beaker. Then a moderate air flow was employed to evaporate most of the water with stirring. After about 1 h of stirring, a wet yellow solid containing about 60 wt% water was obtained. The wet solid was immersed into 10 ml of decane located in a glass tube. Further, the tube was sealed with a rubber septum and Ar was bubbled through the liquid for 5 min with ultrasonic degassing. The polymerization was carried out in a water bath at 60–70°C for 12 h. The obtained quaternized silica was purified by extraction with chloroform for 20 h in a Soxhlet extractor, since both the VEAC and the unbound P(VEAC) can be washed out with $CHCl_3$. Finally, after vacuum drying, 2.6 g of surface quaternized silica were obtained {denoted as Si–P(VEAC)-(1)}. Several quat-

TABLE 1

Deposition of Poly(Quaternary Onium Chloride) on the Surface of Silica

Entry	Q^+Cl^- [a] (g, mmol)	$Q^+ Cl^-$ (mmol) / $Q^+ Cl^-$ (g) + Silica (g)[b]	Content of $Q^+ Cl^-$ [c] in the quaternized silica (mmol of $Q^+ Cl^-$/g)	Percentage of $Q^+ Cl^-$ grafted to silica (%)
Si–P(VEAC)-(1)	VEAC (0.6, 2.37)	0.91	0.86	94.5
Si–P(VEAC)-(2)	VEAC (0.92, 3.64)	1.25	1.14	91.2
Si–P(VEAC)-(3)	VEAC (1.22, 4.82)	1.49	1.23	82.6
Si–P(VEPC)-(E)	VEPC (0.65, 2.41)	0.91	0.91	100
Si–P(VBPC)-(B)	VBPC (0.85, 2.40)	0.84	0.56	66.7

[a] VEAC is 3,4-vinylbenzyl triethyl ammonium chloride; VEPC is 3,4-vinylbenzyl triethyl phosphonium chloride; VBPC is 3,4-vinylbenzyl tributyl phosphonium chloride; E, ethyl; B, butyl; and Q^+ is the onium cation.

[b] Two g of silica were used in all the systems.

[c] Elemental analysis results.

ernized silica have been prepared with various amounts of the quaternary onium monomer (see Table 1).

Binding $Cl(CO)Rh(DPM)_2$ and DPM to the Quaternized Surface of Silica

The procedure is similar to that described in Ref. (*19*): a white powder of DPM (0.19 g, 0.47 mmol) was dissolved in 5 ml of an aqueous solution of formaldehyde. To this solution, 5 ml of a reddish aqueous solution of $RhCl_3$ (9.45×10^{-3} *M*) (the molar ratio of $RhCl_3$/DPM is 0.1) were added with stirring, followed by the addition of a 1 ml solution of NaOH (0.5 *M*). After 15 min, an orange-yellow solution containing the complex $Cl(CO)Rh(DPM)_2$ was obtained. Then 0.5 g of Si–P(VEAC)-(1) (containing 0.44 mmol of VEAC) were introduced into the solution with stirring. The color gradually changed from orange-yellow to yellowish. An air flow was employed to evaporate most of the water. To eliminate the unreacted formaldehyde the yellowish solid was washed twice, each time with 30 ml of ethanol–THF ($v/v = 1$). After filtration and vacuum drying at room temperature, 0.62 g of a yellowish powder were obtained. One gram of prepared supported catalyst contains 0.54 mmol of the DMP, 0.068 mmol of $Cl(CO)Rh(DPM)_2$, and 0.64 mmol of VEAC units.

Hydroformylation of Styrene

The supported catalyst powder (0.3 g containing 0.16 mmol of DMP, 0.021 mmol of $Cl(CO)Rh(DPM)_2$, and 0.19 mmol of VEAC units) was blended with 0.16 g of water. The wet catalyst powder, 10 ml styrene (9.1 g, 88 mmol), 40 ml cyclohexane, and 0.5 ml ethanol were added to the high-pressure reactor. The system was purged with syn-gas four times (to about 10 psi each time) with slow stirring, before reaction. Then the syn-gas was introduced until a pressure of 750 psi was achieved, which increased to 810 psi when the temperature of the reaction system became 85°C. The reactor was charged again with syn-gas as soon as the pressure fell to 650 psi. After 15 h of reaction with mechanical stirring, a colorless liquid phase was taken out from the reactor with a liquid removal tube. The yellow catalyst powder remained precipitated at the bottom of the reactor. The reaction mixture was vacuum concentrated in a rotary evaporator for 2 h at 40–50°C to remove the cyclohexane and the unreacted styrene. Finally, 8.0 g of oil-like liquid were obtained, from

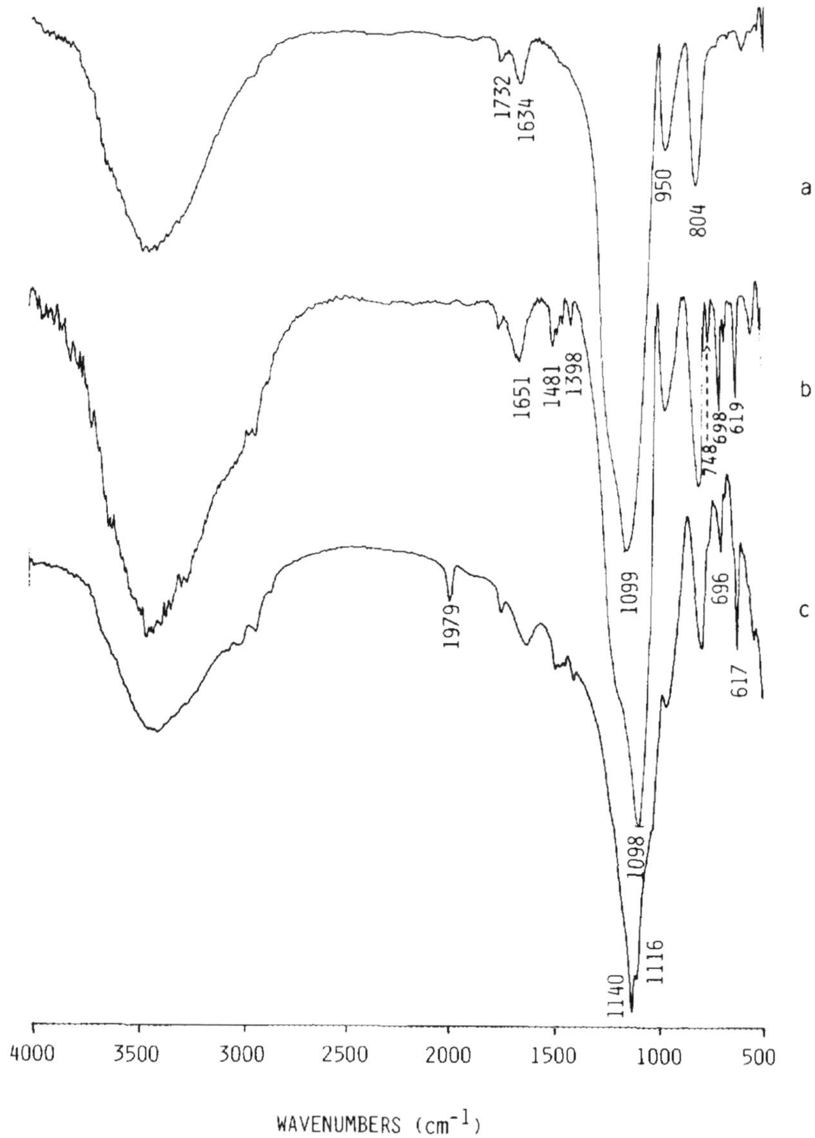

FIG. 1. (a) FT-IR spectrum of the silica substrate; (b) FT-IR spectrum of Si–P(VEAC)-(1) (the amount of deposited polymer is given in Table 1); and (c) FT-IR spectrum of the supported catalyst: Si–P(VEAC)-(1)/Cl(CO)Rh(DPM)$_2$-DPM.

which 1.4 g of polystyrene oligomers could be separated when introduced in ethanol. The ^{1}H-NMR analysis of the mixture of normal and iso phenylpropanals indicated: δ (ppm) = 2.78 (t, -$\underline{CH_2}$CHO); 3.02 (t, Ph$\underline{CH_2}$-); 7.2–7.3 (m, Ph-); 9.82 (s, -C$\underline{H}$O) for normal-phenylpropanal; and δ (ppm) = 1.5 (d, -CH_3); 3.64 (q, -PhC$\underline{H}$<); 7.2–7.3 (m, Ph-); 9.75 (s, -C$\underline{H}$O) for iso-phenylpropanal. The ratio of normal and iso aldehydes was evaluated by the integration of the aldehyde proton signals of the n and i aldehydes.

Hydroformylation of Ethyl 10-Undecenoate

Ethyl 10-undecenoate (10.6 g, 50 mmol), 5 ml of ethanol and Si-P(VEACA)-(2) supported catalyst (0.3 g containing 0.16 mmol of DMP, 0.021 mmol of Cl(CO)Rh(DPM)$_2$, and 0.24 mmol of VEAC units) were used,

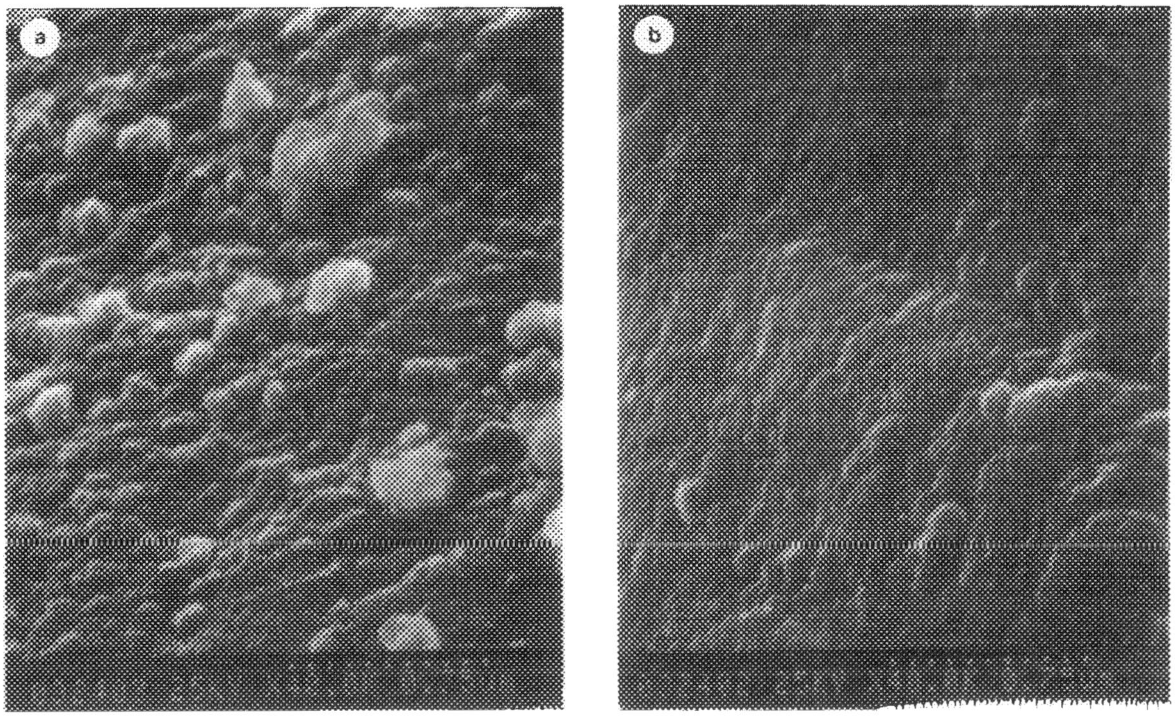

FIG. 2. Scanning electron micrographs of (a) silica and (b) quaternized silica (Si–P(VEAC)-(2) of Table 1).

the other amounts and conditions being the same as for the hydroformylation of styrene. After reaction, a colorless oil (10.7 g) was obtained. The aldehydes were separated from the unreacted olefin by vacuum liquid chromatography (*20*) using TLC silica gel as the stationary phase and mixtures of petroleum ether (30–60°C) and ethyl ether as grade eluant. 5.12 g of a mixture of aldehydes were separated, with a ratio of normal to iso products of 8, which was evaluated by both TLC and ^{1}H-NMR (comparing the integrated signals of -C$\underline{H}$O). The chemical shifts are: δ (ppm) = 1.28 (m, -$(CH_2)_8$-); 1.62 (t, -$OCH_2\underline{CH_3}$); 2.26 (t, -$\underline{CH_2}CO_2Et$); 2.42 (t, -$\underline{CH_2}CHO$); 4.12 (q, -$O\underline{CH_2}CH_3$); 9.76 (s, -C$\underline{H}$O) for $OHCCH_2(CH_2)_8CH_2CO_2CH_2CH_3$, and δ (ppm) = 1.08 (d, -$CH(\underline{CH_3})CHO$); 9.60 (s, -C$\underline{H}$O) for the iso aldehyde. The resonance absorption of the proton -$\underline{CH}(CH_3)CHO$ is overlapped with the CH_2 absorption; the other proton resonance absorptions of $OHCCH(CH_3)(CH_2)_7CH_2CO_2CH_2CH_3$ coincide with those of the normal product.

RESULTS AND DISCUSSION

Polymerization in the Aqueous Layer Adhering to the Surface of Silica

Most of the previous investigations which used silica as substrate for immobilization of functional groups converted the surface Si–OH groups to suitable anchors through several reactions steps (*21*). In this paper, hydrophilic polymers of quaternary ammonium chloride or phosphonium chloride have been attached to the surface of silica via the polymerization of monomers in the adhering aqueous layer on silica. Silica was immersed into an aqueous solution of a quaternary onium monomer and initiator, and the suspension was concentrated by evaporating most of the water until the system remained with about 60 wt% water. This wet solid was introduced in decane, which plays the role of a heat-transfer medium and

TABLE 2

Specific Surface Area of the Silica with Surface Grafted Poly (Q^+ Cl^-)

Sample	Silica support	Si–P(VEAC)-(1)	Si–P(VEAC)-(2)	Si–P(VEPC)-(E)
Specific surface area (m^2/g)	347	215	150	250

also prevents the escape of water. Preserving a water layer on the surface of silica facilitates the polymerization process during the subsequent heating at 60–70°C for 12 h. The quaternized silica powder is finally extracted with chloroform in a Soxhlet extractor to eliminate the unreacted monomer and some of the water. A high percentage of the monomer was attached as polymer chains to the silica, as demonstrated by the elemental analysis given in Table 1. The comparison of the IR spectra (Fig. 1a) of silica and Si–P(VEAC)-(1) (Table 1) indicates that P(VEAC) is present in the porous silica (Fig. 1b). The electron micrographs (Fig. 2) allow to compare the surface morphologies of silica and quaternized silica after the deposition of P(VEAC) on its surface. One can observe that the roughness is higher on silica, which implies that the P(VEAC) chains fill the cavities of silica. This occurs because most of the water and monomer can penetrate in these cavities with the assistance of ultrasonic degassing before polymerization. The specific surface area measurements show that the inclusion of the polymer in the porous silica reduces, as expected, the specific surface area (Table 2).

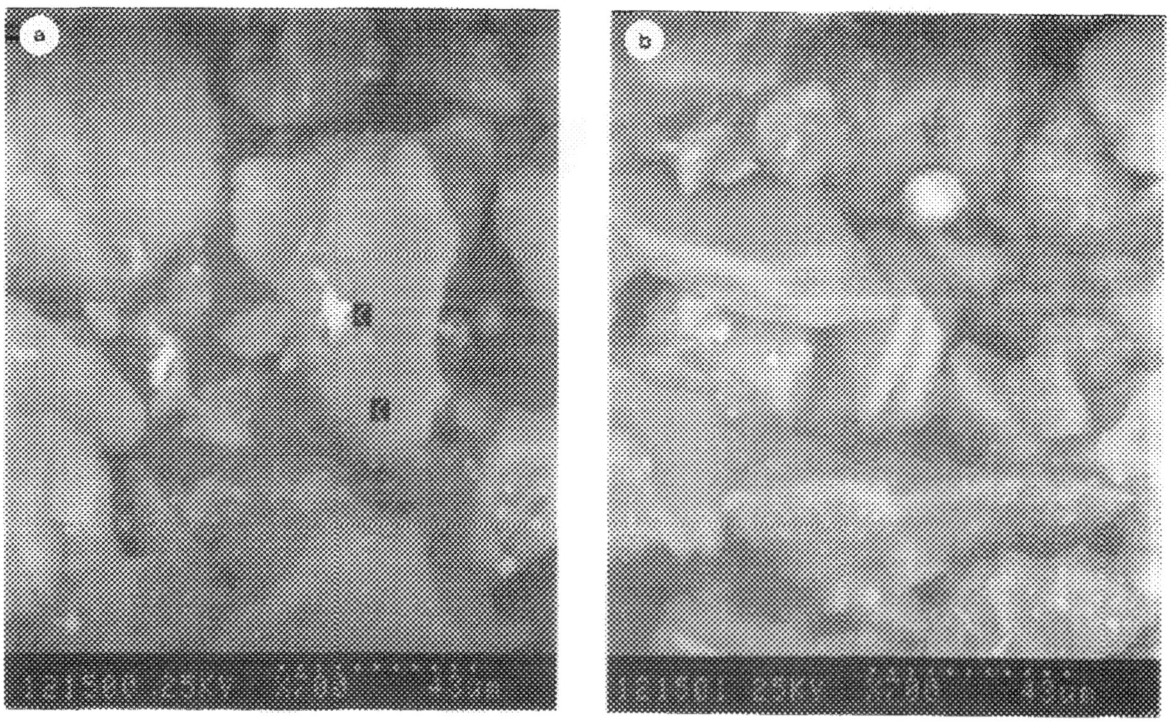

FIG. 3. (a) The backscattered electron image (BEI) of the supported catalyst: Si–P(VEAC)-(1)/Cl(CO)Rh(DPM)$_2$-DPM. (b) The BEI of the same supported catalyst after it was reused three times in the hydroformylation of styrene.

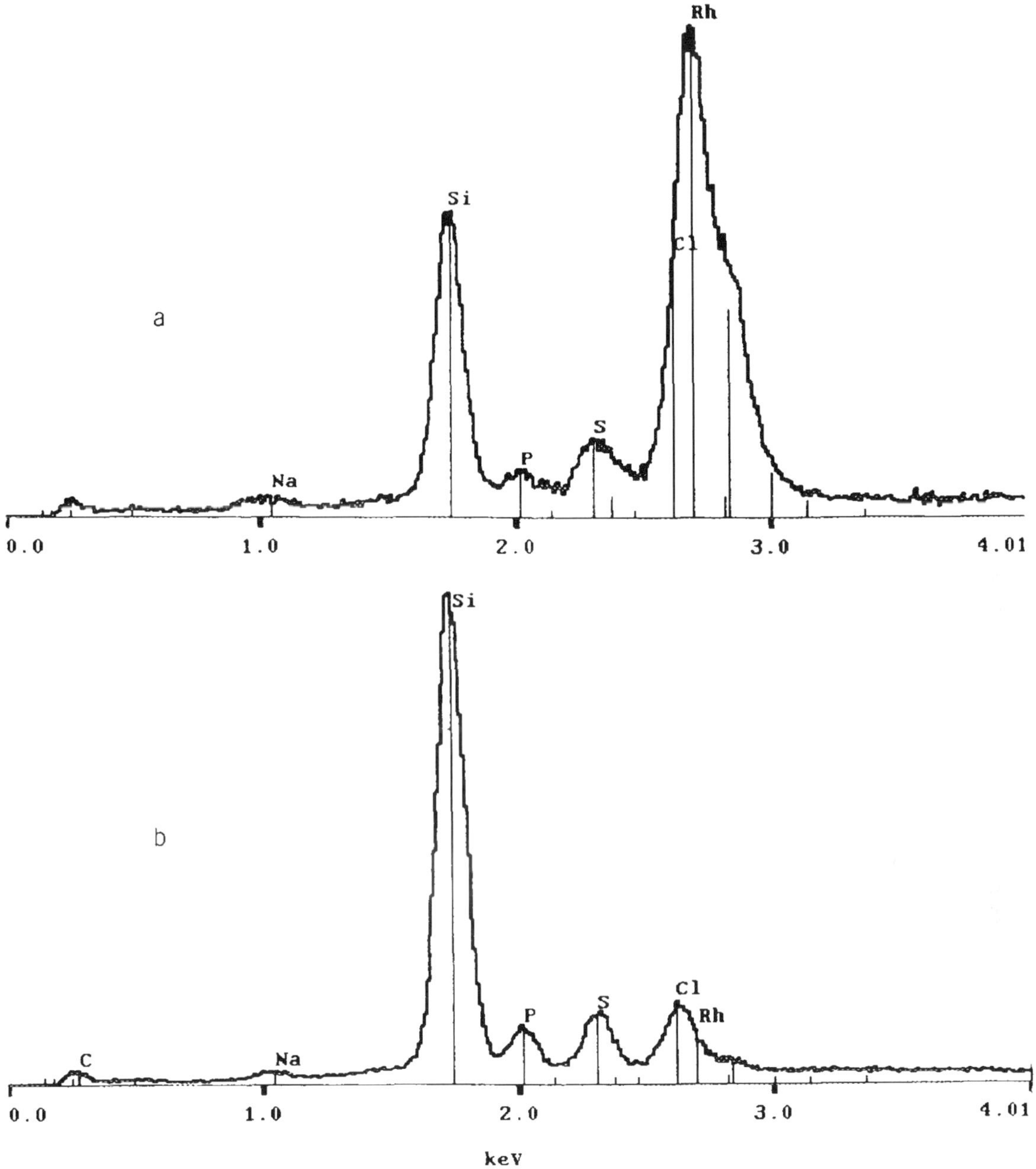

FIG. 4. (a) EDS surface elemental analysis of the white spots shown in Fig. 2a; (b) EDS surface elemental analysis of the area outside of the white spots.

Characterization of the Supported Catalyst

The synthesis of the complex H(CO) Rh(DPM)$_3$ crystals was previously (*22*) carried out by first preparing the oil-soluble complex H(CO)Rh(PPh$_3$)$_3$ and by treating subsequently in THF this complex with an excess amount of DPM for a few days in a refrigerator. In this paper, a single step procedure was employed to prepare the supported catalysts. The rhodium complex was synthesized in an aqueous medium, as described in the experimental section, by a procedure analogous to that used for Cl(CO)Rh(PPh$_3$)$_2$ (*19*). The molar ratio between DPM and RhCl$_3$ was 10. Its structure is expected to be similar to that of

TABLE 3

Hydroformylation of Styrene Catalysed Using Si–P(VEAC)-(1) Supported $Cl(CO)Rh(DPM)_2$

Reuse of the catalyst	Supported catalyst system			Styrene (mmol)	Yield of aldehyde (%)	i/n ratio
	Ligand DPM (mmol)	Complex $Cl(CO)Rh(DPM)_2$ (mmol)	Pendant Q^+Cl^- VEAC (mmol)			
1st	0.16	2.1×10^{-2}	0.19	87.7	56.5	6.2
2nd					55.2	6.6
3rd					53.8	6.3

Note. Reactions have been carried out at 85°C, 800–650 Psi, for 15 h; other conditions are listed in the Experimental section.

$Cl(CO)Rh(PPh_3)_2$. The quaternized silica powder was treated with the above aqueous solution. For Si–P(VEAC)-(1) supported $Cl(CO)Rh(DPM)_2$ and DPM, the total amount of DPM was almost equivalent to the number of pendant onium units. The DPM molecules form via their sulfonic groups ion pairs with the onium units of the polymer. Some of these ion pairs are coordinated to Rh through the bound DPM. The morphologies and compositions of the immobilized systems were investigated with SEM and EDS. The backscattered electron image (Fig. 3a) shows a number of white islands located on the silica particles. The EDS surface elemental analysis (Fig. 4) indicates that these white islands contain high concentrations of Rh complexes, while the regions outside the white spots contain much lower concentrations of such complexes. It is likely that the white islands are generated on P(VEAC) chains that are located on the cracks or larger pores of silica, since they are more easily swollen by water and thus provide a larger number of sites for ion-exchange. This leads to the formation of a large number of ion-pairs between the DPM anion and the onium cation. The FT-IR spectrum (Fig. 1c) shows a week characteristic absorption of the coordinated carbonyl group (at 1979 cm^{-1}) in the bound complex $Cl(DPM)_2RhCO$. Comparing with the corresponding IR absorption (at 1970 cm^{-1}) of the carbonyl group in the complex $Cl(PPh_3)_2RhCO$ (*19*), one can attribute this increase of the wavenumber from 1970 to 1979 cm^{-1} mainly to a weaker coordination bonding between Rh and CO in the immobilized complex than in the latter one. This is probably caused by the electron-withdrawing effect of the *meta*-substituted sulfonate group of one of the benzene rings of triphenylphosphine ligand, which reduces the σ electron donation ability of the phosphorus atom. In conclusion, there are two kinds of

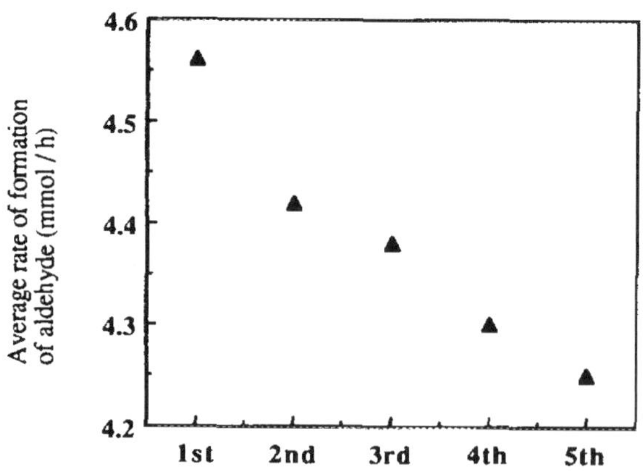

FIG. 5. Reaction conditions: (1) 0.5 g of Si–P(VEAC)-(2) as support containing 0.57 mmol of pendant VEAC units, on which 5.3×10^{-2} mmol of the rhodium complex and 0.42 mmol of DPM are immobilized. The catalytic system contains 50 wt% water; (2) 900 psi ($CO/H_2 = 1$), 85°C, 15 h for each run; (3) styrene (88 mmol) in 45 ml of cyclohexane and 0.5 ml of ethanol for each feedstock.

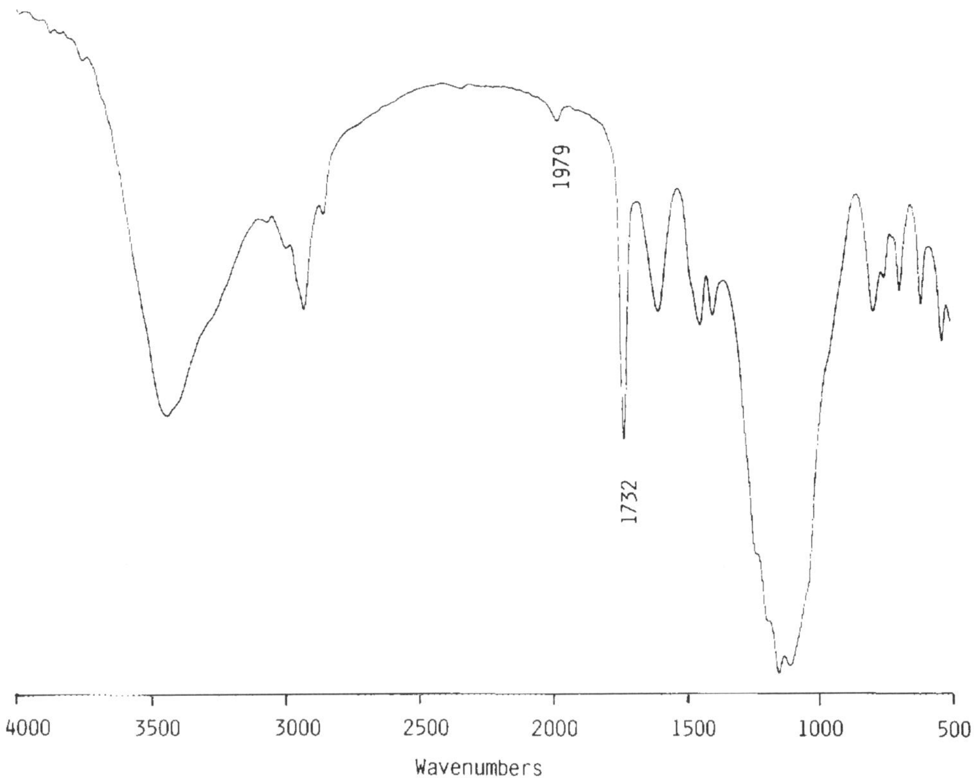

FIG. 6. FT-IR spectrum for the catalyst reused five times (the strong carbonyl absorption band at ν = 1732 cm^{-1} is due to the adsorbed aldehyde compounds).

regions on silica particles: one is covered with catalytic polymer clusters containing higher concentrations of rhodium complex, and another is covered by P(VEAC) and DPM with a much lower concentration of rhodium complex.

Hydroformylation of Functional Olefins

Styrene and ethyl 10-undecenoate are used for illustration purposes, to gain insight into the oxo reaction of aromatic and aliphatic olefins with this type of immobilized catalyst. In the hydroformylation of styrene, the ratio of the iso to normal aldehydes is greater than 6, and the turnover frequency is 156 (mol of olefin/h · mol of catalyst) (Table 3). The phosphine ligand {Ph_2 P(m-$C_6H_5SO_3^-$ N^+Et_3Bzl--)} being large, the complex it forms with rhodium contains only two phosphine ligands (*23*). It appears that the free volume around Rh can preferentially accommodate the secondary alkyl derivative intermediate in which the coordinated H atom attaches to the terminal carbon atom and the center Rh atom to the other carbon of the original double bond. This may be a result of the fact that the above intermediate is shorter than the primary one, but is branched.

The silica supported catalyst Si–P (VEAC)-(1) of Table 3 was hydrated before use to a 35 wt% of water-content. This water forms a thin adherent aqueous layer on the hydrophilic support. After it was reused three times, the catalyst maintained a yellow color, which indicates the presence of the complex, and the SEM micrograph (Fig. 3b) remained very similar to that before the catalyst was used (Fig. 3a). The adherent water layer allows the migration of DPM on the surface, thus providing a ligand-excess environment which prevents the decomposition of the rhodium complex. In addition, the presence of a small amount of ethanol (1.2% volume fraction of the solvent) enhances the yield of the aldehydes, which otherwise is

TABLE 4

Hydroformylation of 10-Ethyl Undecylenoate Catalysed Using Various Supported Catalysts

Entry[a]	Supported catalyst system		Olefin (mmol)	Solvent/additive[b] (ml)	Yield of aldehyde (%)	i/n[c]
	Complex Cl(CO)Rh(DPM)$_2$/DPM (mmol)	Pendant Q^+Cl^- (mmol)				
1	2.1 × 10^{-2}/0.16	VEAC (0.24)	50	Ch (40)/Et (0.5)	11.2	11.2
2	As above	VEPC (0.22)		Tol (40)/Et (0.5)	15	10.6
3	As above	As above		Ch (40)/Et (1.0)	32	9.2
4	As above	VEAC (0.24)		Ch (40)/Et (5.0)	42	8.0
5	Reused the above catalyst (2nd run)			As above	40	8.1
6	Reused the above catalyst (3rd run)			As above	37.5	7.9
7	2.1 × 10^{-2}/0.16	VEAC (0.30)		Ch (40)/Et (5.0)	19	5.8
8	As above	Unquarternized silica		As above	7.2	3.5

[a] The reaction conditions for entries 1–6 are described in the Experimental section. In entries 1 and 4–6, Si–P(VEAC)-(2) was used as substrate; in entries 2 and 3, Si–P(VEPC)-(E) was used; in entry 7, Si–P(VEAC)-(3) was used; in entry 8, silica support was used. The reaction conditions for entries 7 and 8 are 640 Psi, 6 h of reaction time; other conditions are mentioned in the experimental section.

[b] Ch = cyclohexane; Et = ethanol; Tol = toluene.

[c] The n/i ratios were evaluated from the areas of the spots of the corresponding aldehyde in TLC using a mixture of 20 ml petroleum and 5 ml ether as eluant, $R_f(n) = 0.56$; $R_f(i) = 0.66$.

only about 2%. One can speculate that the surface attached polymer coils that contain pendant catalytic ion-pairs, are swollen with ethanol molecules, thus generating sites with suitable polarity and size, able to accommodate the styrene molecules. Subsequently, the generated phenylpropanals play the role of ethanol because they can themselves swell the polymer coils. We found that less than 5% of the styrene converted to oligomer, with an IR spectrum strongly resembling that of polystyrene, terminated however with a monocarboxyl group. The absorption band of the carbonyl group is at 1726 cm^{-1}, as in an aldehyde. No hydrogenation products such as ethylbenzene were detected by GC-MS which was equipped with a fused silica PoraPLOT wide bore capillary column (0.53-mm ID, 28-m length), a 5A molecular sieve, and a high-performance capillary column (HP-1, cross-linked methyl silicone, 0.2-mm ID, 12-m length). Alcohols such as phenylpropanol were also not detected by ^{1}H-NMR. The supported Si–P(VEAC)-(2) catalyst was reused five runs in the oxo reaction of styrene, and the total yield of aldehyde divided by time is plotted in Fig. 5. The infrared spectrum of the catalyst after five runs (Fig. 6) presents an absorption band of Rh–C═O at $\nu = 1979$ cm^{-1}. This indicates that the rhodium complexes survive after 5 reaction runs.

The hydroformylation of ethyl 10-undecenoate led to the high ratio of about 8 of normal to iso aldehydes. In contrast to the hydroformylation of styrene, in which the yield of the iso- aldehyde is higher than that of the linear one, a completely reversed regioselectivity ratio in the oxo reaction of ethyl 10-undecenoate (Table 4) occurs. The free volume around the Rh atom can in this case preferentially accommodate the primary alkyl derivative intermediate. This differs from what occurs in the case of styrene, and a possible explanation can be related to the planarity of the benzene ring in styrene, in contrast to the flexibility of the long hydrocarbon chain in the case of ethyl 10-undecenoate. From Table 4, one can observe that with increasing ethanol content in the reaction medium the yield of the linear aldehyde decreases, but the total yield of the aldehydes increases. This probably occurs because the ethanol molecules decrease via a solvation effect the steric constraints around the coordination center, thus increasing the free volume. From entries 7 and 8 of Table 4, one can see that if the silica surface is not covered with polymer chains of quaternary onium chloride, the yield of the oxo reaction is very low. This is probably a result of the difficulty for the olefin molecules to contact the catalytic sites. The presence of the polymer increases the compatibility between catalyst and olefin.

REFERENCES

1. (a) Hartley, F. R., "Supported Metal Complexes." Reidal, Dordrecht, 1985; (b) Pittman, C. U., Jr., *in* "Comprehensive Organometallic Chemistry" G. Wilkinson, F. G. A. Stone, and E. W. Abel, Eds., Vol. 8, p. 533. Pergamon, Elmsford, NY, 1981; (c) Pomogailo, A. D., *Russ. Chem. Rev.* **61**(2), 133 (1992).
2. (a) Cusumano, J. A., *in* "Perspectives in Catalysis" (J. M. Thomas and K. I. Zamaraev, Eds.), Oxford Blackwell, London, 1992; (b) Kalck, P., Peres, Y., and Jenck, J., *Adv. Organomet. Chem.* **32,** 121 (1991); (c) Russel, M. J. H., *Platinum Met. Rev.* **32,** 179 (1988).
3. (a) Sinou, D., and Amrani, Y., *J. Mol. Catal.* **22,** 319 (1983); (b) Alario, F., Amrani, Y., Colleuille, Y., Dang, T. P., Jenck, J., Morel, D. and Sinou, D., *J. Chem. Soc. Chem. Commun.*, 202 (1986).
4. (a) Panicheva, L. P., Tret'yakov, N. Y., Yalovleva, S. A., and Yuffa, A. Y., *Kinet. Katal.* **31,** 96 (1990); (b) Vigh, L., Joo', F., and Van Hasselt, P. R., *J. Mol. Catal.* **22,** 15 (1983); (c) Fendler, J. H., *Chem. Rev.* **87,** 877 (1987).
5. Lisichkin, G., and Yuffa, Y. Y., *Russ. Chem. Rev.* **59** (12) (1990).
6. (a) Kuntz, E. G., *Chemtech* **Sept.,** 570 (1987); (b) Fr. Patent 2314910 (1975) to Rhône–Poulenc Industries.
7. (a) Haggin, J., *C & EN* **May 21,** 30 (1991); (b) Safi, M., and Sinou, D., *Tetrahedron Lett.* **32** (18), 2025 (1991).
8. (a) Larpent, C., and Patin, H., *J. Mol. Catal.*, 44 (1988) 191; (b) Alario, F., Amrani, Y., Colleuille, Y., Dang, T. P., Jenck, J., Morel, D. and Sinou, D., *J. Chem. Soc. Chem. Commun.*, 202 (1986).
9. Deming, T. J., and Novak, B. M., *Macromolecules* **24**(1), 326 (1991).
10. Oishi, S., *J. Mol. Catal.* **39,** 225 (1987).
11. (a) Siegel, H., and Himmele, W., *Angew. Chem. Int. Ed. Engl.* **19,** 178 (1980); (b) Botteghi, C., Ganzerla, R., Lenarda, M., and Moretti, G., *J. Mol. Catal.* **40,** 129 (1987).
12. Bach, H., Gick, W., Wiebus, E., and Cornils, B. *in "Proceedings, 8th International Congress on Catalysis,* Berlin, 1984," p. 11–417. Verlag Chemie, Weinheim, 1984.
13. Herold, F., and Schneller, A., *Adv. Mater.* **4** (3), 143 (1992).
14. Davis, M. E., *Chemtech* **Aug.,** 499 (1992).
15. Aphancet, J. P., Davis, M. E., Merola, J. S., and Hanson, B. E., *J. Catal.* **121,** 327 (1990).
16. Sheldon, R. A., *Chem. Ind.* **23,** 903 (1992).
17. Nishikubo, T., Uchida, J., Matsui, K., and Iizawa, T., *Macromolecules* **21,** 1583 (1988).
18. Ahrland, S., Chatt, J., Davies, N. R., and Williams, A. A., *J. Chem. Soc.*, 276 (1958).
19. Evans, D., Osborn, J. A., and Wilkinson, G., *Inorg. Synth.* **11,** 99 (1968).
20. Targett, N. M., Kiloyne, J. P., and Green, B., *J. Org. Chem.* **44,** 4962 (1979).
21. (a) Sohn, J. R., *J. Mol. Catal.*, **62,** L1-L4 (1990); (b) Choudary, B. M., and Sarma, M. R. *Tetrahedron Lett.* **31**(10), 1495 (1990).
22. Borowski, A. F., Cole-Hamilton, D. J., and Wilkinson, G., *Nouv. J. Chim.*, 137 (1978).
23. Oswald, A. A., Handriken, D. E., Kastrup, R. V., and Mozeleski, E. J., *in* "Homogeneous Transition Metal Catalyzed Reactions" (W. R. Moser and D. W. Slocum, Eds.,). American Chemical Society, Washington, DC, 1992.

Liquid Polymer Catalyst Immobilized on Polymer-coated Silica: Application to Hydroformylation

Liang Hong and Eli Ruckenstein*

Department of Chemical Engineering, State University of New York at Buffalo, Box 60, Buffalo, New York 14260-4200, USA

α,ω-Bis(diphenylphosphino)-poly(ethylene glycol-400) and its rhodium carbonyl complex have been attached to a poly(2-hydroxyethyl methacrylate) network coating on porous silica; the supported catalytic liquid polymer phase thus obtained has been used in the hydroformylation of ethyl undec-10-enoate.

A recent paper[1] reports the deposition of polymer chains of quaternary onium salts on the surface of hydrophilic silica by the polymerization of water-soluble quaternary onium monomers present in a water medium that is absorbed by the porous silica. This polymer covered silica surface, to which a small amount of water was added, constitutes a suitable catalytic substrate since it provides cation anchors for negatively charged catalytic species. It was employed by us to immobilize monosulfonated triphenylphosphine and its rhodium carbonyl complex. This system was used for the hydroformylation of ethyl undec-10-enoate, using cyclohexane as the reaction medium. The reaction rate was, however, relatively low because this high molecular mass olefin has a low solubility in the supported aqueous phase. For this reason, a different kind of supported catalyst, namely a liquid polymer, has been attached to a silica surface coated with a suitable polymer network. Because of a higher compatibility of the high molecular mass olefin with the liquid polymer phase a higher reaction rate is expected.

Poly(ethylene glycols) (PEGs) or the macrocyclic crowns derived from them have been widely used as encapsulating hosts for various metal ions in phase-transfer catalysis and extraction processes.[2] It will be demonstrated here that the macrocyclic-crown analogues obtained by substituting the hydrogen or the hydroxy substituents at the ends (α,ω-positions) of the PEG with phosphine groups, and then coordinating the two ends to transition metal atoms constitute a supramolecular catalytic system. The main advantage of such a system is that it can be easily immobilized by anchoring the polyether chains to a substrate, either by hydrogen bonding or coordinative complexing, thus generating a catalytic liquid polyether layer. For this purpose we selected silica particles covered with a poly(2-hydroxyethyl methacrylate) network, to which the phosphino-polyether chains had been attached *via* the hydrogen bonding of the OH groups of the network with the glycol ether units of the latter chains. Compared to the supported aqueous phase catalytic systems,[1] the supported liquid polyether layer is generally more compatible with high molecular mass organic compounds because of its suitable polarity.

The polyether ligand α,ω-bis(diphenylphosphino)poly-(ethylene glycol-400) (**1**, Scheme 1) was prepared starting from PEG-400 ($\bar{M}_w$ 400, Aldrich). Diphenylphosphinous chloride (23 g, 100 mmol) in tetrahydrofuran (THF) (20 ml) was dropped into a stirred solution of PEG-400 (20 g, 50 mmol) and pyridine (Py) (8.0 g, 100 mmol) in THF (30 ml) at room temperature. The reaction mixture was stirred for 3 h, then the precipitated Py·HCl was filtered out. The THF solution was passed through a silica-gel column to remove the dissolved Py·HCl. After the solvent THF had been evaporated off, 32 g of α,ω-bis(diphenylphosphino)-PEG-400 **1** was obtained. The molecular mass distribution of **1** was determined by fast atom bombardment (FAB) mass spectrometry. The yellow–green rhodium carbonyl complex (Wilkinson type complex) was prepared by the conventional method.[3] Elemental analyses indicated P : Rh = 3 : 1; FTIR: ν(Rh–H) 2171 cm^{-1} and ν(CO) = 2004 cm^{-1}; the ^{31}P NMR signals changed from δ 19.5 (t) and 22.5 (t) before coordination to δ 80.14 (m) and 80.92 (m) after coordination. The structure of this complex is not yet clear. It may have the structure **2** in Scheme 1, but other possibilities include a coordinative polymeric structure involving successive PEG bridges. Structure **2** seems to be most likely because the complex can be easily dissolved in THF without a major change in viscosity. In addition, Alcock *et al.*[4] have determined by X-ray diffraction a similar structure for phosphino-ether carbonyl complexes which had, however, a shorter PEG chain than ours. Since both the bisphosphino-PEG ligand **1** and the Wilkinson complex **2** have low solubilities in non-polar solvents, it is possible to support them (as a solution of **2** in **1**) on a substrate in order to catalyse organic conversions in which the reactants and products are soluble in non-polar media. The immobilization of the catalytic species is hence a result of their hydrogen bonding to the polymer coated silica and of their insolubility in the non-polar reaction medium.

In this paper, a hydrophilic silica powder (600 m^2 g^{-1}; pore volume 1.1 ml g^{-1}; average particle size 10 μm; Aldrich) was selected as the substrate. To increase the compatibility between the substrate and the immobilized liquid polymer,

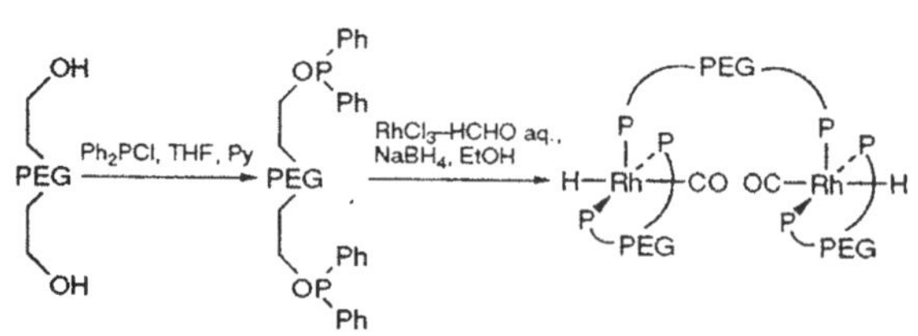

1 **2**

PEG = –[OCH$_2$CH$_2$]$_n$–; P in complex **2** is OP(Ph)$_2$

n = 6 (19.5%), 7 (18.2%), 8 (29.8%), 9 (22.1%), 10 (10.4%)

Scheme 1 Synthesis of polyether phosphine ligand **1** and of its Wilkinson complex **2**

Table 1 Hydroformylation of ethyl undec-10-enoate during six reuses of the same supported liquid polyether catalyst[a]

Run	Reaction time/h (*i*)	Reaction time/h (*ii*)	Total yield of aldehydes 3 and 4 (%)[b] (*i*)	Total yield of aldehydes 3 and 4 (%)[b] (*ii*)	3/4 for (*i*)
1	2.0	2.5	67	34	1.0
2	2.0	2.5	64	35	1.2
3	2.5	2.5	60	33	1.3
4	2.5	2.5	55	34	1.6
5	4.0	5.0	48	45	1.8
6	4.0	5.0	42	43	2.0

[a] Conditions: 800 psi CO–H$_2$ (1 : 1); 85 °C; olefin, 15 g (71 mmol); reaction medium (cyclohexane), 40 ml. Molar ratio olefin/Rh *ca.* 1000. The supported-catalyst system (*i*) contained 0.5 g of substrate and (*ii*) 2.0 g of substrate. [b] The aldehydes were separated from the unreacted olefin by silica-gel column chromatography, using gradient elution with light petroleum–diethyl ether.

This paper, which was originally published in **Journal of the Chemical Society, Chemical Communications**: 1486–1487 (1993), is republished with permission.

poly(hydroxyethyl methacrylate) crosslinked with *N,N*-methylene bisacrylamide (MBAA) was first deposited on silica, by the radical polymerization of the water-soluble monomers 2-hydroxyethyl methacrylate (HEMA) and MBAA present in a water medium that is absorbed by the pores of the silica particles. After polymerization, the coated substrate was washed with THF by Soxhlet extraction in order to remove the unreacted monomers. Elemental analyses indicated that the polymer-coated substrate contained 2.72 mmol of HEMA and 0.34 mmol of MBAA per gram of substrate. The supported liquid polyether catalyst was obtained by the reaction of $RhCl_3 \cdot 3H_2O$ (0.02 g, 0.075 mmol) with an excess of **1** (0.72 g, 1 mmol, calculated from the average molecular mass) in ethanol (15 ml), and then introducing the polymer-coated silica powder (either 0.5 g or 2.0 g) into the ethanol solution. The ethanol solvent was evaporated *in vacuo* with magnetic stirring, and the liquid polymer remained deposited on the PHEMA network coating the silica particles.

In order to examine the hydroformylation of a high molecular mass olefin by this supported catalyst, ethyl undec-10-enoate was selected [eqn. (1)]. Table 1 lists the

$$RCH{=}CH_2 + CO + H_2 \rightarrow \underset{\mathbf{3}}{RCH_2CH_2CHO} + \underset{\mathbf{4}}{RCH(CHO)Me} \quad (1)$$

$$R = EtO_2C[CH_2]_8$$

yields for this reaction for six successive uses of the same catalytic system. The reaction mixture was easily removed by suction from the supported catalytic powder remaining at the bottom of the pressure reactor. Two kinds of catalysts, (*i*) and (*ii*), were used, which had the same amount of catalytic species but were loaded with different amounts of substrate. For comparison, we note that the supported aqueous catalyst previously used by us[1] to catalyse the hydroformylation of ethyl undec-10-enoate (dissolved in cyclohexane containing some ethanol) provided a 40% yield of aldehyde after 15 h of reaction. Higher hydroformylation rates were achieved with the supported liquid polyether catalyst system employed here, because ethyl undec-10-enoate is much less soluble in the former aqueous than in the present liquid polyether layer. Table 1 also shows that catalyst system (*i*) leads to a higher initial reaction rate than (*ii*), and that the yield of the aldehydes decreases for (*i*) during six successive hydroformylations, but remains almost the same for (*ii*). The mass of the recovered supported-catalyst powder was, after washing with cyclohexane, 70% of its original value after the six reuses in case (*i*), but catalyst (*ii*) was recovered almost completely. It is likely that the generation of aldehydes increases the polarity of the solution, thus promoting the dissolution of the liquid polyether molecules that are not bound sufficiently strongly to the (PHEMA–MBAA) network. Larger amounts of substrate are likely to increase the strength of the binding of the catalytic species to the substrate. This explains why in case (*ii*) the recovery of supported-catalyst after six reuses was almost complete. From the 70% recovery after a total of 17 h of reaction for catalyst (*i*), one can estimate that 1 g of substrate possesses a maximum accommodation capacity of about 0.6 g of polyether ligand **1** and its rhodium complex **2**. For the unsupported catalyst under similar conditions after reaction for 5 h, the yield of aldehydes was 47%. The unsupported catalyst is insoluble in cyclohexane, and so was first dissolved in 6 ml of THF. The yields for heterogeneous conditions are thus comparable to those for homogeneous conditions.

Received, 24th May 1993; Com. 3/02967K

References

1 L. Hong and E. Ruckenstein, *J. Catal.*, 1993, **143**, 52.
2 E. Weber, J. L. Toner, I. Goldberg, F. Vögtle, D. A. Laidler, J. F. Stoddart, R. A. Bartsch and C. L. Liotta, *Crown Ethers and Analogs*, ed. S. Patai and Z. Rappoport, Wiley, New York, 1989.
3 D. Evans, G. Yagupsky and G. Wilkinson, *J. Chem. Soc. A*, 1968, 2660.
4 N. W. Alcock, J. M. Brown and J. C. Jeffery, *J. Chem. Soc., Chem. Commun.*, 1974, 829.

Immobilization of alkoxylated phosphine ligands and their Rh complexes to a silica surface coated with an organic mono- or multilayer

Liang Hong, Eli Ruckenstein*

Department of Chemical Engineering, State University of New York at Buffalo, Buffalo, NY 14260, USA

(Received October 13, 1993, accepted December 28, 1993)

Abstract

A new methodology for the immobilization of homogeneous catalysts on a modified silica surface is suggested. Phosphine ligands with polar substituted groups as well as their rhodium carbonyl complexes are attached to modified silica substrates. The modified silica substrates were prepared by either depositing a crosslinked poly(hydroxyethyl methacrylate) network on the surface of porous silica, or by transforming the surface-pendant hydroxyl groups of silica into [3-(siloxysilyl)propyl]-ethylenediamine groups. They are denoted as P(HEMA–MBAM)/SiO_2 and SSPEDA/SiO_2, respectively. The following phosphine ligands were attached to the modified silica: (1) α,ω-bis(diphenylphosphino)-poly(ethylene glycol) (DPP–PEG); (2) 2-methoxy-methylenephenyl-diphenylphosphine (MPDPP); and (3) 1,2-bis{[(2-methoxyacetylphenyl)-phenyl-(diphenyl)]phosphino}ethane (MAPDPPE). The FT-IR investigations of these immobilized catalysts revealed hydrogen bonding between the surface-pendant hydroxyl or amino groups and the oxygen atoms of the phosphine ligands. The SEM and EDS surface analyses showed that the phosphine rhodium carbonyl complexes were uniformly distributed over the polymer coated silica. Hydroformylation reactions of several olefins were employed for the assessment of the efficiency of immobilization, and both hydrophobic (cyclohexane) and hydrophilic (water) reaction media have been employed The time dependence of the hydroformylation of 2-($\pm$)-ethylhexyl acrylate (EHA) in a hydrophobic medium indicated that the modified silica surfaces are more compatible with the hydrophobic reactant molecules than the pure silica surface, and that SSPEDA/SiO_2 is more effective than P(HEMA–MBAM)/SiO_2. The much higher hydroformylation rate for sodium 10-undecenoate (UDNa) than for 1-vinylimidazole when water was used as reaction medium is due to the higher ability of the former surface active molecules to be adsorbed on the catalyst surface. Successive reuses of the immobilized catalysts in the hydroformylation of ethyl 10-undecenoate in cyclohexane have demonstrated that MAPDPPE provides stronger immobilization than MPDPP.

*Corresponding author, fax. (+1-716)6453822.

This paper, which was originally published in **Journal of Molecular Catalysis**: 90, 303–321 (1994), is republished with permission.

Keywords. alkoxylated derivatives; hydroformylation, immobilized catalysts, organic layer coating, phosphine ligands; rhodium, silica

1. Introduction

Organic reactions with heterogenized transition metal complexes as catalysts have been one of the interesting directions of research in the field of homogeneous catalysis [1]. Until now three main methods have been employed to immobilize homogeneous catalysts: (i) Immobilization of transition metal complexes on polymers [2]; in this method, coordination ligands are bound to polymer matrixes either via the copolymerization of a ligand containing monomer with another monomer(s) or via several organic functional group transformations. (ii) Water-soluble ligands bound to a transition metal allow the complex to dissolve in the water phase and the catalytic reactions of oil-soluble high carbon molecules occur at the oil/water interface [3]. (iii) Supported liquid-phase catalysts [4] in which the liquid phase catalyst, which consists of a non-volatile organic solvent and a soluble transition metal complex, coats as a film the internal surface of a porous support.

It is often expensive to bind a homogeneous catalyst to a polymer substrate by employing methods (i). Methods (ii) generally are not efficient for oil/water interface reactions unless a surface active ligand is employed [5]. For the catalytic reactions involving small molecules, such as ethylene or propene hydroformylation, they can, however, become effective if their solubility in the aqueous phase is increased by using high pressures. Methods (iii) seem to offer a simple way to directly immobilize the metal complex catalyst, but they appear to be suitable only for reactions involving gaseous reactants and products, since the presence of flowing organic liquids can entrain the supported liquid out from the system. High reaction temperatures are, therefore, needed, which can, however, bring about also the volatilization of the solvent involved in the liquid phase catalyst. To avoid this, membrane reactors were employed [6]. The transition metal complex and the high boiling point organic solvent are located between two membranes which allow only the passage of the much smaller molecules of the reactants and products.

In a previous paper [7], a supported liquid polymer phase catalyst was prepared in which (1) poly(hydroxyethyl methacrylate) crosslinked with N,N'-methylenebisacrylamide was deposited on a porous silica support (this modified silica is denoted as P(HEMA–MBAM)/SiO_2), and (2) a liquid polymer phase catalyst consisting of the oligomeric ligand α,ω-bis(diphenylphosphino)-poly(ethyl glycol) (denoted as DPP–PEG) and its Rh(I) carbonyl complex was coated on the P(HEMA–MBAM) network. The hydrogen bonding between the pendant hydroxyl groups of the polymer network P(HEMA–MBAM) and the oxygen atoms of the polyether, as well as the low solubility of DPP–PEG in cyclohexane, which was used as reaction medium, ensured the immobilization of the catalyst. Compared with the immobilized aqueous phase catalyst [8,9], which was used for the same reaction, this system exploited in a more suitable manner the difference between the polarities of the catalyst and reaction medium. In this manner, it can simultaneously satisfy two opposite requirements: the access of the hydrophobic olefin to the catalyst and the prevention of catalyst leakage to the reaction medium. Since DPP–PEG possesses less stereospecificity around the phosphorus atom than the commonly used phosphine ligands, such as triphenylphosphine and 1,2-bis(diphenylphosphino)ethane, it cannot achieve high regioselectivity in most reactions. It is, accordingly, of interest to develop methods which involve the immobilization of coordination ligands with higher stereospecificity.

In this paper, the above ideas are exploited by using, in addition to the liquid polyether ligand DPP–PEG supported on a polymer network that coats a porous silica support, P(HEMA–MBAM)/SiO_2, the following modifications: (1) Instead of DPP–PEG, either the triphenylphosphine ligand in which a methoxymethylene group was substituted to one of the benzene rings, or the 1,2-bis(diphenyl phosphino)ethane ligand in which a methoxyacetyl group was substituted to one of the benzene rings was employed. The above substitutions increase the polarity of the molecules thus increasing their insolubility in a hydrophobic reaction medium. In addition, they provide oxygen atoms which allow hydrogen bonding to the modified silica substrate. (2) The porous silica was either covered with the P(HEMA–MBAM) network or was silylated with N-[3-(trimethoxysilyl) propyl]ethylenediamine in an aqueous solution. As a result of silylation, a monolayer of [3-(siloxysilyl)propyl]ethylenediamine covers the surface of the porous silica, generating a modified substrate denoted as SSPEDA/SiO_2. A number of hydroformylation reactions are employed for illustration.

2. Experimental

2.1. Reagents

Most of the chemicals were purchased from Aldrich. 2-Hydroxyethyl methacrylate (97%) was passed through an inhibitor removal column before use. N,N′-Methylenebis-acrylamide (99%), N-[3-(trimethoxysilyl)propyl]ethylenediamine (90%), chlorodiphenyl-phosphine (95%), poly(ethylene glycol) (average MW = 400 or 600), pyridine (anhydrous, 99%), triphenylphosphine (99%), 1,2-bis(diphenylphosphino)ethane (99%), methoxyacetyl chloride (99%), aluminium chloride (Fisher Sci., anhydrous, analysis grade), sodium hydride (80% dispersion in mineral oil), rhodium chloride hydrate, sodium borohydride (99%), formaldehyde (37 wt.% aq.), syngas (CS Gases Inc. 50% CO/H_2), 2-(±)-ethylhexyl acrylate (99%), ethyl 10-undecenoate (98%), cyclohexane (99% A.C.S. reagent), 10-undecylenoic acid (98%), 1-vinylimidazole (99+%), sodium hydroxide (97%) and organic solvents were used without further purification. Silica catalyst support (grade 951), silica gel (TLC high-purity grade, without binder) and TLC plates (silica gel on glass, 5 × 20 cm) were used as support and for the chromatographic separation, respectively.

2.2. Instruments

A 150-ml high pressure reactor (Parr 4841) was employed in the hydroformylation reaction. Energy dispersive spectroscopy surface analysis (EDS) was performed with a PGT/IMIX field emission electron microscopy equipment. The functionalized silica surface was investigated with a scanning electron microscope (SEM) (Hitachi S-800). The infrared absorption spectra of the ligands, the substrates and the immobilized catalysts were obtained with a Mattson Alpha Centauri FT-IR instrument using a KBr cell. ^{1}H-NMR spectra of the synthesized ligand and of the hydroformylation products were recorded on a VXR-400 spectrometer using chloroform-d or methylsulfoxide-d_6 as solvent. The time dependence study of the hydroformylation of 2-($\pm$)-ethylhexyl acrylate was performed using for analysis a HPLC [Isco-2350 pump/Isco-V_4 (UV-254 nm)] containing a Hypersil ENV column (I.D. 4.6 mm, length 150 mm), the mobile phase being 1% acetic acid in methanol. The elemental analysis was carried out by Quantitative Technologies Inc. Whitehouse, NJ.

2.3. Synthesis of the ligands

2.3.1. α,ω-Bis(diphenylphosphino)-poly(ethylene glycol) (DPP–PEG)

Diphenyl phosphine chloride (11.5 g, 50 mmol) in THF (10 ml) was introduced into a stirred solution of PEG-400 (10 g, 25 mmol) and pyridine (Py) (4.0 g, 50 mmol) in THF (20 ml) at room temperature. The reaction mixture was stirred 3 h, then the precipitated Py · HCl was filtered. The THF solution was passed through a silica gel column to remove the soluble Py · HCl. After the solvent THF was evaporated, about 16 g of α,ω-bis(diphenylphosphino)–PEG-400 was obtained.

2.3.2. 2-(Methoxymethylene)phenyl-diphenylphosphine (MPDPP)

Ph_3P (2.45 g, 9.3 mmol) and $AlCl_3$ (2.9 g, 21 mmol) were added to a 50 ml round bottom flask, the flask was sealed with a rubber septum and immersed in a cold water bath. Under magnetic stirring, 7 ml of nitromethane were introduced with a syringe. A red solution was formed accompanied by an intense heat release (Ph_3P is insoluble in nitromethane in the absence of $AlCl_3$; the combination of $AlCl_3$ and Ph_3P generates, however, soluble species). Subsequently, the flask was filled with Ar, and methoxyacetyl chloride (1 ml, 10.8 mmol) was slowly introduced with a syringe into the solution. A large amount of carbon monoxide evolved rapidly [10].

$$(MeOCH_2COCl + AlCl_3 \rightarrow MeO^{(+)}{=}CH_2 + AlCl_4^{(-)} + CO).$$

The flask was placed in a 60°C oil bath and the mixture stirred for 5 h to allow the formation of benzyl methyl ether.

$$(MeO^{(+)}{=}CH_2 + AlCl_4^{(-)} + Ph_3P \cdot AlCl_3 \\ \rightarrow MeOCH_2C_6H_4PPh_2 \cdot AlCl_3 + HAlCl_4)$$

A deep brown–red solution was thus obtained. After cooling, the reaction mixture was introduced into an icy aqueous hydrochloric acid solution (about 12 wt.%), and the mixture was extracted five times with benzene (5×20 ml). The small amount of water contained in the benzene extracted layer was dried by shaking with anhydrous $MgSO_4$ powder and then separated by filtration. After the benzene was removed by vacuum evaporation, 2.14 g of a yellowish solid was obtained. The solid was dissolved in 15 ml of dry THF and added to a THF solution of sodium tetraethylene glycol, which was prepared by treating 1.4 g tetraethylene glycol (7.2 mmol) with 0.44 g NaH (14.6 mmol) in 20 ml THF. The mixture was heated at 60°C for 3 h, and the THF was removed by vacuum evaporation. A candle-like solid remained in the flask, which quickly changed into a white powder when washed with water; the powder was stirred in water and filtered in a Buchner funnel several times. Finally, after drying, 1.5 g of a white powder product of 2-$MeOCH_2C_6H_4PPh_2$ was obtained. The elemental analysis indicated that it contained 77.64 wt.% C and 10.20 wt.% P. The FT-IR spectrum provided the following characteristic absorptions: 3057 cm^{-1}, m (υ_s Ph–H); 2922 cm^{-1}, m (υ_{as} C–H); 2854 cm^{-1}, w (υ_s C–H); 1591 cm^{-1}, m (υ_s Ph ring C=C); 1190 cm^{-1}, vs (υ_s C–O–C); 1120 cm^{-1}, vs (υ_{as} C–O–C); 1095 cm^{-1}, m; 1026 cm^{-1}, w (υ_s P–Ph); 754 cm^{-1}, vs (υ_s out-of-plane=C–H, for the 1,2-disubstituted ring); 721 cm^{-1}, vs, 698 cm^{-1}, vs (υ_s out-of-plane=C–H, for the monosubstituted ring). The ^{1}H-NMR analysis (chloroform-d) identified the chemical shifts: δ (ppm) = 3.59 (t, J=4.9 Hz) and 3.72 (t, J=4.8 Hz) ($Ph_2PC_6H_4$-*o*-$\underline{CH_2}OCH_3$); 3.65 (d, J=2.9 Hz, $Ph_2PC_6H_4$-*o*-$CH_2O\underline{CH_3}$), 7.32 (s, $\underline{Ph}$–P); [7.43–7.48 (q, J=3.0 Hz), and 7.64–7.69 (q, J=7.3 Hz) for -*o*-$\underline{C_6H_4}$-].

2.3.3. 1,2-Bis{[(2-methoxyacetylphenyl)phenyl-(diphenyl)]phosphino}ethane (MAPDPPE)

After $Ph_2PCH_2CH_2PPh_2$ (3.84 g, 9.3 mmol) and $AlCl_3$ (2.9 g, 21 mmol) were added to a 50 ml round bottom flask, the flask was sealed with a rubber septum and immersed in a cold water bath. Under magnetic stirring, 7 ml of nitromethane was introduced with a syringe, and a red solution was formed accompanied by an intense heat release. The flask was further filled with Ar, and methoxyacetyl chloride (1 ml, 10.8 mmol) was added with a syringe to the solution. Since no carbon monoxide evolved, this implies that the entire $AlCl_3$ had combined with the phosphine ligand. After the system was heated to 60°C in an oil bath, a solution of 1.4 g (1.1 mmol) of $AlCl_3$ in 3 ml CH_3NO_2 was introduced intermittently into the stirred solution over a period of 3 h.

$$(MeOCH_2CO^{(+)} + Ph_2PCH_2CH_2PPh_2 + AlCl_4^{(-)}$$
$$\rightarrow (2\text{-}MeOCH_2COC_6H_4)PhPCH_2CH_2PPh_2 + HAlCl_4)$$

The subsequent purification procedures were similar to those employed in the synthesis of 2-(methoxymethylene)-phenyl-diphenylphosphine. Finally, 1.20 g of a white powder were obtained. Elemental analysis indicated that it contained 71.86 wt.% C and 13.90 wt.% P. The FT-IR spectrum gave the following characteristic absorptions: 3055 cm^{-1}, m (υ_s Ph–H); 2941 cm^{-1}, m (υ_{as} C–H); 2850 cm^{-1}, w (υ_s C–H); 1730 cm^{-1}, s (υ_s C=O); 1483 cm^{-1}, m (υ_s Ph ring C=C); 1420 cm^{-1}, m (υ_s P–CH_2); 1176 cm^{-1}, vs (υ_s C–O–C); 1122 cm^{-1}, s (υ_{as} C–O–C); 1103 cm^{-1}, m; 1076 cm^{-1}, w (υ_s P–Ph); 754 cm^{-1}, vs (υ_s out-of-plane=C–H, for the 1,2-disubstituted ring), 731 cm^{-1}, vs and 694 cm^{-1}, vs (υ_s out-of-plane=C–H, for the monosubstituted ring). The ^{1}H-NMR analysis (chloroform-d) identified the chemical shifts: δ (ppm) = 1.71 (s, broad peak, –P$\underline{CH_2CH_2}$P–); 2.51 (s, –CO$\underline{CH_2}$O$\underline{CH_3}$); 7.44–7.46 (d, J=7.8 Hz, $\underline{Ph}$–P); [7.48–7.53 (q, J=7.8 Hz) and 7.67–7.71 (q, J=6.8 Hz) for -*o*-$\underline{C_6H_4}$-].

2.4. Modification of the silica substrate

2.4.1. Deposition of P(HEMA–BMAM) on porous silica support

Porous silica support powder (5.8 g) was slowly introduced in a solution of 2-hydroxyethyl methacrylate (3.5 g, 27 mmol), N,N′-methylene bisacrylamide (0.5 g, 3.2 mmol) and $K_2S_2O_8$ (10 mg) in 10 ml of water with continuous stirring. A wet solid with a small amount of extra liquid phase was thus obtained. The extra water was evaporated by vigorously stirring the mixture with an air flow. The wet solid containing about 6 ml of aqueous solution in the pores of silica was immersed into 10 ml decane held in a glass tube. Further, the tube was sealed with a rubber septum and Ar was bubbled through the liquid for 40 min with ultrasonic degassing. The polymerization was carried out in a water bath at 60°C for 12 h. The P(HEMA–MBAM) deposited on silica was purified by extraction with ethanol for 20 h in a Soxhlet extractor. Finally, after vacuum drying, 10.24 g of polymer modified silica substrate powder (denoted as P(HEMA–MBAM)/SiO_2) were obtained. The elemental analysis indicated that the substrate contained per gram 2.72 mmol of HEMA and 0.34 mmol of MBAM.

2.4.2. Formation of a monolayer of N-[3-(siloxysilyl)propyl]ethylenediamine on the porous silica support

Porous silica support powder (6.0 g) was introduced into 125 ml of dilute hydrochloric acid (0.4 M), and N-[3-(trimethoxysilyl)propyl]ethylenediamine (3.0 g, 12 mmol) was added to the suspension. After the suspension was stirred at 60°C for 72 h, it was filtered in a Buchner funnel. The powder was washed by stirring once in 20 ml sodium carbonate solution (0.5 N), four times in water, and two times in acetone. The substrate was finally vacuum dried, and 6.45 g of white powder were obtained (denoted as SSPEDA/SiO_2). The elemental analysis indicated that the substrate contained 1.1 mmol of SSPEDA surface-pendant groups per gram.

2.5. Coating phosphinorhodium(I) carbonyl hydride complexes and the ligand on the modified surface of silica support

2.5.1. Synthesis of the Wilkinson type rhodium complexes

Four kinds of phosphine ligands have been prepared, they are: α,ω-bis (diphenylphosphino)-poly(ethylene glycol), (DPP–PEG); 2-(methoxymethylene)-phenyl-diphenylphosphine (MPDPP); 1,2-bis{[(2-methoxyacetylphenyl)phenyl-(diphenyl)]phosphino}ethane (MAPDPPE) and 1,2-bis(diphenylphosphino)ethane (DPE). The methods employed for the preparation of Wilkinson type complexes have been described in the literature [11]. Since there is an equilibrium between the five-coordinated complex $P_3Rh(CO)H$ and the four-coordinated complex $P_2Rh(CO)H$ when any of the phosphine ligands MPDPP and MAPDPPE is used, two subscripts for the monodentated ligand MPDPP in the complex $(MPDPP)_{2-3}Rh(CO)H$ and for the bidentated ligands MAPDPPE in the complex $(MAPDPPE)_{1-1.5}Rh(CO)H$ are used.

Table 1
Modified silica-supported phosphine ligands and their rhodium complexes

Supported catalyst	Rh (mmol/g)	P (mmol/g)
{(DPP–PEG-400)$_{1.5}$Rh(CO)H + DPP–PEG-400}/P(HEMA–MBAM)/SiO_2	2.6×10^{-2}	0.70
{(DPP–PEG-400)$_{1.5}$Rh(CO)H + DPP–PEG-400}/SSPEDA/SiO	2.6×10^{-2}	0.70
{$(MPDPP)_{2-3}$Rh(CO)H + MPDPP}/SSPEDA/SiO_2	2.7×10^{-2}	0.87
{$(MAPDPPE)_{1-1.5}$Rh(CO)H + MAPDPPE}/SSPEDA/SiO_2	2.5×10^{-2}	0.90
{$(DPE)_{1-1.5}$Rh(CO)H + DPE}/SSPEDA/SiO_2	2.6×10^{-2}	0.11

2.5.2. Coating the modified substrate with Rh complex and ligand

The surface-modified silica (2.0 g), either P(HEMA–MBAM)/SiO_2 or SSPEDA/SiO_2, was introduced into an ethanol solution (15 ml) of the rhodium complex (0.075, mmol) and the phosphine ligand [DPP–PEG (1.1 mmol), MPDPP (2.45 mmol), MAPDPPE (1.28 mmol) or DPE (0.23 mmol)]. The suspension was kept under stirring at 75°C for about 5–10 min, after which the ethanol was removed from the system by vacuum evaporation under stirring. A yellow powder consisting of P(HEMA–MBAM)/SiO_2 or SSPEDA/SiO_2 with the surface coated by the complex and by the excess ligand was thus obtained (see Table 1).

2.6. Hydroformylation of higher olefins

2.6.1. Conversion of 2-(±)-ethylhexyl acrylate (EHA) as a function of time

Several immobilized catalysts were employed for this reaction; a typical process is described below. The immobilized catalyst powder of {$(MPDPP)_{2-3}$Rh(CO)H + MPDPP} + SSPEDA/SiO_2 (2.8 g), containing 7.6×10^{-2} mmol of Rh, 2.45 mmol of MPDPP and 2.2 mmol of surface-bound SSPEDA groups, together with cyclohexane (40 ml) and 2-(±)-ethylhexyl acrylate (18.6 g, 100 mmol) were loaded into a high-pressure reactor. The system was purged at room temperature before reaction with syngas (5–10 psi) four times with slow stirring. Then, the syngas was introduced until a pressure of 800 psi was achieved. This pressure increased to 910 psi when the system was heated from the room temperature to the reaction temperature of 85°C. The reactor was charged again with syngas as soon as the pressure fell to 840 psi. Samples of 0.1–0.2 ml were withdrawn at one hour intervals through an outlet valve for HPLC analysis. The amount of unreacted olefin was determined on the basis of a calibration curve of EHA against the peak area of the UV absorption at 254 nm. Two major peaks appeared on the HPLC charts of the samples: one for EHA and the other for the mixture of the two enolic forms of (±)-2-ethylhexoxycarbonyl-2′-methyl-acetaldehyde. The latter two were separated from the former on a silica gel column. The ^{1}H-NMR analysis (chloroform-d) identified: δ(ppm) = 0.92 (t, $-CH_2\underline{CH_3}$); 1.36 (m, $-\underline{CH_2}-$); 1.65 (s, $\underline{CH_3}-C{=}CH-OH\cdots O{=}C(OR)-$); 4.22 (m, $-C({=}O)O-\underline{CH_2}-$); 6.4 (d, $CH_3-C{=}\underline{CH}-OH\cdots O{=}C(OR)-$), 7.0 (d, $CH_3-C{=}CH-O\underline{H}\cdots O{=}C(OR)-$) for the aldehyde enolic form; and δ(ppm) = 1.78 (s, $\underline{CH_3}C{=}C(OR)OH\cdots O{=}CH-$); 7.62 (d, $CH_3C{=}C(OR)OH\cdots O{=}\underline{CH}-$); 11.35 (d, $CH_3-C{=}C(OR)O\underline{H}\cdots O{=}CH-$) for the ester enolic form, the other proton resonance absorptions coinciding with those of the aldehyde enolic form. The reaction mixture also contained a large amount of the aldol condensation oligomer of 2-(±)-ethylhexoxycarbonyl-2′-methyl-acetaldehyde–[–CH(OH)–(CH_3)C(COOR)–]$_n$–, which was precipitated and subsequently washed several times with ethanol. After vacuum drying, a highly sticky gel was obtained. Compared with the FT-IR spectrum of poly(EHA), the FT-IR characteristic absorption bands of the gel indicated that the oligomer is a product of aldol condensation rather than poly(EHA).

2.6.2. Hydroformylations of ethyl 10-undecenoate (EUD) in cyclohexane and sodium 10-undecenoate (UDNa) in water

Two of the immobilized catalytic systems already used for the hydroformylation of EHA, namely {(DPP–PEG-400)$_{1.5}$Rh(CO)H + DP–PEG-400}/SSPEDA/SiO_2 and {(MPDPP)$_{2-3}$Rh(CO)H + MPDPP}/SSPEDA/SiO_2, were subsequently reused to catalyse the hydroformylation of EUD and UDNa in cyclohexane and water, respectively. Before reaction the catalysts were washed twice in cyclohexane and recovered by centrifugation.

(a). EUD (15 g, 70 mmol) in 40 ml cyclohexane together with the catalyst were introduced in the pressure reactor, the other conditions being maintained as those for EHA. The normal and iso aldehydes were separated together from the olefin by vacuum liquid chromatography using gradient eluants consisting of petroleum ether and ethyl ether. The ratio of normal to iso aldehydes was determined by comparing the integrated signals of –C$\underline{H}$O of ^{1}H-NMR. Details regarding the separation and ^{1}H-NMR structural assignments of the aldehydes are given in our previous paper [9].

(b). An aqueous solution of NaOH (3.2 g, 80 mmol) in 45 ml H_2O was added to 10-undecenoic acid (15.1 g, 80 mmol) with stirring. A light brown solution (pH = 8.0–8.5) was thus formed. The hydroformylation reaction conditions are as for EHA and EUD. After 6 h of reaction, the catalyst was separated by centrifugation and the liquid phase was acidified by introducing about 6 ml of hydrochloric acid (37%). The separated organic phase together with the aqueous phase (pH = 3.0–4.0) were extracted with 20 ml benzene, and the aqueous phase was again extracted twice with benzene. The benzene solution was dried by shaking with 2–3 g of anhydrous $MgSO_4$ powder. After filtration, the benzene was removed by vacuum evaporation. The remaining oil was mixed with 50–60 ml of petroleum ether (30–60°C) at room temperature, and the mixture was refrigerated overnight to allow the crystallization of the aldehydes (the unreacted olefin is soluble in petroleum ether even at low temperatures). The white aldehydes crystals were further washed twice with cold petroleum ether and vacuum dried. The ratio of normal to iso aldehydes was determined by the ^{1}H-NMR method as described above. The main chemical shifts (in chloroform-d) are: δ(ppm) = 2.36 (t, –$\underline{CH_2}CO_2H$); 2.42(t, –$\underline{CH_2}$CHO); 9.36 (s, –$CO_2\underline{H}$ · · · OHC–, intramolecular hydrogen bonding); 9.77 (s, –C$\underline{H}$O) for $OHCCH_2(CH_2)_8CH_2COOH$, and δ(ppm) = 1.08 (d, –CH($\underline{CH_3}$)CHO); 9.61 (s, –C$\underline{H}$O) for the iso aldehyde.

2.6.3. Testing the strength of immobilization by recycling the catalysts

[(MAPDPPE)$_{1-2}$Rh(CO)H + MAPDPPE}]/SSPEDA/SiO_2 (2.8 g), which contains 2.5×10^{-2} mmol of Rh/g of catalyst and 0.45 mmol of MAPDPPE/g of catalyst was used to catalyse the hydroformylation of EUD (70 mmol) in cyclohexane (40 ml) for five successive runs. Each run of the reaction was carried out at 85°C and under the same initial pressure of CO/H_2 (910 psi), but for different reaction times: 1st (6 h); 2nd (8 h); 3rd (10 h); 4th (12 h); 5th (6 h). The aldehydes were separated from the olefin as described above. After each run, the immobilized catalyst was washed with cyclohexane and separated by centrifugation for the next run of reaction. After five runs, the catalyst powder contained 2.2×10^{-2} mmol of Rh/g and 0.31 mmol of MAPDPPE/g, which were determined by elemental analysis.

3. Results and discussion

3.1. Characterization of the organic layer coating the surface of silica and its effect on hydroformylation

There are two basic requirements for the organic layer attached to the surface of the modified silica substrates: (i) it should bind the catalyst and be compatible enough with the hydrophobic organic reactants in order to allow them to approach the catalyst, and (ii) it should be insoluble in the reaction medium. Two kinds of surface-modified silica substrates were prepared: (1) In one of them, poly(2-hydroxyethyl methacrylate crosslinked with N,N'-methylenebisacrylamide) was attached as a polymer network to the surface of porous silica, [P(HEMA–MBAM)/SiO_2]. (2) In the other, a silylated silica SSPEDA/SiO_2 was employed, in which N-[3-(siloxysilyl)propyl]ethylenediamine groups form a monolayer on the surface after the methoxy groups of the N-[3-(trimethoxysilyl)propyl]ethylenediamine react with the surface-pendant OH groups of silica with elimination of methanol (see Fig. 1a). Based on SEM observations, both modified substrates have similar surface morphologies, the modified surface being smoother than that of silica.

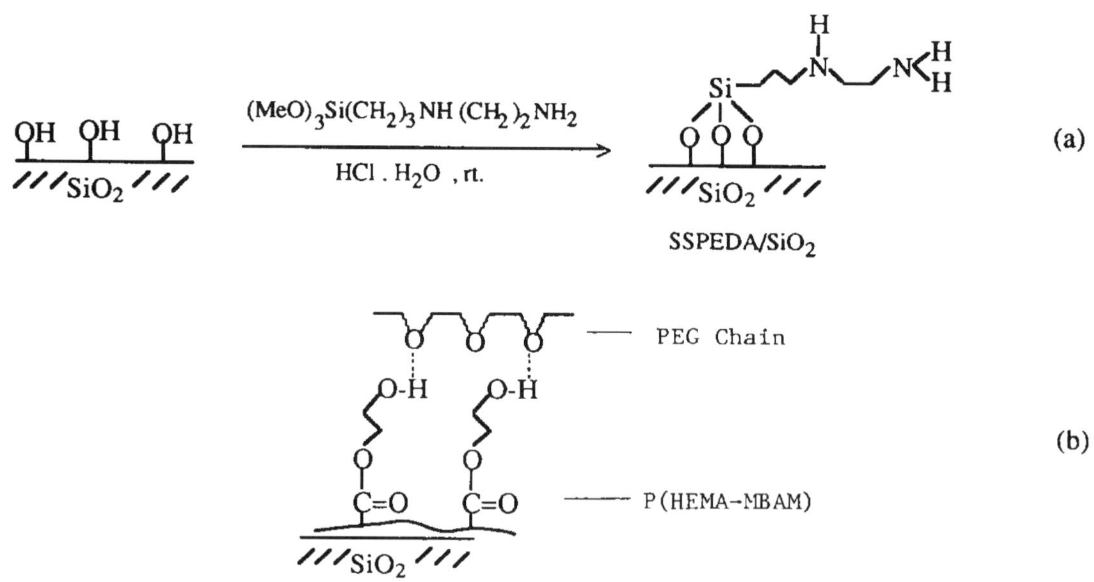

Fig. 1. (a) Transformation of the surface Si–OH groups into N-[3-(siloxysilyl)-propyl]ethylenediamine (SSPEDA) groups; (b) The occurrence of hydrogen bonding between the surface-pendant hydroxyl groups of P(HEMA–MBAM) network and the organic layer of polyether chains.

In order to demonstrate the formation of hydrogen bonding between the ligand and the surface-pendant hydroxyl or amino groups, FT-IR investigations of the two kinds of surface-modified silica substrates and of the corresponding supported catalysts were carried out.

(a) In Fig. 2, the FT-IR spectra of SiO_2 (a), P(HEMA–MBAM)/SiO_2 (b), {(DPP–PEG-400)$_{1.5}$Rh(CO)H + DPP–PEG-400}/SiO_2 (c) and {(DPP–PEG-400)$_{1.5}$Rh (CO)H

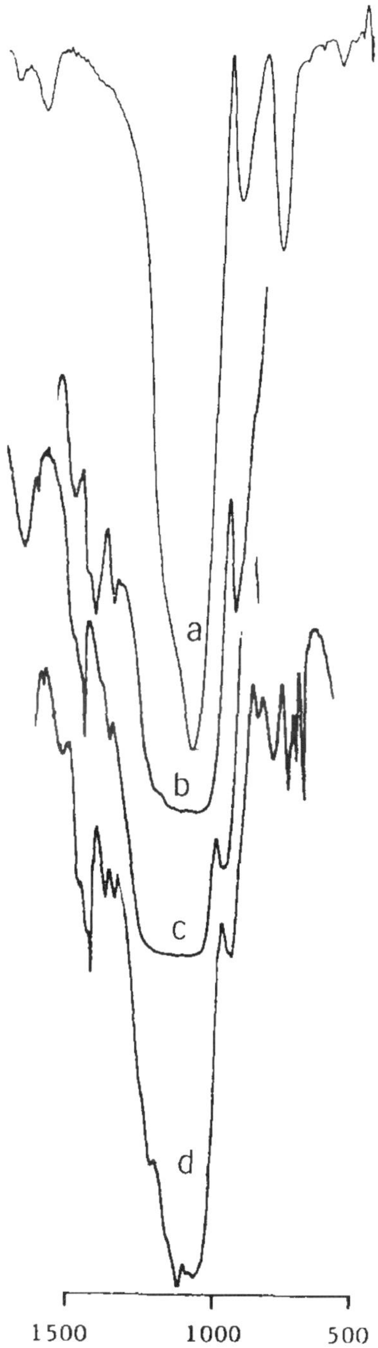

Fig. 2. The FT-IR (cm^{-1}) spectra of: (a) pure silica; (b) P(HEMA–MBAM)/SiO_2; (c) {(DPP–PEG-400)$_{1.5}$Rh(CO)H + DPP–PEG-400}/SiO_2, (d) {(DPP-PEG-400)$_{1.5}$Rh(CO)H + DPP–PEG-400}/P(HEMA–MBAM)/SiO_2.

+ DPP–PEG-400}/P(HEMA–MBAM)/SiO_2 (d) are presented. Spectrum (a) of the pure silica has a relatively narrow and sharp absorption band at 1099 cm^{-1} (ν(Si–O)), whereas the spectra of (b) and (c) exhibit wide and blunt absorption bands in the range 1050–1250 cm^{-1}. This is a result of the partial overlapping of the ν(Si–O) and ν(C–O) absorption bands. For the sample with spectrum (b) there are two types of C–O bonds which belong to the ester and the alcohol, respectively. For the sample with spectrum (c) there is only the C–O bond of the ether. In contrast to (b) and (c), in spectrum (d), the absorption band located between 1050–1250 cm^{-1} is narrower. Since in the latter case DPP–PEG-400 coats the substrate of P(HEMA–MBAM)/SiO_2, a likely possibility for the narrowing of the absorption band is the hydrogen bonding. The hydrogen bonding between the hydroxyl group of the HEMA unit and the oxygen of ethylene glycol unit (Fig. 1b) decreases the C–O bond strength of both of them, and for this reason their ν(C–O) absorption band is weakened. As a result, the contribution of the C–O absorption band to the overlapping is decreased.

(b) In Fig. 3, the FT-IR spectra of SSPEDA/SiO_2 (a), {$(DPP–PEG-400)_{1.5}$Rh(CO)H + DPP–PEG-400}/SSPEDA/SiO_2 (b) and {$(MAPDPPE)_{1-1.5}$Rh(CO)H + MAPDPPE}/SSPEDA/SiO_2 (c) are presented. Since, in general, the formation of hydrogen bonds X–H···X′ (X,X′ = O, N) decreases the bond strength of X–H, the IR absorption of the X–H stretching vibration is weakened. Indeed, the absorption band at 3465 cm^{-1} in (a), which is due to the ν (N–H) of the bound SSPEDA groups, becomes wider and weaker in (b). Since the sample with spectrum (b) was prepared by coating the SSPEDA/SiO_2 with a layer containing the phosphino polyether ligand and its rhodium complex, the surface-pendant amino groups of the substrate can readily interact with the polyether chains through hydrogen bonding [12]. Compared with the corresponding band in (a), an additional shoulder band at about 3275 cm^{-1} appears in (b), which can be assigned to the >N–H···$O(CH_2CH_2O–)_n$– hydrogen bonding [12]. Similarly, the absorption band of ν(N–H) at around 3500 cm^{-1} in (a) is widened and weakened in (c), probably because of the hydrogen bonding of the surface amino groups with the oxygen atoms of the groups (CH_3OCH_2CO–) present into the phosphine ligand of 1,2-bis(diphenylphosphino)ethane.

The backscattered electron images of the supported catalysts showed that there are no bright spots characteristic to rhodium clusters. However, the EDS investigations, indicated the presence of the Rh characteristic peak. This indicates that Rh(I) ions are uniformly distributed over the silica surface.

The effect of the modified organic layer, either P(HEMA–MBAM) or SSPEDA, on the catalytic reaction has been investigated by immobilizing the same amount of polyether liquid catalyst {$(DPP–PEG-400)_{1.5}$Rh(CO)H + DPP–PEG-400} on the following three substrates (2.0 g for each substrate): SiO_2, P(HEMA–MBAM)/SiO_2 and SSPEDA/SiO_2. The hydroformylation of 2-(±)-ethylhexyl acrylate (EHA) was carried out to assess the above three catalytic systems. The conversions of EHA are plotted against reaction time in Fig. 4, which shows that they are in the order SSPEDA/SiO_2 > P(HEMA–MBAM)/SiO_2 > SiO_2.

It is important to note that the comparison between the three catalysts has been made for the same amount of modified silica (2 g). The amount of silica contained by them, determined by elemental analysis, has been 2 g for SiO_2, 1.2 g for P(HEMA–MBAM)/SiO_2 and 1.72 g for SSPEDA/SiO_2. Since the same amount of {(DPP–PEG-

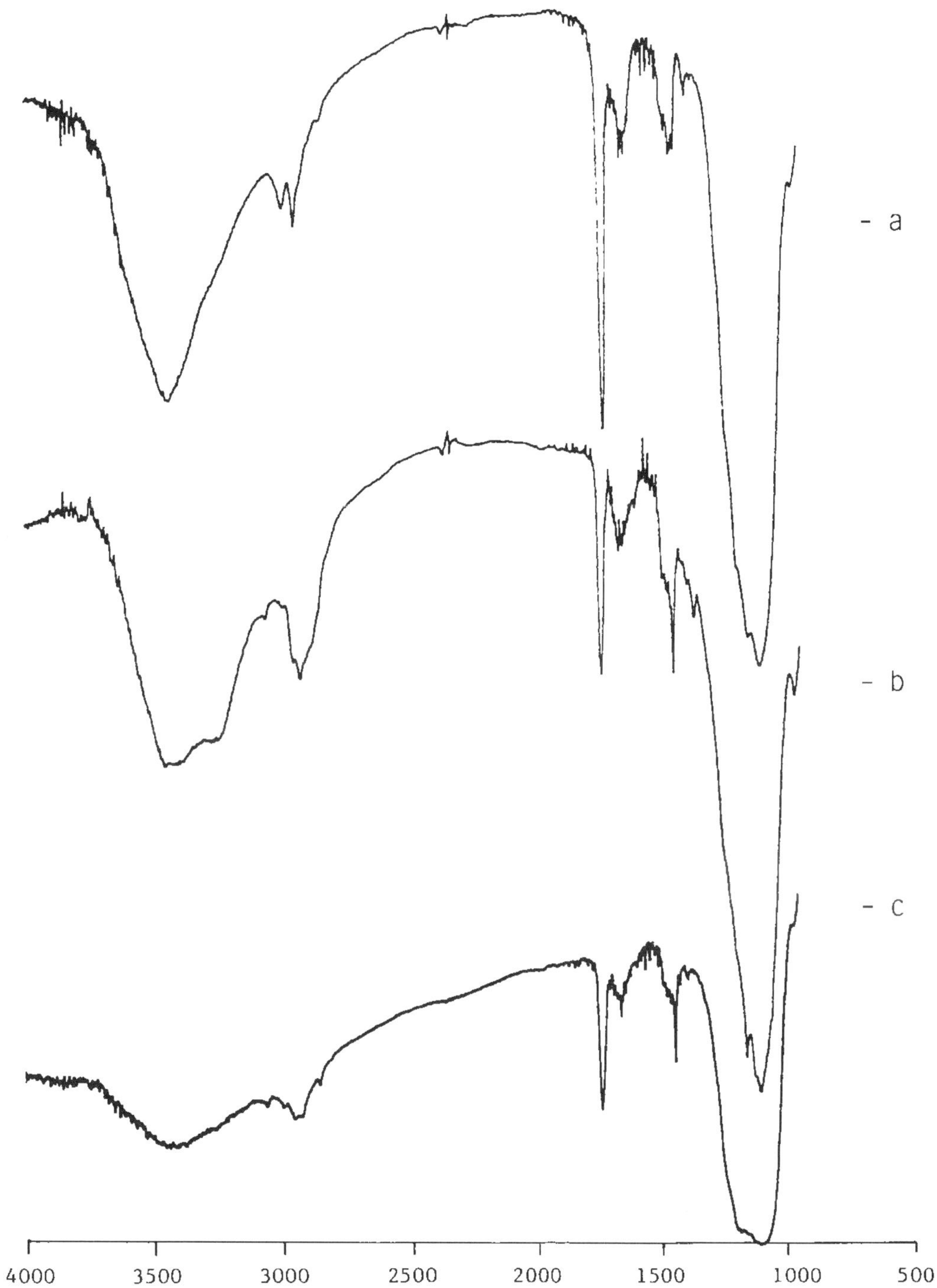

Fig. 3. The FT-IR (cm^{-1}) spectra of: (a) SSPEDA/SiO_2; (b) {(DPP–PEG-400)$_{1.5}$Rh(CO)H + DPP–PEG-400}/SSPEDA/SiO_2; (c) {(MAPDPPE)$_{1\sim1.5}$Rh(CO)H + MAPDPPE}/SSPEDA/SiO_2.

400)$_{1.5}$Rh(CO)H + DPP–PEG-400} is located on the modified substrate, the surface concentration of the catalyst is expected to follow the sequence P(HEMA–MBAM)/SiO_2 > SSPEDA/SiO_2 > SiO_2. The EDS surface elemental analysis (Fig. 5) indicates that

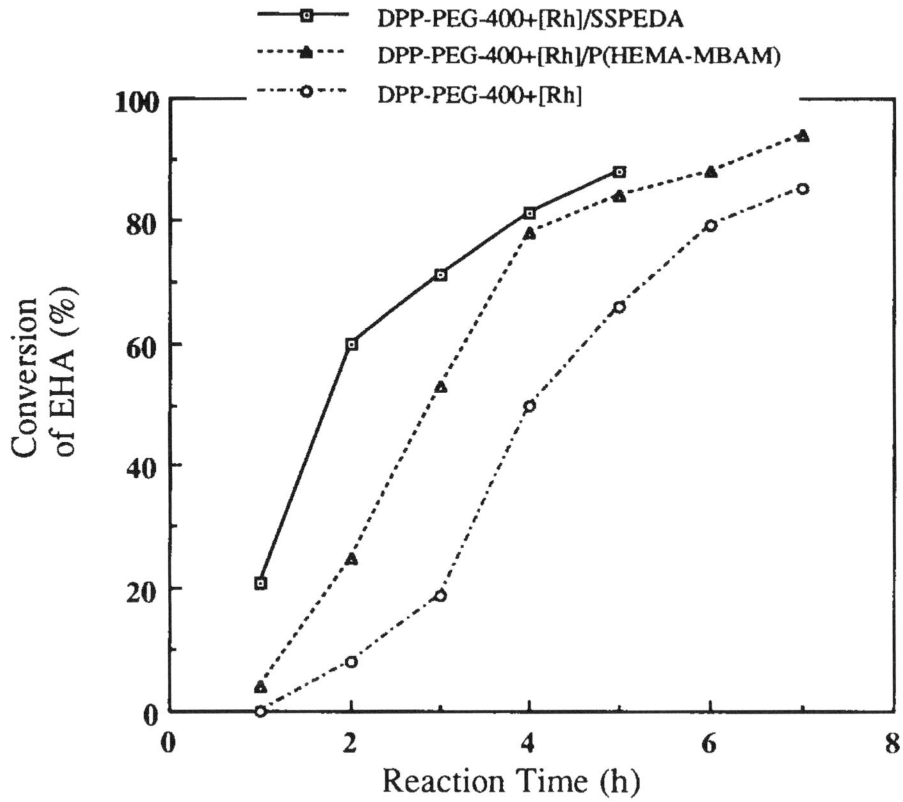

Fig. 4 The conversion against reaction time of EHA in the hydroformylation catalysed by the three catalytic systems indicated in the figure (85°C, 910 psi initial pressure).

this is indeed the case, since the intensities of P and Rh follow the above sequence. From Fig. 4 one can conclude that even though the catalyst support SiO_2 is better exposed, because of the higher surface area, the other two catalysts are more active. This means that the organic layer plays a part in the activity of the catalyst. The sequence of the surface concentration of the catalysts differs from the sequence of activities because the latter depend both upon the amount of catalyst exposed to the reaction medium and the compatibility of the reactant molecules with the catalyst. The organic liquid of the catalyst is responsible for the compatibility. While the system SSPEDA/SiO_2 has an exposed area smaller than that of SiO_2 it has a much better compatibility with the reacting molecules. The former system has, however, a greater exposed area than P(HEMA–MBAM)/SiO_2 and for this reason is more reactive than the latter one.

3.2. Some reaction features of the immobilized catalysts

Three reaction features of the present immobilization systems have been investigated: (1) the effect of the average molecular weight of the oligomeric polyether ligand DPP–PEG on the catalytic activity of the P(HEMA–MBAM) modified silica; (2) the cooperations between the phosphine ligand and the surface-pendant amine groups; and (3) the effect of the hydrophobic accessibility of the surface of the immobilized systems.

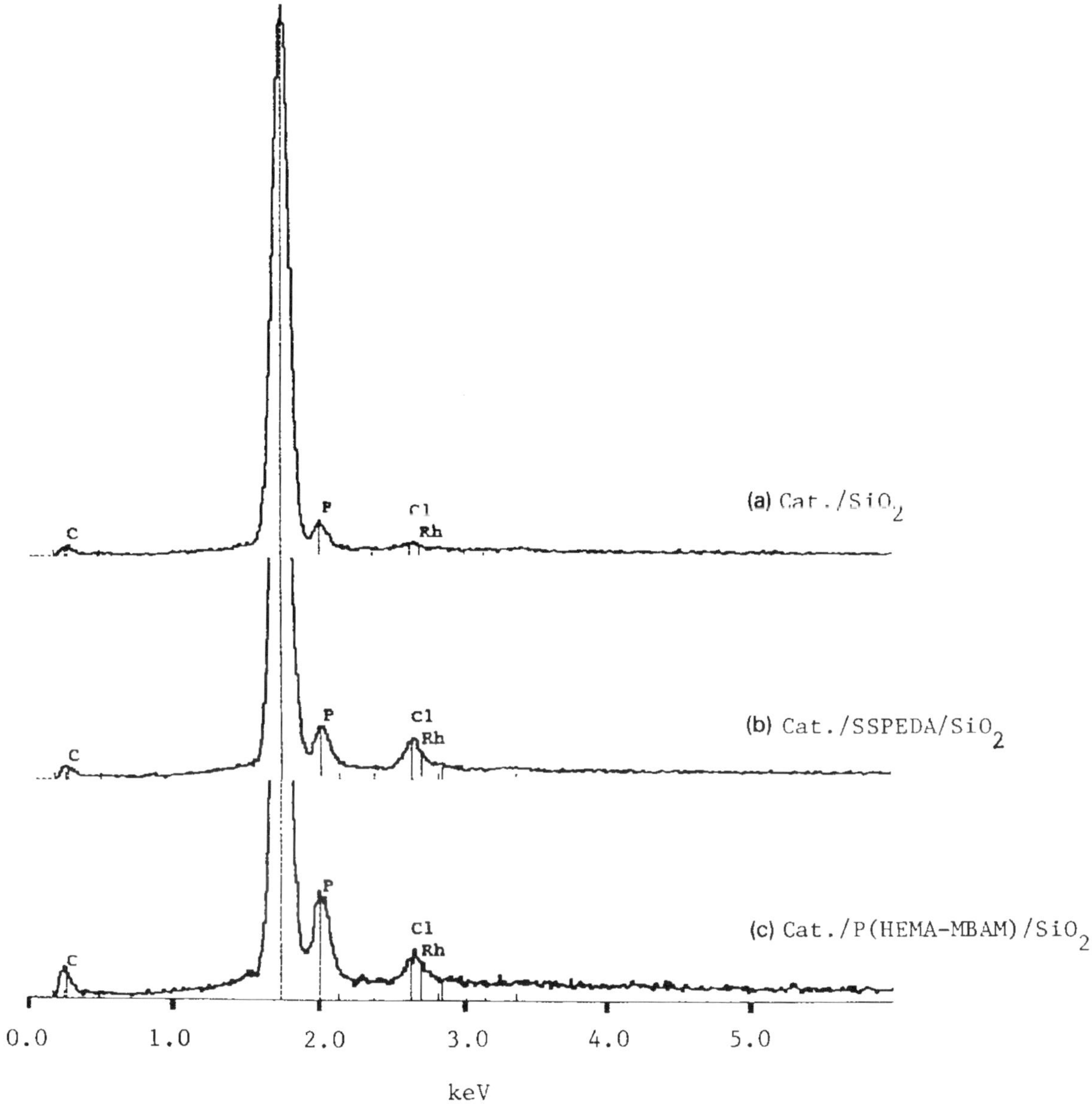

Fig 5. The surface elemental analysis of the supported catalysts by energy dispersive spectroscopy (EDS): (a) {(DPP–PEG-400)$_{1.5}$Rh(CO)H + DPP–PEG-400}/SSPEDA/SiO_2, (b) {(DPP–PEG-400)$_{1.5}$Rh(CO)H + DPP–PEG-400}/P(HEMA–MBAM)/SiO_2, and (c) {(DPP–PEG-400)$_{1.5}$Rh(CO)H + DPP–PEG-400}/SiO_2, which were employed in the reaction of Fig 4

The P(HEMA–MBAM)/SiO_2 substrate was coated with one of the following two oligomeric polyether ligands: DPP–PEG-400 and DPP–PEG-600 (the average molecular weight of the polyethylene glycol moieties being 400 and 600), and their rhodium complexes. The conversions as a function of time in the hydroformylation of EHA with these catalysts, presented in Fig. 6, show that the catalyst based on the lower molecular weight ligand exhibits a higher reactivity. This may be explained by observing that increasing the

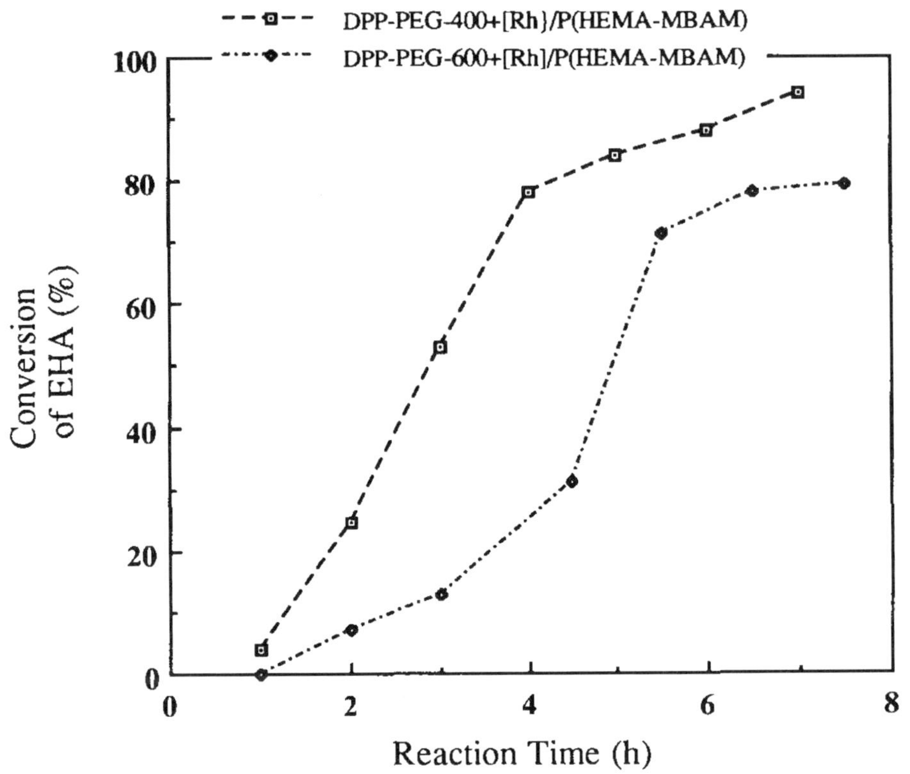

Fig 6 Conversion against reaction time of EHA in the hydroformylation catalysed by {(DPP–PEG)$_{1.5}$Rh(CO)H + DPP–PEG}/P(HEMA–MBAM)/SiO_2. Two DPP–PEG ligands with different average molecular weights were used for comparison. (85°C, 910 psi initial pressure)

length of the PEG chain the molecule will generate a greater number of configurations per unit time, and hence the probability for the Rh coordination center to become entangled and less accessible to the reactants increases.

The surface-pendant ethylenediamine groups of the SSPEDA/SiO_2 substrate may also constitute coordination ligands, even though, in contrast to the phosphine ligands, they do not provide d-orbitals which participate in a major way in the coordination with transition metals [13]. An attempt to take advantage of the cooperation between both ligands (phosphine and diamine) has been made by us in the hydroformylation of EHA. The conversions as a function of time when the catalysts {(MPDPP)$_{2-3}$Rh(CO)H + MPDPP}/SSPEDA/SiO_2 and SSPEDA–Rh/SiO_2 have been employed are presented in Fig. 7. Since the double bond of EHA is conjugated with its electron withdrawing ester group, the 2-(2-ethylhexoxycarbonyl)propylaldehyde, that contains an active hydrogen atom at a tertiary carbon, constitutes the main product of the hydroformylation reaction catalysed by the phosphino–rhodium carbonyl complex. Because of its active hydrogen, this aldehyde can be easily oligomerized by aldol condensation under the catalytic action of the amine groups, and this increases the conversion of EHA. Indeed, a sticky material precipitates in ethanol, and the FT-IR analysis indicated (see the Experimental section) that this material is a product of the aldol condensation. As additional support for this explanation one can raise the fact that only a very weak ^{1}H-NMR peak of –C$\underline{H}$O was observed in the spectrum of the products. The more simple catalyst SSPEDA–Rh/SiO_2 also exhibits a high activity for the above reaction; the activity is, however, much lower than that of {(MPDPP)$_{2-3}$Rh(CO)H + MPDPP}/SSPEDA/SiO_2.

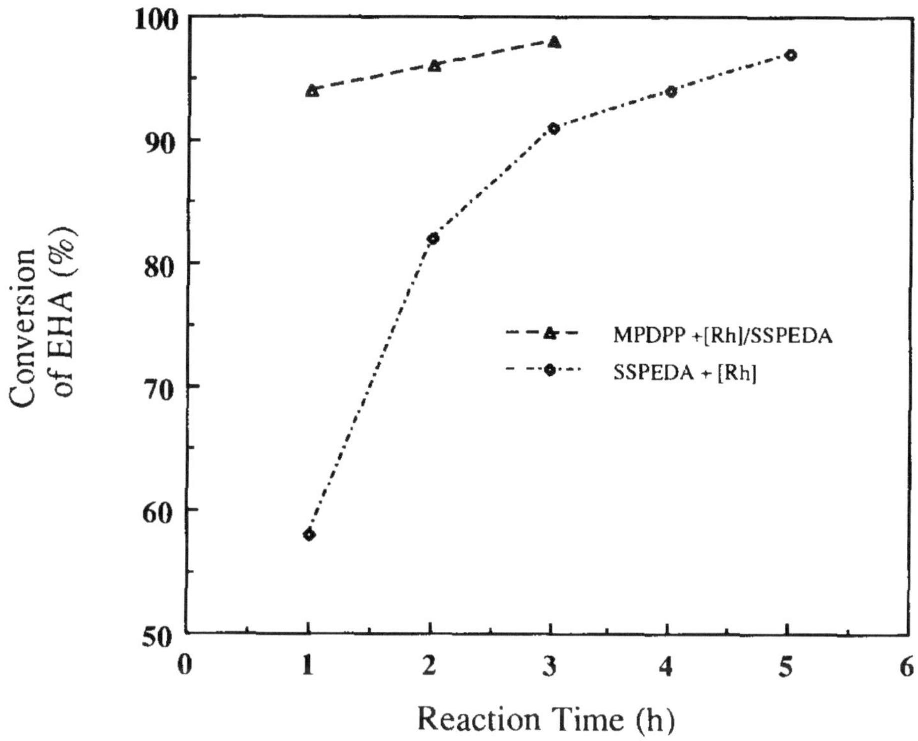

Fig. 7. The effect of the ligand on the conversion of EHA in the hydroformylation catalysed by their supported rhodium carbonyl complexes. (85°C, 910 psi initial pressure).

Table 2
Hydroformylation of EUD in an organic medium and of UDNa and 1-vinylimidazole in an aqueous medium[a]

Entry	Supported catalysts	Yield (%) of aldehydes from olefins[b]			
		EHA[c] (1st run)	EUD (2nd run)	UDNa (3rd run)	1-vinylimidazole (1st run)
1	DPP–PEG-400+[Rh]/ SSPEDA/SiO_2	traces	60 (1 2)	31 (2.4)	negligible
2	MPDPP+[Rh]/ SSPEDA/SiO_2	traces	34 (1 1)	40 (3.8)	negligible

[a]Details are described in the experimental section; the ratio of normal to iso-aldehydes is based on the ratios of the integrated ^{1}H-NMR signals of the corrsponding aldehyde hydrogen.
[b]Their normal/iso ratios after 6 h are in parentheses.
[c]The yield of Aldol condensation of the aldehyde formed from EHA is large

The hydrophobic/hydrophilic accessibility of the immobilized catalysts {(DPP–PEG-400)$_{1.5}$Rh(CO)H+DPP–PEG-400}/SSPEDA/SiO_2 and {(MPDPP)$_{2-3}$Rh(CO)H + MPDPP}/SSPEDA/SiO_2 was examined by performing the hydroformylation of both oil- and water-soluble olefins (Table 2). These catalysts were first employed in the hydroformylation of EHA (see Figs. 4 and 7) in a cyclohexane medium, and then reused to catalyse the hydroformylation of ethyl 10-undecenoate (EUD) in the same medium, and further, in third runs, to catalyse the hydroformylation of sodium 10-undecenoate (UDNa) in a water medium (Table 2).

The conversions of EUD (Table 2) are lower than those of EHA (Figs. 4 and 7) for both catalytic systems. In order to explain this behavior, one can note that while both EHA and EUD are hydrophobic olefins, in EHA the double bond is adjacent to its polar ester group and the hydrophobic carbon chain is at the other end, while in EUD the double bond is at the end of the hydrophobic carbon chain. Consequently, when the reaction medium is more hydrophobic than the surface of the supported catalyst, the double bond of EHA has a higher probability than the double bond of EUD to contact the surface bound catalyst. For this reason, both supported catalysts exhibit higher activities with EHA than with EUD.

Subsequently, in third runs, the above two immobilized systems {(DPP–PEG-400)$_{1.5}$Rh(CO)H + DPP–PEG-400}/SSPEDA/SiO_2 and {(MPDPP)$_{2-3}$Rh(CO)H + MPDPP}/SSPEDA/SiO_2 were used to catalyse the oxo reaction of UDNa using water as reaction medium. The UDNa molecules are soluble in water but are also surface active. Since the surface of the supported catalysts is more hydrophobic than the reaction medium, the adsorption of the UDNa molecules on the surface occurs with the ionic head groups stretched in water. The double bond being present in the hydrophobic medium, the hydroformylation reaction occurs at reasonable rates (as indicated by the consumption of syngas). An interesting result of this reaction is that the ratios of normal to iso aldehyde for both catalysts are higher than those obtained in the hydroformylation of EUD in cyclohexane (Table 2). Since UDNa and EUD have similar structures, the increase of the ratio normal to iso (n/i) may be due to the adsorption of the UDNa surface active molecules at the water/solid interface. To demonstrate that the surface activity plays an important role, the hydroformylation of an olefin, highly soluble in water – 1-vinylimidazole, which is not, however, surface active was also investigated. The results are presented in Table 2 and show that the conversion of this highly soluble olefin is negligible.

3.3. Recycling the catalysts

The scope of this paper is to propose a method to immobilize homogeneous catalysts. The difference between the polarities of the modified ligand MPDPP and the reaction solvent is not large enough to avoid the leakage of the catalyst into the solvent. Indeed MPDPP has some solubility in cyclohexane at temperatures higher than 50°C. For this reason, another

Table 3
Recycling the supported catalysts in the hydroformylation of EUD

The supported catalytic system	Yield (%) of aldehydes				
	1st (6 h)[a]	2nd (8 h)	3rd (10 h)	4th (12 h)	5th (6 h)
MAPDPPE + [Rh]/SSPEDA/SiO_2[b]	36	46	54	56	32
DPE + [Rh]/SSPEDA/SiO_2	62	24			

[a]The ratio of the normal to iso aldehydes is 2.6 in the product of the 1st run of hydroformylation
[b][Rh] represents the complex (MAPDPPE)$_{1.5}$Rh(CO)H

modified phosphine ligand MAPDPPE (see the Experimental section) was also investigated. MAPDPPE is insoluble in cyclohexane even at the boiling temperature.

The catalyst $\{(MAPDPPE)_{1-2}Rh(CO)H+MAPDPPE\}/SSPEDA/SiO_2$ based on one of the above modified phosphine ligands, (containing 2.5×10^{-2} mmol of Rh/g of catalyst and 0.90 mmol of P/g of catalyst), was used in five successive runs to catalyse the hydroformylation of EUD in cyclohexane (Table 3). After five runs, the elemental analysis of the used catalyst indicated that 88% of rhodium and 68% of phosphorus were still retained. For comparison purposes, an unmodified ligand with a similar structure 1,2-bis(dipenylphosphino) ethane (DPE) and its rhodium complex supported on SSPEDA/SiO_2 leading to the catalyst $\{(DPE)_{1-2}Rh(CO)H+DPE\}/SSPEDA/SiO_2$ (containing 2.47×10^{-2} mmol of Rh/g of catalyst and 0.11 mmol of P/g of catalyst) was employed in two successive runs in the hydroformylation of EUD using cyclohexane as reaction medium. The EDS surface elemental analysis indicated that 42% of Rh was retained but almost all DPE had been lost after the two runs. From the above comparison one can conclude that the methoxyacetyl group, which was introduced via the Friedel–Crafts reaction into DPE (which is soluble in cyclohexane) to form MAPDPPE (which is insoluble in cyclohexane), can dramatically improve the immobilization ability of this bidentated phosphine on the SSPEDA/SiO_2 substrate. In addition, the elemental analyses indicate that the surface pendant ethylenediamine groups of the SSPEDA/SiO_2 substrate bind rhodium stronger than the phosphine ligands.

4. Conclusions

An immobilization method is proposed which consists of (1) depositing an organic layer, either poly(hydroxyethyl acrylate crosslinked with N,N′-methylenebisacrylamide) P(HEMA–MBAM) or [3-(siloxysilyl)propyl]ethylenediamine (SSPEDA), over the surface of silica; (2) coating this modified silica with suitable phosphine ligands and their Rh complexes.

The organic layer provides active hydrogen atoms which form hydrogen bonds with the oxygen atoms of the immobilized phosphine ligand; the pendant diamine groups of SSPEDA can also bind Rh(I) ions.

In order to assess the strength of the immobilization methodology, the hydroformylation of 2-(±)-ethylhexyl acrylate (EHA), ethyl 10-undecenoate (EUD) in cyclohexane and sodium 10-undecenoate (UDNa) in water medium were carried out. The main results are: (a) Since the conversion of EHA is higher when the modified instead of pure silica is used as substrate, one can conclude that the organic layer attached to silica facilitates the approach of the hydrophobic reactant molecules to the catalytic sites. (b) The difference between the polarities of the catalyst and reaction medium constitutes an important factor since it affects the adsorption of the reactant molecules on the catalyst. (c) The mono-organic layer (SSPEDA) has a higher activity than the multiple layer P(HEMA-MBAM). (d) The polarities of the synthesized phosphine ligands and the strength of immobilization have the sequence: MAPDPPE (1,2-bis{[(2-methoxyacetylphenyl)-phenyl-(diphenyl)] phosphine}ethane) > MPDPP (2-methoxymethylene-diphenylphosphine).

Acknowledgement

This work was supported by the National Science Foundation.

References

[1] David C. Sherrington, Chemistry and Industry, January 7 (1991), P. Laszlo, (Ed.), Preparative Chemistry Using Supported Reagents, Academic Press, Orlando, FL, 1987; G A Ozin and C Gil, Chem. Rev. (1989) 1749, A.D. Ponogailo, Russ Chem. Rev., 61 (2) (1992) 133.

[2] W.T. Ford, (Ed.), Polymeric Reagents and Catalysts, American Chemical Society, Washington DC, 1986; D.C. Sherrington and P. Hodge, (Eds), Syntheses and Separations Using Functional Polymers, Wiley, Chichester, 1988.

[3] P. Kalck and F. Monteil, Adv Organomet. Chem, 34 (1992) 219,.

[4] P.R. Rony and J.F Roth, J Mol. Catal., 1 (1975) 13, R Datta and R G. Rinker, J. Catal., 95 (1985) 181.

[5] H Bach, W. Gick, E. Wiebus and B Cornlis, Proc. 8th Int Congr. Catal, Verlag Chemie, Weinheim, 1984, V-417

[6] J.S. Kim and R. Datta, AIchE J., 37 (11) (1991) 1675

[7] L. Hong and E. Ruckenstein, J. Chem. Soc., Chem. Commun., (1993) 1487.

[8] M.E Davis, CHEMTECH, August (1992) 489.

[9] L. Hong and E Ruckenstein, J Catal., 143 (1993) 52

[10] A McKillop, F.A. Madjdabadi and D.A Long, Tetrahedron Lett., 24 (18) (1983) 1933

[11] D Evans, G. Yagupsky and G Wilkinson, J Chem Soc. A, (1968) 2660.

[12] E. Weber, J.L Toner, I Goldberg, F Vogtle, D A Laidler, J.F. Stoddart, R.A Bartsch and C.L. Liotta, Crown Ethers and Analogs, John Wiley and Sons, New York, 1989

[13] G H. Olive and S. Olive, Coordination and Catalysis, Verlag Chemie, Weinhein, 1977.

7

METAL SINTERING DURING HEATING IN VARIOUS ATMOSPHERES

INTRODUCTION

Supported metal catalysts contain small crystallites dispersed over the large internal surface areas of refractory metal oxides. These crystallites can coexist with a two-dimensional mono- or multilayer of catalyst atoms or molecules. The degree of dispersion of a metal on a support affects the activity and selectivity of the supported metal catalyst. While during the preparation of the catalyst, a large number of small crystallites are formed, the agglomeration of these small metal crystallites often takes place during heating and particularly during the catalytic reaction. Such a metal particle agglomeration, termed sintering, along with other forms of deactivation such as coking, causes the decrease in the surface area of metal exposed to the chemical atmosphere and ultimately results in the deactivation (loss of activity and selectivity) of supported metal catalysts. Once this occurs, the reaction has to be stopped so as to replace or regenerate the deactivated catalyst. On the other hand, under some other conditions, this metal surface area can be made to increase. This process is termed redispersion. Rejuvenation by redispersion of the crystallites of the aged catalyst has been reported [1–10].

In supported metal catalysts, the commonly used refractory oxide supports, such as silica and alumina, ensure not only a high dispersion of the metal, but also play a role in the reactivity of the catalyst. The physical and chemical interactions between the active catalyst, the oxide support, and the environment affect the morphology of the crystallites and their particle size distribution, and also control the changes that occur in the supported metal during heating and also during the catalytic process. Sintering and redispersion take place in order to minimize the free energy of the system. These processes are therefore affected by the interfacial free energies involved. Since the physical and chemical interactions between crystallites, substrate, and atmosphere affect the values of the earlier free energies, they play a major role in the free energy of a dispersion of small particles.

Parameters that have been found to affect the extent of sintering and redispersion include time, temperature, gas phase composition, the presence of other species on the substrate surface, and the properties of the substrate and metal. Substrate properties include its surface free energy, and the number, strength, and distribution of acidic and basic sites on the support. Active catalyst properties, including the loading (the total amount present per unit surface area of support), the size distribution, the surface free energies (with the gas atmosphere and with the substrate), the surface diffusion coefficients (of metal atoms and of clusters of metal

Heterogeneous Catalysis: Experimental and Theoretical Studies, First Edition. Edited by Eli Ruckenstein and Haiyou Wang.

atoms), and the two-dimensional solubility of the crystallites (the crystallites can coexist with a two-dimensional mono- or multilayer of catalyst atoms). While all this information is rarely available, there are numerous theoretical and experimental approaches that can be used to obtain at least a qualitative picture about the phenomena that control the sintering and redispersion processes.

Our work is mainly focused on metal sintering and redispersion during heating in various atmospheres. Both theoretical and experimental approaches were applied in our studies. In brief, Chapter 7 consists of four parts: the first is concerned with models describing crystallite sintering (**7.1**), ripening (**7.2**), and both sintering and splitting (**7.3**); the second with supported Pd catalysts (**7.4**, **7.5**, and **7.6**); the third with supported Pt catalysts (**7.7–7.8**); and the fourth with supported Fe catalysts (**7.9**).

The degree of dispersion of a metal on a support affects the activity and selectivity of the supported metal catalyst. In a freshly prepared catalyst, the metal is generally highly dispersed and a large number of metal atoms are accessible to the reactants. Deactivation, which occurs after excessive heating, is associated with the aggregation and sintering of the metal crystallites. Models were developed to describe growth kinetics and the size distributions of supported metal crystallites, in which the nonhomogeneity of the surface of the support was taken into account and the time evolutions of the size distribution and of the exposed surface area were examined (**7.1**). The model additionally allows one to analyze the effects of temperature and pressure on the aging process of metal crystallites. A stochastic approach was used to examine the phenomenon of Ostwald ripening for crystallites on a substrate (**7.2**). Two cases were considered, in which the ripening particles were assumed to be (a) three-dimensional hemispherical droplets and (b) two-dimensional circular disks. Furthermore, a general model for the behavior of supported metal crystallites on a substrate in terms of emission, migration, and capture of single atoms and clusters of atoms was suggested (**7.3**). Depending upon the relative magnitudes of the various contributions, the latter model enables one to treat both sintering and splitting in a unitary manner.

Supported metal (Pd, Pt, Fe, Co, Ni, etc) catalysts are used in a wide variety of reactions, including hydrogenation, dehydrogenation, and oxidation of olefins. Substantial sintering of metal crystallites, however, often takes place during heat treatment, resulting in the loss of catalyst activity. The elucidation of the behaviors and characteristics of metal crystallites in various chemical atmospheres is helpful in the improvement of the stability of catalyst. In this regard, various model catalysts were prepared and studied. Model catalysts containing very small Pd particles dispersed on a thin alumina film were heated in H_2 (**7.4**), in O_2 (**7.5**), and alternately in O_2 and H_2 (**7.6**) for various time intervals at various temperatures, and the same region of a specimen was examined by electron microscopy to follow the changes in size, shape, and position of each crystallite. It was found that heating in a hydrogen atmosphere generally results in sintering of palladium crystallites via various mechanisms: (i) crystallite migration and coalescence, (ii) direct ripening, and (iii) Ostwald ripening. Heating in an oxygen atmosphere leads to the formation of pits with the subsequent formation of large cavities via the coalescence of these pits, and the spreading of the oxidized crystallites over the surface of alumina. A variety of events were identified during heating alternately in O_2 and H_2: (i) most Pd crystallites expanded by spreading on the alumina surface during heating in O_2 and recontracted during heating in H_2, (ii) some large crystallites grew in size during heating in H_2, (iii) some small crystallites disappeared during heating in either O_2 or H_2, and (iv) pits were formed on the crystallites during heating in O_2 which disappeared during heating in H_2.

Alternations in the size of the particles on alternate heating in H_2 and O_2 of model Pt/Al_2O_3 catalyst in the range of 300–750°C were observed, in which the particles became smaller on heating in oxygen and larger and extended on heating in hydrogen (**7.7**). This behavior is apparently in contrast to the normally observed extension of particles on oxidation and contraction on reduction. The alternation in size can be explained as a result of the formation of a detectable film (in the form of a halo), or an undetectable film, around each particle due to the strong interactions between crystallites and support in an oxygen atmosphere, and its withdrawal to merge with the respective particle in a hydrogen atmosphere. In another case, the sintering behavior of a model Pt/Al_2O_3 catalysts when heated in hydrogen, and alternately in H_2 and O_2 was investigated at various temperatures between 300 and 800°C (**7.8**).

Traditionally, iron is deposited on oxide supports in the form of ferric salts, prior to calcination and reduction at high temperatures. When present as Fe^{3+} ions, iron exhibits strong reactivity with the support, especially with alumina due to the chemical similarity between Fe^{3+} and Al^{3+}. The strong chemical interaction in turn affects the reducibility of Fe^{3+}, the metal particle size, and ultimately the activity and selectivity of the catalyst. In this regard, the behavior and shape of crystallites and the chemical compounds formed in a model Fe/Al_2O_3 catalyst were investigated. Treatments include heating in H_2 containing trace amounts of oxygen and/or moisture, in additionally purified H_2, and alternately in H_2 and O_2 with observations by transmission electron microscopy and electron diffraction (**7.9**). Electron diffraction indicated the formation of solid solutions of iron oxide with alumina and of the stoichiometric aluminates $FeO \cdot Al_2O_3$ and $Fe_2O_3 \cdot Al_2O_3$ when heated in hydrogen containing trace amounts of oxygen and/or moisture. Such chemical interactions led to enhanced surface spreading phenomena and changes in shape alternately from a torus to a core-and-ring structure. Such alternations were not observed when hydrogen was additionally purified to minimize its oxygen and moisture contents. In addition, the additionally purified H_2 ensured better reducibility to the zero-valent metal. On heating in oxygen, the crystallites were observed to extend considerably, the extension being enhanced with increasing temperatures. After prolonged alternate heating in H_2 and O_2, the crystallites were observed to split into a number of smaller particles, probably as a result of mechanical fatigue. The strong metal–support interactions, the sintering and redispersion process, and the changes in the shape of crystallites are, however, all interlinked as parts of a complex surface phenomenon that cannot be easily categorized as belonging to a simple "either or" phenomenon.

REFERENCES

[1] Adler, S.F., and Keavney, J.J., *J. Phys. Chem.*, **64**, 208 (1960).

[2] Johnson, F.L., and Keith, C.D., *J. Phys. Chem.*, **67**, 200 (1963).

[3] Kearby, K.K., Thorn, J.P., and Hinlicky, J.A., U.S. Patent, 3,134,732 (1964).

[4] Brennan, H.M., Seelig, H.S., and Vander Harr, R.W., U.S. Patent, 3,117,076 (1964).

[5] Coe, R.H., and Randaltt, H.E., U.S. Patent, 3,278,419 (1965).

[6] Jawarska-Galas, Z., and Wrzyszcz, J., *Int. Chem. Eng.*, **6**, 604 (1966).

[7] Kraft, M., and Spindler, H., *4th Int. Cong. Catal.*, Moscow, 1252 (1968).

[8] Schwarzenbek, E.F., and Turkevich, J., U.S. Patent, 3,400,073 (1968).

[9] Weller, S.W., and Montagns, A.A., *J. Catal.*, **20**, 394 (1971).

[10] Ruckenstein, E., and Malhotra, M.L., *J. Catal.*, **41**, 303 (1976).

Growth Kinetics and the Size Distributions of Supported Metal Crystallites

E. RUCKENSTEIN AND B. PULVERMACHER

Department of Chemical Engineering, University of Delaware, Newark, Delaware 19711

Received September 22, 1972

Aging kinetics of supported metal catalysts are represented by a model accounting for diffusion of crystallites on the support and sintering of the colliding crystallites. Equations are established for the time dependence of the distribution of crystallite sizes and for the decay of the exposed surface area of metal for both homogeneous and nonhomogeneous surfaces. For homogeneous surfaces the size distribution becomes after a certain time almost independent of the initial distribution and can be represented by a universal curve. On the basis of the universal curve simple decay equations for the exposed surface area of metal (S) are obtained. For nonhomogeneous surfaces having some sites which interact strongly with crystallites an equilibrium size distribution is reached.

The effect of temperature is accounted for by interactions between the metal and support. At high temperatures, the mobility of the crystallites on the support is relatively great and the sintering process is rate determining. At lower temperatures the aging process is diffusion controlled.

The atmosphere affects the aging process by influencing the surface free energies and hence the work of adhesion and the wetting angle between the crystallites and support. If the wetting angle is decreased the mobility of the crystallites is decreased.

Considerations based on surface chemistry are used to suggest possibilities for increasing the stability of the highly dispersed state of metal. For example, stability may be increased if certain alloys are used rather than pure metals.

INTRODUCTION

The degree of dispersion of a metal on a support affects both the activity and selectivity of the supported metal catalyst. In freshly prepared catalyst the metal is generally highly dispersed so that a large fraction of the metal atoms are accessible to reactants. Deactivation and aging, observed after excessive heating (*1–7*) or use under industrial conditions (*5, 7, 8*), are associated with aggregation and sintering of the small crystallites. For example, Herrmann *et al.* (*6*) and Maat and Moscou (*1*) observed that heat treatment of a highly dispersed supported platinum catalyst caused the formation and growth of platinum crystallites and thus a decay of the exposed surface area of the metal.

Analyzing the available experimental data (*1, 3, 6, 7*), described below in some detail, we have observed that the rate of decay of exposed surface area of metal is given by an equation of the form

$$\frac{dS}{dt} = -KS^n, \qquad (1)$$

where the exponent n has a lower bound of about 2 and an upper bound as large as 8. The large range in the value of the exponent suggests that more than one process is responsible for the decay of the exposed surface area of metal.

Several models have been developed by

This paper, which was originally published in **Journal of Catalysis**: 29, 224–245 (1973), is republished with permission.

the authors (*9*) to describe the aging process caused by thermal treatment. The models assume (a) diffusion of the crystallites upon the surface of the support and (b) sintering of the colliding particles. Quantitative descriptions of the decay of the exposed surface area have been obtained in particular in the two limiting cases of diffusion and sintering control. For sintering control the value of n is about 2, while for diffusion control it may be as large as 8.

The goals of the present research are: (a) from the proposed models to obtain information concerning the size distribution of the metal crystallites for various possible mechanisms and to compare the distribution with available experimental data; (b) to obtain additional information concerning the rate of decay of the exposed metal surface area and to relate the exponent n from Eq. (1) to the mechanism of the aging process; (c) to suggest explanations for the effects of the atmosphere on the aging process; (d) to explain the effect of the method of preparation of the supported metal catalyst on the aging process; (e) to suggest possible ways to increase the life (stability) of the small crystallites.

In the first section of the paper the main results of our previous paper are reviewed and further arguments supporting the physical models are presented. Some modification of the earlier models are added which take into account the nonhomogeneity of the surface of the support. Subsequently the time evolution of the size distribution and of the exposed surface area are analyzed. There follows a detailed comparison between theory and experiment concerning both the time decay of the exposed surface area of metal and the size distribution of the metal crystallites. Finally possible improvements of the stability of the highly dispersed state of the metal are suggested.

Physical Model and Kinetic Equations

The physical model is based upon the assumptions of (a) diffusion of the crystallites upon the surface of the support and (b) sintering of the colliding particles.

We consider an initially highly dispersed state of metal on the surface of the support, assumed for simplicity to be planar and energetically homogeneous. For sufficiently high temperatures the agglomerates of metal atoms have some mobility on the surface of the support. Indeed, Bassett (*10, 11*), using electron microscopy and a high-vacuum system, observed translation of copper and silver islands on amorphous carbon and graphite and rotation as well as translation of silver islands on a molybdenite support at temperatures as low as 525 K. Skofronick *et al.* (*12, 13*) observed motion and coalescence of gold islands on amorphous supports of carbon and silicon in the temperature range 500 to 700 K. They found that the number of islands per unit surface area decreased with time and that their mean radius increased.

Crystallites of supported catalysts are of the order of 10 to 100 Å in diameter, about the same size as the agglomerates in the above-mentioned experiments. Consequently the assumption that the crystallites migrate on the surface of the support has experimental confirmation.

Alternative mechanisms of mass transfer by which growth of the crystallites might occur are (a) atomic breakup of the crystallites with consecutive migration of the atoms over the support and (b) evaporation of the metal with consecutive vapor phase transport. Since the bonding energies of metal atoms to metal crystallites are much larger than the bonding energies of metal atoms to nonmetallic supports, dissociation of metal atoms is unlikely (*13, 18*).* Evaporation rates of metal have been measured and found to be several orders of magnitude too low to account for the observed growth of the metal crystallites (*5, 13*).

The following treatment emphasizes situations in which sintering of the support

* There are special conditions under which breakup of crystallites with consecutive migration of the formed particles can take place. This problem, of particular interest in the regeneration of catalysts, is discussed briefly in the last section of the paper and will be examined more fully in another paper.

is of secondary importance. That such an assumption can be made, at least in a dry atmosphere, is shown for instance in references (*14, 15*).

The decay process is thus a consequence of the migration of crystallites on the support surface. Though the detailed mechanism of this migration is not known, it is clear that it is induced by the thermal motion of the atoms at the metal–support interface. Because of the random character of these thermal motions it is reasonable to represent the motion of the crystallites upon the surface of the support by a diffusional model.

When two crystallites collide, two limiting situations can occur. In one of them the interaction between the colliding particles is so strong that they form a single unit within a time which is short compared to the diffusional time. The rate of sintering is in this case diffusion controlled. If the time of the merging process of the colliding particles into a single unit is long compared to the diffusional time, then the merging process is the rate-determining step. For such a situation the name "sintering control" will be used.

Only the main results of the analysis (*9*) based on the above-mentioned assumptions will be given here. Because only binary collisions are important, the number of crystallites per unit surface area of support composed of k metal units, N_{v_k}, increases by collisions of particles composed of i and $k - i$ metal units and decreases by collisions of the particles composed of k units with the other particles. Consequently, for a homogeneous surface,

$$\frac{dN_{v_k}}{dt} = \frac{1}{2} \sum_{i+j=k} K_{ij} N_{v_i} N_{v_j} - N_{v_k} \sum_{i=1}^{\infty} K_{ik} N_{v_i} \qquad (2)$$

where K_{ij} are second-order rate constants dependent upon the mobility of the particles on the support and on the nature of the interaction between the particles (*9*).

For the two limiting situations of "diffusion control" and "sintering control" the rate constants K_{ij} are given by (*9*)

$$K_{ij} = \frac{8D_{ij}}{\pi} \int_0^\infty e^{-D_{ij}u^2\theta'} \times \frac{du}{[J_0^2(uR_{ij}) + Y_0^2(uR_{ij})]u} \quad \text{(diffusion control)} \qquad (3)$$

and

$$K_{ij} = 2\pi R_{ij} \alpha_{ij} \quad \text{(sintering control).} \qquad (4)$$

α_{ij} is a reaction-rate constant for the merging process, $R_{ij} = R_i + R_j$ is the radius of interaction of the two colliding particles, and D_{ij} is the diffusion coefficient of particle i with respect to particle j.

As pointed out previously (*9*), the present approach involves two time scales: a large scale of time t, introduced in Eq. (2), which coincides with the time of the process, and a small scale of time θ', used in the calculation of the rate of collisions and thus of the rate constants K_{ij}. In the case of sintering control, the rate constants are independent of the small-scale time [Eq. (4)], because the small time affects only the rate of diffusion of the particles towards a selected particle. In this situation the diffusion is not the rate-determining step. In the case of "diffusion control," the rate constants depend on the small-scale time, but in certain situations this dependence is after a very short time very weak (see Appendix I) and

$$K_{ij} = \frac{4\pi D_{ij}}{\ln 4T}, \qquad (3''')$$

(where $T = D_{ij}\theta'/R_{ij}^2$) becomes a good approximation for Eq. (3). Equation (2) together with Eqs. (3''' and 4) describes the sintering process in the two limiting situations. These equations will be used to obtain both the time dependence of the particle-size spectrum and the time evolution of the exposed surface area.

The surface of the support is in reality nonhomogeneous from a topographic and energetic point of view. Some sites can interact so strongly with the crystallites that they are immobilized; also some regions of the surface can entrap crystallites.

Besides the quantities N_{v_k} representing the number of moving crystallites per unit surface area of support, the quantities F_{v_k} representing the number of fixed crystallites per unit surface area of support composed of k metal units are now of significance. Two probabilities are introduced. The probability p that a particle formed by the collision of two moving particles will form a mobile particle and the probability q that a trapped particle will become mobile after a collision with a moving particle. The probability p is a consequence of the fact that some of the collisions take place, either over sites which can interact strongly with the crystallites, or over trapping regions. For simplicity sake p and q are assumed size independent. The moving particles can encounter trapping regions becoming fixed. This is equivalent to introduce in the equations a first-order chemical reaction of the type $\beta_k N_{v_k}$.

The concentration N_{v_k} increases by the collisions between moving particles composed of i and $k - i$ metal units which lead to mobile particles, and by the collisions between moving particles of i units and fixed particles of $k - i$ units which generate moving particles. It decreases by trapping and by collisions between the moving particles of k units and all the other particles.

The concentration F_{v_k} increases by trapping of mobile particles, by the collisions between moving particles of i metal units and fixed particles of $k - i$ units which generate immobilized particles, and by the collisions between moving particles of i and $k - i$ units which form fixed particles. It decreases by collisions between fixed particles of k units and all the moving particles.

Therefore, the rate of change of N_{v_k} and F_{v_k} can be written as

$$\frac{dN_{v_k}}{dt} = \frac{1}{2} p \sum_{i+j=k} K_{ij} N_{v_i} N_{v_j} - N_{v_k} \sum_{i=1}^{\infty} K_{ik} N_{v_i} - N_{v_k} \sum_{i=1}^{\infty} K^*_{ki} F_{v_i} + q \sum_{i+j=k} K^*_{ij} N_{v_i} F_{v_j} - \beta_k N_{v_k} \quad \text{(5a)}$$

and

$$\frac{dF_{v_k}}{dt} = (1 - q) \sum_{i+j=k} K^*_{ij} N_{v_i} F_{v_j} + \frac{1}{2}(1 - p) \sum_{i+j=k} K_{ij} N_{v_i} N_{v_j} - F_{v_k} \sum_{i=1}^{\infty} K^*_{ik} N_{v_i} + \beta_k N_{v_k} \quad \text{(5b)}$$

Equations for the rate constants K^*_{ij} can be obtained as in our previous paper (*9*) taking however into account the trapping and replacing the diffusion coefficient D_{ij} of particle i with respect to particle j ($D_{ij} = D_i + D_j$) by the diffusion coefficient D_i of the moving particle. The rate constants K_{ij} must also be recalculated.

Effect of the Rate Determining Step of the Sintering Process on the Particle Size Distribution for Homogeneous Surfaces

The equations for the discrete spectrum of particle sizes were solved numerically for the initially unisized distribution: $N_{v_{i,0}} = N_0$ for $i = 1$ and $N_{v_{i,0}} = 0$ for $i \neq 1$. A fourth-order Runge–Kutta method was used and as many equations (up to 800) considered as were needed to ensure that no particle was lost in the calculations. Errors due to the underflow of the computing machine were negligible.

The two limiting cases of "sintering control" [Eqs. (2 and 4)] and "diffusion control" [Eqs. (2 and 3‴)] are considered for various size dependencies of the diffusion coefficients and reaction rate constants (see Table 1).

In Figs. 1 to 5 are plotted the results obtained for the size distribution at different times and for different mechanisms. For "diffusion control" (Figs. 1–3) the particle-size distributions develop a maximum because the rate constants K_{ij} decrease with increasing particle size, and therefore the smallest particles in the system disappear the fastest (see also Fig. 6). For "sintering control" a monotone decreasing particle size distribution is obtained (Figs. 4 and

TABLE 1
TIME EVOLUTION OF EXPOSED METAL SURFACE AREA IN THE TWO LIMITING CASES (9)

Assumed size dependence of D_{ij}	Rate constants	Rate equation for exposed surface area
Diffusion control		
$D_{ij} = C_1\left(\frac{1}{r_i^2} + \frac{1}{r_j^2}\right)$	$K_{ij} = C'_1\left(\frac{1}{r_i^2} + \frac{1}{r_j^2}\right)$	$\frac{dS}{dt} = -C''_1 S^6$ (I-1′)
$D_{ij} = C_2\left(\frac{1}{r_i} + \frac{1}{r_j}\right)$	$K_{ij} = C'_2\left(\frac{1}{r_i} + \frac{1}{r_j}\right)$	$\frac{dS}{dt} = -C''_2 S^5$ (I-2′)
$D_{ij} = C_3$	$K_{ij} = C'_3$	$\frac{dS}{dt} = -C''_3 S^4$ (I-3′)

Assumed size dependence of α_{ij}	Rate constants	Rate equation for exposed surface area
Kinetic control		
$\alpha_{ij} = C_4$	$K_{ij} = C'_4(r_i + r_j)$	$\frac{dS}{dt} = -C''_4 S^3$ (I-4′)
$\alpha_{ij} = C_5 \frac{r_i^2 + r_j^2}{r_i + r_j}$	$K_{ij} = C'_5(r_i^2 + r_j^2)$	$\frac{dS}{dt} = -C''_5 S^2$ (I-5′)

5) because the rate constants K_{ij} increase with increasing particle size, and therefore the smallest particles in the system disappear most slowly (Fig. 7).

In Fig. 8 the results obtained for the size distributions for the same value of the ratio S_0/S and for different mechanisms are compared. The particle-size distributions for the diffusion-controlled cases are significantly narrower than those obtained for

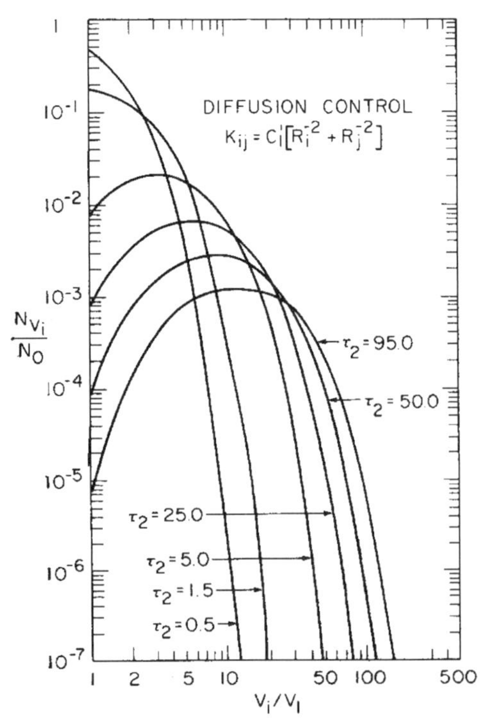

FIG. 1. Particle size distribution from numerical solution of kinetic equations; unisized initial distribution.

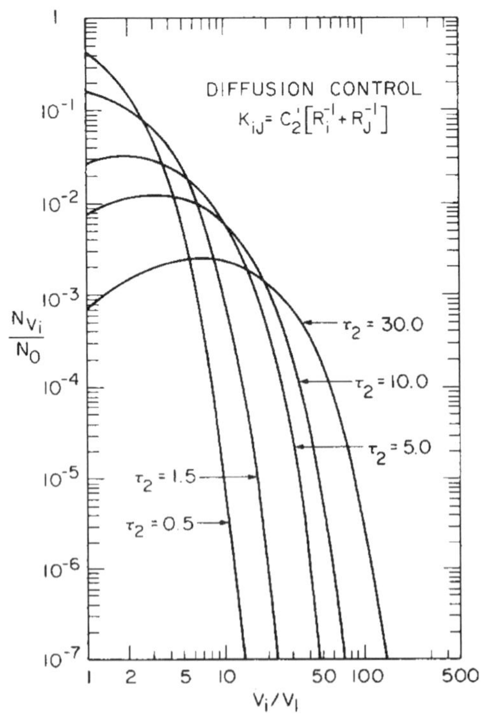

FIG. 2. Particle size distribution from numerical solution of kinetic equations; unisized initial distribution.

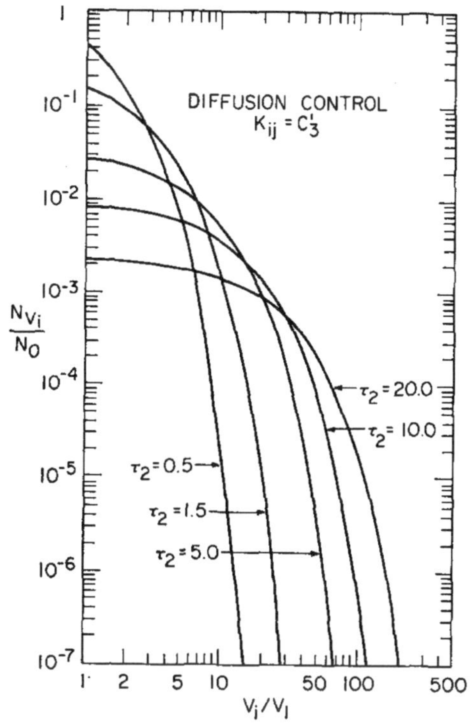

FIG. 3. Particle size distribution from numerical solution of kinetic equations; unisized initial distribution.

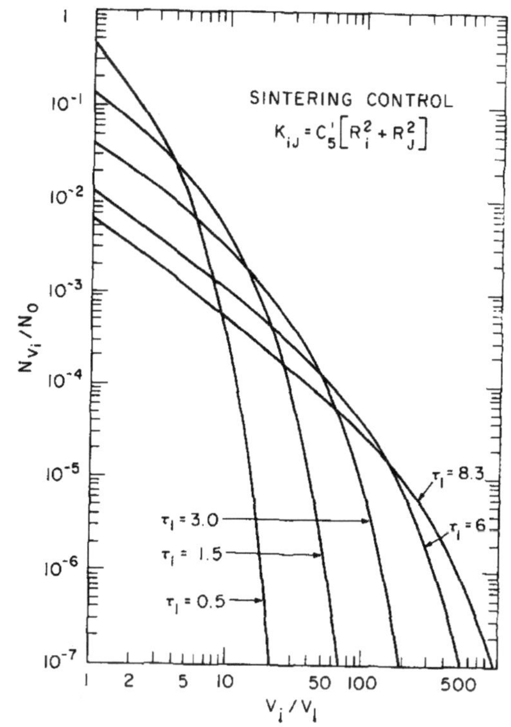

FIG. 5. Particle size distribution from numerical solution of kinetic equations; unisized initial distribution.

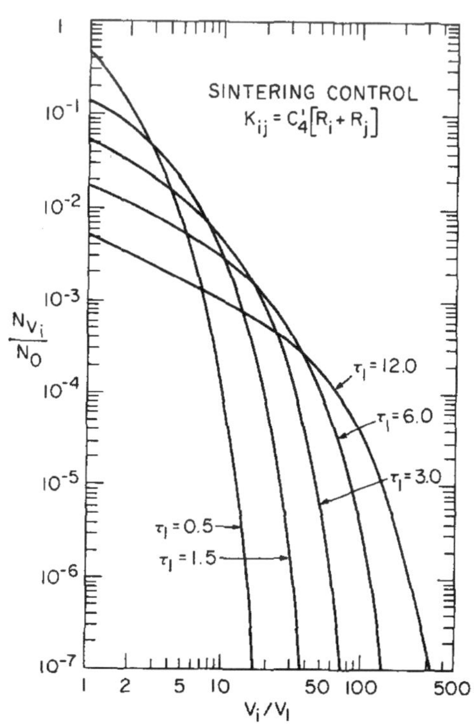

FIG. 4. Particle size distribution from numerical solution of kinetic equations; unisized initial distribution.

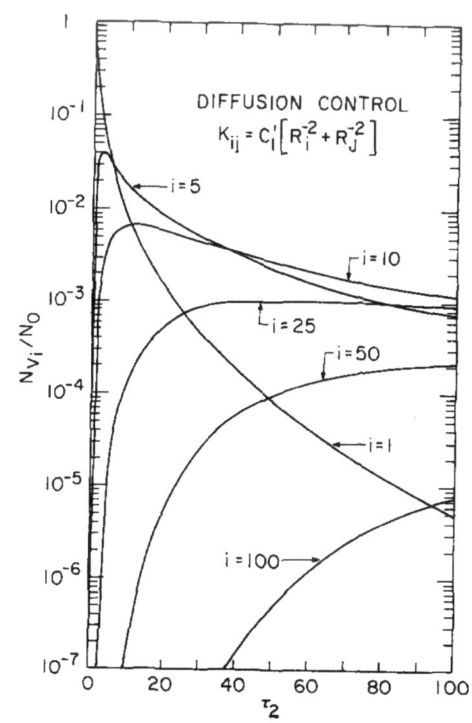

FIG. 6. Number of particles of volume v_i as a function of the dimensionless time τ_2; unisized initial distribution.

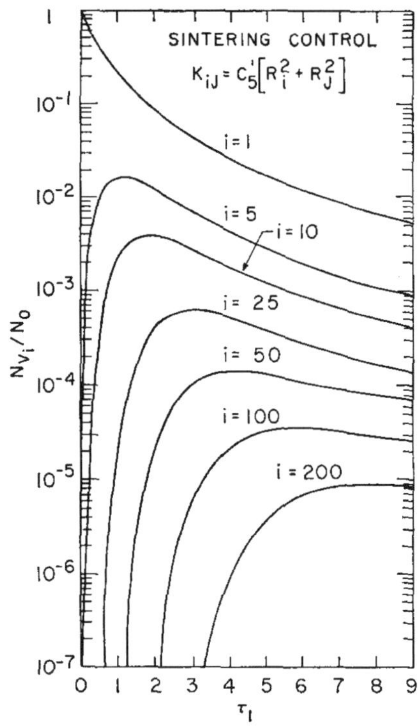

FIG. 7. Number of particles of volume v_i as a function of the dimensionless time τ_1; unisized initial distribution.

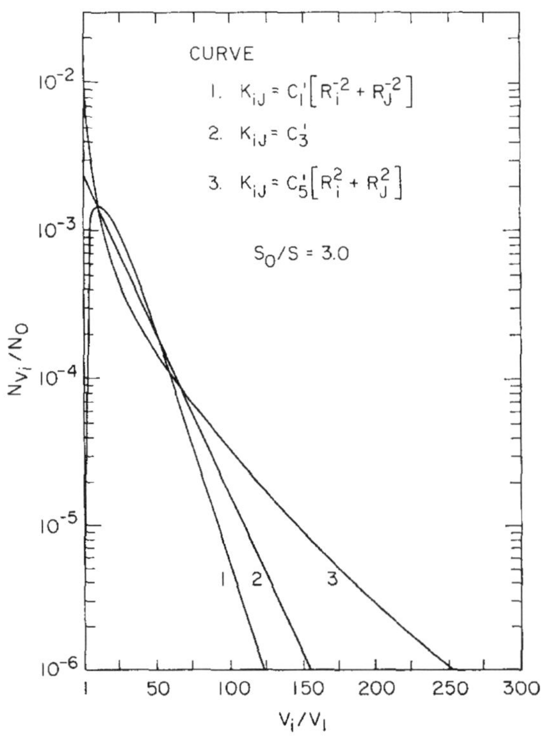

FIG. 8. Particle size distribution for "diffusion control" and "sintering control" for the ratio $S_0/S = 3$, unisized initial distribution.

the sintering-controlled cases. This result suggests a criterion to differentiate between the two mechanisms.

For initial size distributions different from the unisized, our calculations show that for sufficiently long times the size distributions are very similar, approaching the curves obtained for an initially unisized distribution. This result raises the question of the possibility of superposing on a unique curve the time-dependent size spectra. It is shown in Appendix II that such a unique spectrum is obtained if the dimensionless groups $N_{v_i}\phi/v_1N^2$ and v_iN/ϕ are used as variables, where N is the total number of particles per unit surface area of support at a given time t, and ϕ is the total volume of metal per unit surface area. Such a unique spectrum was first used in a different context (size distribution of aerosols) by Friedlander (*34, 35*).

In Fig. 9 the size distributions for the diffusion-controlled case are plotted using these dimensionless groups. Two different initial distributions are considered. For both, the initially monodispersed and exponentially dispersed spectrum, the unique

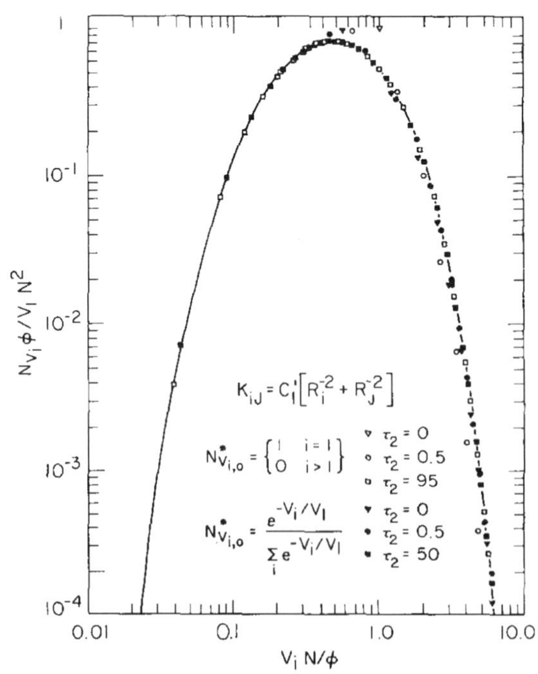

FIG. 9. Dimensionless size distribution for "diffusion control" and different initial distributions; the solid curve represents the similarity solution.

spectrum (solid line) is reached within a short time ($\tau_2 \simeq 2.5$).

Time Evolution of Exposed Surface Area for Homogeneous Surfaces

It will be assumed that the general shape of the particles is preserved during the growth process. Consequently the exposed surface area of a particle containing k units is given by

$$S_k = S_1 k^{2/3}, \tag{6}$$

where S_1 is the exposed surface area of a particle containing one unit. The time evolution of the rate of change of the exposed surface area may be calculated from

$$\frac{dS}{dt} = S_1 \sum_{k=1}^{\infty} k^{2/3} \frac{dN_{v_k}}{dt}. \tag{7}$$

In our previous paper (*9*) the equations for the decay of the surface area [Eqs. (2 and 7)] were solved numerically for an initially unisized distribution. The two limiting cases of "sintering control" and "diffusion control" were considered, taking various size dependencies of the diffusion coefficients and the reaction rate constants into account. The results obtained are summarized in Table 1.

One may observe that the rate equation of the exposed surface area is very sensitive to the rate determining step of the process. In the case of "sintering control" exponents of 3 and smaller are obtained for n [Eq. (1)] while for "diffusion control" the exponent is 4 or larger.

These results, obtained previously for an initially unisized distribution, are of more general validity, because, as shown here, after a relatively short time the size distribution can be represented by the similarity variables. In Appendix II an equation for the decay of the exposed surface area of the metal is derived on the basis of the similarity variables. The resulting equation [Eq. (19) from Appendix II] is indeed equivalent to those established on the basis of numerical computations for an initially unisized distribution (Table 1). The exponent of S in the rate equation for the decay of the exposed surface area of metal [Eq. (1)] is thus dependent only on the assumed mechanism of the process and on the size dependence of the diffusion coefficient and reaction-rate constants, and it is almost independent of the initial distribution.

Size Distribution and Exposed Surface Area for Nonhomogeneous Surfaces with Some Sites which Interact Strongly with Crystallites

In this case, an equilibrium size distribution is reached. Because the strong interacting sites have been active during the preparation process favoring the formation of crystallites, one may expect that a large fraction of these sites are covered by the immobile crystallites from the beginning. Neglecting the trapping regions, one may assume that $p = 1$ and $\beta_k = 0$. Figure 10 shows the size distribution, close to equilibrium, obtained for different values of the initial ratio between moving and fixed islands and for an initially unisized size distribution. By "close to equilibrium" we understand here that less than 1/10 of the particles present are still mobile. Figures 11–13 show the time evolution of the size distribution of the moving particles, fixed

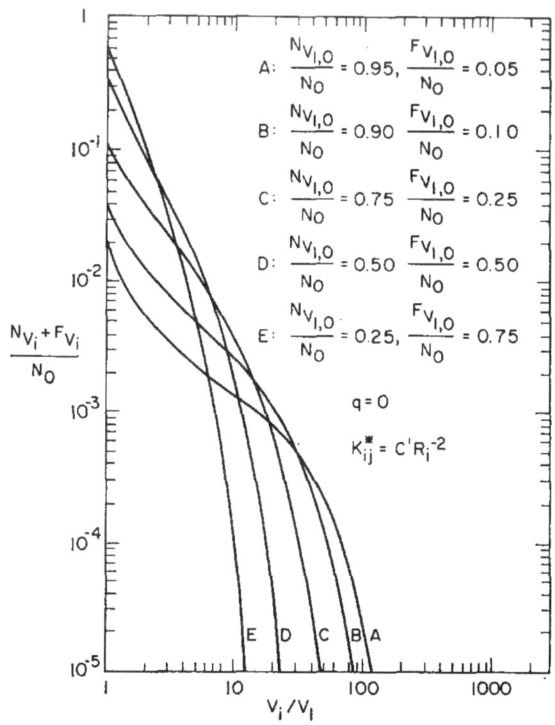

FIG. 10. Particle size distribution in the nonhomogeneous case, close to equilibrium.

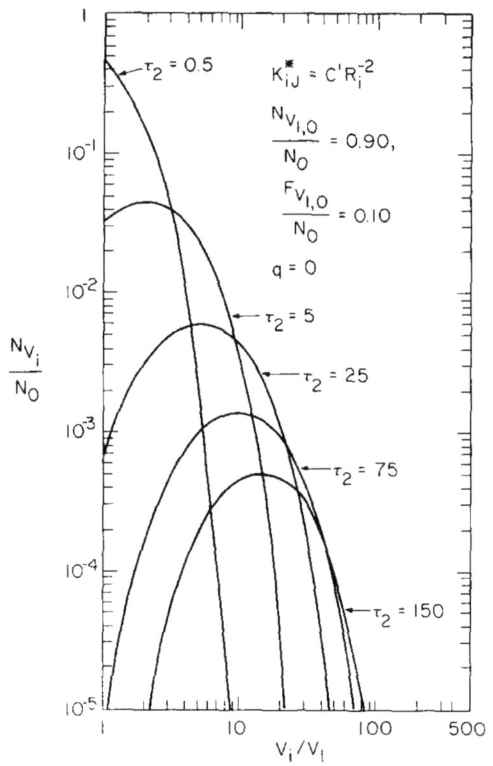

FIG. 11. Time dependence of the size distribution of the moving particles, in the nonhomogeneous case.

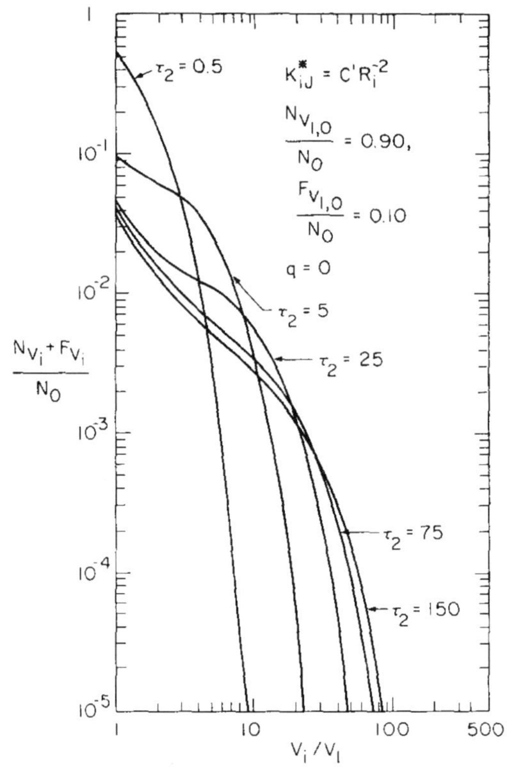

FIG. 13. Time dependence of the size distribution of the total number of particles, in the nonhomogeneous case.

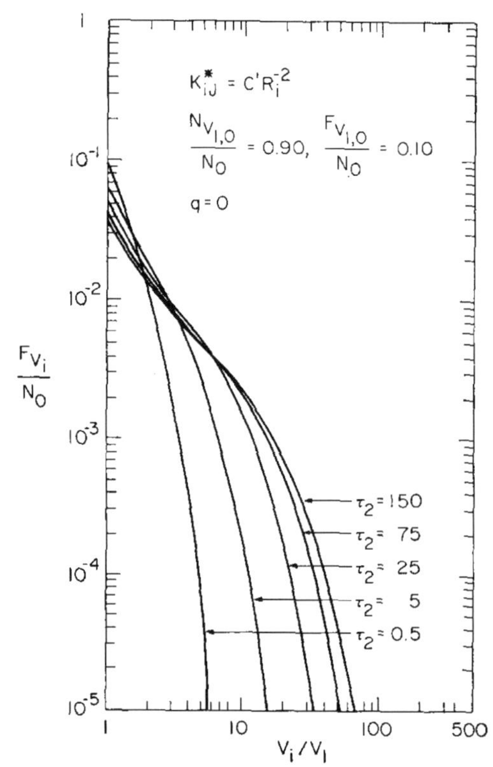

FIG. 12. Time dependence of the size distribution of the fixed particles, in the nonhomogeneous case.

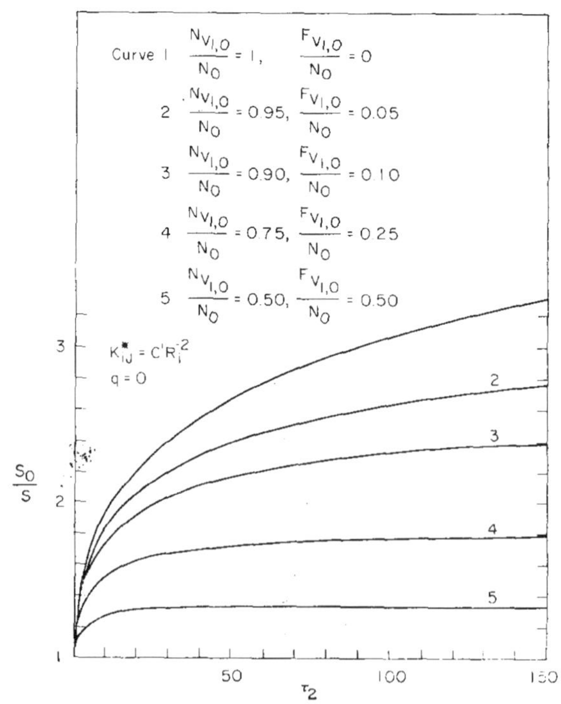

FIG. 14. Effect of fixed islands on the time dependence of S_0/S.

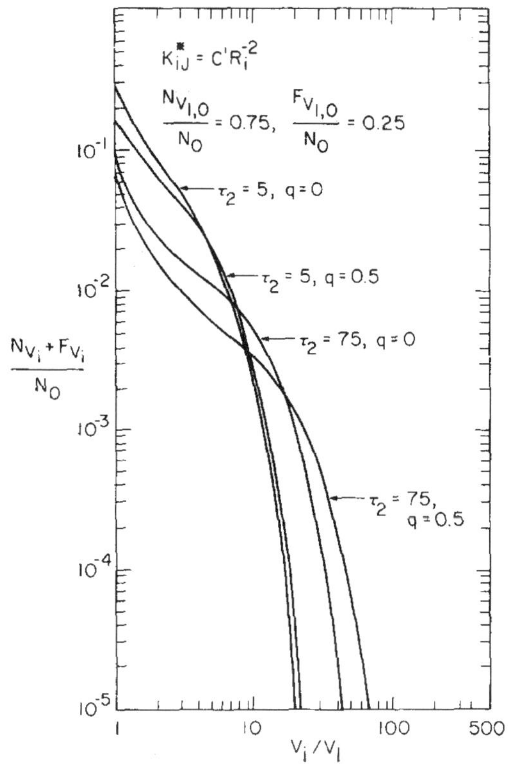

FIG. 15. Effect of the probability q on the time dependence of the size distribution of the total number of particles.

particles, and total number of particles for a specific example. Figure 14 shows the effect of fixed islands on S_0/S for $q = 0$ and various initial values of the ratio N_{v_1}/F_{v_1}. Figure 15 shows the effect of the probability q on the distribution of sizes.

The main conclusions are: (a) a narrow equilibrium size distribution is reached from an initially unisized distribution even if the ratio between mobile and fixed particles is initially as large as 20; (b) the equilibrium size distribution is strongly dependent on the initial distribution (as opposed to the similarity solution obtained previously); (c) an equilibrium value is reached for S; and (d) the size distribution is somewhat broadened when $q = 0.5$. Because small values of q are expected, its influence appears to be of secondary importance.

DISCUSSION AND COMPARISON OF MODEL PREDICTIONS WITH EXPERIMENTAL RESULTS

Table 2 is a list of the experimental data available for the exposed metal surface area of Pt-on-alumina catalysts during heat treatment. One may observe that the exponent in the decay equation of the exposed surface area of metal is in the range 2–8, thus covering both the cases of "sintering control" and "diffusion control."

The results obtained in the particular cases can be interpreted as follows: Maat and Moscou (*1*) measured the rate of sintering of a platinum reforming catalyst at a temperature of 1053 K (Table 2, row 1), probably in an oxidizing atmosphere. They found that the decay of the exposed platinum area followed second order kinetics, which, according to the present considerations, shows a sintering controlled decay. By means of electron microscopy they determined a large dispersion of the particle size distribution. The initial size distribution was between 10 and 50 Å, with an average particle size close to 10 Å. After 17 hr of sintering the distribution was between 10 and 500 Å. Both the observation of a large dispersion of the particle sizes after sintering and the observation that the smallest particles existing initially were still present after excessive sintering (~15-fold decrease in surface area after 17 hr of sintering) are in qualitative agreement with a sintering controlled mechanism.

Herrmann *et al.* (*6*) reported data on the sintering kinetics of Pt–Al_2O_3(Cl) and Pt–Al_2O_3(F, Cl) reforming catalysts in the temperature range 837–898 K (Table 2, row 2) and a dry air atmosphere. The exposed surface area of metal was measured by hydrogen chemisorption. They found that the rate of decrease of the metal surface area—especially at 898 K—was second-order with respect to the metal surface area, thus indicating a sintering-controlled decay.

Gruber (3) reported data on the sintering behavior of Pt-on-alumina catalyst prepared by tetrammino platinum hydroxide impregnation. The sintering experiment was carried out at 773 K (Table 2, row 3) in a reducing atmosphere. Their data can be represented by a sixth-order reaction, which according to the present considerations, is an indication of diffusion controlled decay.

TABLE 2
TIME EVOLUTION OF EXPOSED SURFACE AREA; EXPERIMENTAL RESULTS AND THEIR INTERPRETATION

	Reference	Experimental results			Model interpretation	
		Type of catalyst	Temperature and atmosphere	Rate of decay of exposed metal surface	Controlling	Diffusion coefficient or reaction rate constant
1	(*1*)	Pt on γ-Alumina 0.5–0.7% Cloride 0.6% Pt 140–180 m²/g support	1053 K Probably dry air	$\frac{dS}{dt} = -K_1S^2$	Sintering	$\alpha_{11} = 4. \times 10^{-9}$ cm/sec based on particle 10 Å in size
2	(*6*)	Pt on γ-Alumina a) 0.83% Cl, 0.774% Pt 225 m²/g support b) 0.35% Cl, 0.35% F 0.375% Pt 176 m²/g support	898 K Dry air	$\frac{dS}{dt} = -K_2S^2$	Sintering	$\alpha_{11} = 1. \times 10^{-9}$ cm/sec based on particle 10 Å in size
3	(*3*)	Pt on γ-Alumine 1.1% Pt 210 m²/g support	773 K H_2	$\frac{dS}{dt} = -K_3S^6$	Diffusion	$D_1 \simeq 8. \times 10^{-18}$ cm²/sec based on particle 10 Å in size
4	(*7*)	Pt on Alumina	755 K 811 K H_2	$\frac{dS}{dt} \simeq -K_4S^8$ $\frac{dS}{dt} \simeq -K_5S^{7.8}$	Diffusion	

Hughes *et al.* (*7*) (Table 2, row 4) obtained results similar to those of Gruber (*3*). They measured the exposed surface area of metal as a function of time at 755 and 811 K in a reducing atmosphere. The experimental rate of decay of the exposed surface area of metal indicates a diffusion mechanism, however, with a stronger size-dependent diffusion coefficient than in Gruber's experiments.

We stress that even a qualitative discussion of the above-mentioned data leads to the conclusion that two different mechanisms occur. In some of the cases the decay of the exposed metal surface area is rapid, being for instance a 15-fold decay after 17 hr of heating in an oxidizing atmosphere at 1053 K (*1*) and a 12.5-fold decay after 40 hr at 898 K (*6*). As opposed to these situations, Gruber (*3*) reports that at 773 K and in a reducing atmosphere a heat treatment of 1200 hr leads to only a 2.5-fold decay of the exposed surface area of metal, while Hughes *et al.* (*7*) observed a 3-fold decay after 1000 hr at 811 K also in a reducing atmosphere. Consequently, a decrease in temperature from 1053 to 898 K and in an oxidizing atmosphere results in a decrease of the rate of sintering by about a factor of 4. A further drop in temperature by another 100 K together with a change of the chemical atmosphere to a reducing atmosphere slows down the sintering by more than two orders of magnitude, and changes the exponent n in the rate equation [Eq. (1)] substantially. Obviously the decay of the surface area of metal is controlled in those situations by different processes.

As suggested by the present treatment, the rate of the process is either controlled by the random motion of the crystallites on the surface of the support or by the merging of two colliding particles into a single particle. At sufficiently high temperatures the mobility of the crystallites upon the support is high and one may expect that the merging process is the rate-determining step. For not too high temperatures, the mobility of the crystallites is lower and consequently the diffusion of the crystallites on the support prevails. The data presented above are compatible with this explanation showing that at high temperatures the process is indeed controlled by the merging process, while at lower temperatures the process is diffusion controlled.

The rate of decay depends also on the chemical atmosphere (*5, 7, 8, 17*). At the same temperature the rate of decay is much higher in an oxidizing than in a reducing atmosphere (*5, 17*). Our theoretical approach can explain such behavior, because the diffusion coefficients of the crystallites and the reaction-rate constants of the merging process depend upon the surface properties of the system. The chemisorption process essentially changes the surface free energies and the surface diffusion of the metal on the metal. If the wetting angle of the crystallite on the support becomes larger, due to the changes in the surface free energies, then the surface of crystallite–support contact becomes smaller and consequently their interaction smaller.† In this case the mobility (diffusion coefficient) of the crystallite is increased. The merging process also depends upon the surface free energy and surface diffusion of the metal atoms on the metal surface (*18*). For instance, the surface diffusion coefficient of copper atoms on copper is increased by a factor of about 5 in the presence of oxygen compared to a reducing atmosphere (*19*).

No data are yet available to establish the effect of chemical atmosphere on the wetting angles between metals and supports used in supported metal catalysts. Some data, however, for the wetting angle of drops of $Na_2Si_2O_5$-glass on several metals show an important effect of the atmosphere on the wetting angle (*20, 21*).

Some authors have suggested that a possible explanation for the increase in the rate of sintering in an oxygen atmosphere is the mass transfer caused by formation and vaporization of PtO_2 (*5, 17*). Information available in the literature concerning platinum wires or bands shows that the losses of platinum in an oxygen atmosphere are negligible at temperature below 1050 K (*22–24*). Though for the small crystallites of interest here the vapor pressure is increased due to the curvature effect, our evaluations show that the rate of vaporization [limited in the present case by the diffusion of PtO_2 molecules away from the surface (*25*)] for the temperature range of interest remains negligibly small.

The diffusion coefficient computed by us are very small, of the order of 10^{-17} cm^2/sec at 773 K for a crystallite 10 Å in diameter (Table 2). The values measured for the diffusion coefficient of aluminum islands of the same size on KCl at even lower temperatures are larger by several orders of magnitude, 10^{-12}–10^{-14} cm^2/sec (*26, 27*). The reason for this difference probably lies in the nature of the interactions between metal and support, interactions which depend upon the roughness of the surface of the support and the nature of the bonds between the metal and support. Obviously in the cases in which the pores are not sufficiently large compared to the size of the crystallites, another restriction appears.

Concerning the interactions between metal and support, let us observe that for small aluminum islands evaporated onto the KCl-substrate and also for metal crystallites of a completely reduced supported metal catalyst, relatively weak interactions of the van der Waals type are dominant (*18*). The preparation history of the supported catalyst, on the other hand (for example preparation by adsorption (or ion exchange) as opposed to preparation by impregnation) might influence the nature of the bonds at the metal–substrate interface: If the catalyst is prepared by adsorption from a solution containing for

† For very small crystallites the surface free energies and the wetting angle depend upon their dimensions. Our discussions imply the assumption that their qualitative behavior is as for large objects.

example $(Pt(NH_3)_4)Cl_2$, then during preparation strong ionic bonds are formed between the acidic groups of the support and $Pt(NH_3)_4^{2+}$ (*4*). In the impregnation procedure no such strong interaction between the platinum salt and the support exists. Perhaps even after reduction the strong ionic metal–substrate bonds are partly preserved in the catalyst prepared by adsorption, thus leading to decreased mobility of the particles. Consequently, a higher stability of the highly dispersed state of metal is expected if the support is more acidic and an adsorption procedure is used for the preparation of the supported metal catalyst.

During preparation of the supported metal catalyst, high temperatures are often used, in particular in the reduction period. Thus migration and sintering of the unreduced platinum can also occur. The extent of sintering depends on the temperature, on the duration of the reduction period, and as discussed above, on the method of preparation. Consequently a higher degree of sintering is expected during the reduction period for an acidic support when the impregnation rather than an adsorption method is used. There is experimental confirmation (*4, 29*) of this observation.

It was shown above that for diffusion control the particle-size distribution can be represented by a unique spectrum after a dimensionless time $\tau \simeq 2.5$. (For the catalyst used by Maat and Moscou and for a diffusion coefficient of about 10^{-14} cm^2/sec, one obtains a time of the order of 500 sec.) It is natural to expect that the size distribution for a freshly prepared catalyst, when reduced at high temperatures, can be represented by the unique spectrum. In Fig. 16, the experimental results of Wilson and Hall (*28, 29*) are compared with the theoretical predictions. The ratio

$$\sum_{i}^{\infty} N_{v_k} \Big/ \sum_{1}^{\infty} N_{v_k}$$

rather than ψ^* is selected as a variable both because it is convenient and because, like ψ^* it is dependent only on η_i^*. Theory and experiment are in good agreement in showing that the size distribution of

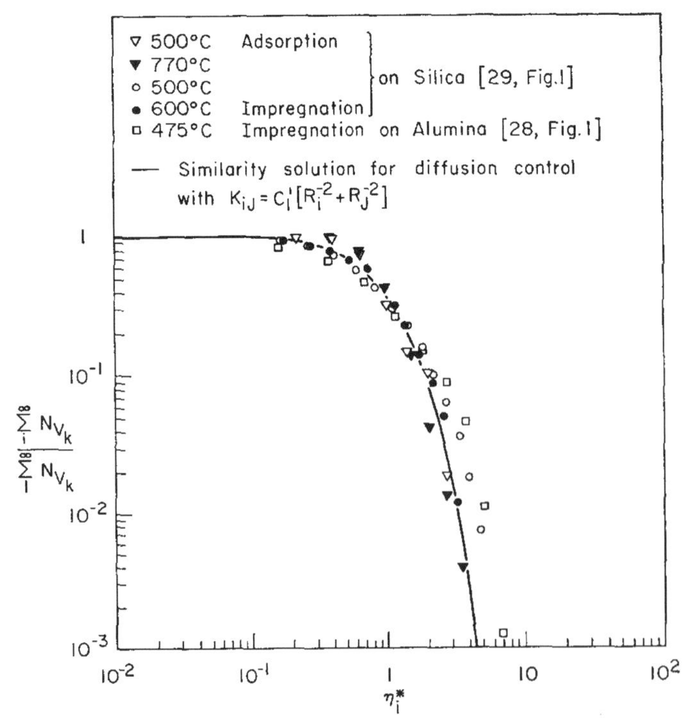

FIG. 16. Comparison between the theoretical similarity solution and Wilson and Hall experimental data for size distribution in fresh catalyst prepared at high reduction temperature.

freshly prepared catalysts reduced at high temperatures already follows the unique spectrum closely.

There are some data in the literature (*2*) which appear to show that after a sufficiently long time of heat treatment sintering no longer occurs. Probably that in those cases the surface was nonhomogeneous and the equilibrium distribution was reached.

Possible Improvements of the Stability of a Highly Dispersed State of Metal

From the previous discussion it results that the stability of a highly dispersed state of metal may be increased by controlling (a) the wetting angle of the metal crystallites, (b) the interactions between the metal and support, (c) the merging behavior of two colliding particles into one particle, (d) the number of fixed islands.

(a) The wetting angle can be changed if, instead of using a pure metal, surface active metals are also present as impurities. Some information concerning the variation of the wetting angle with the concentration c of a soluble impurity can be obtained from Young's equation

$$\sigma_{sg} - \sigma_{ss'} = \sigma_{s'g} \cos\theta \tag{8}$$

Taking the derivative with respect to c, one obtains

$$\frac{d\sigma_{sg}}{dc} - \frac{d\sigma_{ss'}}{dc} = \cos\theta \frac{d\sigma_{s'g}}{dc} - \sin\theta\, \sigma_{s'g} \frac{d\theta}{dc} \tag{9}$$

Gibbs' equation allows us to write

$$\frac{d\sigma_{sg}}{d\mu} = -\Gamma_{sg}; \quad \frac{d\sigma_{ss'}}{d\mu} = -\Gamma_{ss'};$$

$$\frac{d\sigma_{s'g}}{d\mu} = -\Gamma_{s'g} \tag{10}$$

where Γ_{sg} is the surface excess concentration of the impurity on the support–gas interface, $\Gamma_{ss'}$ is the surface excess concentration of the impurity on the support–metal interface, $\Gamma_{s'g}$ is the surface excess concentration on the metal–gas interface, and μ is the chemical potential. For simplicity it is assumed that the solid solution is ideal. Consequently,

$$d\mu = RT d \ln c \tag{11}$$

Equation (9) becomes

$$\sigma_{s'g} \sin\theta \frac{d\theta}{d \ln c} = RT(\Gamma_{sg} - \Gamma_{ss'} - \cos\theta\Gamma_{s'g}). \tag{12}$$

In the situations in which $\Gamma_{sg} - \Gamma_{ss'} - \cos\theta\Gamma_{s'g} < 0$, the impurity contributes to the decrease of the wetting angle and consequently to a lower mobility of the crystallites. If no adsorption occurs on the support–gas interface near the leading edge of the crystallite, then a decrease in the wetting angle will always take place when the surface active metal is preferentially adsorbed on the solid–solid interface.

(b) If the impurity is soluble in the support, then the mobility of the crystallite will decrease because a stronger interaction between the crystallite and support is induced. In terms of concepts from surface chemistry, the stronger interaction has as a consequence the decrease of the surface tension solid–solid and consequently of the wetting angle. Indeed, the surface tension solid–solid, $\sigma_{ss'}$, is related to the surface tensions σ_{sg} and $\sigma_{s'g}$ by the equation**

$$\sigma_{ss'} = \sigma_{sg} + \sigma_{s'g} - U_{ss'} + U_{str'} \tag{13}$$

where $U_{ss'}$ is the interaction energy per unit surface area between the atoms on one side of the interface and the atoms on the other side of the interface, and U_{str} is the strain plus the dislocation or fracture energies per unit surface area, induced by the disregistry between the lattices of the support and of the metal. Consequently $\sigma_{ss'}$ decreases when $U_{ss'}$ increases. Equation (13) can be understood if one considers the limiting situation of two parts of the same solid body separated by an imaginary plane. In this case $\sigma_{ss'}$ must be zero.

However, too much impurity dissolved in the support may have a destructive effect upon the support structure, increasing much more the term U_{str} than the term $U_{ss'}$. In this situation the surface

** The quantity $W = \sigma_{sg} + \sigma_{s'g} - \sigma_{ss'} = U_{ss'} - U_{str}$ represents the work of adhesion.

tension $\sigma_{ss'}$ is increased and consequently (assuming that the other two surface tensions are not affected much) the wetting angle and the mobility of the crystallite increase. One thus may expect that the mobility will be decreased for small concentration of the impurity and increased for large concentrations. Of course the surface active impurities also influence the values of $\sigma_{s'g}$ and σ_{sg}.

There are some experimental results in the literature (*30*) concerning the effect of interfacially active metals on the wetting and adhesion of liquid nickel drops to α-Al_2O_3. They can be interpreted on the basis of the above considerations. For high-purity nickel at 1773 K and 5×10^{-6} Torr the contact angle $\theta = 100.7$ degrees, the surface tension at the Ni–Al_2O_3 interface is equal to 1290 ergs/cm^2 and the work of adhesion to 1390 ergs/cm^2. Addition of 0.99 atomic% titanium to the nickel resulted in a decrease of the wetting angle to 94.5 degrees, a decrease of the surface tension at the Ni–Al_2O_3 interface to 1090 ergs/cm^2 and an increase of the work of adhesion to 1640 ergs/cm^2. Using electron beam X-ray microanalysis, it was found that the titanium was concentrated in the vicinity of the nickel–support interface, resulting in an increase of the adhesion of the drop to the support.

Addition of 1 atomic% zirconium to the high purity nickel resulted in an increase of the wetting angle to 136.2 degrees; in an increase of the surface tension at the Ni–Al_2O_3 interface to 2220 ergs/cm^2; and in a decrease of the work of adhesion to 500 ergs/cm^2. Electron photomicrographs revealed that the zirconium was concentrated in the vicinity of the nickel–substrate interface, chemically reducing the Al_2O_3 support and weakening its structure. The energy U_{str} was thus increased, resulting in a decreased adhesion of the drop to the support. One may conclude that the stability of the highly dispersed state of metal on a support may be substantially increased if certain alloys are used rather than pure metals. The higher stability of reforming catalysts obtained by Sinfelt (*31*) and his group can probably be explained on the basis of the above considerations.

(c) The rate of sintering of two spheres in contact increases with increasing surface diffusion and surface tension (*32*). The sintering of two spherical caps of different radii on a support has not yet been studied from a theoretical point of view. In the last case the surface tension at the metal–support interface will also play a role and the rate of sintering will be decreased with increasing metal–support interactions.

(d) As shown in Section 4 the stability of the catalyst can be much improved if some strongly interacting sites can be generated. Perhaps this can be done using a double adsorption or double impregnation procedure in which first a small amount of metal which interacts strongly both with the active metal and support is introduced.

Regeneration of Catalyst

Regeneration by heating in an oxygen atmosphere of the catalyst containing sintered crystallites (*3, 36, 37, 38*) can be also explained in terms of wetting. Indeed, the new chemical atmosphere can modify the surface free energies in such a manner that the spreading coefficient χ becomes larger than zero

$$\chi \equiv \sigma_{sg} - \sigma_{ss'} - \sigma_{s'g} > 0. \tag{14}$$

For a crystallite in equilibrium

$$\frac{\sigma_{sg} - \sigma_{ss'}}{\sigma_{s'g}} = \cos\theta \leq 1 \tag{15}$$

A spreading coefficient larger than zero is not compatible, however, with $\cos\theta \leq 1$. Consequently if $\chi > 0$ the initial crystallite will no longer be stable, but will split and redisperse. The kinetics of splitting will be analyzed in a subsequent paper.

Conclusion

The aging kinetics of supported metal catalysts are determined by the diffusion of the crystallites upon the surface of the support and by the rate of sintering of the colliding crystallites. The equations obtained on the basis of this model describe the time dependence of the size distribution and the decay of the exposed surface area of metal.

The theory can explain (a) the influence of the preparation procedure, (b) the influence of the atmosphere, and (c) the influence of the temperature upon the rate of decay of the exposed surface area of the metal.

Considerations based upon the wetting (adhesion properties) of the metal on the support suggest possibilities for improving the stability of the highly dispersed state of the metal on the surface of the support.

Appendix I. Time Dependence of the Rate Constants in the Diffusion Model

In the case of diffusion control the present approach involves two time scales, a long time scale t, introduced in Eq. (2), which coincides with the time of the process, and a small time scale θ', used in the calculation of the rate constants K_{ij}. The present treatment can be used if the rate constants are only weakly time-dependent after a time which is sufficiently short compared to the time of the process (*9*). Values for the small scale time are not provided by the theory, but it is reasonable to assume that they are of the order of the time needed for sufficiently small variations of the concentrations N_{v_i} to occur.

Equation (3), valid for "diffusion control," can be approximated by:

$$K_{ij} = 2D_{ij}\pi\left[(\pi T)^{-1/2} + \frac{1}{2} - \frac{1}{4}(T/\pi)^{1/2} + \frac{1}{8}T\cdots\right], \quad (3')$$

for small values of $T = D_{ij}\theta'/R_{ij}^2$, and by

$$K_{ij} = 4D_{ij}\pi\left[\frac{1}{\ln(4T) - 2\gamma} - \frac{\gamma}{[\ln(4T) - 2\gamma]^2}\right]\cdots, \quad (3'')$$

for large values of T. Here $\gamma = 0.5772$ is Euler's constant. If $\ln 4T \gg 2\gamma \simeq 1$, then Eq. (3″) may be further simplified to give

$$K_{ij} = \frac{4\pi D_{ij}}{\ln T4}. \quad (3''')$$

Let us first assume that D_{ij} is of the order of a surface diffusion coefficient, 10^{-4} to 10^{-10} cm^2/sec (*33*). The collision radius R_{ij} is of the order of the particle sizes present in the system (10^{-6} to 10^{-7} cm). Consequently, the dimensionless time T is of the order of (10^2 to 10^{10}) $\times\ \theta'$ and hence is long after a very short time θ'. Since for such large values of T the strong inequality $\ln 4T \gg 1$ is satisfied, Eq. (3′) can be approximated by Eq. (3‴). Although the rate constants depend on the small scale time, this time-dependence is after a short time very weak (Fig. 17). Consequently, K_{ij} from (3‴) can be considered constant and the present treatment in which no possibility for determining θ' exists can be used.

The mobility of the crystallites on the support is, however, much smaller than that of the molecules due both to their larger sizes and to stronger metal–support interactions. For this reason their diffusion coefficient becomes much smaller than a usual surface diffusion coefficient. In this case T is small for sufficiently small values of θ' and consequently the rate constants K_{ij} are initially strongly dependent upon time [Eq. (3′)]. For further clarification we represent in Fig. 18 the dimensionless rate constant $K_{11}/(4\pi D_{11})$ as a function of θ' for $D_{11} = 10^{-16}$ cm^2/sec and $R_{11} = 10^{-7}$ cm. One may observe a rapid variation until $\theta' \simeq 10^3$ sec, and a nearly constant value for $\theta' > 10^3$ sec. Only if the critical value $\theta'_c \simeq 10^3$ sec is small compared to the time t of the process it is reasonable to consider the rate constants independent of the small scale time θ'. In the cases

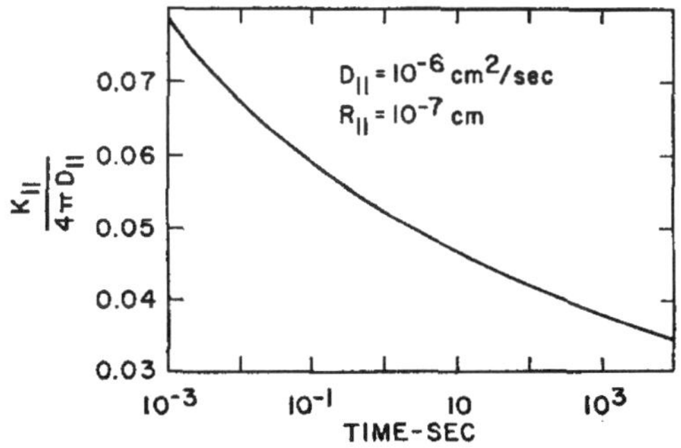

FIG. 17. Dimensionless rate constant in the diffusion controlled case vs time.

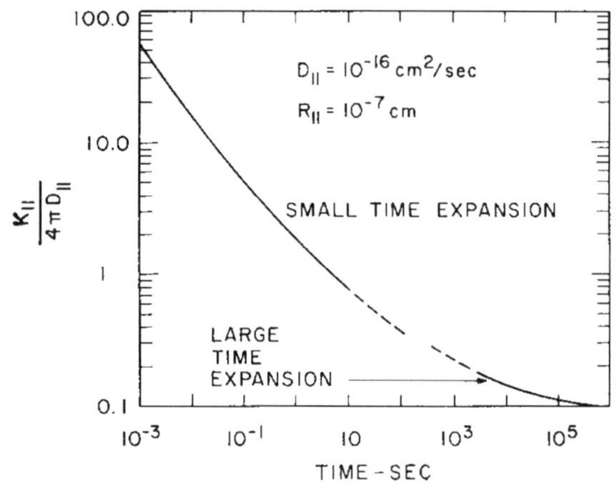

FIG. 18. Dimensionless rate constant in the diffusion controlled case vs time for decreased mobility.

considered here, the time of the process is larger by several orders of magnitude than the critical time.

The present approach can be used if θ' simultaneously satisfies the following conditions: (a) θ' is sufficiently long and as a consequence T large enough for equation (3‴) to be valid, (b) θ' is sufficiently small compared to the time in which an appreciable modification of the concentrations N_{v_k} occurs.

APPENDIX II. SIMILARITY THEORY FOR THE SIZE DISTRIBUTION OF THE SINTERING PROCESS (9)

The dimensionless groups used to account for the similarities in the shapes of the particle-size spectra at different times of the process, can be derived by means of a similarity solution.

First a continuous distribution function $n(v, t)$ is introduced such that $n(v, t)\,dv$ is the number of particles per unit surface area having a volume in the range v to $v + dv$. Equations (2) may be written in the form:

$$\frac{\partial n(v,t)}{\partial t} = \frac{1}{2}\int_0^v K(\tilde{v}, v-\tilde{v})n(\tilde{v},t)n(v-\tilde{v},t)\,d\tilde{v} - n(v,t)\int_0^\infty K(\tilde{v},v)n(\tilde{v},t)\,d\tilde{v} \quad \text{(II-1)}$$

The rate constants $K(\tilde{v}, v)$ are given for the two limiting cases of "diffusion control" and sintering control by Eqs. (3‴ and 4), respectively. If the size dependence of the diffusion coefficient D_{ij} and reaction rate constants for the merging process α_{ij} is taken into account as listed in Table 1, the rate constants $K(\tilde{v}, v)$ can be written as:

$$K(\tilde{v}, v) = K^*(\tilde{v}^m + v^m). \quad \text{(II-2)}$$

Here K^* is a constant and the exponent m depends on the mechanism and on the size dependence of either the diffusion coefficients or the reaction rate constants (see Table 1).

One may observe that the distribution function depends on two variables, v and t. The aim of the similarity solution is to find a change of variables, able to transform Eq. (II-1) into an ordinary integro-differential equation and thus to reduce the number of variables.

The similarity solution will be constructed by means of two parameters: the total number of particles per unit surface area of support $N(t)$, and the total volume of particles per unit area of support ϕ. This choice is arbitrary. Other moments of the distribution function $n(v, t)$, as for instance the total exposed surface area of metal, could be used as well. The results obtained would be equivalent.

$N(t)$ and ϕ are chosen because of their simple physical significance. They are given by

$$N(t) = \int_0^\infty n(v,t)\,dv \quad \text{(II-3)}$$

$$\phi = \int_0^\infty vn(v,t)\,dv. \quad \text{(II-4)}$$

New variables are now introduced

$$\eta = v/v^*(t) \quad \text{(II-5)}$$

and

$$n(v,t) = h(t)\psi(\eta), \quad \text{(II-6)}$$

where $h(t)$ and $v^*(t)$ are functions of time. The form of these functions is determined from Eqs. (II-3) and (II-4). Introducing Eqs. (II-5 and II-6) into (II-3 and II-4), one obtains

$$h(t) = \frac{B}{A^2}\frac{N^2}{\phi} \quad \text{(II-7)}$$

$$v^*(t) = \frac{A}{B}\frac{\phi}{N}, \quad \text{(II-8)}$$

where

$$A \equiv \int_0^\infty \psi(\eta)\, d\eta \quad \text{(II-9)}$$

and

$$B \equiv \int_0^\infty \eta\psi(\eta)\, d\eta \quad \text{(II-10)}$$

are universal constants. The similarity variables $\psi(\eta)$ and η, given by Eqs. (II-5 and II-6), can now be written as

$$\eta = \frac{B}{A}\frac{N(t)}{\phi}v \quad \text{(II-5}')$$

$$n^* = \frac{\phi n(v,t)}{N(t)^2} = \psi(\eta)\frac{B}{A^2}. \quad \text{(II-6}')$$

Such a type of similarity transformation was first suggested, in a different problem, by Friedlander (*34, 35*).

Before transforming Eq. (II-1) into an ordinary integro-differential equation, the rate of change of the total number of particles, $dN(t)/dt$, has to be known. By integrating Eq. (II-1) over all particle sizes one obtains

$$\frac{\partial}{\partial t}\int_0^\infty n(v,t)\, dv = \frac{dN(t)}{dt}$$
$$= -\frac{K^*}{2}\int^\infty \int_0^\infty n(v,t)n(\tilde{v},t)\,[\tilde{v}^m + v^m]\, dv\, d\tilde{v}. \quad \text{(II-11)}$$

Inserting Eqs. (II-5′ and II-6′) into (II-11), yields

$$\frac{dN}{dt} = -b_1N^{2-m}, \quad \text{(II-12)}$$

where

$$b_1 = \frac{K^*}{2}\phi^m\frac{A^{m-2}}{B^m}\int_0^\infty\int_0^\infty \times \psi(\eta)\psi(\tilde{\eta})(\eta^m + \tilde{\eta}^m)\, d\eta\, d\tilde{\eta}. \quad \text{(II-13)}$$

Equations (II-1), (II-2), (II-5′), (II-6′), and (II-12) lead to the following ordinary integro-differential equation

$$\left[2\psi(\eta) + \eta\frac{d\psi(\eta)}{d\eta}\right]\int_0^\infty\int_0^\infty \times \psi(\eta)\psi(\tilde{\eta})(\eta^m + \tilde{\eta}^m)\, d\eta\, d\tilde{\eta}$$
$$+ \int_0^\infty \psi(\eta)\, d\eta \int_0^\eta \times \psi(\tilde{\eta})\psi(\eta - \tilde{\eta})[\tilde{\eta}^m + (\eta - \tilde{\eta})^m]\, d\tilde{\eta}$$
$$- 2\psi(\eta)\int_0^\infty \psi(\eta)\, d\eta \int_0^\infty \times \psi(\tilde{\eta})(\eta^m + \tilde{\eta}^m)\, d\tilde{\eta} = 0 \quad \text{(II-14)}$$

Of course the solution of Eq. (II-14) represents only one of the possible particular solutions of Eq. (II-1). The main problem is whether this particular solution which is compatible only with special types of initial conditions has physical meaning. In what follows an answer to this question is given. A solution of Eq. (II-14) in closed form was obtained only for constant K_{ij} [$m = 0$, Eq. (II-2)], namely

$$\psi(\eta) = e^{-\eta}. \quad \text{(II-15)}$$

For size dependent reaction rate constants the solution of Eq. (II-14) was obtained by replotting the numerical results of the discrete spectrum using the similarity variables. The similarity variables for the discrete case were taken as:

$$\eta^*_i = iv_1N/\phi \quad \text{(II-5}'')$$
$$\psi^*(\eta^*_i) = N_{v_i}\phi/N^2v_1 \quad \text{(II-6}'')$$

Figures 19 and 20 show the obtained results. For the "diffusion control" mechanism the numerical solution after a short time becomes independent of the initial distribution of sizes and can be represented after this short time by the similarity solution (Fig. 19). For "sintering control" a family of curves characterized by a supplementary dimensionless time τ_1, is obtained (Fig. 20). These curves are, however, close to each other. For this reason information concerning some cumulative distributions as are the total number of particles per unit surface area of support and the total exposed surface area of the metal may be obtained from the similarity transformation even in the sintering controlled case.

In what follows an equation for the decay of the exposed surface area of metal

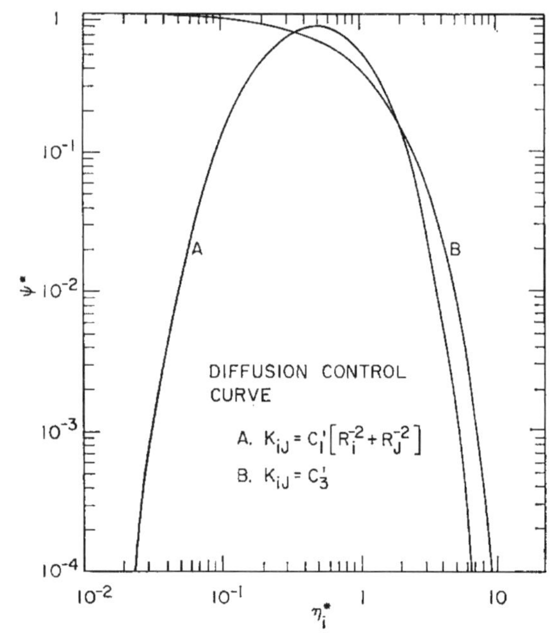

FIG. 19. Similarity solution for diffusion controlled cases.

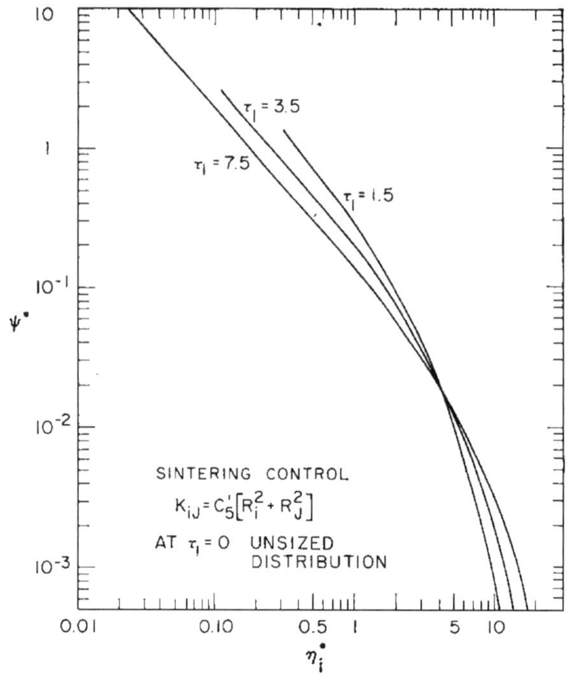

FIG. 20. Similarity solution for sintering controlled case.

is established using the similarity variables. A continuous formulation of the exposed surface area may be written as:

$$S(t) = b_2 \int_0^\infty v^{2/3}\, n(v, t)\, dv \quad \text{(II-18)}$$

where b_2 is a geometric factor, dependent on the shape of the particles. Introducing the similarity variables, Eqs. (II-5′) and (II-6′), into Eq. (II-18), one finds that the rate of decay of the exposed surface area of metal is given by:

$$\frac{dS(t)}{dt} = -b_3 S(t)^{4-3m}, \quad \text{(II-19)}$$

where

$$b_3 = \frac{K^*}{6\phi^{2-3m} A B^{3m-2} b_2^{3-3m}} \times \int_0^\infty \int_0^\infty \psi(\eta)\psi(\tilde{\eta})(\eta^m + \tilde{\eta}^m)\, d\eta\, d\tilde{\eta} \Big/ \left[\int_0^\infty \eta^{2/3}\psi(\eta)\, d\eta\right]^{3-3m}. \quad \text{(II-20)}$$

Eq. (II-19) is equivalent to those established on the basis of numerical computations (see Table 1).

NOMENCLATURE

A	Universal constant defined by Eq. (II-9).
a	Actual area of the solid–solid interface of crystallite and support.
a_{ap}	Apparent area of the solid–solid interface of crystallite and support.
B	Universal constant defined by Eq. (II-10).
b_1, b_2, b_3	Constants in Eqs. (II-12), (II-18), and (II-19).
c	Concentration of the surface active metal within the metal crystallite.
C_i, C'_i, C''_i	Constants independent of time.
D_{ij}	Diffusion coefficient of particle i with respect to particle j.
$F_{v_{k,0}}$, F_{v_k}	Number of fixed particles per unit surface area of support composed of k metal units at $t = 0$, and $t = t$, respectively.
$h(t)$	Function of time defined by Eq. (II-7).

J_0	Bessel function of the first kind and zero order.
K, K_i	Rate constant independent of time.
K^*	Constant independent of particle size.
K_{ij}, K^*_{ij}	Second order rate constant used in the discrete representation.
$K(v, \bar{v})$	Second order rate constant used in the continuous representation.
m	Exponent in Eq. (II-2).
N, N_0	Total number of particles per unit surface area of support for $t = t$ and $t = 0$, respectively.
$N_{v_{k,0}}, N_{v_k}$	Number of moving particles per unit surface area of support composed of k metal units at time $t = 0$ and $t = t$, respectively.
n	Exponent in Eq. (1).
$n(v, t)$	Continuous density distribution function.
p	Probability that a particle formed from two colliding moving particle will move.
q	Probability that a fixed particle will move after collided by a moving particle.
R	Universal gas constant.
R_{ij}	Radius of interaction between two colliding particles i and j.
$r = a/a_{ap}$	Wenzel's ratio, actual area of interface divided by apparent area of interface.
r_i	Radius of particle containing i metal atoms.
S	Total exposed surface area of the metal per unit surface area of support.
S_0	Value of S at the initial moment.
S_k	Exposed surface area of a particle containing k metal units.
$T = D_{ij}\theta'/R_{ij}^2$	Dimensionless time.
t	Time of the process.
$U_{ss'}$	Interaction energy per unit surface area between the atoms on one side of the interface and the atoms on the other side of the interface.
U_{str}	Strain plus dislocation or fracture energy per unit surface area.
u	Dummy variable.
v	Volume of a particle.
v_i	Volume of a particle containing i metal units.
$v^*(t)$	Function of time defined by Eq. (II-8).
W	Work of adhesion.
Y_0	Bessel function of the second kind and zero order.
α_{ij}	Reaction rate constant for the merging process of two particles containing i and j units.
β_k	Trapping rate constants.
Γ_{sg}	Surface excess concentration of the surface active impurity on the support–gas interface.
$\Gamma_{s'g}$	Surface excess concentration of the surface active impurity on the metal–gas interface.
$\Gamma_{ss'}$	Surface excess concentration of the surface active metal on the support–metal interface.
$\gamma = 0.5772$	Euler's constant.
χ	Spreading coefficient.
$\sigma_{sg}, \sigma_{s'g}$	Metal–gas surface tension and support–gas surface tension.
$\sigma_{ss'}$	Solid–solid surface tension.
η	Similarity variable in the continuous representation, dimensionless volume.
η^*_i	Similarity variable in the discrete representation.
θ	Contact angle.

θ'	Small scale time in the diffusion model.
θ^*	$= 1$ sec.
μ	Chemical potential.
$\tau_1 = t\alpha_{11}2\pi R_1 N_0$	Dimensionless time in the sintering-controlled case.
$\tau_2 = \dfrac{t4\pi D_1 N_0}{\ln [4D_{11}\theta^*/(R_{11})^2]}$	Dimensionless time in the diffusion-controlled case.
ϕ	Total volume of metal per unit surface area of support.
ψ	Similarity variable in the continuous representation, dimensionless distribution function.
ψ^*	Similarity variable in the discrete representation.

References

1. Maat, H. J., and Moscou, L., *Proc. 3rd Int. Congr. Catal.* **II,** 1277 (1965).
2. Zaidman, N. M., Dzisko, V. A., Karnautkhov, A. P., Kefeli, L. M., Krasilenko, N. P., Koroleva, N. G., and Ratner, I. D., *Kinet. Katal.* **10,** 386 (1969).
3. Gruber, H. L., *J. Phys. Chem.* **66,** 48 (1962).
4. Dorling, T. A., Lynch, W. J., and Moss, R. L., *J. Catal.* **20,** 190 (1971).
5. Mills, G. A., Weller, S., and Cornelius, E. B., *Proc. 2nd Int. Congr. Catal. Paris,* 2221 (1960).
6. Herrmann, R. A., Adler, S. F., Goldstein, M. S., and DeBaun, R. M., *J. Phys. Chem.* **65,** 2189 (1961).
7. Hughes, T. R., Houston, R. J., and Sieg, R. P., *Ind. Eng. Chem. (Proc. Des. Dev.)* **1,** 96 (1962).
8. Adler, S. F., and Keavney, J. J., *J. Phys. Chem.* **64,** 208 (1960).
9. Ruckenstein, E., and Pulvermacher, B., *AIChE Journal* **19** March, 1973.
10. Bassett, G. A., *in* "Proceedings of the European Regional Conference on Electron Microscopy" (A. L. Houwink and B. J. Spit, eds.), I, De Nederlandse Vereniging voor Elektronen microscopie, Delft (1960).
11. Bassett, G. A., *in* "Proceedings of the International Symposium on Condensation and Evaporation of Solids," (E. Rutner, P. Goldfinger, and J. P. Hirth, eds.), Gordon and Breach, New York (1964).
12. Skofronick, J. G., and Phillips, W. B., *J. Appl. Phys.* **38,** 4791 (1967).
13. Phillips, W. B., Desloge, E. A., and Skofronick, J. G., *J. Appl. Phys.* **39,** 3210 (1968).
14. Ries, H. E., "Advances in Catalysis" (W. G. Frankenburg, V. I. Komarewsky, and R. K. Rideal, eds.), IV, Academic Press, New York (1952).
15. Shapiro, I., and Kolthoff, I. M., *J. Amer. Chem. Soc.* **72,** 776 (1950).
16. Williams, A., Butler, G. A., and Hammonds, J., *J. Catal.* **24,** 352 (1972).
17. Smorjai, G. A., Powell, R. E., Montgomery, P. W., and Jura, G., *in* "Small-Angle X-Ray Scattering," (H. Brumberger, ed.), Gordon and Breach, New York (1965).
18. Geus, J. W., *in* "Chemisorption and Reactions on Metallic Films" (J. R. Anderson, ed.), Academic Press, London and New York (1971).
19. Bradshaw, F. J., Brandon, R. H., and Wheeler, C., *Acta Met.* **12,** 1057 (1964).
20. Zackay, V. F., Mitchell, D. W., Mitoff, S. P., and Pask, J. A., *J. Amer. Ceram. Soc.* **36,** 84 (1953).
21. Pask, J. A., and Fulrath, R. M., *J. Amer. Ceram. Soc.* **45,** 592 (1962).
22. Schäfer, H., and Tebben, A., *Z. Anorg. Allgem. Chem.* **304,** 317 (1960).
23. Raub, E., and Plate, W., *Z. Metallkunde* **48,** 529 (1957).
24. Fryburg, G. C., *Trans. Met. Soc. AIME* **233,** 1986 (1965).
25. Bartlett, R. W., *J. Electrochem. Soc.* **114,** 547 (1967).
26. Kern, R., Masson, A., and Métois, J. J., *Surface Sci.* **27,** 483 (1971).
27. Métois, J. J., Gauch, M., Masson, A., and Kern, R., *Surface Sci.* **30,** 43 (1972).
28. Wilson, G. R., and Hall, W. K., *J. Catal.* **17,** 190 (1970).
29. Wilson, G. R., and Hall, W. K., *J. Catal.* **24,** 306 (1972).
30. Sutton, H., and Feingold, E., *in* "Materials Science Research," (W. W. Kriegel and H. Palmour III, eds.), Vol. 3. Plenum Press, New York (1966).
31. Sinfelt, J. H., *Chem. Eng. News,* June 1972.
32. Kuczynski, G. C., *Trans. Met. Soc. AIME* **185,** 169 (1949).
33. Satterfield, C. N., "Mass Transfer in Heterogeneous Catalysis," M. I. T. Press, Cambridge, MA (1970).
34. Friedlander, S. K., *J. Meteorol.* **18,** 753 (1961).

35. Swift, D. L., and Friedlander, S. K., *J. Colloid Sci.* **19**, 621 (1964).
36. Jaworska-Galas, Z., Wrzyszcz, J., *Int. Chem. Eng.* **6**, 604 (1966).
37. Weller, S. W., and Montagna, A. A., *J. Catal.* **20**, 394 (1971).
38. Johnson, M. F. L., and Keith, C. D., *J. Phys. Chem.* **67**, 200 (1963).

Ostwald ripening: A stochastic approach

Ashok Bhakta and Eli Ruckenstein[a)]
Department of Chemical Engineering, State University of New York at Buffalo, Buffalo, New York 14260

(Received 20 March 1995; accepted 21 July 1995)

The phenomenon of Ostwald ripening is examined theoretically for precipitates in solution, crystallites on a substrate, and islands of insoluble surfactant molecules at the liquid–air interface. The classical theory of Ostwald ripening, developed by Lifshitz and Slyozhov and independently by Wagner, employs a continuum approach in which the size distribution is a continuous function of particle size and the conservation equation is formulated as in hydrodynamics with the particle growth rate serving as a velocity in size space. It was derived for the case in which the diffusion fields of the individual particles do not overlap and provides asymptotic expressions at large times for the two extreme cases of diffusion controlled and kinetic controlled growth. In this paper, the validity of this continuum (hydrodynamic) approach is examined using a more rigorous stochastic microscopic approach (MCE). A detailed comparison is made between the numerical and asymptotic results obtained using the Lifshitz, Slyozhov, Wagner theory and the stochastic microscopic continuity equation (MCE). It is shown that with the MCE, the limiting case of "diffusion controlled Ostwald ripening" ceases to have significance, since the kinetics at the interface play a role even when the transport of molecules from the bulk to the particle is diffusion controlled. For the case of interface kinetic controlled growth in precipitates, the asymptotic trends, which are independent of the initial conditions in both cases, are seen much earlier with the MCE. Besides, for this case (precipitates), the MCE predicts a mean radius which is about 25% larger at any given time. While the scaled (with respect to the maximum) asymptotic distributions obtained from the two approaches are similar, the absolute (not scaled) particle size distributions obtained using the MCE are much broader and the maxima are almost a third of those obtained using the Lifshitz, Slyozhov, Wagner approach. Similar differences in the results are obtained for the two dimensional Ostwald ripening of crystallites on a substrate and islands of insoluble surfactant.

INTRODUCTION

Consider a polydispersed system of particles in a supersaturated solution. Particles larger than a certain critical size gain molecules from the solution and grow, while those smaller than this critical size shrink. This phenomenon is called Ostwald ripening. The higher solubility of smaller particles induces a net transport of monomers through the solution from the shrinking particles to the growing ones resulting in the coarsening of the particle size distribution with time. Ostwald ripening has been the focus of intense investigation since the first theoretical model proposed independently by Lifshitz and Slyozhov[1] and Wagner.[2]

In this classical theory, the size distribution is taken to be a continuous function of the particle radius r. A particle of radius r has a velocity in r space given by its growth rate dr/dt and the continuity equation is written as in hydrodynamics in the form

$$\frac{\partial c_r}{\partial t}+\frac{\partial}{\partial r}\left[c_r\frac{dr}{dt}\right]=0. \tag{1}$$

Here, $c_r\,dr$ represents the number of particles per unit volume having a radius between r and $r+dr$. The Lifshitz, Slyozhov, Wagner theory provides asymptotic expressions at large times for the evolution of the mean particle size and the scaled (with respect to the maximum concentration) size distribution which are independent of the initial conditions. These conclusions, which have been the subject of much debate have been verified numerically.[3,4] Marquesee and Ross[5,6] have shown that the asymptotic solution is actually the lowest order term in a series expansion which is unique for a given growth mechanism. Various extensions of the Lifshitz, Slyozhov, Wagner theory have been made to account for the effect of volume fraction[7–9] and more complicated growth laws.[10] The theory has also been applied to two-dimensional systems[11] in which crystallites on a substrate have been seen to undergo Ostwald ripening. Two-dimensional Ostwald ripening plays a role in the sintering of supported catalysts[12] and in the growth of solid domains in monolayer and bilayer films[13] at the air–water interface.

An alternative to the continuum approach of the Lifshitz, Slyozhov, Wagner theory has been proposed by Dadyburjor and Ruckenstein.[14] They have used more rigorous stochastic balance equations and have shown that these equations reduce to the Lifshitz, Slyozhov, Wagner equations only under certain extreme conditions. Binder and Stauffer have formulated general stochastic equations[15,16] and Binder[16] has derived the Lifshitz, Slyozhov, Wagner equations on their basis. The numerical calculations of Dadyburjor and Ruckenstein,[14] restricted to kinetic control and short times, have shown that the evolution of the particle size distribution predicted by their approach differs significantly from that predicted by the Lifshitz, Slyzhov, Wagner equations.

The scope of the present paper is to examine how good the continuum approach is by making a comparison with the more rigorous stochastic approach. The comparison will be

[a)]Author to whom correspondence should be addressed.

This paper, which was originally published in **Journal of Chemical Physics**: 103(16), 7120–7135 (1995), is republished with permission.

made for both kinetic and diffusion controlled growth by solving the conservation equations of the two approaches numerically.

It is shown that there are significant differences in the asymptotic size distributions and mean size. We also show that with the microscopic approach the term "diffusion controlled Ostwald ripening" loses its significance, since even when the growth of individual particles is diffusion controlled, Ostwald ripening is still affected by the kinetics at the interface.

We shall first consider Ostwald Ripening of precipitates in solution and follow it up with a similar analysis of two-dimensional Ostwald ripening.

PRECIPITATES IN SOLUTION

We assume in the following discussion that the precipitate consists of spherical particles. The concentration (c_{s_j}) of monomers in solution (molecules per unit volume) in equilibrium with a single particle of radius r_j containing j molecules is given by the well known Gibbs–Thompson equation

$$c_{s_j}=c_{s_\infty}\exp\left(\frac{\delta}{r_j}\right), \tag{2}$$

where c_{s_∞} is the solute concentration in equilibrium with a sufficiently large particle ($r_j\to\infty$) and δ is given by

$$\delta\equiv\frac{2\sigma v_m}{k_BT}. \tag{3}$$

Here, σ is the interfacial tension between the solid and liquid phase, v_m is the molecular volume of the solute, k_B is the Boltzmann constant and T is the temperature (K). The radius r_j can be computed using the relation

$$r_j=\left[\frac{3v_mj}{4\pi}\right]^{1/3}. \tag{4}$$

If $\delta\ll r_j$, Eq. (2) can be written as

$$c_{s_j}=c_{s_\infty}\left(1+\frac{\delta}{r_j}\right). \tag{5}$$

Apart from some exceptions,[17] practically all models proposed for Ostwald ripening have made use of Eq. (5). This assumption is justified for cases in which the supersaturation is small.

In the following sections, the microscopic continuity equation (MCE) proposed by Dadyburjor and Ruckenstein[14] is examined in detail. We write the Lifshitz, Slyozhov, Wagner continuity equation analogous to Eq. (1) in j space and proceed to compare the results of our numerical calculations for the two approaches.

Microscopic continuity equation (MCE)

We assume in the following discussion that the growth or shrinkage of a particle occurs only due to the attachment or detachment of monomers. Let c_j be the number of particles per unit volume containing j molecules. The average concentration (molecules per unit volume) of monomers in solution is denoted by c_1. Let c_{0_j} be the actual concentration of monomers at the interface of a j-molecule particle. The rate at which a j-molecule particle gains molecules (L_j^+) is proportional to its surface area and the monomer concentration at the interface and is given by

$$L_j^+=4\pi r_j^2kc_{0_j}. \tag{6}$$

On the other hand, the rate at which molecules are detached from the particle surface is proportional to the saturation concentration at the interface and is given by

$$L_j^-=4\pi r_j^2kc_{s_j}, \tag{7}$$

where the rate constant k is assumed to be independent of the particle size. The equality of the rate constants in Eqs. (6) and (7) results from the conditions at equilibrium

$$L_j^+=L_j^-;\quad c_{s_j}=c_{0_j}. \tag{8}$$

The concentration of j-molecule particles increases due to the loss of molecules by $(j+1)$-size particles and the addition of molecules to $(j-1)$-size particles, and decreases due to the gain and loss of molecules by j-size particles. Thus

$$\frac{dc_j}{dt}=L_{j-1}^+c_{j-1}+L_{j+1}^-c_{j+1}-L_j^+c_j-L_j^-c_j\quad\forall j\geq 2. \tag{9}$$

Combining Eqs. (6), (7), and (9) one obtains

$$\frac{dc_j}{dt}=4\pi k[r_{j-1}^2c_{0_{j-1}}c_{j-1}+r_{j+1}^2c_{s_{j+1}}c_{j+1}$$
$$-r_j^2c_{0_j}c_j-r_j^2c_{s_j}c_j]\quad\forall j\geq 2. \tag{10}$$

We assume that there is no agglomeration of monomers, i.e.,

$$L_1^+=0. \tag{11}$$

This assumption is justified by the fact that the interaction between two monomers is much weaker than between a monomer and a particle. Therefore, the probability that a collision will result in the formation of a dimer is small. If this assumption is not made, the equations show that the formation of dimers dominates the process, since the number of monomers is very large. Of course, in reality, the rate constant is not independent of size as assumed.

Calculation of c_{0_j}

In order to solve Eq. (10), an expression for c_{0_j} is needed. The steady state solution of the diffusion equation in spherical coordinates is given by

$$c=\frac{B_1}{r}+B_2, \tag{12}$$

where c represents the monomer concentration at a distance r from the center of the particle and B_1 and B_2 are constants of integration. We assume that the diffusion fields of particles do not overlap. Thus

$$r\to\infty\Rightarrow c=c_1. \tag{13}$$

The boundary condition at the interface is

$$r=r_j\Rightarrow c=c_{0_j}. \tag{14}$$

If we consider a quasisteady state material balance at the interface, the net rate at which a particle takes up molecules equals the diffusive flux and is given by

$$k(c_{0_j}-c_{s_j})=D\left[\frac{\partial c}{\partial r}\right]_{r=r_j}, \tag{15}$$

where D is the diffusion coefficient of the solute. Using Eqs. (12)–(15) one obtains

$$c_{0_j}=c_{s_j}+\frac{c_1-c_{s_j}}{1+\dfrac{kr_j}{D}}. \tag{16}$$

Introducing the dimensionless number

$$P=\frac{kr_{m_0}}{D}, \tag{17}$$

where r_{m_0} is the initial mean particle radius, Eq. (15) can be written as

$$c_{0_j}=c_{s_j}+\frac{c_1-c_{s_j}}{1+\dfrac{Pr_j}{r_{m_0}}}. \tag{18}$$

Thus when $P\ll 1$, Eq. (18) becomes

$$c_{0_j}-c_{s_j}\approx c_1-c_{s_j} \tag{19}$$

and the transport is controlled by the kinetics of attachment and detachment at the interface. On the other hand, if $P\gg 1$

$$c_{0_j}-c_{s_j}\approx\frac{c_1-c_{s_j}}{\dfrac{Pr_j}{r_{m_0}}}=\frac{D(c_1-c_{s_j})}{kr_j} \tag{20}$$

and diffusion is the rate controlling mechanism.

Finally, since the number of molecules in the system must be conserved

$$N_0=c_1+\sum_2 jc_j, \tag{21}$$

where N_0 is the total number of molecules present in the system per unit volume.

Equation (10) represents a system of initial value problems which can be solved to obtain the evolution of c_j with time for a given initial distribution.

Derivation of the Lifshitz, Slyozhov, Wagner equation in *j* space from the MCE

When Eq. (10) is expanded about j using a Taylor series expansion we get

$$\frac{dc_j}{dt}=4\pi k\left[r_j^2c_{0_j}c_j-\frac{\partial}{\partial j}(r_j^2c_{0_j}c_j)+\frac{1}{2}\frac{\partial^2}{\partial j^2}(r_j^2c_{0_j}c_j)+r_j^2c_{s_j}c_j+\frac{\partial}{\partial j}(r_j^2c_{s_j}c_j)+\frac{1}{2}\frac{\partial^2}{\partial j^2}(r_j^2c_{s_j}c_j)-r_j^2c_{0_j}c_j-r_j^2c_{s_j}c_j\right]. \tag{22}$$

Note that we take $\Delta j=1$ and neglect terms higher than second order. Rearranging Eq. (22), we obtain

$$\frac{dc_j}{dt}=4\pi k\left[-\frac{\partial}{\partial j}r_j^2(c_{0_j}-c_{s_j})c_j\right]+4\pi k\left[\frac{1}{2}\frac{\partial^2}{\partial j^2}r_j^2(c_{0_j}+c_{s_j})c_j\right]. \tag{23}$$

If $2\pi k(\partial^2/\partial j^2)[r_j^2(c_{0_j}+c_{s_j})c_j]\ll 4\pi k(\partial/\partial j)[r_j^2(c_{0_j}-c_{s_j})c_j]$, Eq. (23) becomes

$$\frac{\partial c_j}{\partial t}=-\frac{\partial}{\partial j}[k4\pi r_j^2(c_{0_j}-c_{s_j})c_j]. \tag{24}$$

Since $k4\pi r_j^2(c_{0_j}-c_{s_j})$ represents the net rate at which a j-size particle takes up molecules,

$$\frac{dj}{dt}=k4\pi r_j^2(c_{0_j}-c_{s_j}), \tag{25}$$

and Eq. (24) becomes

$$\frac{\partial c_j}{\partial t}+\frac{\partial}{\partial j}\left(c_j\frac{dj}{dt}\right)=0. \tag{26}$$

Equation (26) represents the Lifshitz, Slyozhov, Wagner continuity equation in j space. The equations of the Lifshitz, Slyozhov, Wagner theory are therefore approximations to the MCE equations. The former are valid when the second term in the right-hand side of Eq. (23) is negligible.

Substitution of the expression for $c_{0_j}-c_{s_j}$ from Eq. (18) into Eq. (25) yields

$$\frac{dj}{dt}=k4\pi r_j^2\frac{c_1-c_{s_j}}{1+\dfrac{Pr_j}{r_{m_0}}}. \tag{27}$$

Thus if $c_1-c_{s_j}>0$, the particle grows and if $c_1-c_{s_j}<0$ the particle shrinks. When $c_1=c_{s_j}$, the particle is said to have the critical size ($j=j_c$) and neither grows nor shrinks. For the numerical calculations, Eq. (26) was discretized as follows:

$$\frac{dc_j}{dt}=4\pi kr_{j+1}^2(c_{s_{j+1}}-c_{0_{j+1}})c_{j+1}-4\pi kr_j^2(c_{s_j}-c_{0_j})c_j\quad\forall j<j_c, \tag{28a}$$

$$\frac{dc_j}{dt}=4\pi kr_{j-1}^2(c_{0_{j-1}}-c_{s_{j-1}})c_{j-1}-4\pi kr_j^2(c_{0_j}-c_{s_j})c_j\quad\forall j>j_c, \tag{28b}$$

$$\frac{dc_j}{dt}=0\quad j=j_c. \tag{28c}$$

This is equivalent to using upwind differencing for Eq. (26) which is a common practice in computational fluid dynamics used to preserve the stability of the system. Thus backward differencing is used when the "velocity" (dj/dt) is positive and forward differencing is used if it is negative. Clearly, this scheme effectively captures the physics of the process, viz., particles larger than the critical size grow and those smaller than this size shrink. Equations (28) will henceforth be referred to as LSE. Thus for $j<j_c$, an increase in c_j is only due to the loss of molecules from particles of size $j+1$, while for $j>j_c$ the increase is due only to the addition of molecules to $(j-1)$-size particles. Since particles of size (j_c+1) gain molecules and particles of size (j_c-1) lose molecules, c_{j_c} is unchanged. A comparison of Eq. (10) and Eq. (28) reveals the essential difference between the two formulations. While in the former, the concentration of particles of size j depends on the concentration of particles of size $j-1$ as well as particles of size $j+1$, in the latter, particles of size j are formed either from particles of size $j-1$ for $j>j_c$ or only from particles of size $j+1$ for $j<j_c$.

Our calculations also serve the additional purpose of verifying the conclusions of the Lifshitz, Slyozhov, Wagner theory regarding the asymptotic behavior at large times. Some results of the Lifshitz, Slyozhov, Wagner theory are presented in brief in the following section.

Asymptotic results of the Lifshitz, Slyozhov, Wagner theory

The Lifshitz, Slyozhov, Wagner theory[1,2,11] provides asymptotic expressions for the size distribution and mean size, depending on the rate laws governing the growth of individual particles. It would therefore be pertinent to discuss the growth laws for the case considered here.

The growth rate (dr_j/dt) is given by

$$\frac{dr_j}{dt}=v_m k(c_{0_j}-c_{s_j}). \tag{29}$$

Substitution of the expression for c_{0_j} from Eq. (18) into Eq. (29) yields

$$\frac{dr_j}{dt}=v_m k\,\frac{(c_1-c_{s_j})}{1+P\dfrac{r_j}{r_{m_0}}}. \tag{30}$$

Thus for $c_1>c_{s_j}$, the j-molecule particle grows, for $c_1<c_{s_j}$ the particle shrinks and for $c_1=c_{s_j}$ the particle neither grows nor shrinks, i.e., it corresponds to the critical size ($j=j_c$ or $r_j=r_c$). The critical radius r_c is therefore given by

$$\frac{c_1}{c_{s_\infty}}=\exp\left(\frac{\delta}{r_c}\right)\approx 1+\frac{\delta}{r_c}. \tag{31}$$

Using Eqs. (5) and (31), c_1 and c_{s_j} can be expressed in terms of r_c and r_j, and Eq. (30) becomes

$$\frac{dr_j}{dt}=-(v_m k c_{s_\infty}\delta)\left(\frac{1}{r_j}\right)\frac{1-\dfrac{r_j}{r_c}}{1+P\dfrac{r_j}{r_{m_0}}}. \tag{32}$$

For interfacial kinetic control $(P\approx 0)$, Eq. (32) becomes

$$\frac{dr_j}{dt}=-\frac{A_i}{r_j}\left(1-\frac{r_j}{r_c}\right), \tag{33}$$

where

$$A_i\equiv v_m k c_{s_\infty}\delta. \tag{34}$$

For the diffusion controlled case $(P\gg 1)$, Eq. (32) leads to

$$\frac{dr_j}{dt}\approx-\frac{A_d}{r_j^2}\left(1-\frac{r_j}{r_c}\right), \tag{35}$$

where

$$A_d\equiv v_m c_{s_\infty}\delta D. \tag{36}$$

According to the Lifshitz, Slyozhov, Wagner theory, for a power law growth equation of the form

$$\frac{dr_j}{dt}=-\frac{A}{r_j^m}\left(1-\frac{r_j}{r_c}\right), \tag{37}$$

the solution of Eq. (1) shows the following asymptotic behavior:

(a) If $G(\rho_j,t)$ represents the number of particles per unit volume between ρ_j and $\rho_j+d\rho_j$, where $\rho_j=(r_j/r_c)$, then $G(\rho_j,t)$ is asymptotically given by

$$G(t,\rho_j)=g(\rho_j)F(t). \tag{38}$$

As a result, $(G)/(G_{\max})=(g)/(g_{\max})$ is independent of time. The subscript "max" refers to the maximum values of G and g.

(b) $\lambda\equiv[A(m+1)/(dr_c^{m+1}/dt)]$ [where m is the exponent in Eq. (37)] is constant. This means that r_c^{m+1} changes linearly with time.

(c) The mean value of $\rho,\bar{\rho}$ is a constant, i.e.,

$$\bar{\rho}=\frac{\int_0^\infty G\rho_j\,d\rho_j}{\int_0^\infty G\,d\rho_j}=\text{constant}. \tag{39}$$

In other words

$$\frac{\bar{r}}{r_c}=\text{constant}. \tag{40}$$

Thus from (b) we get

$$\frac{d(\bar{r})^{m+1}}{dt}=\text{constant}. \tag{41}$$

(d) The total number of particles per unit volume (N_p) is given by

$$N_p\equiv\sum_{j=2}^{j\to\infty}c_j=\int_0^\infty G\,d\rho_j=(\text{constant})t^{-(m_1)/(m+1)}, \tag{42}$$

where $m_1=3$.

Thus $\ln(N_p)$ will change linearly with $\ln(t)$.

The numerical calculations provide the size distribution as a function of the number of molecules (j) in the particles. To compare them with those of Eq. (38) which gives the distribution in terms of ρ_j, one can write

$$c_j\,dj=G(\rho_j,t)d\rho_j. \tag{43}$$

Since $j\propto\rho_j^3$, we get

$$\frac{\rho_j^2 c_j}{(\rho_j^2 c_j)_{\max}}=\frac{g(\rho_j)}{(g(\rho_j))_{\max}}. \tag{44}$$

Nondimensionalization of the equations

For the numerical calculations, Eq. (10) and Eq. (28) were converted to dimensionless form. Let $c_{j_{0_{\max}}}$ be the maximum value of c_j initially ($t=0$). The following dimensionless variables are introduced:

$$c_j^* \equiv \frac{c_j}{c_{j_{0_{\max}}}} \quad j \geq 2 \quad c_1^* \equiv \frac{c_1}{c_{s_\infty}} \quad c_{s_j}^* = \frac{c_{s_j}}{c_{s_\infty}}, \tag{45a}$$

$$c_{0_j}^* = \frac{c_{0_j}}{c_{s_\infty}}, \quad r_j^* \equiv \frac{r_j}{\delta}, \tag{45b}$$

$$t^* \equiv t\,\frac{4\pi\delta^2 c_{s_\infty} k}{1+P}. \tag{45c}$$

Thus for $P \ll 1$

$$t^* = t(4\pi\delta^2 c_{s_\infty} k), \tag{45d}$$

and for $P \gg 1$

$$t^* = t\,\frac{4\pi c_{s_\infty}\delta^2 D}{r_{m_0}}. \tag{45e}$$

The MCE equation [Eq. (10)] therefore becomes

$$\frac{dc_j^*}{dt^*} = (1+P)[r_{j-1}^{*2} c_{0_{j-1}}^* c_{j-1}^* + r_{j+1}^{*2} c_{s_{j+1}}^* c_{j+1}^* - r_j^{*2} c_{0_j}^* c_j^* - r_j^{*2} c_{s_j}^* c_j^*] \quad \forall j \geq 2, \tag{46}$$

and the LSE equation [Eq. (28)] takes the form

$$\frac{dc_j^*}{dt^*} = (1+P)[r_{j+1}^{*2}(c_{s_{j+1}}^* - c_{0_{j+1}}^*)c_{j+1}^* - r_j^{*2}(c_{s_j}^* - c_{0_j}^*)c_j^*] \quad \forall j < j_c, \tag{47a}$$

$$\frac{dc_j^*}{dt^*} = (1+P)[r_{j-1}^{*2}(c_{0_{j-1}}^* - c_{s_{j-1}}^*)c_{j-1}^* - r_j^{*2}(c_{0_j}^* - c_{s_j}^*)c_j^*] \quad \forall j > j_c, \tag{47b}$$

$$\frac{dc_j^*}{dt^*} = 0 \quad j = j_c. \tag{47c}$$

DISCUSSION AND RESULTS

The initial size distribution was assumed to be Gaussian and was computed using the relation

$$\frac{c_j}{c_{j_{0,\max}}} = \exp\left[-\frac{(j-j_{0,\max})^2}{2s^2}\right]. \tag{48}$$

Here, s is the standard deviation and $j_{0,\max}$ is the number of molecules in those particles whose concentration is the highest. Except when the effects of the initial distribution were investigated, all simulations were carried out with $j_{0,\max}=100$ and $s=5$. The value of δ was chosen so that $(\delta)/(r_{m_0}) \leq 0.1$. This was done so as to enable effective comparison with the asymptotic analytical expressions of the Lifshitz, Slyozhov, Wagner theory which use the linearized form of the Gibbs–Thompson equation. The initial supersaturation was chosen so that the initial critical radius was equal to the mean radius ($r_{c_0} \approx r_{m_0}$). The choice of the initial supersaturation is not of much significance since it affects only a short initial transient period within which r_c approaches the mean radius ($\bar{r}$). Indeed, Fig. 1 shows the evolution of ($\bar{r}/r_c$) obtained using Eq. (28) for three values of the initial supersaturation corresponding to $j_c=50$, $j_c=100$ and $j_c=200$ for the interface kinetic controlled regime. The following values were taken for the parameters used:

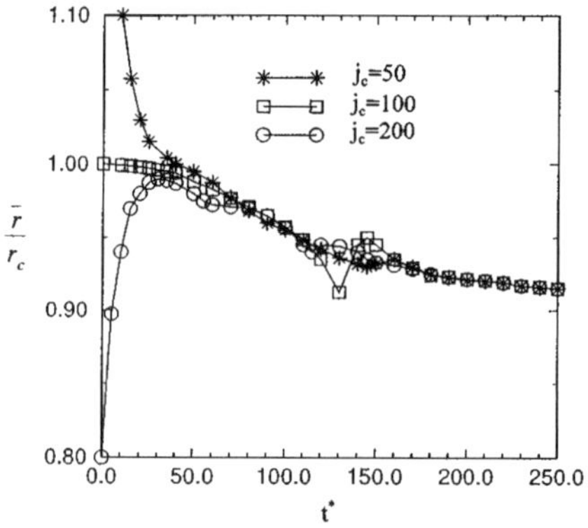

FIG. 1. Profiles of $(\bar{r})/(r_c)$ vs t^* obtained from the numerical solution of the Lifshitz, Slyozhov, Wagner equations [LSE, Eq. (28)] for the kinetic controlled Ostwald ripening of spherical particles. The initial distribution is Gaussian with a mean particle size corresponding to $j=100$ and a standard deviation of 5.

$$v_m = 1.51\times10^{-29}\,\frac{m^3}{\text{molecule}} \quad \text{and} \quad N_0 = 1.25c_{s_\infty}.$$

Note that in the rest of the paper MCE refers to Eq. (10), LSE refers to Eq. (28), and LSW refers to the asymptotic expressions of the Lifshitz, Slyozhov, Wagner theory.

Interface kinetic controlled growth

In this case $m=1$ [see Eqs. (33) and (37)] and the asymptotic expressions of the Lifshitz, Slyozhov, Wagner theory are given by

$$\lambda_i \equiv \frac{2kv_m\delta c_{s_\infty}}{\frac{dr_c^2}{dt}} = 4, \tag{49}$$

$$\frac{c_j\rho_j^2}{(c_j\rho_j^2)_{\max}} = \frac{\rho_j\exp\left(-\frac{6}{2-\rho_j}\right)}{0.00248(2-\rho_j)^5}, \tag{50}$$

$$(\bar{r})^2 \propto t. \tag{51}$$

Figure 2 shows the numerically computed plots of $\lambda_i/4$ vs t^* for Eq. (10) (MCE), Eq. (28) (LSE) along with the analytically predicted value of Eq. (49) (LSW). The computed asymptote based on the LSE is somewhat lower than that predicted by Eq. (49). This is in agreement with similar calculations[4] carried out for ripening of precipitates from solution. The λ_i computed using the MCE tends to a constant

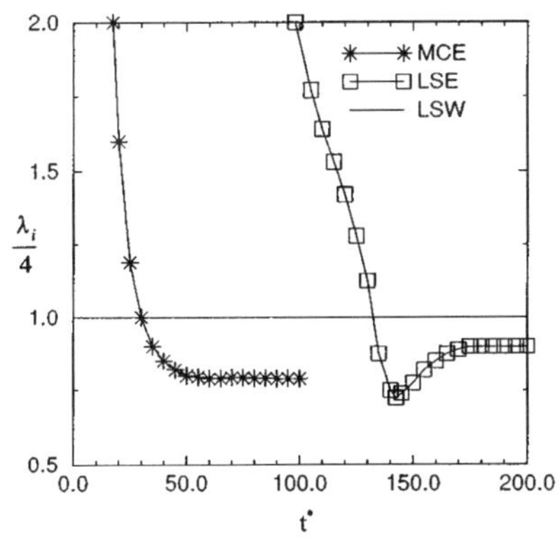

FIG. 2. $\lambda_i/4$ vs t^* for the kinetic controlled ($P=0$) Ostwald ripening of spheres. MCE stands for the microscopic continuity equation [Eq. (10)], LSE stands for the numerical solution of the Lifshitz, Slyozhov, Wagner equations [Eq. (28)], while LSW stands for analytic asymptotic results of the Lifshitz, Slyozhov, Wagner theory. The initial particle size distribution is Gaussian with a standard deviation of 5 and a mean size corresponding to $j=100$. The initial supersaturation corresponds to $j_c|_{t=0}=100$.

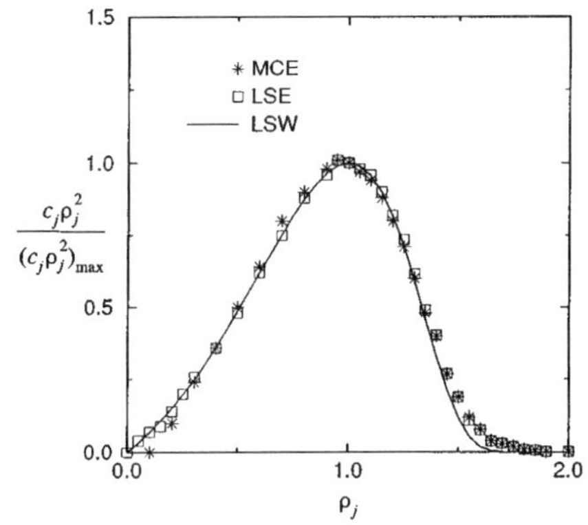

FIG. 4. Normalized (with respect to the maximum) asymptotic particle size distributions for the kinetic controlled Ostwald ripening of spheres for the system of Fig. 2.

much earlier and has a value 10% lower than the LSE value. Figure 3 shows the numerically computed plots of $(\overline{r^*})^2$ vs t^*. Both curves become asymptotically linear with similar slopes. However, the LSE predicts a longer "lag" after which the curve becomes linear. Thus at any given time the MCE predicts a mean particle radius which is about 25% larger. Figure 4 shows the computed normalized (with respect to the maximum) asymptotic distributions

$$\left(\frac{c_j\rho_j^2}{(c_j\rho_j^2)_{\max}}\right)$$

for the LSE and MCE. The continuous curve is obtained from the asymptotic expressions of the Lifshitz, Slyozhov, Wagner theory [Eq. (50)]. The curves correspond to $t^*=215$. These curves overlap with distributions obtained at $t^*=300$, indicating that the asymptote had been reached. The effect of the initial conditions on the asymptotic distribution was investigated by changing $j_{0,\max}$ and s [see Eq. (48)] and was found to be insignificant. Figure 4 shows that except for very small particles, the scaled asymptotic distributions obtained from the LSE and MCE show fair agreement. However, when the distributions are not normalized (see Fig. 5), the difference is extremely significant. The MCE curve is broader and has a peak which is almost one-third that predicted by the LSE.

Some physical feel of the difference may be obtained by considering the hypothetical situation in which at $t=0$ there are particles of only one size ($j_0>j_c$), i.e., $c_j=0 \quad \forall j\neq j_0$. In this case, at $t=0$ we have for the LSE

$$\frac{dc_{j_0}}{dt}=-4\pi k r_{j_0}^2(c_{0_{j_0}}-c_{s_{j_0}})c_{j_0}, \tag{52a}$$

$$\frac{dc_{j_0+1}}{dt}=4\pi k r_{j_0}^2(c_{0_{j_0}}-c_{s_{j_0}})c_{j_0}, \tag{52b}$$

$$\frac{dc_{j_0-1}}{dt}=0. \tag{52c}$$

Finally, for the MCE, we have

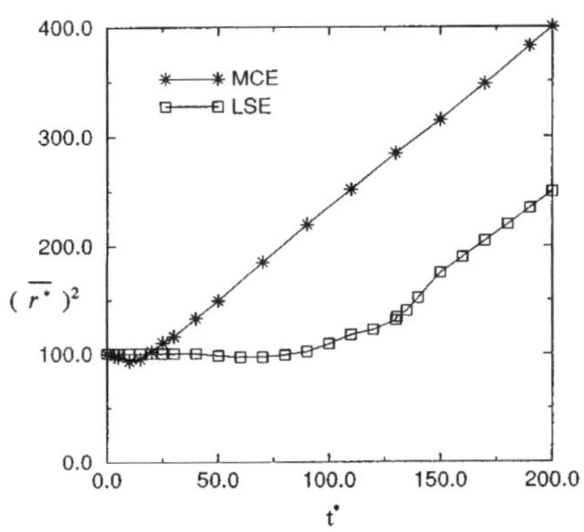

FIG. 3. $(\overline{r^*})^2$ vs t^* for kinetic controlled Ostwald ripening of spheres for the system of Fig. 2.

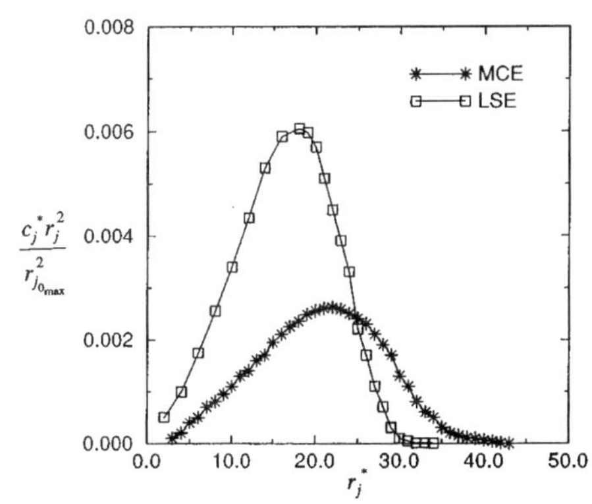

FIG. 5. Absolute size distributions for the kinetic controlled Ostwald ripening of spheres for $t^*=215$ for the system of Fig. 2.

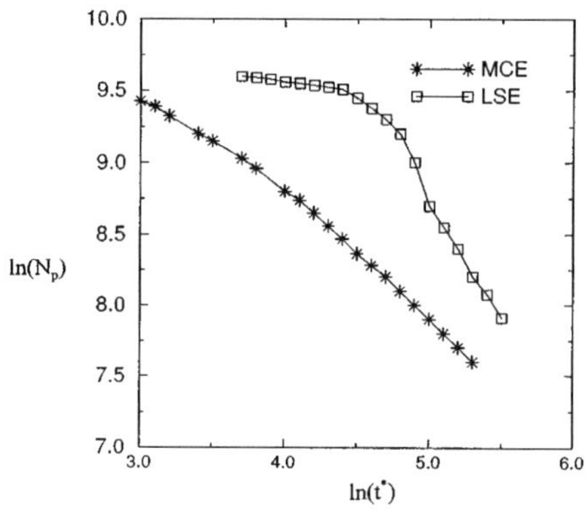

FIG. 6. Evolution of the number of particles per unit volume (N_p) with time for the kinetic controlled Ostwald ripening of spheres for the system of Fig. 2.

$$\frac{dc_{j_0}}{dt} = -4\pi k r_{j_0}^2 (c_{0_{j_0}} + c_{s_{j_0}}) c_{j_0}, \tag{53a}$$

$$\frac{dc_{j_0+1}}{dt} = 4\pi k r_{j_0}^2 c_{0_{j_0}} c_{j_0}, \tag{53b}$$

$$\frac{dc_{j_0-1}}{dt} = 4\pi k r_{j_0}^2 c_{s_{j_0}} c_{j_0}. \tag{53c}$$

Thus according to the LSE, for $j_0 > j_c$ particles of size smaller than j_0 will not form until j_c has increased above j_0. On the other hand, the MCE predicts that particles larger as well as smaller than j_0 will be formed. Thus the MCE tends to flatten the curve to a greater extent resulting in a broader size distribution. Some insight into the difference in the heights of the maxima predicted by the two approaches can be obtained by comparing Eq. (52) and Eq. (53). With the LSE, the decrease in c_{j_0} is proportional to the difference $c_{0_{j_0}} - c_{s_{j_0}}$, while the MCE takes into account the fall in c_{j_0} due to the loss as well as gain of molecules and the decrease is proportional to $c_{0_{j_0}} + c_{s_{j_0}}$. Thus the LSE significantly underpredicts the decrease in c_{j_0} resulting in a higher peak. Figure 6 presents plots of $\ln(N_p)$ vs $\ln(t^*)$ obtained using the LSE and the MCE. While in both cases, the curves become asymptotically linear, the MCE predicts a smaller number of particles at any time.

Diffusion controlled growth

In this case, $P \gg 1$ and we have $m=2$ [see Eqs. (35) and (37)]. The asymptotic expressions of the Lifshitz, Slyozhov, Wagner theory are then given by

$$\lambda_d = \frac{3A_d}{\frac{dr_c^3}{dt}} = \frac{27}{4}, \tag{54}$$

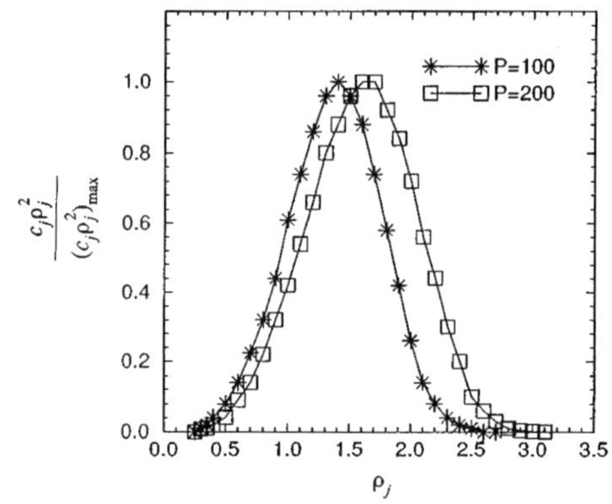

FIG. 7. Normalized (with respect to the maximum) particle size distributions obtained using the MCE [Eq. (10)] for Ostwald ripening of spheres at $t^*=2.4$ with $P=100$ and $P=200$. The initial distribution was Gaussian with a standard deviation of 5 and a mean size corresponding to $j=100$. The initial supersaturation corresponds to $j_c=100$.

$$\frac{c_j\rho_j^2}{(c_j\rho_j^2)_{\max}} = 32.2326\rho_j^2(3+\rho_j)^{-7/3}(1.5-\rho_j)^{-11/3} \times \exp\left[-\left(1-\frac{2\rho_j}{3}\right)^{-1}\right], \tag{55}$$

$$(\bar{r})^3 \propto t. \tag{56}$$

When $P \gg 1$ we have $P+1 \approx P$ and

$$1 + \frac{Pr_j^*}{r_{m_0}^*} \approx \frac{Pr_j^*}{r_{m_0}^*},$$

and the dimensionless LSE equations become

$$\frac{dc_j^*}{dt^*} = r_{j+1}^* r_{m_0}^* (c_{s_{j+1}}^* - c_1^*) c_{j+1}^* - r_j^* r_{m_0}^* (c_{s_j}^* - c_1^*) c_j^* \quad \forall j < j_c, \tag{57a}$$

$$\frac{dc_j^*}{dt^*} = r_{j-1}^* r_{m_0}^* (c_1^* - c_{s_{j-1}}^*) c_{j-1}^* - r_j^* r_{m_0}^* (c_1^* - c_{s_j}^*) c_j^* \quad \forall j > j_c, \tag{57b}$$

$$\frac{dc_j^*}{dt^*} = 0 \quad j = j_c. \tag{57c}$$

Similarly, for the MCE we have

$$\frac{dc_j^*}{dt^*} \approx P[r_{j-1}^{*2} c_{s_{j-1}}^* c_{j-1}^* + r_{j+1}^{*2} c_{s_{j+1}}^* c_{j+1}^* - 2r_j^{*2} c_{s_j}^* c_j^*] + [r_{j-1}^* r_{m_0}^* c_{j-1}^* (c_1^* - c_{s_{j-1}}^*) - r_j^* r_{m_0}^* c_j^* (c_1^* - c_{s_j}^*)] \quad \forall j > 2. \tag{58}$$

A comparison of Eqs. (57) and (58) reveals a very fundamental difference. While the LSE is independent of P, P still appears explicitly in the MCE. Figures 7 and 8 show plots of the normalized (with respect to the maximum) distributions and $(\overline{r^*})^3$ for $P=100$ and $P=200$ obtained using the MCE.

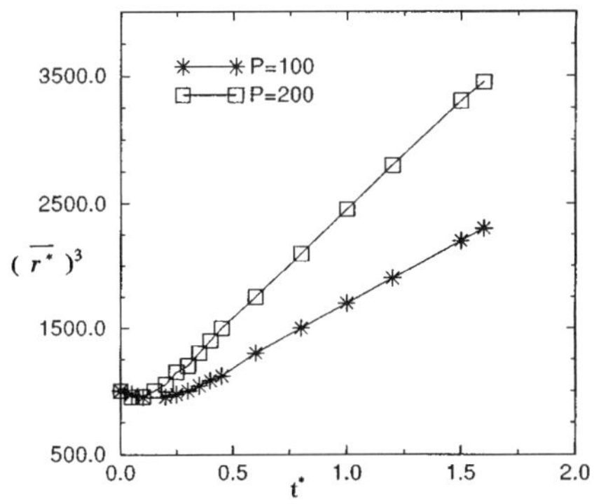

FIG. 8. $(\overline{r^*})^3$ vs t^* obtained using the MCE for $P=100$ and $P=200$ for the system of Fig. 7.

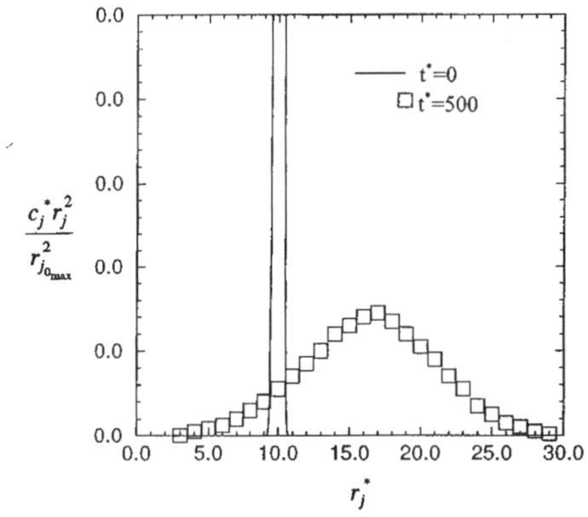

FIG. 9. Absolute particle size distributions (spheres) predicted by the MCE [Eq. (10)] for $P \to \infty$ at $t=0$ and $t^*=500$. The initial distribution is the same as that for Fig. 7.

It is clear from the figures that the results obtained from the MCE do not become independent of P at large values of P. However, as expected from the basic equations, the corresponding results from the LSE were found to be independent of P.

Equations (57) can be rewritten as

$$\frac{dc_j}{dt} = 4\pi D[r_{j+1}(c_{s_{j+1}} - c_1)c_{j+1} - r_j(c_{s_j} - c_1)c_j] \quad \forall j < j_c, \tag{59a}$$

$$\frac{dc_j}{dt} = 4\pi D[r_{j-1}(c_1 - c_{s_{j-1}})c_{j-1} - r_j(c_1 - c_{s_j})c_j] \quad \forall j > j_c, \tag{59b}$$

$$\frac{dc_j}{dt} = 0 \quad \forall j = j_c \tag{59c}$$

and Eq. (58) as

$$\frac{dc_j}{dt} \approx 4\pi k[r_{j-1}^2 c_{s_{j-1}} c_{j-1} + r_{j+1}^2 c_{s_{j+1}} c_{j+1} - 2r_j^2 c_{s_j} c_j] + 4\pi D[r_{j-1}c_{j-1}(c_1 - c_{s_{j-1}}) - r_j c_j(c_1 - c_{s_j})] \quad \forall j \geq 2. \tag{60}$$

Thus in the theoretical limit as $D \to 0$ $(P \to \infty)$, the LSE reduces to

$$\frac{dc_j}{dt} \approx 0 \tag{61}$$

while for the MCE we get

$$\frac{dc_j}{dt} \approx 4\pi k[r_{j-1}^2 c_{s_{j-1}} c_{j-1} + r_{j+1}^2 c_{s_{j+1}} c_{j+1} - 2r_j^2 c_{s_j} c_j] \quad j \geq 2. \tag{62}$$

So even when the diffusion coefficient tends to 0, the MCE still predicts that the size distribution will evolve with time, i.e., c_j increases if $r_{j+1}^2 c_{s_{j+1}} c_{j+1} + r_{j-1}^2 c_{s_{j-1}} c_{j-1} > 2r_j^2 c_{s_j} c_j$ and vice versa. Physically, this corresponds to a scenario in which the monomer concentration close to the particle surface increases or decreases due to attachment or detachment from the particle, but there is no diffusion of monomers to or from the bulk. The results for this situation are illustrated in Fig. 9 which shows the absolute distributions predicted by the MCE for this case. The continuous curve represents the initial distribution. Thus, when the equations are written in microscopic form, even when the growth of a particle is diffusion controlled, the Ostwald ripening is still affected by the kinetics of attachment and detachment at the interface. It is therefore appropriate to say that there is no such thing as diffusion controlled Ostwald ripening.

TWO-DIMENSIONAL OSTWALD RIPENING

Ostwald ripening plays an important role in the growth of metal crystallites on supports such as Pt/alumina (sintering)[12] which results in a reduction in the activity of metal supported catalysts. It has also been shown to be responsible for the growth of solid domains on monolayer and bilayer films.[13] A study of Ostwald ripening in two dimensions is therefore relevant. We shall consider two cases in which the ripening particles are assumed to be (a) three-dimensional hemispherical droplets (b) two-dimensional circular disks.

Hemispherical droplets

In this section we consider the ripening particles to be hemispherical. The surface concentration (n_{s_j}) of monomers in equilibrium with a hemispherical particle of radius r_j containing j molecules is given by

$$n_{s_j} = n_{s_\infty} \exp\left(\frac{\delta_1}{r_j}\right) \approx n_{s_\infty}\left(1 + \frac{\delta_1}{r_j}\right), \tag{63}$$

where n_{s_∞} is the concentration of monomers in equilibrium with a sufficiently large particle and δ_1 is given by

$$\delta_1 \equiv \frac{\sigma_1 A_m}{k_B T}. \tag{64}$$

Here, σ_1 is the line tension and A_m is the molecular area. The radius r_j can be computed using the relation

$$r_j=\left[\frac{3v_m j}{2\pi}\right]^{1/3}. \quad (65)$$

Microscopic continuity equation (MCE)

Let n_j be the number of particles per unit area containing j molecules. The average concentration (molecules per unit area) of monomers on the substrate is denoted by n_1. Let n_{0_j} be the actual concentration of monomers at the periphery of j-molecule particles. If we assume that there is no transport of molecules through the vapor phase, the rate at which a j-molecule particle gains molecules (L_j^+) is proportional to the periphery of the particle and is given by

$$L_j^+=2\pi r_j k_1 n_{0_j} \quad (66)$$

and the rate at which molecules are lost is

$$L_j^-=2\pi r_j k_1 n_{s_j}, \quad (67)$$

where the rate constant k_1 is assumed to be independent of the particle size. The change in the concentration of j-molecule particles is then given by

$$\frac{dn_j}{dt}=L_{j-1}^+ n_{j-1}+L_{j+1}^- n_{j+1}-L_j^+ n_j-L_j^- n_j \quad j\geqslant 2. \quad (68)$$

Combining Eqs. (66)–(68) we get

$$\frac{dn_j}{dt}=2\pi k_1[r_{j-1}n_{0_{j-1}}n_{j-1}+r_{j+1}n_{s_{j+1}}n_{j+1}-r_j n_{0_j}n_j-r_j n_{s_j}n_j] \quad j\geqslant 2. \quad (69)$$

Assuming no agglomeration of monomers

$$L_1^+=0. \quad (70)$$

Evaluation of n_{0_j}

The steady state solution of the diffusion equation in cylindrical coordinates is given by

$$n=C_1 \ln r+C_2, \quad (71)$$

where n represents the monomer concentration at a distance r from the center of the particle. We assume the diffusion field of a j-molecule particle to be confined to a circle of radius A_j. Thus

$$r=A_j\Rightarrow n=n_1. \quad (72)$$

If we assume the area of this circle to be proportional to the area occupied by the particle, A_j is given by

$$A_j=\frac{r_j}{\sqrt{\Sigma n_j \pi r_j^2}}. \quad (73)$$

The boundary condition at the periphery is

$$r=r_j\Rightarrow n_j=n_{0_j}. \quad (74)$$

Assuming a quasisteady state, the diffusive flux at the interface ($r=r_j$) equals the net rate at which the molecules are attached and we have

$$k_1(n_{0_j}-n_{s_j})=D_s\left[\frac{\partial n}{\partial r}\right]_{r=r_j}, \quad (75)$$

where D_s is the surface diffusion coefficient of the monomer. Using Eqs. (72), (74), and (75), one obtains

$$n_{0_j}=n_{s_j}+\frac{n_1-n_{s_j}}{1+\cfrac{\cfrac{k_1 r_j}{D_s}}{\ln \cfrac{A_j}{r_j}}}. \quad (76)$$

Let us define a dimensionless number

$$P_1=\frac{\cfrac{k_1 r_{m_0}}{D_s}}{\left[\ln \cfrac{A_j}{r_j}\right]_0}, \quad (77)$$

where r_{m_0} is the initial mean particle radius and $[\ln(A_j/r_j)]_0$ refers to the initial value. Equation (76) can therefore be written as

$$n_{0_j}=n_{s_j}+\frac{n_1-n_{s_j}}{1+\cfrac{P_1 r_j \ln\left(\cfrac{A_j}{r_j}\right)}{r_{m_0}\left[\ln\left(\cfrac{A_j}{r_j}\right)\right]_0}}. \quad (78)$$

Thus when $P_1\ll 1$, Eq. (78) becomes

$$n_{0_j}-n_{s_j}\approx n_1-n_{s_j} \quad (79)$$

and the transport is controlled by the kinetics of attachment and detachment at the interface. On the other hand, if $P_1\gg 1$

$$n_{0_j}-n_{s_j}\approx\frac{n_1-n_{s_j}}{\cfrac{P_1 r_j \ln\left(\cfrac{A_j}{r_j}\right)}{\left[\ln\left(\cfrac{A_j}{r_j}\right)\right]_0}}=\frac{D_s(n_1-n_{s_j})}{k_1 r_j \ln\left[\cfrac{A_j}{r_j}\right]} \quad (80)$$

and surface diffusion is the rate controlling mechanism.

Finally, since the number of molecules in the system must be conserved

$$N_t=n_1+\sum_2 jn_j, \quad (81)$$

where N_t is the total number of molecules present in the system per unit area.

Equation (69) represents a system of initial value problems which can be solved to obtain the evolution of n_j with time for a given initial distribution.

Lifshitz, Slyozhov, Wagner formulation (analytical asymptote)

The growth equation in this case has the form

$$\frac{d}{dt}\left(\frac{2}{3}\pi r_j^3\right)=v_m 2\pi k_1 r_j(n_{0_j}-n_{s_j}), \tag{82}$$

i.e.,

$$\frac{dr_j}{dt}=\frac{v_m k_1}{r_j}(n_{0_j}-n_{s_j}). \tag{83}$$

Substitution of the expression for n_{0_j} from Eq. (78) into Eq. (83) yields

$$\frac{dr_j}{dt}=\frac{v_m k_1}{r_j}\frac{(n_1-n_{s_j})}{1+P\dfrac{r_j}{r_{m_0}}\dfrac{\ln\dfrac{A_j}{r_j}}{\left[\ln\dfrac{A_j}{r_j}\right]_0}}. \tag{84}$$

The critical radius r_c is given by

$$\frac{n_1}{n_{s_\infty}}=\exp\left(\frac{\delta_1}{r_c}\right)\approx 1+\frac{\delta_1}{r_c}. \tag{85}$$

Using Eq. (85), Eq. (84) can be written as

$$\frac{dr_j}{dt}=-\left(\frac{v_m k_1 n_{s_\infty}\delta_1}{r_j}\right)\left(\frac{1}{r_j}\right)\frac{1-\dfrac{r_j}{r_c}}{1+P_1\dfrac{r_j}{r_{m_0}}\dfrac{\ln\left(\dfrac{A_j}{r_j}\right)}{\left[\ln\left(\dfrac{A_j}{r_j}\right)\right]_0}}. \tag{86}$$

The asymptotic results are essentially the same as for spheres, the only differences being that $G(\rho_j,t)d\rho_j$ now denotes the number of particles per unit area.

Numerical solution (LSE)

The LSE equations in this case are analogous to Eqs. (28) and are written as

$$\frac{dn_j}{dt}=2\pi k_1 r_{j+1}(n_{s_{j+1}}-n_{0_{j+1}})n_{j+1} - 2\pi k_1 r_j(n_{s_j}-n_{0_j})n_j \quad \forall j<j_c, \tag{87a}$$

$$\frac{dn_j}{dt}=2\pi k_1 r_{j-1}(n_{0_{j-1}}-n_{s_{j-1}})n_{j-1} - 2\pi k_1 r_j(n_{0_j}-n_{s_j})n_j \quad \forall j>j_c, \tag{87b}$$

$$\frac{dn_j}{dt}=0 \quad j=j_c. \tag{87c}$$

The equations are nondimensionalized using the following dimensionless variables:

$$n_j^*\equiv\frac{n_j}{n_{j_{\max_0}}} \quad j\geqslant 2, \quad n_1^*\equiv\frac{n_1}{n_{s_\infty}}, \quad n_{s_j}^*\equiv\frac{n_{s_j}}{n_{s_\infty}}, \tag{88a}$$

$$n_{0_j}^*=\frac{n_{0_j}}{n_{s_\infty}}, \quad r_j^*\equiv\frac{r_j}{\delta_1}, \tag{88b}$$

$$t^*\equiv t\,\frac{4\pi\delta_1 n_{s_\infty}k_1}{1+P_1}. \tag{88c}$$

Thus for $P_1\ll 1$ we have

$$t^*=t(2\pi\delta_1 n_{s_\infty}k_1), \tag{88d}$$

and for $P_1\gg 1$ we have

$$t^*=t\,\frac{2\pi n_{s_\infty}\delta_1}{r_{m_0}}\frac{D_s}{\left[\ln\dfrac{A_j}{r_j}\right]_0}. \tag{88e}$$

Equations (87) can therefore be rewritten as

$$\frac{dn_j^*}{dt^*}=(1+P_1)[r_{j+1}^*(n_{s_{j+1}}^*-n_{0_{j+1}}^*)n_{j+1}^* - r_j^*(n_{s_j}^*-n_{0_j}^*)n_j^*] \quad \forall j<j_c, \tag{89a}$$

$$\frac{dn_j^*}{dt^*}=(1+P_1)[r_{j-1}^*(n_{0_{j-1}}^*-n_{s_{j-1}}^*)n_{j-1}^* - r_j^*(n_{0_j}^*-n_{s_j}^*)n_j^*] \quad \forall j>j_c, \tag{89b}$$

$$\frac{dn_j^*}{dt^*}=0 \quad j=j_c, \tag{89c}$$

and the MCE [Eq. (69)] as

$$\frac{dn_j^*}{dt^*}=(1+P_1)[r_{j-1}^*n_{0_{j-1}}^*n_{j-1}^*+r_{j+1}^*n_{s_{j+1}}^*n_{j+1}^* - r_j^*n_{0_j}^*n_j^*-r_j^*n_{s_j}^*n_j^*] \quad \forall j\geqslant 2. \tag{90}$$

Discussion and results

The initial distribution was computed using Eq. (48), with $s=5$ and $j_{\max,0}=100$. As in the earlier case the value of δ_1 was chosen so that $(\delta_1)/(r_{m_0})\leqslant 0.1$ and the initial supersaturation was chosen so that the initial critical radius was equal to the mean radius ($r_{c_0}\approx r_{m_0}$). The following values were taken for the parameters used: $v_m=1.51\times 10^{-29}$ (m^3/molecule) and $N_t=1.25n_{s_\infty}$. The results of the numerical calculations are discussed in the following sections.

INTERFACE KINETIC CONTROLLED GROWTH

For the interface kinetic controlling case ($P_1\approx 0$), Eq. (86) becomes

$$\frac{dr_j}{dt}\approx-\frac{A_k}{r_j^2}\left(1-\frac{r_j}{r_c}\right), \tag{91}$$

where

$$A_k\equiv v_m k_1 n_{s_\infty}\delta_1. \tag{92}$$

Thus $m=2$ [see Eq. (37)] and the asymptotic expressions are given by

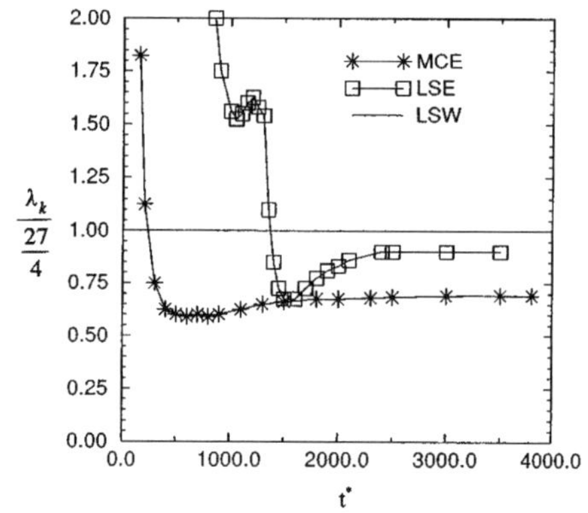

FIG. 10. $\lambda_k/(27/4)$ vs t^* for the kinetic controlled ($P_1=0$) Ostwald ripening of hemispheres. The initial distribution is Gaussian with a mean size corresponding to $j=100$ and a standard deviation(s) of 5. The initial supersaturation corresponds to $j_c=100$. MCE stands for the microscopic continuity equation [Eq. (69)], LSE stands for the numerical solution of the Lifshitz, Slyozhov, Wagner theory [Eq. (87)], and LSW stands for the analytical asymptotes predicted by the Lifshitz, Slyozhov, Wagner theory.

$$\lambda_k \equiv \frac{3k_1 v_m \delta_1 n_{s_\infty}}{\dfrac{dr_c^3}{dt}} = \frac{27}{4}, \tag{93}$$

$$\frac{n_j\rho_j^2}{(n_j\rho_j^2)_{\max}} = 32.2326\rho_j^2(3+\rho_j)^{-7/3}(1.5-\rho_j)^{-11/3} \times \exp\left[-\left(1-\frac{2\rho_j}{3}\right)^{-1}\right], \tag{94}$$

$$(\bar{r})^3 \propto t. \tag{95}$$

Figure 10 shows the numerically computed plots of

$$\frac{\lambda_k}{\frac{27}{4}}$$

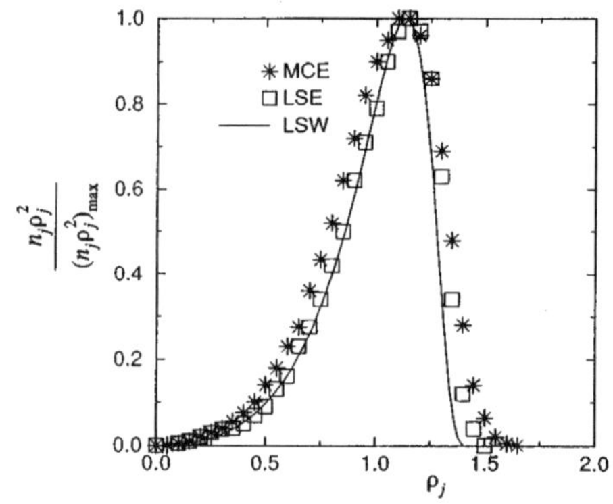

FIG. 12. Normalized (with respect to the maximum) asymptotic particle size distributions for the kinetic controlled Ostwald ripening of hemispheres for the system of Fig. 10.

vs t^* for Eq. (69) (MCE) and Eq. (89) (LSE) along with the analytically predicted value of Eq. (93) (LSW). The computed asymptote from the LSE is somewhat lower than that predicted by Eq. (93). The λ_k computed using the MCE tends much earlier to a constant which is about 20% lower than the LSE value. Figure 11 shows the numerically computed plots of $(\overline{r^*})^3$ vs t^*. While both curves become asymptotically linear with similar slopes, the MCE predicts a mean particle radius which is about 15% larger at any given time. Figure 12 shows the computed scaled (with respect to the maximum) asymptotic distributions

$$\left(\frac{n_j\rho_j^2}{(n_j\rho_j^2)_{\max}}\right)$$

for the LSE, MCE, and LSW equations. These curves correspond to $t^*=4000$. As before, tests to confirm the asymptotic nature of the curves were carried out by comparing with distributions obtained at $t^*=5000$. These curves were also found to be independent of the initial distribution. It is clear from Fig. 12 that while the maxima of the curves coincide, the MCE distribution is broader. Figure 13 shows the distributions when they are not normalized. Again the difference

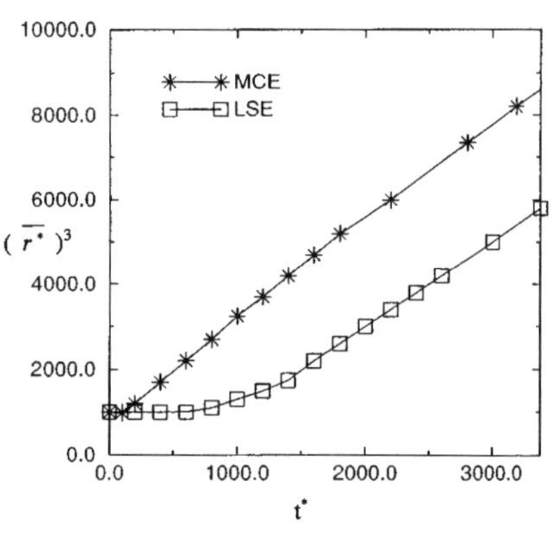

FIG. 11. $(\overline{r^*})^3$ vs t^* for the kinetic controlled Ostwald ripening of hemispheres for the system of Fig. 10.

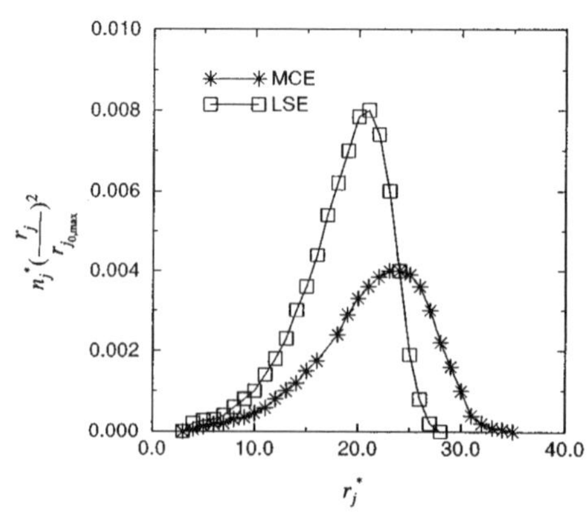

FIG. 13. Absolute particle size distributions for the kinetic controlled Ostwald ripening of hemispheres for $t^*=4000$ for the system of Fig. 10.

is magnified and the LSE predicts a narrower curve with a peak value that is double that obtained from the MCE. The evolution of the number of particles per unit area is similar to that obtained for precipitates (Fig. 6), i.e., the MCE predicts a smaller number of particles per unit area at any time.

SURFACE DIFFUSION CONTROLLED GROWTH

In this case ($P_1 \gg 1$) Eq. (86) becomes

$$\frac{dr_j}{dt} \approx -\frac{A_s}{r_j^3}\left(1-\frac{r_j}{r_c}\right), \tag{96}$$

where

$$A_s = \frac{v_m n_{s_\infty} \delta_1 D_s}{\ln\left(\frac{A_j}{r_j}\right)}.$$

Thus we have $m=3$ and the asymptotic expressions of the LSW theory are then given by

$$\lambda_s = \frac{4A_s}{\frac{dr_c^4}{dt}} = 9.5, \tag{97}$$

$$\frac{n_j\rho_j^2}{(n_j\rho_j)_{\max}} = \frac{10.07\rho_j \exp\left[\frac{-2}{3(1.33-\rho_j)}\right]\exp\left[-\frac{1}{6\sqrt{2}}\tan^{-1}\left(\frac{1.33+\rho_j}{1.33\sqrt{2}}\right)\right]}{(1.33-\rho_j)^{-19/6}(\rho_j^2+2.66\rho_j+5.32)^{23/12}} \quad \forall \rho_j \leq 1.33, \tag{98a}$$

$$\frac{n_j\rho_j^2}{(n_j\rho_j^2)_{\max}} = 0 \quad \forall \rho_j \geq 1.33, \tag{98b}$$

$$\bar{r}^4 \propto t. \tag{99}$$

As for spheres, the dimensionless LSE equations for diffusion control are

$$\frac{dn_j^*}{dt^*} = \frac{1}{\ln\left[\frac{A_j}{r_j}\right]}[r_{m_0}^*(n_{s_{j+1}}^* - n_1^*)n_{j+1}^* - r_{m_0}^*(n_{s_j}^* - n_1^*)n_j^*], \tag{100a}$$

$$\frac{dn_j^*}{dt^*} = \frac{1}{\ln\left[\frac{A_j}{r_j}\right]}[r_{m_0}^*(n_1^* - n_{s_{j-1}}^*)n_{j-1}^* - r_{m_0}^*(n_1^* - n_{s_j}^*)n_j^*], \tag{100b}$$

$$\frac{dn_j^*}{dt^*} = 0 \quad \forall j = j_c. \tag{100c}$$

And for the MCE

$$\frac{dn_j^*}{dt^*} \approx P_1[r_{j+1}^* n_{s_{j+1}}^* n_{j+1}^* + r_{j-1}^* n_{s_{j-1}}^* n_{j-1}^* - 2r_j^* n_{s_j}^* n_j^*] + \frac{1}{\ln\left[\frac{A_j}{r_j}\right]}[r_{m_0}^*(n_1^* - n_{s_{j-1}}^*)n_j^* - r_{m_0}^*(n_1^* - n_{s_{j-1}}^*)n_j^* - r_{m_0}^*(n_1^* - n_{s_{j-1}}^*)n_j^* - r_{m_0}^*(n_1^* - n_{s_j}^*)n_j^*]. \tag{101}$$

It is clear that unlike the LSE, the MCE equation for diffusion control still depends on P_1. Figure 14 which shows the evolution of $(\overline{r^*})^4$ predicted by the MCE for $P_1=10$ and $P_1=50$ illustrates this point. In the extreme case ($D=0$, $P_1 \to \infty$) we have for the LSE

$$\frac{dn_j}{dt} = 0, \tag{102}$$

and for the MCE

$$\frac{dn_j}{dt} \approx 2\pi k[r_{j+1}n_{s_{j+1}}n_{j+1} + r_{j-1}n_{s_{j-1}}n_{j-1} - 2r_j n_{s_j} n_j]. \tag{103}$$

Figure 15 shows the absolute distributions for this case for $t(2\pi k_1 n_{s_\infty}\delta_1) = 4000$. Thus the conclusions for this case are similar to those obtained in the three-dimensional case.

TWO-DIMENSIONAL CIRCULAR DISKS

When the particles are approximated by circular monomolecular disks the linearized Gibbs–Thompson equation is written as

$$n_s = n_{s_\infty}\left(1+\frac{\delta_2}{r}\right), \tag{104}$$

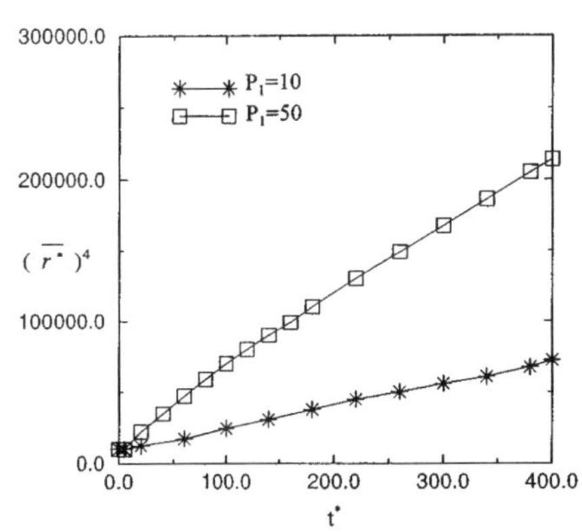

FIG. 14. $(\overline{r^*})^4$ vs t^* predicted by the MCE for hemispheres with $P_1=10$ and $P_1=50$. The initial conditions are the same as that for Fig. 10

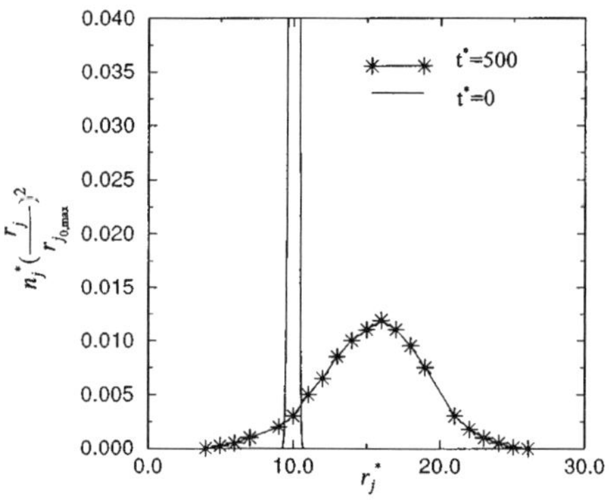

FIG. 15. Absolute particle size distributions (hemispheres) predicted by the MCE for $P_1 \to \infty$ at $t=0$ and $t^*=500$ for the system of Fig. 14.

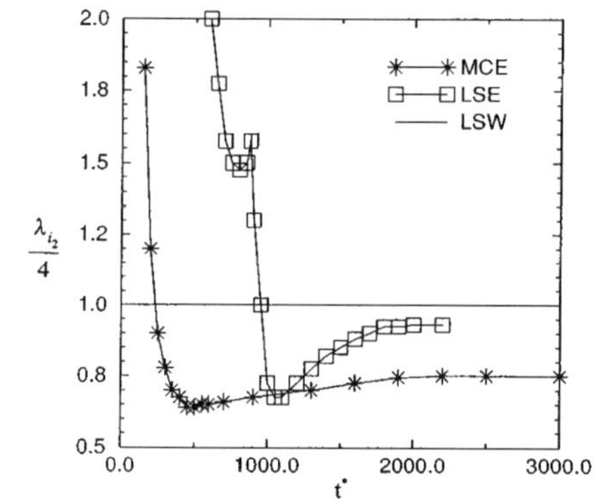

FIG. 16. $\lambda_{i_2}/4$ vs t^* for the kinetic controlled ($P_1=0$) Ostwald ripening of monomolecular disks. The initial distribution is Gaussian with a standard deviation of 5 and a mean size corresponding to $j=100$. The initial supersaturation corresponds to $j_c=100$. MCE stands for the microscopic continuity equation [Eq. (69)], LSE stands for the numerical solution of the Lifshitz, Slyozhov, Wagner theory [Eq. (87)], and LSW stands for the analytical asymptotes predicted by the Lifshitz, Slyozhov, Wagner theory.

where $\delta_2 \equiv [(\sigma_1 A_m)/(k_B T)]$ and σ_1 and A_m represent the line tension and molecular area, respectively. The basic balance equations are essentially the same as in the previous case, the only difference being that the radius of a particle containing j molecules is given by $r_j = \sqrt{A_m j/\pi}$ and the growth relation is

$$\frac{d\pi r_j^2}{dt} = 2\pi r_j A_m k_2 (n_{0_j} - n_{s_j}), \tag{105}$$

i.e.,

$$\frac{dr_j}{dt} = k_2 A_m \frac{(n_1 - n_{s_j})}{1 + \dfrac{P_1 r_j \ln\left(\dfrac{A_j}{r_j}\right)}{r_{m_0}\left[\ln \dfrac{A_j}{r_j}\right]_0}}. \tag{106}$$

The asymptotic expressions are again given by Eqs. (37)–(42). However, in this case, since the area occupied by the molecules is conserved we have $m_1=2$ in Eq. (42).

DISCUSSION AND RESULTS

Interface kinetic controlled growth

Putting $P_1=0$ in Eq. (106), we have

$$\frac{dr_j}{dt} = -\frac{A_{i_2}}{r_j}\left[1 - \frac{r_j}{r_c}\right], \tag{107}$$

where $A_{i_2} = k_2 A_m n_{s_\infty} \delta_2$ and the asymptotic LSW relations are

$$\lambda_{i_2} = \frac{2A_{i_2}}{\dfrac{dr_c^2}{dt}} = 4, \tag{108}$$

$$\frac{n_j \rho_j}{(n_j \rho_j)_{\max}} = 50.197 \frac{\rho_j}{(2-\rho_j)^4} \exp\left[\frac{-4}{2-\rho_j}\right], \tag{109}$$

$$(\bar{r})^2 \propto t. \tag{110}$$

Figures 16–19 show the results for this case. As in the earlier cases the MCE predicts a lower asymptote for λ_{i_2}. The $\bar{r}^2$ obtained from the MCE becomes linear earlier and has a higher value than that predicted by the LSE at any given time. The normalized (with respect to the maximum) and absolute distributions are shown in Figs. 17 and 18 and show the same characteristics as the curves for hemispheres, i.e., the normalized MCE curve is somewhat broader and the absolute distributions from the LSE are significantly higher.

Diffusion controlled growth

In this case we have

$$\frac{dr_j}{dt} = -\frac{A_{s_2}}{r_j^2}\left(1 - \frac{r_j}{r_c}\right), \tag{111}$$

where

$$A_{s_2} = \frac{D_s \delta_2 n_{s_\infty}}{\ln \dfrac{A_j}{r_j}}.$$

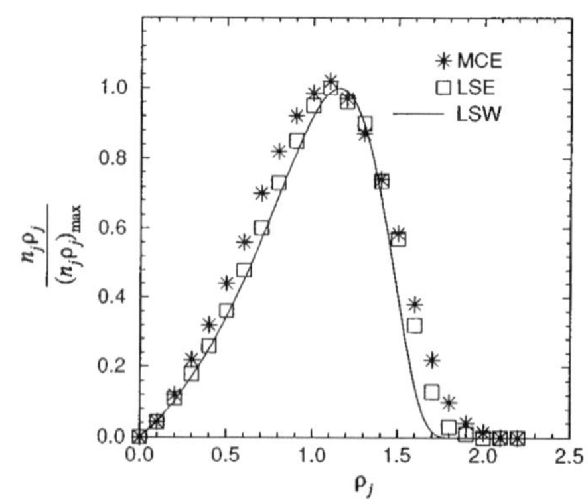

FIG. 17. Normalized (with respect to the maximum) asymptotic distributions for kinetic controlled Ostwald ripening of disks for the system of Fig. 16.

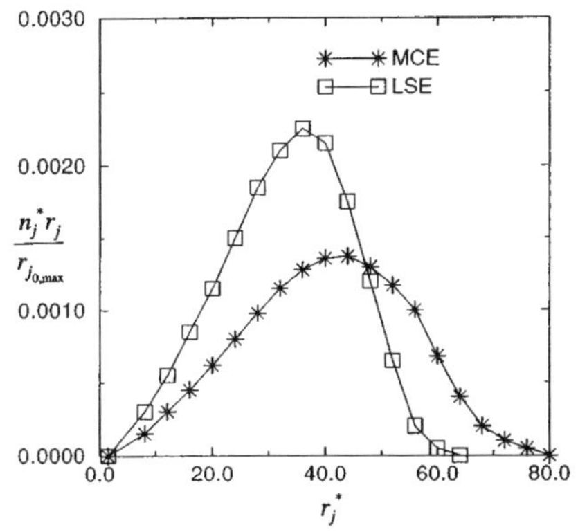

FIG. 18. Absolute distributions for the kinetic controlled Ostwald ripening of disks at $t^*=215$ for the system of Fig. 16.

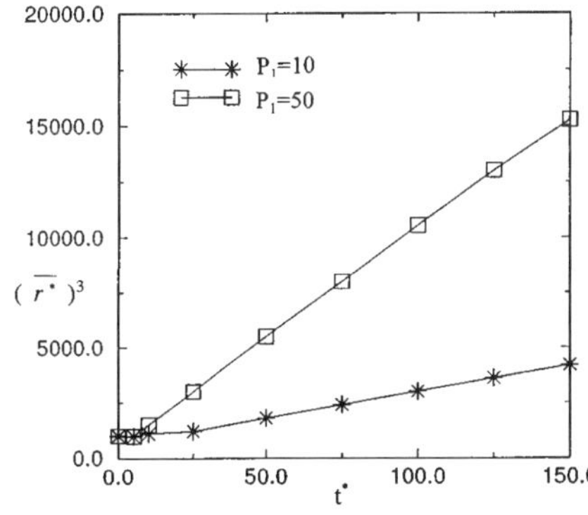

FIG. 20. $(\overline{r^*})^3$ vs t^* obtained using the MCE for growth of monomolecular disks with $P_1=10$ and $P_1=50$. The initial conditions are the same as Fig. 16.

Figure 20 shows a plot of $(\overline{r^*})^3$ vs t^*. The result for the extreme case ($P_1 \rightarrow \infty$) is shown in Fig. 21.

There is a paradox when we take $D=0$ for a system containing particles of a single size and no monomers. Details of this paradox and its resolution are given in Appendix A.

CONCLUSIONS

A theoretical study of Ostwald ripening was carried out. An attempt was made to investigate the appropriateness of the continuum approximation used in the traditional theory of Lifshitz, Slyozhov, and Wagner by making a comparison with numerically obtained results from a more rigorous stochastic microscopic approach. While in the latter, the concentration of particles of size j depends on the concentration of particles of size $j-1$ as well as particles of size $j+1$, in the former particles of size j are formed either from particles of size $j-1$ for $j>j_c$ or only from particles of size $j+1$ for $j<j_c$.

Ostwald ripening of precipitates in supersaturated solutions and two-dimensional Ostwald ripening in which the particles are considered to be hemispherical droplets or monomolecular circular disks are discussed in detail.

The following conclusions emerge from this study:

(a) When the stochastic microscopic equations are employed, the limiting case of "diffusion controlled Ostwald ripening" ceases to have significance, since the kinetics at the interface play a role even when the transport of molecules from the bulk to a particle is diffusion controlled. The conclusions of the Lifshitz, Slyozhov, Wagner theory for this extreme case are therefore invalid.

(b) Similar to the Lifshitz, Slyozhov, Wagner theory, the time evolution of the mean particle radius, the number of particles per unit volume and the critical radius based on the MCE have asymptotic behaviors which are independent of the initial conditions.

(c) With the MCE, the asymptotic trends are observed much earlier than with the Lifshitz, Slyozhov, Wagner approach.

(d) The MCE predicts a larger mean radius (25% larger for precipitates and about 15% larger for hemispheres and monomolecular disks) at any given time than the Lifshitz, Slyozhov, Wagner theory.

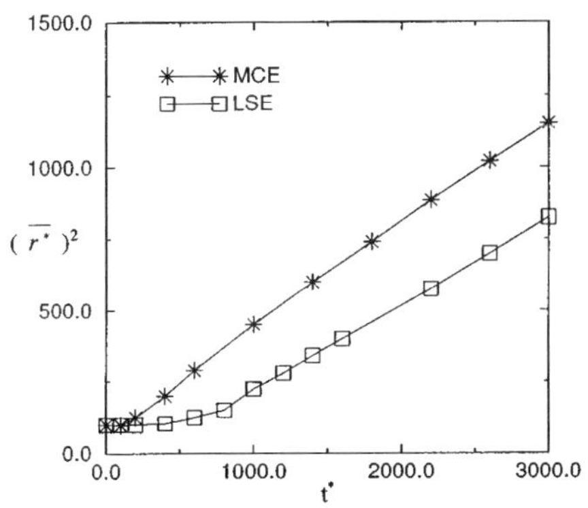

FIG. 19. $(\overline{r^*})^2$ vs t^* obtained for the kinetic controlled growth ($P_1=0$) of monomolecular disks for the system of Fig. 16.

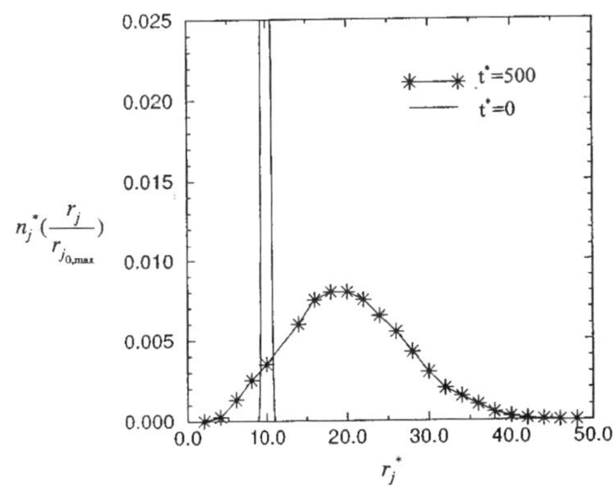

FIG. 21. Absolute particle size distributions (monomolecular disks) predicted by the MCE for $P_1 \rightarrow \infty$ at $t=0$ and $t^*=500$ for the system of Fig. 20.

(e) The particle size distributions obtained using the MCE differ significantly from those obtained using the Lifshitz, Slyozhov, Wagner theory. The maxima of the MCE distributions are much lower (approximately one-third for precipitates and about half for hemispheres and monomolecular disks) and the widths much larger than those predicted by the Lifshitz, Slyozhov, Wagner theory.

APPENDIX A: A PARADOX

As mentioned in the text, when $D=0$, there is a paradox when an extreme case is considered. Suppose, that initially there are particles of only one size and there are no monomers in solution. If the diffusion coefficient is zero, monomers detaching from a particle cannot move away. Thus the concentration of monomers around a particle will keep on increasing and one would expect that all the particles will decrease in size. Thus if only particles containing "p" molecules are initially present, we have

$$c_j|_{t=0}=0 \quad \text{for } j\neq p. \tag{A1}$$

Intuitively, one would expect that no particles containing "$p+1$" molecules will be formed. However, from Eq. (60) we have

$$\frac{\partial c_{p+1}}{\partial t}=4\pi k[r_p^2 c_{s_p} c_p] \quad \text{at } t=0. \tag{A2}$$

In other words, the MCE predicts that particles of size $p+1$ will be formed. However, if we consider the MCE in its original form [Eq. (10)], i.e., before the c_{0_j} is computed using the transport equations, we have

$$\frac{dc_{p+1}}{dt}=4\pi k[r_p^2 c_{0_p} c_p]=0 \quad \text{at } t=0, \tag{A3}$$

which is acceptable since $c_{0_p}=0$. Consequently, the above paradox occurs because of the incompatibility between the transport equations and the expressions for the kinetics at the interface. A minor modification of the transport equations eliminates, however, the paradox. At quasisteady state, the rate at which single atoms diffuse from a particle containing j atoms is equal to the rate at which a $(j+1)$-atom particle emits atoms minus the rate at which a j-atom particle captures single atoms; thus, using the quasisteady state approximation, the flux of monomers $[J(j)]$, towards a j sized particle is given by

$$4\pi r_j^2 J(j)=k_a r_{j+1}^2 c_{s_{j+1}}-k_b r_j^2 c_{0_j}. \tag{A4}$$

Since at equilibrium, one obtains

$$c_{0_j}=c_{s_{j+1}}, \tag{A5}$$

we have

$$k_a r_{j+1}^2=k_b r_j^2. \tag{A6}$$

If instead of Eq. (8), we employ Eqs. (A4) and (A6), Eq. (62) is replaced by

$$\frac{dc_j}{dt}\approx 4\pi k_b[r_{j-1}^2 c_{s_j} c_{j-1}+r_j^2 c_{s_{j+1}} c_{j+1} -(r_{j-1}^2 c_{s_j}+r_j^2 c_{s_{j+1}})c_j], \tag{A7}$$

and Eq. (A2) by

$$\frac{\partial c_{p+1}}{\partial t}=4\pi k_b[r_p^2 c_{s_{p+1}} c_p]. \tag{A8}$$

Since $c_{p+1}=0$, it is clear that $c_{s_{p+1}}$ does not exist and $(\partial c_{p+1}/\partial t)=0$ as it should. While this modification eliminates the paradox, the modification of the basic equations on this basis does not affect any of the other calculations in the paper.

APPENDIX B: NOMENCLATURE

A_j=The radius of the circle within which the diffusion field of a particle (hemisphere or disk) is confined
$A_i\equiv v_m k c_{s_\infty}\delta, A_d\equiv v_m c_{s_\infty}\delta D$ = Constants arising in the interface kinetic controlled growth and diffusion controlled growth of spheres

$$A_k\equiv v_m k_1 n_{s_\infty}\delta_1, \quad A_s\equiv \frac{v_m n_{s_\infty}\delta_1 D_s}{\ln\left(\frac{A_j}{r_j}\right)}$$

are quantities in the expressions for the kinetic controlled and diffusion controlled growth of hemispheres
$A_{i_2}\equiv k_2 A_m n_{s_\infty}\delta_2$ = Constant arising in the kinetic controlled growth of monomolecular disks

$$A_{s_2}\equiv \frac{D_s\delta_2 n_{s_\infty}}{\ln \dfrac{A_j}{r_j}}$$

is the quantity arising in the diffusion controlled growth of monomolecular disks.
c=Monomer concentration (molecules per unit volume)
c_1=Monomer concentration in the bulk
c_j=Number of particles of size j per unit volume
$c_{j_{0\max}}$ = Maximum value of c_j at $t=0$
c_{s_∞} = Saturation concentration of monomers in equilibrium with a large particle
c_{s_j} = Monomer concentration in equilibrium with a particle having j molecules
$c_j^*\equiv c_j/c_{j_{0\max}}$
$c_1^*\equiv c_1/c_{s_\infty}$
$c_{s_j}^*\equiv c_{s_j}/c_{s_\infty}$
r=Particle radius
c_r=Number density of particles having a radius between r and $r+dr$
c_{0_j} = Concentration of monomers near the surface of a particle containing j molecules
D=Diffusion coefficient
D_s=Surface diffusion coefficient
j_c=Number of molecules in a particle of critical size.
k_B=Boltzmann constant
k=Kinetic rate constant for the attachment and detachment

of molecules at the surface of a sphere
k_1=Kinetic rate constant for hemispheres
k_2=Rate constant for disks
L_j^+=Rate at which a j-sized particle gains molecules
L_j^-=Rate at which a j-sized particle loses molecules
N_0=Total number of molecules present in the system per unit volume
n_j=Number of particles per unit area having j molecules
$n_{j_{0\max}}$ = Maximum value of n_j at $t=0$
n_{s_∞} = Saturation concentration of monomers at the periphery of a large particle (molecules per unit area)
N_p=Total number of particles per unit volume
n_{0_j} = Monomer concentration (molecules per unit area) at the periphery of the particle
N_t=Total number of molecules in the system per unit area
n_{s_j} = Saturation monomer concentration at the periphery of a particle (molecules per unit area)
P=Dimensionless number given by $(kr_{m_0})/(D)$
P_1=Dimensionless number given by

$$\frac{\dfrac{k_1 r_{m_0}}{D_s}}{\left[\ln \dfrac{A_j}{r_j}\right]_0}$$

r_c=Critical radius
r_{m_0} = Initial mean particle radius
r_j=Radius of a particle containing j molecules
$\bar{r}$=Mean radius
r_j^* = Dimensionless radius of a particle containing j molecules
s=Standard deviation of the initial Gaussian distribution
t=Time
t^*=Dimensionless time
v_m=Molecular volume

Greek symbols

$\delta \equiv 2\sigma v_m / k_B T$
$\delta_1 \equiv \sigma_1 A_m / k_B T$
$\delta_2 \equiv \sigma_1 A_m / k_B T$
$\lambda_s \equiv 4A_s / dr_c^4 / dt$
$\lambda_k \equiv 3A_k / dr_c^3 / dt$
$\lambda_{i_2} \equiv 2A_{i_2} / dr_c^2 / dt$
$\lambda_i \equiv 2A_i / dr_c^2 / dt$
$\lambda_d \equiv 3A_d / dr_c^3 / dt$
$\rho_j = r_j / r_c$
σ=Interfacial tension between the solid and liquid phase
σ_1=Line tension

[1] I. M. Lifshitz and J. Slyozhov, J. Phys. Chem. Solids **19**, 35 (1961).
[2] C. Wagner, Z. Elektrochim. **65**, 243 (1961).
[3] G. Venzl, Ber. Bunsenges Phys. Chem. **87**, 976 (1984).
[4] M. P. Shepilov, J. Non-Cryst. Solids **146**, 1 (1992).
[5] J. A. Marquesee and J. Ross, J. Chem. Phys. **79**, 373 (1983).
[6] J. A. Marquesee, J. Chem. Phys. **81**, 976 (1984).
[7] J. A. Marquesee and J. Ross, J. Chem. Phys. **80**, 536 (1984).
[8] K. Tsumuraya and Y. Miyata, Acta Metall. **31**, 437 (1983).
[9] J. H. Yao, K. R. Elder, H. Guo, and M. Grant, Phys. Rev. B **45**, 8173 (1992).
[10] T. Ahn and J. K. Tien, J. Phys. Chem. Solids **37**, 771 (1976).
[11] B. K. Chakraverty, J. Chem. Solids **28**, 2401 (1967).
[12] E. Ruckenstein, in *Metal-Support Interactions in Catalysis, Sintering and Redispersion*, edited by S. A. Stevenson, J. A. Dumesic, R. T. K. Baker, and E. Ruckenstein (Van Nostrand Reinhold, New York, 1987).
[13] L. L. Evert, D. Leckband, and J. Israelachvili, Langmuir **10**, (1994).
[14] D. B. Dadyburjor and E. Ruckenstein, J. Cryst. Growth **40**, 279 (1977).
[15] K. Binder and D. Stauffer, Adv. Phys. **25**, 343 (1976).
[16] K. Binder, Phys. Rev. B **15**, 4425 (1977).
[17] H. H. Lee, J. Catalysis **63**, 129 (1980).

Mechanisms of Aging of Supported Metal Catalysts

ELI RUCKENSTEIN AND DADY B. DADYBURJOR

Faculty of Engineering and Applied Sciences, State University of New York at Buffalo, Buffalo, New York 14214

Received August 31, 1976; revised February 28, 1977

The behavior of a population of small metal crystallites on a substrate is accounted for by the splitting of single metal atoms and multiatom particles from larger particles, their diffusion along the substrate and the coalescence of particles (or of a particle and an atom) that collide. The equations are stochastic, allowing for the treatment of a large number of particles. The rate at which one particle is captured by another is determined by the rate of diffusion of particles from the bulk to the interface of the other particles and the kinetics of the capture process at the interface. The rate of emission is proportional to the equilibrium concentration, on the original particle interface, of particles having the size of the particle to be emitted. If emission is an activated process, the probability of multiatom emission is negligibly small compared to single atom emission. Hence, even if particles can lose single atoms to the substrate by activated breaking of bonds the loss of aggregates of two or more atoms must occur via another mechanism. This alternate process is not entirely clear but is probably related to the existence of cracks on the crystallite and their propagation due to internal stresses caused either thermally or by the formation of a chemical compound. The present model can account for an increase or a decrease in the total surface area with time, as well as an initial increase followed by a decrease or an initial decrease followed by an increase. In any particular case the path followed depends upon the metal loading, the saturation concentration of metal atoms, the diffusion rate and the emission rate.

NOMENCLATURE

A_0	Preexponential factor in capture, emission coefficients, Eqs. (12a), (15), (17)
A_j	Radius of the area "governed" by a j-atom particle, Eqs. (18a) and (22)
a	Lattice parameter
C_j, C_j'	Creation rate of j-atom particles from particles of larger, smaller, size, Eqs. (5) and (8)
$C_{j,1}$	Rate of creation of j-atom particles by capture or emission of single atoms, Eq. (3)
$\bar{C}_p$	Concentration of metal atoms on facet of nucleating crystal
$D_{j,i}$	Diffusion coefficient of i-atom particle near a j-atom particle, Eq. (18a)
$E_{a,j,i}$, $E_{b,j,i}$	Activation energies for emission, capture of an i-atom particle from a j-atom particle, Eqs. (12), (15), (17)
E_{mm}	Energy required to remove a metal atom surrounded by other metal atoms, Eq. (12b)
E_{ms}	Energy required to remove metal atom surrounded by atoms of substrate species, Eq. (12b)
G	Growth rate of a crystal
ΔG_c	Free energy of formation of a crystal nucleus
ΔH_c	Migration energy of an atom over a crystal

This paper, which was originally published in **Journal of Catalysis**: 48, 73–86 (1977), is republished with permission.

i, j	Number of atoms in a particle
$J_{j,i}$	Diffusion flux of i-atom particles from j-atom particles, Eqs. (18)
K	Rate constant of sintering equation, Eq. (1)
k	Boltzman constant
$k_{a,j,i}$	Coefficient of emission of an i-atom particle from a j-atom particle, Eq. (15)
$k_{b,j,i}$	Coefficient of capture of an i-atom particle by a j-atom particle, Eq. (17)
L_j, L_j'	Loss rate of j-atom particles to form particles of larger, smaller size, Eqs. (6) and (7)
$L_{j,1}$	Rate of loss of j-atom particles by capture or emission of single atoms, Eq. (2)
m	Exponent of the size dependence of the diffusion coefficient on size, Eq. (23)
$n_j(t)$, $n(j, t)$	Concentration of j-atom particle at time t
$n_{1\infty}^s$	Equilibrium concentration of single atoms at the interface of an infinitely large crystallite
$n_{i,j}^s$	Equilibrium concentration of i-atom particles at the interface of a j-atom particle
$n_{i,j}^0$	Concentration of i-atom particles at the interface of a j-atom particle
N_0	Total number of metal atoms present per unit area of substrate, Eq. (10)
P^+	Probability of atoms arriving at the surface of a crystal
P_q^-	Probability of an atom leaving a crystal surface with q occupied sites
p	Index of sintering equation, Eq. (1)
p_q	Ratio of P_q^- to P^+
Q	Maximum number of sites on crystal surface
q	Number of sites occupied on crystal surface
R	Gas constant
r_j	Radius of a j-atom particle
r_{crit}	Critical radius of metal species, Eq. (14c)
S	Total exposed surface area of metal at time t
S_0	Total exposed surface area of metal at $t = 0$
t	Time
T	Temperature
V_m	Molar volume of metal
$\alpha_{j,i}$	Emission rate; number of i-atom particles emitted from a j-atom particle per unit time
$\beta_{j,i}$	Capture rate; number of i-atom particles captured by a j-atom particle per unit time
δ	Constant in Eq. (14), $2\sigma V_m/RT$
δ_{ij}	Kronecker delta; $=1$ if $i = j$, $=0$ otherwise
ν_p	Vibration of an atom on a crystal
π	3.14159
ρ_c	Radius of crystal nucleus
σ	Interfacial free energy between metal and substrate per unit area

INTRODUCTION

Supported metal catalysts consist of metal crystallites 10 to 100 Å in size dispersed upon the internal surface of a porous substrate. The catalyst loses its activity after excessive heating and during the catalytic process. In the absence of chemical "poisoning," this aging phenomenon is generally recognized as being due to sintering or loss of active surface area by coalescence of the metal crystallites to form larger particles. Experimental evidence for sintering has been reviewed in a number of recent papers (*1–3*). Under certain conditions, however, sintered crystallites have been shown to split and redisperse, thus regaining the activity lost by coalescence. The results of the splitting process have been observed directly by using transmission electron microscopy (*4*).

Some indirect experimental evidence exists (*5–9*) that small metal crystallites in a model system migrate over the surface of substrate. The model system contained a high density of metal islands.

Assuming that sintering occurs because of the migration of the islands, an effective diffusion coefficient D of 10^{-15} cm^2/sec was obtained (*2*) for 10 Å gold particles on a silicon substrate. For a commercial catalyst containing a total of N_0 metal atoms per surface area of the substrate, uniformly distributed as j-atom particles, the distance between particles can be written as $(j/N_0\pi)^{\frac{1}{2}}$. Assuming that metal particles in a commercial catalyst diffuse in a manner similar to that of particles in the model system, the time t taken for two particles to diffuse towards one another can be obtained by equating this distance to $2(Dt)^{\frac{1}{2}}$. For a metal loading N_0 equal to 1×10^{13} atoms/cm^2 of substrate and uniform particles containing 16 atoms (corresponding to 0.5% Pt/Al_2O_3 containing particles of 10 Å) the distance is approximately 70 Å. If the diffusion coefficient equals that above for the model system, the time involved is only a few minutes. Hence, diffusion of metal crystallites may play a part in the overall sintering of the supported metal catalysts.

A crystallite has many bonds between metal and substrate atoms that would have to be broken before the crystallite can diffuse along the surface, while a free metal atom requires fewer bonds to be broken. Hence, it might be supposed that crystallite migration would be slow compared to the motion of free atoms. However, the spacings of metal and substrate atoms are such that only a fraction of metal atoms on the crystallite base are in an energy well between atoms of the substrate. On the other hand, free metal atoms can almost always be placed without hindrance between substrate atoms. Hence, although the number of metal–substrate bonds may be much greater in the case of a large crystallite, the intensity of each bond is generally less than that for a single atom. This makes possible some motion of metal crystallites on the substrate. Further, by virtue of their size, crystallites may be able to overcome obstacles such as valleys on the irregular substrates that might trap and immobilize the smaller single atoms.

Of course migration is size dependent and will be very small for particles larger than about 50 Å. Assuming that the diffusion coefficient is proportional to the base area of the crystallite, then for a 0.5% Pt/Al_2O_3 system containing 10^2 Å particles uniformly distributed, the distance between particles is of the order of 10^3 Å. In this case, the time for particles to diffuse towards each other is of the order of 10^2 hr (as opposed to a few minutes calculated above for 10 Å particles). However, for systems containing both large and small metal crystallites, the larger crystallites, despite their own lack of mobility, would nevertheless increase in size when smaller particles collided against them and coalesced. Recent controlled atmosphere electron microscopy (CAEM) experiments cannot observe particles smaller than approximately 25 Å (*10, 11*); the fact that no migration could be observed here is perhaps due to this size limitation.

A phenomenological model was developed (*1*) based upon migration of particles and coalescence of particles that collide. Solution of the kinetic equations led to a family of rate equations for the total exposed surface area S of metal of the form,

$$\frac{dS}{dt} = -KS^p. \tag{1}$$

The index p was found to have values dependent upon the relative magnitudes of the diffusion coefficient and the coalescence constant as well as the particular size dependencies used for these parameters. Analysis of experimental work from other laboratories yielded values of p from 2 to 8 (*12–15*). These values correspond to different, but feasible, size dependencies for the model parameters. For example, $p = 2$ corresponds to coalescence controlled process with the coalescence constant proportional to particle radius. On the other hand, $p = 6$ corresponds to a diffusion controlled process with the diffusion coefficient inversely proportional to the square of the particle radius. If the diffusion coefficient is inversely proportional to higher powers of the radius then larger values of p can be obtained.

Wynblatt and Gjostein (*3*) comment that such a model is valid for particles smaller than about 50 Å but invalid for larger particles because the latter cannot move. We disagree with this. As mentioned above, the size dependence of the diffusion coefficient in the diffusional model accounts for the fact that the larger particles have a negligible migration. However, the larger particles grow because the smaller particles, which migrate much faster, would coalesce with them.

Wynblatt and Gjostein (*16*) obtained values of p equal to about 13 in their sintering experiments on supported Pt. They attempted to explain this by starting with an alternate mechanism, whereby large particles grow while smaller particles decay. This occurs via capture and emission of single atoms which are the only mobile species. However, this Ostwald ripening-like phenomenon could yield only values of p that were even smaller ($p = 3$ to 5) than those of the migration and coalescence model (*1*). In later work (*17*) they have attempted to resolve the differences between experiment and theory by considering an additional, slow, nucleation step in the growth of the larger particles during the ripening process.

Flynn and Wanke (*18*) also considered only single atom emission and transfer, but the formulation of their equations was different. They used a model system containing a relatively small number of particles. They calculated the individual rates of single atom emission and capture of each particle. The initial absence of single atoms in their systems restricted their results to predicting either total redispersion, i.e., a continuous increase in the exposed surface area (for metal loadings less than the saturation concentration of the metal atoms), or an initial increase followed by a decrease in the exposed surface area.

Allowing only single atoms to move on the substrate would seem like too strong a restriction. If monomers can move on the surface, it might be argued that dimers could also be expected to move, albeit at a marginally slower rate, and so on. Flynn and Wanke (*18*) assume that all particles of less than 14 atoms are inherently unstable and reduce instantaneously to single atoms, but this comes about by assuming that all particles are in the form of fcc crystal structures. Certainly small clusters of atoms will not have the same structure as the bulk metal, but this does not mean that they cannot exist in some other form. Further, Geus (*19*) observes that many experimental results indicate that dimers are stable on a substrate.

In this work we suggest a model for the behavior of metal crystallites supported on a substrate that entails the emission, migration and coalescence of particles of every size, including single atoms. In this manner we have an alternate mechanism to explain large values of p without having to postulate what in the model of Ref. (*1*) might be unfeasible size dependences of the diffusion coefficient. Instead, these large values of p are explained by allowing the emission of particles as a mechanism that offsets the capture of particles. In contrast to the model of Ref. (*1*), the present model will also predict particles of smaller size than the smallest initial particle. We show later that the present model can predict a continuous decrease in surface area with time and a decrease followed by an increase in surface area, as well as continuous increase in surface area with time and an increase followed by a decrease. The behavior depends upon the relative magnitudes of emission, diffusion, and capture, although the final state depends upon the metal loading and the saturation concentrations of the metal species on the substrate.

METHODS

1. Single Atom Emission and Capture

For simplicity we consider first the equations of emission and capture of single atoms. They will be extended to the more general case of multiatom transfer below.

The total rate at which particles containing j atoms are lost by capture or emission is

$$L_{j,1} = (\beta_{j,1} + \alpha_{j,1})n_j. \qquad (2)$$

Here the first term represents the number of j-mers per unit time that capture an atom to become $(j + 1)$-mers, and the second term represents the number of j-mers per unit time that become $(j - 1)$-mers by the loss of an atom. The creation rate of j-mers is given by

$$C_{j,1} = \beta_{j-1,1}n_{j-1} + \alpha_{j+1,1}n_{j+1}. \qquad (3)$$

Now since the net rate of change of j-mers can be written as

$$\frac{dn_j}{dt} = C_{j,1} - L_{j,1},$$

we get for the continuity equation for the j-atom particles

$$\frac{dn_j}{dt} = (\alpha_{j+1,1}n_{j+1} - \beta_{j,1}n_j) - (\alpha_{j,1}n_j - \beta_{j-1,1}n_{j-1}). \qquad (4)$$

In the Appendix we derive the differential form of Eq. (4) and note that it is different from the continuity equation of Lifshitz and Slyosov (*20*) frequently used in the treatment of Ostwald ripening [see for example Ref. (*21*)].

It is important to realize that Eq. (4), considering as it does solely the emission and capture of single atoms, is a more rigorous formulation of the mechanism of Wynblatt and Gjostein (*16*) or that of Flynn and Wanke (*18*) that leads in supersaturated situations to Ostwald ripening.

2. The Migration–Coalescence–Emission Model

We now extend each term in Eq. (4) to include multiatom processes. The first term in Eq. (4) represents the rate at which the concentration of j-atom particles is increased by emission of atoms from larger particles. The corresponding term for multiatom transfer is

$$C_j = \sum_{i=1}^{\infty} (1 + \delta_{ij})\alpha_{j+i,i}n_{j+i}, \qquad (5)$$

where the Kronecker delta, δ_{ij}, accounts for the formation of two j-atom particles if $i = j$. The second term in Eq. (4) is the rate at which j-atom particles are lost when they capture atoms. Noting that two j-atom particles are lost when $i = j$, we write

$$L_j = \sum_{i=1}^{\infty} (1 + \delta_{ij})\beta_{j,i}n_j, \qquad (6)$$

as the loss rate of j-atom particles to larger particles. The emission of i atoms by j-atom particles, described for single atom transfer by the third term in Eq. (4), is equivalent to the emission by the same particle of $(j - i)$ atoms. To avoid duplication we write the corresponding term as

$$L_j' = \tfrac{1}{2} \sum_{i=1}^{j-1} \alpha_{j,i}n_j. \qquad (7)$$

The final term in Eq. (4) represents the increase in the concentration when smaller particles capture the required number of atoms. Since the capture of an i-atom particle by a $(j-i)$-atom particle is identical to the capture of a $(j-i)$-atom particle by an i-atom particle we write, following Eq. (7),

$$C_j' = \tfrac{1}{2} \sum_{i=1}^{j-1} \beta_{j-i,i} n_{j-i}. \quad (8)$$

From Eqs. (5) through (8), the rate of change of the concentration of j-atom particles is given by

$$\frac{dn_j}{dt} = \sum_{i=1}^{\infty} (1 + \delta_{ij}) \alpha_{j+i,i} n_{j+i} + \tfrac{1}{2} \sum_{i=1}^{j-1} \beta_{j-i,i} n_{j-i} - \sum_{i=1}^{\infty} (1 + \delta_{ij}) \beta_{j,i} n_j - \tfrac{1}{2} \sum_{i=1}^{j-1} \alpha_{j,i} n_j. \quad (9)$$

We have implicitly assumed that this form of the continuity equation for the j-atom particle is valid only for aggregates of atoms and not for single atoms; that is, for $j \geq 2$. The concentration of single atoms emerges from the additional equation,

$$\sum j n_j(t) = N_0, \quad (10)$$

where N_0 is the metal loading, i.e., the total number of metal atoms present per unit area of the substrate. Equation (10) ensures the conservation of metal atoms.

In reality the surface of the substrate is nonhomogeneous from a geographic and energetic point of view. These nonhomogeneities trap some of the mobile particles, for example those caught in a grain boundary of the substrate, or those trapped by acid sites or impurities on the substrate. A treatment has been developed (*1*) to take into account the behavior of these fixed particles by modifying the rate equation to consider the probabilities of formation of fixed and mobile particles. However, we do not consider this in the present paper.

3. Evaluation of Capture and Emission Rates

Metal crystallites on a substrate, like crystals in a solution, may be subjected to a number of complicated processes which lead to increases or decreases in the crystallite size. In the previous section we have simply written the rate of emission of i atoms from a j-atom particle as $\alpha_{j,i}$ and the rate of capture of i atoms by a j-atom particle as $\beta_{j,i}$; however, the evaluation of these terms will depend upon the processes by which capture and loss take place.

A. Single Atom Transfer

Let us first consider some mechanisms by which crystallites may grow. Jackson's model (*22*) involved a crystal containing a maximum of Q sites on its surface. A crystal with, say, q sites occupied gains and loses atoms until Q sites are occupied. This is, of course, identical to having no occupied sites on the next surface, and so the process continues in this manner. If the probability, P^+, of atoms arriving on the surface is independent of the number of sites occupied, the steady state volume rate of growth of the crystal is

$$G = QP^+(1 - p_0 p_1 \cdots p_{Q-1}) / \left\{ Q + \sum_{q=0}^{Q-1} (p_q) + \sum_{q=0}^{Q-1} (p_q p_{q+1}) + \cdots + \sum_{q=0}^{Q-1} (p_q p_{q+1} \cdots p_{q-2}) \right\},$$

where $p_q \equiv P_q^- / P^+$ is the ratio of the rates at which a surface with q occupied sites loses and gains atoms. Hence, the rate at which the crystal grows by capturing single atoms is proportional to the rate at which atoms approach the surface and hence to the concentration of single atoms near the surface.

For a crystal with facets, growth may be slow until a certain number of sites are occupied (a nucleus is formed), after which there is rapid formation of a monolayer surrounding the original crystal. Growth will then be reduced until a nucleus is formed again on the new crystal. If nucleation is the rate determining step, the volume rate of growth may be written as (*17, 23*)

$$G = 4\pi r_j^2 \left(\frac{\Delta G_c}{\pi RT}\right)^{\frac{1}{2}} \exp\left[-\frac{\Delta G_c}{RT}\right] \frac{V_m}{\rho_c a} \nu_p \times \exp\left[-\frac{\Delta H_p}{RT}\right] \tilde{C}_p,$$

where ΔG_c and ρ_c are the free energy and radius of a nucleus, a is the lattice parameter of the substrate, ν_p and ΔH_p are the vibration frequency and migration energy of an atom over the crystal and $\tilde{C}_p$ is the concentration of single atoms on the facet.

In either of these cases, the volume growth rate of the crystal, i.e., the rate at which it captures single atoms, depends upon the concentration of single atoms. In a formal sense, therefore, no rigor is lost if the usual condensation equation is used to obtain $\beta_{j,1}$. Here $\beta_{j,1}$ depends upon the perimeter of the j-atom particle and the interfacial concentration of single atoms, i.e.,

$$\beta_{j,1} = k_{b,j,1} 2\pi r_j n_{1,j}^0, \qquad (11)$$

where $k_{b,j,1}$ is a rate constant and $n_{1,j}^0$ is the monomer concentration at the interface of the particle. In Eq. (11) the perimeter of the single atom is ignored with respect to that of the larger particle. The rate constant can be written

$$k_{b,j,1} = A_0 \exp[-E_{b,j,1}/kT]. \qquad (12a)$$

The activation energy $E_{b,j,1}$ can be expressed in terms of metal–metal and metal–substrate energies, E_{mm} and E_{ms}, as

$$E_{act,j,1} = \tfrac{1}{2}(E_{mm} + E_{ms}). \qquad (12b)$$

The factor $\frac{1}{2}$ is considered because the number of nearest neighbors is one half of those corresponding to the bulk solid.

When the diffusion of metal atoms on the substrate is very fast, the concentration of single atoms is uniform so that $n_{1,j}^0 = n_1$, the bulk concentration of single atoms.

Analogous to the condensation relation, Eq. (11), we can write the evaporation equation for the loss of single atoms from a j-atom particle to the substrate as

$$\alpha_{j,1} = k_{a,j,1} 2\pi r_j n_{1,j}^s, \qquad (13)$$

where $n_{1,j}^s$ is the concentration of single atoms in equilibrium with a j-atom particle. This concentration is given by the well known Gibbs–Thomson relation

$$n_{1,j}^s = n_{1\infty}^s \exp\left[\frac{2\sigma V_m}{RTr_j}\right] \qquad (14a)$$

$$\simeq n_{1\infty}^s \left(1 + \frac{\delta}{r_j}\right). \qquad (14b)$$

Here r_j is the particle radius and $n_{1\infty}^s$ refers to the equilibrium solubility of the monomer species, i.e., monomer concentration in equilibrium with an infinitely large crystallite.

From Eqs. (11) and (13), with $n_{1,j}^0 = n_1$, when the bulk monomer concentration equals the equilibrium concentration at the interface of some particle, then growth of the particle by addition of single atoms is exactly countered by the decay of the particle by loss of single atoms. Such a particle, which suffers no change in size at this point, is said to have a critical size. The critical radius is given by

$$r_{crit} \equiv \delta n_{1\infty}^s/(n_1 - n_{1\infty}^s). \qquad (14c)$$

The critical radius is significant only when $n_1 > n_{1\infty}^s$, i.e., in supersaturated systems. Under these conditions, particles of radius smaller than r_{crit} will undergo a net loss by monomer transfer while particles larger than r_{crit} will undergo a net gain by monomer transfer. This is of course the mechanism of Ostwald ripening.

B. Multiatom Transfer

The rate of emission of multiatoms from larger particles to the substrate can perhaps be written similar to Eq. (13), viz,

$$\alpha_{j,i} = k_{a,j,i}2\pi r_j n_{i,j}^s, \quad (15a)$$

where $k_{a,j,i}$ is the rate constant of the activated step,

$$k_{a,j,i} = A_0 \exp[-E_{a,j,i}/kT], \quad (15b)$$

and $E_{a,j,i}$ is the activation energy. Chen (*24*) has computed values of $E_{a,j,i}$ for various values of i and j. The activation energy for emission of dimers, for instance, from j-atom particles was found to be about twice that for single atom emission; since $E_{a,j,i}$ is exponentially related to $\alpha_{j,i}$, the rate of emission of dimers is many orders of magnitude smaller than the corresponding rate for single atoms. Hence, even when emission of single atoms occurs by an activated process, emission of $i \geq 2$ atom particles by a similar process is probably not to be expected. A more plausible mechanism of multiatom emission is one associated with fracture of the particles because of the propagation of some preexisting cracks. The detailed mechanism by which this fracture takes place has not been completely understood but we suspect that in some conditions some compounds are formed in the metal crystallites. This generates stresses in the crystallites which consequently fracture. It is not yet possible to model the dynamics of this process. Since there may be conditions in which no fracture occurs, we consider here the emission of only single atoms.

Analogous to the condensation equation for the capture of single atoms, Eq. (11), the rate of capture of an i-atom particle by a j-atom particle can be written

$$\beta_{j,i} = k_{b,j,i}2\pi(r_j + r_i)n_{i,j}^0, \quad (16)$$

where $k_{b,j,i}$ is the rate constant of the kinetics of combination of two adjacent particles, given by

$$k_{b,j,i} = A_0 \exp[-E_{b,j,i}/kT]. \quad (17)$$

In Eq. (16) $n_{i,j}^0$ is the concentration of i-atom particles at the interface of the j-atom particle. Hence, n_{ij}^0 depends upon the rate at which particles diffuse towards one another as well as the rate at which touching particles combine.

To obtain an expression for $n_{i,j}^0$, consider the steady state diffusion flux of i-atom particles from a j-atom particle, given in Ref. (*1*) as

$$J_{j,i} = \frac{2\pi D_{ji}}{\ln[A_j/(r_i + r_j)]}(n_{i,j}^0 - n_i). \quad (18a)$$

Here A_j is a measure of the area "governed" by a particle of size j. Under quasi steady state conditions, $J_{j,i}$ is equal to the rate at which $(j + i)$-atom particles lose i atoms, less the rate at which j-atom particles capture i-atom particles. Hence

$$J_{j,i} = \alpha_{j+i,i} - \beta_{j,i}. \quad (18b)$$

At equilibrium $J_{j,i} = 0$ and, by definition, the interface concentration of i-atom particles is the equilibrium concentration. Equations (15), (16) and (18b) now yield

$$k_{a,j+i,i}r_{j+i} = k_{b,j,i}(r_i + r_j). \quad (18c)$$

This is of course the principle of microscopic reversibility and must be valid even far from equilibrium.

Using Eqs. (15a) and (16) and equating Eqs. (18a) and (18b), the concentration of i-atom particles at the interface of a j-atom particle is

$$n_{ij}^0 = \frac{\{k_{a,j+i,i}r_{j+i}\}n_{i,j+i}^s + \{D_{j,i}/\ln[A_j/(r_i + r_j)]\}n_i}{k_{b,j,i}(r_i + r_j) + D_{j,i}/\ln[A_j/(r_i + r_j)]}, \quad (19a)$$

or if $j \gg i$

$$= \frac{\left\{\frac{\ln [A_j/(r_i + r_j)]}{D_{j,i}}\right\} n_{i,j}{}^s + \left\{\frac{1}{k_{b,j,i}(r_i + r_j)}\right\} n_i}{\frac{\ln [A_j/(r_i + r_j)]}{D_{j,i}} + \frac{1}{k_{b,j,i}(r_i + r_j)}}. \tag{19b}$$

Equation (19) reduces to the two special cases $n_{i,j}{}^0 \simeq n_i$ when the kinetics at the interface controls the process (fast diffusion), and $n_{i,j}{}^0 \simeq n_{i,j}{}^s$ when diffusion controls.

Substituting Eq. (19a) into Eq. (13a) yields

$$\beta_{j,i} \sim 2\pi k_{b,j,i}(r_i + r_j)n_i, \tag{20a}$$

when the kinetics at the interface control the capture process. Now, if the activation energy for emission, $E_{a,j,i}$ in Eq. (15), is such that multiatom emission is not possible and if n_i is not very much larger than $n_{i,j+i}{}^s$ then, from microscopic reversibility [Eq. (18c)] and Eq. (20a), there can be no capture of multiatom particles when the interface kinetics control the capture process. Hence, there exists only single atom emission and capture in these circumstances (Ostwald ripening if the system is supersaturated). When diffusion controls the capture process, i.e., when

$$D_{j,i}/\ln [A_j/(r_i + r_j)] \ll k_{b,j,i}(r_i + r_j), \tag{20b}$$

the capture rate is given by

$$\beta_{j,i} \sim \frac{2\pi D_{j,i}}{\ln [A_j/(r_i + r_j)]} n_i + k_{a,j+i,i} 2\pi r_{j+i} n_{i,j+i}{}^s. \tag{20c}$$

Further, if the system is highly supersaturated, i.e., if $n_i \ggg n_{i,j+i}{}^s$, then in spite of inequality (20b) and microscopic reversibility, Eq. (18c), the capture rate can be given by

$$\beta_{j,i} \sim \frac{2\pi D_{j,i}}{\ln [A_j/(r_i + r_j)]} n_i.$$

The emission rates are negligible due to the small values of $n_{i,j+i}{}^s$. In such circumstances we have only the migration and capture of atoms and particles, i.e., the model of Ref. (*1*).

In many cases we may have the single atom concentration, n_1, of the same order of magnitude, or smaller, than the saturation, although n_i is still much greater than $n_{i,j+i}{}^s$ for $i \geq 2$. In such circumstances we have the occurrence of single and multiatom capture and single atom emission.

4. Illustrative Example

The differential-algebraic equations (9) and (10) can be solved numerically for given initial conditions to obtain concentrations $n_j(t)$ of j-atom particles at various times t. Since the surface area of all particles containing j atoms is proportional to $n_j j^{\frac{2}{3}}$, the total surface area at time t can be written as a fraction of the initial surface area

$$\frac{S(t)}{S_0} = \sum_j j^{\frac{2}{3}} n_j(t) / \sum_j j^{\frac{2}{3}} n_j(0), \tag{21}$$

from a knowledge of the size distribution. This fractional surface area is of interest here, rather than the size distribution, since the former affects the catalytic activity more directly and can be more easily measured.

For our example we consider the behavior of a system consisting initially of particles containing 50 atoms each. Such a system may be unrealistic both because of the smallness of the particles and the fact that they are all of the same size. However, we intend here to demonstrate only the inter-

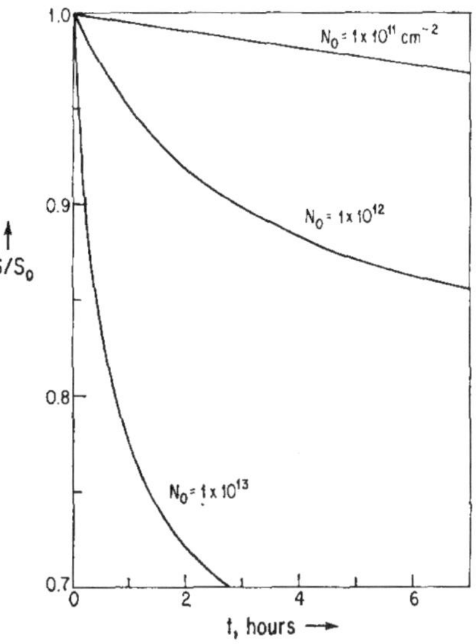

FIG. 1. Sintering curves for various values of N_0 (atom/cm^2) of metal crystallites on a substrate. The initial system contains only 50-atom particles of metal. Capture of all sized particles is allowed but only single atoms are emitted. The capture process is controlled by diffusion, $D_{j,i} = D_j + D_i$, where $D_k \alpha r_k^{-2}$. Particles are hemispherical, $r_j \alpha j^{\frac{1}{3}}$. The parameters used are $D_{16} = 2 \times 10^{-15}$ cm^2/sec, $\delta = 12$ Å, $V_m = 9$ cm^3, $n_{1\infty}{}^s = 10^3$ cm^{-2}, $T = 10^3$°K, $A_0 = 10^4$ cm/sec, $E_{mm} = 130$ kcal/mole, $E_{ms} = 90$ kcal/mole. It is not clear whether the curves will follow the trends shown here at very large times.

relationships between terms in Sec. 3, and these are perhaps better shown on a simple system. The nature of Eqs. (9) and (10) is such that every size (i.e., every value of j) must be accounted for; hence, small enough particles are essential for a reasonable amount of computer time. No significant differences are expected if a larger particle size is used initially. Further, the formulation of Eq. (10) is stochastic, differing from those in Refs. (*16*, *18*). For a system of particles initially of one size, the latter formulations yield changes in particle sizes with time, but at any one time the systems always contain identical particles. In contrast, Eq. (10) allows us to start with a system of identical particles and to obtain a spectrum of sizes after a period of time. Since our formulation does not lead to singular results when starting with a system containing particles of only one size, it is advantageous to use such a system for the sake of simplicity and ease in analyzing the results.

All particles in the model system are assumed to be perfect hemispheres. In this way the radius of a j-atom particle is proportional to $j^{\frac{1}{3}}$.

The effective diffusion coefficient for two particles $D_{j,i}$ is the sum of the coefficients for each particle, $D_i + D_j$. It is reasonable to expect that the diffusion coefficient D_i is inversely proportional to the base area of a particle of i atoms, hence $D_{j,i}$ is set proportional to $r_i^{-2} + r_j^{-2}$.

The parameters A_j may be obtained by assuming that the area πA_j^2 of the substrate associated with each particle of j atoms is proportional to the area of the base of the particle. Then A_j is the equivalent radius of the corresponding subdivision so that

$$A_j^2 = r^2{}_j/\pi \sum_k n_k r_k^2. \tag{22}$$

Figure 1 illustrates the behavior of our sample system as plots of the fractional change in surface area of crystallites, S/S_0, against time, t, at a temperature of 1000°K. If the metal–metal bond energy can be related to the heat of sublimation, then E_{mm} is of the order of 100 kcal/gmole. The values used in Fig. 1 are 130 kcal/gmole for E_{mm} and 90 kcal/gmole for E_{ms}. The value of δ was taken as 12 Å, consistent with values for the Pt/Al_2O_3 system. The diffusion coefficients $D_{j,i}$ were obtained from the earlier size dependence assuming that $D_i = 2 \times 10^{-15}$ cm^2/sec for i corresponding to a particle of 10 Å diameter. The pre-exponential terms A_0 in Eqs. (15) and (17) were taken to be 1×10^4 cm/sec. The equilibrium concentration of single atoms at the interface of an infinitely large particle, $n_{1\infty}{}^s$, was taken as 10^3 cm^{-2}. The molar volume, V_m, of Pt metal is 9 cm^3 (*25*), so that $r_1 = 2$ Å. These values yield diffusion as the controlling step in the overall diffusion and capture process. Single and multiatom particles are captured but there is emission of only single atoms.

In Fig. 1 the sintering is relatively small for smaller values of N_0, the metal loading, but increases with larger values of N_0. The values of the sintering index p [Eq. (1)], obtained by the best fit of the curves of Fig. 1 to the form of Eq. (1), changes with time. The initial value of p and the values after approximately 3 and 5 hr are shown in Table 1 for the three curves. Interestingly, though the curves of Fig. 1 are so disparate, the initial values of p are approximately the same in two cases. This shows that N_0 affects K, the rate constant of Eq. (1), to a significantly greater extent than it affects p.

TABLE 1

Value of the Sintering Index p from Fig. 1[a]

N_0 (atoms/cm²)	p		
	$t = 0$	$t = 3$	$t = 5$
1×10^{11}	10.5	10.3	10.3
1×10^{12}	10.6	18.3	24.5
1×10^{13}	7.6		

[a] The time t is in hours.

Consider now the process of Ref. (*1*), which entails only migration and capture and does not involve any form of emission. Under these conditions, and with diffusion controlling the overall process, a relationship was derived between p and m, the exponent of the size dependence of the diffusion coefficient

$$p = 4 - 3m. \tag{23}$$

If m is set equal to $-\frac{2}{3}$, the value used in the present computations, the value of p is 6. The increase in the value of p between this value and those of Table 1 demonstrates the improvement that may be expected in the fitting of experimental data when the emission of single atoms is added to the model of migration and capture of single atoms and multiatom particles.

DISCUSSION

At least initially, the sign of the change of surface area depends upon the relative rates of two processes. The first is the formation of 100-atom particles by collision of two 50-atom particles. The second is the splitting of 50-atom particles into monoatoms and 49-atom particles. Small enough values of N_0 result in large average distances between the initial particles and small probabilities that one particle approaches another. Since diffusion controls the capture process, there will be only a small rate of formation of the 100-atom particles so that there is a lesser decrease of surface area by capture. On the other hand, N_0 and the $D_{j,i}$ have virtually no effect on the initial emission of single atoms. The net result is an increase in the total surface area with time. Larger values of N_0 or the $D_{j,i}$ may increase the rate of the capture process sufficiently to lead to a decrease in the total surface area with time.

The present model and those models involving only monoatomic transfer differ in the paths available to the final (equilibrium) states. If $N_0 > n_{1\infty}^s$, the final state consists of a single large particle of radius r surrounded by single atoms having a concentration $n_1^s(r)$ as given in Eq. (5); if $N_0 < n_{1\infty}^s$, the final state consists of only single atoms. If only single atoms are transferred, and if the initial distribution contains no single atoms, the case of Flynn and Wanke (*18*), the possible paths correspond in Fig. 2 to curves A ($N_0 < n_{1\infty}^s$) and C ($N_0 > n_{1\infty}^s$).

In contrast, if multiple atom transfer is allowed, an additional element is considered in the overall process. In this case, assuming diffusion controls the capture process, the relative magnitudes of the diffusion coefficients and the emission coefficients determine the path followed by the system, although of course only $n_{1\infty}^s$ and N_0 determine the final state. Thus, curves A and C correspond to the paths if the

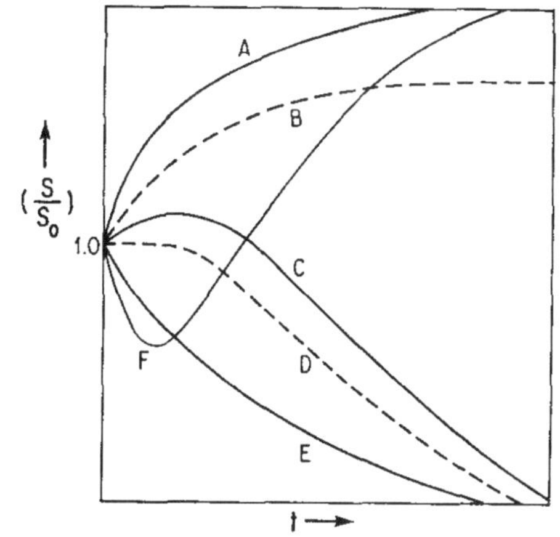

FIG. 2. Possible paths for sintering or splitting supported metal crystallites depend at small times upon the relative magnitudes of emission, migration and capture of metal species, and at large times upon the relative magnitudes of loading and saturation concentration of metal species. (A) The initial increase in surface area by emission is greater than the initial decrease by migration and capture, metal loading is less than saturation concentration. (B) Emission overrides capture initially, metal loading equals saturation concentration. (C) Emission overrides capture initially, metal loading is greater than saturation concentration. (D) Emission balances capture initially, metal loading is greater than saturation concentration. (E) Capture overrides emission initially, metal loading is greater than saturation concentration. (F) Capture overrides emission initially, metal loading is less than saturation concentration.

decrease in surface area by diffusion of particles and their coalescence is overshadowed by the increase in surface area by emission of particles. For the case of metal loading less than saturation concentration, this situation continues until finally only single atoms are left on the substrate (curve A). For larger metal loadings, emission causes an initial increase in surface area. This is followed by a decrease as concentrations of smaller particles increase, leading to an increase in the diffusion process and hence an increase in the capture rate, as shown in curve C. Curves E and F correspond to the paths for large diffusion coefficients. In curve E, where the metal loading is greater than the saturation, the large diffusion rates cause rapid agglomeration and the final state consists of one large particle and the equilibrium concentration of single atoms. If, however, the metal loading is small, the diffusion and capture process eventually slows down as particles grow larger and are further apart. Emission will gradually take over, as in curve F, and the final state should consist of only single atoms on the substrate.

Even though the final state is dictated by the metal loadings, the time required for this state to be approached may be very large. Hence, a supported metal catalyst corresponding to curve F may only sinter during its effective life. Similarly the catalyst of curve C may effectively not sinter at all.

CONCLUSIONS

This work presents a general model for the behavior of supported metal crystallites on a substrate in terms of the emission, migration and capture of single atoms and groups of atoms. It was possible to define conditions under which only single atoms are emitted and captured; there is no emission and particles of all sizes are captured; or single atom emission and multiatom capture takes place. The latter case was discussed here in detail.

We assumed that the process for the emission of a multiatom particle is analogous to the evaporation of a single particle. However, consideration of the number of bonds that need to be broken for multiatom emission leads to rates of particle emission that are many orders of magnitude smaller than rates of emission of single atoms. To account satisfactorily for emission of particles, rate equations for particle fracture must be developed.

In all the present calculations, diffusion controlled the overall migration and capture process. Depending upon the relative magnitudes of the diffusion and emission coefficients of all particles, the overall metal loading, and the saturation concentration, the model predicts a continuous increase in available surface area of the metal, an increase followed by a decrease, a decrease followed by an increase, or a continuous decrease. Hence, the present model enables us to treat both sintering and splitting in a unitary manner.

APPENDIX

Derivation of a Modified Lifshitz–Slyosov Continuity Equation

We start with the continuity equation for the j-mers, Eq. (4),

$$\frac{dn_j}{dt} = (\alpha_{j+1,1}n_{j+1} - \beta_{j,1}n_j) - (\alpha_{j,1}n_j - \beta_{j-1,1}n_{j-1}). \quad \text{(A1)}$$

If we consider n to vary continuously with j, Eq. (A1) can be rewritten for n a function of two variables, j and t, as

$$\frac{\partial n(j,t)}{\partial t} = \frac{\partial}{\partial j}[\{\alpha(j) - \beta(j)\}n(j,t) + \beta(j)n(j,t) - \beta(j-1)n(j-1,t)]. \quad \text{(A2)}$$

Substituting the first two terms of the Taylor series of $\beta(j-1)n(j-1,t)$ around $\beta(j)n(j,t)$ for the last term in brackets yields

$$\frac{\partial n(j,t)}{\partial t} = \frac{\partial}{\partial j}[\{\alpha(j) - \beta(j)\}n(j,t)] + \frac{\partial^2}{\partial j^2}[\beta(j)n(j,t)]. \quad \text{(A3)}$$

Of course this is not the only continuous variable approximation that can be made to Eq. (A1). Expanding α around $(j+\frac{1}{2})$ and β around $(j-\frac{1}{2})$, then expanding the result around j we obtain

$$\frac{\partial n(j,t)}{\partial t} = \frac{\partial}{\partial j}[\{\alpha(j) - \beta(j)\}n(j,t)] + \frac{1}{2}\frac{\partial^2}{\partial j^2}[\{\alpha(j) + \beta(j)\}n(j,t)]. \quad \text{(A4)}$$

The quantitative treatment of Ostwald ripening is based on the Lifshitz–Slyosov equation which is, in the present notation (*20*)

$$\frac{\partial n}{\partial t} + \frac{\partial}{\partial j}[n\{\beta(j) - \alpha(j)\}] = 0. \quad \text{(A5)}$$

Comparing Eqs. (A3) and (A4) we see that both contain a second derivative. However, Eq. (A5) does not. We suggest therefore, that either of the Eqs. (A3) or (A4) be used in the treatment of Ostwald ripening in supersaturated solutions. Which of these is a better choice depends upon the forms postulated for α and β. For example, if $\alpha = \beta$ Eqs. (A3) and (A4) are trivially identical.

ACKNOWLEDGMENT

We gratefully acknowledge the useful comments of Dr. S. E. Wanke.

REFERENCES

1. Ruckenstein, E., and Pulvermacher, B., *J. Catal.* **29**, 224 (1973).
2. Pulvermacher, B., and Ruckenstein, E., *J. Catal.* **35**, 115 (1974).
3. Wynblatt, P., and Gjostein, N. A., *Progr. Solid State Chem.* **9**, (1975).
4. Ruckenstein, E., and Malhotra, M. L., *J. Catal.* **41**, 303 (1976).
5. Basset, G. A. *in* "Proceedings of the European Regional Conference on Electron Microscopy" (A. L. Houwink and B. J Spit, Eds.), I. De Nederlandse Vereniging voor Electronen Microscopie, Delft, 1960.
6. Skofronick, J. G., and Phillips, W. B., *J. Appl. Phys.* **38**, 4791 (1967).
7. Masson, A., Métois, J. J., and Kern, R., *Surface Sci.* **27**, 463 (1971).
8. Kern, R., Masson, A., and Métois, J. J., *Surface Sci.* **27**, 483 (1971).

9. Chu, Y. F., and Ruckenstein, E. *Surface Sci.*, in press 1977.

10. Baker, R. T. K., Thomas, C., and Thomas, R. B., *J. Catal.* **38**, 510 (1975).

11. Baker, R. T. K., and France, J. A., *J. Catal.* **39**, 481 (1975).

12. Hughes, T. R., Houston, R. J., and Sieg, R. P., *Ind. Eng. Chem. Proc. Des. Develop.* **1**, 96 (1962).

13. Maat, H. J., and Moscou, L., *Proc. Int. Congr. Catal., 3rd, 1964* **2**, 1277 (1965).

14. Gruber, H. L., *J. Phys. Chem.* **66**, 48 (1962).

15. Hermann, R. A., Adler, S. F., Goldstein, M. S., and DeBaum, R. M., *J. Phys. Chem.* **65**, 2189 (1961).

16. Wynblatt, P., and Gjostein, N. A., *Scripta Met.* **7**, 969 (1973).

17. Wynblatt, P., and Gjostein, N. A., *Acta Met.* **24**, 1165 (1976).

18. Flynn, P. C., and Wanke, S. E., *J. Catal.* **34**, 390 (1974); **34**, 400 (1974).

19. Geus, J. W., *in* "Chemisorption and Reactions on Metallic Films" (J. R. Anderson, Ed.), Vol. 1. Academic Press, New York, 1971.

20. Lifshitz, I. M., and Slyosov, V. V., *J. Phys. Chem. Solids* **19**, 35 (1961).

21. Chakraverty, B. K., *J. Phys. Chem. Solids* **28**, 2401 (1967).

22. Jackson, K. A., *J. Crystal Growth* **5**, 13 (1969).

23. Hirth, J. P., and Pound, G. M., *in* "Progress in Materials Science" (B. Chalmers, Ed.), Vol. 11. Macmillan, New York, 1963.

24. Chen, J., personal communication.

25. "Handbook of Chemistry and Physics" (R. C. Weast, Ed.), 51st ed., p. B198. Chem. Rubber Co., Cleveland, Ohio 1970.

Sintering of Palladium on Alumina Model Catalyst in a Hydrogen Atmosphere

JAMES J. CHEN AND ELI RUCKENSTEIN[1]

Department of Chemical Engineering, State University of New York at Buffalo, Amherst, New York 14260

Received June 26, 1980; revised December 23, 1980

The sintering behavior of Pd crystallites supported on γ-alumina in 1 atm flowing hydrogen as a function of time and temperature is reported. Model catalysts consisting of Pd crystallites supported on thin films of alumina are heated at 650, 725, and 800°C for various time intervals. The same region of a specimen is examined with a transmission electron microscope after each heat treatment to follow the changes of each Pd crystallite. The various events which occur during short, intermediate, and long durations of heating, are identified. The possible mechanisms are outlined and discussed.

INTRODUCTION

The supported metal catalysts are widely used in petrochemical, chemical, and automotive industries. They consist of small metal crystallites dispersed on a support of high surface area such as alumina or silica. Sintering of metal particles which occurs at high temperature results in a decrease of the active metal surface area. This affects the activity and sometimes the selectivity of the catalyst.

Supported palladium catalysts are mainly used as petroleum-reforming catalysts and automotive exhaust catalysts. Substantial sintering of Pd supported on alumina during heating in hydrogen has been found (*1*, *2*). Just 1 hr of heating at 650°C was enough to decrease the dispersion by 60% (*1*). In contrast to Pt/Al_2O_3 catalysts (*3–9*), the sintering kinetics of Pd/Al_2O_3 catalysts have not been systematically studied.

The purpose of this paper is to obtain information about the behavior of Pd crystallites as a function of time and temperature. Model catalysts containing very small Pd particles dispersed on a thin alumina film were heated in hydrogen for various time intervals at 650, 725, and 800°C. Samples were examined in a JEOL 100U transmission electron microscope (TEM). The same region of a specimen was observed after each heat treatment to follow the change in size, shape, and position of each particle. About 1000 to 1500 crystallites were measured to obtain the size distribution. Numerous events have been examined at various stages of heating with the goal of identifying possible mechanisms by which sintering occurs. Crystallite migration and coalescence seem to play a role in the sintering process. Direct ripening which involves atomic migration between pairs or a few neighboring particles with the smaller particle decreasing in size and then disappearing also has been observed occasionally. The disappearance of a large number of small crystallites all over the surface could be explained either by crystallite migration and coalescence or by Ostwald ripening. In contrast to the local direct ripening, Ostwald ripening is a global process involving a large number of particles and a two-dimensional phase of single metal atoms on the substrate.

EXPERIMENTAL PROCEDURES

Preparation of Thin Alumina Films

A thin (0.025 mm) high-purity (99.999%) aluminum foil was first chemically cleaned

[1] Correspondence concerning this paper should be addressed to E. Ruckenstein.

This paper, which was originally published in **Journal of Catalysis**: 69, 254–273 (1981), is republished with permission.

in a polishing solution of orthophosphoric acid and nitric acid and then anodized at 20 V for 30 sec in a 3%wt tartaric acid solution at a pH of 5.5. The details of the anodic oxidation procedure to prepare thin films of alumina have been described in previous papers of this laboratory (*9, 10*). The nonporous and amorphous oxide film thus built up on aluminum was about 300 Å thick. The alumina film was separated from the unoxidized portion by floating small squares of foil on a dilute mercuric chloride solution in which aluminum dissolved. The films were then washed in distilled water. Finally, they were picked up on gold microscope grids and dried in air. The samples were calcinated in air at 800°C for 72 hr to transform amorphous alumina into crystalline γ-alumina (confirmed by electron diffraction). This prolonged heat treatment ensured that no further phase transformation and no displacements of the grains of the substrate would occur during the sintering experiments.

Preparation of the Model Catalyst

A thin Pd film about 15 Å thick was vacuum evaporated on to the support in an Edwards 306 vacuum coater at pressure below 10^{-6} Torr. The support was kept at room temperature during deposition. Pd wire of 99.999% purity wound on tungsten filament was the source. The film thickness could be controlled by the amount of Pd and the distance between source and target. The Pd/Al_2O_3 samples were heated at 500°C in flowing hydrogen inside a quartz tube for 6 hr to produce small Pd crystallites. This initial size distribution had crystallite diameters in the range of 10 to 70 Å.

Heat Treatment

All the sintering experiments were carried out at 1 atm in a quartz tube 5 ft. long and 1.5 in. in diameter in a 2-ft. tubular furnace. The following procedures were adopted to reduce the time span of heating-up and cooling-down. The quartz tube was first evacuated to 10^{-2} Torr and then nitrogen was introduced to flow through. The middle section of the tube was heated up to the desired temperatures in about 2 to 4 hr. In the meantime, samples sitting in a ceramic boat and being placed into the tube before evacuation were staying at room temperature at the end of the tube where nitrogen was flowing in. The ceramic boat was then displaced by a stainless-steel rod to the middle of the tube where its temperature reached the desired temperature in 3 to 5 min. Then hydrogen, replacing nitrogen, started to flow into the tube. As soon as the prespecified time interval was reached, the furnace was turned off and the samples were cooled down in a nitrogen stream. Compressed laboratory air was allowed to flow through the annular region between furnace and tube such that the tube temperature dropped to 500°C in 1.5 to 2.5 min. Preliminary experiments showed that no sintering took place at 500°C. Then the air flow rate was reduced so that the system slowly cooled to room temperature in about 1.5 hr to avoid the generation of strain in the crystallites. Nitrogen and hydrogen gases were provided by Linde Division, Union Carbide Corporation, and have a minimum purity of 99.999%. Before entering the tube, the gas passed through a column of 5A molecular sieves to remove any trace of water. The flow rate of the gases was 30 ml/min.

The same region of a specimen was examined after each heat treatment to follow the change in size, shape, and position of each particle. In each temperature series several regions of two or three similar specimens were examined. Heating effects due to electron bombardment in the microscope were largely eliminated by keeping the illumination intensity at about one-tenth of the level at which modifications in the samples were noticed. Samples were stored in a vacuum desiccator between each of the steps of the experiment to avoid contamination.

RESULTS

Figure 1 and Table 1 summarize various events observed during the sintering process on electron micrographs. It includes the disappearance of small crystallites, crystallite migration, crystallite migration and coalescence, the appearance of new crystallites, direct ripening, and faceting.

The size distribution of Pd crystallites at every stage was measured. In order for the size distribution to represent the whole population as closely as possible, the total number of crystallites examined at every stage was between 1000 and 1500. Two kinds of average sizes are reported here. They are defined as number-average diameter:

$$\bar{D}_n = \Sigma N_i \bar{D}_i / \Sigma N_i \qquad (1)$$

and surface-average diameter:

$$\bar{D}_s = \Sigma N_i \bar{D}_i^3 / \Sigma N_i \bar{D}_i^2. \qquad (2)$$

By assuming hemispherical crystallites, the specific surface area, defined as the surface

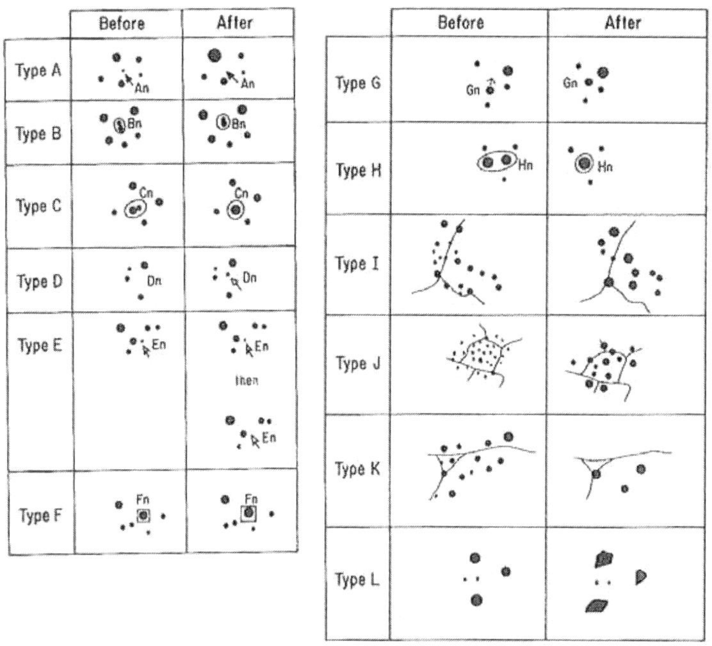

FIG. 1. Schematic illustration of various events observed during the sintering process: (Type A) Small crystallite disappears. Surrounding larger crystallites are stationary and either do not change or grow in size. (Type B) Two neighboring crystallites come in contact with each other and either do not change or grow in their sizes. (Type C) Two touching crystallites coalesce into a larger one. (Type D) A new crystallite appears at a place where no crystallite was located before. The previous location of this new crystallite is unknown. (Type E) A small crystallite very close to a larger one decreases in size and then disappears. (Type F) Crystallite is stationary and grows in size. (Type G) A crystallite migrates to a new position and grows or remains unchanged in size. (Type H) Two or more neighboring crystallites coalesce into a new crystallite somewhere between their original positions. (Type I) (The schemes I, J, and K contain grain boundaries. These events can, however, occur everywhere and not just near grain boundaries.) This is a global event which involves many crystallites. The former situation occurs after 30 min of heating at 650°C. The latter situation occurs after an additional 30 min of heating at the same temperature. The process leads to a new distribution of crystallites with larger diameters. (Type J) This event occurs during the first 10 min of heating of fresh catalyst at 725 and 800°C. The process is similar to a Type I event. (Type K) This event occurs during the additional 20 min of heating at 725°C after the first 10 min. The process is similar to that of a Type I event. (Type L) The shape of large crystallites, with diameter of about 125 Å or beyond, changes from spherical to faceted while those of the small ones remain unchanged.

TABLE 1

Possible Mechanisms and Examples of the Events of Fig. 1 Observed on the Electron Micrographs

Type	Possible mechanisms	Examples
A	There are two possible mechanisms. In the first, the small crystallite migrates and coalesces with one of the large ones. In the second, the small crystallite emits single atoms onto the substrate surface and either disappears or becomes smaller than the resolution of the electron microscope. The large crystallites capture the emitted single atoms.	A_1 to A_5 of Figs. 2a (before) and b (after); A_6 to A_{10} of Figs. 2b (before) and c (after); A_{11} to A_{13} of Figs. 2d (before) and e (after).
B	The process implies some displacement of particles.	B_1 to B_4 of Figs. 2a (before) and b (after).
C	Crystallite coalescence.	C_1 of Figs. 2a (before) and b (after).
D	The migration and coalescence of very small particles with sizes below the resolution of the electron microscope (unresolved particles) is one of the possible mechanisms. The new crystallite could also form through direct ripening of the unresolved particles.	D_1 and D_2 of Figs. 2a (before) and b (after); D_3 to D_7 of Figs. 2d (before) and e (after).
E	The small crystallite transfers single atoms directly to the larger one. This phenomenon is called direct ripening and will be discussed in detail later.	E_1 and E_2 of Figs. 2a (before), b, and c (after); E_3 of Figs. 2d (before), e, and f (after).
F	The migration of unresolved particles and coalescence with the large one is one of the possible mechanisms. The unresolved particles could also be captured by the large one through direct ripening. Spreading of crystallites is also a possible mechanism.	F_1 and F_2 of Figs. 2b (before) and c (after).
G	The growth of the moving crystallite could possibly be attributed to its capture of small unresolved particles on the way to its new position.	G_1 to G_5 of Figs. 2d (before) and e (after).
H	This process may involve the migration and coalescence of crystallites and/or direct ripening.	H_1 to H_5 of Figs. 2d (before) and e (after); H_6 to H_8 of Figs. 2e (before) and f (after).
I	Because of the appearance of crystallites at places where no crystallites were located before, it seems that crystallite migration and coalescence are at least in part responsible for this process.	Figs. 2c (before) and d (after).
J	Similar to a Type I event.	Figs. 5a (before) and b (after).
K	Similar to a Type I event.	Figs. 5b (before) and c (after).
L	Faceting which occurs after extended heating on sufficiently large crystallites because this shape is thermodynamically more stable.	Figs. 5d (before) and e (after); Figs. 7a (before) and b (after).

area per gram of metal, is given by[2]

$$S = \left(\frac{\pi}{2}\right)(\Sigma N_i\bar{D}_i^2) \Big/ \left(\frac{\rho\pi}{12}\right)(\Sigma N_i\bar{D}_i^3) = 6/\rho\bar{D}_s, \quad (3)$$

where N_i is the number of crystallites with diameter between $\bar{D}_i - \Delta D_i/2$ and $\bar{D}_i + \Delta D_i/2$, $\bar{D}_i$ is the average diameter in each diameter interval, and ρ is the density of Pd.

Effect of Time and Temperature

Figures 2a–f show the time sequence of the same region heated at 650°C. As time progressed, the population density of crystallites decreased and the size of crystallites increased. For the first 30 min (Figs. 2a–c) all the changes occurred locally and the events which took place are of Types A to F of Fig. 1 and Table 1. The most drastic changes occurred between the third and the fourth stages (Figs. 2c and d), from $t = 30$ min to $t = 1$ hr. The event of Type I of Fig. 1 and Table 1 is the best description of these changes. Almost all the crystallites were involved in the process. The result was a new distribution of the crystallites. For comparison, the upper right part of Fig. 2c is overlapped in Fig. 3 over the corresponding part of Fig. 2d. The particles marked by solid lines and those marked by broken lines indicate the positions of the crystallites at the third and the fourth stages, respectively. Crystallite 2 probably originated from crystallite 1 which seems to migrate about 80 Å and to grow from 60 to 85 Å in diameter. Its growth in size probably took place either by the capturing of small particles with sizes below the resolution of the electron microscope and/or by spreading. Crystallite 5 might be the combination of crystallites 3 and 4. Crystallite 6 could have moved toward crystallite 7 to form crystallite 8. Crystallite 10 may have originated from crystallite 9 and the small crystallites (including some unresolved ones) nearby. Crystallite 13 probably formed by the migration of crystallite 11 across the grain boundary and the coalescence with crystallite 12. Similarly, migration and coalescence of crystallites 14 and 15 could have occurred and led to the formation of crystallite 16.

Since so many new crystallites appear in Fig. 2d at places where no crystallites have been located before (Fig. 2c) and the disappearance of crystallites in Fig. 2c suggest that a large-scale crystallite migration and coalescence might have occurred during this 30-min interval. An alternative explanation is atomic migration associated with nucleation of new particles at some sites where the interactions between metal particles and substrate are strong. Figure 3 seems to suggest that the latter explanation is less likely.

The sintering rate slowed down with further heating. All the events involved at most a few neighboring crystallites. Type G and H events of Fig. 1 and Table 1 seemed to dominate the latter stages of the process. Crystallites as large as 60 to 75 Å in diameter (events G_1 to G_5 of Figs. 2d and e) were observed to migrate. Events H_1 to H_5 of Figs. 2d and e involved two to five crystallites which coalesced into new crystallites. The appearance of new small crystallites without any evidence about their previous locations, events of Type D, indicates that some small crystallites might be generated through the coalescence or direct ripening of smaller particles and not necessarily through the decrease in size of crystallites.

Table 2 lists the events observed on the electron micrographs of all the regions examined after each treatment. Events B and C were found only during the early stages because they involve small crystallites and take place in a short time interval. Probably a Type H event replaced them during the latter stages.

[2] The actual shape of the crystallites is not hemispherical. Tilting experiments can provide information about the thickness of the crystallites. However, tilting capabilities are not available in this laboratory.

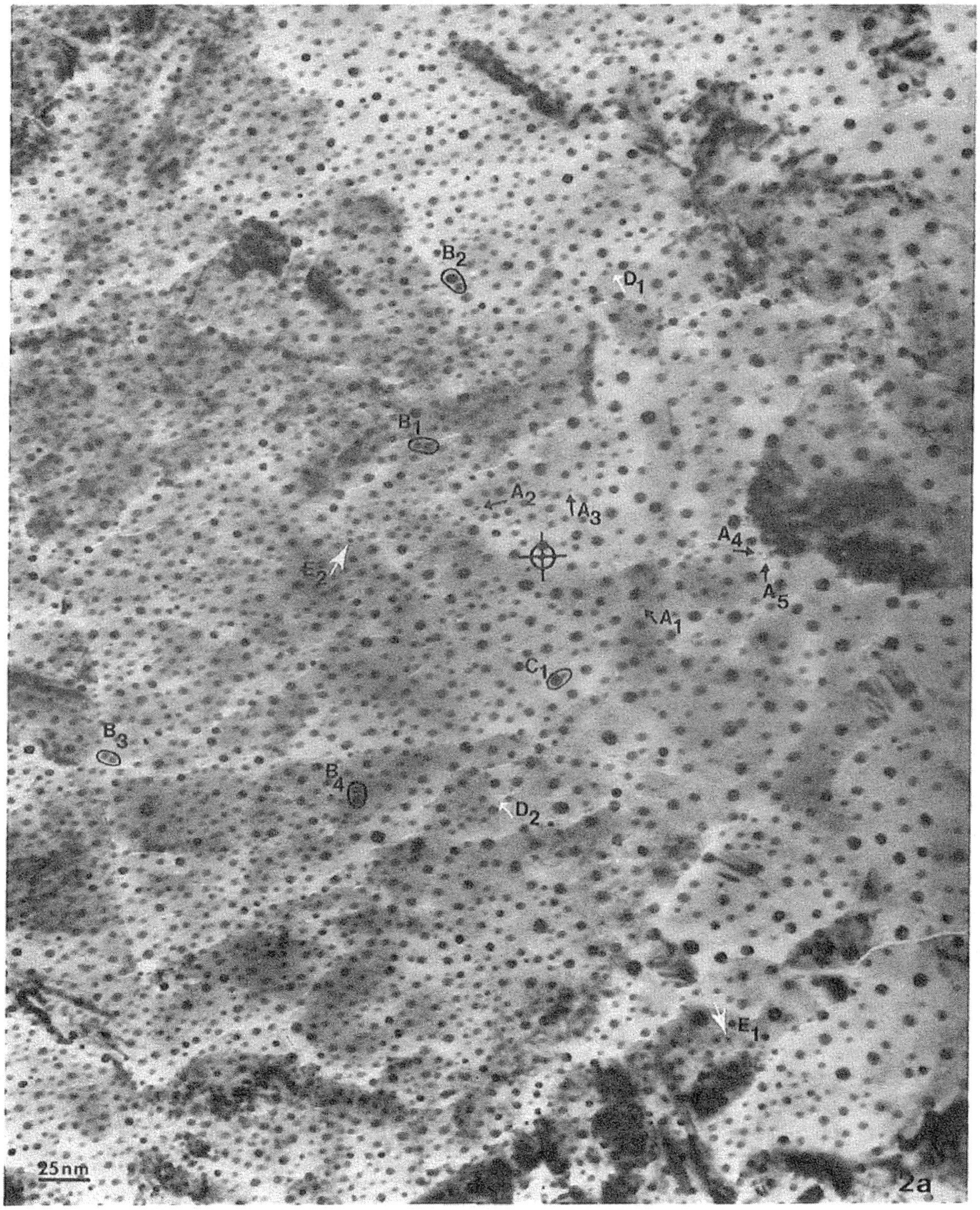

FIG. 2. Time sequence of the same region of Pd/Al_2O_3 model catalyst: (a) fresh catalyst; heated in 1 atm H_2 at 650°C for (b) 10 min; (c) 30 min; (d) 1 hr; (e) 4 hr; (f) 8 hr.

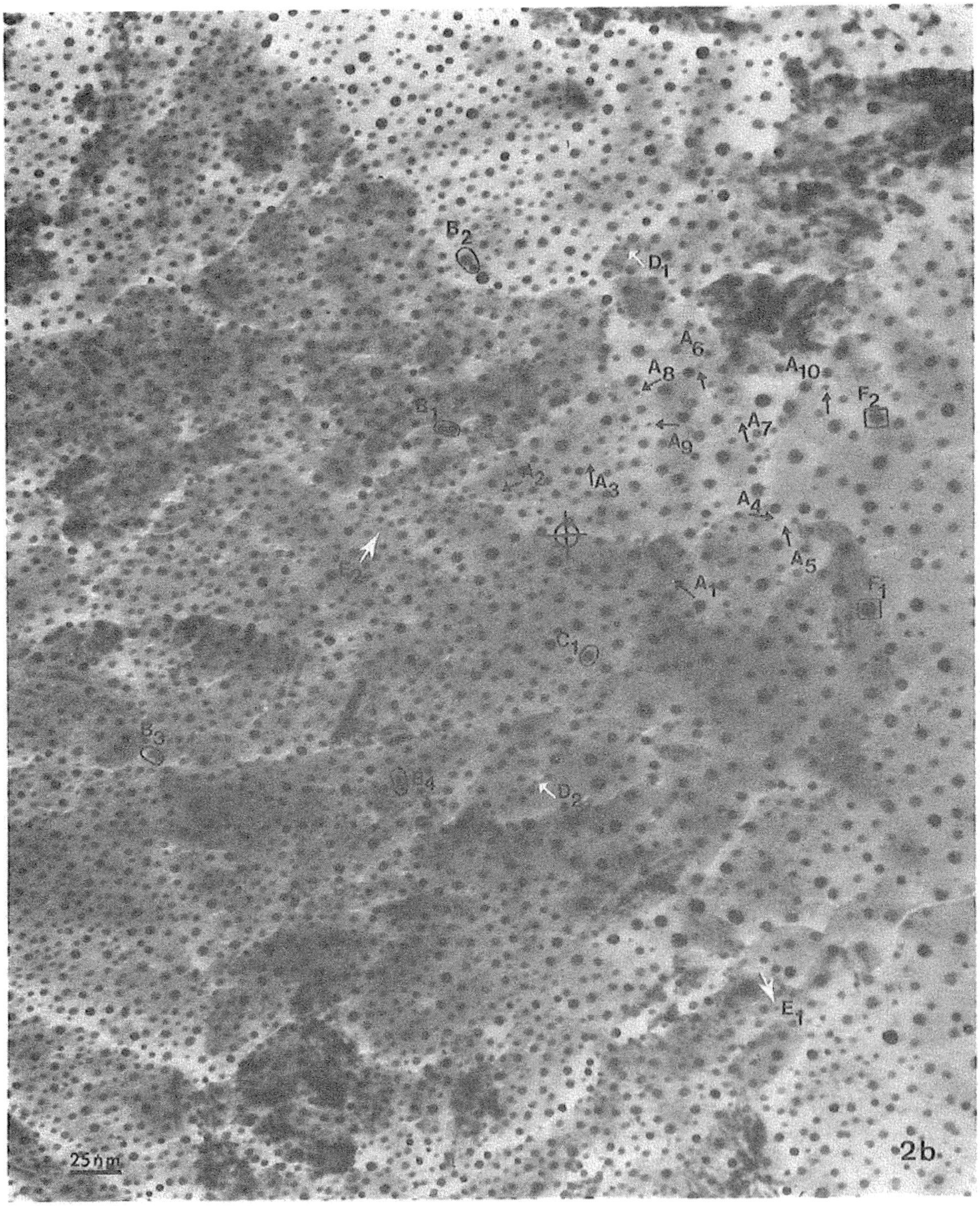

FIG. 2—*Continued.*

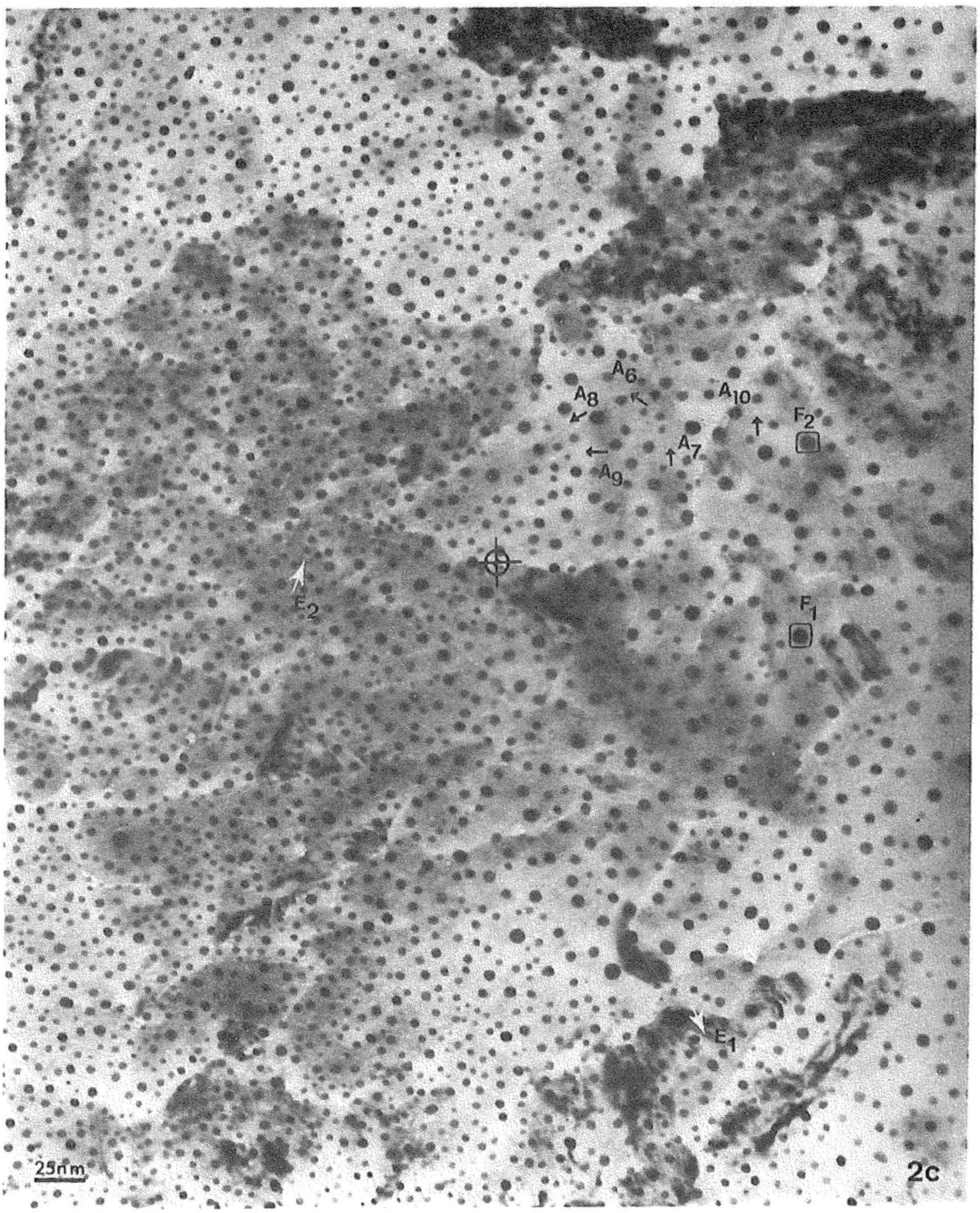

FIG. 2.—*Continued.*

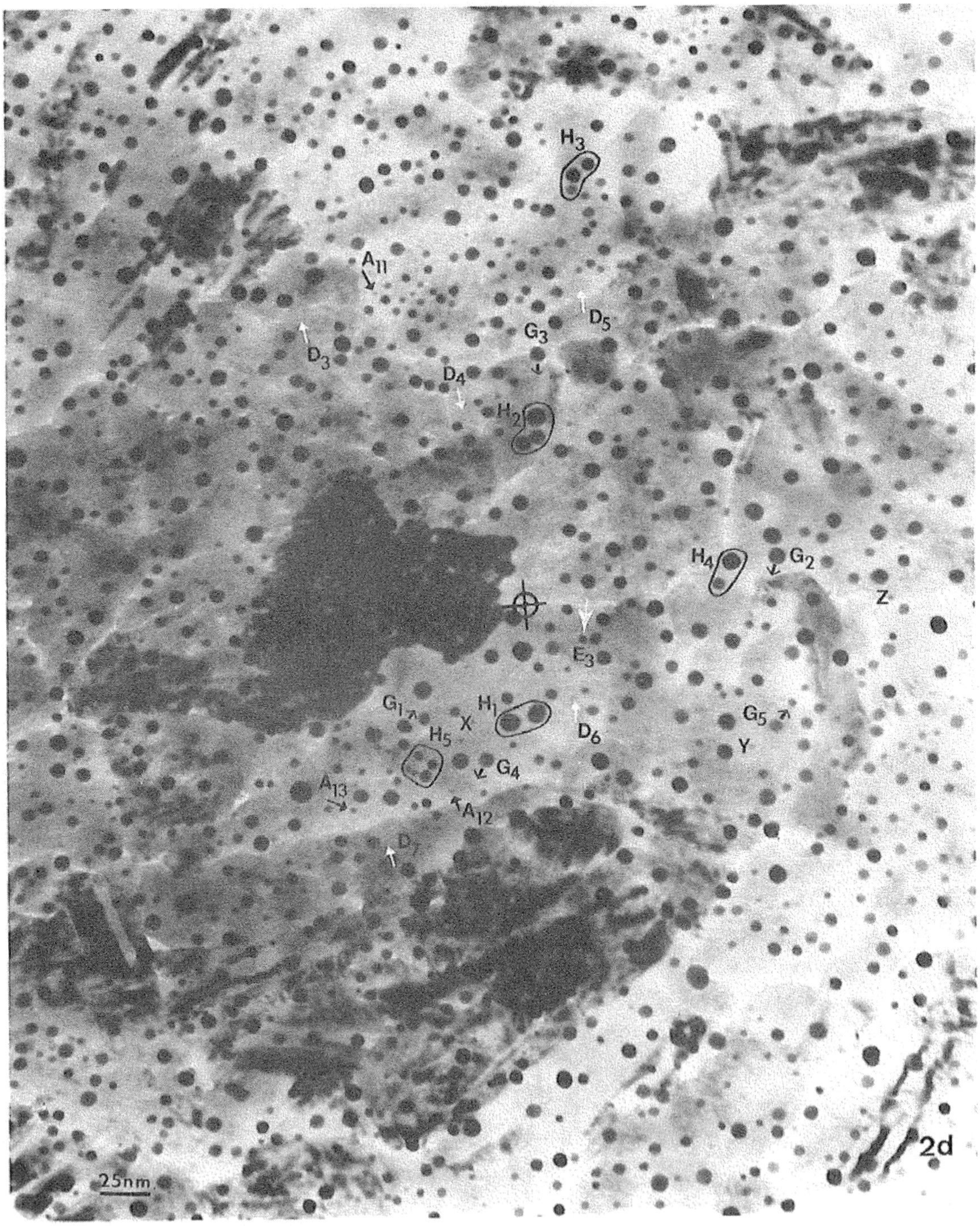

FIG. 2.—*Continued.*

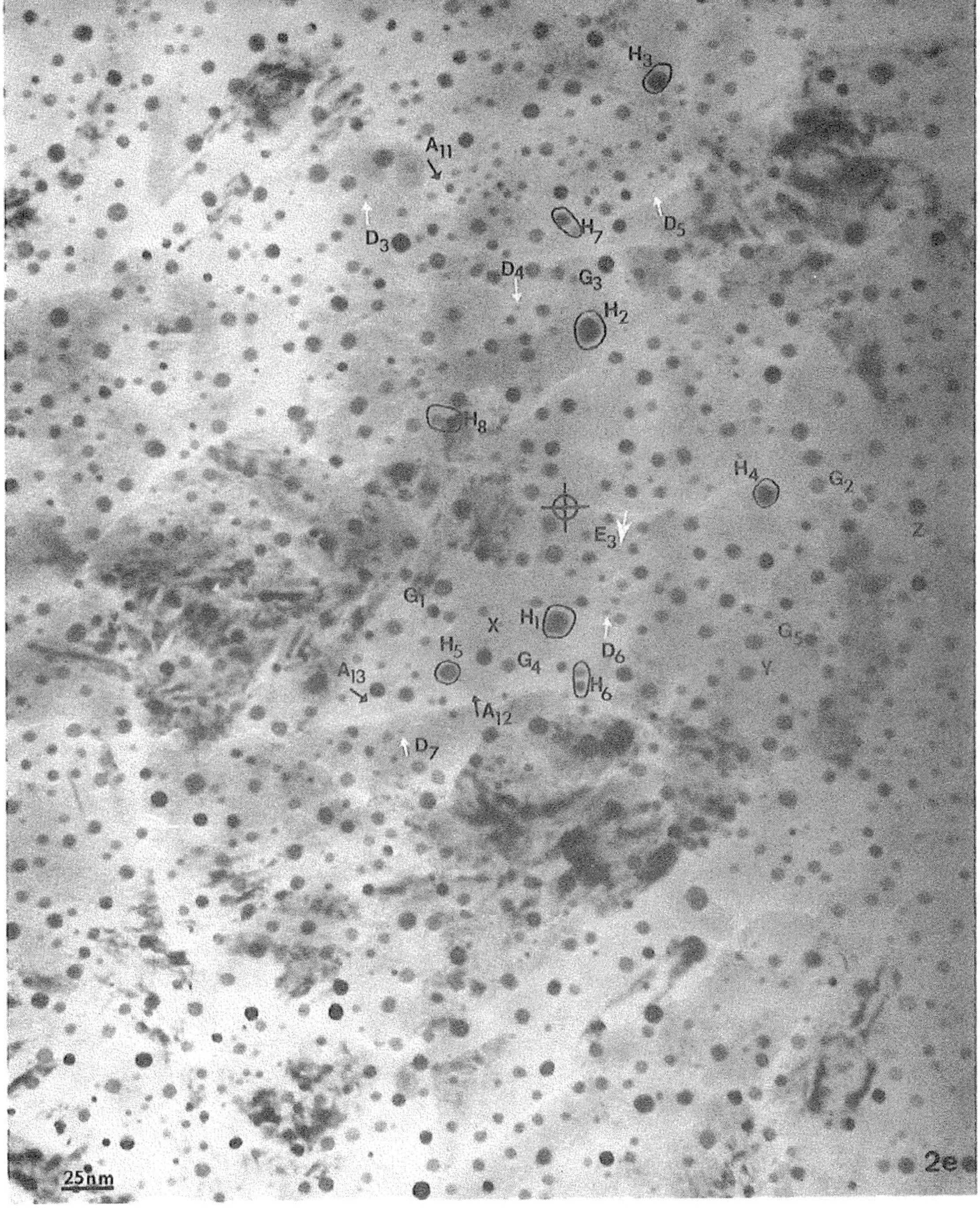

FIG. 2.—*Continued.*

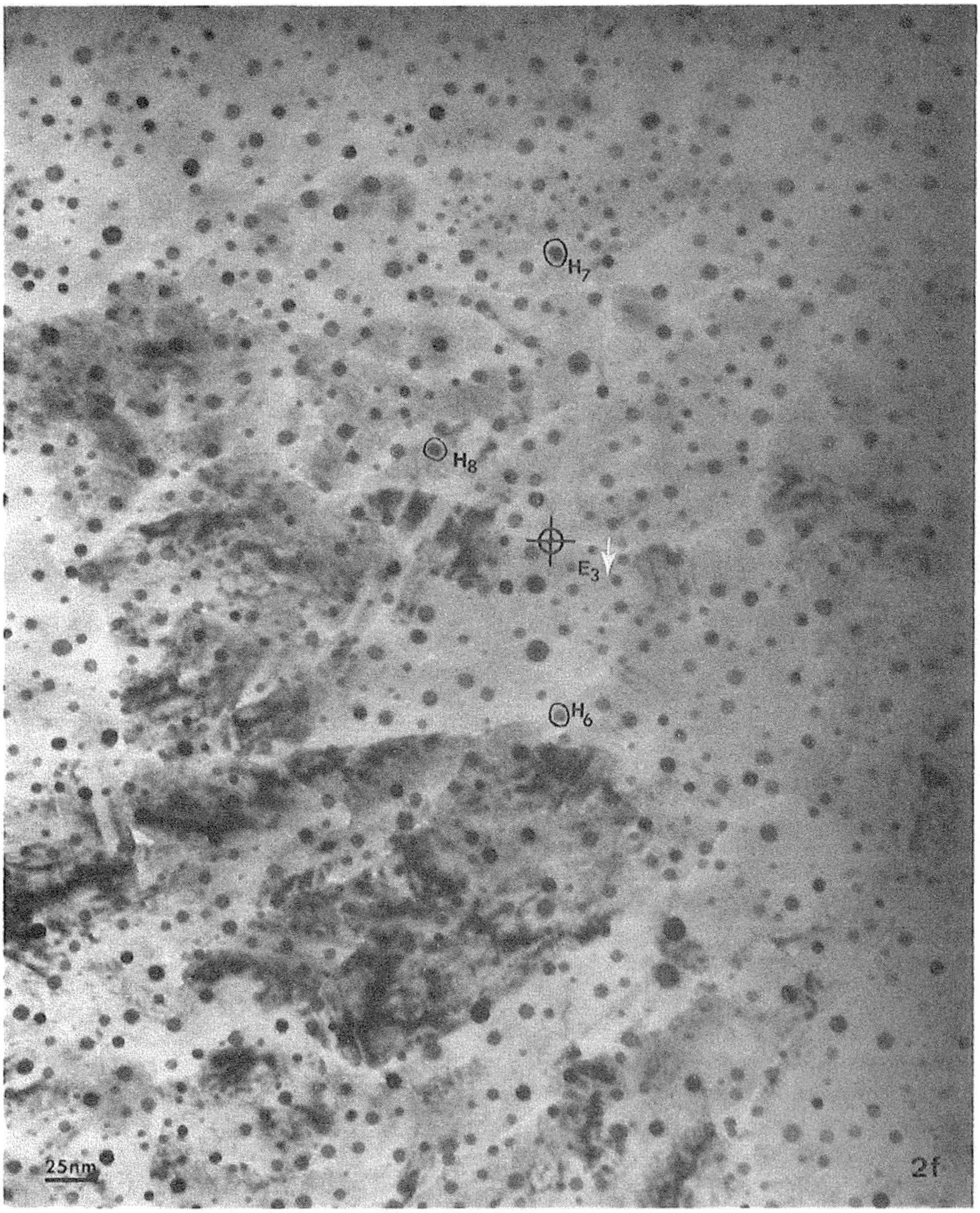

FIG. 2.—*Continued.*

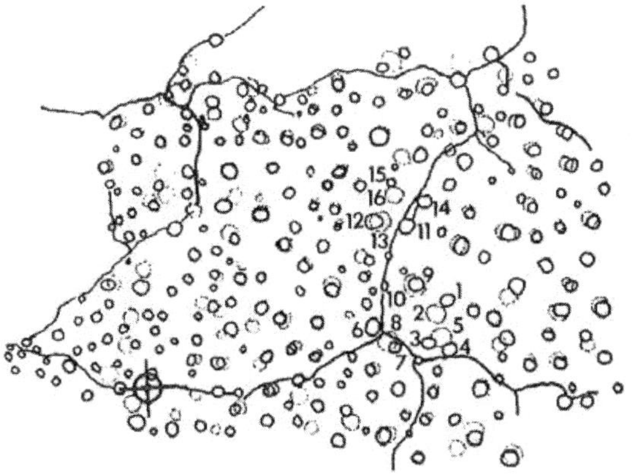

FIG. 3. Position distribution of the crystallites from Fig. 2c, solid lines, and Fig. 2d, broken lines.

Figures 4a–d give the size distribution of more than 1000 Pd crystallites at four different stages. As sintering proceeded, the percentage of small crystallites decreased, the curve broadened and developed tails toward the large-diameter side of the curve.

The time sequence of average crystallite size and surface area based on the size distribution is shown in Table 3. The drastic changes taking place between the third and the fourth stages are also reflected in the variation of the average diameter and surface area.

Figures 5a–e show the same region of the same specimen heated at 725°C for various lengths of time. It is impossible to follow the changes which occurred on each crystallite during the first 30 min, Figs. 5a–c, because the rate of sintering was extremely high. Types J and K events of Fig. 1 and Table 1 described the process. During further heating, the rate of sintering was much slower. Faceting of the crystallites larger than 125 Å in diameter occurred after 20 hr of heating (Fig. 5e).

Table 4 lists the events observed on all the regions examined. They are similar to the latter stages of the 650°C series. The size distributions shown in Figs. 6a–c are similar in shape to those at 650°C. However, a multimodal distribution starts to

TABLE 2

Events Observed during Sintering at 650°C

Time interval (hr)	Events
0–0.167	A[a], B, C, D, E[b], F
0.167–0.5	A[a], B, C, D, E[b], F
0.5–1	I
1–4	A, D, E[b], F, G, H[a]
4–8	A, D, E[b], F[a], G, H
8–20	A, D[b], F[a], G, H

[a] Most frequently observed event.
[b] Least frequently observed event.

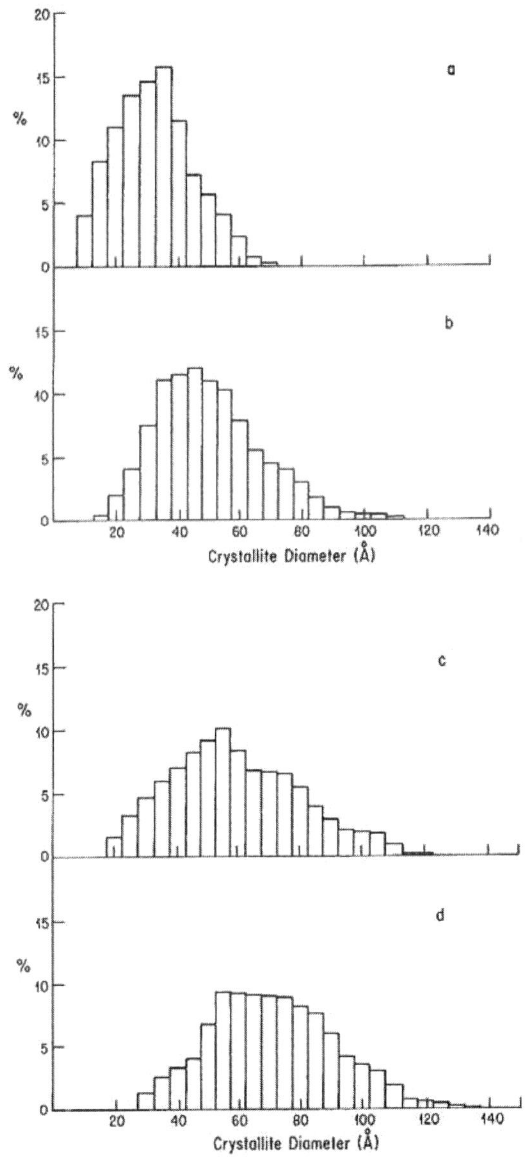

FIG. 4. Size distribution of Pd crystallites: (a) initial distribution; heated in 1 atm H_2 at 650°C for (b) 1 hr; (c) 4 hr; (d) 20 hr.

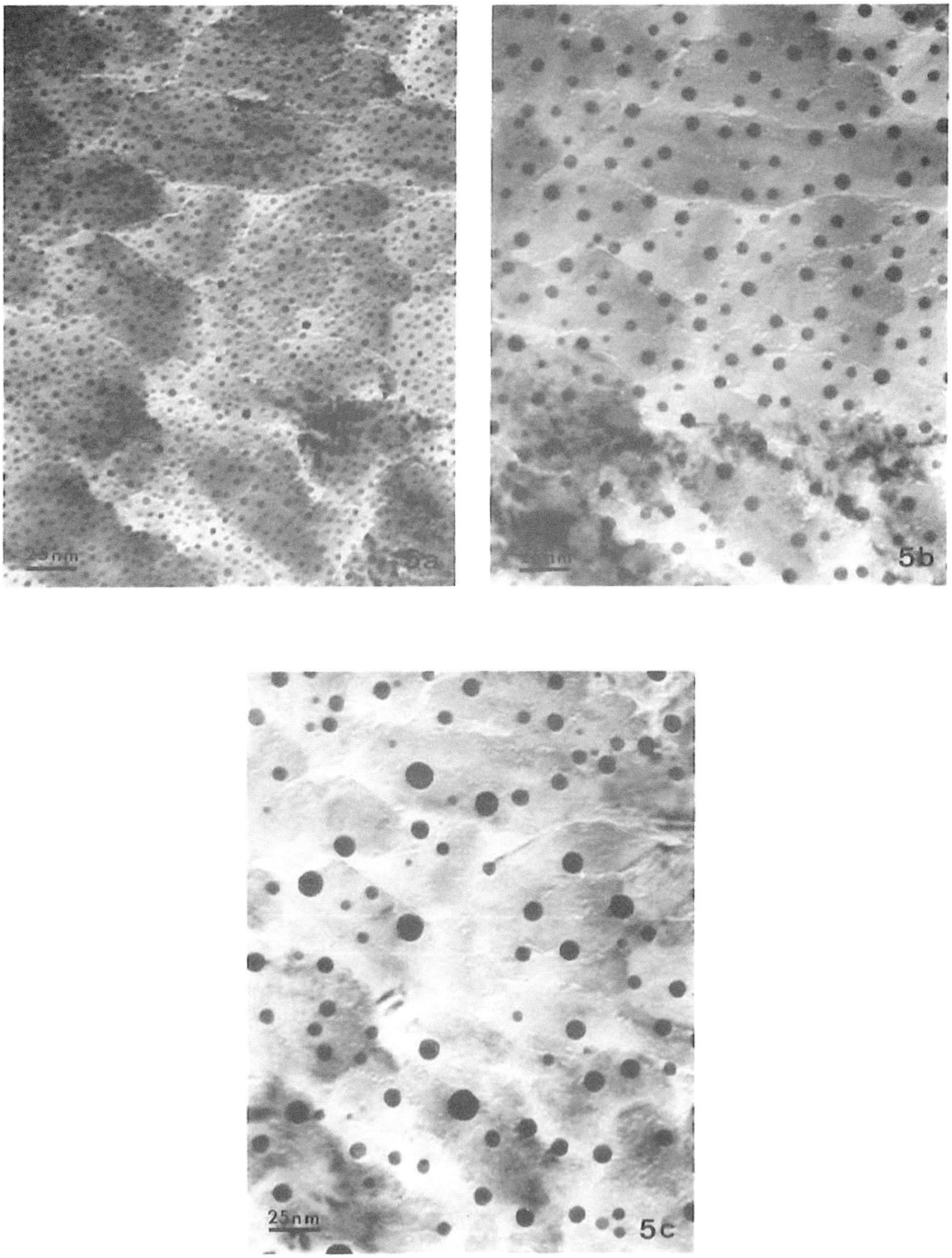

FIG. 5. Time sequence of the same region of Pd/Al_2O_3 model catalyst: (a) fresh catalyst; heated in 1 atm H_2 at 725°C for (b) 10 min; (c) 30 min; (d) 8 hr; (e) 20 hr.

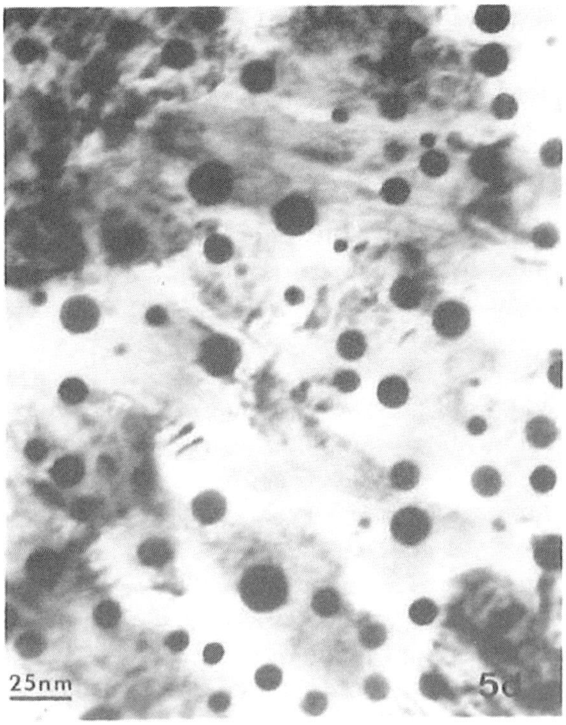

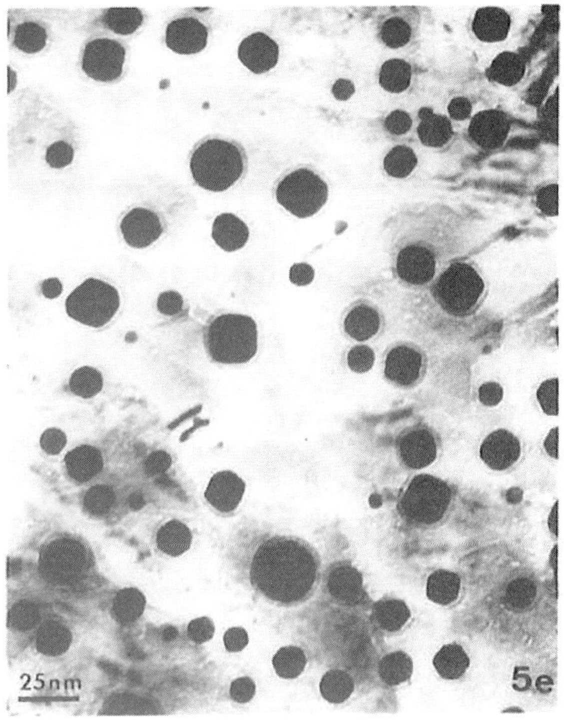

FIG. 5—*Continued.*

develop in Fig. 6d. This seems to be associated with the formation of faceted crystallites. The average crystallite size and surface area are listed in Table 5. The surface area dropped to 66% of the initial value during the first 10 min and reached 30% after 20 hr.

The events occurring during heating at 800°C are summarized in Table 6. (For average crystallite size and surface area see Table 7). In this case, a few faceted crystallites started to form just after 2 hr of treatment (Fig. 7a) and, almost all the large crystallites changed their shapes from spherical to faceted after 4 hr (Fig. 7b). At this stage, t = 4 hr, most crystallites had square, tetragonal, pentagonal, or hexagonal shape. Further sintering of these faceted crystallites led to the appearance of particles with more irregular shapes which show different contrast areas in a crystallite (Fig. 7c). Twinned crystallites might have

TABLE 3

Average Crystallite Size and Specific Surface Area of Pd/γ-Al_2O_3 at 650°C

Cumulative time (hr)	Number-average diameter (Å)	Surface-average diameter (Å)	Specific surface area (m^2/g Pd)	$\left(\frac{S}{S_0}\right)$
0	32.5	41.8	119.402	1.00
0.167 (10 min)	35.3	43.1	115.759	0.97
0.5	39.3	46.3	107.707	0.90
1	50.3	62.4	80.033	0.67
4	59.7	74.3	67.203	0.56
8	61.3	76.4	65.366	0.55
20	71.2	82.3	60.670	0.51

S: specific surface area at time t; S_0: initial specific surface area.

TABLE 4

Events Observed during Sintering at 725°C

Time interval (hr)	Events
0–0.167	J
0.167–0.5	K
0.5–1	A, D, E[a], F, G, H[b]
1–2	A, D, E[a], F, G, H[b]
2–4	A[a], D, F[b], G, H
4–8	A[a], D, F[b], G, H
8–20	A[a], D, F[b], G, H, L

[a] Least frequently observed event.
[b] Most frequently observed event.

formed (*11*) probably by the coalescence of neighboring crystallites.

The size distributions, Figs. 8a–d, started developing into a bimodal curve at t = 4 hr and became well-developed multimodal at t = 20 hr. The time at which the bimodal distribution appears seems to be related to the formation of faceted crystallites.

DISCUSSION

Two sintering mechanisms have been proposed. Ruckenstein and Pulvermacher (*12, 13*) developed a scheme of migration and coalescence of metal crystallites over the support surface. Chakraverty (*14*) and Flynn and Wanke (*7, 15*) have used the Ostwald ripening mechanism to account for sintering. In Ostwald ripening the large crystallites increase in size while the small ones decrease in size. Since these two mechanisms are not exclusive, a comprehensive treatment has been developed by Ruckenstein and Dadyburjor (*16*), which combines emission, migration, and coalescence of particles of every size, including single atoms.

Crystallite Migration

In Table 1 crystallite migration is listed as one of the possible mechanisms for various events. Crystallites as large as 75 Å in diameter were observed to migrate at 650°C (Figs. 2d and e). However, small crystallites such as X, Y, and Z of the same figures were found to remain stationary. This fact suggests that the mobility of the crystallites is not just a function of their size. Strictly speaking, the substrate surface is not homogeneous from a microscopic point of view. The local surface structure and the local surface irregularities affect the interaction force between crystallites and substrate. Therefore, the migration of crystallites on the surface of substrate may be a direction-selective process rather than a random walk-type diffusion process. Particular crystallites may migrate with preference in one direction but not in another. A small crystallite trapped in a valley will migrate with more difficulty than a larger

TABLE 5

Average Crystallite Size and Specific Surface Area of $Pd/\gamma\text{-}Al_2O_3$ at 725°C

Cumulative time (hr)	Number-average diameter (Å)	Surface-average diameter (Å)	Specific surface area (m^2/g Pd)	$\left(\frac{S}{S_0}\right)$
0	31.8	40.4	123.556	1.00
0.67 (10 min)	52.0	61.2	81.563	0.66
0.5	63.8	74.3	67.183	0.54
1	67.3	88.1	56.682	0.46
2	72.1	97.2	51.375	0.42
4	80.9	113.9	43.831	0.36
8	84.3	119.2	41.881	0.34
20	98.6	134.3	37.164	0.30

S: specific surface area at time t; S_0: initial specific surface area.

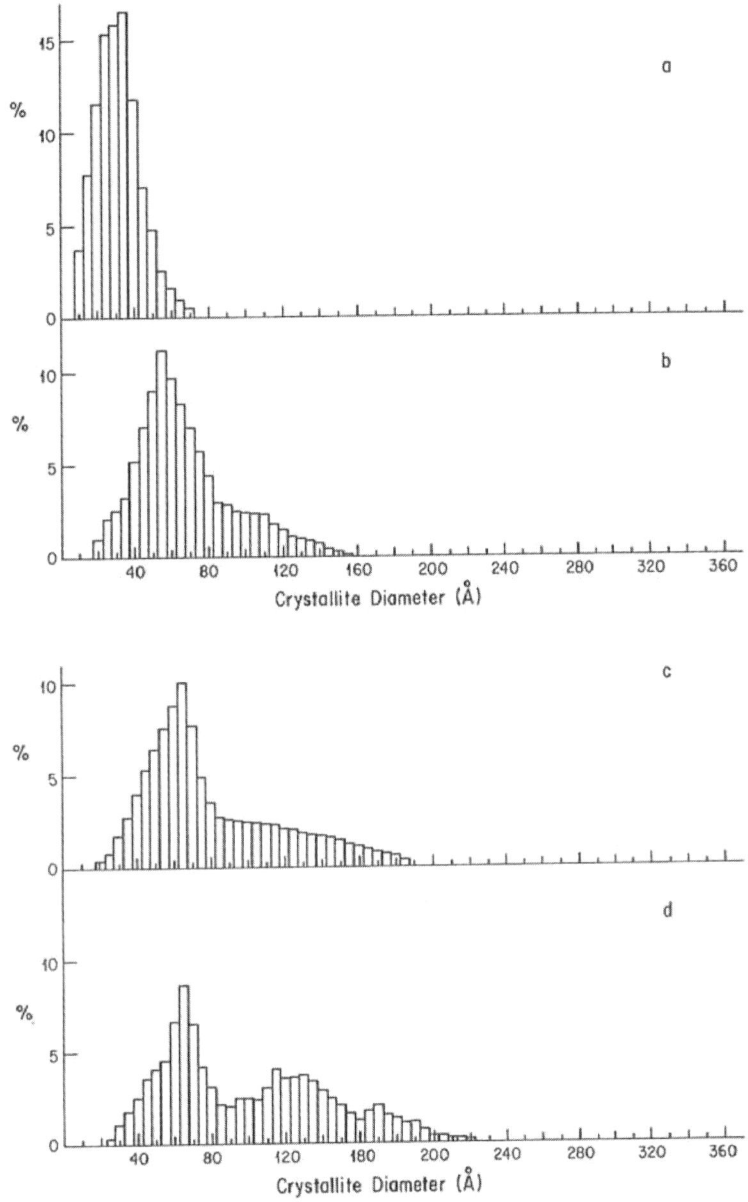

FIG. 6. Size distribution of Pd crystallites: (a) initial distribution; heated in 1 atm H_2 at 725°C for (b) 1 hr; (c) 4 hr; (d) 20 hr.

one which contacts only the tips of the hills of the surface.

Movements of the grain boundaries were not observed; nevertheless, local movements of the alumina substrate might have occurred, stimulating the migration of crystallites.

Ripening

Ripening is another possible mechanism for the sintering process. The equilibrium concentration of single atoms at the interface of a smaller particle is larger than that of a larger one. The equilibrium interface concentration of single atoms (expressed as atoms per unit area of substrate) for a particle of infinite size is the solubility of the system at that temperature. In a supersaturated system, i.e., a system in which the concentration of single atoms per unit area of substrate is greater than the solubility, the particles which have a radius

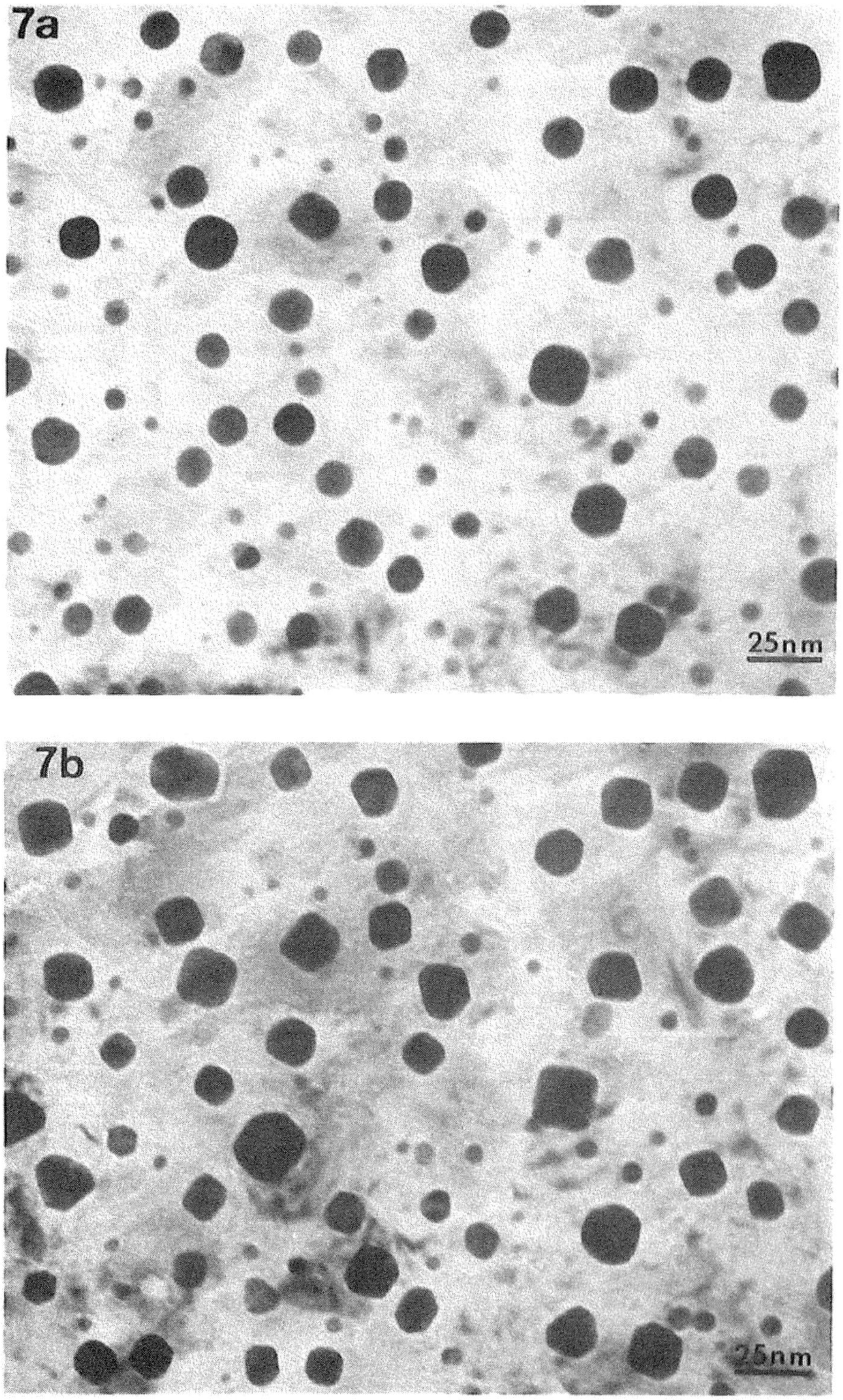

FIG. 7. Time sequence of the same region of Pd/Al_2O_3 model catalyst after heating in 1 atm H_2 at 800°C for (a) 2 hr; (b) 4 hr; (c) 20 hr.

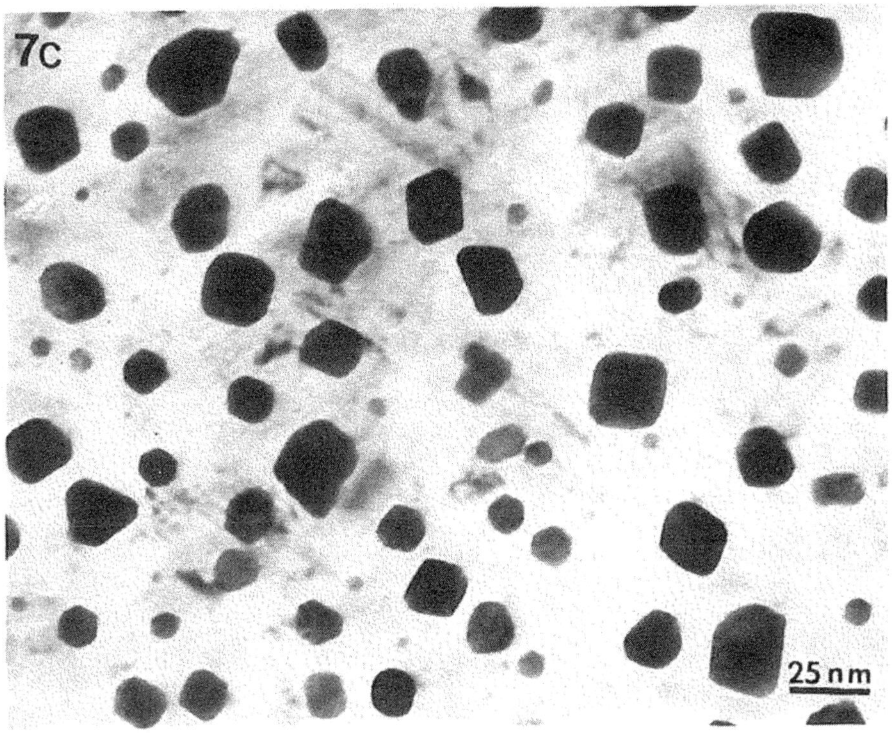

FIG. 7—*Continued.*

smaller than a critical value will release single atoms into the bulk and decrease in size. On the other hand, particles larger than the critical radius will gain single atoms from the bulk and increase in size. This occurs because for a supersaturated system, the equilibrium concentration at the leading edge of the smaller particles is larger than the bulk concentration, while that at the leading edge of the larger particles is smaller than the bulk concentration. The result is the growth of the larger particles at the expense of the smaller ones. This phenomenon is called Ostwald ripening and is a global process involving a two-dimensional phase of single atoms and a large number of particles. During the experiment, a large number of small crystallites were observed to disappear (Type A event) between consecutive stages, e.g., Figs. 2a (before) and b (after). Although there is no direct evidence to prove that they did first decrease in size and then disappear, the possibility of Ostwald ripening cannot be ruled out. Of course, the disappearance of these crystallites could also be explained via a migration–coalescence process.

A few small crystallites very close to a larger one were observed to decrease in size and then disappear (Type E event). This phenomenon which occurred in localized regions of the surface and involved only neighboring crystallites could possibly be explained by direct ripening (*17*). In contrast to Ostwald ripening which can

TABLE 6

Events Observed during Sintering at 800°C

Time interval (hr)	Events
0–0.167	J
0.167–0.5	A, D, E[a], F, G, H[b]
0.5–1	A, D, E[a], F, G, H[b]
1–2	A, D, F, G, H[b], L[a]
2–4	A[a], D, F[b], G, H, L
4–8	A, D, E[a], F, G, H[b]
8–20	A[a], D, F[b], G, H

[a] Least frequently observed event.
[b] Most frequently observed event.

TABLE 7

Average Crystallite Size and Specific Surface Area of $Pd/\gamma\text{-}Al_2O_3$ at 800°C

Cumulative time (hr)	Number-average diameter (Å)	Surface-average diameter (Å)	Specific surface area (m^2/g Pd)	$\left(\frac{S}{S_0}\right)$
0	31.7	41.0	121.748	1.00
0.167 (10 min.)	57.3	65.2	76.560	0.63
0.5	68.5	79.6	62.700	0.51
1	70.9	91.4	54.645	0.45
2	74.6	101.4	49.234	0.41
4	87.4	126.3	39.509	0.33
8	90.4	129.9	38.438	0.32
20	112.2	164.0	30.440	0.25

S: specific surface area at time t; S_0: initial specific surface area.

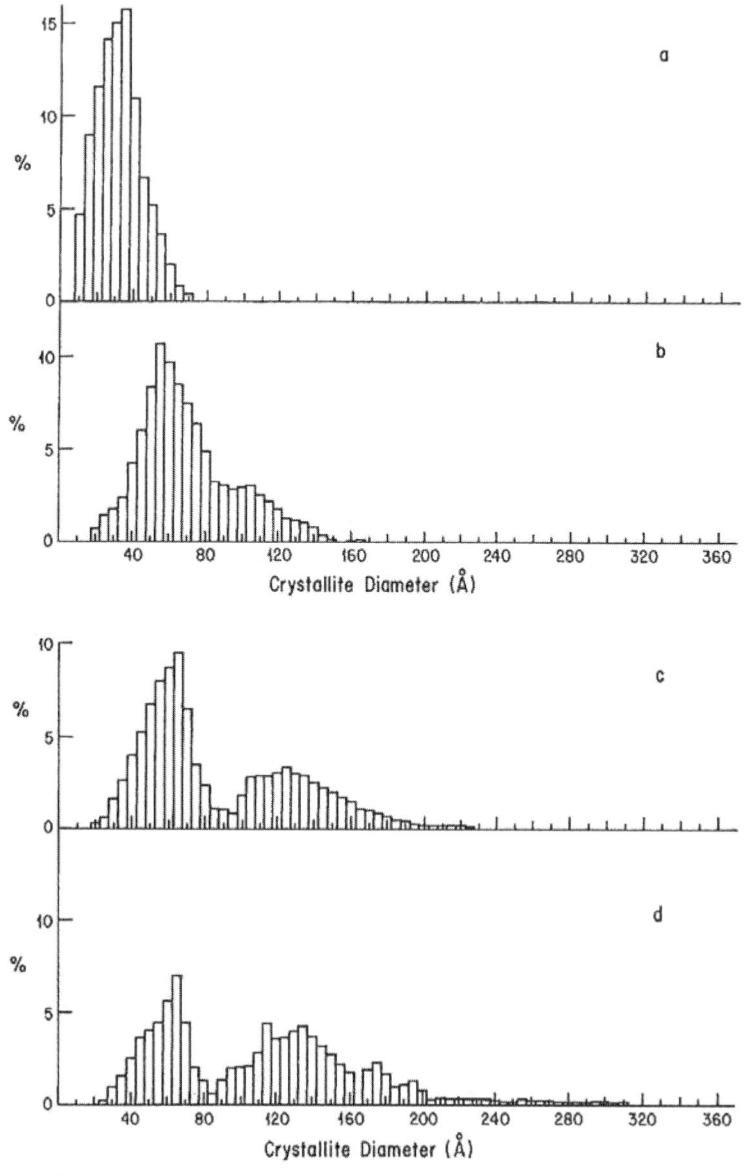

FIG. 8. Size distribution of Pd crystallites: (a) initial distribution; heated in 1 atm H_2 at 800°C for (b) 1 hr; (c) 4 hr; (d) 20 hr.

only occur in a supersaturated system, the direct ripening can occur even if the whole system is unsaturated. As long as two or more particles are close enough to one another, single atoms may transfer directly from the smaller to the larger one since the former has a higher surface pressure than the latter. It is a local event rather than a global process. The localized character of direct ripening probably stems from the heterogeneity of the (alumina) substrate surface, as well as from the particularities of the interactions at the leading edge of the crystallites.

CONCLUSION

The sintering of palladium on alumina during heating in a hydrogen atmosphere is a complex phenomenon which may occur via various mechanisms: (a) crystallite migration and coalescence; (b) direct ripening; and (c) Ostwald ripening. It seems that crystallite migration is not only dependent on the crystallite size and temperature but also on the irregularities of the substrate since the latter affect the interaction of crystallite with the substrate. In some cases, the location of the crystallite rather than its size affects the mobility. Evidence for direct ripening which involves atomic migration between a small and a larger particle is also brought. Ostwald ripening or migration and coalescence are two possible mechanisms which could explain the disappearance of a large number of small crystallites that was observed between two consecutive stages of heating.

REFERENCES

1. Klimisch, R. L., Summers, J. C., and Schlatter, J. C., *in* "Catalysts for the Control of Automotive Pollutants" (J. E. McEvoy, Ed.). Advances in Chemistry Series No. 143, Amer. Chem. Soc., Washington, D.C., 1975.
2. Aben, P. C., *J. Catal.* **10**, 224 (1968).
3. Herrmann, R., Adler, S. F., Goldstein, M. S., and DeBaun, R. M., *J. Phys. Chem.* **65**, 2189 (1961).
4. Graber, H. L., *J. Phys. Chem.* **66**, 48 (1962).
5. Somorjai, G., *Anal. Chem.* **1**, 101 (1968).
6. Maat, H. J., and Moscou, L., *in* "Proceedings, 3rd International Congress on Catalysis, Amsterdam, 1964," Vol. 11, p. 1277. Wiley, New York, 1965.
7. Flynn, P. C., and Wanke, S. E., *J. Catal.* **37**, 432 (1975).
8. Baker, R. T. K., Thomas, C., and Thomas, R. B., *J. Catal.* **38**, 510 (1975).
9. Chu, Y. F., and Ruckenstein, E., *J. Catal.* **55**, 281 (1978).
10. Ruckenstein, E., and Malhotra, M. L., *J. Catal.* **41**, 303 (1976).
11. Avery, N. R., and Sanders, J. V., *J. Catal.* **18**, 129 (1970).
12. Ruckenstein, E., and Pulvermacher, B., *AIChE J.* **19**, 356 (1973).
13. Ruckenstein, E., and Pulvermacher, B., *J. Catal.* **29**, 224 (1973).
14. Chakraverty, B. K., *J. Phys. Chem. Solids* **28**, 2401 (1967).
15. Flynn, P. C., and Wanke, S. E., *J. Catal.* **33**, 233 (1974).
16. Ruckenstein, E., and Dadyburjor, D., *J. Catal.* **48**, 73 (1977).
17. Ruckenstein, E., and Dadyburjor, D. B., *Thin Solid Films* **55**, 89 (1978).

Role of Interfacial Phenomena in the Behavior of Alumina-Supported Palladium Crystallites in Oxygen

James J. Chen† **and Eli Ruckenstein***

Department of Chemical Engineering, State University of New York at Buffalo, Buffalo, New York 14260

(Received: December 12, 1980; In Final Form: March 4, 1981)

Palladium crystallites with sizes in the range of 30–350 Å supported on thin alumina films have been heated in 1 atm of oxygen at 350, 500, 750, and 920 °C. Formation of pits and their coalescscence into large cavities have been identified on the crystallites by transmission electron microscopy. In addition, because the wetting angle of the oxide is smaller than that of the metal, some spreading of the oxidized crystallites over the surface of alumina was observed within the temperature range in which PdO is stable (below 870 °C). At 350 °C, even though cavities form on all the crystallites, only the smaller crystallites spread appreciably on the substrate. While at 500 °C, the spreading of the oxidized crystallites is very extensive and spreading-induced rupture and fragmentation were also observed, at 750 °C spreading is not as extensive. At 920 °C, where the oxide is unstable, the particles sinter into faceted crystallites just as observed previously during heating in hydrogen. The spreading of crystallites over the surface of alumina is explained in terms of wetting. A possible explanation for the formation of pits and cavities is as follows: The oxidation takes place at some active sites. The PdO thus formed has a lower interfacial tension than the unoxidized metal and in order to decrease the free energy of the system spreads over the metal thus generating pits on the crystallites. The stress generated by the surface tension gradient constitutes the driving force for this motion. The pits coalesce into large cavities as oxidation proceeds. The stresses generated because of the spreading of the crystallites over the substrate can lead to the rupture and even to the fragmentation of the crystallites. The stresses induced by the surface tension gradient on the surface of the crystallites and by the spreading of the crystallites as well as the oxidation process may act in cooperation thus accelerating the formation of pits and/or the rupture.

Introduction

Supported palladium catalysts are used in a wide variety of reactions such as hydrogenation, dehydrogenation, and oxidation of olefins. However, their behavior and characteristics in different chemical environments have not been well investigated. In a previous paper of this laboratory,[1] the sintering during heating in hydrogen of palladium crystallites supported on alumina has been systematically studied. The present work deals with the behavior of palladium crystallites when the heating takes place in an oxygen atmosphere.

Palladium crystallites supported on alumina films of about 300 Å thickness were heated in oxygen at 350, 500, 750, and 920 °C for various time intervals. To follow the changes in size, shape, and position of each crystallite, we examined the same region of a specimen with a JEOL 100U transmission electron microscope after repeated heating at the same temperature for various time intervals. The formation of pits on the crystallites, the coalescence of these pits into cavities, and the spreading of the oxidized crystallites over the alumina surface have been observed during the experiments. The formation of pits is tentatively explained by the oxidation which starts at some active sites on the crystallite surface and by the spreading of the formed palladium oxide, which has a lower surface tension than palladium, over the nonoxidized metal. The pits developed during the early stages of oxidation coalesce to form large cavities. The spreading of crystallites is dependent on the extent of oxidation and can be explained in terms of wetting.

Experimental Section

The details of the experimental procedure have been described in a previous paper.[1] A thin and highly pure aluminum foil was first chemically polished and then anodized at 20 V for 30 s in a 3 wt.% tartaric acid solution at pH 5.5. The nonporous and amorphous oxide film, about 300 Å thick, was separated from the unoxidized aluminum foil by amalgamation. The films were then transferred to and washed in a large amount of distilled water. Finally, they were picked up on gold microscope grids and dried in air. Phase transformation of amorphous alumina samples into crystalline γ-alumina was carried out by calcination in air at 800 or 920 °C for 72 h. This pretreatment ensured that, during heating in oxygen, no further phase transformation and no motions of the grains of the substrate would occur.

A Pd film of about 10 Å thickness was deposited onto the γ-alumina substrate by vacuum evaporation of Pd wire wound on a tungsten filament. The Pd/Al_2O_3 samples were heated from room temperature to 800 °C in flowing hydrogen inside a quartz tube and then cooled to room temperature in flowing helium. This treatment broke up the Pd film into Pd crystallites. The heating in oxygen was carried out at 1 atm in a quartz tube placed in a tubular furnace. The quartz tube was first preheated to the desired temperature with oxygen flowing through and then samples contained in a ceramic boat were displaced to the middle of the tube. As soon as the prespecified time interval was reached, the furnace was turned off and the samples were cooled to room temperature in a helium stream. Compressed laboratory air was allowed to flow through the annular region between furnace and tube to facilitate cooling. High-purity oxygen and helium gases, purchased from Union Carbide Corp., were passed through a column of 5A molecular sieves before entering the tube to remove any trace of water. The flow rate of the gases was about 30 mL/min.

Results

Figure 1 (a–d) shows the time sequence of the same region containing some large Pd crystallites. These large crystallites, with diameter of about 150 Å or larger, became

† J. J. Chen is now associated with Exxon Corporation, New Jersey.

(1) Chen, J. J.; Ruckenstein, E. *J. Catal.* In press.

This paper, which was originally published in **Journal of Physical Chemistry**: 85, 1606–1612 (1981), is republished with permission.

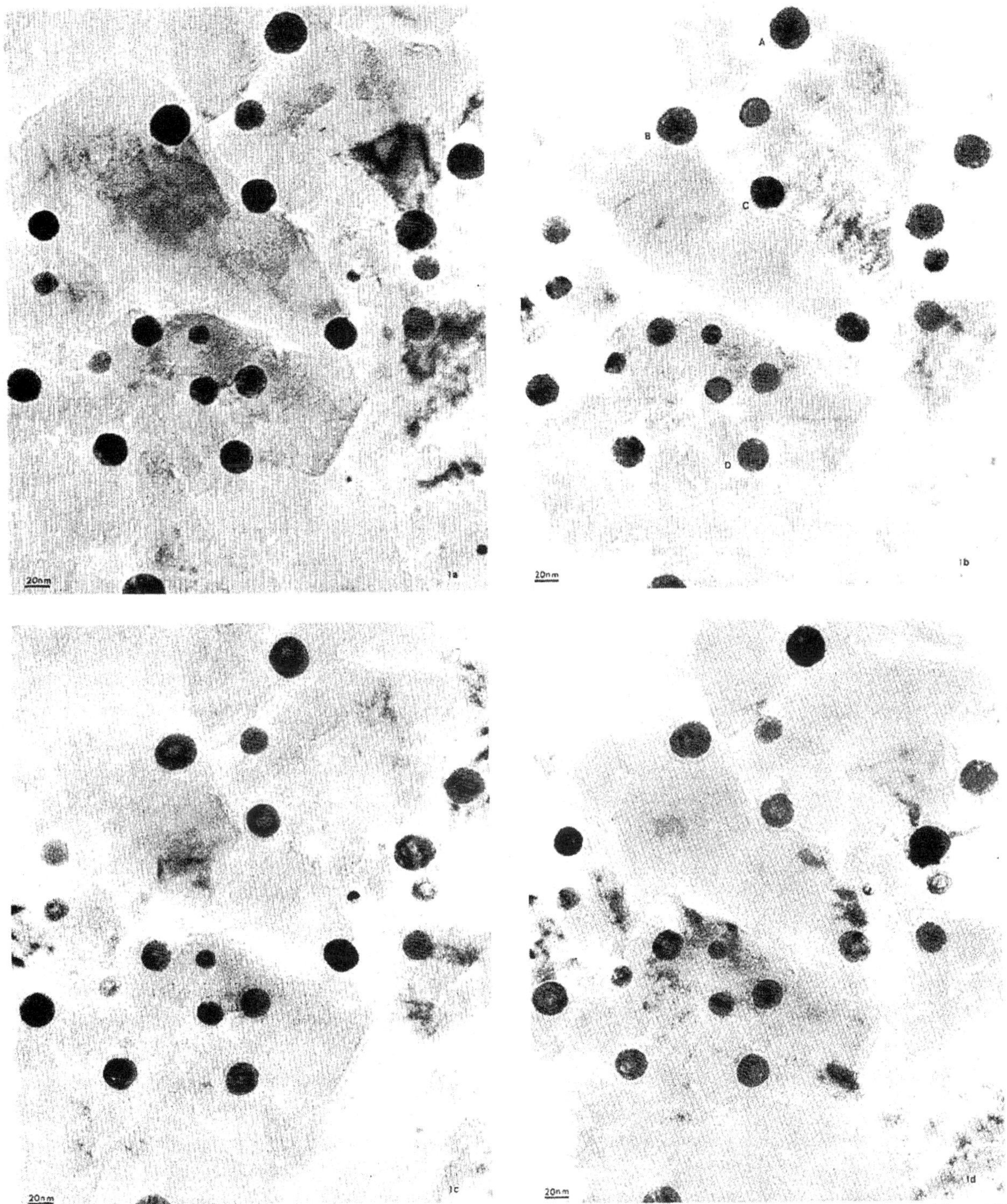

Figure 1. Time sequence of the same region of Pd/Al_2O_3 specimen (large crystallites): (a) unoxidized specimen; heated in 1 atm of O_2 at 350 °C for (b) 2 h; (c) 4 h; (d) 24 h.

porous after heating in O_2 at 350 °C for 2 h (crystallites A, B, C, and D of Figure 1b). The porous structure is more obvious when similar samples were heated at higher temperature (Figure 4a). Additional heating for 2 h at 350 °C led to the formation of toruslike crystallites (Figure 1c). The crystallites showed a boundary in their central part, while their size and leading edge remained largely unchanged perhaps because the height of the crystallites has increased. The toruslike shape is stable at 350 °C, since further heating for up to 20 h did not cause any significant change in shape and size (Figure 1d).

The smaller crystallites, on the other hand, show more drastic changes at 350 °C (Figure 2a–d) than the larger ones do. During the first 2 h of heating, they acquired a toruslike shape as well (Figure 2b). However, the crystallites could no longer preserve their circular leading edge

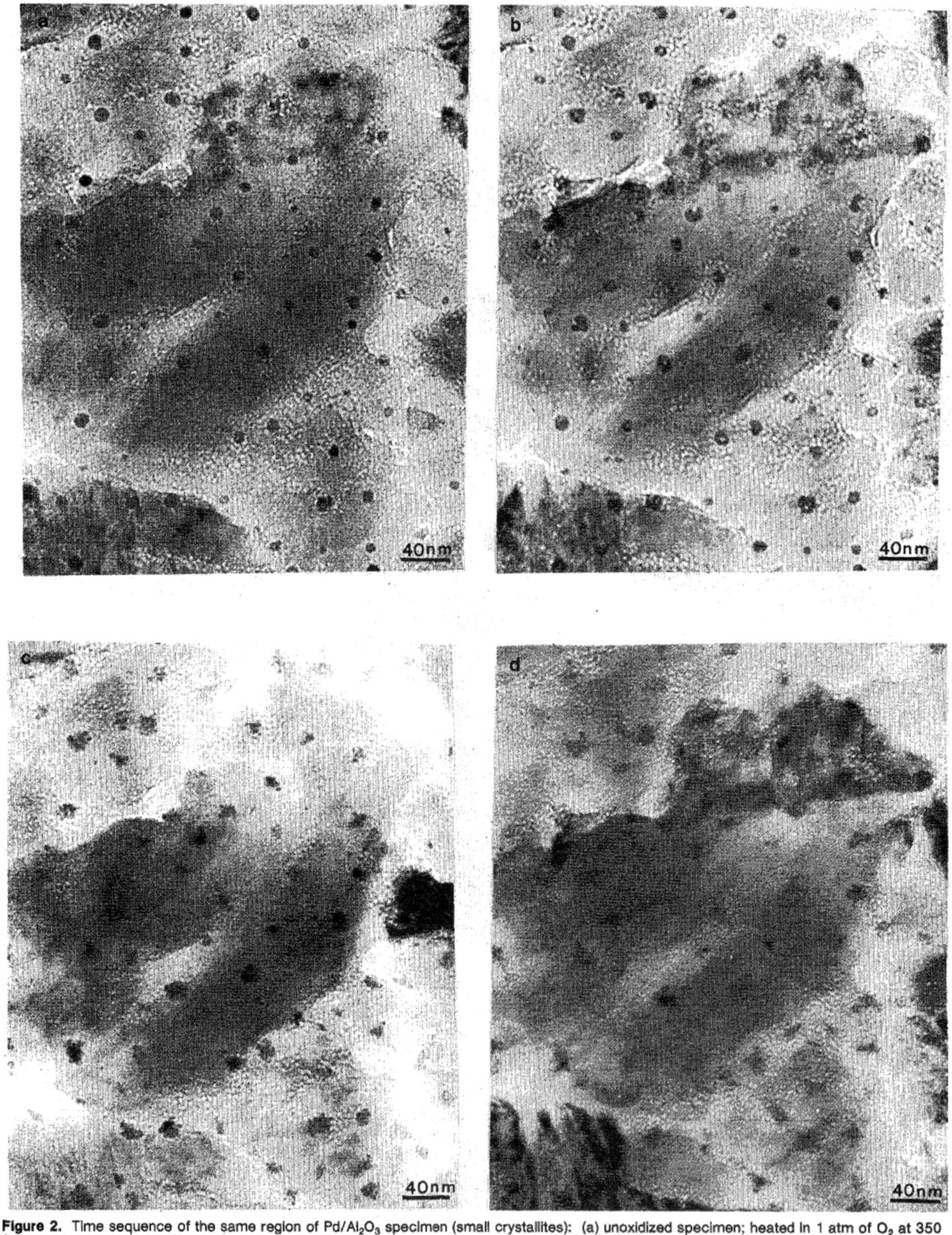

Figure 2. Time sequence of the same region of Pd/Al_2O_3 specimen (small crystallites): (a) unoxidized specimen; heated in 1 atm of O_2 at 350 °C for (b) 2 h; (c) 3 h; (d) 24 h.

after additional heating for 1 h. The crystallites spread during this period of heating over the surface of the substrate and their shapes became irregular (Figure 2c). Further spreading and more irregular shapes of the crystallites were observed as the oxidation progressed (Figure 2d).

The size of the crystallites plays a role in the oxidation at 350 °C but does not have such a major effect at higher temperature. Figure 3a shows a region of a sample heated at 500 °C for 1.5 h. Almost all the crystallites have irregular shapes and contain cavities of different sizes. The dumbbell shape of some crystallites is the result of spreading of neighboring crystallites. Further heating for an additional 5 h made the crystallites spread even more extensively over the surface of the crystallites (Figure 3b). The leading edges of crystallites became ambiguous and ruptures in the crystallites occurred. More spreading and rupture, and even some fragmentation, were observed as the oxidation proceeded further (Figure 3c).

The palladium crystallites showed irregular shapes and porous structures after just 20 min of heating at 750 °C (Figure 4a). As during oxidation at 500 °C, torus-shape cavities and spreading of crystallites have been observed. Figure 4b shows the same region as Figure 4a after 24 h of heating at 750 °C. Spreading has led the neighboring crystallites in areas A and B to form two large crystallites. However, the crystallites did not spread over the surface of the substrate to the same extent as those heated at 500 °C (Figure 3c).

As the temperature was raised to 920 °C, which is 50 °C above the decomposition temperature of the bulk palladium oxide in 1 atm of oxygen,[2] palladium particles sintered and formed faceted crystallites (Figure 5). No pits, cavities, or spreading have been detected during the heating at 920 °C.

Discussion

The formation of pits on the crystallites and the spreading of crystallites are associated with the oxidation of palladium to palladium oxide. Indeed, these two phenomena were not observed when similar samples were heated in hydrogen[1] and in oxygen but at temperatures higher than the decomposition temperature of the bulk palladium oxide. The decomposition temperatures of PdO in 1 atm of air and 1 atm of oxygen are 700 and 870 °C, respectively.[2] Small Pd particles supported on silica were found to be completely oxidized to PdO at temperatures between 538 and 580 K when heating in oxygen.[3] For these reasons we expect Pd to be oxidized to PdO if the temperature is not too high.

Formation of Pits and Cavities. Generally speaking, the actual surface of a crystallite is not homogeneous. Some locations on the surface are more reactive than others. At those locations, PdO forms. The stress generated by PdO which has a volume about 1.5 times that of Pd (calculated from the density of PdO and Pd at 20 °C)[4] could cause the crystallite surface to crack. This kind of cracking was observed, for instance, in a nickel film undergoing oxidation at 350 °C.[5] In addition, PdO has a lower surface tension than Pd. As a consequence, in order to decrease the free energy of the system, the PdO has the tendency to move on the crystallite surface to cover the unreacted regions of the crystallite. This phenomenon, driven by the surface tension gradient, was previously identified for liquid-fluid interfaces and called the Marangoni effect.[6] The motion is caused by the stress induced by the surface tension gradient and takes place in the direction of increasing surface tension. Both cracking and migration of PdO molecules out of their positions expose fresh (hence more reactive) Pd atoms in deeper layers to oxygen. Therefore as the process proceeds, a pit might develop in the crystallite. These pits could coalesce and large cavities thus form.

Spreading. Spreading of metal particles over the surface of the substrate was used in ref 7 and 8 to explain the redispersion of platinum crystallites on alumina in an oxygen atmosphere. Of course, the spreading caused by a decrease in the wetting angle increases the number of metal atoms on the free surface, thus increasing the dispersion of the metal. However, two more important possible mechanisms for redispersion based on spreading have been emphasized in the above references. The first mechanism implies that no wetting angle exists between the oxidized metal crystallites and substrate. This occurs when the surface tension of the metal oxide formed in the oxygen atmosphere and the interfacial tension between the crystallite and substrate become sufficiently low so that no equilibrium wetting angle is possible. In this case the crystallites will spread, rupture, and fragment to form ever smaller particles. In the second, metal crystallites coexist with a two-dimensional phase of metal oxide. Redispersion occurs in this case via the emission of metal oxide molecules by the crystallites (of course, the capture of these molecules by the large crystallites because of Ostwald ripening must be either nonexistent or less important than the emission by the smaller ones). (An additional mechanism of redispersion is the splitting of crystallites.[8] Direct evidence for this mechanism is available in ref 8, where it is shown that planar and thin crystallites of Pt on alumina can be split by applying a perturbation (by changing, for instance, the chemical atmosphere), and in ref 9, where it is shown that thin crystallites of Pt on graphite split upon cooling.) These considerations have also been extended[10] to explain the behavior of platinum on titanium oxide when heated in oxygen and hydrogen.[11]

In the present experiments, the oxidized crystallites spread over the surface of alumina to a new internal equilibrium wetting angle which is smaller for the oxide than for the metal. The spreading was found to be dependent on temperature and size, since the smaller crystallites oxidized more completely and thus the interfacial tension of the crystallite–substrate is smaller. It is possible that a two-dimensional phase of PdO molecules coexists with the crystallites. However, our experiments cannot provide any evidence in this direction.

At 350 °C, the smaller crystallites, Figure 2 (a–d), spread over the surface of alumina, while the larger ones, Figure 1 (a–d), were able to preserve their size and circular leading edge. From the fact that the large crystallites remained largely unchanged after they acquired a torus shape, one may conclude that these large crystallites were not completely oxidized. The pure metal part of the large crystallites might be covered by a layer of palladium oxide such that further oxidation was prevented, while the smaller crystallites were nearly completely oxidized. The irregular shape of the crystallites is due to the nonuniform oxidation and resulting nonuniform spreading.

The extent of spreading is higher at 500 °C than at 350 °C since almost all the crystallites have been completely oxidized at the higher temperature (Figure 3 (a–c).) At 750 °C, the crystallites show less spreading than at 500 °C,

(2) Samsonov, G. V., Ed. "The Oxide Handbook" (C. N. Turton and T. L. Turton, Transl.); Plenum: New York, 1973.
(3) Lam, Y. L., Boudart, M. *J. Catal.* **1977**, *47*, 393.
(4) "Handbook of Chemistry and Physics"; CRC Press: Cleveland, 1974; 55th ed.
(5) Anderson, J. R., Ed.; "Chemisorption and Reactions on Metallic Films", Academic Press: New York, 1971; Vol. 2, p 287.
(6) Adamson, A. W., "Physical Chemistry of Surfaces"; Interscience: New York, 1967; 2nd ed.
(7) Ruckenstein, E.; Pulvermacher, B. *J. Catal.* **1973**, *29*, 224.
(8) Ruckenstein, E.; Chu, Y. F. *J. Catal.* **1979**, *59*, 109.
(9) Chu, Y. F.,; Ruckenstein, E. *Surf. Sci.* **1977**, *67*, 517.
(10) Baker, R. T. K. *J. Catal.* **1980**, *63*, 523.
(11) Baker, R. T. K.; Prestridge, E. B.; Garten, R. L. *J. Catal.* **1979**, *59*, 293.

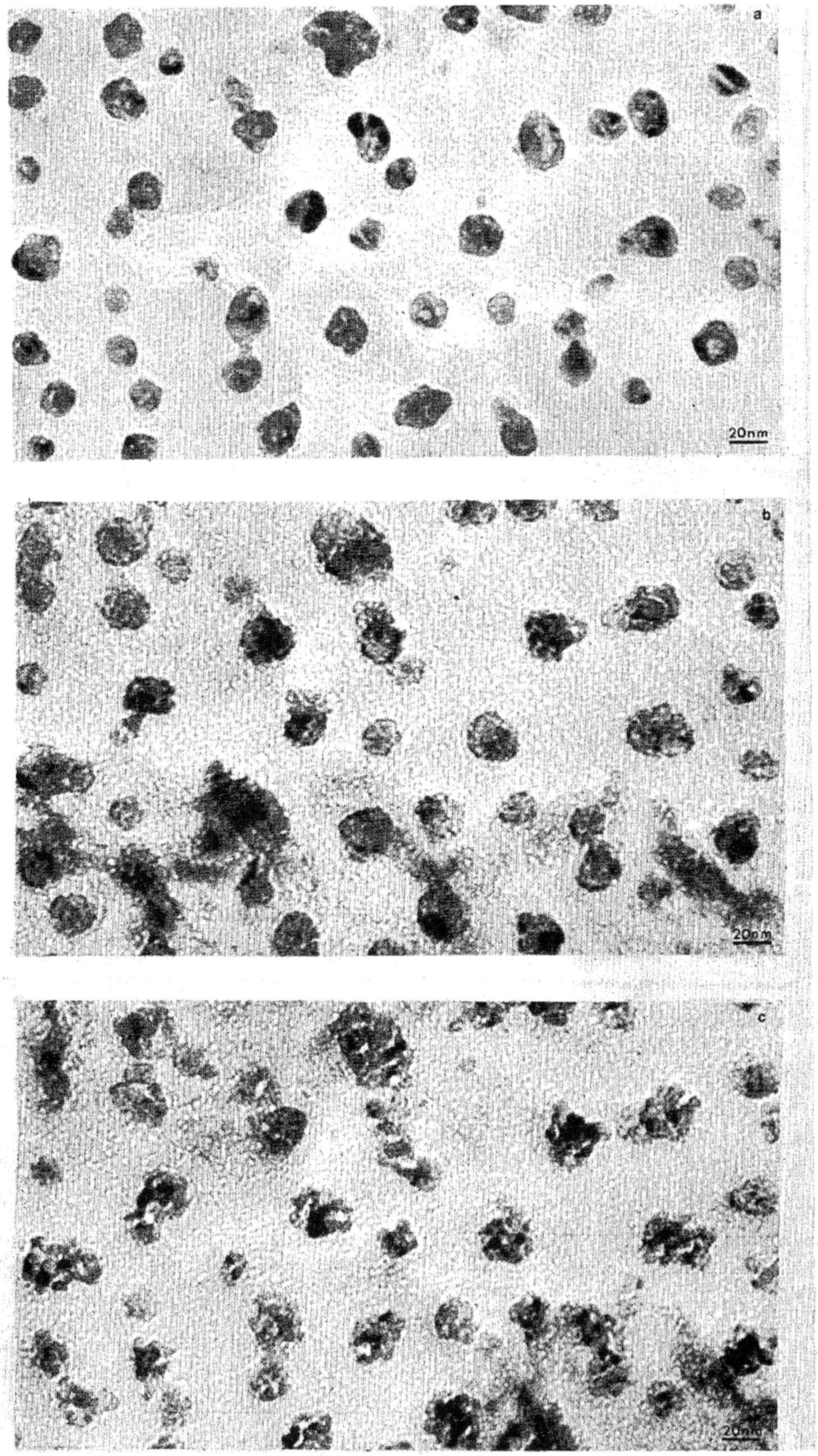

Figure 3. Time sequence of the same region of Pd/Al_2O_3 specimen after being heated in 1 atm of O_2 at 500 °C for (a) 1.5 h; (b) 6.5 h; (c) 24 h.

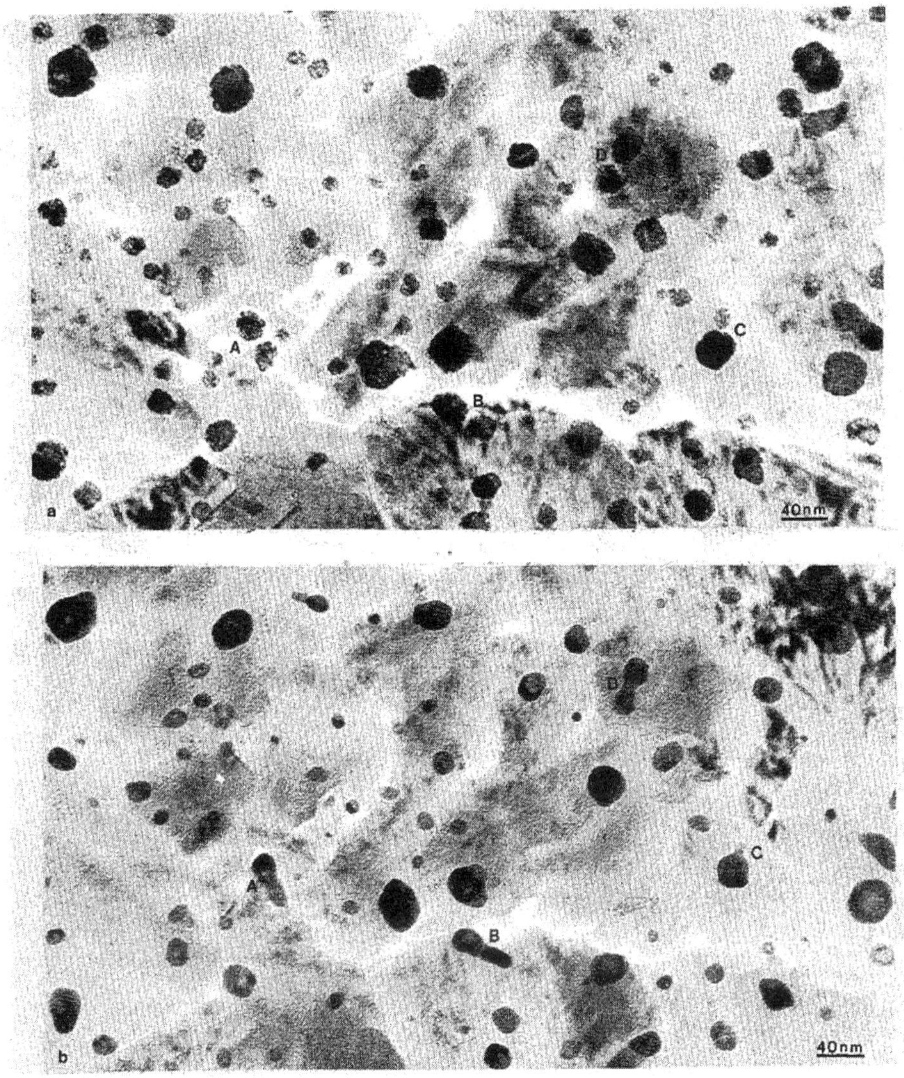

Figure 4. The same region of a Pd/Al_2O_3 specimen after being heated in 1 atm of O_2 at 750 °C for (a) 20 min; (b) 24 h.

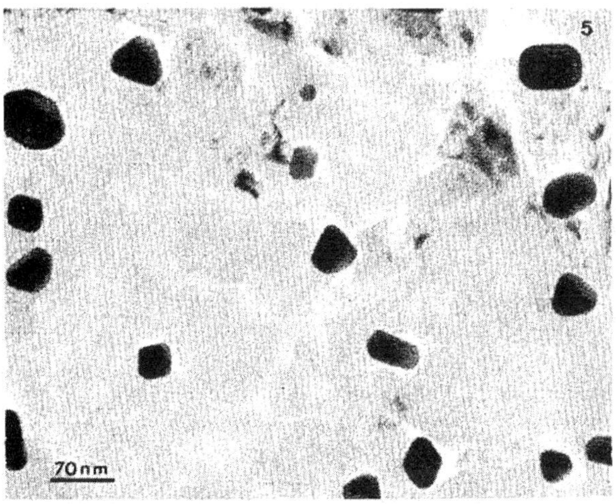

Figure 5. Transmission electron micrograph of a Pd/Al_2O_3 specimen after being heated in O_2 at 920 °C for 6 h.

probably because the wetting angle is larger at the former temperature or because the oxide is less stable at 750 °C.

The micrographs of Figure 3 indicate that rupture and fragmentation of the crystallites occur during spreading. These are probably caused by the stresses generated by spreading which may be large enough to cause the propagation of existing small cracks on the surface of the crystallites. The continuous oxidation at the tip of the cracks contributes to and might be essential for this process (see Appendix). The stresses induced by the surface tension gradient and by spreading as well as the oxidation process may act in cooperation thus accelerating the formation of pits and/or rupture.

Summary

Along with the oxidation to PdO, Pd crystallites develop first a porous structure which is followed by the formation of cavities. In addition, spreading of the crystallites over the surface of alumina takes place because the wetting angle of the oxide differs from that of the metal. The development of pits is probably due to the oxidation occurring at some particular active sites and the motion of PdO which has a lower surface tension, over the unoxidized metal surface with a higher surface tension. The pits coalesce to form large cavities on the crystallites. The extent of spreading is dependent on the extent of oxidation of the crystallites. The stresses induced by spreading probably play a role in the rupture and fragmentation of the crystallites and in the pit formation.

Appendix

It is well-known that the stress needed to separate two atomic layers is very large and that in the presence of cracks it can be appreciably decreased. Griffith[12] has derived the following expression for the critical stress σ_c needed to initiate crack growth:

$$\sigma_c = \left(\frac{2E\gamma}{\pi r}\right)^{1/2}$$

where E is the Young elastic modulus, γ is the surface tension, and r is a length characterizing the crack. Using $E \simeq 10^{12}$ dyn/cm^2, $\gamma = 10^3$ dyn/cm, and $r \simeq 10^{-7}$ cm, one obtains $\sigma_c \simeq 10^{11}$ dyn/cm^2. For a crystallite of 10^{-6} cm size the stress generated by wetting cannot be larger than about $\gamma/10^{-6} \simeq 10^9$ dyn/cm^2 which is small compared with $\sigma_c \simeq 10^{11}$ dyn/cm^2. For this reason we are inclined to believe that the oxidation occurring at the tip of the crack, a process which is not accounted for in Griffith's derivation, plays an essential role in this case.

(12) Griffith, A. A., *Phil. Trans. R. Soc., Ser. A.* **1920**, *221*, 163.

Wetting Phenomena during Alternating Heating in O_2 and H_2 of Supported Metal Crystallites

E. RUCKENSTEIN[1] AND J. J. CHEN

Department of Chemical Engineering, State University of New York, Buffalo, New York 14260

Received February 16, 1981; accepted June 1, 1981

Specimens of Pd crystallites (20 to 200 Å in size) supported on thin alumina films have been heated alternately in O_2 and H_2. The changes in size, shape, and position of each crystallite after each heat treatment have been followed by examining the same areas of the specimens with a transmission electron microscope. During heating in O_2, most of the crystallites extended by spreading on the alumina surface and became irregular in shape. Pits and cavities developed on many of the crystallites. During heating in H_2, most of the crystallites which extended during the previous heating in O_2 contracted to a more circular form. Some large crystallites, however, grew in size. Some small crystallites disappeared during heating in O_2 or H_2. The extension and recontraction are explained in terms of wetting. Pit formation is explained as either a surface tension gradient driven phenomenon enhanced by a spreading-produced stress, and/or as a crack propagation phenomenon induced by the spreading-generated tensile stress and enhanced by the oxidation of the freshly exposed atoms at the tip of the crack.

INTRODUCTION

The behavior of Al_2O_3 supported Pd crystallites during heating in hydrogen or oxygen has been investigated and reported recently by the authors (1, 2). Severe sintering of the Pd crystallites was found to occur during heating in H_2 above 650°C. During heating in O_2, the formation of pits on the crystallites, the coalescence of these pits into cavities, the extension by spreading of the oxidized crystallites on the alumina surface and fragmentation induced by spreading have been observed.

The scope of the present paper is to examine the behavior of Pd crystallites (20 to 200 Å in initial size) supported on alumina, under heat treatment in alternating oxidizing and reducing atmospheres. These experiments have been carried out because the alternation between the two extreme conditions can better emphasize the phenomena which occur in each of them and the effect of the chemical atmosphere. Palladium crystallites supported on thin alumina films have been heated alternately in 1 atm O_2 at 750°C and 1 atm H_2 at 550 or 600°C, for various time intervals. A JEOL 100U transmission electron microscope was used to examine the same areas of the specimens after each treatment in order to follow the changes in size, shape, and position of each crystallite. A variety of events has been identified during the experiments. Most Pd crystallites expanded by spreading on the alumina surface during heating in O_2 and recontracted during heating in H_2. Some large crystallites grew in size during heating in H_2. During heating in either O_2 or H_2, some small crystallites disappeared. Pits formed on the crystallites during heating in O_2 and disappeared during heating in H_2.

While extensive literature is available concerning the sintering of supported metal crystallites and its mechanisms [some references are (3–17)], no evidence seems to be available on the spreading of the crystal-

[1] To whom correspondence should be addressed.

This paper, which was originally published in **Journal of Colloid and Interface Science**: 86(1), 1–11 (1982), is republished with permission.

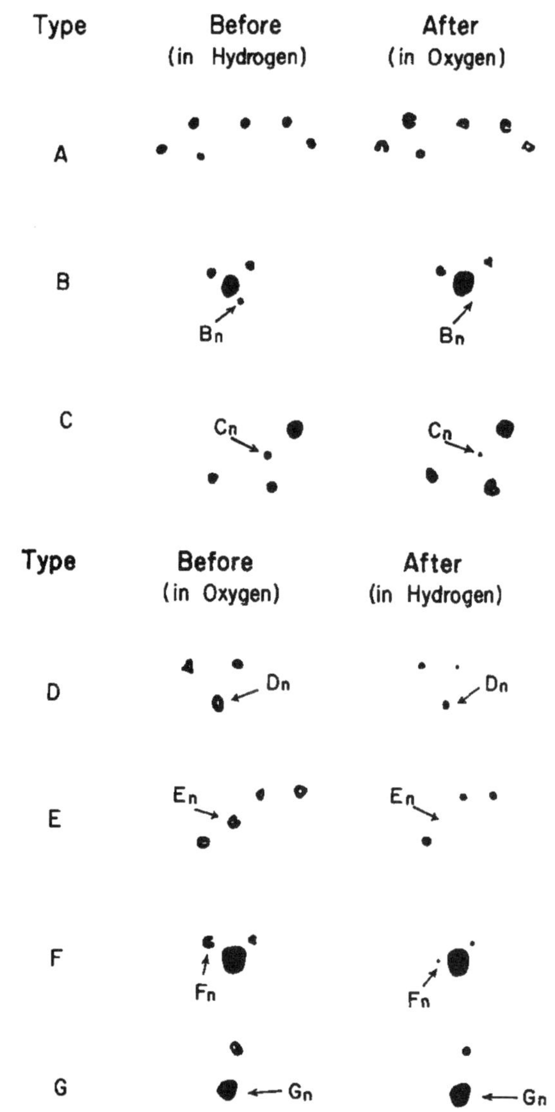

FIG. 1. Schematic illustration of various events observed during heating of Pd/γ-Al_2O_3 specimens in alternating O_2 and H_2 atmospheres. (Type A) During heating in oxygen, the crystallites change their shape from circular to irregular (with or without cavities) and extend by spreading on the alumina surface. (Type B) A small crystallite disappears while others change their shape during heating in oxygen. (Type C) A small crystallite decreases in size while others change their shape during heating in oxygen. (Type D) Previously oxidized crystallites contract back to more circular shapes and decrease in size during heating in hydrogen. (Type E) A previously oxidized crystallite disappears while others contract back to a more circular shape during heating in hydrogen. (Type F) A previously oxidized crystallite contracts and migrates to a new position during heating in hydrogen. (Type G) Previously large oxidized crystallites grow in size while the smaller ones contract during heating in hydrogen.

lites during heating in O_2 and their contraction after an additional heating in H_2, as well as on pit formation during heating in O_2.

EXPERIMENTAL PROCEDURES

Thin alumina films, about 300 Å thick, were used as supports for the palladium crystallites. They were prepared, as suggested in Ref. (18), by anodization of high-purity aluminum foil followed by amalgamation to separate the alumina film from the unoxidized portion of the aluminum foil. The films were then washed in a large amount of distilled water, picked up on gold microscope grids, and dried in air. Before the deposition of palladium, the nonporous and amorphous alumina films were calcined in air at 800°C for 72 hr to transform amorphous alumina into crystalline γ-alumina. The purpose of this prolonged heat treatment was to ensure that no phase transformation and no motions of the grains of the substrate would occur during further heating.

A thin Pd film about 10 Å thick was vacuum-evaporated onto the support from a 99.999% purity Pd wire wound on a tungsten filament. The Pd/Al_2O_3 specimens were heated inside a quartz tube from room temperature to 800°C in a 1-atm H_2 stream and cooled to room temperature in a He stream. This treatment generated Pd crystallites.

The heating of the specimens in H_2 or O_2 was carried out at 1 atm in a quartz tube that is surrounded by a tubular furnace. The specimens contained in a ceramic boat were displaced to the middle of the tube after the tube had been heated up to the desired temperature with He flowing through. Then, H_2 or O_2 was introduced into the tube to replace He. The specimens were cooled down in a He stream after the prespecified time duration was reached. The time span of cooling down was reduced by letting compressed laboratory air to flow through the annular region between furnace and tube. All the gases, H_2, O_2, and He were of high-purity grade and were purchased from Union Carbide Corporation. The gases were passed through a column of 5A molecular sieves to remove any trace of water before entering the tube. The flow rate of gas was about 30 ml/min.

After each step of heat treatment, the same regions of the specimens were examined by a transmission electron microscope, to follow the changes in size, shape, and position of each crystallite.

RESULTS

Figure 1 lists schematically various kinds of events observed on the electron micrographs taken after each stage of alternate heating of the Pd/Al_2O_3 specimens in 1 atm O_2 or H_2. During heating in O_2, most of the crystallites have expanded by spreading on the alumina surface and became irregular in shape. Many of them contain pits or cavities. Some small crystallites, below about 65 Å in diameter, either disappeared or decreased in size. During heating in H_2, most of the crystallites which have extended during the previous heating in O_2 contracted back to a more circular shape with a decrease in size. The pits formed during the previous heating in O_2 disappeared. Disappearance of crystallites and crystallite migration have also been observed. Some large crystallites, with diameters greater than 90 Å before the previous oxidation grew in size upon heating in H_2. Micrographs 2a to g show the time sequence of the same region of a Pd/Al_2O_3 specimen and provide the examples listed in Table I for various kinds of events. Besides the events listed, occasionally new crystallites appeared during heating in H_2, e.g., crystallite N of micrograph 2e.

Figures 3a to e give the size distribution of (about 1000) crystallites measured after each reduction step. Due to the irregular shapes of the oxidized crystallites, the size distribution of crystallites after the oxidation steps was not measured. The decrease in size of many small crystallites during run 2 and the growth in size of some large crystallites shift the minimum diameter to the smaller sizes and lead to a bimodal distribution (Fig. 3b). During later stages many small crystallites disappeared during heating in O_2 and some disappeared during heating in H_2, while the large ones grew gradually during heat treatment in both O_2 as well as H_2. The size distributions of Figs. 3c to e reflect this development.

Table II lists the experimental conditions for each heat treatment, the events observed after each treatment on the electron micrographs of all the regions examined, and two average crystallite sizes after each reduction stage. These two average sizes are: the number-average diameter defined by

$$\bar{D} = \sum N_i\bar{D}_i/\sum N_i \qquad [1]$$

and the surface-average diameter defined as

$$\bar{D}_s = \sum N_i\bar{D}_i^3/\sum N_i\bar{D}_i^2. \qquad [2]$$

Here N_i is the number of crystallites with diameters between $\bar{D}_i - \Delta D_i/2$ and $\bar{D}_i + \Delta D_i/2$, and $\bar{D}_i$ is the average diameter in each interval, ΔD_i.

DISCUSSION

Phenomena Observed during Heating in O_2

The phenomena observed during heating in O_2 include the formation of cavities on the crystallites, the extension by spreading of crystallites on the alumina surface, the disappearance of some small crystallites, and the decrease in size of some other small crystallites. The formation of pits as well as the spreading of the crystallites are associated with the oxidation of Pd to PdO. Indeed, these two phenomena have not been observed during heating in H_2 (1) and during heating in 1 atm O_2 at temperatures larger than 870°C, temperature at which the oxide becomes unstable.

The following mechanisms possibly explain pit (cavity) formation:

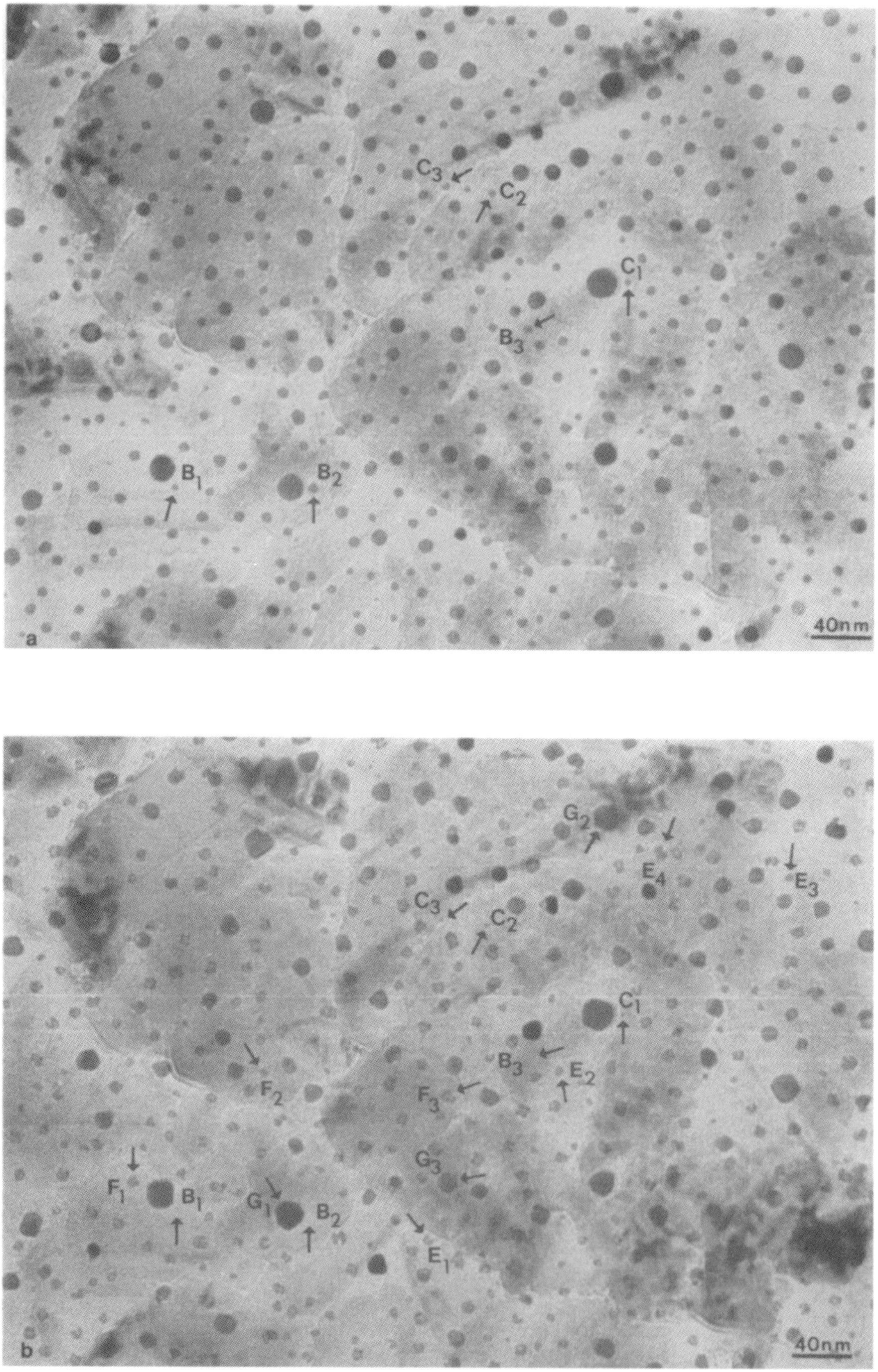

FIG. 2. Transmission electron micrographs showing the same region of a $Pd/\gamma\text{-}Al_2O_3$ specimen in time sequence (a) Fresh specimen; (b) after run 1; (c) after run 2; (d) after run 3; (e) after run 4; (f) after run 5; (g) after run 6. Table II contains the characteristics of the runs.

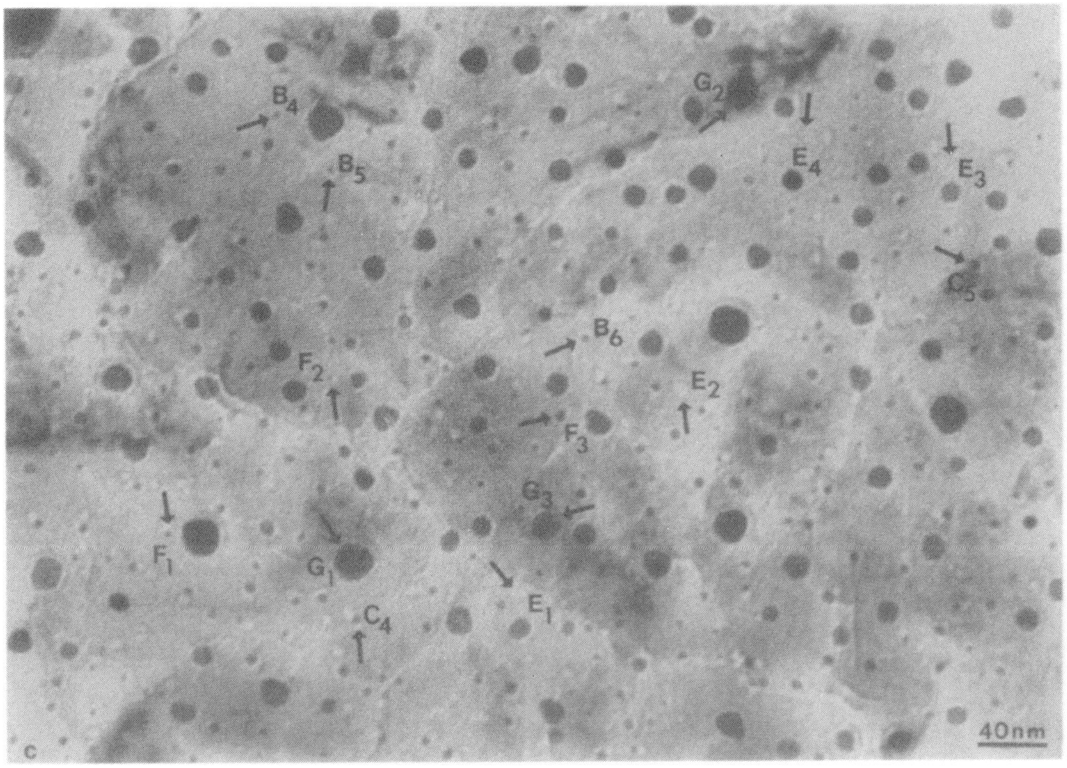

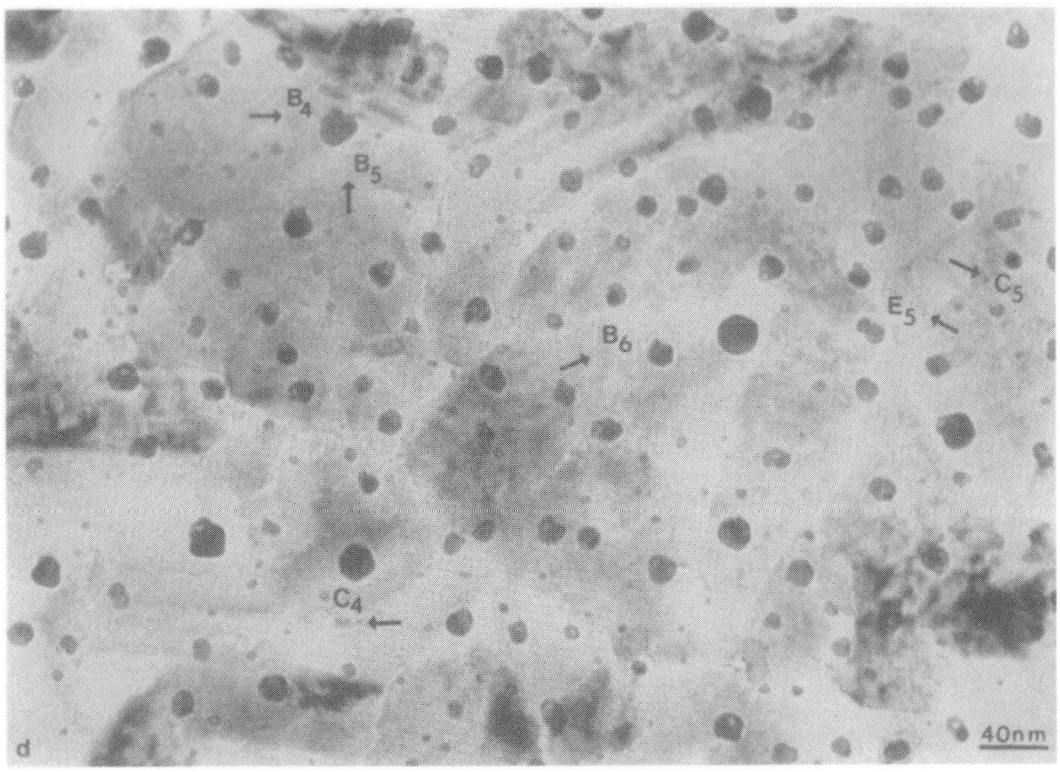

FIG. 2—*Continued.*

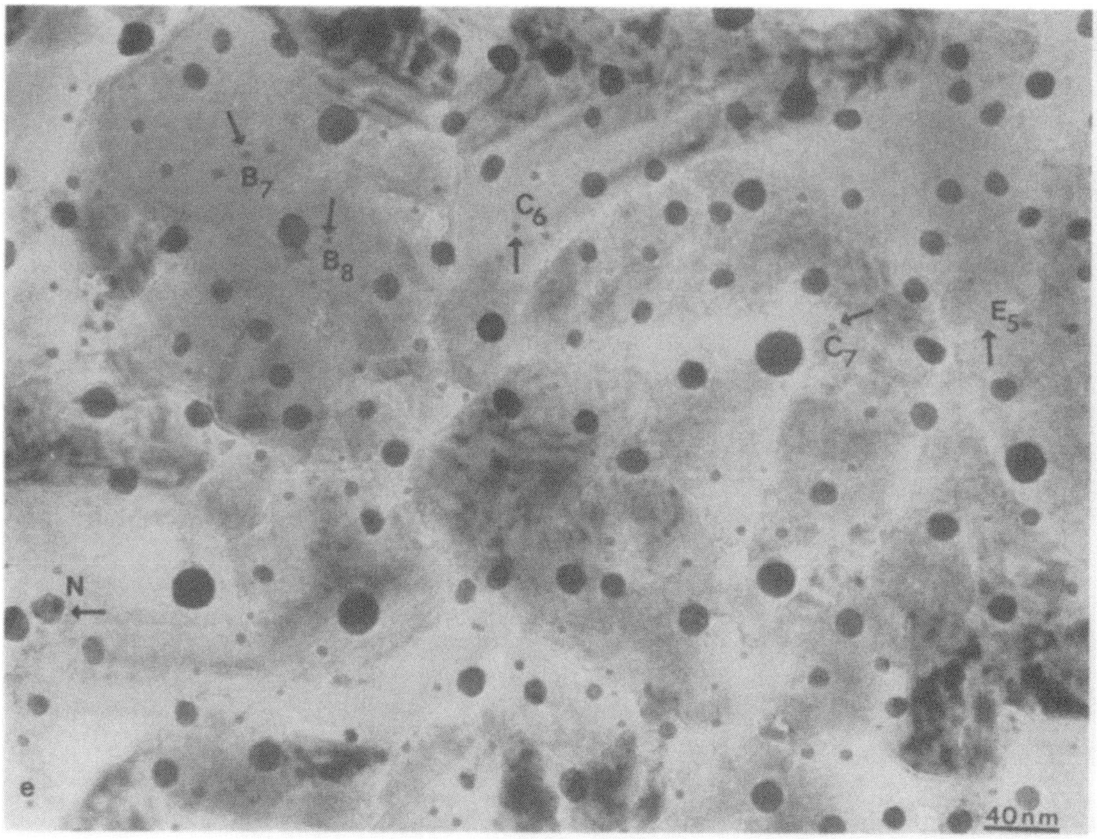

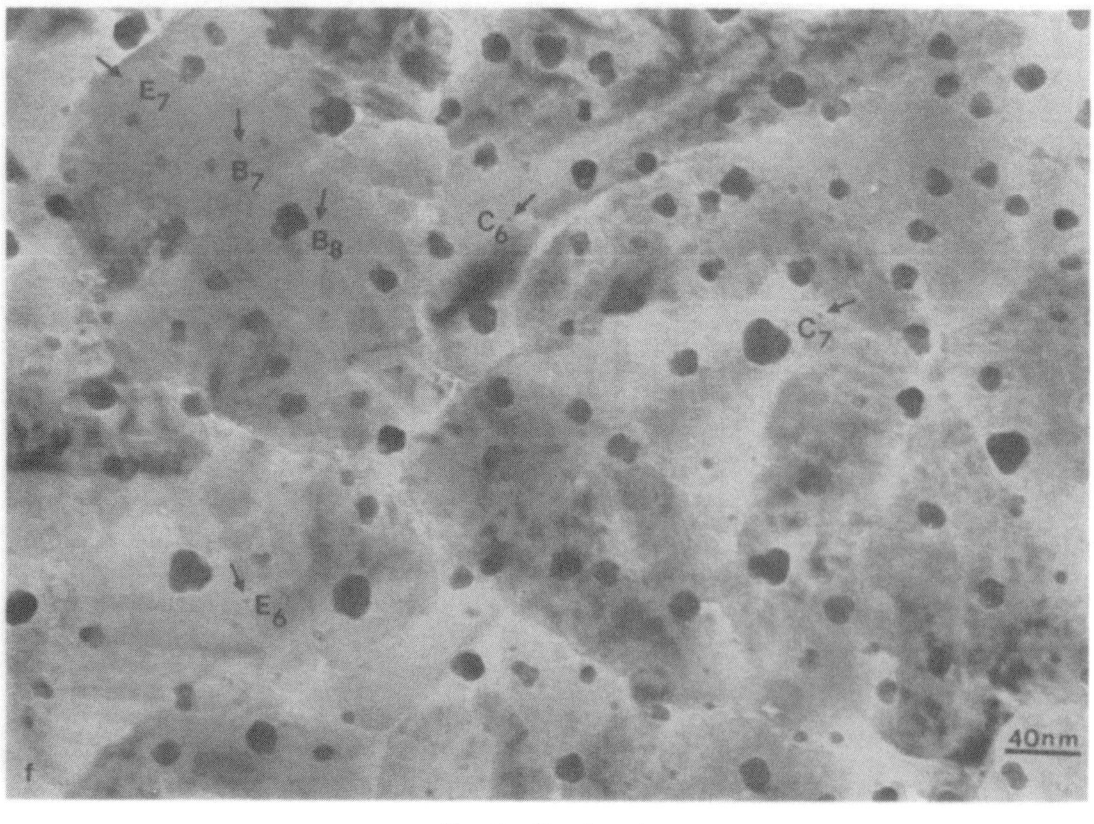

FIG. 2—*Continued.*

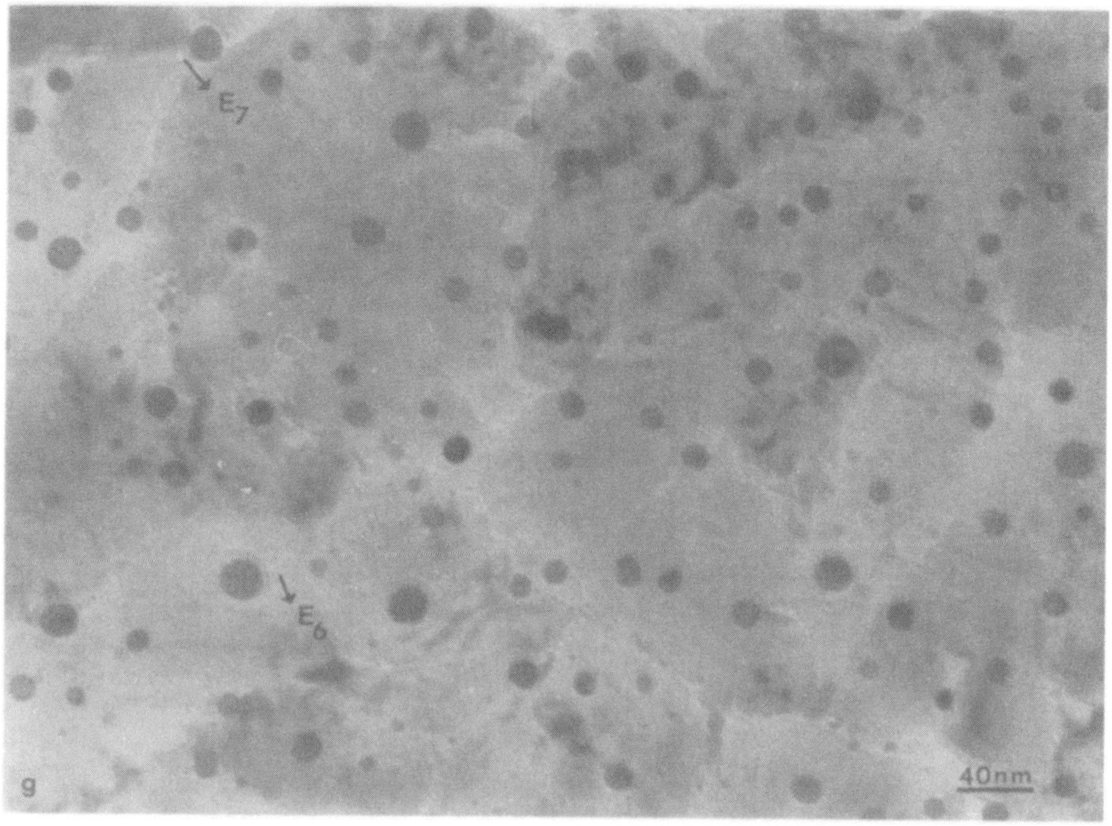

FIG. 2—*Continued.*

(a) The oxidation starts at some particular (active) sites on the heterogeneous crystallite surface. The lower surface tension of PdO molecules induces their migration toward the unoxidized areas of Pd since the metal has a higher surface tension than PdO. This process, driven by the surface tension gradient toward regions with higher surface tensions, is similar to the Marangoni effect in liquids (19). Migration of PdO molecules exposes fresh inner layers of Pd atoms to O_2. Thus the oxidation continues and a pit forms. The large cavities are probably generated by the coalescence of these pits. Since, as explained below, the crystallites extend, during heating in O_2, to a smaller equilibrium wetting angle, the stresses generated during this process may enhance the rate of growth of the pits.

(b) The spreading generated stresses may be sufficiently large to lead to the propagation of small cracks existing on the surface of the oxidized crystallites and thus to generate pits. This process is enhanced by the oxidation of the freshly exposed atoms at the tip of the crack.

(c) A combination of mechanisms (a) and (b).

The extension by spreading of the oxidized crystallites can be explained in terms of wetting. Generally speaking, when a crystallite on the substrate surface is at thermodynamic equilibrium, the wetting angle θ is related to σ_{gs}, σ_{cg}, and σ_{cs}, the interfacial tensions between gas–substrate, crystallite–gas, and crystallite–substrate, via Young's equation[2]:

$$\sigma_{gs} = \sigma_{cs} + \sigma_{cg} \cos \theta \qquad [3]$$

As the crystallite is oxidized, σ_{cg} decreases substantially and σ_{cs} also decreases. Under these conditions, the wetting angle decreases

[2] For small crystallites the equilibrium problem is more complicated and has to be examined along the lines developed in Ref. (7). Because the present discussion is largely qualitative, Young's equation, which is valid for relatively large droplets, is used as the starting point.

TABLE I

Examples of the Events of Fig. 1

Type	Examples
A	Figs. 2a (before) and 2b (after) Figs. 2c (before) and 2d (after) Figs. 2e (before) and 2f (after)
B	B_1 to B_3 of Figs. 2a (before) and 2b (after) B_4 to B_6 of Figs. 2c (before) and 2d (after) B_7 and B_8 of Figs. 2e (before) and 2f (after)
C	C_1 to C_3 of Figs. 2a (before) and 2b (after) C_4 and C_5 of Figs. 2c (before) and 2d (after) C_6 and C_7 of Figs. 2e (before) and 2f (after)
D	Figs. 2b (before) and 2c (after) Figs. 2d (before) and 2e (after) Figs. 2f (before) and 2g (after)
E	E_1 to E_4 of Figs. 2b (before) and 2c (after) E_5 of Figs. 2d (before) and 2e (after) E_6 and E_7 of Figs. 2f (before) and 2g (after)
F	F_1 to F_3 of Figs. 2b (before) and 2c (after)
G	G_1 to G_3 of Figs. 2b (before) and 2c (after)

and hence, the crystallite has the tendency to extend over the surface of the substrate. The irregular shape of the extended crystallites during the heating in O_2 is probably due to the nonuniform oxidation and heterogeneity of the substrate.

Emission of PdO, Ostwald Ripening

Since some small crystallites decrease in size during heating in O_2, it is reasonable to conclude that emission of PdO to the free surface of the substrate occurs. This can generate a two-dimensional phase of PdO on the surface of the substrate (5, 21).

The dissolution surface pressure of the crystallite, P_{sr}, is related to the radius r of the crystallite–substrate interface by a Kelvin-type equation

$$P_{sr} = P_{s\infty} \exp\left(\frac{\delta}{r}\right), \quad [4]$$

where $P_{s\infty}$ is the dissolution surface pressure for a crystallite of infinite size, and

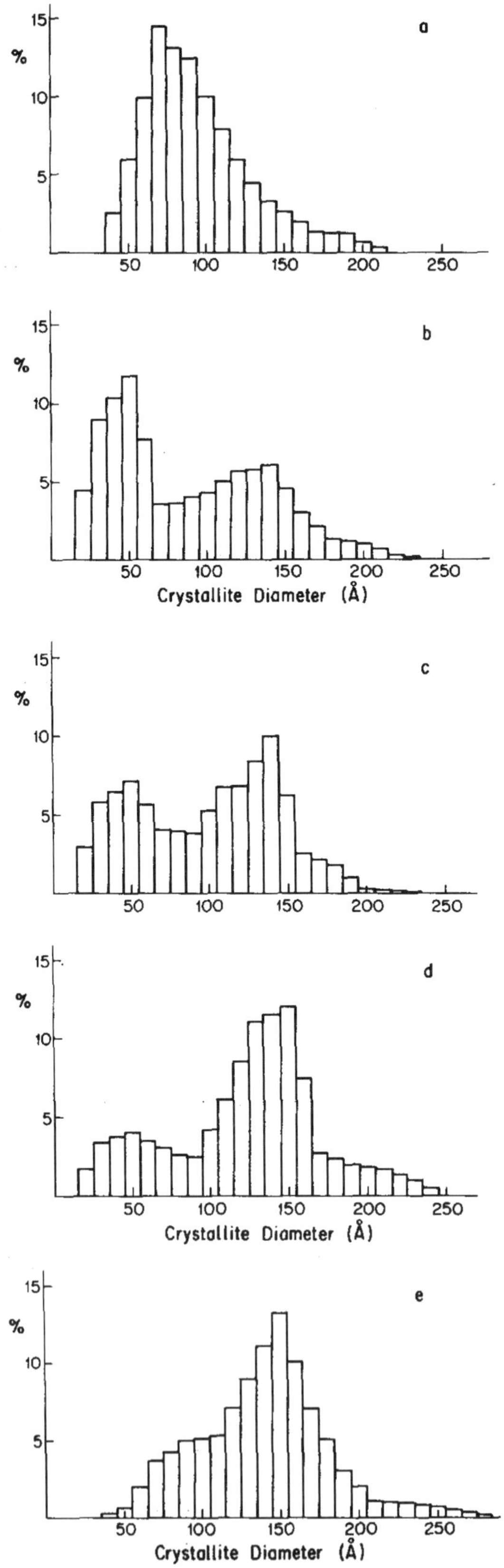

FIG. 3. Size distribution of Pd crystallites on Al_2O_3 substrate. (a) Fresh specimen; (b) after run 2; (c) after run 4; (d) after run 6; (e) after run 14.

TABLE II

Conditions of Heat Treatment, Average Crystallite Sizes, and Events Observed during Experiment

Run	Atmosphere (1 atm)	Temperature (°C)	Time interval (hr)	Events observed	Number-average diameter (Å)	Surface-average diameter (Å)
Fresh specimen					94.30	119.42
1	O_2	750	0.5	A, B, C		
2	H_2	550	1.5	D, E, F, G	92.68	143.97
3	O_2	750	1.0	A, B, C		
4	H_2	550	3.0	D, E, F, G	105.58	145.58
5	O_2	750	2.0	A, B, C		
6	H_2	550	6.0	D, E, G	122.69	153.45
7	O_2	750	4.0	A, B		
8	H_2	550	12.0	D, E	130.75	156.89
9	O_2	750	4.0	A, B		
10	H_2	550	12.0	D, E	135.03	159.51
11	O_2	750	4.0	A, B		
12	H_2	550	12.0	D, G	137.24	161.02
13	O_2	750	10.0	A, B		
14	H_2	600	18.0	D	139.07	162.07

$$\delta = \sigma_l S_l / kT. \qquad [5]$$

Here, σ_l is the line tension (erg/cm) at the leading edge of the crystallite, S_l is the surface area per molecule at the crystallite–substrate interface, k is the Boltzmann constant, and T is the temperature in degrees Kelvin. With the simplified assumption of a two-dimensional ideal gas,

$$P_{s\infty} = n_{s\infty} kT, \qquad [6]$$

where $n_{s\infty}$ is the concentration of single PdO molecules in a two-dimensional phase in equilibrium with a crystallite of infinite size, expressed as molecules per unit area of substrate. Thus,

$$P_{sr} = n_{s\infty} kT \exp\left(\frac{\delta}{r}\right). \qquad [7]$$

Obviously, the dissolution surface pressure is larger when r is smaller. The two-dimensional phase of PdO molecules generates, near the leading edge of the crystallite, a surface pressure, P_s, which, assuming the two-dimensional phase to be an ideal gas, is given by

$$P_s = nkT. \qquad [8]$$

Here n is the concentration, at a given moment, of the molecules in the two-dimensional phase, near the leading edge of the crystallite.

Of course, if the equilibrium concentration of molecules at the leading edge of the crystallite, $n_{s\infty} \exp(\delta/r)$, is larger than n, or in other words, if $P_{sr} > P_s$, single molecules would be emitted from the crystallite onto the substrate. The opposite happens when $P_{sr} < P_s$. Ostwald ripening occurs when $P_{sr} > P_{sb}$ for the small crystallites and $P_{sr} < P_{sb}$ for the large ones, where P_{sb} is the surface pressure in the bulk of the two-dimensional phase. In such cases, the small crystallites decrease in size, while the large ones increase [Ostwald ripening (5)].

The disappearance of some small crystallites during heating in oxygen (event B, Fig. 1) can be explained as follows: A small crystallite has a much larger dissolution pressure than a large one does and thus its emission of molecules per unit length of leading edge and unit time is also higher. Somewhat larger crystallites will only decrease in size (event C, Fig. 1). Since the dissolution pressure is negligible for large crystallites, they may increase their radius by: (1) the spreading caused by the oxidation of Pd; (2) the coalescence with small migrating crystallites; and, if the surface pressure P_{sb} is sufficiently large, (3) by the capture of molecules of the two-dimensional phase of PdO molecules.

Phenomena Observed during Heating in H_2

Upon heating in H_2, most of the crystallites which extended during heating in O_2 contracted back to more circular shapes because the oxide is reduced to metal. The sizes of the reduced crystallites are usually smaller than those of the oxidized ones (event D). This is because the equilibrium wetting angle between Pd and Al_2O_3 is larger than that between PdO and Al_2O_3. The sizes of the reduced crystallites are even smaller than those of the corresponding crystallites before oxidation. This suggests that the crystallites might have lost some of their atoms during the oxidation step.

One of the two possible mechanisms for event E is that the emission of single PdO molecules from the crystallite during heating in O_2 is so substantial that the size of the crystallite resulting during heating in H_2 becomes smaller than the resolution of the electron microscope. An alternative mechanism is the migration of the crystallite and its capture by a neighboring large crystallite. The migration of resolved and unresolved particles and coalescence with the large one is a possible mechanism for the growth in size of a crystallite during heating in H_2.

The appearance of a new crystallite N of Fig. 2e could be due to the migration and coalescence of unresolved particles.

CONCLUSION

The behavior of small Pd crystallites is quite different from that of the larger ones during alternating heating in O_2 and H_2. Many small crystallites disappear during heating in O_2 and some disappear during heating in H_2. While the disappearance of the small crystallites during heating in O_2 is probably associated with the emission of PdO by the particles to a two-dimensional phase, their disappearance during heating in H_2 is probably caused by migration and coalescence. Somewhat larger crystallites extend by spreading on the surface of the substrate during heating in O_2 and contract after an additional heating in H_2. Pits and cavities are generated in some of the crystallites during heating in O_2 and disappear after an additional heating in H_2. The large crystallites grow further during heating in H_2 probably because of the capture of the smaller, migrating, resolved, and unresolved particles. The extension as well as the contraction can be explained in terms of wetting. Pit formation is explained as either a surface tension gradient driven phenomenon enhanced by a spreading generated stress, or as a crack propagation phenomenon induced by the stress produced by spreading and enhanced by the oxidation of the freshly exposed atoms at the tip of the crack.

REFERENCES

1. Chen, J. J., and Ruckenstein, E., *J. Catal.* **9**, 254 (1981).
2. Chen, J. J., and Ruckenstein, E., *J. Phys. Chem.* **85**, 1606 (1981).
3. Skofronick, J. G., and Phyllips, W. B., *J. Appl. Phys.* **38**, 4791 (1967).
4. Phillips, W. B., Desloge, E. A., and Skofronick, J. G., *J. Appl. Phys.* **39**, 3210 (1968).
5. Chakraverty, B. K., *J. Phys. Chem. Solids* **28**, 2401 (1967).
6. Kern, R., Masson, A., and Metois, J. J., *Surface Sci.* **27**, 483 (1971).
7. Metois, J. J., Gauch, M., Masson, A., and Kern, R., *Surface Sci.* **30**, 43 (1972).
8. Ruckenstein, E., and Pulvermacher, B., *AIChE J.* **19**, 356 (1973).

9. Ruckenstein, E., and Pulvermacher, B., *J. Catal.* **29,** 224 (1973).
10. Flynn, P. C., and Wanke, S. E., *J. Catal.* **33,** 233 (1974).
11. Zanghi, J. C., Metois, J. J., and Kern, R., *Philos. Magn.* **29,** 1213 (1974).
12. Heinemann, K., and Poppa, H., *Thin Solid Films* **33,** 237 (1976).
13. Chu, Y. F., and Ruckenstein, E., *Surface Sci.* **67,** 517 (1977).
14. Metois, J. J., Heineman, K., and Poppa, H., *Philos. Mag.* **35,** 1413 (1977).
15. Baker, R. T. K., Thomas, C., and Thomas, R. B., *J. Catal.* **38,** 510 (1975).
16. Ruckenstein, E., and Chu, Y. F., *J. Catal.* **59,** 109 (1979).
17. Baker, R. T. K., Prestridge, E. B., and Garten, R. L., *J. Catal.* **59,** 293 (1979).
18. Ruckenstein, E., and Malhotra, L., *J. Catal.* **41,** 303 (1976).
19. Adamson, A. W., "Physical Chemistry of Surfaces," 2nd ed. Interscience, New York (1967).
20. Ruckenstein, E., and Lee, P. S., *Surface Sci.* **52,** 298 (1975).
21. Ruckenstein, E., *in* "Studies in Surface Science and Catalysis" (J. Bourdon, Ed.), Vol. 4, p. 57. Elsevier, Amsterdam (1980).

Redispersion of Pt/Alumina via Film Formation

I. SUSHUMNA AND E. RUCKENSTEIN

Department of Chemical Engineering, State University of New York, Buffalo, New York 14260

Received April 7, 1987; revised June 9, 1987

Alternations in the size of the particles on alternate heating in H_2 and O_2 of model Pt/Al_2O_3 catalysts is reported. The catalysts have been investigated by electron microscopy and electron diffraction. The particles appear smaller on heating in oxygen and larger and extended on heating in hydrogen (on heating in hydrogen they regain more or less the size they had prior to the oxygen treatment). This behavior is apparently in contrast to the normally observed extension of particles on oxidation and contraction on reduction. The alternation in size is explained on the basis of the formation of a detectable film (in the form of a halo), or an undetectable film, around each particle due to the strong interactions between crystallites and support in an oxygen atmosphere, and its withdrawal to merge with the respective particle in a hydrogen atmosphere. The films remain in the immediate vicinity of and in contact with the parent particles and do not interconnect with one another to form a contiguous surface film. It is suggested that one of the possible mechanisms of redispersion (regeneration) of Pt/Al_2O_3 often observed in oxidizing atmospheres is by extension and film formation.

INTRODUCTION

Pt/Al_2O_3 catalysts have been widely used in a number of industries. These supported catalysts, however, gradually lose their activity. This is attributed in part to the gradual decline in the exposed metal surface area due to sintering. Sintering of supported metal catalysts has been extensively investigated with both model and industrial catalysts (see (*1–3*) for reviews). The approach to the more significant problem of the regeneration of the sintered industrial catalyst has been more or less empirical. However, especially in the case of supported Pt catalysts, a number of investigations (*1–17*) regarding regeneration have been carried out and the mechanisms involved have been debated. Ruckenstein and Malhotra (*4*) suggested that the catalyst redispersion was a result of particle splitting due to a buildup of strain in the particles. Subsequently Stulga *et al.* (*9*) reported that (in the case of a similar system under similar conditions) they did not observe particle redispersion. Gollob and Dadyburjor (*10*), on the other hand, observed redispersion which they attributed to particle splitting. Ruckenstein and Pulvermacher (*11*) had suggested earlier that redispersion could occur by emission of particles, and Fiedorow and Wanke (*5*) attributed the observed increase in dispersion (measured by H_2 chemisorption) to a similar cause, namely emission of atoms or molecules. Dautzenberg and Wolters (*12*) suggested that an oxygen treatment could recover only the fraction of platinum complexed with alumina during prior high-temperature H_2 treatment and that it could not actually redisperse the sintered large platinum crystallites. Lieske *et al.* (*13*) also argued that oxygen by itself cannot lead to redispersion and that redispersion is possible only in the presence of chloride ions. Straguzzi *et al.* (*14*) and Bournonville and Martino (*15*) also suggested that chlorine plays a role in the redispersion of alumina-supported platinum catalysts. Ruckenstein and Chu (*16*) reported that in the case of Pt/Al_2O_3, following a few cycles of alternate heating in H_2 and O_2 at 750°C, sintering in H_2 and redispersion in O_2 were observed. They suggested that the redisper-

This paper, which was originally published in **Journal of Catalysis**: 108, 77–96 (1987), is republished with permission.

sion in O_2 may be due to the coexistence of a two-dimensional phase of single atoms, or of a multilayer film, with the crystallites. The concept of the coexistence of single atoms and a particulate phase was used also by Yao *et al.* (*17*) to explain sintering or redispersion. Evidence for the redispersion of large crystallites via extension, film formation, and its breakup as well as by splitting of crystallites was reported recently in the case of Fe/Al_2O_3 (*18, 19*) and Ni/Al_2O_3 (*20*) model catalysts.

We report here results with model Pt/Al_2O_3 catalysts heated alternately in H_2 and O_2 at temperatures in the range 300–750°C, which show that there is an alternation in the size, namely an apparent decrease in the size of the particles on heating in O_2 and a recovery of more or less the original size on subsequent heating in H_2. This size alternation continued over a number of cycles and was observed at temperatures between 500 and 700°C. The results are discussed in terms of the redispersion of the crystallites via the formation of disconnected individual films around the particles, which is a result of the strong interactions between crystallites and support.

EXPERIMENTAL

The method of preparation of model Pt/Al_2O_3 catalysts and the treatment procedure are presented in detail in Refs. (*16, 21*) and they are therefore not elaborated here. The term hydrogen refers to the as-received ultra-high-purity hydrogen (≤ 1 ppm O_2 and ≤ 3 ppm moisture) from Linde Division, Union Carbide Corp., and "purified hydrogen" refers to the above hydrogen purified further by passing it through a Deoxo unit (Engelhard Industries), followed by a column of silica gel, a bed of 15 wt% MnO on silica, and finally, through a column of 13× and a column of 5-A molecular sieves immersed in liquid nitrogen.

RESULTS

The results presented here have been obtained with the same samples for which the sintering results were described in Ref. (*21*). The experiments have been carried out with a number of specimens. The results obtained with five of them are presented in some detail.

1. A 2-nm initial film thickness sample was heated at 500°C for 8 h in purified hydrogen; this generated small particles of about 1.6 nm average size. In order to more easily identify the processes that occur, the size of the particles was increased further by heating the sample in H_2 at 600°C for 4 h followed by heating at 700°C for 5 h, at 800°C for 2 h, and finally again at 600°C for 2 h. This heat treatment led to relatively large particles of about 7 nm average size. On subsequent heating of the sample in O_2 at 300°C, there was a decrease in the size of all the particles. Additional heating in O_2 at progressively higher temperatures in steps of 50°C up to 600°C (2-h duration at each temperature) did not result in significant changes in either the size distribution or the shape of the particles. However, when the sample was subsequently heated in H_2 at 500°C, all the particles, small and large, increased significantly in size. Few particles disappeared, indicating that neither ripening nor migration and coalescence involving the detectable particles were responsible for the growth. Further heating in O_2 at 500°C resulted again in a decrease in the size of all the particles. Subsequently, when the sample was heated in H_2 at 500°C, the particles again increased in size and regained more or less the size they had prior to the oxygen treatment. Alternate heating in H_2 and O_2 at 500°C was carried out over six cycles and the above alternation in size was observed in each cycle (Fig. 1). A decrease in size on heating in O_2 and an increase in size on heating in H_2 were also observed when the same specimen was subsequently heated at 700°C.

2. To verify whether the observed alternations in size were caused by the gradual and prolonged heat treatment in O_2 between 300 and 600°C prior to the cycling in H_2 and O_2 at 500°C, another sample with a

similar prehistory, but without the extended oxygen treatment between the initial size distribution and the cycling in H_2 and O_2 at 500°C, was investigated. During the first two cycles, the particles decreased in size only marginally on heating in O_2 and increased in size on heating in H_2. However, during the subsequent four cycles, the changes in size were more pronounced, even though the extent of size reduction was smaller than that for Sample 1. Perhaps the gradual heating in O_2 at progressively higher temperatures prior to the cycling in H_2 and O_2 enhanced the wetting of the substrate by the particles in the case of Sample 1. It is also to be noted that following the heating in O_2 at 500°C, the smallest of the detectable particles were still present on the substrate and did not appear to have disintegrated via molecular emission. In addition, following the heating in O_2, detectable films in the form of a halo around each particle could be seen (Fig. 2a). Subsequent to the above six cycles, the treatment in oxygen was carried out at higher temperatures. On heating in O_2 at 600°C few particles disappeared and the films around the particles could be seen even better than at 500°C (particles marked with large arrows in Fig. 2c). Even the smallest particles remained (particles marked with small arrows in Figs. 2a–2d) and regained more or less their previous sizes on subsequent heating in H_2 at 500°C. A large number of crystallites considerably extended to form thin, light patches on the substrate on subsequent heat treatment in O_2 at 700°C (Regions 1, 2, and 3 in Fig. 2e, for example). Some particles are no longer detectable. Not all the particles present before the last O_2 treatment were regained on subsequent heating in H_2 at 500°C. Smaller particles were generated in regions where the particles had extended considerably before (Regions 1–3 in Figs. 2e and 2f). Films around some particles could still be seen (particles marked with large arrows). A decrease in the number of detectable particles on the substrate surface during heating in O_2 may suggest the possibility of material loss via evaporation. Evaporation of Pt as PtO_2 in oxygen atmospheres at temperatures greater than 700°C has been suggested before (*22–24*). However, in light of the observations indicating the presence of detectable films around the particles on heating in O_2 and the generation of smaller particles in some regions where the particles had considerably extended before, it is more likely that a part of the material (Pt or PtO_x) diffused into or spread over the surface of the substrate. These probably happen due to the strong interactions between particles and substrate. Further, particles 0.8 nm in size or even smaller remained without disappearing through the various cycles of alternate heating in H_2 and O_2 at 500°C and also following the high-temperature (up to 700°C) O_2 treatments (particles marked with small arrows in Figs. 2a–2f), suggesting that molecular emission and/or evaporation played a minor role, if any.

3. A third sample of 2 nm initial film thickness and from the same batch as the preceding sample was heated in a slightly different sequence. Particles were generated as before by heating the Pt/alumina sample in purified H_2 at 500°C for 12 h and subsequently at 600°C for 5 h. Before the alternate heating in H_2 and O_2, Sample 1 described above was heated in O_2 between 300 and 600°C in steps of 50°C, without hydrogen treatments in between. If the particles were already oxidized at the lower temperature of 300°C, subsequent heat treatment of the oxidized sample in O_2 at higher temperatures would cause little change as observed up to 600°C with Sample 1. In order to verify whether there is a threshold temperature beyond which film formation is detected following heating in O_2, it is appropriate to start with a reduced sample prior to the oxidation step at each temperature. Therefore, the present sample was also heated in O_2 in the range 300–600°C in steps of 50°C, but unlike in Sample 1, in between each oxidation step, this sample was heated in H_2 at 500°C in

order to reduce the oxide to metal prior to the next oxidation. As reported earlier (*21*), except for some marginal sintering, mostly by the coalescence of neighboring particles, there was very little change in the size or shape of the particles up to the 500°C O_2 treatment. No detectable decrease in the particle size or the formation of films around the particles could be detected following the 300 to 450°C O_2 treatments. Particles as small as 0.8 nm in size or even smaller remained essentially unaffected following the above treatments in H_2 and O_2 as observed also with Sample 2 on cycling at 500°C. This indicates that even though an oxide might have formed at or below 500°C, no significant molecular emission was involved. (It is to be noted that oxidizing atmospheres are more likely to induce molecular/atomic emission.) At 500°C, changes regarding particle size and film formation begin to be noticed. As for Sample 2, the decrease in the size of the particles following heating in O_2 at 500°C was marginal during the first two cycles of heating but more pronounced during the subsequent cycles; the extent of size reduction was again smaller than that for Sample 1.

The behavior of this sample on subsequent heating in O_2 at higher temperatures was slightly different. On heating in O_2 at 600°C a number of particles disappeared, as observed before (Fig. 3). Some large particles have decreased in size considerably or disappeared, while nearby smaller particles have only decreased marginally in size (Regions 1–5 in Fig. 3). Subsequent heating in O_2 at 700°C resulted in an almost complete disappearance of the existing particles on the substrate. Instead, a number of very large, thin, irregular hexagonal patches could be observed. These patches can be a result of either a redistribution of the parti-

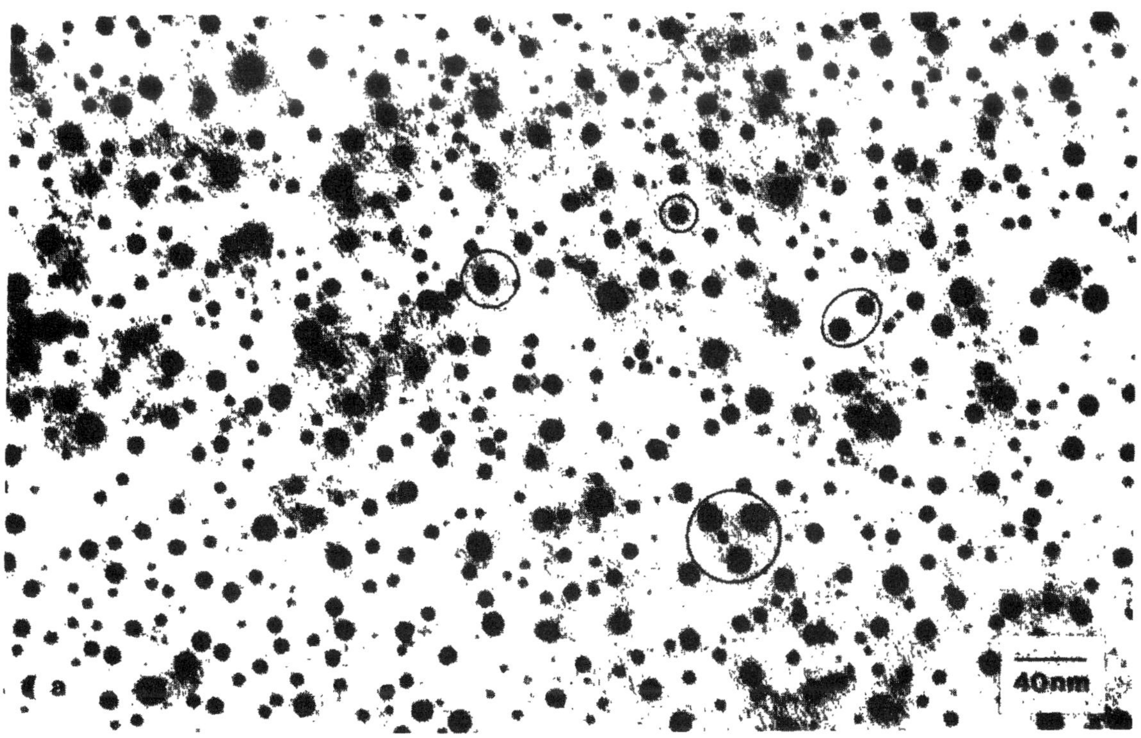

FIG. 1. Changes in particle size of a 2 nm initial film thickness Pt/Al_2O_3 sample on alternate heating in H_2 and O_2 at 500°C. The sample has had a long heating history prior to Fig. 1a (see text). (a) H_2, 500°C, 2 h; (b) O_2, 500°c, 2 h (or 16 h); (c) H_2, 500°C, 4 h.

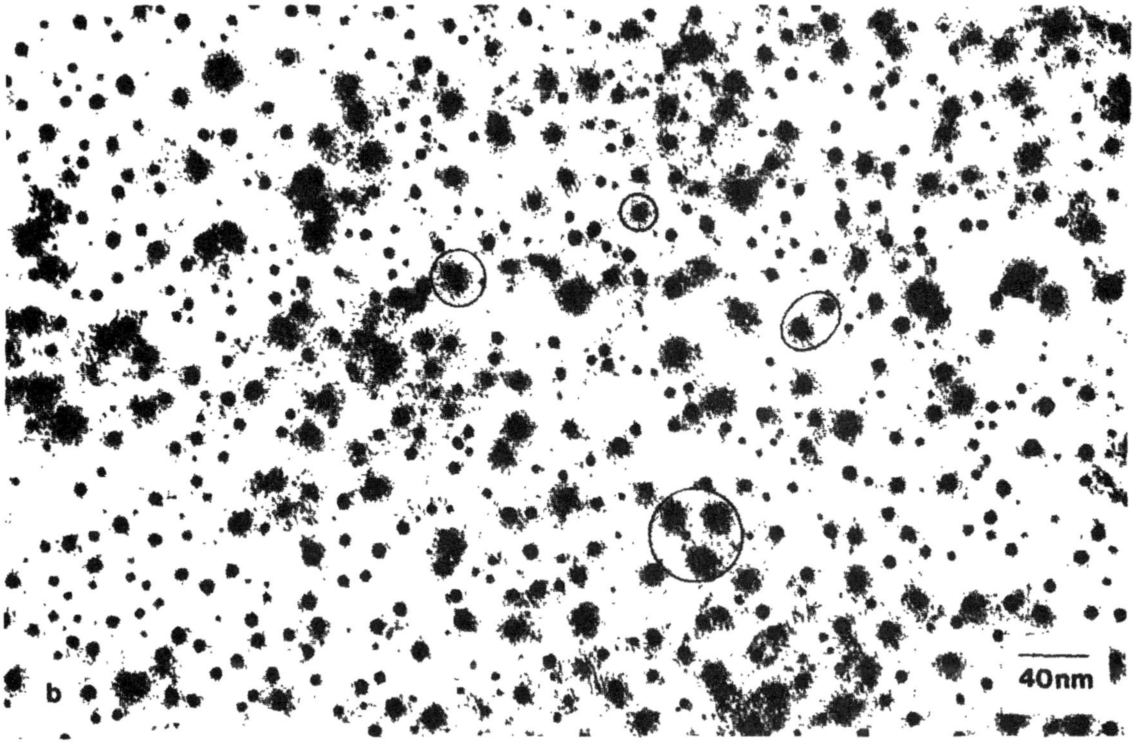

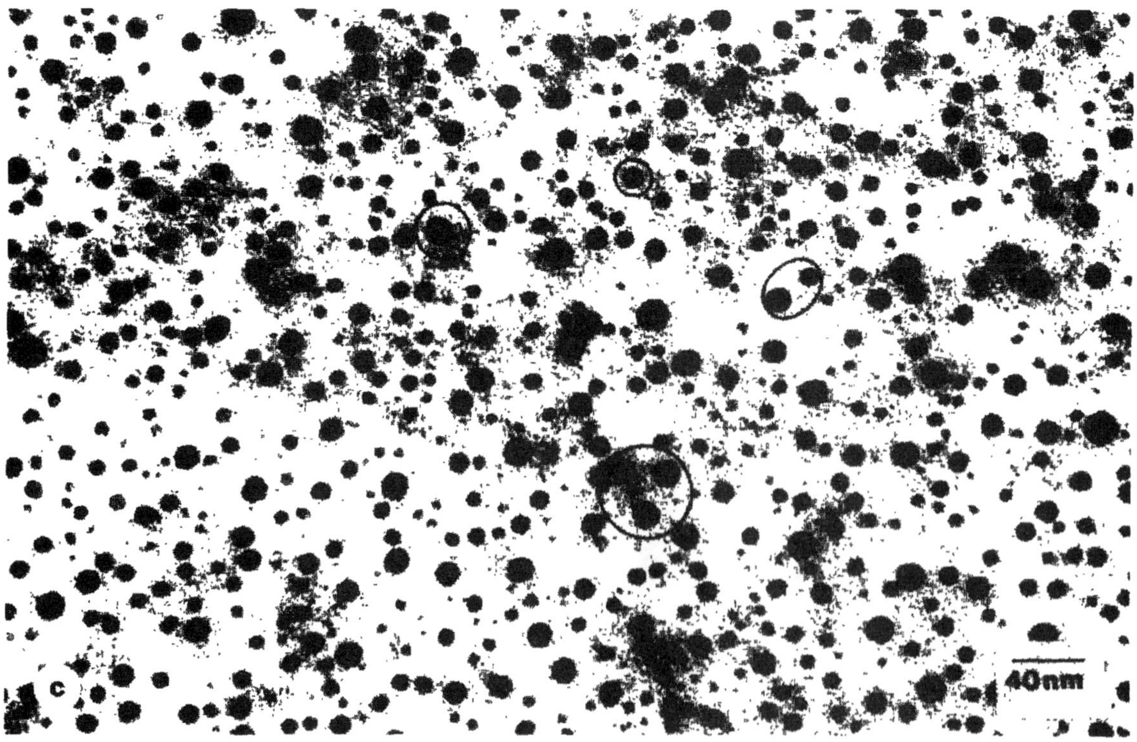

FIG. 1—*Continued.*

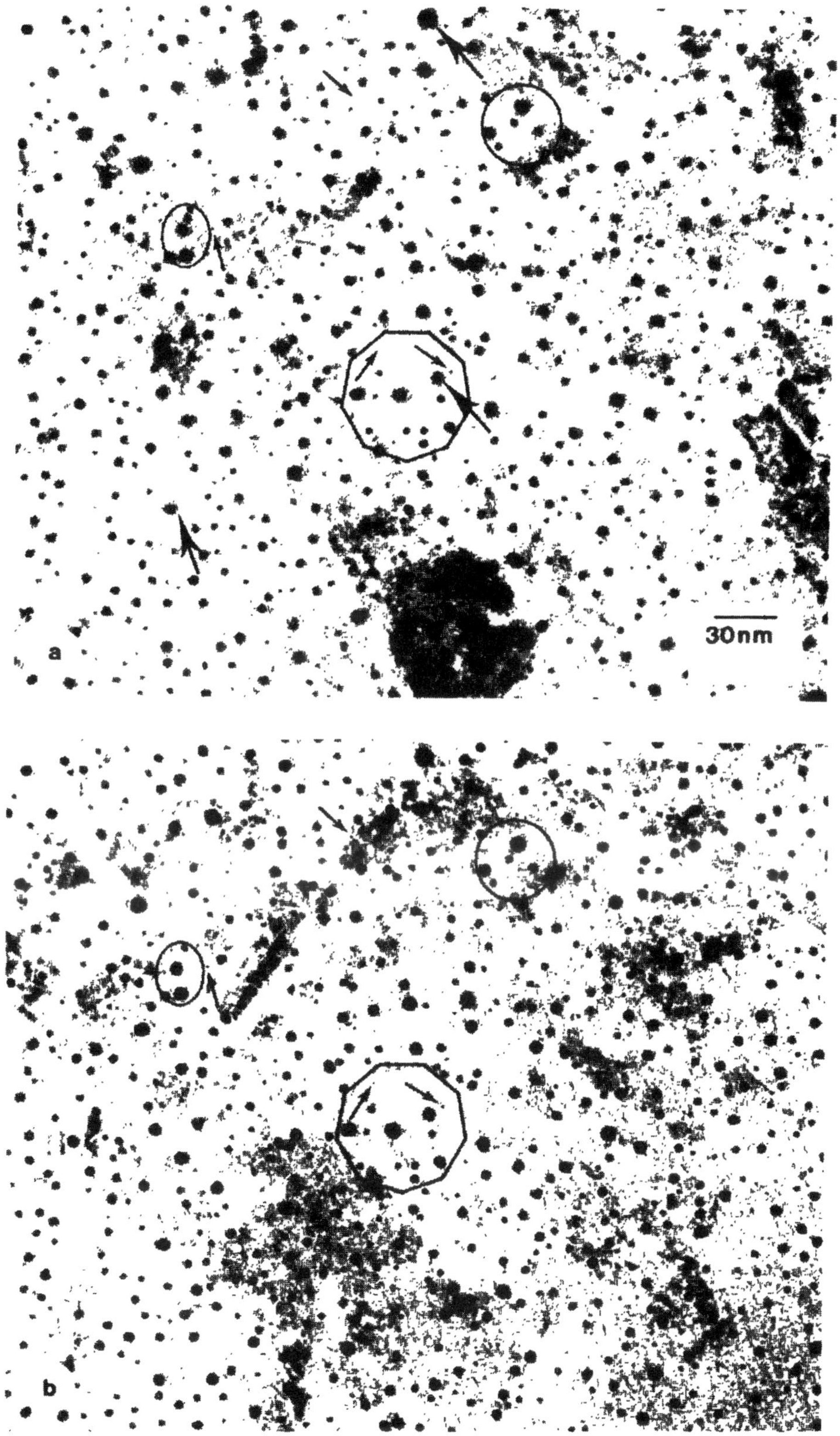

FIG. 2. Changes on heating a 2 nm initial film thickness Pt/Al_2O_3 sample alternately in H_2 and O_2. Micrographs for only a few cycles of alternate heating are shown. Prior to the following, the sample had undergone a few cycles of alternate heating in H_2 and O_2 (see text). (a) O_2, 500°C, 5 h; (b) H_2, 500°C, 2 h; (c) O_2, 600°C, 5 h; (d) H_2, 500°C, 2 h; (e) O_2, 700°C, 5 h; (f) H_2, 500°C, 2 h.

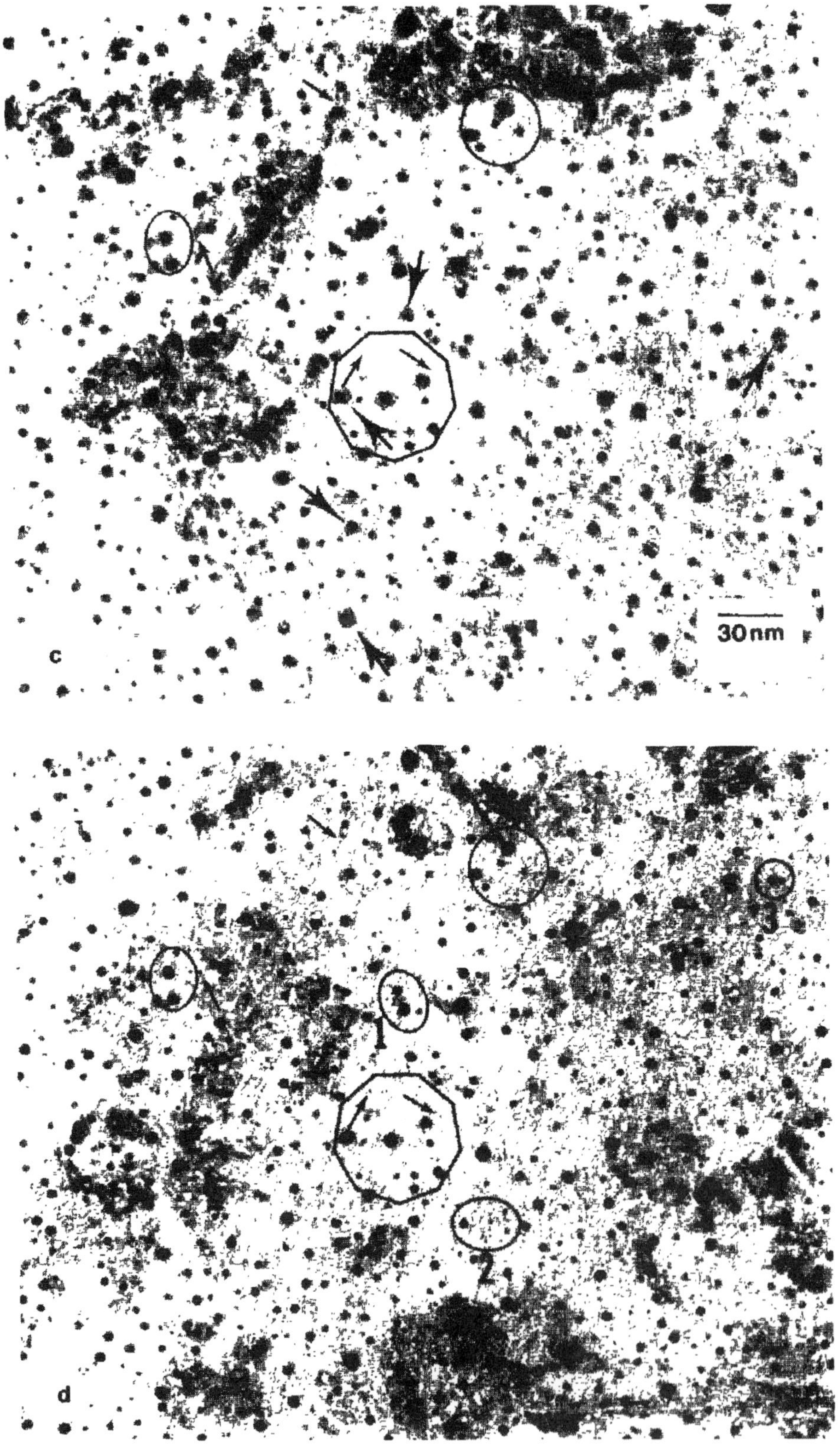

FIG. 2—*Continued.*

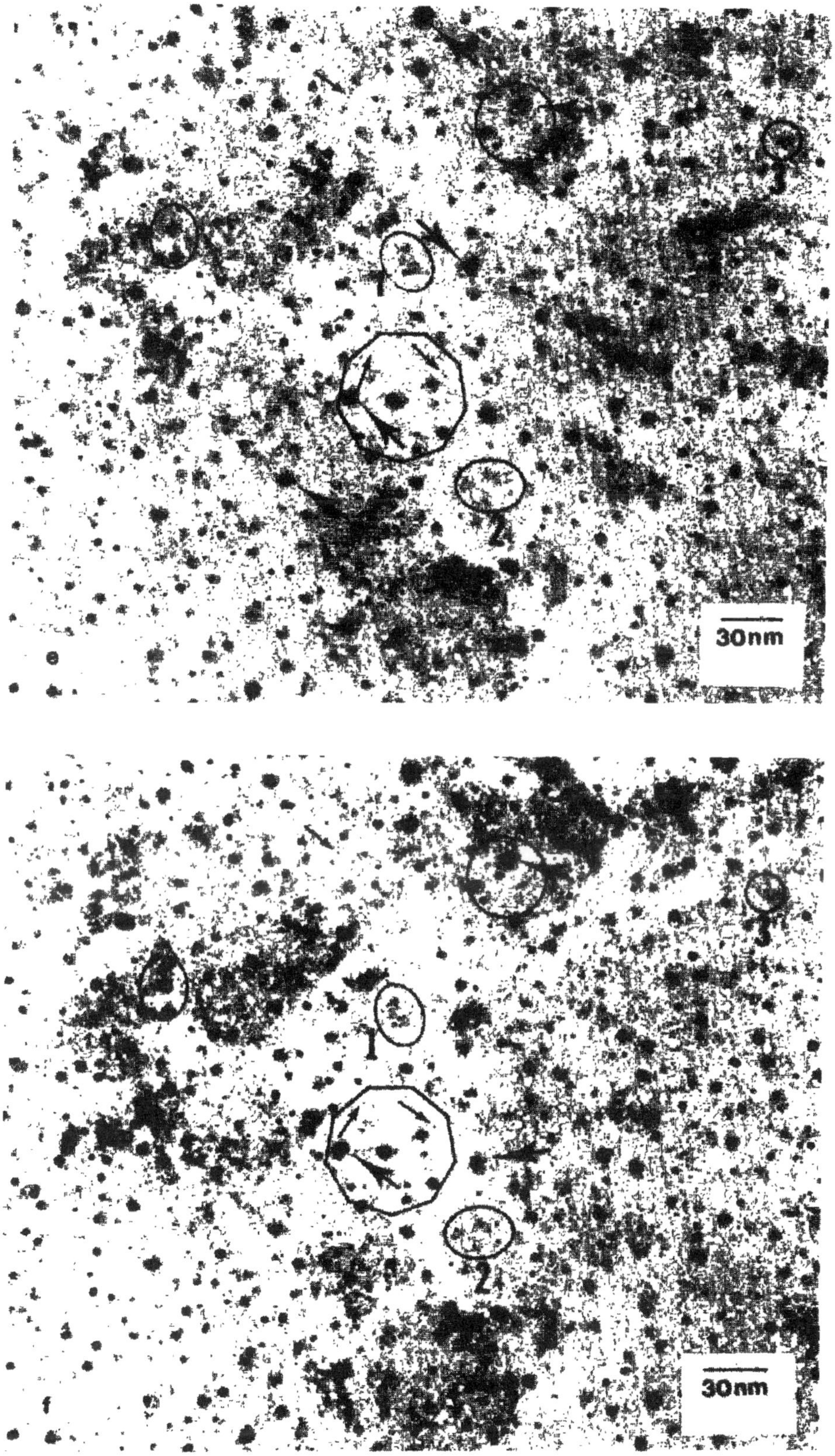

FIG. 2—*Continued.*

cles (due to the strong interactions between particles and substrate) or contamination. These patches remain, however, and even increase in size, on heating in O_2 at 800°C. One would expect that they would disappear on heating in O_2 if they were hydrocarbon or carbon impurities. The patches have sharp edges and are facetted in irregular, hexagonal shapes as seen in Figs. 3d and 3e. (Fig. 3e shows a larger area of the sample.) Some patches overlap (particles marked (a) and (b)). The patches remain essentially unchanged on subsequent heating in H_2 at 500 or 700°C. From the micrographs, it appears that the patches represent large extended particles, which are formed as a result of the strong interactions between particles and support (which in turn are enhanced by the high-temperature, 700 or 800°C, oxygen treatment). The interactions are probably so strong that little change in their morphology occurs on subsequent heat treatment in H_2 or O_2. It is to be noted that a thin, hexagonal, pill box morphology of Pt particles observed on heating in H_2 at $\geq$500°C was previously attributed to the so-called "SMSI," which was reported to be restricted to particles on reducible oxide supports, such as TiO_2 (*28*).

4. In order to verify whether the observed alternation in size was also exhibited by samples with smaller average sizes (since the average particle size in industrial Pt/Al_2O_3 catalysts is usually small), another sample with a smaller loading (0.75 nm initial film thickness) was heated in H_2 and O_2 at 500°C. Since this specimen was not heated in H_2 or O_2 at >500°C at any time during either the pretreatment or the cycling in H_2 and O_2, the particles were small. Any marginal decrease in size of such small particles on heating in O_2 will not be easily detectable unless the particles spread completely. Consequently, very little change, other than that due to sintering, was observed. After heating the sample for a total of 80 h, which incidentally generated sufficiently large particles, clearly detectable alternations in size were subsequently observed on cyclic heating in H_2 and O_2 at 500°C. These results indicate that the films around the particles, or an apparent decrease in the particle size as a result of the film formation, become detectable only when the particles are sufficiently large. The inability to detect such changes by electron microscopy does not imply, however, that they do not occur for small particles as well.

5. Whether the alternation in size on cyclic heating in H_2 and O_2 and the formation of films around particles in O_2 were restricted to temperatures below the decomposition temperature of PtO_2, namely $\simeq$580°C, or it is possible also at higher temperatures was verified by heating another sample at 750°C. The formation of films around the particles following heating in O_2 was in fact more obviously exhibited by a sample (of 1.5 nm initial film thickness) heated at 750°C. After Pt deposition, the sample was heated in as-received ultra-high-purity H_2 at 750°C for 15 h. This generated very large particles in the range 5–15 nm (Fig. 4). When this specimen was subsequently heated in O_2 at 750°C, a large number of small and large particles disappeared without any kind of preference or pattern (Figs. 4a and 4b). All the remaining particles, including the small ones decreased in size. It could be surmised that because of the higher temperature, emission of atoms/molecules to the substrate or material loss via evaporation was involved. However, on subsequent heating in H_2 at 750°C for 5 h, the remaining particles considerably increased in size while some grew even larger than their size prior to the O_2 treatment. There was very little decrease in the number of detectable particles. The increase in average particle size on switching to H_2 from O_2 was greater than could possibly be explained by the very small decrease in the number of detectable particles (Figs. 4b and 4c). This indicates that the decrease in size during prior heating in O_2 could not have been solely due to the loss of material by evaporation. Two of the

possibilities are (i) the material was present on the substrate surface as single atoms/molecules following the O_2 treatment, which were subsequently captured by the particles on H_2 treatment, and (ii) following the O_2 treatment the material was present as individual unconnected films around each particle which subsequently contracted and merged with the respective particle on heating in H_2, to result in the observed increase in size.

A subsequent prolonged heat treatment of this sample for 12 h in H_2 at 750°C led to the appearance of films around the particles (Fig. 4d). This behavior is reminiscent of a similar observation in the case of Fe/Al_2O_3 on prolonged heat treatment in as-received H_2 (*18*). There it was suggested that the extension of the particles was a result of the oxidation of iron by the residual O_2 and H_2O present in the H_2 and the subsequent strong interaction between the iron oxide

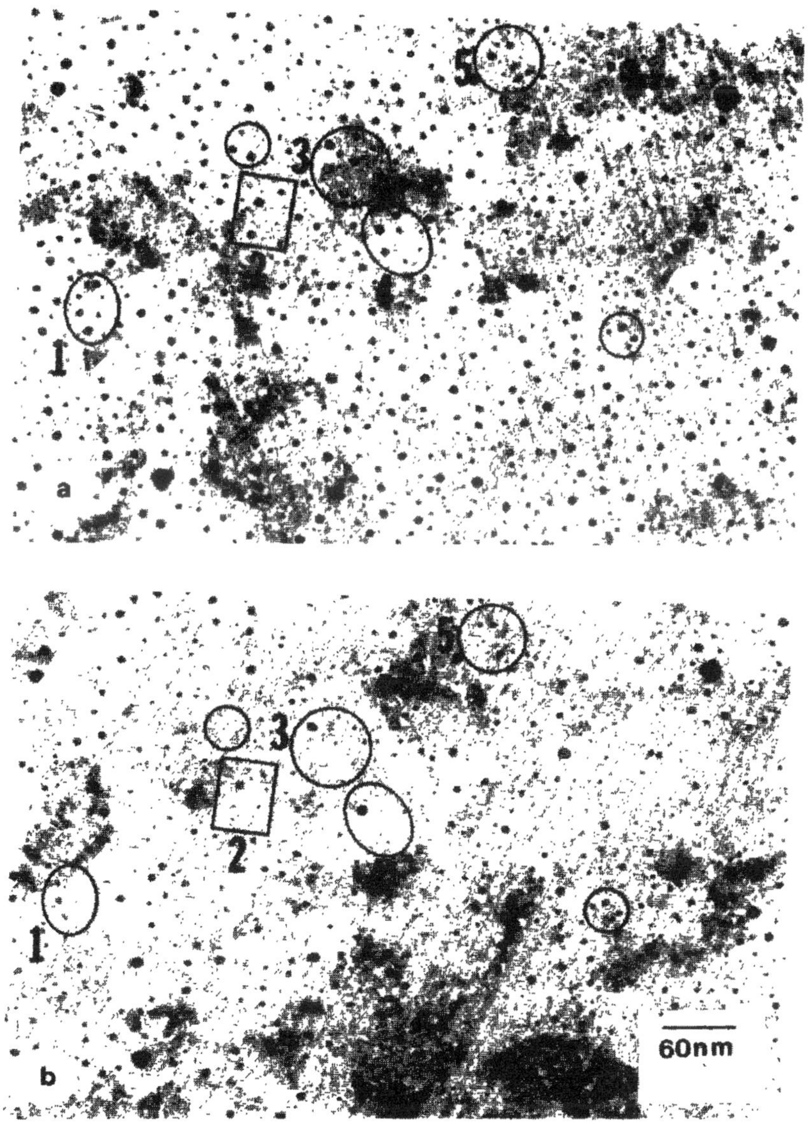

FIG. 3. Micrographs showing changes in a 2 nm initial film thickness Pt/Al_2O_3 sample subsequent to a few cycles of alternate heating in H_2 and O_2 (see text). (a) H_2, 500°C, 2 h; (b) O_2, 600°C, 5 h; (c) (O_2, 700°C, 5 h) + H_2, 500°C, 2 h; (d) O_2, 800°C, 3 h; (e) same as (d), a larger area shown at a lower magnification.

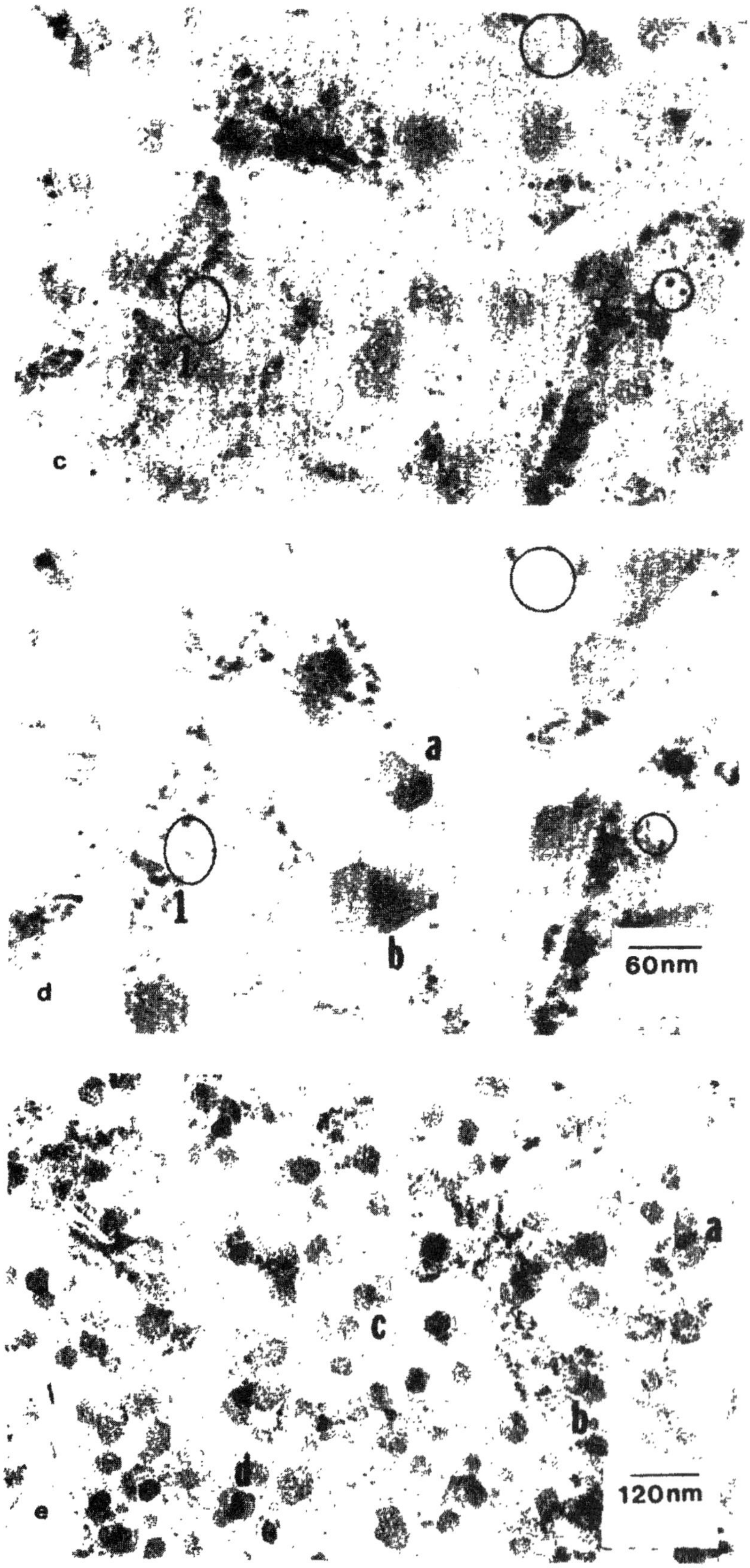

FIG. 3—*Continued.*

and the alumina. In the present case even though a stoichiometric platinum oxide (PtO_2) is not stable at 750°C, transient phases of platinum oxide might have formed which might have reacted with alumina. Even otherwise, because of the high temperature, an interaction compound between Pt and Al_2O_3 might have formed leading to the extension of the particles. In this regard, it is noted in the next section that the presence of an interaction compound is suggested in the electron diffraction patterns in both H_2 and O_2. Subsequently, an apparent decrease in size of the solid core and the appearance of films around the particles on heating in O_2 and an apparent increase in size on heating in H_2 were observed following three more cycles of alternate heating of this sample in H_2 and O_2 at 750°C (Fig. 5).

It follows from the above experimental results that an apparent decrease in size on heating in O_2 and a recovery in size on heating in H_2 is exhibited by model Pt/Al_2O_3 catalysts over a range of temperatures (500–750°C used in the present experiments) and initial metal loadings (0.75–2 nm initial film thickness) even though there are some differences perhaps due to the differences in the history of the samples.

DISCUSSION

A few mechanisms have been suggested in the literature for the redispersion of sintered supported metal catalysts in an oxidizing atmosphere. The sintered crystallites can emit atoms or molecules to the substrate which can then coexist with the crystallites as a two-dimensional dispersed phase, thereby increasing the dispersion

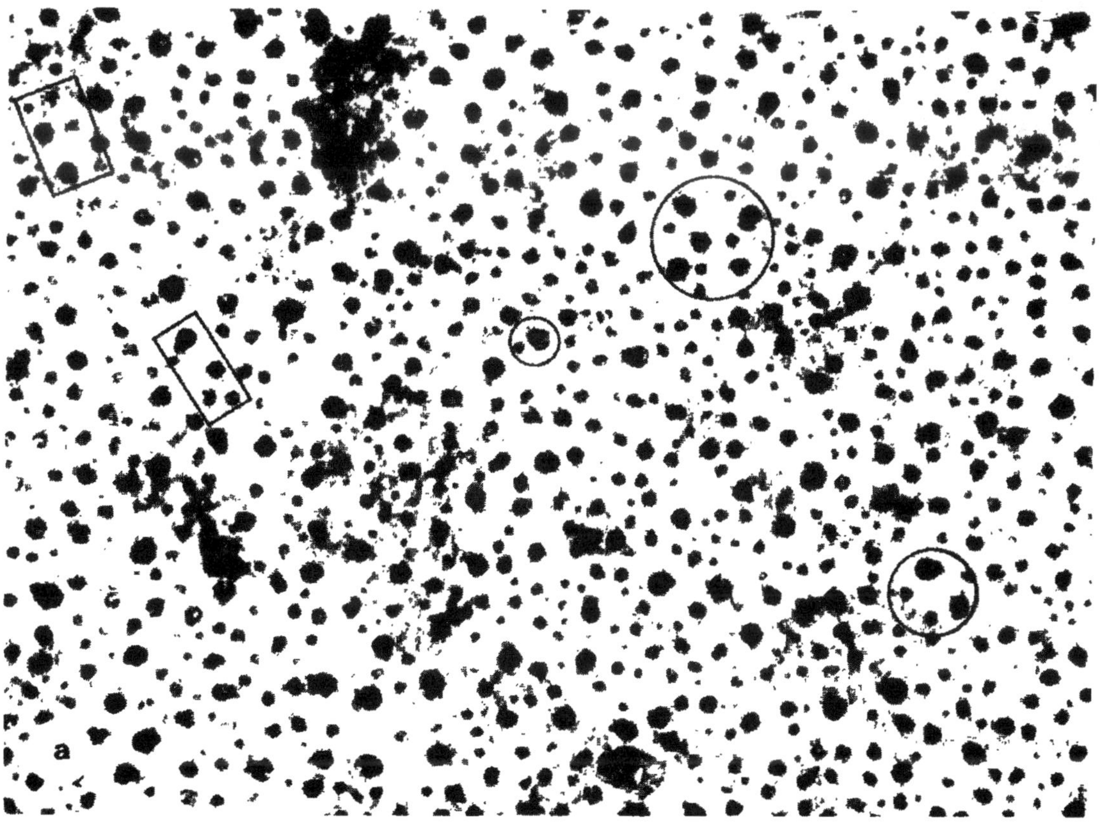

FIG. 4. Electron micrographs showing sequence of changes in a 1.5 nm initial film thickness Pt/Al_2O_3 specimen heated at 750°C. (a) Initial (H_2, 750°C, 15 h); (b) O_2, 750°C, 5 h; (c) H_2, 750°C, 5 h; (d) H_2, 750°C, + 12 h (17 h); (e) H_2, 750°C, + 6 h (23 h).

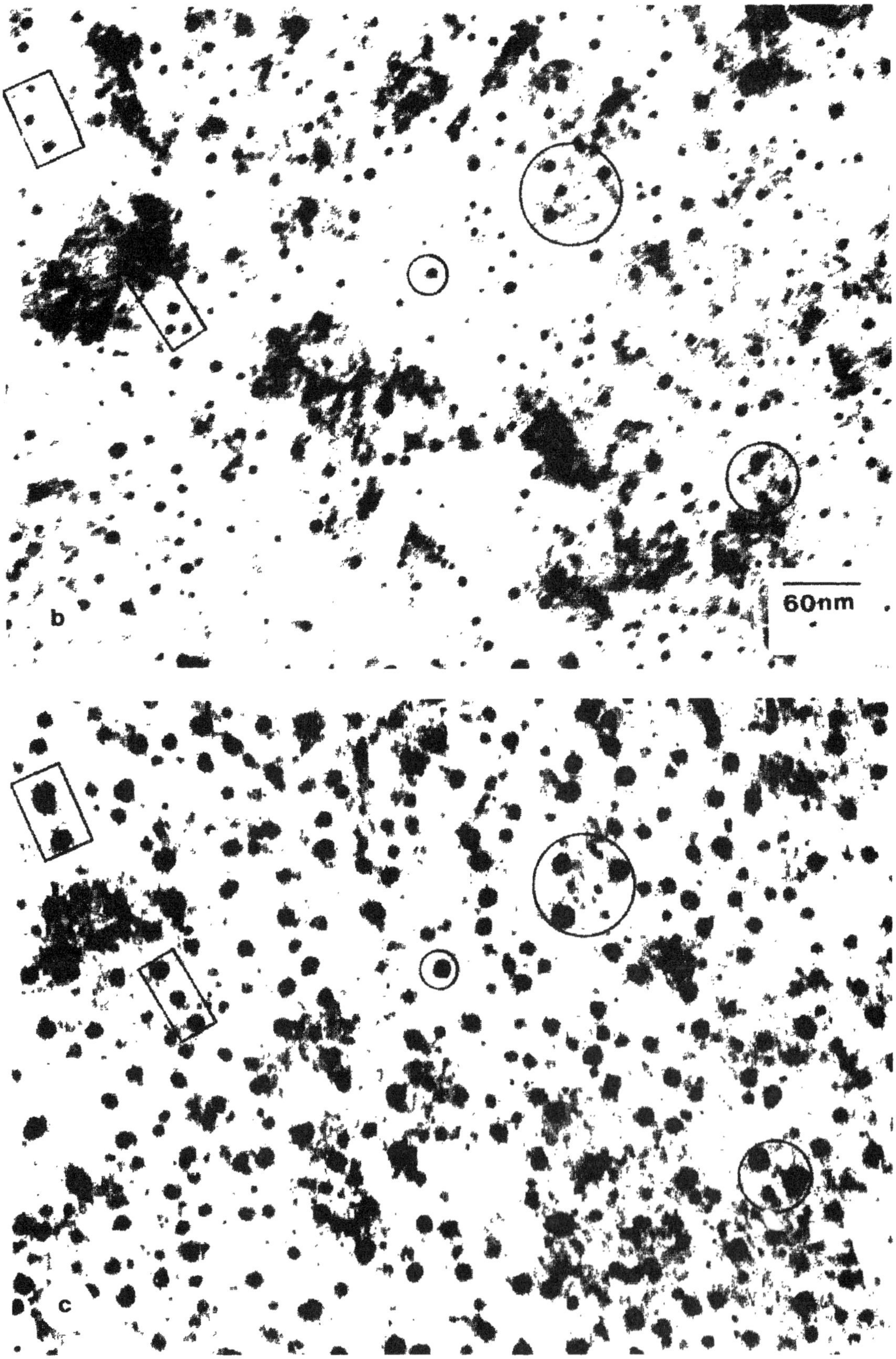

FIG. 4—*Continued.*

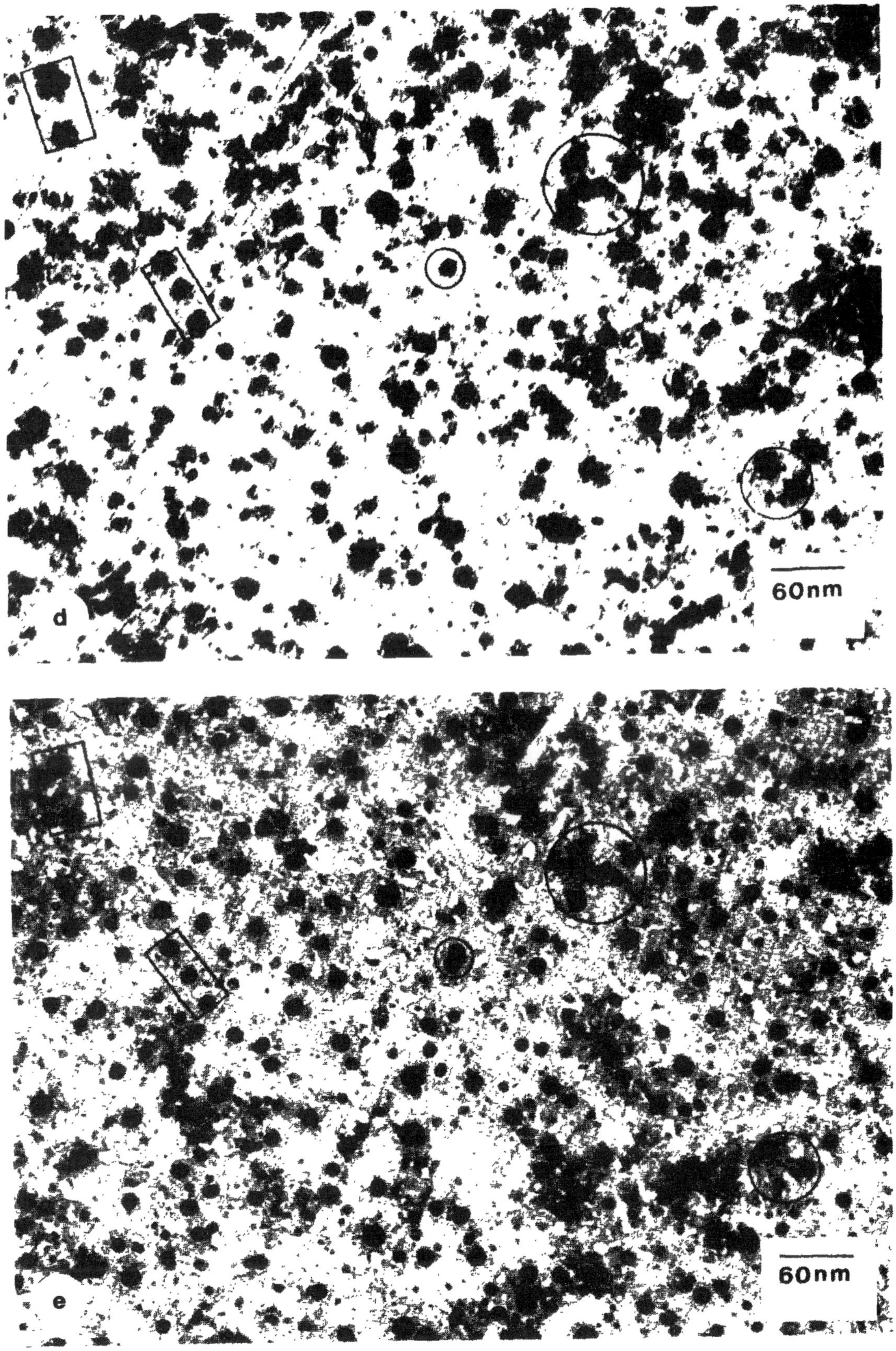

FIG. 4—*Continued.*

(*5, 16, 17*). Such a molecular emission mechanism is expected to be favored at high temperatures and in an oxidizing atmosphere rather than in a reducing one, since in oxidizing atmospheres the possibility of a decrease in the free energy of the system via the strong interactions between the platinum oxide molecules and the alumina substrate creates a driving force for the emission of molecules from the particles to the substrate. On subsequent heating in H_2, the dispersed oxide molecules and the small crystallites are reduced (to metal), the interactions with the support become weaker, and subsequently two possibilities can arise. (i) Either new small particles are formed (not necessarily in the same locations where the particles might have disintegrated by emission of atoms or molecules to the substrate), and/or (ii) the atoms (reduced molecules) are captured by the existing three-dimensional crystallites. Since new particles have not been generated in our experiments either following heating in O_2 or on subsequent heating in H_2, one could suppose that on heating in O_2, dispersed single molecules were present on the substrate and that they were captured by the existing crystallites on subsequent heating in H_2. However, the extent of capture of atoms by various crystallites will be different because of the differences in their radii of curvature. Since their rate of emission is smaller, the large crystallites are expected to have a greater growth than the smaller ones (*1, 3*). Consequently, a size distribution different from the one before is expected. However, as noted under Results, especially at 500°C, all the detectable particles increase in size and attain more or less the size they had prior to the O_2 treatment. Therefore, it appears that a mechanism based on emission and capture of single atoms from a two-dimensional phase of single atoms located on the substrate surface is unlikely to be responsible for the observed behavior.

Another possible, and in the present case more likely, mechanism for redispersion in an oxygen atmosphere is by the spreading of a film in the form of a halo around each particle. As discussed in detail in (*18, 25*), the surface free energy (γ_{cg}) of a metal is larger than the surface free energy of the corresponding oxide. Also, since, in general, a metal oxide (platinum oxide in the present case) interacts more strongly with the substrate (which is another oxide) than a metal does, the interfacial free energy γ_{cs} between the crystallite and substrate is also smaller in an oxidizing than in a reducing atmosphere. Since γ_{cg} and γ_{cs} are both smaller in an oxygen atmosphere, the contact angle θ is also smaller as can be seen from Young's equation shown below and the particle would be more extended in an O_2 atmosphere than in a H_2 atmosphere:

$$\cos\theta = \frac{\gamma_{sg} - \gamma_{cs}}{\gamma_{cg}}.$$

The interfacial free energy γ_{cs} is related to the net interaction energy U_{cs} between crystallite and substrate via the expression

$$\gamma_{cs} = \gamma_{cg} + \gamma_{sg} - (U_{int} - U_{str}) = \gamma_{cg} + \gamma_{sg} - U_{cs},$$

where U_{int} is the interaction energy per unit contact area between substrate and supported material and U_{str} is the strain energy per unit contact area due to the mismatch of their lattices. The net interaction energy is larger when the support and the supported material are similar in structure (two oxides, for instance, rather than an oxide and a metal), or when there are covalent interactions between the two. In the present case, if the interactions between the platinum oxide and the alumina are strong enough, the small particles (which will be more completely oxidized than the larger ones) could spread out completely (because of the large driving force for spreading when U_{cs} is very large or γ_{cs} is very small) while the larger particles would only extend. Therefore, the small particles would no longer be detected in the micrographs following heating in O_2, while the larger particles would appear smaller if the films extending from the leading edge of the particles are so thin as not to be detectable by electron microscopy. If the above interactions are not very strong (i.e., if the driving force for spreading is not very large), or if the kinetics of wetting is slow, even the smallest particles would only extend and could therefore still be detected in the micrographs (of couse, with a smaller apparent size and an undetectable film around). In some cases, for example when the particle size is large, the extended film around the particles may be thick enough to be detected in the micrographs (particles marked with large arrows in Figs. 2 and 5). In these cases, the particles do not spread completely, but only acquire detectable or undetectable films around them. These films would contract and merge with the respective crystallites on subsequent heating in H_2 and , depending upon the reconstruction that occurs, the particles can regain the sizes they had prior to the oxygen treatment. This scenario appears to explain the observed results better than the two-dimensional phase of single atoms or molecules that covers the entire substrate can.

It is to be noted that at 500°C, the films around the particles are not likely to be interconnected to form a contiguous patch as observed in the case of Fe/Al_2O_3 at high temperatures (*19*). If the films were to be interconnected, the particles would be pulled closer together and sinter when the interconnecting film contracts on subsequent heating in H_2. Since little sintering is observed, it appears that the film constitutes an atmosphere around the particle (in

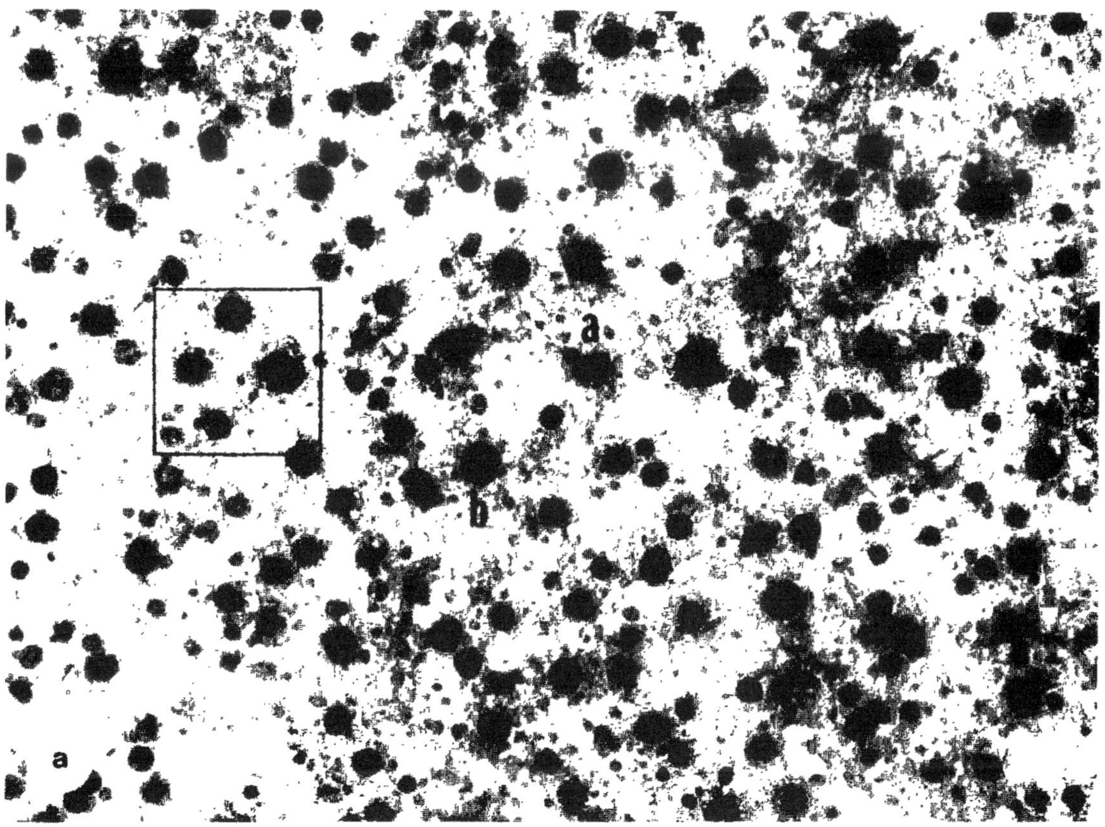

FIG. 5. Same sample as that of Fig. 4. Micrographs shown are from a different region, at a different magnification, and for treatments subsequent to those in Fig. 4e. (a) H_2, 750°C, 32 h; (b) O_2, 750°C, 12 h; (c) H_2, 750°C, 12 h.

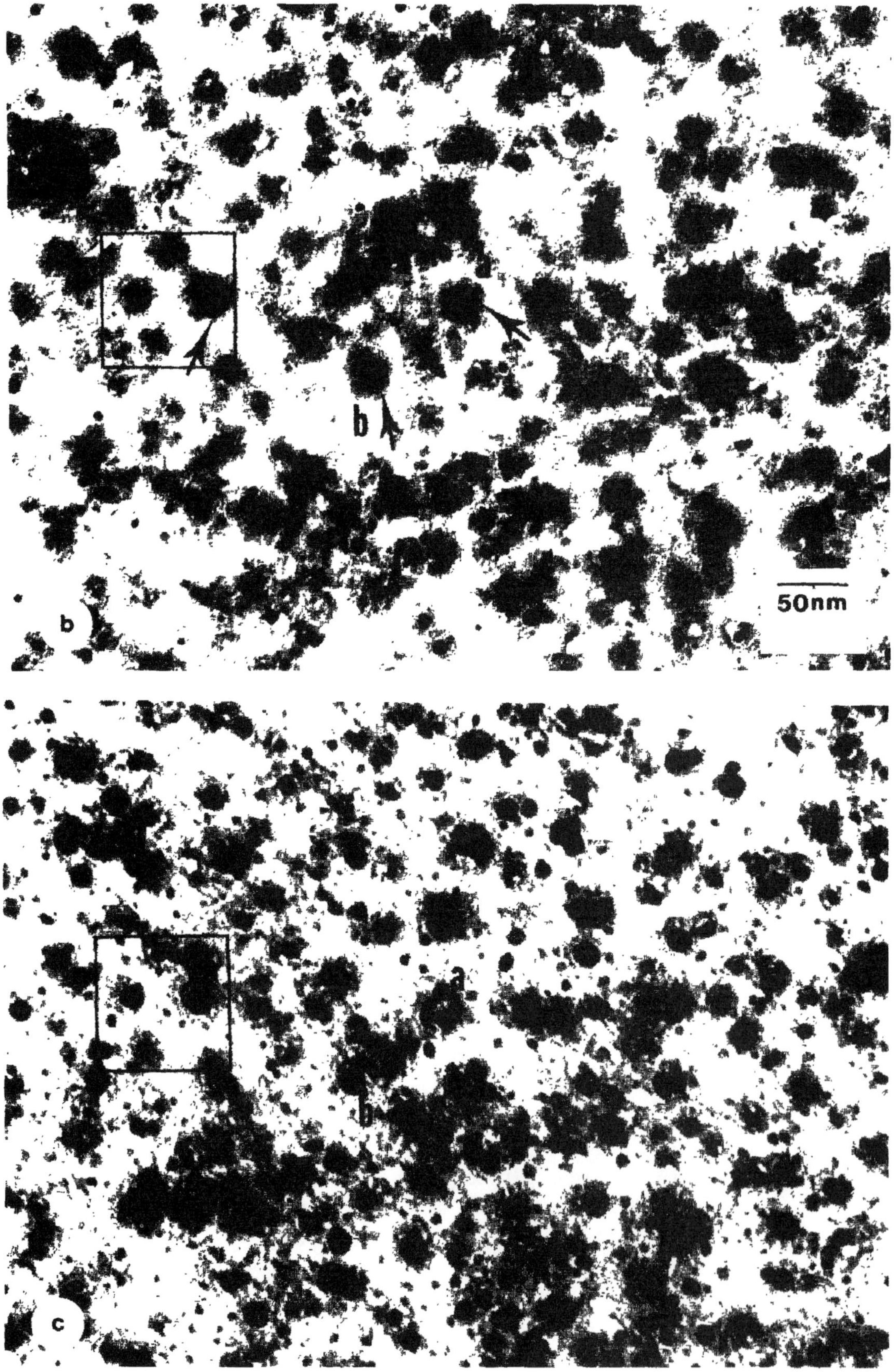

FIG. 5—*Continued.*

the form of a halo), extended only in its immediate vicinity, so that on subsequent heating in H_2 it contracts and merges entirely with the parent particle. However, it is likely that there exists a critical metal loading beyond which the films would extend farther away from the particles, interconnect with those from nearby particles, and form a contiguous film covering the substrate surface. Similarly, when two or more particles are close to one another, the films around them may overlap and become interconnected.

Another alternative could also be envisioned. In a H_2 atmosphere, the Pt crystallites can catalyze the partial reduction of alumina. Similar suggestions have already been made (*26*). The partially reduced substrate is relatively more reactive. Also, because of its nonstoichiometric nature, the surface free energy γ_{sg} of the partially reduced substrate is larger than that of Al_2O_3. In addition, as suggested for Pt/TiO_2 (*27, 29*), the reduced substrate species may migrate onto the crystallites and because of their interaction with the crystallites they may decrease γ_{cg}. Similarly, because of its reactive nature, the net interaction energy U_{cs} between the substrate and the crystallite at the base of the crystallites is increased and therefore γ_{cs} may be decreased. As a result of the increase in γ_{sg} and decreases in γ_{cg} and γ_{cs}, the contact angle will decrease and the crystallite would extend on the substrate in a reducing atmosphere, as in the case of Pt/TiO_2 (*28*). That is, in the present case, the larger size of the crystallite, observed following heating in H_2, is because of its extended form. On subsequent heating in O_2, the partially reduced substrate can be reoxidized to Al_2O_3 and γ_{sg} will therefore decrease. If the interactions between platinum oxide and Al_2O_3 are weaker than those between Pt and the reduced alumina, γ_{cs} may increase compared to its value in the reducing atmosphere. Also, because now the interactions between the particles and the partially reduced substrate species migrated onto them are absent, γ_{cg} may also increase although it may be smaller compared to the initial γ_{cg} of the metal. As a result of the increases in γ_{cg} and γ_{cs} and the decrease in γ_{sg}, the contact angle will increase and the particles would appear contracted and smaller. That is, the smaller size of the particles, observed following heating in O_2, is a result of their contraction to a larger wetting angle. However, this mechanism is not likely to be responsible for the observed behavior especially since in a few micrographs a kind of film is detected around the particles following heating in O_2. It is more likely that stronger interactions and extension occur in an oxygen atmosphere rather than in a reducing one.

Apart from the indirect evidence, via film formation, for strong interactions between the crystallites and support, there is also some direct evidence for such interactions from electron diffraction results, even though the compounds formed could not be identified with certainty. Electron diffraction patterns were recorded for all the samples following each heat treatment. Following heating in oxygen, especially above 500°C, additional rings, a few of which corresponded to Pt_3O_4 and Al_2Pt, appeared in the diffraction patterns. The rings corresponding to the compound between Pt and Al did not disappear on subsequent heating in H_2, only their *d* values changed marginally. This indicates that once an interaction compound is formed it is not easily decomposed. This compound formation may involve the diffusion of Pt species into the bulk of the alumina substrate and it may be difficult to be reversed. However, at the interface between the particle and the substrate, the compound may be relatively easily reduced (even if only partly) and lead to the observed contraction of the particles. It is to be noted that especially on heating in O_2 at temperatures greater than 600°C a large number of particles disappeared, suggesting possible loss of material by evaporation. In this regard, Chen and Schmidt (*24*) have reported that there is considerable loss of Pt as PtO_2 when Pt/SiO_2 is heated in air or O_2 at ≥700°C. However Leitz *et al.* (*30*) have noted that only in the case of chlorine-containing samples and at ≥800°C was some loss of Pt observed. It was pointed out in the previous section that only a small fraction, if any at all, of the material must have been lost via evaporation, since substantial recovery of the material in the form of crystallites occurred on subsequent heating in H_2 (Figs. 4a–4c). As suggested before, the apparent "loss" of material might have been due to considerable extension of the particles (formation of thin undetectable films on the substrate) as well as to the diffusion of material into the substrate to form an interaction compound with alumina.

Even though the identity of the compounds formed could not be ascertained, the appearance of additional rings in the electron diffraction pattern (some of which correspond to some known Pt–Al compounds) nonetheless suggests compound formation between PtO_x and Al_2O_3 to yield $Pt_pAl_qO_r$, where r can also be zero. In this regard it is to be noted that den Otter and Dautzenberg (*31*) proposed the formation of Pt_3Al to explain the suppression of room temperature H_2 adsorption activity following high-temperature (≥500°C) reduction. Lagarde *et al.* (*32*) suggested that a Pt–O bonding between crystallites and Al_2O_3 is more likely than a Pt–Al bonding. But they were inclined to the view that chlorine atoms are also present in the complex. As pointed out in the Introduction, Leiske *et al.* (*13*) argued that oxidative redispersion is possible only in the presence of chlorine or chloride ions, and that oxygen alone does not lead to redispersion. Even though the presence of chlorine or chloride ions in our samples is very unlikely, it cannot be ruled out completely. It is worth noting that in the procedure followed to obtain electron transparent alumina films, subsequent to the anodic oxidation of an aluminum foil, the alumina films are stripped of their aluminum backing by dissolving the latter in mercuric chloride solution. Even though the alumina films are washed repeatedly in distilled water, there may be strongly bound chloride ions left behind and, therefore, we cannot rule out its absence or confirm its presence. A number of other authors have postulated complex formation between crystallites and support in Pt/Al_2O_3 samples (*6, 32–36*) and the exact nature and mechanism of formation of the complex have been debated. The present results have shown, both via electron diffraction and via film formation, that there are strong interactions between the crystallites and support in Pt/Al_2O_3 model catalysts and that the oxidative redispersion (regeneration), routinely achieved in industries by oxidative treatments between 500 and 600°C, occurs more via spreading and film formation than by emission of molecules to a two-dimensional phase of single atoms or molecules. It can be further noted that while Pt/Al_2O_3 is differentiated as being a "non-SMSI" system, the results indicate that there are strong interactions in this system as well, similar to those exhibited by the SMSI systems.

CONCLUSION

Pt/Al_2O_3 model catalysts are observed to undergo alternate changes in particle size when heated alternately in H_2 and O_2 in the range 500–700°C. When a sample heated previously in H_2 at $T \geq 500$°C is heated in O_2 at $500 \leq T \leq 700$°C, there is an apparent decrease in the particle size. On subsequent heating in H_2, approximately the previous sizes of the particles are recovered. Even the smallest of the particles exhibit this behavior and (particles down to about 0.8 nm are observed to) remain on the substrate through the various cycles of heating, suggesting that the molecular emission–capture mechanism is unlikely to be the cause of the alternations in size. In some cases, especially at 600 or 700°C, on heating in O_2, a kind of film, in the form of a halo, is observed around each particle and some particles extend out considerably into detectable or undetectable patches. In light of these observations, the alternation in size at 500°C is explained on the basis of the formation in O_2 of a thin, undetectable film around each particle which results in the appearance of a smaller size of the particles since only the solid core is detected. It is surmised that the film is in the immediate vicinity of and in contact with the parent particle and that the films around the particles do not interconnect to form a contiguous patch. On subsequent heating in H_2, the films contract and merge with the respective parent particle. Depending upon the reconstruction that follows, the particles can recover approximately their previous sizes. Such extension in O_2 and contraction in H_2 are enhanced by prolonged pretreatment in O_2 and/or by a few initial cycles in H_2 and O_2 which perhaps improve the wetting characteristics. It is suggested that extension and film formation are possible and likely mechanisms for the routinely observed oxidative redispersion of supported platinum catalysts.

REFERENCES

1. Wynblatt, P., and Gjostein, N. A., *Prog. Solid State Chem.* **9,** 21 (1975).
2. Burch, R., *in* "Catalysis," (G. C. Bond and G. Webb, Senior Reporters), Vol. 7 (a specialist periodical report). The Royal Society of Chemistry, London, 1985.
3. Ruckenstein, E., and Dadyburjor, D. B., *Rev. Chem. Eng.* **1**(3), 251 (1983).
4. Ruckenstein, E., and Malhotra, M. L., *J. Catal.* **41,** 303 (1976).
5. Fiedorow, R. M. J., and Wanke, S. E., *J. Catal.* **43,** 34 (1976).
6. Johnson, F. L., and Keith, C. D., *J. Phys. Chem.* **67,** 200 (1963).
7. Kraft, M., and Spindler, H., "Proceedings, 4th International Congress on Catalysis, Moscow, 1968" (B. A. Kazansky, Ed.), p. 1252. Adler, New York, 1968.
8. Yates, D. J. C., U.S. Patent 1433864 (1976).
9. Stulga, J. E., Wynblatt, P., and Tien, J. K., *J. Catal.* **62,** 59 (1980).
10. Gollob, R., and Dadyburjor, D. B., *J. Catal.* **68,** 473 (1981).
11. Ruckenstein, E., and Pulvermacher, B., *J. Catal.* **29,** 224 (1973).
12. Dautzenberg, F. M., and Wolters, H. B. M., *J. Catal.* **51,** 26 (1978).
13. Lieske, H., Leitz, G., Spindler, H., and Volter, J., *J. Catal.* **81,** 8 (1983).
14. Straguzzi, G. I., Aduriz, H. R., and Gigola, C. E., *J. Catal.* **66,** 171 (1980).
15. Bournonville, J. P., and Martino, G., *in* "Studies in Surface Science and Catalysis" (B. Delmon and G. Froment,Eds.), Vol. 6, p. 159. Elsevier, Amsterdam, 1980.
16. Ruckenstein, E., and Chu, Y. F., *J. Catal,* **59,** 109 (1979).
17. Yao, H. C., Sieg, M., and Plummer, H. K., Jr., *J. Catal.* **59,** 365 (1979).
18. Sushumna, I., and Ruckenstein, E., *J. Catal.* **94,** 239 (1985).
19. Ruckenstein, E., and Sushumna, I., *J. Catal.* **97,** 1 (1986).
20. Ruckenstein, E., and Lee, S. H., *J. Catal.* **86,** 457 (1984).
21. Sushumna, I., and Ruckenstein, E., *J. Catal.*, in press.
22. Nowak, E. J., *Chem. Eng. Sci.* **21,** 19 (1966).
23. Schmidt, L. D., and Luss, D., *J. Catal.* **22,** 269 (1971).
24. Chen, M., and Schmidt, L. D., *J. Catal.* **55,** 348 (1978); **56,** 198 (1979).
25. Ruckenstein, E., and Sushumna, I., *in* "Hydrogen Effects in Catalysis" (Z. Paal and P. G. Menon, Eds.). Dekker, New York, 1987, in press.
26. Weller, S. W., and Montagna, A. A., *J. Catal.* **21,** 303 (1971).
27. Mériaudeau, D., Dutel, J. F., Dufaux, M., and Naccache, C., *Stud. Surf. Sci. Catal.* **11,** 95 (1982).
28. Baker, R. T. K., *J. Catal.* **63,** 523 (1980).
29. Ruckenstein, E., *in* "Strong Metal Support Interactions" (R. T. K. Baker, S. J. Tauster, and J. A. Dumesic, Eds.). ACS Symp. Ser. No. 292, Washington, DC, 1986.
30. Leitz, G., Lieske, H., Spindler, H., Hanke, W., and Walter, J., *J. Catal.* **81,** 17 (1983).
31. den Otter, G. J., and Dautzenberg, F. M., *J. Catal.* **53,** 116 (1978).
32. Lagarde, P., Murata T., Vlaic, G., Freund, E., Dexpert, H., and Bournonville, J. P., *J. Catal.* **84,** 333 (1983).
33. McHenry, K. W., Bertolacini, R. J., Brennan, H. M., Wilson, J. L., and Seelig, H. S., *in* "Proceedings, 2nd International Congress on Catalysis, Paris, 1960," Paper 117 (1960), Vol. **2,** p. 2295, Technip, Paris, 1961.
34. Ushakov, V. A., Moroz, E. M., Zhdan, P. A., Boronin, A. I., Bursian, N. R., Kogan, S. B., and Levitskii, E. A., *Kinet. Katal.* **19,** 744 (1978).
35. Kunimori, K., Ideda, Y., Soma, M., and Uchijima, T., *J. Catal.* **79,** 185 (1983).
36. Mansour, A. N., Cook, J. W., Jr., Sayers, D. E., Enrich, R. J., and Katzer, J. R., *J. Catal.* **89,** 462 (1984).
37. Margitfalvi, J., Kern-Talas, E., and Szedlacsek, P., *J. Catal.* **92,** 193 (1985).

Events Observed and Evidence for Crystallite Migration in Pt/Al_2O_3 Catalysts

I. Sushumna and E. Ruckenstein

Department of Chemical Engineering, State University of New York at Buffalo, Buffalo, New York 14260

Received February 3, 1987; revised July 28, 1987

The sintering behavior of model Pt/Al_2O_3 catalysts when heated in hydrogen, and alternately in H_2 and O_2 was investigated at different temperatures (between 300 and 800°C) by means of transmission electron microscopy. At both relatively low (500°C) and high temperatures (750°C), short-distance migration (usually 1–3 particle diameters) and coalescence of particles was observed to contribute considerably toward sintering. At 500°C and following two or three cycles of alternate heating in H_2 and O_2, significant migration of particles (up to 8-nm particles migrating over 25 nm) were observed on exposure of the oxidized particles to H_2 and/or of the reduced particles to O_2. Sintering is fast and pronounced in the initial 4 to 6 h of heating of a fresh sample; then it becomes slow. However, further continuous heating for extended periods of 12 h or more causes additional significant sintering by particle migration and coalescence among other mechanisms. This suggests that investigations based on only 1 or 2 h of heating, as often reported in the literature, yield incomplete information. The present results indicate that at both 500 and 750°C, sintering by ripening (apparently only of a localized kind) also occurs. The present results are evidence of the occurrence of a variety of phenomena such as short-distance migration and coalescence of particles, migration toward or away from another particle, decrease in size and/or disappearance of larger particles near unaffected smaller particles, decrease in size and/or disappearance of both small and larger particles, decrease in size and/or disappearance of small particles near larger particles (ripening), decrease in size and subsequent migration of particles or vice versa, collision without coalescence, and collision–coalescence–separation into two particles again. The particles seem to feel the presence of other nearby particles via long-range interparticle forces that induce the migration of a particle toward another or emission of atoms toward a nearby particle. The two major mechanisms of sintering of supported metal crystallites appear to be (i) short-distance, direction-selective (in contrast to random) migration of particles followed by (a) collision and coalescence or (b) direct transfer of atoms between the two approaching particles, or (ii) localized ripening (direct ripening) between a few stationary, adjacent particles.

INTRODUCTION

Alumina-supported Pt catalysts are relatively thermostable in comparison to supported Ag, Fe, Ni, etc. Nevertheless, Pt/Al_2O_3 catalysts still lose their activity gradually on use in reducing atmospheres (such as in catalytic reforming). A part of the activity loss is attributed to the sintering of the metal particles (*1–3*). The supported catalysts lose exposed surface area also during heating in oxidizing atmospheres (such as in automotive exhaust oxidation and during coke burn off) (*1, 4*). That is, coarsening of the supported crystallites occurs in both reducing and oxidizing atmospheres, though its extent may vary, depending on the catalyst system and the treatment conditions employed. In specific cases, such as Pt/Al_2O_3 heated in O_2 between 500 and 580°C, redispersion of the catalysts has also been observed (*5–10*). Basically, two physical models have frequently been employed to explain the growth kinetics of the supported crystallites (*11*). One, known as the crystallite migration model, considers that the growth occurs via random migration, collision, and coalescence of crystallites (*12, 13*). The other, known as the atomic migration

This paper, which was originally published in **Journal of Catalysis**: 109, 433–462 (1988), is republished with permission.

model or Ostwald ripening, assumes that the growth occurs via emission of single atoms or molecules by the small crystallites and their capture by the larger crystallites (*14–16*). For single-atom emission and capture, two possibilities have been noted. In one of them, which is global, the small crystallites lose atoms to a surface phase of single atoms dispersed over the substrate, while the large ones capture atoms from this phase. In the other case, called direct ripening (*17*), atoms released by a small particle move directly to a neighboring large crystallite.

A number of investigations with supported catalysts including Pt/Al_2O_3 have been carried out in an attempt to identify the mechanisms involved in, and the factors responsible for, the growth of the crystallites, as well as for the regeneration of the sintered catalysts (*3, 11, 18–41*). Both model and industrial-type catalysts have been investigated, predominantly by selective chemisorption of gases, transmission electron microscopy, and X-ray line broadening (for reviews, see Refs. (*11, 14, 42–44*)). As a result of the anomalous chemisorption behavior (such as the suppression of room-temperature H_2 chemisorption following high-temperature (500°C) H_2 treatment) exhibited by certain catalyst systems such as Pt/Al_2O_3 (*45, 46*) and Pt/TiO_2 (*47–49*), it is now recognized that chemisorption of gases may require the availability of adsorption sites on the crystallites (which may be covered and hence prevent chemisorption, even though the dispersion is high). Consequently, there is not necessarily a direct correlation between the extent of chemisorption and the catalyst dispersion. Caution is needed therefore in drawing conclusions on the basis of dispersion values obtained from chemisorption measurements alone (*50*). A number of investigations of the mechanism of sintering of supported metal catalysts have relied on such chemisorption measurements. Selective chemisorption of gases is, in general, indeed a very valuable method of measuring catalyst dispersion. However, especially in cases where ambiguities and anomalies of the kind mentioned above are involved, it is appropriate to use a complementary technique such as electron microscopy. As is well known, transmission electron microscopy (TEM), though tedious, provides direct information on the crystallite shape, size, size distribution, and, to a large extent, on the mechanisms involved in sintering and redispersion. Therefore, TEM has been employed by a number of researchers to investigate the mechanism of sintering of supported metal catalysts.

The results have been discussed in terms of both Ostwald ripening (*8, 11, 14–18, 26–30, 37*) and crystallite migration models (*32–40*). On the basis of their electron microscopy or chemisorption results, a few authors (*26, 28, 29*) reported that they had never observed crystallite migration or dumbbell-shaped particles on the supports with the catalysts they had investigated and that atomic or molecular migration is the only mechanism of sintering and redispersion. Recent results, from this laboratory as well as from others, of direct TEM observation of changes in the same region of the samples following successive heat treatments have shown, however, that crystallites can migrate and that crystallite migration and coalescence contribute significantly to the growth of supported crystallites. Besides the early electron microscopy observations of supported crystallite migration by Bassett (*51*) in the case of copper and silver islands on carbon, graphite, or molybdenite substrates and by Skofronick *et al.* (*52*) in the case of gold islands on carbon and silicon substrates, there are more recent observations of supported crystallite migration and coalescence. Chu and Ruckenstein (*24*) observed significant migration of crystallites larger than 20 nm in the case of Pt/C heated in O_2 atmospheres. They observed migration of particles larger than 10 nm also in the case of Pt/Al_2O_3 at 750°C (*25*). Heinemann and Poppa (*53*) presented direct *in situ* electron microscopic evidence for the simultaneous occurrence of short-distance crystallite migration and coalescence as well as ripening in the sintering of Ag/graphite system in the temperature range 25–450°C. They noted that though slow ripening occurred over the entire temperature range, the overriding surface transport mechanism was short-distance (<10 nm) cluster mobility. Chen and Ruckenstein (*36*) in the case of Pd/Al_2O_3, Ruckenstein and Lee (*40*) in the case of Ni/Al_2O_3, and Sushumna and Ruckenstein (*39*) in the case of Fe/Al_2O_3 observed crystallite migration among a large number of other events when the samples were heated in H_2 and O_2. Arai *et al.* (*37*) reported *in situ* TEM results of sintering of Ni, Pt, and Ag on amorphous carbon, SiO_2, and Al_2O_3 and noted that in the case of Pt/Al_2O_3 and above 600°C, only particles larger than 10 nm grew through abrupt surface movement and subsequent coalescence of adjacent pairs. Richardson and Crump (*32*) in the case of Ni/SiO_2 concluded, on the basis of magnetic measurements of the particle size distributions, that especially above 450°C particle migration and coalescence were responsible for the sintering in a He atmosphere. Granquist and Buhrman (*54, 55*), analyzing various previous experimental results on sintering, argued on the basis of the shape of the particle size distributions that crystallite (migration and) coalescence is more likely than ripening to be responsible for the coarsening of the supported particles. Abundance of experimental results notwithstanding, the mechanism(s) of sintering has yet to be clarified; in general, some authors support one mechanism, while others support the other mechanism. However, recent results (*36, 39*) seem to suggest that the phenomena of sintering and regeneration involve complex interrelated processes that cannot be restricted to one or even to both of the mechanisms noted above.

In the past, investigations of supported metal catalysts, including Pt/Al_2O_3 have been based in general on either short (1 or 2 h) (*26–30, 41*) or long (24 or 48 h) durations of heating (*9, 25, 29*). However, no systematic investigations of short intervals of heating, especially of a fresh catalyst, over extended time periods, at various temperatures, have been carried out. Especially on the basis of 1 or 2 h of heating, a few authors have reported that they never observed crystallite migration on the supports with the catalysts they had investigated (*26–30*). Evidence for migration and coalescence with Pd/Al_2O_3 (*36*), Pt/Al_2O_3 (*25*), Ni/Al_2O_3 (*40*), and especially Fe/Al_2O_3 (*39*) have been presented from this laboratory among others in the past. Here we present additional transmission electron microscopic evidence for crystallite migration among other phenomena in Pt/Al_2O_3 model catalysts heated at different temperatures in H_2 and O_2.

EXPERIMENTAL

Preparation of Alumina Supports

The method of preparation of electron-transparent films of γ-Al_2O_3 has been described in detail previously (*25, 39*). Thin aluminum foils (99.999% pure, Alfa Products, Inc.) were cleaned thoroughly by chemical polishing and were subsequently washed in distilled water. An amorphous aluminum oxide layer about 30 nm thick was built up on the cleaned aluminum foil by anodic oxidation at 20 V in a 3 wt% tartaric acid solution. The oxidized foil was subsequently cleaned thoroughly in distilled water. The oxide layer was separated from aluminum by floating small pieces of the oxidized foil in mercuric chloride solution, in which the unoxidized aluminum dissolved. The aluminum oxide films were subsequently washed repeatedly in distilled water and eventually picked up on gold electron microscope grids. The oxide films on the grids were dried and subsequently heated at 800°C in laboratory air for about 72 h and then slowly cooled to the room temperature. The prolonged heat treatment is carried out to convert the amorphous films to crystalline γ-Al_2O_3 and to ensure that no further changes would occur on subsequent heat treatments. It is to be noted that after the above heat treatment and prior to or after the deposition of the metal, these alumina samples were not brought into contact with any solution or liquid and therefore no artifacts of the kind discussed by Glassl *et al.* (56) were likely to have been generated.

Platinum films 0.75, 1.5, and 2 nm thick were deposited onto the alumina films in three different batches, by evaporating the corresponding amounts of Pt wire (99.999% pure, Alfa Products, Inc.) from a tungsten filament in an Edwards vacuum evaporator. The vacuum in the chamber was better than 10^{-6} Torr. The substrate was maintained at room temperature during deposition.

It should be pointed out that the industrial supported Pt catalysts are relatively thermostable. The very low metal loading of Pt/Al_2O_3 usually used in industry contributes to the slow growth rate. The model catalysts in general have a relatively high loading, which enhances sintering. However, even though the metal loadings of the model catalysts may not correspond to those of the industrial catalysts, the model systems provide at least a qualitative indication of the processes that occur in supported catalyst systems. In this regard it is worth noting that Smith *et al.* (*41*) and Baker *et al.* (*27*) have noted previously that the behaviors of the model catalysts of the kind described here are in general similar to those of the industrial catalysts.

The samples on which metal had been deposited were heated in a quartz boat inside a quartz tube. Ultrapure hydrogen, both as-received (99.999% pure with <1 ppm O_2 and <3 ppm moisture) and further purified by being passed through a Deoxo unit (Engelhard Industries) followed by a column of 3-Å molecular sieves immersed in liquid nitrogen, were used in the experiments. Helium, 99.999% pure with <3 ppm moisture, was also passed through the above column of molecular sieves immersed in liquid nitrogen. Ultrahigh-purity-grade oxygen was used as received. All the gases were supplied by Linde Division, Union Carbide Corporation. For each heat treatment, the following procedure was adopted. With the sample inside, the tube was flushed with He for at least 5 minutes before the power to the furnace was turned on. The sample was heated in He until the preset temperature was reached, at which point He was turned off and the desired gas was let in. The sample was then heated for the predetermined length of time, at the end of which the furnace and the gas were turned off and He was allowed to flow through again. The sample was cooled slowly to the room temperature in He before it was exposed to the atmosphere.

The samples were observed in a JEOL 100U transmission electron microscope operated at 80 kV. Each sample was scanned thoroughly over the entire specimen area to check for uniformity in behavior, and pictures were taken at a few regions. Following each heat treatment, the same regions were photographed at the same magnification (in the 60–100-K range). Electron diffraction patterns were also recorded following each heat treatment. A liquid nitrogen trap was used to reduce specimen contamination in the microscope. Also, the intensity of the electron beam and the specimen's exposure to the beam were kept low to minimize the contamination of the specimen. The alumina film moved very little on the supporting grid, and, especially following the initial one or two observations, the same regions could be located relatively easily after successive heat treatments. The particle migrations shown in the next section were followed with reference to certain fixed grain boundaries or other "landmarks" on the support.

RESULTS

Table 1 lists some of the events observed on Pt/Al_2O_3 model catalysts heated in H_2 and O_2. Figures 1, 1′, and 1″ show some micrographs of the time sequence of a sam-

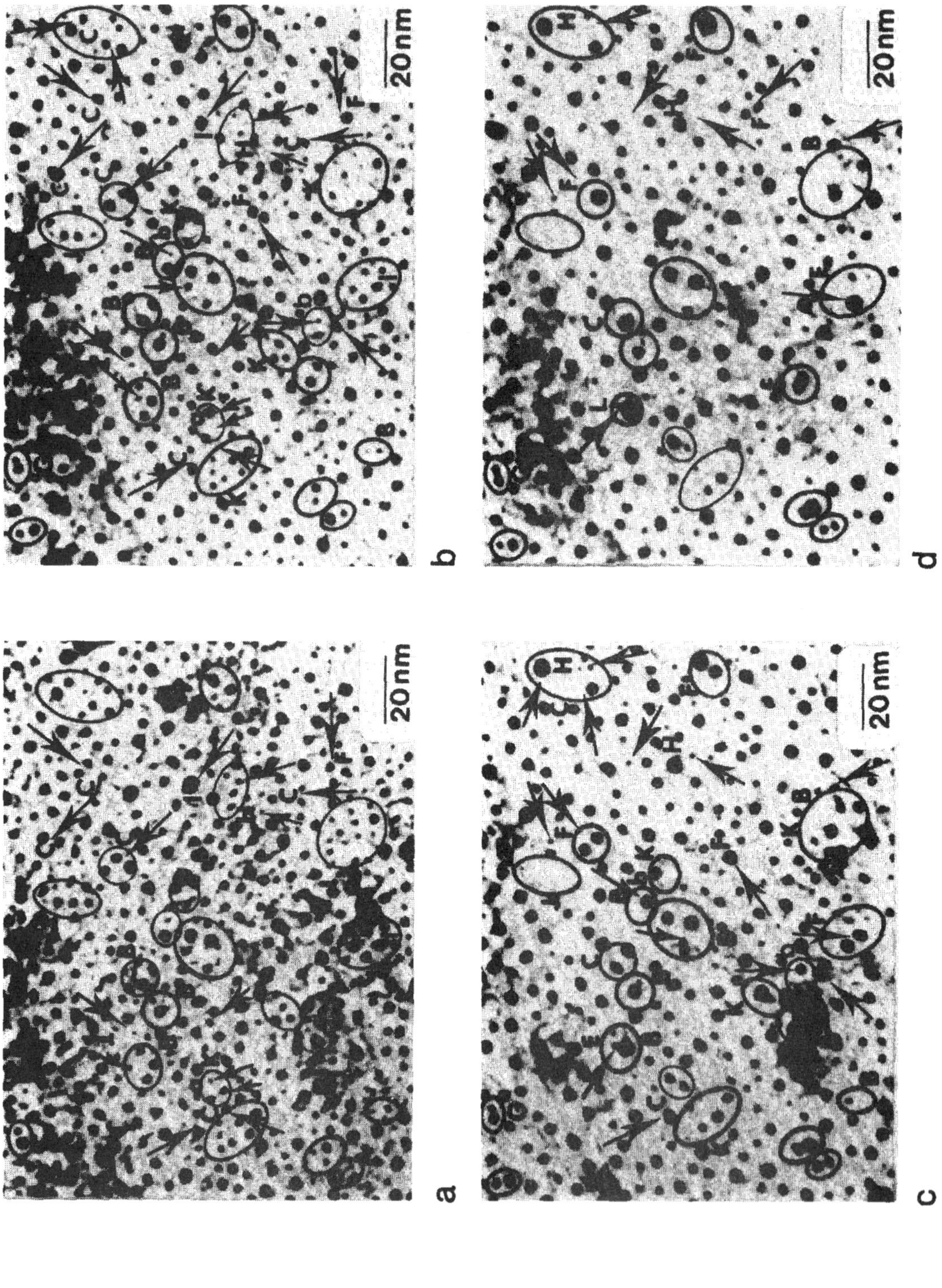
20 nm
20 nm
20 nm
20 nm
a
b
c
d

ple with an initial film thickness of 1.5 nm heated in as-received ultrapure hydrogen. The letters on all the micrographs correspond to the respective events listed in Table 1. When the metal-deposited sample was heated in H_2 for 6 h at 500°C, detectable particles in the range 0.8–5 nm formed, with the average around 2 nm. The sample with this initial distribution was heated subsequently in the following cumulative sequence: 1, 2, 3, 4, 6, 8, 12, 17, 29, 43, and 65 h. Following the initial 1 h heating, a number of neighboring particles, especially the larger ones, coalesced, decreasing the particle number density (B, C, F in Figs. 1a–1d). A number of small particles 0.8–1.5 nm in size remained unaffected while a number of other small particles decreased in size or disappeared (regions 1 and H in Figs. 1a–1d and 1′a–1′d, for example). All the large particles also appear to have decreased slightly in size (I in Figs. 1a and 1b, for example). Since, as just mentioned, a number of small particles on the substrate remained unaffected, the decrease in the size of the large particles may be a result of emission of atoms to the substrate, due not to a global ripening mechanism, but more likely to local variations in curvature along the periphery of the large particles. Or, it may be a result of extension of an undetectable film from around the particles or of diffusion of material into the substrate. It may also be a result of mere reconstruction with time to a more compact shape. During the next three 1-h intervals of heating, the number of particles decreased further. Both the decrease and disappearance of smaller particles (H, J) and the coalescence of nearby particles in the vicinity of unaffected or decreasing smaller particles were observed (C, F in Figs. 1b–1d). Also, a number of nearby

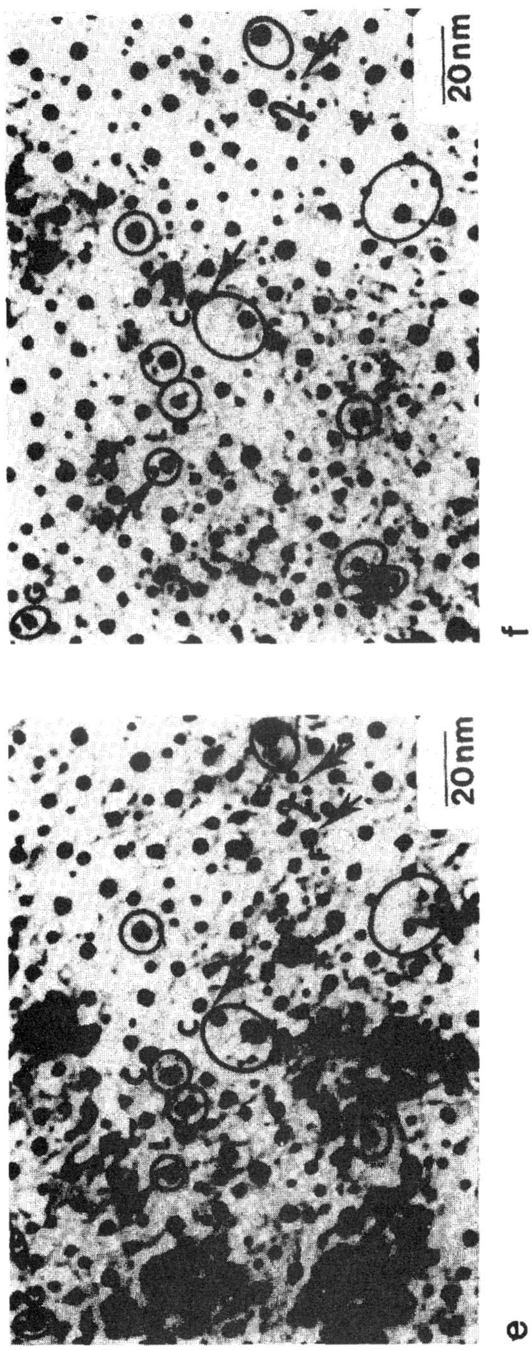

FIG. 1. Sequence of changes in the same region of a Pt/Al_2O_3 sample of 1.5 nm initial film thickness on heating in as-supplied hydrogen at 500°C. The following durations of heating, starting with (b), are cumulative. (a) Initial (6 h H_2); (b) 1 h; (c) 2 h; (d) 4 h; (e) 17 h; (f) (65 h H_2 +) 16 h O_2.

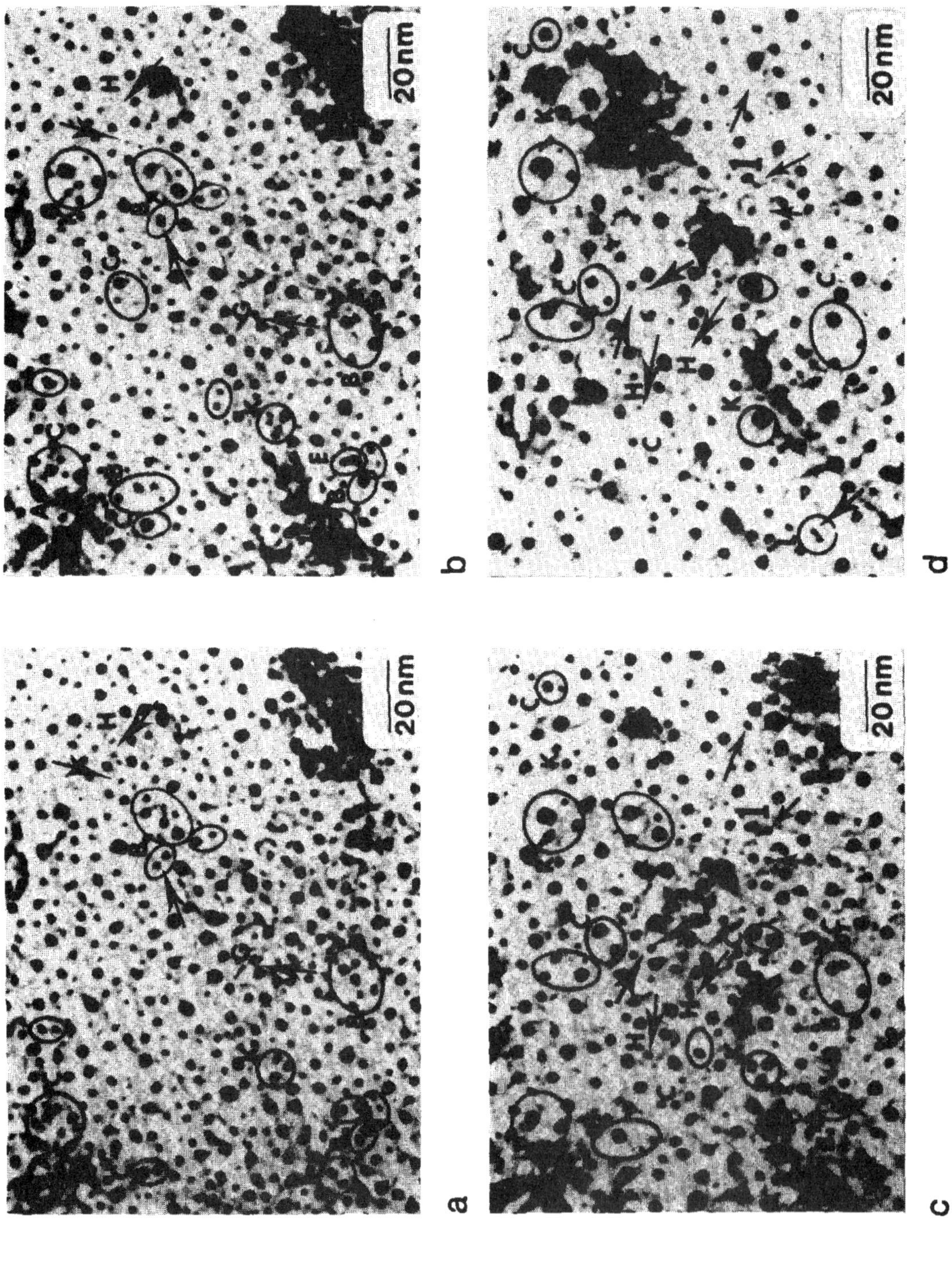
20 nm
a
20 nm
b
20 nm
c
20 nm
d

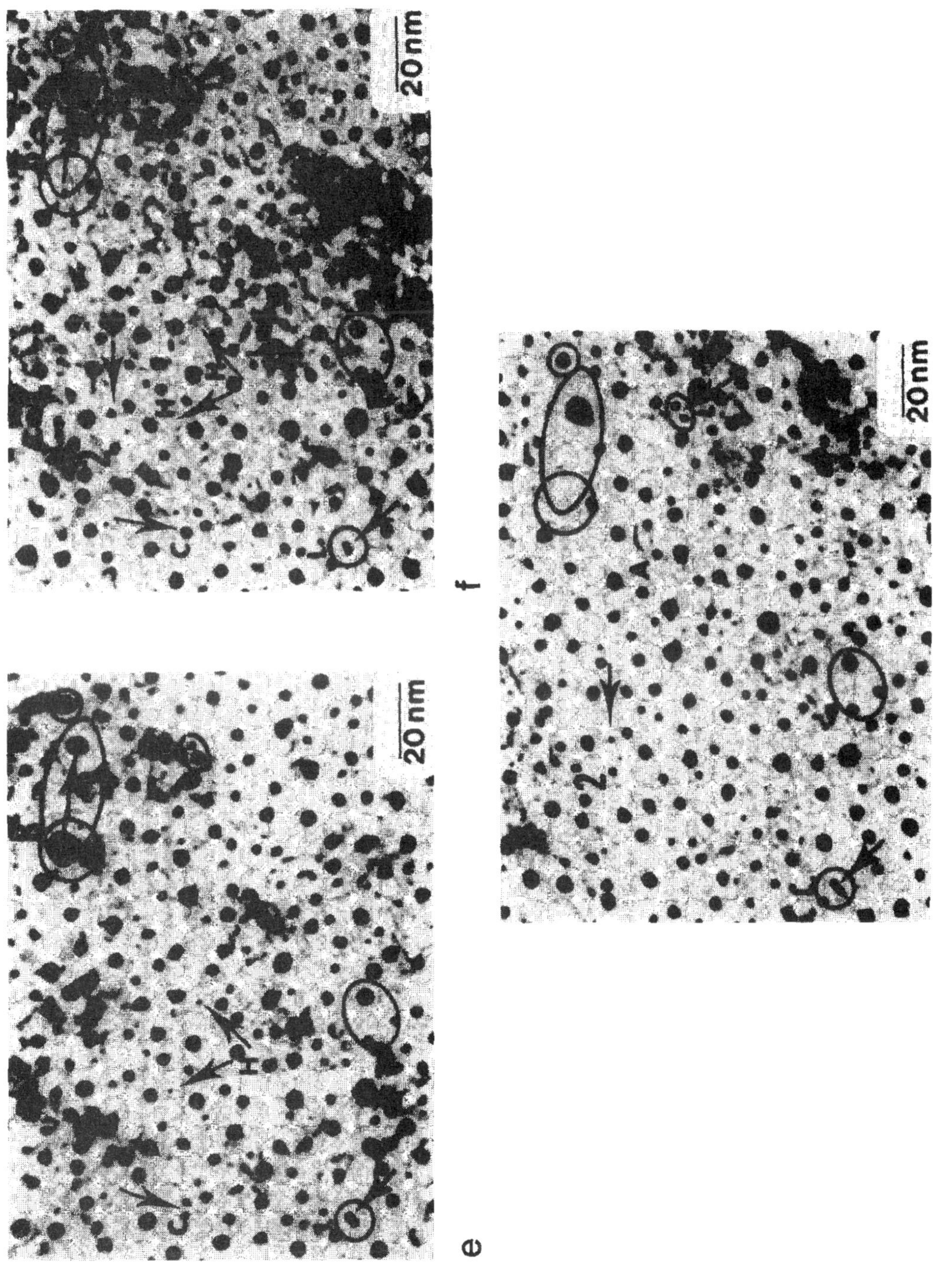

FIG. 1′. Same sample as in Fig. 1. The micrographs are of a different region. (a) Initial (6 h H_2); (b) 1 h; (c) 2 h; (d) 12 h; (e) 29 h; (f) 65 h; (g) 16 h O_2.

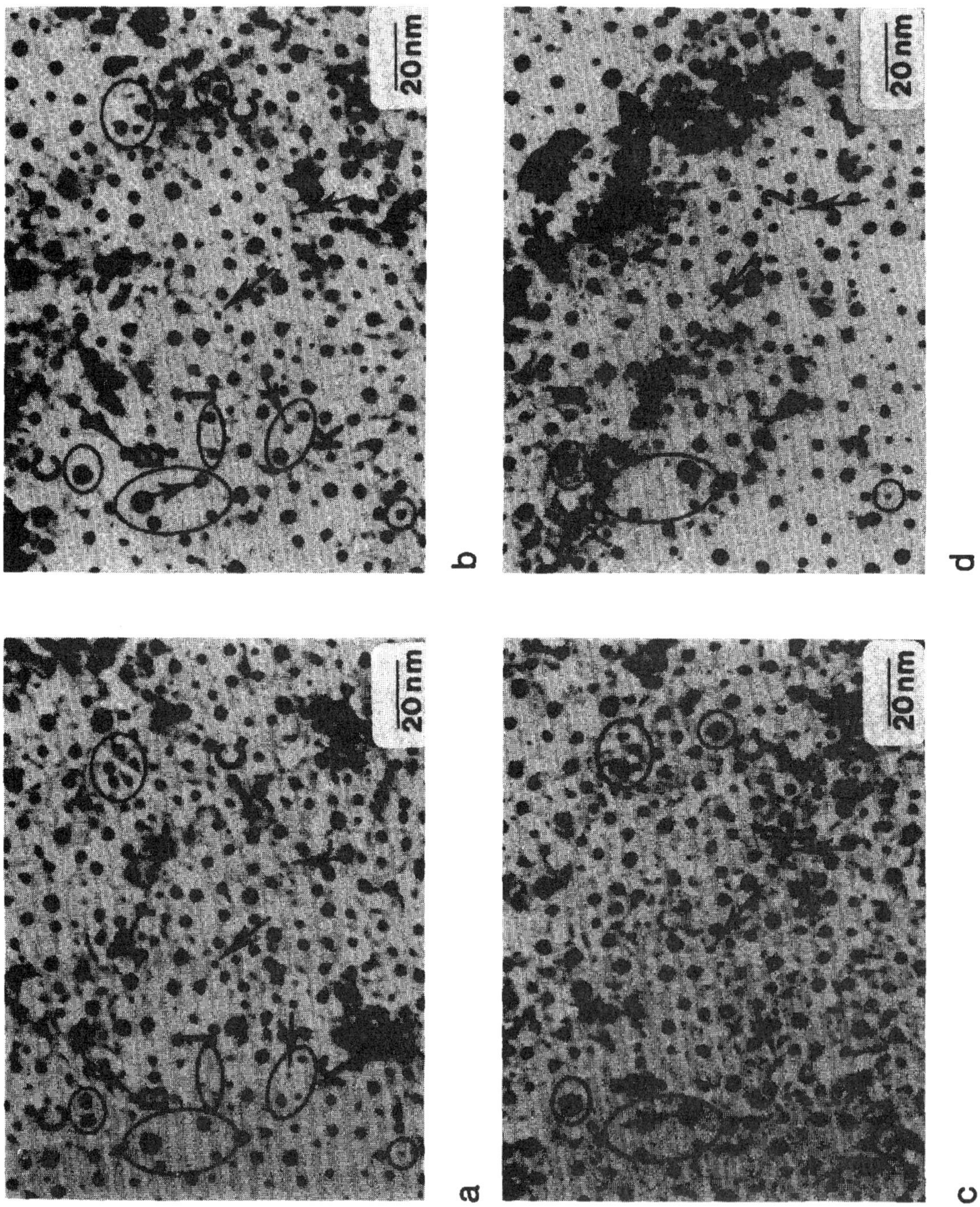

FIG. 1″. Same sample as in Fig. 1. The micrographs are of yet another region. (a) 6 h H_2; (b) 29 h H_2; (c) 65 h H_2; (d) 16 h O_2.

small particles appear to have coalesced to form larger particles (K); this growth occurs either by migration and coalescence of the crystallites or by direct ripening between nearby crystallites, more likely by the former, as the particles are of about the same size and small, ~2 nm (K in Figs. 1b, 1c and 1′b, 1′c). In fact, in the region marked "a" in Figs. 1a and 1b, a small particle (about 0.8 nm) appears to have grown to about 2.5 nm, very likely as a result of the migration and coalescence with it of a larger particle from nearby. A similar event is visible at "b" in Figs. 1b and 1c. The subsequent heat treatment for up to a total of 8 h brought very little change (not all the micrographs for the entire heating sequence are shown here). At the end of a total of 12 h of heating only a few additional pairs of nearby particles had coalesced. However, there was a marginal increase in the size of all the particles, both small and large. Small particles, as small as 1.5 to 2 nm still remained on the substrate. It is possible that all the particles over the entire observable size range could have increased in size marginally at the expense of particles much smaller than about 0.8 nm, either by their coalescence with the migrating crystallites or by a global ripening mechanism. On the other hand, the increase in size may also be a result of extension and/or reconstruction of the particles. Following 17, 29, and 43 h total heating, a number of pairs of large particles coalesced. Very few small particles disappeared. Particles were formed in some places where no particles had been apparent previously (region 1 in Figs. 1″a, 1″b). After heating for additional 22 h (to 65 h of total heating), a number of large crystallites (up to about 8 nm) migrated large distances (up to about 30 nm). Some of these crystallites coalesced with other particles (B in Figs. 1′e, 1′f and in 1″b, 1″c), while some stopped short of colliding (not shown in figures). A number of other particles near each other coalesced, most likely by migration and collision. It is to be noted that long durations of heating lead to significant sintering, especially when the loading is high, though it is not as rapid and pronounced as in the initial stages. During this sequence of heating it was also observed that sometimes two nearby particles made contact via a faint, narrow neck or a whiskerlike bridge. The particles then either sintered via a transfer of atoms along the bridge or separated again as the interconnecting bridge was broken (G in Figs. 1b–1f and 1′a, 1′b). In some cases the two particles more or less completely coalesced and then tended to separate again (L in Figs. 1d–1f and 1′b–1′g). Additional events observed are marked on the micrographs by various letters corresponding to the events listed in Table 1.

When the above sample was subsequently heated in O_2 at 500°C for up to 28 h, no particle migration or detectable ripening of particles was observed. However, most of the particles, small and large, decreased in size (region 2 in Figs. 1f, 1′g, and 1″d) as a result of either emission of atoms/molecules to the substrate or of the extension and formation of films undetectable by electron microscopy around the particles. This phenomenon is discussed further in Refs. (*57, 58*). Two additional cycles of heating in H_2 and O_2 for a total of 50 h resulted in marginal particle coarsening mostly via coalescence of neighboring particles (C in Figs. 2a–2b). Particle migration without coalescence (A in Figs. 2a, 2b) and elongation of some particles (O in 2a, 2b) could also be observed. Almost all the detectable particles of about 0.8 nm and even less remained (some of them are marked with arrows in Figs. 2a, 2b). Following subsequent cycles of alternate heating in H_2 and O_2, a large number of events of migration and coalescence of particles (A, B in Figs. 2c–2f and 2′a–2′e) and dumbbell-shaped particles (E in Figs. 2e–2f and 2′a–2′e) could be observed in both O_2 and H_2, as shown in Figs. 2 and 2′. As seen on the micrographs (Figs. 2 and 2′), there is, in addition, a marginal decrease in the size of all the particles on heating in oxygen and an increase in size on

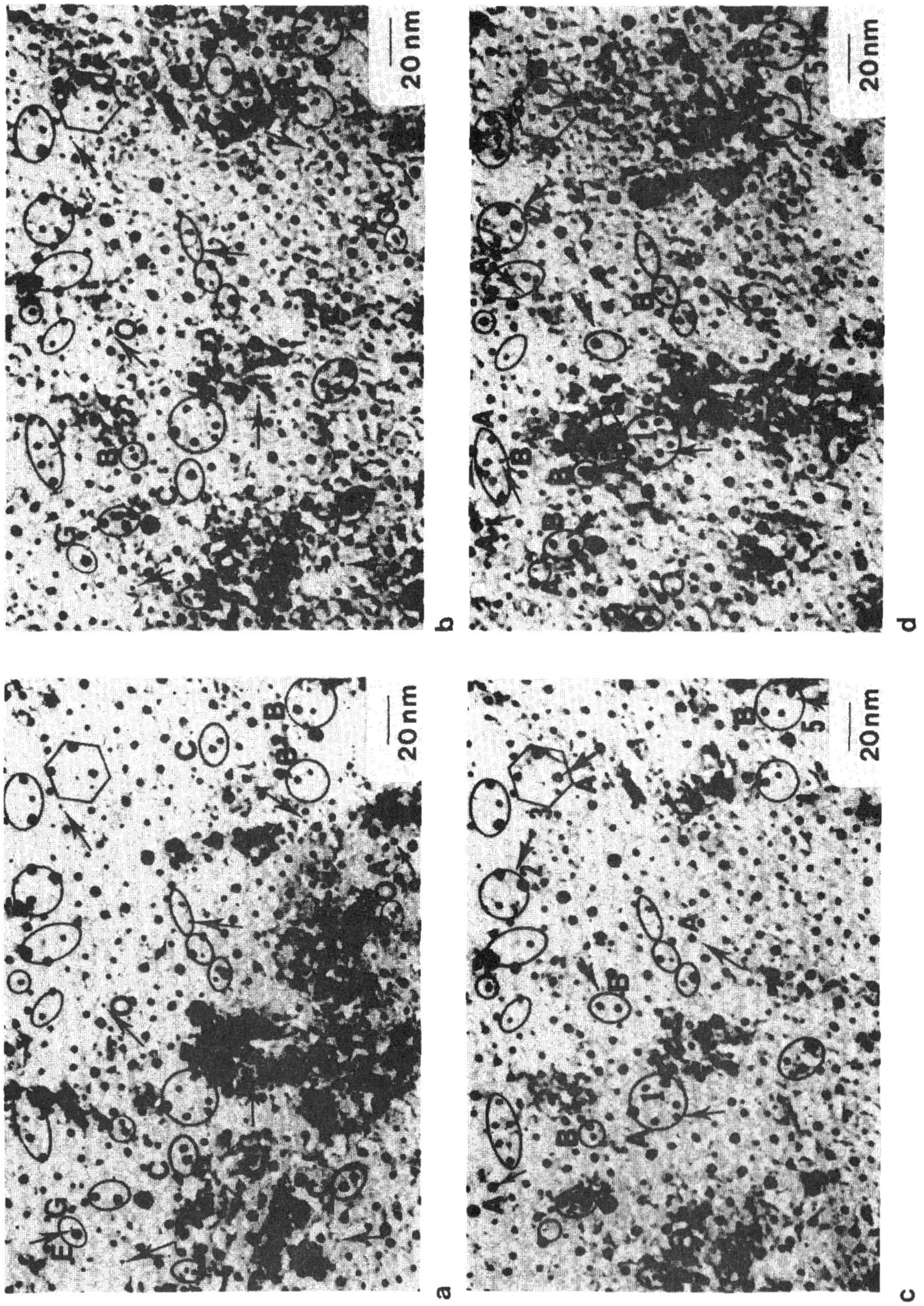
20 nm
20 nm
20 nm
20 nm
a
b
c
d

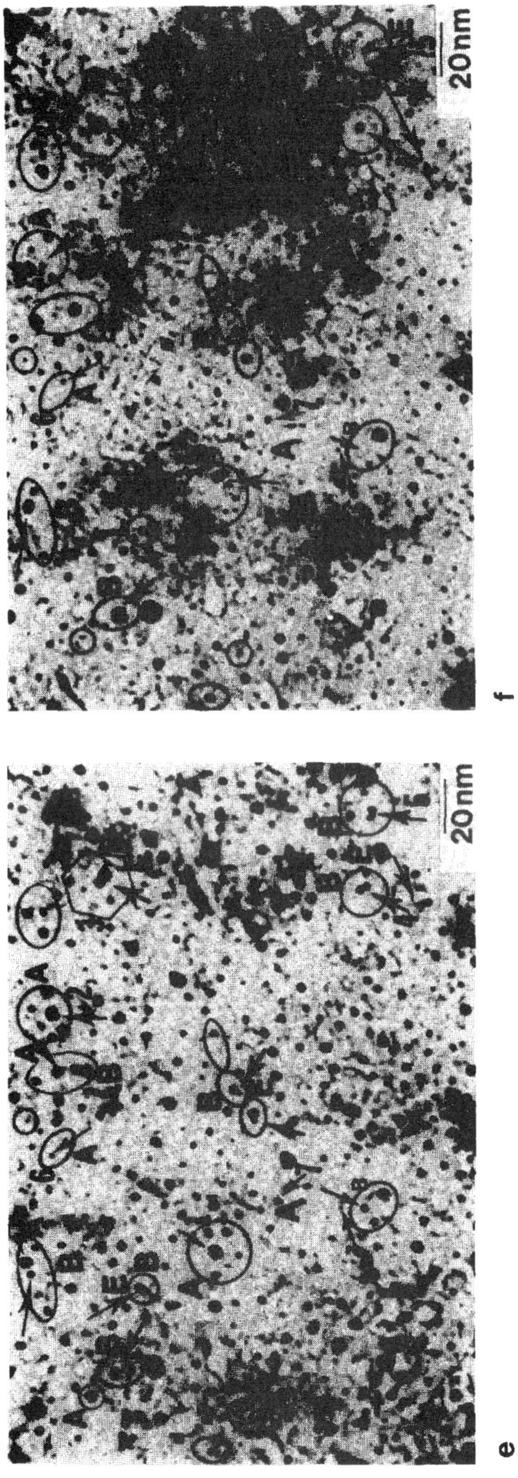

heating in hydrogen (Figs. 2c and 2d), which is discussed in detail in Refs. (*57*, *58*). It can be seen that while there is very little ripening, especially of the global kind, large particles (regions marked 1–5 in Figs. 2c–2f and regions 1 and 2 in Figs. 2′a–2′e), as large as 8 nm (region 1 in Figs. 2′a–2′e), migrate gradually over large distances (25 nm) and coalesce with other particles. At region 2 in Figs. 2e and 2f, the large particle appears to have migrated over a smaller particle. The smaller particle is probably trapped in a valley on the substrate surface and does not hinder the migration of the larger particle over it. On the other hand, at region 1 in Fig. 2′, for example, the large particle incorporates into itself the smaller particles it encounters while migrating. The numerous events marked A, B, and E in the micrographs provide evidence for more than isolated incidents of crystallite migration in supported model catalysts. Also, dumbbell-shaped particles (E in Figs. 2 and 2′) remain over extended periods of heat treatment, suggesting that coalescence can be a rate-limiting step. This is in contrast with the suggestion that coalescence must be fast (*14*). In addition, as noted before, sometimes two neighboring particles collide, partially coalesce, and then separate into two again (L in Figs. 2e, 2f and 2′d, 2′e).

It is clear that migration and coalescence play a role in the coarsening of Pt/Al_2O_3 catalysts not merely during the initial stages of sintering of a fresh sample but also after prolonged heat treatment of the sample. In contrast to the opinion that especially in an oxidizing atmosphere only atomic/molecular migration takes place, the present results show that migration and coalescence of crystallites also occur.

From the above results one can note that when an already sintered sample is heated

FIG. 2. Same sample as in Fig. 1 heated subsequently in H_2 and O_2 alternately at 500°C. (a) Initial (65 h H_2 + 28 h O_2 + 18 h H_2 + 14 h O_2); (b) 6 h H_2; (c) 12 h O_2; (d) 14 h H_2; (e) 12 h O_2; (f) 10 h H_2 + 10 h O_2.

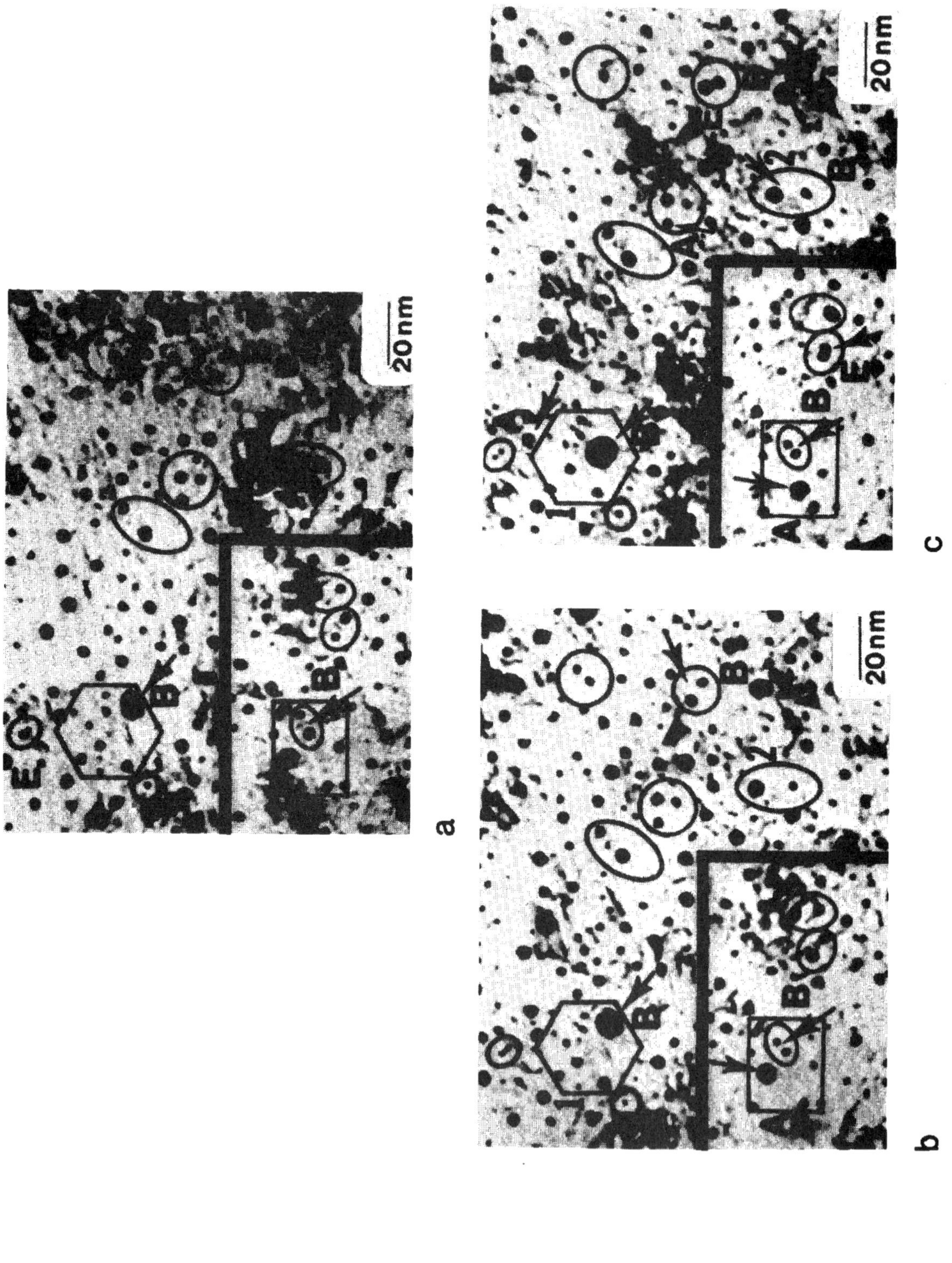
20nm
a
20nm
b
20nm
c

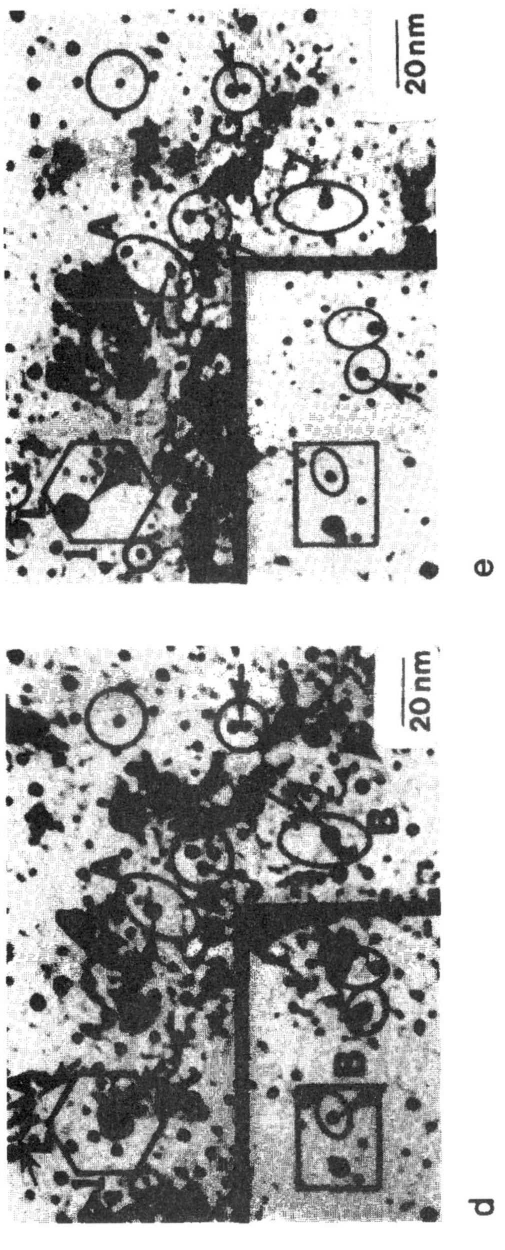

further, changes begin to occur after two or three cycles of alternate heating in H_2 and O_2. Very little change with regard to sintering is noticed on extended heating in the same atmosphere. A change in exposure, either of an oxidized sample to a reducing atmosphere or of reduced particles to O_2, brings about particle motion, coalescence, etc. This is perhaps due to the heat of reaction involved, as suggested also by Glassl *et al.* (*33*). The above results show that migration of particles in Pt/Al_2O_3 catalysts occurs at the relatively low temperature of 500°C, both with a fresh sample in H_2 and a "sintered" sample in both O_2 and H_2.

A different sample of 2-nm initial film thickness was heated in additionally purified H_2 at 500°C for 12 h and subsequently at 600°C for 5 h to generate particles. Following heating at 500°C, a large number of particles of about 1.6 nm average size were formed. The specimen employed in the experiments described in the previous paragraphs, which had a lower metal loading of 1.5-nm initial film thickness, yielded slightly larger particles following only 6 h of heating at 500°C. The difference in the sintering behaviors of these two specimens points to the role of the type of hydrogen used. Sintering is faster and relatively more pronounced in the as-received hydrogen (which contains traces of water) than in the further purified H_2. Similar trends were also observed by Smith *et al.* (*41*). The additional heating at 600°C for 5 h caused severe sintering. Dumbbell-shaped particles could be observed following both the 500 and the 600°C heat treatments (particles marked with arrows in Figs. 3a and 3b). Even though the sintering mechanism could not be followed during these heat treatments because of the severity of sintering, the presence of dumbbell-shaped particles

FIG. 2′. Same sample as in Fig. 2. The micrographs are of a different region. (a) Initial (same as in Fig. 2); (b) 6 h H_2; (c) 12 h O_2 + 14 h H_2; (d) 12 h O_2; (e) 10 h H_2 + 10 h O_2. The regions bounded by the black margins are from a different area.

suggests that at both temperatures crystallite migration and coalescence were involved. This sample was subsequently heated alternately in O_2 and purified H_2. A few select micrographs are shown in Figs. 3′ and 3″. The heat treatment in O_2 was carried out from 300 to 500°C, with a 50°C increment in successive cycles. But the heat treatment in H_2 in each cycle was carried out at 500°C. (The incremental heating in O_2 was carried out in an attempt to identify the lowest temperature at which film formation around the particles, or a decrease in size of all the particles, occurs. Film formation around particles when Pt/Al_2O_3 model catalysts are heated in O_2 is discussed in Refs. (*57, 58*).) Throughout this treatment and especially up to the 500°C O_2/H_2 cycling, the growth was slow and gradual and the marginal growth occurred mostly by coalescence of nearby particles (C in Figs. 3′b–3′f). However, at 500°C, as with the other sample reported in the previous paragraphs, significant migration and coalescence were observed on alternate heating in H_2 and O_2. A number of particles as small as 1 nm remained with very little change throughout the various cycles of heating for up to a total of about 100 h (some are marked with arrows in Figs. 3′a–3′f and 3″a–3″h), while a few others disappeared. This suggests that at low temperatures very little global ripening, involving a large number of particles, takes place (in spite of the alternating oxidizing and reducing treatments). Particle migration (A), migration and coalescence (B and C in Figs. 3′ and 3″), and direct ripening between a small and a larger particle via a narrow whiskerlike bridge (G, region 1 in Fig. 3″) were all observed. In other cases nearby particles made and broke contact a few times without any change in the size of the particles (region 2 in Fig. 3″). These events indicate that in some cases the coalescence is retarded or inhibited altogether and that coalescence may also be a rate-limiting step, as stated before.

Another sample from the same batch as the preceding one with a 2-nm initial film thickness of Pt was heated similarly in purified H_2 at 500°C for 12 h and subsequently at 600°C for 5 h to obtain an initial crystallite distribution (Fig. 4a). Heating this sample further in purified H_2 at 700°C for only $\frac{1}{2}$ h led to severe sintering (Fig. 4b). A large number of small and relatively large particles disappeared, other particles of the same size and even smaller remained unaffected, while some of the smaller particles also decreased in size (particles marked with arrows in Figs. 4a, 4b). In addition, a large number of neighboring pairs of crystallites coalesced, as indicated by the elongated and/or dumbbell-shaped particles in the micrographs (C, E in Figs. 4a, 4b). In one location, out of the two small particles adjacent to a much larger particle, the seemingly larger of the two decreased in size while the smaller grew or remained unaffected (regions 1, 2 in Figs. 4a–4d). In another place, two large particles of about the same size coalesced into one, while the nearby very small particles decreased only slightly in size (F in Figs. 4a– 4d). This indicates that both local ripening and short-distance migration and coalescence occur adjacent to each other. Probably because of the high temperature, most of the particles were faceted and rectangular. Subsequent heat treatment at the same temperature for up to a total of 5 h resulted in gradual sintering, and similar events such as coalescence of neighboring particles (C), decrease in size of small particles (H), decrease in size and migration (D), gradual migration and coalescence (B), and disappearance of small particles (particles marked with arrows) could be observed (Figs. 4b–4d). Also, the particles became relatively circular. Heating this sample further in H_2 at 800°C for 2 h led to a drastic decrease in the number of particles and an increase in the average particle size. A large number of both small and large particles disappeared (J and particles marked with arrows in Figs. 4d, 4e). Some relatively large particles decreased in size while considerably smaller

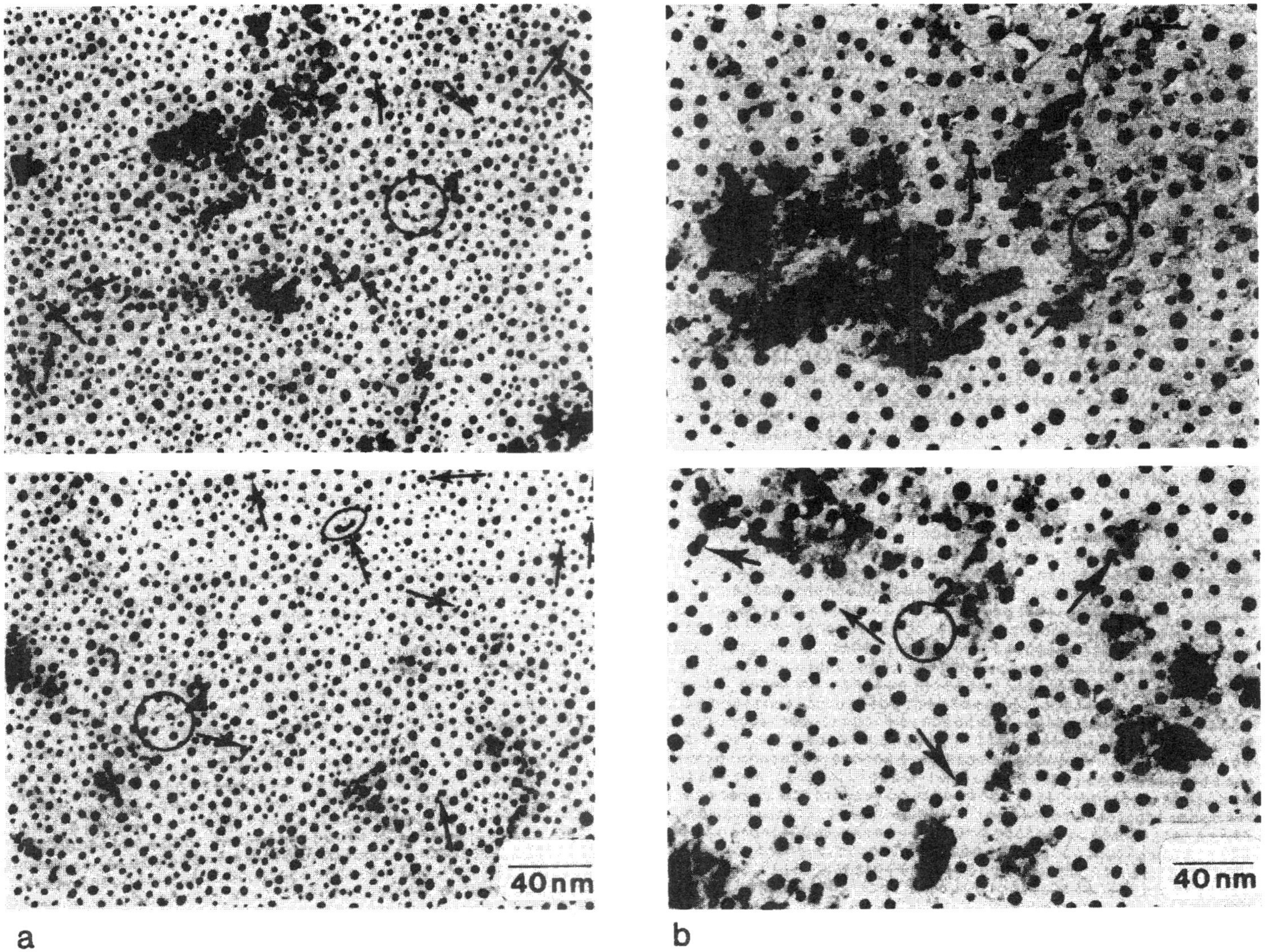

FIG. 3. Changes on heating a Pt/Al_2O_3 sample of 2 nm initial film thickness in purified hydrogen. The micrographs are of the same regions. (a) 12 h H_2, 500°C; (b) 5 h H_2, 600°C.

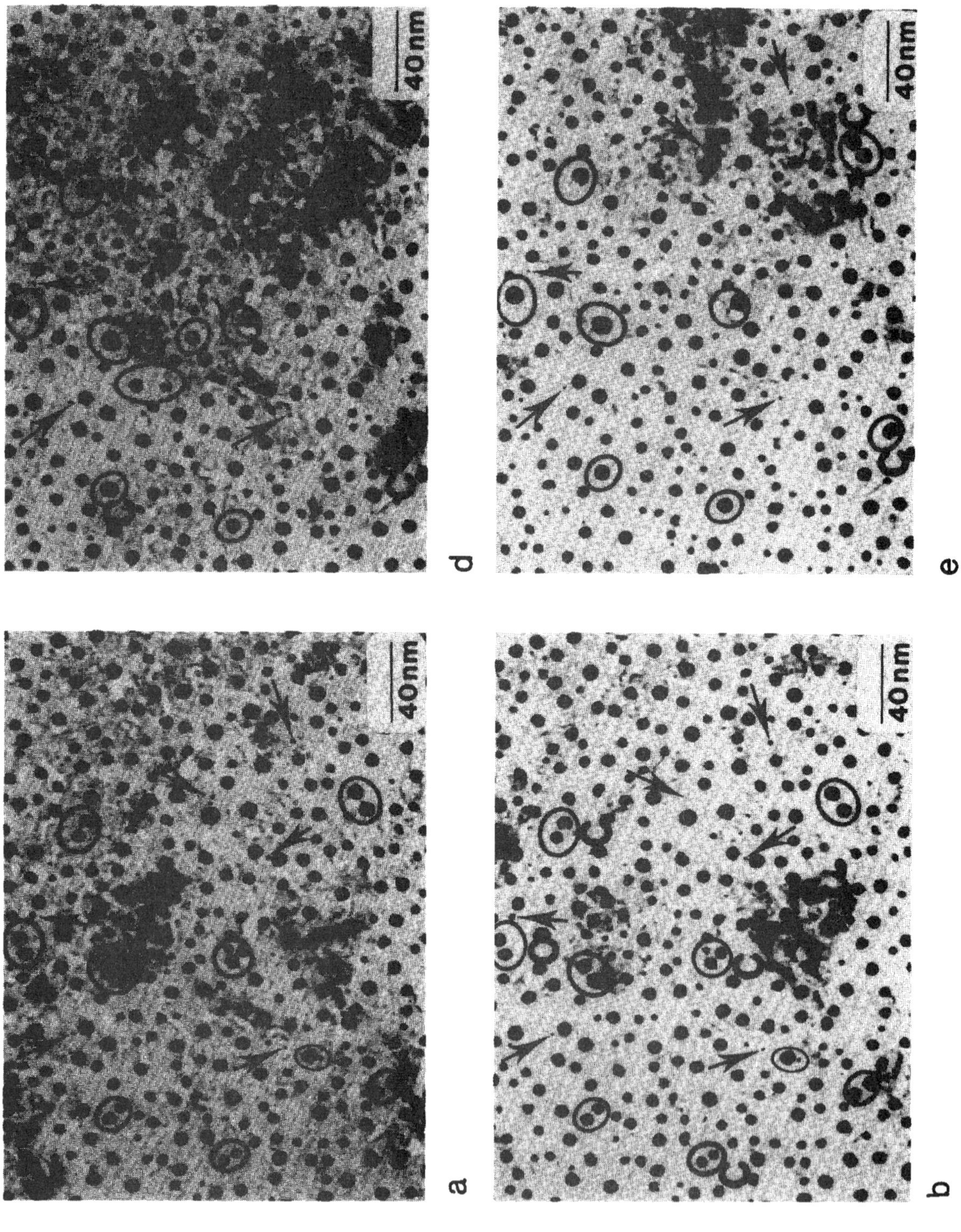
d
40nm
e
40nm
a
40nm
b
40nm

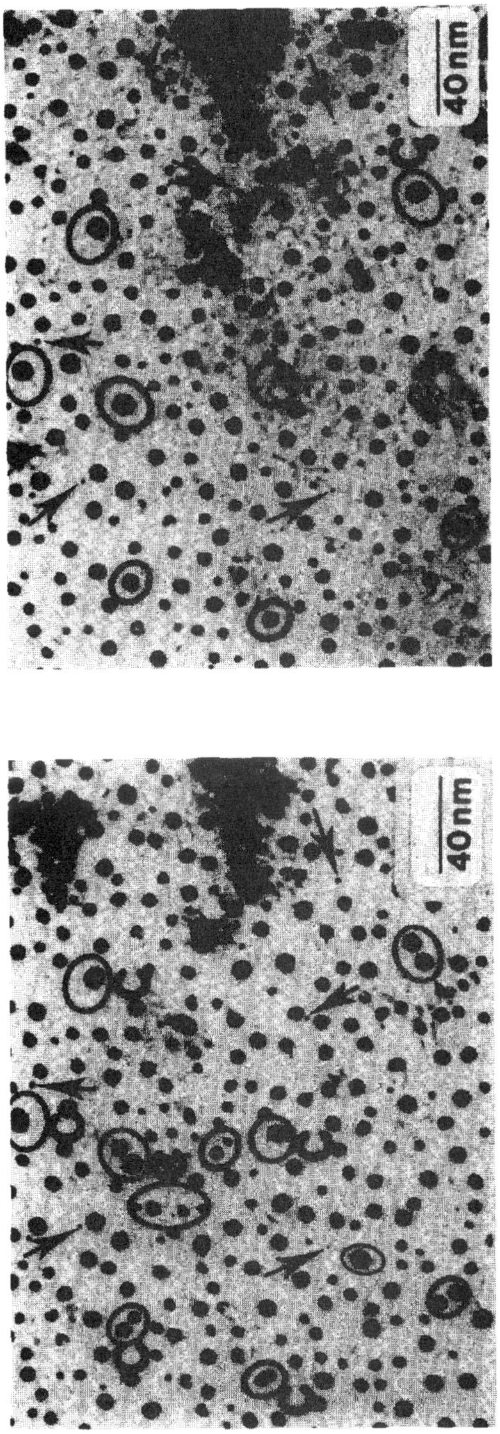

f

c

particles nearby remained unchanged (I). The sintering was very fast and it can only be inferred that both short-distance migration and subsequent coalescence of particles (B, C) and localized ripening (H) and migration of smaller particles, which decreased in size as they migrated (regions 1, 3 in Figs. 4d, 4e), contributed to the growth of the particles.

Another sample of 1.5-nm initial film thickness was heated in purified H_2 at 750°C for 1 h to generate crystallites. Only 1 h of this high-temperature treatment of the fresh sample with a lower loading (1.5 nm) yielded particles considerably larger than those obtained with a sample of higher loading (2 nm) after 12 h of heating at 500°C. As expected, the sintering is faster and more pronounced at the higher temperatures. Further heating in purified H_2 at 750°C for $\frac{1}{2}$ h led to a drastic decrease in the number of particles. The average size of the particles also decreased. The decrease in the number of particles and also an overall decrease in their size cannot possibly be attributed to sintering by a global ripening mechanism. A part of the crystallites is lost via evaporation, or more likely has diffused onto or into the substrate, or has spread as a thin undetectable film. Figures 5a, 5b, 5′a and 5′b show some of the events. A large number of small and large particles disappeared completely without significant growth of the remaining particles (J). In fact, in some locations, the large particles decreased in size significantly while much smaller particles adjacent to them were affected only marginally or even remained unchanged (I). Larger particles were detected in regions where there were smaller particles before, suggesting growth by collision and coalescence of smaller particles (K). Small particles could be seen in re-

FIG. 3′. Sequence of changes in a different region of the same sample as in Fig. 3. (a) Initial (12 h H_2, 500°C + 5 h H_2, 600°C); (b) 4 h O_2, 300°C; (c) 2 h H_2, 500°C; (d) 4 h O_2, 350°C + 4 h H_2, 500°C; (e) 3 h O_2, 400°C + 4 h H_2, 500°C; (f) 5 h O_2, 450°C.

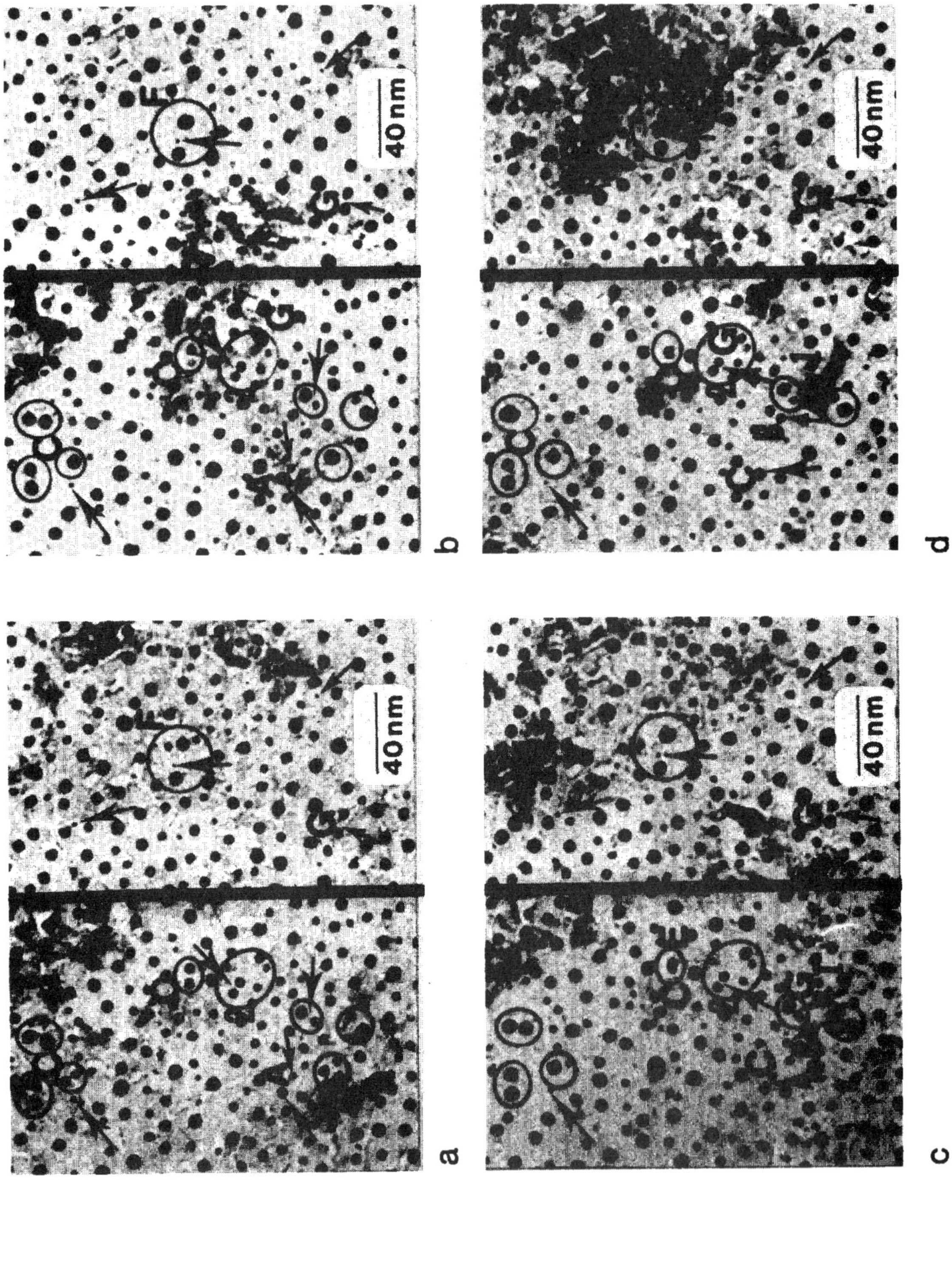
a
b
c
d
40nm
40nm
40nm
40nm

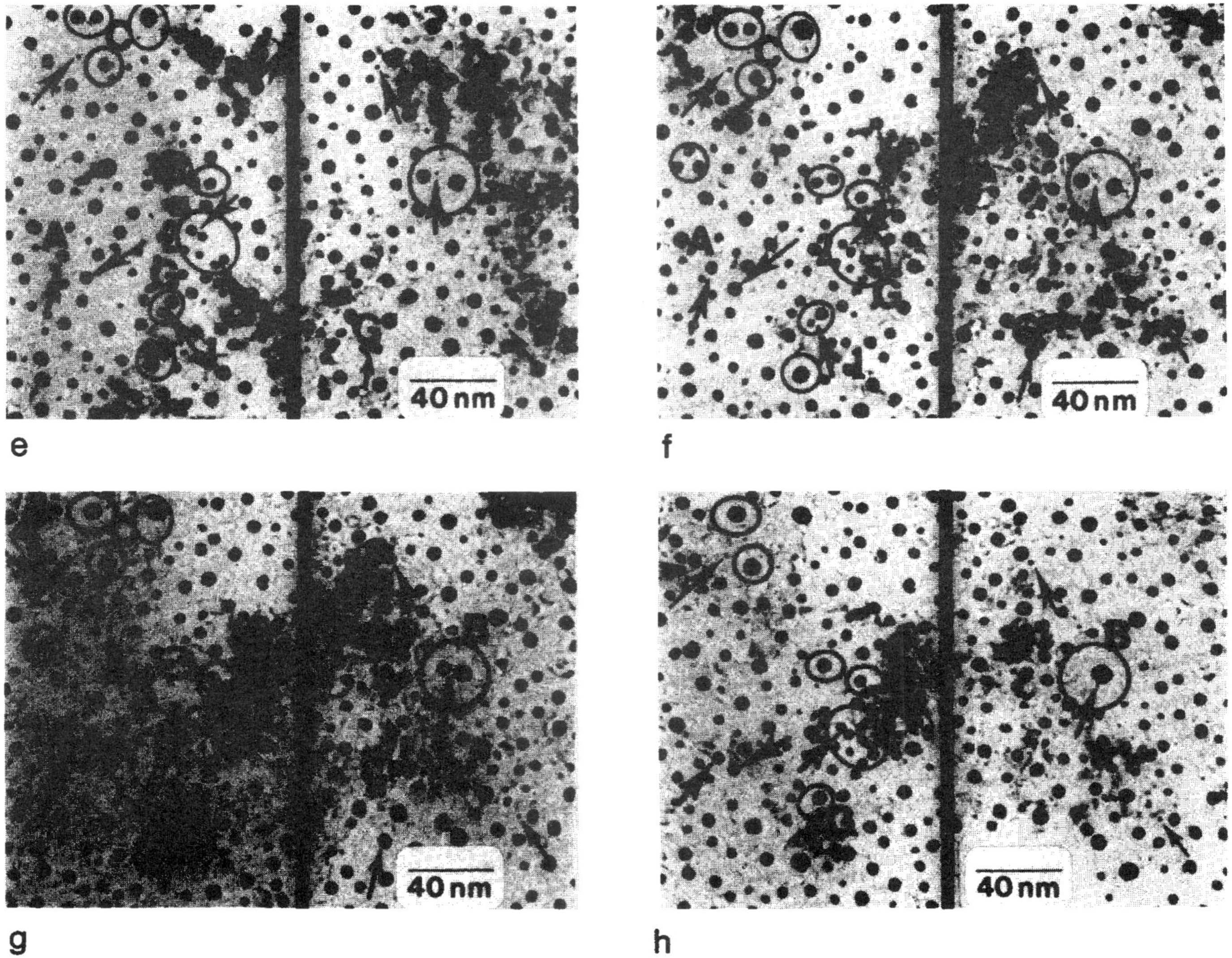

FIG. 3″. Same sample as in Fig. 3. The micrographs are of a different region. (a) Initial (12 h H_2, 500°C + 5 h H_2, 600°C); (b) 4 h O_2, 300°C + 2 h H_2, 500°C; (c) 4 h O_2, 350°C; (d) 4 h H_2, 500°C; (e) 3 h O_2, 400°C; (f) 4 h H_2, 500°C; (g) 4 h O_2, 450°C; (h) 2 h H_2, 500°C.

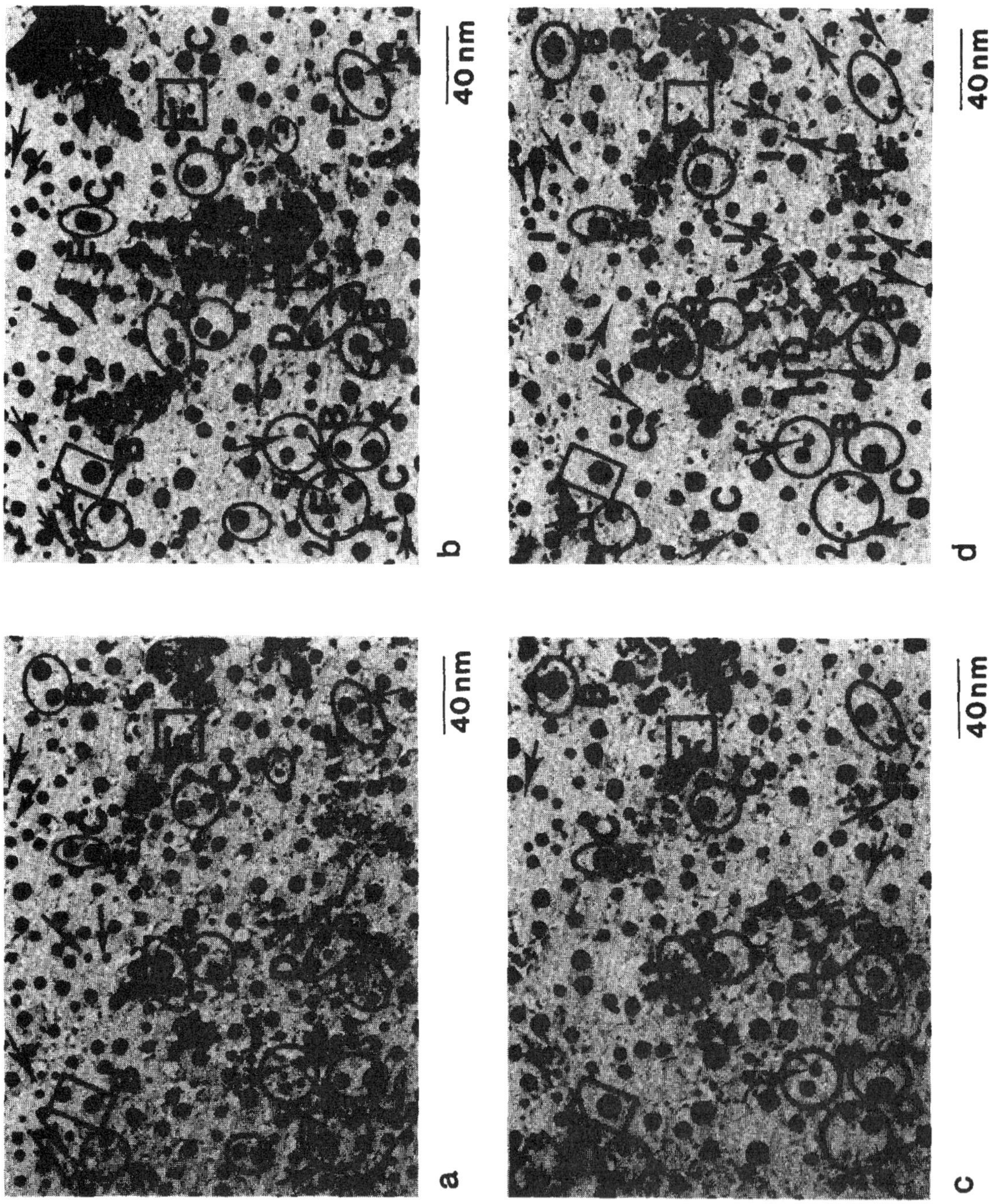
a
b
c
d
40 nm
40 nm
40 nm
40 nm

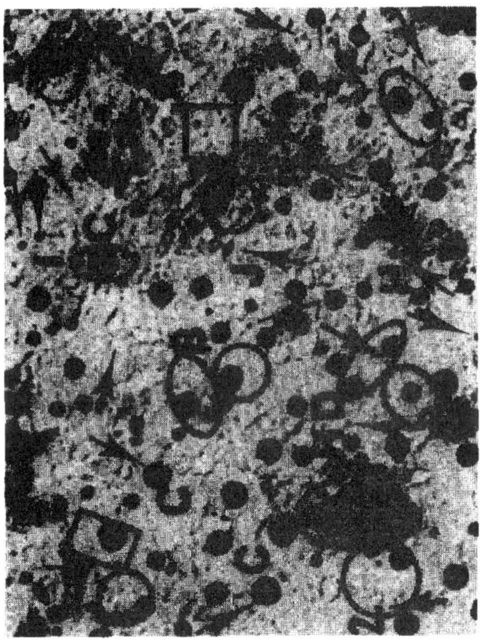

gions adjacent to places where larger particles had disappeared (D), indicating decrease in size and subsequent migration of particles or vice versa. It is unlikely that the new small particles have grown in size from undetectable particles via atom capture because considerably larger particles could be seen to decrease in size or disappear nearby, which is not compatible with the growth of smaller particles from capture of single atoms. It is likely that these particles are migrating remnants of dissolving larger particles. Short distance migration of small particles (~1.5 nm) without coalescence (A) and a few instances of migration with subsequent coalescence (B) could be observed following heating for an additional ½ h at 750°C (Figs. 5′b, 5′c). However, the decrease in the number of particles was very small. Subsequent heating at ½- and 1-h intervals resulted in migration and coalescence of nearby particles (B) (Figs. 5′c, 5′d). A few small particles also disappeared (particles marked with arrows). It is likely that these small particles collided with nearby larger particles and merged with them, since other small particles of about the same size or larger migrated (A) and a few nearby small particles grew (region 1 in Figs. 5′c, 5′d). Beyond 4 h, when longer heating intervals were employed, particle coarsening continued up to a total of 31 h of heating in H_2 (Figs. 5′e–5′g). In addition to coalescence of nearby particles, the decrease in size and/or disappearance of a number of small particles were also observed.

Figure 6 shows another sample heated at 750°C. The micrographs show the events observed on heating in H_2 at 750°C, subsequent to one cycle of heating in H_2 and O_2 at the same temperature. The micrographs show migration of particles (A), particles

FIG. 4. Sequence changes on heating a Pt/Al_2O_3 sample of 2 nm initial film thickness in purified H_2. (a) Initial (12 h H_2, 500°C + 5 h H_2, 600°C); (b) ½ h H_2, 700°C; (c) 2 h H_2, 700°C; (d) 5 h H_2, 700°C; (e) 2 h H_2, 800°C.

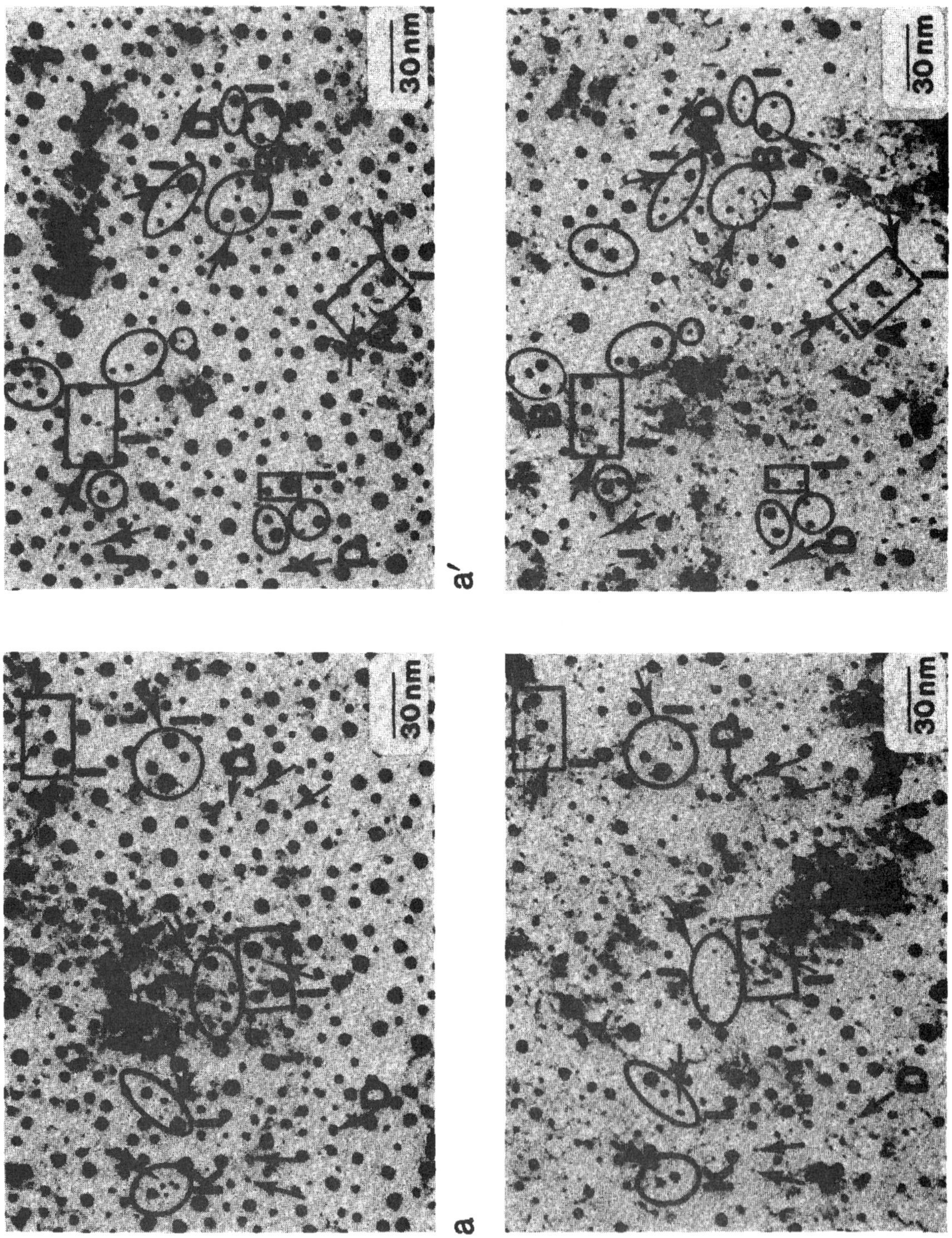

FIG. 5. Micrographs showing the sequence of changes in a Pt/Al_2O_3 sample of 1.5 nm initial film thickness on heating in purified hydrogen at 750°C. (a, a′) Initial (1 h H_2, 750°C); (b, b′) $\frac{1}{2}$ h H_2, 750°C.

that were initially apart but later contacted each other, indicating migration and collision (B), dumbbell-shaped particles (E), separation into two of contacting particles (L), and generation of two particles from one (M). Also, a number of instances of overlapping particles (N) could be seen.

DISCUSSION

The results reported here have provided further evidence for crystallite migration and coalescence and their contribution to sintering of Pt/Al_2O_3 catalysts in H_2 and on alternate heating in H_2 and O_2. However, as reported before in the case of Pd/Al_2O_3 (*36*) and Fe/Al_2O_3 (*39*), ripening, especially of the localized kind, also often takes place concurrently. In fact, a variety of events takes place during sintering of supported metal catalysts and a few of them observed here in the case of Pt/Al_2O_3 are summarized schematically in Table 1. Even though short-distance migration and coalescence of nearby particles (particles that are one to a few diameters apart) is probably the major mechanism of particle growth, especially in the initial stages, a decrease in size of smaller particles because of localized ripening has also been observed. A number of small particles of about 1.5 to 2 nm in diameter or smaller disappeared. Since, as mentioned in the previous section, a number of particles of similar size remained unaffected or migrated and since larger particles have also been observed to migrate on the same micrograph, there appears to be no compelling reason to attribute the disappearance of such small particles to ripening only. It is possible and quite probable that these particles migrate relatively fast, especially because of their small sizes, and coalesce with the larger particles. Some small particles that remain stationary and unchanged might have been trapped in the valleys of the heterogeneous substrate surface, as also noted previously in the case of Pd/Al_2O_3 (*36*).

It is worth noting that in some cases, two particles approach each other to a kind of distance of minimum approach and growth subsequently occurs via transfer of atoms or molecules from one particle to the other through a very narrow bridge established between the particles (G in Figs. 1′a, 1′b and in Figs. 3″a–3″h). Such a growth by ripening between two adjacent particles via a whiskerlike bridge contact has also been observed with Fe/Al_2O_3 (*39*). It is also possible that the particles coalesce when pulled together because of the contraction of the connecting bridge. Arai *et al.* (*37*) have also postulated that the growth of Pt particles occurs via formation of a bridge between two adjacent particles and coalescence by an abrupt movement that brings the particles together, but only above 600°C and only with particles larger than 10 nm. It appears that when the particles are close to each other, growth via ripening is more of a localized nature, aided by long-range interparticle interactions (which apparently generate a localized curvature that facilitates the removal of atoms or molecules from one particle to the other). In fact, in addition, the micrographs suggest that the migration of crystallites is not diffusional (random) but directional, guided by long-range interparticle interactions. The particles seem to feel the presence of other particles nearby and, in general, migrate toward each other. Of course, migration of particles away from each other has also been observed, though only occasionally. It should be noted that there are also other kinds of interactions affecting the migration of the particles, such as the interactions between particles and support and the surface roughness of the support. Direction-selective migration of particles has been suggested before (*36*). When the interparticle force is strong, it is possible that even if a particle is trapped in a valley or is too large to migrate, it may still rotate and rearrange to orient its more highly curved protrusions along the periphery (and in extreme cases even form long whiskerlike protrusions) in the direction of a nearby particle to emit atoms or molecules toward it (G in Figs.

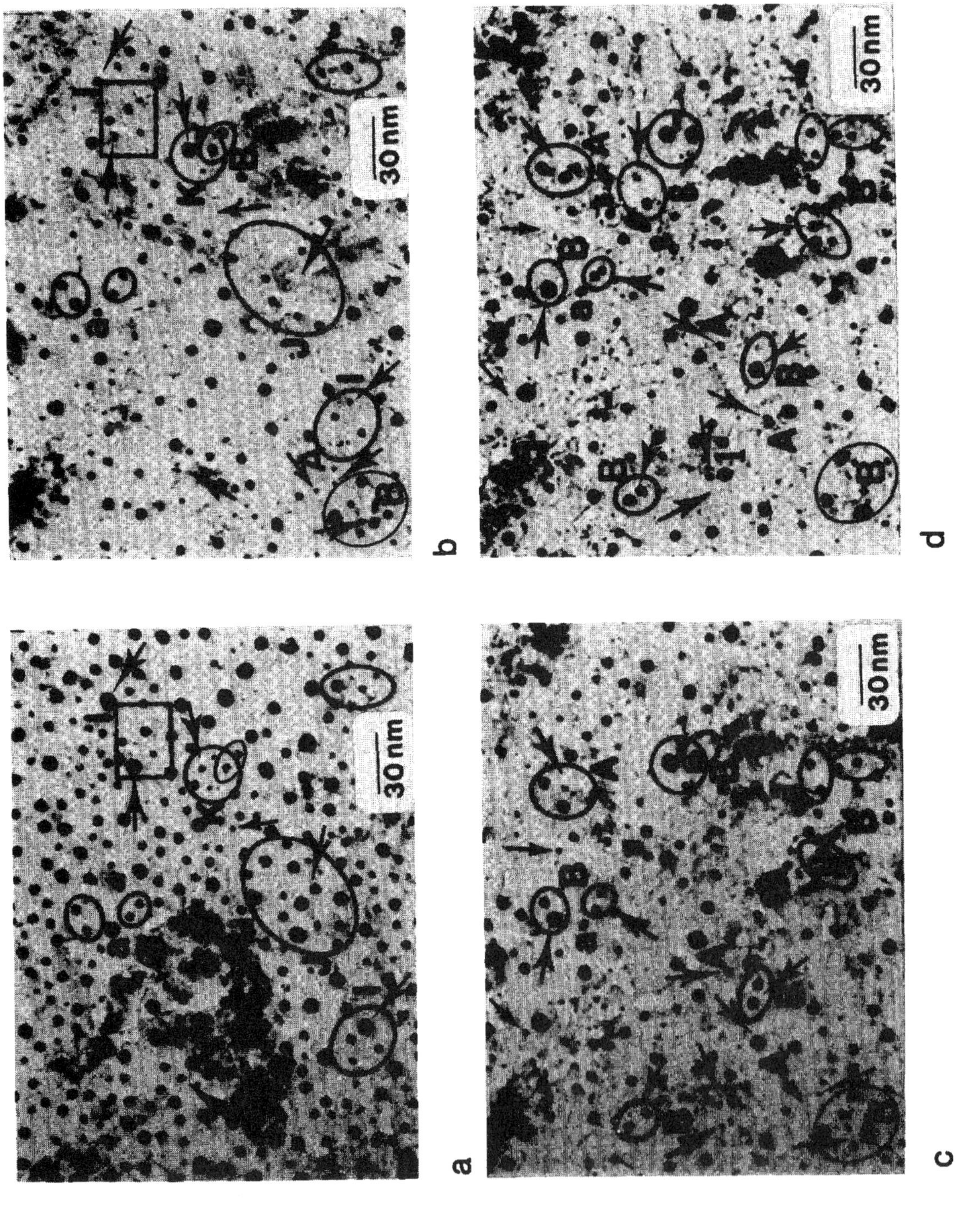
30 nm
a
b
c
d

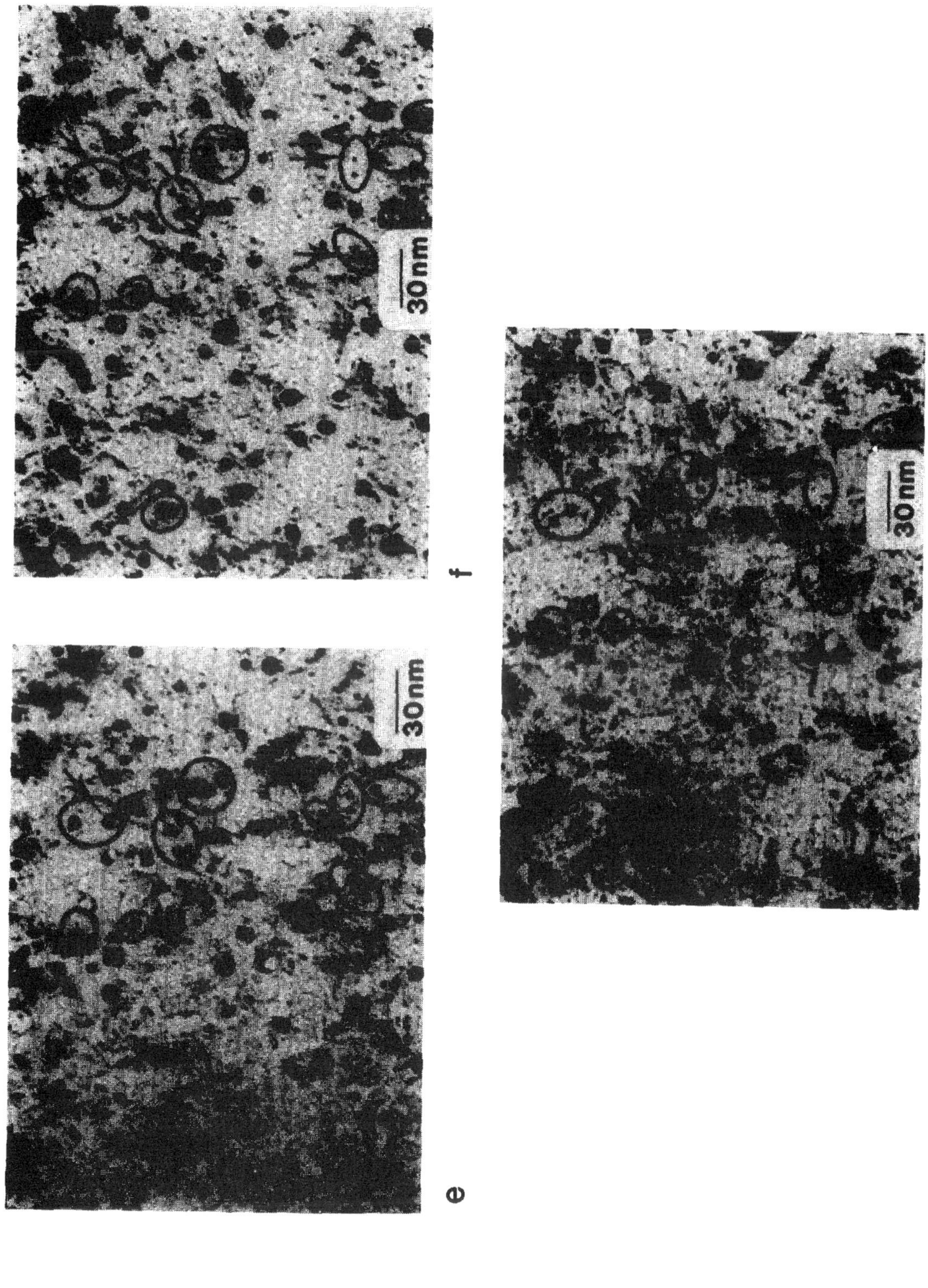

FIG. 5′. Same sample as in Fig. 5. The micrographs are from a region different from those in Fig. 5. The following durations of heating are cumulative. (a) Initial (1 h H_2, 750°C); (b) $+\frac{1}{2}$ h H_2; (c) 1 h H_2; (d) 3 h H_2; (e) 6 h H_2; (f) 9 h H_2; (g) 13 h H_2.

TABLE 1

Events Observed on Pt/Al_2O_3 Model Catalysts Heated in H_2 and O_2

Type	Event	Schematics	Figure Nos.
A	Migration, without coalescence, of small or large particles toward or away from another particle		1, 1′, 2, 2′, 3″, 5, 5
B	Migration, collision, and coalescence; gradual migration and coalescence		1, 1′, 1″, 2, 2′, 3″, 5, 5′, 6
C	Coalescence of nearby particles		1, 1′, 1″, 2, 3′, 3″
D	Migration followed by a decrease in size or vice versa		4, 5
E	Dumbbell-shaped particles, particles in fusion		1, 1′, 2, 2′, 3, 3″, 4
F	Coalescence of nearby larger particles adjacent to unaffected/decreasing smaller particles		1, 1′, 3″, 4
G	Transfer of atoms between two nearby particles via a neck or narrow whiskerlike bridge; formation and breakup of a contact between two particles via a narrow whiskerlike bridge		1, 1′, 3″
H	Decrease and/or disappearance of a small particle near a larger particle that grows in size or appears unaffected (ripening); decrease in size and/or disappearance of one particle amid particles of equal size		1, 1′, 4
I	Decrease of a large particle nearby smaller particles		4, 5, 5′
J	Disappearance of large and small particles		4, 5, 5′
K	Growth of smaller particles or coalescence of small particles to yield larger particles		1, 1′, 1″, 5, 5′
L	Collision, coalescence, and subsequent separation of two particles; separation of particles in contact		1, 1, 2, 6
M	Appearance of two particles in place of one (splitting, or wetting and reconstruction)		6
N	Overlapping particles		6

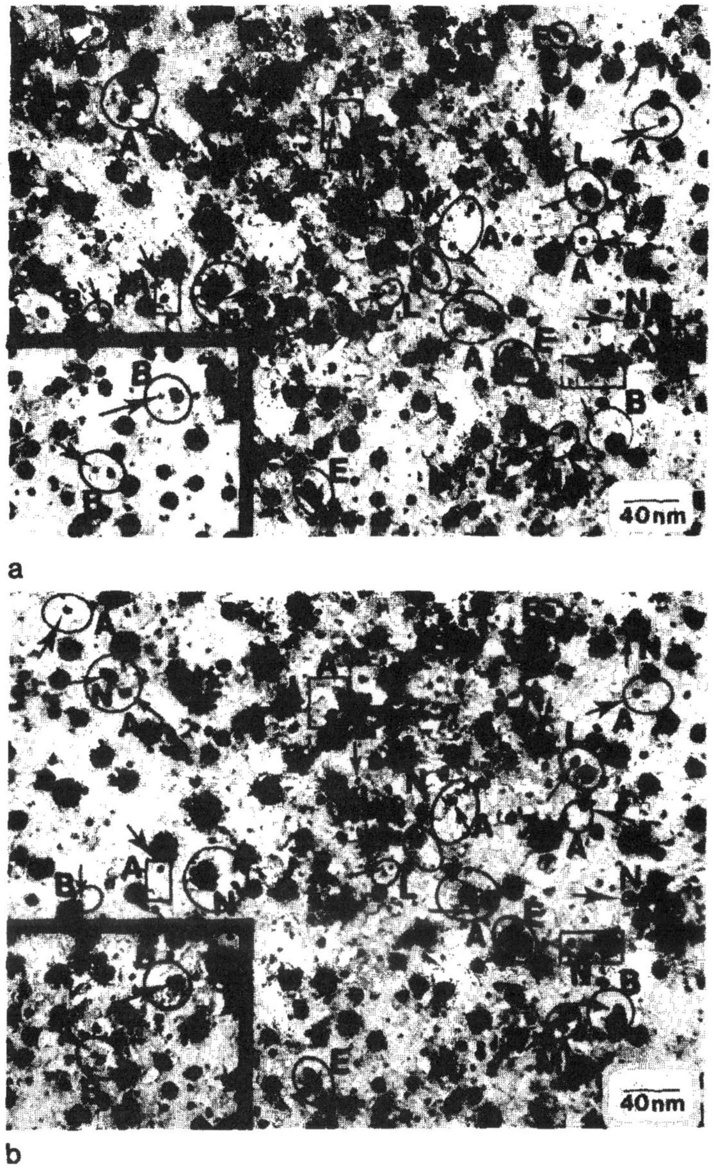

FIG. 6. Sequence of changes in a Pt/Al_2O_3 sample of 1.5 nm initial film thickness heated in purified H_2 and O_2 at 750°C. (a) Initial (15 h H_2, 750°C + 5 h O_2, 750°C + 17 h H_2, 750°C + 6 h H_2, 750°C); (b) 9 h H_2, 750°C.

3″a–3″h). Such events, though isolated, have been observed with Fe/Al_2O_3 previously (*39*). Of course, in the case of supported iron particles the interparticle interaction forces may be stronger because of the ferromagnetic or superparamagnetic nature of the particles.

Finally, there are at least two possible reasons for failure to detect crystallite migration even in systems where it occurs. Since the particles in a fresh sample are, in general, small (less than 1 or 2 nm), they migrate relatively fast and even 1 or 2 h of observation (as is often used) may be too long to identify such a mechanism. Shorter heating intervals, especially in the initial stages, may be helpful. Even in the case of 1 or 2 h of heating, particle migration and coalescence may be detected or inferred if considerable care is taken to follow the same particles in various regions covering a large area of the sample. Second, if the observation is not continued for longer than 1 or 2 h of heating, evidently it is not possible to detect the slower migration of the now relatively larger particles and/or other phenomena that may occur. Also in this case the observation made above regarding careful scanning of a large area might be useful.

In the case of model Pt/Al_2O_3 catalysts, as with other traditional supported metal catalysts, sintering is pronounced at higher temperatures, higher loadings, and at any temperature in the initial few (4–6) hours of heating a fresh catalyst. Sintering in as-received ultrapure hydrogen (which, however, contains traces of moisture and O_2) is enhanced relative to that in the same hydrogen further purified to eliminate/reduce the trace moisture and O_2. In a hydrogen atmosphere and below about 700°C, sintering seems to occur primarily by short-distance migration and coalescence of nearby particles as well as by localized ripening between a few neighboring particles. Significant migration of larger particles occurs around 500°C when H_2-treated samples are heated in O_2 or vice versa. At $T < 500$°C sintering is very slow in H_2 and on alternate heating in H_2 and O_2 and even slower in O_2.

CONCLUSION

Transmission electron microscopic evidence for the role of crystallite migration and coalescence among other mechanisms in the sintering of model Pt/Al_2O_3 catalysts is provided. Results of model Pt/Al_2O_3 catalysts heated in H_2 and alternately in H_2 and O_2 at temperatures in the range 500–750°C indicate that a large number of phenomena such as coalescence of nearby particles, crystallite migration (of both small, 1.5 nm, and large, about 8 nm, particles), migration followed by coalescence, decrease in size and disappearance of small particles near larger particles, disappearance of small and larger particles, decrease in size of large particles near unaffected smaller particles, decrease in size and subsequent migration of particles or vice versa, collision and inhibited coalescence of particles, and collision–coalescence–separation of particles occur. It appears that the particles on the support feel the presence of nearby particles probably via the interaction forces between them and migrate or emit atoms toward them, and consequently the migration is directional and the ripening is local.

REFERENCES

1. Mills, G. A., Weller, S., and Cornelius, E. B., *Proc. 2nd Int. Congr. Catal., Paris*, p. 2221 (1960).
2. Adler, S. F., and Keavney, J. J., *J. Phys. Chem.* **64,** 208 (1960).
3. Maat, H. J., and Moscou, L., *Proc. 3rd Int. Congr. Catal., Amsterdam* **2,** 1277 (1965).
4. Dalla Betta, R. A., McCune, R. C., and Sprys, J. W., *Ind. Eng. Chem. Prod. Res. Dev.* **15,** 169 (1976).
5. Johnson, F. L., and Keith, C. D., *J. Phys. Chem.* **67,** 200 (1963).
6. Kraft, M., and Spindler, H., *4th Int. Congr. Catal., Moscow,* p. 1252 (1968).
7. Kearby, K. K., Thorn, J. P., and Hinlicky, J. A., U.S. Patent 3,134,732 (1964).
8. Fiedorow, R. M. J., and Wanke, S. E., *J. Catal.* **43,** 34 (1976).
9. Ruckenstein, E., and Malhotra, M. L., *J. Catal.* **41,** 303 (1976).
10. Gollob, R., and Dadyburjor, D. B., *J. Catal.* **68,** 473 (1981).
11. Ruckenstein, E., and Dadyburjor, D. B., *Rev. Chem. Eng.* **1**(3), 251 (1983).
12. Ruckenstein, E., and Pulvermacher, B., *AIChE J.* **19**(2), 356 (1973).
13. Ruckenstein, E., and Pulvermacher, B., *J. Catal.* **29,** 224 (1973).
14. Wynblatt, P., and Gjostein, N. A., *Prog. Solid State Chem.* **9,** 21 (1975).
15. Flynn, P. C., and Wanke, S. E., *J. Catal.* **34,** 390 (1974).
16. Ruckenstein, E., and Dadyburjor, D. B., *J. Catal.* **48,** 73 (1977).
17. Ruckenstein, E., and Dadyburjor, D. B., *Thin Solid Films* **55,** 89 (1978).
18. Flynn, P. C., and Wanke, S. E., *J. Catal.* **37,** 432 (1975).
19. Baker, R. T. K., Thomas, C., and Thomas, R. B., *J. Catal.* **38,** 510 (1975).
20. Hermann, R., Adler, S. F., Goldstein, M. S., and Debaum, R. M., *J. Phys. Chem.* **65,** 2189 (1961).
21. Hughes, T. R., Houston, R. J., and Sieg, R. P., *Ind. Eng. Chem. Process Des. Dev.* **1,** 96 (1962).
22. Gruber, H. L., *J. Phys. Chem.* **66,** 48 (1962).
23. Somorjai, G., *Anal. Chem.* **1,** 101 (1968).
24. Chu, Y. F., and Ruckenstein, E., *Surf. Sci.* **67,** 517 (1977).

25. Chu, Y. F., and Ruckenstein, E., *J. Catal.* **55,** 281 (1978).
26. Chen, M., and Schmidt, L. D., *J. Catal.* **55,** 348 (1978).
27. McVicker, G. B., Garten, R. L., and Baker, R. T. K., *J. Catal.* **54,** 129 (1978).
28. Wang, T., and Schmidt, L. D., *J. Catal.* **66,** 301 (1980).
29. Fiedorow, R. M. J., Chahar, B. S., and Wanke, S. E., *J. Catal.* **51,** 193 (1978); Wanke, S. E., *in* "Sintering and Heterogeneous Catalysis" (G. C. Kuczynski, A. E. Miller, and G. A. Sargent, Eds.), *Mat. Sci. Res.,* Vol. 16, p. 223. Plenum, New York, 1984.
30. Baker, R. T. K., Prestridge, E. B., and Garten, R. L., *J. Catal.* **56,** 390 (1979); Baker, R. T. K., *Catal. Rev. Sci. Eng.* **19**(2), 161 (1979).
31. Yao, H. C., Sieg, M., and Plummer, H. K., Jr., *J. Catal.* **59,** 365 (1979).
32. Richardson, J. T., and Crump, J. G., *J. Catal.* **417** (1979).
33. Glassl, H., Kramer, R., and Hayek, K., *J. Catal.* **64,** 303 (1980).
34. Kuo, H. K., Ganesan, P., and DeAngelis, R. J., *J. Catal.* **66,** 171 (1980).
35. Straguzzi, G. I., Aduriz, H. R., and Gigola, C. E., *J. Catal.* **66,** 171 (1980).
36. Chen, J. J., and Ruckenstein, E., *J. Catal.* **69,** 254 (1981).
37. Arai, M., Ishikawa, T., Nakayama, T., and Nishiyama, Y., *J. Colloid Interface Sci.* **97,** 254 (1984).
38. Kim, K. T., and Ihm, S. K., *J. Catal.* **96,** 12 (1985).
39. Sushumna, I., and Ruckenstein, E., *J. Catal.* **94,** 239 (1985).
40. Ruckenstein, E., and Lee, S. H., *J. Catal.* **86,** 457 (1984).
41. Smith, D. J., White, D., Baird, T., and Fryer, J. R., *J. Catal.* **81,** 107 (1983).
42. Moss, R. L., *in* "Catalysis" (A Specialist Periodical Report) (C. Kemball and D. A. Dowden, Eds.), Vol. 4, p. 31. R. Soc. Chem., London, 1980.
43. Baird, T., *in* "Catalysis" (A Specialist Periodical Report) G. C. Bond and G. Webb, Eds.), Vol. 5, p. 172. R. Soc. Chem., London, 1981.
44. Burch, R. *in* "Catalysis" (A Specialist Periodical Report) (G. C. Bond and G. Webb, Eds.), Vol. 7, p. 149. R. Soc. Chem., London, 1983.
45. Dautzenberg, F. M., and Wolters, H. B. M., *J. Catal.* **51,** 26 (1978).
46. Menon, P. G., and Froment, G. F., *J. Catal.* **51,** 26 (1978).
47. Tauster, S. J., Fung, S. C., and Garten, R. L., *J. Amer. Chem. Soc.* **100,** 170 (1978).
48. References in "Metal-Support and Metal-Additive Effects in Catalysis" (B. Imelik, C. Naccache, G. Coudourier, H. Praliaud, P. Meriaudeau, P. Gallezot, G. A. Martin and J. Vedrine, Eds.). Elsevier, Amsterdam, 1982.
49. Bond, G. C., and Burch, R., *in* "Catalysis" (A Specialist Periodical Report) (G. C. Bond and G. Webb, Eds.), Vol. 6, p. 27. R. Soc. Chem., London, 1983.
50. Kunimori, K., Ikeda, Y., Soma, M., and Uchijima, T., *J. Catal.* **79,** 185 (1983).
51. Bassett, G. A., *Proc. Eur. Reg. Congr. Electron Microsc.*, Delft, **1,** 270 (1960).
52. Skofronick, J. G., and Phillips, W. B., *J. Appl. Phys.* **38,** 4791 (1967).
53. Heinemann, K., and Poppa, H., *Thin Solid Films* **33,** 237 (1976).
54. Granquist, C. G., and Buhrman, R. A., *J. Catal.* **42,** 477 (1976).
55. Granquist, C. G., and Buhrman, R. A., *J. Catal.* **46,** 238 (1977).
56. Glassl, H., Kramer, R., and Hayek, K., *J. Catal.* **63,** 167 (1980).
57. Sushumna, I., and Ruckenstein, E., *in* "Hydrogen Effects in Catalysis" (Z. Paal and P. G. Menon, Eds.). Dekker, New York, 1988.
58. Sushumna, I., and Ruckenstein, E., *J. Catal.* **108,** 77 (1987).

Role of Physical and Chemical Interactions in the Behavior of Supported Metal Catalysts: Iron on Alumina—A Case Study

I. SUSHUMNA AND E. RUCKENSTEIN[1]

Department of Chemical Engineering, State University of New York, Buffalo, New York 14260

Received November 1, 1984; revised January 22, 1985

Iron, being the most reactive of the group VIII metals, has been chosen to examine the role of strong chemical interactions between gas, crystallite, and substrate in the behavior of supported metal crystallites. Model catalysts of iron deposited by vacuum evaporation onto thin films of alumina have been heated in ultrahigh pure hydrogen (which, however, contains traces of O_2 and moisture), in additionally purified hydrogen (to eliminate or minimize the above impurities), in oxygen, and alternately in H_2 and O_2. Transmission electron micrographs provide information on the behavior and shape of the crystallites and electron diffraction provides information on the chemical compounds formed. It is shown that iron oxide which forms even in hydrogen containing traces of oxygen and/or moisture undergoes strong chemical interaction with alumina to form solid solutions or/and stoichiometric aluminates. Such chemical interactions lead to enhanced surface spreading phenomena and to unusual shapes of the crystallites. The role of strong chemical interactions in the wetting behavior of the crystallites is examined in some detail. It is suggested that because of these interactions the interfacial tension between crystallite and substrate can be decreased enormously (it may even become negative under nonequilibrium conditions). This then leads to a rapid spreading of the crystallite over the substrate, to torus formation, and/or fragmentation of the crystallite. At high temperatures, the smaller crystallites spread completely over the substrate and, as inferred from the electron micrographs, a multilayer film appears to coexist with the crystallites. A variety of mechanisms of sintering could be identified, such as migration and coalescence (even crystallites over 400 Å in size were observed to migrate occasionally), direct ripening, and Ostwald ripening. However, the micrographs indicate that these processes, as well as redispersion, are not simple, but parts of complex surface phenomena involving chemical interactions, wetting, shape changes, splitting etc.

INTRODUCTION

Currently, metal catalysts are being prepared mostly in a highly dispersed, supported form replacing the old fused or unsupported powder types. While the commonly used refractory oxide supports, silica and alumina, increase the metal dispersion, they are not inert especially toward the non-noble metals (*1*). The chemical reactivity between the active metal and the oxide support in turn affects the particle size distribution and consequently the activity and selectivity of the catalyst. Iron, the most electropositive of the group VIII elements and consequently a very reactive metal, is employed extensively as a catalyst in such important industrial processes as ammonia synthesis, the recently rejuvenated Fischer–Tropsch synthesis, and other hydrogenation and dehydrogenation reactions (*2*). Traditionally, iron is deposited onto the oxide supports in the form of ferric salts, prior to calcination and reduction at high temperatures (*3, 12*). When present as Fe^{3+} ions, iron exhibits strong reactivity with the support, especially with alumina because of the chemical similarity between Fe^{3+} and Al^{3+} ions (*4, 5*). Consequently, at low loadings on alumina, iron is not reduced easily to the zero-valent state, because a large fraction of the iron ions interact strongly with the substrate. The strength of the interactions between the ac-

[1] To whom correspondence should be addressed.

This paper, which was originally published in **Journal of Catalysis**: 94, 239–288 (1985), is republished with permission.

tive-metal ion and the support and consequently the reducibility and the particle size of the supported iron catalyst depend upon the chemical precursor used, the catalyst pretreatment (particularly the temperature of decomposition), and the metal loading. For example, Garten and Ollis (*6*) have shown, for samples prepared by impregnation, that when the metal loading on alumina is low (0.05%), the ferric ions can be reduced only to the ferrous state, even when the reduction with hydrogen occurs at temperatures as high as 700°C. On the other hand, Brenner and Hucul (*7*) have observed that the decomposition at low temperatures (<150°C) of zero-valent iron complexes, such as the iron carbonyls, adsorbed on alumina, yields highly dispersed iron catalysts. In this case the initial dispersion can be more than an order of magnitude higher than that of the catalyst prepared by the more traditional method of impregnation with ferric ions, but, because of the weaker interactions with the support, the zero valent particles may sinter more easily. Raupp and Delgass (*8*) have shown that mild pretreatment of the impregnated catalysts by slow vacuum drying, prior to reduction, produces significantly smaller particles of iron in the case of Fe/SiO_2. The nonreducibility and/or the smaller particle sizes arising as a result of the interactions between metal and support can have important consequences on the activity and selectivity of the supported iron catalysts in the Fischer–Tropsch reaction (*8–11*).

The literature abounds in Mössbauer spectroscopic investigations of supported iron catalysts, especially regarding their role in the Fischer–Tropsch reaction (*11–14*). Mössbauer spectroscopy is a powerful tool to characterize catalysts involving ferromagnetic metals. Nonetheless, when a strong interaction between metal and support exists, the interpretation of the spectra becomes difficult regarding both the particle size distribution and the exact chemical compounds formed (*15*). In addition, information regarding the microstructure and the shape of the particles cannot be obtained (*16*). A complementary technique such as electron microscopy can provide both the size distribution and the shape of the particles and, via electron diffraction, can also identify the chemical compounds formed. Electron microscopy and electron diffraction have been used in the present paper to investigate the behavior of a model Fe/Al_2O_3 catalyst and results are presented regarding the physicochemical changes accompanying their heating at different temperatures in a reducing (H_2) atmosphere, in an oxidizing (O_2) atmosphere, and during an alternate heating in these atmospheres. The present investigation emphasizes the role of the surface phenomena and of the chemical interactions in the complex processes observed with model Fe/Al_2O_3 catalysts. Accordingly, the electron micrographs, which provide physical information on shape and size, are supplemented with chemical infomation on the compounds formed via electron diffraction.

EXPERIMENTAL

Preparation of model alumina supports. Electron-transparent and nonporous films of amorphous Al_2O_3 were prepared by anodically building up the oxide on a clean, thin, high-purity aluminum foil (99.999%, Alfa Products Inc.) and stripping the oxide film off by dissolving the unoxidized aluminum in a dilute mercuric chloride solution (*21*). The oxide films were washed in distilled water and picked up on gold electron-microscope grids. They were subsequently heated in air at 800°C for 72 h to transform the amorphous alumina into the γ or η form and also to ensure that no further changes would occur during its subsequent use.

Deposition of the active metal. Iron films of different thicknesses were deposited onto the alumina substrates by evaporating the corresponding amounts of 99.998% pure iron wire (Alfa Products Inc.) from a tungsten basket in an Edwards vacuum evaporator under a vacuum of better than 2×10^{-6} Torr. The thickness of the metal film deposited was estimated from the amount of metal evaporated and the distance between the source and the target. The substrate was kept at room temperature during deposition.

Heat treatment. Samples were heated in a quartz boat inside a 3.8-cm-diameter and 120-cm-long quartz tube. A predetermined gas flowed through the quartz tube during heating and its flow rate of 150 cm^3/min was maintained constant during the heat treatment. The gases used in the experiments were all ultrahigh purity (UHP) grade, supplied by Linde Division, Union Carbide Corporation. Hydrogen, 99.999% pure, contained less than 1 ppm O_2 and less than 2 ppm moisture. Helium, also 99.999% pure, contained less than 3 ppm moisture. For some experiments, the hydrogen was further purified by passing it through a Deoxo unit (Englehard Industries) followed by a silica gel column and then through a 5A molecular sieve bed immersed in liquid nitrogen. Helium was also purified by passing it through the liquid-nitrogen trap.

After introducing the boat containing the sample into the reactor, each heat treatment followed the sequence: flushing the reactor with helium for at least 5 min; heating the specimen from room temperature to the desired value in a helium atmosphere, switching over the gas stream to hydrogen or oxygen as desired and heating for the predetermined length of time, and cooling down the specimen slowly to room temperature in a helium atmosphere before it is withdrawn for observation. Following each heat treatment, the same regions of the sample were photographed using a JEOL 100U transmission electron microscope. During each observation, the entire specimen area was scanned to check if the behavior was uniform throughout. Electron diffraction patterns were obtained for all the specimens after each treatment.

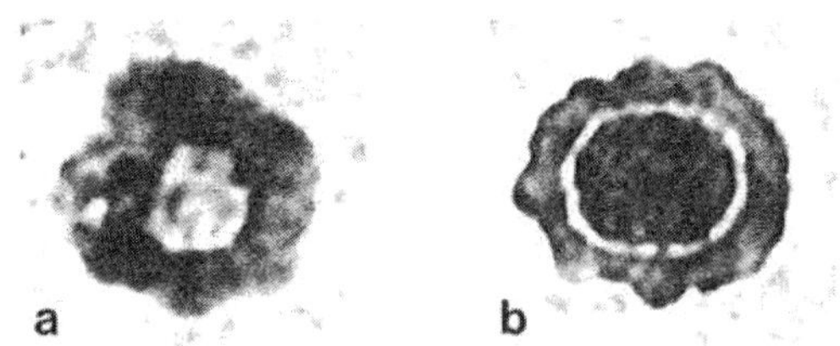

FIG. 1. Representation of (a) torus and (b) core-and-ring structures.

RESULTS AND DISCUSSION

Several samples have been investigated at each of the following loadings corresponding to film thicknesses of 5, 6, 7.5, 10, and 12.5 Å. Except for the cases in which the behavior was different, the results are reported for either 6- or 7.5-Å loadings. A large number of events have been observed at each of the temperatures investigated regarding the sintering and the wetting behavior as well as the changes in the shape of crystallites following heating in hydrogen, oxygen, and alternate heating in these two atmospheres. Table 1 summarizes the various events observed. Tables 2 to 5 provide the events observed under various conditions, the chemical compounds identified, and the corresponding micrograph numbers. The torus and core-and-ring structures, to which we refer frequently later in the text, are represented in Figs. 1a and b.

A. Behavior with as-Supplied Hydrogen

A.1. Heating in Hydrogen. It is seen from Table 2 that distinct crystallites do not form easily at 300°C, whereas at 400°C they form after heating for only 1 h. At both temperatures, a compound whose *d* values lie between those of γ-Fe_2O_3 (Fe_3O_4) and the higher aluminate, $Al_2Fe_2O_6$, is detected, in addition to other compounds. (γ-Fe_2O_3 and Fe_3O_4 have the same lattice constant and therefore cannot be easily differentiated by electron diffraction). It is to be noted that the major *d* values of $Al_2Fe_2O_6$ are greater than those of γ-Fe_2O_3 (Fe_3O_4). Therefore, an increase in the major *d* values from those of γ-Fe_2O_3 toward those of $Al_2Fe_2O_6$ would indicate the formation of solid solutions of iron oxide in alumina ultimately leading to

TABLE 1

Types of Events Observed

Type	Event and figure number	Schematic of the events (Before → After)
A	Coalescence of nearby particles. Appearance of dumbbell-shape particles (2f–2i; 3a–c; 4a, 4b; 8a–c; 13b–d)	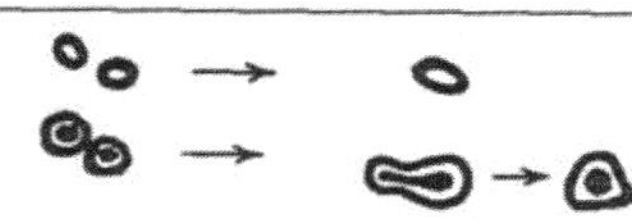
B	Migration with subsequent coalescence (3a–c; 7b, c; 13a–d) (or, coalescence followed by migration)	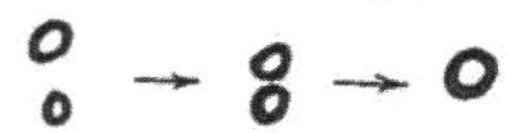
C	Migration without coalescence (4′c, d; 6d, e; 7b, c; 11a, b; 13b, c)	
D.a	Decrease and disappearance of a few smaller particles (Localized ripening between a few particles. Direct ripening) (3a, b; 4′a–c)	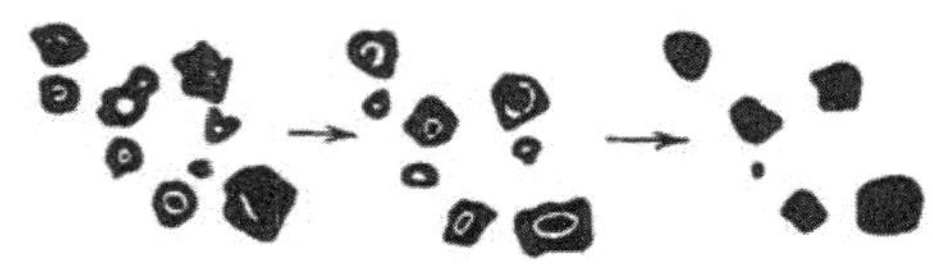
D.b	Ripening between particles of almost the same size. Probably due to local curvature difference (3a, b; 4′a–c)	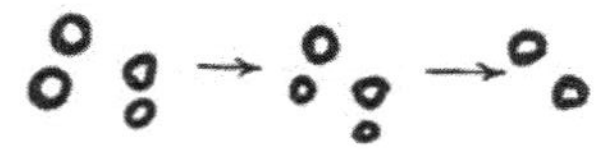
D.c	Transfer of molecules via a narrow, whisker-like bridge or a neck formed between two particles (3a, b; 4′a–c)	
E	Decrease and disappearance of a large number of small particles. (Global ripening, Ostwald ripening; break-up to unresolved crystallites and their migration; extension to undetectable islands.) (4a–d; 12a–c)	
F	Considerable sintering with redistribution of particles over some areas of the specimen (7b, c)	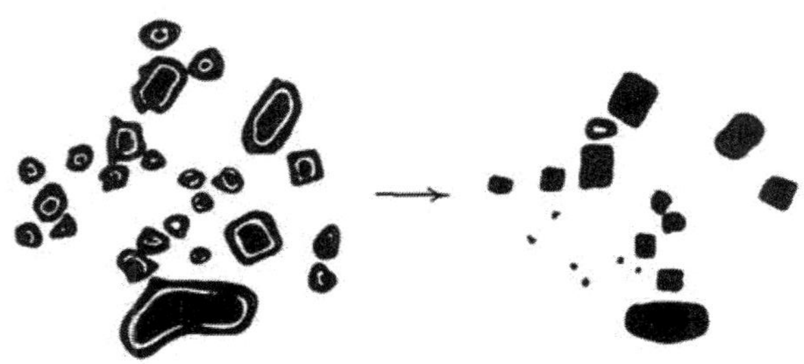

G	Decrease in size or disappearance of a particle with no obvious growth of nearby particles (3a, b; 4a, b; 6d, e)	
H	Disappearance of large particles without decreasing in size first; A large particle nearby a smaller particle decreases in size and/or disappears or coalesces with the latter. The smaller particle either increases in size or remains unaffected (3a–c; 13a–d)	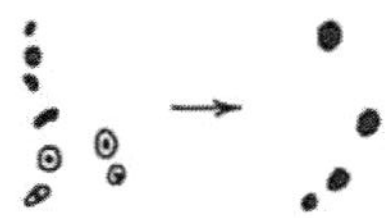
I	Disappearance of small as well as large particles (3a, b; 4a, b; 13a, b)	
J	Growth of some small particles (3b, c)	
K	Increase in contrast (darkening) of all particles. (Growth of all particles small and large); Light and thin particles grow bigger and darker. (6b, c; 7a, b)	 
L	Appearance of new particles. (6e, f; 12a–d; 13a–c)	
M	Splitting of particles without complete separation of the units. (2g–i; 13b–d)	
N	Splitting of particles with separation and migration. (6d–f; 13a, b)	
O	Extension of small particles; recontraction of the extended particles. (2b–e; 13c, d)	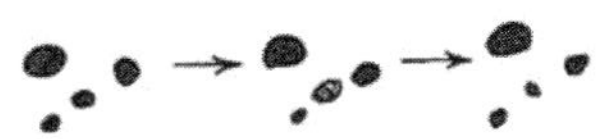
P	Appearance of periphery marks around particles. (7c; 9b, c)	
Q	Appearance of a film extending out from around the particles. (12a, b)	
R	Substrate grain boundaries appear less distinct on heating in oxygen. A film appears to cover the substrate. The grain boundaries appear sharp again on heating in hydrogen subsequently. (12a, c, d; 15c, d, e)	
S.a	Considerable extension of the crystallite to result in a torus shape. (4e; 9a; 10b)	
S.b	Considerable extension and appearance of a torus shape with a small remnant particle in the cavity. (14b, d; 15a)	

TABLE 1—*Continued*

Type	Event and figure number	Schematic of the events (Before → After)
T	Considerable extension to a kind of thick film with a remnant particle on top. (14b, d; 15d, e)	
U	Rupture of crystallites. The crystallite is composed of very small particles held close-together making it appear porous. (8c, d; 9d–f; 15d, e)	
V	Break-up of the crystallites with the subunits interconnected. The sub-units coalesce subsequently. (9a, b; 10b–d; 15a, b)	
W	Considerable break-up of the crystallites to very small pieces. (9g)	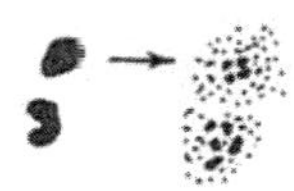
X	Contraction of periphery marks to form small particles. (11c)	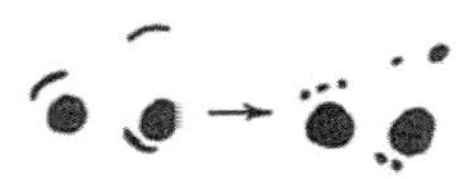
Y	Faceting of particles. (11c)	

TABLE 2

Heating in As-Supplied Hydrogen

Total heating time	Event observed	Compound identified
Table $2_{6,300}$[a]		
1 h	No distinct large particle formation	γ-Al_2O_3
Up to 9 h	No detectable change after heating in a few steps, except for changes in chemical composition.	$FeO \cdot Al_2O_3$ ($FeAl_2O_4$) and solid solution of Fe_2O_3 in Al_2O_3 approaching $Al_2Fe_2O_6$
Table $2_{6,400}$		
½ h	Scattered, light, small, and large particles with torus shape observed. (2a)[b]	$FeO \cdot Al_2O_3$ and traces of Fe_3O_4 (γ-Fe_2O_3)
1 h	An increase in number of large and distinct crystallites. Some particles acquire the core-and-ring structure. (2b)	
2 h	Increase in size and darkening of small and large crystallites. Particles have a torus shape and are of uniform size. No change in the number of particles. (2c)	An increase in the intensity of $FeAl_2O_4$ rings and a decrease in the intensity of Fe_3O_4 rings
Up to 36 h	Considerable extension of some particles and their recontraction (O in 2c–e). Alternation in the shape of the crystallites (2e–i). Coalescence of nearby pairs of particles (A in 2f–i). Tendency to split and subsequent coalescence of these particles (M in 2g–i).	$FeAl_2O_4$ and a solid solution with *d*-values close to those of $Al_2Fe_2O_6$ when torus shape is observed. $FeAl_2O_4$ and a solid solution with *d*-values decreased from those corresponding to the case above and trace α-Fe when core-and-ring structure is observed
Table $2_{6,500}$		
1 h	Uniform size particles with torus shape (3a).	$FeAl_2O_4$, traces of $Al_2Fe_2O_6$
2.5 h	Increase in size of particles. Torus shape is maintained still. Very small cores are present in cavities of particles (1.b in 3c). Disappearance of small and large particles (I in 3b, c). Disappearance of a large particle nearby a smaller particle (H in 3b, c) coalescence of nearby particles (A in 3a–c), dumbbell-shape particles (A in 3a, c), decrease in size and disappearance of some particles (D.b in 3b, c), migration and coalescence of particles (B in 3a–c), growth of some small particles (J in 3b, c) and extension and/or recontraction of particles are observed.	Increase in the intensity of $Al_2Fe_2O_6$
Up to 27 h	Marginal extension and contraction.	An increase in $Al_2Fe_2O_6$ and a decrease in $FeAl_2O_4$
Table $2_{12.5,500}$		
2 h	Particles with sharp, core-and-ring structure form. Small particles have a torus shape. (4a)	Fe_3O_4(γ-Fe_2O_3), α-Fe

TABLE 2—*Continued*

Total heating time	Event observed	Compound identified
3 h	All particles acquire torus shape. Some extension, and coalescence following contact. A few small particles decrease in size or disappear.	
4 h	Particles contract. Particles acquire a core-and-ring structure. A number of small particles disappear (E in 4a, b). Coalescence of a few nearby pairs of particles (A in 4a, b).	
5 h	Formation of compact particles. Decrease in size and/or disappearance of small particles, some of which coalesce with neighbors (A, E in 4b, c). Ripening between a few particles (D.a, D.b, D.c, in 4′a–c).	Mostly α-Fe
6 h	Disappearance of some more small particles.	Almost only α-Fe
Up to 24 h in a few steps	Compact, dark particles are formed. No other change. (4d)	Almost only α-Fe
46 h (22 in one step)	Particles are considerably extended and each particle encloses a cavity. The particles appear porous. Small fragments seen in the cavities of some particles (S.a in 4e).	Solid solution of Fe_2O_3 and Al_2O_3 with *d*-values slightly larger than those of Fe_3O_4 (γ-Fe_2O_3)
Table $2_{6,600}$		
2 h	Particles with core-and-ring structure form. The core is smaller and the annular gap is wider than at lower temperature. (5a, c)	More of $FeAl_2O_4$ and less of Fe_3O_4 (γ-Fe_2O_3)
4 h	The particles contract and tend to form compact particles. Very little change in the number of particles.	
Up to 41 h	Alternation in shape between a core-and-ring structure and a 3-dimensional compact structure. (5b, d)	More of Fe_3O_4 (γ-Fe_2O_3), traces of $Al_2Fe_2O_6$ and α-Fe
Table $2_{6,700}$		
1 h	Very few distinct crystallites are detected. (6a)	Al_2O_3
5 h	Light (low contrast) and mostly small particles are detected. (6b)	Faint rings of α-Fe, Al_2O_3
8 h	Increase in size and darkening of all particles. (6c)	$FeAl_2O_4$, faint rings of α-Fe and Fe_3O_4 (γ-Fe_2O_3)
18 h	Growth of particles. Crystallites are dark and compact. Very little change in the number of resolved particles. (6d)	$FeAl_2O_4$
Up to 34 h	Migration (C), splitting (N), coalescence (A) and appearance of new particles (L) are detected. Traces of particle peripheries (P) and film around particles (Q) are also observed. (6e, f)	
Table $2_{6,800}$		
	(Sample previously heated at 600°C for 41 h). (7a)	

TABLE 2—*Continued*

Total heating time	Event observed	Compound identified
$2\frac{1}{2}$ h	Very dark, compact particles are formed (K). No change in the number and position of the particles. (7b)	A compound with *d*-values slightly greater than those of Fe_3O_4 (γ-Fe_2O_3)
4 h	Considerable sintering and rearrangement of particles is observed (F). Migration of some large crystallites detected (C). Marks of particle peripheries seen (P). (7c)	Fe_3O_4 (γ-Fe_2O_3) and $Al_2Fe_2O_6$

[a] The first subscript refers to the loading in Å and the second to the temperature in °C.
[b] The numbers in parentheses refer to the figure numbers.

$Al_2Fe_2O_6$. Thus an increase in the *d* values toward those of $Al_2Fe_2O_6$ indicates the formation of Fe_2O_3 and its subsequent dissolution in Al_2O_3; conversely, a decrease in the *d* values of one such solid solution toward those of γ-Fe_2O_3 (Fe_3O_4) indicates a partial precipitation of Fe_2O_3 from the solid solution and its eventual reduction to the metal and/or a nonstoichiometric lower oxide. In comparison to the former case, the latter can be considered a relatively reduced (less oxidized) state. The major *d* values of Al_2O_3 and of the lower aluminate, $FeAl_2O_4$, are in the order $FeO > FeAl_2O_4 > Al_2O_3$ and therefore, the formation of solid solutions between FeO and Al_2O_3, ultimately leading to $FeAl_2O_4$, can also be inferred from the increase or decrease in the *d* values. One may note that FeO is not stable below 560°C as a separate phase. The stoichiometric aluminates, $FeAl_2O_4$ and $Al_2Fe_2O_6$, have been detected in some cases. When the *d* values have not corresponded to the stoichiometric aluminates or oxides, solid solution formation has been inferred for reasons discussed above.

At 400°C, the total number of crystallites remained almost constant for heating times greater than 1 h. A few nearby pairs of particles coalesced following their contact because of extension (A in Figs. 2f–i). The crystallites, however, underwent alternate changes in shape from a torus enclosing a cavity to one in which an annular ring separates the torus from a core in the cavity (Figs. 2e–i). The torus shape was found to be associated with a more oxidized state and the core-and-ring structure with a relatively less oxidized state as discussed in detail in Ref. (*22*).

At 500°C, after heating in hydrogen for 6 h, both 6- and 12.5-Å specimens exhibited stability with respect to further growth (Tables $2_{6,500}$ and $2_{12.5,500}$).[2] However, whereas at the lower loading almost no metal was detected in the diffraction pattern even after a total of 27 h of heating in the as-supplied hydrogen, at the higher loading almost only metal was detected after only 6 h of heating in the same as-supplied hydrogen. This shows that only samples with high metal loadings can be reduced to the zero-valent state with hydrogen. As seen from the tables, the events observed and the compounds formed indicate that the chemical interactions between the crystallites and the substrate are very strong and result in the dissolution of the metal oxide in alumina and/or of alumina in the metal oxide. It appears that for the low loading used, it is almost impossible at 400 or 500°C to obtain the zero-valent metal with the as-supplied hydrogen. Even though there is some reducibility which causes the shape alternations (*22*), it is not sufficiently large to yield detectable α-Fe. The higher loading (12.5 Å) yielded, however, compact metallic par-

[2] The first subscript refers to the initial film thickness in Å and the second to the temperature in °C.

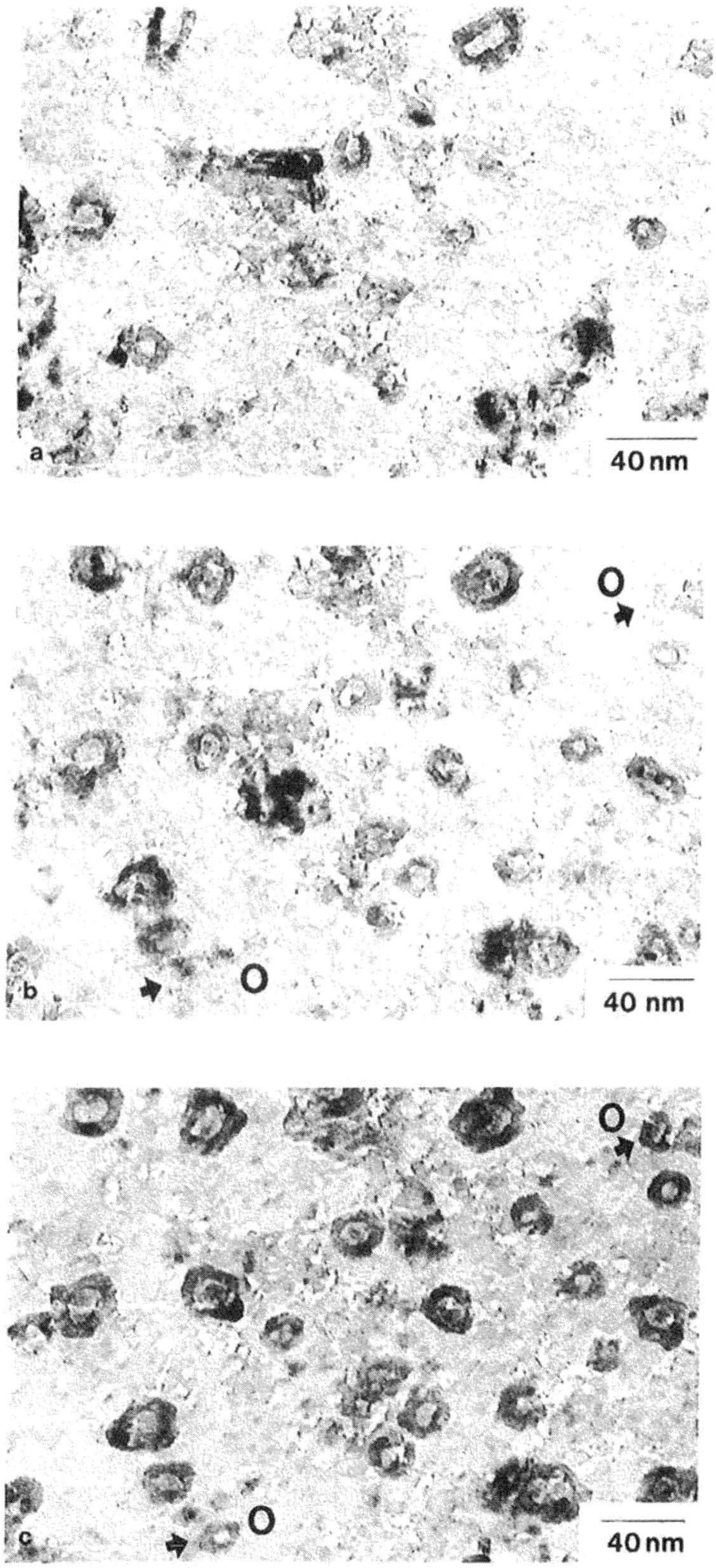

FIG. 2. Sequence of changes of a specimen heated at 400°C in as-supplied hydrogen. Initial loading, 6 Å. The micrographs show the same region after (a) $\frac{1}{2}$ h, (b) 1 h, (c), 2 h, (d) 3 h, (e) 6 h, (f) 12 h, (g) 21 h, (h) 33 h, (i) 36 h.

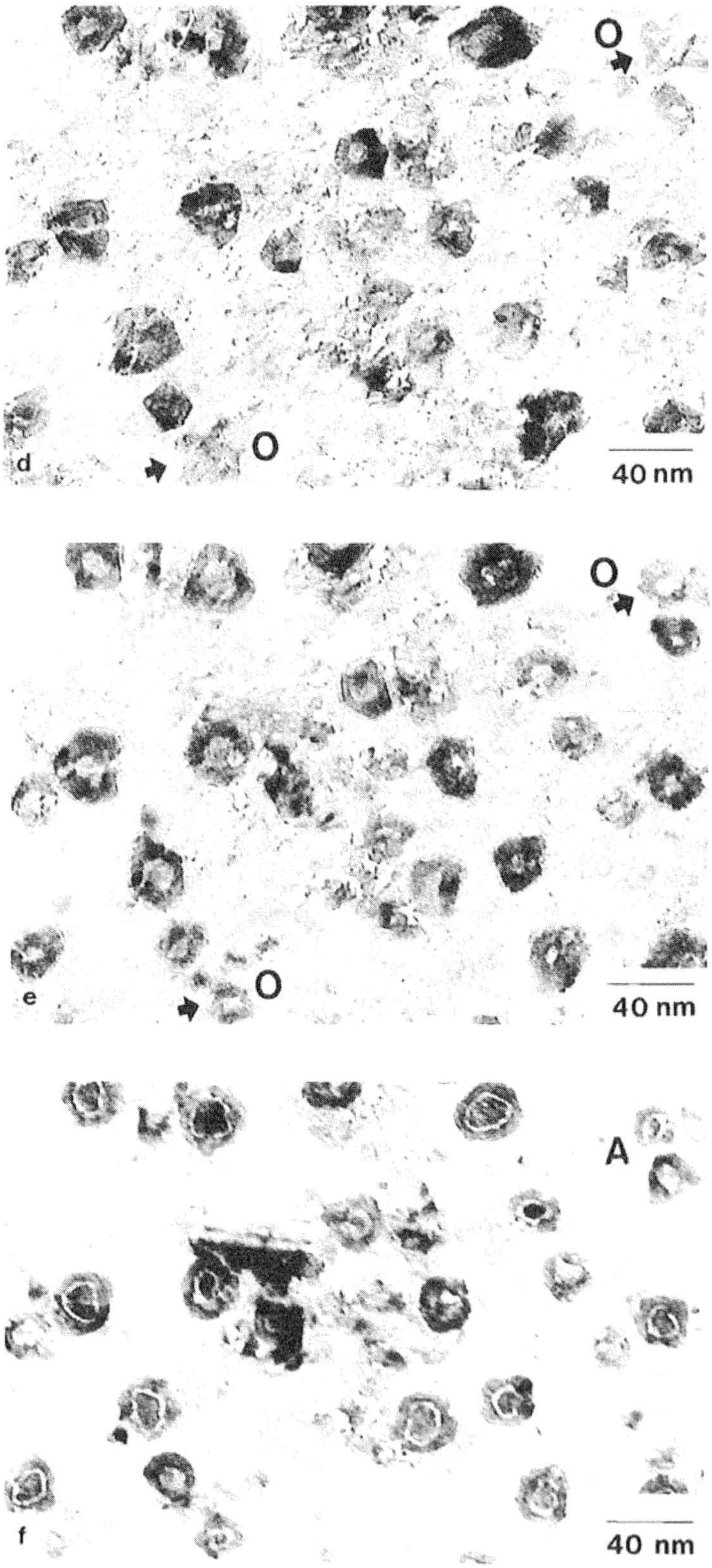

FIG. 2—*Continued*.

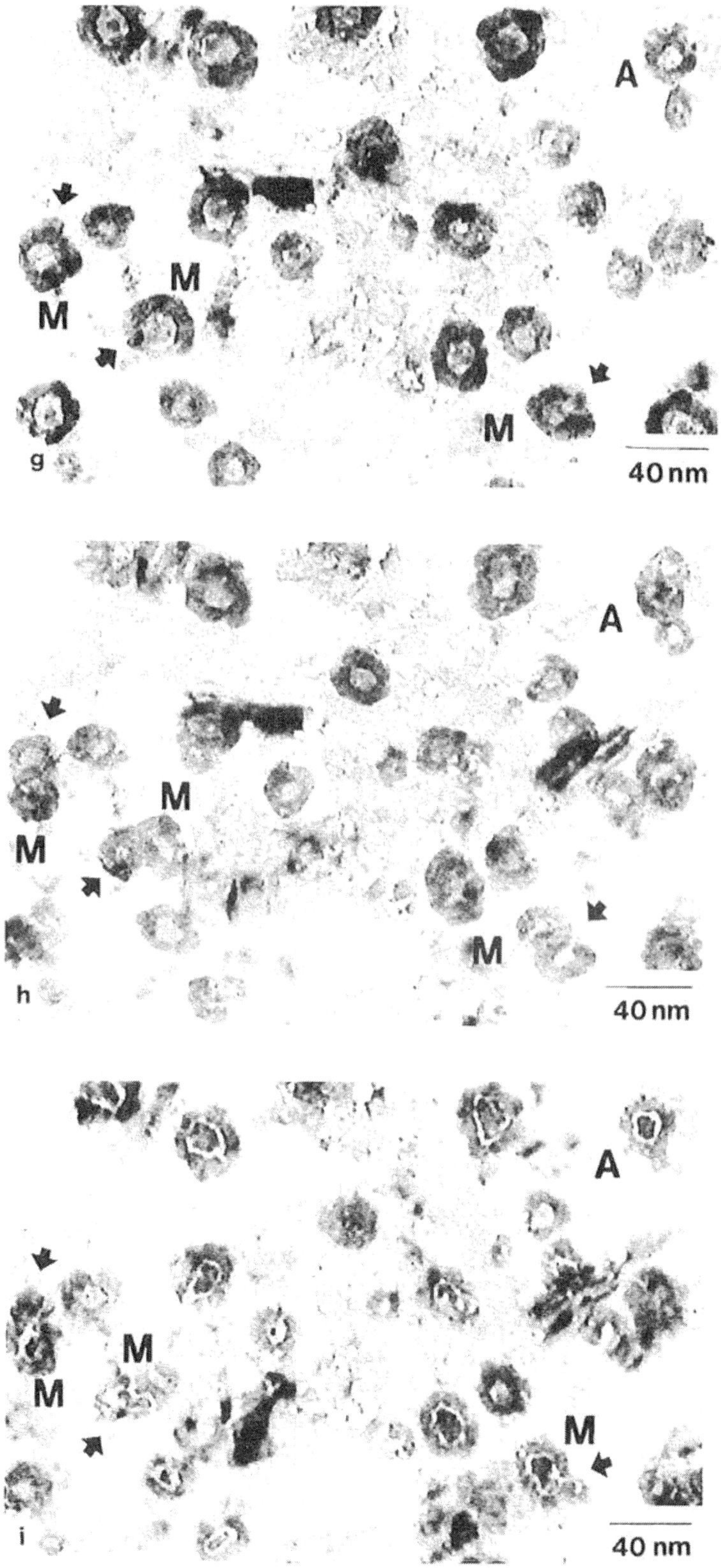

FIG. 2—*Continued.*

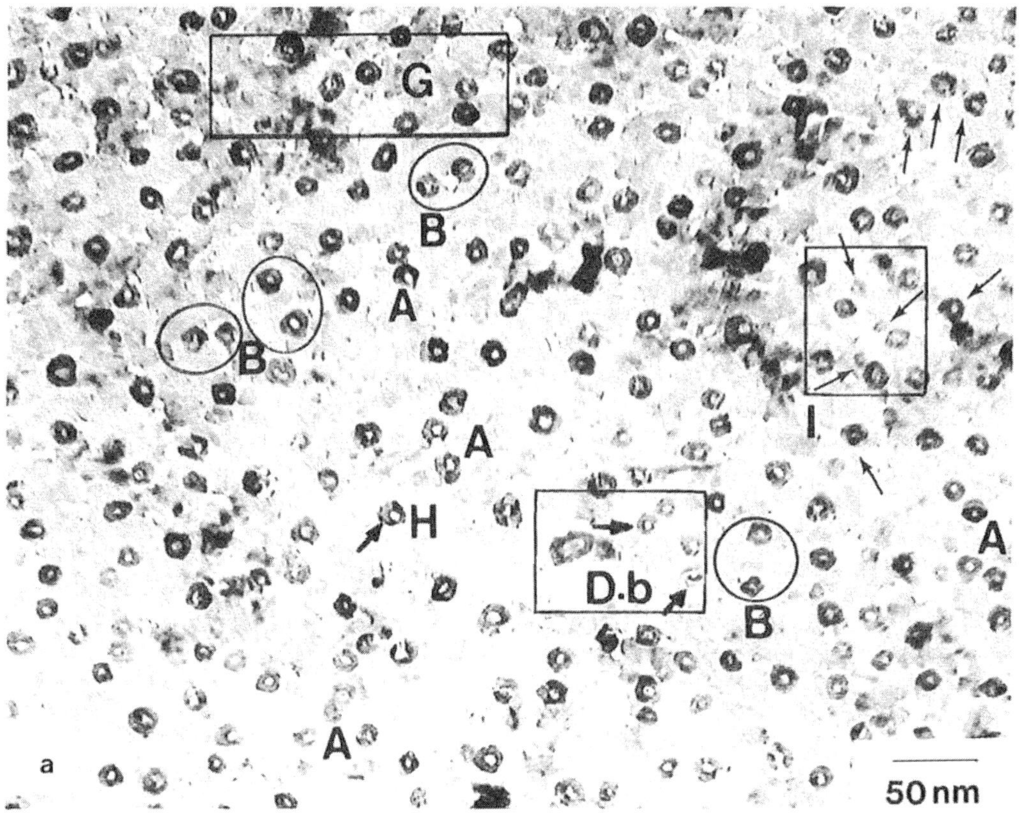

FIG. 3. Time sequence of the same region of the specimen heated previously at 300°C on subsequent heating in as-supplied hydrogen at 500°C. (a) 1 h, (b) 2 h, and (c) 5 h.

ticles of α-Fe which subsequently sintered considerably.

The results at 600°C for a loading of 6 Å are listed in Table $2_{6,600}$. Unlike at lower temperatures, the crystallites alternated in shape between a core-and-ring structure and a three-dimensional, compact structure, instead of a torus (Figs. 5a–d). At this temperature and loading, the compact structure is associated with $FeAl_2O_4$, Fe_2O_3, and only traces of α-Fe. In contrast, at 500°C and 12.5-Å loading, the compact particles were associated mostly with α-Fe.

Distinct crystallites formed more slowly at 700°C (Table $2_{6,700}$) than at 400 to 600°C (Table $2_{6,400}$–$2_{6,600}$). Distinct crystallites could not be detected after an initial heating for 1 h (Fig. 6a). In addition, only rings of Al_2O_3 were present in the diffraction pattern. Possibly, very thin particles were present in an oxidized extended form on the substrate, undetectable by TEM and electron diffraction. The same observation is also valid for the 300°C treatment. At the lower temperature of 300°C, after about 2 h of heating the iron oxide molecules of the small (not well formed) clusters probably diffused into the substrate to form aluminate. At the higher temperature of 700°C, they are perhaps present initially as large, extended, thin islands of oxide, or even aluminate, too thin to be detected by electron diffraction. During the initial few hours of heating, the rate of aluminate formation is probably faster at 300°C than at 700°C because a larger fraction of molecules is in contact with alumina in the former than in the latter case. (In addition, since the clusters may not be well formed at 300°C, the molecules may be loosely held to one another.) The subsequent increase in size and contrast, with the number of crystallites remaining almost the same (Table $2_{6,700}$, Fig. 6c), occurs probably as a result of coales-

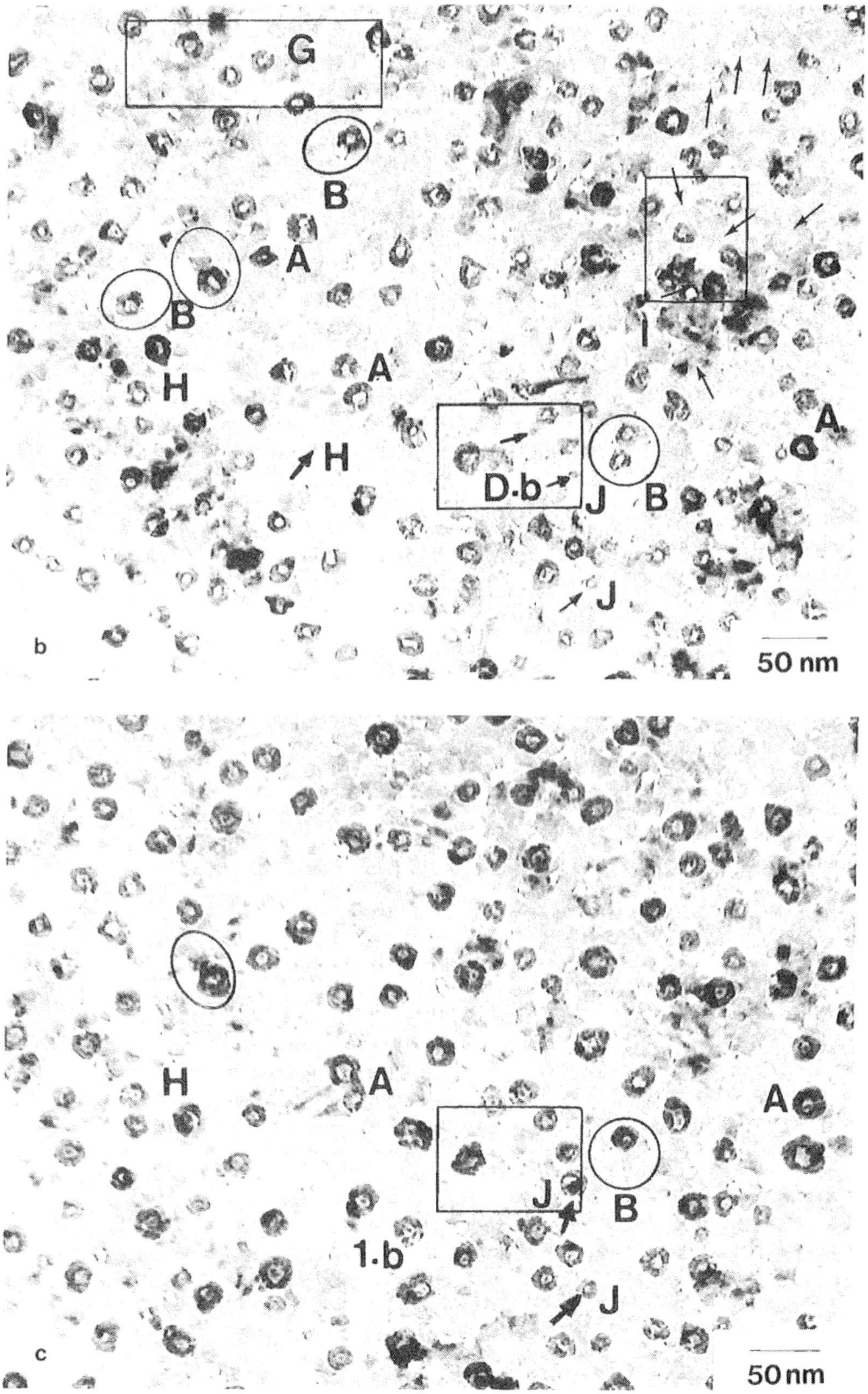

FIG. 3—*Continued.*

TABLE 3

Alternate Heating in Hydrogen and Oxygen

Heating atmosphere	Event observed	Compound identified
Table $3_{6,400}$		
O_2	Torus shape particles with a tendency to fill in the cavity of the torus are observed.	Fe_3O_4 (γ-Fe_2O_3), $Al_2Fe_2O_6$
H_2	Particles acquire the core-and-ring structure. Coalescence of nearby particles and rupture of some particles are seen.	$FeAl_2O_4$, α-Fe
Table $3_{6,500}$		
O_2	Considerable filling in of the cavities of the torus. Some pairs of neighboring particles make contact following extension. (8c)	$Al_2Fe_2O_6$, Fe_3O_4 (γ-Fe_2O_3), trace $FeAl_2O_4$
H_2	Particles acquire a core-and-ring structure initially but on continued heating assume a torus shape. Particles in contact coalesce. (8a, b, d, e)	Fe_3O_4 (γ-Fe_2O_3), trace $FeAl_2O_4$
Table $3_{12.5,500}$		
O_2	Extension. Tendency to fill in the cavity. Coalescence of nearby particles. Particles composed of loosely interconnected units. (9a, d, e, f; 10b)	*d*-Values approaching those of $Al_2Fe_2O_6$ and $FeAl_2O_4$
H_2	Splitting of particles into a few interconnected sub-units, a few of which coalesced subsequently (V). Periphery marks seen around particles (P). Fragmentation of particles (W). (9b, c, g; 10a, c, d)	α-Fe, traces of Fe_3O_4 (-Fe_2O_3)
Table $3_{6,600}$		
O_2	(Sample was previously heated in hydrogen at 800°C for 4 h, Fig. 7c). Migration of some particles (C). A film appears to extend out from underneath and around the particles. (11a, b)	Fe_3O_4 (γ-Fe_2O_3), $Al_2Fe_2O_6$, traces of $FeAl_2O_4$
H_2	Contraction and formation of more circular particles. Periphery marks, observed before, contract to form small particles (X). (11c)	Mostly Fe_3O_4 (γ-Fe_2O_3), traces of $FeAl_2O_4$, traces of α-Fe
Table $3_{6,700}$		
O_2	A thick film extends from the periphery of the particles. Considerable extension or disappearance of small particles (E). Subtrate grain boundaries appear less sharp (R). Very little extension of large particles. (12b, c)	Mostly $FeAl_2O_4$ and some Fe_3O_4 (γ-Fe_2O_3)
H_2	Appearance of a few, new, small particles (L). (12d)	$FeAl_2O_4$

cence of the unresolved extended islands with the nearby resolved particles or by ripening. After a total of 18 h of heating, almost only $FeAl_2O_4$ is detected. The fact that only aluminate is detected indicates that, perhaps, it is present not only in the substrate, but on its surface and also in as well as on the particles.

A.2. Alternate heating in oxygen and as-supplied hydrogen. Heating in oxygen caused, in general, extension of the crystallites over the substrate, as seen from Table 3. In addition, in the oxidized and extended state, the crystallites appear to be composed of a number of very small units held together (Figs. 8c, 9e, f). Except for the coalescence of a few nearby pairs of particles following their contact, as a result of extension on oxidation, the number of crystallites remained, in general, constant during alternate heating in oxygen and hydrogen. Heating in hydrogen following heating

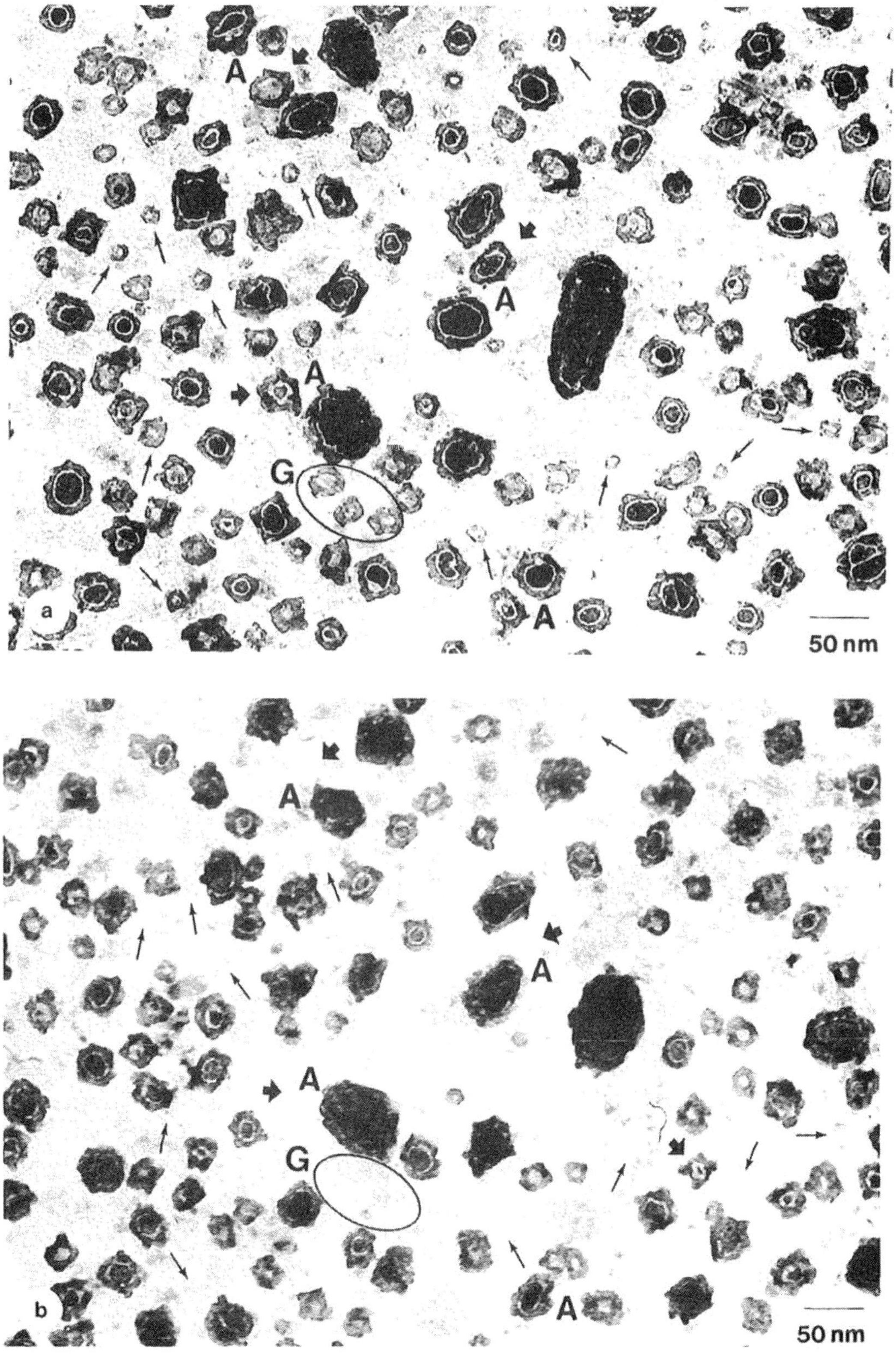

FIG. 4. Sequence of changes at 500°C in a specimen with higher loading. Initial loading, 12.5 Å. Two adjacent regions (Figs. 4 and 4′) are shown after (a) 2 h, (b) 4 h, (c) 5 h, (d) 24 h, and (e) 46 h.

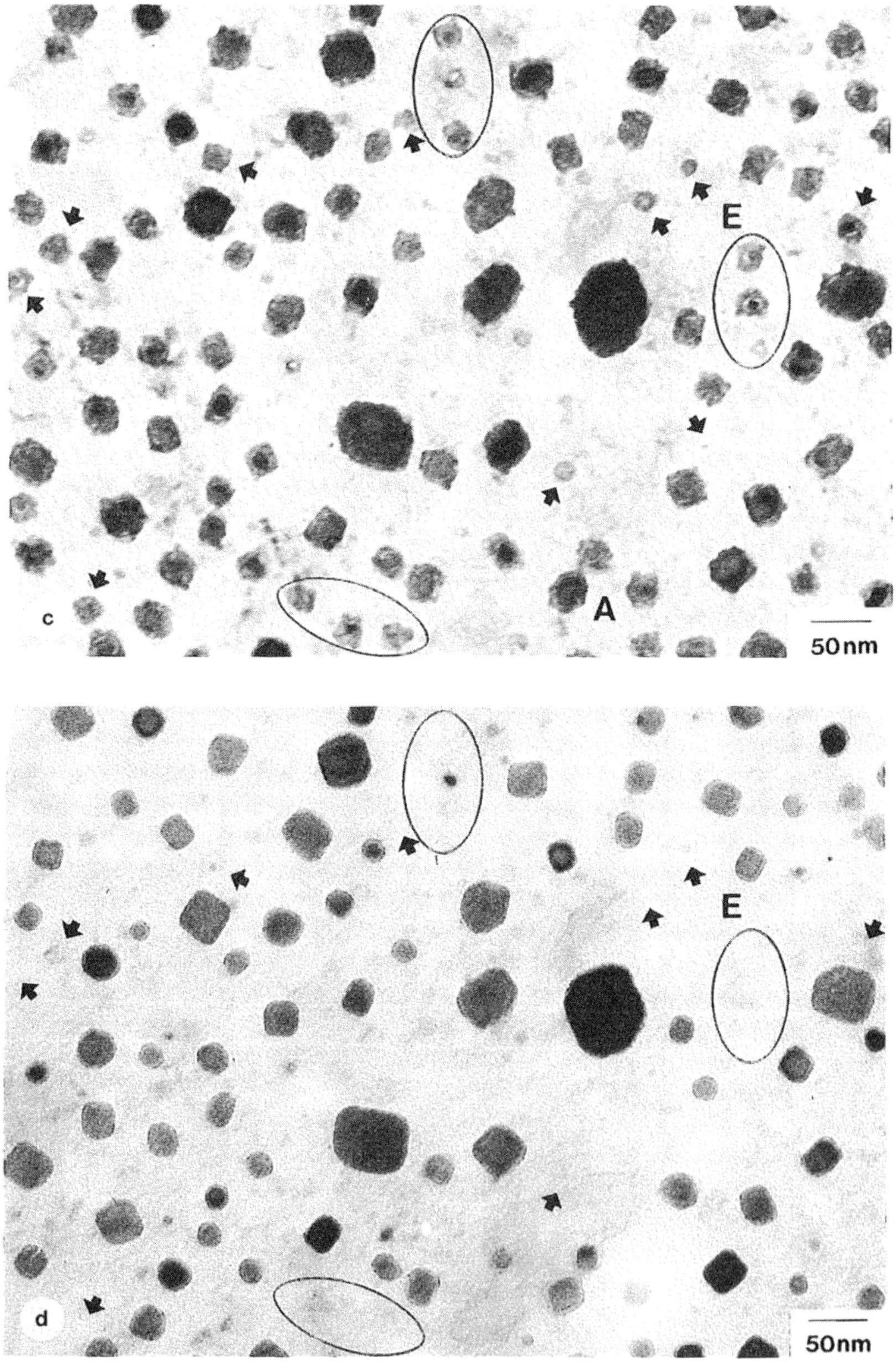

FIG. 4—*Continued.*

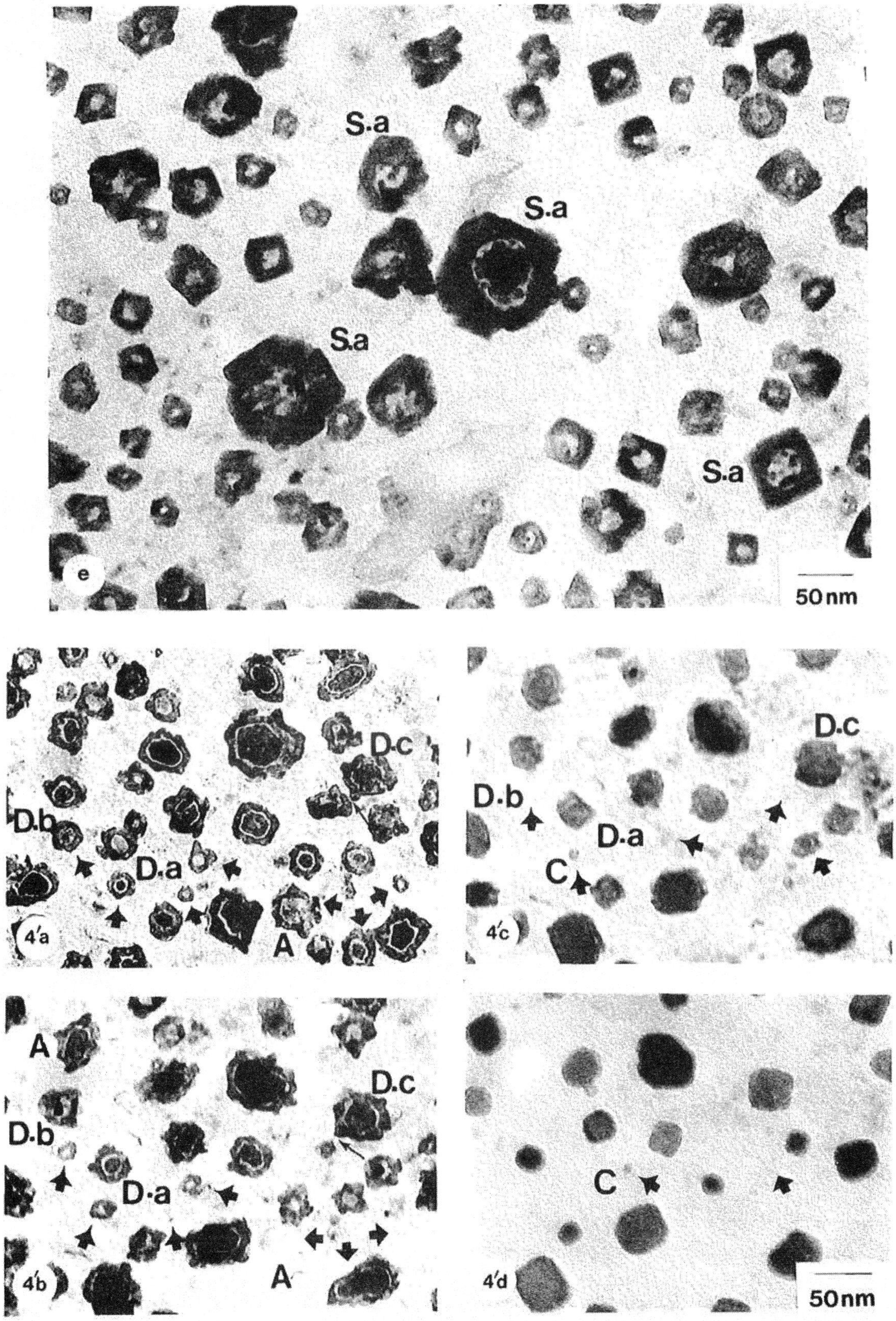

FIG. 4—*Continued.*

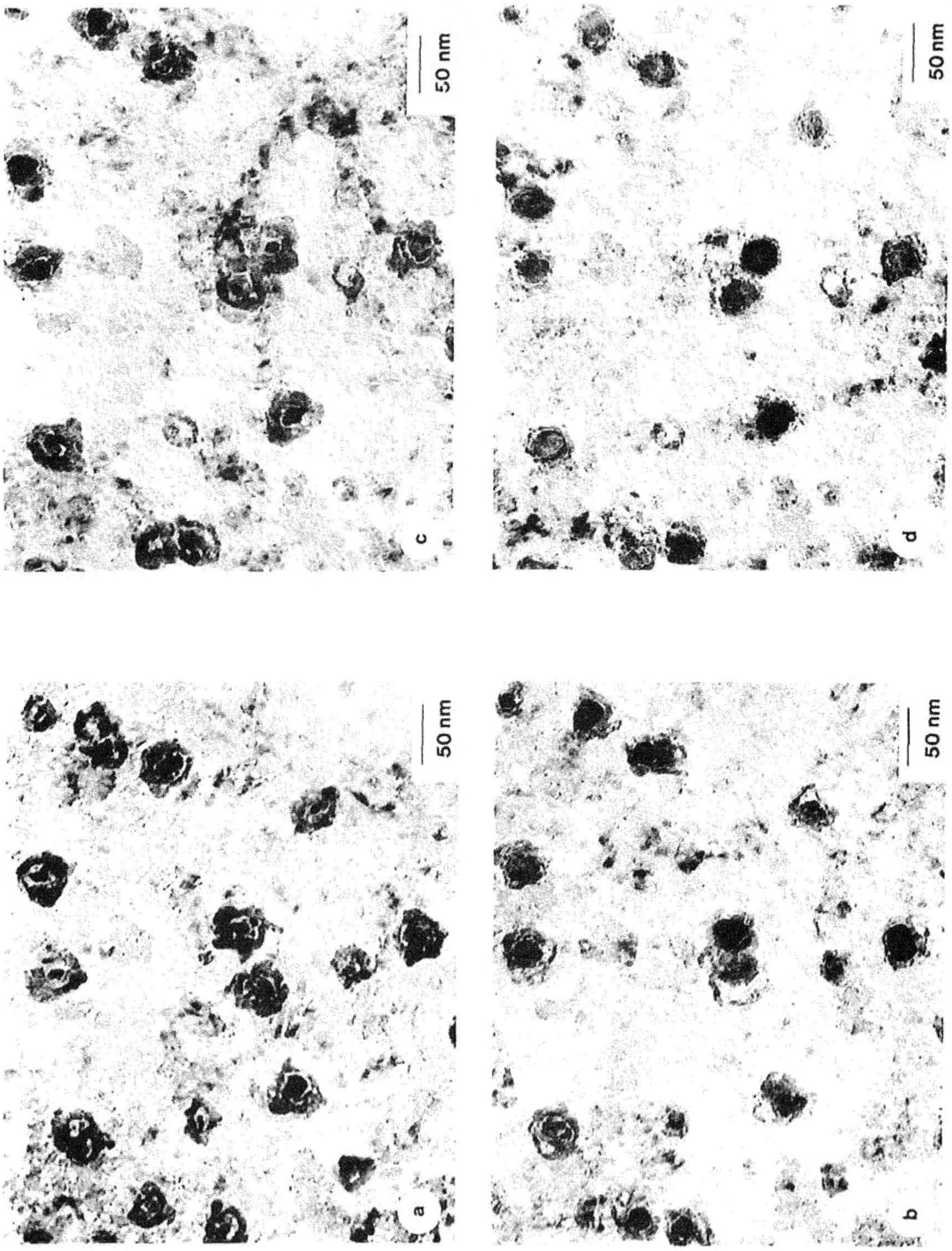

FIG. 5. Sequence of changes on heating a 6-Å initial loading specimen in as-supplied hydrogen at 600°C. (a) 2 h, (b) 6 h, (c) 14 h, (d) 30 h.

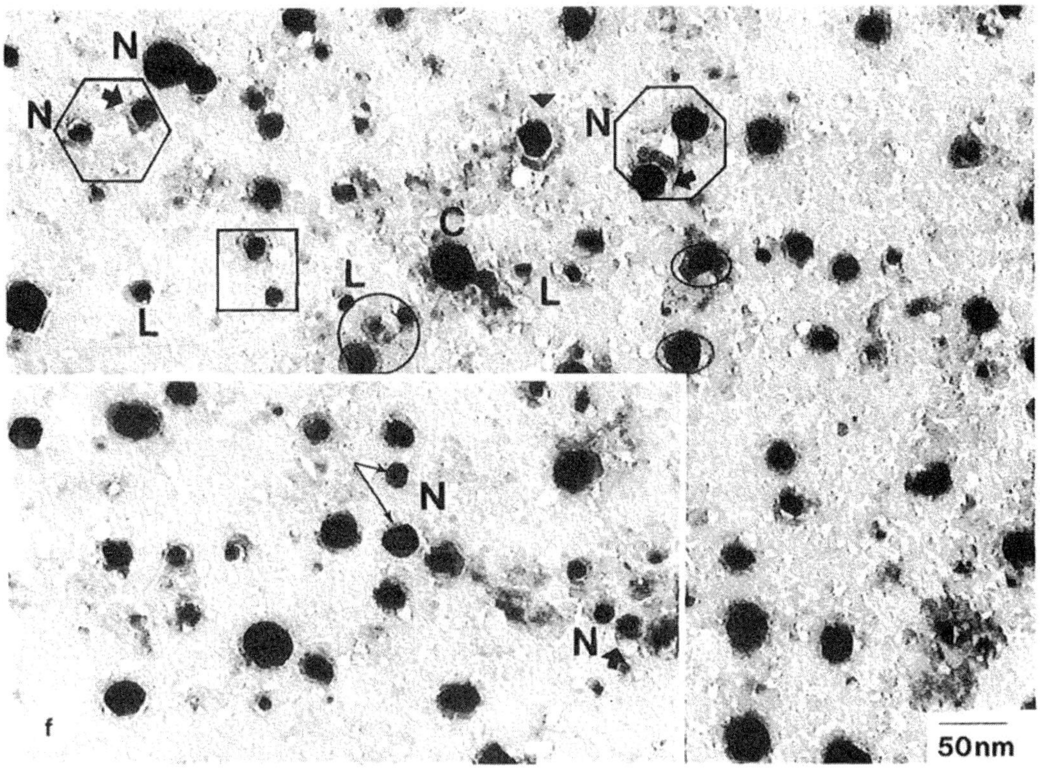

FIG. 6—*Continued.*

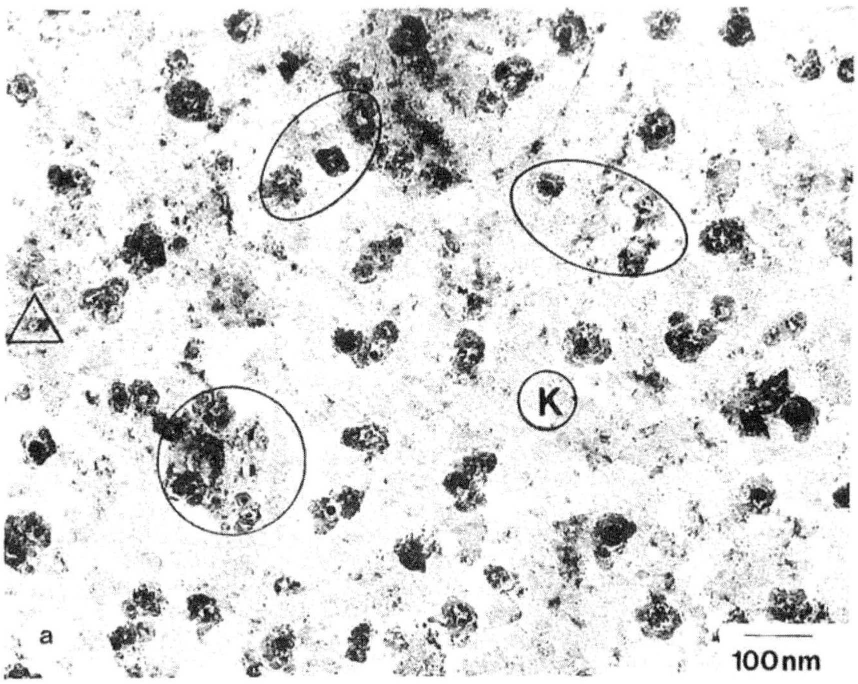

FIG. 7. Sequence of changes on heating the specimen of Fig. 5 at 800°C. The micrographs correspond to the same region: (a) 41 h, 600°C (included for reference) (b) 2 h 30 min, 800°C; (c) 4 h, 800°C.

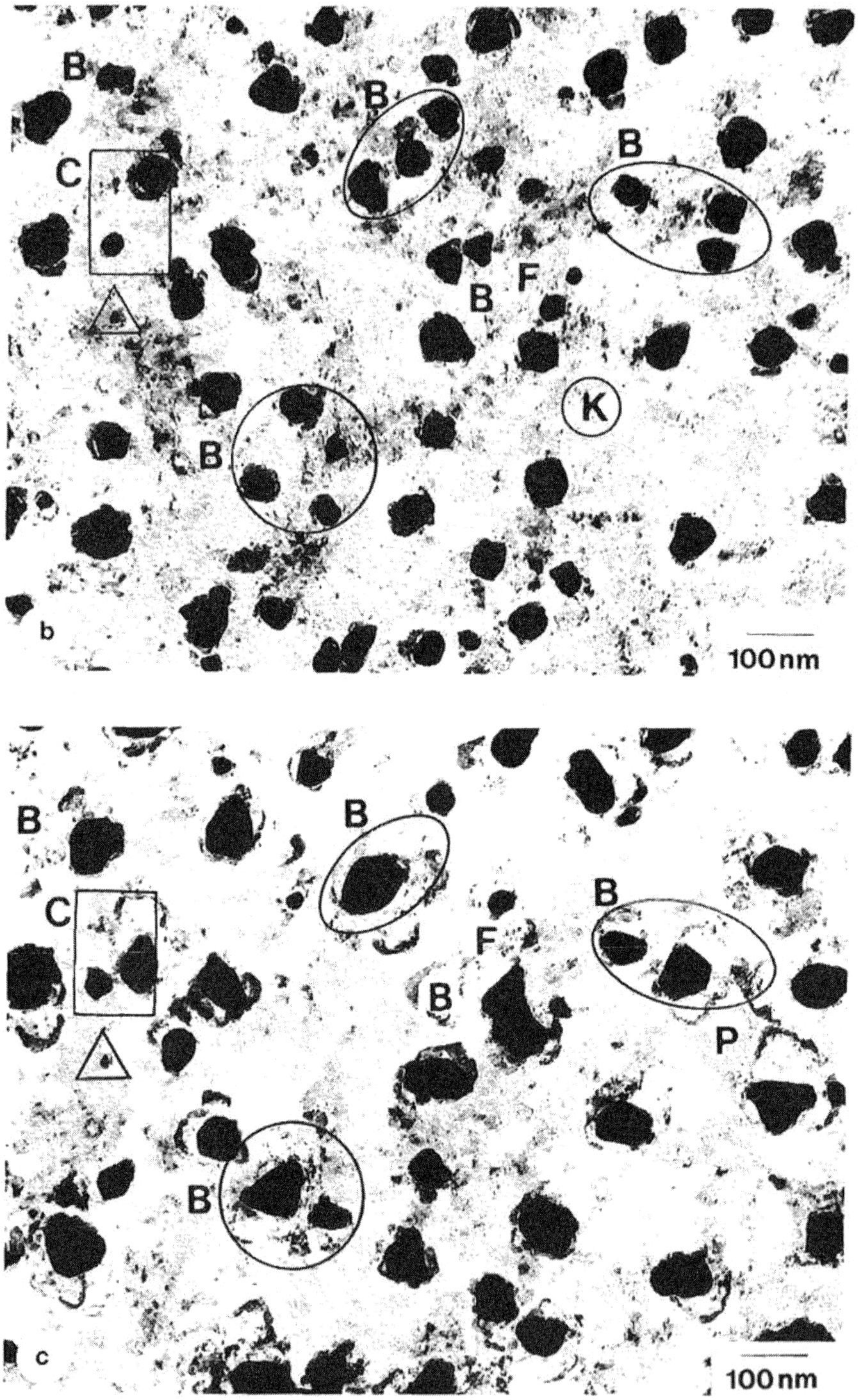

FIG. 7—*Continued.*

rous and are composed of a large number of smaller particles held together. Also, the substrate is covered with a very large number of small and thin crystallites seen clearly in a magnified micrograph (Fig. 9f). Subsequent heating in hydrogen causes the particles to contract and appear distinct, accentuating the fragmentation. The micrograph appears as if a kind of explosive shattering of the particles had occurred (Fig. 9g). In fact, the breakup occurs during heating in oxygen, as a result of the consider-

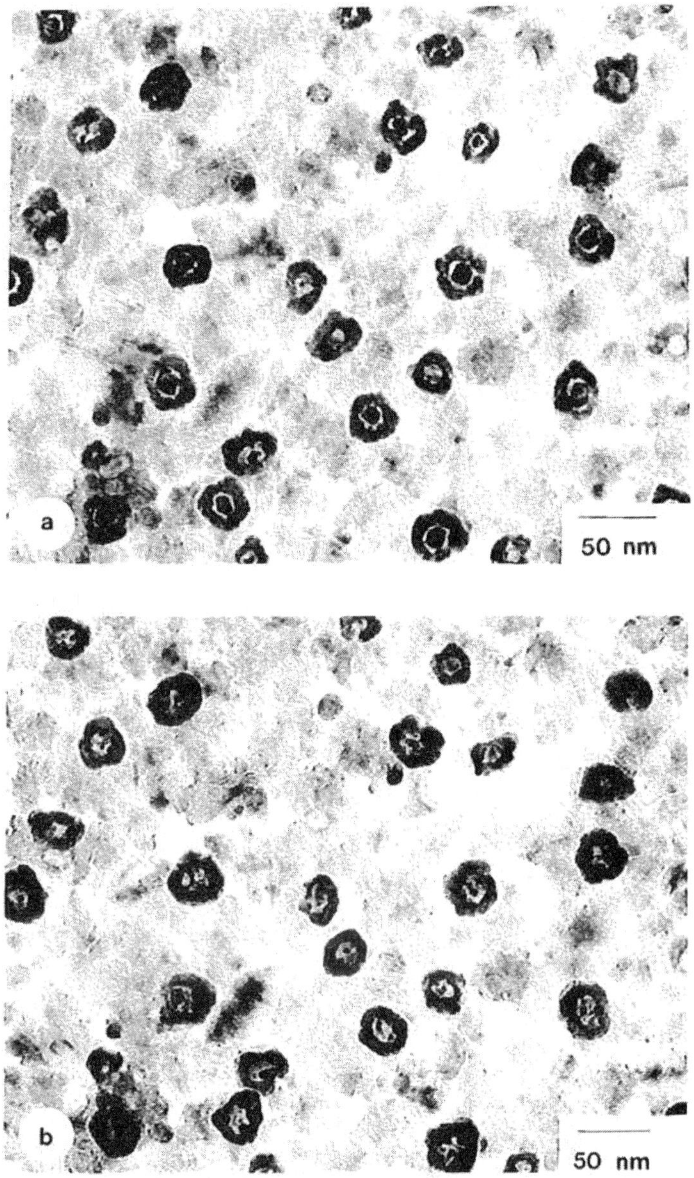

FIG. 8. Sequence of changes in the specimen of Fig. 3 on subsequent alternate heating in oxygen and hydrogen at 500°C. The region shown here is from the same specimen but of a different region than that of Fig. 3. The substrate is thicker in this region probably as a result of tearing and folding upon itself, initially. The particles are larger and the changes are sharper. The micrographs included show the changes subsequent to an oxidation step. (a) 5 h, H_2; (b) 10 h, H_2; (c) 10 h, O_2; (d) 10 min, H_2; (e) 16 h, H_2. The behavior is similar at 400°C.

able extension which, in addition, probably occurs very rapidly causing the rupture. Mechanical fatigue might have also contributed to the fragmentation since the specimen had already been heated for about 200 h.

On heating in oxygen at 700°C (Table $3_{6,700}$), a thick film extends out from the periphery of the larger particles and also the grain boundaries of the substrate appear less distinct (Figs. 12b, c). The substrate might be covered by a film. In the same micrograph, a number of small particles have disappeared (E). It is possible that both the crystallites that disappeared as well as some unresolved ones are present considerably extended on the surface covering the substrate as a thin film which co-

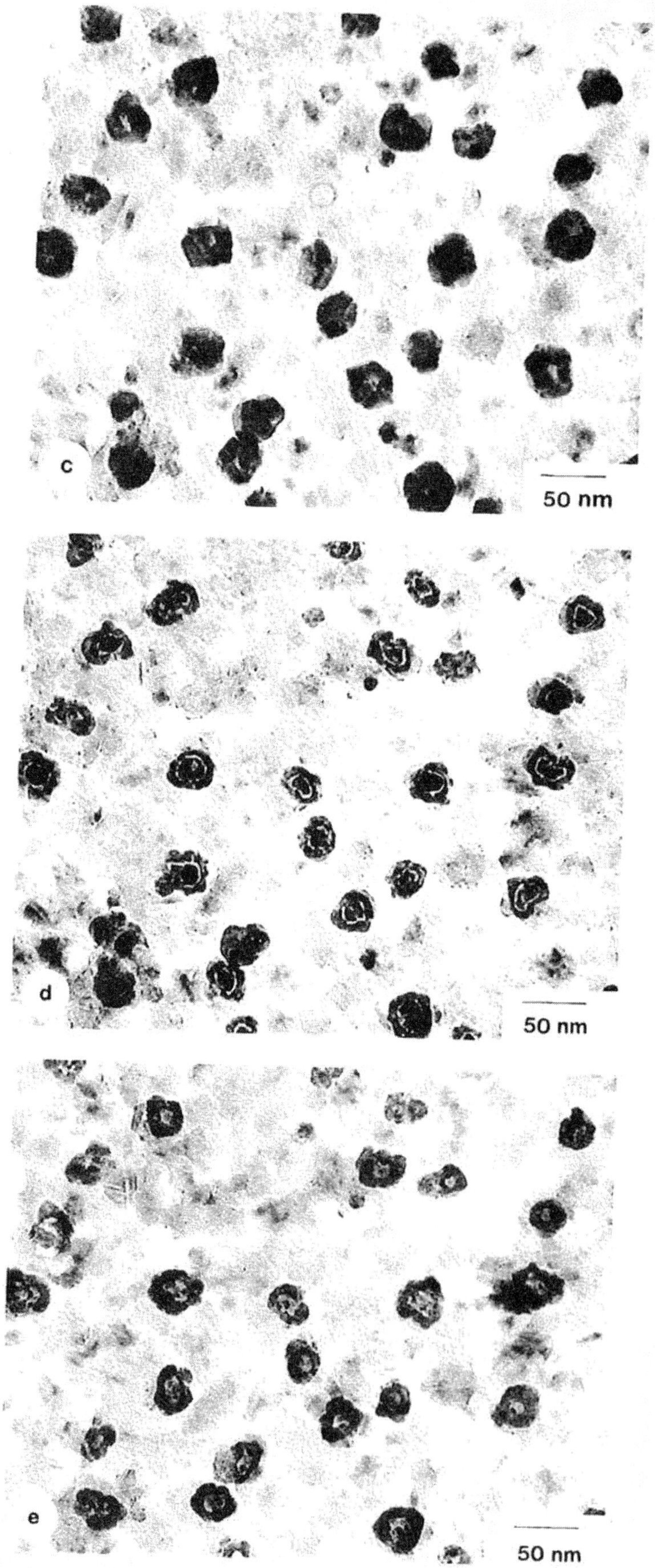

FIG. 8—*Continued.*

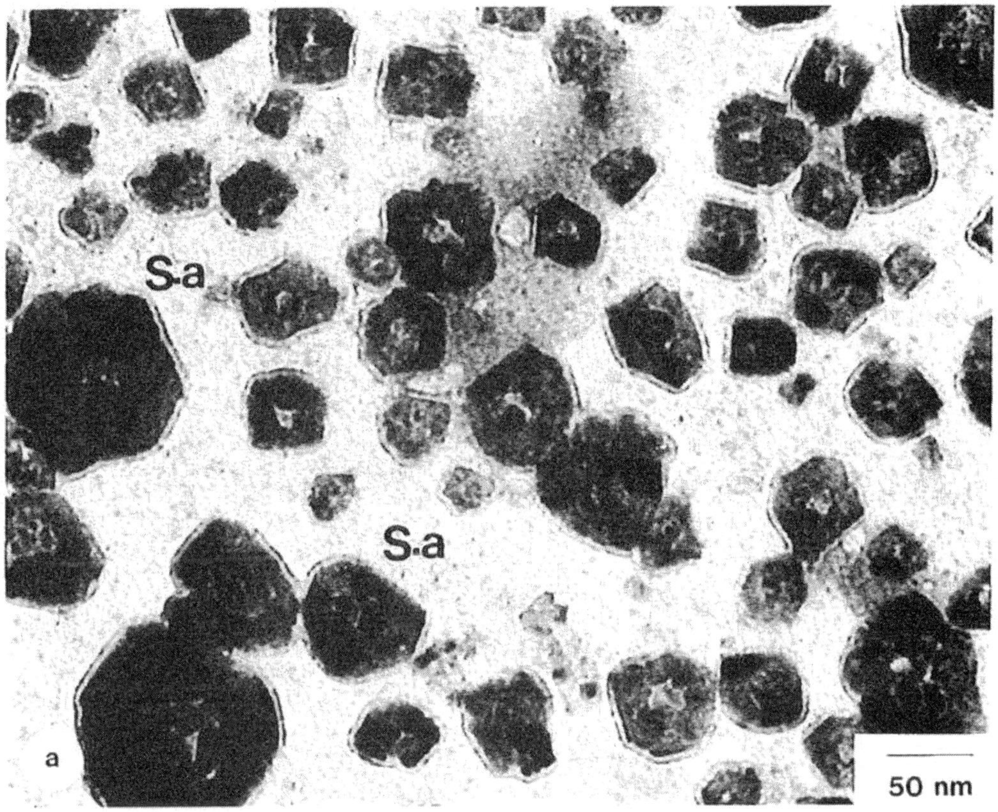

FIG. 9. Sequence of changes on heating the specimen of Fig. 4e alternately in oxygen and hydrogen at 500°C. (a) 2 h, O_2; (b) 2 h, H_2; (c) 66 h, H_2; (d) 1 h, O_2; (e) 40 h, O_2, an adjacent region; (f) magnified micrograph of (e); (g) 45 min, H_2.

TABLE 4

Heating in Purified Hydrogen

Total heating time[a]	Event observed	Compound identified
400°C, 4 h	Particles with very sharp core-and-ring structure and a broad size distribution are formed. Cores in some big particles have fringes indicating twinning (x, y in Fig. 13a). Smaller crystallites have a torus shape.	Fe_3O_4 (γ-Fe_2O_3) α-Fe
500°C, 2 h	Dark, compact particles are formed. Extremely narrow annular rings are still detected. Considerable sintering has occurred. A number of small and large particles disappeared (I). Also, a number of smaller crystallites appeared in other places (L). Coalescence of nearby particles (A) and disappearance of a large particle nearby an unaffected smaller particle (H) are detected (13a, b).	α-Fe
600°C, 4 h	Particles contract. Circular and restructured particles are formed. (13c)	α-Fe, traces of Fe_3O_4 (α-Fe_2O_3)
700°C, 2 h	Disappearance of some small particles, appearance of new ones (L), splitting (M), coalescence (A) and migration (B, C) are observed. (13d)	α-Fe

[a] 7.5 Å, progressive heating.

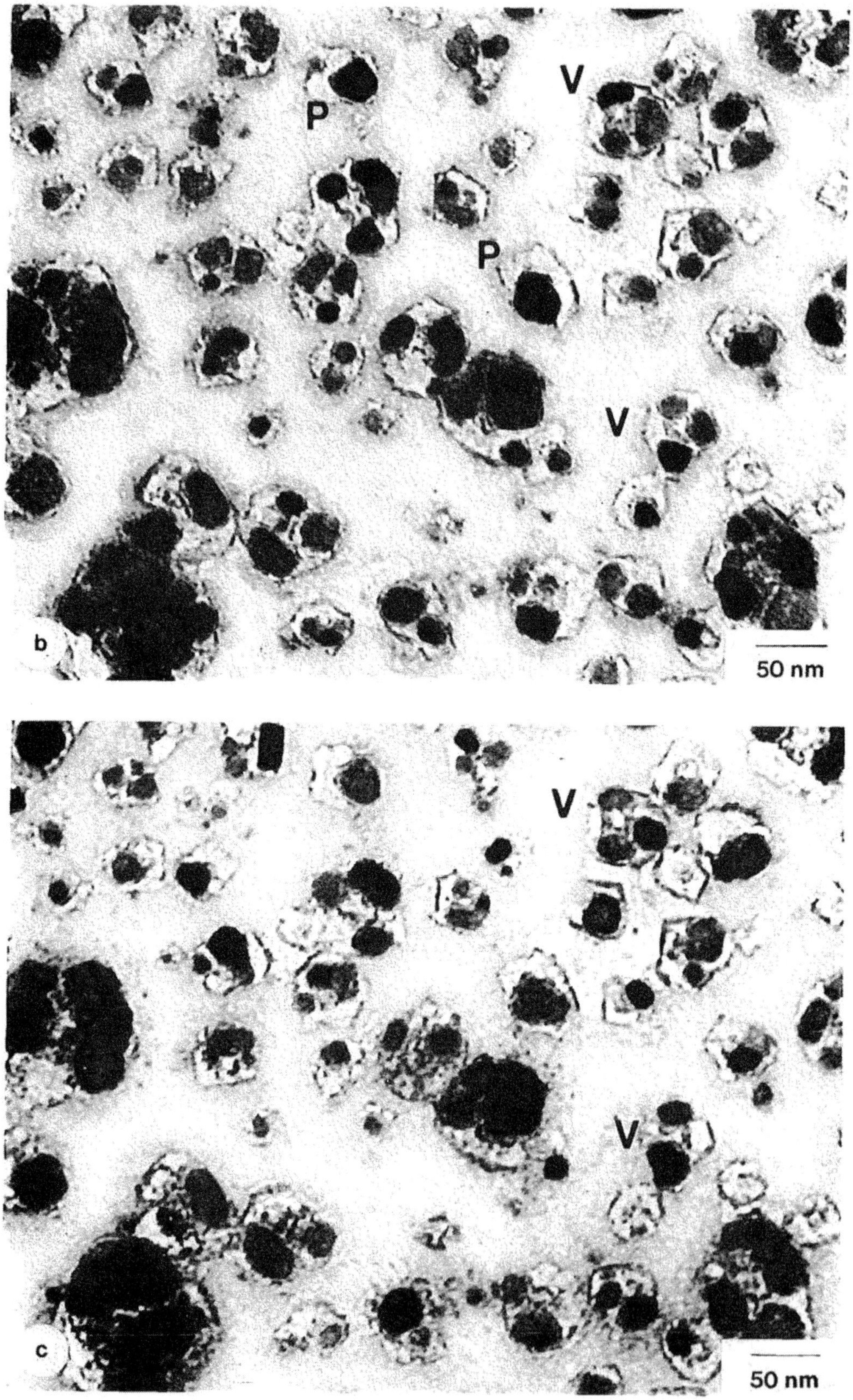

FIG. 9—*Continued.*

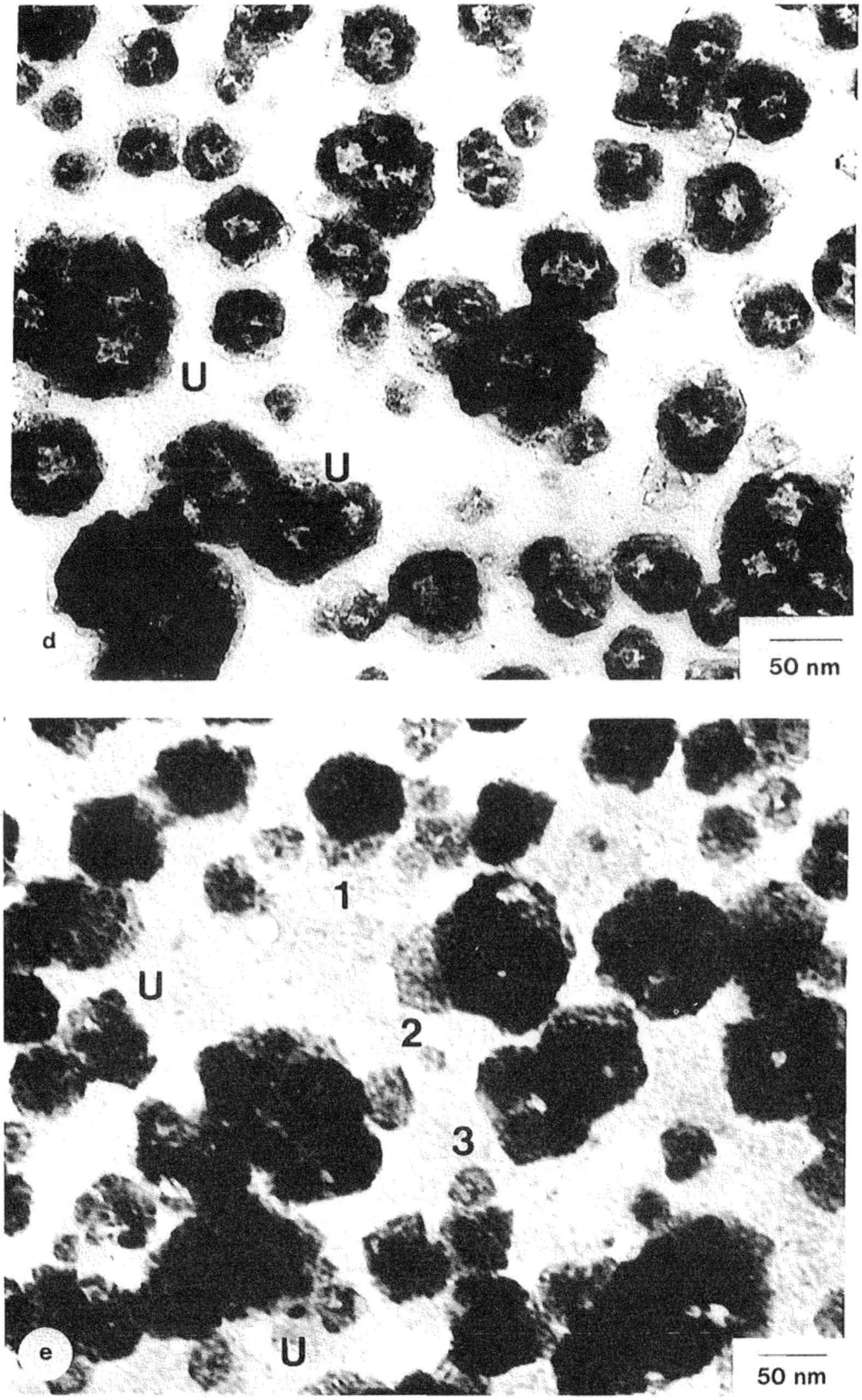

FIG. 9—*Continued.*

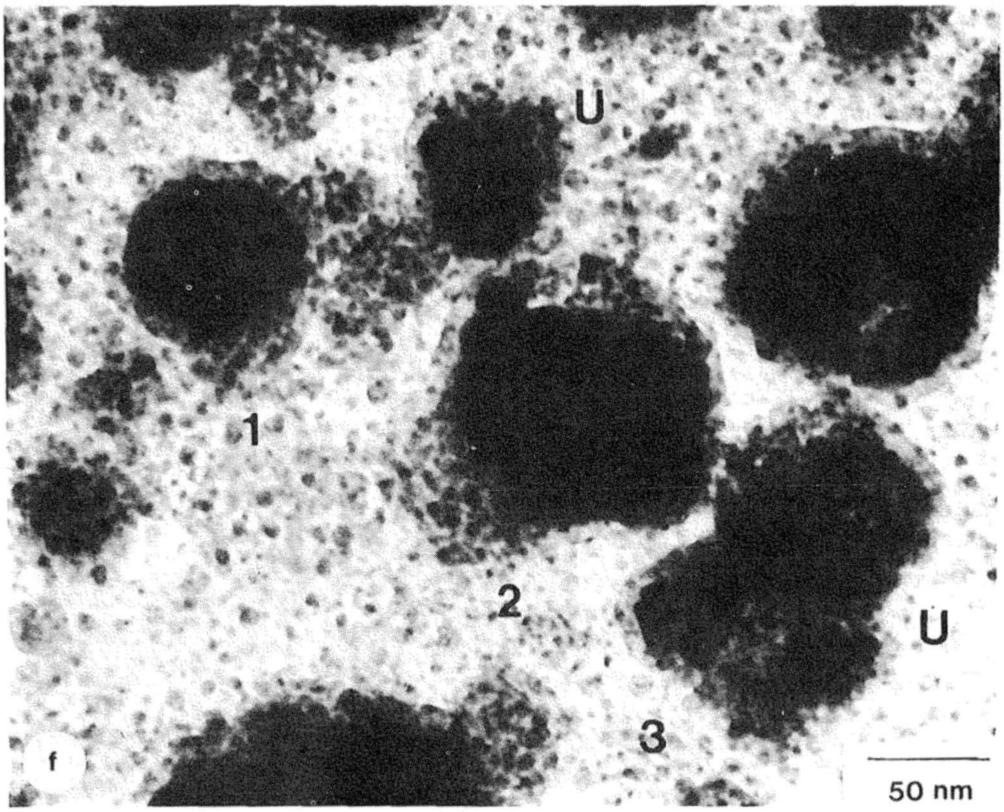

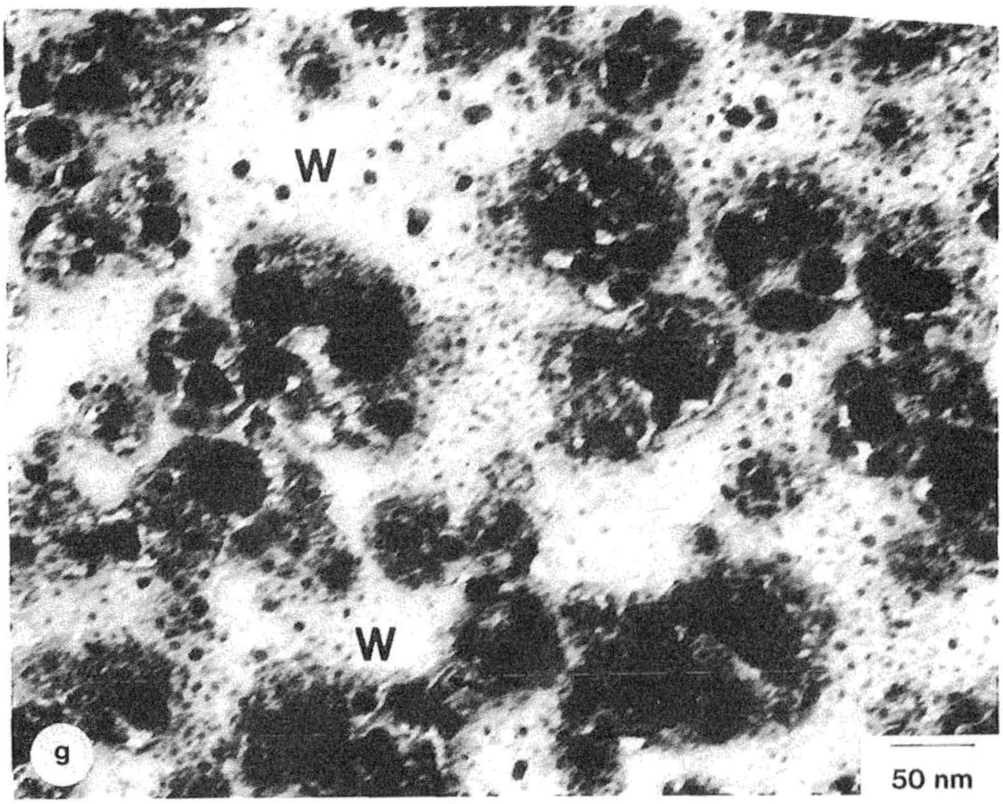

FIG. 9—*Continued.*

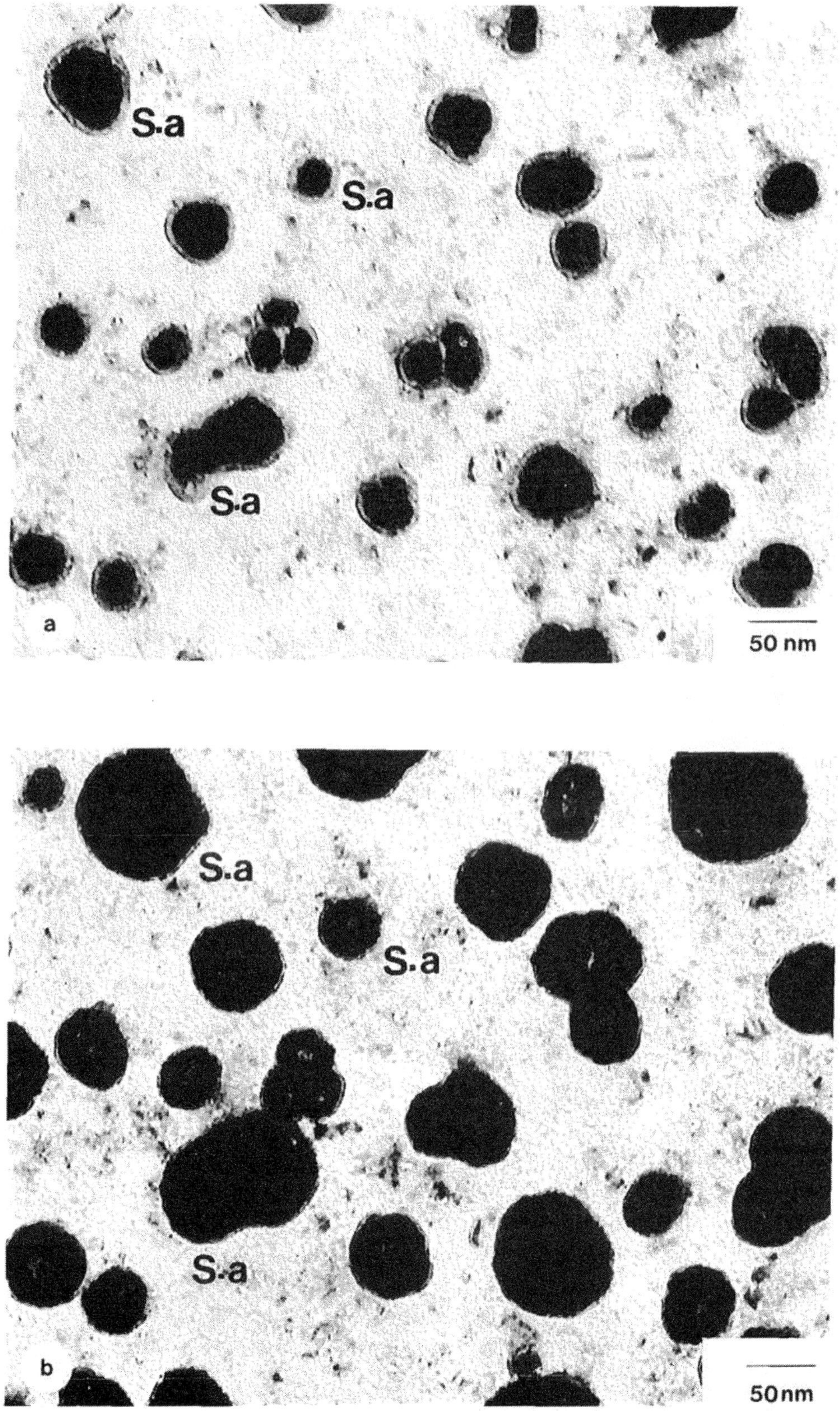

FIG. 10. As Fig. 9 but a different region of the same specimen better indicating the crystallite splitting. (a) 2 h, H_2; (b) 1 h, O_2; (c) 1 h, H_2; (d) 2 h, H_2.

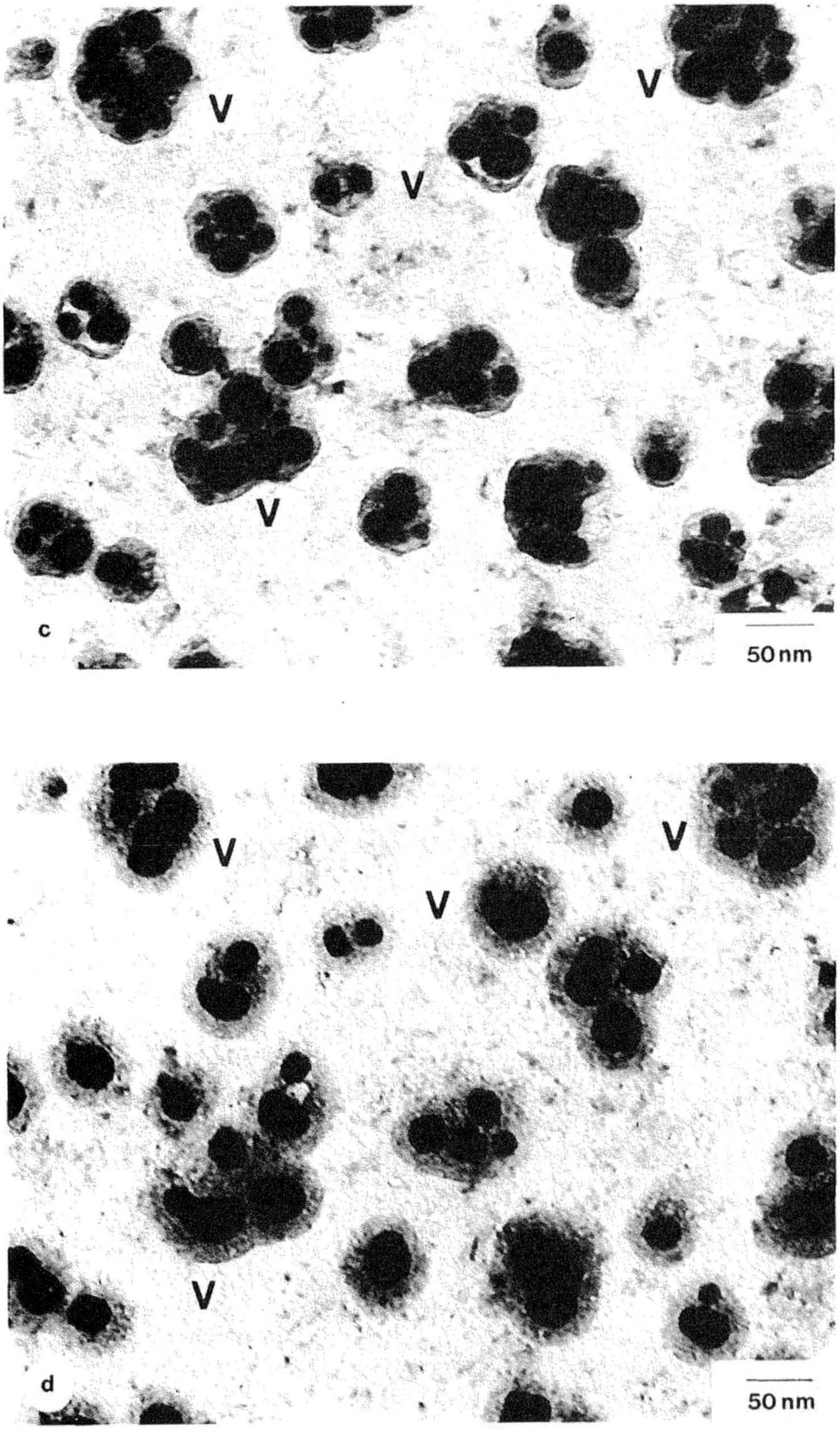

FIG. 10—*Continued.*

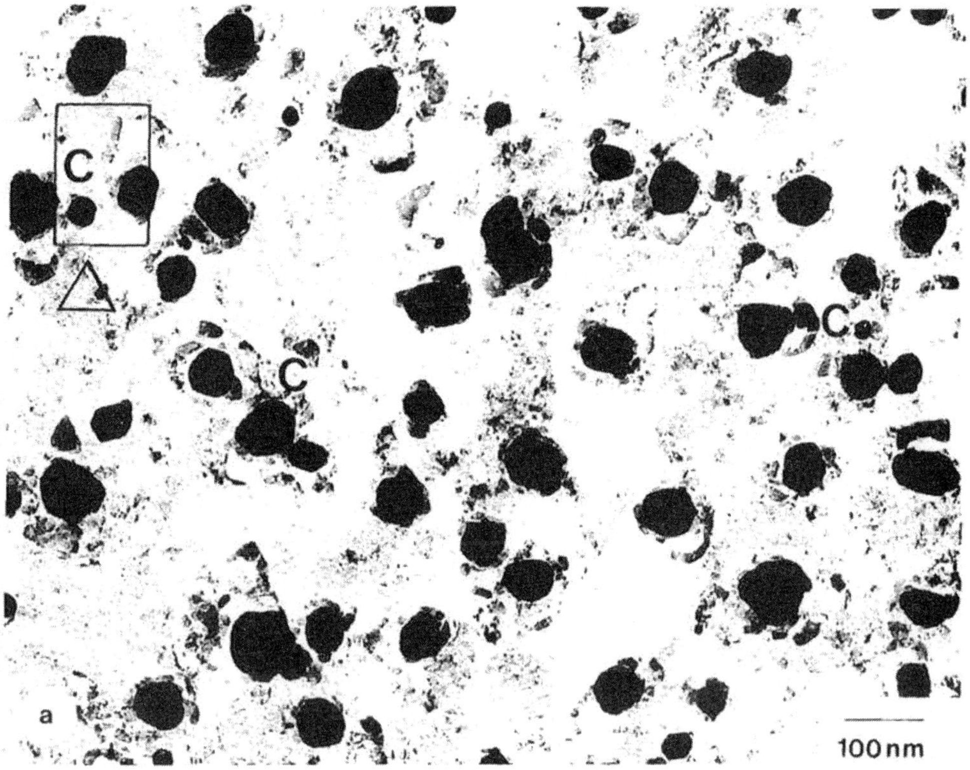

FIG. 11. Time sequence of changes on heating the specimen of Fig. 7c alternately in oxygen and hydrogen at 600°C. (a) ½ h, O_2; (b) 1½ h, O_2; (c) 1 h, H_2.

exists with the three-dimensional crystallites. The thin film which appears to cover the substrate might be an oxide, or more likely, an aluminate, since almost only $FeAl_2O_4$ is detected in the diffraction pattern.

B. Behavior in Purified Hydrogen

B.1. Heating in hydrogen. The behavior of the specimens when heated in purified hydrogen was different in a few respects from that observed with as-supplied hydrogen (Sect. A.1). Unlike the previous case, the initially formed crystallites are typically larger and have a bimodal distribution. Some of these initial particles are so thin that the substrate grain boundaries can be seen through them (a–c in Fig. 13a). The particles, especially the large ones, seem to prefer predominantly rectangular shapes to the usual circular ones. The events observed on heating a 7.5-Å initial film in purified hydrogen at progressively higher temperatures are listed in Table 4. Metallic iron is very clearly detected even at 400°C. At 500°C and higher temperatures almost only iron is detected, with trace quantities of oxide. In contrast, with the as-supplied hydrogen the metal could not be detected even at 700°C. In addition, at 500°C and even at low loadings, relatively compact particles formed, unlike heating in the as-supplied hydrogen which led to torus-shape particles. These results point out that the difficulty in reducing the iron catalysts to the metallic state is clearly a result of the chemical interactions with the support which are caused to a considerable extent by the oxygen and/or moisture present in the reducing chamber. The micrographs (Figs. 13a, b) show that a number of particles of about 100 Å and also a few particles considerably larger (about 350 Å) disappear (H) and, instead, a number of small crystallites of about 30 Å or less appear in all the regions, including locations where no particles were

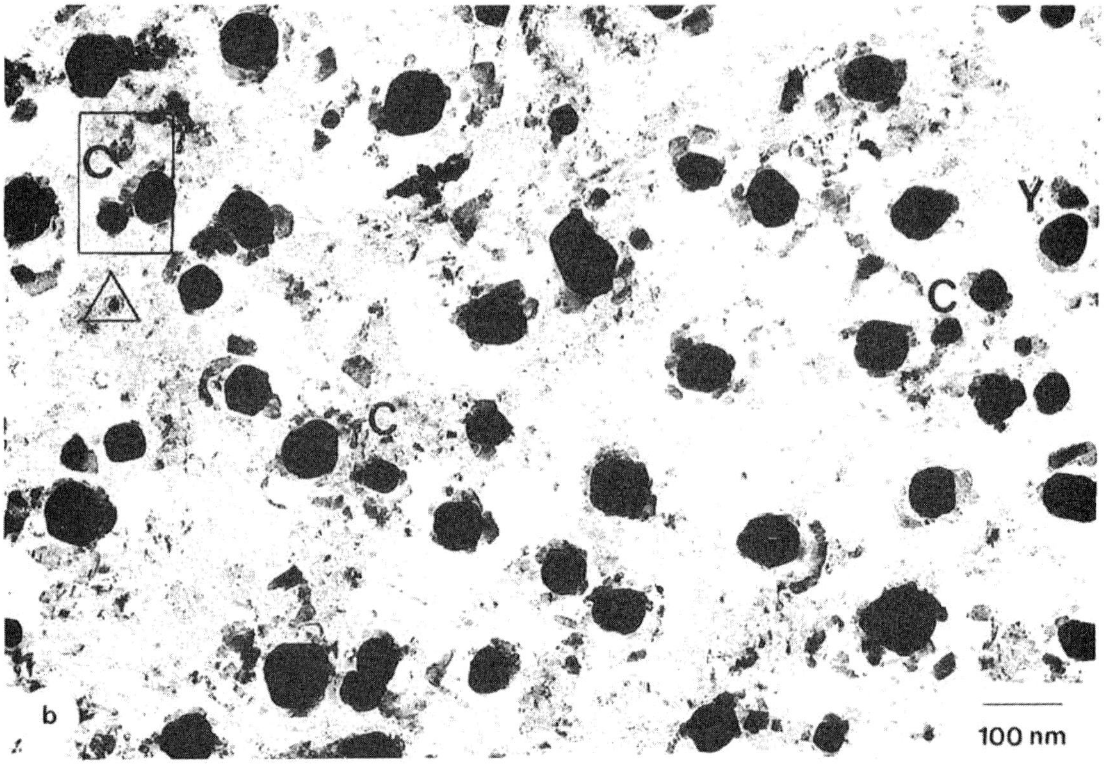

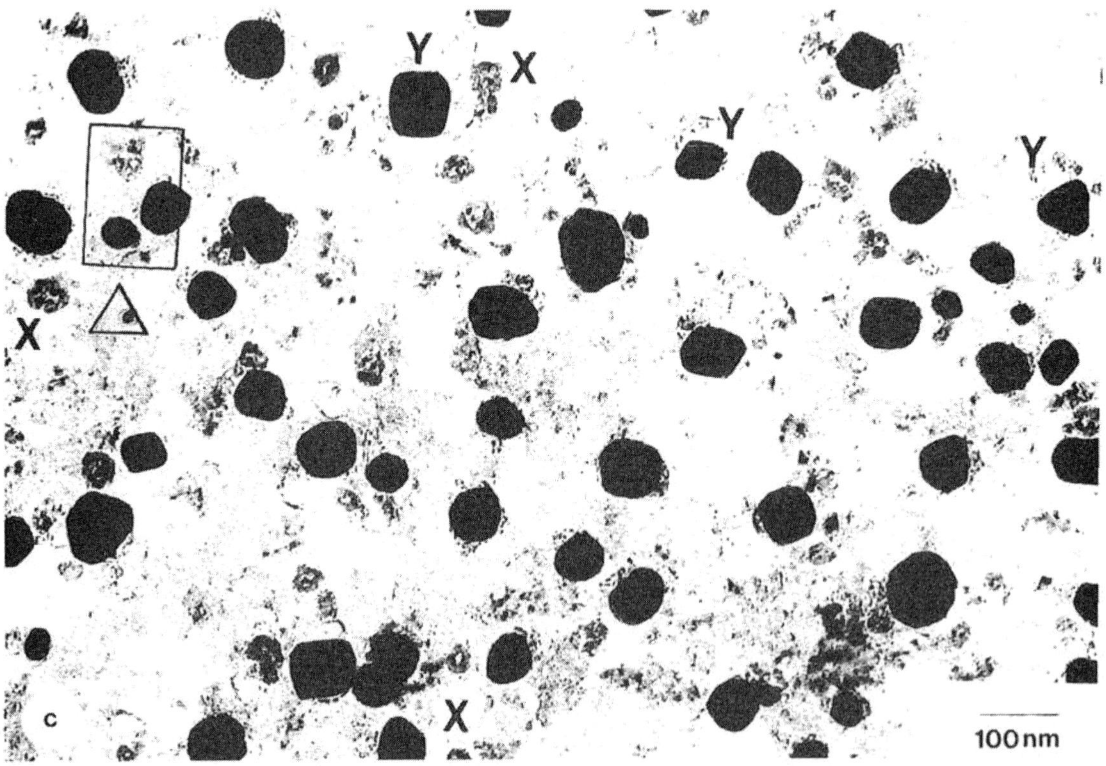

FIG. 11—*Continued.*

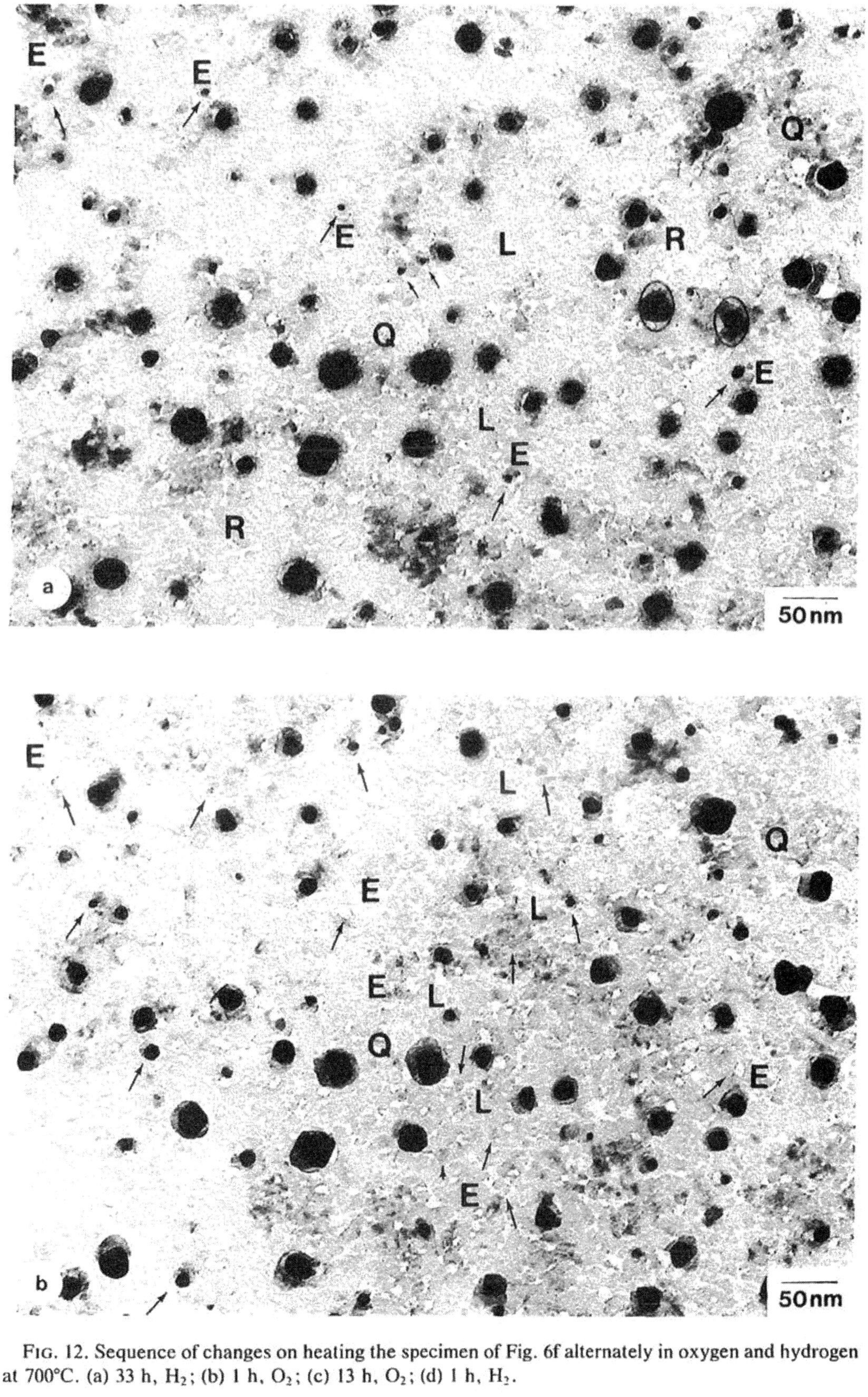

FIG. 12. Sequence of changes on heating the specimen of Fig. 6f alternately in oxygen and hydrogen at 700°C. (a) 33 h, H_2; (b) 1 h, O_2; (c) 13 h, O_2; (d) 1 h, H_2.

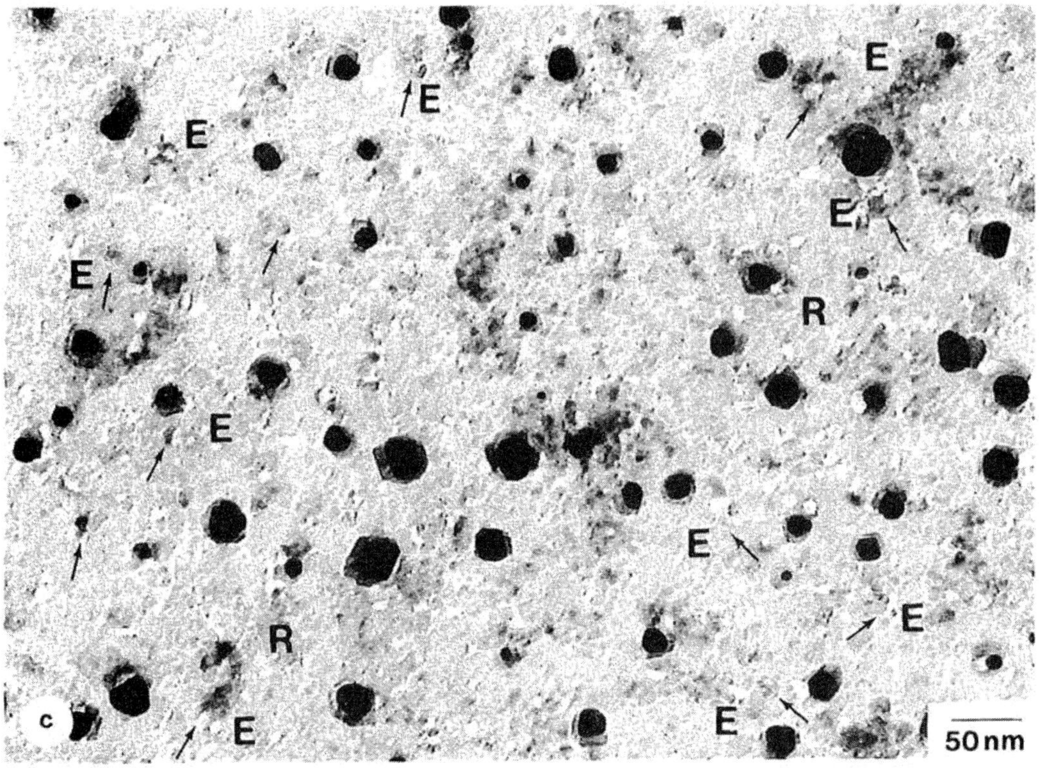

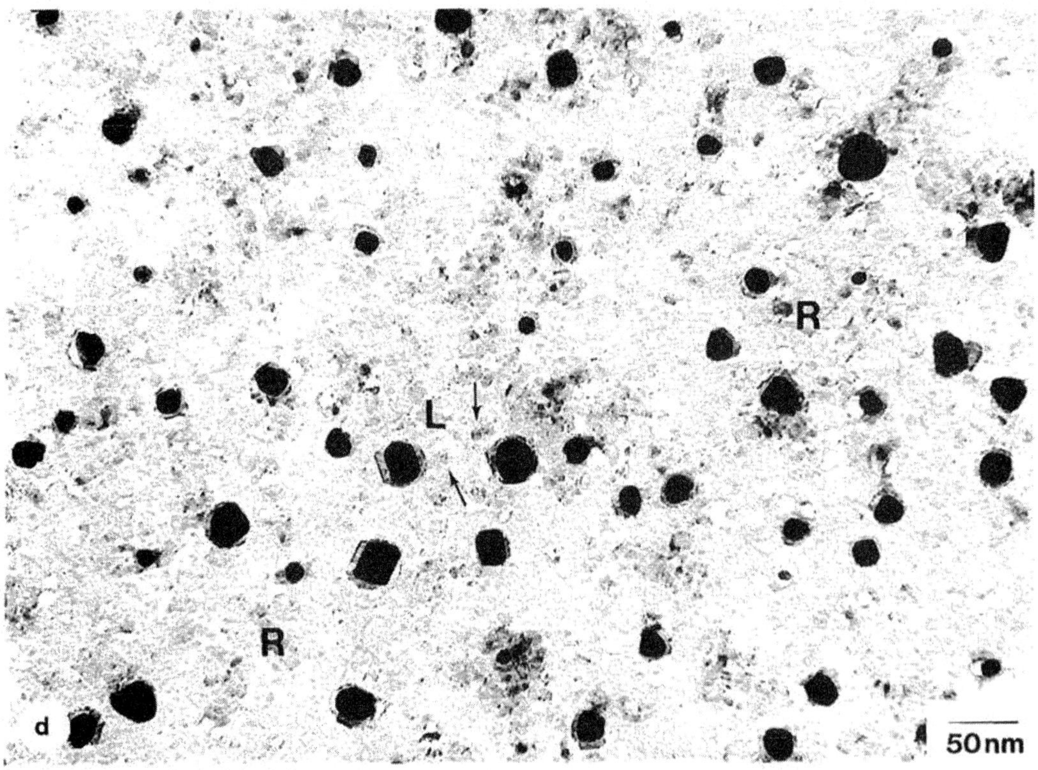

FIG. 12—*Continued.*

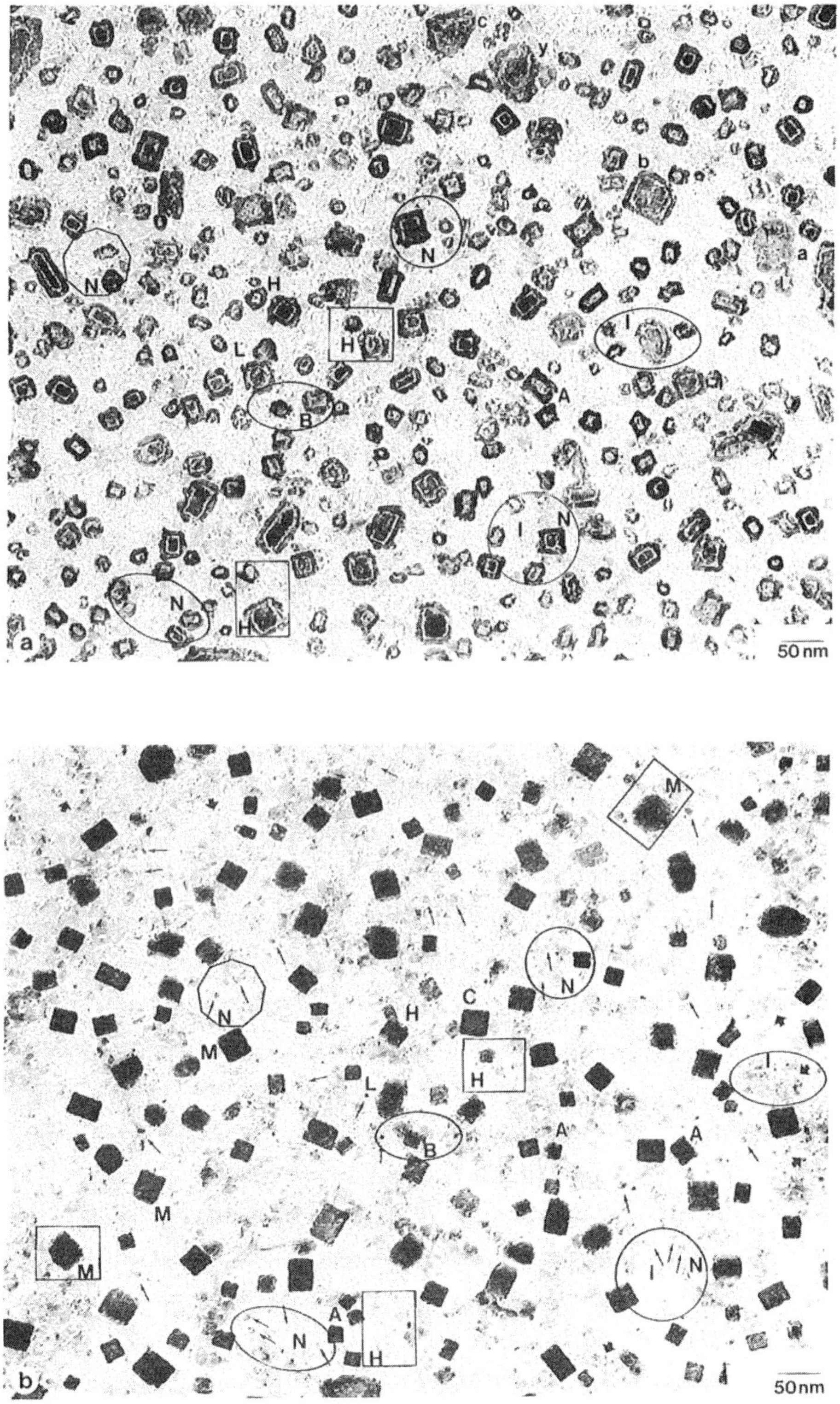

FIG. 13. Sequence of changes on heating a model Fe/Al_2O_3 specimen in purified hydrogen at progressively higher temperatures. Initial loading, 7.5 Å. The same region is shown after (a) 4 h, 400°C; (b) 2 h, 500°C; (c) 4 h, 600°C; (d) 2 h, 700°C.

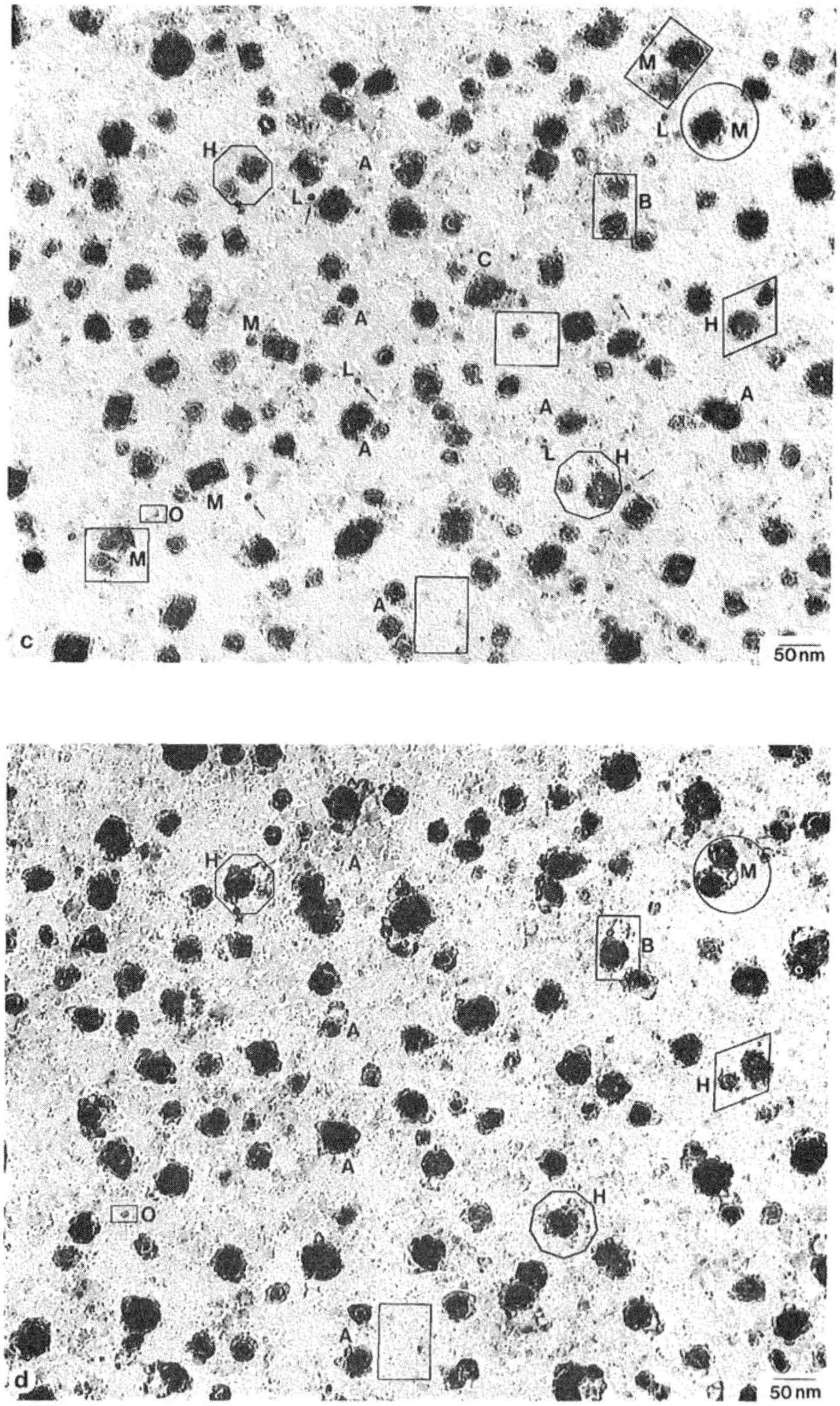

FIG. 13—*Continued.*

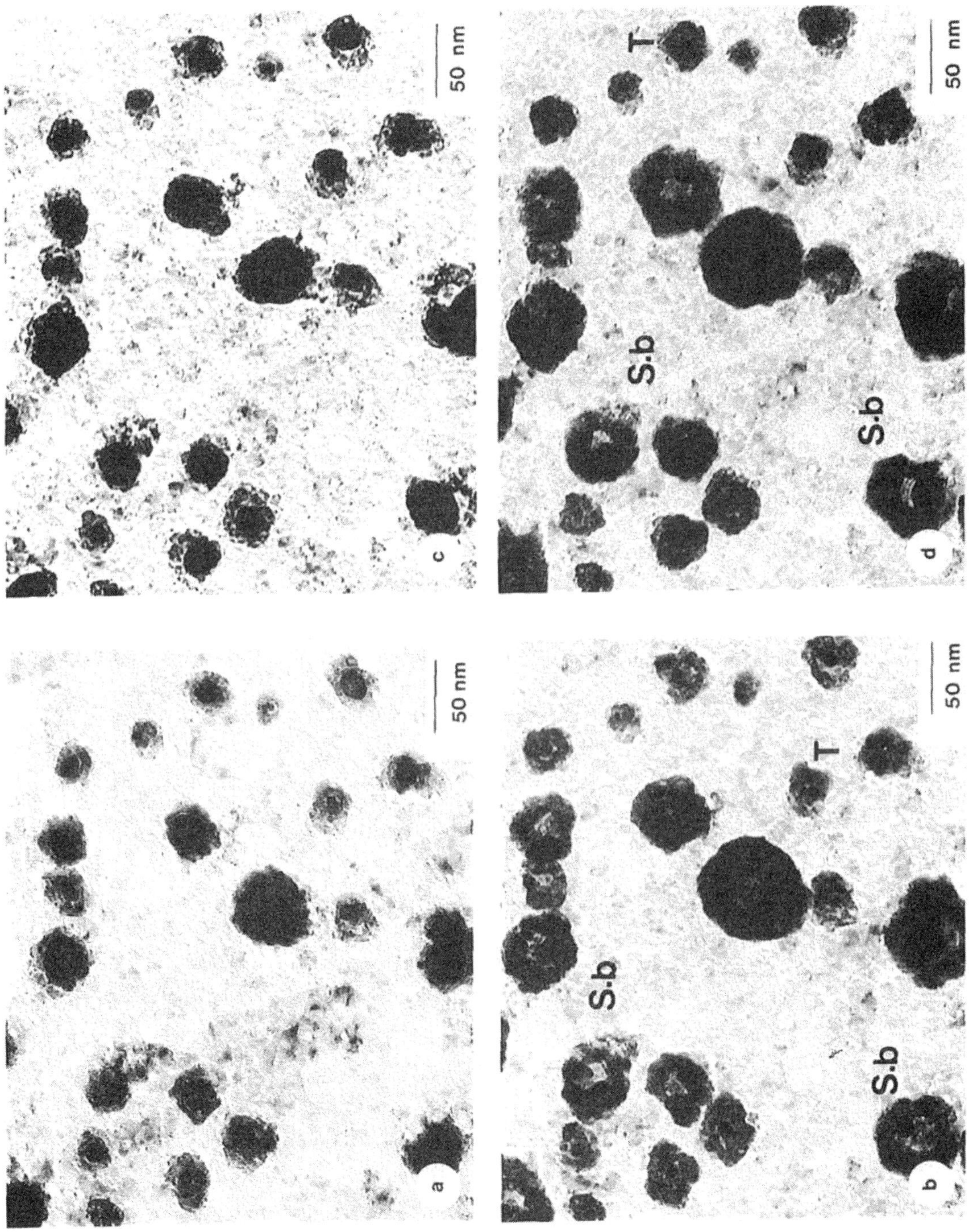

FIG. 14. Sequence of changes on heating a specimen alternately in oxygen and purified hydrogen. (a) 6 h, H_2, 600°C; (b) 1 h, O_2, 500°C; (c) 5 h, H_2, 600°C; (d) 1 h, O_2, 600°C.

TABLE 5

Alternate Heating in Oxygen and Purified Hydrogen

Heating atmosphere	Event observed	Compound identified
Table $5_{6,500}$		
O_2	Particle extension. Particles have a torus shape and a film appears to fill in the cavity. (14b)	*d*-values approaching those of $FeAl_2O_4$ and $Al_2Fe_2O_6$
H_2	Particle contraction. Faint annular ring is observed initially and compact particles are formed on continued heating.	$FeAl_2O_4$ decreases in amount. α-Fe and traces of Fe_3O_4 (γ-Fe_2O_3)
Table $5_{6,600}$		
O_2	Crystallites extend and assume a torus shape, enclosing a large cavity and having a small residual particle in the cavity. (14d)	Fe_3O_4 (γ-Fe_2O_3), $FeAl_2O_4$
H_2	Contraction and compact particle formation. Core-and-ring structure is not observed. (14a, c)	
Table $5_{7.5,500}$		
O_2	(Heated before in hydrogen at 700°C, Fig. 13d). Considerable extension. Large cavity formation in the crystallite (S.b.). A small residual particle is present in the cavity of each crystallite. No change on prolonged heating. (15a)	$FeAl_2O_4$, Fe_3O_4 (γ-Fe_2O_3)
H_2	Splitting of some crystallites into a few interlinked particles, some having a core-and-ring structure (V). Sintering of the sub-units on further heating. (15b)	$FeAl_2O_4$, α-Fe, traces Fe_3O_4 (γ-Fe_2O_3)
Table $5_{7.5,700}$		
H_2	Contraction and formation of more compact particles. Marks of particle peripheries (P). (15c)	$FeAl_2O_4$, Fe_3O_4 (γ-Fe_2O_3)
O_2	Considerable extension. No cavities in particles. A small residual particle is present on the extended parent crystallite (T). The substrate appears covered by a film (R). (15d)	$FeAl_2O_4$, $Al_2Fe_2O_6$
H_2	Contraction and formation of compact particles. No core-and-ring structure. (15e)	$FeAl_2O_4$

present before (L, N). The crystallites might have fragmented into smaller particles (including single atoms) which subsequently migrated on the substrate before colliding and coalescing with other particles. Of course, the migrating crystallites might have simultaneously lost some atoms too, thus decreasing further in size.

B.2. Alternate heating in oxygen and hydrogen. The extension following oxidation is more pronounced when purified hydrogen is used in the reduction step instead of the as-supplied hydrogen (Sect. A.2). In addition, a few differences in the behavior of the two cases have been observed, as listed in Tables 3 and 5. The events observed in Table $5_{7.5,700}$ are of particular interest. After 1 h of oxidation at 500°C, each crystallite extended considerably acquiring a torus shape (Fig. 15a). However, a small residual particle remained randomly located within the cavity. The leading edge of the torus has an undulating form and the torus appears to be composed of interlinked

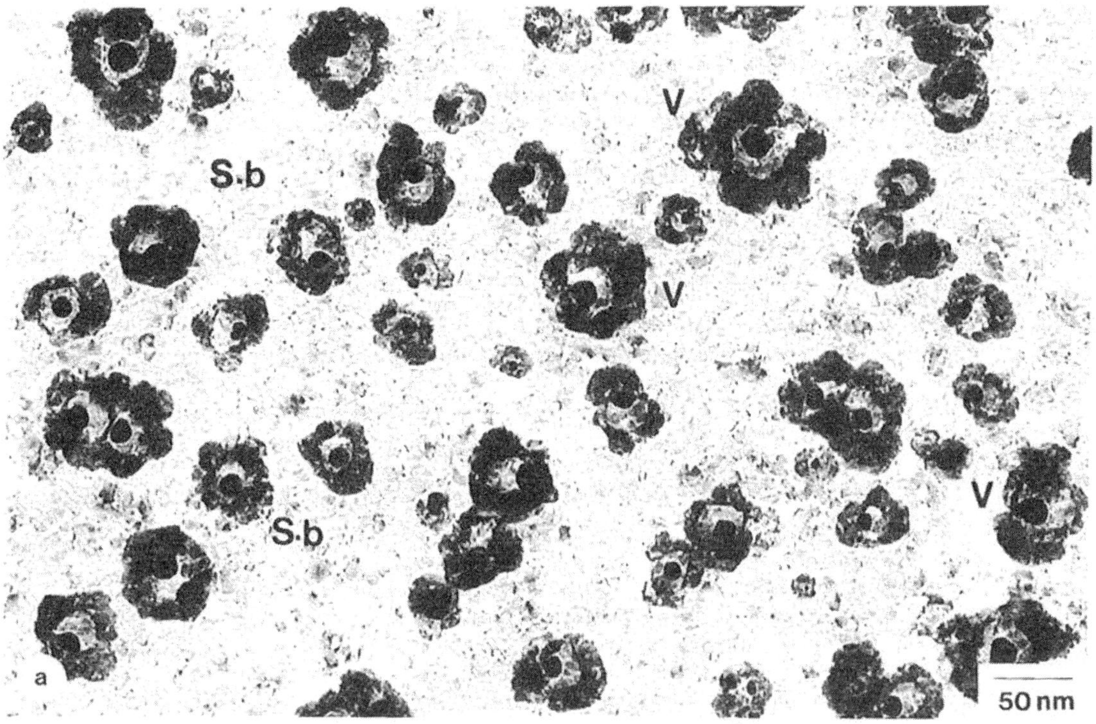

FIG. 15. Sequence of changes on heating the specimen of Fig. 13d alternately in oxygen and hydrogen. (a) 1 h, O_2, 500°C; (b) 30 min, H_2, 500°C (reduction for a longer duration followed by oxidation at the same temperature was carried out between (b) and (c)); 2 h, H_2, 700°C; (d) 12 h, O_2, 700°C (there is very little difference between 1 and 12 h oxidation); (e) 1 h, H_2, 700°C.

subunits. Heating this specimen subsequently in hydrogen at 500°C for only $\frac{1}{2}$ h caused splitting, especially of the larger crystallites, into a few interconnected subunits, some of which have a core-and-ring structure (Fig. 15b). Subsequently, the subunits sintered to form compact particles (Fig. 15c). Heating in oxygen at 700°C caused pronounced wetting and extension of the crystallites. However, unlike the previous cases, cavities were not formed (Fig. 15d). Instead, each crystallite extended as a uniformly thick film with a residual particle on top. The substrate grain boundaries also appear less distinct, probably because of a film covering the substrate, as observed also in Section A.2 on oxidation at 700°C. Subsequent heating in hydrogen at 700°C caused each particle to contract to form a single compact particle, instead of splitting into a few as observed at 500°C.

C. Effect of Moisture and Residual Oxygen Present in the as-Supplied Hydrogen

The effect of moisture on the crystallites and their composition was investigated by first purifying the as-supplied hydrogen and then bubbling it through distilled water. The effect of moisture appears to be similar to that of pure flowing oxygen. At 500°C, the crystallites extended and formed torus-shape particles. Solid solutions leading to $FeAl_2O_4$ were detected in the diffraction pattern. The effect of residual oxygen in the absence of moisture was examined by passing the as-supplied hydrogen through a desiccant. In contrast to the residual oxygen, which causes shape changes and some extension, the moisture causes much greater extension and easier aluminate formation. In other words, moisture considerably en-

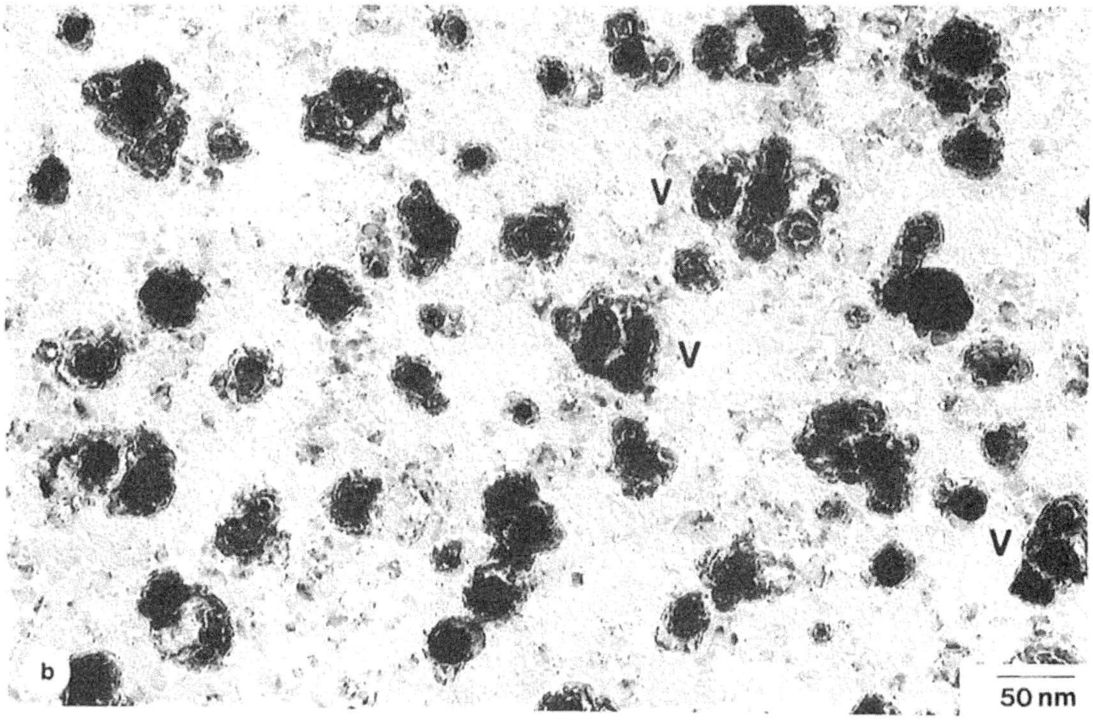

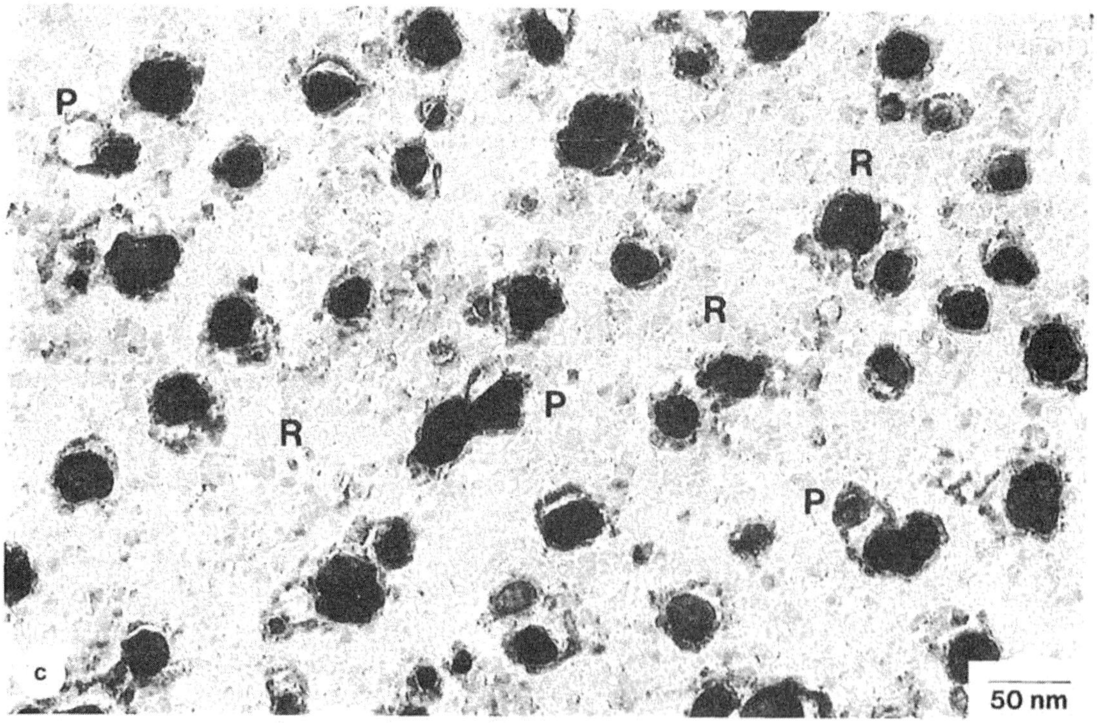

FIG. 15—*Continued.*

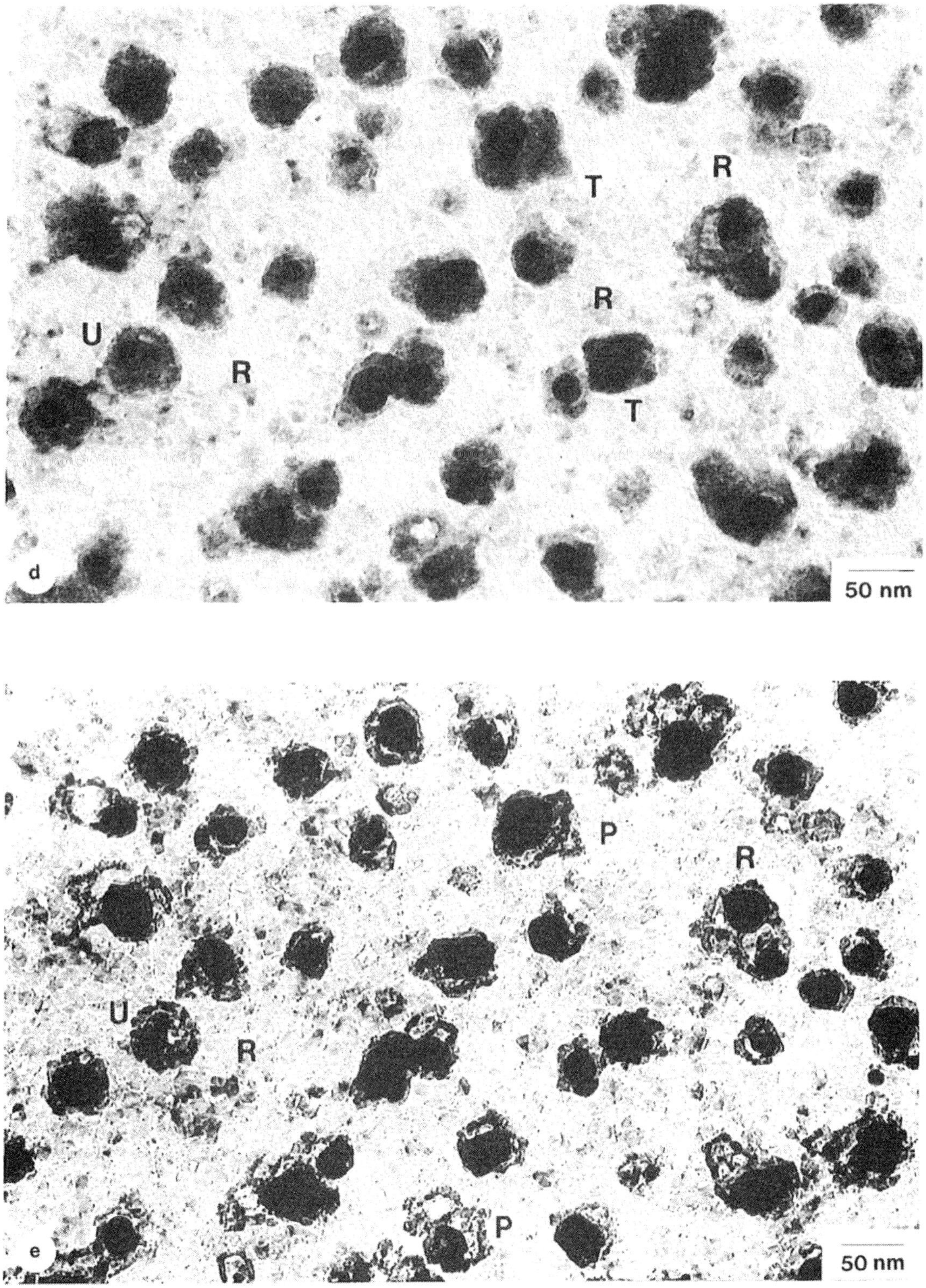

FIG. 15—*Continued.*

hances the effects caused by the residual oxygen.

DISCUSSION

The results presented in the previous section bring evidence for a variety of phenomena such as sintering, wetting and extension, shape changes, splitting, etc., which occur during heating of iron-on-alumina model catalysts. In what follows, some of these phenomena are discussed in more detail.

Sintering Behavior

The process of sintering can occur by several mechanisms (*17*), such as migration of crystallites and their coalescence (*23*), emission of single atoms by the small crystallites and their capture by the large ones (*24–27*), and a combination of the two (*27*). Two possibilities have been suggested for sintering by single-atom emission and capture. In one of them, a large number of crystallites are involved and the small crystallites lose atoms to a surface phase of single atoms dispersed over the substrate, while the large ones capture atoms from this phase. This process, known as Ostwald ripening, occurs when the substrate surface phase of single atoms is supersaturated with respect to the large crystallites and undersaturated with respect to the small ones.

In the other case, called direct ripening (*28*), atoms released by a small particle move directly to a neighboring large crystallite, even though the surface phase of single atoms is, on the average, undersaturated with respect to all the crystallites present. In contrast to Ostwald ripening, which is global, the latter is local and can be considered as a fluctuation from the average behavior, determined by the local surface pressures generated by the particles involved.

As seen from Table 1, migration and coalescence of crystallites (Figs. 3a–c, 7b, c, 13a–d), disappearance of a large number of small particles (Ostwald ripening, Figs. 4a–d, 12a–c) and decrease and disappearance of a few particles near larger particles (direct ripening, Figs. 3a, b, 4′a–c) could all be detected in the present experiments. Interestingly, large crystallites appear to have migrated over large distances. For example, crystallites greater than 90 Å in size have migrated at 700°C (Figs. 6a–d). Also, crystallites of about 480 Å migrated over distances of about 600 Å at 800°C (Figs. 7b, c). Even though the number of particles whose migration could be detected is small, the few instances of migration of large particles over large distances, the redistribution of particles sometimes observed (Figs. 6d–f, 7b, c), the appearance of dumbbell-shape particles (Fig. 3c), and the appearance of small particles in regions unoccupied before (Figs. 6e, f; 12a–d; 13a–c) indicate that migration, more likely of the smaller crystallites, plays a role in the sintering phenomenon. Similarly, a large number of small as well as large crystallites disappear at temperatures as low as 400 or 500°C. In addition to the ripening mechanism, the disappearance of these small particles may be caused by their considerable extension to an undetectably thin film which subsequently coalesces with the nearby particles, or by their migration following fragmentation. In fact, in some cases, a surface film (instead of a surface phase of single atoms) appears to coexist with the three-dimensional crystallites. As pointed out in a previous section (Sect. B.1), the disappearance of the small particles from some regions and the appearance of other, new particles in regions far removed from the previous indicate that mechanisms other than just migration and coalescence (which is expected to occur over only shorter distances), or ripening (since contrary to what is predicted by ripening, large particles have vanished while smaller ones remained (H in Figs. 3b, c, 15a, b)) probably occur. Other mechanisms, such as fragmentation followed by migration of the smaller fragments (N in Figs. 6d–f), emission of atoms from the migrating crystallites, considerable extension to an undetectable film which splits and contracts to form new particles, etc., might have also occurred. The last process occurs, either because of alternate oxidation and reduction in the as-supplied hydrogen, and/or because of the heterogeneities of the substrate which interacts strongly with the particles in some regions and weakly in other regions.

Wetting Behavior

It is important to recognize that the wetting characteristics of the crystallites on the support affect the sintering and redispersion behavior, the thermostability of the size distribution, and the shape of the crystallites. The role of wetting in the redispersion of supported metal crystallites and in the changes of their shape has been emphasized in Refs. (*19, 20, 23, 29*).

Wetting angle. The ability of a crystallite to wet a support and its extent are determined by the equilibrium of the three interfacial tensions, substrate–gas (σ_{sg}), crystallite–gas (σ_{cg}), and crystallite–substrate (σ_{cs}), as expressed by Young's equation

$$\sigma_{sg} - \sigma_{cs} = \sigma_{cg} \cos \theta, \qquad (1)$$

where θ is the equilibrium wetting or contact angle. When, under some conditions such as oxidizing,

$$\sigma_{sg} - \sigma_{cs} > \sigma_{cg} \qquad (1a)$$

is satisfied, no wetting angle can exist, since $\cos \theta > 1$, and the crystallite spreads over the substrate. On the other hand, when

$$\sigma_{sg} - \sigma_{cs} < \sigma_{cg} \qquad (1b)$$

is satisfied, the crystallite forms an equilibrium contact angle, θ, with the substrate, whose value lies in the range $0° < \theta < 180°$, depending upon both the difference ($\sigma_{sg} - \sigma_{cs}$) as well as σ_{cg}. Under vacuum, in inert, or reducing atmospheres, the metals which are commonly used as supported catalysts have, in general, a high surface tension σ_{cg}, as well as a high interfacial tension, σ_{cs}, with silica and alumina supports (*30, 31*). These lead to contact angles greater than 90° and the crystallites do not wet well the support. However, in an oxidizing environment both σ_{cg} and σ_{cs} are smaller, since these values are smaller for the oxide compared to those of the corresponding metal, and the contact angle can decrease. Accordingly, the crystallite will extend and wet the support better.

Interaction energy and wetting. When two phases are brought into contact to form an interface, and there are no molecular interactions between them, the interfacial tension between the two should be expressed by the sum

$$\sigma_{cs} = \sigma_c + \sigma_s. \qquad (2)$$

However, there are always attractive molecular interactions between two phases brought into contact and because of this, there is a corresponding decrease in the interfacial tension. These interactions are strong when they are of a chemical nature, for instance, the formation of compounds, or weak when they are of a physical nature, such as dispersion or polar interactions. The interfacial tension between crystallite and substrate is therefore given by

$$\sigma_{cs} = \sigma_c + \sigma_s - (U_{int} - U_{str}) = \sigma_c + \sigma_s - U_{cs}, \qquad (3)$$

where, U_{int} is the interaction energy per unit area of crystallite–substrate interface between the atoms or molecules of the crystallite and those of the substrate and U_{str} is the strain energy per unit area arising as a result of the mismatch of the two lattices.

When a chemical interaction takes place at the crystallite–substrate interface (leading to the formation of a compound), as in the case of iron on alumina in an oxidizing environment, U_{cs} will become large, thereby appreciably decreasing σ_{cs}. In fact, if the chemical bonding at the interface is very strong, U_{cs} will become so large that σ_{cs} can decrease to zero or (under non-equilibrium conditions) even below zero. Such a large decrease in the dynamic interfacial tension, σ_{cs}, and also a concomitant decrease in the value of σ_c (in an oxidizing environment the metal forms an oxide, and σ_c of the oxide is smaller than σ_c of the metal) lead to a rapid spreading of the crystallite over the substrate, since the driving force for spreading ($\sigma_{sg} - \sigma_{cs} - \sigma_{cg} \cos \theta$) is considerably increased.[3] Initially, the reaction occurs only at the interface and then it proceeds in the bulk of either the substrate or the crystallite, more likely the former. Bulk reaction occurs because it decreases the free energy of the system and it continues until the rate of dissolution of the oxide molecules into the substrate is considerably limited by the diffusional resistance. Immediately following the reaction at the interface, the crystallite is probably present in a highly extended and unstable configuration because of the large driving force for spreading. Any thinning produced by perturbations (such as diffusion into the substrate, thermal fluctuations, etc.) will tend to grow, leading finally to the torus which, under the conditions of the experiment, might be the thermodynamically stable configuration. The rapid extension following the initial surface reaction is soon arrested since, subsequently, U_{cs} is relatively reduced. Indeed, after some of the oxide has already dissolved into the substrate, the subsequent surface interaction energy, U_{cs}, between the aluminate and the oxide is no longer as large as that between alumina and oxide, since, unlike the latter, the former does not lead to a chemical compound at the interface. Consequently, the driving force for spreading $\{\sigma_{sg} - (\sigma_s + \sigma_c - U_{cs}) - \sigma_{cg} \cos \theta\}$ becomes smaller. To summarize, σ_{cs} has a large decrease initially, even possibly becoming negative, and then increases. Correspondingly, the initial large driving force for spreading also decreases subsequently. Therefore, the considerably extended crystallite formed by the initial rapid extension subsequently reaches a new equilibrium wetting angle with the new layer of substrate (reaction product) at the (crystallite–substrate) interface. The time evolution of U_{cs} and σ_{cs} are shown schematically in Fig. 16.

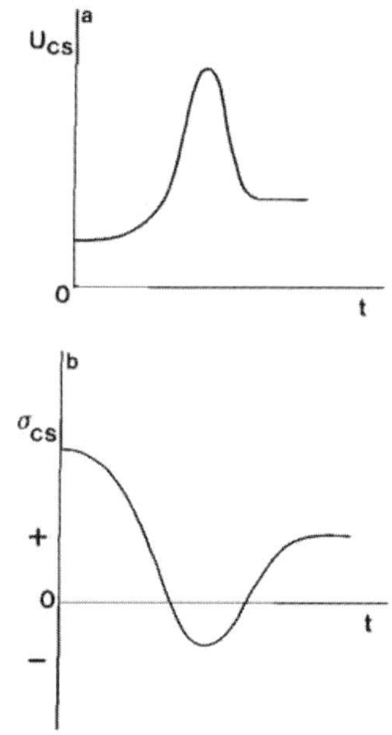

FIG. 16. Schematic of the variation with time of (a) the interaction energy, U_{cs}, and (b) the crystallite–substrate interfacial tension, σ_{cs}, when there is a chemical reaction between the crystallite and the substrate.

If a small cavity had initially nucleated in the extended crystallite because of the rapid extension and the dissolution of a part into the substrate, the subsequent tendency to form a larger wetting angle, for reasons just described above, would make both the inner and the outer leading edges of the cavity-containing crystallite to withdraw. This causes the expansion of the cavity and leads to the observed torus shape. It should be noted, however, that depending upon the substrate material (aluminate or alumina) ahead of each of the two leading edges being the same or different, the final equilibrium angles at the outer and inner edges of the torus may be the same or different. In addition, as a result of the rapid extension, the crystallite may undergo fragmentation and the torus may appear to be composed of interconnected smaller units (Figs. 9d, 9f). On the other hand, if the interactions at the leading edge with the substrate are very strong so as to bind the crystallite to the substrate and if for some reasons the initial cavity does not form in

[3] Here θ is the instantaneous nonequilibrium wetting angle.

the extended crystallite, then the subsequent tendency to contract following reduction would generate opposing forces, since the crystallite tends to contract while its periphery is constrained from following through. This generates the annular gap and gives rise to the core-and-ring structure (*32*). This tendency is enhanced by temperatures which are sufficiently low so as not to break the binding at the crystallite periphery. If the temperature is high, however, the binding at the periphery can be broken and the particle will contract as a single unit to form a compact structure. Compact particles and marks of their previous particle peripheries on the substrate were indeed observed at the high temperatures of 700 and 800°C (Figs. 6d; 7c; 11a, b) (and also at 500°C but at higher loading or when pure hydrogen was used for reduction, Figs. 9c, 15c). From the above discussion it results that depending upon the kinetics of the different processes (such as dissolution, reaction, wetting etc.) either a torus shape or a core-and-ring structure may form initially. Subsequent alternation between these shapes, observed with Fe/Al_2O_3 heated in as-supplied hydrogen, is a result of alternate oxidation and relative reduction and is discussed in Refs. (*22, 32*). The events observed in the present experiments can be explained on the basis of the above strong chemical interaction and wetting considerations as follows.

Iron does not wet alumina in reducing or inert atmospheres when there is no reaction or mass transfer at the interface, its wetting angle near the melting point being 141° (*31, 33*). However, in either partially or fully oxidizing environments, such as hydrogen contaminated with oxygen and/or moisture (even only in traces) or flowing oxygen, the present results indicate that the crystallites are extended. Electron diffraction indicates that in both the oxidizing environments, in addition to some oxide (γ-Fe_2O_3 (Fe_3O_4)), an interaction product, probably a solid solution of iron oxide in alumina and/or vice versa, is formed. The reactivity of iron, in general, with oxygen is very high. In the case of supported iron (iron supported on alumina), oxygen, even present only in traces, is preferentially adsorbed at the interface (*34*), decreasing both σ_{cg} and σ_{cs}. This decreases θ and the crystallite extends on the alumina surface. An additional decrease in σ_{cs} occurs as a result of the interaction of the iron ions with the alumina support. Alumina, especially the γ form, possesses a defect structure and a dehydroxylated surface (*35*) and is therefore very reactive. At low loadings, the iron ions move into the vacant sites of γ-Al_2O_3. The solubility of Fe^{3+} ions in γ-Al_2O_3 can be appreciable because of the chemical similarity of the cations and also the availability of defects in γ-Al_2O_3. As a result of this metal oxide reaction with alumina, a strong chemical bonding results between the crystallite and the support, and U_{cs} is tremendously increased. Such strong interactions lead to the observed extended forms and the torus or core-and-ring shapes as described in the previous paragraphs. However, although they prevent subsequent sintering, such strong interactions with the support cause difficulties in reducing the iron ions to the zero-valent metal. The reaction products, either the solid solutions of iron oxide with alumina or the stoichiometric aluminates, $FeAl_2O_4$ and $Al_2Fe_2O_6$, are very difficult to be reduced (as seen in Tables 2–5) compared to the pure oxide of iron. In fact, iron oxide itself requires only 300°C for reduction to the metallic state (*12*). The literature (*6, 16*) also indicates that ferric ions supported on Al_2O_3 or silica in low concentrations (about 0.1 wt%) can be only partially reduced by hydrogen to the ferrous state, even at 700°C. Direct evidence for the formation, at low loadings, of the difficult-to-reduce aluminates or solid solutions in the presence of even traces of oxygen and/or moisture has been indicated by electron diffraction in the present experiments (Tables 2–4). Their presence could explain the difficulty encountered in the literature to reduce Fe^{3+} ions below the Fe^{2+} state at temperatures as high as 700°C, if the reducing hydrogen contained even traces of oxygen or moisture. Other factors in the catalyst preparation procedure, such as the precursor used, pH, etc., may facilitate the formation of such compounds. On the other hand, if the loading is large, initially large particles form, which have a relatively smaller fraction of their atoms in contact with the substrate at the interface. The interaction compounds (aluminates or solid solutions) formed between the two will be in relatively small amounts compared to the oxide. Therefore they are not easily detected by electron diffraction and they do not pose difficulties in reducing the oxide to the metal, even at the relatively low temperature of 500°C. Hence, in such cases the oxide particles are reduced easily (α-Fe is detected by electron diffraction) and the reduced crystallites will have relatively weak interactions with the substrate. As a result, considerable sintering occurs and dense, compact particles are formed (Fig. 4d). In fact, when further purified hydrogen is used during the reduction step, formation of compact particles and reduction to α-Fe occur even for relatively low loadings even at the low temperatures of 500°C (Table 13). This occurs because of decreased interaction between crystallite and support in the absence of oxygen and/or moisture.

A thick film extends from around the particles on heating in oxygen at 700°C and only aluminate is detected in the diffraction pattern. A thick film of aluminate may cover the alumina substrate if $\sigma_{\text{aluminate-gas}} + \sigma_{\text{aluminate-alumina}} < \sigma_{\text{alumina-gas}}$. Similarly, a thick film of aluminate may cover the oxide particles if $\sigma_{\text{aluminate-gas}} + \sigma_{\text{aluminate-iron oxide}} < \sigma_{\text{iron oxide-gas}}$.

As seen in Figs. 14 and 15 the degree of extension of the crystallites, when they are heated in flowing oxygen, is higher at 700°C than at 600 and 500°C. The fact that the temperature coefficients of surface and interfacial tensions are negative for most materials (σ_{cg}, σ_{cs}, and σ_{sg} decrease with increasing temperatures) and the possibility of an enhanced reaction between the gas, the crystallite, and the substrate at higher temperatures may explain this observation. (Of course, this implies sufficient decreases with temperature of σ_{cs} and/or σ_{cg} compared to that of σ_{sg}.) In fact, it is possible that heating the crystallites in oxygen at temperatures higher than 700°C will cause even more extension, and eventually complete spreading. This can happen if the inequality (1a) is satisfied but will be observed only if the time needed for complete spreading is shorter than the heating time.

SMSI

The SMSI (strong metal–support interactions) referred to frequently in the recent literature (*36*) is in reality (as already noted by Ruckenstein and Pulvermacher (*23*)) a reflection of the quantity U_{cs}, which in turn is related via Eq. (3) to the interfacial tension σ_{cs}. Large values of U_{cs}, hence strong interactions (*23*) lead to small values of σ_{cs} (even to negative values of σ_{cs} under nonequilibrium conditions). As a result, a variety of surface phenomena associated with wetting and spreading, of the kind discussed in Ref. (*23*) and in this paper, occur.

Splitting Behavior

Splitting of crystallites has been observed under different conditions and in different forms (Tables 1–3, 5). When heated in as-supplied hydrogen at 400°C, some crystallites were observed to split into two and to merge again subsequently (M in Figs. 2g–i). The crystallites had a torus shape both before as well as after the splitting. Such splitting may be a result of the extension and the instability of the extended toroidal shape to local variations in curvature along the periphery. Following heating in oxygen, in the majority of cases, the crystallites appear to be composed of a number of smaller subunits held together (Figs. 9d, 9f). Such a breakup is more likely a result of the spreading generated stresses and rupture (*20, 28*). Palladium crystallites on alumina were also observed to become porous and develop cavities when heated in oxygen (*18–20*). It was shown that if the crystallites were small or if the temperature was high, the crystallites extended irregularly, exhibiting a kind of tearing phenomenon.

The extension of the crystallites following oxidation generates stresses in the crystallite and leads to the growth of existing cracks if these stresses exceed the critical stress for crack propagation given by Griffith's expression

$$\tau_c = \left(\frac{2E\gamma}{\pi r}\right)^{1/2},$$

where γ is the surface tension, r is a characteristic length of the crack, and E is Young's modulus of elasticity. Further oxidation, especially at the tip of the cracks, facilitates the crack propagation and the crystallite is ruptured and appears to be made up of disjointed smaller units. However, because iron oxide has a tendency to wet alumina, the individual units of the crystallites may be very near to one another. Splitting is, therefore, not observed clearly in oxidizing atmospheres, but the crystallites appear porous. Subsequent heating in hydrogen tends to contract the particle and the splitting becomes evident as each individual unit tends to contract to a higher equilibrium wetting angle. Of course, on continued heating in hydrogen the individual crystallites coalesce and sinter together.

Redisperion of the crystallites may also occur as a result of rupture and contraction of an extended surface film. As seen in Figs. 12c and 15d a kind of thick film either of iron oxide or of aluminate may spread on alumina from the periphery of the crystallites. In an oxygen atmosphere there may even be a multilayer surface film on the surface of alumina, coexisting with the three-dimensional crystallites. The film may not be continuous, but ruptured. When heated subsequently in hydrogen, the islands will contract to a higher equilibrium angle. During this process of contraction a part of the film may coalesce with the adjacent particles and a part may form new, smaller crystallites, effectively causing redispersion.

The fragmentation of the crystallites to a large number of smaller particles observed after repeated alternate heating in hydrogen and oxygen for a total of about 200 h (Figs. 9f and g) is again a result of the spreading generated stresses and rupture. If a chemical interaction occurs between the crystallites and the substrate, repeated alternations may modify the support and affect the wetting behavior of the crystallites as a result of the repeated diffusion of the iron oxide molecules in and out of the substrate. In addition, the fragmentation may be influenced by the mechanical fatigue that sets in the crystallites due to repeated extensions and contractions.

CONCLUSION

Several phenomena have been observed to occur during heating model iron/alumina catalysts in hydrogen containing trace amounts of oxygen and/or moisture, in additionally purified hydrogen, and in oxygen. Particle growth by migration and coalescence, by emission and capture of atoms involving only a few neighboring crystallites (direct ripening) as well as a large number of the crystallites (Ostwald ripening) have been observed. These are, however, only a part of the complex surface phenomena that occur. Coalescence of nearby particles, extension and disappearance of particles, appearance of new particles in regions unoccupied previously, and splitting of particles also occur. In addition, ppm quantities of oxygen and/or moisture in hydrogen are shown to affect the reducibility of the iron ions and also to cause stability with respect to growth of the crystallites. Electron diffraction indicated considerable chemical interaction of the crystallites with the support and indicated the formation of solid solutions of iron oxide with alumina or of the stoichiometric aluminates $FeO \cdot Al_2O_3$ and $Fe_2O_3 \cdot Al_2O_3$. As a result of these chemical interactions, the crystallites have been observed to undergo changes in shape alternately from a torus to a core-and-ring structure when heated, especially at low temperatures of 400 or 500°C, in ultrahigh pure hydrogen containing trace quantities of oxygen and/or moisture. Such alternations were not observed when hydrogen was further purified to minimize oxygen and moisture contents. The additionally purified hydrogen caused better reducibility to the zero-valent metal. On heating in as-supplied hydrogen, at the low loadings of 6 or 7.5 Å, very little Fe^0 forms even at 700°C, whereas almost complete reduction to metallic iron is achieved even at 500°C when the loading is high, about 12.5 Å. In contrast, with additionally purified hydrogen, metallic iron could be detected even at 400°C for a loading of only 7.5 Å. On heating in oxygen, the crystallites were observed to extend considerably, the extension being enhanced with increasing temperatures. Toroidal shapes with a small remnant particle in the cavity were observed on heating in oxygen at 500 and 600°C. The extent of wetting increased at 700°C. In addition, it was inferred from the electron micrographs that on heating in oxygen at 700°C, a film whose thickness should be greater than monomolecular, coexists with three-dimensional crystallites. After prolonged alternate heating in hydrogen and oxygen, the crystallites were observed to split into a number of smaller particles, probably as a result of mechanical fatigue. The results are discussed in terms of the interactions between metal, gas, and support and the consequent wetting characteristics of the metal–support system. It is suggested that such strong chemical interactions may lead to a considerable decrease in the interfacial tension between crystallite and substrate and, therefore, to a tendency for the crystallites to spread out. It is stressed that the strong metal–support interaction, the sintering and redispersion phenomena, and the changes in the shape of the crystallites are all interlinked parts of a complex surface phenomenon that cannot be easily categorized as belonging to a simple "either or" phenomenon.

REFERENCES

1. Anderson, J. R., "Structure of Metallic Catalysts." Academic Press, New York, 1975.
2. Satterfield, C. N., "Heterogeneous Catalysis in Practice." McGraw–Hill, New York, 1980.
3. Amelse, J. A. Butt, J. B., and Schwartz, L. H., *J. Phys. Chem.* **82**(5), 558 (1978).
4. Guczi, L., *Catal. Rev.-Sci. Eng.* **23**(3), 329, (1981).
5. Yoshioka, T., Koezuka, J., and Ikoma, H., *J. Catal.* **16**, 264 (1970).
6. Garten, R. L., and Ollis, D. F., *J. Catal.* **35**, 232 (1974).
7. Brenner, A., and Hucul, D. A., *Inorg. Chem.* **18**(10), 2836 (1979).
8. Raupp, G. B., and Delgass, W. N., *J. Catal.* **58**, 337 (1979).
9. Raupp, G. B., and Delgass, W. N., *J. Catal.* **58**, 348 (1979).
10. Vannice, M. A., *J. Catal.* **37**, 462 (1975).
11. Perrichon, V., Charcosset, H., Barrault, J., and Forquy, C., *Appl. Catal.* **7**, 21 (1983).
12. Topsøe, H., Dumesic, J. A., and Morup, S., *in* "Applications of Mössbauer Spectroscopy" (R. L. Cohen, Ed.), Vol. 2. Academic Press, New York, 1980.
13. Raupp, G. B., and Delgass, W. N., *J. Catal.* **58**, 361 (1979).
14. Amelse, J. A., Grynkevich, G., Butt, J. B., and Schwartz, L. H., *J. Phys. Chem.* **85**, 2484 (1981).
15. Hobson, M. C., Jr., and Gager, H. M., *J. Catal.* **16**, 254 (1970).
16. Garten, R. L., *J. Catal.* **43**, 18 (1976).
17. Ruckenstein, E., and Dadyburjor, D. B., *Reviews in Chem. Eng.* **1**, 251 (1983).
18. Chen, J. J., and Ruckenstein, E., *J. Catal.* **69**, 254 (1981).
19. Ruckenstein, E., and Chen, J. J., *J. Colloid Interface Sci.* **86**(1), 1 (1982).
20. Chen, J. J., and Ruckenstein, E., *J. Phys. Chem.* **85**, 1696 (1981).
21. Chu, Y. F., and Ruckenstein, E., *J. Catal.* **41**, 385 (1976).
22. Sushumna, I., and Ruckenstein, E., *J. Catal.* **90**, 241 (1984).
23. Ruckenstein, E., and Pulvermacher, B., *J. Catal.* **29**, 224 (1973).
24. Chakraverty, B. K., *J. Phys. Chem. Solids* **28**, 2401 (1967).
25. Wynblatt, P., and Gjostein, N. A., *Progr. Solid State Chem.* **9**, 21 (1975).
26. Flynn, P. C., and Wanke, S. E., *J. Catal.* **34**, 390 (1974).
27. Ruckenstein, E., and Dadyburjor, D. B., *J. Catal.* **33**, 233 (1977).
28. Ruckenstein, E., and Dadyburjor, D. B., *Thin Solid Films* **55**, 89 (1978).
29. Ruckenstein, E., and Chu, Y. F., *J. Catal.* **59**, 109 (1979).

30. Overbury, S. H., Bertrand, P. A., and Somorjai, G. A., *Chem. Rev.* **75**(5), 547 (1975).
31. Beruto, D., Barco, L., and Passerone, A., *in* "Oxides and Oxide Films" (Ashok K. Vijh, Ed.), Vol. 6. Dekker, New York, 1981.
32. Ruckenstein, E., and Sushumna, I., "Proceedings, 9th Iberoamerican Symposium on Catalysis," Vol. II, p. 1074, July (1984).
33. Naidich, Ju. V., *in* "Progress in Surface and Membrane Science" (D. A. Cadenhed and J. F. Danielli, Eds.), Vol. 14, p. 354. Academic Press, New York, 1981.
34. Halden, F. A., and Kingery, W. D., *J. Amer. Ceram. Soc.* **59**, 557 (1955).
35. Maciver, D. S., Tobin, H. H., and Barth, R. T., *J. Catal.* **2**, 485 (1963); **3**, 502 (1964).
36. Imelik, B., *et al.* (Eds.), "Studies in Surface Science and Catalysis," Vol. 11. Elsevier, Amsterdam, 1982.

8

HETEROGENEOUS CATALYSIS: A THEORETICAL APPROACH

INTRODUCTION

Chapter 8 is devoted to some theoretical approaches to heterogeneous catalysis. In brief, this chapter consists of three parts: the first is concerned with the mechanisms by which the chemisorbed H, O, Cl, or S atoms affect the stability of metal clusters (**8.1**) as well as those by which certain additives, such as Cu, Ag, Cl, and S, affect the catalytic behavior of metal clusters (**8.2**); the second with the role of multiple landing area in determining adsorption and catalytic performance (activity and selectivity) of a heterogeneous catalyst (**8.3–8.4**); the third with the migration and removal of oxygen in a bismuth molybdate lattice during catalytic selective oxidation (**8.5–8.6**).

The importance of metal cluster size and structure as well as of the support is well known for the stability, activity, and selectivity of supported metal catalysts. It is also well known that chemical reagents and/or additives can cause changes in the catalyst surface layer thus affecting its catalytic behavior. For example, small amounts of S adsorbed on the surface of a Pt catalyst render it ineffective for the ring opening of cyclopentanes. The (111) crystal face of nickel changes to the (100) face in the presence of H_2S, C_2H_4 or benzene. Halogens can increase the surface diffusion of silver atoms. A copper-treated Pt/Al_2O_3 catalyst, prereduced in a H_2 atmosphere, results in a significant increase in the activity for CO oxidation and to a more moderate increase in the activity for ethylene oxidation in comparison with a copper-free catalyst. In contrast, the presence of sulfur in the feed decreases the activity of the Pt/Al_2O_3 catalyst both for the oxidation of CO and for the oxidation of ethylene. To elucidate the mechanisms behind these phenomena, the extended Hückel method was used to investigate how H, O, S or Cl atoms chemisorbed on Pt(100) or Pt(111) planes change the interaction energy and bond population between Pt atoms of small clusters (**8.1**). Such calculations were also used to study how the electronic properties of the Pt and Ag clusters change under the action of preadsorbed gases, or by the addition of a second metal, and what are the effects of these changes on the activity and selectivity of the Pt and Ag cluster catalyst (**8.2**).

It has been experimentally proven that small molecules such as CO and H_2 adsorb on a surface in more than one configuration. For example, IR studies show that CO can adsorb on Pd in two forms: one adsorbed form requires one Pd site per molecule, while the other requires two. Different forms of adsorbed hydrogen have also been postulated based on experimental results of temperature-programmed desorption (TPD), tracer studies, and inelastic neutron scattering. For larger, more complicated molecules, it must be the same. Different configurations of a reactant adsorbed on a catalyst surface, that is, adsorbates on different landing areas, might react differently and would form different products. Therefore, the catalytic performance (activity and

Heterogeneous Catalysis: Experimental and Theoretical Studies, First Edition. Edited by Eli Ruckenstein and Haiyou Wang.

selectivity) of a catalyst can be related to the relative amounts of the reactant adsorbed in different configurations, that is, to the landing-area distributions. In general, only a fraction of the surface may be available to a particular landing area, which is termed the availability factor for that landing area. The variation of the landing area distribution with temperature and partial pressure of the gas phase species was examined, and so was the kinetics of metal-catalyzed skeletal isomerization and hydrogenolysis of hydrocarbons on supported metal catalyst on "uniform" surfaces (on which all sites are identical) in the framework of the landing area (**8.3**). A model for the landing areas expected in the skeletal isomerization of hydrocarbons by noble metals, originally introduced for a "uniform" surface (**8.3**), was extended to apply to the general case of a "nonuniform" surface, which consists of two types of crystallites: cubic and icosahedral (**8.4**). Changing the relative amounts of the two types of crystallites changes the average dispersion of the surface and the availability factor for each landing area and ultimately the catalytic activity and selectivity.

Bismuth molybdate (Bi_2MoO_6) is known for its excellent activity and selectivity toward the partial oxidation of hydrocarbons. Oxygen species on the surface and in the bulk of lattices of bismuth molybdate play important roles during partial oxidation. Different arguments have been presented regarding whether the oxygen ion from the lattice near a bismuth site or a molybdate site is responsible for the catalytic oxidation. The method of minimum energy paths was used to estimate the relative ease of the motion of O^{-2} ions into and inside an idealized lattice of bismuth molybdate, which is completely filled except for the single anion vacancy to be filled by the mobile O^{-2} ion (**8.5**). It was shown that the following path is energetically favorable: O^{-2} ion outside the lattice $\rightarrow$ O^{-2} ion affiliated with Mo^{6+} $\rightarrow$ O^{-2} ion in intermediate layer $\rightarrow$ O^{-2} ion affiliated with Bi^{3+}. The pathway described earlier reinforces the view that the bismuth-bound O^{-2} oxidizes the hydrocarbon, the anionic vacancy thus formed is replenished via the intermediate layer by a molybdate-bound O^{-2} moving through the lattice, and the reactant oxygen from the gas phase finally fills the molybdate-bound anionic vacancy. The removal of a bismuth-bound O^{-2} ion out of the lattice of a Bi_2MoO_6 catalyst is an important step in the catalytic selective oxidation–reduction cycle. Nevertheless, a real catalyst, particularly after prereduction, will possess a lattice with appreciable vacancy concentrations or even nonstoichiometries. Furthermore, during the catalytic process, charge imbalances (local nonelectroneutrality) may take place. All these play a significant role in the performance of the catalyst and must be considered in the evaluation of the energy paths. In this regard, the activation energies for the removal of oxygen from bismuth molybdate during prereduction step and during the actual oxidation process were calculated (**8.6**). It was found that the interaction energies between ion and lattice are significantly lower for a nonelectroneutral, vacancy-containing lattice (encountered during reaction) than for an electroneutral lattice containing vacancies (encountered before and after prereduction). As a result, a Bi-bound O^{2-} ion in the "original" lattice, that is, before prereduction, would be able to leave the lattice during the actual reaction, whereas it was prevented from doing so before and after prereduction. In other words, the steady-state loss of O^{2-} ions with a Bi_2MoO_6 lattice containing anion vacancies is energetically favorable.

EFFECTS OF CHEMISORPTION OF ELECTRON ACCEPTOR ELEMENTS ON THE STABILITY OF PLATINUM CLUSTERS

T.D. HALACHEV * and E. RUCKENSTEIN **

Department of Chemical Engineering, State University of New York at Buffalo, Amherst, New York 14260, USA

Received 18 November 1980; accepted for publication 24 February 1981

The extended Hückel method (EHM) is used here to investigate how the bond strengths of Pt(100) and Pt(111) clusters, containing 9 and 10 atoms, respectively, are affected by the chemisorbed H, O, Cl, or S atoms. Three adsorption sites are considered on each cluster. For most adsorption sites Cl and S weaken all the Pt–Pt bonds, while H and O strengthen the bonds between some Pt atoms and weaken the bonds between some other Pt atoms. The enhanced mobility of the Pt atoms, that occurs upon adsorption of the above elements, is proposed as the mechanism behind the first step in corrosive chemisorption and in the redispersion of supported Pt catalysts. The strong destabilizing effect of S and Cl is attributed to the empty d orbitals of these elements. An attempt is made to explain the poisoning by Sulfur on the basis of long range structural and electronic changes that occur in the Pt clusters upon chemisorption of S.

1. Introduction

Theoretical and experimental work carried out in the last decade has shown the importance of cluster size and structure as well as of the support, for the stability, activity and selectivity of supported metal catalysts. It is also well known that chemical reagents can bring about changes in the catalyst surface layer thus affecting its catalytic behavior.

Silica and alumina supports contain sulphur as impurity and this leads to selective or nonselective poisoning of certain reactions [1]. Also, small amounts of S adsorbed on the surface of a Pt catalyst render it ineffective for the ring opening of cyclopentanes [2]. The (111) crystal face of nickel changes to the (100) orientation in the presence of H_2S, C_2H_4 or benzene [3]. Similarly, experiments show [4] that halogens can increase the surface diffusion of silver atoms. If an alumina supported Pt catalyst is heated alternatively in O_2 and H_2 atmospheres, a sufficiently large number of times, redispersion of the metal may occur during heating in oxygen [5].

* On leave from the Institute of Organic Chemistry, Bulgarian Academy of Sciences, Sofia 1113, Bulgaria.

** Address correspondence to E. Ruckenstein.

This paper, which was originally published in **Surface Science**: 108, 292–310 (1981), is republished with permission.

In addition, adsorbed species can affect sintering of metal crystallites.

The mechanisms behind the above phenomena are not well understood. The first step towards their understanding probably rests in the recognition that adsorbed species may reduce the interaction energy and hence decrease the bond population between the Pt atoms of the cluster. This in turn would enhance the mobility of the Pt atoms thus increasing the distance between them and possibly leading to a reconstruction of the surface layer(s).

The scope of the present paper is to investigate how H, O, S or Cl atoms chemisorbed on Pt(100) or Pt(111) planes change the interaction energy and bond population between Pt atoms of small clusters. The mentioned adsorbates have been chosen, since they are often present in typical reaction atmospheres. In addition, the choice enables one to assess how adsorbates with different electronic structures affect the properties of the Pt clusters. For example, despite the fact that sulfur is isoelectronic with oxygen, these two elements affect the catalytic behavior of Pt clusters in different ways. S or Cl have stronger effects on Pt clusters, when compared with H or O atoms, because they have complex electronic structures, especially empty d orbitals, that interact with the orbitals of the Pt clusters.

2. Method of calculation

The Extended Hückel Method (EHM) [6] employed in the present calculations has previously proved successful in predicting adsorption characteristics [7–10]. Because of its simplicity and the established limits of its validity, it is still a valuable tool in theoretical catalytic studies. A complete ab initio quantum mechanical treatment of transition metal clusters with adsorbates is presently unavailable. The CNDO method which has also been used frequently to investigate chemisorption uses empirical parameters which have not been evaluated for the fifth and sixth row transition metals.

The wave function of the system in the Extended Hückel Method is composed of a linear combination of atomic orbitals (LCAO):

$$\psi_i = \sum_j c_{ij}\chi_j \,, \tag{1}$$

where ψ_i is the ith molecular orbital, c_{ij} is the coefficient of the atomic orbital and χ_j is the Slater type atomic orbital (STO).

Minimization of the energy of the system with respect to the coefficients c_{ij} gives the secular matrix equation:

$$|H - ES| = 0, \tag{2}$$

where H is the Hamiltonian matrix, and S and E are the overlap and energy matrices, respectively. The elements H_{ii} are given by:

$$-H_{ii} = \text{VSIP} \quad \text{(valence state ionization potential of AO}_i\text{)}, \tag{3}$$

whereas the off-diagonal elements are calculated by the Wolfsberg–Helmholtz formula:

$$H_{ij} = \tfrac{1}{2}kS_{ij}(H_{ii} + H_{jj}). \tag{4}$$

Here k is a constant (for which the value 1.75 is generally taken) and S_{ij} is the overlap integral between orbitals i and j:

$$S_{ij} = \int \psi_i \psi_j \, \mathrm{d}\tau \, . \tag{5}$$

The energy levels of the system are computed after diagonalization of the matrix equation (2), and the total energy is given by

$$E = \sum_i \epsilon_i n_i \, , \tag{6}$$

where ϵ_i is the energy of the ith molecular orbital (MO) and n_i is its occupation number. The Wolfsberg–Helmholtz formula is applicable to molecules, such as organic molecules, for which the orbitals in the basis set are of similar "diffuseness", large occupation numbers and close energies. Transition metal clusters do not satisfy these conditions because of the existence of unoccupied, high energy s and p basis functions. Therefore, for the latter case, one uses the so-called "weighted" expression:

$$H_{ij} = \tfrac{1}{2}k' S_{ij}(H_{ii} + H_{jj}) \, , \tag{7}$$

where

$$k' = k + \Delta^2 + \Delta^4(1 - k) \, , \tag{8}$$

$$\Delta = (H_{ii} - H_{jj})/(H_{ii} + H_{jj}) \, . \tag{9}$$

This formula and the more diffuse "double zeta" functions for the d orbitals of platinum were employed in the calculations. Mulliken analysis [11] is used to compute the atomic bond population between atom A and atom B:

$$P_{\mathrm{AB}} = \sum_i^{\mathrm{A}} \sum_j^{\mathrm{B}} P_{ij} S_{ij} \, . \tag{11}$$

Here S_{ij} is an element of the overlap matrix and P_{ij} is an element of the bond-density matrix defined by

$$P_{ij} = \sum_\mu n_\mu c_{\mu i} c_{\mu j} \, . \tag{12}$$

The values of the parameters used in the calculations are listed in table 1.

Table 1
Parameters used in the calculations

Atom	r [Å]	H_{ii} (d)	H_{ii} (s)	H_{ii} (p)	μ (s)	μ (p)	μ_1 (d)	μ_2 (d)	c_1	c_2
Pt	0.94	−10.989 [a)]	−8.091 [a)]	−6.3 [a)]	2.554	2.535	6.013	2.696	0.633	0.551
H	0.529		−13.6		1.3					
O	0.73		−32.3	−14.8	2.275	2.275				
Cl	0.99	−9.0	−30.0	−15.0	2.033	2.033	2.033			
S	1.04	−8.0	−20.0	−13.3	1.817	1.817	1.5			

a) The IP values for Pt are taken from ref. [15]. All other IP and exponents are from QCPE N344. Here μ are the exponents in the STO and c_1 and c_2 are the coefficients in the double zeta d orbitals of Pt. The H_{ii} are expressed in eV.

3. Structure of the clusters used in the calculations

The calculations have been carried out for clusters containing 10 and 9 Pt atoms. The 10 atoms are arranged as in the (111) configuration (fig. 1A) and the 9 atoms as in the (100) configuration (fig. 1B). The numbering of the atoms is indicated in the figure. The plane in which atom 1 lies will be referred to as the first layer for both crystallographic structures, and the plane formed by three (fig. 1A) or four (fig. 1B) Pt atoms constitutes the second layer. The value of the nearest neighbor distance between the Pt atoms is taken equal to that in the bulk metal (2.775 Å). Three types of adsorption sites are considered: (a) adsorption of a single atom above the central atom 1 in both structures; (b) adsorption of a single atom below the gap formed by three (fig. 1A) or four (fig. 1B) Pt atoms in the second layer; (c) adsorption of two atoms below two opposite Pt–Pt bonds of the second layer of fig. 1B and of a single chemisorbed atom below a Pt–Pt bond of the second layer of fig. 1A.

The latter adsorption site is especially meaningful for the (100) configuration since adsorption of two oxygen atoms below two opposite bonds of the second layer simulates the PtO structure [12]. With one exception to be mentioned below, the distance between the cluster and the adsorbed atom is taken equal to the sum of the atomic radii of Pt and the particular chemisorbed atom. For oxygen chemisorbed at the third site of fig. 1B, a distance, smaller than the above sum, and equal to the distance in the bulk PtO, is also considered.

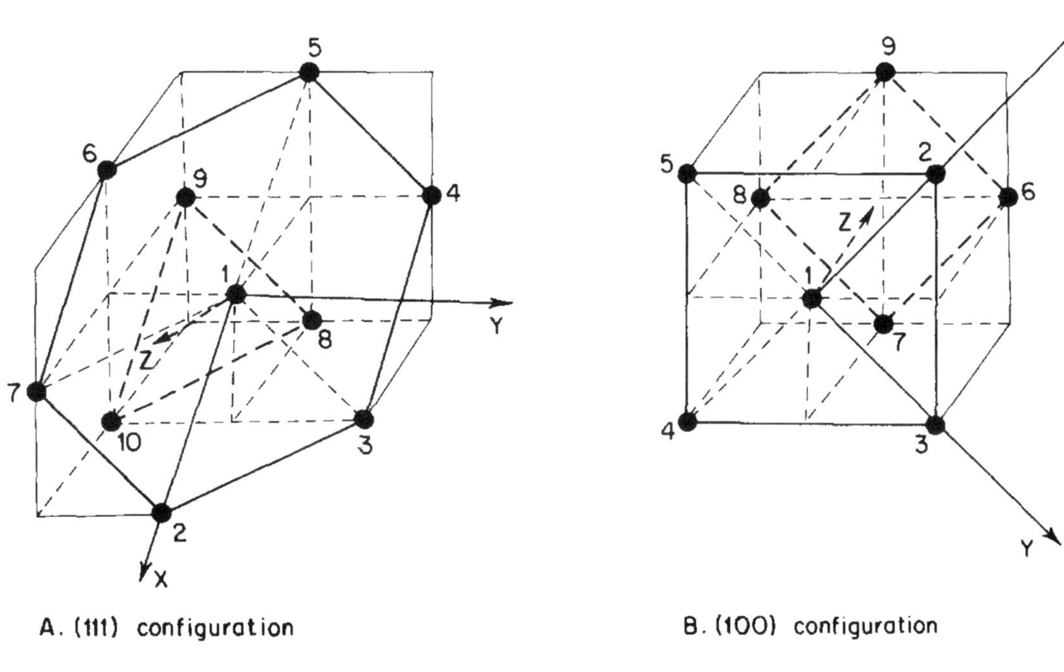

Fig. 1. Clusters used in the calculations.

All the calculations have been carried out with and without adsorbed species and the results for both cases have been compared.

4. Results and discussion

4.1. Adsorption of an H or O atom on the Pt(100) plane

Calculated values of the bond population in a Pt(100) plane before and after adsorption of a H or O atom, are presented in table 2. Because of symmetry, the Pt atoms can be recognized to belong to three types, represented by atoms 1, 2 and 6. The results show that the chemisorption of a H or O atom below the gap between the four Pt atoms (fig. 1B) reduces all the bond-populations among the Pt atoms. In addition, the interaction energy between atoms 6 and 8 (and between 7 and 9) becomes positive (i.e., the interaction becomes repulsive) and the bond population negative. This major reduction in bond strength destabilizes the system and may facilitate the break-away of some surface atoms. For instance, the chemisorbed H atom decreases the bond population between atoms 6 and 7 by 20%. In the case of the chemisorbed O atom, the decrease is greater than 23%. The destabilization may lead to a widening of the distances between Pt atoms and hence, to a reconstruction of the surface layer. This, of course, would change the selectivity and activity of the catalyst. It is difficult, however, to predict exactly how the rearrangements occurs.

Table 2 also contains results corresponding to adsorption above the central Pt atom (atom 1). Chemisorption of H increases by a small amount the bond population between the central atom and the surrounding atoms in the same layer; all other interactions among the cluster atoms decrease. In the case of O, the bond strength between the central Pt atom 1 and the neighboring atoms in the same layer is decreased, while the bond strength among atoms of different layers increases. This indicates a possible decrease of the distance between the two layers, in the case of O chemisorption.

Consider now two H or O atoms adsorbed below two opposite Pt–Pt bonds in the second layer (below the bond between atoms 6 and 9 and between atoms 7 and 8). When the adsorbed atom is O, the structure is close to that of PtO. The calculations have been carried out with two different distances between the oxygen atom and the Pt–Pt bond. The first distance was chosen to be equal to the sum of the atomic radii of Pt and O (1.67 Å), while the second one was taken equal to the distance in the bulk PtO (1.43 Å). The results are summarized in table 3. Destabilization is observed, in particular, for the Pt atoms nearest to the adsorbed H or O atoms. The destabilization produced by O is greater when the distance between Pt and O is smaller. This decrease of the bond strength should lead to a reduction in the surface tension of the particles. Because of its better wetting ability [13], the PtO that is eventually formed spreads over the nonoxidized part of the Pt surface, or in some cases over the support.

Table 2
Bond population matrix before and after H or O adsorption on (100) plane

System	Bond population matrix					
	1		2		6	
9 Pt atoms in (100)	2	0.1936	1	0.1936	1	0.1770
configuration without	3	0.1936	3	0.0048	2	0.2023
adsorbed atoms	4	0.1936	4	0.0001	3	0.2023
	5	0.1936	5	0.0048	4	0.0006
	6	0.1770	6	0.2023	5	0.0006
	7	0.1770	7	0.0006	7	0.2118
	8	0.1770	8	0.0006	8	0.0053
	9	0.1770	9	0.2023	9	0.2118
9 Pt atoms in (100)	2	0.1759	1	0.1759	1	0.1552
configuration and a	3	0.1759	3	0.0026	2	0.1782
H atom below the gap	4	0.1759	4	0.0001	3	0.1782
of the four Pt atoms	5	0.1759	5	0.0026	4	0.0002
lying in the second	6	0.1552	6	0.1782	5	0.0002
layer	7	0.1552	7	0.0002	7	0.1696
	8	0.1552	8	0.0002	8	−0.0018
	9	0.1552	9	0.1782	9	0.1696
	10	0.0006	10	0.0001	10	0.0878
9 Pt atoms in (100)	2	0.1770	1	0.1770	1	0.1570
configuration and	3	0.1770	3	0.0028	2	0.1840
an O atom below	4	0.1770	4	0.0001	3	0.1840
the gap of the four	5	0.1770	5	0.0028	4	0.0002
Pt atoms lying in	6	0.1570	6	0.1840	5	0.0002
the second layer	7	0.1570	7	0.0002	7	0.1626
	8	0.1570	8	0.0002	8	−0.0024
	9	0.1570	9	0.1840	9	0.1626
	10	−0.0006	10	0.0000	10	0.0491
9 Pt atoms in (100)	2	0.1960	1	0.1960	1	0.1757
configuration and a	3	0.1960	3	0.0041	2	0.1953
H atom above the	4	0.1960	4	0.0001	3	0.1953
central atom 1 of the	5	0.1960	5	0.0041	4	0.0005
first layer	6	0.1757	6	0.1953	5	0.0005
	7	0.1757	7	0.0005	7	0.1967
	8	0.1757	8	0.0005	8	0.0047
	9	0.1757	9	0.1953	9	0.1967
	10	0.0593	10	0.0023	10	−0.0001
9 Pt atoms in (100)	2	0.1909	1	0.1909	1	0.1887
configuration and an	3	0.1909	3	0.0045	2	0.2031
O atom above the	4	0.1909	4	0.0001	3	0.2031
central atom 1 of	5	0.1909	5	0.0045	4	0.0005
the first layer	6	0.1887	6	0.2031	5	0.0005
	7	0.1887	7	0.0005	7	0.2065
	8	0.1887	8	0.0005	8	0.0045
	9	0.1887	9	0.2031	9	0.2065
	10	0.4404	10	−0.0004	10	0.0001

Table 3
Bond population matrix after adsorption of H or O atoms below two opposite Pt–Pt bonds of the second-layer

System	Bond population matrix							
		1		2		3		6
9 Pt atoms in (100) configuration with O atoms below the bonds between atoms 6 and 9 and 7 and 8. Distance between the Pt–Pt bond and the oxygen atom equal to 1.67 Å	2	0.1800	2	0.1800	1	0.2029	1	0.1722
	3	0.2029	3	0.0036	2	0.0036	2	0.1889
	4	0.1800	4	0.0001	4	0.0036	3	0.2142
	5	0.2029	5	0.0036	5	0.0001	4	0.0004
	6	0.1722	6	0.1889	6	0.2141	5	0.0004
	7	0.1722	7	0.0004	7	0.2141	7	0.2056
	8	0.1722	8	0.0004	8	0.0003	8	0.0015
	9	0.1722	9	0.1889	9	0.0003	9	0.1027
	10	−0.0001	10	0.0000	10	0.0000	10	0.1954
	11	0.0000	11	0.0000	11	0.0000	11	0.0000
9 Pt atoms in (100) configuration with O atoms below the bonds between atoms 6 and 9 and 7 and 8; distance between the Pt–Pt bond and the oxygen atoms equal to 1.43 Å	2	0.1794	1	0.1794	1	0.1977	1	0.1712
	3	0.1977	3	0.0033	2	0.0033	2	0.1980
	4	0.1794	4	0.0001	4	0.0033	3	0.2083
	5	0.1977	5	0.0033	5	0.0001	4	0.0004
	6	0.1712	6	0.1870	6	0.2083	5	0.0003
	7	0.1712	7	0.0004	7	0.2083	7	0.2025
	8	0.1712	8	0.0004	8	0.0003	8	0.0005
	9	0.1712	9	0.1780	9	0.0003	9	0.0765
	10	−0.0002	10	0.0000	10	0.0000	10	0.2732
	11	−0.0002	11	0.0000	11	0.0000	11	−0.0008
9 Pt atoms in (100) configuration with H atoms below the bond between atoms 6 and 9 and atoms 7 and 8	2	0.1770	1	0.1770	1	0.1971	1	0.1674
	3	0.1971	3	0.0029	2	0.0029	2	0.1790
	4	0.1770	4	0.0001	4	0.0029	3	0.1941
	5	0.1971	5	0.0029	5	0.0001	4	0.0004
	6	0.1674	6	0.1790	6	0.1941	5	0.0002
	7	0.1674	7	0.0004	7	0.1941	7	0.1892
	8	0.1674	8	0.0004	8	0.0002	8	0.0007
	9	0.1674	9	0.1790	9	0.0002	9	0.1384
	10	0.0002	10	0.0004	10	−0.0001	10	0.2535
	11	0.0002	11	0.0000	11	−0.0001	11	−0.0005

4.2. Adsorption of H and O on the Pt(111) plane

The bond population matrix between 10 Pt atoms arranged as in the (111) plane, before and after adsorption of H or O atoms, is presented in table 4. The adsorption of a H atom, below the gap formed by the three Pt atoms of the second layer, alters little the bond population between the atoms of the first layer of the cluster. The greatest reduction in the bond population, about 18%, occurs between

Table 4
Bond population matrix before and after adsorption of H or O atom on (111) plane

System	Bond population matrix					
	1		2		8	
10 Pt atoms in (111) configuration without adsorbed species	1	0.1825	1	0.1825	1	0.1735
	3	0.1825	3	0.2076	2	0.0041
	4	0.1825	4	0.0005	3	0.2018
	5	0.1825	5	0.0001	4	0.2018
	6	0.1825	6	0.0005	5	0.0041
	7	0.1825	7	0.1969	6	0.0005
	8	0.1735	8	0.0041	7	0.0005
	9	0.1735	9	0.0005	9	0.2091
	10	0.1735	10	0.2018	10	0.2091
10 Pt atoms in (111) configuration with a H atom below the gap between the three Pt atoms of the second layer	2	0.1853	1	0.1853	1	0.1714
	3	0.1853	3	0.2080	2	0.0031
	4	0.1853	4	0.0004	3	0.2040
	5	0.1853	5	0.0001	4	0.2040
	6	0.1853	6	0.0005	5	0.0031
	7	0.1853	7	0.1992	6	0.0004
	8	0.1714	8	0.0031	7	0.0004
	9	0.1714	9	0.0004	9	0.1709
	10	0.1714	10	0.2040	10	0.1709
	11	−0.0004	11	−0.0001	11	0.1530
10 Pt atoms in (111) configuration with an O atom below the gap between the three Pt atoms of the second layer	2	0.1855	1	0.1855	1	0.1790
	3	0.1855	3	0.2053	2	0.0032
	4	0.1855	4	0.0004	3	0.2049
	5	0.1855	5	0.0001	4	0.2049
	6	0.1855	6	0.0004	5	0.0032
	7	0.1855	7	0.1971	6	0.0004
	8	0.1790	8	0.0032	7	0.0004
	9	0.1790	9	0.0004	9	0.1450
	10	0.1790	10	0.2049	10	0.1450
	11	0.0000	11	0.0000	11	0.1142
10 Pt atoms in (111) configuration with a H atom above the central Pt atom 1 of the first layer	2	0.1804	1	0.1804	1	0.1731
	3	0.1804	3	0.2003	2	0.0034
	4	0.1804	4	0.0004	3	0.1962
	5	0.1804	5	0.0001	4	0.1962
	6	0.1804	6	0.0004	5	0.0034
	7	0.1804	7	0.1896	6	0.0005
	8	0.1731	8	0.0034	7	0.0005
	9	0.1731	9	0.0005	9	0.1910
	10	0.1731	10	0.1962	10	0.1910
	11	0.6585	11	0.0012	11	−0.0002

Table 4 (continued)

System	Bond population matrix											
		1		2		8						
10 Pt atoms in (111)	2	0.1759	1	0.1759	1	0.1845						
configuration with	3	0.1759	3	0.2047	2	0.0036						
an O atom above the	4	0.1759	4	0.0004	3	0.2040						
central Pt atom 1	5	0.1759	5	0.0001	4	0.2040						
of the first layer	6	0.1759	6	0.0004	5	0.0036						
	7	0.1759	7	0.1950	6	0.0005						
	8	0.1845	8	0.0036	7	0.0005						
	9	0.1845	9	0.0005	9	0.1982						
	10	0.1845	10	0.2040	10	0.1982						
	11	0.4322	11	−0.0009	11	0.0000						
		1		2		3		6		8		9
10 Pt atoms in (111)	2	0.2001	1	0.2001	1	0.1808	1	0.1813	1	0.1724	1	0.1811
configuration an O	3	0.1808	3	0.2079	2	0.2079	2	0.0004	2	0.0033	2	0.0004
atom below the bond	4	0.1808	4	0.0004	4	0.1927	3	0.0001	3	0.1998	3	0.0005
between atoms 9 and	5	0.2001	5	0.0001	5	0.0004	4	0.0005	4	0.1998	4	0.0038
10 of the second	6	0.1813	6	0.0004	6	0.0001	5	0.1992	5	0.0033	5	0.2254
layer	7	0.1813	7	0.1992	7	0.0005	7	0.2061	6	0.0005	6	0.2004
	8	0.1724	8	0.0033	8	0.1998	8	0.0005	7	0.0005	7	0.0034
	9	0.1811	9	0.0004	9	0.0005	9	0.2004	9	0.1904	8	0.1904
	10	0.1811	10	0.2254	10	0.0038	10	0.0034	10	0.1904	10	0.1107
	11	0.0000	11	−0.0000	11	−0.0000	11	0.0000	11	0.0013	11	0.1909

the three Pt atoms of the second layer. When O is adsorbed on site (b), strengthening of the bond between atom 1 and all other atoms is observed. Nevertheless, a decrease of about 30%, in the interaction between the three Pt atoms in the second layer is observed which is much greater than that for chemisorbed H. It seems that O chemisorption on this site decreases mainly the interaction between the surface atoms while the atoms in the lower layer become more strongly bonded to the atoms of the upper layer. The results for adsorption of a H or O atom above the central atom 1 are also included in table 4. The H chemisorption on this site decreases the bond population between all Pt atoms. If an O atom is adsorbed, the picture is different. The interactions between Pt atom 1 and its neighbors in the same layer decrease but those between the atoms of the first layer and the three Pt atoms in the second layer are increased. This is similar to the effect on the (100) plane and indicates a possible decrease of the distance between the two layers.

When O is adsorbed below the bond between atoms 9 and 10 (fig. 1A), (Table 4) a decrease in the bond population between the Pt atoms of the second layer occurs. The decrease in the bond population between atoms 9 and 10 is larger than 47%.

For both crystallographic structures, the chemisorption of H or O atoms at different sites results in the destabilization of the cluster, thus providing the possibility of greater mobility for the Pt atoms.

4.3. Adsorption of a sulphur or chlorine atom on the (111) and (100) plane of Pt clusters

The bond population matrices between the atoms of the Pt clusters after adsorption of a sulphur or chlorine atom are presented in tables 5 and 6. It seems that when a S or Cl atom is adsorbed below a gap of the second layer, its destabilizing effect (decrease in the strength of the Pt–Pt bonds) on that layer is greater than that of a chemisorbed H or O atom. The destabilizing effect of chemisorbed S or Cl atom above Pt atom 1 is much stronger than that of a H or O atom, especially for the (111) plane. Comparing the results for both configurations, it is easily seen that the destabilization of the (100) configuration is greater than that of the (111) configuration when the adsorption of S or Cl atom is below the gap formed by the Pt atoms of the second layer. However, if the adsorption takes place above the central atom 1 of the first layer, the destabilization of the (111) configuration is the greater. One should note that adsorption of a S or Cl atom above the atom 1 leads, for both crystallographic structures, to far greater changes than those caused by oxygen, even though oxygen is isoelectronic with sulphur. Generally, the adsorption of a S or Cl atom results in the decrease of the bond population among all Pt atoms. This effect can be attributed to the empty d orbitals of the S and Cl atoms.

If should be noted that the d orbitals of S and Cl have to be carefully used. Wrong parameters can lead to very high occupancy of these orbitals and overestimation of their contribution to the electron redistribution in the system cluster-adsorbed atom. Table 7 contains the atomic orbital occupation numbers for S and

Table 5
Bond population matrix after adsorption of Cl or S atom on (111) plane

System	Bond population matrix					
	1		2		8	
10 Pt along in (111) configuration with a Cl atom below the gap of the three Pt atoms of the second layer	2	0.1854	1	0.1854	1	0.1726
	3	0.1854	3	0.2079	2	0.0030
	4	0.1854	4	0.0005	3	0.2052
	5	0.1854	5	0.0001	4	0.2052
	6	0.1854	6	0.0005	5	0.0030
	7	0.1854	7	0.1988	6	0.0003
	8	0.1726	8	0.0030	7	0.0003
	9	0.1726	9	0.0003	9	0.1415
	10	0.1726	10	0.2052	10	0.1415
	11	−0.0007	11	0.0000	11	0.3683
10 Pt atoms in (111) configuration with a Cl atom above the central Pt atom 1 of the first layer	2	0.1584	1	0.1584	1	0.1709
	3	0.1584	3	0.1937	2	0.0031
	4	0.1584	4	0.0003	3	0.1932
	5	0.1584	5	0.0001	4	0.1932
	6	0.1584	6	0.0003	5	0.0031
	7	0.1584	7	0.1829	6	0.0004
	8	0.1709	8	0.0031	7	0.0004
	9	0.1709	9	0.0004	9	0.1899
	10	0.1709	10	0.1932	10	0.1899
	11	1.1899	11	0.0102	11	−0.0002
10 Pt atoms in (111) configuration with a S atom below the gap of the three Pt atoms of the second layer	2	0.1837	1	0.1837	1	0.1779
	3	0.1837	3	0.2041	2	0.0027
	4	0.1837	4	0.0004	3	0.2000
	5	0.1837	5	0.0001	4	0.2000
	6	0.1837	6	0.0004	5	0.0027
	7	0.1837	7	0.1959	6	0.0003
	8	0.1779	8	0.0027	7	0.0003
	9	0.1779	9	0.0003	9	0.1418
	10	0.1779	10	0.2000	10	0.1418
	11	−0.0021	11	0.0003	11	0.4997
10 Pt atoms in (111) configuration with a S atom above the central Pt atom 1 of the first layer	2	0.1583	1	0.1583	1	0.1695
	3	0.1583	3	0.1970	2	0.0032
	4	0.1583	4	0.0003	3	0.1991
	5	0.1583	5	0.0001	4	0.1991
	6	0.1583	6	0.0003	5	0.0032
	7	0.1583	7	0.1866	6	0.0004
	8	0.1695	8	0.0032	7	0.0004
	9	0.1695	9	0.0004	9	0.1967
	10	0.1695	10	0.1991	10	0.1967
	11	1.2832	11	0.0349	11	−0.0004

Table 6
Bond population matrix after adsorption of Cl or S atom on (100) plane

System	Bond population matrix					
		1		2		6
9 Pt atoms in (100) configuration with a Cl atom below the gap of the four Pt atoms of the second layer	2	0.1776	1	0.1776	1	0.1576
	3	0.1776	3	0.0027	2	0.1793
	4	0.1776	4	0.0001	3	0.1793
	5	0.1776	5	0.0027	4	0.0002
	6	0.1576	6	0.1793	5	0.0002
	7	0.1576	7	0.0002	7	0.1504
	8	0.1576	8	0.0002	8	−0.0035
	9	0.1576	9	0.1793	9	0.1504
	10	−0.0011	10	0.0000	10	0.2049
9 Pt atoms in (100) configuration with a Cl above the Pt atom 1 of the first layer	1	0.1729	1	0.1729	1	0.1715
	3	0.1729	3	0.0032	2	0.1914
	4	0.1729	4	0.0001	3	0.1914
	5	0.1729	5	0.0032	4	0.0004
	6	0.1715	6	0.1914	5	0.0004
	7	0.1715	7	0.0004	7	0.1941
	8	0.1715	8	0.0004	8	0.0045
	9	0.1715	9	0.1914	9	0.1941
	10	1.2000	10	0.0125	10	−0.0003
9 Pt atoms in (100) configuration with a S atom below the gap of the four Pt atoms of the second layer	2	0.1781	1	0.1781	1	0.1571
	3	0.1781	3	0.0029	2	0.1835
	4	0.1781	4	0.0001	3	0.1835
	5	0.1781	5	0.0029	4	0.0002
	6	0.1571	6	0.1835	5	0.0002
	7	0.1571	7	0.0002	7	0.1606
	8	0.1571	8	0.0002	8	−0.0033
	9	0.1571	9	0.1835	9	0.1606
	10	0.0011	10	−0.0001	10	0.3212
9 Pt atoms in (100) configuration with a S atom above the central Pt atoms of the first layer	2	0.1758	1	0.1758	1	0.1736
	3	0.1758	3	0.0034	2	0.1978
	4	0.1758	4	0.0001	3	0.1978
	5	0.1758	5	0.0034	4	0.0004
	6	0.1736	6	0.1978	5	0.0004
	7	0.1736	7	0.0004	7	0.2026
	8	0.1736	8	0.0004	8	0.0046
	9	0.1736	9	0.1978	9	0.2026
	10	1.3003	10	0.0390	10	−0.0007

Cl d orbitals. These occupation numbers and the coefficients in the LCAO-MO representation show that the participation of sulfur and chlorine d orbitals in the formation of the molecular orbitals of the system is not unreasonably high.

Sulphur is a well known inhibitor for a large variety of surface reactions catal-

Table 7
Occupation of S and Cl d orbitals at different adsorption sites

Orbital	S above Pt atom 1	S below the gap	Cl above Pt atom 1	Cl below the gap
Pt (100) cluster				
$x^2 - y^2$	0.09856	0.08428	0.07852	0.05563
z^2	0.03579	0.11801	0.11400	0.08817
xy	0.06146	0.15609	0.06094	0.15797
xz	0.34544	0.18118	0.51373	0.20760
yz	0.34544	0.18118	0.51373	0.20760
Pt(111) cluster				
$x^2 - y^2$	0.08786	0.11812	0.07217	0.12150
z^2	0.04527	0.16144	0.10207	0.18239
xy	0.08786	0.11812	0.07217	0.12150
xz	0.35370	0.21493	0.51600	0.30621
yz	0.35370	0.21493	0.51600	0.30621

yzed by supported Pt catalysts. A possible explanation of this inhibition is the restructuring of the (111) to the (100) configuration, produced by its adsorption [14]. This involves the assumption that before adsorption the (111) plane is more stable than (100), since it has a lower surface tension, while after adsorption the order of stability is inverted. Our computations show that the adsorption of sulphur produces major electronic changes in practically all the surrounding atoms. Even when this electron redistribution does not produce structural changes it would still affect catalytic activity and selectivity. Most probably electron redistribution, sometimes accompanied by restructuring, is responsible for the inhibition caused by sulphur.

4.4. Atomic charge distribution

The atomic charge distribution is given in tables 8 and 9. The acceptor character of the adsorbed H atom is not strong enough for either crystallographic structures to lead to positive charges on all Pt atoms. In contrast, the adsorption of oxygen or sulfur atom generates positive charges on all Pt atoms. Also, when Cl is adsorbed above the central Pt atom, the electron transfer from this atom is very high and all Pt atoms become positively charged. Most probably the charges generated when adsorption of Cl is considered on the central Pt atom are overestimated because of the shortcomings of the method. Chlorine adsorbed below the gap of the Pt(100) plane does not generate positive charges on the Pt atoms constituting the first layer. Even though the calculated charges may outline the trend, they may be unrealistically large because of the shortcomings of the method.

Table 8
Charge distribution in the Pt (100) clusters before and after adsorption of H, O, S and Cl atoms (atom 10 is the adsorbed atom)

System: atoms	Atomic charges								
	Pure cluster	H above Pt atom 1	H below the gap	O above Pt atom 1	O below the gap	S above Pt atom 1	S below the gap	Cl above Pt above 1	Cl below the gap
1	−0.30814	0.28289	0.01249	0.79585	0.10961	0.73221	0.05145	1.16050	0.07108
2	−0.03823	−0.02200	−0.04148	0.03903	0.05087	0.09925	0.04561	0.03171	−0.03623
3	−0.03823	−0.02200	−0.04148	0.03903	0.05087	0.09925	0.04561	0.03171	−0.03623
4	−0.03823	−0.02200	−0.04148	0.03903	0.05087	0.09925	0.04561	0.03171	−0.03623
5	−0.03823	−0.02200	−0.04148	0.03903	0.05087	0.09925	0.04561	0.03171	−0.03623
6	0.11527	0.05162	0.18779	0.16660	0.37368	0.23746	0.43401	0.10585	0.33821
7	0.11527	0.05162	0.18779	0.16660	0.37368	0.23746	0.43401	0.10585	0.33821
8	0.11527	0.05162	0.18779	0.16660	0.37368	0.23746	0.43401	0.10585	0.33821
9	0.11527	0.05162	0.18779	0.16660	0.37368	0.23746	0.43401	0.10585	0.33821
10		−0.40134	−0.59775	−1.61835	−1.80779	−2.07904	−1.96993	−1.71073	−1.27900

Table 9
Charge distribution in the Pt (111) cluster before and after adsorption of H, O, S and Cl atoms (atom 11 is the adsorbed atom)

System: atoms	Atomic charges								
	Pure cluster	H above Pt atom 1	H below the gap	O above Pt atom 1	O below the gap	S above Pt atom 1	S below the gap	Cl above Pt atom 1	Cl below the gap
1	−0.28662	0.27789	0.00821	0.72562	0.46344	0.77995	0.49442	1.24370	0.07039
2	−0.00642	−0.02770	−0.00041	0.04256	0.02793	0.09134	0.04581	0.00878	0.02517
3	−0.00642	−0.02770	−0.00041	0.04256	0.02793	0.09134	0.04581	0.00878	0.02517
4	−0.00642	−0.02770	−0.00041	0.04256	0.02793	0.09134	0.04581	0.00878	0.02517
5	−0.00642	−0.02770	−0.00041	0.04256	0.02793	0.09134	0.04581	0.00878	0.02517
6	−0.00642	−0.02770	−0.00041	0.04256	0.02793	0.09134	0.04581	0.00878	0.02517
7	−0.00642	−0.02770	−0.00041	0.04256	0.02793	0.09134	0.04581	0.00878	0.02517
8	0.10838	0.09584	0.17446	0.21535	0.35259	0.26069	0.40617	0.13689	0.42406
9	0.10838	0.09584	0.17446	0.21535	0.35259	0.26069	0.40617	0.13689	0.42406
10	0.10838	0.09584	0.17446	0.21535	0.35259	0.26069	0.40617	0.13689	0.42406
11		−0.39923	−0.52912	−1.62701	−1.68877	−2.11004	−1.98781	−1.70704	−1.49360

4.5. Origin of the destabilization of clusters by adsorbed electron acceptor species

Generally, if two orbitals with different energies interact, splitting occurs and the lower energy orbital is stabilized and the upper energy orbital is destabilized. As a result of the interaction between the cluster and the adsorbate two kinds of cluster–adsorbate orbitals are formed, with a relatively large energy gap of a few eV between them. The first have low energies and consist of the s and p orbitals of the adsorbate (which play the dominating role) mixed with some orbitals of the metal. These predominantly adsorbate orbitals have lower energies than the pure adsorbate orbitals, because they are generated after the interaction with the higher energy orbitals of the metal. The second kind of orbitals has generally higher energies and is composed of orbitals of the metal, which now play the dominating role, mixed with some s and p orbitals of the adsorbate. Mixing with d orbitals of the adsorbate also takes place if the adsorbed species are sulfur or chlorine. If the d orbitals of the adsorbate have higher energies than the orbitals of the metal, some of the second kind of orbitals can be stabilized in comparison with the pure metal orbitals.

The filling of the two kinds of cluster–adsorbate orbitals with electrons has the following effects on the stability of the system: (1) Electrons belonging to the highest occupied antibonding orbitals of pure Pt fill the empty orbitals of the first kind which, as mentioned above, have low energies, thus stabilizing the system; (2) The filling with electrons of the second kind of cluster–adsorbate orbitals causes stabilization or destabilization of the system depending on whether these orbitals have lower or higher energies in comparison with the pure metal orbitals. Of course, when the destabilization overcomes the stabilization effects, an overall destabilization results.

Now, we will examine, in some detail each of the cluster–adsorbate systems. When a H atom is adsorbed, the lowest molecular orbital has predominantly s character (because of the high contribution of the s orbital of H), mixed with s and p orbitals from the metal. This orbital is of the first kind. There is a gap of about 3 eV between this orbital and the next occupied orbitals which are of the second kind. The higher energy orbitals have predominantly Pt d character and only a few of them are mixed with the hydrogen s orbital. If a comparison is made between the orbitals above the lowest orbital and the corresponding orbitals in the pure Pt cluster, one can conclude that the molecular orbitals after H adsorption have energies which are either higher, or equal to those of the molecular orbitals of the pure cluster. The coefficients of the molecular orbitals whose energies remain unchanged after adsorption indicate that there is no participation of the hydrogen s orbital in these cluster–adsorbate orbitals. For the cluster–adsorbate orbitals of the second kind that have higher energies, some mixing with the s orbital of H exists. The destabilization of the molecular orbitals of the cluster by the hydrogen s orbital decreases the bond strength among certain Pt atoms since it is able to overcome the stabilizing effect of the removal of an electron from the highest occupied antibonding orbital of platinum. Of course, the bonds between some other Pt atoms may be strengthened.

The situation is similar when oxygen is adsorbed. The lowest occupied orbital of the system has predominantly s character (resulting from the s orbital of O), mixed with some d orbitals of the metal. The next three occupied orbitals have p character (resulting from the oxygen p orbitals), mixed with d orbitals of the metal. These four orbitals are of the first kind. If a comparison is made between the energies of the occupied orbitals of the second kind following the lowest four orbitals and the corresponding orbitals in the pure cluster, it is found again, that the cluster–adsorbate orbitals have energies, which are either equal or higher than those of the pure cluster. Destabilization occurs always for molecular orbitals mixed with s or p orbitals of the oxygen atom. If the energies are equal, the coefficients of the molecular orbitals show, that no mixing of orbitals from cluster and adsorbate exists. Again, because the destabilizing effects overcome the stabilizing effect caused by the transfer of electrons from the highest occupied, antibonding, levels of platinum to the empty, predominantly oxygen p orbitals of the first kind, destabilization occurs for some Pt–Pt bonds.

When S or Cl atom is adsorbed the picture is somewhat more complicated. The lowest four occupied orbitals are similar to those after oxygen adsorption, but some of the higher occupied molecular orbitals of the second kind are slightly shifted towards lower energies, when compared to the corresponding orbitals of the pure cluster. The reason is that they have predominantly Pt d character and as their coefficients show, they are mixed with high lying d orbitals of S or Cl. However, this shift towards lower energies is not enough to compensate for the destabilization caused by the mixing with the s and p orbitals of the adsorbed atoms.

5. Conclusions

(1) The adsorption of H, O, Cl or S atom below the gap of the (100) plane causes a decrease of the bond strength between the Pt atoms and affects practically all the bonds in the cluster. The destabilizing effect is in the order $Cl \simeq S > O > H$.

(2) The adsorption of Cl or S atom above the central atom 1 of the (100) configuration decreases the bond strength among all Pt atoms. The destabilizing effect is in this case stronger than that of a H or O atom. The adsorption of a H atom on the same site results in a decrease of the bond strength among all Pt atoms with the exception of that between atom 1 and its neighbors in the same layer. The adsorption of oxygen on this site results in a decrease in the interaction within the layers, but the interaction increases between the atoms of the two layers. This may result in a decrease of the distance between the two layers. The destabilizing effect is in the order $Cl > S > O > H$.

(3) The adsorption of a H, O, Cl or S atom below the gap of the (111) plane affects mainly the layer upon which the adsorption takes place. The destabilizing

effect of a Cl or S atom upon this layer is stronger than that of H or O atom, but is weaker than that of a Cl or S atom adsorbed below the gap of the (100) plane.

(4) The adsorption of a Cl or S atom above atom 1 of the (111) configuration decreases the bond population between all Pt atoms and its destabilizing effect is stronger than that of a H or O atoms. It is also stronger than that of a S or Cl atoms adsorbed above atom 1 in the (100) configuration. A chemisorbed O atom on the same site causes an increase in the interaction between the atoms of the two layers thus, possibly, decreasing the distance between the layers. The destabilizing effect is in the order $Cl \simeq S > O > H$.

(5) The adsorption of O atoms on site (c) for both crystallographic structures results in a strong decrease of the bond strength between the Pt atoms directly connected with the oxygen atoms. Therefore, the surface tension of the clusters decreases and the oxide eventually formed may spread over the non-oxidized part of the cluster or eventually over the support because of its better wetting ability.

(6) Generally, the destabilizing effect of Cl and S atoms is stronger than that of H or O atoms and affects all Pt bonds. The major changes of the electron structure of the Pt clusters after S adsorption is probably responsible for the inhibition produced by this elemet in catalytic processes. This strong effect can be attributed to the empty d orbitals of sulfur.

(7) The calculations show that the adsorption of atoms with high occupancy and with lower ionization potentials than that of Pt (e.g., electron acceptors) on different adsorption sites of Pt (100) and (111) cluster configurations results in destabilization of the clusters, in a decrease of the surface tension and greater mobility of the surface atoms.

References

[1] R. Maurel, G. Leclercq and J. Barbier, J. Catalysis 37 (1975) 324.
[2] R.L. Smith, P.A. Naro and A.J. Silvertry, J. Catalysis 20 (1971) 359.
[3] J.J. McCaroll, T. Edmonds and R.C. Pitkethly, Nature 223 (1969) 1260.
[4] F. Delamare and G.E. Rhead, Surface Sci. 28 (1971) 267.
[5] E. Ruckenstein and Y.F. Chu, J. Catalysis 59 (1979) 309.
[6] R. Hoffmann, J. Chem. Phys. 39 (1963) 1397.
[7] D.J. Fassaert, H. Verbeek and A. van der Avoird, Surface Sci. 29 (1972) 501.
[8] R.C. Baetzold, Surface Sci. 36 (1972) 123.
[9] R.C. Baetzold, J. Catalysis 29 (1973) 129.
[10] K. Takahashi, J. Coloid Interface Sci. 73 (1980) 162.
[11] R.S. Mulliken, J. Chem. Phys. 23 (1955) 1833.
[12] W.J. Moore and L. Pauling, J. Am. Chem. Soc. 63 (1941) 1392.
[13] E. Ruckenstein, J. Crystal Growth 47 (1979) 666.
[14] G.A. Somorjai, J. Catalysis 27 (1972) 453.
[15] C.C. Lu, T.A. Carlson, F.B. Malik, T.C. Tucker and C.W. Nestor, Atomic Data 3 (1971) 1-131.

Poisoning and Promoting Effects of Additives on the Catalytic Behavior of Metal Clusters

T. HALACHEV[1] AND E. RUCKENSTEIN[2]

Department of Chemical Engineering, State University of New York, Buffalo, New York 14260

Received May 22, 1981; revised October 1, 1981

Extended Hückel calculations are used to explain the mechanisms by which additives such as Cu, Ag, Cl, and S affect the catalytic behavior of Pt or Ag clusters. The complete oxidation of ethylene over Pt/Al_2O_3, the oxidation of CO over Pt/Al_2O_3, and the ethylene epoxidation over supported silver catalysts are used as examples. It is found that Cu or Ag, unlike S, change mainly the net atomic population (the localized electrons around each atom) and very slightly the bond population among the atoms of the Pt clusters. The increased electron density at the surface of Pt clusters due to the electron transfer from Cu or Ag facilitates the electron transfer from the cluster to the π^* antibonding MO of adsorbed O_2. As a result, the dissociative chemisorption of O_2 over Cu- or Ag-treated Pt catalyst is enhanced and the reaction rates for the complete oxidation of C_2H_4 and the oxidation of CO are increased. The electronic effect of Cu on the electron redistribution in the clusters is stronger than that of Ag. Sulfur inhibits these reactions because it is a strong electron acceptor and the competition for electrons, as well as the destabilization of the MOs of the cluster, reduces the dissociative chemisorption of O_2. The inhibition effect of sulfur overcomes the promoting effect of Cu or Ag. Two forms of chemisorbed O_2 molecule, normal and parallel to the surface of a Ag (111) cluster, have been assumed in the calculations. For both forms, the coadsorption of chlorine results in a significant energy shift of the O_2 dissociation curves toward higher energies, thus inhibiting the chemisorption of O_2 and its dissociation. As a consequence, a loss in overall activity occurs, while the selectivity for epoxidation is higher. In the presence of chlorine, the O_2 adsorbed normal to the surface seems to be more stable than the O_2 adsorbed parallel to the surface. In the absence of chlorine, the former form becomes less stable than the latter if the elongation of the O–O bond is larger than 1.4 Å.

INTRODUCTION

This paper is concerned with the mechanisms by which certain additives, such as Cu, Ag, Cl, and S affect the catalytic behavior of metal clusters. For illustrative purposes, we consider the oxidation over supported platinum of both ethylene to CO_2 and H_2O, and of CO to CO_2, and the ethylene epoxidation over supported silver catalysts. The extended Hückel method is employed in the investigation.

A copper-treated Pt/Al_2O_3 catalyst (that is used in the pollution abatement of automobile exhausts), prereduced in a H_2 atmosphere to avoid the presence of oxides, leads to a significant increase in the activity for CO oxidation and to a more moderate increase in the activity for ethylene oxidation (*1*) in comparison with a copper-free catalyst. The addition of Ag, instead of Cu, has a similar, but less strong, effect. The presence of sulfur in the feed decreases the activity of the Pt/Al_2O_3 catalyst both for the oxidation of CO and for the oxidation of ethylene (*1*). The addition of Cu to the Pt/Al_2O_3 catalyst compensates in part for the negative effects of sulfur.

The first objective of the present paper is to explain the mechanisms by which these promoting and inhibiting effects take place and in particular, to understand why the inhibition effect of S is stronger than the promoting effects of Cu or Ag. Effects similar to those described above have been observed in the epoxidation of ethylene over Ag/Al_2O_3 catalysts, when the substrate was

[1] On leave from the Institute of Organic Chemistry, Bulgarian Academy of Sciences, Sofia 1113, Bulgaria.

[2] Address correspondence to Eli Ruckenstein.

This paper, which was originally published in **Journal of Catalysis**: 73, 171–186 (1982), is republished with permission.

TABLE 1

Parameters Used in the Calculations

Pt	Ag	Cu	S	Cl	O
$\mu(6s) = 2.554$[a]	$\mu(5s) = 2.244$	$\mu(4s) = 1.98$	$\mu(3s) = 1.817$	$\mu(3s) = 2.033$	$\mu(2s) = 2.275$
$\mu(6p) = 2.535$	$\mu(5p) = 2.244$	$\mu(4p) = 1.20$	$\mu(3p) = 1.817$	$\mu(3p) = 2.033$	$\mu(2p) = 2.275$
$\mu_1(5d) = 6.013$	$\mu_1(4d) = 6.07$	$\mu_1(3d) = 5.95$	$\mu(3d) = 1.5$		
$\mu_2(5d) = 2.696$	$\mu_2(4d) = 2.663$	$\mu_2(3d) = 2.10$			
$C_1(5d) = 0.633$[b]	$C_1(4d) = 0.589$	$C_1(3d) = 0.58$			
$C_2(5d) = 0.551$	$C_2(4d) = 0.637$	$C_2(3d) = 0.62$			
$H_{ii}(6s) = -8.091$ eV[c]	$H_{ii}(5s) = -7.56$ eV	$H_{ii}(4s) = -7.66$ eV	$H_{ii}(3s) = -20.00$ eV	$H_{ii}(3s) = -30.00$ eV	$H_{ii}(2s) = -32.3$ eV
$H_{ii}(6p) = -6.36$ eV	$H_{ii}(5p) = -3.83$ eV	$H_{ii}(4p) = -3.96$ eV	$H_{ii}(3p) = -13.30$ eV	$H_{ii}(3p) = -15.00$ eV	$H_{ii}(2p) = -14.8$ eV
$H_{ii}(5d) = -10.989$ eV	$H_{ii}(4d) = -11.58$ eV	$H_{ii}(3d) = -10.62$ eV	$H_{ii}(3d) = -8.00$ eV		

[a] μ is the orbital exponent.
[b] C_1 and C_2 are the coefficients of the double-zeta orbitals.
[c] H_{ii} is the ionization potential.

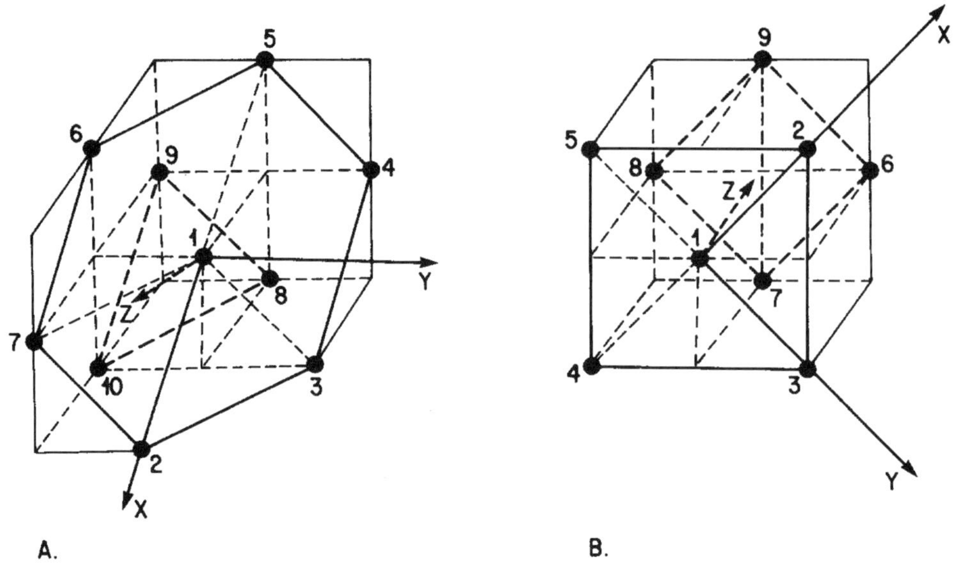

FIG. 1. Clusters used in the calculations.

doped with germanium oxide or magnesium oxide (*2*). The overall activity and the selectivity for epoxidation are enhanced when the catalyst is doped with magnesium oxide, which is a p-type semiconductor, but are decreased when germanium oxide, which is an n-type semiconductor, is used. The same authors observe that the addition of chlorine to the feed promotes the selectivity of the catalyst for epoxidation (*3*). For this reason, as a second objective, we examine the two types of adsorption of O_2 on the clusters (normal and parallel to the surface) and the effect of chlorine on their dissociative chemisorption. In essence, we are interested in investigating how the electronic properties of the clusters change under the action of preadsorbed gases, or by the addition of a second metal, and what are the effects of these changes on the activity and selectivity of the cluster catalysts.

METHOD OF CALCULATION AND STRUCTURE OF THE CLUSTERS

The familiar Extended Hückel Method (EHM) (*4*) is employed in the computations. The method accurately predicts trends (and eventually values) for some quantities which are significant in catalysis. Although the method is not as sophisticated as other quantum-chemical methods (CNDO, X_α, ab initio), its application is useful, if one is interested in trends and not in precise calculated values. Mulliken's expressions (*5*) are used to calculate the bond population and the atomic charges of the system. The parameters used in the calculations are listed in Table 1. The ionization potentials of Pt are taken from Ref. (*6*) and those for Cu and Ag from Ref. (*7*). The exponents of the Slater orbitals for the double-zeta-type orbitals are taken from Ref. (*8*). All the parameters for sulfur, oxygen, and chlorine are from Quantum Chemistry Program Exchange N344.

The calculations have been carried out for clusters containing 10 or 9 platinum or silver atoms. The 10 atoms are arranged as in the (111) fcc plane (Fig. 1A) and the 9 atoms as in the (100) fcc plane (Fig. 1B).[3] The numbering of the atoms is indicated in the figure. The plane in which atom 1 lies will be referred to as the first layer for both crystallographic structures and the plane below, formed by the 3 (Fig. 1A) or 4 (Fig.

[3] The coordinate systems shown in Figs. 1A and B are connected with the calculations and not with the crystallographic orientation of the clusters.

1B) metal atoms, constitutes the second layer. The distance between the metal atoms in the clusters is taken equal to the distance in the bulk metals (2.775 Å for platinum and 2.984 Å for silver). The distance between the cluster and the adsorbed

TABLE 2

Bond Population Matrices and Charge Distribution in Pt (111) Cluster after Adsorption of Single Cu, Ag, or S Atoms

System	Atom	Bond population matrix			Atomic charges
		1	2	8	
Ten Pt atoms in (111) plane without adsorbed species	1	9.4788	0.1825	0.1735	−0.28662
	2	0.1825	9.6092	0.0041	−0.00642
	3	0.1825	0.2076	0.2018	−0.00642
	4	0.1825	0.0005	0.2018	−0.00642
	5	0.1825	0.0001	0.0041	−0.00642
	6	0.1825	0.0005	0.0005	−0.00642
	7	0.1825	0.1969	0.0005	−0.00642
	8	0.1735	0.0041	9.3894	0.10838
	9	0.1735	0.0005	0.2091	0.10838
	10	0.1735	0.2018	0.2091	0.10838
Ten Pt atoms in (111) plane and a Cu atom adsorbed above the central Pt atom 1	1	8.9917	0.1877	0.1805	0.01420
	2	0.1877	9.7909	0.0036	−0.18631
	3	0.1877	0.2004	0.1958	−0.18631
	4	0.1877	0.0005	0.1958	−0.18631
	5	0.1877	0.0001	0.0036	−0.18631
	6	0.1877	0.0005	0.0004	−0.18631
	7	0.1877	0.1906	0.0004	−0.18631
	8	0.1805	0.0036	9.7096	−0.18814
	9	0.1805	0.0004	0.1888	−0.18814
	10	0.1805	0.1958	0.1888	−0.18814
	11	0.3208	0.0112	−0.0008	1.66810
Ten Pt atoms in (111) plane and a Cu atom adsorbed above the gap formed by the three Pt atoms of the second layer	1	9.4303	0.1821	0.1753	−0.23847
	2	0.1821	9.7088	0.0039	−0.09870
	3	0.1821	0.2031	0.1964	−0.09870
	4	0.1821	0.0004	0.1964	−0.09870
	5	0.1821	0.0001	0.0039	−0.09870
	6	0.1821	0.0004	0.0005	−0.09870
	7	0.1821	0.1925	0.0005	−0.09870
	8	0.1753	0.0039	9.5211	−0.06012
	9	0.1753	0.0005	0.1947	−0.06012
	10	0.1753	0.1964	0.1947	−0.06012
	11	−0.0023	0.0003	0.1117	1.01106
Ten Pt atoms in (111) plane and a Ag atom adsorbed above the central Pt atom 1	1	9.3354	0.1823	0.1729	−0.26187
	2	0.1823	9.6972	0.0037	−0.08910
	3	0.1823	0.2030	0.1962	−0.08910
	4	0.1823	0.0005	0.1962	−0.08910
	5	0.1823	0.0001	0.0037	−0.08910
	6	0.1823	0.0005	0.0005	−0.08910
	7	0.1823	0.1923	0.0005	−0.08910
	8	0.1729	0.0037	9.5517	−0.03747
	9	0.1729	0.0005	0.1988	−0.03747
	10	0.1729	0.1962	0.1988	−0.03747
	11	0.2405	0.0047	0.0004	0.90885

TABLE 2—*Continued*

System	Atom	Bond population matrix			Atomic charges
		1	2	8	
Ten Pt atoms in (111) plane and a Ag atom adsorbed above the gap formed by the three Pt atoms of the second layer	1	9.4338	0.1818	0.1742	−0.24067
	2	0.1818	9.7046	0.0039	−0.09348
	3	0.1818	0.2029	0.1956	−0.09348
	4	0.1818	0.0004	0.1956	−0.09348
	5	0.1818	0.0001	0.0039	−0.09348
	6	0.1818	0.0004	0.0005	−0.09348
	7	0.1818	0.1920	0.0005	−0.09348
	8	0.1742	0.0039	9.5196	−0.04355
	9	0.1742	0.0005	0.1976	−0.04355
	10	0.1742	0.1956	0.1976	−0.04355
	11	0.0002	0.0001	0.0784	0.93218
Ten Pt atoms in (111) plane and a S atom adsorbed above the central Pt atom 1	1	7.8495	0.1583	0.1695	0.77995
	2	0.1583	9.5186	0.0032	0.09134
	3	0.1583	0.1970	0.1991	0.09134
	4	0.1583	0.0003	0.1991	0.09134
	5	0.1583	0.0001	0.0032	0.09134
	6	0.1583	0.0003	0.0004	0.09134
	7	0.1583	0.1866	0.0004	0.09134
	8	0.1695	0.0032	9.2554	0.26069
	9	0.1695	0.0004	0.1967	0.26069
	10	0.1695	0.1991	0.1967	0.26069
	11	1.2832	0.0349	−0.0004	−2.11004
Ten Pt atoms in (111) plane and a S atom adsorbed above the gap formed by the three Pt atoms of the second layer	1	8.6886	0.1837	0.1779	0.49442
	2	0.1837	9.5602	0.0027	0.04581
	3	0.1837	0.2041	0.2000	0.04581
	4	0.1837	0.0004	0.2000	0.04581
	5	0.1837	0.0001	0.0027	0.04581
	6	0.1837	0.0004	0.0003	0.04581
	7	0.1837	0.1959	0.0003	0.04581
	8	0.1779	0.0027	8.9102	0.40617
	9	0.1779	0.0003	0.1418	0.40617
	10	0.1779	0.2000	0.1418	0.40617
	11	−0.0021	0.0000	0.4997	−1.98781

atom is assumed equal to the sum of their atomic radii. The dissociation energy of O_2 over the Pt or Ag catalysts is calculated keeping the metal–oxygen distance constant and then allowing the O–O distance to elongate. The adsorption sites considered in the calculations are indicated in the tables containing the numerical results.

RESULTS

1. Adsorption of Cu, Ag, Ag and S, or Cu and S Atoms on the Pt (111) Plane

In this section, the results for adsorption of a single Ag or Cu atom are compared to those obtained when simultaneous adsorption of Ag and S or Cu and S atoms occurs. Table 2 contains the bond population matrices (the electron density between the atoms) and the atomic charges of the cluster atoms (charges of each individual atom) after adsorption of a single Cu, Ag, or S atom on different adsorption sites. The values of the charges in Table 2 are probably higher than the real ones because of the approximations involved in the method. There is a great difference in the effect of sulfur as compared with the effect of Ag or

Cu. The numerical values show that the effect of sulfur on the cluster is opposite to that of copper or silver atoms. Sulfur decreases considerably the bond population among the atoms of the cluster. In addition, the diagonal elements of the bond population matrix which provide the localized electron density around each atom (Mulliken's net atomic population) also decrease considerably. In contrast, the effect of copper or silver atom on the bond population is negligible and the shift toward lower energies of the cluster-adsorbate MOs dominated by the AOs of Pt is negligible. The electron redistribution in the latter case affects mainly the net atomic populations. Copper has a more pronounced effect on the electronic properties of the cluster than silver.

Calculations have also been carried out for simultaneous adsorption of Cu and S or Ag and S atoms. Several adsorption sites have been considered. Table 3 contains the computed charges of all atoms. For comparison, the results for adsorption of single Cu, Ag, or S atoms on the peripheral atoms of the cluster are also included in the table. It is obvious that for any of the considered adsorption sites, sulfur largely (but not totally) compensates the negative charges caused by copper. The negative charges of the platinum atoms for the simultaneous copper–sulfur adsorption are larger than those of the pure platinum clusters. We will return to this point when we discuss the inhibition by sulfur.

The situation is somewhat different when the preadsorbed metal is silver. Sulfur overcomes in this case the electron donor effect of silver and thus generates positive or negative charges (but less negative than in a pure cluster) on different Pt atoms.

TABLE 3

Atomic Charge Distribution after Simultaneous Copper–Sulfur or Silver–Sulfur Adsorption on Pt (111) Plane

Cu over Pt atom 2		S over Pt atom 2		S over Pt atom 1 Cu over Pt atom 2		S over Pt atom 2 Cu over Pt atom 1		S over Pt atom 5 Cu over Pt atom 2		Ag over Pt atom 2		S over Pt atom 1 Ag over Pt atom 2		S over Pt atom 2 Ag over Pt atom 1		S over Pt atom 2 Ag over Pt atom 5	
Pt 1	−0.25977	Pt 1	0.39730	Pt 1	0.75955	Pt 1	0.57056	Pt 1	−0.12189	Pt 1	−0.27555	Pt 1	0.79657	Pt 1	0.19304	Pt 1	0.10324
Pt 2	−0.02426	Pt 2	0.86187	Pt 2	0.18012	Pt 2	0.33969	Pt 2	0.41485	Pt 2	−0.07531	Pt 2	0.07281	Pt 2	0.72180	Pt 2	0.73608
Pt 3	−0.13532	Pt 3	0.02380	Pt 3	−0.05814	Pt 3	−0.16121	Pt 3	−0.07724	Pt 3	−0.08705	Pt 3	−0.00657	Pt 3	−0.03107	Pt 3	−0.02170
Pt 4	−0.15751	Pt 4	0.07647	Pt 4	−0.08799	Pt 4	−0.18356	Pt 4	−0.12734	Pt 4	−0.09769	Pt 4	−0.02377	Pt 4	0.01396	Pt 4	0.00048
Pt 5	−0.16723	Pt 5	0.06424	Pt 5	−0.10315	Pt 5	−0.20425	Pt 5	0.40694	Pt 5	−0.10155	Pt 5	−0.01673	Pt 6	−0.03208	Pt 5	0.04793
Pt 6	−0.16424	Pt 6	0.8410	Pt 6	−0.09196	Pt 6	−0.17759	Pt 6	−0.10504	Pt 6	−0.10079	Pt 6	−0.02923	Pt 6	0.00301	Pt 6	0.00139
Pt 7	−0.13199	Pt 7	0.17053	Pt 7	−0.07016	Pt 7	−0.15651	Pt 7	−0.09472	Pt 7	−0.08781	Pt 7	−0.00462	Pt 7	0.08682	Pt 7	0.04928
Pt 8	−0.14895	Pt 8	0.14265	Pt 8	−0.04431	Pt 8	−0.12045	Pt 8	−0.10005	Pt 8	−0.05230	Pt 8	0.07099	Pt 8	0.03727	Pt 8	0.04794
Pt 9	−0.16070	Pt 9	0.15571	Pt 9	−0.04008	Pt 9	−0.10154	Pt 9	−0.02525	Pt 9	−0.05538	Pt 9	0.08203	Pt 9	0.05683	Pt 9	0.06990
Pt 10	−0.08553	Pt 10	0.15843	Pt 10	0.05315	Pt 10	−0.05881	Pt 10	0.00211	Pt 10	−0.03339	Pt 10	0.12399	Pt 10	0.12292	Pt 10	0.13844
Cu 11	1.43550	S 11	−2.13511	S11	−2.00257	S 11	−2.04297	S 11	−2.17599	Ag 11	0.96680	S 11	−2.02246	S 11	−2.08411	S 11	−2.14213
				Cu 12	1.50552	Cu 12	2.29673	Cu 12	2.00361			Ag 12	0.95705	Ag 12	0.91162	Ag 12	0.96915

2. *Adsorption of Cu, Ag, Ag and S, or Cu and S Atoms on the Pt (100) Plane*

The trends observed for the Pt (111) plane are preserved for the Pt (100) plane as well. The calculated results are presented in Tables 4 and 5. Sulfur again decreases considerably both the bond population among the Pt atoms and their net atomic charges. Copper again causes a greater electron redistribution than silver. As in the previous case, sulfur compensates largely the electron donor effect of copper, but the negative charge of the Pt atoms for simultaneous copper–sulfur adsorption is greater than that without adsorbed species. The coadsorption of silver and sulfur results in positive or negative (but less negative than in a pure cluster) charges on different Pt atoms. Since from a qualitative point of view, the results for the (100) plane do not differ from those for the (111) plane, they are no longer discussed in what follows.

3. Dissociation of O_2 over Pt Clusters

The dissociation of O_2 over Pt (111) clusters in the presence or absence of adsorbed Cu or Ag atoms was investigated by keeping the Pt–O distance constant and increasing the O–O distance. The binding energies for O_2 adsorbed normal to the surface above Pt atom 2 in the presence and absence of a Cu atom (adsorbed above the central Pt atom 1) are shown in Fig. 2. In the presence of Cu the energy curve is shifted by about 0.1 eV, toward the lower energies. The energy shift between the two curves increases with the O–O distance. The same relationship is obtained for the bond population between the oxygen atoms in the presence of Cu.[4] In this case the oxygen–oxygen bond population values are smaller than those which exist in the absence of copper. Obviously the Cu atom enhances the dissociative chemisorption of O_2 over Pt clusters.

Calculations have also been carried out for other adsorption sites, such as: Cu above Pt atom 2 and O_2 above Pt atom 1, Cu above Pt atom 2 and O_2 above Pt atom 5. Even though in the latter case there are no direct interactions between O_2 and Cu, there are indirect interactions between them via the cluster. The results of these calculations confirmed that the dissociation of O_2 takes place easier in the presence of Cu. The calculations carried out in the present paper allow one to propose the following explanation: The molecular orbitals of O_2 before adsorption have much lower energy levels than the molecular orbitals of the Cu-containing Pt clusters. For this reason after the adsorption one can distinguish the group of low-energy orbitals in which the weight of the atomic orbitals of oxygen is dominant and the group of high energies which are dominated by the Pt and/or Cu atomic orbitals. The adsorption shifts slightly the low-energy levels to even lower ones (stabilizes these orbitals). Because of the complex mixing of atomic orbitals, with close energies, in the group dominated by Pt and Cu, some of the high-energy molecular orbitals are slightly stabilized while others are slightly destabilized. While it is difficult to identify all the stabilization and destabilization details of these high-energy orbitals, the overall effect of the low and high occupied MOs is a stabilization of the system. This produces a shift of the dissociation energy curve toward lower (more negative) energies in comparison with Cu-free Pt clusters and enhances the dissociation of O_2.

The main process behind this energetic description is the electron transfer from the cluster to the π^* antibonding orbital of O_2 which decreases the strength of the O–O bond. The mechanism of this electron transfer can be described as follows: The adsorption of Cu results in a charge transfer from the s and d_{z^2} orbitals of copper (and to a lesser extent from the d_{xz} and d_{yz} orbitals) mainly to the d_{z^2} atomic orbitals of the Pt atoms of the first layer surrounding the central Pt atom, above which the Cu atom is adsorbed. The atomic-orbital occupation values for the occupied MOs (Table 6)

[4] For example, the O–O bond population for an O_2 molecule is 0.6780, for O_2 adsorbed above Pt 2 it is 0.4366, for O_2 above Pt 2 and Cu above Pt 1 it is 0.4299, and for O_2 above Pt 2 and Cu above Pt 5 it is 0.4291. The above values are calculated before any elongation of the O–O bond takes place.

show that after the adsorption of O_2 the main electron transfer takes place from the electron-enriched (enriched because of the presence of Cu) d_{z^2} (and to a lesser extent from) d_{xy}, d_{yz}, and d_{xz} atomic orbitals of the Pt atom above which O_2 is adsorbed, to the other Pt atoms and to the antibonding π^* MO of O_2. (For the latter they are not coming from d_{z^2} because of symmetry considerations.) The main electron density changes occur on the Pt atom above which the adsorption takes place and its immediate

TABLE 4

Bond Population Matrices and Charge Distribution in Pt (100) Cluster after Adsorption of Single Cu, Ag, or S atoms

System	Atom	Bond population matrix			Atomic charge
		1	2	6	
Nine Pt atoms in (100) planc without adsorbed atoms	1	9.5668	0.1930	0.1770	−0.30814
	2	0.1936	9.7337	0.2023	−0.03823
	3	0.1936	0.0048	0.2023	−0.03823
	4	0.1936	0.0001	0.0006	−0.03823
	5	0.1936	0.0048	0.0006	−0.03823
	6	0.1770	0.2023	9.3789	0.11527
	7	0.1770	0.0006	0.2118	0.11527
	8	0.1770	0.0006	0.0053	0.11527
	9	0.1770	0.2023	0.2118	0.11527
Nine Pt atoms in (100) plane and a Cu atom adsorbed above the central Pt atom 1	1	9.0133	0.2009	0.1874	0.03854
	2	0.2009	9.8830	0.1976	−0.19392
	3	0.2009	0.0045	0.1976	−0.19392
	4	0.2009	0.0001	0.0005	−0.19392
	5	0.2009	0.0045	0.0005	−0.19392
	6	0.1874	0.1976	9.7441	−0.23217
	7	0.1874	0.0005	0.1949	−0.23217
	8	0.1874	0.0005	0.0038	−0.23217
	9	0.1874	0.1976	0.1949	−0.23217
	10	0.3431	0.0155	−0.0011	1.66585
Nine Pt atoms in (100) plane and a Cu atom adsorbed above the gap formed by the four Pt atoms in the second layer	1	9.5577	0.1917	0.1759	−0.29133
	2	0.1917	9.8185	0.1961	−0.11534
	3	0.1917	0.0043	0.1961	−0.11534
	4	0.1917	0.0001	0.0005	−0.11534
	5	0.1917	0.0043	0.0005	−0.11534
	6	0.1759	0.1961	9.5315	−0.05679
	7	0.1759	0.0005	0.1994	−0.05679
	8	0.1759	0.0005	0.0041	−0.05679
	9	0.1759	0.1961	0.1994	−0.05679
	10	−0.0032	−0.0001	0.0785	0.97987
Nine Pt atoms in (100) plane and a Ag atom adsorbed above the central Pt atom 1	1	9.3985	0.1947	0.1777	−0.26640
	2	0.1947	9.8100	0.1976	0.11299
	3	0.1947	0.0047	0.1976	−0.11299
	4	0.1947	0.0001	0.0006	−0.11299
	5	0.1947	0.0047	0.0006	−0.11299
	6	0.1777	0.1976	9.5567	−0.04822
	7	0.1777	0.0006	0.2020	−0.04822
	8	0.1777	0.0006	0.0046	−0.04822
	9	0.1777	0.1976	0.2020	−0.04822
	10	0.2460	0.0055	0.0004	0.91126

TABLE 4—*Continued*

System	Atom	Bond population matrix			Atomic charge
		1	2	6	
Nine Pt atoms in (100) plane and a Ag atom adsorbed above the gap formed by the four Pt atoms in the second layer	1	9.5510	0.1917	0.1757	−0.28606
	2	0.1917	9.8189	0.1960	−0.11582
	3	0.1917	0.0044	0.1960	−0.11582
	4	0.1917	0.0001	0.0006	−0.11582
	5	0.1917	0.0044	0.0006	−0.11582
	6	0.1757	0.1960	9.5271	−0.04039
	7	0.1757	0.0006	0.2009	−0.04039
	8	0.1757	0.0006	0.0047	−0.04039
	9	0.1757	0.1960	0.2009	−0.04039
	10	0.0004	0.0001	0.0512	0.91092
Nine Pt atoms in (100) plane and a S atom adsorbed above Pt atom 1	1	7.9190	0.1758	0.1736	0.73221
	2	0.1758	9.5917	0.1978	0.09925
	3	0.1758	0.0034	0.1978	0.09925
	4	0.1758	0.0001	0.0004	0.09925
	5	0.1758	0.0034	0.0004	0.09925
	6	0.1736	0.1978	9.2730	0.23746
	7	0.1736	0.0004	0.2026	0.23746
	8	0.1736	0.0004	0.0046	0.23746
	9	0.1736	0.1978	0.2026	0.23746
	10	1.3003	0.0390	−0.0007	−2.07904
Nine Pt atoms in (100) plane and a S atom adsorbed above the gap formed by the four Pt atoms in the second layer	1	9.2776	0.1781	0.1571	0.05145
	2	0.1781	9.6788	0.1835	0.04561
	3	0.1781	0.0029	0.1835	0.04561
	4	0.1781	0.0001	0.0002	0.04561
	5	0.1781	0.0029	0.0002	0.04561
	6	0.1571	0.1835	8.9842	0.43401
	7	0.1571	0.0002	0.1606	0.43401
	8	0.1571	0.0002	−0.0033	0.43401
	9	0.1571	0.1835	0.1606	0.43401
	10	0.0011	−0.0001	0.3212	−1.96993

neighbors. This justifies the assumption on which all cluster model calculations are based, namely, that the chemical interactions during adsorption are largely localized to a small number of atoms. One should note that, in the presence of O_2, the electron transfer from Cu to the Pt cluster is somewhat reduced compared to the transfer in the absence of O_2.

The atomic orbitals involved in the charge transfer, when the adsorbed metal is Ag, are the same as those for Cu. However, the interactions among orbitals are weaker and the electron transfer from the silver atom to the cluster and from the cluster to the O_2 molecule is reduced in comparison with Cu. As a result, the effect of silver on the dissociative chemisorption of O_2 is weaker than that of copper. The shift toward lower energies of the dissociation energy curve of O_2 in the presence of Ag, is about half of that obtained in the presence of Cu.

Figure 2 also contains the calculated energy curve for the dissociation of O_2 adsorbed normal to the surface of a Pt (111) cluster above Pt atom 2, in the presence of a S atom adsorbed above Pt atom 1. As can be seen, the dissociative chemisorption of O_2 is inhibited by the presence of sulfur, since the curve is shifted toward higher energies by about 0.3 eV. This shift is larger

TABLE 5

Atomic Charge Distribution after Simultaneous Copper–Sulfur or Silver–Sulfur Adsorption on Pt (100) Plane

Cu over Pt atom 2		S over Pt atom 2		S over Pt atom 1 Cu over Pt atom 2		S over Pt atom 2 Cu over Pt atom 1		S over Pt atom 4 Cu over Pt atom 2		Ag over Pt atom 2		S over Pt atom 1 Ag over Pt atom 2		S over Pt atom 2 Ag over Pt atom 1		S over Pt atom 2 Ag over Pt atom 4	
Pt 1	−0.29230	Pt 1	0.25026	Pt 1	0.67082	Pt 1	0.63449	Pt 1	−0.05953	Pt 1	−0.29621	Pt 1	0.74111	Pt 1	0.32964	Pt 1	0.25744
Pt 2	−0.05029	Pt 2	0.71988	Pt 2	0.23315	Pt 2	0.13407	Pt 2	0.46546	Pt 2	−0.08971	Pt 2	0.06792	Pt 2	0.61567	Pt 2	0.65214
Pt 3	−0.18306	Pt 3	0.05060	Pt 3	−0.07143	Pt 3	−0.15680	Pt 3	−0.17258	Pt 3	−0.12982	Pt 3	0.00509	Pt 3	−0.01701	Pt 3	−0.04894
Pt 4	−0.18569	Pt 4	0.08150	Pt 4	−0.10247	Pt 4	−0.18882	Pt 4	0.20785	Pt 4	−0.12298	Pt 4	−0.01037	Pt 4	−0.00705	Pt 4	0.04062
Pt 5	−0.18306	Pt 5	0.05060	Pt 5	−0.07143	Pt 5	−0.15680	Pt 5	−0.17258	Pt 5	−0.12982	Pt 5	0.00509	Pt 5	−0.01701	Pt 5	−0.04894
Pt 6	−0.10170	Pt 6	0.22904	Pt 6	−0.00248	Pt 6	−0.11087	Pt 6	0.09681	Pt 6	−0.04447	Pt 6	0.07982	Pt 6	0.06701	Pt 6	0.05806
Pt 7	−0.16776	Pt 7	0.22760	Pt 7	−0.09823	Pt 7	−0.16652	Pt 7	−0.10863	Pt 7	−0.05704	Pt 7	0.04273	Pt 7	0.07068	Pt 7	0.06663
Pt 8	−0.16776	Pt 8	0.22760	Pt 8	−0.09823	Pt 8	−0.16652	Pt 8	−0.10863	Pt 8	−0.05704	Pt 8	0.04273	Pt 8	0.07068	Pt 8	0.06663
Pt 9	−0.10170	Pt 9	0.22904	Pt 9	−0.00248	Pt 9	−0.11087	Pt 9	−0.09681	Pt 9	−0.04447	Pt 9	0.07982	Pt 9	0.06701	Pt 9	0.05806
Cu 10	1.43331	S 10	−2.06615	S 10	−1.99160	S 10	−2.03857	S 10	−2.15991	Ag 10	0.97156	S 10	−1.99944	S 10	−2.05908	S 10	−2.10907
				Cu 11	1.53438	Cu 11	2.32722	Cu 11	2.30216			Ag 11	0.94549	Ag 11	0.87966	Ag 11	1.00737

TABLE 6

Atomic-Orbital Population on Pt (111) Cluster in the Presence or Absence of Adsorbed Species

Orbital	Pure cluster		O_2 above Pt 2				Cu above Pt 1			Cu above Pt 1 and O_2 above Pt 2					S above Pt 1			S above Pt 1 and O_2 above Pt 2				
	Pt 1	Pt 2	Pt 1	Pt 2	O 1	O 2	Pt 1	Pt 2	Cu	Pt 1	Pt 2	Cu	O 1	O 2	Pt 1	Pt 2	S	Pt 1	Pt 2	S	O 1	O 2
s	0.55323	0.30177	0.53215	0.36106	1.66839	1.82368	0.58931	0.31464	0.16824	0.56760	0.36147	0.20885	1.66975	1.81869	0.51246	0.27924	1.74479	0.50176	0.33606	1.74805	1.66816	1.82147
P_x	0.17539	0.8610	0.17833	0.11989	1.90312	1.99012	0.20531	0.08721	0.01563	0.19327	0.11762	0.02698	1.94188	1.98530	0.20671	0.08118	1.86452	0.20878	0.11372	1.85793	1.88048	1.98254
P_y	0.17539	0.09149	0.17639	0.14244	1.91853	1.99246	0.20531	0.09526	0.01563	0.19106	0.14303	0.01951	1.94024	1.99530	0.20671	0.09309	1.86452	0.20734	0.14273	1.86555	1.92037	1.99194
P_z	0.10310	0.05150	0.10264	0.25224	0.924	1.23048	0.17847	0.05662	−0.00876	0.16661	0.28717	0.00592	0.92955	1.22454	0.34805	0.04985	1.70464	0.35334	0.24383	1.70475	0.92619	1.22930
$d_{x^2-y^2}$	1.81214	1.86461	1.58314	1.81303			1.81417	1.90832	1.99721	1.77252	1.90677	1.99802			1.75227	1.85004	0.08786	1.56889	1.79646	0.08372		
d_{z^2}	1.82650	1.83612	1.79822	0.89599			1.47183	1.92002	1.43771	1.39923	0.99838	1.38738			1.13932	1.82163	0.04527	1.10186	0.86459	0.04142		
d_{xy}	1.81214	1.86513	1.74394	1.83843			1.81417	1.87931	1.99721	1.81939	1.85944	1.99888			1.75227	1.87555	0.08786	1.17843	1.85477	0.08577		
d_{xz}	1.91437	1.94945	1.86400	1.49220			1.85362	1.94874	1.85452	1.87050	1.92113	1.95079			1.65112	1.90135	0.35370	1.59407	1.38413	0.32509		
d_{yz}	1.91437	1.96025	1.91082	1.68877			1.85362	1.97613	1.85452	1.88063	1.92113	1.93333			1.65112	1.95674	0.35370	1.64830	1.71283	0.35546		

Note. Pt 1 and Pt 2 are the platinum atoms 1 and 2 while O 1 and O 2 are the oxygen atom which is adsorbed and the second atom, respectively.

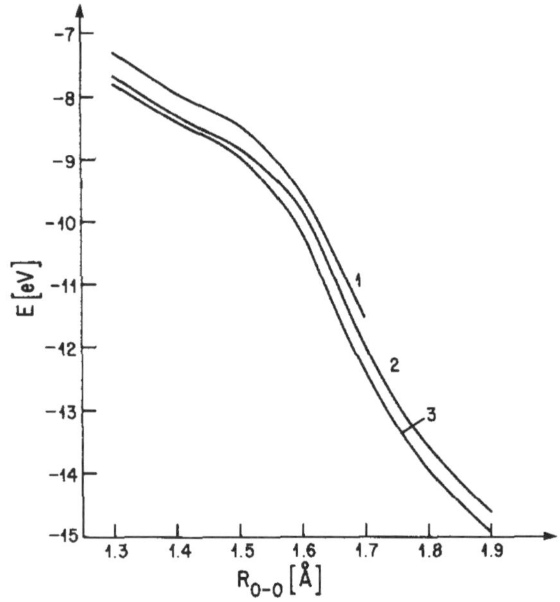

FIG. 2. Potential energy curves for the dissociation of O_2 adsorbed normal to the surface of a Pt (111) cluster above Pt atom 2 vs the bond lengths R_{O-O}: (1) dissociation of O_2 in the presence of S adsorbed above Pt atom 1; (2) dissociation of O_2 over a pure cluster; (3) dissociation of O_2 in the presence of Cu adsorbed above Pt atom 1.

than the shift in the opposite direction caused by the presence of a Cu or Ag atom. Therefore, one might expect that the inhibiting effect caused by a single S atom prevails over the promoting one caused by a single Cu or Ag atom. This has been, indeed, observed experimentally (*1*).

4. Dissociation of O_2 over Ag Clusters in the Presence and Absence of Chlorine

Calculations have been carried out to investigate the effect of adsorbed Cl on the ability of silver (111) clusters to dissociate adsorbed oxygen molecules. The *d* orbital functions of chlorine have been used in some of the calculations, but led to unrealistically high charge numbers and very high occupation numbers for these orbitals. For this reason, in the calculations presented here, the basis set of wavefunctions does not include *d* orbitals. Two configurations of the O_2 molecule have been assumed in the computation: (a) O_2 is adsorbed normal to the cluster surface and above silver atom 2; (b) the O_2 molecule is adsorbed parallel to the cluster surface and is parallel to and above the bond between silver atoms 5 and 6. These two possibilities have been chosen because they are often discussed in literature. Kilty and Sachtler (*9*) favor the former, while Kenson and Lapkin (*10*) prefer the latter configuration.

It is generally accepted that the nondissociatively adsorbed O_2 is responsible for the epoxidation reaction and therefore, if the presence of Cl in the feed improves the selectivity of the catalyst, this means that Cl should decrease the rate of dissociation of O_2. The dissociation energy curves for the parallel adsorption of O_2 are plotted in Fig. 3, which show that the presence of Cl adsorbed above Ag atom 1, shifts the curve toward higher energies by about 1.5 eV. The change of the oxygen–oxygen bond population value follows the same trend, e.g., they are slightly larger than those in the absence of Cl. Obviously, chlorine

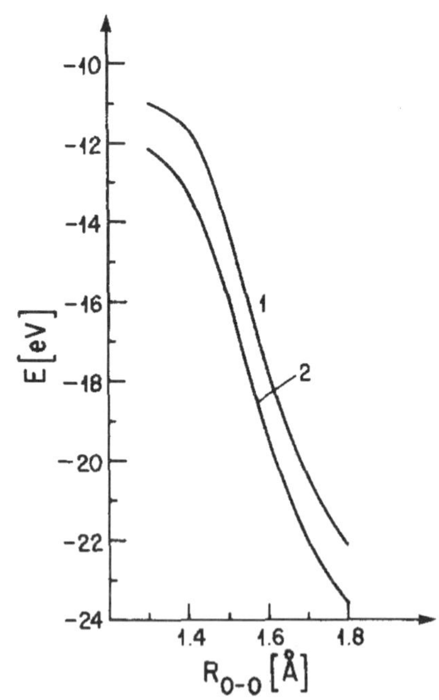

FIG. 3. Potential energy curve for the dissociation of O_2 adsorbed parallel to the surface of a Ag (111) cluster above the bond between Ag atom 5 and Ag atom 6: (1) dissociation of O_2 in the presence of Cl adsorbed above Ag atom 1; (2) dissociation of O_2 over a pure cluster.

makes both the adsorption and dissociation of O_2 more unfavorable.

The present calculations allow us to suggest the following explanation for the effect of chlorine. The *s* orbitals of silver are the main orbitals that interact with the O_2 molecule. Electrons from the *s* orbitals are transferred to the π^* antibonding MO of adsorbed O_2. Chlorine interacts with the silver cluster in a similar way, electrons being transferred mainly from the silver *s* orbitals to the *p* orbitals of chlorine. Because of the competition for electrons between O_2 and Cl, the electron transfer from the cluster to the oxygen molecule is reduced. As a consequence, the bond strength between the oxygen atoms is increased and the dissociation rate of adsorbed O_2 is decreased. In addition, the molecular orbitals of oxygen are stabilized by the interaction with the orbitals of Ag and Cl, while the molecular orbitals of the cluster are destabilized. However, the destabilization overcomes the stabilization and leads to a lower heat of adsorption of O_2 in the presence of chlorine, reducing the amount which is adsorbed. For the above reasons, the overall activity is reduced while the selectivity for epoxidation is increased.

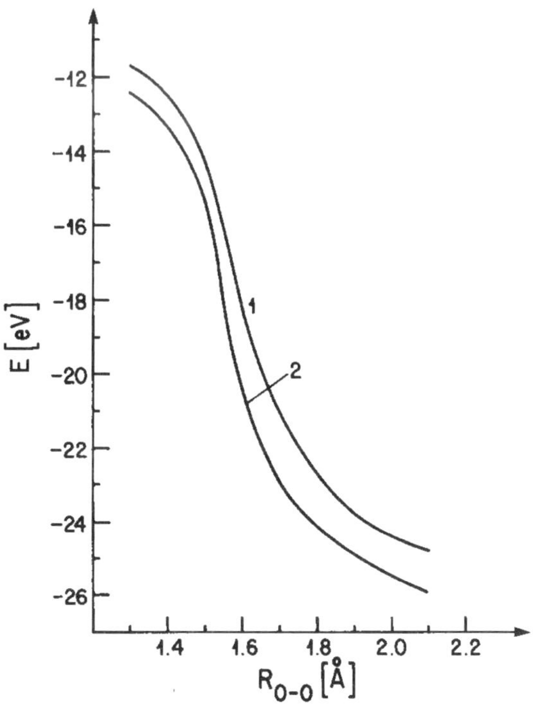

FIG. 4. Potential energy curves for the dissociation of O_2 adsorbed normal to the surface of a Ag (111) cluster above Ag atom 2: (1) dissociation of O_2 in the presence of Cl adsorbed above Ag atom 1; (2) dissociation of O_2 over a pure cluster.

The results of the calculations for the adsorption of O_2 normal to the surface are shown in Fig. 4. The results are similar to those for the parallel adsorption. However, the effect of Cl adsorbed above Ag atom 1 is now weaker, since the energy shift of the potential energy curve is about 0.9 eV instead of 1.5 eV. Calculations have also been carried out with O_2 adsorbed normal to the surface above Ag atom 2 and Cl adsorbed above Ag atom 5. They show that, in this case, there is no direct interaction between O_2 and Cl. They interact, however, via the cluster atoms. The results are similar to those obtained for the other adsorption sites.

5. *Effect of cluster Size*

Previous calculations based on the EHM (*17*) carried out for Ag and Pd clusters indicate a dependence of the ionization potential, IP [identified with the highest occupied MO (HOMO)], and of the electron affinity, EA [identified with the lowest unoccupied MO (LUMO)], on the size of the cluster. These two quantities affect the charge transfer from or to the cluster. They have practically identical values for bulk metals, but they differ significantly when the number of atoms in the cluster is small. Increasing the number of atoms results in a rapid decrease of the energy gap between them. The above-mentioned calculations (*17*) show a rapid change of the IP and the EA up to 10–15 atoms and a much slower change for larger sizes. Therefore, the charge transfer to or from the cluster is not expected to differ drastically for clusters between 10 and 20 atoms. Furthermore, EH calculations (*18*) carried out for the dissociative chemisorption of H_2 on Pd clusters of different sizes show that the adsorption characteristics of H_2 depend insignificantly on the cluster size, but depend on the adsorption site. The latter observation is also confirmed by our calculations. The change of the adsorption site changes, however, only the numerical values, but does not change the observed trend of the effect of the coadsorbed atoms on the dissociation of O_2. We note, that the increased number of cluster atoms could result in a larger charge transfer from the Cu or Ag atom to the cluster, since the LUMO shifts toward lower energies with the increase of the size of the cluster. This could contribute to an increased compensation of the inhibition effect of S. However, it is difficult to believe that the orbital interaction between the cluster orbitals and the orbitals of Cu and Ag would become so strong with increasing cluster size, that the resulting shift of the dissociation energy curve of O_2 toward lower energy would overcome the opposite shift caused by the strong cluster–sulfur interaction and, thus, change the observed trends.

6. Catalytic Implications

Let us consider first the complete oxidation of C_2H_4 to CO_2 and H_2O. Ethylene is a well-known nucleophilic reagent. Therefore, one might expect that the more positive the surface of the catalyst is, the easier the adsorption of C_2H_4 will take place and the larger the reaction rate will be. If the copper- or silver-treated Pt/Al_2O_3 catalyst is used instead of pure Pt/Al_2O_3, our calculations indicate that the catalyst surface is more negatively charged. Hence, the catalytic activity should decrease. If, however, the rate-determining step is the interaction between the adsorbed oxygen atoms and the gas-phase ethylene, the reaction rate should increase, since, as our calculations show, the dissociation of oxygen molecules on the more negatively charged surface of the catalyst is easier. The experimental results in (*1*) seem to support the latter conclusion. This point of view is compatible with the explanation given in Ref. (*2*), where the decreased selectivity of a Ag catalyst, used for ethylene epoxidation, supported on a germanium oxide- (n-type semiconductor) doped alumina, is explained in terms of electron transfer from the support to the silver clusters. This decreases the fraction of nondissociatively adsorbed oxygen molecules, which are probably responsible for the selective epoxidation, and increases the fraction of dissociatively adsorbed oxygens which participate in the complete oxidation reaction.

The addition of Cu or Ag to the Pt/Al_2O_3 catalyst has similar effects on the reaction rate in the oxidation of CO. According to the Blyholder mechanism (*11, 12*), the adsorption of CO on the surface of transition metals involves electron transfer from and to the metal surface. The electron transfer to the metal atoms is from the nonbonding orbital of CO. This does not affect the bond strength between the C and O atoms. On the contrary, the electrons coming from the metal fill the antibonding orbitals of CO thus changing its reactivity. Therefore, the increased electron density at the surface of the Cu- or Ag-treated Pt/Al_2O_3 catalyst should increase the catalytic activity in the CO oxidation. In addition, as the curves in Fig. 2 show, the bimetallic catalyst enhances the dissociation of O_2 thus further contributing to the increase of the reaction rate. These theoretical considerations are again supported by the experiments of Ref. (*1*).

Let us now consider the poisoning effect of sulfur. Somorjai (*13*) has suggested that the structural change from (111) to (100) planes, which occur on the surface of Pt crystallites after adsorption of sulfur, constitutes the main reason for the inhibition of platinum catalyst by sulfur. The results of Sivasanker and Ravaswamy (*14*) and Maurel *et al.* (*15*) show that this interpretation cannot always explain the experimental facts. While it is obvious that any structural change will bring changes in the electronic properties of the catalyst, the adsorption of contaminants can change the electronic properties of the atoms at the surface without causing major structural rearrangements of the surface atoms. Previous quantum-chemical calculations (*16*) show that structural changes of the (111) and (100) Pt planes can, indeed, occur after adsorption of electron acceptor elements, such as sulfur or chlorine, because of the significant decrease in the bond strength among the atoms of the cluster. However, the same calculations indicate that the inhibiting effect of sulfur is also caused by a change in the electronic properties of the platinum atoms themselves. It should be noted that other mechanisms of the inhibition by sulfur include: the blocking of the active sites and/or its reaction with the reactants to produce undesirable reaction pathways. Without excluding the latter possibilities, we will emphasize the electronic changes brought on the platinum atoms by the adsorption of sulfur.

To explain the inhibiting effect of sulfur, we use the calculated atomic-orbital population values of a Pt (111) cluster (Table 6) in the presence and absence of sulfur. These values suggest that sulfur adsorbed above the central Pt atom 1, accepts electrons, mainly, from the d_{z^2} (and to a lesser extent from the d_{xz} and d_{yz} atomic orbitals) of that atom. Electrons are also taken from the s, d_{z^2}, d_{xz}, and d_{yz} atomic orbitals of the surrounding Pt atoms. Similar results have been obtained (Table 6) when O_2 is absorbed above Pt atom 2 of a sulfur-free Pt (111) cluster. The O_2 molecule accepts electrons, mainly, from the d_{z^2}, d_{xz}, and d_{yz} orbitals of that atom. However, in the presence of sulfur, the electron population of these orbitals is diminished. Hence, the electron transfer to the antibonding MO of the adsorbed O_2 is impeded and the dissociative chemisorption of O_2 is inhibited. The chemisorption of O_2 on a S-free Pt (111) cluster is accompanied by the following energy changes: The low-energy cluster-adsorbate MOs are composed, mainly, of AO of oxygen. These orbitals are, however, stabilized (shifted toward lower energies) in comparison with the MOs of a free O_2 molecule. The cause of this stabilization is the interaction of the low-energy orbitals of O_2 with the high-energy orbitals of the cluster. The stabilization of these orbitals is accompanied by the destabilization of the higher-energy MOs of the cluster-adsorbate, which are composed, mainly, of Pt AO. Concerning the effect of S, we mention that the ionization potential of the 3*s* AO of S is smaller than that of the 2*s* AO of O, but higher than that of the 5*d* AO of Pt. That is why the simultaneous chemisorption of O_2 and S results in further stabilization of the MOs composed mainly of oxygen AO and further destabilization of the high-energy MOs, composed mainly of Pt AO. This resultant destabilization caused by O_2 and S overcomes the stabilization of the lowest (cluster-adsorbate) MOs and thus decreases the heat of adsorption of O_2 over the sulfur-contaminated cluster by about 0.3 eV (Fig. 2).

It is worth noting that in contrast to the energy curves for the dissociation of O_2 (Fig. 2), the numerical values of the atomic charges of the Pt atoms (Table 3) seem to suggest that the promoting effect of a Cu atom is stronger than the inhibiting effect of a S atom. However, as we have mentioned before, the adsorption of Cu above the Pt cluster results in a major change of the net atomic population, while the bond population values are only slightly affected. This means that even though the charges of the atoms are high, the shifts toward lower energies of the cluster-adsorbate MOs dominated by the AOs of Pt, are small. These shifts, as well as the slight stabilization of the cluster-adsorbate MOs dominated by the AOs of oxygen, contribute to the shift of the O_2 dissociation curve toward lower energies. Sulfur stabilizes slightly the cluster-adsorbate MOs dominated by the AOs of oxygen, but destabilizes the MOs dominated by the AOs of Pt. This destabilization is higher than the stabilization caused by Cu, although the atomic charges generated in the presence of sulfur are smaller in absolute value than those caused by Cu.

The promoting effect of small amounts of chlorine on the selectivity of the silver catalysts in the reaction of ethylene epoxidation (*3*) can be explained in a similar way. As the calculations indicate, chlorine adsorption increases the probability for nondissociative chemisorption of O_2 molecules adsorbed in any of the two adsorption forms and decreases the heat of adsorption of O_2. Concerning the two chemisorbed forms of oxygen molecules, it seems that in the absence of chlorine and at a bond length between the oxygen atoms equal to that in the free molecule, the form adsorbed normal to the surface is favored energetically. Indeed, in this case, this form has a heat of adsorption about 0.25 eV larger than that of form b, adsorbed parallel to the surface. However, the distance between the dissociation curves decreases with the elongation of the O_2 molecule and they cross one another at an oxygen–oxygen distance of about 1.4 Å (Fig. 5). Therefore, one can tentatively conclude that during the dissociation process form a may transform into form b before the breaking of the oxygen–oxygen bond, which takes place at a bond length of about 1.5 Å for both adsorption forms. It seems that in the absence of Cl both forms may exist. In the presence of Cl, the two curves (Fig. 6) meet at a bond length of about 1.5 Å without crossing one another. This means that in the presence of Cl, the adsorbed form a is favored energetically to form b. It is difficult to conclude whether one of the two adsorption forms, or both of them, take part in the epoxidation reaction since ethylene itself may play a part in this choice. Most probably both forms exist simultaneously, even though the normal configuration seems to be slightly prevailing energetically.

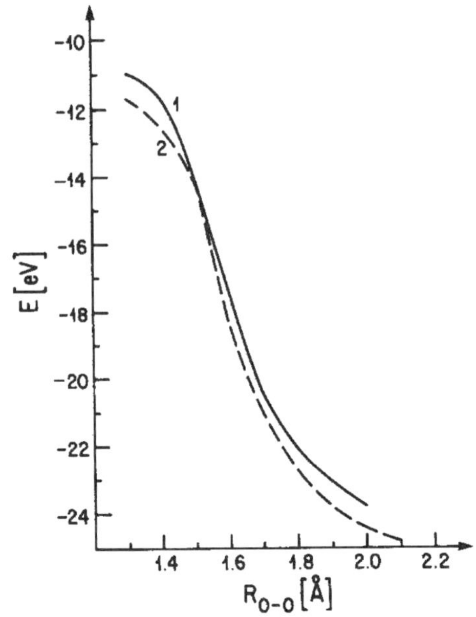

FIG. 6. Potential energy curves for the dissociation of O_2 over a Ag (111) cluster in the presence of Cl adsorbed above Ag atom 1: (1) O_2 adsorbed parallel to the cluster above the bond between Ag atom 5 and Ag atom 6; (2) O_2 adsorbed normal to the cluster above Ag atom 2.

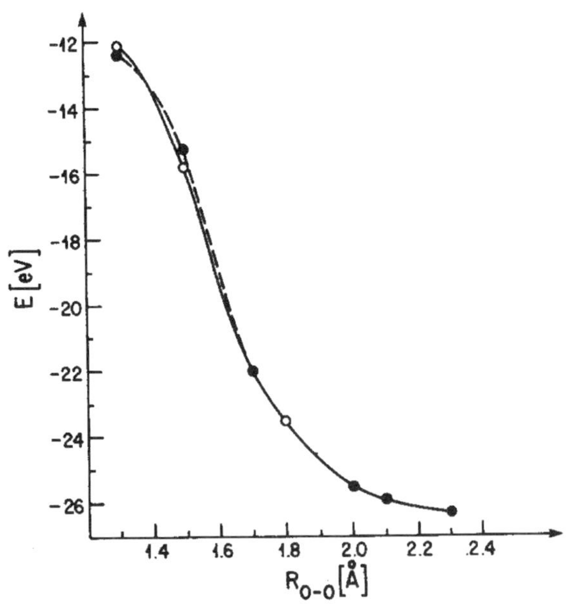

FIG. 5. Potential energy curves for the dissociation of O_2 over a pure Ag (111) cluster: (●) O_2 adsorbed normal to the cluster above Ag atom 2; (○) O_2 adsorbed parallel to the cluster above the bond between Ag atom 5 and Ag atom 6.

CONCLUSIONS

The EH method has been applied to study the promoter and poisoning effects of different additives to Pt and Ag cluster catalysts and the calculated results have been correlated with the catalytic activity for ethylene oxidation, CO oxidation, and ethylene epoxidation. Additives such as Cu or Ag cause a shift toward lower energies of the O_2 dissociation curve and slightly increase the population of the π^* antibonding MO of O_2. These effects favor the dissociative chemisorption of O_2 and constitute the main reason for the increased catalytic activity in reactions such as CO oxidation and the complete oxidation of ethylene. For the latter reaction, the computations favor a mechanism in which C_2H_4 from the gas phase interacts with adsorbed O atoms. The inhibition by sulfur for the above reactions is due to the shift of the dissociation curve of O_2 toward higher energies and the decrease of the population of the π^* antibonding MO of O_2. Since sulfur affects both the adsorption of CO and the dissociation of O_2, the reaction of CO oxidation is more sensitive to sulfur poisoning than the complete oxidation of C_2H_4. The poisoning effect of sulfur is stronger than the promoter effect of Cu or Ag, because the electronic changes of the catalyst caused by sulfur are much greater than those caused by Cu or Ag.

The effect of Cl on the dissociation of O_2 over Ag clusters is similar to the effect of S on Pt clusters and the origin of this effect is the same for both cases. Concerning the epoxidation reaction, Cl increases the selectivity because it inhibits the dissociative chemisorption and decreases the overall activity since it decreases the heat of adsorption.

Our calculations suggest that most probably two types of O_2 chemisorbed on Ag may exist—normal to the surface and parallel to it. The normal form eventually transforms into the parallel one with the elongation of the O–O bond before the dissociation occurs. The presence of Cl changes this behavior favoring, under all conditions, the normal configuration.

REFERENCES

1. Kummer, J. T., *J. Catal.* **38,** 166 (1975).
2. Verikios, X. E., Stein, F. P., and Coughlin, R. W., *J. Catal.* **66,** 147 (1980).
3. Verikios, X. E., Stein, F. P., and Coughlin, R. W., *Catal. Rev. Sci. Eng.* **22**(2), 1987 (1980).
4. Hoffman, R., *J. Chem. Phys.* **39,** 1397 (1963).
5. Mulliken, R. S., *J. Chem. Phys.* **23,** 1833 (1955).
6. Lu, C. C., Carlson, T. A., Malik, F. B., Tucker, T. C., and Nestor, C. W., *At. Data* **3,** 1–131 (1971).
7. Moore, C. E., *Natl. Bur. Stand (U.S.) Circ.,* 1–3 (1949).
8. Basch, H., and Gray, H. B., *Theor. Chim. Acta* **4,** 367 (1966).
9. Kilty, P. A., and Sachtler, W. M. H., *Catal. Rev. Sci. Eng.* **10**(1), 1 (1974).
10. Kenson, R. E., and Lapkin, M., *J. Phys. Chem.* **74,** 1493 (1970).
11. Blyholder, G., *J. Phys. Chem.* **68,** 2772 (1974).
12. Doyen, G., and Ertl, G., *Surf. Sci.* **43,** 197 (1974).
13. Somorjai, G. A., *J. Catal.* **27,** 453 (1972).
14. Sivasanker, S., and Ravaswamy, A. V., *J. Catal.* **37,** 553 (1975).
15. Maurel, R., Leclercq, G., and Barbier, J., *J. Catal.* **37,** 324 (1975).
16. Halachev, T., and Ruckenstein, E., *Surf. Sci.* **108,** 292 (1981).
17. Baetzold, R., *Surf. Sci.* **51,** 1 (1975).
18. Baetzold, R., *J. Catal.* **29,** 129 (1973).

ADSORPTION ISOTHERMS AND CATALYST SELECTIVITY FOR A UNIFORM SURFACE: THE ROLE OF MULTIPLE LANDING AREAS

ELI RUCKENSTEIN

Department of Chemical Engineering
State University of New York at Buffalo
Amherst, NY 14260

DADY B. DADYBURJOR

Department of Chemical Engineering and Environmental Engineering
Rensselaer Polytechnic Institute
Troy, NY 12181

(Received June 26, 1981; in final form October 20, 1981)

In general a gas phase species can be present simultaneously in more than one adsorbed form on a uniform solid surface. The different adsorbed forms tie up different combinations of surface sites. Each such combination is termed a landing area. The different adsorbed forms present on the surface of a catalyst may form different products. (In the more general case, each different adsorbed form yields a different distribution of a set of products.) Hence the relative amount of each adsorbed form and landing area present on the surface determines the selectivity of the solid catalyst. The variation of the landing area distribution with temperature and partial pressure of the gas phase species is examined in the first part of this work. In general, increasing the pressure or decreasing the temperature leads to the relative decrease of molecules adsorbed on landing areas consisting of larger number of sites and the relative increase of molecules adsorbed on smaller landing areas. The kinetics of metal catalyzed skeletal isomerization and hydrogenolysis of hydrocarbons on supported metal catalysts is examined in the framework of the landing area in the second part of this work. A simple model for the variation of reaction order with pressure agrees with experimental data from other laboratories.

1. INTRODUCTION[1]

It is axiomatic that the first steps of a solid-catalyzed reaction include the (chemical) adsorption of at least one of the reactants on to the solid surface. In the simple case of a single reactant, the adsorbed species may undergo one or more reactions on the surface and the product eventually desorbs back to the bulk fluid phase. For a single reaction on the surface, the process may be written as:

$$A(g) + nS \rightleftharpoons A(a) \rightarrow B(a) \rightleftharpoons B(g) + nS$$

The Langmuir-Hinshelwood-Hougen-Watson (LHHW) analysis for chemisorption and reaction assumes a uniform distribution of surface sites for adsorption, an absence of interaction between adsorbate molecules, no more than monolayer coverage of adsorbate molecules on the surface, and a coverage-independent heat of adsorption.

[1]The basic idea of this work was presented by us at a meeting in Namur (Belgium) in December 1977 during the discussion following a presentation by Professor Frennet.

This paper, which was originally published in **Chemical Engineering Communications**: 14, 59–89 (1982), is republished with permission.

The inhomogeneity of the surface has been taken into account by considering a distribution of energies among the sites on the surface (see for example Brunauer *et al*, 1942). Interactions between adsorbate species have also been taken into account (see for example Thomas and Thomas, 1967, Young and Crowell, 1962, and the references cited therein).

An implicit assumption in the above analysis is that each molecule that adsorbs on the surface does so in only one form. A considerable amount of experimental evidence indicates that in many cases a unique adsorbate is not found. For example the IR work of Eischens *et al*. (1956) showed that molecules of CO could adsorb on Pd in two forms. One adsorbed form requires one Pd site per molecule, while the other requires two. Low Energy Electron Diffraction (LEED) (Ertl *et al*, 1977) and Electron Energy Loss Spectroscopy (EELS) (Froitzheim *et al*, 1977) confirm these findings. Different forms of adsorbed hydrogen have also been postulated, from the results of temperature programmed desorption (TPD) experiments of Tsuchiya *et al* (1970), tracer studies of Paál and Thomson (1973), and inelastic neutron scattering data from Renouprez *et al* (1980).

The present analysis takes into account the simultaneous presence of various adsorbed forms of a given species. The different combinations of surface sites used by the different possible adsorbed forms of a single species we term the different landing areas available on the solid surface. Then the CO species above can be considered to adsorb in two landing area configurations; one tying up one site, the other tying up two.

For more complicated molecules, the number of landing possibilities would be much greater. The landing areas could differ from one another either in the number of the sites on which the molecule is chemisorbed, or in the arrangement of these sites (linear, cyclic, etc.) More subtle differences may also be possible. For example, two landing areas may consist of identical sites at which strong (chemisorptive) bonds exist with the adsorbate, but different sites at which weaker bonds (physisorptive, in the limiting case) exist with the adsorbate.

In this work, we show that the distribution of an adsorbing species on different landing areas depends upon properties of the fluid phase, such as temperature and pressure. Here we consider a large uniform solid surface such as that of a single face of a crystal. Our primary interest is centered on conditions where adsorption is followed by surface reaction. Then each adsorbed form may react to form a different product or, in the general case, to form a set of products with different distributions. Figure 1 illustrates this. Here reactant A may adsorb on any one of three different groups of sites, forming any one of three adsorbed states, C, D or E. Each adsorbed state could form each of three products, F, H, and I, but the relative amount of a particular product formed will vary from one adsorbed state to the next. In a special case, product F alone could be formed from adsorbed state C, product H from adsorbate D, and product I from adsorbed state E. Then the overall selectivity and activity of the catalyst is related to the distribution of the adsorbed forms on the different landing areas, and through this to fluid and solid properties. Hence, the driving force behind this paper is the expectation that catalyst activity and selectivity can be related in a rational way to the properties of the fluid phase. In companion papers (Ruckenstein

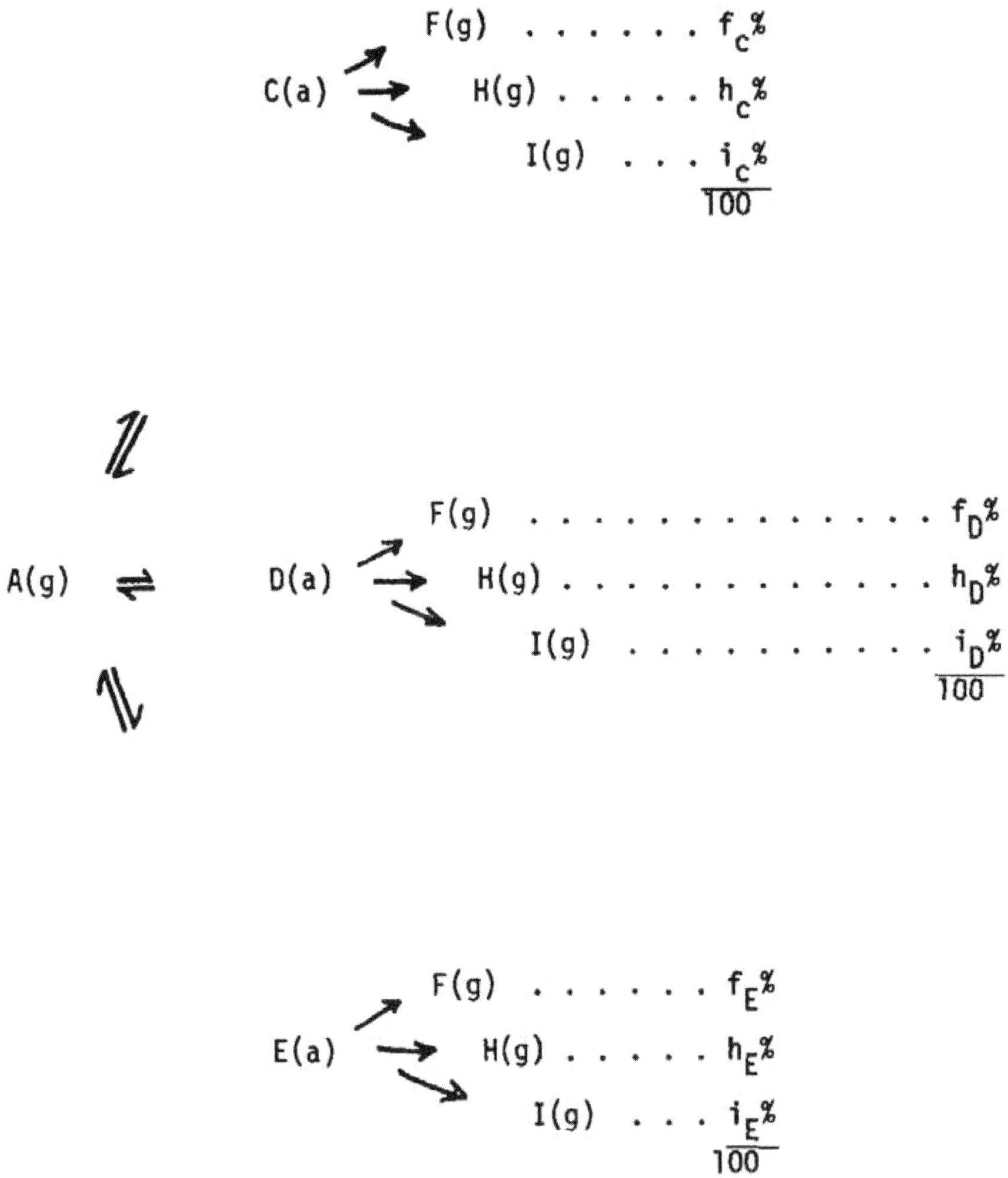

FIGURE 1 Schematic of a possible catalytic reaction of gas phase A. Three landing area conformations are hypothesized, leading to adsorbed states C, D, and E. Each adsorbed state can form products F, H and I, but in differing proportions f_j, h_j, and i_j, $j = C, D, E$. In a special case, each landing area conformation could lead to only one, unique, product. Then $f_C = h_D = i_E = 100$, all others $= 0$.

and Dadyburjor, 1980, Dadyburjor and Ruckenstein, 1981), we show the effect of properties of the solid surface, such as particle size of crystallites, on the landing area distribution.

Clear distinction must be made between the work reported here, the "landing site" concept of Frennet *et al.* (1978), the analysis of Zeldowitsch (1934) and Halsey (1952), and the results of Monte Carlo techniques reported by Callahan and Grasselli (1963), Butt and coworkers (1974), and Schwarz and Smith (1980). In all these cases, every reactant is adsorbed in a single form only, i.e., there is only one landing area per species. In the work of Frennet *et al.* (1978), experimental measurements are used to obtain an "average" landing area, consisting of a suitably defined mean number of sites. We return to this later. Zeldowitsch (1934) and Halsey (1952) quantify the inhomogeneity of the surface by a distribution of heats of adsorption. For one such distribution, i.e., a particular inhomogeneous surface, the empirical Freundlich isotherm arises naturally when the LHHW analysis is applied to each portion of the surface. The empirical Temkin isotherm can be similarly explained *a posteriori*. Callahan and Grasselli (1963), Butt and coworkers (1974), and Schwartz and Smith (1980) use Monte Carlo techniques to find the distribution of groups of adsorbed species, i.e., how many adsorbates are isolated, how many exist in groups of two, three, four, and so on, and how many surface sites are unoccupied. The distribution of different sized groups of (completely identical) adsorbate species is then related to the selectivity. This type of analysis is of value where only one type of adsorbed form/landing area configuration is expected to predominate.

The approach used in this work incorporates a thermodynamic equilibrium constant for each type of adsorbed form of each species. Of course a molecule adsorbed as one form, on a particular landing area, may change to another form involving a different collection of surface sites, i.e. another type of landing area. Even so, at equilibrium the distribution of adsorbate forms on landing areas is fixed. Throughout this work, it is assumed that adsorption equilibrium is instantaneously established on the catalytic surface. Then the values of the equilibrium constants, K_i for the adsorbed form of type i, characterize the distribution of adsorbate molecules among the possible landing area configurations.

In the present work, the simplest possible model for the solid surface is used. We assume that all sites on the surface are identical, in other words, any surface site would be as likely to be a part of one landing area configuration as any other configuration possible on that surface. We term such a surface as "uniform." A uniform surface is clearly not realistic in general but is approached, for example, by a sufficiently large single face of a crystal. The utility of our assumption lies in the fact that even for this most simple surface, the selectivity of the catalyst can be demonstrated to change with changes in temperature and pressure. The additional changes in selectivity with process conditions when the surface is not uniform are described by us elsewhere (Ruckenstein and Dadyburjor, 1980, Dadyburjor and Ruckenstein, 1981).

In Section 2, an adsorption isotherm is established which allows for a distribution of landing areas as a function of pressure and temperature. The surface is taken to be energetically and topographically uniform. The derived adsorption isotherm is then used to represent experimental adsorption data. Adsorbed forms hypothesized from TPD and other experimental measurements are consistent with the temperature and pressure dependence of the landing area distribution.

If the adsorbed species undergo reactions while on the uniform solid surface, different products may be formed from adsorbates on different kinds of landing area. Associating a given kind of landing area with a given product we suggest in Section 3 a possible effect of the pressure on selectivity. Finally we examine the role of the spectrum of the landing area in the kinetics of metal catalyzed isomerization and hydrogenolysis of hydrocarbons.

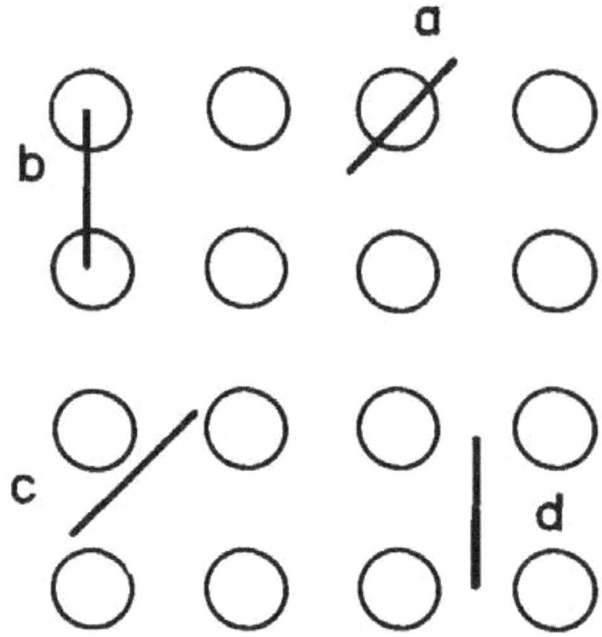

FIGURE 2 Examples of landing area choices available to a molecule for its initial adsorption on a solid surface. Circles denote sites or surface atoms. Lines at locations (a) through (d) denote possible initial locations of the adsorbate tying up 1 through 4 sites on the surface.

2. ADSORPTION

2A. The Multiple Landing Area Isotherm

In accordance with the premise of the previous section, consider a gas-solid system in which gas molecules have the option of landing on the surface in one of several ways. A molecule may adsorb on a landing area comprising only one surface site, or on a landing area of two sites, and so on. In general all such adsorbed forms will be found. Some landing area possibilities are illustrated in Fig. 2. In this regard, note that dissociative single-site adsorption is not equivalent to adsorption on a two-site landing area. In the latter case, the adsorbed species is shared by two surface sites. In the former case, each adsorbed moiety is restricted to one surface site. Further, two landing areas may require the same number of sites, but differ in the relative location of these sites. The treatment below can be extended easily to cover these cases, but they will not be explicitly considered in this work.

A central assumption is that the surface is uniform; in other words, one site could as likely host one adsorbed form as another adsorbed form. This provides us with a simple base case upon which the complexities of a realistic surface can later be added. It is further assumed that the adsorbed forms are in equilibrium with the gas phase, and that the net rate at which molecules adsorbed as one form change to other adsorbed forms is zero. Then the rate at which a molecule of species A adsorbs on a landing area of i unoccupied sites can be written as

$$r_{a,i} = k_{a,i}P_A f_a(i,\theta_{tot})$$

Here P_A is the partial pressure of A in the gas phase, and f_a is the probability of finding the appropriate number of empty sites for adsorption to occur. The latter term depends upon the number of sites required in the landing area, i, and the number of empty sites available, which is expressed in terms of the fraction of sites tied up by the adsorbate on all landing areas (total fractional coverage) θ_{tot}. The interaction forces between the adsorbed molecules and the surface, and the relative orientation of sites required for this landing area are included in the adsorption rate constant, $k_{a,i}$. The rate at which a molecule adsorbed on a landing area of i sites desorbs from the surface can be written as

$$r_{d,i} = k_{d,i} f_d(i,\theta_i)$$

Here f_d is the probability of finding the molecule adsorbed on the particular landing area, and may be considered to depend upon i and the fraction θ_i of sites on the surface that are tied up by the adsorbate on this type of landing area.

At equilibrium, adsorption and desorption rates are equal, and so

$$k_{a,i} P_A f_a(i,\theta_{tot}) = k_{d,i} f_d(i,\theta_i) \tag{1a}$$

If n different landing areas are possible, Eq. (1a) yields n relations for $(n + 1)$ unknowns. Finally, the total fractional coverage can be related to the individual landing area coverages by

$$\theta_{tot} = \theta_1 + \theta_2 + \ldots + \theta_i + \ldots + \theta_n \tag{1b}$$

To find the relative amounts of the surface covered by adsorbates on each type of landing area from Eq. (1) requires that the functional form of f_a and f_d be known. This in turn requires some knowledge of the nature of the landing area and the adsorbed species.

Consider the situation when the adsorbed species does not dissociate. Then the probability of finding a single unoccupied surface site is $(1 - \theta_{tot})$. If the unoccupied sites are randomly distributed, the probability of finding a second unoccupied site is also $(1 - \theta_{tot})$, and the probability of finding two adjacent unoccupied sites is $(1 - \theta_{tot})^2$. Consequently, the probability of finding i adjacent empty sites can be written as

$$f_a = (1 - \theta_{tot})^i$$

The functional form of f_d is more involved. The probability of finding a single site occupied by a molecule on an i-site landing area is θ_i. However, the remaining $(i - 1)$ sites occupied by the same molecule must be adjacent to the first site, and are not randomly distributed throughout the surface. It is difficult to derive an expression for f_d. However upper and lower bounds for this quantity can be obtained. Observe that the probability of finding each of the $(i - 1)$ sites lies between a lower bound of θ_i, corresponding to randomness, and an upper bound of unity corresponding to the hypothetical situation in which every site surrounding the given site is occupied by the same molecule. Hence a lower bound for f_d is

$$f_{dl} = (\theta_i)^i$$

and an upper bound for f_d is given by

$$f_{du} = \theta_i$$

Substituting the values of f_a and the lower bound of f_d into Eq. (1a) yields n equations of the type

$$k_{a,i}\mathrm{P_A}(1 - \theta_{\mathrm{tot}})^i = k_{d,i}(\theta_i)^i \tag{2a}$$

where $i = 1, \ldots, n$. The $(n + 1)$th equation is given by Eq. (1b), reproduced below

$$\theta_{\mathrm{tot}} = \theta_1 + \ldots + \theta_i + \ldots + \theta_n \tag{2b}$$

For the upper bound of f_d, the analogs of Eq. (2) are

$$k_{a,i}\mathrm{P_A}(1 - \theta_{\mathrm{tot}})^i = k_{d,i}\theta_i \tag{3a}$$

and

$$\theta_{\mathrm{tot}} = \theta_1 + \ldots + \theta_i + \ldots + \theta_n \tag{3b}$$

Now consider the adsorbing molecule dissociating into two entities. Assume that the two entities are identical (e.g. atoms from a diatomic molecule), and that each entity occupies a landing area requiring i sites. Then the relations corresponding to f_{dl} would be

$$k_{a,i}\mathrm{P_A}(1 - \theta_{\mathrm{tot}})^{2i} = k_{d,i}(\theta_i)^{2i} \tag{4a}$$

and

$$\theta_{\mathrm{tot}} = \theta_1 + \ldots + \theta_i + \ldots + \theta_n \tag{4b}$$

Similarly the relation using the upper bound of f_d would be

$$k_{a,i}\mathrm{P_A}(1 - \theta_{\mathrm{tot}})^{2i} = k_{d,i}\theta_i^2 \tag{5a}$$

and

$$\theta_{\mathrm{tot}} = \theta_1 + \ldots + \theta_i + \ldots + \theta_n \tag{5b}$$

Note that here we assume that both entities from a dissociating molecule require the same number of sites, and that two adjacent adsorbed forms are required to be similar for desorption to occur. Though these assumptions are not necessarily valid, more general expressions for adsorption and desorption rates would only increase the level of mathematical complexity without changing the qualitative picture.

The cases represented by Eq. (2) or (4) generate a closed form solution for the landing area coverages θ_i. Thus for an undissociating molecule, Eq. (2) yield

$$\theta_{\text{tot}} = \frac{\sum_{j=1}^{n} K_j^{1/j} P_A^{1/j}}{1 + \sum_{j=1}^{n} K_j^{1/j} P_A^{1/j}} \tag{6a}$$

and

$$\theta_i = \frac{K_i^{1/i} P_A^{1/i}}{1 + \sum_{j=1}^{n} K_j^{1/j} P_A^{1/j}} \tag{6b}$$

where

$$K_i \equiv k_{a,i}/k_{d,i} \tag{7}$$

represents a thermodynamic equilibrium constant for the ith landing area/adsorbed form. For a dissociating molecule, Eq. (4) yield

$$\theta_{\text{tot}} = \frac{\sum_{j=1}^{n} K_j^{1/2j} P_A^{1/2j}}{1 + \sum_{j=1}^{n} K_j^{1/2j} P_A^{1/2j}} \tag{8a}$$

and

$$\theta_i = \frac{K_i^{1/2i} P_A^{1/2i}}{1 + \sum_{j=1}^{n} K_j^{1/2j} P_A^{1/2j}} \tag{8b}$$

When the upper bound of f_d is used, Eq. (3) requires the solution of the transcendental equation

$$x(1 + K_1P_A) + K_2P_Ax^2 + \ldots + K_iP_Ax^i + \ldots + K_nP_Ax^n = 1 \tag{9a}$$

for x to enable the coverages to be obtained. The relations corresponding to Eq. (5) are

$$\theta_{\text{tot}} = 1 - x \tag{9b}$$

and

$$\theta_i = K_i P_A x^i \tag{9c}$$

The case of the dissociating molecule using f_{du}, as given by Eq. (5), yields results similar to Eq. (9) except that now

$$\theta_i = K_i^{1/2} P_A^{1/2} x^i \tag{10a}$$

and x is given by the solution of

$$x(1 + K_1^{1/2} P_A^{1/2}) + K_2^{1/2} P_A^{1/2} x^2 + \ldots + K_i^{1/2} P_A^{1/2} x^i + \ldots + K_n^{1/2} P_A^{1/2} x^n = 1 \tag{10b}$$

The relation between x and θ_{tot} is unchanged, as

$$\theta_{tot} = 1 - x \tag{10c}$$

In some cases, experimental results lie closer to those expected from one bound than the other. (See for example Section 3E). However, both models yield qualitatively similar results for the effects of pressure and temperature on the relative importance of each adsorbed form. For simplicity, therefore, these effects are demonstrated below using only the model corresponding to Eq. (6).

At very large partial pressures of A, P_A is very much greater than $P_A^{1/2}$ so that $\theta_1 \gg \theta_2$. Similarly, $\theta_2 \gg \theta_3$, and so on. Consequently, most adsorbed sites at large P_A's could be expected to contain A in the single site adsorbed form, few in the two-site form, and decreasing amounts in higher-site forms. At low values of P_A, the difference between P_A and $P_A^{1/2}$ is smaller than at high P_A's; consequently a relatively larger fraction of sites could be expected to be used for two-site adsorption, or in general for multi-site adsorption. Of course, $P_A \gg P_A^{1/2}$ implies $\theta_1 > \theta_2$ only if $(k_{a,2}/k_{d,2})^{1/2}$ is not very much greater than $k_{a,1}/k_{d,1}$. However the ratio (θ_2/θ_1) will decrease with increasing P_A irrespective of the relative magnitude of $K_2^{1/2}$ and K_1.

The obvious generalization is that increasing pressures lead to decreasing adsorption on landing areas requiring greater numbers of surface sites per molecule. The mathematical reasoning can be reinforced by a simple physical argument. At sufficiently low pressures, a large number of unoccupied sites exist per molecule of gas, hence an adsorbed species has the opportunity to tie up as many sites as it desires. As the pressure increases, so does the number of molecules in the gas phase per unit surface. Consequently, a species adsorbed on a landing area occupying a large number of sites is increasingly likely to be forced to assume a landing area which occupies fewer sites, in order to allow a gas phase molecule to adsorb on the sites relinquished. At sufficiently high pressures, of course, there will be a sufficiently large number of adsorbed species so that only single-site adsorption is possible.

The temperature variation of θ_i is enclosed in the constant K_i which can be written from Eq. (7) as

$$K_i = k_{a,i}/k_{d,i} = K_i^{\circ}\exp\left[\frac{Q_i}{RT}\right] \tag{11}$$

where Q_i is the heat of adsorption on an i-site landing area. Note that Q_i is independent of coverage (θ_i or θ_{tot}), but it will vary from one landing area type to the next, i.e. it is dependent on i. From Eq. (6), the ratio (θ_{i+1}/θ_i) increases with temperature, provided (Q_i/i) decreases with an increase in i. In general we expect Q_i to increase with i, the number of sites in the landing area, but to a small enough extent that (Q_i/i) would decrease with i. Then it can be seen that higher temperatures favor surface coverage as landing areas of larger numbers of sites.

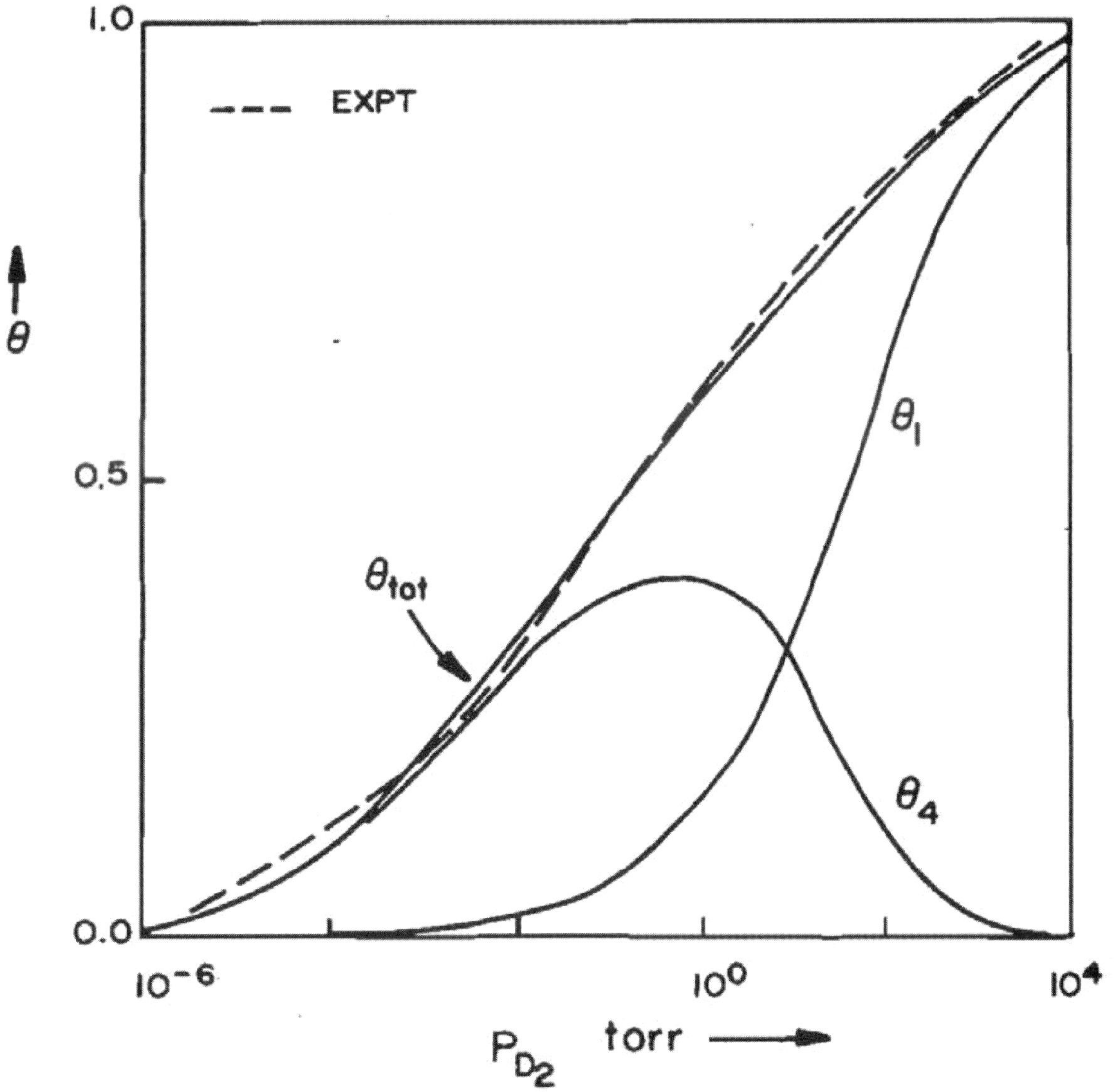

FIGURE 3 Comparison of the experimental data of Frennet *et al.* (1978) with a model incorporating several landing area possibilities on the surface. The dissociated species may adsorb on landing areas tying up from 1 through 4 sites. The experimental result is shown by the dashed curve. Contributions to the total fractional coverages from landing areas incorporating 1 and 4 sites are labelled θ_1 and θ_4 respectively. The total fractional coverage is labelled θ_{tot}.

Consider now adsorption of a single species on three different types of landing areas, occupying either a single site (characterized by a fractional number of sites θ_1), two sites (θ_2), or three sites (θ_3). Summarizing the contents of the last few paragraphs, θ_1 will be larger than θ_2 or θ_3 at high pressures and low temperatures. On the other hand, θ_3 will be a major contributor at low pressures and high temperatures. The two-site adsorbed form may or may not ever be very much larger than the other two, depending upon the values of K_i^0 and Q_i, but it will reach some maximum value at intermediate values of temperature and pressure.

2B. Comparison with Experiment—Effect of Pressure on Adsorption

Both qualitative and quantitative (fitting) comparisons are possible for the variation with pressure of the relative coverages of different landing areas.

Qualitatively, a variety of experimental tools—LEED (Ertl *et al.*, 1977), EELS (Froitzheim *et al.*, 1977), Infrared Reflection Spectroscopy (Krebs and Luth, 1977)—applied to CO adsorption on Pt (111) surfaces show that three sites are tied up by an adsorbing molecule at low coverages, and two sites at higher coverages. Although this can only be taken as an indication that the average, or most abundant, adsorbed form ties up a decreasing number of sites at increasing coverages, it does confirm that increasing concentrations in the fluid phase favor landing areas which tie up fewer surface sites.

A simple system and data over a wide range of pressures are required if we are to obtain quantitative information on landing areas from experiment. The data of Frennet *et al.* (1978) for the adsorption of D_2 on a rhodium film cover pressures from 10^{-6} to 10^4 torr, ten orders of magnitude. Here both conditions appear to have been met. In this regard it should be recalled that Renouprez *et al.* (1980) have suggested up to four-site adsorption of hydrogen atoms on small crystallites of platinum, based on the results of their inelastic neutron scattering experiments. Hence a dissociative model with landing areas requiring up to four sites can be used to analyze the data of Frennet. Fitting Eq. (10) to the experimental isotherm yields the curves shown in Fig. 3, with

$$K_1^{1/2} = 0.38 \qquad \text{torr}^{-1/2}$$

$$K_2^{1/2} = 0.31 \qquad \text{torr}^{-1/2}$$

$$K_3^{1/2} = 0.42 \times 10^{-3}\ \text{torr}^{-1/2}$$

$$K_4^{1/2} = 15.1 \qquad \text{torr}^{-1/2}$$

The four site landing area contributes most of the total fractional coverage at low values of the partial pressure. Landing areas tying up fewer sites make increasing contributions at increasing partial pressures, and the single site landing area dominates at large partial pressures. This is as expected.

It is to be expected that a model with what amounts to four adjustable parameters will fit the experimental curve very well. However it should be noted that the four parameters evolve from *a priori* experiments, and not *a posteriori* convenience. Of course, assuming that only one type of landing area exists for adsorption of the gas would also lead to an isotherm. However, it should be noted that no single integral value of i can result in an isotherm which fits the experimental data to anything near the experimental accuracy throughout the range of the pressures used in the measurements. Of course an acceptable value of a single non-zero equilibrium constant could be obtained by force-fitting data over a limited range of pressures, say one order of magnitude. But this value would be an apparent adsorption coefficient, valid only inside that particular pressure range. Clearly a combination of landing areas must be assumed to explain the experimental data.

There is a similarity between the mathematical analysis performed by Frennet *et al.* (1978) and our treatment of the data. However, an important distinction exists between the two analyses. Frennet *et al.* have tried to find one single form of adsorption, a single (noninteger) value of i, which would best fit the data. In our analysis, several types of landing areas are possible, leading to different adsorption configurations being present simultaneously on the surface. Varying conditions change the relative amounts of these configurations, and we wish to use the experimental data to illustrate and quantify these changes.

2C. Comparison with Experiment—Effect of Temperature on Adsorption

The TPD work of Tsuchiya *et al.* (1970) indicates that at least 4 types of adsorption of hydrogen are possible over a Pt surface for adsorption temperatures ranging from −196°C to 50°C. Types α and β are postulated to represent undissociated adsorption of a H_2 molecule on a single Pt site and on two Pt sites respectively. Types γ and δ are postulated to represent dissociative adsorption of hydrogen, each atom requiring one and two sites respectively. Comparing runs where the β, γ, and δ forms can be unequivocally assigned, it can be seen that when H_2 is adsorbed at lower temperatures, the β peaks are higher and broader than the same peaks for the higher temperature adsorption. Hence the area under the β peak, a measure of the amount of hydrogen present in that form, is decreased at higher temperatures. Further, the areas under the δ peaks are much more pronounced for the case of the higher temperature adsorption measurements. These qualitative observations are in line with the predictions of the temperature dependence of landing area conformations, viz. that higher temperatures favor adsorption on landing areas of larger numbers of sites.

The recent TPD measurements of Menon and Froment (1979) for H_2 adsorbed on Pt black at constant temperatures between 200°C and 500°C are also in agreement with the model. The measurements were performed under conditions such that the α and β forms cannot be detected. The γ and δ forms are detected, as well as small amounts of another form that could not be detected by the procedure of Tsuchiya *et al.* (1970). Extrapolating from the assignment of landing areas for forms γ and δ, this other form (ϵ?) could be considered to tie up three (or more) Pt sites. Comparing the surface forms present during adsorption at 200°C and 500°C, the lower temperature yields more of the γ and δ types and significantly less of the ϵ type, compared with adsorption at 500°C. Again, this is as expected.

3. SELECTIVITY

3A. Pressure Selectivity

If the solid catalyzes chemical reactions on the surface, it is reasonable to suppose that each adsorbed form may give rise to a different distribution of products after the reaction. Herein lies the reason for our attention to the different θ_i. Clearly the selectivity of a given product can be related to the fractional number of sites occupied by the corresponding adsorbed species. Hence, by adjusting the parameters (T,P) of the process to maximize the θ_i of interest, we can obtain the highest possible selectivity from the catalyst. Conversely, in many simple cases, we can analyze experimental relationships between selectivity and process parameters, and explain these in terms of landing area considerations.

In the simplest case, assume that only one reactant, A_2, is present. The molecule dissociates on the surface, and each of the entities may adsorb on landing areas requiring one, two, or three sites in the following manner

$$A_2 + 2S \rightleftharpoons 2[AS_1] \tag{12a}$$

$$A_2 + 4S \rightleftharpoons 2[AS_2] \tag{12b}$$

$$A_2 + 6S \rightleftharpoons 2[AS_3] \tag{12c}$$

Here, terms in square brackets indicate surface adsorbed forms. Assume that each adsorbed form reacts to form only one product, distinct from those obtained via the other two adsorbed forms. Thus

$$[AS_1] \rightarrow P_1 + S \tag{12d}$$

$$[AS_2] \rightarrow P_2 + 2S \tag{12e}$$

$$[AS_3] \rightarrow P_3 + 3S \tag{12f}$$

Note that Eq. (12) quantify a special case of Fig. 1, with one product only per landing area.

If adsorption and desorption are in equilibrium, then following the approximation that gives rise to Eq. (5), we can write

$$K_1P_A(1 - \theta_{tot})^2 = \theta_1^2 \tag{13a}$$

$$K_2P_A(1 - \theta_{tot})^4 = \theta_2^2 \tag{13b}$$

$$K_3P_A(1 - \theta_{tot})^6 = \theta_3^2 \tag{13c}$$

$$\theta_{tot} = \theta_1 + \theta_2 + \theta_3 \tag{13d}$$

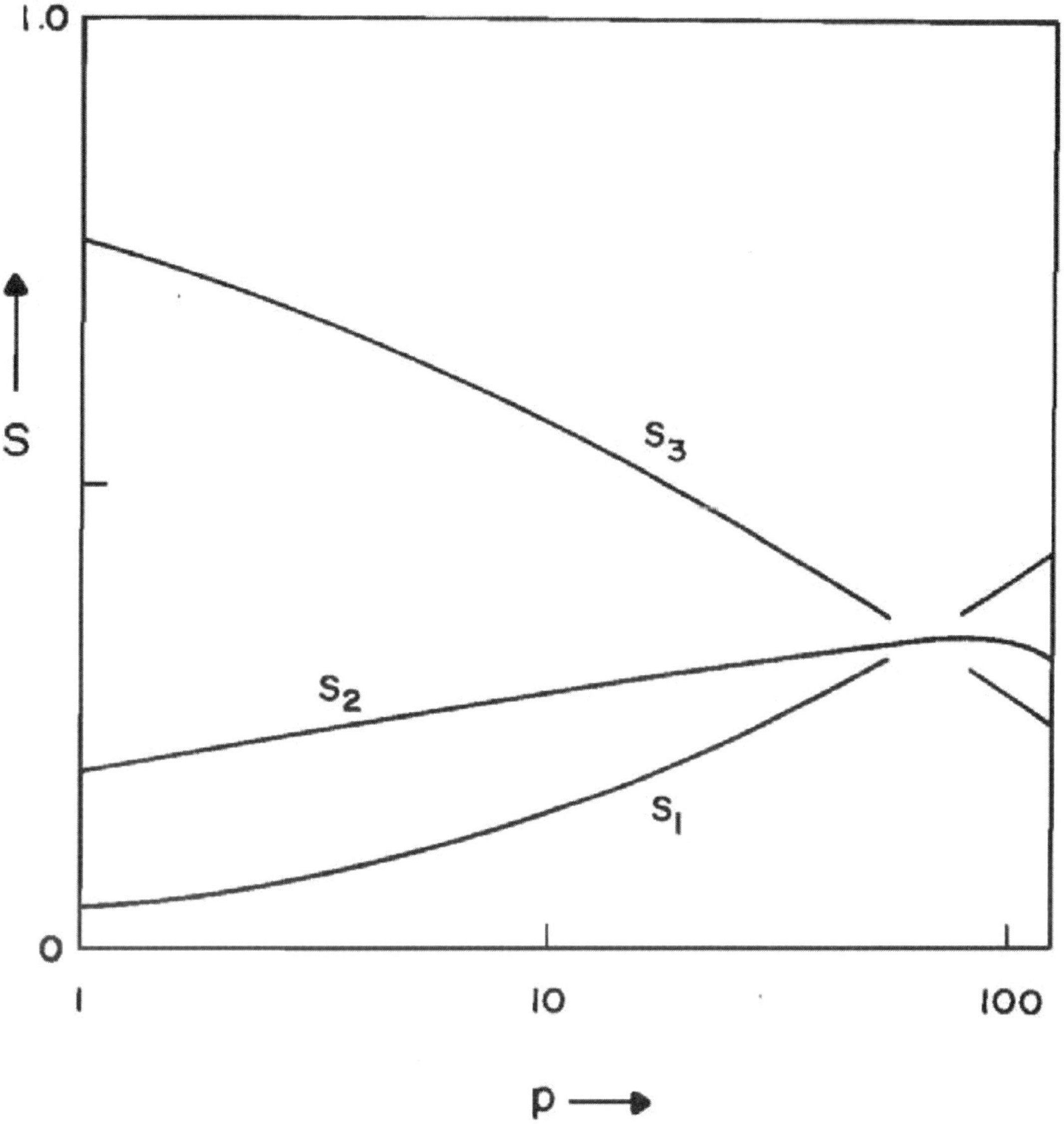

FIGURE 4 Variation of selectivity with respect to pressure. The curves refer to the products obtained from adsorbed species of types $i = 1$ through 3. Eqs. (12) and (13) are used, with $k_{r,1} = 0.1$, $k_{r,2} = 1.0$, $k_{r,3} = 10$, and $k_{a,i}/k_{d,i} = 1$ for all i.

where θ_i is the coverage of $[AS_i]$. If the surface reactions are the rate determining steps, the rate of formation of product P_i may be written as

$$r_i = k_{r,i}\theta_i \tag{13e}$$

where $k_{r,i}$ is the i-th reaction rate constant. Define the instantaneous fractional yield of product P_i as

$$s_i = r_i/r \tag{13f}$$

where r is the rate of formation of all products

$$r = r_1 + r_2 + r_3 \tag{13g}$$

In Fig. 4 we plot the variation in the instantaneous fractional yields, s_1, s_2, s_3, as a function of the total pressure for the model of Eq. (12) by solving Eq. (13). Note that the parameters used yield a maximum in s_2 at a pressure of approximately 80 units. Hence, if product P_2 was of primary interest, the optimum pressure of operation at the temperature of Fig. 4 would be at this maximum point. As might be expected, P_1 is formed at high pressures, and P_3 predominates at the lower pressures.

3B. Metal Catalyzed Isomerization/Hydrogenolysis Reactions

The experimental examples chosen so far to illustrate the role of different landing areas have involved relatively simple molecules. Evidence of the existence of more than one adsorbed form for these kinds of molecules, though indirect, appears firm. It is therefore not unrealistic to suppose that molecules that are not as simple as H_2 and CO would also have multiple landing areas possible on the surface.

We devote the rest of this work to a relatively complicated situation, one of perhaps more industrial importance. The case considered is the set of reactions, including scrambling, isomerization and hydrogenolysis, between a straight chain hydrocarbon (specifically n-pentane) and hydrogen, catalyzed by the metal component of a Pt/Al_2O_3 catalyst. The scrambling reaction refers to the process whereby the carbon atoms interchange position in the molecule, but the overall structure of the molecule is unchanged. The arguments of this sub-section are mainly qualitative, and are

TABLE I

Product distributions from Garin and Gault (1975)

Temperature, °C	283	269
Run numbers	3 & 21	5 & 24
Distributions, moles/100 mole of n-pentane-2-^{13}C:		
	3.1	0.8
	87.4	94.0
	1.6	0.4
CH_4 and C_4H_{10}	1.3	0.95
C_2H_6 and C_3H_8	2.9	1.9
	0.32	0.20
	0.45	0.20
	1.45	0.70
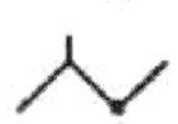 and 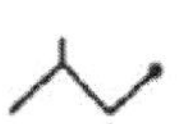	1.7	0.85

Each set of data has been calculated from two separate runs under similar conditions; one run yielding isopentane distributions, and the other the remaining products. Conversions range from 5 to 8% of *n*-pentane-2-^{13}C. The partial pressure of H_2 is 760 mm Hg in all cases.

quantified in sub-section 3C. The careful ^{13}C-tracer work of Garin and Gault (1975) furnish the data for our analysis, and some of these data are summarized in Table I.

Metal-catalyzed isomerization and hydrogenolysis reactions for n-pentane are known to occur by two mechanisms. The so-called cyclic mechanism postulates the formation of a five-membered ring intermediate on the catalyst surface. The bond shift mechanism postulates the formation of one of several different chain-type intermediates whereby two adjacent or nonadjacent carbon atoms are adsorbed on the metal surface. We postulate here that each of these intermediates has a characteristic landing area, i.e. requires a distinct set of surface sites.

Intermediates and the reaction network for the labelled compound n-pentane-2-^{13}C are indicated in Fig. 5. For simplicity, only three sets of landing areas are identified, and the first two of these are further subdivided.

The intermediate for the X-type of landing area corresponds to adsorption bonds on two nonadjacent C atoms of the molecule. Both points of attachment to the surface, indicated by arrows in Fig. 5, are at non-terminal C atoms for the X_1 set. One point of attachment for the X_2 set is a terminal C atom, and the other is a non-terminal C atom. The presence of a ^{13}C atom unsymmetrically placed in the molecule leads to two different intermediates produced on the X_2 set of landing area with equal probability. Each intermediate produces one of two isomers, again with equal probability. (Changes in the probabilities due to isotope effects are not expected to be significant for our present purposes.)

The intermediate for the Y type of landing area corresponds to adsorption of two adjacent C atoms, again either including two non-terminal C atoms (Y_1) or one terminal and one non-terminal C atom (Y_2). Again, the unsymmetrically placed ^{13}C atom leads to two different adsorbed forms for each intermediate on the Y_1 and Y_2 sets, as shown in Fig. 5. The bond shift mechanism can allow two different isomers from each intermediate to be formed. Lower molecular weight hydrogenolysis products are also formed.

Finally, the intermediate corresponding to the Z type of landing area is the adsorbed ring structure. Desorption of the intermediate yields cyclopentane. The random breaking of any one bond yields n-pentane again, but now with the ^{13}C atom randomly placed in the molecule. This is the so-called scrambling reaction.

There is, of course, no theoretical reason to restrict the landing area possibilities to five in number (X_1, X_2, Y_1, Y_2, Z). Other intermediates are possible, qualitatively different from these discussed above, and requiring qualitatively different landing areas. Alternatively, what is considered to be one intermediate, requiring one landing area, may in reality be an average of various intermediates. For example, the intermediate indicated on the X_1 type of landing area could be an average of various intermediates, each of which could contain additional points of attachment to the surface. These additional bonds would be perhaps less strong than the chemisorption bonds that are indicated by the location of the arrows. Each different point of attachment could then contribute to a marginally different landing area requirement. Although all forms should be included in the spectrum of possible landing areas to be considered, they may be neglected or lumped with one of the above five already considered for the sake of simplicity.

Many combinations of adsorbed species formed on the X- and Y-types of landing areas have common initial adsorption forms, or precursors. For example, in the case

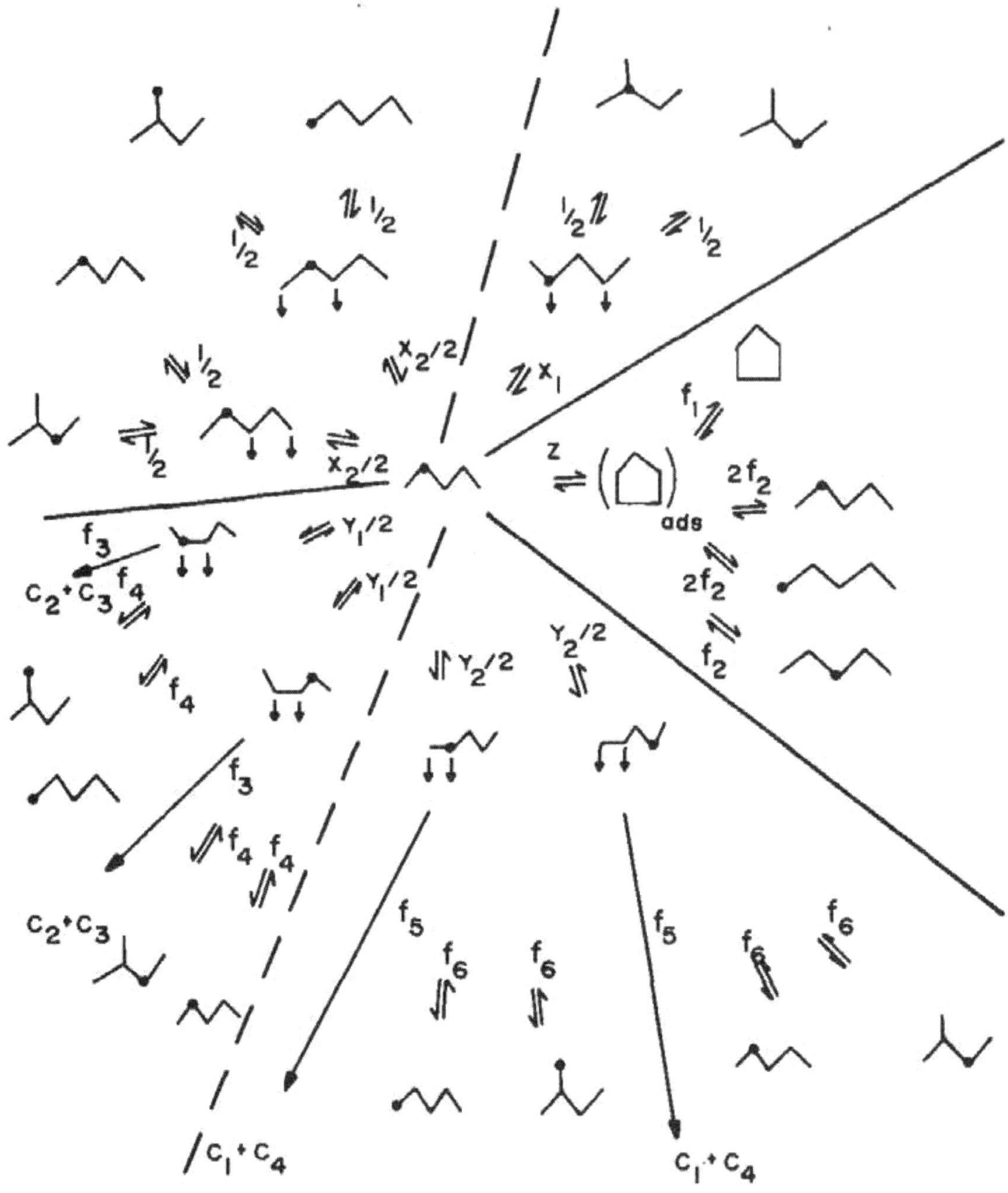

FIGURE 5 The set of reactions that can occur over metal catalyst for *n*-pentane-2-^{13}C. The dot denotes the location of the ^{13}C species. Arrows denote points of attachment to the surface. See text for explanation of symbols.

where *n*-pentane-2-^{13}C is the reactant, the species 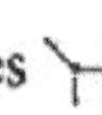 and  could both be formed from the precursor . Here arrows denote points of attachment to the surface, as before, and the dot defines the location of the ^{13}C. The former intermediate requires a Y_1 type landing area, while the second intermediate is characterized by an X_1 type of landing area. Further, note that the Y_1 intermediate species could be formed from either the precursor shown above, or from 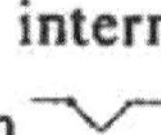, which could itself form intermediates requiring the X_2 landing area types, and so on.

The reactions for the formation of precursors, intermediates and products can be written in a manner similar to that leading to Eq. (1). The difficulty lies in the fact that the number of sites required for each type of landing area is not known, since it is not clear that this number can be simply equated to the number of points of attachment of the species to the surface. For example, it is doubtful whether the X- and Y-type areas require sites that are identical, in number and orientation, simply because both are doubly attached to the surface. Two sites perhaps represent the minimum number required in both the X- and Y- landing areas, since this number of sites "anchors" the intermediates in both cases. This may be considered the core, or "umbra," of the landing area. However, the "non-anchored" portions of the adsorbed species may make difficult adsorption of other molecules on sites near to the anchor sites. Hence, these sites too are "tied up" by the landing area configuration, but in a looser sense than the "umbra" sites. They may be considered the periphery, or "penumbra," of the landing area. Unlike the umbra, the penumbra would not necessarily be a whole number of sites. The number of the penumbra sites would depend upon the ability of the non-anchored portions to move, i.e. upon the dimensions of the molecule and the location of the points of adsorption. The range of the interaction force between two adsorbate species also plays a role. For example, the Y_2-type landing area, with intermediate 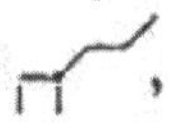, and the X_1-type, with , would have the same number of umbra sites, but different penumbra sites. The Z-type landing area, with its 5 membered ring intermediate, would probably have more umbra sites than the two areas above, but fewer penumbra sites.

Some of the steps in the reactions of *n*-pentane-2-^{13}C with hydrogen on a metal catalyst are described below in Appendix I. If a complete set of reactions could be written out, a reaction rate could be obtained. This would be expressed in terms of reactant concentrations, forward and reverse rate constants, and the site numbers of landing areas. In principle, experimental kinetic data on the formation of the reaction products could yield information on the numbers of sites tied up by the various precursor and intermediate landing areas. In practice, this determination would be extremely cumbersome. In the present case, the available experimental data is simply too sparse to allow a fitting procedure to be even attempted.

To obtain a better understanding of the essentials behind the complex hydrocarbon-hydrogen reactions catalyzed by the metal, we consider in the following section a model of a simpler (hypothetical) process that has many of the features of this reaction set.

3C. Model for the Metal-Catalyzed Reactions

Our model considers the reactants R and H_2 in the gas phase above a metal catalyst surface. The hydrogen species is also present in the molecule of R. The reactions of R and H_2 are shown in Table II. The H_2 molecule dissociatively adsorbs on 2 vacant sites S.[2] Reactant R can adsorb in several different configurations. We consider three

[2]In principle we should allow the dissociated hydrogen atoms to land on more than one type of landing area as well. However the hydrocarbon–hydrogen reactions are generally carried out in large excess of hydrogen. Hence it is reasonable to assume that only landing areas consistent with high partial pressures of hydrogen would be used. The use of only the one-site landing area for hydrogen results in a considerable simplification in the mathematics with little loss of rigor.

TABLE II

Reactions of the model system. See text for explanation.

$H_2 + 2S \rightleftharpoons 2[HS]$
$R + 4S \rightleftharpoons [RS_2] + 2[HS]$
$R + 5S \rightleftharpoons [RS_3] + 2[HS]$
$R + 8S \rightleftharpoons [RS_5] + 3[HS]$
$[RS_2] + 4[HS] \rightarrow P_h + 6S$
$[RS_3] + 2[HS] \rightarrow P_i + 5S$
$[RS_5] + 3[HS] \rightarrow P_s + 8S$

configurations in the present work. (This is the smallest number of configuration which yields useful results. Additional configurations are of course not impossible but will not change the results qualitatively.) Each configuration dissociates different numbers of H atoms in the process, and each dissociated remnant of R ties up different numbers of surface sites. The R species adsorbed on the 2-site landing area loses two H atoms and forms two surface bonds. This intermediate then reacts with more adsorbed H species to form hydrogenolysis products, collectively termed P_h. The $[RS_3]$ intermediate also loses two H atoms and forms two C-S bonds, but ties up three surface sites because of the relative placing of the bonds. This intermediate yields an isomerization product, P_i. The third adsorbate loses two H atoms to form a cyclic species, and an additional H atom to form a surface bond. (More than one C-S bond is possible, of course, but this does not change the qualitative results). The cyclic nature of the species causes it to tie up a larger number, say 5, surface sites. Ring breakage of this intermediate yields a product, denoted by P_s, which is similar overall to reactant R but which contains the C atoms in a different order from those in R. This is the "scrambled" product. The similarity of these three equations in Table II to equations (AI-8), (AI-5), and (AI-6), respectively, should be noted.

For simplicity, assume that the rates of formation of products P_h, P_i, and P_s is governed by the rate at which the corresponding adsorbed form ($[RS_2]$ in the case of P_h, and so on) combines with a single [HS] species. Adsorption and desorption phenomena are assumed to reach equilibrium rapidly. Then the rate of formation of P_h is

$$r_h = k_h \theta_2 \theta_H \tag{14a}$$

Similar relations can be written for products P_i and P_s as

$$r_i = k_i \theta_3 \theta_H \tag{14b}$$

$$r_s = k_s \theta_5 \theta_H \tag{14c}$$

Here θ_j denotes the fractional area of the surface taken up by j-th landing area conformation, and k_j is the rate constant for the reaction of type j.

Further, the coverages of the various landing areas follow the equilibrium relations:

$$K_5 C_R \theta_S^8 = \theta_5 \theta_H^3 \tag{14d}$$

$$K_3 C_R \theta_S^5 = \theta_3 \theta_H^2 \tag{14e}$$

$$K_2 C_R \theta_S^4 = \theta_2 \theta_H^2 \tag{14f}$$

$$K_H C_H \theta_S^2 = \theta_H^2 \tag{14g}$$

where θ_S represents the fractional area of the unoccupied surface, and K_i denote the various equilibrium coefficients. C_R and C_H are of course the concentrations of R and H_2 in the gas phase respectively. These four relations are not the only way that equilibrium could be postulated. They correspond to Eq. (5) in Section 2; the other case, involving the lower bound for f_d, is considered (and discarded) in sub-section 3E.

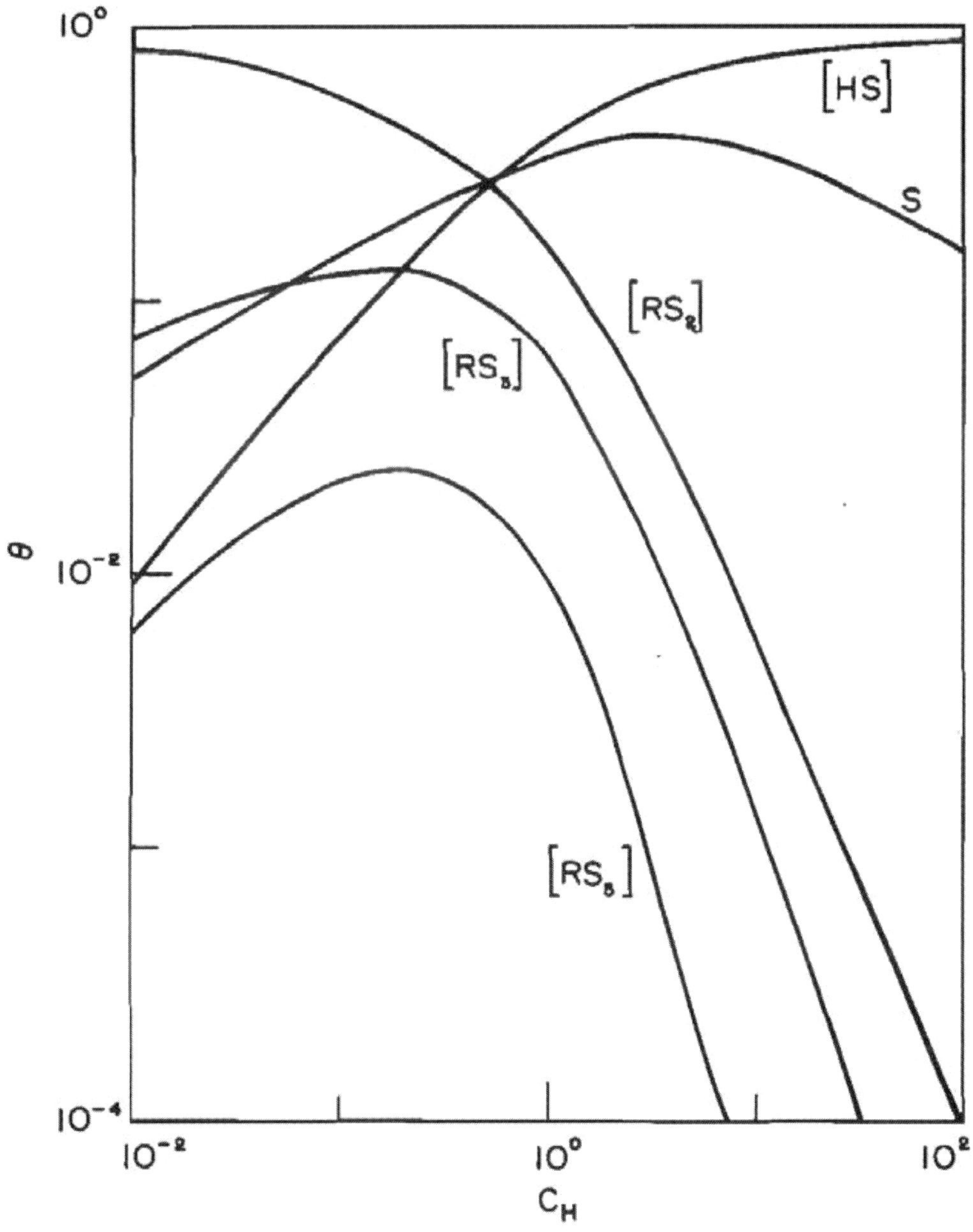

FIGURE 6 Variation of surface coverage with the concentration of H_2 in arbitrary units. The model of Table II and Eq. (14) is used, with $C_R = 1$, $K_H = K_2 = K_3 = K_5 = 1$, all in arbitrary units.

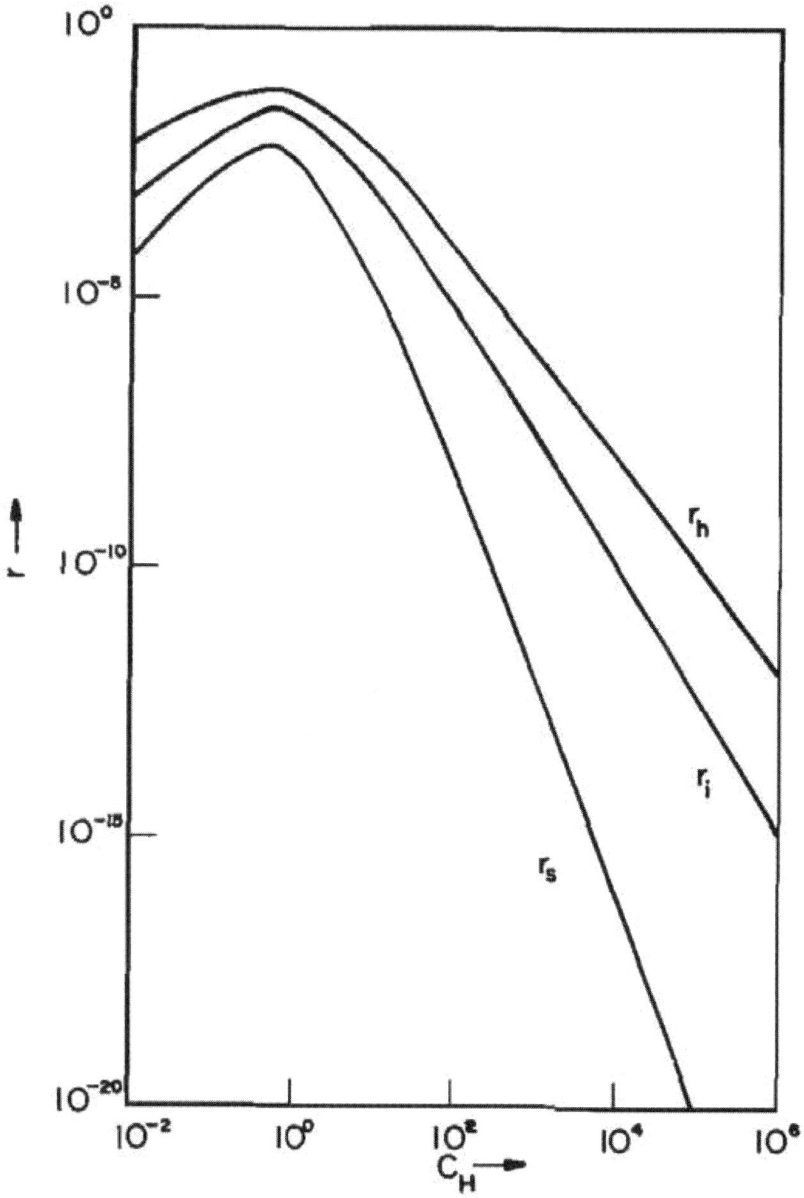

FIGURE 7 Reaction rates of Table II and Eq. (14) as functions of the concentration of H_2. Values of the reaction rate constants are $k_h = k_i = k_s = 1$ in arbitrary units. Values of other model parameters are those of Fig. 6.

Finally, by definition

$$\theta_H + \theta_2 + \theta_3 + \theta_5 + \theta_S = 1 \tag{14h}$$

Figures 6 and 7 illustrate the results of this model when C_R is kept constant and C_H is increased over several orders of magnitude.

Figure 6 is a log-log plot of the fractional coverages for the different types of landing area conformations for R and for H_2. When C_R is much greater than C_H, the surface is mainly covered with adsorbed R. The most common conformation of R on the surface is $[RS_2]$, as this uses the smallest number of sites. As C_H is increased, the fraction of the surface covered with [HS] increases, while the coverage of [RS_2] decreases. The increase in the (relatively low) coverages of [RS_5], which contains the largest number of surface sites, can be explained by noting that the relative amount of R in the gas phase decreases as C_H increases. For $C_H < C_R$, the slope of the [RS_5] curve is greater than that for [RS_3], and consequently the relative rate of increase of coverage is the greatest for the largest site conformation. The total amount of R adsorbed in all landing areas of course decreases. As C_H approaches and exceeds C_R, the coverages of [RS_3] and [RS_5] in turn reach a maximum value and then decrease. The fraction of sites that are unoccupied also reaches a maximum value and then decreases as C_H is increased.

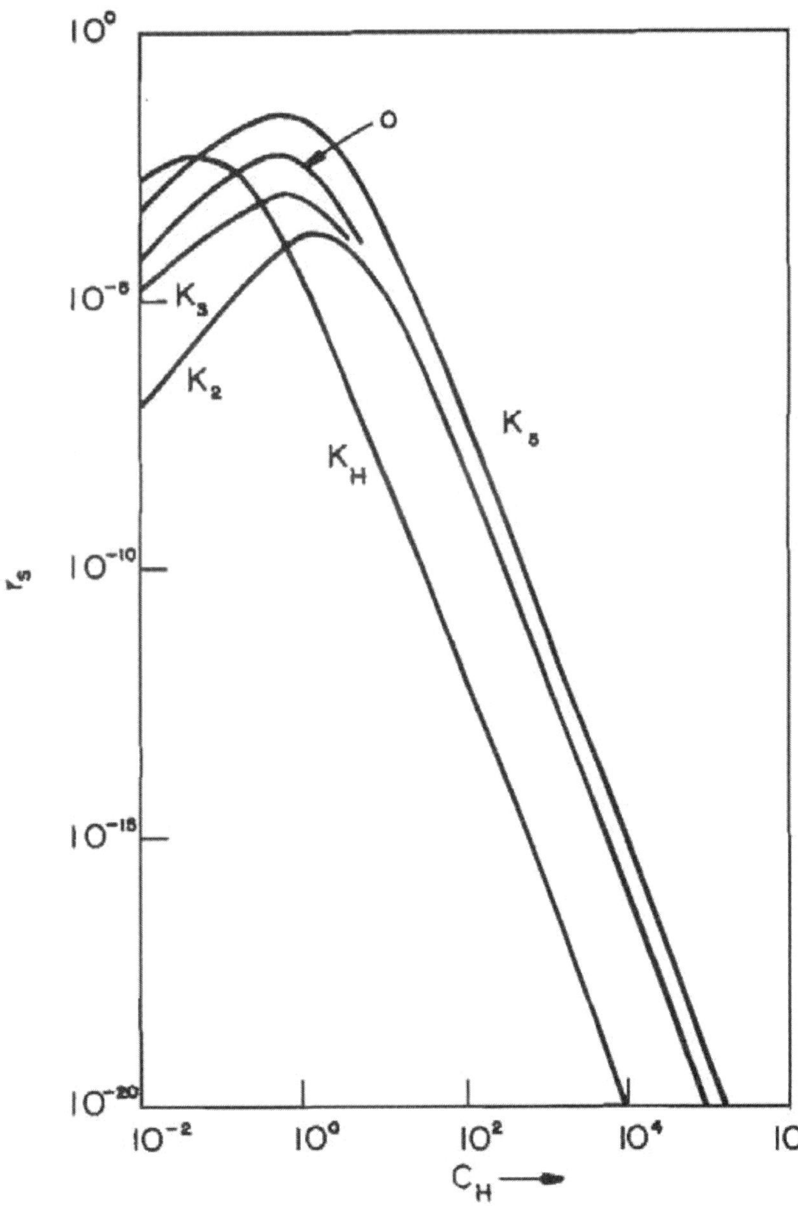

FIGURE 8 Variation in (a) r_s (b) r_i (c) r_a with changing equilibrium constants K_H, K_2, K_3, K_5. Parameter values are those of Fig. 7 except that the equilibrium constant used to label each curve is set equal to 10 rather than 1. The curves labelled 0 correspond to those in Fig. 7, i.e. with all parameters set equal to 1. The smaller the value of C_H at which each curve reaches its maximum rate (corresponding to a zero slope), the smaller the algebraic value of b at a fixed C_H for this set of parameter values.

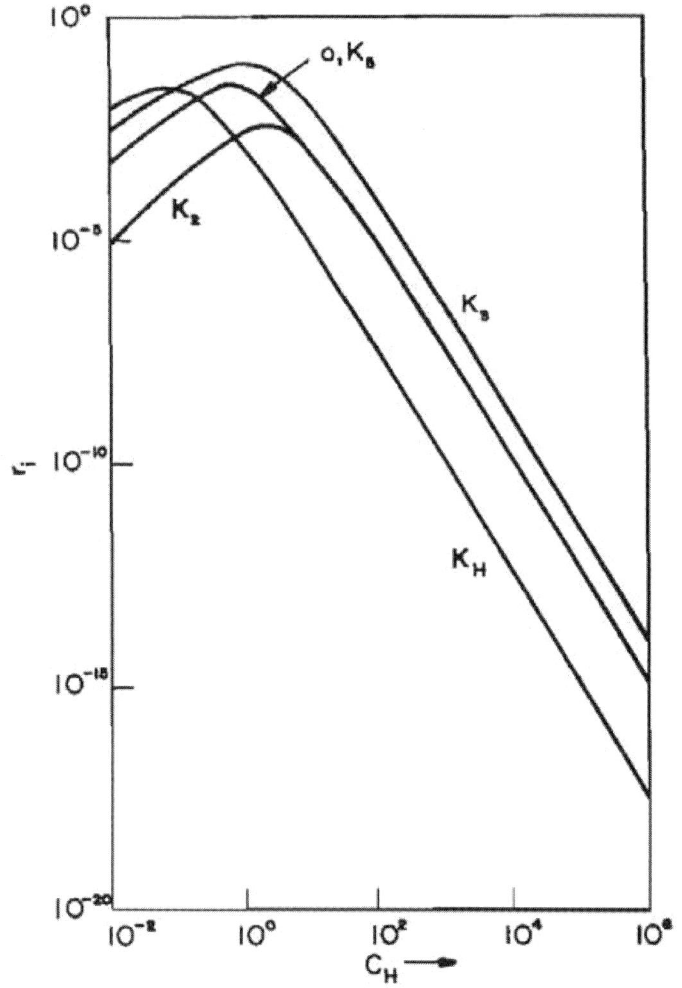
10⁰
10⁻⁵
10⁻¹⁰
10⁻¹⁵
10⁻²⁰
r_i
0, K_S
K_2
K_S
K_H
10⁻²
10⁰
10²
10⁴
10⁶
C_H

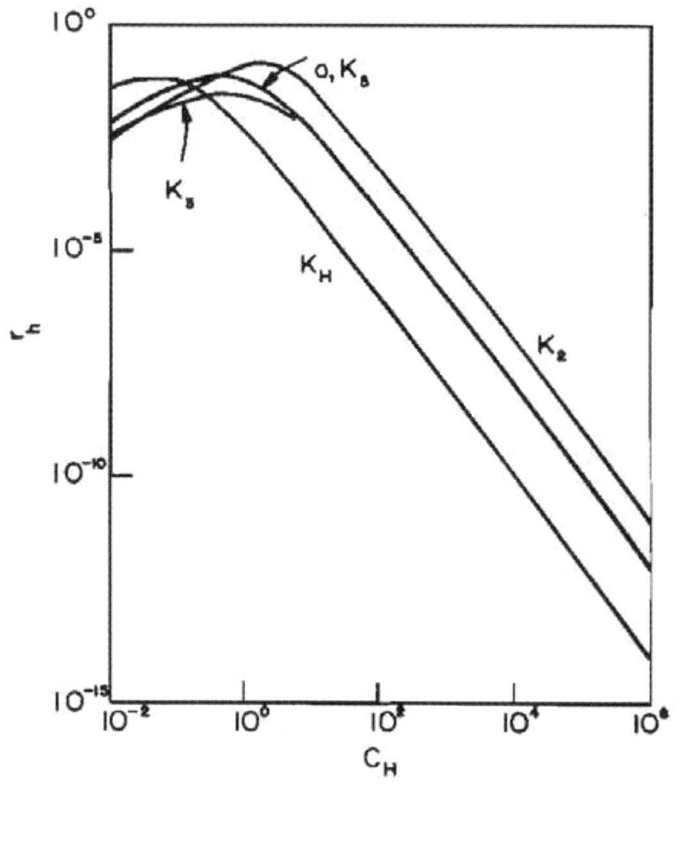
10⁰
10⁻⁵
10⁻¹⁰
10⁻¹⁵
r_h
0, K_S
K_S
K_H
K_2
10⁻²
10⁰
10²
10⁴
10⁶
C_H

In Fig. 7, the rates of formation of P_h, P_i, and P_s are plotted as functions of C_H in logarithmic coordinates. This enables the H_2 concentration dependency of the rates to be obtained, since the order of the reaction rate with respect to H_2 is the slope of the log rate vs. log C_H curve. In all three cases, the slope varies with C_H; is positive for small values of C_H, passes through zero and then becomes negative. At large excess of H_2, the slope apparently tends to a constant value, i.e. the curve becomes a straight line.

The effects on r_h, r_i, and r_s of increasing the equilibrium constants K_H, K_2, K_3, and K_5 are shown in Fig. 8. As expected, when H_2 is present in large excess, the asymptotic slopes do not change from those in Fig. 7. However, the curves are shifted on changing the equilibrium values, so that both the maximum value of the rate and the corresponding value of C_H are altered. Hence, at low and intermediate concentrations of H_2, the orders of reaction with respect to H_2 depend not only on the concentration of H_2 but also on the equilibrium constants.

From Fig. 8a the value of C_H for a maximum value of r_s decreases by about an order of magnitude when K_H is increased. Hence, the algebraic value of the slope decreases with an increase in K_H. On the other hand, the algebraic value of the slope does not change appreciably with an increase in K_5, and increases with an increase in K_3 and K_2, with the greatest change occuring when K_2 is increased.

Similar comments can be made regarding the effect of equilibrium constants on the slopes (orders of reaction with respect to H_2) of reaction rates r_i and r_h from Figs. 8b and 8c respectively. However, note that in the last two figures, the changes on increasing K_5 are not discernable. From Eq. (14) it is not surprising that the effect of K_5 is most felt in the rate r_s.

3D. Comparison of Model With Experiment

The present model is postulated to exhibit many of the salient features of hydrocarbon–hydrogen reactions over a noble metal catalyst. Consequently, the slopes of the logarithmic rates for P_h, P_i, and P_s formation with respect to C_H calculated for the model above should correspond respectively to the rate orders with respect to hydrogen observed for hydrogenolysis, isomerization with the bond shift mechanism, and the scrambling reactions for the cyclic mechanism.

The rates for various such reactions have been measured at constant temperature by Garin and Gault (1975). The experimental conditions encompass a relatively small range of partial pressures of hydrogen at relatively high ratios of hydrogen to hydrocarbon partial pressure. Hence it is possible to compare the observed values of the rate orders with respect to hydrogen, denoted by b, with the values of the asymptotic slopes of Fig. 7 and 8. This is done in Table III. Note that in all cases the observed values of b are smaller in absolute magnitude than the asymptotic slopes calculated from the model. Apparently the partial pressure of H_2 under the experimental conditions is sufficient for the slopes to be negative, but not large enough for the asymptotic portion of the curve in Fig. 7 to be used.

This conclusion is reinforced by the fact that different products formed in the same type of reaction have different observed values of b. For example, the two sets of hydrogenolysis products ($C_3 + C_2$) and ($C_4 + C_1$) formed from n-C_5 have orders

TABLE III

Comparison between experiments and the theoretical asymptote of Fig. 7. The parameter b is the exponent of the hydrogen partial pressure in the rate equation $r_j = k_j(p_{C_5})^a (p_{H_2})^b$ for the formation of product j.

Reactions	$(-b)$obsd	$(-b)$calcd.	Model product
	3.4 ± 0.1	<4.0	P_s
	1.8 ± 0.1	<2.5	P_i
	1.3 ± 0.2	<2.0	P_h
	0.6 ± 0.2	<2.0	P_h
	2.3 ± 0.15	<2.5	P_i
	1.9 ± 0.2	<4.0	P_s
	1.65 ± 0.15	<2.5	P_i
	1.65 ± 0.2	<2.0	P_h
	0.7 ± 0.1	<2.0	P_h

of -1.3 and -0.6 respectively, with respect to H_2. Our simplified model cannot predict this becasue it assumes only one product is formed from each reaction type; in this case, P_h with a slope greater than or equal to -2. However, recall that changing the values of K_H, etc., in Fig. 8c changes the slopes at intermediate values of the H_2 concentration. Hence, it is reasonable to consider that the differences in reaction order stem from different equilibrium constants K_2 for the different landing area configurations leading to the different hydrogenolysis products. From Fig. 5 (which, although containing a more complete picture than that given in Table II, is itself a simplified version of the process), it is the Y_1 landing area configuration that yields $(C_3 + C_2)$ from n-C_5, while the Y_2 landing area yields $(C_1 + C_4)$. Since the observed value of b for $(C_3 + C_2)$ is algebraically less than that for $(C_1 + C_4)$, the value of K_2 for the Y_1 landing area configuration must be less than that for the Y_2 landing area configuration.

When i-C_5 is used as the reactant in place of n-C_5, the intermediates for the hydrogenolysis products are not those in Fig. 5. The intermediate leading to the formation of $(C_3 + C_2)$ would probably be 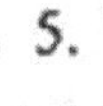 , while 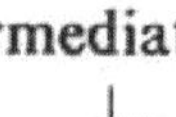 would result in $(C_1 + C_4)$. The landing areas for hydrogenolysis products from i-C_5 may then differ from Y_1 and Y_2, which correspond to n-C_5. However, note from Table III that the observed value of b for $(C_3 + C_2)$ is greater than that for $(C_1 + C_4)$, using either i-C_5 or n-C_5 as reactant. Hence, even when i-C_5 is the reactant, the value of K_2 for adsorption and desorption on the landing area leading to $(C_3 + C_2)$ must be greater than the value corresponding to $(C_1 + C_4)$ as the product.

It is not clear whether comparisons can be made between values of K_2 for landing areas forming the same product but using different reactants. If Y_1 and Y_2 are not too different from the landing areas for the two intermediates from i-C_5, then the comparisons are meaningful. Assuming that this is so, note that the observed values of b indicate that for either ($C_2 + C_3$) or ($C_1 + C_4$) as products, the value of K_2 using i-C_5 as reactant is smaller than that using n-C_5 as reactant.

Now consider the isomerization reaction forming P_i in Table II. The different values of b obtained experimentally as shown in Table III for the different isomerization products and reactants can be considered to be due to different values of the corresponding equilibrium constant, in this case K_3. From the results of Fig. 8b, recall that the algebraic value of b increases with increasing K_3. Then the largest value of K_3 is that for adsorption/desorption of n-C_5 by the landing areas that yield i-C_5. (Note from Fig. 5 that one or both of the two different X type landing areas (X_1, X_2) may be involved. This is not considered in the model of Table II.) The smallest value of K_3 will be that corresponding to the reverse isomerization, from i-C_5 to n-C_5.

From Fig. 8a, K_5 does not appreciably alter the value of the slope. Hence the significant increase in the value of b for the scrambling reaction of i-C_5 compared to that for n-C_5 must be due to a qualitative change in the landing area for the former reaction compared to the Z type of landing area for the latter.

As a summary of this subsection, note that experimental values of b, the exponent of the hydrogen partial pressure in the rate expression, correlate quantitatively with those predicted from the simple model assuming a distribution of landing areas and adsorbed forms for the hydrocarbon reactant. Further, the changes in b corresponding to different products from the same reactant and similar but not identical landing area are compared with changes expected in the model when the adsorption equilibrium constants K are changed. This allows the K values for such landing areas to be compared. A similar procedure allows the comparison of K values for landing areas forming the same products but starting with different reactants.

3E. Other Models of Adsorption/Desorption and Reaction

Equations (14) represent only one of the possible ways of expressing the relationship between adsorption, desorption and reaction on the three landing areas. Specifically we assume that the rates of reaction on the surface are first order in adsorbed hydrogen, and we use the upper bound for f_d to describe the adsorption/desorption equilibrium for the hydrocarbon. It is worthwhile to discuss alternatives to these assumptions, so as to judge how well they fit the physical picture and the agreement between the experimental and predicted reaction order of hydrogen.

A power of greater than unity could be considered for the dependence of r_s, r_h and r_i on the coverage of H, θ_H, in Eq. (14). From Table II the rates could be considered as

$$r_s = k_s \theta_5 \theta_H^a \tag{15a}$$

$$r_h = k_h \theta_2 \theta_H^a \tag{15b}$$

and

$$r_i = k_i \theta_3 \theta_H^a \tag{15c}$$

where a could be as high as 3 in Eq. (15a), 4 in Eq. (15b) and 2 in Eq. (15c). These equations can be used, in place of Eq. (14a–c,) together with Eq. (14d) through (14h) to determine the rates of reaction. If this is done, it can be seen that the asymptotic shape of the rate curves will not change. This is as expected. Since the asymptotic values are obtained at large concentrations of H_2 in the bulk vapor phase, the value of θ_H is approximately unity, virtually independent of C_H. Hence $\theta_H^a \approx \theta_H \approx 1$, regardless of the value of a. Even at low and intermediate values of C_H, the shapes of the rate curves are similar to those of Fig. 6.

Equations (14a–c) are perhaps closer to physical reality than the higher order relations above. For the rate of formation of P_l, Eq. (14b) is what might be expected if the following sequence was followed:

$$[RS_3] + [HS] \rightarrow [RS_3'] + S$$

$$[RS_3'] + [HS] \rightarrow P_l + 4S$$

and if the net rate of formation of $[RS_3']$ was zero, i.e. if $[RS_3']$ was a reaction intermediate. Similar paths can be written to explain Eq. (14a) and (14c). Note that other paths are also possible to explain Eq. (14a–c).

As mentioned previously, Eq. (14d,e,f) are based on the upper bound for f_d in the absorption/desorption relationship for the hydrocarbon. A second possibility corresponds to the lower bound of f_d. Here

$$K_5\, C_R\, \theta_S^8 = \theta_5^5\, \theta_H^3 \tag{16a}$$

$$K_3\, C_R\, \theta_S^5 = \theta_3^3\, \theta_H^2 \tag{16b}$$

$$K_2\, C_R\, \theta_S^4 = \theta_2^2\, \theta_H^2 \tag{16c}$$

Certainly Eq. (14d,e,f) are no more physically meaningful than Eq. (16), so no choice can be made between them on that score. Using Eq. (16) in place of Eq. (14d,e,f) to solve the rest of Eq. (14), we get the following closed-form expressions

$$r_h' = k_h\,(K_2\, C_R)^{1/2} \,/\, (\text{den})^2 \tag{17a}$$

$$r_l' = k_l\,(K_3\, C_R)^{1/3}\,(K_H\, C_H)^{1/6} \,/\, (\text{den})^2 \tag{17b}$$

and

$$r_s' = k_s\,(K_5\, C_R)^{1/5}\,(K_H\, C_H)^{1/5} \,/\, (\text{den})^2 \tag{17c}$$

where (den) is used as short hand notation for

$$(\text{den}) = 1 + (K_H\, C_H)^{1/2} + (K_2\, C_R/K_H\, C_H)^{1/2} + (K_3\, C_R/K_H\, C_H)^{1/3} + (K_5\, C_R)^{1/5}/(K_H\, C_H)^{3/10} \tag{17d}$$

Equations (17) predict that the experimental values of b for P_h, P_l, and P_s-type products should be no less than -1, $-5/6$, and $-4/5$ respectively. From Table III, this is clearly not so. Since Eq. (16) and (17) are contradicted by the experimental data while Eq. (14) and Fig. 6 are not, Eq. (16) should not be used to model the present system.

3F. Other Theoretical Treatments

It is of interest to compare the reactions of the model system as given in Table II with hydrogenolysis and isomerization schemes cited by Garin and Gault (1975) based upon the treatment of Cimino *et al.* (1954) or of Frennet and Lienard (1971). The former scheme can be rewritten in the present notation as

$$H_2 + 2S \rightleftharpoons 2\,[H - S] \tag{18a}$$

$$n\text{-}C_5H_{12} + 2S \rightleftharpoons [C_5H_{11} - S] + [H - S] \tag{18b}$$

$$[C_5H_{11} - S] + 2xS \rightleftharpoons [C_5H_{11-x} - S_{1+x}] + x[H - S] \tag{18c}$$

$$[C_5H_{11-x}S_{1+x}] + H_2 \rightarrow [C_3H_{8-y} - S_y] + [C_2H_{6-z} - S_z] \tag{18d}$$

and the latter as

$$H_2 + 2S \rightleftharpoons 2\,[H - S] \tag{19a}$$

$$n\text{-}C_5H_{12} + [H-S] \rightleftharpoons [C_5H_{11}-S] + H_2 \tag{19b}$$

$$[C_5H_{12-x+1}-S_{x-1}] + [H-S] \rightleftharpoons [C_5H_{12-x}-S_x] + H_2 \tag{19c}$$

$$[C_5H_{12-x}-S_x] \rightarrow [i\text{-}C_5H_{12-x}-S_x] \tag{19d}$$

Equations (18) and (19) differ fundamentally from one another and from the equations in Table II for hydrogenolysis and isomerization. In Eq. (18), hydrogenolysis occurs by successsive dehydrogenation of the adsorbed species and the creation of surface-adsorbate bonds together with formation of adsorbed hydrogen atoms. In Eq. (19), dehydrogenation of the adsorbate and formation of C–S bonds leads to formation of gas phase hydrogen molecules. Consequently the two types of postulated mechanisms are essentially different. On the other hand, the hydrogenolysis and isomerization reactions of the present work (Table II) are treated in a unitary fashion. Both involve dissociated adsorption of reactant and of hydrogen. However, the reactant is adsorbed in different forms, with different patterns of C-S bonds, and on different landing areas. The different adsorbed forms undergo different reactions, forming isomers and hydrogenolysis products respectively.

A significant quantitative difference in the present approach lies in the fact that the formation of P_h and P_i involves the loss of no more than two hydrogen atoms and the formation of two C-S bonds during the dissociative adsorption of the hydrocarbon.[3] On

[3]Cimino *et al.* (1954) and Frennet and Lienard (1971) do not consider the scrambling reaction, and so we exclude the formation of the adsorbate leading to the formation of P_r from the comparison. Of course the formation of $[RS_3]$ involves the loss of three hydrogen species from R. However, two of these species are lost to give rise to C-C bonds in the cyclic adsorbate, and there is only one C-surface atom bond.

the other hand, for the model of Eq. (18) to match the experimental value of the order with respect to hydrogen, four to six hydrogen atoms must be lost, and an equal number of bonds with the surface formed. Similarly, from two to four hydrogen atoms must be lost and a like number of surface bonds formed in the model corresponding to Eq. (19).

4. SUMMARY AND CONCLUSIONS

Since multiple adsorption conformations (multiple landing areas) occur for relatively simple molecules adsorbing on a surface, their existence cannot be denied when more complicated molecules adsorb on a surface. In most cases, the landing area possibilities would reduce to a small number; in some cases all adsorbed configurations could react equivalently so that they would not need to be distinguished. An analysis of selectivity considerations is presented here in a landing area framework, considering the absorbed states on different landing areas to have different adsorption equilibrium coefficients and to lead to different reaction products or product distributions.

In this work the nature of the catalyst surface is not altered. For simplicity the surface is assumed to be uniform, i.e., a surface site can be involved in any landing area combination. Even for this simple case, changes in the temperature and pressure of the gas phase can be shown to change catalyst selectivity through the change in the distribution of landing configurations. In general, landing areas corresponding to larger numbers of surface sites are favored at higher temperatures and at lower partial pressures of the corresponding gas phase species.

Calculations are made on the basis of a model for the reactions of hydrocarbons with hydrogen over the metal component of supported metal catalysts. The landing area framework upon which this model is based allows the unitary treatment of hydrogenolysis, isomerization and scrambling reactions. No more than two carbon-surface bonds need to be formed from the hydrocarbon using this analysis; compared to this number, other models require the formation of up to 6 surface bonds. The reaction orders with respect to the hydrogen component for the reactions above are shown to be positive for low hydrogen concentrations, pass through zero, and then tend to an asymptotic negative value at high concentrations of hydrogen. All the reaction orders obtained from experiment lie at or above the corresponding lower bounds of the calculated reaction orders.

The simple model does not allow for more than one product for each of the hydrogenolysis, isomerization and scrambling reactions. Nevertheless it is shown that the different reaction orders for different products from the same type of reaction with the same reactant can be related to the different equilibrium constants for the corresponding adsorbed forms. Similarly, the change in the reaction orders when the same product is formed from different reactants is related to the different landing area equilibrium constants.

The effect of altering the surface characteristics of the catalyst has not been considered here. In another paper, it is shown how such a change, say by changing the average diameter of the crystallites of the supported catalyst on the substrate, can lead to a change in catalyst selectivity. Facile and demanding selectivity can thus be explained from a landing area point of view.

REFERENCES

Brunauer, S., Love, K.S., and Keenan, R.S., *J. Amer. Chem. Soc.* **64**, 751 (1942).

Butt, J.B., and Starzec, L.T., *J. Catal.* **32**, 99 (1974). See also Welsch, R.J., and Butt, J.B., *Chem. Eng. Comm.* **1**, 155 (1973); Dabrowski, J.E., Butt, J.B., and Bliss, H., *J. Catal.* **18**, 297 (1970).

Callahan, J.L., and Grasselli, R.K., *A.I.Ch.E. J.* **9**, 755 (1963).

Cimino, A., Boudart, M., and Taylor, H., *J. Phys. Chem.* **58**, 796 (1954).

Dadyburjor, D.B., and Ruckenstein, E., *J. Phys. Chem.*, **85**, 3396 (1981).

Eischens, R.P., Francis, S.A., and Pliskin, W.A., *J. Phys. Chem.* **60**, 194 (1956).

Ertl, G., Neumann, M., and Streit, K.M., *Surf. Sci.* **64**, 393 (1977).

Frennet, A., and Lienard, G., *J. Chim. Phys. Phys.-Chim. Biol.* **68**, 1526 (1971).

Frennet, A., Lienard, G., Crucq, A., and Degos, L., *J. Catal.* **53**, 150 (1978).

Froitzheim, H., Hopster, H., Ibach, H. and Lehwald, S., *Appl. Phys.* **13**, 147 (1977).

Garin, F., and Gault, F.G., *J. Amer. Chem. Soc.* **97**, 4466 (1975).

Halsey, G.D., *Adv. Catal.* **4**, 259 (1952).

Krebs, H.-J. and Lüth, H., *Appl. Phys.* **14**, 337 (1977).

Menon, P.G., and Froment, G.F., *J. Catal.* **59**, 138 (1979).

Paál, Z., and Thomson, S.J., *J. Catal.* **30**, 96 (1973).

Renouprez, A., Fouilloux, P., and Moraweck, B., in *Growth and Properties of Metal Clusters*, Bourdon, J., ed., Elsevier, Amsterdam (1980) p. 421.

Ruckenstein, E. and Dadyburjor, D.B., in *Growth and Properties of Metal Clusters*, Bourdon, J., ed., Elsevier, Amsterdam, (1980), p. 435.

Schwarz, J.A., and Smith, R.P., *J. Catal.* **62**, 176 (1980).

Thomas, J.M. and Thomas, W.J., *Introduction to the Principles of Heterogeneous Catalysis*, Academic, New York (1967).

Tsuchiya, S., Anemomiya, J., and Cvetanovic, R.J., *J. Catal.* **19**, 245(1970).

Young, D.M., and Crowell, A.D., *Physical Adsorption of Gases*, Butterworths, London (1962).

Zeldowitsch, J., *Acta Physicochim.* USSR **1**, 961 (1934).

APPENDIX I

Some of the steps in the reactions of *n*-pentane-2-^{13}C with H_2 on a metal catalyst are described below. Here x_1, x_2, y_1, y_2, and z represent the number of sites tied up with the corresponding landing area, and w_i represents the number of sites required to form a singly adsorbed precursor species, where the point of attachment to the surface is on the *i*-th C atom. Although we have shown earlier that several simultaneous forms of adsorbed hydrogen exist, with their own site requirements, we assume here that the hydrogen landing area can be represented by a single form, with h sites tied up for each dissociated hydrogen atom.

The formation of precursors involves the reversible removal of atomic hydrogen through a reaction of the type

 $+ (w_2 - h)S \rightleftharpoons [$ 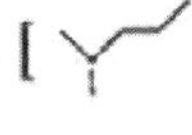 $- S_{w_2}] + [H - S_h]$ (AI-1)

Here square brackets denote an adsorbed species, dots denote the location of the ^{13}C, arrows denote the points of attachment to the surface, and $-S_i$ denotes that the adsorbed species involves i surface sites. The presence of the unsymmetrically placed ^{13}C species results in five, rather than three, such reactions being distinct, each corresponding to a different point of attachment to the surface. From these precursors, intermediate species form, also reversibly, similar to the following example

$[$ $- S_{w_2}] + (x_1 - w_2 + h)S \rightleftharpoons [$ $- S_{x_1}] + [H - S_h]$ (AI-2)

From Fig. 5 this precursor can be seen to form intermediates requiring Y_1- and Y_2-type landing areas, in addition to the intermediate on an X_1-type landing area shown in Eq. (AI-2). Other precursors form other X- and Y-type landing area intermediates. The formation of the cyclic intermediate on the Z-type landing area may be written as

(AI-3)

The hydrogen adsorption/desorption reactions are

$$H_2 + 2hS \rightleftharpoons 2[H - S_h] \qquad \text{(AI-4)}$$

Formation of products from intermediates follows reactions of the type

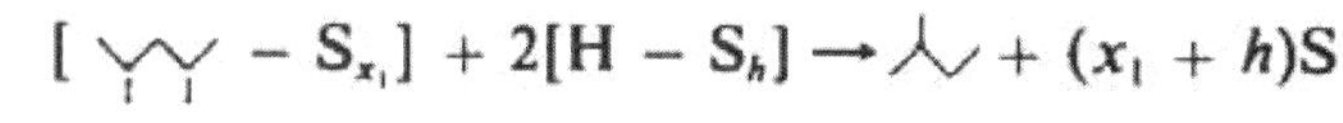

(AI-5)

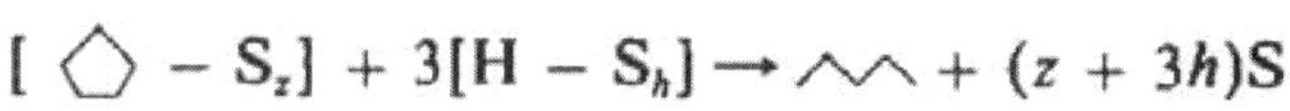

(AI-6)

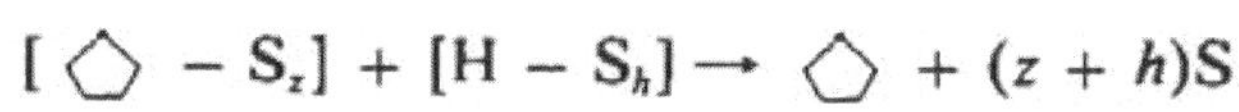

(AI-7)

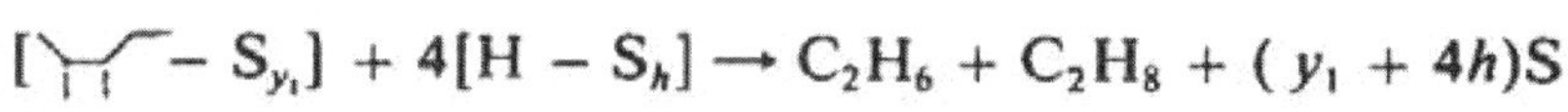

(AI-8)

Here Eq. (AI-5) is an example of the bond shift mechanism. Several other isomers are of course possible; see Fig. 5. The "scrambling" isomerization reactions which occur when the cyclic intermediate suffers arbitrary bond breakage, and of which Eq. (AI-6) is an example, yields *n*-pentane with the tracer ^{13}C at any point on the chain. Only *n*-pentane-2-^{13}C is shown here. Eq. (AI-7) forms the other product from the cyclic intermediate, cyclopentane. An example of a hydrogenolysis product is shown in Eq. (AI-8).

Role of Multiple Landing Areas for Adsorption in Facile or Demanding Activity and Selectivity of a Heterogeneous Catalyst

Dady B. Dadyburjor

Department of Chemical Engineering and Environmental Engineering, Rensselaer Polytechnic Institute, Troy, New York 12181

and Eli Ruckenstein*

Department of Chemical Engineering, State University of New York at Buffalo, Amherst, New York 14260

(Received: December 16, 1980; In Final Form: June 22, 1981)

A reactant can be adsorbed in different forms on the surface of a heterogeneous catalyst, and each adsorbed form may give rise to a different product. The different adsorbed forms tie up different combinations of surface sites; i.e., each adsorbed form has a characteristic landing area. Hence, the relative amounts of these adsorbed forms, or the landing-area distribution, inter alia determines the catalyst selectivity. In general, only a fraction of the surface may be available to a particular landing area and is termed the availability factor for that landing area. Since a particular site may be available to one or more (or none) of the landing areas, the sum of the availability factors over all landing areas is not unity. A change in the nature of the surface can change the availability factors for one or more (or none) of the landing areas, and in this manner the selectivity can change. For supported metal catalysts, the nature of the surface could change by altering the average particle size, i.e., the dispersion. Hence, the variation or constancy in selectivity with dispersion (i.e., the structure sensitivity or insensitivity, the demanding or facile nature of the catalyst, with respect to selectivity) can be explained from a landing-area framework. Analogously the demanding or facile nature of the catalyst with respect to activity is also discussed in a landing-area framework. The example used is the metal-catalyzed isomerization, scrambling, and hydrogenolysis reactions of C_6 hydrocarbons. A simple model is developed for the landing areas of this reaction set, and for the variation in availability factors with dispersion of the catalyst. The results agree with experimental data.

1. Introduction

It has been experimentally verified that small molecules such as CO and H_2 adsorb on a surface in more than one configuration.[1] It is therefore reasonable that the same must be true for larger, more complicated molecules. Previous work[2] quantified the premise that different configurations of a reactant adsorbed on a catalytic surface, i.e., adsorbates on different landing areas, might react differently and would form different products. We related the selectivity of a catalyst to the relative amounts of the reactant adsorbed in different configurations, i.e., to the landing-area distribution.

Even on a surface where all sites are equivalent, i.e., where a particular surface site could as easily be a part of one type of collection of sites (landing area) as another type, it was shown that the distribution of the different landing areas for the same single molecule would depend upon the temperature and the pressure of the gas phase. In fact, increasing the pressure leads to the relative decrease of molecules adsorbed on landing areas consisting of larger numbers of sites, and the relative increase of the molecules adsorbed on smaller landing areas. The total number of adsorbate molecules would also increase. Decreasing the temperature has the same effect on the landing-area distribution as increasing the pressure. Hence, even for this so-called "uniform" surface, the change in selectivity with temperature and/or pressure can be explained.

TABLE I: Reactions of the Model System

$H_2 + 2S \rightleftharpoons 2[HS]$
$R + 4S \rightleftharpoons [RS_2] + 2[HS]$
$R + 5S \rightleftharpoons [RS_3] + 2[HS]$
$R + 8S \rightleftharpoons [RS_5] + 3[HS]$
$[RS_2] + 4[HS] \rightarrow P_h + 6S$
$[RS_3] + 2[HS] \rightarrow P_i + 5S$
$[RS_5] + 3[HS] \rightarrow P_s + 8S$

The present work will focus on the effect of the catalytic surface on selectivity using the landing-area approach. For a so-called nonuniform surface, i.e., where not all sites can be a part of every possible landing area, the distribution of landing area may change as the characteristics of the surface are changed at the same pressure and temperature. Hence, the selectivity of the catalyst will be altered. Characteristics of the catalytic surface that could affect the landing-area distribution include the crystallographic orientation and, for supported metal catalysts in particular,

(1) Eischens, R. P.; Francis, S. A.; Pliskin, W. A. *J. Phys. Chem.* **1956**, *60*, 194. Tsuchiya, S.; Anemomiya, J.; Cvetanovic, R. J. *J. Catal.* **1970**, *19*, 245.
(2) Ruckenstein, E.; Dadyburjor, D. B. *Chem. Eng. Commun.* In press.

This paper, which was originally published in **The Journal of Physical Chemistry**: 85(23), 3396–3406 (1981), is republished with permission.

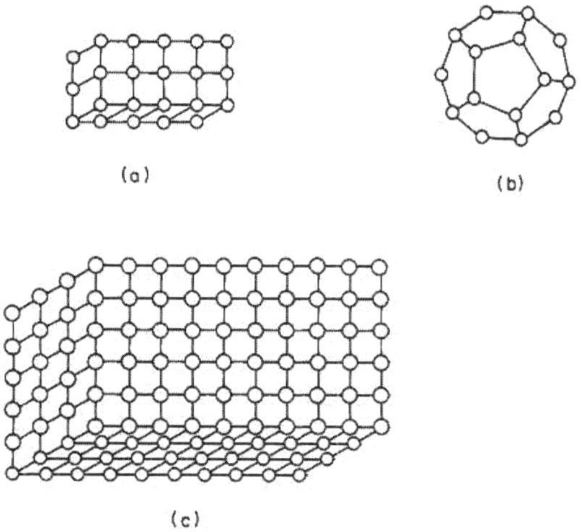

Figure 1. Various supported metal crystallites used to demonstrate the effect of particle size and shape on the availability factor ϕ and selectivity dispersion a_s for different landing areas. (a) A simple cubic structure containing 30 metal atoms. The unseen 5 × 3 face lies on the support. (b) Icosahedral structure, also containing 30 metal atoms. Note the 5-fold symmetry of the planes. For small crystallites, such a structure is probably more stable than structure a.[3] (c) A simple cubic structure similar in shape to structure a, but now every dimension is twice as long (240 metal atoms). The unseen 10 × 6 face lies on the support. For such large sizes, the cubic structure is probably most stable.

the size and shape of the metal crystallites. Some examples of these changes are shown in section 2, using the model developed in ref 2 for isomerization and hydrogenolysis of a hydrocarbon. The concept of availability factors is introduced here as a measure of the fraction of the surface capable of supporting one landing-area configuration. This parameter is used to explain in section 3 whether the selectivity and the specific activity of the catalyst depend upon the particle size of the metal crystallite, i.e., whether the catalyst is demanding or facile in selectivity and activity in a landing-area framework. The kinetics of demanding selectivity are developed in sections 4 and 5.

2. Availability Factor

Here and in what follows we consider supported metal catalysts almost exclusively. However, it should be noted that the landing-area approach is a general one and can be used for other catalysts as well.

The skeletal isomerization and hydrogenolysis of hydrocarbons in the presence of hydrogen is modeled in Table I. The chemistry is described in a later section. Compounds R and H_2 are considered, and hydrogen is present also inside the molecule of R. Three landing areas are taken into account. In one of these, two surface sites are tied up by two carbon–surface bonds with adsorbate [RS_2], which leads to hydrogenolysis products, P_h. Another landing area ties up three sites and is bonded to adsorbate [RS_3] with two nonadjacent C–S bonds. The last landing area ties up five sites to form a cyclic intermediate [RS_5] with one C–S bond. [RS_3] leads to isomerization products P_i, while [RS_5] leads to a "scrambled" product, P_s, where the carbon atoms in the reactant are moved around without changing the nature of the molecule. Here [RS_x] denotes a surface species where x surface sites are incorporated in the landing area of adsorbed R. Hydrogen is considered to adsorb dissociatively on only one type of landing area, incorporating a single surface site per H atom. The adsorbed species [HS] is involved in the surface reactions forming P_h, P_i, and P_s as shown in Table I.

TABLE II: Variation of Availability ϕ and Selectivity Dispersion a_s Corresponding to Each of the Four Landing Areas for Each of the Three Crystallites of Figure 1[a]

crystallite (dispersion)	parameter	[RS_2]	[RS_3]	[RS_5]	[LLA]
Figure 1a (a = 0.9)	ϕ	1.0	5/9	0	5/9
	a_s	0.9	0.5	0	0.5
Figure 1b (a = 1.0)	ϕ	1.0	0	1.0	0
	a_s	1.0	0	1.0	0
Figure 1c (a = 0.6)	ϕ	1.0	1.0	0	1.0
	a_s	0.6	0.6	0	0.6

[a] Descriptions of the four landing-area possibilities are given in the text.

In this section we describe the demands imposed upon the distribution of these three landing areas by the nature of the catalyst surface. Only geometric constraints are considered here; other chemical (electronic) constraints are contained in the kinetic and thermodynamic constants. We describe in turn the effects of the number and the orientation of sites in the landing area, and the shape and size of the crystallite.

Consider a metal crystallite as shown in Figure 1a. It consists of 30 atoms in a simple cubic structure. A landing area for the [RS_2] adsorbate can be found at any place on any of the five exposed faces of the crystallite, assuming that such a landing area would require two adjacent sites only. Further assume that the landing area for the [RS_3] adsorbate requires three adjacent sites on a straight line and that steric effects prevent adsorption parallel to the surface on the bottom layer of the crystallite. Then not every surface site of the crystallite is capable of being part of the [RS_3] landing area. In other words, the catalytic surface is "nonuniform". In fact, only the top face of this crystallite can provide this type of landing area; i.e., 15 atoms out of 30. Finally, if the landing area for the [RS_5] adsorbate requires the surface sites to be at the corners of a regular pentagon, then this landing area will not be found anywhere on this crystallite.

To quantify this approach, define an availability factor ϕ for each adsorbed form as the fractional number of catalyst atoms on the adsorbing surface that can support the corresponding landing area. Then values of ϕ for [RS_2], [RS_3], and [RS_5] for the crystallite of Figure 1a are given in Table II. Note that 27 atoms of this crystallite are on the adsorbing surface, i.e., its dispersion a is 0.9.

For a particular adsorbate, the relative orientation of sites constituting the landing area also plays a role in determining the availability factor. Consider a landing area requiring five sites, not in a cyclic pattern as for [RS_5], but instead in a straight line. The value of ϕ for this linear landing-area adsorbate, [LLA], is now $^5/_9$ for the crystallite of Figure 1a, using the same steric constraint for the bottom layer of the crystallite. This is compared in Table II with a value of ϕ = 0 for [RS_5] over the same crystallite.

Changing the shape of the crystallite may dramatically change the availability factor for each landing area. The icosahedron of Figure 1b also contains 30 atoms, and the dispersion is unity. However, since it contains 16 exposed pentagonal faces, any of its atoms can be part of the landing area for the [RS_5] adsorbate. The [RS_3] adsorbate cannot be accommodated if the landing area requires the three surface sites to be linear. As before, any surface atom can be a part of the two-site landing area for the [RS_2] adsorbate. The corresponding values of ϕ are collected in Table II. As shown by Hoare and Pal[3a] and by Burton,[3b] crystallites of small numbers of atoms organize differently from large crystallites. According to their results, a particle of 30 atoms would probably be more stable in the form of Figure 1b than in the form of Figure 1a.

The size of the crystallite is also of importance in determining the availability of landing areas. Consider Figure 1c, where the crystallite is doubled in every dimension from that of Figure 1a. This increase for Figure 1c does not affect the availability of the surface atoms for landing areas of [RS_2], or the lack thereof for [RS_5], compared with that for Figure 1a. However, for [RS_3], all surface atoms in Figure 1c are now capable of being a part of the landing area for [RS_3]. Hence, ϕ for this landing area is now unity. However, the increase in crystallite size also changes the fraction of metal atoms present on the surface of the crystallite, i.e., the dispersion, a. Only 0.6 of the atoms are now on the surface. This decreases the total landing areas available for an adsorbate such as [RS_2], even though the corresponding values of ϕ are identical for both crystallites.

Clearly the availability factor for a particular landing area is of value in determining the specific reaction rate (rate per unit surface site) of the corresponding adsorbate. On the other hand, a measure of the landing areas available to the adsorbate as a fraction of the total amount of catalyst present would be of value in determining an overall reaction rate. One such quantity is the product of the dispersion a and the availability factor ϕ for the adsorbate. This parameter we term the selectivity dispersion, a_s, for the corresponding landing area. It is the number of catalyst atoms that can support the corresponding landing area of the adsorbate, expressed as a fraction of the total number of catalyst atoms present. In Table II the value of a_s is also tabulated for each combination of the three types of particles and four types of adsorbates.

Of course, a physical system would contain supported metal particles of a variety of sizes and shapes. As mentioned earlier, smaller particles, those having large values of the dispersion a, are more likely to exist in the form of Figure 1b, whereas large particles, corresponding to small dispersions, exist in the form of Figure 1c. Consequently, a system that has a large value of the average dispersion would consist mainly of particles similar to those of Figure 1b. Such a system would have a marked propensity toward allowing the adsorption of the cyclic [RS_5] and [RS_2] but would allow the adsorption of only a small amount of [RS_3]. In other words, such a system has a relatively large value for the average selectivity dispersion a_s of [RS_5] or [RS_2] and a relatively small value for a_s of [RS_3]. On the other hand, a system with a small value of the average dispersion a yields relatively small values for a_s of [RS_5], large values for a_s of [RS_2], and relatively large values for a_s of [RS_3]. As a quantitative example, see the results plotted in Figure 2 for a system that contains only crystallites of the type shown in parts b and c of Figure 1, in varying proportions.

In assigning availability factors and selectivity dispersions to various landing areas for different crystallites and systems of crystallites, for the sake of simplicity, we have omitted any special consideration of the relationships required between carbon–carbon bond lengths of the adsorbing species and distances between active sites on the catalyst. These relationships have been discussed by Balandin.[4] In the present work, we assume that the intersite distances are suitable for all landing-area adsor-

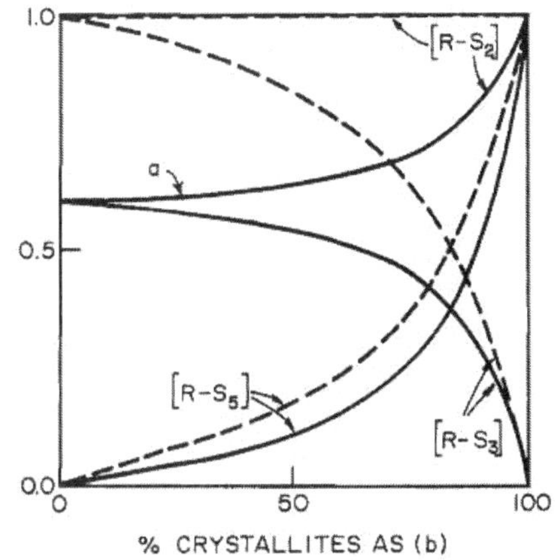

Figure 2. Parameters of a hypothetical system of metal crystallites consisting only of particles as shown in Figure 1, b and c. The independent variable is the percentage of crystallites that are of the type of Figure 1b. The variation of the availability factors ϕ and the selectivity dispersion a_s for the different landing areas in such a system is indicated by broken lines (ϕ) and unbroken curves (a_s). The overall dispersion a is also shown.

bates. This yields upper bounds for all values of ϕ and a_s.

3. Landing Areas in Facile/Demanding Selectivity

Chambellan et al.[5] modified the terms "facile" and "demanding", originally suggested for specific reaction rate (i.e., rate per surface catalyst atom) by Boudart, to apply to selectivity. Thus a "facile", or structure-insensitive, selectivity for a supported metal catalyst is independent of the size of the catalyst particle, while a size dependence exists for a catalyst that is "demanding", or structure sensitive, with respect to selectivity.

In the previous section, it was seen how changing the size and shape of the crystallites could change the total number of surface sites (through the dispersion a) and/or the ability of a reactant to be adsorbed in one or more landing-area conformations (through the fraction ϕ of sites available for that landing area). The availability factors ϕ for each landing area permit a discussion of facile or demanding activity and selectivity using the landing-area approach.

First, we consider the case where the crystallite change alters the dispersion, but the distribution of the relative surface available for each type of landing area does not change. As an example, consider the crystallites of parts a and c of Figure 1, if the possible intermediates were only [RS_2] and [RS_5]. Then, although a changes from 0.9 to 0.6 as the crystallite dimensions are doubled, the value of ϕ for [RS_2] remains unity, and for [RS_5] the value of ϕ remains zero. Clearly, the selectivity will not change. The global reaction rate will change, but the specific rate will be constant. Consequently, such a reaction would be facile, or structure insensitive, both in selectivity and in activity.

If both the dispersion and the availability of sites were to change, the situation is more complex. As an example, consider again parts a and c of Figure 1, now with the possibilities on the surface as [RS_2] and the intermediate [LLA] requiring five sites on a straight line. The value of a changes, as before, from 0.9 to 0.6. While ϕ remains at unity for [RS_2], it increases from $^5/_9$ to 1.0 for [LLA]. The selectivity of the process would then be altered.

(3) Hoare, M. R.; Pal, P. *J. Cryst. Growth* **1972**, *17*, 77. Burton, J. J. *Catal. Rev.–Sci. Eng.* **1974**, *9*, 209.

(4) Balandin, A. A. *Adv. Catal.* **1958**, *10*, 96; **1969**, *19*, 1.

(5) Chambellan, A.; Dartigues, J. M.; Corolleur, C.; Gault, F. G. *Nouv. J. Chim.* **1977**, *1*, 41.

Further, since the landing area of [LLA] ties up more of the surface than that of the other, the total number of intermediates formed per unit surface area of the metal would then decrease. Consequently, even if the intermediates were to desorb as products with the same ease as before, the rate of disappearance of R per unit area of the metal surface would then decrease. In such a situation, the process would be structure sensitive, or demanding, in both selectivity and activity.

If the new concentration of [LLA] were large relative to the original concentration of this intermediate, but much smaller than the concentration of [RS_2], the selectivity would change with the dispersion, but the specific reaction rate may remain virtually unaltered. Consequently, the process is demanding in selectivity while being facile in activity.

Finally, consider the change in dispersion to be accompanied by ϕ values altered in equal proportions for all landing-area possibilities. Again consider parts a and c of Figure 1, now using only [RS_3] and [LLA] as possible adsorbates. Then the total number of adsorbed intermediates per unit area of metal surface area may increase on changing from Figure 1a to Figure 1c; however, the relative distribution of these intermediates would be more or less the same. This implies a change in the specific reaction rate but no change in the selectivity. Such a process would be demanding in activity while being facile in selectivity.

4. Landing Areas from Initial Product Distribution

We return now to the specifics of the reactions of hydrocarbons in the presence of hydrogen catalyzed by the metal component of a supported metal catalyst. These reactions are used as the basis of the model of Table I and include isomerization, hydrogenolysis, and the so-called scrambling reaction. Metal-catalyzed isomerization and hydrogenolysis reactions occur by two mechanisms. The bond-shift mechanism postulates the formation of one of several different chain-type intermediates whereby two carbon atoms are bonded to the metal surface. The two carbon atoms may be adjacent or nonadjacent. The cyclic mechanism postulates the formation of a five-membered-ring intermediate on the metal surface. In the scrambling reaction, the carbon atoms in the hydrocarbon interchange positions in the molecule, but the overall structure of the molecule is unchanged. This requires the formation of the cyclic intermediate. We postulate that each intermediate has a characteristic landing area and requires a distinct set of surface sites. As the nature of the surface is changed, the availability factors for one or more of the landing areas may change, and this will alter the product distribution.

In this section, experimental information[5] on initial product distribution is used to obtain changes in the landing-area distribution due to changes in the nature of the surface obtained by changing metal dispersions, metal loadings, and supports. For any given example of a surface, actual landing-area distributions can be obtained from product distributions only if the rates for all reactions are of the same order of magnitude. However, different surfaces can be compared by this procedure irrespective of the relative reaction rates.

The relations between product distribution and landing-area distribution are shown in Figure 3. The hydrocarbon is 2-methylpentane-2-^{13}C, the reactant used in ref 5. Five possible landing areas are considered. Landing areas X_1 and X_2 correspond to a species adsorbed on two nonadjacent C atoms, one C atom being an end atom for X_2, and both C atoms being interior atoms of the molecule for the X_1 landing area. Bond-shift isomerization occurs for adsorbates on these landing areas. The Y_1 and Y_2 types of landing area correspond to adsorption on two adjacent C atoms, including one or both atoms in the interior of the molecule for Y_2 and Y_1, respectively. Both hydrogenolysis and bond-shift isomerization are posulated to occur on these landing areas. The Z type of landing area corresponds to cyclic adsorption. Adsorbates on this landing area undergo cyclic isomerization, including scrambling, and cyclization reactions. Note that some adsorbates are twice as likely as others for the same landing area, e.g., those characterized by $^2/_3X_2$ and $^1/_3X_2$ in Figure 3.

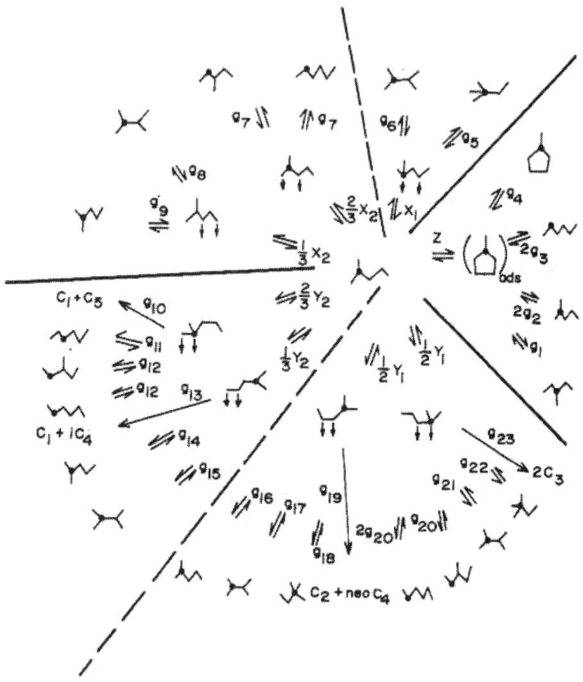

Figure 3. Set of reactions that can occur over a metal catalyst for 2-methylpentane-*2*-^{13}C. The dot denotes the location of the ^{13}C species. Arrows denote points of attachment to the surface. See the text and Appendix I for an explanation of the symbols.

The relationships between the amount of a particular product formed initially and the amount of reactant adsorbed on the different landing areas are discussed in Appendix I. The experimental data on the initial product distribution are listed in Table III from various places in ref 5. Five different supported metal catalysts are considered, containing different total dispersions, amounts of metal, and types of support.

The experimental data are insufficient to enable an unambiguous solution to be obtained as to the moles of reactant adsorbed on each of the landing areas. One possible set of values is given in Table IV.

Note that the values corresponding to landing area X_1 are all set equal to zero in Table IV. This is because none of the two products possible from Figure 3 was observed (or noted) in ref 5. Steric difficulties could be a possible reason for the absence of these products. Further, to obtain the moles of 2-methylpentane adsorbed on the Y_1 landing area in Table IV, it is assumed that only hydrogenolysis, and not bond-shift isomerization, occurs here. Finally, the number of moles adsorbed on the X_2 and Y_2 landing areas cannot be independently obtained. The numbers in Table IV corresponding to X_2 (or Y_2) are average values, i.e., one-half the maximum values if Y_2 (or X_2) was set equal to zero. For details on the calculations, see Appendix I.

TABLE III: Initial Product Distributions from Ref 5 for the Reactions of Hydrogen and 2-Methylpentane-2-^{13}C at 254 °C over a Number of Pt/Al_2O_3 Catalysts of Various Dispersions, a[a]

	Catalyst					
a = H/Pt	0.05	0.07	0.12	0.55	0.70	1.00
% Pt	8.5	8.4	8.4	4.1	2.5	0.3
support[b]	A′	B'_{10}	B'_{10}	B'_{210}	B'_{210}	B'_{210}
	Products					
conversion, %	15	6		18	15	12
$6C_1$	0	6		0	0	1
$C_1 + C_5$	27	21		24	29	20
$C_2 + C_4$	14	12		11	14	14
$2C_3$	9	7		8	8	6
3MP	34.6	32.6		23.6	20.5	20
2-^{13}C	29.1[d]		26.1	16.5	7.8	3.2
3-^{13}C	5.5[d]		6.5	7.1	12.7	16.8
n-hexane	12	12		30	26	34
2-^{13}C	7.7		7	17	13.4	17.3
3-^{13}C	4.3		5	13	12.6	16.7
methylcyclopentane	4.4	9.5		2	3	6.5

[a] Partial pressure of the hydrocarbon is 3.9 torr. The product distribution is in units of moles of reactant hydrocarbon forming given product per 100 mol reacting to form all products. The dot indicates the location of the ^{13}C atom. [b] A and A′ are Al_2O_3 that have different extents of exchangeable hydroxyl groups; B'_x consists of A′ calcined at 600 °C for x hours. [c] Isotopic distribution obtained with 2-MP-4-^{13}C as reactant, not 2-MP-2-^{13}C. [d] Isotopic distribution obtained with support A, not A′.

TABLE IV: Landing-Area Distribution from the Data of Table III Obtained by Using Appendix I[a]

landing area	8.5% A′	8.4% B'_{10}	4.1% B'_{210}	2.5% B'_{210}	0.3% B'_{210}
X_1	0	0	0	0	0
X_2[b]	40.8	33.0	31.0	23.7	11.9
Y_1[c]	28.0	24.0	22.0	28.0	28.0
Y_2[b]	27.3	22.1	20.8	15.8	7.9
Z	29.5	39.0	49.3	66.3	90.3
% X	33	28	25	18	9
% Y	44	39	35	33	26
% Z	24	33	39	50	65

[a] The distribution is in units of moles of reactant adsorbed on the given landing area per 100 mol of reactant forming all products. Adsorbates corresponding to the various landing areas are as in Figure 3. [b] X_2 and Y_2 values cannot be independently determined. Values given here are average. See text for details. [c] Lower limit of Y_1, assuming only hydrogenolysis occurs for adsorbates on this landing area. See text for details.

The results of Table IV bear out the trends expected. There is a dramatic increase in the percent of reactant molecules adsorbed on the Z type of landing area, from 24% to 65%, as the average dispersion is increased from 0.05 to 1.00. This mirrors the large increase in the selective dispersion for [RS_5] expected with the increasing number of smaller particles in the system. Note that a similar observation can be made from the simple system of Figure 2.

Further, there is more than a 3-fold decrease in the percentage of reactant adsorbed on the X type of landing areas as the average dispersion increases. This mirrors the decrease in the selective dispersion of [RS_3] for the simple system of Figure 2.

Finally, there is a moderate decrease in the percentage of reactant adsorbed on the Y type of landing area as the dispersion increases, even though the selective dispersion for [RS_2] increases in Figure 2. This illustrates the fact that the value of ϕ alone does not control the landing-area distribution. In this case, given the option of adsorbing on a Z or Y type of landing area, reactant molecules tend to favor the former. This may be related to the low partial pressure of the reactant; under these conditions the landing area tying up a greater number of sites is preferred.[2]

To obtain more insights into this system requires more information on the landing areas. In the following section, assumptions are made about possible landing areas for the hypothetical reactions that were used earlier in section 2 and in ref 2 and that closely parallel the present situation. This then allows an investigation into the effects of altering the availability of one or more landing areas.

5. Kinetics for Demanding Selectivity

The model system used is described in section 2 and is given in Table I. It is of interest to examine the selectivity of this system under conditions when the selective dispersion a_s for one or more landing areas is changed.

In section 2 the selective dispersion was shown to consist of two factors, the dispersion a and the availability ϕ of the given landing area. Section 3 indicated that, if the values of ϕ stayed constant as a was varied, the catalyst would be facile in activity and selectivity. This situation is analogous to the case treated in ref 2 and will not be considered further. If the values of ϕ for the different landing areas were changed by different amounts, the catalyst would be demanding in selectivity. In this section we consider the rates of formation of the three model products P_i, P_s, and P_h of Table I as the availability of the landing areas changes. Recall that P_i and P_h represent products of isomerization and hydrogenolysis reactions, respectively. P_s represents the "scrambled" product, whose molecule has the same overall structure as that of reactant R but where the carbon atoms are no longer found in the same order as in R.

We denote the availability for the landing areas of adsorbates [RS_5], [RS_3], and [RS_2] as ϕ_5, ϕ_3, and ϕ_2, respectively. For the single adsorbate [HS] of the other species, it is assumed that all areas of the surface are available, i.e., ϕ_H is unity. Since a given site may be available to more than one type of landing-area conformation, it is necessary to define not only θ_i, the number of sites occupied by landing-area-i adsorbates as a fraction of all sites on the surface (the fractional site coverage), but also $\theta_{i,j}$, the number of sites that are occupied by landing-area-i adsorbates and that would also be available, if unoccupied, to landing-area-j adsorbates as a fraction of all sites on the surface (the cross-coverage). Then for sites available to [RS_5]

$$\theta_5 + \theta_{3,5} + \theta_{2,5} + \theta_{H,5} + \theta_{S,5} = \phi_5 \quad (1a)$$

Similarly for [RS_3]

$$\theta_{5,3} + \theta_3 + \theta_{2,3} + \theta_{H,3} + \theta_{S,3} = \phi_3 \quad (1b)$$

and

$$\theta_{5,2} + \theta_{3,2} + \theta_2 + \theta_{H,2} + \theta_{S,2} = \phi_2 \quad (1c)$$

for sites available to [RS_2]. A total site balance yields

$$\theta_5 + \theta_3 + \theta_2 + \theta_H + \theta_S = 1 \quad (1d)$$

In eq 1, θ_S is the fractional number of unoccupied sites, and $\theta_{S,j}$ is the number of unoccupied sites available to the j-site landing area as a fraction of all sites on the surface.

Further, an adsorbate corresponding to a particular landing area is not affected by whether the sites that it is adsorbed on are available or not to other landing-area conformations. In other words, the ratio $\theta_{i,j}/\theta_i$ is independent of pressure and temperature. (Note that θ_i is a function of pressure and temperature, as shown in ref 2.)

The ratio is denoted by $\phi_{i,j}$ and termed the cross-availability. In general the cross-availability $\phi_{i,j}$ depends on the nature of landing areas i and j rather than on the pressure or temperature of the gas phase or the nature of the surface.

As in ref 2, it is assumed that adsorption and desorption rapidly achieve equilibrium and that reactions on the surface are the rate-determining steps. Then, equating the rates of adsorption and desorption, we can obtain the equilibrium fractional coverage for each landing-area conformation. For species H which adsorbs on only one type of landing area

$$k_{a,H}C_H f_{a,[HS]} = k_{d,H} f_{d,[HS]}$$

where $k_{a,H}$ and $k_{d,H}$ are rate constants of adsorption and desorption, respectively, C_H is the concentration of H_2 in the gas phase, $f_{a,[HS]}$ is the probability of finding empty sites suitable for the adsorption of H_2, and $f_{d,[HS]}$ is the probability of finding adsorbed species which could desorb as H_2. Clearly, $f_{d,[HS]_2}$ can be set equal to θ_H^2, and $f_{a,[HS]}$ can be set equal to $\theta_{S,H}^2$ where $\theta_{S,H}$ is the number of empty sites available to the [HS] landing area as a fraction of the total number of sites on the surface. Then

$$K_H C_H \theta_{S,H}^2 = \theta_H^2 \tag{2a}$$

where

$$K_H \equiv k_{a,H}/k_{d,H} \tag{2b}$$

is the thermodynamic equilibrium constant for adsorption of H_2.

For the three landing areas of species R, similar relations can be obtained so that

$$K_5 C_R f_{a,[RS5]} = f_{d,[RS5]} \tag{3a}$$

$$K_3 C_R f_{a,[RS3]} = f_{d,[RS3]} \tag{3b}$$

$$K_2 C_R f_{a,[RS2]} = f_{d,[RS2]} \tag{3c}$$

Here C_R is the concentration of R in the gas phase, and the probabilities f are similar to those defined above.

The thermodynamic equilibrium constants K_H, K_2, K_3, and K_5 incorporate information on the chemistry of the adsorption and desorption process, and on the relative orientation of sites (e.g., cyclic, linear) of a landing area.

Table I indicates that eight adjacent empty sites are required for the adsorption of R on the five-site landing area. Of these sites, five must be available to [RS_5] and the others, for [HS], need not be available to [RS_5]. Clearly two extreme cases are possible for the evaluation of $f_{a,[RS5]}$, depending upon whether the sites available for particular landing areas are grouped together or are randomly scattered. For our particular case, a supported metal catalyst, it is reasonable to believe that the sites available for particular landing areas are grouped together, following the corresponding crystallographic orientation of a particular crystallite size. Hence, all of the sites on which are adsorbed the hydrogen atoms around a five-site landing area are also available for the [RS_5] adsorbate. Below, we evaluate the probabilities f_a and f_d for both extreme cases, but further analysis in this work will be restricted to the extreme case of grouped-together landing areas.

For the former extreme case

$$f_{a,[RS5]} = \theta_{S,5}^8$$

In the other extreme case, where the available sites are randomly distributed throughout the surface, the probability of finding five empty sites available for [RS_5] and three adjacent empty sites available for [HS] would be

$$f_{a,[RS5]} = \theta_{S,5}^5\theta_S^3$$

Recall that all sites are available for the adsorption of H.

Similarly, the expression for $f_{d,[RS5]}$ can be written for the first extreme case as

$$f_{d,[RS5]} = \theta_5\theta_{H,5}^3$$

while it would be

$$f_{d,[RS5]} = \theta_5\theta_H^3$$

for the second extreme case.

Note that in f_d we are considering the probability of finding an occupied five-site landing area as θ_5. In fact the value chosen is an upper bound for this probability. This value is close to physical reality and probably closer than the lower bound. Certainly the probability of finding one site of the landing area containing a [RS_5] adsorbate can be given by θ_5. The probability of finding each of the remaining sites of the landing area lies between a lower bound of θ_5, corresponding to randomness, and an upper bound equal to 1, corresponding to the hypothetical situation in which every site surrounding the given site is occupied by the same molecule. Hence, the probability of finding an occupied five-site landing area lies between θ_5 and θ_5^5. The actual value is closer to the upper bound in our particular case since the five sites are always adjacent to one another. (For the case of dissociative adsorption of H_2, where each dissociated moiety can move on the surface, the probability is given by the lower bound.)

When the available landing areas are grouped together, and when the upper bound in the expression for f_d is used, then eq 3a can be written as

$$K_5 C_R \theta_{S,5}^8 = \theta_5\theta_{H,5}^3 \tag{4a}$$

Similarly, eq 3b and 3c reduce to

$$K_3 C_R \theta_{S,3}^5 = \theta_3\theta_{H,3}^2 \tag{4b}$$

$$K_2 C_R \theta_{S,2}^4 = \theta_2\theta_{H,2}^2 \tag{4c}$$

The rates of formation of products P_h, P_i, and P_s per unit surface area, assuming the rate-determining step is the reaction between intermediate and one H, can be written

$$r_h = k_h\theta_2\theta_{H,2} \tag{5a}$$

$$r_i = k_i\theta_3\theta_{H,3} \tag{5b}$$

$$r_s = k_s\theta_5\theta_{H,5} \tag{5c}$$

As before, it is assumed that sites available to a particular landing-area conformation are grouped together. In the other extreme case, for a random distribution of available sites, eq 5c would be replaced by

$$r_s = k_s\theta_5\theta_H$$

and similar expressions could be written for eq 5a and 5b.

Equations 5a–c relate reaction rates to the coverages (θ_S, θ_H, θ_2, θ_3, and θ_5), and to the cross-coverages ($\theta_{i,j}$, 12 in all). The coverages and cross-coverages are related by the definition of the cross-availability factors $\phi_{i,j}$. Recall that they depend upon the nature of landing areas i and j. The cross-availability factors are specified in Appendix II for the surface of Figure 2. Then the site balances of eq 1 and the equilibrium relations of eq 2 and 4 can be used to obtain values of the θ_i and $\theta_{i,j}$, from which the reactions rates can be calculated by way of eq 5. This is done below.

By inspection, the entire surface is available for [HS] and [RS_2]. Hence

$$\phi_H = 1 \tag{6a}$$

$$\phi_2 = 1 \tag{6b}$$

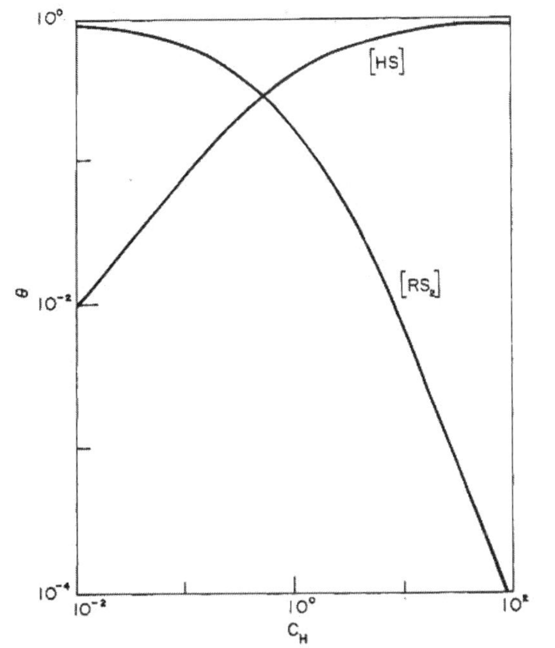

Figure 4. Variation of surface coverage of [HS] and [RS_2] with the concentration of H_2 in arbitrary units. The model of Table I is used, as quantified by eq 4–7. The availability of sites for [HS] is unity, and so is that for [RS_2] following eq 6. The coverages of these two species is not affected significantly by ϕ_3 or ϕ_5. $C_R = 1$, $K_H = K_2 = K_3 = K_5 = 1$, all in consistent units.

Since all of the surface in our present system is available to either [RS_3] or [RS_5] and since the two landing areas are mutually exclusive, then ϕ_3 and ϕ_5 are related by

$$\phi_3 + \phi_5 = 1 \tag{6c}$$

From Appendix II we note the following values for the cross-availability factors $\phi_{i,j}$ in our present system:

$$\phi_{2,H} = \phi_{3,H} = \phi_{5,H} = 1 \tag{7a}$$

$$\phi_{H,2} = \phi_{3,2} = \phi_{5,2} = 1 \tag{7b}$$

$$\phi_{H,3} = \phi_{2,3} = \phi_3 \tag{7c}$$

$$\phi_{H,5} = \phi_{2,5} = \phi_5 \tag{7d}$$

$$\phi_{5,3} = \phi_{3,5} = 0 \tag{7e}$$

Recall that the values of ϕ shown in Figure 2 are upper bounds for the availability factors. For the sake of simplicity, it has been assumed that the distances between catalytic sites are suitable for all three kinds of landing-area adsorbates. In reality, of course, this need not be so. In the present work and in ref 2, analogies have been drawn between the X type of landing area of Figure 3 and that corresponding to the three-site adsorbate [RS_3], between the Y sites and those for [RS_2], and between the Z sites and those for [RS_5]. If this is so, then note that nonadjacent carbon atoms are bonded to the surface at the three-site landing area [RS_3]. The resulting slack may make the distance requirement between neighboring catalyst sites considerably less stringent a restriction for the three-site landing area than it is for the other two landing areas. Hence, ϕ_5 and ϕ_2 for a given value of ϕ_3 may be less than the values obtained from eq 6. Since this will lead to quantitative, rather than qualitative, changes in the results, we use eq 6 and 7 in what follows.

Figures 4–6 show the variation in the fractional coverage by the different landing areas as a function of the amount

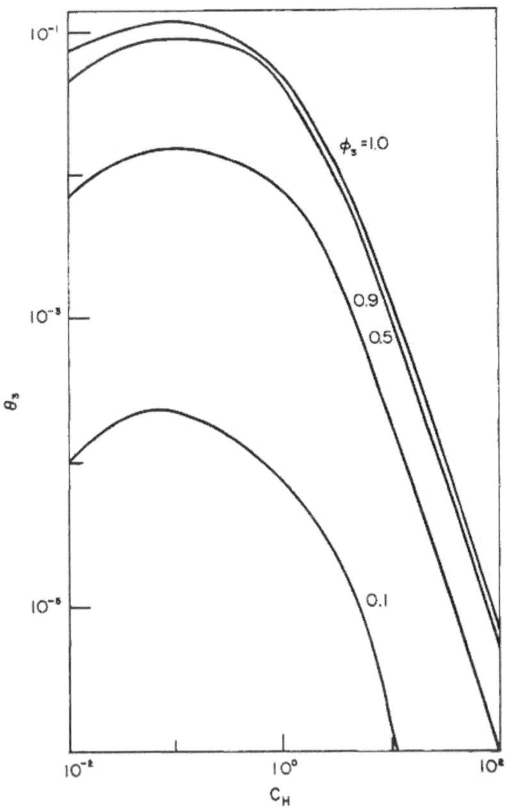

Figure 5. Variation in surface coverage of [RS_3] with concentration of H_2 for various values of ϕ_3. The corresponding values of ϕ_5 are given by eq 6c for the hypothetical catalyst described in Figure 2. Other parameters are as in Figure 4.

of H_2 present in the gas phase at a constant concentration of R when the values of the availability factors ϕ_3 and ϕ_5 are changed according to eq 6c. Following eq 6a and eq 6b, ϕ_2 and ϕ_H are kept unchanged at unity. As a result, the fractional coverages of [RS_2] and of [HS] change only marginally with changes in ϕ_3 and ϕ_5, and they are shown in Figure 4.

Shown in Figure 5 are values of the fractional coverage of [RS_3]. As expected, these values increase with increasing values of ϕ_3. The relative rate of the increase is not as rapid for larger values of ϕ_3. In Figure 6 is illustrated the behavior of the fractional coverage of [RS_5]. Note that the values of ϕ_5 used here complement the ϕ_3 values used in Figure 5; e.g., $\phi_3 = 0.1$ corresponds to $\phi_5 = 0.9$.

For a particular set of values of ϕ_2, ϕ_3, and ϕ_5, the rates of formation of products P_i, P_s, and P_h are shown in Figure 7 as functions of the concentration of H_2. Also shown is the ratio r_s/r_i. This parameter describes the relative rates of formation of the scrambled product P_s via the adsorbed form [RS_5], and the isomerization product P_i via the adsorbed form [RS_3]. Hence, this ratio is analogous to the ratio of isomerization by the cyclic mechanism and by the bond-shift mechanism. The latter ratio can be obtained experimentally from tracer measurements. We will return to this later.

In Figure 8 is illustrated the effect of changing the availability factors ϕ_3 and ϕ_5 following eq 7 for fixed values of C_H and C_R. As expected, r_i increases as larger fractions of the surface became available to the [RS_3] landing area. At the same time, r_s decreases as less of the surface is available to the [RS_5] landing area. Note that r_h is not significantly altered, since all of the surface is always

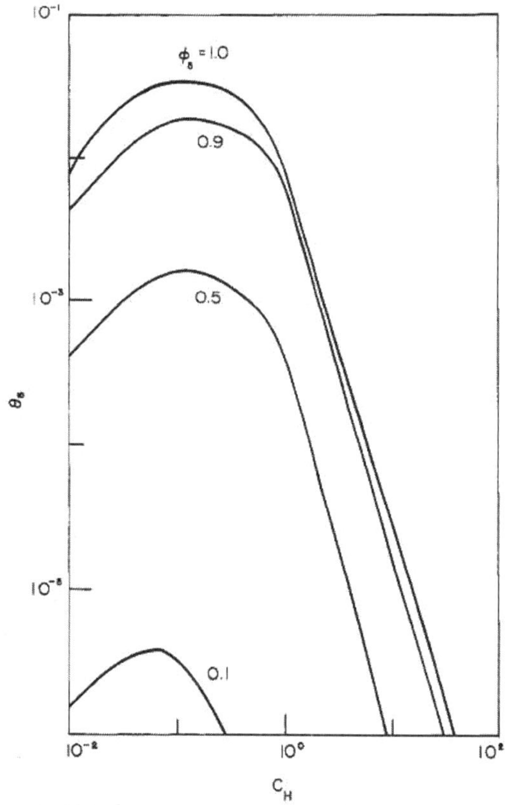

Figure 6. Variation in surface coverage of $[RS_5]$ with concentration of H_2 for various values of ϕ_5. The corresponding values of ϕ_3 are given by eq 6c for the hypothetical catalyst of Figure 2. The same values of ϕ_3 are used in Figure 5. Other parameters are as in Figure 4.

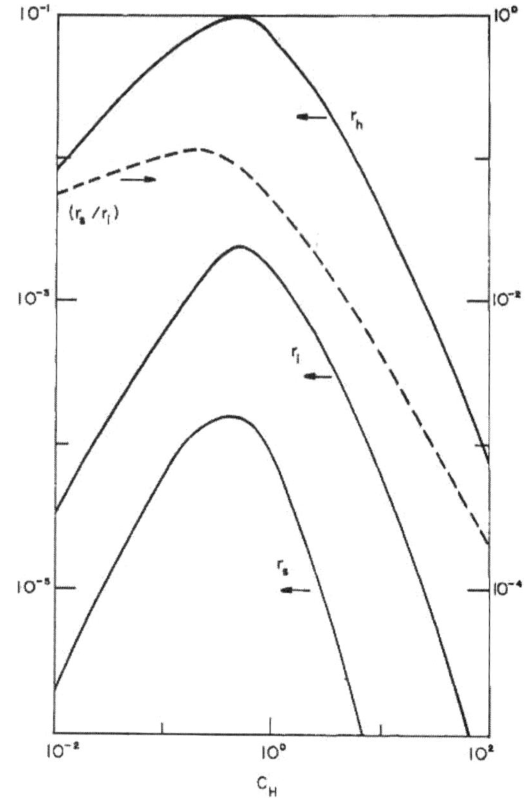

Figure 7. Reaction rates r_s, r_i, and r_h for the products P_s, P_i, P_h of Table I as a function of H_2 concentration for $\phi_2 = 1$, $\phi_3 = 0.5$, and $\phi_5 = 0.5$. The ratio (r_s/r_i) is a measure of the contribution of the cyclic and bond-shift mechanisms. The model of eq 4–7 is used with $k_h = k_i = k_s = 1$ in consistent units. Other parameters are as in Figure 4.

available to the $[RS_2]$ landing area in the simple model of Figure 2. The values of the average dispersion a from Figure 2 are also indicated in Figure 8. Note that for this particular case the values of a range from 0.6 to 1.0.

According to Figure 8, the ratio of the rates (r_s/r_i) varies from 0 at $a = 0.6$ to ∞ at $a = 1.0$ for the simple hypothetical catalyst of Figure 2. As mentioned earlier, this ratio, obtained from the model of Table I, is analogous to the ratio of the contribution of the cyclic mechanism to that of the bond-shift mechanism, obtained from the experimentally determined formation of isomers. The cyclic mechanism[6] involves a cyclopentane intermediate, similar to that formed on a Z-type landing area and characterized in our simple model of Table I as $[RS_5]$. The bond-shift mechanism[7] involves a carbon–carbon bond displacement, similar to that occurring at adsorbate $[RS_3]$ in Table I corresponding to landing areas of types X_1 and X_2. For isomerizations of various C_6 hydrocarbons, the contribution of the cyclic mechanism to the overall isomerization has been measured by using isotopic tracers by Dartigues et al.[8] as a function of the average dispersion a. In Table V results of these measurements are tabulated for the series of catalysts considered in Tables III and IV, both for the isomerization of 2-methylpentane and for the isomerization of n-hexane. The real catalyst surface is so much more complicated than the simple system used in Figure 2 (for instance, a in Figure 2 cannot be less than 0.6) that quantitative comparisons between Table V and Figure 8 would be meaningless. Qualitatively, however, note that both sets of results are similar: an increase in the dispersion yields an increase in the cyclic/bond shift ratio in Table V and an increase in r_s/r_i in Figure 8.

The present analysis of Dartigues's results in the landing-area framework is of course not inconsistent with that advanced by Dartigues et al.[8] However, there is an important difference. They specify the formation of particular geometrical planes corresponding to given particle sizes, e.g., the set of B_5 sites in the incomplete cubeocta-

TABLE V: Results of Dartigues et al.[8] on the Relative Contribution of Cyclic and Bond-Shift Mechanisms to the Initial Rates of Isomerization of C_6 Hydrocarbons to 3-Methylpentane at 254 °C[a]

catalyst Pt/Al_2O_3	av dispersion (a)	cyclic mechanism/bond-shift mechanism	
		A[b]	B
8.5% A′	0.05	0.33	0.92
8.4% B'_{10}	0.07	0.19	1.2
4.1% B'_{210}	0.55	0.43	
2.5% B'_{210}	0.70	1.6	4.9
0.3% B'_{210}	1.00	5.3	>200

[a] Catalysts correspond to those of Table III.

[b] A = , B =
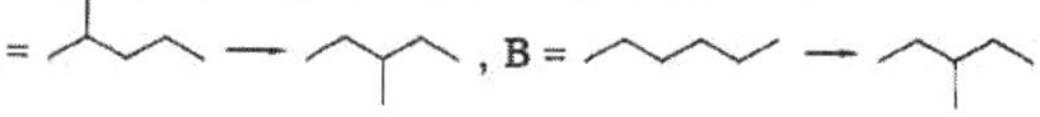

(6) Barron, J.; Maire, G.; Cornet, D.; Muller, J. M.; Gault, F. G. *J. Catal.* **1963**, *2*, 152.
(7) Anderson, J. R.; Baker, B. G. *Proc. R. Soc. London, Ser. A* **1963**, *271*, 402.
(8) Dartigues, J. M.; Chambellan, A.; Gault, F. G. *J. Am. Chem. Soc.* **1976**, *98*, 856. Dartigues, J. M.; Chambellan, A.; Corolleur, S.; Gault, F. G.; Renouprez, A.; Moraweck, B.; Bosch-Giral, P.; Dalmai-Imelik, G. *Nouv. J. Chim.* **1979**, *3*, 591.

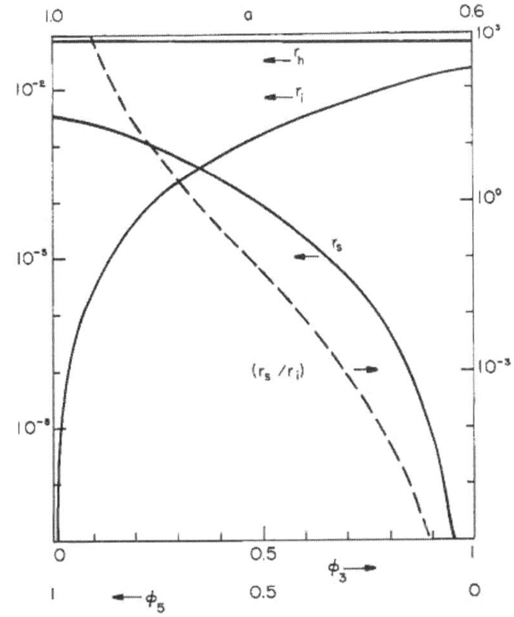

Figure 8. Reaction rates and r_s/r_i as a function of the nature of the surface of the hypothetical catalyst of Figure 2. The nature of the surface can be described in terms of the dispersion a, related to ϕ_3 and ϕ_5 by Figure 2 and eq 6. The value of C_H used is 1.0. Other parameters are as in Figure 7.

hedron. A molecule adsorbed on that collection of sites is hypothesized by them to desorb only as a particular product. On the other hand, in the present work, a site itself does not determine uniquely the product that will be formed—successive adsorbed species on the same site (together with different other sites) could form different products. What fixes the product formed is the orientation of the reactant while adsorbing, i.e., which other sites are involved at the same time. Certain orientations may be more likely assumed by the reactant, and certain orientations may be unacceptable by the surface, and it is in the changing of these likelihoods that the selectivity is altered.

6. Summary and Conclusion

Multiple adsorption or landing-area conformation possibilities must be seriously considered. In most cases, of course, the possibilities in fact reduce to a small number, and in some cases only one landing area may be possible. Nevertheless, the multiplicities in adsorption are present even in relatively simple systems such as hydrogen over a metal film and must exist in more complex situations.

A particular site on the surface is not, in general, restricted to being capable of being a part of only one landing-area conformation. Exactly which conformation it is in at any time would depend upon the nature of the surface, the temperature and pressure of the system, the relative amounts of the reactants, and the amounts of the other landing areas present on the surface. In most cases of interest, landing areas corresponding to larger numbers of surface sites are favored at higher temperatures and at lower partial pressures of the gas-phase species corresponding to the adsorbate. The relation between the nature of the surface and the distribution of adsorbed forms/landing areas is less quantitative and less general and is considered in this work.

In the simplest situation, all surface sites can equally well accommodate all landing-area possibilities that are actually present. The surface is then defined as being "uniform". Such a surface was considered earlier.[2]

If some landing areas are not possible in certain parts of the surface, the surface may be considered to contain specific adsorption sites. Such a situation is considered in the present work. Note, however, that we do not, in general, expect the surface to be divided into regions with each region being "allowed" only one landing area. Put in mathematical form, the availability ϕ of a landing area, the fraction of the surface in which a particular landing area can form, when summed over all landing-area possibilities is not, in general, expected to equal unity.

The changes, if any, in specific reaction rate (activity) and selectivity with the metal crystallite size are considered in terms of the fractional areas available for each of the possible landing-area conformations. A selectivity dispersion factor is defined for each landing area in terms of the dispersion and the availability factor for the landing area. Facile and demanding reactions with respect to activity and selectivity are related to the relative changes in dispersion and availability factors for each landing area or, equivalently, to the changes in the selectivity dispersion for each landing area.

A model for the landing areas expected in the skeletal isomerization of hydrocarbons by noble metals, originally introduced for a "uniform" surface in ref 2, is extended to apply to the general case of a nonuniform surface. A simple model of a "nonuniform" surface is used, consisting of two types of crystallite: cubic and icosahedral. The latter is expected for small crystallites, and the former for larger crystallites. Changing the relative amounts of the two types of crystallites changes the average dispersion of the surface and the availability factor for each landing area. (Hence, the selectivity dispersion also changes.) For each such surface the relative coverages of the landing areas can be obtained, and consequently so can the specific rates of formation of all of the products of the model. The predictions of the model with respect to selectivity as a function of average dispersion compare well with experimental results.

Appendix I

Define $\langle I \rangle$ as the number of moles of the reactant, 2-methylpentane-2-^{13}C, that adsorbs (and reacts) on the landing area of type I per 100 mol of the reactant lost. Then the number of moles of each product formed, $\langle\langle J \rangle\rangle$ for product J, can be related as follows to the various landing-area contributions $\langle I \rangle$ and to fractions g_1–g_{23} defined in Figure 3. For product 3-methylpentane-*3*-^{13}C, the relation is

$\langle\langle$ 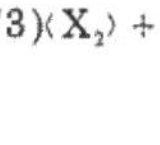 $\rangle\rangle = g_1\langle Z \rangle$ (A-1)

since this isotopic isomer is formed only from the Z-type landing area. For other products with 2-methylpentane-2-^{13}C as reactant, analogous relationships are presented in eq A-2–A-8.

$\langle\langle$ 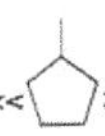 $\rangle\rangle = g_4\langle Z \rangle$ (A-2)

$\langle\langle$  $\rangle\rangle = g_7(2/3)\langle X_2 \rangle + g_{12}(2/3)\langle Y_2 \rangle + 2g_{20}(1/2)\langle Y_1 \rangle$ (A-3)

$\langle\langle$ 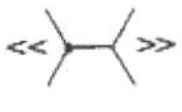 $\rangle\rangle = g_8(1/3)\langle X_2 \rangle + g_6\langle X_1 \rangle + g_{15}(1/3)\langle Y_2 \rangle + 2g_{17}(1/2)\langle Y_1 \rangle + g_{21}(1/2)\langle Y_1 \rangle$ (A-4)

$\langle\langle$ 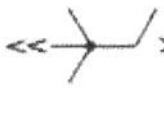 $\rangle\rangle = g_5\langle X_1 \rangle + g_{18}(1/2)\langle Y_1 \rangle + g_{22}(1/2)\langle Y_1 \rangle$ (A-5)

$$\langle\langle 2C_3\rangle\rangle = g_{23}(1/2)\langle Y_1\rangle \quad \text{(A-6)}$$

$$\langle\langle C_2 + C_4\rangle\rangle = g_{19}(1/3)\langle Y_1\rangle \quad \text{(A-7)}$$

$$\langle\langle C_1 + C_5\rangle\rangle = g_{10}(2/3)\langle Y_2\rangle + g_{13}(1/3)\langle Y_2\rangle \quad \text{(A-8)}$$

 $= g_{11}(2/3)\langle Y_2\rangle$

 $=$
$2g_3\langle Z\rangle + g_7(2/3)\langle X_2\rangle + g_{12}(2/3)\langle Y_2\rangle + 2g_{20}(1/2)\langle Y_1\rangle$

For the first eight of these products, the isotopic distribution of the compounds formed from 2-methylpentane-2-^{13}C is shown in Table III from ref 5. For the last two products, the isotopically separated n-hexanes, however, the reactant used in ref 5 was rather 2-methylpentane-*4*-^{13}C. The number of moles of these two products per 100 mol of this reactant are denoted by

with the open circles serving to remind us that the reactant is different. These product distributions can be obtained from a diagram analogous to Figure 3 and can be seen to be

 $= g_3\langle Z\rangle$ (A-9)

 $= g_3\langle Z\rangle + g_7(2/3)\langle X_2\rangle + g_{11}(2/3)\langle Y_2\rangle + g_{12}(2/3)\langle Y_2\rangle + 2g_{20}(1/2)\langle Y_1\rangle$ (A-10)

The definitions of fractions g_1–g_{23} yield the following additional relations:

$$g_1 + 2g_2 + 2g_3 + g_4 = 1 \quad \text{(A-11)}$$

$$g_5 + g_6 = 1 \quad \text{(A-12)}$$

$$2g_7 = 1 \quad \text{(A-13)}$$

$$g_8 + g_9 = 1 \quad \text{(A-14)}$$

$$g_{10} + g_{11} + 2g_{12} = 1 \quad \text{(A-15)}$$

$$g_{13} + g_{14} + g_{15} = 1 \quad \text{(A-16)}$$

$$g_{16} + 2g_{17} + g_{18} + g_{19} = 1 \quad \text{(A-17)}$$

$$4g_{20} + g_{21} + g_{22} + g_{23} = 1 \quad \text{(A-18)}$$

Equations A-1–A-18 need to be solved by using the data of Table III. In principle this allows us to obtain the various values of $\langle I\rangle$, which are measures of the contributions of the various landing areas, and the fractions g_1–g_{23}. Clearly, there is insufficient information to determine these parameters rigorously. One possible set of values for $\langle X_1\rangle$, $\langle X_2\rangle$, $\langle Y_1\rangle$, and $\langle Z\rangle$ is given for each catalyst in Table IV. Below we indicate how these values were obtained.

Note first that ref 5 makes no mention of the presence of either of the following:

 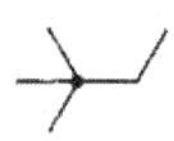

Assuming that this means that these compounds were not formed to any appreciable extent, then each contributor to the right-hand side of eq A-4 and A-5 can be set equal to zero. Consequently

$$\langle X_1\rangle = 0 \quad \text{(A-19)}$$

eq A-14 reduces to

$$g_9 = 1 \quad \text{(A-20)}$$

and eq A-16–A-18 are now

$$g_{13} + g_{14} = 1 \quad \text{(A-21)}$$

$$g_{16} + g_{19} = 1 \quad \text{(A-22)}$$

$$4g_{20} + g_{23} = 1 \quad \text{(A-23)}$$

Further, it is reasonable to suppose that g_1 and g_2 are equal. Hence, eq A-11 is now

$$3g_1 + 2g_3 + g_4 = 1 \quad \text{(A-24)}$$

From eq A-1, A-2, A-9, and A-24, we have

$\langle Z\rangle = 3$  $+ 2$  $+$ 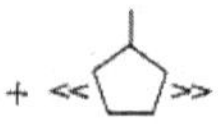 (A-25)

Equation A-7 yields

$$\langle Y_1\rangle = 2\langle\langle C_2 + C_4\rangle\rangle / g_{19}$$

Since g_{19} is less than or equal to 1, a lower bound for $\langle Y_1\rangle$ is given by

$$\langle Y_1\rangle \geq 2\langle\langle C_2 + C_4\rangle\rangle \quad \text{(A-26)}$$

Note that the equality would hold if only hydrogenolysis reactions proceeded on the Y_1 landing area. Using the lower bound for $\langle Y_1\rangle$, eq A-6 yields

$$g_{23} = \langle\langle 2C_3\rangle\rangle / \langle\langle C_2 + C_4\rangle\rangle$$

and therefore, from eq A-23

$$g_{20} = \{\langle\langle C_2 + C_4\rangle\rangle - \langle\langle 2C_3\rangle\rangle\} / 4\langle\langle C_2 + C_4\rangle\rangle$$

Substituting this into eq A-3 and using eq A-13 gives

$g_{12}(2/3)\langle Y_2\rangle =$
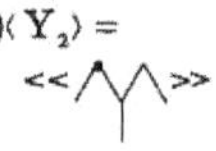 $- (1/3)\langle X_2\rangle - (1/2)\{\langle\langle C_2 + C_4\rangle\rangle - \langle\langle 2C_3\rangle\rangle\}$

From eq A-3, A-9, and A-10

$g_{11}(2/3)\langle Y_2\rangle =$  $-$  $-$ 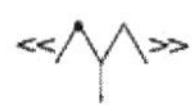

while eq A-8 yields

$$g_{10}(2/3)\langle Y_2\rangle = \langle\langle C_1 + C_5\rangle\rangle - (1/3)\langle Y_2\rangle$$

provided g_{13} is set equal to unity. (Note that this simplification is consistent with setting g_{19} equal to unity, as was used to obtain the lower bound of eq A-26.) Summing the last three equations yields

$\langle Y_2\rangle + (2/3)\langle X_2\rangle = \langle\langle C_1 + C_5\rangle\rangle - \langle\langle C_2 + C_4\rangle\rangle + \langle\langle 2C_3\rangle\rangle +$
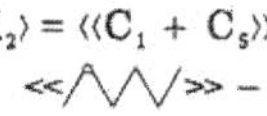 $-$ 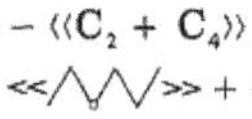 $+$ 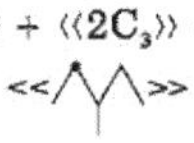 (A-27)

Because of the lack of information, values of $\langle Y_2\rangle$ and $\langle X_2\rangle$ cannot be obtained independent of each other. However, eq A-27 does allow upper bounds of $\langle Y_2\rangle$ and $\langle X_2\rangle$ to be obtained, by neglecting the other contributor to the left-hand side of eq A-27.

The values of $\langle X_2\rangle$ and $\langle Y_2\rangle$ shown in Table IV are one-half of these upper bounds:

$$\langle X_2\rangle = \tfrac{3}{4}\{\text{rhs}\} \quad \text{(A-28)}$$

$$\langle Y_2\rangle = \tfrac{1}{2}\{\text{rhs}\} \quad \text{(A-29)}$$

where {rhs} denotes the right-hand side of eq A-29. The value of $\langle X_1\rangle$ in Table IV is zero, in accordance with eq A-19. The value of $\langle Y_1\rangle$ in Table IV is the lower bound of eq A-26. The value of $\langle Z\rangle$ in Table IV is taken from eq A-25.

Appendix II

Evaluation of Cross-Availability Factors $\phi_{i,j}$.[9] Recall that the cross-availability $\phi_{i,j}$ is defined as the ratio $\theta_{i,j}/\theta_i$,

(9) We are grateful to one of the reviewers for suggesting such an analysis.

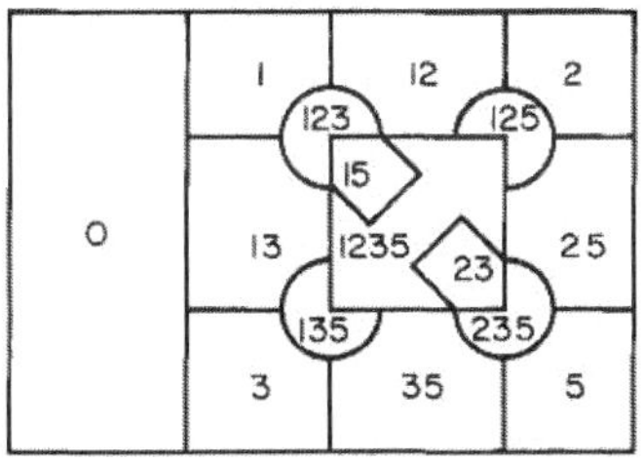

Figure 9. Schematic of surface of hypothetical catalyst showing the most general distribution of possibilities for the spectrum of landing areas of Table I. See Appendix II for an explanation of region notation.

where $\theta_{i,j}$ is the fractional surface covered as landing area i and which is also amenable to landing area j. The factors $\phi_{i,j}$ are first evaluated for the most general situation involving the present spectrum of four landing-area types (corresponding to [HS], [RS_2], [RS_3], and [RS_5]) on the surface, as shown in Table I. For the surface used in the present work, described in Figure 2, a number of simplifications are possible, and these are described below. In brief, we first show that, if all sites on the surface are available for a type-k adsorbate, i.e., if $\phi_k = 1$, then $\phi_{k,j} \equiv \phi_j$ and $\phi_{i,k} \equiv 1$. We finally show that, if sites for landing areas k and m are mutually exclusive, then $\phi_{k,m} = \phi_{m,k} = 0$. These relations allow us to obtain values for all of the $\phi_{i,j}$.

Consider Figure 9, which represents schematically and in general the surface available for each type of landing area considered in this work. Here region 0 is the fractional surface not available for any landing area. Region 1 represents the fractional surface available only for the [HS] type of landing area; i.e., this surface is not available for [RS_2], [RS_3], or [RS_5]. Further, regions 2, 3, and 5 represent fractional surfaces exclusively for [RS_2] alone, [RS_3] alone, and [RS_5] alone, respectively. Region 13 represents the fractional area available to either [HS] or [RS_3], but not to [RS_2] or [RS_5]; similarly, region 123 is unavailable only to [RS_5]. Other regions may be analogously assigned.

Central to this development is the assumption that the probability of tying up a site as part of a type-i landing-area adsorbate does not depend upon whether the site would also be available, if unoccupied, for a type-j landing-area conformation. To prove that $\phi_{i,k} = 1$ if $\phi_k = 1$, we consider for definiteness the cross-coverage $\theta_{H,2}$. From Figure 9 and the assumption above, $\theta_{H,2}$ is related to θ_H by

$$\frac{\theta_{H,2}}{A_{12} + A_{123} + A_{125} + A_{1235}} = \frac{\theta_H}{A_1 + A_{12} + A_{13} + A_{15} + A_{123} + A_{125} + A_{1235}} \quad \text{(A-30)}$$

where A_k is the fractional area of region k in Figure 9. Then by definition

$$\phi_{H,2} = \frac{A_{12} + A_{123} + A_{125} + A_{1235}}{A_1 + A_{12} + A_{13} + A_{15} + A_{123} + A_{125} + A_{1235}} \quad \text{(A-31)}$$

in general. Now from Figure 9 the availability factor for [RS_2] can be written

$$\phi_2 = A_2 + A_{12} + A_{23} + A_{25} + A_{123} + A_{125} + A_{235} + A_{1235} \quad \text{(A-32)}$$

Since the sum of all of the fractional areas in Figure 9 is unity, then setting $\phi_2 \equiv 1$ is equivalent to

$$A_0 + A_1 + A_3 + A_5 + A_{13} + A_{15} + A_{35} + A_{135} = 0 \quad \text{(A-33)}$$

Since the A_k values are either positive or zero, eq A-33 requires that each term on the left-hand side be zero. Substituting this into eq A-31, we have the required relation

$$\phi_{H,2} = 1 \quad \text{(A-34)}$$

Similarly

$$\phi_{3,2} = \phi_{5,2} = 1 \quad \text{(A-35)}$$

Since ϕ_H is also equal to unity, we can write

$$\phi_{2,H} = \phi_{3,H} = \phi_{5,H} = 1 \quad \text{(A-36)}$$

We now show that $\phi_{k,j} \equiv \phi_j$ if $\phi_k = 1$ by considering the case of $\phi_{2,3}$. Similar to eq A-31 we can write $\phi_{2,3}$ as

$$\phi_{2,3} = \frac{\theta_{2,3}}{\theta_2} = \frac{A_{23} + A_{123} + A_{235} + A_{1235}}{A_2 + A_{12} + A_{23} + A_{25} + A_{123} + A_{125} + A_{235} + A_{1235}} \quad \text{(A-37)}$$

and similar to eq A-32, ϕ_3 is given by

$$\phi_3 = A_3 + A_{13} + A_{23} + A_{35} + A_{123} + A_{135} + A_{235} + A_{1235} \quad \text{(A-38)}$$

Since each term on the left-hand side of eq A-33 is zero, and since $\phi_2 = 1$, then eq A-32, A-37, and A-38 yield

$$\phi_{2,3} \equiv \phi_3 \quad \text{(A-39)}$$

Similarly

$$\phi_{2,5} \equiv \phi_5 \quad \text{(A-40)}$$

and since ϕ_H is also unity

$$\phi_{H,3} = \phi_3 \quad \text{(A-41)}$$

$$\phi_{H,5} = \phi_5 \quad \text{(A-42)}$$

Finally, we show that for mutually exclusive sites $\phi_{k,m} = \phi_{m,k} = 0$. In the present spectrum of landing areas, [RS_3] cannot be adsorbed on sites suitable for the landing area of [RS_5]. To evaluate $\phi_{3,5}$ and $\phi_{5,3}$, note that in general

$$\phi_{3,5} = \frac{A_{35} + A_{135} + A_{235} + A_{1235}}{A_3 + A_{13} + A_{23} + A_{35} + A_{123} + A_{135} + A_{235} + A_{1235}} \quad \text{(A-43)}$$

$$\phi_{5,3} = \frac{A_{35} + A_{135} + A_{235} + A_{1235}}{A_5 + A_{15} + A_{25} + A_{35} + A_{125} + A_{135} + A_{235} + A_{1235}} \quad \text{(A-44)}$$

Because no fraction of the present surface can accommodate both [RS_3] and [RS_5], each term in the numerators of eq A-43 and A-44 is identically zero. Then

$$\phi_{3,5} = 0 \quad \text{(A-45)}$$

$$\phi_{5,3} = 0 \quad \text{(A-46)}$$

Equations A-34–A-36, A-39–A-42, A-45, and A-46 are the required values of the cross-availability factors $\phi_{i,j}$ reproduced in eq 7.

Path of Oxygen in a Bismuth Molybdate Lattice During Selective Oxidation

Dady B. Dadyburjor† **and Eli Ruckenstein***

Faculty of Engineering and Applied Sciences, State University of New York at Buffalo, Buffalo, New York 14214 *(Received June 23, 1977; Revised Manuscript Received February 17, 1978)*

Interaction potentials are calculated between an oxygen ion and either a bismuth or a molybdenum ion. Paths are calculated for the motion of oxygen ions at the interface of or inside a 2:1 bismuth molybdate lattice and thus under the influence of bismuth, molybdenum, and oxygen ions. The minimum energy requirements for these paths show that if an oxygen bound to bismuth is removed in oxidation, a lattice oxygen bound to a nearby molybdenum ion would occupy the vacancy near the bismuth while an external oxygen ion would later transfer to fill the position bound to the molybdenum. It is suggested that this pumping behavior of the molybdenum ions plays a role in the superior activity and selectivity of the mixed oxides bismuth molybdate as oxidation catalysts, compared with bismuth oxide or molybdenum oxide.

1. Introduction

Compounds consisting of two or more oxides often possess catalytic properties that are far superior to those of their components. For example, bismuth oxide Bi_2O_3 catalyzes the oxidation of hydrocarbons to completion so that carbon dioxide and water are the only products. Conversely, when molybdenum oxide MoO_3 is used, the reaction hardly occurs at all but almost all the products are unsaturated organic compounds of industrial importance. Various bismuth molybdates can be formed by combination of these two oxides in certain proportions under prescribed conditions.[1] One such compound, the 2:1 bismuth molybdate (Bi_2MoO_6), is known[2] to combine the favorable activity of the bismuth oxide component with the favorable selectivity of the molybdenum oxide component. Hence, this mixed oxide is often used as a catalyst for the partial oxidation of hydrocarbons.

In practice, the oxidation reaction occurs in the presence of sufficient oxygen gas. However, even in the absence of the reactant oxygen when only bismuth molybdate and the hydrocarbon are present there is, at least initially, some formation of the unsaturated compounds of interest. Clearly the bismuth molybdate itself yields oxygen from its lattice.

† D. B. Dadyburjor is now with the Department of Chemical and Environmental Engineering, Rensselaer Polytechnic Institute, Troy, N.Y. 12181.

Therefore, in the presence of reactant oxygen, the role of the bismuth molybdate is to furnish the oxygen necessary for the partial oxidation of the hydrocarbon, while the role of the external oxygen is to replenish the bismuth molybdate lattice. Furthermore, since the mixed oxide bismuth molybdate contains the advantages of both of its constituent oxides and the disadvantages of neither, there would have to be some interplay between the lattice oxygen ions associated with molybdenum and those associated with bismuth.

In a series of papers, Peacock et al.[3] argue that the external oxygen replaces the lattice ion near a bismuth site, while an oxygen ion from the lattice near a molybdenum site plays a role in the oxidation of the hydrocarbon. According to these workers the reactions for the oxidation of propylene can be written as

$$C_3H_6 + Mo^{6+} + O^{2-} \rightarrow C_3H_5 + Mo^{5+} + OH^-$$

$$Mo^{5+} + Bi^{3+} \rightarrow Mo^{6+} + Bi^{2+}$$

$$2Bi^{2+} + {}^1/_2O_2 \rightarrow 2Bi^{3+} + O^{2-}$$

Thus, in the overall reaction, an O^{2-} ion near Mo^{6+} is removed and an O^{2-} ion is placed near Bi^{3+}. Further, the Bi^{3+} ion in bismuth molybdate serves merely to reoxidize the Mo^{5+} to Mo^{6+}. In the process of losing an electron to return to the Bi^{3+} state, it creates the replacement O^{2-} lattice ion from the reactant oxygen. The assumption of ref 3 is that the olefin is attracted near the Mo^{6+}, rather

This paper, which was originally published in **The Journal of Physical Chemistry**: 82(13), 1563–1575 (1978), is republished with permission.

than near the Bi^{3+}, because of the higher charge associated with the former ion.

Haber and co-workers[4] have the opposite point of view, i.e., that it is the bismuth region that loses the initial O^{2-}. They recall that Swift et al.[5] observed that, in the absence of oxygen, bismuth oxide reacts with propylene to yield 1,5-hexadiene. Repeating the experiment with bismuth molybdate instead of bismuth oxide, Haber et al. found that acrolein is the product formed. Haber explained the results of both reactions in a unitary manner. He postulated that the first step in both reactions is the loss of an oxygen ion from near the Bi^{3+} ion and a hydrogen ion from the olefin to yield $CH_2{=}CH{-}CH_2^-$. In the absence of molybdenum, i.e., for Swift's experiments, two such species combine to yield 1,5-hexadiene. If molybdenum is present, another hydrogen ion is lost from the species and oxygen is added from the oxide catalyst to the species to yield acrolein. Haber believes that the $CH_2{=}CH{-}CH_2^-$ moves to near the $O^{2-}{-}Mo^{6+}{-}O^{2-}$ portion of the bismuth molybdate lattice for the last hydrogen–oxygen transfer; we will return to this shortly.

Otsubo et al.[6] performed tracer experiments with 2:1 bismuth molybdate samples prepared in three different ways. In one method of preparation, bismuth oxide containing 9% ^{18}O atoms was combined with molybdenum oxide containing the ambient isotopic mixture (sample B), while in the second, yielding sample M, the usual bismuth oxide was combined with molybdenum oxide containing 9% ^{18}O atoms. Lastly, sample BM was obtained by reoxidizing with $^{36}O_2$ a partially reduced bismuth molybdate initially containing no added ^{18}O. All three samples finally contained ^{18}O in 4.5% of the oxygen ions in the lattice. The samples were individually reduced with hydrogen at 400 °C, and the water produced was analyzed for ^{18}O at various times. The water from sample B initially contained over 6% of ^{18}O but the concentration dropped with increasing time, levelling off to about 3% after 5 min. Sample M on the other hand yielded water initially containing about 1% of ^{18}O and the concentration increased with time, levelling off after 5 min to the same value as for sample B, about 3%. These results indicate that the lattice oxygen ions lost are associated with bismuth ions, and the lost ions are replaced with those associated with molybdenum. After only 5 min the oxygen ions initially associated with molybdenum exist with equal probability at molybdenum or bismuth sites so that the ions leaving the lattices of both samples B and M contain approximately the average proportion of ^{18}O. The water from sample BM followed the same trend as that from sample M. This indicates that after bismuth molybdate is reduced, the external oxygen enters the lattice at oxygen vacancies associated with molybdenum. Hence the scheme of oxygen transfer in the lattice during the oxidation of the olefin seems to be that an O^{2-} associated with a Bi^{3+} ion is initially taken up by the olefin. This O^{2-} is replaced by one associated with Mo^{6+}, and the vacancy thus created near Mo^{6+} is filled by external oxygen. We note that this process is consistent with that of Haber mentioned above, provided one assumes that the second hydrogen abstraction also occurs near the Bi^{3+} site, the Mo^{6+} site supplying the oxygen required.

In this work we bring arguments in favor of Otsubo's sequence of steps. We thus hope to clarify a portion of the overall scheme of selective oxidation by bismuth molybdates.

Based on the pair potentials $O^{2-}{-}O^{2-}$, $O^{2-}{-}Bi^{3+}$, and $O^{2-}{-}Mo^{6+}$, we calculate the total potential of an oxygen ion under the influence of bismuth, molybdenum, and oxygen ions in the lattice of the 2:1 bismuth molybdate. The assumption of pairwise additivity is used. More complex summations, made elsewhere for other systems, do not lead to significant changes in the total energy.[7,8] One particular O^{2-} ion, either from the lattice or from the reactant, is considered mobile while other ions in the known 2:1 bismuth molybdate structure are considered stationary. Then the mobile ion has a preferred path along which the total potential, as calculated above, is always a minimum. Although the actual catalytic reaction occurs at temperatures above 400 °C, we neglect any effects the vibration of the components of the lattice may have upon the path of a moving O^{2-} ion. The energy along the preferred path may have maxima and/or minima. Consider now two such paths with the same initial position but occurring under different lattice conditions and having different final positions. Let the total energy along the first path decrease monotonically from the initial to the final stage while the energy along the second path passes over a maximum, i.e., an energy barrier. Clearly the first path is favorable while the second is at best less so. Similarly for two paths with different energy barriers to overcome, the path with the smaller difference between initial energy and maximum energy will be favored. We use this approach to estimate the relative ease with which a mobile O^{2-} ion moves to various empty O^{2-} sites in the bismuth molybdate lattice. We show that an O^{2-} ion removed from near the Bi^{3+} by the oxidation reaction is most easily replaced by an O^{2-} ion associated with Mo^{6+}, while an external O^{2-} ion has an energetically favorable path in replacing the latter O^{2-} ion. We should point out at this time that we do not consider here the process by which the reactant oxygen molecule is converted to external O^{2-} ions.

2. Interaction Potentials between Isolated Ion Pairs

The energy of interaction between two ions, of the same or different species, can be considered as the sum of Coulombic interactions, polarization contributions from permanent, induced, and instantaneous (van der Waals) dipoles and higher multipoles, the short-range electron shell repulsion terms, and the contribution from covalent bonding. Because the relations for some of these terms have not been given explicitly in the literature, they are presented in the Appendix. A comprehensive list of the parameters in the interaction potentials between the pairs $O^{2-}{-}O^{2-}$, $O^{2-}{-}Bi^{3+}$, and $O^{2-}{-}Mo^{6+}$ is given in Table I. The interaction energies for the three cases are shown in Figure 1. At small separation distances, the total energy passes over a maximum and then tends to $-\infty$ as the distance tends to 0. This is not realistic, as it is expected that the energy will steadily increase with decreasing separation. The maximum occurs only because of the mathematical nature of the forms used for the various contributions to the total interaction energy, and is known as the "artificial maximum". The interaction energy is set at $+\infty$ when the separation distance is less than the position of the artificial maximum, r_{max}. This is most easily seen in the case of the ion pair $O^{2-}{-}O^{2-}$ where r_{max} = 1.27 Å.

3. Oxygen Paths

3.1. *The Solid Lattice.* This work is concerned with the mechanism by which an external O^{2-} ion is incorporated into the lattice of the mixed oxide 2:1 bismuth molybdate, Bi_2MoO_6. We are not dealing with the motion of the ion in the bulk of the solid lattice, i.e., in the subsurface unit cells of the lattice, but only in the solid near the surface. Hence we consider the lattice of the solid to be doubly infinite in the two directions along the surface and singly

TABLE I: Parameters for the Interaction Energy, E_{ij}, between Ions of Species i and j, Separated by a Distance r_{ij}[a]

i	Range of r	A_{ij}	B_{ij}	G_{ij}	C_{ij}	D_{ij}	F_{ij}
O^{2-}	$r < 0.28$	1.35×10^4	10.5	0	15.6	45.5	57.6
	$0.28 < r < 0.6$	3.8×10^3	6.0				
	$0.6 < r < 2.5$	5.7×10^2	3.04				
	$r > 2.5$	8.68	1.37				
Bi^{3+}	$r < 0.3$	1.24×10^5	11.2	6.17	43.9	291.0	−86.4
	$r > 0.3$	4.5×10^4	7.6				
Mo^{6+}	$r < 0.24$	7.41×10^4	12.0	2.60	46.4	312.0	−172.8
	$0.24 < r < 0.7$	1.62×10^4	5.7				
	$0.7 < r < 1.3$	1.58×10^3	2.4				
	$1.3 < r < 1.7$	4.02×10^2	1.33				
	$1.7 < r < 1.95$	1.8×10^2	0.85				
	$1.95 < r < 2.35$	1.07×10^2	0.58				
	$r > 2.35$	69.0	0.4				

[a] Here the subscript j always represents the O^{2-} species. The value of E_{ij} is obtained from the parameters A_{ij} through G_{ij} and the relationship $E_{ij} = A_{ij} \exp[-B_{ij}(|r_{ij} - (G_{ij}/r_{ij}^2)|)] - C_{ij}/r_{ij}^6 - D_{ij}/r_{ij}^4 + F_{ij}/r_{ij}$ where E_{ij} is in eV, r_{ij} is in Å. See Nomenclature for an explanation of symbols.

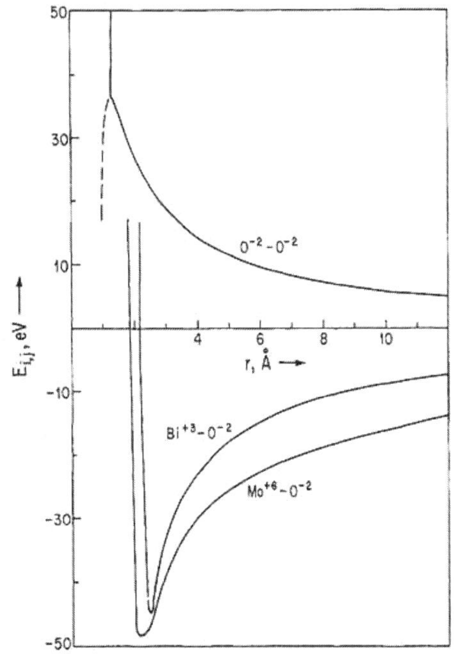

Figure 1. Interaction potentials O^{2-}–O^{2-}, Mo^{6+}–O^{2-}, Bi^{3+}–O^{2-}. The separation distance r_{ij} is in Å, the potentials E in eV. For O^{2-}–O^{2-}, the potential is assumed to be infinite at distances less than the artificial maximum as shown with dashed lines.

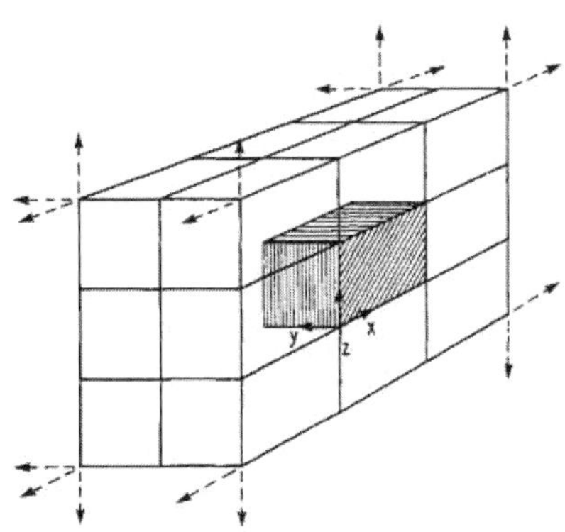

Figure 2. Truncated lattice used in the computations of the total interaction energy for a mobile O^{2-} ion. The unit cells considered are symmetric in the $\pm x$, $+y$, and $\pm z$ directions around the center unit cell, shaded, which lies on the external surface of the total lattice at $y = 0$. The dashed arrows show the directions in which the truncated lattice can be extended to form the full lattice.

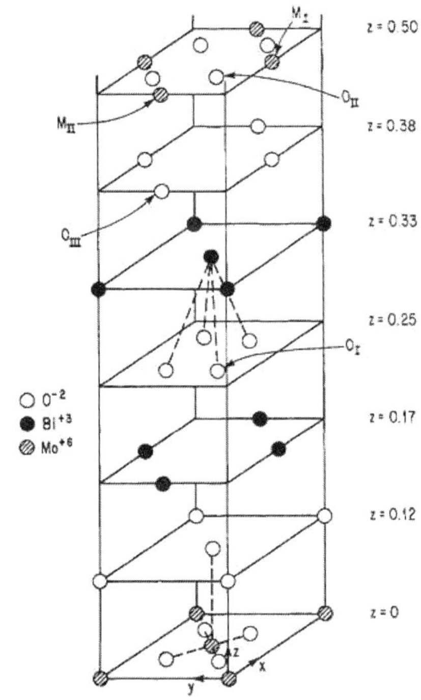

Figure 3. Location of ions in the lower half of a unit cell of 2:1 bismuth molybdate Bi_2MoO_6. The upper half is a mirror image through $z = 0.5$. The vertical coordinate has been exaggerated to show the position of ions in the various horizontal planes. The broken lines indicate how O^{2-} ions are considered affiliated with Bi^{3+} and Mo^{6+} ions. O_I represents an O^{2-} ion affiliated with Bi^{3+}, O_{II} and O_{III} are affiliated with a Mo^{6+} ion at the base and apex of the MoO octahedron, respectively. M_I represents a Mo^{6+} ion at the external surface of the lattice, at $y = 0$. M_{II} is a Mo^{6+} ion inside the cell.

infinite perpendicular to the surface. However, for simplicity we will compute the interactions of the mobile O^{2-} only with those Mo^{6+}, Bi^{3+}, and O^{2-} ions present in a truncated lattice shown in Figure 2. This lattice consists of all unit cells that are one lattice parameter away in the $\pm x$, $+y$, and $\pm z$ direction from the center unit cell, shaded in Figure 2. The $-y$ direction is excluded because the shaded cell is taken to be on the external surface of the lattice at $y = 0$. The inclusion of more unit cells to the truncated lattice, symmetric about the shaded cell, will change the total interaction energy of the mobile ion but will not have an appreciable effect upon the path it follows. Although surface imperfections such as dislocations are known to play a role in heterogenous catalysis, we make the simplifying assumption that the surface we consider here does not contain such imperfections.

3.2. *The Bi_2MoO_6 Unit Cell.* Figure 3 represents the lower half of the ideal unit cell of the 2:1 bismuth molybdate, Bi_2MoO_6, and is based upon ref 9. The upper half is a mirror image of the lower half through the xy plane at $z = 0.5$. The dimensions of the unit cell, a, b, and c in the x, y, and z directions, respectively, are $a = b = 5.5$ Å, $c = 16.1$ Å. The structure of Figure 3 may be thought to consist of layers of $(MoO_4)^{2-}$ and $(Bi_2O_2)^{2+}$, or alternatively O^{2-}, $(MoO_2)^{2+}$, O^{2-}, and $(Bi_2O_2)^{2+}$. Adoption of one or the other assignment of layers makes it easy to identify different types of oxygen ions in the lattice. For example, the former assignment is used to draw dashed lines in Figure 3 identifying O^{2-} ions on the base of a molybdenum–oxygen octahedron (at $z = 0$), at the apex of the octahedron ($z = 0.12$), and affiliated to Bi^{3+} ($z = 0.25$). Therefore while O_I in Figure 3 is a Bi^{3+} bound ion, O_{II} is a "base ion", and O_{III} is an "apex ion". With the latter assignment of layers, the "apex ions" are not unambiguously affiliated to Mo^{6+}, hence we may write of Bi^{3+}-bound ions (O_I), Mo^{6+}-bound ions (O_{II}), and intermediate ions (O_{III}). The intermediate ions are sometimes said to lie in the mobile O^{2-} layer. We will use the former notation in most of this work.

3.3. *Total Energy of a Mobile O^{2-} Ion in a Lattice.* In the rest of this section we use the pair potentials $E_{Bi,O}$, $E_{Mo,O}$, and $E_{O,O}$ of section 2 to calculate the energy of an O^{2-} ion at some known position interacting with ions of Bi^{3+} and Mo^{6+} and other O^{2-} ions at fixed positions in the lattice of Bi_2MoO_6. If (x_O, y_O, z_O) describes the position of the O^{2-} ion of interest with respect to some coordinate axes, then the energy of the ion is

$$E_O = \sum_l E_{Bi,O}(r_{l,O}) + \sum_m E_{Mo,O}(r_{m,O}) + \sum_n E_{O,O}(r_{n,O}) \quad (1)$$

The summations over l, m, and n are over all Bi^{3+}, Mo^{6+}, and O^{2-} ions in the lattice. If the coordinates of the lth Bi^{3+} ion are (x_l, y_l, z_l) then $r_{l,O}$, the distance between this Bi^{3+} ion and the O^{2-} ion of interest, is

$$r_{l,O} = \{(x_O - x_l)^2 + (y_O - y_l)^2 + (z_O - z_l)^2\}^{1/2} \quad (2)$$

A similar relation can be obtained for $r_{m,O}$, the distance between the mth Mo^{6+} ion and the O^{2-} ion of interest, and for $r_{n,O}$, the distance between the nth O^{2-} ion in the lattice and the O^{2-} ion of interest.

The energy E_O, as a function of (x_O, y_O, z_O), yields paths the ion will follow under various circumstances. If one coordinate, e.g., $x_O{}^*$, is specified, the position of the ion must be such that $E_O{}^* = E_O(x_O{}^*, y_O{}^*, z_O{}^*)$ is at least a local minimum. In this work, the values of the other two coordinates were obtained by the method of steepest descent. These two coordinates as a function of the specified coordinate define the path followed by the mobile ion in the direction of the specified coordinate. The corresponding values of the total energy, E_O at the local minima, as a function of the specified coordinate determine whether the path is energetically favored or not. In later sections paths will be calculated for a mobile O^{2-} ion when certain ions are removed from the solid lattice. Under some conditions the energy along one path decreases monotonically until the moving ion is in the vacant position, after which the energy drastically increases. Under different circumstances, another path of the same ion may be such that the energy goes through a maximum before the vacant position is reached. Clearly, the first type of path is more favorable than the second.

3.4. *Path of External Oxygen to the Lattice Surface.* In this section we calculate the path of an external O^{2-} ion to the surface of the solid lattice, truncated as in Figure 2, consisting of perfect unit cells containing Bi^{3+}, Mo^{6+}, and O^{2-} ions in positions as shown in Figure 3. We do not consider the process by which the reactant gaseous molecule O_2 is converted to external ions O^{2-}. We assume that this is done at some time prior to our interest in the O^{2-} ion, and without essentially disturbing the structure of the solid lattice. It may well be that the process of gaining electrons is closely related to the process of approaching the surface; however we ignore this coupling for the sake of simplicity. The coordinate system of the rest of this paper is that of Figure 2, with x, y, and z nondimensionalized with the unit cell parameters a, b, and c, respectively.

The position of an external O^{2-} ion outside the $y = 0$ surface of the lattice is computed by considering the path followed by such an ion as its y coordinate y_O changes from -0.2 to 0.0. We are not implying that the ion is formed a large distance from the external surface of the lattice. We suggest rather that, wherever the ion is formed, it will seek a position in this path such that its energy is at an absolute minimum. The initial guess for the position of the ion at $y_O{}^* = -0.2$ corresponds to the smallest value of $E_O(x_O, y_O{}^*, z_O)$ obtained from a series of calculations, using eq 1 and 2 and Table I, for x_O and z_O values between 0 and 1, i.e., above the shaded unit cell in Figure 2. The method of steepest descent is then used to vary x_O and z_O until the position of the minimum is obtained for this $y_O{}^*$ with sufficient accuracy, i.e., $x_O{}^*$ and $z_O{}^*$ to within $\pm 1 \times 10^{-5}$ in absolute value. These values of $x_O{}^*$ and $z_O{}^*$ are used as the initial guess for the next value of $y_O{}^*$, which is obtained by increasing $y_O{}^*$ by $dy = 0.02$. The computations are continued up to the surface of the lattice, i.e., $y_O{}^* = 0$.

The results are illustrated in Figure 4. The upper portion represents the energy of the path as $y_O{}^*$ is increased from -0.2 to 0.0, representing the surface of the lattice. The energy initially decreases as the O^{2-} ion approaches the lattice, due to the increasing (long range) attraction of the Bi^{3+} and Mo^{6+} ions. After a while, the energy path passes through a minimum and then increases as the (short range) repulsions predominate. The external O^{2-} ion will therefore lie at the position of the minimum along the path, i.e., at $y_O{}^* \simeq -0.166$. The coordinates of the path, plotted in the lower portion of Figure 4, show that the minimum occurs at $x_O{}^* \simeq 0.045$ and $z_O{}^* \simeq 0.470$. Comparing the x and z coordinates of the position of the minimum with those of the fixed ions in Figure 3 we see that the external O^{2-} ion lies roughly between two Mo^{6+} ions of the type labeled M_I, and above the Mo^{6+} ion labeled M_{II}, in Figure 3. Hence these three ions probably influence the external O^{2-} ion to the greatest extent.

3.5. *Path of External Oxygen to a Bismuth-Bonded Oxygen Vacancy.* In this section we obtain the path followed by the external O^{2-} ion if one of the O^{2-} ions associated with a Bi^{3+} ion in the lattice is missing. For definiteness we assume that the ion labeled O_I in Figure 3 is missing from the shaded center unit cell of the truncated lattice of Figure 2. The loss is presumably due to the oxidation reaction. We emphasize here that only one ion is missing, not also the mirror image to O_I through $z = 0.5$ in Figure 3. The starting point of the path is the position of the minimum, obtained in section 3.4 as (0.045, -0.166, 0.470). Based on energy paths calculated in both the y and the z directions, we assert that the minimum energy path will be one where the motion is first down the

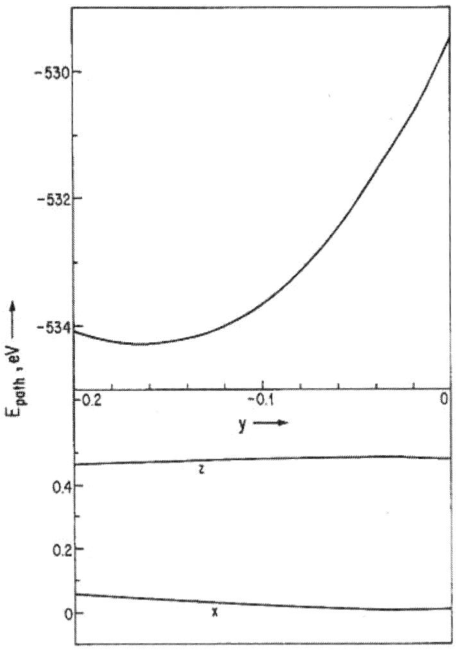

Figure 4. Motion of an external O^{2-} ion to the surface ($y = 0$) of the solid. The truncated lattice of Figure 2 is used for the computation of the total interaction energies. The external ion will lie at the position corresponding to the minimum of the curve of energy E_{path} in the upper figure. The minimum occurs at $y \simeq -0.166$. The x and z coordinates of the path are given in the lower figure. The position of the minimum of E_{path} corresponds to $x \simeq 0.045$, $z \simeq 0.47$.

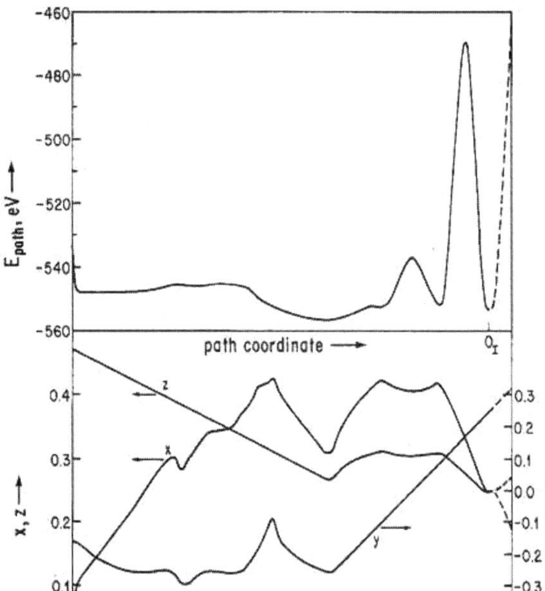

Figure 5. Motion of an external O^{2-} ion through the solid surface at $y = 0$ if the lattice ion at O_I in Figure 3 is missing. The ion at O_I is affiliated with a Bi^{3+} ion. Neither y nor z is an independent variable for the entire path, since the external ion first moves parallel to the z axis and then parallel to the y axis. Hence the abscissa is a path coordinate which is a measure of progress along the path. The portion of the path after the vacancy at O_I is shown as dashed lines.

z axis and then along the y axis. This motion is illustrated in Figure 5, where the abscissae are "path coordinates".

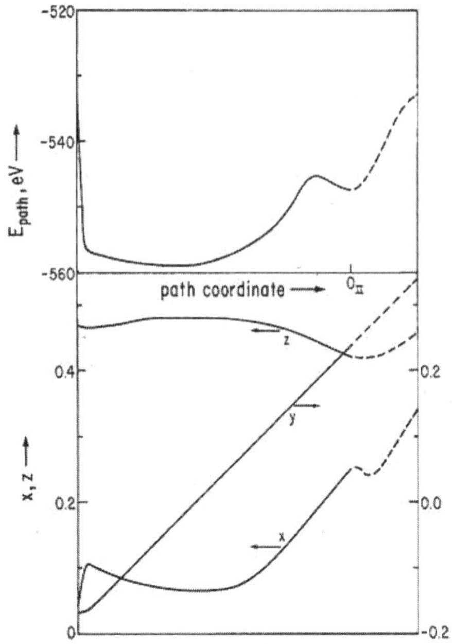

Figure 6. Motion of an external O^{2-} ion through the solid surface at $y = 0$ if the lattice ion at O_{II} in Figure 3 is missing. The ion at O_{II} is affiliated with a Mo^{6+} ion at the base of the MoO octahedron. The path is parallel to the y axis. The portion of the path after the vacancy at O_{II} is shown as dashed lines.

Several maxima can be seen in the energy curve of Figure 5. The first of these occurs when the mobile O^{2-} ion is proceeding down the z axis outside the surface of the lattice. This increase in energy is due to the presence of the oxygen layer at $z \simeq 0.38$ in the unit cell. Note that at this time the mobile ion is farthest from the lattice surface ($y_O = -0.3$). The mobile ion would probably not have to overcome the small energy barrier at this point if the energy (required to be a minimum) to define the original position of the ion had been calculated taking into account the vacancy at O_I instead of being calculated on the basis of a full lattice. In other words, if the removal of O_I were to occur before the arrival of the external ion, the initial position of the external ion would probably not be as in section 3.4. Rather it would correspond to the first broad minimum in the curve of Figure 5, having coordinates (0.25, −0.27, 0.26), i.e., directly above the vacancy O_I.

Of course there is no easy way of predicting the real sequence. However, the energy required to overcome the barrier at $z = 0.38$ is in any case minimal compared to the energy required to enter into the lattice, at $y = 0$, or to occupy the vacancy, at $y = 0.25$. The latter activation energy is in fact enormous and chracterizes the energy path as very unfavorable.

3.6. *Path of External Oxygen to Molybdenum-Bonded Oxygen Vacancies.* a. *Base Oxygen.* In this section we consider paths to vacancies of various O^{2-} ions affiliated with a Mo^{6+} ion in the lattice. We first consider a vacancy in an O^{2-} position at the base of a molybdenum–oxygen octahedron, for definiteness O_{II} in Figure 3. We will later consider an O^{2-} vacancy at the apex of the octahedron.

For the vacancy at O_{II}, as before, the starting coordinates are those of the external O^{2-} ion determined in section 3.4, viz. (0.045, −0.166, 0.470). The energy path is clearly in the y direction and is shown in Figure 6. The coordinates of O_{II} appear in the path as corresponding to a local energy minimum. Though the vacancy does not behave as a position of absolute minimum, the path to the vacancy is

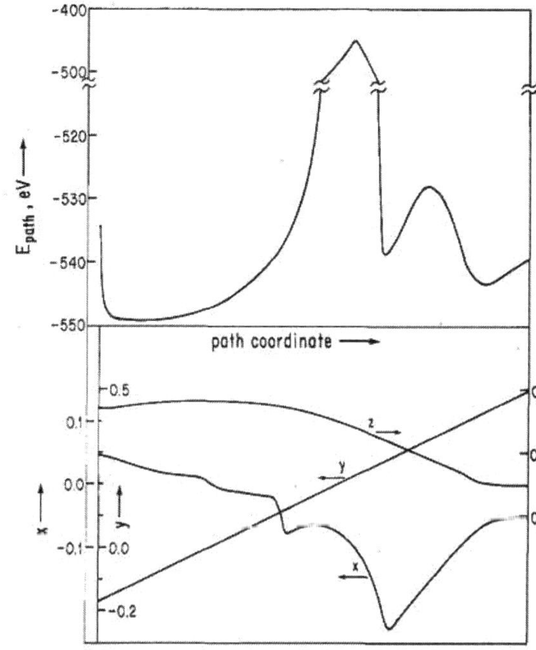

Figure 7. Motion of an external O^{2-} ion through the solid surface at $y = 0$. The lattice ion at O_{III} in Figure 3, i.e., at the apex of the MoO octahedron, is missing. The path is along the y axis.

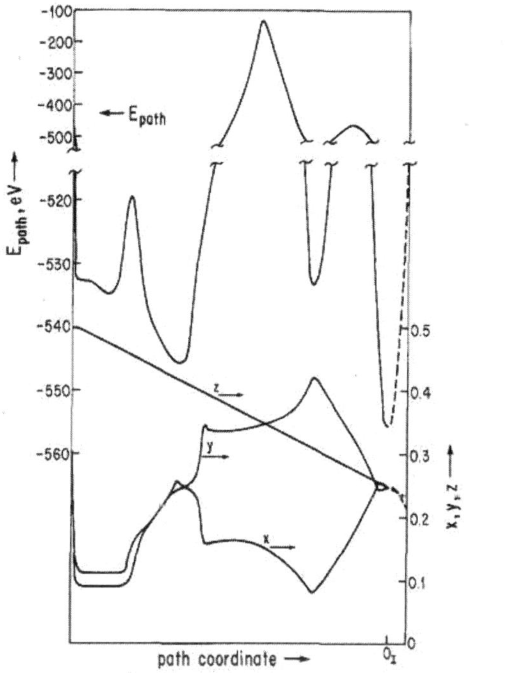

Figure 8. Motion of O^{2-} ion originally at the base of a MoO octahedron (O_{II}) if an O^{2-} ion bound to a Bi^{3+} ion (O_I) is removed. The path is calculated along the z axis. The portion of the path after O_I is shown as dashed lines.

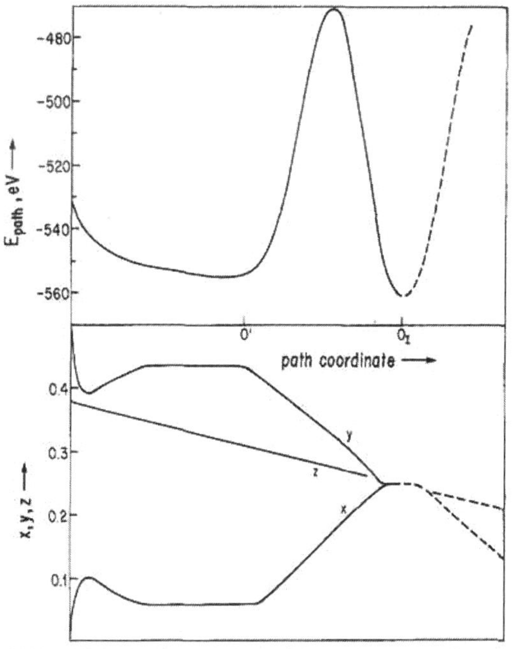

Figure 9. Motion of O^{2-} ion originally at the apex of a MoO octahedron (O_{III}) if an O^{2-} ion bound to a Bi^{3+} ion (O_I) is removed. The path is in the direction of the z axis. The portion of the path after O_I is shown as dashed lines. O′ denotes the probable final location of the moving ion, near O_I.

not entirely unfavorable. This is because the local maximum (at $y \simeq 0.18$) that the ion has to overcome to reach the vacancy (at $y \simeq 0.25$) occurs at an energy level that is lower than that corresponding to the starting point of the path.

b. *Apex Oxygen.* An oxygen ion at an apex of a molybdenum–oxygen octahedron is labeled O_{III} in Figure 3. The energy required for the path of the external O^{2-} ion to a vacancy at O_{III} is shown in Figure 7. The ion enters the lattice approximately at the level of the Mo^{6+} ions M_I and M_{II}. However, the presence of the ion O_{II} in the lattice causes a steep rise in the energy required at approximately $y = 0.25$ before the external O^{2-} ion can reach $y = 0.50$ corresponding to the vacancy at O_{III}. Hence, the path of the external ion directly to the vacancy at the apex site is not favorable.

3.7. *Path of Molybdenum-Bonded Oxygen Ions to a Bismuth-Bonded Oxygen Vacancy.* a. *Base Oxygen.* In the nomenclature of Figure 3, we first consider the motion of O^{2-} ion O_{II} to a vacancy at position O_I. We later consider the motion of O_{III} to the vacancy at O_I. In both instances the presence of an external O^{2-} ion at (0.045, −0.166, 0.470) is accounted for in the calculations.

Figure 8 illustrates the characteristics of the path of O_{II} to O_I. The layer of O^{2-} ions at $z = 0.38$ causes a large increase in the energy required for the path, which is very unfavorable energetically.

b. *Apex Oxygen.* The path along the z axis corresponding to the motion of ion O_{III} to a vacancy in position O_I is shown in Figure 9. The minimum energy of the path occurs not exactly at the vacancy O_I at (0.25, 0.25, 0.25) but at approximately (0.1, 0.4, 0.3) which is relatively close. The path to the latter point is obviously favored from an energetic point of view.

3.8. *Path of a Molybdenum-Bonded Oxygen Ion at the Base of the Octahedron to a Vacancy at the Apex.* In this section we consider the path of the ion originally at O_{II} when the position at O_{III} is vacant. This path is described in Figure 10. The absolute minimum of the energy occurs at $z = 0.41$ rather than at $z = 0.38$ but this difference is small. The path to near the O_{III} vacancy is obviously favored energetically.

4. Results

The selective oxidation of a hydrocarbon in the presence of bismuth molybdate catalyst is considered to be a

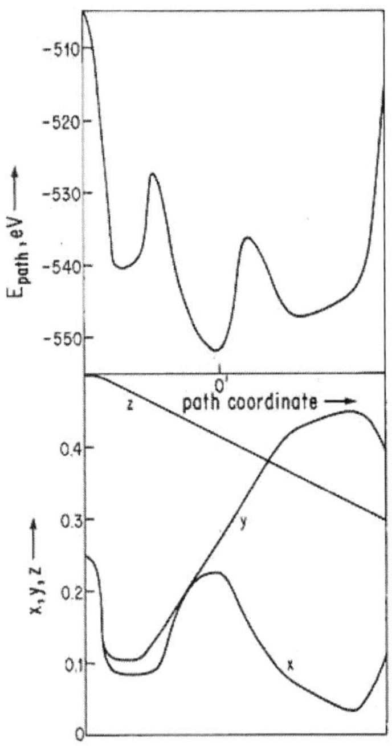

Figure 10. Motion of O^{2-} ion originally at the base of a MoO octahedron (O_{II}) if an O^{2-} ion at the apex (O_{III}) is removed. The path is in the direction of the z axis. O′ denotes the probable final location of the moving ion, near O_{III}.

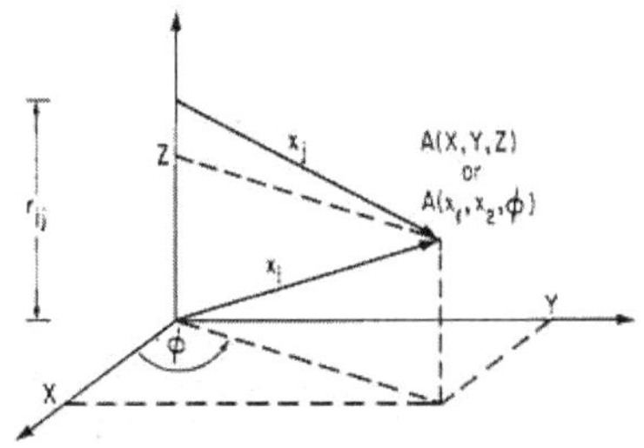

Figure A1. Coordinates (x_i, x_j, ϕ) of the two-center integration compared with Cartesian coordinates (X, Y, Z).

two-step process. In the first step the hydrocarbon is oxidized by the bismuth molybdate. In the second step the reactant oxygen reoxidizes the reduced bismuth molybdate. The calculated paths of section 3 may help in understanding the process by which oxygen travels through the bismuth molybdate lattice. The scheme used here is simplified in that only one ion at a time is allowed to move while the others are considered to be fixed. This is clearly artificial because we expect some accommodation to be made by all ions in the unit cell; also because more than one such process may be occurring at the same time. Further, we consider the unit cell to be completely filled, and do not take into account the presence of surface imperfections, vacancies, and localized nonstoichiometries that may influence the motion of the O^{2-} ion. Lastly we assume that the structure of unit cells on the surface of the catalyst is identical with that of cells in the bulk of the catalyst.

Consider first the hypothesis that the oxygen ion removed from the lattice of the catalyst by the hydrocarbon is in one of the bismuth layers. It seems apparent that the external oxygen ion O^{2-}, somehow derived from the gaseous oxygen, finds it extremely difficult to enter the lattice of any point near the bismuth layers (Figure 5). The minimum energy for this path lies outside the lattice, at $y = -0.26$. The ion must pass through an energy barrier to enter the lattice and another, much larger, barrier to approach the vacant position.

On the other hand, an O^{2-} ion at the apex of a MoO octahedron can readily move through the lattice (Figure 9). No energy barrier is required to be crossed to move the ion to a position near the vacancy created near the Bi^{3+}. Apparently an energy barrier must be crossed to reach the actual vacancy assuming that all of the other ions in the lattice are fixed. While the vacancy at the apex position cannot directly be filled by the external O^{2-} (see Figure 7), Figure 10 shows that an O^{2-} ion at the base of the MoO octahedron can easily approach very near to the vacancy. That the ion will not exactly fill the vacancy is perhaps related to our neglecting the motion of all other ions in the lattice and to the approximations in the computation of the interaction potential. The external O^{2-} ion can now move as in Figure 6 either to the absolute minimum on the $y = 0$ surface or to the local minimum at the vacancy near Bi^{3+}. It should be noted that the energy required to surmount the local maximum between the minima at $y = 0$ and at the vacancy near Bi^{3+} is considerably less than the energy of the ion at the start of the path. This implies that the path to the vacancy itself is not totally unfavorable.

Now consider the hypothesis that an oxygen ion at the molybdenum layer is removed by the hydrocarbon. If the processes shown in Figures 8 and 9 are reversed, we see that an ion affiliated to a Bi^{3+} ion will have great difficulty in approaching the vacancy near Mo^{6+}. On the other hand, the external O^{2-} could fill this vacancy. The substitution would be direct if a base O^{2-} ion on the MoO octahedron was lost. If an apex O^{2-} ion was lost, a base O^{2-} ion could replace it and the external O^{2-} would then fill the vacancy at the base.

The nature of the present calculations is such that these two hypotheses cannot be distinguished here. However, the former hypothesis implies that both bismuth and molybdenum layers play a role in the replenishment of the lattice oxygen, while the latter hypothesis allows from an energetic point of view only the molybdenum layers to play a role in this process. It is reasonable to believe that the superior catalytic performance of the 2:1 bismuth molybdate over those of bismuth oxide and molybdenum oxide individually is due to the cooperative behavior of the two components of the mixed oxide (see section 1). Hence the process involving only the molybdenum layers probably takes place at best to a very small extent in the bismuth molybdate. Cooperation between the layers by replacement of a molybdenum-bound O^{2-} by a bismuth-bound O^{2-} has been shown to be energetically very unfavorable and will probably never occur. However, replacement of a bismuth-bound O^{2-}, removed by reaction, by a molybdenum-bound O^{2-} which is then replaced by O^{2-} from reactant oxygen is energetically favorable and is probably the dominant process.

5. Conclusion

Interaction potentials for the ion pairs O^{2-}–O^{2-}, Bi^{3+}–O^{2-}, and Mo^{6+}–O^{2-} have been calculated almost exclusively from first principles. These interaction potentials are used to define a set of paths of oxygen into the 2:1 bismuth molybdate.

The results of the present computation indicate that an oxygen ion associated with the bismuth is probably the one removed by the hydrocarbon when it is oxidized. This oxygen is replaced by oxygen ion in the intermediate, or mobile, layer between $(Bi_2O_2)^{2+}$ and $(MoO_2)^{2+}$ layers. This in turn is replaced by an oxygen ion associated with the

TABLE AI: Steps in the Calculation of Electron Density ρ of O^{2-} as a Function of Distance x from Center of Ion[a]

n	1		2		2		$\rho =$
l	s		s		p		$\sum_{n,l} \{\omega_{nl} \cdot$
ω_{nl}	2		2		6		$(P_{nl}^{HS})^2\}$
x^{HS}	P_{nl}^{HS}	$\omega_{nl}(P_{nl}^{HS})^2$	P_{nl}^{HS}	$\omega_{nl}(P_{nl}^{HS})^2$	P_{nl}^{HS}	$\omega_{nl}(P_{nl}^{HS})^2$	
0.0933	1.2947	3.3535	0.3055	0.1867	0.0240	0.0036	3.5428
0.49	1.7283	5.9740	0.1905	0.0726	0.3482	0.7275	6.7740
1.21	0.3994	0.3190	−0.7333	1.0755	0.7922	3.7656	5.1601
2.27	0.0269	0.0014	−0.9597	1.8420	0.8319	4.1526	5.9960
3.67	0.0010	0	−0.5720	0.6544	0.5817	2.0301	2.6845
5.43	0	0	−0.2345	0.1100	0.3206	0.6168	0.7268
7.57	0	0	−0.0734	0.0108	0.1504	0.1356	0.1464
10.12	0	0	−0.0170	0.0006	0.0588	0.0209	0.0215
13.13	0	0	−0.0030	0	0.0190	0.0022	0.0022
16.65	0	0	−0.0004	0	0.0046	0.0001	0.0001
20.78	0	0	0	0	0.0009	0	0
25.62	0	0	0	0	0.0001	0	0
>26	0	0	0	0	0	0	0

[a] Reduced wave functions P_{nl}^{HS} are interpolated from the values of ref 19 at various reduced distances, x^{HS}. The values of x^{HS} chosen are those required for numerical integration by the Gauss-Legendre quadrature method. See Nomenclature for definition of n, l, and ω_{nl}.

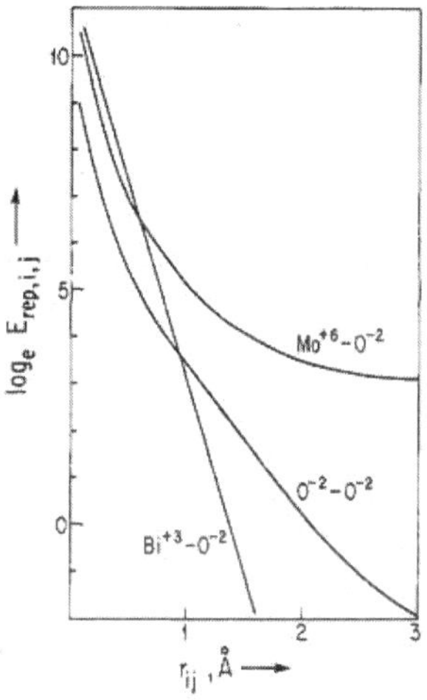

Figure A2. Short-range repulsion interactions O^{2-}–O^{2-}, Mo^{6+}–O^{2-}, Bi^{3+}–O^{2-} to enable A_{ij} and B_{ij} to be calculated.

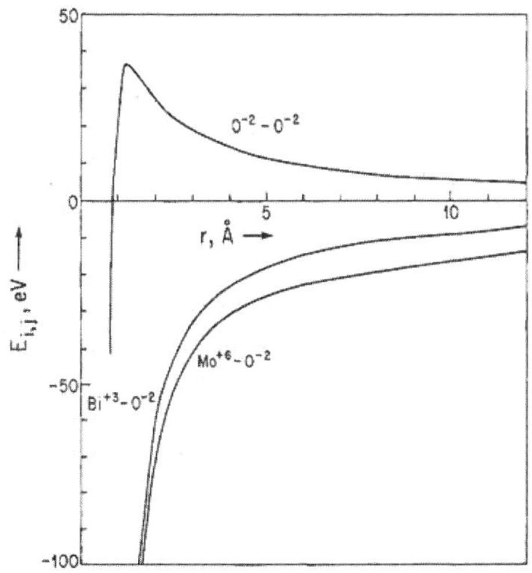

Figure A3. Form of the interactions if the polarization catastrophe is not corrected for. The separation distance r is in Å, the potentials E in eV.

molybdenum and this is eventually replaced by an oxygen ion derived from reactant oxygen in the gas phase. We can therefore infer that one of the uses of molybdenum in the bismuth molybdate catalysts is to deliver its oxygen to the bismuth and receive oxygen from the chemical atmosphere to take its place. Thus, the molybdenum can be thought of as a "pumping station", receiving external oxygen and transporting it to the bismuth site for oxidation of the hydrocarbon. This may be a reason for the superior performance of the mixed oxides bismuth molybdates as catalysts, compared to the behavior of bismuth oxide or molybdenum oxide, in oxidation. We further suggest that the best combination of the two components in the mixed oxide requires, inter alia, the presence of just enough molybdenum to "pump" oxygen to the bismuth site at just the right rate.

Acknowledgment. We acknowledge useful discussions with Dr. D. O. Welch.

Appendix

AI. *Form of the Interaction Potentials.* We use the term ion as a general form to include a neutral atom as a special case. An ion of species k is characterized by (i) its atomic number Z_k, which is the total number of protons in the nucleus, (ii) the total number of electrons N_k it contains, and (iii) its ionicity I_K or net charge integer, given by $Z_k - N_k$. The energy of interaction between two ions, of the same or different species, is considered as the sum of Coulombic interactions, polarization contributions from permanent, induced, and instantaneous (van der Waals) dipoles and higher multipoles, and the short-range electron shell repulsion terms.

In this section we mention only the expected dependence of the ionic energy terms upon the distance r_{ij} separating the two ions. The Coulombic energy is inversely proportional to the separation distance and may be written as

$$E_{\mathrm{coul},i,j} = F_{ij}/r_{ij} \tag{A1}$$

The dependence of F_{ij} on ion properties is given in section AII.

It is assumed that no permanent dipoles exist in any of the species of interest here. The induced dipole interaction is given by

$$E_{\mathrm{ind},i,j} = -D_{ij}/r_{ij}^4 \tag{A2}$$

where the negative sign indicates that induced polarization terms are always attractive in nature. The expression for

D_{ij} is derived in section AIII. The energy due to instantaneous electron–nucleic dipoles can be given by

$$E_{\text{inst},i,j} = -C_{ij}/r_{ij}^6 \quad \text{(A3)}$$

where C_{ij} values are derived in section AIV.

From quantum-mechanical considerations, the short-range repulsion contribution is of the form

$$E_{\text{rep},i,j} = A_{ij} \exp[-B_{ij}r_{ij}] \quad \text{(A4a)}$$

A_{ij} and B_{ij} may be dependent on r_{ij} and are derived in section AV. However, the shell model[10,11] recognizes that the outermost electrons which are primarily responsible for the repulsion are affected most by the induced polarization of the ions. This coupling between the polarization and repulsion terms is considered by using an effective separation distance r_{eff} in place of r_{ij} in eq A4a. We show in section AVI that

$$r_{\text{eff}} = r_{ij} - G_{ij}/r_{ij}^2 \quad \text{(A4b)}$$

where G_{ij} depends upon the dipole induced in each of the two ions.

AII. *Coulombic Energy.* The Coulombic energy between two particles of charge I_ie and I_je is well known to be

$$E_{\text{coul},i,j} = \left(\frac{I_iI_je^2}{4\pi\epsilon_0}\right)\frac{1}{r_{ij}} \quad \text{(A5a)}$$

where $4\pi\epsilon_0$ is the energy conversion factor, so that F_{ij} in eq A1 is

$$F_{ij} \equiv 14.403 I_iI_j \text{ eV Å} \quad \text{(A5b)}$$

Note that F_{ij} is positive if the ionicities of the two species are of the same sign and negative if the signs are different.

AIII. *Induced Polarization Energy.* Even though there are no permanent dipoles in the systems of interest, a dipole may be induced in one ion due to the charge of the other ion. The magnitude of the induced dipole is proportional to the field of the charge, the proportionality constant being the polarizability of the species of ion in which the dipole is induced. The total energy due to interactions between each ion and the induced dipole in the other ion may be written

$$E_{\text{ind},i,j} = -\frac{1}{2}\left(\frac{e^2}{4\pi\epsilon_0}\right)\{\alpha_iI_j^2 + \alpha_jI_i^2\}\frac{1}{r_{ij}^4} \quad \text{(A6a)}$$

so that, if the polarizabilities are in units of [Å^3], D_{ij} in eq A2 is

$$D_{ij} \equiv 7.2\{\alpha_iI_j^2 + \alpha_jI_i^2\} \text{ eV Å}^4 \quad \text{(A6b)}$$

To eq A6a should also be added terms corresponding to the interaction between the two dipoles themselves, and to the interaction due to the self-induced dipole of each species. However, these are expected to be small in comparison with eq A6a and are neglected here.

Equations A6 assume constant polarizabilities that are independent of the separation of the ions. The values for the polarizabilities are obtained from values of the gram-atom refractivity collected by Rich.[12]

AIV. *Instantaneous Polarization Energy.* Due to symmetry, the net dipole moment between the positive nucleus of an ion and its electron cloud is identically zero. However, at any time there exists an instantaneous dipole (and instantaneous higher multipoles) in the ion and this dipole can polarize the other ion. The interaction energy for these instantaneous multipoles averaged over the electron cloud is not zero. The energy for the instantaneous dipoles is the van der Waals energy and has been calculated using variational and perturbational techniques. For two unlike ions having N_{0i} and N_{0j} electrons in the outermost shells, the energy averaged over all electron states is[13]

$$E_{\text{inst},i,j} = -\frac{6}{(4\pi\epsilon_0)^2 r_{ij}^6}\sum_{n,m}\frac{|\mu_z|_{in}^2 + |\mu_z|_{jm}^2}{h\nu_{in} + h\nu_{jm}}$$

where h is Planck's constant, $|\mu_z|_{kl}$ is the matrix element of the z component dipole moment of the kth species between the ground and the lth excited state, and ν_{kl} is the frequency of the lth excited state of the kth species. Averaging over n and m

$$E_{\text{inst},i,j} = -\frac{3}{2}\frac{e\hbar}{(4\pi\epsilon_0)^2}m_e^{-1/2} \times \frac{\alpha_i\alpha_j}{(\alpha_i/N_{0i})^{1/2} + (\alpha_j/N_{0j})^{1/2}}\frac{1}{r_{ij}^6} \quad \text{(A7a)}$$

Hence in eq A3, the value of C_{ij} is

$$C_{ij} = 15.7\frac{\alpha_i\alpha_j}{(\alpha_i/N_{0i})^{1/2} + (\alpha_j/N_{0j})^{1/2}} \text{ eV Å}^6 \quad \text{(A7b)}$$

if, as before, the polarizabilities α are in units of [Å^3].

AV. *Short-Range Repulsion.* The exponential nature of the short-range repulsion forces was probably first pointed out by Pauling.[14] In the case of a few compounds, notably alkyl halides, the interaction parameters A_{ij} and B_{ij} have been obtained by comparison of experimentally determined properties with theoretical relationships of the perfect crystal. Unfortunately, no such analyses exist for the ionic species of interest here and calculations must be carried out from first principles. We use below the approach of Wedepohl[15] as modified by Wilson and Bisson.[16]

In brief, the total interaction energy between two ions, less the polarization terms, is calculated as the sum of electron–electron, electron–nucleus, and nucleus–nucleus interaction terms, to which are added a kinetic correction and an exchange correction. The interaction terms are obtained from the general electrostatic relations using the square of the wave function as a measure of the electron density. The shell repulsion energy is obtained by subtracting the Coulomb interaction energy from the total interaction energy. It might be thought that the subtraction of the Coulomb energy from the total energy and its later addition would be a wasted effort, but this is not so. Since the repulsion energy can be well fitted by an exponential curve, the calculations can be performed once for a given ion pair to obtain values of A_{ij} and B_{ij} in eq A4a; using these in all future calculations of the interaction energy frees us from laborious numerical integrations and results in a considerable saving of computational time. Because the relationships describing each of the contributions to the total repulsion energy have not been given explicitly in the literature, we briefly describe these calculations below.

a. *Electron Shell–Electron Shell Interaction.* Consider two elementary spherical shells of radii x_i and x_j with centers a distance r_{ij} apart. The interaction energy between the two shells can be written

$$\mathrm{d}E_{\text{ee}}(r_{ij}) = \int_{r_{ij}}^{\infty}\mathrm{d}F_{\text{ee}}(r)\,\mathrm{d}r \quad \text{(A8)}$$

where $\mathrm{d}F_{\text{ee}}$ represents the elemental force between the two shells when they are a distance r apart. Where the two shells do not overlap, i.e., where r is such that $0 < x_j < r$

$-x_i$, and $0 < x_i < r$, the elemental force is

$$dF_{ee}(r) = \frac{dq_i dq_j}{(4\pi\epsilon_0)}\,\frac{1}{r^2} \tag{A9a}$$

where dq_i and dq_j are the elemental charges of the shells. Where there is partial overlap, i.e., if $|x_i - r| < x_j < (x_i + r)$, and $0 < x_i < \infty$, we have

$$dF_{ee}(r) = \frac{dq_i dq_j}{16\pi\epsilon_0}\left[\frac{r^2 - (x_i - x_j)^2}{r^2 x_i x_j}\right] \tag{A9b}$$

Where one shell is totally inside the other, i.e., if $0 < x_j < x_i - r$ and $x_i > r$ or if $x_j > x_i + r$ and $x_i > 0$

$$dF_{ee}(r) = 0 \tag{A9c}$$

Equations A8 and A9 can be applied to calculate the differential energy dE_{ee} between all electrons a distance x_i from the center of ion i and all electrons a distance x_j from the center of ion j. The elemental charges dq_i and dq_j are related to the electron densities in the electron clouds by

$$dq_k = 4\pi x_k{}^2\, dx_k\, e\rho_k(x_k) \tag{A10}$$

Here ρ_k represents the electron density (number per unit volume) of species k at a radial distance x_k from the center of the ion of species k, and can be obtained from the appropriate wave function. For simplicity we assume that the electron density depends only upon the radial coordinate x_k and not upon the angular coordinates θ and ϕ. In other words the wave function to be considered must be one that is spherically averaged.

The total electron–electron interaction can be written by integrating dE_{ee} assuming that the electron cloud of each ion is infinitely large.

$$E_{ee}(r_{ij}) = \int_{x_j=0}^{\infty}\int_{x_i=0}^{\infty} dE_{ee}$$

From eq A8–A10

$$E_{ee}(r_{ij}) = \frac{\pi e^2}{\epsilon_0}\int_{x_i=0}^{\infty}\int_{x_j=0}^{\infty} x_i x_j\, \rho_i(x_i)\, \rho_j(x_j) \times \mathrm{X}(x_i, x_j, r_{ij})\, dx_j\, dx_i \tag{A11}$$

where

$$\mathrm{X}(x_i, x_j, r_{ij}) \begin{cases} = 2\{x_i + x_j - |x_i - x_j|\} & r_{ij} < |x_i - x_j| \\ = 4x_i - \dfrac{(r_{ij} + x_j - x_i)^2}{r_{ij}} & x_i + x_j > r_{ij} > |x_i - x_j| \\ = 4x_i x_j / r_{ij} & r_{ij} > x_i + x_j \end{cases}$$

Further evaluation of eq A11 requires values of the spherically averaged electron density $\rho(x)$ for each of the two ions.

b. *Electron–Nucleus Interactions.* Consider first the interaction between the electron cloud of ion i and the nucleus of ion j, $E_{e_in_j}$. We can write

$$E_{e_in_j} = \int_{x_i=0}^{\infty} dE_{e_in_j}$$

where $dE_{e_in_j}$ is the interaction energy between a point charge of magnitude Z_je and a hollow shell of charge density $\rho_i(x_i)$, radius x_i, and thickness dx_i, whose center is a distance r_{ij} from the point charge. Again the relationship between force and energy may be written

$$dE_{e_in_j} = \int_{r_{ij}}^{\infty} dF_{e_in_j}\, dr$$

where

$$dF_{e_in_j} = 0 \qquad r < x_i$$

i.e., for shells far enough from the i nucleus that they completely overlap the j nucleus. If the shell does not overlap the other nucleus

$$dF_{e_in_j} = -\frac{Z_j e^2 x_i{}^2 \rho_i(x_i)}{\epsilon_0 r^2}\, dx_i \qquad r \geqslant x_i$$

Since the nucleus is a point charge, partial overlap need not be considered. If the integration over r is carried out, we have

$$E_{e_in_j} = -\left(\frac{Z_j e^2}{\epsilon_0}\right)\int_{x_i=0}^{\infty} x_i{}^2 \rho_i(x_i)\, \mathrm{Min}\left[\frac{1}{x_i}, \frac{1}{r_{ij}}\right] dx_i$$

where $\mathrm{Min}[A_1, A_2]$ is the smaller of the two values A_1 and A_2. The total electron–nucleus interaction consists of the sum of $E_{e_in_j}$ and a similar expression for the interaction of the nucleus of ion i with the electron cloud of ion j. Hence

$$E_{en} = -\frac{e^2}{\epsilon_0}\int_{x=0}^{\infty}\{Z_j\rho_i(x) + Z_i\rho_j(x)\}x^2 \cdot \mathrm{Min}\left[\frac{1}{x}, \frac{1}{r_{ij}}\right] dx \tag{A12}$$

c. *Nucleus–Nucleus Interaction.* This can be written as the interaction of two point charges of magnitude Z_ie and Z_je. Hence

$$E_{nn}(r_{ij}) = \frac{Z_i Z_j e^2}{(4\pi\epsilon_0) r_{ij}} \tag{A13}$$

This seems an appropriate place to make a comparison between the Coulomb energy, as given by eq A5, and the sum $E_{ee} + E_{en} + E_{nn}$ from eq A11–A13. From the definition of the ionicities

$$E_{coul,i,j} = \left(\frac{e^2}{4\pi\epsilon_0}\right)\left(\frac{1}{r_{ij}}\right)\{(N_iN_j) - (N_iZ_j + N_jZ_i) + (Z_iZ_j)\}$$

The identity of the third term in parentheses and eq A13 is obvious. Comparing the other two terms with eq A11 and A12 and noting from the definition of ρ_i and ρ_j that

$$\frac{N_k}{4\pi} \equiv \int_{x=0}^{\infty} x^2 \rho_k(x)\, dx \qquad k = i, j$$

we can say that the difference between eq A5 and the sum of eq A11–A13 is a measure of the electron distribution. Clearly if $\rho(x) \boxminus \delta(0)$, i.e., if all the electrons were in the center ($x = 0$), then

$$E_{coul} \equiv E_{ee} + E_{en} + E_{nn}$$

d. *Kinetic Correction.* The kinetic energy of an electron cloud of density $\rho(x)$ in a volume v is[17]

$$\mathrm{KE}(v) = \int_v \left(\frac{3}{5}\right)\frac{\hbar^2\pi^2}{2m_e}\left(\frac{3}{\pi}\rho\right)^{2/3}\rho\, dv$$

Due to the nonlinear relation between the kinetic energy and the electron density, the kinetic energy of the electron cloud of the two ions in the region of overlap is not the sum of the kinetic energies of the two electron clouds measured separately. The difference contributes to the energy of interaction between the two ions and is known as the kinetic correction, E_k. Hence

$$E_k = \left(\frac{3}{5}\right)\frac{\hbar^2\pi^2}{2m_e}\left(\frac{3}{\pi}\right)^{2/3} \int_V (\{\rho_i(x_i) + \rho_j(x_j)\}^{5/3} - \{\rho_i(x_i)\}^{5/3} - \{\rho_j(x_j)\}^{5/3})\,dV$$

where V represents the region of overlap, x_i and x_j represent the radial distances of the volume element dV from the centers of the two ions a distance r_{ij} apart. The coordinate system used in evaluating this two-center, multiple integral is that of Abrahamson[18] and is shown in Figure A1. The elemental overlap volume can be written in terms of these coordinates as

$$dV = (x_i x_j / r_{ij})\,dx_i\,dx_j\,d\phi \tag{A14a}$$

and the overlap region V is defined by

$$|r_{ij} - x_i| \leqslant x_j \leqslant r_{ij} + x_i \tag{A14b}$$

$$0 \leqslant x_i \leqslant \infty \tag{A14c}$$

$$0 \leqslant \phi \leqslant 2\pi \tag{A14d}$$

Since the electron densities ρ are spherically averaged, and therefore independent of ϕ, the integration over ϕ may be trivially performed to give

$$E_k = \frac{2\pi}{r_{ij}}\left(\frac{3^{5/3}\pi^{4/3}\hbar^2}{10m_e}\right)\int_{x_i=0}^{\infty}\int_{x_j=|r_{ij}-x_i|}^{x_j=r_{ij}+x_i} x_i x_j \times (\{\rho_i(x_i) + \rho_j(x_j)\}^{5/3} - \{\rho_i(x_i)\}^{5/3} - \{\rho_j(x_j)\}^{5/3})\,dx_i\,dx_j \tag{A15}$$

e. *Exchange Correction.* For a single ion, overlapping wave functions of pairs of electrons in a volume v contribute the so-called exchange term to the total energy. The exchange energy

$$\mathrm{EE}(v) = \frac{3}{4}\left(\frac{e^2}{4\pi\epsilon_0}\right)\int_v \left(\frac{3}{\pi}\rho\right)^{1/3} \rho\,dv$$

is a quantum mechanical contribution to the total energy of a single ion, arising when pairs of electrons interchange orbitals. For an ion pair, the exchange energy is given by substituting the sum $\rho_i + \rho_j$ for the single term ρ. This then gives the exchange energy when two electrons of the same ion or different ions exchange orbitals. If from this energy is subtracted the sum of the exchange energy of the two ions acting independently, what remains is the contribution when pairs of electrons corresponding to different ions exchange orbitals. The difference is an indication of the extent of sharing of electrons by the two ions, i.e., a measure of covalent bonding. This energy is the exchange correction, E_a, defined by

$$E_a = \frac{3}{4}\left(\frac{3}{\pi}\right)^{1/3}\left(\frac{e^2}{4\pi\epsilon_0}\right)\int_V (\{\rho_i(x_i) + \rho_j(x_j)\}^{4/3} - \{\rho_i(x_i)\}^{4/3} - \{\rho_j(x_j)\}^{4/3})\,dV$$

The elemental overlap volume dV and the limits of the overlap region V can again be transformed as in eq A14 so that, after integration over ϕ, we have

$$E_a = \left(\frac{3\pi}{2r_{ij}}\right)\left(\frac{e^2}{4\pi\epsilon_0}\right)\left(\frac{3}{\pi}\right)^{1/3}\int_{x_i=0}^{x_i=\infty}\int_{x_j=|r_{ij}-x_i|}^{x_j=r_{ij}+x_i} x_i x_j \times (\{\rho_i(x_i) + \rho_j(x_j)\}^{4/3} - \{\rho_i(x_i)\}^{4/3} - \{\rho_j(x_j)\}^{4/3})\,dx_i\,dx_j \tag{A16}$$

f. *The Electron Density.* In this work, the spherically averaged density $\rho(x)$ represents the number of electrons per unit volume at a radial distance x from the center of the ion. From quantum mechanics

$$\rho(x) \equiv \sum_{n,l} \{R_{nl}(x)\}^2 \omega_{nl}$$

where R_{nl} is the radial component of the time-independent eigenfunction having principal quantum number n and azimuthal quantum number l, and ω_{nl} is the occupation number of the (n, l)th orbital of the ion, i.e., the number of electrons in that particular orbital. Values of $R_{nl}(x)$ can be obtained by solution of the appropriate Schroedinger equation. In this work we use the very comprehensive tabulation of Herman and Skillman[19] based on the Hartree–Fock–Slater approximations. The following relationships exist between values used in the previous sections and the reduced values of ref 19, denoted by the superscript HS.

$$\rho_i(x_i) \equiv \frac{Z_i^{2/3}}{4\pi K^2 a_{\mathrm{H}}}(x^{\mathrm{HS}})^{-2} \sum_{n,l} \omega_{nl}(P_{nl}(x^{\mathrm{HS}}))^2 \tag{A17a}$$

$$x_i = (K/Z_i^{1/3})x^{\mathrm{HS}} \tag{A17b}$$

where

$$K \equiv 0.88534138a_{\mathrm{H}} \tag{A17c}$$

and a_{H} is the classical radius of the hydrogen orbit. Similar relations exist for ρ_j and x_j.

g. *Evaluation of E_{rep}.* A 15-point Gauss–Legendre quadrature was used for numerical integration to the infinite limits of eq A11 and A12. The inner integrals of the kinetic and exchange corrections were obtained by using a Gauss–Chebyshev scheme. The values of P_{nl} were interpolated from those available in ref 19 for the corresponding atoms although our concern here is with ions. However, calculations carried out[19] indicate that this approximation is acceptable. Of course, the ionic values for the occupation numbers ω_{nl} are used in the present calculations. Table AI illustrates the steps in the calculation of $\sum_{n,l}\omega_{nl}\{P_{nl}(x^{\mathrm{HS}})\}^2$ for oxygen.

The values of the repulsion energy E_{rep} are calculated from

$$E_{\mathrm{rep}}(r_{ij}) = E_{\mathrm{ee}}(r_{ij}) + E_{\mathrm{en}}(r_{ij}) + E_{\mathrm{nn}}(r_{ij}) + E_k(r_{ij}) + E_a(r_{ij}) - E_{\mathrm{coul}}(r_{ij})$$

They are plotted in Figure A2 as a function of the interionic separation distance r_{ij} for each of the pairs O^{2-}–O^{2-}, O^{2-}–Bi^{3+} and O^{2-}–Mo^{6+}. Values of A_{ij} and B_{ij} are obtained from these curves as functions of r_{ij}.

AVI. *Polarization Catastrophe.* The interaction energies for the three cases, as calculated up to this point, are shown in Figure A3. For the case of bismuth and molybdenum, the calculated repulsion energy seems never quite strong enough to overcome the attraction contribution, thus leading to the impossible situation that the equilibrium separation of each of the two pairs is apparently zero. This is the so-called "polarization catastrophe" that occurs if the coupling between repulsion and polarization terms is ignored. The coupling occurs because the outermost electrons influence both the repulsion and the induced polarization. Following the suggestion of ref 10 and 11, we account for the coupling and consequent distortion of the electron clouds by substituting an effective separation distance r_{eff} for the actual interionic distance r_{ij} in the exponential relation of eq A4a. The effective separation distance for repulsion is related to $\mu_{\mathrm{ind},i}$ and $\mu_{\mathrm{ind},j}$, the dipoles induced in each of the ions of the pair, by

$$r_{\mathrm{eff}} = \left| r_{ij} + \frac{\mu_{\mathrm{ind},i}}{Q_i} - \frac{\mu_{\mathrm{ind},j}}{Q_j} \right|$$

where $|A_1|$ is the absolute value of A_1, and Q_i and Q_j are constants for species i and j. Using average polarizabilities α_i and α_j in the induced dipole moments, the effective separation can be written

$$r_{eff} = \left| r_{ij} - \left(\frac{\alpha_j I_i}{Q_j} - \frac{\alpha_i I_j}{Q_i}\right)\left(\frac{e}{4\pi\epsilon_0}\right)\frac{1}{r_{ij}^2}\right| \quad (A18a)$$

$$= |r_{ij} - (G_{ij}/r_{ij}^2)| \quad (A18b)$$

Dienes et al.[11] calculated values of Q_i and Q_j for Al_2O_3 by matching theoretical and experimental values of the high-frequency dielectric constant. We suggest instead that the value of G_{ij} be obtained for each ion pair such that the interaction energy of the ion pair is at a minimum at the experimentally observed value of the separation distance of the ions in the respective oxides. The separation distance for Bi^{3+} and O^{2-} is taken as[20]

$$r_{min} = 2.5\ \text{Å}$$

and that for Mo^{6+} and O^{2-} is[21]

$$r_{min} = 2.0\ \text{Å}$$

The interaction energy E_{ij} for each ion pair Bi^{3+}–O^{2-} and Mo^{6+}–O^{2-} is obtained from the sum of eq A1–A4a, using the effective radius r_{eff} defined in eq A4b in place of r_{ij} in eq A4a. Then the value of G_{ij} for each ion pair is that value for which

$$(\mathrm{d}E_{ij}/\mathrm{d}r_{ij})_{r_{ij}=r_{min}} = 0$$

The corresponding values of G_{ij} are

$$G_{ij} = 6.17\ \text{Å}^3 \text{ for } Bi^{3+}\text{–}O^{2-} \quad (A19a)$$

and

$$G_{ij} = 2.60\ \text{Å}^3 \text{ for } Mo^{6+}\text{–}O^{2-} \quad (A19b)$$

It may be noted here that setting

$$r_{min} = \infty$$

for the minimum energy separation of two O^{2-} ions will yield a value of

$$G_{ij} = 0 \text{ for } O^{2-}\text{–}O^{2-} \quad (A19c)$$

This is consistent with the definition of G_{ij} in eq A18, since $Q_i = Q_j$, $\alpha_i = \alpha_j$, and $I_i = I_j$.

AVII. *Interaction Potentials*. The interaction energies for the three cases are shown in Figure 1, using the results of the previous sections. We repeat that, at small separation distances, the "artificial maximum" occurs only because of the mathematical nature of the forms used for the various contributions to the total interaction energy. The interaction energy is set at $+\infty$ when the separation distance is less than the position of the artificial maximum, r_{max}. A comprehensive list of all the parameters in the interaction potentials has been given in Table I.

Nomenclature

a_H	radius of hydrogen
a, b, c	lattice parameters of Bi_2MoO_6
A_{ij}, B_{ij}	parameters of the short-range repulsion energy between ions i and j
C_{ij}	parameter of the instantaneous dipole interaction energy
D_{ij}	parameter of the induced dipole interaction energy
e	electron charge (absolute magnitude)
E_a	measure of covalent energy between two ions
$E_{coul,i,j}$	Coulombic interaction energy between ions i and j
E_{ee}	contribution toward short-range repulsion energy by interaction between electron clouds of ions i and j
E_{en}	contribution toward short-range repulsion energy by interaction between electron cloud and nucleus of other ion
$E_{e_i n_j}$	contribution toward E_{en} by electron cloud of ion i and nucleus of ion j
$E_{i,j}$	interaction energy between ions i and j
$E_{ind,i,j}$	induced dipole interaction energy between ions i and j
$E_{inst,i,j}$	instantaneous dipole interaction energy between ions i and j
E_k	kinetic correction for the interaction between two ions
E_{nn}	contribution toward short-range repulsion energy by interaction between nuclei of ions i and j
E_O	energy of mobile O^{2-} ion at (x_O, y_O, z_O) interacting with bismuth molybdate lattice
E_O^*	energy of mobile O^{2-} ion along path at (x_O^*, y_O^*, z_O^*)
$E_{rep,i,j}$	short-range repulsion energy between ions i and j
F_{ij}	parameter of the Coulombic interaction energy
G_{ij}	shell model parameter for species i and j
h	Planck's constant
$\hbar$	$h/2\pi$
I_i, I_j	charge integer of ion i, j
K	conversion factor between actual radial distance x_k of an electron from the center of an ion of species k and the corresponding reduced distance x_k^{HS}
l	azimuthal quantum number of electron orbital
m_e	electron mass
n	principal quantum number of electron orbital
N_i, N_j	total number of electrons in species i, j
N_{Oi}, N_{Oj}	number of electrons in outermost shell of species i, j
P_{nl}^{HS}	reduced wave function of an electron in an orbit characterized by n and l
Q_i, Q_j	shell model constant for species i, j
r_{eff}	effective separation distance, used for calculation of apparent repulsion energy
r_{ij}	distance between ion i and ion j
r_{max}	position of artificial maximum of interaction between two ions
r_{min}	distance of minimum energy between two ions
R_{nl}	wave function of principal quantum number n and azimuthal quantum number l
V	overlap region of electron clouds corresponding to two ions
x, y, z	coordinate axes of the lattice
x_O, y_O, z_O	position of mobile ion
x_O^*, y_O^*, z_O^*	position of mobile ion along minimum energy path
x_i, x_j	radial distance of electron from the center of the ion of species i, j
x_k^{HS}	reduced distance of an electron from the center of the ion of species k
Z_i, Z_j	atomic number of species i, j
α_i, α_j	polarizability of species i, j
$\delta(x)$	Dirac delta function
ϵ_O	conversion factor

$\mu_{ind,i}$, $\mu_{ind,j}$	induced dipole on ions i, j
π	3.14159
ρ_i, ρ_j	electron density of ions i, j
ω_{nl}	number of electrons in an orbital of principal quantum number n and azimuthal quantum number l

References and Notes

(1) P. A. Batist, A. H. W. M. Der Kinderen, Y. Leeuwenburgh, F. A. M. G. Metz, and G. C. A. Schuit, *J. Catal.*, **12**, 45 (1968).
(2) D. J. Hucknall, "Selective Oxidation of Hydrocarbons", Academic Press, London, 1974, Chapter 3.
(3) J. M. Peacock, A. J. Parker, P. G. Ashmore, and J. A. Hockey, *J. Catal.*, **15**, 373, 387, 398 (1969); J. M. Peacock, M. J. Sharp, A. J. Parker, P. G. Ashmore, and J. A. Hockey, *ibid.*, **15**, 379 (1969).
(4) J. Haber, *Int. Chem. Eng.*, **15**, 21 (1975); K. Brückmann, J. Haber, and J. Janas, *Bull. Acad. Polon. Sci., Ser. Sci. Chim.*, **20**, 813 (1972).
(5) H. E. Swift, J. E. Bozik, and J. A. Ondrey, *J. Catal.*, **21**, 212 (1971).
(6) O. Otsubo, H. Miura, Y. Morikawa, and T. Shirasaki, *J. Catal.*, **36**, 240 (1975).
(7) I. N. Schmit, *Surf. Sci.*, **55**, 589 (1976).
(8) A. J. Bennett, *Phys. Rev. B*, **9**, 741 (1974).
(9) A. F. van der Elzen and G. D. Rieck, *Acta Crystallogr., Sect. B*, **29**, 2436 (1973).
(10) D. O. Welch and G. J. Dienes, in "Defects and Transport in Oxides: Battelle Institute Materials Science Colloquia, 1973", M. S. Seltzer and R. I. Jaffee, Ed., Plenum Press, New York, N.Y., 1974.
(11) G. J. Dienes, D. O. Welch, C. R. Fischer, R. D. Hatcher, O. Lazareth, and M. Samberg, *Phys. Rev. B*, **11**, 3060 (1975).
(12) R. Rich, "Periodic Correlations", W. A. Benjamin, New York, N.Y., 1965.
(13) J. C. Slater, "Insulators, Semi Conductors and Metals", Vol. 3, "Quantum Theory of Molecules and Solids", McGraw-Hill, New York, N.Y., 1967.
(14) L. Pauling, *Acta Kristallogr.*, **67**, 377 (1928).
(15) P. T. Wedepohl, *Proc. Phys. Soc.*, **92**, 79 (1967).
(16) W. D. Wilson and C. L. Bisson, *Phys. Rev. B*, **3**, 3984 (1971).
(17) H. A. Bethe, "Intermediate Quantum Mechanics", W. A. Benjamin, New York, N.Y., 1964.
(18) A. A. Abrahamson, *Phys. Rev.*, **130**, 693 (1963); **133**, 990 (1964).
(19) H. Herman and S. Skillman, "Atomic Structure Calculations", Prentice-Hall, Englewood Cliffs, N.J., 1963.
(20) G. Sillen, *Z. Kristallogr.*, **103**, 274 (1941), as reported in *Structure Reports*, **8**, 142 (1940–1941).
(21) A. Magnelli, *Acta Chem. Scand.*, **4**, 793 (1950), as reported in *Structure Reports*, **13**, 220 (1950).

Removal of Oxygen from Bismuth Molybdate during Prereduction and Catalytic Oxidation. Initiation and Steady-State Steps

Eli Ruckenstein

Faculty of Engineering and Applied Science, State University of New York at Buffalo, Buffalo, New York 14214

and Dady B. Dadyburjor*

Department of Chemical and Environmental Engineering, Rensselaer Polytechnic Institute, Troy, New York 12181
(Received: January 21, 1980; In Final Form: June 9, 1980)

The loss of a bismuth-bound O^{2-} ion out of the lattice of a Bi_2MoO_6 catalyst is an important step in the mechanism of selective oxidation. Here it is suggested that two kinds of such a process exist. The "initiation" process refers to the loss of an ion that was initially in the lattice, i.e., before the start of the oxidation process. The lattice vacancy created by this process may be filled by an ion originally outside the lattice, i.e., as gas-phase O_2. The loss of such an ion is termed the "steady-state process". The energy path method indicates that difficulties may arise in the initiation loss of a Bi-bound O^{2-} ion, due to a high potential barrier to its removal from a perfect lattice. The potential barrier can be significantly decreased by the loss of local electroneutrality added to the removal of certain labile Mo-bound and intermediate-layer O^{2-} ions. The latter phenomenon occurs at prereduction, and the former is ascribed to the presence of the reactant hydrocarbon. This decrease in potential barrier allows the bismuth-bound O^{2-} ion to leave the lattice at the start of the selective oxidation reaction. The steady-state loss occurs more easily. The path of an O^{2-} ion into and out of a partially vacant (i.e., imperfect) Bi_2MoO_6 lattice is found to be qualitatively similar to the corresponding path in a perfect lattice. Further, at no point in this path is the energy of the O^{2-} ion greater than at the start of the path. Hence the steady-state loss of O^{2-} ions with a Bi_2MoO_6 lattice containing anion vacancies can be considered energetically favorable.

1. Introduction

In the bismuth molybdate lattice, oxygen ions are found in three types of locations: in a bismuth layer, affiliated with molybdenum, and in a layer intermediate between the other two. It would appear that the bismuth–oxygen neighborhood in the bismuth molybdate lattice is particularly relevant in the selective oxidation of hydrocarbons. For example, Haber and co-workers[1] extrapolated from work on bismuth oxide to indicate that, over the bismuth molybdate, the site where the hydrocarbon is oxidized is around the bismuth ion. Haber found that, in the absence of oxygen in the chemical atmosphere, propylene oxidized over bismuth oxide yields a significant amount of 1,5-hexadiene in the product. This compound is postulated to be formed by the dimerization of the propyl species. The propyl intermediate is also found when propylene is oxidized to acrolein in the presence of oxygen over bismuth molybdate. Since the chemical intermediate is the same in both cases, it seems reasonable that over bismuth molybdate, as over bismuth oxide, the catalytic site is at the bismuth layer. The role of the molybdenum layer in the bismuth molybdate would appear to be to provide the bismuth layer with oxygen ions, replenishing its own supply from the chemical atmosphere. Other experimental work confirming this mechanism has been collected in ref 2.

Dadyburjor and Ruckenstein[3] used the computational technique of minimum energy paths[4] to describe the

This paper, which was originally published in **The Journal of Physical Chemistry**: 84(26), 3487–3495 (1980), is republished with permission.

motion of O^{2-} ions into and inside the lattice of bismuth molybdate. An idealized lattice was considered, completely filled except for the single anion vacancy to be filled by the mobile O^{2-} ion. By comparing the activation energies required for motion of an O^{2-} ion along various paths, it was shown that the following path is energetically favorable:

O^{2-} ion outside the lattice →
O^{2-} ion affiliated with Mo^{6+} →
O^{2-} ion in intermediate layer →
O^{2-} ion affiliated with Bi^{3+} (1)

The pathway described above reinforces the view that the bismuth-bound O^{2-} oxidizes the hydrocarbon, that the anionic vacancy thus formed is replenished via the intermediate layer by a molybdenum-bound O^{2-} moving through the lattice, and that reactant oxygen from the gas phase finally fills the molybdenum-bound anionic vacancy.

We recently started to consider[5] the process that logically succeeds the pathway described in eq 1, i.e., the motion of an O^{2-} ion, affiliated with bismuth, out of (but near) the Bi_2MoO_6 lattice. An activation energy of removal, ΔE, was defined as the difference between the maximum energy in the ion path and the energy at the start of the path. For a completely filled, perfect, Bi_2MoO_6 lattice, it was found that the minimum energy path for such a motion requires an activation energy that is exceedingly large. On the other hand, it was found that O^{2-} ions in the intermediate layer, and those affiliated with molybdenum, require relatively small activation energies for similar paths leaving an otherwise fully occupied lattice.

These results appear to indicate that it is much more difficult for a bismuth-bound O^{2-} ion to leave the lattice (and presumably oxidize the hydrocarbon) than for a molybdenum-bound O^{2-} ion to perform the same function. Consequently it would appear that there is a discrepancy between these results and the experimental results described earlier (and the results of ref 3 as summarized by eq 1), which indicate that it is the bismuth-bound O^{2-} ion which leaves the lattice and is responsible for the catalytic oxidation.

However, note that the lattice of ref 5 was taken as a completely occupied, perfect structure. Particularly after prereduction, a real catalyst will possess a lattice with appreciable vacancy concentrations or even nonstoichiometries. Further, during the catalytic process charge imbalances (local nonelectroneutrality) may occur. All these play a significant part in the performance of the catalyst and must be considered in the evaluation of the energy paths.

This is done in the present work. We consider that the prereduction step introduces vacancies in anion positions associated with the molybdenum layer and intermediate layers in the Bi_2MoO_6, since these O^{2-} ions are relatively easily removed, even from a completely filled lattice.[5] Concurrent with the formation of anion vacancies is the corresponding loss of positive charge in surrounding cations, in order to keep the lattice electrically neutral. We calculate the activation energy of removal, ΔE, for the Bi-bound O^{2-} ion from a lattice containing increasing numbers of vacancies. We find that, at the end of the prereduction step, i.e. for an electroneutral lattice containing vacancies, the tendency of the Bi-bound O^{2-} ion to leave the lattice is not significantly altered.

During the actual reaction process, however, charge transfers occur between reactants, catalyst, and products. Hence we expect that a number of cations which lost some positive charges during prereduction could return to their normal charged state. Of course this would imply that small portions of the lattice have a net charge (the condition of local nonelectroneutrality); however the reactant–catalyst–product system would be electrically neutral. Repeating the calculations of ΔE under conditions now approximating those found during reaction, i.e., for a nonelectroneutral, vacancy-containing lattice, we find that the interaction energies between ion and lattice are significantly lowered. Consequently, a Bi-bound O^{2-} ion in the "original" lattice, i.e., before prereduction, would be able to leave the lattice during the actual reaction, whereas it was prevented from doing so before and after prereduction. The process whereby an "original" Bi-bound O^{2-} ion leaves the lattice we term the initiation loss.

Replacements for the ions lost in this initiation step must come eventually from oxygen in the gas phase. We define the steady-state loss as the transfer of O^{2-} ions from the gas phase, through the lattice, to just outside the lattice. It is found that the energy path for this process requires no activation energy, when a nonelectroneutral, vacancy-containing lattice (similar to the lattice used for a favorable initiation step) is used. The energy path for the steady-state loss from the present lattice is consistent with that obtained earlier[3] for a perfect lattice, viz., that motion of the external ion occurs via the molybdenum and intermediate layers to the bismuth layer vacancy, and hence to the surface of the lattice.

The lattices used to model the presence of various levels of anionic vacancy and local nonelectroneutrality are described quantitatively in section 2. In sections 3–5 are shown the energy paths corresponding to the initiation step for prereduced lattices, and for prereduced lattices under reaction conditions, and the energy path for the steady-state step under reaction conditions. The levels of vacancy formation and nonelectroneutrality required for the initiation step are discussed in section 6, along with mechanisms for the prereduction process, the initiation loss, and the steady-state loss.

2. Lattices Considered

The lattices correspond to the "working lattice" described in ref 3, with the modifications described below. Briefly, the lattices contain a central unit cell, surrounded by unit cells one lattice parameter away in the $\pm x$, $+y$, and $\pm z$ direction, as shown in Figure 1. The coordinates x, y, and z are nondimensionalized with respect to c_1 (=5.5 Å), c_2 (=c_1), and c_3 (=16.1 Å), respectively. Figure 2 describes the idealized unit cell.

In this work, there are vacancies in intermediate-layer and molybdenum-bound positions of oxygen ions in the central unit cell of the working lattice. If the working lattice must be electrically neutral, as is expected during prereduction of the oxide catalyst, the absence of an O^{2-} ion implies that one or more ions of the two types of positively charged species (Bi^{3+} or Mo^{6+}) must lose a total of two electronic charge units. Corresponding to each O^{2-} vacancy in the molybdenum layer, two nearby Mo ions lose one unit of electronic charge each (e.g., from Mo^{6+} to Mo^{5+}). Corresponding to each O^{2-} vacancy in the intermediate layer, one nearby Bi ion and one nearby Mo ion each lose one electronic charge unit. (It is shown later that it is not critical which ions lose the charges.) The energy path of a Bi-bound O^{2-} ion leaving a lattice such as that described above is treated in section 3 and is discussed in section 6.

If electrons liberated by the removal of O^{2-} ions can be used elsewhere, the Bi and Mo ions would return to their former charges of 3+ and 6+, respectively. When one notes that electron transfer occurs in the Mars–van Krevelen

TABLE I: Interaction Potentials E_{jk} between an O^{2-} Ion (j) and Various Ions (k) as Functions of Interionic Distances r[a]

k	range of r	A_{jk}	B_{jk}	C_{jk}	D_{jk}	F_{jk}	G_{jk}
Bi^{2+}	$r < 0.28$	1.33×10^5	1.18				
	$0.28 < r < 0.75$	2.97×10^4	6.45	−52.1	−262.4	−57.6	6.17
	$r > 0.75$	1.26×10^6	11.5				
Mo^{5+}	$r < 0.3$	5.99×10^4	10.5				
	$0.3 < r < 0.84$	1.09×10^3	4.80	−20.6	−249.3	−144.0	2.60
	$0.84 < r < 1.35$	1.45×10^3	2.38				
	$1.35 < r < 2.06$	2.35×10^2	1.02				
	$r > 2.06$	55.7	0.305				
Mo^{4+}	$r < 0.45$	5.42×10^4	9.30				
	$0.45 < r < 1.25$	3.29×10^3	3.10	−20.57	−198.1	−115.2	2.60
	$1.25 < r < 1.84$	293	1.18				
	$r > 1.84$	7.0	0.40				
Mo^{2+}	$r < 0.40$	5.42×10^4	9.40				
	$0.4 < r < 0.90$	6.84×10^3	4.22	−20.57	−129.9	−57.6	2.60
	$0.9 < r < 1.7$	7.13×10^2	1.74				
	$r > 1.7$	77.0	0.44				

[a] The constants A_{jk} to G_{jk} are used in the equation $E_{jk}(r) = A_{jk} \exp[-B_{jk}|r - (G_{jk}/r^2)|] - C_{jk}/r^6 - D_{jk}/r^4 + F_{jk}/r$ with E_{jk} in eV and r in Å.

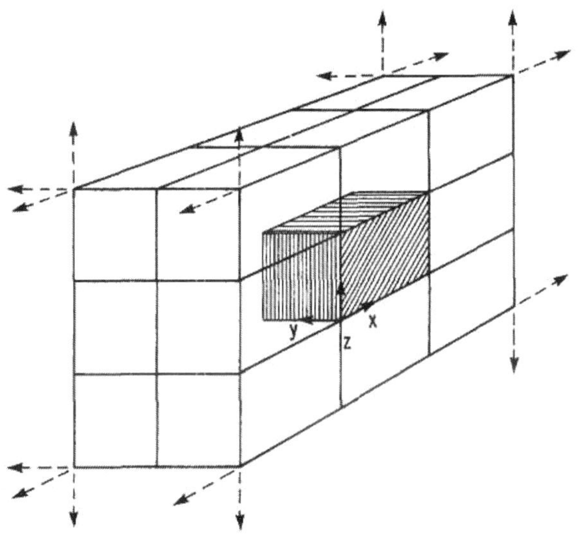

Figure 1. Truncated lattice used in the computation of the interaction energy of a mobile O^{2-} ion. The central unit cell, shaded, lies on the external surface of the solid lattice at $y = 0$. Dashed arrows show the directions that the truncated lattice can be extended to form the infinite lattice.

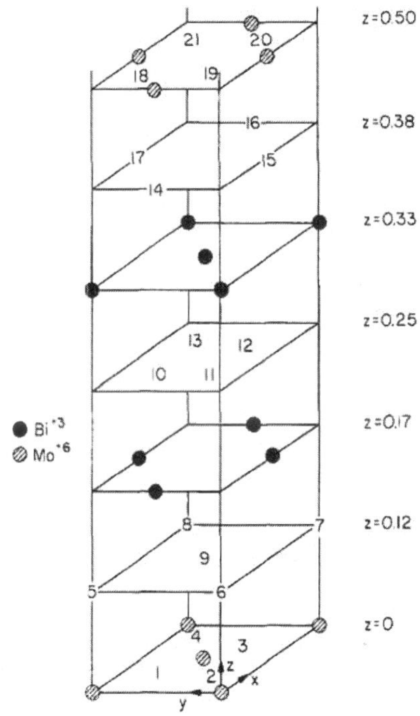

Figure 2. Lower half of an idealized unit cell of Bi_2MoO_6. The upper half is a mirror image through the plane $z = 0.5$. Locations of Bi^{3+} and Mo^{6+} ions are indicated by filled circles and shaded circles, respectively. Positions of O^{2-} ions are indicated by numbers 1–21, which are used to identify the locations in the text.

model[6] of selective oxidation, such a locally nonelectroneutral lattice is likely to be formed during the catalytic process itself. The charge imbalance in the working lattice can however be assumed corrected by charges of the adsorbed and reacting hydrocarbon molecules somewhere in the appropriately infinite lattice. The difference between the central unit cell of such a working lattice after the transfer of electrons and Figure 2 is the existence of vacancies at the appropriate Mo-bound and intermediate-layer oxygen ion positions. The energy path of a Bi-bound O^{2-} ion leaving this lattice is treated in section 4 and is discussed in section 6.

3. Energy Paths of Bi-Bound O^{2-} Ions Leaving the Working Lattice after Prereduction

Interaction Potentials. The calculation of energy paths in such a lattice requires the use of interaction potentials of ion pairs not considered in ref 3. These potentials are calculated in a manner similar to that used to generate Figure 1 of ref 3. In brief, Coulombic, induced-dipole, instantaneous-dipole, and electron-shell repulsion terms are obtained as described by us earlier.[3] These terms are gathered in Table I for the ion pairs of present interest.

Energy Paths. Figure 3 illustrates energy paths of a bismuth-bound O^{2-} ion leaving the lattice through the $y = 0$ face. The ion corresponds to number 11 of Figure 2. The lower curve represents the y coordinate of the ion corresponding to the path coordinate. The upper curves O, A, B, and C are the energy paths, describing the energy of the O^{2-} ion as a function of its progress along the path as given by the path coordinate. Curve O describes the energy path of the O^{2-} ion leaving a full Bi_2MoO_6 lattice and is taken from ref 5. As first noted in that work, the energy barrier for motion of this ion is formidable under these circumstances. Curve A represents the energy path of the O^{2-} ion when the central unit cell has no O^{2-} ions

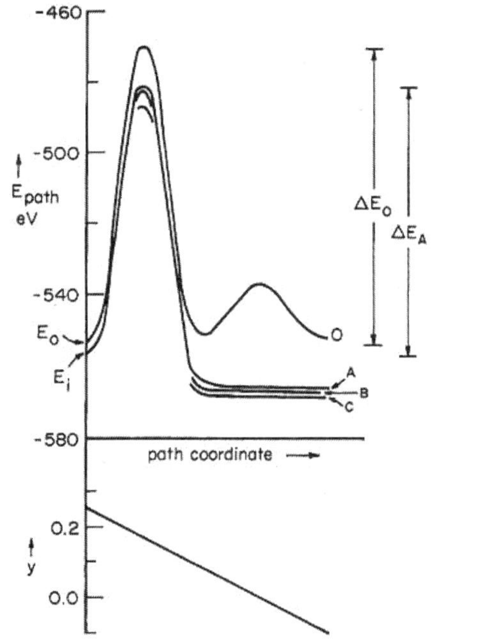

Figure 3. Energy paths of ion 11 leaving electroneutral lattices. Curve O describes the energy path from the fully occupied lattice and corresponds to Figure 6.[5] The independent variable is an arbitrarily defined "path coordinate". The motion is in the $-y$ direction, and the y coordinate of the moving ion is shown as a function of the path coordinate in the bottommost curve. Curves A, B, and C correspond to energy paths out of lattices which are electroneutral and after different numbers of O^{2-} ions have been removed from the central unit cell. Curve A corresponds to the lattice when vacancies are present in three intermediate-layer O^{2-} ion positions. As described in the text, one Bi ion and one Mo ion lose one unit of positive charge each for each such vacancy, in order to maintain electroneutrality. For energy path B, six intermediate-layer O^{2-} ion vacancies exist in the central unit cell. In addition to the vacancies of curve A at positions 6, 7, and 15, there are vacancies at positions that are mirror images of these three through $z = 0.5$. Bismuth ions at (0,0.5,0.17), (1,0.5,0.17), (1,0,0.33), (1,0,0.67), (0,0.5,0.83), and (1,0.5,0.83) are Bi^{2+}; there is an Mo^{4+} ion at (0.5,0,0.5), and Mo^{5+} ions exist at (0,0,0), (1,0,0), (0,0,1), and (1,0,1). A total of 12 vacancies exist in the central unit cell for curve C. In addition to the six vacancies in curve B, molybdenum-bound O^{2-} ions numbers 2, 3, 19, and 20, and the mirror images of numbers 2 and 3 through $z = 0.5$ are removed. Six Bi^{2+} ions exist, identical with those of curve B. Mo^{5+} ions exist at (0,0.5,0.5) and (1,0.5,0.5); the ions at (0,0,0), (1,0,0), (0.5,0.5,0), (0,0,1), (1,0,1), and (0.5,0.5,1) are Mo^{4+}; and there is a Mo^{2+} ion at (0.5,0,0.5).

corresponding to locations 6, 7, and 15 of Figure 2. These three ions are in the intermediate layer; hence, in accordance with the electroneutrality condition, three bismuth and three molybdenum ions each lose one unit of positive charge. The three bismuth ions so altered are those at (1,0,0.33), (0,0.5,0.17), and (1,0.5,0.17); the Mo^{5+} ions are at (0,0,0), (1,0,0), and (0.5,0,0.5). Such a lattice was shown in a previous section to be consistent with conditions after prereduction of the catalyst.

Note that the energy level at the start of curve A, given by point E_i, is somewhat less than that at the start of curve O, given by point E_0. The assumption here is that the number-11 ion is immobile during the removal of the intermediate-layer anions (although its energy level changes from E_0 to E_i) and attemps its passage to the surface of the lattice only after the removal of the three labile O^{2-} ions is complete. The removal of the labile O^{2-} ions is of course not instantaneous and probably occurs concurrent with small changes in the position of the number-11 ion. However, the treatment of only one mobile ion at a time is a computationally convenient device and is not likely

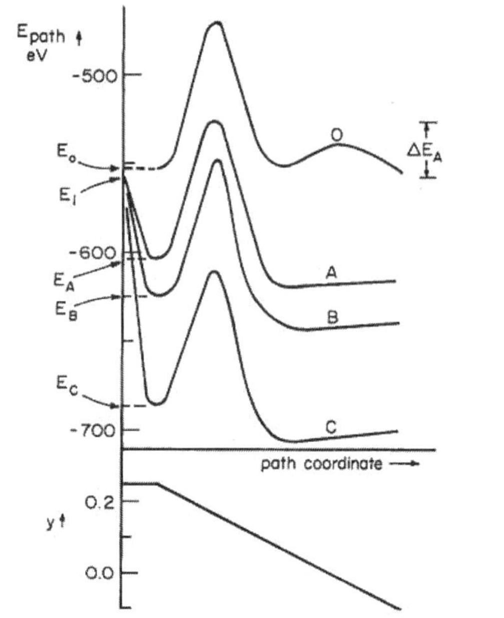

Figure 4. Curves A, B, and C represent energy paths of Bi-bound O^{2-} ion number 11 leaving lattices containing 3, 6, and 12 O^{2-} vacancies in the central unit cell, respectively. The O^{2-} ion positions that are vacant are the same as those in the corresponding curves of Figure 3. The initial portion of the curves (to E_A, E_B, E_C) represent the decreasing energy levels of the number-11 ion in the vacancy-containing lattice as bismuth and molybdenum ions return to Bi^{3+} and Mo^{6+} and the lattice turns locally nonelectroneutral at the start of the reaction.

to result in a significant error in the path. Curves O and A and the corresponding energy barriers to the motion of the number-11 ion under these conditions are discussed in section 6.

To consider more extreme cases of prereduction, first 6 and then 12 O^{2-} ions are removed from the central unit cell of the working lattice. The O^{2-} ions removed are all either in the intermediate layer or bound to molybdenum ions; according to ref 5 such ions are relatively easily removed even from a full lattice. As before, bismuth and molybdenum ions lose units of positive charge to maintain electroneutrality. The corresponding energy paths are shown in Figure 3, curves B and C. The initial energy level in these curves is not identically E_i, the corresponding value for curve A, but is very nearly this value. Curves B and C are also discussed in section 6.

4. Energy Paths of Bi-Bound O^{2-} Ions Leaving the Lattice at the Start of Reaction Conditions

In this section are presented energy paths for O^{2-} ion number 11, Bi-bound, leaving the lattice with O^{2-} ions previously removed from the central unit cell. Three, six, and twelve O^{2-} ions are successively removed, and those removed are identical with those of section 3. However, the lattice is not electrically neutral; during the process all the bismuth ions return to the Bi^{3+} form, and all the molybdenum ions to Mo^{6+}. Except for the O^{2-} ions removed, therefore, the central unit cell is identical with that of Figure 2. It was shown in section 2 that such a working lattice is consistent with conditions when the reactants are first introduced, i.e., at the start of the reaction.

The energy paths can be seen in Figure 4. As in the previous section, curves A, B, and C indicate that 3, 6, and 12 O^{2-} ions are missing from the central unit cell of the working lattice when the energy path of O^{2-} ion number 11 is computed leaving the lattice. The position of the ion in the path, as given by the y coordinate, is shown in the

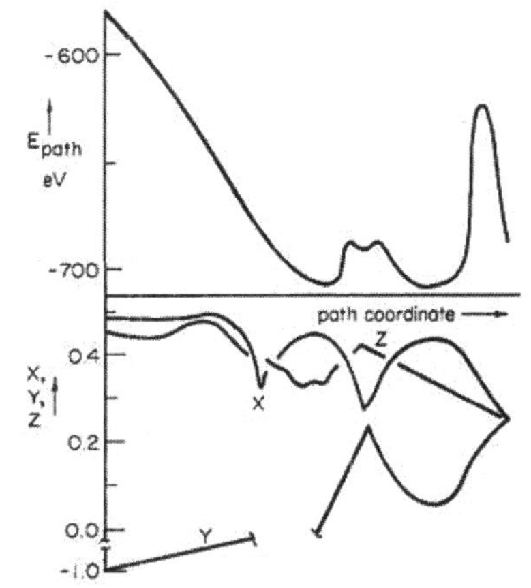

Figure 5. Energy paths and coordinates for an external O^{2-} ion entering the lattice in the $+y$ direction and then moving inside the lattice in the $-z$ direction toward the bismuth-bound site. The central unit cell is nonelectroneutral and corresponds to case C in Figure 4. In addition, there is a vacancy corresponding to O^{2-} ion 11, at the bismuth-bound site.

lower figure. As in Figure 3 the computations are carried out by assuming no change in position of the number-11 ion until the electrons have been removed from the bismuth and molybdenum ions, i.e., until the condition of local nonelectroneutrality has been met (and after the intermediate- and Mo-layer O^{2-} ions have been removed as in section 3). Accordingly, the first portion of curve A is a decrease in the energy level of the mobile ion from E_i to E_A, the mobile ion being stationary in the number-11 position while electrons are being removed from the bismuth and molybdate ions. After the transfer of electrons is complete, the remainder of curve A characterizes the energy path of the motion of the number-11 ion toward the surface.

Similar observations can be made for curves B and C. The differences between Figures 3 and 4 are discussed in section 6.

5. Energy Paths of External O^{2-} Ions Entering a Bi-Bound Site during the Reaction Process

Figure 5 describes the energy path (above) and corresponding coordinates of the trajectory (below) for an O^{2-} ion originally outside the lattice (extreme left of the figure), entering the lattice, and moving inside it toward a Bi-bound anion vacancy corresponding to number 11 (extreme right of figure). The lattice is equivalent to that of the previous section, i.e., is nonelectroneutral with 12 O^{2-} vacancies, and contains in addition the O^{2-} vacancy at position 11.

Figure 5 is of interest when discussing the steady-state replenishment and loss of O^{2-} ions and is considered in the following section.

6. Discussion

Processes Considered. Two types of processes are considered in this work, each characteristic to a particular type of O^{2-} ion. The ions are differentiated not by position in the lattice, as both are at position 11 at some point in their characteristic process; rather the ions have different histories prior to being at position 11. The first type of O^{2-} ion is one which is an "original" member of the lattice of Bi_2MoO_6, i.e., one present in the lattice prior to prereduction and the subsequent reaction. The characteristic process considered for this ion is its removal to the surface of the lattice. This initiates the catalytic oxidation reaction, and hence is termed the "initiation" step. We relate below how changing the number of vacancies and the ionic charges in the lattice alters the energy path of the initiation step, changes the energy barrier of the process, and hence improves the ease with which the process can occur.

The vacancy left when the "original" number-11 ion leaves the lattice is filled by the second type of O^{2-} ion considered here. This ion arrives at this position from being initially in the gas phase. The process considered for this ion is its insertion into the lattice near the molybdenum ion, its movement through the lattice to near the bismuth ion, and its removal to the surface of the lattice. This process, occurring after the initiation step, is termed the "steady-state" step. The steady-state step requires the presence of an adsorbed hydrocarbon and is not considered in detail.

Initiation Step. First consider Figure 3, where curve O represents the energy path of the Bi-bound number-11 O^{2-} ion moving toward the surface of the full working lattice, i.e., one with all other O^{2-} ions present, and curves A, B, and C represent energy paths when the electroneutral lattice contains 3, 6, and 12 O^{2-} vacancies, respectively. Then the energy barrier to the motion of the number-11 ion toward the surface is the difference between the maximum energy level in the path and the initial energy level. For the case of curve O, the initial energy level is E_0, the interaction energy of the ion with the full lattice. It is reasonable to suppose that the motion of the labile O^{2-} ions giving rise to the vacancies is slow. Hence for curve A, with three O^{2-} ion vacancies, the initial energy level will be E_i, the interaction energy of the ion in the number-11 position with the rest of the vacancy-containing, electroneutral lattice. Consequently, the energy barrier for motion of this ion to the surface of the lattice is given by ΔE_A in Figure 3. Note that ΔE_A is not significantly different from the energy barrier for curve O, ΔE_0. For curves B and C, the initial energy is very nearly E_i, and the maximum energy level is only slightly different from that for curve A. Since the energy barriers for curves A, B, and C are not significantly different from that for curve O (approximately 80 eV), it may reasonably be concluded that the formation of anion vacancies while the lattice is kept electrically neutral does little to facilitate the removal of bismuth-bound O^{2-} ions in the initiation step. Hence at the end of the prereduction process, the Bi-bound O^{2-} ion is still unable to leave the lattice.

The reformulating of bismuth and molybdenum ions as Bi^{3+} and Mo^{6+} yields the energy paths shown as curves A, B, and C in Figure 4 for the electron-transfer process and the motion of the number-11 ion toward the surface of the lattice. As in Figure 3, an energy barrier for each of these curves, corresponding to nonelectroneutral lattices with increasing anion vacancies, can be given by the difference between the maximum energy level of the path and the energy of the ion at the start of the path. For curve A the latter energy level will be E_i, the energy of interaction of the number-11 ion with the vacancy-containing, electroneutral lattice. The transfer of electrons out of the working lattice probably occurs much faster than the loss of O^{2-} ions to form the anion vacancies for Figure 3. Hence the drop in potential energy of the mobile ion from E_i to E_A during the first part of the energy path is not dissipated, but instead helps the ion to attempt to move over the energy barrier in its path to the surface of the lattice. For curve A, the maximum energy is approximately 30 eV greater than E_i. Though this still represents a difficult requirement, the drop in ΔE from 80 to 30 eV is significant. For curve B (6 vacancies) the value of ΔE drops to ~8 eV, and for the extreme case of 12 vacancies in curve C, the maximum energy level of the path is below the initial point (approximately E_i). Hence the mobile O^{2-} ion should easily overcome the rise in the portion of the curve between E_C and the maximum energy level and move to the surface of the lattice.

Clearly the condition of local nonelectroneutrality is required, along with the presence of anion vacancies, if the initiation step is to be favorable.

In terms of interaction potentials such as those of Table I, it is not difficult to understand the small difference between curve O (full lattice) and curves A, B, and C in Figure 3 (electroneutral lattice with vacancies) or the large differences between curve O and curves A, B, and C in Figure 4 (nonelectroneutral lattice with vacancies).

Consider Figure 4 first. The difference in the interaction energy of O^{2-} ion 11 with the two types of lattices arises because of the absence of O^{2-}–O^{2-} pair potentials corresponding to the 3, 6, or 12 anion vacancies. These pair potentials are mainly Coulombic repulsion terms, although there are small electron-shell (repulsion) and polarization (mainly attraction) contributions even at the relatively large interionic distances involved between ion 11 and O^{2-} ions to be absent. The absence of 3, 6, or 12 repulsion terms makes the lattice interaction energy of the mobile ion more negative. Consequently, the value at the peak of the energy path curve may approach and even be less (more negative) than the energy of the mobile ion at the start of the path.

Now consider the curves of Figure 3. As in Figure 4, the lattice is missing several O^{2-} ions and so the total interaction energy of the mobile O^{2-} ion at any position in the lattice is decreased. However, in order to maintain electroneutrality, some ions in the working lattice are less positively charged than in Figure 4 or in the full lattice. For each such ion, the pair potential with the mobile O^{2-} ion now carries a weaker (less negative) Coulombic attraction term. For example, as a bismuth ion changes from Bi^{3+} to Bi^{2+}, the product of the charges with respect to O^{2-} changes from 6– to 4–. There are additionally small changes in polarization and electron-shell repulsion. The net effect is an increase in the ion-pair potential. The number of positive charges lost (electrons gained) is equal to the number of negative charges disappearing because of the vacancy formation. Consequently, in the interaction energy between the mobile ion and electroneutral working lattice with vacancies, the algebraic sum of all the Coulombic terms is approximately the same as that in the original, filled, working lattice. The sum is not exactly the same, of course, because the charge compensations occur at different interionic distances from the mobile ion. The small net difference in these two Coulombic terms, together with the changes in the smaller contributors to the interionic potentials—electron-shell, polarization—account for the relatively minor improvement in the ease with which the bismuth-bound O^{2-} ion can be removed from the lattice when anion vacancies are coupled with the electroneutrality condition.

Further, note that the changes in the Coulombic terms of anion–anion and anion–cation potentials will cancel to about the same level of approximation, more or less regardless of their position with respect to the mobile O^{2-} ion, so long as the cations that lose the positive charges are reasonably close to the location of the anion that becomes a vacancy and consequently "loses" negative charges. Hence, curves A, B, and C in Figure 3 are almost independent of which O^{2-} ions become vacancies and which Bi and/or Mo ions lose how many positive charges, so long as the formation of the vacancies and the loss of charges occur at adjacent ionic positions.

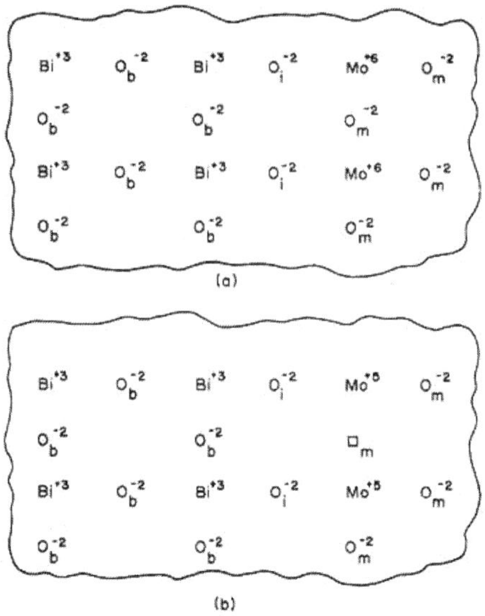

Figure 6. Simplified representation of one possible process during the prereduction of Bi_2MoO_6. (a) The view of the catalyst is facing the top surface. Subscripts i, b, and m denote O^{2-} ions located in an intermediate layer, and bound to bismuth and molybdenum ions, in that order. (b) After prereduction. One anion vacancy (□) is shown, for an O_m^{2-} ion. O_i^{2-} ions can also leave the lattice (not shown) but O_b^{2-} ions cannot leave. Note that electroneutrality is maintained.

Mechanism of Prereduction to Vacuum. Recall now that the electroneutral lattice with vacancies can be expected at the end of a prereduction state and that the electroneutrality condition may not be satisfied in portions of the lattice during the actual oxidation process. (The overall system would, of course, be electroneutral because of the presence of adsorbed intermediates.) Consequently, we can hypothesize the following processes occurring during prereduction and at the start of the oxidation process, leading to the initiation step.

Prereduction occurs either into vacuum or in the presence of hydrogen and serves to increase the number of anion vacancies present in the intermediate and molybdenum layers. In the former case, O^{2-} ions from these layers can be removed relatively easily, even from the ideal, fully occupied lattice. The anions leave as O_2 and the electrons migrate to cations, reducing their positive charge. Prereduction into vacuum is illustrated in Figure 6. If one recalls the large energy barriers to the motion of ion 11 in Figure 3, the Bi-bound O^{2-} ion is not expected to leave the electroneutral lattice at the end of the prereduction step.

Prereduction by hydrogen represents of course a steady-state chemical reaction between the hydrogen and the bismuth molybdate to form water. A discussion of this process is best postponed until after consideration of the steady-state replenishment and loss process below. We discuss first the start of the oxidation process, culminating in the initiation loss of the Bi-bound O^{2-} ion.

Mechanism of the Initiation Step. When the hydrocarbon and oxygen are introduced in the gas phase, the oxidation process may be considered to start. Now electrons leave the cations of lower charge, Bi^{2+} and Mo^{5+}, possibly to help convert gas-phase O_2 to lattice O^{2-} ions elsewhere in the lattice. In this manner Bi^{3+} and Mo^{6+} ions are formed once again near the anionic vacancies (Figure 7a,b). The original bismuth-bound O^{2-} ion is able to leave the (nonelectroneutral, vacancy-containing) lattice easily, helped perhaps by the presence of a hydrocarbon near this

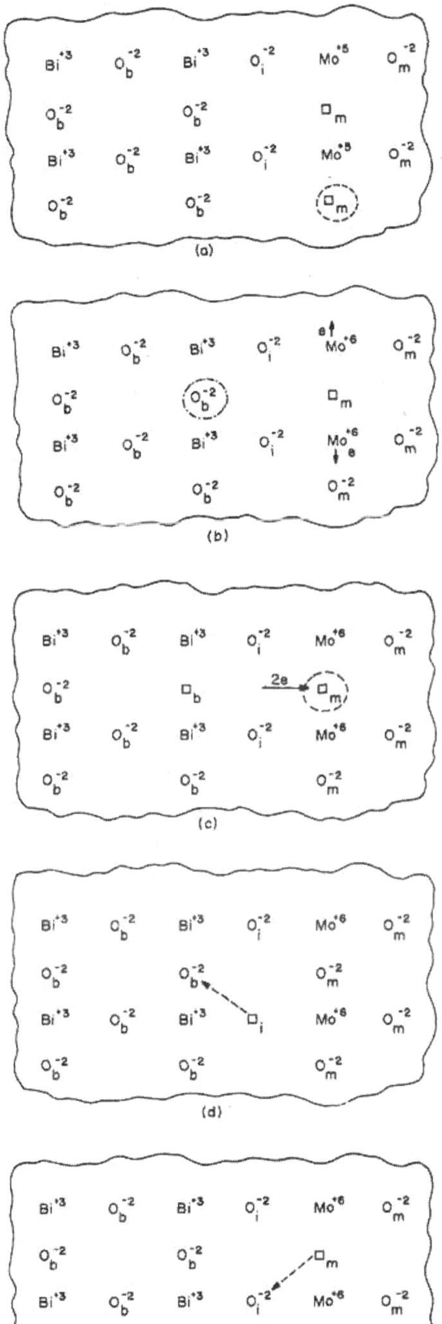

Figure 7. One possible scheme for the oxidation process. Symbols are as in Figure 6. (a) After prereduction, a number of anion vacancies are present corresponding to O_i^{2-} and O_m^{2-} ions. This figure is similar to Figure 6b except that an additional vacancy is shown. Note also, above the catalyst surface, the presence of gas-phase oxygen, denoted by the dashed lines. (b) In the presence of gas-phase oxygen, electrons leave cations, perhaps to enable the oxygen to fill excess anion vacancies. Cations are now Mo^{6+}. The presence of a hydrocarbon above the catalyst surface is shown schematically by dashed-dotted lines. (c) An O_b^{2-} ion is removed from the lattice, leaving $\square_b$ behind. Its electrons move to a molybdenum site, above which gas-phase oxygen is present. (d) Oxygen fills the molybdenum-bound vacancy, and O_2^- migrates to the vacant bismuth-bound position. (e) O_m^{2-} migrates to fill $\square_i$. Parts a–d can now be repeated.

site. The electrons from the leaving O^{2-} ion can be transferred, via a Mars–van Krevelen mechanism,[6] to a molybdenum site.

Mechanism of Prereduction with Hydrogen. Prereduction with hydrogen is in fact a chemical reaction of the hydrogen with bismuth molybdate. Since hydrogen is expected to be adsorbed on the surface, the mechanism of this prereduction probably bears resemblances to Figure 7 as well as 6. In this case local nonelectroneutrality, due to the presence of adsorbed forms of hydrogen on the surface, and the presence of naturally occurring anion vacancies in the lattice allow bismuth-bound O^{2-} ions to be removed. The resulting anion vacancies can be quickly filled with intermediate-layer or molybdenum-bound O^{2-} ions (see ref 3), resulting in vacancies in those positions. In the absence of gas-phase oxygen, the latter vacancies are not filled. The anions now leave as H_2O, and, as in the case of vacuum prereduction, electrons migrate to cations, reducing their positive charge. Whether in vacuum or by hydrogen, therefore, the prereduced lattice contains anion vacancies in the molybdenum and intermediate layers, and remains electroneutral by adjusting the charge of nearby cations. Further, the initial energy of a bismuth-bound O^{2-} ion after hydrogen prereduction, as after vacuum prereduction, is given by E_i of Figures 3 and 4.

Finally, none of the above should be taken to imply that the oxidation process cannot occur in the absence of prereduction of the catalyst. Dissociative adsorption of the hydrocarbon itself leads to the presence of adsorbed hydrogen on the surface, which may have the same effect as prereduction with hydrogen. Further, it is inconceivable that the actual lattice of the bismuth molybdate is perfect, so that O^{2-} vacancies are to be expected even before the reduction process. However, prereduction can reasonably be expected to increase the number of vacancies and thus to accelerate the initiation of the selective oxidation process. This hypothesis is consistent with the recent experimental reports of Yong et al.[7] They found that prereducing the catalyst increases the density of anion vacancies from $\sim 10^{18}$ to more than 10^{19} per gram of catalyst.

Steady-State Loss. Following the loss of the "original" Bi-bound O^{2-} ion from the vacancy-containing lattice to the outside in the initiation step, its place is taken by another O^{2-} ion, which in its turn moves to the outside, and so on. When a steady state is reached, the replacement O^{2-} ion must have originally come from the gas phase. Consequently its initial energy level is not E_i, corresponding to that of the O^{2-} ion inside the electroneutral vacancy-containing lattice, as is that for the "original" O^{2-} ion. Instead the initial energy level is that of an O^{2-} ion outside the nonelectroneutral lattice containing vacancies, and in the presence of a hydrocarbon, perhaps adsorbed on the surface. Similarly the energy path of the leaving O^{2-} ion must also take the hydrocarbon into account. Then, as in the case of the initiation loss, the ease of the steady-state loss will depend upon the difference between the initial energy of the O^{2-} ion outside the lattice and the maximum energy level on the energy path of the moving ion.

Taking into account the presence of a hydrocarbon adsorbed on the surface of the working lattice in computing the energy path of the moving ion would introduce a significantly higher level of complexity to the problem. Instead we consider the working lattice in the absence of the hydrocarbon, and the energy path of interest is one where the O^{2-} ion enters the lattice from outside the surface, moves inside the lattice, and then leaves the lattice. As mentioned earlier, for the sake of definiteness the working lattice is taken to be one with 12 anion vacancies plus a vacancy in position 11 (and is locally nonelectroneutral). Since essentially the same lattice is used to obtain curve C of Figure 4, the last part of the energy path of present interest, where the mobile ion leaves the lattice, is given by the curve of Figure 4C after point E_C. The first two parts of the energy path of interest, where the mobile ion enters the lattice and moves inside it, are shown in Figure 5.

The steady-state process is relatively quick. Hence the potential energy of the mobile ion arriving near position 11 from the outside (right-hand side of Figure 5) is not dissipated before the ion moves to the surface (Figure 4C after E_C). Consequently, the energy level at the start of the path must be compared with the maximum energy level in the paths of Figure 5 and of Figure 4C to determine the energy barrier for the steady-state loss of O^{2-}. For the start of the path as given in Figure 5, there is no energy barrier to the motion of the external ion to the lattice, inside the lattice, and out again. Note that in Figure 5 the path of the external O^{2-} ion is started at $y = -1$, i.e., 5.5 Å outside the surface of the lattice, for the sake of definiteness. The actual y coordinate for the start of the path would depend upon the position and nature of the hydrocarbon and the energetic characteristics of the O_2 molecule and the O^{2-} ions. In fact, it can be seen from Figure 5 that the initial energy would have to correspond to $y \geq -0.5$, i.e., closer than 2.8 Å outside the surface, for any energy barrier to be present for the steady-state loss of O^{2-}. Considering that the van der Waals diameter of an oxygen atom is of that magnitude, it is reasonable to suppose that the formation of the ion would not occur at a distance from the surface that was significantly less than this value. Hence it is likely that there is no barrier to the steady-state loss of O^{2-} from a nonelectroneutral, vacancy-containing lattice.

The role of the hydrocarbon is not evaluated in this analysis. However note that it is probably adsorbed near the bismuth site and consequently has little effect on the energy path of the mobile ion passing through the molybdenum or intermediate positions. From Figure 5, the energy path of the mobile ion near the bismuth site is at its highest peak (excluding the start of the path). The presence of the nearby hydrocarbon may serve to decrease the energy level in this region. This would make even more favorable the motion of the O^{2-} ion through the lattice.

Recall that for the full lattice, i.e., one with no vacancies, it has been shown[3] that energetically favorable paths exist for (a) an external O^{2-} ion to a Mo-bound site, (b) a Mo-bound O^{2-} ion to an intermediate-layer site, and (c) an intermediate-layer O^{2-} ion to a Bi-bound site. An examination of Figure 5 shows some differences in the paths from those described above and in ref 3 for the full lattice. In the present case, the O^{2-} ion enters the lattice between the molybdenum-bound site and the intermediate site and travels to between the intermediate site and the bismuth-bound site. Considering the increased numbers of vacancies in the present lattice, as compared to the full lattice, such modifications in the paths can hardly be unexpected.

7. Conclusion

The initiation step considered here refers to the loss, at the start of catalytic oxidation, of a bismuth-bound O^{2-} ion that was actually present in the lattice before prereduction and catalytic oxidation. The steady-state step refers to the loss of an O^{2-} ion not bound to bismuth originally (i.e., before prereduction) but one that moves to that location from the gas phase, indirectly via a molybdenum-bound site, during the catalytic oxidation itself.

The initiation loss of bismuth-bound O^{2-} ions is contingent on three factors. The first is the presence of a sufficient quantity of intermediate-layer and molybdenum-bound anion vacancies. Figure 4 demonstrates the relative ease of loss of the ion when increasing numbers of vacancies are present in the lattice. It is worth noting that the incremental ease of removal of the bismuth-bound O^{2-} ion decreases with an increased number of vacancies. Consequently, there must be an optimum number of vacancies beyond which it is not worthwhile to proceed.

The second factor is the (local) nonelectroneutrality of the lattice. The anion vacancies must be surrounded by bismuth and molybdenum ions in their "normal" state—Bi^{3+} and Mo^{6+}. Comparing Figure 4 with Figure 3, one sees that the loss of electrons from cations returning to their "normal" state decreases the potential of the mobile ion and the magnitude of the potential barrier, thus facilitating the approach of the ion to the lattice surface.

The final factor influencing the initiation loss process involves the presence of the hydrocarbon near the bismuth site. The exact role of the hydrocarbon is not clear from this analysis. However, it must be significant, because Figure 4 indicates that over six vacancies are required to be present if the initial bismuth-bound O^{2-} ion is to leave the unit cell in the absence of the hydrocarbon. This is probably unrealistic. The number can probably be reduced if the presence of the hydrocarbon can increase the ease of removal of the bismuth-bound O^{2-} ion to the external surface of the lattice and to the hydrocarbon.

The role of prereduction is tied to the first two of these requirements. The end result of prereduction, whether under vacuum or with the use of hydrogen, is to increase the number of intermediate-layer and molybdenum-bound O^{2-} ion vacancies in the lattice, while keeping it electrically neutral. Under vacuum, the motion of the corresponding ions to the external surface of the lattice is found to be relatively easily accomplished. The presence of hydrogen creates these vacancies in a more indirect manner. Reaction of hydrogen with the bismuth-bound O^{2-} ions creates vacancies that are filled by anions moving from the intermediate layer and the molybdenum layer of the bismuth molybdate. At the end of the prereduction stage, the lattice is electrically neutral by adjustment of cationic charges, and bismuth-bound O^{2-} ions cannot be easily removed from the lattice.

The local electroneutrality condition is lifted during the actual oxidation process, and these bismuth-bound O^{2-} ions can now leave, thus initiating the oxidation of the hydrocarbon. This initiation loss of an O^{2-} ion from the lattice results in a vacancy that is filled by an O^{2-} ion originally from the gas phase. In its turn, this O^{2-} ion leaves the lattice too, during the steady-state step.

As in the case of the initiation process, the role of the hydrocarbon is also significant in the steady-state process. Here the O^{2-} ion, initially outside the lattice, enters at the molybdenum layer, moves to the bismuth layer, and leaves from there to oxidize the hydrocarbon. However, Figure 5 indicates that, even in the absence of the hydrocarbon, the energy required for such an ion to enter the (vacancy-containing, nonelectroneutral) lattice, move inside it, and leave is never greater than the energy level of the ion outside the lattice. The presence of the hydrocarbon, adsorbed near a position corresponding to the highest peak of the energy path of Figure 5, probably decreases the energy level at that point and hence decreases the energy barrier even further. Consequently this process and the steady-state replenishment and loss are favorable.

References and Notes

(1) Haber, J.; Grzybowska, G. *J. Catal.* **1973**, *28*, 489. Haber, J. *Int. Chem. Eng.* **1975**, *15*, 21.
(2) Dadyburjor, D. B.; Jewur, S. S.; Ruckenstein, E. *Catal. Rev.—Sci. Eng.* **1979**, *19*, 293.
(3) Dadyburjor, D. B.; Ruckenstein, E. *J. Phys. Chem.* **1978**, *82*, 1563.
(4) Ruckenstein, E.; Dadyburjor, D. B. *AIChE J.* **1976**, *22*, 785.
(5) Dadyburjor, D. B.; Ruckenstein, E. *J. Catal.* **1980**, *63*, 383.
(6) Mars, P.; van Krevelen, D. W. *Chem. Eng. Sci. Suppl.* **1954**, *3*, 41.
(7) Yong, L. F.; Howe, R. F.; Keulks, G. W.; Hall, W. K. *J. Catal.* **1978**, *52*, 544.